Mrinal K. Das
Lennoxville
1994

EMBRYOLOGY

An Introduction to Developmental Biology

STANLEY SHOSTAK

University of Pittsburgh

1817

HarperCollins*Publishers*

This book is dedicated to my mother
and to the memory of my father.

Art Direction: Lucy Krikorian
Text Design: Lee Goldstein
Cover Design: Delgado Design Inc.
Cover Illustration: Scanning electron micrograph of sand rat *Psammomys obesus* spermatozoa bound to egg's zona pellucida *in vitro*. Courtesy of Professor D. M. Phillips, The Population Council, Rockefeller University, New York.
Production: Kewal K. Sharma

EMBRYOLOGY: An Introduction to Developmental Biology

Library of Congress Cataloging-in-Publication Data

Shostak, Stanley.
    Embryology : an introduction to developmental biology / Stanley Shostak.
      p.   cm.
    Includes bibliographical references.
    ISBN 0-06-046126-8
    1. Embryology.   I. Title.
QM601.S46   1991
574.3'3 — dc20

89-26774
CIP

90 91 92 93   9 8 7 6 5 4 3 2 1

BRIEF CONTENTS

Part 7

GASTRULAS AND GASTRULATION

Part 8

EMBRYOGENESIS AND MORPHOGENESIS: ACQUIRING SPECIES CHARACTERISTICS

Part 9

ORGANOGENESIS AND SPECIFICATION

DETAILED CONTENTS

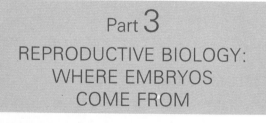

Part 3

REPRODUCTIVE BIOLOGY:
WHERE EMBRYOS
COME FROM

Part 4

FERTILIZATION AND THE ZYGOTE

Part 5

CLEAVAGE AND BLASTOMERES

Part 6

BLASTULATION AND BLASTULAS

Part 7
GASTRULAS AND
GASTRULATION

Part 8

EMBRYOGENESIS AND
MORPHOGENESIS: ACQUIRING
SPECIES CHARACTERISTICS

Part 9

ORGANOGENESIS AND
SPECIFICATION

PREFACE

Twenty-five years ago when I started teaching embryology and developmental biology, I made a commitment to tell my students what they wanted to know about embryos. From listening to their comments and questions, I learned that my students wanted me to explain difficult subject matter, not simplify it. From their term papers, I learned that they were eager to bring the subject matter up to date and grapple with current research. And from their examinations, I learned that I served my students best by placing before them my own concerns with change and growth.

Embryology: An Introduction to Developmental Biology is the composite of what I have told my students. The book is a collection from lectures over the years, rather than a single lecture series, and students and instructors should sort out the material according to their needs and the limitations of their time. The subject matter is organized conceptually, chronologically, and by degrees of difficulty, but the reader should be able to pick and choose desired material among the chapters and utilize the remainder of the text for background information where necessary.

The demands of a printed page are different from those of an academic hour. On the debit side, as a writer, I lose, for the most part, access to the students' comments I found so valuable in creating my lectures. On the credit side, the book allows me to highlight details and data, as well as concepts, and to lay a foundation of knowledge as well as build an edifice of facts. In addition, students gain confidence in the information presented if only because they can return to the printed page without worrying if their notes are correct.

My objective as a teacher is to explain the sources and inspiration of knowledge of embryos and development. Embryology springs from a dynamic view of life. Embryology situates the embryo in a continuum of life cycles; it integrates form with change and function with evolution. The embryological literature contains descriptions and comparisons, analysis and theory. Upgraded and updated, a course in embryology provides a meaningful body of information and a sustaining intellectual challenge.

The subject of embryology receives a rounded and balanced treatment in *Embryology*. Examining organisms' early history, *Embryology* is concerned with origins, causes, and effects. Building on a cellular foundation, the text reaches out toward anatomy and histology and grows toward developmental biology and reproductive physiology.

Addressed to advanced undergraduates, *Embryology* provides biology majors with information on embryos that they will need in other courses. Students with aspirations to careers in health sciences and medicine receive fundamentals for acquiring more precise knowledge of human development, and students on their way to graduate school obtain contemporary information on the known and unknown sides of embryos. *Embryology* also serves as a reference, a lexicon, a source of data, and a window to the literature of embryology.

The first parts of *Embryology* are introductory: Part 1 introduces embryology, while Part 2 introduces developmental biology. Classical embryology begins with reproductive physiology, gametes, and gametogenesis in Part 3, and the embryo's own development begins with fertilization in Part 4. Parts 5–9 are organized chronologically as the embryo unfolds.

The sections present increasingly complex

subject matter and concepts. Description gives way to analysis, and a comparative approach yields to experimentation. Following the adage that "history is the postscript of wisdom," sections end with discussions of the history of the subject. Conceding to the practical limitations, the book focuses on vertebrates for its final chapters and more or less ends at the point where the mammalian embryo transits into a fetus.

Each part is followed by Summary and Highlights, Questions for Discussion, and Recommended Readings. These are all intended to aid study by pointing the way backward or forward to information or by jogging memory. Key words and species are listed in the Index. Numbers in boldface identify pages in the text containing definitions. Additional information may be pursued through citations dotting the text and listed alphabetically by the first author's last name in the References.

Ultimately, the success of *Embryology* will not be measured by how far the text has taken students but by how much further students take the text. Friedrich Seidel, who ranks among the parents of modern developmental biology, makes this point while explaining his motivation for studying development[1]:

I have been very fortunate to participate at a Congress of the German Zoological Society in Göttingen 60 years ago (1920) and to listen to the talk of Hans Spemann on his transplantation experiments with Triton eggs. This has been a truly incisive event in my life. I have given a detailed report of these investigations to many of my student friends . . . . Then I have gathered all available publications of Spemann and my decision was

made: "Something like that you ought to be doing!"

My goal for *Embryology* was to distill Seidel's "something like that" for contemporary students. If I succeeded, it was entirely due to the scientists who labor in embryology.

In addition, I would like to thank all those who helped me bring this book to fruition: Professor Marcia Landy, whose belief in the project and effort on its behalf never flagged; Claudia Wilson, college editor, who told me at the beginning to "write your own book" and meant it; Albert Chung, Chairperson of the Department of Biological Sciences, and Peter Koehler, Dean of the Faculty of Arts and Sciences, who granted me sabbatical leave for the purpose of completing the manuscript; Drynda Lee Johnston, departmental reference librarian, and her able assistants who located and procured the literature I required; Ruth Ann Schulte, who patiently gathered all the permissions, and the following reviewers, whose comments were a tremendous help: Dr. Klaus Kalthoff, University of Texas; Dr. Brian Kay, University of North Carolina; Dr. Gerald Bergtron, University of Wisconsin; Dr. Hugh Stanley, Utah State University; Professor Helmut Sauer, Texas A&M University; Professor Howard Nornes, Colorado State University; Albert Harris, University of North Carolina at Chapel Hill; James Fowler, State University of New York; Betty L. Black, North Carolina State University; Herbert Phillips, University of Texas, Arlington; James Asher, Jr., Michigan State University; Dr. William Forbes, Indiana University of Pennsylvania; Dr. Ronald Harris, University of Southern California. Finally, let me thank the publishers and authors who gave permission to use illustrations, especially those who contributed photographs from original work.

STANLEY SHOSTAK

---

[1]F. Seidel, 1981. Introductory address to the conference on progress in developmental biology, Mainz, March 25–28, 1980. In H. W. Sauer, ed., *Progress in developmental biology.* Fischer Verlag, Stuttgart, pp. 1–5.

Part 1

INTRODUCTION

*Let us a little permit
Nature to take her own way; she better understands her own affairs than we.*

MICHEL DE MONTAIGNE, Works

*I would, therefore,
have you, gentle reader, to take nothing on
trust from me concerning the generation of
animals; I appeal to your own eyes as my
witnesses and judge.*

WILLIAM HARVEY, Anatomical Exercises on the
Generation of Animals

*E*mbryology is probably the most natural of the natural sciences. You do not have to be told that a developing organism is fascinating; you only have to see one. It is an organism in the process of becoming, and a generation in the process of rejuvenation. Embryos are an endless source of wonder, and embryology is the scientific expression of that wonder.

Part 1 has two objectives: to introduce embryos as seen by embryologists and to situate the embryo both as a stage in a lifetime and a link between life cycles. Embryologists view the embryo as both possessing and acquiring a variety of qualities, from the general properties of life and features of a phylum to the narrow properties of species and individual variation. Embryologists are concerned with the development of all these qualities.

EMBRYOS AND DEVELOPMENT

*The study of development, or **embryology,** because it offers the possibility of finding out how the most fundamental characteristic of living things, their organization, comes into being, has always been of compelling interest to everyone who has been concerned with the position of living things in the general philosophical scheme.*

CONRAD H. WADDINGTON (1962, p. 16)

An **embryo** (Gk. *embryon* from *em-* in + -*bryein* to swell, referring to the organism within the swelling of pregnancy) is a sexually reproduced organism near the beginning of its lifetime (Fig. 1.1). Embryos are the subject of embryology (Gk. *embryon* + -*logia* from *logos* word, discourse, hence the doctrine or theory of embryos), and the discovery of general principles of development is the objective. Historically, embryology has advanced the concept that all embryos are related; they represent a comparable stage (or series of stages) in a lifetime; they confront and cope with similar problems, and their development represents a comparable process (or set of processes).

Figuratively speaking, all organisms are confronted with problems they must solve to survive. For example, organisms must have resources for metabolism, defenses against adversity, and dump sites for toxic wastes. Unlike adult organisms that solve their problems by utilizing existing structures and adjusting their physiology and behavior, embryos solve their problems by developing new structures and adding novel functions.

Development is the embryo's way of life, but development also precedes and follows the embryo. Embryos are, after all, only part of a lifetime, and embryology is only part of the larger study of developmental biology.

While embryology considers development within the confines of sexual reproduction, developmental biology extends to asexual reproduction, regeneration, repair, maintenance, and so on. While embryology is traditionally limited to animal embryos, developmental biology extends to bacteria, protoctists (protistans), fungi, and plants. Moreover, while embryologists tend to think synthetically about how embryos are put together, developmental biologists tend to think analytically about how individual parts work within the whole.

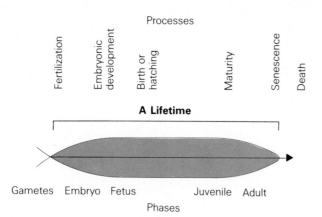

FIGURE 1.1. The lifetime of a typical multicellular organism. While the arrow of time leads to death in the case of most multicellular organisms, the loop illustrates the possibility that an adult will produce gametes. Some processes or events in a lifetime are listed above the arrow, and several stages in a lifetime are listed below.

EMBRYOLOGY AS A SCIENCE

Two aspects of differentiation must be recognized, the spatial aspect . . . where one region becomes different from another, and the temporal aspect, when qualitative changes occur within a particular unit in the course of time.

A. E. Needham (1964, p. 3)

Embryology's Questions

Like any other science, embryology is identified with a set of questions. While dynamic and evolving throughout history, embryology's questions focus on change or, at least, the perception of change in space and time.

What develops? Development is progressive, constructive change. It is an attribute applied to embryos and even to the gametes or sex cells preceding the embryo. A larger ovum (sing., ova pl.) or egg is prepared in the female parent, and a smaller spermatozoon (sing., spermatozoa or sperm pl.) is prepared in the male parent. Fusing at fertilization, the gametes produce a zygote or fertilized egg which begins embryonic development (Fig. 1.2).

Although a mere single cell, the zygote develops rapidly and dramatically, becoming increasingly heterogeneous in structure and complex in function. Organized structures appear and grow (i.e., increase in mass). Organs emerge, tissues diversify, and cells differentiate as they produce characteristic intracellular and extracellular materials.

Ultimately, an adult is formed, much like the adults who furnished the earlier gametes.

Development thus links the past and the future. While adults are parents of the embryo through reproduction, the embryo is the source of the adult through development, or, as the poet reminds us, "The child is father of the man" (William Wordsworth, *My heart leaps up when I behold*).

Development erases the major differences between the zygote and the adult as it creates similarities between offspring and parents. The similarities extend from general characteristics of the organism's phylum to those of species, subspecies, varieties, and parents. All these similarities constitute heredity.

While development proceeds horizontally through a lifetime, heredity proceeds vertically between succeeding generations and links lifetimes (Fig. 1.3). Heredity is transmitted to the zygote by the gametes, especially by their genes, or deoxyribonucleic acid (DNA), present in the gametes' nuclei. The organism's genotype, its genome or collection of genes, is largely established at fertilization. The organism's phenotype, its traits or manifest heredity, develops as genetic information unfolds under environmental auspices.

When does development occur? Development may go on throughout a lifetime, and different kinds of development may occur at various points in a lifetime. Ideally, stages of development correspond to conspicuously different developmental activities taking place at different times (Fig. 1.4), but, in general, development is continuous, and stages overlap one another rather than begin and end at discrete borders (for mammals and chick, see Butler and Juurlink, 1987).

Initially enclosed within egg membranes, the young organism breaks through the membranes and hatches into another stage of development. Most animals pass through some form of morphological and physiological transition as they leave their embryonic existence and enter the external environment as free instruments of independent life. Birds, for example, hatch as immature versions of adults (i.e., peeps or chicks), but many other animals hatch as larvae which do not resemble adults. Larvae and adults may occupy different niches and play different roles in the life cycle.

In mammals, including human beings, hatching occurs quite early, prior to implantation in the uterus, and the hatched organism is still considered an embryo. It becomes a fetus when it has

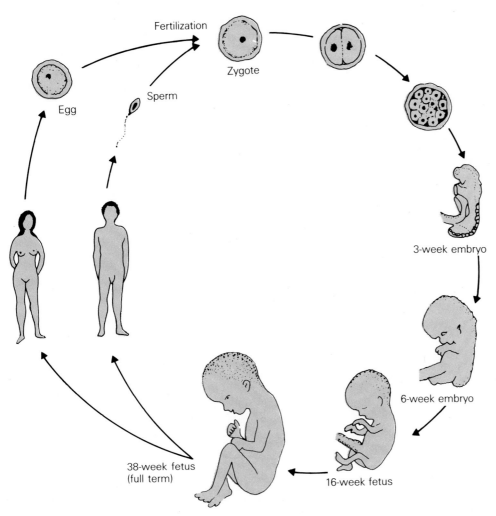

FIGURE 1.2. Development's phases for human beings. Egg and sperm develop in adults and produce a zygote at fertilization which develops into an adult. Adapted from D. L. Hartl, *Our uncertain heritage,* 2nd ed. Harper & Row, New York, 1985. Used by permission.

acquired rudimentary forms of the structures and system that will later be part of the adult, and it becomes a neonate or infant at birth. It grows rap-

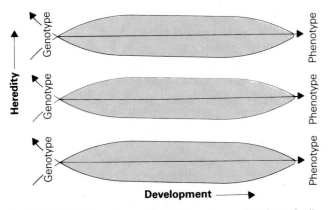

FIGURE 1.3. Hereditary information transmitted vertically to succeeding generations and developmental information translated horizontally in the course of an individual's lifetime. Each organism receives its genotype as a consequence of fertilization and its phenotype as a consequence of complex interactions occurring during development.

idly at the juvenile or preadolescent stage and then matures into the adult when gametes are formed. The sex cells may take part in another act of fertilization, but senescence and death bring the mammal's lifetime to an end (Fig. 1.1).

Early and late development are conspicuously different. In the embryo and fetus, the rate of constructive change exceeds the rate of degenerative change for a net gain. In the young adult, these rates are nearly equal. Adult structures may be in a dynamic or steady state of apparent (but not real) stasis in which rates of replacement equal the rates of "wear and tear." Later, at senescence, the rate of degenerative change exceeds the rate of constructive change for a net loss. Finally, during the last phase of a lifetime, development may occur, but not necessarily of a healthy variety. Cancers or malignant tumors, after all, also develop.

What causes development? The premise of most embryologists and developmental biologists is that development results from ordinary proper-

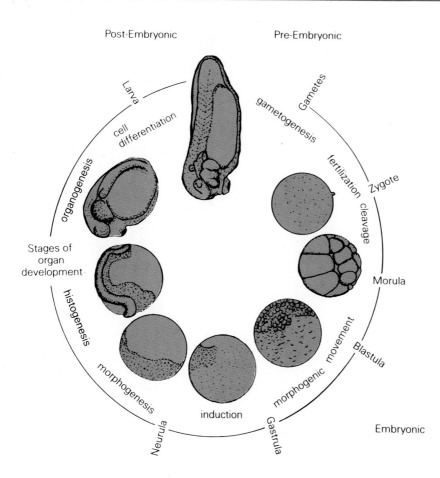

Post-Embryonic · Pre-Embryonic

Larva · cell differentiation · gametogenesis · Gametes

organogenesis · fertilization cleavage · Zygote

Stages of organ development

histogenesis · Morula

morphogenesis · morphogenic movement · Blastula

Neurula · induction · Gastrula

Embryonic

FIGURE 1.4. An embryological clock for amphibians: the chronology of phases and stages of the development.

ties of material, energy, and information according to the same chemical, thermodynamic, and cybernetics rules that govern other things in our world. But while these properties ordinarily dissipate energy, in a developing organism, they concentrate energy and increase temporal and spatial organization (Schrödinger, 1967).

An embryo develops because a chance event sets in motion a chain of reactive events. Fertilization is ordinarily the chance event. Chance is involved in the sense that the precise gametes that combine are not chosen in advance. These gametes meet because they are available at the same time and in the same place. As long as everything required by the developmental chain reaction is available, and nothing stops it, the reaction proceeds to construct the adult organism.

The developmental explosion that follows fertilization is interpreted across a broad spectrum of views. At one extreme, a preformationist model portrays development as a single explosion. At the other extreme, an epigenic model (Gk. *epi-* upon or after + *-genesis* from *gignesthai* to be born, hence progressive development) suggests that development is a series of interactive explosions.

According to the preformationist view, the embryo, like any bomb, must have its parts assembled beforehand, and, once triggered, the explosion cannot be called back. Within limits, development is viewed as invulnerable to influences arising from the environment.

The embryonic "bombshell" is thought to be assembled from cytoplasmic determinants of development and preformed coded information in nuclear DNA. The preformationist model of development implies that an embryo develops because it has been "loaded." It has all the right materials, energy, and information packaged and deliverable at precisely the right places and at exactly the right times as soon as the "fuse" is lit by fertilization.

In contrast, according to the epigenic view, embryonic development is subject to controls and environmental influences in the intervals between its several "explosions." Epigenesis implies that the sequence of developmental events and the changing of embryonic materials influence further events and the production of other materials. In the epigenic view, the embryo develops not because it has been switched on but because one thing has led to another.

Looking at the whole picture, development seems to defy both purely preformationist and epigenic models. The epigenic model works for sequences of developmental events, while the preformationist model works for consequences of these events, but neither works equally well with sequences and consequences.

As embryologists continue to seek the underlying impetus for development, they adopt both preformationist and epigenic views. The gene is central in both views. Preformationist theories looking in the direction of genes are called neopreformationist, while epigenic theories taking off from genes are called neoepigenic or neoepigenetic. Someday, both types of theory may merge in a truly synthetic theory of development.

Why does the embryo develop? "Why" questions raise issues of remote or ultimate causes. Although troubled by a long history of seeking answers in ancestors, embryologists now ask why a particular feature of the embryo develops as it does instead of why it develops at all.

Answers are sought in terms of function or adaptation in the context of evolutionary relationships. For example, the question "Why do embryos of terrestrial vertebrates have branchial (i.e., gill) arches?" is answered by the arches' adaptive value for vertebrate embryonic circulation (i.e., in conveying blood from the ventral heart to the dorsal aorta).

Answers to embryology's "why" questions must pass two acid tests. They must avoid restating premises while ascribing functions to things, and they must not violate the law of parsimony, that the simplest explanation for a phenomenon must be given priority. Statements containing phrases such as "in order to develop" cannot be offered as answers, since they fail both acid tests. Not only are such phrases circular, but they create a false dichotomy between embryonic parts that are developing and those that are functioning in the maintenance of the embryo.

The embryo and its parts do not lie about while developing but contribute to the life of the embryo. Life is not interrupted to allow an embryo to develop, and explanations of development must not separate the life of an embryo from its development.

How is the embryo constructed? "How" questions raise the issue of proximate or immediate causes. Since the embryo is built by its cells, the immediate causes of development involve the synthesis of materials in accordance with the coded instructions of nuclear genes and all the other self-perpetuating cytoplasmic constituents transmitted from one generation to the next (Fig. 1.5). Each constituent may have its own form of information waiting to be interpreted in the course of development.

Genes constitute a molecular "memory," a coded blueprint of hereditary information. Written in the language of nucleotides in DNA, this information is transcribed to ribonucleic acid (RNA), refined or processed, transported to the cytoplasm, and translated into protein. Whether as enzymes or structural material, as intracellular or extracellular proteins, the products of translation provide the core materials for embryonic construction.

Many, if not all, parts of embryonic cells also convey developmental information. From the machinery for synthesizing proteins to the structure of the cortex or surface of the cell, developmental information both instructs and allows cellular constituents to interpret hereditary information in the course of producing the organism's phenotype.

The cells of some embryos build from the ground up. Starting with only the minimum of differences among them, embryonic cells gradually create differences through synthetic activities. The cells in other embryos contain a variety of prefabricated materials (e.g., ribosomes and structural proteins) "ready-to-be-assembled" for intracellular development. In addition, embryonic cells may also contain stored "directions" (e.g., messenger RNA) capable of guiding the early course of embryonic synthesis.

The separate cells, or blastomeres, formed in the embryo may be capable of moving and performing a variety of morphogenic activities. Mass movements of embryonic cells, or morphogenic movements, bring cells together from diverse places and allow cells to interact. The development of composite organs (i.e., those consisting of more than one tissue or part) is especially affected by movement. Different embryonic cells brought together frequently induce each other to perform new activities.

Embryonic cells release a variety of extracellular materials, some of which may be especially important for cell movement and induction. In addition, the direction and rate of cell movement may be determined by internally controlled changes in the cell's size and shape, and cells may induce each other through points of intimate cell-to-cell contact.

The embryo is not constructed at one time, by one process, or of one thing. As the construction

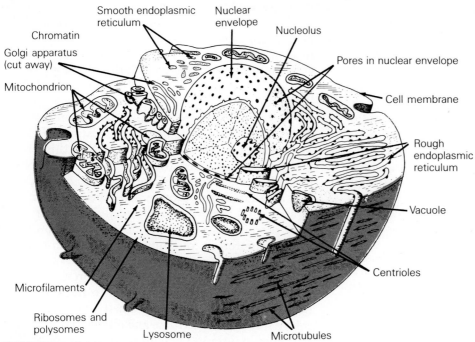

Chromatin
Golgi apparatus (cut away)
Mitochondrion
Smooth endoplasmic reticulum
Nuclear envelope
Nucleolus
Pores in nuclear envelope
Cell membrane
Rough endoplasmic reticulum
Vacuole
Centrioles
Microtubules
Lysosome
Ribosomes and polysomes
Microfilaments

FIGURE 1.5. Cut-away diagram of a generalized animal cell showing typical cell organelles and subcellular components. Adapted from D. L. Hartl, *Our uncertain heritage,* 2nd ed. Harper & Row, New York, 1985. Used by permission.

of the embryo progresses, new processes and entities come into play.

Embryology's Methods

Embryologists seek answers to their questions in fundamentally different ways, and embryology is divided correspondingly into different fields: descriptive, comparative, experimental, and, recently, theoretical embryology. The information gathered in each field is nevertheless additive and serves as a cross-reference and check on information gathered in any other ways.

> To a great extent the success of biological investigations depends on the skillful combination of studies of living material with studies of dead material.
>
> Ernst Mayr (1963, p. 82)
>
> We murder to dissect.
>
> William Wordsworth, *The Tables Turned*

Descriptive embryology establishes embryology's facts. Descriptive embryology documents spatial and temporal changes during development. Descriptive embryologists ask "what, where, and when" questions. The answers are sought through verifiable and repeatable observations and are pre-

sented as description. This method of deriving knowledge from observation or sensory perception, known as **empiricism** or the **empirical method,** provides embryology with its core of facts (e.g., for human beings, see Moore, 1988).

Spatial changes in the course of development frequently come into focus at different **levels of organization** or levels of integration. These levels, like steps of a pyramid (Fig. 1.6), proceed upward from a base of basics to a summit of gross structures.

Levels of organization also correspond to levels of observations dictated by the different research instruments used to make observations. The light microscope lies at one end of the spectrum; the mass spectroscope lies at the other end, and many other instruments lie between.

Paul A. Weiss pioneered applying the concept of levels of organization to embryology. He put it this way: "the body appears as a *complicated system of hierarchies of different orders of magnitude each level of which has its own mode of organization*" (Weiss, 1939, p. 104, italics original).

Recently, Francois Jacob, a microbiologist turned embryologist, reiterated Weiss' concept: "Construction in successive stages is the principle governing the formation of all living systems, whatever their degree of organization. . . . It is thus

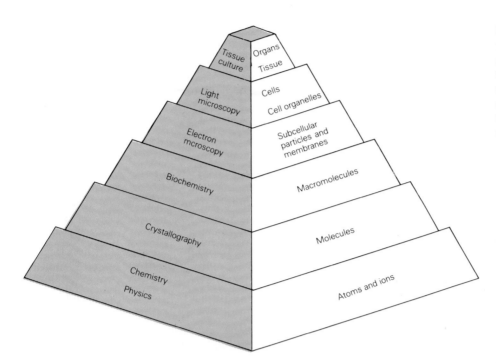

FIGURE 1.6. Levels of organization. One side of the pyramid shows levels in terms of the sizes of components (i.e., from atoms to organs). The other side shows the kinds of instrumentation used in research for the corresponding components.

by combining more and more elaborate elements, by fitting subordinate structures into one another, that complexity is born in living systems" (Jacob, 1976, p. 302).

As convenient as levels of organization are for embryology, they are also problematic, because of the limitations of the instruments employed for investigations at different levels. In practice, relating observations at one level to observations at another is often a matter of speculation. Rare successes in finding relationships between levels, like the relationship of DNA's structure to replication and transcription, are hailed as "breakthroughs."

Levels of organization are also problematic for theorizing about development. One can take an embryo apart by levels, but how do you put it back together again (Fig. 1.7)?

Temporal change is a difficult concept for everyone used to thinking of time as constant. Embryologists measure time on an "embryological clock" (e.g., Fig. 1.4) and describe the chronology of development in terms of conspicuous achievements or plateaus reached by the organism during its development. For the most part, embryology is concerned with three phases of development (preembryonic, embryonic, and postembryonic), each of which is further broken down into **stages.**

Like levels of organization, stages are not without their problems, and like stratifying embryos spatially, staging embryos temporally is not without its hazards. Stages should not be used to take the embryo apart temporally. They are chosen arbitrarily by embryologists for convenience or be-

cause of conspicuous features and should not become obstacles to thinking about developmental continuity or stumbling blocks to appreciating sequences in development.

> By this [comparative] method it becomes possible with greater or less certainty to distinguish the secondary from the primary or ancestral embryonic characters, to determine the relative value to be

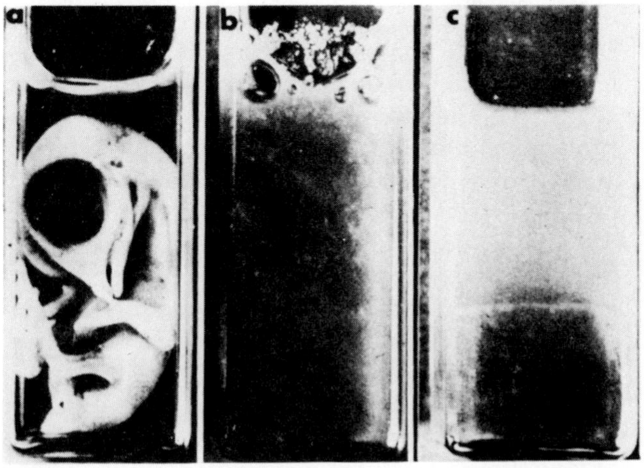

FIGURE 1.7. Tubes with chicks in various states of preparation for research. (a) Preserved fetal chick; (b) after homogenization (pureed in blender); (c) nomogenate after centrifugation for differential sedimentation of cell components. Reducing the chick to components is easy. Putting the components back into a chick is presently impossible, except for the egg. From P. A. Weiss, *Dynamics of development.* Academic Press, Orlando, 1968. Used by permission.

attached to the results of isolated observations, and generally to construct a science out of the rough mass of collected facts.

Francis M. Balfour (1880, pp. 3–4).

If the time ever comes when every step in the normal development of a single individual is known, the cause of development will not be far to seek.

Edwin Grant Conklin (1896, reprinted in Maienschein, 1986, p. 151)

Comparative embryology searches for remote links among embryos. Like the comparative anatomist, the comparative embryologist asks the "why" question and often performs analyses of similarities, especially those similarities that can be attributed to heredity.

That even remotely related organisms can inherit and share genes through the course of evolution suggests that similarities in development (e.g., Fig. 1.8) can be based on shared genes. Critics of this concept disparagingly call it the argument of biological inertia. The difficulty is that the roles of most genes in determining normal developmental pathways are almost totally unknown. Similarities in embryonic development therefore cannot be used as evidence for similarities in genes.

Supporters of the comparative method, on the other hand, are quick to show that one type of similarity is frequently related to another type. Biological generalities, after all, are not based on examining whole species but on sampling a portion of the whole. To the extent that genes are the cognates of structures, structures are the cognates of genes. Whether the extension of genes to structures is valid or a case of mixing "apples and oranges" must be determined specifically for every case.

Similarities in embryos can be both qualitative and quantitative, based on degree or number. Differences may represent ends of a spectrum filled by similarities rather than disconnected opposites. Comparative embryologists attempt to fill in intermediate cases or missing links and place different appearing embryos in a continuum of similar embryonic types.

Comparative embryologists also have an additional criterion for evaluating similarities: "Do structures seen in embryos of different species become comparable structures in the adults?" Only when structures in different embryos share similar developmental fates is the comparative embryologist likely to conclude that the organisms possess similar genes.

Development is synonymous with gene action.

John Tyler Bonner (1962, p. 77)

Experimental embryology searches for direct causes of development. It is the analytical branches of embryology devoted to ascertaining the mechanisms and controls of development (e.g., for the mouse, see Bürki, 1986).

The **experimental method** is a way of deducing the mechanical causes of events from the consequences of manipulating things or their parts. Causes can be mechanisms (i.e., how parts work together to produce an effect or function) or controls (e.g., how mechanisms are switched on and off, tuned in [i.e., modulated], or speeded up and slowed down).

An experiment is a manipulation, and data (pl., datum sing.) obtained by observing the results of a manipulation are experimental results. Educated guesses or plausible explanations about how a thing works are called hypotheses. A **working hypothesis** is a prediction about an experimental result.[1] The process of making predictions and then determining if they are borne out experimentally is called **testing a hypothesis.**

Although an embryologist may want to test a single favorite hypothesis, more often than not, several hypotheses must be tested at the same time. This is because more than one variable usually affects an embryo, and all the variables that can have a bearing on the results have to be considered in the complete experiment.

Ideally, each variable is tested separately (a situation that is hardly ever realized). The variables that one hopes are unimportant are called control variables or simply **controls,** and experiments done to test them are called control experiments. The variable of primary interest is the **experimental variable,** and the experiment done to test it, while sometimes considered the actual experiment, is really no more or less important than the control experiment.

When all the experiments are finally done and all the data gathered, the analysis begins. Frequently, rigorous statistical criteria are applied for testing hypotheses, and **negation,** or showing that the hypothesis is false (also known as falsification or disproof), is the route to an unambiguous conclusion.

Negation provides the objective and the efficient part of the experimental method, since to disprove a hypothesis is to do so forever. Hypotheses

---

[1]The term "working hypothesis" is sometimes used incorrectly to denote an overall plan of research.

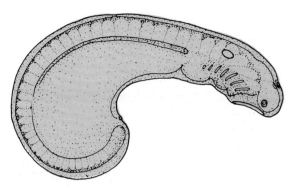

A. CYCLOSTOME (PETROMYZON)

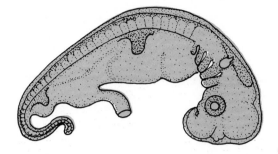

E. REPTILE (ANOLIS)

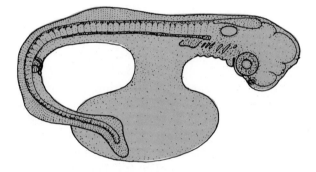

B. ELASMOBRANCH (SQUALUS)

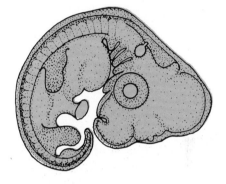

F. BIRD (GALLUS)

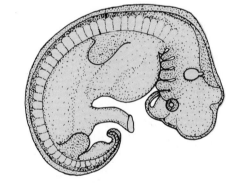

C. OSTEICHTHYES (FUNDULUS)

G. MAMMAL (SUS)

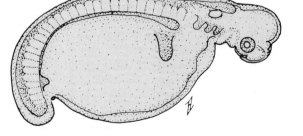

D. AMPHIBIAN (CRYPTOBRANCHUS)

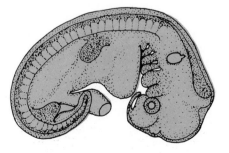

H. MAMMAL (HOMO)

FIGURE 1.8. Various vertebrate embryos at the organogenesis stage of development illustrating similarity of gross structure. From H. E. Lehman, *Chordate development,* 3rd ed. H. E. Lehman & Hunter Textbooks, Inc., Winston-Salem, 1987. Used by permission.

that are not eliminated by the results of an experiment are not proved, since they remain vulnerable to disproof in the next experiment. Even many results that consistently fail to disprove a hypothesis over many years do not prove it.

Consistency is not proof, but the importance of causal analysis is not a metaphysical assertion about proof and disproof. Deductions based on analysis are of value because they bestow the power to predict, and through that power science makes its contribution to society.

> Instead of searching for an illusory genetic program, developmental biologists must try to fathom the complex network of causal relations to which the sequence of developmental events owes its regularity.
>
> G. S. Stent (1985, p. 1)

Theoretical embryology searches for the overriding rules that govern development. In general, theories are rounded principles which incorporate diverse facts. Theories have explanatory value if they explain new or unrelated facts, and they have heuristic value if they provide incentives for further research. The cell theory, for example, states that organisms are made by cells. It led directly to an explanation of cleavage and inspired research on pathology.

Ideally, the theoretical method is comparable to the experimental method. Experimentalists state their hypotheses in terms of the physics and chemistry of things and test their hypotheses in the laboratory. Theoreticians state their hypotheses as quantitatively precise models or equations and test them with pencil and paper or with the help of computers.

Many embryological theories have catalyzed and spurred new investigations, inspired debate, and galvanized opinion, but they may be difficult to test or untestable. Rather than being vulnerable to negation, these theories tend to be amended when facts fail to conform with predictions.

Theoreticians interested in development usually use more or less qualitative or verbal models and derive quantitative or mathematical models secondarily. Ideally, testing a model involves determining how well it simulates development and mimics the results of experiments. At present, most models are still qualitative, and even quantitative models are not sufficiently precise to be eliminated or negated by the results of computation. Hopefully, this situation will soon change.

SEXUAL REPRODUCTION

> The problem is . . . how, while retaining the total potentialities of the egg, regional differentiation takes place to produce the different parts of the body. It is this latter problem that engages the serious attention of the student of embryology.
>
> T. H. Morgan (1927, pp. 6–7)

Situated between eggs and adults, embryos typically owe their past, present, and future to sexual reproduction. While in the conventional view of sexual reproduction, the embryo is seen as a new organism, in the embryological view, the embryo is also seen as a vital link in the chain of life.

Clarifying Terms

In ordinary speech, a generation is the time elapsing before children supersede parents. Biologically, a **generation** is a complete rotation of the life cycle (see Calow, 1978).

In contrast to a lifetime (Fig. 1.1) with its beginning, middle, and end, a **life cycle** is continuous. It connects the gametes producing the organism with organisms producing the gametes (Fig. 1.9). It consists of a haploid phase with cells containing a single set of unique chromosomes in their nuclei and a diploid phase with cells containing two sets of chromosomes in their nuclei. The connections, or nodes between the phases, are meiosis and fertilization.

While each chromosome in a haploid set is unique, each chromosome in a diploid set is similar to another chromosome. Pairs of similar chromosomes are called homologues (or homologs), each complete set of pairs representing the contribution of both gametes and hence both parents.

Reproduction within a phase of the life cycle is called **asexual reproduction.** Since the life cycle does not undergo a revolution, asexual offspring are members of the same generation as their parents.

Ordinarily, sex (attributed to Lat. *secare* to cut, referring to separate sexes) is identified with female and male organisms and with behavior related to sexual reproduction. Biologically, **sex** is the production of an organism by cells moving through the life cycle, and **sexual reproduction** is procreation requiring meiosis and fertilization (with certain exceptions, e.g., parthenogenesis found in some animals and plants).

Biologically speaking, sex is not necessarily reproductive. It might even be "deproductive," since,

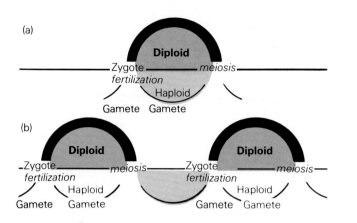

FIGURE 1.9. Generalized life cycle: (a) viewed as a cycle; (b) viewed as a wave. Diploid and haploid phases are connected by fertilization and meiosis.

at fertilization, two cells fuse to make one (i.e., spermatozoon and egg combine in the zygote).

Sex is only reproductive for two reasons. (1) Life cycles turn faster than lifetimes, allowing generations to accumulate. (2) A pair of sexually reproducing individuals may generate more gametes and more offspring than those required to replace the parental pair.

Differences between the common and biological definitions of sex and sexual reproduction are not really as great as they may seem. Biologists reduce the idea of female and male organisms to female and male sex cells, namely, to egg and sperm, or, more generally, to gametes. Sex is identified with meiosis and with the fusion of gametes at fertilization. Sexual reproduction is equated with procreation via gametes.

Sexual Reproduction: Turning the Life Cycle

Phases of the life cycle are defined by the number of sets of chromosomes present in a cell's nucleus. The number of chromosomes in a single set is conventionally designated n, and their unreplicated content of DNA is designated C. Cells in the **haploid** phase of the life cycle have n chromosomes with C DNA, while cells in the **diploid** phase have twice the haploid number of chromosomes or $2n$ chromosomes and twice the unreplicated haploid amount of DNA or $2C$ DNA. For example, for human beings, n is 23; $2n$ is 46. C is 3.2×10^{-12} grams (picograms, pg), and $2C$ is 6.4 pg.

More generally, a **hemizygoid** (or monoploid) number designates the number of chromosomes present after meiosis, and a **zygoid** number designates the number of chromosomes present after fertilization. In **euploid** species, the minimum number of chromosomes present in a single chromosomal set is equal to the haploid number, and

the number of chromosomes present in body cells or in germ cells prior to meiosis equals the diploid number. In **polyploid** species, the zygoid number is a multiple of the haploid number greater than 2, and in **aneuploid** species, the zygoid number is not a multiple of the haploid number.

Mitosis (*Gk.* mit- *thread* + -otic *adjectival suffix signifying action, referring to chromosomal movement accompanying nuclear division*) *or karyokinesis* (*Gk.* karyo- *kernel* + kinein *to move, referring to the division of nuclei*) **is the chromosomal and nuclear part of cell division.** **Cytokinesis** is the cytoplasmic part of the cell's division and ordinarily follows mitosis. Cell division is ordinarily called **mitotic division** to distinguish it from other forms of division not involving chromosomes.

The two cells produced as a result of mitotic division are called **sibling cells** (formerly daughter cells). These cells may divide in turn to produce two more sibling cells and so on. Cells dividing repeatedly exhibit a **division** or **cell cycle.**

At comparable points in a division cycle, sibling cells have the same n and C as the cell that gave rise to them. Chromosomal constancy is maintained, because the cells stick slavishly to the formula: one act of DNA replication and chromosomal duplication for each act of mitotic division (Fig. 1.10).

During the interval of the cycle called interphase, the cell's nuclear DNA, and hence the cell's genes, **replicate,** that is, are synthesized as exact copies. At the beginning of mitosis, chromosomes appear in the cell's nucleus. They consist of pairs of chromatids having identical molecules of DNA.

During mitosis, one chromatid from each chromosome is delivered to each sibling cell. The new cells' nuclei (their chromosomes, content of DNA, and genes) are qualitatively and quantitatively

Cell Division

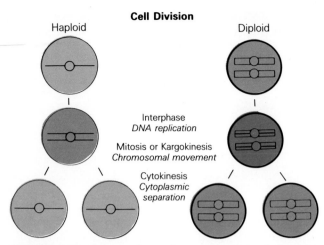

Diploid

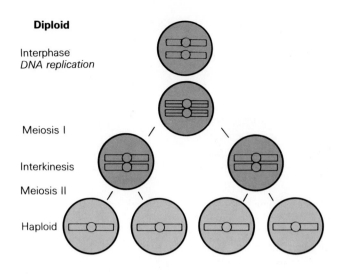

FIGURE 1.10. Scheme for ordinary cell division in haploid and diploid cells. One round of synthesis for each division. DNA synthesis (replication) occurs in interphase, and chromosomes containing duplicate chromatids undergo mitosis followed by cytokinesis. The circles are nucleated cells. The lines with dots are chromosomes consisting of one or two chromatids. Cells with two sets of unreplicated chromosomes are diploid. Those with one set of unreplicated chromosomes are haploid.

FIGURE 1.11. Scheme for meiosis. One round of synthesis for two rounds of division. DNA synthesis (replication) occurs in interphase, and chromosomes containing duplicate chromatids appear as meiosis I begins. Homologous chromosomes first pair and then separate at the first division. Interkinesis follows, and, at meiosis II, the chromatids of each remaining chromosome separate. The circles are nucleated cells. The lines with dots are chromosomes consisting of one to four chromatids. Cells with two sets of chromosomes are diploid. Those with one set are haploid.

equivalent to each other and to the nucleus of the parental cell prior to replication (Fig. 1.10).

Meiosis is the chromosomal and nuclear counterpart to fertilization. At meiosis, $2n$ and $2C$ are halved to n and C in contrast to fertilization where n and C are doubled to $2n$ and $2C$. Meiotic cells have their own formula for halving their chromosomal number and DNA content: one act of replication for two acts of division (Fig. 1.11).

The two meiotic divisions are called meiosis I and meiosis II. The interval between the divisions is called **interkinesis** (i.e., between movement). Replication does not take place during interkinesis, and, in some organisms such as the round worm, *Ascaris*, the chromosomes may retain their integrity during this period rather than become diffuse as they usually would during ordinary interphase.

Pairs of homologous chromosomes associate temporarily during meiosis I, and portions of chromosomes may be exchanged via crossing over. As a result of the recombination of genes, meiotic chromosomes represent new structures unlike those present prior to meiosis.

Representatives of each pair of chromosomes move to the cells formed by meiosis I, and chromatids from each chromosome move to the cells formed by meiosis II. After the second meiotic division, a single chromatid represents each chromosome in the nucleus of a haploid cell.

The nomenclature of mitosis, meiosis, and fertilization is rooted in history. Before the cytological events of meiosis were established at the close of the 19th century, August Weismann had predicted that a **reduction division** of chromosomes would take place in germ cells on their way to becoming gametes. Embryologists soon showed that, in round worms and sea urchins, meiosis was two divisions, not one, and controversy soared over which division should be designated the reduction division.

The retention of the term reduction division today is a tribute to the shrewdness of Weismann's deduction. It is used most often for the first meiotic division, since homologous chromosomes combine at this time and the number of physical chromosomes is reduced (Fig. 1.11).

The second meiotic division is often called the **equational division,** since the homologous chromosomes are said to split equally. Meiosis I and II are preferable to reduction and equational divisions, since numbers only identify sequence and avoid undue implications about the events.

Cells involved in meiosis are called **meiocytes** in general but oocytes and spermatocytes in animals. Cells produced following meiosis are called **spores** in algae and fungi and megaspores or microspores in plants. Spores are generally involved in asexual reproduction, and both megaspores and microspores undergo mitotic division. Animal cells produced by meiosis might be called **meiomeres** or spore-equivalent cells, although they typically do not undergo further division.

The differentiation of spores is called **sporogenesis.** Except in the case of conjugation (or fertilization) in some protozoans, nothing equivalent to sporogenesis occurs in animals. Instead, meiomeres go right on to differentiate as gametes.

Cells capable of undergoing fertilization are **gametes** or **sex cells,** and, for the botanist and mycologist, **gametogenesis** refers to the differentiation of gametes from spores or related cells already having a haploid nucleus. For animals, gametogenesis is defined more broadly to include all the processes leading to the production of gametes, from ordinary cell division in diploid progenitor cells, through meiosis, and finally to the differentiation of haploid cells into gametes.

Contemporary views of fertilization and sexual reproduction in multicellular animals have evolved from the work of Frank R. Lillie who directed the Marine Biological Laboratory at Woods Hole early in this century. In most male animals, each of the four haploid cells resulting from meiosis differentiates as a spermatozoon. In female animals, the meiotic divisions are grossly unequal, and only the largest of the four haploid cells goes on to form an egg. The remaining cells are abortive eggs called polar bodies.

The formation of the zygote (Gk. *zygon* yoked or paired referring to the fusion of gametes into a single cell) is dominated by two processes: (1) plasmogamy (the fusion of gametes' plasmalemmas) (Austin, 1974) or syngamy (Gk. *syn-* by means of + *-gamia* marriage) and (2) karyogamy[2] or amphimixis[3] (Gk. *amphi-* both + Lat. *miscere* to mix or breed; also called zygogenesis, meaning the production of the zygote but referring to its nucleus). The nuclei that fuse are called pronuclei, since they contribute to the formation of single diploid nuclei, or synkaryons, in later embryonic cells.

Sexual Reproduction of Multicellular Organisms

Multicellularity occurs in either or both the haploid or diploid phases of the life cycle. It refers

[2]Karyogamy is borrowed from developmental biologists who use it to identify the process of nuclear fusion following experimental cell fusion.

[3]Syngamy and amphimixis are often used interchangeably.

to the presence of multiple (i.e., upward from four) nuclei within an organism usually separated by cell membranes.

In most multicellular species, development depends on cells remaining together after division. Exceptions occur in the Acrasiomycetes, the cellular slime mold, such as *Dictyostelium*, and in the Myxomycetes, the plasmodial slime molds, such as *Physarum*. In these "fungal animals" amoeboid cells (myxamoebae) converge into organisms (pseudoplasmodia or plasmodia, respectively) (see Bold et al., 1987).

Typically, haploid cells of multicellular algae, fungi, and plants remain together after division to form **gametophytes,** a haploid multicellular organism that produces gametes (e.g., Figs. 1.12 and 1.13). Nothing comparable to a gametophyte occurs in multicellular animals, but a multinucleate form of some protozoans represents an equivalent to a gametophyte among unicellular animals.

Gametogenesis may occur in all the cells of a gametophyte or be limited to those cells present in gametangia (pl., gametangium sing.). In some algae and fungi, all the cells in a gametangium may be capable of becoming gametes, but, in others and in the flowering plants, gametangia include sterile cells that do not themselves undergo gametogenesis.

The diploid cells of multicellular algae and plants form a **sporophyte,** a diploid organism that produces spores (Fig. 1.13). In the case of large multicellular plants, such as the flowering plants, the

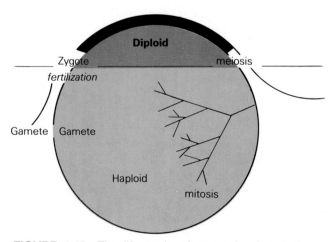

FIGURE 1.12. The life cycle of some fungi and algae (sometimes called the haplontic life cycle). Meiosis occurs in the zygote and mitosis in haploid cells. The products of meiosis undergo cell division to produce a haploid multicellular organism (a gametophyte). Haploid spores produced through sporogenesis reproduce the species asexually.

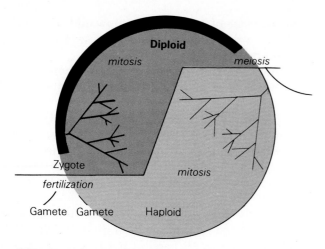

FIGURE 1.13. The life cycle of multicellular plants called the alternation of generation. Mitosis occurs in the zygote and spore. Gametogenesis is delayed, while a haploid organism (a gametophyte) is produced, and meiosis is delayed, while a diploid organism (a sporophyte) is produced. The life cycle is diplobiontic when both types of organism are independent or monobiontic when one type is permanently a part of the other.

sporophyte consists of leaves, branches, stems, and roots, but in algae, the sporophyte body or thallus (Gk. *thallos* stem or sprout) is usually simpler. No convenient term is available for the diploid, multicellular animal's equivalent of the sporophyte.

Multicellular diploid algae and plants develop from an embryo defined as the stage in the life cycle lacking morphologically distinct organs.[4] The word "embryo" is not applied to the early stage of multicellular haploid organisms developing from the spore. The early haploid organism may be called a **protonema** (e.g., as in ferns) for its stringlike appearance or a **prothallus** before gametangia appear.

Since, with few exceptions (e.g., the parthenogenesis of some insects), multicellular animals are not formed from originally haploid cells, animals have only one developing form that qualifies as an embryo, namely, the diploid embryo formed from the zygote. Animal embryology is the development of the multicellular animal via sexual reproduction and the zygote. **Normogenesis**[5] is the all-inclusive term for the development of multicellular organisms in either phase of a life cycle, whether via a

[4]Terrestrial plants are sometimes called embryophytes, since they develop from embryos. The part of their early embryo giving rise to the plant proper is called a proembryo.

[5]Normogenesis is also defined as the average developmental process in a species (*Stedman's Medical Dictionary,* 23 edition).

zygote or not, and whether via asexual or sexual reproduction.

Different Types of Life Cycle

In the course of evolution of different species, either the haploid or diploid phase of the life cycle became dominant. As a rule, the diploid phase became dominant for larger organisms and the haploid phase for smaller organisms.

This relationship of ploidy to organismic size is also reflected in cell size (Commoner, 1964; Szarski, 1976). The quantity of DNA in cells might have channeled or inclined species toward one or the other evolutionary path.

Multicellular plants and algae exhibit the greatest variation in their styles of life cycles. The chief variable is the occurrence of mitosis. When mitosis occurs exclusively in haploid cells, the haploid phase is extended (sometimes called the haplontic life cycle, Fig. 1.12). The zygote undergoes **zygotic meiosis** immediately after fertilization or, commonly, after a period of dormancy.

The **alternation of generation** life cycle has mitosis in both phases and a multicellular sporophyte as well as a multicellular gametophyte (Fig. 1.13). In **diplobiontic species** (two organisms), haploid and diploid organisms are separate, while in **monobiontic species** (one organism), the sporophyte and gametophyte are permanently fused.

The morphology of sporophyte and gametophyte in diplobiontic species may be similar (e.g., *Ulva*, sea lettuce) or different (ferns, brown and red algae). Sporophyte and gametophyte are always different in monobiontic species. In mosses, the gametophyte is the conspicuous plant and the sporophyte is a mere stalk, while in trachaeophytes, the gametophyte is reduced to a vestige of a multicellular organism contained within the sporophyte.

The trachaeophyte plants include the longest living and largest organisms ever to develop. These plants represent the sporophyte almost entirely. Meiosis is tightly linked to the formation of gametes (i.e., gametic meiosis), but a period of sporogenesis follows meiosis and leads to the development of **megaspores** and **microspores.**

The megaspore of angiosperms forms the female gametophyte, known as the **embryo sac,** and the microspore forms the immature male gametophyte, a **pollen grain** or pollen. An egg is present in the embryo sac, and a sperm nucleus is produced in a pollen tube developing from the pollen grain. A zygote is formed by the union of a sperm nucleus and the egg, while a nutritive, triploid endosperm

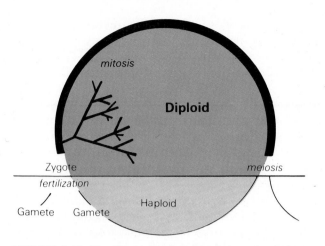

FIGURE 1.14. The life cycle of multicellular animals (sometimes called the diplontic life cycle). Mitosis occurs in the zygote, while meiosis occurs in specialized diploid cells belonging to the germ line. The products of cell division produce a diploid embryo which becomes a multicellular organism. Groups of diploid cells may reproduce the species asexually. Meiosis is limited to cells in specialized reproductive organs and is followed immediately by the differentiation of gametes.

is formed by the fusion of a second sperm nucleus and the nuclei of two polar cells belonging to the embryo sac. The fusion of sperm nuclei with two parts of the embryo sac constitutes the **double fertilization** of angiosperms, but only the zygote forms the embryo and the future sporophyte.

The life cycles of multicellular animals resemble the monobiontic life cycles of algae and plants with two exceptions. (1) The animal equivalent of the gametophyte is a single cell. (2) Animal cells preserving the ability to perform meiosis are often limited to **germ cells** and frequently belong to an exclusive **germ line.** Meiosis is delayed and occurs only in oocytes and spermatocytes destined to become the respective gametes (i.e., gametic meiosis, Fig. 1.14). Gametic differentiation accompanies or follows immediately after meiosis.

Mitosis in multicellular animals is found exclusively in cells of the diploid phase. Following fertilization, mitotic divisions provide the cells with which the embryo builds the adult. Meiosis will typically not proceed until sexual maturity and then only in cells of the germ line.

EMBRYOLOGY'S HISTORY

How, then, are the other parts formed? Either they are all formed simultaneously—heart, lung, liver, eye, and the rest of them—or successively, as we read in the poems ascribed to Orpheus, where

he says that the process by which an animal is formed resembles the plaiting of a net. As for simultaneous formation of the parts, our senses tell us plainly that this does not happen: some of the parts are clearly to be seen present in the embryo while others are not.

Aristotle, *Generation of Animals*

Embryology only became a recognized branch of the natural sciences in the second half of the 18th century. *Embryologie* was admitted into the French language by the Académie in 1762, and "embryology" entered English in the 19th century (Singer, 1959). Nevertheless, embryology's roots lie in antiquity.

Empirical Method and Progenitors of Embryology

Aristotle (384–322 B.C.) bequeathed embryology his empirical method and his epigenic vision. He introduced diagrams to document development and descriptions to narrate it, and from observations on embryos of the chick, cuttlefish, squid, and octopus, he showed that development was cumulative.

Naively, as it turned out, Aristotle also proposed that embryos arose *de novo* from totally unformed matter in each generation, and he was guilty of other errors, but embryology would have progressed rapidly had it proceeded in the direction Aristotle gave it. This was not to be. While research on reproductive physiology continued in Moslem cities in Asia, Africa, and Spain, elsewhere, embryology was uprooted and did not find fertile soil again for two millennia.

The return to Aristotelian empiricism and progress in embryology coincided with the development of the microscope (named by Johannes Faber [1574–1629]). The new instrument was seriously flawed and did not improve rapidly, but by the end of the 17th century, scarcely anything through which light could pass escaped microscopic examination. Moreover, the early practitioners of microscopy advanced what became a central premise of embryology—that macroscopic events can be explained by microscopic events.

In their advocacy of minute structures as instruments for the development of larger structures, the early microscopists spawned the preformationist model of embryology. Ultimately, the concept of preformed parts evolved into the present concept of genes bearing coded hereditary information, and preformation became neopreformationism, one of the pillars of modern embryological theory. In the

18th century, however, preformation was largely a matter of speculation and thus an impediment to empirical studies of epigenesis.

In the 19th century, preformationism was adopted by practitioners of two divergent philosophies. **Natural theology** and **Naturphilosophie** (nature philosophy), which were unrelated otherwise, derived their preformationist bias from Platonic **essentialism,** the belief that the essence of an organism was a type or theme (*Urbildung* or archetypes). Essences could be arranged in an Aristotelian-like *scala naturae,* and development proceeded along rungs of the *scala* according to a preordained (i.e., preformed) plan (Fig. 1.15).

Natural theologists were uncompromisingly fundamentalist and ardent creationists. They believed in an abundant, static world and a creative, all-powerful God. The job of the natural theologian was to reveal God's handiwork in nature. For example, embryos were seen as developing from lower to higher rungs of the *scala* as they moved as close to God as their species could take them.

In America, natural theology was best represented by the Swiss-born Louis Agassiz (1835–

FIGURE 1.15. ''Ladder'' development: development as progress toward perfection.

1910), who proposed from empirical evidence that the themes developing in organisms were also revealed among fossils. Intending to show God's plan through repetition, Agassiz interpreted the fossil record in terms of fetal stages of modern organisms.

Nature philosophy, on the other hand, did not invoke God to explain nature. A branch of the German romantic movement, nature philosophy was concerned with revealing nature's own general laws.

Organisms reproduced from a natural urge to return to origins and developed according to an equally natural urge to progress. Johann Friedrich Meckel (1781–1838) proposed that every animal began life at the same low level of existence (i.e., the egg) and climbed the rungs of the ladder toward a stable level, in the case of human beings, to the rung of near perfection at the top. According to Meckel and, later, the French nature philosopher, Étienne Serrès, the development of ''higher'' organisms paralleled the *scala naturae,* a parallelism sometimes known as the **Meckel–Serrès law.**

The nature philosopher and poet, Johann Wolfgang von Goethe (1749–1832), who coined the term **morphology** for the aesthetics of scientific essentialism, pondered how a limited number of natural themes could generate the almost endless variation of different species. Goethe sought to find basically constant parts that became modified by reproduction and was successful with his discovery of the intermaxillary bone in human beings. Previously thought to be absent, the discovery broke down the barrier and demonstrated continuity between *Homo sapiens* and the anthropoid apes.

The foremost nature philosopher, Lorenz Oken (1779–1851), reversed the process of comparing structures in different organisms to comparing different structures in the same organism. His concern with segmentally arranged structures led him to theorize that the skull was derived from fused vertebrae. While incorrect in detail, Oken's concerns are echoed today by research on vertebrate somites (see Bellairs et al., 1986) and arthropod segmentation (see Scott and O'Farrell, 1986).

Nature philosophers tended to broaden embryological investigations and correct the static views of preformationists with a more dynamic portrayal of development. Moreover, nature philosophers contributed the concept that organisms on any rung of the *scala* were naturally related and thereby prepared later 19th century scientists to accept common descent and hence evolution (see Mayr, 1982).

On the other hand, much of the work of the

nature philosophers was fantastic if not ludicrous. Prone to draw farfetched analogies, nature philosophers continued to elevate generality above observation and to lead embryology into still more years of sterile debate.

> The greatest progressive minds of embryology have not searched *for* hypotheses; they have looked *at* embryos.
>
> Jane M. Oppenheimer (1968, p. 168, emphasis original)

Embryology's Growing Pains and the Comparative Method

Embryologists were slow to gather in defense of Aristotelian empiricism, but skirmishes occasionally took place between individual embryologists practicing the empirical approach and those engaged in speculation. Not surprisingly, the embryologists who returned to empiricism also turned away from preformation and toward Aristotle's epigenesis.

In the 19th century, epigenesis was the biological term for change, and embryologists studying epigenesis paralleled other scientists studying change in nature and society. They were all swept up by the great intellectual ferment of the time, to understand change, and their efforts produced the classic studies of both biological and social change.

In general, these studies relied on the comparative method to draw conclusions: if something was observed repeatedly where one looked, it would probably be observed under comparable circumstances where one did not look. The trick was to find similarities within the body of empirical data which could serve as the basis of generalities.

Deemed the parent of comparative embryology, Estonian-born Karl Ernst von Baer (1791–1876) was one of the most avid defenders of empiricism and epigenesis in biology. Among von Baer's accomplishments was the discovery of the mammalian egg. After his first sighting of the egg in the dog, von Baer extended his observations to cows, sheep, rabbits, deer, porpoise, dolphin, pigs, and human beings (von Baer, 1827, 1828). To say that he discovered the mammalian egg is not to say that he discovered eggs in every mammal, but he found eggs throughout his sample of mammals and concluded that they would be found in every other mammal as well.

Von Baer also saw the germ layers and the notochord in early chick embryos and found similar structures in mammalian embryos. He demonstrated the generality of germ-layer development in vertebrate embryos and guided embryology back to embryos and away from speculation.

Von Baer was not entirely liberated from nature philosophy, however, and remained interested in natural laws (or guiding metaphors, as the case may be). On the basis of his own detailed microscopic studies of embryos, von Baer argued that embryos of different species resembled each other at comparable stages of development. Species spread apart, like spines of a fan, as epigenic changes accumulated (Fig. 1.16).

Von Baer's version of epigenesis shifted the basis of comparison from vertical to horizontal, from progress in development to comparisons at stages of development. Although the amount of change occurring between an embryo and the adult varied in different species, the adult of one species never resembled the embryo of another species more than the embryos of the two species resembled each other (e.g., Fig. 1.17).

Embryology might have been aided had embryologists adopted von Baer's model, but, in the competition for general acceptance, the "fan" lost out to the "ladder." Von Baer was subject to personal attacks, and by the end of the century, his version of epigenesis was deemed "impossible" by such leading embryologists as August Weismann.

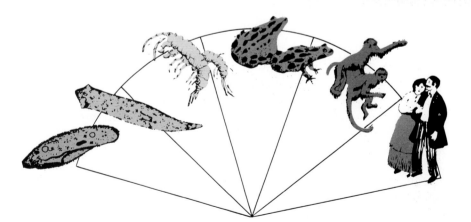

FIGURE 1.16. Von Baer's "fan": development as the accumulation of differences.

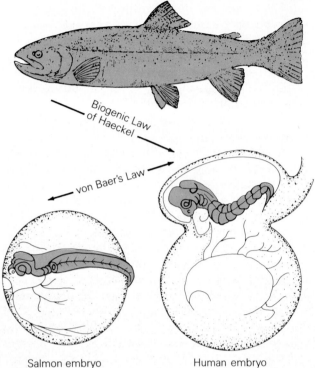

FIGURE 1.17. Comparing the embryos of a salmon and a human according to von Baer and the adult salmon and a human embryo according to recapitulation.

Of course, the pendulum also swings back, and today the comparative method is used widely and epigenesis has new advocates in molecular biology. Even the Nobelist, Francois Jacob, referring to changes in surface antigens in human and mouse embryos, has acknowledged von Baer's priority:

> These findings, in a way, are a mere confirmation, at the molecular level, of Karl von Baer's observation that, among related organisms, early stages of development remain similar, differences showing up at later stages. (Francois Jacob, 1979, p. 134)

Embryology's Adolescence and Evolution

Many theories about change emerged in the mid-19th century, and several of these had a great deal in common with epigenesis. Most conspicuously, Charles Darwin's (1809–1882) concepts of fortuitous variations and natural selection paralleled epigenesis.

Darwin argued in *The Origin of Species by Means of Natural Selection or the Preservation of Favoured Races in the Struggle for Life* that different forms of life were produced by blind luck coupled with selection. His thesis would ultimately

change how human beings thought about life, but it was unacceptable at the time in large segments of staid English society (see Bowler, 1988).

Convinced of God's largess and devoted to a sense of purposefulness in life, few of Darwin's compatriots adopted his view that chance was the motor of change in nature. The Lamarckian view—that evolution was progressive—was much more congenial. It fit the individualistic and competitive character of Victorian society and justified imperialism and capitalism. While Darwin might have wavered, the best efforts of his friend and admirer, Thomas Henry Huxley (1825–1895), could not turn Darwin around on the issue of progress in human society and ethics (Helfand, 1977).

Darwinism was also weakened by its lack of theoretical underpinnings. Without the gene as it is known today and mutation as an instrument for changing genes, Darwinism could not explain the origins of variation. As a consequence, by the end of the century, Darwinism was in retreat throughout the English-speaking world.

In the revolutionary centers of central Europe, however, accidents frequently provided opportunities for progress, and chance seemed as likely a wellspring for new species as for better societies. For many Europeans, Darwin held up a beacon that cast a natural light on revolutionary aims for society. Among these Europeans were biologists, especially those in the new fields of embryology, histology, and cytology. They rallied under the banners of republicanism and evolution.

Ernst Heinrich Phillip Haeckel (1834–1919) led the defense of evolution in Europe (especially, Haeckel, 1892). Borrowing the term **biogenesis** from Huxley (who coined it to mean the origin of living things from previously living things), Haeckel drafted the **biogenic** (or **biogenetic**) **law.** In its succinct and catchy form, the law states that, "ontogeny epitomizes phylogeny." "Ontogeny" (Gk. *onta-* the thing or individual which exists + *-genesis*) refers to the development of the individual; "phylogeny" (Gk. *phylon-* race + *-genesis*) refers to the evolution of races or higher levels of classification (e.g., phyla), while "epitomizes" refers to an abridgment or condensation of the whole.

The word "recapitulates" is sometimes substituted for "epitomizes," and the law is referred to as Haeckel's law of recapitulation. In any case, the law finds in an individual's development a shortened version of its species' evolution.

Haeckel's contribution to biology in general and to embryology in particular was enormous. He was a prodigious coiner of terms, and much of what passes as traditional biology was actually of Haeckelian origin.

Haeckel's view of organisms as historical beings continues to provide the explanation for the retention (sometimes called conservation) of primitive traits within the embryos of advanced organisms. Above all, his emphasis on embryos as sources of systematic and evolutionary information is completely vindicated.

Other parts of Haeckel's work are more controversial. Haeckel realized that the Meckel–Serrès law of parallelism was tailor-made for evolution. With a stroke of the pen, he substituted plateaus of evolution for rungs on the *scala* and provided pedigrees for species in place of paths of progress.

According to Haeckel, the evolution of ancestral adults was the cause for the development of today's embryos. Only the most skillful scientific sleight of hand could pull adult ancestors out of embryos (e.g., see Fig. 1.17), but Haeckel's skills as artist, author, and orator were more than adequate. His success removed evolution from the "endangered concepts" list but placed embryology in jeopardy as a science (see Nordenskiöld, 1929).

With respect to the proposition that evolution is the cause of development, Haeckel's biogenesis is not a law but a theory. It had enormous heuristic value, and, like a theory, it could be amended to suit data (rather than negated by contrary data). It launched many embryologists on careers of describing the details of development but failed to be overturned despite the accumulation of evidence against it.

Even Darwin, whose standing as a scientist was most affected, did not support the biogenic law. Tenaciously, he rejected the comparison of embryos and adults and even deleted a reference to Agassiz, the author of the first comparison of embryos and fossils, from the last edition of *Origin*, thus removing the taint of recapitulationism from the seminal work.

In this respect, Darwin was far ahead of his time. In fact, most embryologists did not join him in renouncing the biogenic law until the mid-20th century (Churchill, 1984).

The recognition of the advantage of applying the method of experiment to problems of development may be said to have begun with Wilhelm Roux in 1883. Today the need of this procedure seems so obvious that we are apt to forget the strong opposition that the movement at first met from embryologists of the old school, who had already

developed a philosophy dealing with the developmental process as a series of historical events.

Thomas Hunt Morgan (1927, p. 1)

Maturity and Emergence of Experimental Embryology

The idea of testing hypotheses went back to the English Franciscan monk, Roger Bacon (1214–1294), but did not come into widespread use until the late 16th century when scientific societies appeared and issued journals for publishing the results of experiments. In the 17th century, Francis Bacon Lord Verulam (1561–1626) promoted the "let's see what happens if" version of the experimental method, and under his tutelage, if not his example, experimentation and mechanistic models changed how scientists thought and worked.

Embryology was influenced by no less a mechanist than René Descartes (1596–1650). Although not personally familiar with embryological material, Descartes championed the hypothesis that embryos worked like machines, and, like machines, embryos could be analyzed. Descartes failed, however, to establish embryology as a mechanistic science.

The question of when the embryo acquired its soul blocked the advance of mechanism in embryology. The soul was placed outside the scope of science by religious authorities, and the 17th century was not, in general, a healthy time to challenge established doctrine. It was the century in which the great mechanist, Galileo (1564–1642), chose recantation to torture, and other mechanists chose silence to censure. Even Descartes agreed to leave the soul outside the laboratory.

By the end of the 17th century, the gulf between mechanistic thinking and embryological theory had widened. Preformation and nature philosophy fell into the void, but, by the late 19th century, embryology was ready to emerge.

The lifeline that came to embryology's rescue was thrown by Wilhelm Roux (1850–1924). The experiments that Roux performed were not especially difficult or done particularly well, but the results he published in the last decades of the 19th century showed that "the embryo could be grappled with experimentally" (Oppenheimer, 1967, p. 163). His real accomplishment was to demonstrate the efficacy of experimentation for answering questions about development.

Roux coined the word *Entwicklungsmechanik*, or "developmental mechanics," to embody the mechanist approach he pioneered, and, in 1894, he founded a new journal to ensure the publication of reports on mechanisms of development and the experimental analysis of embryos and other developing systems. Roux' journal, _Archiv für Entwicklungsmechanik der Organismen_ (_Archives for Developmental Mechanics of Organisms_), now known as _Roux' Archives of Developmental Biology_, is today one of the longest continuously published journals in biology. For all these accomplishments, Roux became known as the parent of experimental embryology and the first of the "new" generation of embryologists.

In the early days of the 20th century, disillusioned with squabbles over recapitulationism, embryologists turned to experimentation with its prospect for objective and verifiable explanations of development. Interest in mechanistic models of living things increased, and experimental embryology became the springboard for experimental biology.

_____

It will be necessary . . . _to make as much or even more use of hypotheses, as physicists and chemists are compelled to do_ when they cope with the fundamental processes of their respective sciences. And just as in these sciences, we shall have to regard those assumptions as approximating most nearly to the truth which explain the most facts and permit of the successful prediction of new facts; and _ceteris paribus_ we shall prefer that explanation which appears to be the "simplest," not forgetting, however, that we may easily fall into error on this point. . . ."

> Wilhelm Roux (1894, translated by Wheeler in Maienschein, 1986, p. 126, emphasis in original)

Embryology is, in essence, a physiological science; it has not only to describe the building up of every single form from the egg, according to its different phases, but to trace it back in such a way that every stage of development with all its peculiarities appears as the necessary result of those immediately preceding.

> Wilhelm His (1878, translated by Russell, 1930, p. 95)

Embryology's Adulthood: Reductionism and Developmental Biology

The Swiss embryologist and microscopist, Wilhelm His (1831–1904), like Roux, was dedicated to mechanism. Self-consciously imitating the methods of chemists, engineers, and physicists in his study of development, His brought **reductionism** to embryology.

Reductionism, or the application of an exclusively mechanistic approach to science, focuses experimental analysis on crucial or control points in complex operations. By narrowing the sights of mechanism, reductionism increases its efficiency, and by focusing on specific solutions to particular analytical problems, reductionism increases mechanism's utility.

When embryologists adopt reductionism, they are often accused of dealing with oversimplifications, producing auxiliary work, or merely building apparatus (Platt, 1964), but their choice of methods has been vindicated many times by accomplishments. After the microscope was brought to its present state of optical excellence, reductionists were the first to recognize the chromosomes as vehicles for developmental and hereditary factors. These scientists were also among the first to appreciate the vast implications of the gene concept for development, and, in pursuing implications of the gene, they created developmental biology.

Developmental biology is a chimera, a science in which biochemistry, molecular biology, and microbiology are grafted to experimental embryology. Today, developmental biology reigns supreme, chalking up one success after another.

Genes, above all, have been fruitfully analyzed by developmental biologists. The discovery of the structure of DNA and the promulgation of the "central dogma" that DNA codes for RNA and RNA codes for protein turned the gene into a reductionist's dream come true.

The chief concept of developmental biology is a molecular "house that Jack built" called self-assembly in which events at one level of organization determine events at the next higher level. The amino acid sequence or primary structure of polypeptide chains produces the helices and pleats of secondary structure, which in turn produce the twists and bends of tertiary structure, and even the conformational surfaces of quarternary structure that allow one polypeptide chain to interact with another and determine function, whether structural or enzymatic.

The successes of developmental biologists are not limited to the molecular level of organization. Success has also been achieved in research at the cellular level of organization. Moreover, tissue culture and other _in vitro_ research techniques make it possible to reduce questions about the supercellular level (e.g., about the adhesiveness and movement of embryonic cells) to experiments.

The triumphs of reductionism were not achieved without losses, the greatest of which was to the breadth of embryology's vision. In the "old days," wonder was a sufficient reason to look at developing eggs. For some embryologists, asking "who, what, where, and when" questions provided the motivation for making painstaking and detailed observations on development.

For other embryologists, "why" questions were the only ones worth asking, and ultimate causes were the only ones worth pursuing. The embryologist's role was to solve the mysteries of development not the technical problems of performing experiments. Depending on the answers and on one's point of view, the embryo might "justify God's ways to man" or free the spirit from the grip of religious tyranny.

In the 20th century, curiosity about development itself is no longer sufficient justification for embryological research. The day of peering through a microscope for the sheer joy of watching an embryo develop is gone. Reductionism does not have a place for this kind of "research," but an alternative to reductionism is on the horizon.

The alternative is **holism,** an approach to science focusing on the whole object in its setting. For the present, holism's successes are in the realm of theory.

Conrad H. Waddington first pushed embryology toward a modern theoretical position. He transformed **epigenesis** to **epigenetics** and placed the gene squarely in the center of forces shaping the embryo, but he erased the lines that separated genes from each other and other cellular entities. From his point of view, development operated through switches, like those in a railroad marshaling yard or forks in the valleys of an **epigenetic landscape.** Access to any developmental path depended on paths already taken and available branching points.

Imagine that genes lie under every developmental switch or path and that biochemical products of genes determine the direction taken by cells, tissues, organs, and ultimately the organism throughout its development. The cells moving down a path find themselves in a biochemical environment in which the future is **channelized** by multiple decisions. At each junction, embryonic cells move closer to particular final destinations and, at the same time, farther from alternative destinations. Thus, development proceeds both toward and away from different paths, and cells undergo restriction as well as progress.

According to Waddington, equations for the dynamic aspects of development could describe the sequence and timing of switches that determine each cell's movement down one of the many tracks of cellular differentiation. None of these tracks can be totally isolated, and each is influenced by many switches. In Waddington's words (1962, p. 45):

> any general theory of development should envisage each pathway of histogenesis, or creode, as the resultant of essentially all the available gene-action systems whose intensities are mutually adjusted by interlocking control systems.

Today, many conceptual or qualitative models of development share features of the epigenetic landscape, although with less emphasis on the gene. The future will tell whether any of these models is an accurate way of describing embryonic development holistically.

Chapter

2

EMBRYOS AND ADULTS

*T*here are three reasons for looking at adult organisms before examining embryos in detail: (1) to establish the multiple, intimate, and complex relationships connecting embryos and adults through the life cycle; (2) to provide a retrospective view of the embryo in terms of what it will become; and (3) because the longest journey seems shorter when the destination is known. Chapter 2 examines these aspects of embryo–adult relationships at three levels of organization: the organism, the tissue, and the cell.

THE ORGANISM

Embryos are sometimes portrayed merely as part of the developmental history of adults, but the embryo's role in producing the adult may be much broader. Likewise, adults are sometimes portrayed merely as sources of gametes for the embryo, but adults play much larger roles in the histories of offspring. Continuity between adults and embryos dictate that relationships extend over the range of living activities from behavior and ways of making a living to organization and ways of structuring an organism.

Behavior

Biologically, behavior refers to an organism's activities. Behaviors are characterized activities, frequently divisible into more refined behavioral items (e.g., aggressive behavior in chimpanzees include piloerection and swaying as well as lunge and bite). An ethogram is the complete list of an organism's behaviors.

Generally, behaviors are directly relevant to the survival of an individual at the time the behavior is performed (e.g., appetitive behaviors leading to drinking and eating). Play behaviors and behaviors involved in reproduction, on the other hand, do not seem to benefit the individual at the time and may actually pose a threat to the individual's survival. These behaviors seem to anticipate requirements at some other stage of life: in the case of play, a later stage; in the case of reproduction, an earlier stage not occurring until the next generation.

Several larval and embryonic behaviors also seem to function not so much in the survival of the larva or embryo but in the survival of the adult. The excessive eating or gorging behavior of caterpillars, for example, exposes caterpillars to enormous risk from predatory birds. The caterpillars hardly need all they eat. They exhaust themselves eating. The payoff for the behavior only comes later, when, equipped with requisite bulk and resources, the caterpillars metamorphose into reproductive adults.

Larvae may also play a role in habitat selection for the adult. The mobile ascidian tadpole (see Fig. 5.15), for example, selects the home for the sessile adult.

Equipped with sensory cells, a rudimentary nervous system, and muscle bands arrayed along a notochordal rod, the ascidian tadpole is capable of taking cues from its environment and moving directionally. Soon after hatching in the morning, it swims toward light, while later the same day, it swims away from the light. This change in behavior tends to bring the tadpole to the shaded undersurface of fixed objects where it attaches by adhesive papillae and proceeds to metamorphose into the adult.

Read backward, the behavior of the tadpole makes eminently good sense. The object selected by the tadpole for attachment shields the adult against falling detritus and excess sunlight while providing a stationary substratum in the midst of floating food. What does not make sense is why the tadpole exposes itself to the hazards of life in the plankton.

Adult behaviors also bear on the survival of offspring. For example, dispersal behavior, or movements away from members of the same species, as seen among adult anemones, relieves competition in areas and offers the space required by offspring. Similarly, brooding, nesting, retrieval, and nursing behaviors fall into the category of parental care behaviors. Each of these parental investments in offspring involves a risk or loss on the balance sheet of energy and individual survival.

The most obvious example of behavior with little or no advantage to the adult is reproductive behavior or behavior that brings potential mates together and allows them to copulate or release gametes under conditions likely to result in fertilization. Adults frequently invest enormous amounts of energy and make heroic sacrifices, including exposing themselves to hazards and death, while performing reproductive behaviors (e.g., the nuptial migration of Pacific salmon).

On the surface, none of these behaviors makes sense, since they do not promote the survival of the individual. Life is not an individual, however. It is a continuity of cycles, and organisms in the various stages of life are only parts of that continuity. Adaptation therefore cannot be measured by the fitness between the individual and the environment. It must be measured by the fitness of life cycles and the environment. Many behaviors associated with reproduction diminish an individual's chance of survival but augment its **inclusive fitness** or likelihood of introducing its share of the gene pool into the next generation.

Organization

The first thing one notices about almost all organisms is that they are not amorphous but have characteristic form. They have symmetry, structure, and a body plan.

**Morphological symmetry specifies correspondence in the profile of an organism and its parts.** Symmetry is defined operationally by rotation rather than by sets of values. If an object does not appear to change when rotated on every axis, it has **spherical symmetry** (e.g., a ball). If the object does not change when rotated 180 around two axes but changes when rotated around all other axes, it has **biradial symmetry** (an American football); and if the object changes when rotated along all but one axis, it has **radial symmetry** (a shuttlecock).

An object has **bilateral symmetry** (i.e., two-sided symmetry, see Figs. 2.1 and 2.2) when a median longitudinal or midsagittal plane separates lateral right and left sides. Other longitudinal or sagittal planes run lengthwise through either the right or left sides. Frontal sections (called coronal in humans) divide right and left sides into dorsal and ventral regions. Transverse sections (also called cross sections) are perpendicular to both sagittal and frontal planes and divide right and left sides into anterior and posterior parts.

In general, anterior means leading, while posterior means following. Ventral refers to the surface bearing the mouth and dorsal to the backside or surface opposite ventral. Neither the anterior and posterior nor the dorsal and ventral parts are equal in bilaterally symmetrical organisms the way right and left sides are equal.

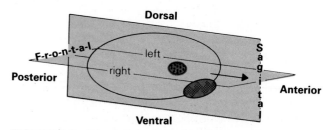

FIGURE 2.1. Drawing to illustrate planes, sides, and aspects of a bilaterally symmetrical organism.

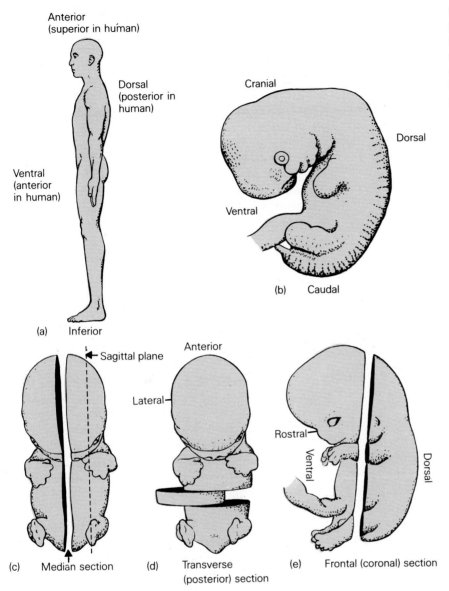

FIGURE 2.2. Drawings of adult (a), embryonic (b), and fetal (c–e) human being illustrating and naming axes and planes of symmetry. From K. L. Moore, *The developing human,* 4th ed., W. B. Saunders, Philadelphia, 1988. Used by permission.

Traditionally, another set of terms is used to describe the symmetry of bipedal animals such as human beings (Fig. 2.2a). Anterior still refers to leading, but bipedal organisms lead with their ventral surface. Likewise, the dorsal surface trails in bipedal organisms and is called posterior. Superior, or atop, replaces anterior, and inferior, or below, identifies the part of the posterior surface closest to the ground.

Generally, any line between geometrically opposite parts of an object is an axis, and axes around which rotation fails to change an object are **axes of symmetry.** In bilaterally symmetrical objects, a median line along the midsagittal plane running from anterior to posterior (or, in humans, from superior to inferior) extremes is called the longitudinal or anterior–posterior axis.

The right side is the lateral half seen when an object's dorsal region is upward, and its anterior end is toward the right. The left side is the opposite lateral half (Fig. 2.1). The right and left halves are equal but not exchangeable (i.e., not superimposable) and therefore are described as having **mirror-image symmetry.**

An axis or plane is **polarized** when its ends have opposite or contrasting properties. **Polarity** defines the direction along the axis or plane in which properties change.

The spatial organization of a right and a left arm (Fig. 2.3) illustrates polarity and mirror-image symmetry. Each arm has leading and trailing edges (equal to anterior and posterior edges) and dorsal and ventral surfaces paralleling those of the body, and each arm is organized along a polar longitudi-

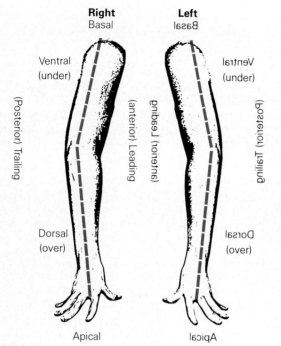

FIGURE 2.3. Schematic drawing of right and left arms illustrating their opposite polarity and mirror-image symmetry.

nal axis or apical–basal axis (apical referring to the farthest point from the body and basal to the closest point). The arms are equal in the sense of length and appearance, but the polarities of the apical–basal axes lie 180° apart, and that makes them nonsuperimposable.

In the course of development, the organism does not produce an arm or even two arms but two polarized arms. Each develops in the correct place, emerges on the opposite side near the anterior end of the organism, and grows outward with anterior and posterior edges and dorsal and ventral surfaces in parallel. Embryology must account for the origins and development of the arms' polarity and symmetry as well as for their tissues and anatomical parts.

Structural anatomy is generally considered the most stable of adult characteristics and is frequently made the basis for classification. Stable anatomical features appear in the embryo at about the gastrula stage and these generally provide the criteria for the larger categories of classification.

Multicellular animals, or Metazoa, are separated into sponges, or Parazoa, and all other metazoans, or Eumetazoa, on the basis of embryonic anatomy. Sponge embryos exhibit a peculiar way of turning inside out or inverting not seen in eumetazoan embryos. In addition, sponges have only vaguely organized tissues and lack specialized nerves.

The eumetazoans are placed into one of two categories depending on the maximum number of cell layers present in gastrulas (Table 2.1, Fig. 2.4). **Diploblastic** (also didermic and biepithelial) animals have gastrulas with no more than two cell layers. Corresponding to Cuvier's Radiata or Zoophyta and Lankester's Enterocoela, diploblastic animals are the coelenterates (Cnidaria and Ctenophora).

The second category encompasses **triploblastic** (sometimes endomesodermal) animals whose gas-

TABLE 2.1. Classification based on postgastrula embryos[a]

	Diploblastic	Triploblastic			
		Protostomes			Deuterostomes
		Acoelomates	Pseudocoelomates	Eucoelomates	
				Schizocoels	Enterocoels
Stomodeum formed first	na	na/+	+	+	−
Proctodeum formed near blastopore or equivalent	na	na	−	−	+
"Coelom" body cavity(ies) between gut and body wall (pseudocoelom) or	−	−	+	−	−
Cavity completely lined by mesoderm (true coelom)	na	−	−	+	+
Coelom formed by cavitation or	na	na	+	+	−
Coelom formed by evaginating gut	na	na	−	−	+

[a] +, Present or yes; −, absent or no; na, not applicable.
Source: Adapted from Hyman (1940).

Animals with diploblastic gastrulas

Animals with triploblastic gastrulas

Acoelomates   Pseudocoelomates   Eucoelomates

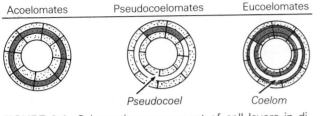

Pseudocoel   *Coelom*

FIGURE 2.4. Schematic arrangement of cell layers in diploblastic and triploblastic animals.

trulas have three cell layers, an outer ectoderm, an inner endoderm, and a middle mesoderm. Corresponding to Hatschek's Bilateria, triploblastic animals include all eumetazoans not classified as diploblastic.

One scheme for classifying triploblastic metazoans (attributed to W. Schimkevitch, see Hyman, 1940) features the cavity or embryonic coelom (also coelome, Gk. *koiloma* cavity) between the embryo's body wall and gut (Fig. 2.4). **Acoelomates** (e.g., Platyhelminthes and Nemertinea) have no coelom. **Pseudocoelomates** (equivalent to Aschelminthes, include Entoprocta, Rotifera, Gastrotricha, Nematoda, and others) have a pseudocoel not completely lined by a layer of mesoderm. True or **eucoelomates** (all other invertebrates and vertebrates) have an embryonic body cavity (not necessarily retained in the adult) completely lined by a layer of mesoderm.

The mode of origin of the coelom provides the grounds for making further distinctions. Named by T. H. Huxley, **schizocoelomates** (including the Mollusca, Echiuroidea, Annelida, and Arthropoda, among others) form coeloms by the cavitation (i.e., splitting) of a mesodermal or mesenchymal mass derived from the embryo's surface. Enterocoelomates (including the Chaetognatha, Echinodermata, and Chordata [Protochordata and Vertebrata]) form coeloms by evaginating (i.e., pocketing), folding, or splitting mesoderm derived from the embryonic gut (Table 2.1). Acoelomates and pseudocoelomates are allied with the schizocoelous eucoelomates by a variety of developmental criteria.

Another basis for classification is the mode of forming the mouth or stomodeum (also stomodaeum, Gk. *stoma* mouth + -*odeum* resembling

[as in odor, having the smell of]) at the beginning of the digestive system and the anus or proctodeum (Gk. *proktos* buttocks + -*odeum*) at the end of the digestive system. **Protostomes** (Gk. *pro* ahead or first + *stoma*) form their stomodeums around an early pocket and their proctodeums around a late pocket, while **deuterostomes** (Gk. *deuteros* second + *stoma*) form their proctodeums in the vicinity of an early pocket and their stomodeums around a late pocket. The deuterostomes correspond to enterocoelomates. The protostomes include all the other invertebrates (Table 2.1).

The body plan is a generalized pattern of internal structures and their external manifestations present throughout a phylum. Applicable to adult organisms at all stages, body plans emerge in embryos after the gastrula stage.

The body plan may emerge directly or indirectly. **Direct development** occurs where embryonic structures merge with adult structures. The unfolding of the body plan is equivalent to the early production of adult structures. **Indirect development** occurs where a post-hatching larval stage intervenes between the embryo and adult stages. The body plan in the intervening stage may not include adult structures.

Internally, the body plan of enterocoels resembles the dorsal–ventral opposite of the body plan of schizocoels and their acoelous and pseudocoelous allies (Fig. 2.5 upper row). For example, in vertebrates, as representatives of enterocoels, the nervous system is organized as a dorsal tube

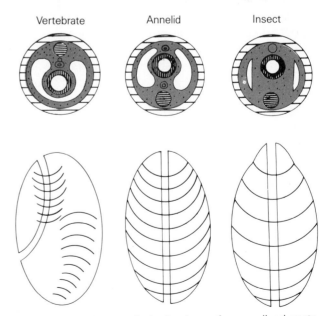
Vertebrate   Annelid   Insect

FIGURE 2.5. Schematic body plans of generalized vertebrate, annelid, and insect.

(neural tube) overlying the major distributive blood vessel (aorta). In contrast, in annelids and insects, as representatives of schizocoels, the nervous system is organized as ventral cords with accompanying ganglia underlying the major longitudinal blood vessel. Major differences in the embryology of enterocoels and schizocoels arise during the development of dorsal–ventral polarity.

Externally, the body plan of many eucoelomates (mollusks being the conspicuous exception) is dominated by segmentation or, in the case of chordates, by partial segmentation (Fig. 2.5, lower row). Beyond the segmental arrangement of appendages, the surface corrugations of larval schizocoels reflect internal partitioning. Muscle is concentrated in discrete bands of somites separated by mesenteries stretching across the entire animal and separating complete segments. Nerve ganglia and nephridia (functioning in fluid balance) are arranged segmentally.

The main part of vertebrates' internal cavity is not segmental and nephrons (functioning in fluid balance) are not grouped by segments. The partial segmentation of vertebrates is mainly a reflection of somites and does not extend to partitions across the body cavity. Anticipated by paraxial mesoderm (on either side of the dorsal midsagittal line) and by somitomeres along the length of the embryo, vertebrate somites emerge in the postgastrula behind the head and give rise to segmentally arranged vertebra, ribs, muscles, and connective tissue.

In addition, the neural tube at the rear of the brain may be partially segmented (i.e., divided into neuromeres), and neural ganglia are arranged segmentally in the neck and thorax. Anteriorly, the branchial or embryonic pharyngeal region is partially segmented into branchial arches some of which develop gills in fish and amphibians.

Developmentally, great differences appear in both segments and partial segments, usually with the greatest amount of differentiation occurring in anterior segments (Fig. 2.5 lower row). In annelids, the first or prostomial segment contains the large cerebral ganglia. Pharyngeal segments enclose the pharynx. Trunk segments follow; the anterior ones may be specialized for reproductive functions, while the posterior ones enclose the anus. Advanced annelids (e.g., oligochaetes) superimpose a variety of structures on the underlying segments, and related phyla (e.g., leeches) add other structures while erasing segmentation.

In insects, the specialization of segments is far greater than in annelids. A procephalon begins with a labrum and continues as the enlarged cephalic lobe with eyes and antennal segments.

The mouth follows accompanied by segments bearing pairs of leglike mandibles, maxillas, and labia comprising the gnathocephalon or eating head. In the adult, all these segments may be enclosed in a head capsule obscuring the original segmentation.

The thorax follows with three segments: prothorax, mesothorax, and metathorax, each with a pair of legs. The mesothoracic and metathoracic segments may also have wings. Eight to 11 abdominal segments follow, ending in the anus.

In vertebrates, such as terrestrial tetrapods, the number of anterior segments is in doubt. Segments may have condensed or divided, but they are obscure in the dorsal and branchial parts of the head. In mammals, four occipital segments seem to compose the bottom of the skull; eight cervical segments compose the throat, and 12 thoracic segments with ribs compose the thorax. The abdomen is supported by five lumbar and five sacral segments followed by five or more postanal, tail, or caudal segments.

TISSUES AS EXEMPLIFIED IN VERTEBRATES

Tissues are defined by both static and dynamic qualities. Statically, their main characteristics include intercellular relationships, extracellular spaces, and synthetic products. Dynamically, tissues differ in the size and activities of proliferating cell populations and in the frequency of cells leaving the tissue.

Classification of Vertebrate Tissues

Presently, four to six basic tissues are recognized, while the number of cell types is estimated at 210 (J. Lewis, cited in Slack, 1983). The four tissues are epithelial, connective, muscular, and nervous tissues. A fifth, vascular tissue (blood and lymph), is usually grouped with connective tissue, while a sixth, germ-line tissue, is lumped (inappropriately) with epithelial tissue.

Epithelial tissues or epithelia (Fig. 2.6) are characterized by maximal intercellular contacts, minimal extracellular material, and the presence of an underlying basal lamina (Fig. 2.7). Epithelia are typically surface tissues. They are said to hate a free edge and will stretch and migrate as if struggling to encompass a surface even when placed in a large tissue culture vessel. Embryonic cells linked together and covering surfaces may lack a

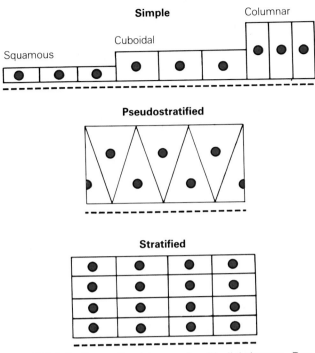

FIGURE 2.6. Generalized types of epithelial tissues. Rectangles and triangles, cells; circles, nuclei, dashed lines, basal lamina.

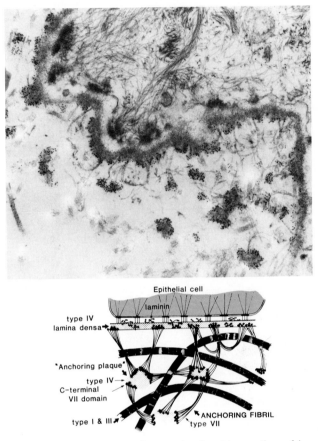

FIGURE 2.7. Electron micrograph of a thin section of human fetal foreskin (a) and schematic drawing showing base of epithelial cell resting on basal lamina and underlying basement membrane (b). The basal lamina consists of the densely stained lamina densa below and the electron-transparent layer above. A loosely organized basal reticulum (or reticular lamina) of thin collagen fibrils immediately below the basal lamina merges imperceptibly with the adjacent loose connective tissue and with more organized collagen fibrils. ×12,800. The locations of unbanded collagen (types IV, VII) and different types of banded collagen (types I, III) are indicated. Anchoring plaques (open arrows) and the lamina densa are stained densely with 5 nm colloidal gold coupled to monoclonal antibodies for an N-terminal globular domain of type VII collagen. Anchoring fibrils of type VII collagen (closed arrows) apparently link plaques to the lamina densa. From R. E. Burgeson, 1988. Reproduced, with permission, from the *Annual Review of Cell Biology*, Volume 4, © 1988 by Annual Reviews Inc. and by courtesy of the author. Electron micrograph courtesy of D. R. Keene. Used by permission.

basal lamina (e.g., the hypoblast of chicks) but still be considered epithelial, if only for want of another term.

The underlying basal lamina (Fig. 2.7), resolved by the electron microscope, and the **ground substance,** or extracellular material among epithelial cells, are amorphous. The basal lamina contains nonfibrillar **type IV collagen** (Table 2.2; also called amorphous and network collagen), which the epithelial cells probably secrete. The basal lamina also contains the glycoprotein **laminin,** which may function in attaching the epithelium to collagen (Terranova et al., 1980). In addition, the ubiquitous glycoprotein **fibronectin** is present.

The basal lamina is not the same as the **basement membrane** seen with the light microscope. Where present, the basement membrane represents the composite of a thin basal lamina and a thick **basal reticulum.** The fibrillar collagens of the basal reticulum are probably secreted by underlying connective tissue, not the epithelium, and are frequently tissue specific (Table 2.2) (Burgeson, 1988).

Epithelial cells (Fig. 2.6) may be **squamous** (scalelike, flattened), **cuboidal** (cubic), or **columnar** (tall, Fig. 2.8) or may lie somewhere in between. Epithelial tissues may be **simple** (consisting of one cell layer, Fig. 2.8), **stratified** (consisting of more than one layer, Fig. 2.15), or **pseudostratified** (having all cells joined to the basal lamina but nuclei

at different levels; see Fig. 26.2 for an example in embryos).

At their apical ends (i.e., opposite the basal lamina), epithelial cells frequently have **terminal bars,** as seen with the light microscope, or bands of **junctional complexes,** as seen with the electron microscope (Fig. 2.8), wrapped around the cells.

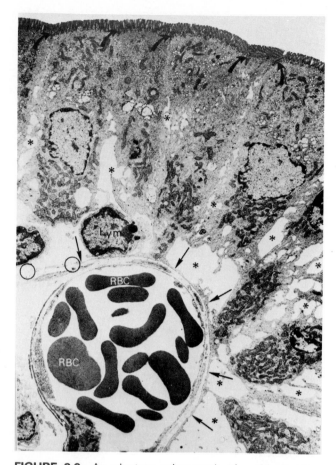

FIGURE 2.8. An electron micrograph of a thin section through a mammalian intestinal villus. Specialized cell contacts or *junctional complexes* appear among columnar epithelial cells (curved arrows). Note the more conspicuous spaces between cells (asterisks) near their bases compared to close contacts near the junctional complexes. Straight arrows point to the cell processes that form a lining over the basal lamina, and the circles show where this lining is broken by pores. × 2450. RBC, red blood cell in a capillary close to the intestinal epithelium; Lym, lymphocyte penetrating the intestinal epithelium. From E. J. Reith and M. H. Ross, *Atlas of descriptive histology.* Harper & Row, New York, 1977. Used by permission.

The complexes are further seen to consist of impermeable **tight junctions** or **occluding zones** (also **zonula** or **zona occludens**) and tenacious **zonula adherens**. Strengthening the bonds between cells are punctate **macula adherens** or **spot desmosomes**. Elsewhere, processes from adjacent cells may interdigitate across the intercellular space (Fig. 2.8), and **gap junctions** (Fig. 2.9), providing electrophysiological coupling, are frequently present.

Tissues in the early embryo are sometimes assigned to the epithelial category because the cells are, or soon become, broadly in contact with each other and, especially at their outer surfaces, form

junctional complexes. The designation of epithelium is a bit problematic, however, until blastomeres form a basal lamina.

When a *bona fide* basal lamina is present (e.g., in a blastula), the cells are legitimately epithelial, and the blastocyst may be considered an epithelial vesicle. Gastrulas generally contain epithelia, but **mesenchymal cells** are also common. Considered an embryonic form of connective tissue, mesenchymal cells have only limited points of contact with each other and are usually spread throughout a gelatinous matrix of extracellular material.

Embryonic tissues are not stable, and cells may move from epithelium to mesenchyme (i.e., de-epithelialization) and even back to epithelium. Gradually, cells acquire more stability. They diversify into tissues and differentiate.

Connective tissue (CT) cells are minimally in contact with each other, do not rest on a discrete extracellular membrane, and typically lie amidst more extracellular material or matrix (ECM) than cells in other tissues. Several types of connective tissue are distinguished in classical histology.

So-called **loose connective tissue** connects blood vessels and nerves to epithelial tissue. The lamina propria is the loose connective tissue layer beneath the basal lamina of mucus-secreting epithelia in moist membranes, and the outer portion of the dermis, or leather, is the loose connective tissue beneath the epidermis. **Dense connective tissue** connects bone to bone or bone to muscle and forms thick, tight bindings around bones (known as the periostium) and cartilage (known as the perichondrium, Fig. 2.10). **Bone** and **cartilage** form the skeletal elements of vertebrates. Fat or **adipose tissue** is also considered a form of connec-

TABLE 2.2. Some types of collagen and their locations[a]

Type	Alpha chains	Location
I	Two I-1 and one I-2	Most abundant fibrillar collagen; typically found in adult and embryonic dermis and bone
II	II	Cartilage; embryonic notochord, vitreous body of eye, corneal stroma (of chicken eye)
III	III	Embryonic tissues, blood vessels, skin, lung, liver, but not dentin (of teeth) or bone
IV	IV	Basal lamina (network forming or amorphous collagen)

[a]*Source*: Adapted from Burgeson, R. E., 1988. *Ann Rev Cell Biol,* 4:551.

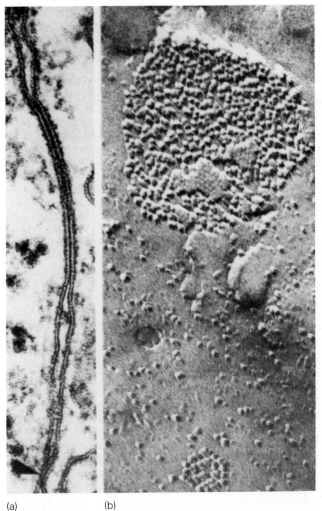

(a) (b)

FIGURE 2.9. Electron micrograph of a thin section (a: ×126,000) and a replica of a freeze-fracture plane (b: ×108,000) showing similar gap junctions (large one, top center; small one, arrows toward bottom). Aggregates of intramembranous particles bring plasma membranes close together and couple cells electrophysiologically. From N. B. Gilula et al., *Nature (London),* 235:262 (1972). Reprinted from *Nature* Vol. 235, pp. 262. Copyright © 1972 Macmillan Magazines, Ltd. and the authors.

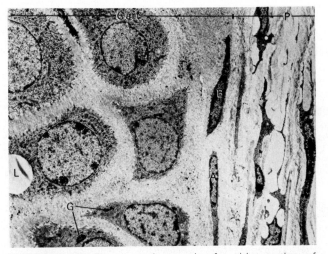

FIGURE 2.10. Electron micrograph of a thin section of mouse cartilage. Cells progress from fibroblasts (a) in dense connective tissue (P, perichondrium) to chondroblasts (b) and chondrocytes (c) embedded in the cartilagenous matrix (Cart). Glycogen (G) and lipid (L) may be present in the cartilage cells, while calcium deposits (Ca) accumulate in the matrix outside the cells. ×3600. From E. J. Reith and M. H. Ross, *Atlas of descriptive histology.* Harper & Row, New York, 1977. Used by permission.

tive tissue, since connective tissue cells frequently possess lipid globules (L in Fig. 2.10) that can accumulate in large amounts.

The most common connective tissue cells are called **fibroblasts.** They produce collagen of different types (Table 2.2): type I in the dermis and bone, type II in cartilage, type III in embryonic mesenchyme, blood vessels, and other extracellular fibrous material, and possibly other types. Fibroblasts involved in making bone become **osteoblasts** or **osteocytes** once they are embedded in bone. Fibroblasts involved in making cartilage are **chondroblasts** and **chondrocytes** when surrounded by a cartilaginous matrix. In addition, large connective

TABLE 2.3. Specialized regions of cell membranes[a]

Name	Type	Function
Desmosome	Macula adherens = spot desmosome Zonula adherens = belt desmosome Hemidesmosome	Mechanical support; adhesion
Gap junctions	Plaques	Intercellular communication; electrophysiological coupling
Tight junctions	Zonula occludens (belt or band); endothelial cell junctions (discontinuous belt or band)	Sealing element; epithelial permeability regulation

[a]See N. B. Gilula, Junctions between cells. In R. P. Cox, ed., *Cell communication.* Wiley, New York, 1974.

tissue **macrophages** capable of ingesting extracellular particles are abundant in loose connective tissue and accumulate following trauma or infection.

The extracellular matrix produced by connective tissue cells tends to polymerize into fibrillar solids, such as collagen and elastin, and into amorphous solids, such as glycosaminoglycans (GAG) or mucopolysaccharides. The matrices of cartilage (Fig. 2.10) and bone are especially dense, and bone is calcified.

Muscle is the contractile tissue. It contains organized arrays of contractile proteins, or mechanoproteins, especially actin and myosin. Vertebrate muscle is further classified as **smooth, striated,** and **cardiac muscle.**

Smooth or **involuntary muscle** (under autonomic nervous control) consists of parallel, spindle-shaped cells. It forms circular and longitudinal layers over blood vessels, the intestine, and other tubular structures, including the uterus. Smooth muscle also forms sphincters at the ends of many tubular structures or where one tubular structure passes into another.

Striated or **skeletal muscle** (Fig. 2.11, sometimes called **voluntary muscle,** since it is controlled by nerves originating in the central nervous system) consists of syncytia or **muscle fibers** containing cylindrical **myofibrils** lying between thin layers of cytoplasm known as **sarcoplasm.** Actin and myosin filaments are arranged in precise parallel arrays within **sarcomeres.** Laying in register in adjacent myofibrils, the striations of sarcomeres give the entire fiber a regularly striped appearance.

Cardiac muscle has much the same striped appearance as striated muscle but consists of individual cells rather than syncytia. Nerves and hormones influence the heart's rate of contraction, but the heart's intrinsic pulsation arises directly from cardiac muscle cells. Even the fibers carrying the impulse for contraction from one part of the heart to another are not nerves but modified cardiac muscle fibers.

Nervous tissue is the conductive tissue. Nerves convey excitatory impulses to other nerves (Fig. 2.12), muscle, and secretory epithelia. The nerve cell body, or **neuron,** is found either in the **central nervous system** (CNS, the brain and spinal chord) or in dense clusters called **ganglia** in the **peripheral nervous system** (PNS, autonomic motor ganglia and sensory ganglia). The cytoplasmic processes (called axons and dendrites) extending from neurons sometimes run for great distances through the organism. Nerve cells stimulate each other through junctions called **synapses** and stim-

FIGURE 2.11. Electron micrograph of longitudinal section through parts of two muscle fibers separated by their cell membranes (CM) and collagen (Col) in the extracellular space. A fibroblast (Fib) is also present. In the syncytial muscle fibers, cylindrical myofibrils (Myf) are separated by sarcoplasm (Sp) containing double membranes of the sarcoplasmic reticulum (a form of smooth surface endoplasmic reticulum) and glycogen granules. The stripes through the muscle identify sarcomeres (S) in which the actin and myosin filaments are in register. ×3900. From E. J. Reith and M. H. Ross, *Atlas of descriptive histology.* Harper & Row, New York, 1977. Used by permission.

ulate muscle or glandular epithelia tissue through **end plates.**

The embryonic precursor of the CNS in vertebrates is a pseudostratified **neural epithelium.** Its epithelial character is lost as nerve cells send out processes. PNS cells are derived from migrating **neural crest** cells.

The cablelike anatomical structures called "nerves" are actually bundles of processes. **Afferent** nerves convey **sensory** impulses from the periphery toward the central nervous system, and **efferent** nerves convey motor impulses from the central nervous system to organs, muscles, and blood vessels.

Nerve cells are not alone in the nervous sys-

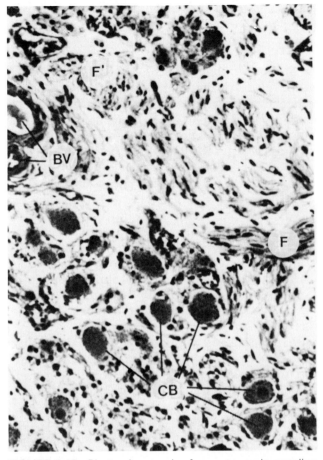

FIGURE 2.12. Photomicrograph of an autonomic ganglion (i.e., sympathetic chain ganglion) showing nerve cell bodies (CB) and fibers (F, cut longitudinally; F', cut across). BV, blood vessels. ×160. From E. J. Reith and M. H. Ross, *Atlas of descriptive histology.* Harper & Row, New York, 1977. Used by permission.

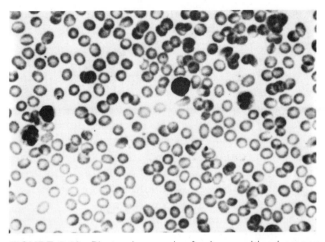

FIGURE 2.13. Photomicrograph of a human blood smear showing occasional white blood cells (with irregularly shaped nuclei) and lymphocytes (with round nuclei) among a large number of red blood cells (erythrocytes). ×400. From E. J. Reith and M. H. Ross, *Atlas of descriptive histology.* Harper & Row, New York, 1977. Used by permission.

tem. Many of the nerve processes in the PNS are wrapped in sheaths of **myelin** formed by **neurolemmocytes** (formerly Schwann cells), while nerve processes in the CNS are wrapped by extensions of **glial cells** (i.e., "glue" cells). Glial cells of several types constitute the supportive cells of the adult CNS and possibly the guiding cells in the developing CNS.

The **sensory cells** transduce environmental stimuli to electrochemical impulses and excite associated nerves. The sensory cells of **special sensory organs,** such as the ear, eye, and nose, and of taste buds are developmentally and physiologically related to nerves but are frequently classified as epithelial cells, because they rest on membranes and are tightly coherent to each other. **General sensory organs** are made of isolated sensory cells and distributed throughout the body.

Some general sensory cells and even free nerve endings in the skin are especially sensitive to pres-

sure (see Fig. 26.19). **Stretch receptors** known as muscle spindles in muscle and Golgi tendon organs within tendons are sensitive to tension.

Vascular tissue includes blood cells, erythrocytes or red blood cells (Figs. 2.8 and 2.13), white blood cells or leukocytes of different types (Fig. 2.13), megakaryoctyes, platelets (Fig. 2.18), and lymphocytes (Fig. 2.8). Precursor cells that give rise to vascular tissue cells are classified as vascular or as connective tissue. Red blood cells and lymphocytes are characterized by their products (e.g., hemoglobin and immunoglobulin, respectively). White blood cells are characterized by the appearance of their nuclei, cytoplasmic granules, and their responses to various traumas and infection.

Cells of the vascular system may be considered as extreme varieties of connective tissue cells for several reasons. Much as connective tissue cells may break intercellular contacts, vascular cells maintain no prolonged cell contacts, and, like extracellular matrix around connective tissue cells, vascular cells are bathed in serum. Moreover, some white blood cells and connective tissue macrophages are interconvertible.

The germ line or germ tissue in animals includes the male and female gametes, sperm and egg, and all the cells leading up to them. The germ tissue is unique in animals by way of its ability to support meiosis, but it is also distinguished from other tissues by its tendency to form **intercellular bridges,** and **germ-line cells** are frequently set aside or isolated as primordial germ cells relatively early in development. The primordial germ cells establish a **germ-cell lineage** which gives rise to all the germ cells the animal will ever produce.

In male vertebrates, the early germ-line cells become sequestered within epithelial cords, while in females, the cells become concentrated under the organ's epithelial covering. The tradition of classifying the germ line as an epithelium is based on the mistaken belief that germ cells originate from these cords or surface epithelia. In light of the origin of the germ line from migrating cells, tradition should be set aside and the germ line recognized as a tissue in its own right.

Among the unique characteristics of the germ line is the formation of **intercellular bridges** between mitotically dividing cells. When the intercellular bridges are later eliminated, germ cells become uniquely independent, having no junctional complexes or gap junctions with other germ-line cells.

Both male and female germ cells are unique in undergoing meiosis, maturation, and, as gametes, fertilization. Differences between male and female germ cells appear mainly during differentiation when sperm shed cytoplasm soon after completing maturation, while eggs accumulate great amounts of yolk and other materials during a prolonged period prior to maturation.

Tissue Dynamics

Tissues differ in population dynamics as well as in their synthetic products. Four types of cell population are distinguished by the presence, absence, and distribution of dividing cells (Leblond and Walker, 1956) (Fig. 2.14): static, expanding, steady-state, and intermediate cell populations.

Constraints on cell division may have evolved as adaptations to the "division of labor" among tissues and organs in multicellular organisms (see Shostak, 1977).

Static cell populations add no new cells after an initial developmental period. In adults, nerve cells and sensory cells are generally members of static cell populations. Adaptation, not an inherent incompatibility between cell division and differentiation, seems to dictate proliferative stasis in static cell populations. For example, given the elaborate circuits achieved through development of the nervous system (e.g., Fig. 2.12), division by nerves in an adult could very well disrupt function.

The germ cells of some female vertebrates (e.g., mammals, reptiles, and birds) are static cell populations. Other vertebrate females (fish, amphibians) produce eggs without constraining premeiotic proliferation. Neither the requirements of meiosis nor the deposition of yolk seem to explain this difference.

Expanding cell populations retain the ability to divide even among their differentiated and functional cells, but cells that once divided rapidly in the embryo may do so only slowly in the adult. The cells produced are kept in the population, which is therefore always expanding even if slowly. Cells in expanding populations may also divide rapidly were the population to undergo a cell loss or experience a physiological stress (Goss, 1978).

The parenchyma of the kidney, liver, and other organs are expanding cell populations, as are the fibroblasts of connective tissue and smooth and cardiac muscle. Even fibroblasts that differentiate into immobile, nonproliferative chondrocytes (e.g., Fig. 2.10, cells A, B, C) may be considered members

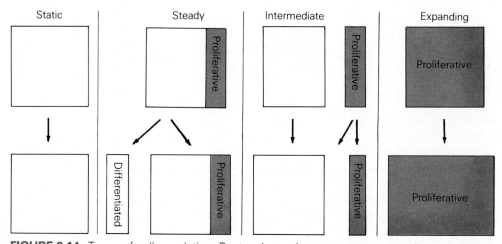

FIGURE 2.14. Types of cell population. Rectangles and squares represent cell populations within tissues.

of expanding cell populations, since, in response to injury, these cells may yet proliferate.

Glial cells of the central nervous system are also members of expanding cell populations. These cells retain the ability to divide and may even do so when the organism learns new tasks.

Steady-state or renewing cell populations preserve the ability to divide in a subpopulation of proliferative cells (sometimes called stem cells). While performing most of the tissue's functions, the proliferative cells occupy special sites where some functions are limited. On the average, half the cells produced by cell division replace cells leaving the overall population as a consequence of **terminal differentiation,** while the other half remains in the proliferative subpopulation. The size of a steady-state cell population is thus held constant, even while the member cells are changing.

The simple columnar epithelium lining the intestine (Fig. 2.8) (Quastler and Sherman, 1959) and the stratified squamous epithelium, or epidermis, covering the organism (Fig. 2.15) are examples of steady-state populations. In the case of the intestinal epithelium, the proliferative subpopulation lies within crypts, while in the epidermis, the proliferative cells lie in the basal layer. In both cases, cells become terminally differentiated as they move apically.

Male germ-cell populations are of the steady-state type as are female germ cell populations with the conspicuous exception of mammals, reptiles, and birds. Red and white blood cells and their proliferative precursor cells also constitute a steady-state population. Lymphocytes of different types with their own proliferative subpopulations likewise constitute steady-state populations. The clearing stations where terminal vascular cells break down are the spleen and liver, and the exit portal where disposal takes place is the intestine.

Intermediate cell populations combine features of expanding and steady-state populations. Functional cells in intermediate cell populations do not divide or leave the population, but a subpopulation of nonfunctioning cells is capable of proliferating and adding new cells to the functional population (Leblond, 1972). This subpopulation of reserve cells or **satellite cells** typically proliferates and differentiates slowly, but under some conditions, especially arising from injury, they provide **reserve cells** capable of rapid proliferation and differentiation.

Striated muscle is an example of an intermediate cell population. This type of muscle does not divide. Even growth following continued exercise is due to accumulation within existing cells rather than the addition of new cells. But under conditions of injury or stress, satellite cells (Fig. 2.16) found on the periphery of muscle fibers differentiate as new muscle.

In general, the nuclei of differentiated syncytia do not undergo mitosis and must be "fed" into syncytia by the fusion of cells from without. Since the fusion of myoblasts is an ordinary part of development in the embryo, muscle satellite cells in adults may be considered "carry-overs" from embryonic development.

The difference between steady-state populations and intermediate cell populations is complicated when reserve cells are attached to an otherwise proliferative population. The male germ-cell population of mammals, for example, is broadly a steady-state population, but it contains reserve stem cells (so-called A_d cells; see Chap. 7) which are generally nonproliferative. Like satellite cells, the reserve stem cells of the male germ line may respond to trauma by dividing and replacing lost cells.

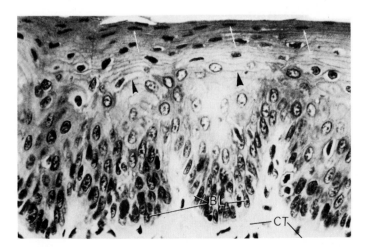

FIGURE 2.15. Photomicrograph of a section through the stratified squamous epithelium (epidermis) of a monkey's tongue showing some mitotic figures (telophase nuclei immediately above CT) in the basal layer (BL) resting on the basal lamina. As the cells acquire keratin, their borders become thicker (arrowheads); nuclei become pycnotic and flattened (white arrows), and cells finally slough at the tissue's outer surface. CT, connective tissue. × 380. From E. J. Reith and M. H. Ross, *Atlas of descriptive histology.* Harper & Row, New York, 1977. Used by permission.

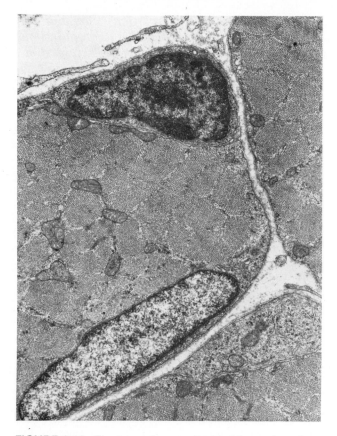

FIGURE 2.16. Electron micrograph of a thin section of rat muscle showing two reserve "satellite" cells in a group of three muscle fibers. A membrane encloses the satellite cells and muscle fibers. The nucleus of the upper satellite cell is much denser than the nucleus of the muscle fiber. The cytoplasm of the lower satellite cell contains numerous free ribosomes. From F. P. Moss and C. P. Leblond, *Anat. Rec.,* 170:421 (1971). Reprinted by permission of Alan R. Liss and courtesy of C. P. Leblond.

Another problem for classification occurs when a proliferative population includes reserve cells. The mammalian epidermis, for example, is by and large a steady-state population, but it contains reserve cells poised on the brink of cell division. In the event of a wound, these reserve cells divide and provide the first line of cells for wound healing. Intermediate populations, whether self-contained or parts of larger populations, seem ideally suited for tissues that occasionally encounter trauma.

Concept of Tissues

Predating the idea of tissues, adult organs were considered composites of **parenchyma** (Gk. *parenchein* to pour in, referring to the belief that the stuff of internal organs was poured into them from the blood) and **stroma** (Gk. *stroma* bed covering,

referring to outer covering of organs). Even today, if mainly for soft organs, cells that perform the main physiological function of an organ are called **parenchymal cells.** The tough covering of organs and the internal support and fibrous network within organs are called the stroma.

The present concept of tissue dates back to the 18th century. "Tissue" is derived from the French *tissu* meaning woven (Lat. *texere* to weave, construct), referring to the texture of parts of organs. The word was introduced by the French anatomist, Marie Francois Xavier Bichat (1771–1802), sometimes considered the parent of **histology** (Gk. *histos* loom or web + *-ology* theory or doctrine of), who had a fundamentally synthetic concept of organs and defined tissues as their finely homogeneous parts.

Like the unseen elementary atoms of chemistry which combined to form compounds, Bichat believed that elementary tissues (he described 21 of them) combined to form organs. The same tissue might be present in different organs, but organs also contained different tissues.

Bichat was strongly influenced by the emerging field of chemistry and followed his many careful dissections of cadavers by preparing material in the manner of chemists: putrefaction, drying, boiling, and macerating in acids, alkalies, and salts. Disdaining the microscope, he drew all his conclusions, as chemists of his day, from observations made with the naked eye.

In the opening years of the 19th century, Bichat's concept of elementary tissues found an appreciative audience among anatomists who wanted to simplify the increasingly complex world revealed by the microscope. Unlike Bichat, these anatomists were enthusiastic microscopists, and microscopy soon prevailed over chemistry as the primary research method of histologists. Soon, with the ascendance of the **cell theory,** tissues were redefined as functionally specialized arrangements of similar cells.

Developmentally, tissues were thought to **diversify** from embryonic germ layers, and finer differences within tissue and among tissue cells were supposed to be acquired secondarily through cellular **differentiation.** Some cells belonging to the same tissue might even be sufficiently different to warrant placing them in separate classes or **cell types** within tissues.

Strictly speaking, tissues do not always consist of entities that meet the classic definition of cells (i.e., a membrane-bound vesicle containing a nucleus within a cytoplasm). Some so-called cells are "trimmed down," while others are "scaled up."

For example, circulating adult mammalian red blood cells and platelets lack nuclei. They also lack common cytoplasmic constituents such as mitochondria (e.g., see platelet, Fig. 2.18). Other so-called cells continue to possess nuclei, but these become **pycnotic** (Gk. *piknoun* to condense) and presumably nonfunctional (Fig. 2.15). Some of the defective cells, like those of the epidermis, die and disappear, while others, like those of the lens, are stable for the lifetime of the organism.

Scaled up tissue entities include **syncytia** (Gk. *syn-* together + *-kyt-* hollow vessel [cell] + *-ium* suffix of diminutive force, hence fused cells) containing many nuclei within a common cytoplasm. Syncytia, such as vertebrate striated muscle, are derived by the fusion of originally single cells. Following fusion, the syncytium's nuclei no longer divide. Similarly, embryonic syncytia, like the syncytial trophoblast of mammals (see Fig. 24.29) contain nondividing nuclei.

Cells or syncytia in which nuclei retain the ability to divide without triggering cytokinesis are sometimes called **plasmodia** (pl., *plasmodium* sing., e.g., the plasmodial slime mold). Regrettably, embryonic tissues with multiple nuclei derived through repeated nuclear division are identified as syncytia (e.g., the syncytial blastoderm of insects, Fig. 11.9, or the yolk syncytial layer of fish, Fig. 18.21), not plasmodia.

CELL BIOLOGY

The elementary parts of most tissues, when traced backwards from their state of complete development to their primary condition, are only developments of cells.

Theodor Schwann (1893)

In many ways, cells are to biology what atoms are to chemistry. They are fundamental units, parts of or whole things that cannot be reduced without the loss of many, if not all, their properties.

Still, the analogy of cells and atoms cannot be carried very far. While molecules are made of atoms and living things are made of cells, molecules do not synthesize other molecules. Cells, on the other hand, are capable of synthesis and always contain products of their own making. Furthermore, while atoms may split into other elements, cells divide without necessarily changing their properties.

Cell biology is the study of cells in all their unique activities and manifestations. Overlapping with all other fields of biology, cell biology offers insights over a range of embryological concerns from gametogenesis to the differentiation of cells in organisms.

Cells

At some remote time, at least a billion and a half years ago, organisms diverged into two great groups composed of fundamentally different cells. Members of one group, the **Monera** of Haeckel or, **prokaryotes** (also procaryotes, Gk. *pro-* before + *karyon* kernel + *ote*, suffix denoting inhabitant, hence cells occupied by a rudimentary nucleus) consist of cells whose DNA, while concentrated in a **nucleoid** and associated with **DNA binding proteins,** lacks the basic proteins known as **histones** and is exposed directly to cytoplasm. Members of the other group, the **eukaryotes** (also eucaryotes, Gk. *eu-* true + *karyon* + *-ote*, hence cells occupied by a true nucleus) consist of cells with chromosomal DNA closeted in nuclei away from the cytoplasm and "moth-balled" in histone.

Prokaryotes are today's bacteria and blue-green algae. They specialize in small size and live as single cells, colonies, and filaments. Aggregates may produce masses of macroscopic dimensions, but these display little of the organized, integrated, and differentiated systems associated with multicellularity. Economy of function seems to come with small size, and the cells utilize all their DNA for structural and control genes.

Blue-green algae may lack sexual processes. Bacteria may be said to have sexual processes, but they lack a completely diploid phase and hence have no zygote and no embryo.

In general, eukaryotic cells (e.g., see Fig. 1.5) are larger and more complex than prokaryotic cells. In contrast to prokaryotic DNA, eukaryote DNA includes nongenic regions and repetitive sequences. Eukaryotes include the **unicellular** protistans, the protozoa and unicellular algae, and the **multicellular** fungi, algae, plants, and animals—thus, all organisms utilizing an embryo in reproduction.

Eukaryotic Cytoplasm

Roughly half solid and half liquid, cytoplasm is anything but simple. Following **homogenization,** or cellular disruption, and **fractionation** by differential sedimentation in a centrifuge (e.g., see Fig. 1.7c), the cytoplasm of eukaryotic cells is separated into a liquid component, called the **cell sap** and a sedimentable component containing **inclusions** (passive products of cell activity, such as starch granules) and **organelles** (little organs, such as the nucleus, centrioles, various membranous structures, mitochondria, and, in plants, plastids).

In the intact cell, the sap is a solid–liquid interphase called the **cytosol.** Inclusions are usually suspended in the cytosol. Some of the organelles are parts of a unified **endomembranous system** (stretching from the surface of the cell to the nucleus), and some (mitochondria and plastids) are independent.

The cytosol represents a dynamic, constantly transforming sol-gel system. Cytosol is a reservoir of dissolved materials and, at the same time, a structure. It is the home of "housekeeping" enzymes doing the work of intermediate metabolism, the site of glycolysis, and the place where polysomes synthesize proteins. It is also the superstructure where these activities are integrated in the life of the cell.

The cytosol contains dissolved components or monomeres of relatively stable filamentous inclusions. The fibers are broadly classified as thin microfilaments (7 nm in diameter), thicker microtubules (24–28 nm in diameter), and intermediate filaments (10 nm in diameter) all of which comprise the cytoskeleton (Fig. 2.17) (Fulton, 1984).

Very fine filaments (2–3 nm in diameter) seem to comprise a **microtrabecular network** and bind other types of filament (Schliwa et al., 1981) while leaving liquid-filled spaces of about 100 nm in diameter. Just about everything dissolved in the cell sap may be connected to some structuring element.

Microfilaments are made of **actin,** polypeptide subunits wound in a double helix. Actin is one of the two main contractile elements of muscle and possibly the chief element involved in changing cell shape. In dividing animal cells, actin filaments form a **contractile ring** responsible for cytokinesis (see Schroeder, 1986), and, in migrating cells, actin is concentrated in **stress fibers** or bands. A separate pool of actin provides the core of the fingerlike projections called **microvilli** protruding from the surface of some cells (e.g., intestinal epithelium).

Microtubules are noncontractile, hollow tubules composed of 13 longitudinally arrayed **protofilaments** made of two polypeptide subunits, alpha and beta **tubulin.** Bundles of microtubules frequently form more or less rigid structures and stabilize cell shape (Fig. 2.18), but they also form the mitotic apparatus of dividing cells and play a role in moving chromosomes.

Microtubules are found in long cellular extensions such as nerve processes. In flagella (e.g.,

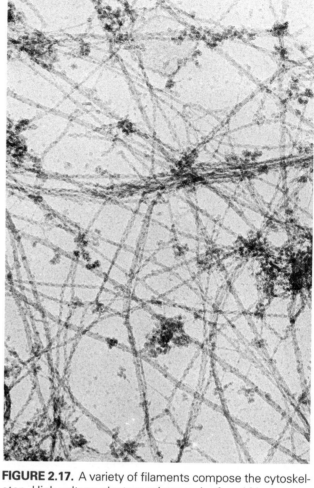

FIGURE 2.17. A variety of filaments compose the cytoskeleton. High-voltage electron micrograph of an African green monkey kidney cell (strain BSC-1) showing a variety of cytoskeletal elements including smooth intermediate filaments, actin filaments, microtubules, and very fine 2–3-nm filaments associated with the microtrabecular network. ×70,000. From M. Schliwa, J. van Blerkom, and K. B. Pryzwansky, *Cold Spring Harbor Symp. Quant. Biol.,* 46:51 (1982) by permission of the Cold Spring Harbor Laboratory of Quantitative Biology and courtesy of the authors.

sperm tails) and cilia (e.g., oviductal epithelium), sometimes collectively called **undulipodia,** microtubules have a characteristic arrangement of nine pairs of fused tubules surrounding two attached tubules (i.e., the "9 + 2" arrangement, Fig. 2.19). In the basal bodies of cilia and flagella and in centrioles, microtubules are generally present in nine sets of triplets (Fig. 2.20) (see Dustin, 1978).

The intermediate filament (IF) is the most abundant and the most stable (i.e., least dynamic) intracellular fiber. The distribution and density of particular kinds of IFs are among the major characteristics of differentiated cells.

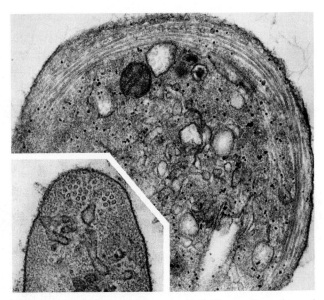

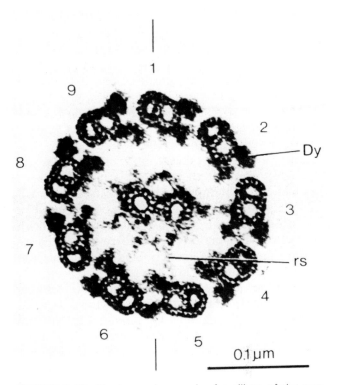

FIGURE 2.18. Electron micrograph of thin sections through the edge of a blood platelet. Microtubules form marginal bands close to the membrane. (Insert) Some of the microtubules contain an unusual central element. From O. Behnke, *Triangle*, Sandoz Journal of Medical Science, 13:1 (1974). Copyright Sandoz Ltd., Basle, Switzerland. Used by permission and courtesy of the author.

FIGURE 2.19. Electron micrograph of a cilium of the sea urchin, *Lytechinus*, showing typical "9 + 2" arrangement of microtubules. Dynein (Dy), a protein, forms arms or dense bodies attached to complete microtubules (tubule A) involved in bending of cilium. Radial spokes (rs) connect peripheral microtubules to globular region in central core. From K. Fujiwara and L. G. Tilney, Substructural analysis of the microtubule and its polymorphic forms. *Ann. N.Y. Acad. Sci.*, 253:27(1975) by permission of the New York Academy of Sciences and the authors.

Cells make the IFs of their cytoskeleton from related polypeptides coded by an IF multigene family. Different types of cells make their IFs from different members of the family. For example, the vertebrate keratin family consists of 20 or so IFs. Specific combinations of these proteins constitute the keratins of different epithelial cells and the appendages (hair and feathers) and tumors derived from these cells (see Sawyer, 1987).

Large numbers of granules are contained in the cytosol or attached to the microtrabecular network and membranes. The carbohydrate storage product, glycogen, is present in granular form as are protein storage products (e.g., the yolk of eggs). Ribosomes and their subunits, messenger RNA, and polyribosomes or polysomes are also present in the cytosol.

The liquid parts of the cytosol (sometimes equated to the cytoplasm) occupy the **extracisternal** space. Possibly confined to narrow channels within the microtrabecular network, the liquid cytosol conveys intracellular messages, both those created in the cell and those imported from outside the cell. For example, specific intracellular receptors in the liquid cytosol bind specific molecules and transport them to the nucleus.

In general, the regulation of protein activity (e.g., via phosphorylation-dephosphorylation, methylation–demethylation, disulfide–sulfhydryl

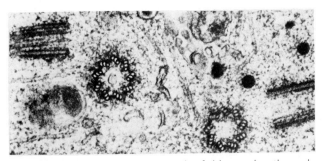

FIGURE 2.20. Electron micrograph of thin section through pairs of centrioles in cultured Chinese hamster fibroblast cell. The centrioles had doubled previously as the cell prepared to divide. Note triplets of microtubules in cross sections forming "pinwheel" centrioles. The centriolar bodies in each centriole are perpendicular to each other. The cell cultures had been treated with the nucleic acid stain, propidium iodide, prior to preparation for electron microscopy. ×60,000. From M. McGill, D. P. Highfield, T. M. Monahan, and B. R. Brinkley, *J. Ultrastruct. Res.*, 57:43 (1976) by permission of Academic Press.

interconversion), and hence the regulation of cellular activity, takes place in the cytosol. The process begins with a two part sequence which, like dominos in a row, triggers a cascade of events. Initially, some new circumstance triggers a change in the concentration of a low molecular weight **secondary messenger.** Proteins dependent on the secondary messenger are then either activated or deactivated by the change in messenger concentration, and these proteins proceed to initiate changes in the **target proteins,** the objects of regulation.

Catalytic subunits of kinases (i.e., phosphorylating enzymes), in particular, may be activated by low molecular weight secondary messengers such as **cyclic adenosine monophosphate** (cAMP). Similarly, other kinases are activated by **free calcium** (Ca^{2+}) in the liquid cytosol.

The response of a cell to changes in cAMP and Ca^{2+} concentrations in the liquid cytosol depends on which target enzymes are activated. cAMP and Ca^{2+} can also alter each other's concentrations by their effects on synthetic and degrading enzymes and on channels or pumps for Ca^{2+} transport.

The plasma membrane or plasmalemma surrounding the cell is made of a unit membrane (Robertson, 1959) containing lipid leaflets or bilayers, typically 4–5 nm in thickness, a variety of proteins, and external carbohydrates. At normal temperatures the lipids are melted, turning the membrane into a **fluid mosaic** of suspended proteins (Singer and Nicolson, 1972).

Proteins traversing the membrane have extracellular, intramembranous, and intracellular **domains,** while peripheral proteins lying on one side of the membrane lack an intramembranous domain. Outside, the membrane may be covered by a carbohydrate sheath or **glycocalyx** through which the cell interacts with its microenvironment including other cells. The plasma membrane also plays an important role as a transducer of extracellular information via **membrane-bound receptors.**

Low molecular weight lipids, such as sex steroids, move from the extracellular side of the plasma membrane into cells by dissolving in the plasma membrane. Other extracellular materials move into cells passively as part of the fluid contents of **endocytic vesicles** (also called pinocytotic vesicles) formed during **constitutive endocytosis** (so called because it occurs continuously, i.e., constitutively) or pinocytosis (cell drinking).

Still other extracellular materials (e.g., from nutrients, hormones, proteins, and viruses) are transported into cells via **receptor-mediated endocytosis** (also called absorptive endocytosis and clathrin-facilitated endocytosis). These extracellular materials are first bound to surface receptors concentrated in **coated pits. Clathrin,** a fibrous protein lining the pits internally, plays a role in sorting and concentrating receptors and, later, in facilitating their transport from one vesicle to another. The pits sink into the surface and move into cells as **coated vesicles.** Similarly, clathrin facilitates the secretion or **exocytosis** of proteins bound in secretory vesicles (see Brodsky, 1988).

An endomembranous or double membrane system of cisternae, tubules, and vesicles extends from the plasma membrane to the nuclear envelope of some cells. The most conspicuous parts of the system form an **endoplasmic reticulum** (ER) that functions independently or through interactions with other cellular organelles. A **rough surface ER** (Palade and Porter, 1954) (see Fig. 1.5) of cisternae and tubules with polysomes on their outer surfaces is involved in the manufacture of proteins that are later transported outside the cell (Siekevitz and Palade, 1958) or stored within the cell (e.g., in lysosomes). A **smooth surface ER** of tubular and vesicular elements may be involved in the manufacture of lipids.

The contents of **intracisternal spaces** of the ER may move during processing and transport out of the cell, or products of cellular synthesis may be stored within the spaces. In addition, calcium ions are stored within vesicular elements and released to the **extracisternal space** under the influence of secondary messengers.

The **Golgi apparatus** is also composed of cisternae or flattened sacs sometimes stacked and molded into a cup shape (Fig. 1.5). Typically found near the nucleus, the Golgi apparatus plays a role in the manufacture and transport of high molecular weight polysaccharides as well as proteins, and vesicles formed at tips of flattened Golgi cisternae may transport and package materials for cellular export (Dalton and Felix, 1953).

Some intracellular vesicles are involved with the treatment of materials entering the cell or made in the cell. **Peroxisomes,** which detoxify wastes, and **lysosomes,** involved with intracellular digestion, converge with Golgi elements and contribute to the movement of materials and to cellular recycling.

Other vesicles are more concerned with the movement of materials into and out of cells. The coated vesicles, for example, formed during receptor-mediated endocytosis lose their clathrin coats and fuse into **endosomes,** which in turn fuse with lysosomes or tubular elements of the ER. Vacuoles, or membrane-bound secretory granules, may also acquire a clathrin undercoat prior to exocytosis (see Brodsky, 1988).

Mitochondria (see Fig. 1.5) *in all eukaryotic cells are also membranous, but their inner membranes are not continuous with the remainder of the cell's membranous system* (see Margulis, 1981). Mitochondria lie in the interstices between microtubules and are oriented by them. The electron transport system and all the enzymes, cytochromes, and cofactors of oxidative phosphorylation are present in mitochondria. Mitochondrial DNA (mtDNA), with its own DNA polymerases, is self-replicating, and mitochondria are replicating organelles with their own division cycle, but most mitochondrial proteins are derived from the cell.

The centriole (Fig. 2.20) (see Dustin, 1978) *is also a replicating organelle but without its own DNA.* Pairs of centrioles, or a **diplosome,** are present in animal cells where they seem to play a role in mitosis. Because centrioles lie close to the limits of resolution with the light microscope (i.e., 0.1–0.2 μm), a single **centrosome** is more likely to be observed than the two centrioles.

Eukaryotic Nucleus

The nucleus is the cell's command post for gene action and the seat of the **nucleolus** (sing., nucleoli pl.), the factory for ribosomal manufacture. Bound by a **nuclear envelope** having an inner and outer membrane with a **perinuclear space** between them, the nucleus is filled with **nuclear matrix** or nucleoplasm and **chromatin** (see Fig. 1.5).

The nucleus cannot be separated entirely from the cytoplasm. The nuclear envelope is itself a double membrane attached to the cell's system of double membranes. The outer nuclear membrane, occasionally studded with ribosomes, is continuous with membranes of the rough surface ER, and the intracisternal space of the ER is continuous with the perinuclear space. Moreover, **nuclear pores,** connecting the two membranes of the nuclear envelope and surrounded by an ordered nuclear pore complex, provide channels between the nuclear matrix and the cytosol. The number of pores and electrical resistance of the envelope change greatly with nuclear activity (Feldherr, 1965).

A matrix of three fibrous, nonhistone proteins, called **nuclear lamina** or the **nuclear cage,** lies within the nuclear envelope (i.e., on the nucleoplasmic side) (see Goldman et al., 1986; Gerace, 1988). Nuclear lamina plays several roles, shaping the nucleus, taking part in breaking it down and building it up during cell division. In addition, particular genes may be attached to specific sites in the nuclear lamina (Cook and Brazell, 1980), and the movement of gene products through nuclear pores may be facilitated by nuclear lamina.

The **nucleolus** is the site within the nucleus where **nucleolar organizer genes** form most of the ribosomal RNA (rRNA) and where ribosomal proteins and rRNA are packaged into ribosomal subunits. A continuous granular component of the nucleolus consists of mature ribosomal precursors. A discontinuous component contains the nucleolar organizer DNA, and a fibrillar component consists of newly synthesized rRNA.

The nucleolus is a large structure in cells actively engaged in protein synthesis but may disappear when protein synthesis is suspended. Eggs engaged in synthesizing ribosomes for storage may contain many nucleoli resulting from the specific amplification of ribosomal organizer genes.

Chromatin (Gk. *chroma* color + -*in* suffix denoting of or belonging to) received its name for the dark color acquired from **nuclear stains,** such as hematoxylin, in standard histological preparations. Only condensed portions of chromatin, known as heterochromatin, actually stain darkly. Loosened or decondensed portions of chromatin, known as **euchromatin,** do not stain.

DNA is a **duplex** of nucleotide chains held together by internal hydrogen bonds between complementary purines and pyrimidines (i.e., base pairing between adenine and thymine, guanine and cytosine). In chromatin, about 1.75 turns of DNA or 146 base pairs (bp) are wrapped around an octamer of four pairs of **core histones,** H2A, H2B, H3, and H4 (molecular weights 10,000–16,000) to form a core **nucleosome** (the "beads" on the DNA "string," Fig. 2.21) (Mirzabekov, 1980). The addition of a fifth histone, H1, or a tissue-specific variant, draws another 15–20 bp of DNA into the structure and completes a second turn. The entire unit (sometimes called a chromatosome) contains 160 bp.

The H1 histone is also associated with a variable amount of **interparticle** or **linker DNA** (e.g., on the order of 60–80 bp) stretching between nucleosomes. The **repeat length** or **nucleosomal spacing** is determined by the linker DNA and changes with the activity and condensation of chromatin. Nucleosomes containing histone H1 interact and, when more than six are present in a row, they become compact, while their DNA loses its ability to undergo transcription (Wolffe and Brown, 1988).

Stacks of nucleosomes with H1 form the **10-nm chromatin fiber** (or type A chromatin fiber) which coils into a **solenoid** or **30-nm chromatin**

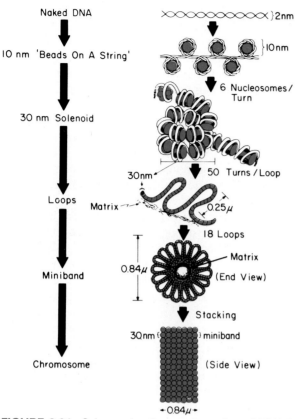

FIGURE 2.21. Scheme for the condensation of DNA into chromosomes through coupling with nucleosomes and the nuclear matrix. From K. J. Pienta and D. S. Coffey, in C. Nicolini and P. O. P. Ts'o, eds., *Structure and function of the genetic apparatus.* Plenum Press and NATO Scientific Affairs Division, 1985. Used by permission of the authors.

fiber (or type B chromatin fiber) with six nucleosomes per turn (Fig. 2.21). The higher order structure and mechanism of coiling are still a matter of speculation, but **folds** occurring in distinct folding or **loop domains** of about 50 turns per loop may correspond to genic units. Anchored at their bases to **scaffolding proteins** of the nuclear lamina, loops may maintain their integrity, both when they are invisible at the light microscopic level during interphase and when they are condensed into chromosomes during division (see Pienta and Coffey, 1985).

At cell divisions, chromatin condenses into **chromosomes** (Gk. *chroma* color + *soma* body) consisting of two **chromatids** (also known as chromonemata). While typically free along **arms,** the chromatids of each chromosome are bound together at (or by) a **centromere.**

Each chromatid contains an astonishingly long loop of chromatin condensed to an amazing degree. For example, human body cells about to divide

contain 3.8 meters (m) of DNA. The loop of DNA present in each of the 92 chromatids is therefore about 4 centimeters (cm) long, but the chromatids are only 2–8 micrometers (μm) in length, 1/10,000th the length of the DNA loop.

Chromatids are sometimes seen to contain rings or bands (chromomeres) of various degrees of fineness. In minibands, 18 loops may radiate from the center (Fig. 2.21), while in less dense regions, euchromatin alternates with more dense regions of heterochromatin to form larger chromomeres. Darkly staining portions on the arms of chromosomes are **facultatively heterochromatic,** since they may or may not remain darkly staining.

The centromere is the strongest point of chromatid attachment and the last firm point of contact prior to separation during mitotic and meiotic divisions. It is **constitutively heterochromatic,** since it is always darkly staining.

Centromeric DNA corresponds to **satellite DNA,** a fraction of DNA that does not sediment during centrifugation with the bulk of eukaryotic nuclear DNA. Its differential density is attributable to highly repeated, simple sequences of relatively dense guanine and cytosine (GC) base pairs. Satellite DNA itself or the proteins of centromeres may hold chromatids together until the end of cell division.

CELL THEORY

Wir haben gesehen, dass alle Organismen aus wesentlich gleichen Theilen, nämlich aus Zellen zusammengesetzt sind.[1]

Theodor Schwann (1839, p. 227)

In what sense is the **cell theory** a theory? Often misrepresented, the cell theory is not concerned with whether living things are made *of* cells and their products but with whether living things are made *by* cells. While the cellular structure of living things (which might be called the "cell law") can be confirmed without exception (or with few exceptions), the issue raised by the cell theory is still very much with us: "What are the smallest units concerned with the construction of organisms, or just how reducible is the process of making multicellular organisms?"

The beginning of the cell theory can be traced to Robert Hooke (1635-1703) who is credited with coining the word "cell" in 1665 to designate the cavities or hollow vessels seen in cork. The word

[1]We have seen, that all organisms are assembled by essentially the same parts, namely by cells.

was also used by other 17th century botanists who examined plant tissue with the aid of primitive microscopes. "Cells" were the minute compartments surrounded by solid walls, not the contents of the compartments. Today, the name is retained but applied to contents and not walls.

During the Age of Enlightenment, the words "utricle" (as in uterus, designating a bag) and "areola" (as in area or open space) came slightly closer to a contemporary concept of cells. A utricle meant any bladder, vesicle, or pustule found in tissue, and areolae[2] were spaces enclosed by fibers seen in animal organs prepared for microscopic examination, not the actual cells embedded among the fibers.

More progress was made by 18th-century microscopists, who described the first real cells as "globules" and "granules" in embryonic tissues. Casper Friedrich Wolff (1738–1794), best known for his contribution to the epigenic concept of development, was among the first to speculate on the importance of "globules."

After studying the cellular development of vessels in plants, Wolff turned his attention to the development of vessels in hen eggs. Among his accomplishments was his discovery of the embryonic nephric duct now called the Wolffian duct. Employing alcohol to harden tissue before dissecting it (a procedure first employed by Robert Boyle [1622–1691] who, along with Robert Hooke, discovered what is now called oxygen), Wolff also discovered that the chick embryo contained globules similar to those of plants. Moreover, the globules present in embryos formed leaflike layers. In the case of the intestine, an embryonic layer built an adult structure through growth and infolding, and globules continued to be present in adult structures.

Today, historians of embryology (e.g., Oppenheimer, 1967; Oppenheim, 1982) see the kernel of the cell theory in Wolff's observations, but he is not cited as one of the theory's early formulators. In fact, Wolff had little impact on the science of his time. He played the role of gadfly, running against the prevailing tide of preformationist prejudices, and found little support among biologists.

The authorities, including Albrecht von Haller (1708–1777), argued against Wolff's globular concept of embryonic tissue. Embryos were expected to contain miniature versions of adults, not globules. At the very least, embryonic globules might have been small versions of adult globules but not globules like those of adult tissues.

[2]"Areolar" is retained today in the term areolar connective tissue, also known as loose connective tissue, referring to a fibrous tissue with small, irregular spaces.

The turning point for the cell theory came with the 19th century. Along with the industrial revolution, with smoke billowing from factory chimneys, and with political liberalism, empiricism burst forth in biology (at least in some quarters). Cells were not only rediscovered, but, like workers, they provided convenient explanations for some of the enigmas of life. The liberal school of nature philosophy soon lent its authority to the idea of cells and saw in them the possibility of resolving long outstanding issues, including a pivotal controversy on the classification of animals.

The controversy centered on the "Infusoria." Today, infusoria refers to ciliated protozoans, but, at the time, it referred to organisms found in "infusions" (i.e., mixtures of water and organic material such as hay) and included organisms as different as rotifers and amoebas.

In 1805, Lorenz Oken (1779–1851), a prominent nature philosopher, recommended limiting the term "infusorian" to animals made of "single mucous vesicles," as opposed to animals of "multiple mucous vesicles" or "agglomerations" of vesicles. Oken's mucous vesicle contained a vague *Urschleim* (or primitive mucus) but it set the stage for thinking about living things as composed of discrete cells.

The discoveries of the nucleus and cytoplasm helped to fill out the idea of the cell. Robert Brown (1773–1858), a botanist, characterized the nucleus as a fundamental part of plant tissues in 1831. He also described the directional movement of material within cells, now called "cytoplasmic streaming," and the random movements, now called "Brownian motion."

Cytoplasm was first described by Felix Dujardin (1801–1862) in 1835. He called it the "sarcode," since he based his description on amoebas (i.e., alias Sarcodina). Finally, **protoplasm,** consisting of cytoplasm and nucleus (but not confined to cells), was named by Johannes Evangelista Purkinje (1782–1869) in 1839.

Priority for the cell theory is sometimes claimed on behalf of Henri Dutrochet (1776–1847) who first asserted in 1824 that all living things are made of cells. Dutrochet atomized the organism into cells and called them fundamental units. He also asserted that the differences among tissues were attributable to substances contained in cells or produced by them, but he did not articulate a theory.

The idea that turned the cell from a "hollow vessel," an "areola," or a "utricle" into a sensible principle of how life was organized appeared for the first time in 1839, when Theodor Schwann

(1810–1882), developing a suggestion by his friend, Mathias Jakob Schleiden (1804–1881), placed the cell at the center of a mechanism for development. According to Schwann, the idea "that there exists one general principle for the formation of all organic productions, and that this principle is the formation of cells . . . may be comprised under the term *cell theory*" (Schwann, 1839, quoted from Libby, 1922, p. 259). This is still the cell theory.

The initial cell theory turned out to be a false start. This is not to say that it did not have enormous heuristic value. It did, but its explanatory value was marred by numerous erroneous notions about cells and their production. These notions required correction, and amending a theory is always the prerogative of theorists.

The most glaring problem was the emphasis prior theory placed on the "wall" of cells rather than on the nucleus and cytoplasm. Attention only shifted from the outside of cells to the inside when embryologists showed that cartilage was a secreted product of developing cells rather than a part of cells.

Another major problem was the idea that cells in multicellular organisms were essentially the same independent units as unicellular organisms. Much verbiage was spent to correct this notion with little impact (Russell, 1930).

The problem was in part overgeneralization and in part semantic. Because the cell theory required all living things to be made by cells, the absence of cells within an infusorian meant that the infusorian itself had to be a cell, a "unicellular" organism. Furthermore, if infusorians were independent and capable of meeting all their needs, the cells of multicellular organisms must also be independent and capable of meeting all their needs.

The problem might have been avoided if unicellular organisms had been called "noncellular organisms" and the word "cell" used exclusively for the cells of multicellular organisms. Regrettably, this convention was never widely adopted. The term "noncellular" seemed to violate the cell theory by implying that some living things were not made by cells.

Today, the possibility of fundamental differences between unicellular organisms and the cells of multicellular organisms is more acceptable. Phenomena in unicellular organisms such as extragenic inheritance (i.e., the perpetuation of the cell's pattern of ciliary rows through the cortex independently of the nucleus) (Sonneborn, 1967) set them apart from multicellular organisms. Moreover, polyphyletic theories for the origins of eukaryotic cells based on endosymbiont theories of

prokaryotic origins for mitochondria and chloroplasts and the spirillian origins for flagella and cilia undermine faith in the unity of cells (see Margulis, 1981). But the problem of overgeneralizing the cell theory continues to daunt contemporary biology.

Still another flaw in the original cell theory concerned the mechanisms of cell production. Some of the theory's early advocates, including Schwann, thought that the nucleus of a cell was a center of preformed parts. Coming into being by a type of crystallization, a "daughter" cell was supposed to pop out of a "mother" nucleus. Today, ideas about the role of the nucleus in cell division are very different, but the connotation of a nucleus of crystallization is still implicit in the name of the egg nucleus, the **germinal vesicle.**

The first correction to the initial cell theory was made by the hard-nosed embryologist, Robert Remak (1815–1865), and the visionary physiologist, Johannes Müller (1801–1858), who observed "cleavage" in the early development of frogs. Others, including the classical microscopist, Jan Swammerdam (1632–1680), also observed amphibian embryos divide, but Remak and Müller appreciated that a cell, rather than a nucleus, produced new cells (i.e., **sibling cells** in current usage).

Rudolf Virchow (1821–1902) soon clarified the role of division in the cell theory and reformulated the cell theory. Indeed, the current version of the cell theory may legitimately be attributed to Virchow.

He introduced the theory to physicians and recast histology and pathology in the mold of cells. He classified tissues by their characteristic cells and showed how pathological states were variations of physiological states. Virchow also described various abnormalities in the fetus and drew attention to the similarities between less developed parts of the fetus and tumors or "neoplasms" in the adult.

In Virchow's version of the cell theory, cells had no other mode of origin other than previous cells: "Wherever there is a cell, there has been one before,"[3] or, more succinctly, *omnis cellula e cellula* (every cell is from a cell). Thus, Virchow pointed all the way back to egg and sperm as progenitor cells and all the way ahead to egg and sperm as ultimate cells in succeeding generations. Embryology's mission was to fill in the gap between the successive rounds of gametes.

Embryologists quickly accepted Virchow's

---

[3]Wo eine Zelle entstent, da muss eine Zelle vorasugegangen sein (Virchow, 1858, *Cellularpathologie*, quoted from Wilson, 1896, p. 45).

challenge. Remak and Müller already recognized the egg as a cell, and embryologists placed embryos of every sort under scrutiny, usually with the same result: cells were made by cells, and embryos were made by masses of cells.

Vindicated at last, the cell theory proved to have enormous explanatory and heuristic value. The fresh breeze of descriptive and mechanistic hypotheses that the theory blew over biology has never grown stale.

Technical progress and increased sophistication in methods for observing cells soon followed. The compound light microscope was vastly improved during the 19th century by the perfection of achromatic lenses and later oil immersion optics. Increasingly reliable techniques for processing tissues and embryos for microscopic observation also became available. The microtome was invented for cutting sections of tissue automatically, and histologists developed many of the histological and cytological stains still used today to visualize tissues in sections.

Yet problems remained, and new prejudices had replaced older ones. Remak's version of cleav-

age, for example, was about as complicated as separating links of sausage by squeezing and twisting. The nucleolus was thought to cleave first, then the nucleus, and finally the cytoplasm.

By the 1870s, Edouard Strasburger (1844–1912) using plants and Walther Flemming (1843–1915) using animals had characterized the events of **karyokinesis** correctly. The term **mitosis** (Gk. *mit-* thread + *osis*) was later introduced for the behavior of the threadlike chromosomes or colored bodies that came into view at the beginning of karyokinesis.

When the relationship of nuclear events to **cytokinesis** was clarified, Remak's version of cleavage became known as "cell fragmentation," "direct division," and finally **amitosis** as distinct from "cell division," "indirect division," or **mitotic division.**

In the last decade of the 19th century, the stages of mitosis (prophase, metaphase, anaphase, and telophase, Fig 2.22 a–h) were established. With the exception of prometaphase, added later to designate events originally assigned to late prophase, the stages remain unchanged to the present.

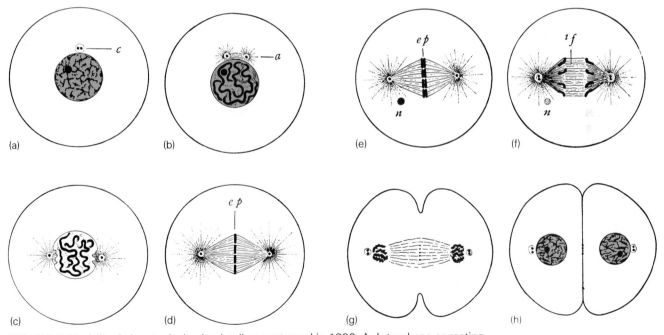

FIGURE 2.22. Mitosis in a typical animal cell as portrayed in 1896. A. Interphase or resting cell. A single nucleus with nucleolus is present but two centrosomes (c) have already appeared. B. Early prophase. The mitotic spindle has begun to form between centrosomes, and chromosomal threads appear within the nucleus. C. Late prophase (prometaphase). Chromosomes have formed from threads, and the nuclear membrane fades. D. Metaphase. Mitotic (or karyokinetic) figure is established with chromosomes at equatorial (or metaphase) plate (ep). E. Chromosomes have split into chromatids. F. Anaphase. Chromatids move to poles of spindle. G. Telophase. Division of cell body (cytokinesis) begins as chromosomes fade and nuclei begin reconstruction. H. Division complete. From E. B. Wilson, *The cell in development and inheritance.* Macmillan, London, 1896. Used by permission.

In the 20th century, cytologists realized that the cell's lifetime was spent rotating between the two phases of a cycle. One phase, the mitotic phase, was associated with changes in the nucleus. These changes resulted in the uniform and equal division of the cell's chromosomes. Generally, equal division of the cell's cytoplasm followed.

The second phase, called **interphase** because the nucleus was between divisions, was widely thought to be associated with changes in the cytoplasm. It was never considered a "resting" period in the life of the cell, but technical difficulties retarded studying the cytoplasmic events of interphase.

Contemporary research finally rounded out the cell theory by showing that cells prepared for division grew and differentiated during interphase. Cells did not build the organism through division alone. They built it with the products of growth and differentiation as well as with new cells

Today, the greatest challenge to the cell theory comes from molecular biologists. They do not question whether cells exist, how they divide, or the importance of interphase. Rather, they question whether cells are ultimately the fundamental units of life. DNA may be more fundamental and cells merely convenient factories for fabricating the organism according to DNA's plan.

If all that the egg and sperm bring to the zygote is DNA, then the molecular biologists are right. If, however, additional information is conveyed to the zygote through the organization of the cytosol, organelles, the membranous system, autonomous mitochondria, and the environment, then the cell theory will survive this new challenge.

PART 1 SUMMARY AND HIGHLIGHTS

Embryos are developing multicellular organisms produced through sexual reproduction. When two sex cells, otherwise on the verge of extinction, fuse to form a zygote, they are part of a process that potentially has no end. Life cycles consisting of haploid or hemizygoid and diploid or zygoid phases are linked by meiosis and fertilization. Meiosis combines two cell divisions with one act of replication and results in the creation of cells with haploid nuclei. Fertilization involves the fusion of gametes and their nuclei in the formation of a zygote with $2n$ chromosomes and $2C$ DNA.

Sexually reproducing species may exist as unicellular or multicellular organisms in one or both phases of the life cycle. Gametogenesis in plants, algae, and fungi refers to the production of gametes and, in animals, to all the events leading to the formation of gamete producing cells as well.

In animals, meiosis occurs only in cells about to differentiate as gametes, and cells in the haploid phase of the life cycle have no mitotic divisions. The embryo is formed by cells produced following mitosis in the zygote.

In some protoctists and fungi, the zygote formed by fertilization undergoes meiosis, and mitosis occurs only in haploid cells which may differentiate as unicellular organisms or undergo sporogenesis to become spores capable of asexual reproduction.

In the alternation of generation occurring in large algae and plants, haploid cells form a multicellular organism known as a gametophyte capable of forming gametes. When the zygote undergoes mitosis and the cells formed remain integrated in an embryo, they form a multicellular diploid organism known in algae and plants as a sporophyte.

Embryologists look at embryos and dissect them, magnify and pulverize them, compare and analyze them. Descriptive embryologists ask "what, who, where, and when" questions and draw answers from empirical observations. "Why" questions are asked by comparative embryologists who commonly seek answers in terms of remote causes and ancestral relationships or adaptations and functions.

Experimental embryologists ask "how" questions and analyze developing systems through levels of organization to find answers in terms of immediate or proximal causes. Theoreticians employ models, equations, and simulations, and developmental biologists attempt to reduce problems of embryology to subcellular levels of organization.

Aristotle's empirical method and concept of epigenesis, or emergent development, provided the basis for descriptive embryology, while comparative embryology was launched when Karl Ernst von Baer tested generalizations about development by searching for similar structures in different embryos.

Theoretical embryology begins with nature philosophy and recapitulationism. The biogenic law of Ernst Haeckel and Thomas Huxley emerged when Darwin's concept of evolution by chance and selection was yet to be integrated fully into biological theory. The idea that ancestry determined development was ultimately replaced by the idea that embryos, like adults, bear similarities to congeners, both past and present.

Preformationism, or the concept of preformed (now programmed) parts directing development, set the stage for the appearance of experimental embryology, and hypothetical minute parts matured into our modern version of genes. Under the guidance of Wilhelm Roux, experimental embryology became the driving force behind embryology's 20th century renaissance and the emergence of developmental biology.

Reductionism is unsurpassed for its efficiency at answering questions within a narrow range of interests, but it sometimes leads to "learning more and more about less and less." The gene is the crowning achievement of reductionists, but cells, their lineages, and subcellular materials have also yielded secrets. Other aspects of development may require more synthetic and holistic approaches and methods.

The end point of embryonic development is a complex mixture of behaviors and organization. Larval behaviors may pass benefits on to adults, and adult behaviors may redound to embryos. Because life cycles evolve, adaptive advantages may be shifted from one stage to another, and individual fitness at any one time may be sacrificed for inclusive fitness in the course of time.

Development turns the overtly structureless egg into the structured embryo and adult. It turns the spherical or radial symmetry of the egg into the biradial and bilateral symmetry of the adult and imposes polarity on structures. It assigns dorsal and ventral, anterior (superior) and posterior (inferior), right and left values to parts, and defines longitudinal axes, sagittal (lateral), frontal (coronal), and transverse (cross sectional) planes of symmetry.

Eumetazoan gastrulas provide major criteria for animal classification. Diploblastic animals have gastrulas with two cell layers, while triploblastic animals have gastrulas with three layers. Among triploblastic animals, protostomes form an early

stomodeum, while deuterostomes form an early proctodeum. Acoelomates fail to develop a coelom; pseudocoelomates develop a coelom incompletely lined by mesoderm, and eucoelomates form a completely lined coelom. Among the eucoelomates, schizocoelomates generally form a coelom by cavitation, while enterocoelomates form a coelom from cells associated with the endoderm.

Development provides the organism with a phylum-specific body plan either directly through the embryo or indirectly through a larva. The body plan of enterocoels is the reverse of the body plan of schizocoels, pseudocoelomates, and acoelomates. Complete segmentation is conspicuous in annelids and arthropods, while partial segmentation is a major feature of vertebrates. Development proceeds through growth and differentiation within and between segments and partial segments.

Tissues and their cell types are populations of similar cells having characteristic relationships. Diversification gives rise to different tissues, and differentiation produces cells of different tissue and cell types. Differentiated tissues are classified primarily by the relationships cells have to each other through specialized areas of their plasma membranes and by the relative amounts and types of synthetic product they accumulate or secrete.

Organs contain a soft parenchyma and a supportive stroma. Mature vertebrates have at least four stable tissues. Epithelial tissues (simple: squamous, cuboidal, columnar; pseudostratified; stratified) are made of cells broadly in contact with each other and resting on a basal lamina containing nonfibrillar collagen secreted by the epithelium and laminin. Epithelia cover surfaces and comprise the parenchyma of many organs.

Connective tissues (loose, dense, cartilage, bone, adipose) have cells with relatively few intercellular contacts and abundant extracellular materials. Many different types of connective tissue cells seem to be derived from fibroblasts while tissue macrophages are interconvertible with vascular white blood cells.

Muscle is contractile tissue characterized by ordered arrays of actin and myosin filaments. Smooth (involuntary) and cardiac muscles are cellular, but skeletal muscle (striated or voluntary) is syncytial (many nuclei joined in a common cytoplasm surrounded by a single plasma membrane).

Nervous tissue contains nerve cells capable of stimulating excitatory impulses to other nerves via synapses and to muscle and glands via end plates. Nerve cell bodies are located in the central nervous system and in ganglia of the peripheral nervous system. Glial cells and neurolemmocytes invest nerve fibers in sheaths and may guide development. Sensory cells act as transducers of environmental stimuli.

Blood cells (RBCs and leukocytes), lymphocytes, and their progenitors constitute a vascular tissue, although these cells have many affinities with connective tissue cells. Germ cells may also be considered a distinct tissue, although they are traditionally considered epithelial. In most animals, germ cells come from a unique cell line isolated relatively early in development and characterized by the ability to perform meiosis and differentiate as gametes. Early embryos have changing populations of epithelial and mesenchymal cells.

Most if not all embryonic cells divide, but cells in adult tissues may suppress division. Steady-state or renewal cell populations occur in tissues exposed to the external environment or subject to wear and tear. Cells either undergo terminal differentiation or remain in a proliferative subpopulation. No dividing cells are found in highly structured static cell populations and in some female germ-cell populations. Few of the cells of expanding cell populations normally divide, but most can do so in the event of injury. Intermediate populations maintain a nonfunctional subpopulation of occasionally proliferative cells adapted for repair.

Embryonic and adult eukaryotic cells typically contain a nucleus suspended in cytoplasm enclosed by a plasma membrane. The liquid content of the cytoplasm or cytosol is a dynamic sol-gel interphase in which sedimentable and nonsedimentable components are exchanged. Secondary messengers and intracellular receptors carry information to various parts of the cell through the cytosol.

A cytoskeleton, consisting of microfilaments (actin), microtubules, and more stable intermediate filaments, influences cell shape, movement, and function. Soluble enzymes of intermediate metabolism involved in cellular "housekeeping" and ribosomes involved in the synthesis of intracellular proteins are structured by a microtrabecular network consisting of fine microfilaments.

Mitochondria are replicating membrane-bound organelles involved in energy metabolism. Centrioles associated with the mitotic apparatus of an-imal cells and basal bodies associated with cilia and flagella are also replicating organelles.

Cellular membranes are unit membranes with a lipid bilayer construction in the form of a fluid mosaic suspending proteins. Beginning with the plasma membrane covered by a glycocalyx, a double-membrane system extends through the cell as cisternae, tubules, and vesicles of the smooth and rough surface endoplasmic reticulum and the nuclear envelope. Ribosomes, linked by messenger RNAs into polysomes, may be bound to membranes of the rough surface endoplasmic reticulum.

Cisternae of the Golgi apparatus communicate with other parts of the cell's membrane system through vesicles. Lysosomes participate in intracellular digestion and recycle components of the plasma membrane back to the surface. Pinocytosis or constitutive endocytosis and receptor-mediated endocytosis involving clathrin-associated coated pits are the chief ways of transporting materials into cells, while exocytosis is the chief way of moving materials out of cells.

Pores through the double-layered nuclear envelope are presumably the main channels of communication between the nuclear sap and the cytoplasm. Within the nucleus, a nucleolus containing nucleolar organizer genes is the site of most ribosomal RNA synthesis and of ribosome construction. Nuclear lamina stabilizes the nucleus, regulates its construction and breakdown during cell division, and influences chromatin condensation and gene action.

Chromatin is tightly bound in heterochromatin and loosely bound in euchromatin. Nucleosomes, consisting of an octameric core of histones wrapped in DNA, are complexed with additional histone and linker DNA, wound into chromatin fibers, folded, bound, and wrung into the higher ordered structure of chromosomes.

The idea of cells and concepts of their production have changed enormously over the last three centuries. Casper Friedrich Wolff anticipated much of the cell theory from his studies of plants and chick embryos, but his concept of a globular organization of tissues was rejected. Cells became acceptable to biologists when nature philosophers adopted cells as a basis of classification. By drawing attention to cell production in multicellular animals, Schwann provided a mechanism for constructing organisms and turned the idea of cells into a "cell theory" of development.

Virchow was largely responsible for rewriting the cell theory in modern terms and extending it to histology and pathology. Misunderstandings about cell production began to be corrected when embryologists described cleavage. Mitosis was linked to cell division in the late 19th century, and, in the 20th century, interest shifted toward the interphase portion of the cell cycle. Questions linger in developmental biology about whether the fundamental unit of development can be reduced beyond the cell to genes.

PART 1 QUESTIONS FOR DISCUSSION

1. What is a generation? A life cycle? Define sex and sexual reproduction for eukaryotes. Contrast and compare fertilization and meiosis.

2. What is the alternation of generation? Define zygotic meiosis and gametic meiosis. What are gametophytes and sporophytes? Gametes and spores? Distinguish between the diplobiontic and monobiontic life cycles.

3. What levels of organization encompass embryological studies, and how are they relevant to development? What is epigenesis and how is it different from preformation?

4. What are the implications of stages of embryonic development to the concept of recapitulation? What is the biogenic law? Have you heard of this law in other courses, such as anthropology and psychology? How is recapitulationism discussed in these courses?

5. Compare experimental and theoretical aspects of embryology. What is reductionism? Which of the branches of embryology do you consider the most interesting and which the most important? Why?

6. Name the main contributions of Aristotle, Karl Ernst von Baer, Ernst Haeckel, and Wilhelm Roux to embryology?

7. What direction would you like embryology to take in the future? What problems would you like to see future embryologist tackle? If embryologists gained greater control over embryonic development, what improvements in human beings would you hope embryologists would bring about?

8. What are the roles of embryos and larvas in the life cycle, and how are these roles fulfilled? How do life cycles explain the evolution of behaviors that are disadvantageous for the organism performing them? What behaviors of larvas aid adults, and what behaviors of adults aid offspring? Give specific examples.

9. How do embryos fulfill their roles in development? What sorts of symmetry do you see in eggs, embryos, larvas, and adults? What would an adult look like without bilateral symmetry? Without polarity?

10. What kinds of adults develop from gastrulas with two cell layers? With three cell layers? What kind of adults develop from gastrulas that fail to form coeloms? That form a coelom without a complete mesodermal lining? That form a coelom by cavitation? That form a coelom by folds associated with endoderm? Distinguish between a stomodeum and proctodeum, and describe their formation in protostomes and deuterostomes.

11. Define tissue diversification and cellular differentiation. List the characteristics of vertebrate tissues, and describe some of their variations. Should germ cells be considered a separate tissue and why? Should vascular tissue be combined with connective tissue and why?

12. Distinguish between steady-state, expanding, intermediate, and static cell populations. Suggest what adaptive advantages each type of tissue dynamics might have in particular situations.

13. Define the cell sap and cytosol. Describe the components and structures of the microtrabecular network and cytoskeleton. Describe the cell's membranous system.

14. Should the nucleus be classified as a cell organelle? Describe the nucleolus and its function. What is chromatin, and how are its components organized?

15. What changes have occurred in the concept of the cell over the last three centuries? How have concepts of cell production changed? What is the cell theory? How did it profit from the description of cleavage by embryologists? Is the cell theory accepted today? In what way is it under attack?

PART 1 RECOMMENDED READING

Bell, G., 1982. *The masterpiece of nature: The evolution and genetics of sexuality.* University of California Press, Berkeley.

Bloom, W. and D. W. Fawcett, 1975. *A textbook of histology,* 10th ed. Saunders, Philadelphia.

Calow, P., 1978. *Life cycles: An evolutionary approach to the physiology of reproduction, development and ageing.* Chapman and Hall, London.

Dustin, P., 1978. *Microtubules.* Springer-Verlag, Berlin.

Fulton, A., 1984. *The cytoskeleton: Cellular architecture and choreography.* Chapman and Hall (Methuen), New York.

Hyman, L. H. 1940. *The invertebrates: Protozoa through Ctenophora.* McGraw-Hill, New York.

Jacob, F., 1976. *The logic of life.* Vintage Books, New York.

Lewin, R. 1984. Why is development so illogical? *Science,* 224:1327–1329.

Marx, J. L. 1984. The riddle of development. *Science,* 226:1406–1408.

Needham, J. 1959. *A history of embryology,* 2nd ed. with A. Hughes. Cambridge University Press, Cambridge.

Stent, G. S., 1985. Thinking in one dimension: the impact of molecular biology on development. *Cell,* 40:1–2.

CONTINUITY AND CHANGE: HEREDITY AND DEVELOPMENT

*H*ow does reproduction result in off-spring resembling parents when the mechanism of development is change? The elucidation of the gene in the present century represents a giant step toward solving the enigma of heredity's relationship to development. Lying at the hub of both heredity and development, the gene in the form of DNA is both a vehicle for heredity information and a path for developmental change.

In Part 2, the movement of information through heredity and development is examined at both the microscopic and molecular levels of organization. Chromosomes are followed through mitosis and meiosis; chromatin is traced through interphase and DNA through replication and gene rearrangement. RNA is followed through transcription, processing, modification, and translation. Finally, heredity and development are reviewed as concepts in genetics and embryology.

Chapter

3

CELL REPRODUCTION AND MOLECULAR REPLICATION

Cell division is an indispensable process for living organisms. It is important in both the qualitative and quantitative senses; the cell has to pass its genetic information on to its offspring, and at the same time, it has to increase its own number and type in order to proliferate and/or make up a multicellular body. The former is carried out by nuclear division and the latter, by cytokinesis.

I. Mabuchi (1986, p. 175)

CELL CYCLE

Most embryonic and some adult eukaryotic cells rotate between periods of mitosis and interphase. The rotation is called the **cell cycle** or the **mitotic cycle** (Fig. 3.1). Events occurring in each part of the cycle are indispensable for events in other parts. The replication of DNA and synthesis of histones during interphase provide the fundamental materials of mitotic chromosomes, and chromatids delivered to sibling cells by mitosis provide the chromatin of interphase. While interphase is absent or truncated in early embryonic cells, it is expanded in adult cells.

Closer Look at Mitotic Division

Mitosis functions to distribute chromatids equally to the sibling cells produced by cell division (see Fig. 1.10). The period of mitosis is delineated by the appearance of chromosomes in the nu-

cleus of a dividing cell and the disappearance of chromatids in the reforming nuclei of sibling cells.

Mitosis has four traditional phases plus prometaphase. During the first phase, **prophase** (Gk. *pro-* before + *phase* appearance), the mitotic apparatus develops outside the nucleus, and chromosomes make their appearance within the nucleus. Although not yet apparent (Fig. 3.2a), the chromosomes are longitudinally doubled or **dyad chromosomes,** consisting of two chromatids joined at a centromere.

Cytoplasmic preparations for mitosis parallel nuclear preparation. In animal cells, the first morphological evidence of impending cell division is seen in the **centrosome** (or diplosome), a pair of **centrioles** and halo of finely granular material (Fig. 3.3). During interphase (at the S period, see below), each centriole becomes a **parent centriole** by forming a **procentriole.** As cell division approaches, a parent centriole and its growing procentriole migrate around the nucleus.

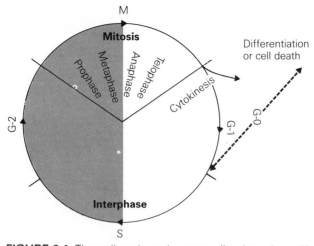

FIGURE 3.1 The cell cycle and a noncycling branch to differentiation. Cells alternating between periods of mitosis and interphase are cycling. Those that have abandoned mitosis enter a G-0 state from which they may not return. M, mitosis; G-1, postmitotic gap; S, synthesis of DNA; G-2, premitotic gap; G-0, noncycling gap.

An **achromatic** (i.e., a colorless) **apparatus** or **spindle** is responsible for the distribution of chromosomes. The **mitotic apparatus** consists of the spindle plus a **chromatic** (colored) **apparatus** of **chromosomes.** In typical animal embryonic cells, the mitotic apparatus is tipped with two star-shaped **asters** (Gk. *aster* star) and is called an **amphiaster** (Gk. *amphi-* around or on both sides + *aster*; Fig. 3.3). In adult animal cells, the asters are smaller than in embryonic cells and may be lost toward the end of mitosis.

Microtubules (MTs) are heavily concentrated in mitotic spindles and asters. These tubules gather at **microtubule organizing centers** (MTOCs; Brinkley, 1985) near the **spindle poles.** A diplosome typically lies within the MTOCs of animal cells (Fig. 3.3) but not always (e.g., early blastomeres of the mouse) (Szollosi et al., 1972). Seed plant cells lack centrioles but contain MTOCs in an area called a centrosphere (Fig. 3.2b).

Only the parent centrioles attach to microtubules in the mitotic apparatus. The new centriole will only attach to microtubules when it becomes a parent centriole in the next cell division.

As outwardly directed **aster rays,** microtubules end in the cytoplasm or against the plasma membrane. As inwardly directed **polar fibers,** microtubules run into the spindle from each MTOC but do not stretch to the other MTOC. Instead, an area at the equator of the spindle contains overlapping microtubules originating at each pole.

Later, when the spindle is fully developed, **kinetochore fibers,** originating at **kinetochores** on

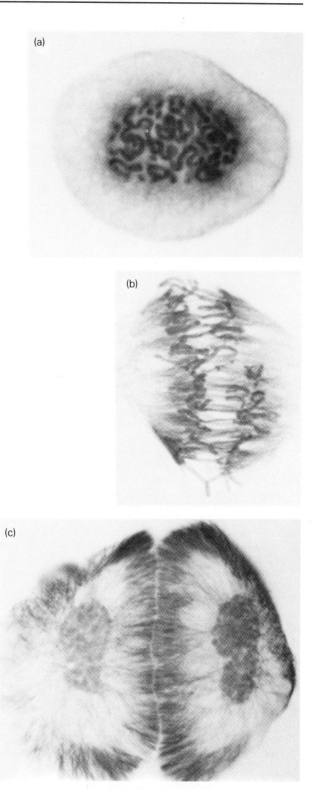

FIGURE 3.2 Light micrographs of endosperm cells from *Haemanthus katherinae* Bak. a: prophase. b: mid-anaphase. c: telphase. The deeply staining thin fibers of the spindle and reforming cells are bundles of microtubules visualized by immunogold staining. The chromosomes are stained with toluidine blue. ×1000. From L. C. Morejohn et al., *Planta,* 172: 252 (1987) by permission of Springer-Verlag and the authors.

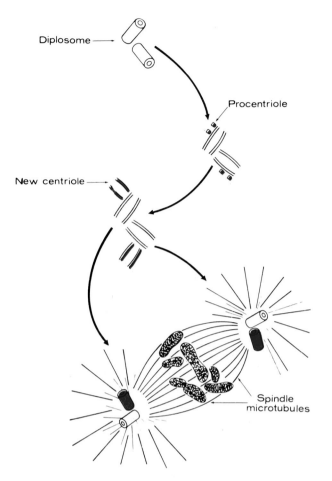

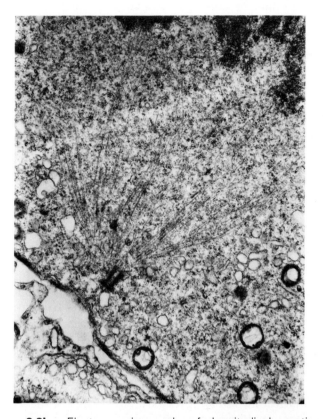

FIGURE 3.3a Scheme for the replication of centrioles and their participation in the mitotic spindle of an animal cell. Originally, two centrioles are present in a diplosome. Procentrioles appear and grow into new centrioles prior to division.

3.3b. Electron micrograph of longitudinal section through half of the spindle of a human cell at metaphase. Microtubular spindle fibers fan out from the centriol at the bottom left, while mitotic or kinetochore fibers stretch from kinetochores of chromosomes at the top right. Bends in the kinetochore fibers suggest that they are under tension. ×11,000. Part A from D. W. Fawcett, in R. A. Beatty and S. Gluecksohn-Waelsch, eds., *The genetics of the spermatozoon.* R. Beatty and S. Gluecksohn-Waelsch and others, Edinburgh, N.Y. (1972). Used by permission. Part B from O. Bennke. *Triangle,* 13: 7 (1974) by permission of Santoz and the courtesy of the author.

either side of the centromeres of dyad chromosomes (see Mitchison, 1988), run into the spindle toward the poles (i.e., opposite the direction of polar fibers). The two types of fiber acquire a **pinnate organization** and form a complex **microtubular "fir tree"** — a trunk of kinetochore fibers rising from the equator and side branches of polar fibers descending from the poles.

Kinetochore fibers aid in locating the chromosomes at the equator of the spindle and in orienting the chromosomes' centromeres toward the poles of the spindle. The pinnate organization of the spindle is also necessary for chromosomal migration at anaphase (Bajer and Molè-Bajer, 1985).

At the same time that kinetochore fibers are forming, the nuclear envelope breaks down. Enzymatic phosphorylation of the nuclear lamina seems to be responsible for separating chromosomes from the nuclear envelope and fracturing

the envelope into double-membrane structures indistinguishable from double membranes of the rough surface endoplasmic reticulum.

The breakdown of the nuclear envelope and the formation of kinetochore fibers identify the **prometaphase** stage of mitosis (Inoué, 1981). In some algae and protozoans, however, the nuclear envelope does not break down, and the spindle develops outside the nucleus without the fibers penetrating into it.

With the disappearance of the nuclear envelope, chromosomes on the spindle constitute the

mitotic figure occupying the former position of the nucleus. Chromosomes move rapidly and erratically as the mitotic apparatus is completed.

When the chromosomes come to rest, they lie on the spindle's **equatorial plane** or **metaphase plate** (Gk. *meta-* [akin to mid-] between + *phase*). Metaphase is the pivotal stage in mitosis. Before metaphase, the spindle is developing, and the chromosomes are forming and jockeying for position. After metaphase, the spindle is disintegrating, and the chromosomes are moving apart and disappearing.

The metaphase chromosomes seem to hover on the brink of separation, but when the chromatids finally move apart at **anaphase** (Gk. *ana-* against + *phase*, referring to the apparent mutual repulsion of chromatids) they do so rapidly. The centromeres on the chromatids disassociate, and single chromatids, or **monad chromosomes,** move toward the poles of the spindle (Fig. 3.2).

The chromatids appear to be only passively involved in their movement, since nonchromosomal objects in the spindle move in the same direction and at the same rate as chromosomes. Various "push-me pull-me" forces have been suggested as the driving force of anaphase chromosomes.

The assembly of microtubules in some part of the spindle ordinarily parallels the disassembly of microtubules in another part. Shortening of the kinetochore fibers may pull the chromosomes, while elongation of polar fibers may push them. Alternatively, a continuous assembly of side branches on the microtubular "fir trees" may be all that is necessary for chromosome migration (Bajer and Molè-Bajer, 1985).

A **dyneinlike** protein, similar to that found among microtubules in cilia and involved in ciliary movement (see Fig. 2.19), may operate in chromosomal movements too. The polar fibers in the region of fiber overlap at the equator may slide past each other through the action of dyneinlike protein and push chromatids apart.

Alternatively, anaphase movements may be due to the relaxation of tension within the spindle. Initially, the "zippering" of microtubules into parallel arrays near the equator may create tension, manifest as bending toward the MTOCs at the poles. Intraspindle tension would reach its maximum during metaphase when the amount of parallelism between microtubules is maximum. Then, breakdown and disorganization near the spindle poles accompanying anaphase would release tension and produce a force capable of moving chromatids (Bajer, 1977).

At **telophase** (Gk. *telo-* end + *phase*), the chromatids reach the spindle poles. The kinetochore fibers and the kinetochores disappear; the nuclear lamina is dephosphorylated, and cytoplasmic vesicles coalesce into new nuclear envelopes. The chromatids **decondense** and chromatin disperses. The nucleolar organizers synthesize rRNA; nucleoli reappear and coalesce, and ribosomes are produced again (Fig. 3.2c).

The duration of the phases of mitosis vary widely (Table 3.1), but anaphase is usually the shortest phase, and metaphase is frequently the longest. Prophase and telophase are difficult to measure, since the beginning of the one and the end of the other are ambiguous.

Cytokinesis, or the division of the cytoplasm and separation of the telophase nuclei in sibling cells, is intimately related to mitosis and sometimes considered the final phase of mitotic division. Not only is the division of the cytoplasm begun during anaphase, but the plane separating sibling cells is the same as the plane of the former metaphase plate.

Still, mitosis and cytokinesis should not be too closely linked, since they seem to depend on separate mechanisms. Mitosis sometimes may occur without cytokinesis (e.g., intralecithal cleavage of insect eggs), and cytokinesis may occur without directly following mitosis (e.g., the cellularization of the insect blastoderm).

In animal cells, the **signal** determining the plane of cell division appears to be given by the mitotic apparatus or, in the case of embryonic cells, by the asters. Actin fibers, present in the cell **cortex** beneath the plasma membrane throughout the cell, become concentrated in a circular band or

TABLE 3.1. Duration in minutes of periods of mitosis in some tumor cells[a]

	Prophase	Metaphase	Anaphase	Telophase
Yoshida sarcoma	14	31	4	21
MTK sarcoma	10	44	5	18
HeLa cells	18	77	9	110
Ehrlich ascites	250	22	14	22

Source: G. G. Steel, *Growth kinetics of tumours.* Clarendon Press, Oxford, 1977.

contractile ring, where they are joined by myosin fibers (although the latter have yet to be visualized in the electron microscope) (see Mabuchi, 1986). Like strings at the mouth of a purse, contraction of the ring draws the opposing surfaces of the dividing cell together.

In many early embryonic cells the total cellular volume does not change as a result of division. New plasma membrane must therefore be added to the cell's original plasma membrane. For example, a spherical egg that divides equally into two hemispherical cells (e.g., see Fig. 2.22) must add half again as much membrane as that present in the original egg (i.e., the surface area of a sphere is $4\pi r^2$, while the surface area of two hemispheres cut out of the same sphere is $4\pi r^2 + 2\pi r^2$, r being the radius of the sphere). Since the membrane hardly stretches this much, large amounts of new membrane are required by rapidly dividing embryonic cells.

New membrane begins to be synthesized before mitosis and accumulates as **blebs** or granules of membrane waiting to be added to the plasma membrane of the dividing cell. The addition of blebs to the membrane may require sharp folds, since the electrostatic repulsive force of the membrane might otherwise prevent contact with blebs and fusion (Pethica, 1961). The shortening of aster rays attached to the internal surface of the plasma membrane may "pucker" the surface and create the requisite folds.

Separating cells frequently remain attached to each other through an **intercellular bridge,** called the **midbody** (or telophasic body), filled with the remnants of the mitotic spindle, especially the overlapping, parallel (i.e., "zippered") polar fibers from the spindle's equator (Fig. 3.4). In some cells, such as dividing germ cells, intercellular bridges remain intact, and many cells are joined until the final stage of their differentiation, but, in most cells, the bridges break before the next round of cell division.

Closer Look at Interphase

Until the mid-20th century, interphase was viewed more or less as a homogeneous period. The possibility of interphase periods arose when new techniques for measuring DNA microscopically (cytophotometric techniques using stained cells, ultraviolet spectroscopy, and radiological methods) (see Caspersson, 1950) revealed regular changes in cellular DNA content.

DNA synthesis followed one of two patterns. In prokaryotes and some protozoans (e.g., *Tetra-*

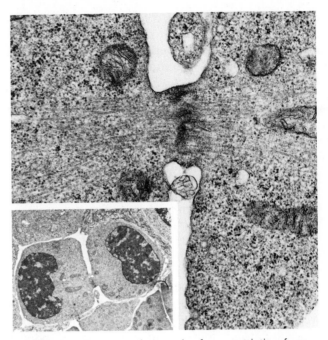

FIGURE 3.4 Electron micrograph of a constriction furrow between sibling cells (insert). Remnants of spindle microtubules pass between cells through the midbody. ×20,000. From O. Behnke, Triangle, *Sandoz Journal of Medical Science,* Vol. 13, No. 1, 1974, 7, Copyright Sandoz Ltd., Basle, Switzerland. Used by permission.

hymena), the amount of DNA present per cell increased throughout interphase. In other protozoans (including *Paramecium*) and eukaryotes in general, the amount of DNA began to increase only some time following mitosis.

Howard and Pelc (1951) devised the nomenclature for the **periods** (or divisions) of interphase and made the first estimates of their durations from the results of experiments on bean sprouts. Soon after exposing the sprouts to radioactive phosphorus, only a few interphase nuclei and no mitotic figures contained labeled DNA, but, several hours later, many of the interphase nuclei and half of the mitotic figures had labeled DNA.

A discrete **S period** (S) of DNA synthesis followed by a **premitotic gap** or **G-2 period** (G-2, Fig. 3.1) was hypothesized to account for the lapse between the appearance of labeled interphase nuclei and labeled mitotic figures. A **postmitotic gap** or **G-1 period** (G-1) was hypothesized to account for the difference between the empirically determined **intermitotic time** (or duration of the cell cycle) and the sum of the durations of S, G-2, and mitosis (M).

The durations of the periods of interphase vary greatly from organism to organism and from one set of conditions to another. For example, S can be as short as 10 minutes in yeast cells (dividing with an intermitotic time of 30 minutes) or require 30–35 hours in mouse ear skin *in vivo*. Estimates of the durations of the periods frequently suggest that S, G-2, and M are more stable and uniform than G-1.

The duration of G-1 in rapidly growing mammalian tissue culture cells of different strains is between 6 and 8 hours, but G-1 can also be indefinitely long. The duration of S is typically between 6 and 9 hours *in vitro*. With the exception of some tissues that preserve G-2 cells, it seems, in anticipation of wounding (e.g., mammalian epidermis), G-2 typically lasts about 5 hours. The duration of mitosis itself, or M, is usually placed at 1 hour.

The relative constancy of the periods in rapidly growing tissue culture cells suggests that cells move through precise events and controls during interphase. The possibility for an indefinite delay in G-1 indicates that the cell reaches a bottleneck step in this period. Satellite cells in intermediate cell populations and cells in expanding cell populations may enter a nondividing or quiescent state by leaving G-1 for a noncycling **G-0 period** (Pederson, 1972) from which they can return at some later time and reenter the cell cycle. In static and steady-state populations, G-0 cells go on to differentiate without the possibility of returning to the cycle (Fig. 3.1).

Mazia (1963) has suggested that a **chromosomal cycle** superimposed on the cell cycle explains the orderly sequence of events beyond G-1 and how the cycle can be interrupted. According to Mazia's hypothesis, a cell makes a decision to launch another round of cell division when a crucial event in **chromosomal synthesis** (possibly the initiation of histone synthesis) takes place in G-1. Thereafter, everything in the cell's storehouse and powerhouse is put at the disposal of cell division. If the decision to enter another round of division is not made, the event does not take place, and the cell remains in G-1 or G-0.

For the most part, if a cell enters S, it will usually go through the rest of the cycle and complete mitosis. The cycle is therefore divisible into two additional periods. A **stable period** or **mitogenic period** (also known as the "B" state) begins late in G-1 and ends early in the next G-1. An **unstable period** (also known as the "A" state) occurs entirely in G-1.

Theoretically, the decision to divide may be stochastic, that is, determined at random. A cell may "decide" to divide and pass a **restriction point** in the "A" state much like a potentially radioactive atom "decides" to emit its radiation (see Pardee et al., 1978). Quiescent or G-0 cells would be those with a low probability of leaving the "A" state. Rapidly cycling cells would be those with a high probability of entering the "B" state. Differences between various sorts of cell populations therefore may reflect differences in the numbers of cells present in each state. More precise molecularly based or cell-factor-based models could help clarify the issue.

MOLECULAR "REPRODUCTION"

Taken all together we may see the problem as that of how chromosomes can multiply or divide their size without appearing to change their genetic character.

C. D. Darlington (1981, p. 2).

In the first half of the 20th century, the idea of biological reproduction was extended to molecules. Heredity, it was thought, could have no foundation other than molecular duplication by complementary molecules (Pauling and Delbrück, 1940). Further understanding of organismic reproduction therefore depended on revealing the detailed molecular structure of hereditary materials.

Complementariness and Chromosomes

Initially, most of the authorities believed that hereditary material was made of protein. Opinion only began to shift toward DNA after 1944 when Oswald Avery and colleagues eliminated the possibility that protein accounted for a heritable shift called **transformation** from one strain of pneumococcal bacteria to another. The alternative—that DNA affected the shift—could not be eliminated.

Soon thereafter, the DNA content of every differentiated body cell in several multicellular organisms was shown to be very nearly the same (Mirsky and Ris, 1949), and the amount of DNA in a body cell's nucleus was found to be twice that in a gametic nucleus (see Vendrely and Vendrely, 1956). DNA therefore met all the essential criteria of a hereditary material: it was reduced by half during meiosis, doubled by fertilization, and passed on equally by mitosis.

By the mid-20th century, DNA was established as the hereditary material. It was known to contain long-chained polymers of nucleotides. The three-dimensional structures of the individual nucleotides were known, and X-ray diffraction photographs suggested that strands of DNA formed a regular helix. When Edwin Chargaff (1950) established that the molar ratios of purines to pyrimidines (adenine to thymine and guanine to cytosine) were approximately one, the possibility arose that DNA existed as a double-stranded structure.

Aided by an X-ray diffraction pattern obtained by Rosalind Franklin, Maurice H. F. Wilkins (Wilkins et al., 1953), James D. Watson, and Francis H. C. Crick (1953a)[1] proposed that the nitrogenous bases of two strands of DNA turned inward (see Judson, 1978; Watson, 1968). The requirements of hydrogen bonding and molecular stacking were met by winding the strands around a common axis in a double helix, and the rest of DNA's structure began to fall into place along with several pieces in the puzzle of molecular "reproduction."

By pairing nitrogenous bases internally, Watson and Crick (1953b) built **complementariness** into their model of the DNA molecule. Double-stranded duplexes synthesized against the format of existing duplexes would be reproductions (in the sense of copies) of each other.

Of the several hypothetical mechanisms for DNA synthesis first suggested to explain complementariness, the Watson–Crick "semiconservative duplication," or today's **semiconservative replication,** ultimately prevailed. In each round of synthesis, one of the two strands, now called the **template strand,** was **conserved** (or preserved), while the second **complementary** strand was made afresh (Fig. 3.5).

[1]The last three shared the Nobel Prize in Physiology or Medicine in 1962 for determining the structure of DNA.

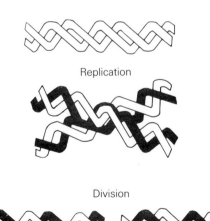

FIGURE 3.5 Replication of a DNA duplex results in the production of two DNA duplexes, each with one new strand (black).

Ironically, the semiconservative hypothesis was initially in danger of foundering on the enormity of the problem it solved so simply. Because of the vast amount of DNA known to be present in eukaryotic chromosomes, the possibility that a chromatid contained one continuous duplex of DNA was unthinkable. Autoradiographs of replicated chromosomes, that today are models of semiconservative replication, were misinterpreted in tortuous ways designed to accommodate complex chromosomal structures (see Prescott and Bender, 1963). The idea that a eukaryotic chromatid contained a single DNA duplex (i.e., a unineme) as opposed to multiple DNA duplexes (a multineme) only became reasonable after semiconservative replication was tested with heavy nitrogen incorporation and a sophisticated density-gradient centrifugation technique in *Escherichia coli* (Meselson and Stahl, 1958). Autoradiographs of tritiated-thymidine labeled DNA from *E. coli* (Cairns, 1963) also showed it to be a single, circular duplex.

Similar results with eukaryotes and the law of parsimony, requiring the adoption of the simplest explanation until it is shown to be inadequate, finally required a reexamination of premises. The unthinkable became thinkable, and semiconservative replication became a tenet of molecular biology for eukaryotes as well as for prokaryotes.

Mechanisms of Semiconservative Replication

The steps of replication (Fig. 3.6) have been garnered from results of experiments with *in vitro* systems involving enzymes from prokaryotes, such as *E. coli* and *Bacillus subtilis,* and from *in vivo* systems, especially with mutant forms of these bacteria. In addition, experiments have been performed with a variety of bacteriophage and animal viruses, the fruit fly, mammalian and avian tissue culture lines, tumors, and normal fetal and adult tissues. Replication may be similar in all the systems studied, but differences are apparent when the activities of particular enzymes are compared (Table 3.2).

Probably the greatest problem encountered by researchers trying to piece together the cell's mechanism for replicating DNA is the loss of intermolecular organization every time components of the system are isolated. In all likelihood, the proteins involved in replication are not individual agents floating in a sap but structured units, some of which probably occupy specific sites on membranes.

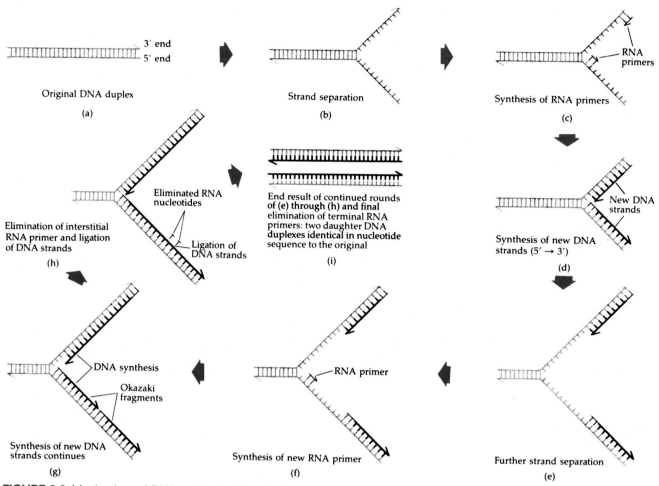

FIGURE 3.6 Mechanism of DNA replication (description in text). From D. L. Hartl, *Human genetics.* Harper & Row, New York, 1983. Used by permission.

TABLE 3.2. Some 5′ → 3′ DNA polymerases and their characteristics

	Function Principally Associated With			Activity Primarily	
	DNA Replication	DNA Repair	Exonuclease Activity	Processive	Distributive
Animal DNA polymerases:					
α					
Holoenzyme	+		−	+	
β		+	−		+
δ			+		
E. coli DNA polymerases:					
Pol I	+		+		+
Pol II			−		
Pol III					
Holoenzyme	+		+	+	

+, Activity especially high or known to occur; −, activity absent under particular conditions; blanks indicate uncertain or incomplete information.

Source: Adapted from Kornberg (1982, Table S6.3, p. S46) and Kornberg (1980, Tables 5.1 and 6.2, pp. 169 and 204).

The DNA origin or **ori** *region (also* **Ori** *region) and autonomous replication sequences are the sites where replication begins.* Its **core component** has a defined region of tightly bound DNA (i.e., supercoiled or containing superhelical turns) capable of being recognized by **initiator proteins** and a region rich in adenine and thymine (AT-rich areas) that "melts" (via the breakage of noncovalent internal bonds) into single-stranded DNA. Catalytic unwinding of the DNA continues bidirectionally from this point, followed by synthesis of complementary strands.

In *Escherichia coli*, the DNA origin (*OriC*) contains groups of tandem repeated (i.e., iterated) base pair sequences. The tightly bound region contains one group of repeated sequences that binds particles of initiator protein, and the nearby AT-rich region contains another group of repeated sequences that are the initial sites of melting. Similar tandem repeated sequences are found in other bacteria and bacteriophage (Bramhill and Kornberg, 1988).

The initiation of synthesis of mammalian viral DNA depends on a **palindrome** (i.e., an inverted repetitive sequence of nitrogenous bases within a single DNA strand) that creates one or two **hairpin** loops (Tattersall and Ward, 1976). In the case of the animal tumor virus SV40, the DNA origin (*ori-SV40*) contains an AT-rich region and binds the viral-encoded initiator factor known as the **T antigen** (Dodson et al., 1987).

Eukaryotic *ori* regions contain one or more **auxiliary components** which resemble *cis*-acting transcription elements (see Chapter 4) in addition to the core component where replication begins. Auxiliary components are related to cell type and determine the efficiency of replication. One type of auxiliary components (i.e., Group I or promoter type) lies near the AT-rich end of the origin and enhances melting. Another type (Group II or enhancer type) lies anywhere with respect to the core component (i.e., independent of distance and orientation) and directs synthesis toward replication rather than transcription (i.e., RNA synthesis) by binding replication proteins to the *ori* core. Still another type of auxiliary component (Group III, found in mitochondrial DNA) utilizes RNA, whose synthesis began upstream, to initiate replication (i.e., acts as a primer for DNA synthesis; see below) (DePamphilis, 1988).

The enzymes of replication are often interacting. An **enzyme complex** or **holoenzyme** (i.e., a multimeric or multisubunit complex) consisting of a polypeptide core and additional **polypeptide subunits** and **accessory proteins** seems to be responsible for replication. Accessory proteins may clamp the enzymes to the DNA, thereby allowing them to remain on track even through hazardous hairpin helices, and polypeptide subunits have some of the activities of the whole, but it is the holoenzyme that replicates the DNA efficiently and does so in long stretches.

Enzymes called **topoisomerases** (Gk. *topo-* surface or place, + *-iso-* same or equal + *mer-* part, and the ending *-ase* specifying an enzyme) change the surface configuration of a molecule. **Helicases** unwind the DNA helix, possibly by causing supercoiling elsewhere. **Swivel enzymes** or **gyrases** relieve the stress produced by unwinding and relax the strands, while **stabilizing proteins** or **single-stranded binding proteins** (ssb proteins) preserve the single strands of DNA and protect them from intracellular digestive enzymes.

$5' \rightarrow 3'$ DNA polymerases polymerize complementary deoxyribose nucleotides into a new DNA strand (Table 3.2). These enzymes catalyze the replacement of the 3' hydroxyl on the end of the growing DNA strand with the phosphate group of the next nucleotide. The hydrolytic reaction is fueled by the cleavage of the diphosphate group off the 5' end of the nucleotide triphosphate substrate (Fig. 3.6d) and the rapid hydrolysis of the diphosphate (pyrophosphate).

The first prokaryotic DNA polymerase to be isolated, known as **DNA pol I**, or the Kornberg enzyme, had the reputation of being a DNA repair enzyme capable of removing a wrong (i.e., noncomplementary) nucleotide from a strand (Watson, 1976). Along with the third DNA polymerase, or **DNA pol III**, DNA pol I is now recognized primarily as a replication enzyme (Kornberg, 1982).

Eukaryotic **DNA polymerase α** or I, isolated from chick embryos and from a variety of normal mammalian tissues (e.g., calf thymus, rat liver) and tumors (e.g., mouse myeloma, rat hepatoma), seems to be the principal enzyme of eukaryotic chromosomal DNA replication. The primary structure of DNA polymerase α from all these sources is remarkably similar or, in the argot of the field, "highly conserved."

A lower molecular weight eukaryotic DNA polymerase, known as β or II, seems to function primarily in DNA repair rather than replication. DNA polymerase β has far greater activity in tissues containing nondividing cells than in tissues containing dividing cells. Moreover, this DNA polymerase operates most efficiently on very short single-stranded regions of DNA rather than on larger regions associated with replication.

Other DNA polymerases (e.g., δ) may function

primarily in related processes such as repair, recombination, and the movement of transpositional elements in prokaryotes or mobile elements in eukaryotic chromosomes. Cells may also produce unique DNA polymerases in response to viral infection. Furthermore, mitochondria have DNA polymerase γ, and chloroplasts have their own DNA polymerases.

Prokaryotic DNA pol III and eukaryotic DNA polymerases α require a DNA template and an **RNA primer** or **RNA initiator** to start the reaction. DNA pol I also requires an RNA primer but prefers short single-stranded regions of DNA for its template.

Enzymes called **primases,** or **RNA polymerases,** manufacture the RNA primer (Fig. 3.6c) required for the initiation of replication. Using a single strand of DNA as a template, and usually beginning with adenine, the primases run up primers of about five nucleotides in prokaryotes, 10 in animal cells.

Why DNA polymerases require an RNA primer is uncertain. Possibly, the RNA primers act like wedges aiding the separation to single-stranded DNA. The possibility that the RNA primer is a relic from a pre-DNA past should be dismissed as another example of arguments from biological inertia.

DNA polymerases remove the bits of **non-proofread** RNA primer between stretches of single-stranded DNA and add the correct DNA nucleotides in place of the primer (Fig. 3.6h). Enzymes known as **DNA ligases** then connect the pieces of DNA (Fig. 3.6i).

The problem of preventing the progressive loss of genes at the 3' terminus through the substitution of terminal deoxyribonucleotides with ribonucleotides in RNA primers seems to have been solved in eukaryotes (e.g., the ciliate *Tetrahymena*) by the addition of a "buffer" of tandem repeated thymine–guanine sequences (sometimes called G-rich repeats) at the very ends of chromosomes.

The ends protrude as **telomeres,** which are regenerated by **telomerase** through the enzymatic addition of thymine–guanine sequences. Telomerase is a large ribonucleoprotein (RNP) particle whose RNA and protein components are essential for recognizing telomere-specific DNA structure and adding the repeats in register. Telomeres may also function in chromosome movement and organization (see Weiner, 1988).

Proofreading and correcting mistakes is the prerogative of the same DNA polymerases that synthesize DNA. Prokaryote DNA pol I and pol III holoenzyme and eukaryotic DNA polymerase δ

(but probably not eukaryotic DNA polymerases α and β; Table 3.2), in addition to polymerizing complementary nucleotides into DNA, function in **proofreading** and correcting mistakes in the new DNA strands.

Acting as a **3' → 5' exonuclease** (and, in addition, 5' → 3' exonuclease in the case of DNA pol I), nucleotides that do not form complementary pairs with the nucleotides in the template strand are eliminated or **cleaved** out and replaced by complementary nucleotides. One possible mechanism (known as V_{max} discrimination) seems to involve the kinetics of lengthening the DNA strand. Because the enzyme moves on faster from a correctly paired nucleotide than from an incorrectly paired nucleotide, the latter is likely to be replaced while the enzyme dawdles over it (see Kunkel, 1988).

The **active sites** for polymerase and exonuclease activities reside in separate **domains** (portions of the polypeptide with distinct properties). DNA pol I, for example, has a large fragment (known as the Klenow fragment) with a polymerase domain at the carboxyl terminus and a 3' → 5' exonuclease proofreading domain at the amine terminus. The 5' → 3' exonuclease activity is found elsewhere on the polypeptide.

In the multimeric enzymes, the active sites may even be on separate polypeptides. In DNA pol III holoenzyme, polymerase activity is in an α subunit and 3' → 5' exonuclease activity is in an ε subunit.

Proofreading and addition enhances the fidelity of replication, 10-fold for DNA pol I, 200-fold for DNA pol III holoenzyme, and about 100-fold for eukaryotic DNA polymerase δ. The system works so well that even when nucleotide bases are added at maximal rates, errors are made in no more than 1 in 10 billion bases!

Replicating forks are the points where the original DNA duplex branches into two arms of single-stranded DNA (Fig. 3.6b). In effect, the topoisomerases of replication turn DNA inside out at the replicating fork, and nitrogenous bases become accessible for matching up with complementary nucleotide triphosphates.

Because the strands of the DNA duplexes are **antiparallel** (i.e., going in opposite directions, like a two-way street, Fig. 3.6a) and the 5' → 3' DNA polymerases add nucleotides in only one direction, replication proceeds in opposite directions on the two arms of the fork. **Processive** (in the sense of a procession or march) or continuous replication proceeds toward the fork on the so-called **leading strand** (reading 3' → 5' toward the fork), while **distributive** (in the sense of apportioned or divided) or discontinuous replication proceeds away from the fork on the **lagging strand** (also called retrograde arm, reading 5′ → 3′ toward the fork, Fig. 3.6e).

Processive replication proceeds uninterruptedly behind one primer. Prokaryotic DNA pol III and eukaryotic DNA polymerase α add nucleotides **processively** or continuously at nearly 1000 nucleotides per second in prokaryotes or approximately 40 nucleotides per second in eukaryotes.

Distributive replication requires multiple primers. After a brief period of replication on an unwound stretch of the lagging strand, a new primer is made on the next unwound stretch (Fig. 3.6f). The DNA polymerase begins again at the primer and quickly lengthens the new strand until it reaches the previous primer (Fig. 3.6g).

The replicated DNA, called Okazaki fragments after their discoverer (Okazaki et al., 1968), with RNA nucleotides at the 5′ end, are short (1000–2000 nucleotides in prokaryotes, 100–200 nucleotides in most eukaryotes, but only 40–50 in *D. melanogaster*). Ultimately (after the primers are removed), DNA ligase brings the fragments together in a single DNA strand. The requirements for ''backstitching'' Okazaki fragments delays the process, and replication of the lagging strand follows replication of the leading strand (hence the names of the strands).

This combination of processive replication of the leading strand and distributive replication of the lagging seems to be universal. RNA also seems to be a nearly universal primer (the exception being some viruses), especially for discontinuous synthesis of the lagging strand (see Alberts, 1986).

Replicating units lie between replication forks. The replication fork in Fig. 3.6 is shown for simplicity as if it appears in isolation along a length of DNA. In prokaryotes, mitochondria, and chloroplasts, the arms of two replication forks typically lengthen in opposite directions as DNA synthesis proceeds around the entire DNA duplex. In eukaryotes, pairs of replication forks surround a **replicon** or **replicating unit** (RU) which may correspond to a chromomere or chromosome band.

Replication occurs simultaneously on both strands and in both directions (see Sheinin et al., 1978). Some part of both strands of newly synthesized DNA therefore is produced distributively and contains Okazaki fragments, while another part of the same strand is produced processively.

The matching up and ligating of Okazaki fragments that have occurred on one side of an RU also occur on the other side, and Okazaki fragments and processively produced pieces on opposite sides of the RU meet to form a **replication bubble.** Repli-con-sized duplexes of DNA are thus made which fuse with other replicon-sized duplexes, thereby enlarging replication bubbles until two chromosome-sized duplexes are completed.

Although in prokaryotes, only one ''bubble'' occurs at a time, in eukaryotes, several hundred RUs are involved in the replication of each chromosome during a typical S period. The massive lengths of the DNA molecules in chromosomes (e.g., about 500 million nucleotide pairs in human chromosome 1) require that many sites of replication proceed at maximum velocity if replication is to be accomplished within the allotted time.

Replication at RUs in the nucleus of most eukaryotes (yeast being an exception) requires concomitant protein synthesis in the cytoplasm, but the event that triggers initiation or determines the speed of replication of particular RUs may not be the synthesis of a particular protein. In cultured cells, different RUs are replicated at characteristic times during S (see Prescott, 1976), and many chromosomes can be classified as **early replicating** or **later replicating** depending on when their DNA is synthesized. In cleaving embryos, on the other hand, replication, which may be 100 times faster than in adult cells, involves large numbers of small RUs operating simultaneously (Callan, 1973).

A model for replication based on Escherichia coli, but generalizable to other prokaryotes, bacteriophage, and possibly eukaryotes, begins at the supercoiled DNA-origin (OriC in E. coli). Protein particles assembled on the DNA provide the complete replicating ''machine'' or **replisome** that moves locomotive style along the DNA duplex track as it synthesizes complementary DNA (see Alberts, 1986). The process is broken into four stages (Table 3.3).

First, the DNA is ''opened.'' This is accomplished by particles of initiator protein (DnaA or dnaA protein) that binds to tandem repeated base pair sequences (e.g., four 9 base pair sequences, called 9-mers or dnaA-boxes in *E. coli*) in the supercoiled region of the DNA origin. Initiator protein melts the AT-rich duplex beginning with other tandem repeated regions (e.g., 13-mers) thereby creating a ''bubble'' or **open complex.**

Second, **prepriming reactions** stabilize the open complex. Additional proteins (dnaC protein in *E. coli*) are directed into the bubble and bind helicases (DnaB helicase or dnaB protein) into **prepriming** or **prepriming complexes.** Gyrases, single-stranded binding proteins, and other proteins must also be present.

Third, by unwinding the DNA in both directions, the preprimosome generates two forks for

TABLE 3.3. Steps of replisome formation and initiation of replication (based on *Escherichia coli*)

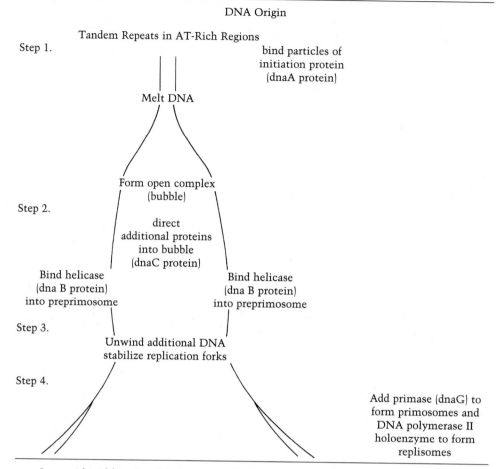

DNA Origin

Tandem Repeats in AT-Rich Regions

Step 1. bind particles of initiation protein (dnaA protein)

Melt DNA

Form open complex (bubble)

Step 2. direct additional proteins into bubble (dnaC protein)

Bind helicase (dna B protein) into preprimosome   Bind helicase (dna B protein) into preprimosome

Step 3. Unwind additional DNA stabilize replication forks

Step 4. Add primase (dnaG) to form primosomes and DNA polymerase II holoenzyme to form replisomes

Source: Adapted from Bramhill and Kornberg (1988).

bidirectional replication. The step is ATP dependent.

Fourth, a primase (dnaG) joins in to form a **primosome,** and DNA polymerase II holoenzyme is added to form the **replisome** (Bramhill and Kornberg, 1988).

Similarly, replication of DNA tumor virus SV40 *in vitro* requires, in addition to the DNA origin (*Ori*SV40 or *ori*SV40), ATP, topoisomerase, and single-stranded binding protein (both from HeLa cells), a DNA polymerase α–primase complex and the viral-encoded T antigen. This antigen operates as both an initiator protein (i.e., a site-specific DNA-binding protein) and a helicase (DNA-unwinding protein) and probably guides a DNA polymerase α–primase complex to the DNA origin (Dodson et al., 1987).

For the moment, neither eukaryotic replisomes nor primosomes have been identified, al-though the holoenzyme form of DNA polymerase α is likely to be part of a multisubunit complex. The starting system for eukaryotic chromosomal DNA is also unknown.

Histones

At the same time that nuclear DNA is replicated in eukaryotic cells, **histones** are being synthesized in the cytoplasm. Core histone synthesis is **replication dependent** (i.e., linked to DNA synthesis), but **tissue-specific variants** of H-1 found in nondividing cells (e.g., H-5 in avian erythrocytes) may be only partially replication dependent or even replication independent.

After transportation to the nucleus, the core histones are fashioned into nucleosomes and added to both strands of replicated DNA. Old nucleosomes also remain associated with both DNA

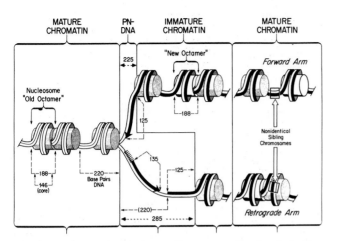

FIGURE 3.7 Chromatin formation at a replicative fork. The original nucleosomes accompany both arms at random, while new nucleosomes are acquired as soon as sufficient base pairs are available. PN, prenucleosomal DNA. From M. L. DePamphilis and P. M. Wassarman, *Annu. Rev. Biol. Chem.* 49:627 (1980) by permission Annual Reviews.

strands (DePamphilis, 1987) (Fig. 3.7). In addition, nonhistone **high-mobility group** proteins (or HMG proteins) and RNA are added to make the chromatin of the G-2 cell. The chromosomes also begin their condensation (see above, Fig. 2.21) during G-2, although dyad chromosomes will not appear until the beginning of the mitotic phase of the cell cycle.

Centromeres

When almost all else is done, centromeric DNA is synthesized. The time of replication in the centromere may be related to the unique role of the centromere in maintaining contact between chromatids until anaphase of the upcoming mitosis.

Earlier in the S period, topoisomerase activity resolves twists or **catenations** caused by unwinding the original DNA duplex. In the case of the centromere, topoisomerase activity is weak, and, when two replication forks meet at the centromere, significant catenations result (see Murray and Szostak, 1985). Each double helical turn is converted into a catenation.

When the DNA is packaged into chromosomes, the catenations between sister chromatids in the centromeric DNA are maintained as **linkages.** After the attachment of microtubules to the chromosomes' kinetochores, DNA topoisomerase seems to be reactivated and catenations in the centromeric DNA are resolved with the release of chromatids from their last point of attachment.

FIDELITY AND MEIOSIS

Although the copying of DNA is generally very faithful, there are two phenomena of great biological importance that bring about changes in the nucleotide sequence of DNA. These are, of course, mutation and genetic recombination.

Matthew Meselson (1965, p. 11)

Replication before meiosis is not entirely the same as replication before mitosis. The sensitivity of organisms produced sexually to mutations in the germ line places a burden on meiosis not carried by mitosis, and meiotic DNA is prepared with an extra measure of care known as **recombinational repair.** Still, mutant genes are replicated faithfully and are passed along through meiosis.

Mutation

Mutants are variant individuals, and mutations are stable variants of genes that cause the mutant to develop. Spontaneous mutations occur without the intervention of scientists, while induced mutations are caused artificially, but mutation as such implies nothing about the mechanism of DNA or chromosomal change.

Many mechanisms account for the appearance of mutants. Problems in chromosomal replication such as breaks, deletions, translocations coupled to position effects, failures to separate (nondisjunction), and even crossing over between sister chromatids (i.e., sister chromatic exchange) can yield mutants. **Transposable** or **mobile DNA sequences** (or **elements**), either short **insertion sequences** or longer **transposons** containing numerous genes (such as *Ac/Ds* in maize, *P* in *Drosophila*), which move into and out of chromosomes, also cause mutations. In certain strains of *Caenorhabditis elegans* the transposable element *Tc1* is responsible for the majority of spontaneous mutations (see Emmons, 1987).

At the molecular level, mutations are lapses in the fidelity of replication. Many kinds of lapses are possible, only some of which are consequential in heredity, and most of which are **repaired** before replication is completed.

Although DNA may be altered in many ways (depurination, depyrimidination, cross-linking), damaged DNA is not always inherited. In bacteria, damage is more likely to create a block to replication, and replication-blocking lesions are readily removed by highly efficient **excision-repair mechanisms.** Similar lesions in the chromatin of some rodent cell lines, however, are readily tolerated.

Other changes in the sequence of base pairs in DNA may be heritable, but, because they may be repaired, they are not necessarily the source of mutations. Mismatched base pairs can result from substitutions, additions, deletions, and rearrangements. The wrong nucleotide may be inserted at a site, for example, as a result of high local concentrations of nucleotides. The $3' \rightarrow 5'$ exonuclease activity of DNA polymerase is inhibited under these circumstances, and the enzyme works against itself.

Studies in the bacterium *Escherichia coli* on the repair of errors in DNA have provided testable models of repair in eukaryotic cells (Hanawalt, 1982). Errors may be repaired by a variety of DNA processing mechanisms. For example, the photoreactivation enzyme photolyase can repair pyrimidine dimers induced by ultraviolet (UV) radiation, and alkyl transferases can remove methyl or ethyl groups from guanine (Fig. 3.8 top line).

In human beings, the failure to perform dimer excision in the absence of a specific UV endonuclease seems to be responsible for the classical variant of xeroderma pigmentosus. Cells from patients with another variant carry out excision repair but fail to perform postreplication repair. Victims of these mutations suffer from great sensitivity to sunlight and proneness to ulcerative and cancerous skin lesions.

Mismatched bases in one DNA strand may be corrected during replication by enzyme-mediated mechanisms using the undamaged DNA strand as a template. Highly specific enzymes such as DNA glycosylase can recognize modified bases including aberrant uracils. A direct base replacement mechanism utilizing insertase for apurinic sites or an incision mechanism by an apyrimidinic endonuclease coupled to excision-resynthesis mechanisms and a joining ligase replace the incorrect base with the correct one. Alternatively, damage may be recognized by damage-specific endonucleases that incise the DNA directly and replace the wrong base.

Another mechanism of DNA repair involves modified replicases with weakened **proofreading** functions. Acting as DNA polymerases, these enzymes read through damaged DNA without performing the usual excision and replacement activities, but if they hit upon particular combinations of bases they correct the damage (i.e., translesion synthesis; Fig. 3.8, bottom line). The mechanism is **error prone,** however, and just as likely to perpetuate the error as it is to repair the error and may even make matters worse. Repair mechanisms such as this are sometimes considered part of a cell's last ditch efforts or **SOS responses** that sustain replication despite damage.

Errors that have reached both strands of DNA in a premeiotic cell are the most likely to affect eukaryotic heredity. Since an error in both strands eliminates the redundant strand that might have acted as a template, **double-stranded errors** are not

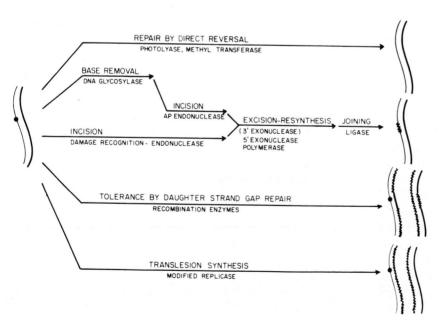

ENZYMATIC PROCESSING OF DAMAGED DNA

FIGURE 3.8 The repair of damaged DNA by enzymatic processing. From P. C. Hanawalt, in R. T. Schimke, ed., *Gene amplification.* Cold Spring Harbor Laboratory, Cold Spring Harbor, NY, 1982, by permission of the author.

readily eliminated during replication. They can, however, be corrected or tolerated by recombinational repair.

Meiosis and Recombinational Repair

The roles of meiosis in heredity, independent assortment, genetic linkages, and recombination have been appreciated for a long time (see Sturtevant, 1965), but several features of meiosis have defied explanation until recently. The most enigmatic features occur during prophase I, and their most provocative explanation is recombinational repair.

Meiotic prophase I distinguishes itself from mitotic prophase by the appearance of chromosomal axes at the electron microscopic level of resolution, by synapsis or the joining of homologous chromosomes, and by the appearance of chiasmata or bridges between homologous chromosomes. Vastly more complex than mitotic prophase, prophase I is divided into five subphases: leptotene, zygotene, pachytene, diplotene, and diakinesis (e.g., Figs. 3.9 and 3.10).[2]

In leptotene (Gk. leptos thin or fine + tene, a descriptive termination), **or the leptonema stage (leptos + nema thread), threadlike chromosomes make their appearance** (Fig. 3.9). These chromosomes are already **dyadic** (i.e., made of two chromatids), but their dual character only becomes apparent later when the chromosomes thicken.

At the electron microscopic level, leptotene chromosomes appear as **single axes** or dense fibers attached to the nuclear envelope at their ends. Nothing similar appears in mitotic chromosomes. Homologous chromosomes in the meiotic cell then associate either at their ends (Fig. 3.11b) or at one or more **convergent regions** elsewhere along their lengths.

At zygotene (Gk. zygon paired or yoked + tene), **or the zygonema stage, homologous chromosomes (sometimes called sister chromosomes) gradually pair or synapse** (Gk. synaptein to come together). The chromosomes are now called **bivalents** because they contain two dyad chromosomes or **tetrad chromosomes** because they contain four chromatids.

At the electron microscopic level, homologous chromosomes become increasingly attached across

a ribbonlike **synaptonemal complex** (Fig. 3.11a and b). The lateral elements of the complex correspond to the single axes of the homologous chromosomes, while the central element does not seem to contain DNA.

Lengthening and twisting during zygotene and the succeeding stage, the complexes may function in aligning the chromosomes or in stabilizing alignment once achieved. Attached to the nuclear envelope at its ends, synaptonemal complexes extend from one end of a bivalent chromosome to the other end.

The pairing of chromosomes is facilitated by **reassociation** or **r protein.** In addition, a heavy lipoprotein complex, associated with r protein, is necessary for synaptonemal complex formation.

Remarkably, some 0.3–0.4% of the DNA of the bivalent chromosomes replicates during zygotene (Stern, 1977). This **zygotene DNA** or zygotene-replicated DNA may be involved in the mechanism of pairing between homologous chromosomes. It is interspersed on all the chromosomes amidst the other DNA as sequences of about 10,000 base pairs. When zygotene-DNA synthesis is prevented by inhibitors of DNA replication, not only is the formation of synaptonemal complexes halted, but meiosis remains incomplete.

At pachytene (Gk. pachys thick), or the pachynema stage, the chromosomes condense. Synaptonemal complexes remain intact, and convergent regions between homologous chromosomes become visible at the light microscopic level as **chiasmata** (pl., chiasma sing.), putative points of actual **crossing over** or exchange between chromatids (Fig. 3.10a).

Following crossing over, **autosomal chromosomes** (i.e., all non-sex-determining chromosomes) no longer correspond purely to either of the original parental chromosomes. Only the noncoding centromeres joining the chromatids can still be considered as representing one or the other of the original parents' chromosomes.

With one chiasma per bivalent chromosome in mammals, chiasmata do not seem to account for all the crossing over occurring between homologous chromosomes. The number of crossing over events detected in genetic experiments is greater than the number of chiasmata observed cytologically. How additional crossing over occurs is unknown. Possibly, crossing over occurs earlier in meiosis and chiasmata formation is the consequence of several collected events.

Not all homologous chromosomes pair (e.g., chromosome 4 of *Drosophila*), however, and the

---

[2]"Lazy packmen die ducking," loosely made up from the sounds of the subphases' first syllables, may be of some assistance as a mnemonic device, an aid for remembering the subphases.

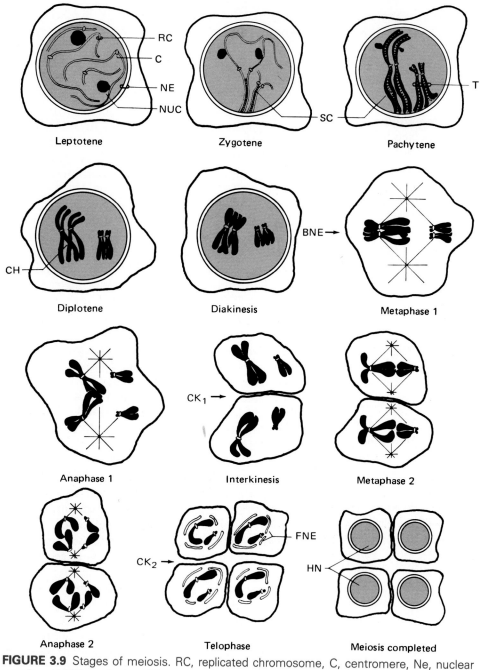

FIGURE 3.9 Stages of meiosis. RC, replicated chromosome, C, centromere, Ne, nuclear envelope, NUC, nucleolus, SC, synaptonemal complex, T, tetrad chromosome, CH, chiasmata, BNE, breakdown nuclear envelope, CK1 and CK2, cytokinesis, FNE, forming nuclear envelope, HN, haploid nuclei. F. Longo and E. Anderson, in J. Lash and J. R. Whittaker, eds., _Concepts in development._ Sinauer Associates, Inc., Sunderland, MA (1974). Used by permission of the authors.

chromosomes of some organisms do not normally form chiasmata or cross over at all (e.g., those of male _Drosophila_). Some chromosomes, especially **sex chromosomes** (e.g., X), are resistant to crossing over and retain their identity through meiosis.

An endonuclease appearing at pachytene is dif-ferent from ordinary endonucleases. Being specific for double-stranded DNA but breaking only one strand, the enzyme operates at points of crossing over. DNA repair synthesis follows, preferentially in regions that are **moderately repetitive** in the complexity of their sequences. At the same time,

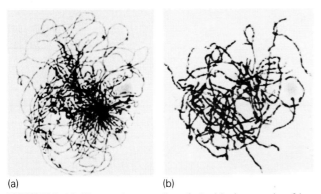

(a)   (b)

FIGURE 3.10 Photomicrographs of air-dried spreads of human chromosomes dropped on slides from fixed and dispersed free oocytes and stained with Giemsa. (a) pachytene, (b) diplotene. ×140. J. M. Luciani, M. Devictor-Vuillet, R. Gagne, and A. Stahl, *J. Reprod. Fertil.,* 36:409 (1974). With permission from the Journal of Reproduction and Fertility and the author.

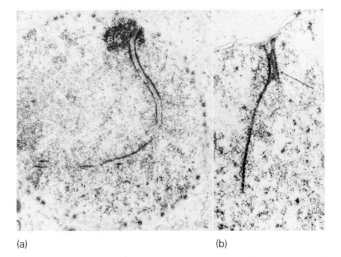

(a)   (b)

FIGURE 3.11 Electro micrographs of mouse spermatocyte showing homologous chromosomes paired in synaptonemal complexes attached at their ends to the nuclear envelop. The ribbonlike, tripartite complex consists of lateral elements or chromosomal axes and a central element formed during synapsis. Synaptonemal complexes run from one end of the meiotic chromosome to the other end. (a) A pachytene nucleus. The basal knob (BK) at one end of this synaptonemal complex is the centromere. ×14,000. (b) A leptotene nucleus. An early synaptonemal complex (arrow). ×26,000. From A. J. Solari, in S. J. Segal et al., eds., *The regulation of mammalian reproduction,* 1973. Courtesy of Charles C. Thomas, Springfield, Illinois.

topoisomerase may mediate site-specific recombinations across synapses and **DNA strand exchange.** The enzymes act processively after forming a complex with the DNA (Wasserman et al., 1985).

The exchange of parts from homologous chromosomes, or **recombination,** in eukaryotic cells, is thought to be achieved by a mechanism similar to that involved in **recombinant DNA** formation in prokaryotes (Fig. 3.12). An endonuclease **nicks** or breaks into one member of the pair of DNA strands. If the same enzyme produces a similar nick on another strand, the unwound nicked strands can form a **heteroduplex** region consisting of nicked strands aligned with unnicked strands. Ligases fuse the molecules, forming a molecular bridge or **chi structure.**

Chi structures are inherently unstable and fall victim to further breaking and ligating. The breaks can occur either along the originally unnicked strands of the DNA molecules or along the originally nicked strands. Different lengths of the original strands of DNA may thus be inserted in the resulting DNA. When larger pieces of DNA including genes are exchanged, **recombinant DNA duplexes** result. Identical heteroduplex regions are

exchanged with no genetic consequences, but the exchange of nonidentical heteroduplexes results in mismatching between the DNA strands.

In the case of meiosis, the mismatching of base pairs can be corrected by **mismatch repair mechanisms.** An endonuclease may nick one of the strands near the mismatched pair, and an exonuclease may remove bases including the mismatched one. Then nucleotides complementary to (i.e., matching) the nucleotides in the unnicked template strand may be inserted, and a perfectly complementary duplex of DNA is restored.

This is the essence of recombinational repair. While not well characterized, it reduces the number of errors in yeast DNA (Cole, 1975) and presumably in other eukaryotes' DNA. As well as restoring the DNA duplex following crossing over, the mechanism is eminently suitable for carrying out repair.

A similar mechanism is found in *Escherichia coli* in the form of **daughter strand gap repair** (Fig. 3.18) or **multiplicity reactivation** found when many bacteriophages infect the same cell. Additional DNA is employed as a template for the resynthesis of damaged DNA. The recombination

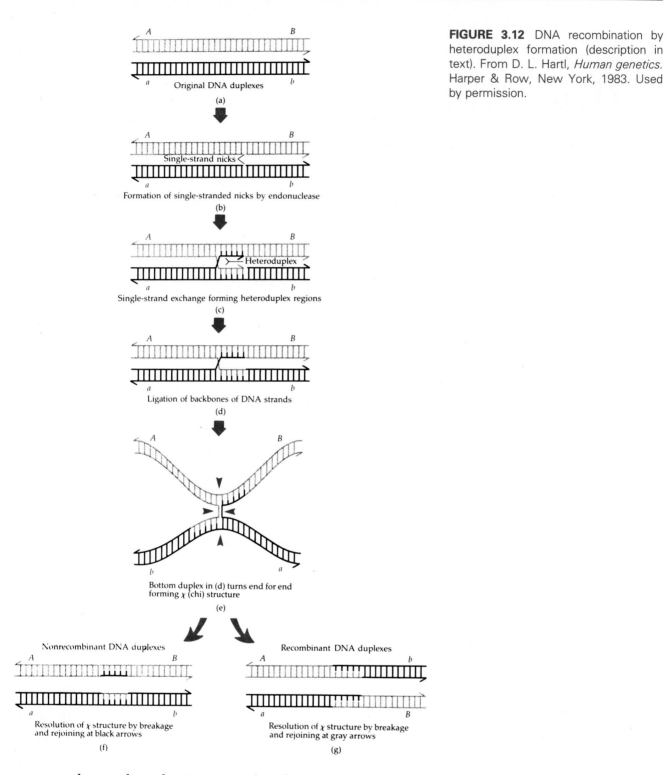

FIGURE 3.12 DNA recombination by heteroduplex formation (description in text). From D. L. Hartl, *Human genetics.* Harper & Row, New York, 1983. Used by permission.

enzymes that produce the DNA strand exchange allow a replication fork to overcome lesions in the DNA that otherwise block chain elongation.

At diplotene (Gk. diploos double), double chromosomes are conspicuously held together at chiasmata. Actually, while some chiasmata have become obvious (Fig. 3.10b), others have disap-

peared, presumably by condensing with other chiasmata, and additional DNA synthesis has occurred. But the most unusual aspect of diplotene is the resumption of RNA synthesis. The diplotene chromosomes, called **lampbrush chromosomes,** seem to sprout "whiskers" as loops of DNA transcribe large amounts of RNA (Fig. 3.13).

Diplotene is prolonged compared to the other

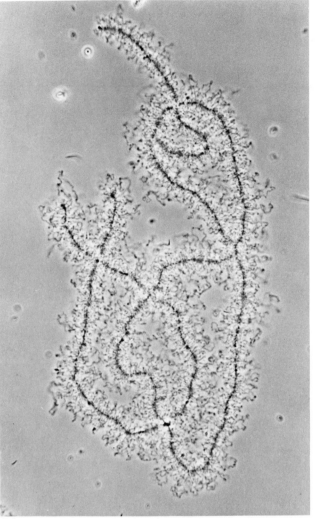

(a)

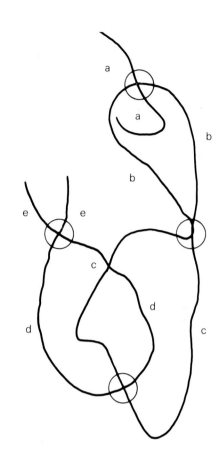

(b)

FIGURE 3.13 (a) Photomicrograph of lampbrush chromosome from an amphibian egg. (b) An interpretative drawing of photomicrograph. a–e, arbitrary points intended to aid in tracing chromatids. Chiasmata are circled. From J. G. Gall, "Chromosomes and Cytodifferentiation," in M. Locke, ed., *Cytodifferentiation and molecular synthesis.* Used by permission of Academic Press, Orlando, 1963. Courtesy of J. G. Gall.

stages. In the case of mammals, eggs remain suspended in diplotene (also known as the dictyate stage) from the time of follicle formation to shortly before ovulation or atrophy. For human beings, this period can be as long as 50 years, from some time before birth until menopause.

In diakinesis (Gk. dia- apart + kinesis motion), the chromosomes undergo saltatory movements (i.e., become jumpy). The movements distribute the chromosomes evenly and widely around the nuclear envelope and continue until the chromosomes find their places on the metaphase I plate (Fig. 3.9).

The nucleolus, or nucleoli, disappears, and later the nuclear envelope breaks down. The bivalent chromosomes become engaged in the spindle with their centromeres facing opposite poles.

Chiasmata now slide down the chromosomes and **terminalize** or move to the chromosomes' ends. The final resolution of these chiasmata will occur during the separation of chromosomes.

Later Meiotic Stages

After reaching the metaphase I plate, homologous centromeres separate, and dyad anaphase I chromosomes move to the spindle poles. This separation, sometimes called **disjunction** or meiotic segregation, is a function of the orientation of homologous centromeres toward opposite spindle poles.

Several alternative devices may be involved in aligning the chromosomes. So-called **exchange pairing** and disjunction between homologous chromosomes requires chiasmata formation and accompanies crossing over. Chiasmata, which add to the rigidity of bivalent chromosomes, stabilize their two-dimensional profile. A single chiasma places the bivalent chromosome's centromeres on opposite sides and, when lined up between spindle fibers on the metaphase plate, facing opposite poles.

Homologous chromosomes, such as X and Y sex chromosomes, that do not form chiasmata do not utilize exchange pairing in aligning centromeres. One mechanism of **nonexchange pairing,** known as distributive disjunction, aligns chromosomes by size (see Grell, 1976). Other devices for nonexchange pairing may involve qualitative characteristics and the number of chromosomes with similar characteristics.

Separation at the level of DNA is similar to separation at the level of chromosomes but with a twist. In the case of meiosis, replication during the preleptotene S period is incomplete. The centromeric DNA of the meiotic cell remains unreplicated, and each tetrad chromosome possesses only two kinetochores. As a result, chromosomes on the first metaphase plate offer only two points of attachment for microtubules (Fig. 3.14a). When homologous centromeres separate at anaphase (Fig. 3.14b), they are attached to dyadic chromosomes (Figure 3.14c) (see Murray and Szostak, 1985).

Interkinesis is generally brief. The nuclear envelope may not even reform, and the dyad chromosomes may congregate on the second metaphase plate without pause (Fig. 3.9). DNA topoisomerase resolves the twists or catenations in

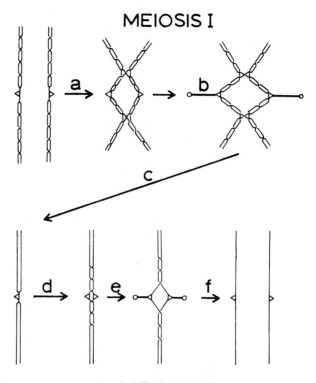

MEIOSIS I

MEIOSIS II

FIGURE 3.14 Schema for meiosis I and II illustrating the role of twists or catenations and centromeric replication in holding chromatids together. From A. W. Murray and J. W. Szostak, *Annu. Rev. Cell Biol.,* 1:289 (1985) by permission of Annual Reviews, and the authors.

the chromatids (Fig. 3.14d), and kinetochores replicate. When microtubules attach to chromosomes (Fig. 3.14e), the situation resembles that of mitosis more than that of meiosis I (Fig. 3.14a). Meiosis II is therefore sometimes called **mitosis-like** and, like mitosis, results in the delivery of one DNA duplex of each chromosome to both sibling cells (Fig. 3.14f).

Chapter

4

THE GENE AND DIFFERENTIATION

The cell is thus a minute factory, bustling with rapid, organized chemical activity. Under suitable molecular controls, enzymes busily synthesize lengths of messenger RNA. A ribosome will jump onto each messenger RNA molecule, moving along it, reading off its base-sequence and stringing together amino acids (carried to it by tRNA molecules) to make a polypeptide chain which, when finished, will fold on itself and become a protein. Nature invented the assembly line some billions of years before Henry Ford. Moreover, this assembly line produces many different highly specific proteins, the machine tools of the cell, which themselves shape and reshape the organic chemical molecules in order to provide raw material for the assembly lines and also all the molecules needed to build the structure of the factory.

FRANCIS CRICK (1981, pp. 70–71)

The great diversity of tissues and cell types in eukaryotes does not necessarily mean that the mechanisms and controls of change are also greatly diverse. Since mid-century, developmental biologists have believed that diversity can be reduced to differences in proteins and differentiation can be reduced to differences in protein synthesis. The basis for the belief is the compelling simplicity of the **central dogma.**

MECHANISMS OF CHANGE

Francis Crick's (1958, 1970) "central dogma" is that DNA makes RNA and RNA makes protein. **Gene expression** begins with **transcription,** or synthesis of an **RNA transcript** on a DNA template, and reaches its fruition with **translation,** or the

synthesis of a protein's polypeptide on a **messenger RNA** (mRNA).

Specificity is transferred down this molecular path (Ochoa, 1963[1]). The central dogma predicts that the sequence of bases in DNA is translated **colinearly** to amino acids in polypeptide chains. The **genetic code** is faithfully transcribed as **codons** in mRNA and translated via **anticodons** in transfer RNA (tRNA) to the amino acid sequence of individual polypeptides (Nirenberg[2] and Leder, 1964).

---

[1]Severo Ochoa and Arthur Kornberg were awarded the Nobel Prize in Physiology or Medicine in 1959 for their work on mechanisms of DNA and RNA synthesis.

[2]Robert Holley, Har Gobind Khorana, and Marshall W. Nirenberg were awarded the Nobel Prize in Physiology or Medicine in 1968 for their contributions to discovering the genetic codes for amino acids.

Currently, the central dogma is thought of more like a constitution than a creed. It has been significantly amended and will no doubt require further amendments. Colinearity has been substantially shaken: not in prokaryotes, where it has been amply confirmed, but in eukaryotes, where **gene rearrangement** and **RNA processing** constrict colinearity to short DNA/amino acid sequences (see Strickberger, 1986).

In addition, important changes occur in DNA itself during differentiation. In the words of Arthur Kornberg (1982, p. S201), "the ways in which DNA is modified, repaired, and recombined, even within a single cell, continue to multiply and impress the observer with the dynamic character of the once staid DNA molecule."

Transcription = DNA-Dependent RNA Synthesis

The subunit structure of RNA differs from that of DNA mainly in the presence of ribose sugars instead of deoxyribose sugars, and in the replacement of U (uridine) for the complementary T (thymine) (Fig. 4.1). In the transcription of a **DNA template** or **sense strand** to a complementary **RNA transcript,** a short stretch of the DNA double helix is unwound, allowing nitrogenous bases of ribonucleotide triphosphates to find their complementary base pairs in the DNA sense strand. Enzymes then catalyze the successive replacement of the 3' hydroxyl group of one ribonucleotide with the phosphate group of the next.

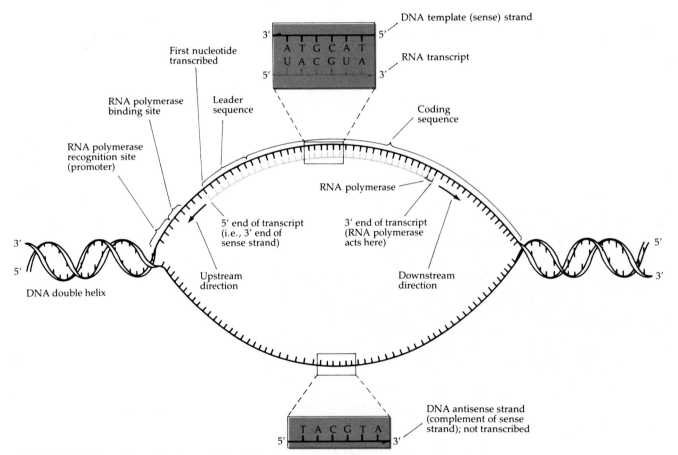

FIGURE 4.1. The scheme for transcription. The *first nucleotide transcribed* on the sense strand would be +1, and subsequent downstream nucleotides would be numbered consecutively. In the *upstream direction,* nucleotides would be numbered consecutively beginning with −1. Transcription proceeds in the *downstream direction* (right arrow) along the sense strand of the double-stranded DNA. The other strand, known as the *antisense strand,* is not transcribed except in the case of some viral DNAs and during the orgy of transcription occurring in some oocytes. From D. L. Hartl, *Human genetics.* Harper & Row, New York, 1983. Used by permission.

Only one **RNA polymerase** catalyzes the synthesis of all the RNAs of prokaryotes, but three RNA polymerases are required to accomplish the same mission in eukaryotes. **RNA polymerase I** (typically called pol I) or **A** produces the precursor RNA for most of the eukaryotic cell's ribosomal RNA (rRNA). **RNA polymerase II** (pol II) or **B** produces the RNA that will later become mRNA. **RNA polymerase III** (pol III) or **C** produces the tRNAs, 5S rRNA found in the large ribosomal subunit, and several other small RNAs including some small viral RNAs found in the nuclei of infected cells. Actually, transcription by RNA polymerase II and III may not be quite as exclusive, at least *in vitro* where the enzymes may synthesize transcripts of the same gene.

Transcription begins at the gene's **transcription initiation site** with the **initiation nucleotide** or the **first nucleotide transcribed.** The initiation nucleotide may not remain the first nucleotide in eukaryotes, however, since the ends of the primary transcript undergo posttranscriptional modifications (see below).

The sequence of nucleotides at the 5′ end of the transcript is a **leader sequence** containing bases necessary for ribosome binding and subsequent translation. A **coding sequence,** beginning with the first 5′-AUG-3′ codon, follows downstream (i.e., toward the 3′ end). This codon identifies an **open reading frame,** a series of codons specifying the amino acids later translated into a polypeptide chain.

Cis-*acting control sequences are sequences of nitrogenous bases on a DNA duplex required for the transcription of a specific mRNA.* **These sequences are primarily responsible for the control (i.e., turning on and off) or regulation (turning up and down) of differential gene transcription. They tend to be **modular, having sequence **domains** more or less specialized for different functions.

One type of RNA polymerase *cis*-acting sequence is the transcriptional **enhancer.** An **octanucleotide sequence motif** (i.e., a sequence of eight nucleotides) is characteristic of several enhancers. Studied primarily in RNA polymerase II transcription (i.e., mRNA transcription), enhancers or enhancer-like elements are also involved in the operation of eukaryotic RNA polymerases I and III.

Found as far as several thousand base pairs (i.e., kilobase pairs or kbp) away from the relevant coding sequence, enhancers can be on either side of a gene, within parts of the transcribed region (known as an intron, see below), and not even on the same strand of DNA as the gene. *Cis*-acting **negative ele-** ments or silencers resemble enhancers by way of these locational properties but have opposite effects on transcription.

Enhancers activate while negative elements tone down the other type of *cis*-acting sequence, the promoter. By regulating the rate of transcription, enhancers and negative elements influence gene expression across a spectrum of parameters, from timing and induction in embryogenesis to hormone responsiveness and cell-type specificity in differentiation (see Atchison, 1988).

Prokaryotic and eukaryotic **promoters** are required for binding RNA polymerase and for the correct initiation of transcription. Promoters are generally **gene-external** DNA sequences (i.e., not part of coding sequences) upstream from the first nucleotide transcribed. They are about 100 bases long but have a modular structure in which particular shorter sequences exercise different aspects of promoter activities.

An **RNA polymerase recognition site** (Figs. 4.1 and 4.11) is a domain that binds proteins helping bacterial RNA polymerase or eukaryotic RNA polymerases I and II recognize suitable stretches of DNA. The sequence 5′-TGTGA-3′ (or a variation thereof), for example, is found at the recognition site in some prokaryotic genes. On the other hand, recognition may be prevented by specific regulatory proteins bound to DNA sequences in the vicinity of recognition sites.

The **Pribnow box** (Fig. 4.11), another domain of the promoter found in prokaryotes, is the nucleotide sequence 5′-TATAATG-3′ (and similar sequences), five to six nucleotides upstream from the first nucleotide transcribed. The sequence 5′-TTGACA-3′ (or something close) frequently follows 17 nucleotides farther upstream (see Queen and Rosenberg, 1981).

Similarly, in eukaryotes, a **TATA box** or **Goldberg–Hogness box** contains a 5′-TATAAAA-3′ sequence (or something like it, ranging from ATA to TATAAATA) beginning 30 to 20 nucleotides upstream from the transcription initiation site. While binding occurs initially upstream on the double-stranded DNA, TATA boxes may secure the polymerase to the sense strand after initial unwinding.

In addition, genes may have **proximal sequence elements** (PSE), or **upstream promoter elements.** A **CAAT box** (often preceded by another C) some 80 to 70 nucleotides upstream from the transcription initiation site is commonly required for RNA II polymerase binding (see Maniatis et al., 1987). Gene far-upstream promoters (more than 200 bp) upstream of the initiation site may share elements with RNA polymerase II promoters and enhancers (from TATA-like sequences to octanucleotides).

In the case of eukaryotic RNA polymerase III, **gene-internal** promoters, or **internal control regions** (ICR), operate as well as gene-external promoters. The enzyme begins transcribing upstream from protein bound to the gene-internal promoter and slips past it without dislodging it during transcription. Actually, different gene-internal promoters or combinations of promoters may be associated with each type of RNA transcribed (box B sequences with tRNA, box C sequences with 5S rRNA, and box A sequences with both but not all) (see Sollner-Webb, 1988).

Several protein factors may be required simultaneously for accurate and efficient transcription of RNA polymerase III transcription products. For example, the intragenic promoter element of 5S rRNA genes of _Xenopus laevis_ binds a **transcriptional factor** (TF)—specifically, TFIIIA (i.e., the first or A transcription factor discovered for RNA polymerase III). The initiation of the 5S rRNA genes' transcription depends on a **transcription complex** of TFIIIA with at least two more TF proteins (TFIIIB and TFIIIC) (Wolffe and Brown, 1988).

**Termination of the transcript is achieved when the RNA polymerase moving down the DNA sense strand encounters a termination signal.** Several factors may facilitate termination, but a fold or group of self-complementary bases typically provides a sufficient termination signal _in vitro_. The actual length of a eukaryotic transcript may be shorter than its length at termination, however, as a result of enzymes cleaving the transcript.

At an average rate of 30 ribonucleotides per second, RNA polymerase produces a typical transcript of 8 kilobases (kb) in 5–7 minutes. When several RNA polymerases follow each other down the same gene, transcripts are produced at a faster rate.

Primary transcripts destined to become messenger RNA compose the **heterogeneous nuclear RNA** (HnRNA) fraction. These transcripts complex with nuclear protein to form **ribonucleoprotein particles** (RNP particles) in which the RNA is stable and protected, much as DNA bound in nucleosomes is stable and protected (see Dreyfuss, 1986).

Modifying and Processing RNA

All types of eukaryotic RNA are synthesized as **primary transcripts** that are longer than final **RNA-product strands**. While still in the nucleus, transcripts are shortened and their ends modified.

**Transfer RNA precursors are modified by the nontranscriptional addition of a unique 5'-CCA-3' triplet at the 3' end of the molecule and by enzymatic methylation and hydroxylation of some of the ribonucleotides.** These modifications and the inclusion of an occasional thymine and other rare nucleotides result in the molecule's characteristic double-stranded regions and loops (Fig. 4.2).

Transfer RNA precursors are processed by trimming to the requisite 73–93 ribonucleotides. **RNA endonuclease** or **RNA ribonuclease** (a multimeric macromolecule containing its own structural RNA) cleaves out some parts, while **RNA ligase** splices other parts together. In the typical tRNA molecule, variation in the number of nucleotides is limited to the **DHU loop** (containing two hydroxyuridines), the small **extra loop,** and the duplexed **stems** leading to these loops. Otherwise,

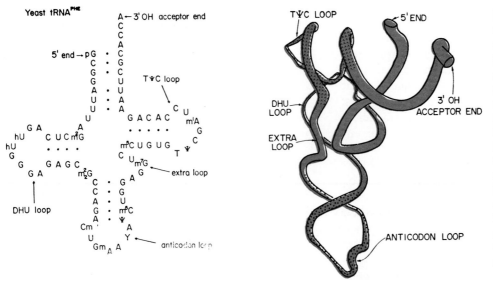

FIGURE 4.2. The tRNA cloverleaf and postulated three-dimensional structure of the yeast tRNA for phenylalanine. A, adenine; C, cytosine; G, guanine; T, thymine; U, uridine; m, methyl; n, hydroxyl; phi, pseudouracil; Y, an ambiguous pyrimidine. Numbers indicate the position of modifications. Dots represent one or more hydrogen bonds. From S. H. Kim, et al., _Science,_ 179:285 (1973). Used by permission of The American Association for the Advancement of Science and the authors.

the number of nucleotides in the parts of the molecule is remarkably constant.

The structure and regularity created through RNA processing is the key to understanding tRNA function. Adenine at the 3' terminus identifies the **3' OH acceptor end,** the site that will later become **charged** with a specific amino acid. The **anticodon** exposed in an **anticodon loop** will complement an mRNA codon during translation, while the rest of the molecule will match slots in a ribosome and bring the amino acid into an appropriate relationship for delivering it to a growing polypeptide.

Ribosomal RNA precursors are processed into the different rRNAs of small and large ribosomal subunits. The **small ribosomal subunit** in mammalian cells is composed of an 18S[3] rRNA and 33 proteins. The **large ribosomal subunit** is composed of a 28S rRNA, a 5.8S rRNA, 5S rRNA, and 49 proteins. The fruit fly, *Drosophila melanogaster,* has an additional 2S rRNA in its large subunit.

In the African clawed frog, *Xenopus laevis,* and in *D. melanogaster,* the greater part of the ribosome's rRNAs is transcribed in single 40S and 34S precursors, respectively. In the mouse, human beings, and other mammals, the similar precursor is 45S (Fig. 4.3).

[3]S is the symbol for a Svedberg unit, named after Theodor Svedberg, and is equal to the sedimentation coefficient (usually ascertained through centrifugation) times 10^{13}. S is directly proportional to molecular weight and buoyancy and inversely proportional to friction.

Synthesized on **nucleolar organizer genes,** the precursors appear in nucleoli already complexed with protein and integrated into ribonucleoprotein particles with ribosomal proteins. During the processing of the precursor, several specific ribose sugars and a few nitrogenous bases are methylated. More dramatically, the precursor is cleaved by **endonucleases** and unused portions are degraded by **exonucleases.** The three rRNA components (28S, 18S, and 5.8S) removed from the precursor are formed at the expense of nearly half the original ribonucleotides.

The 5S rRNA of the eukaryotic large ribosomal subunit is made on genes located outside the ribosomal organizer gene complex. This rRNA joins the other rRNAs in the nucleolus.

All the components of ribosomes are capable of **self-assembly** and of forming the appropriate subunits entirely on their own *in vitro.* Whether they do so *in vivo* remains to be seen. In any case, 60S and 40S subunits leave the nucleus independently and form functional ribosomes in the cytoplasm.

Modifying mRNA precursors alters the resulting mRNA transcripts in several ways. Even while transcription is still in progress, any 5' **preleader sequence** (i.e., an untranslated sequence ahead of the leader) present is trimmed down to the so-called **cap site.** There, a modified nucleotide (typically 7-methylguanosine) is added 5' → 5' by a triphosphate bridge, and both adenine and its attached ribose sugar are methylated. This is the **cap**

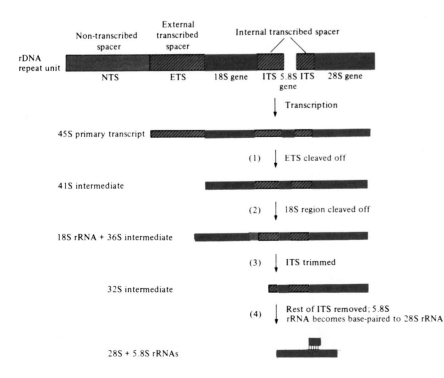

FIGURE 4.3. Ribosomal RNA processing in mammalian cells. Adapted from B. Lewin, *Gene expression,* Vol. II, *Eucaryotic chromosomes,* 2nd ed. Wiley-Interscience, New York, 1980 by permission of the author.

that provides an **initiation signal** that ribosomes recognize as belonging to an mRNA.

The 3' end is also altered. With the exception of histone transcripts, the 3' end of pre-mRNA is trimmed back to a **poly-A addition site.** Trimming occurs independently of whether transcription is ended by a termination signal in the DNA template or by the enzymatic cleavage of the RNA chain. The poly-A addition site is contained in sequences known as **trailers** that follow the **translation terminator sequences** or **stop codons** (UAA, UAG, or UGA, i.e., sequences that will later bring translation to an end).

The enzyme **poly-A polymerase** binds to a poly-A addition-recognition sequence (AAUAAA) above the poly-A addition site and adds a series of 100–200 adenylic acids (i.e., a tail) to the 3' end of the transcript. When accomplished, the transcript has become **poly(A)⁺ mRNA.** The poly(A)⁺ stretch (an isostich) may function to stabilize the mRNA and identify it, upon entering the cytoplasm, as a suitable mRNA for translation. The length of the poly(A)⁺ tail is important in the differential control of translation, and the maintenance of tail length prolongs the useful life of the mRNA.

Processing messenger RNA precursors in eukaryotes can greatly alter the molecule. Premessenger (pre-mRNA) transcripts form part of the nuclear fraction called **heterogeneous nuclear RNA (HnRNA)** because of its wide range of transcript sizes.

Most vertebrate pre-mRNAs have RNA to spare. They are generally 10–30 times longer than their ultimate coding sequences. Polypeptide chains which could be translated from an mRNA of 500–3000 bases are commonly coded in pre-mRNA transcripts of 50 kb and sometimes in transcripts as long as 200 kb. The prize may go to the **Duchenne type muscular dystrophy** (DMD) gene which is more than 2000 kb long and would take a day to transcribe.

The shortening of the pre-mRNA typically involves **cutting and splicing.** Pre-mRNA transcripts encoded by **split genes** contain **expressed sequences,** known as **exons,** and noncoding **intervening sequences,** or **introns.** With the exception of the mRNAs for histones and some interferon genes, vertebrate mRNAs seem to be processed by the splicing of exons and the excision of introns (Fig. 4.4).

Splicing vastly increases the potential of a cell for manufacturing different mRNAs. For example, the C and J domains of the light chains of immunoglobulins (Fig. 4.5) are brought together by cutting and splicing the original transcript.

Introns are eukaryotic structures, but they are especially common in vertebrates. They are vir-

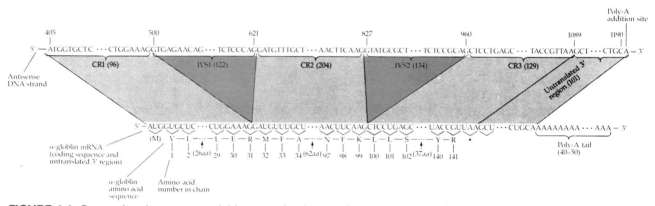

FIGURE 4.4. Processing the mouse α-globin transcript. Intervening sequences, or introns (IVS1 and IVS2), of the primary transcript are removed, and exons or the coding regions (CR1, CR2, and CR3), are spliced together. The in-frame splice takes place after the second base on the 5' donor side and before the third base on the 3' acceptor side. Note: In the figure, the antisense strand of DNA is shown (top) rather than the complementary sense strand in order to facilitate comparison with the final mRNA (bottom). Code letters in DNA and RNA: A, adenine; C, cytosine; G, guanine; T, thymine; U, uridine. Code letters in the amino acid sequence: A, alanine; E, glutamic acid; F, phenylalanine; K, lysine; L, leucine; M, methionine; N, asparagine; R, arginine; S, serine; V, valine; Y, tyrosine. The missing portions of the molecules are indicated by ellipses, and the number of amino acids (aa) in the absent portions are indicated in parentheses beneath arrows. From Y. Nishioka and P. Leder, *Cell,* 18:875 (1979) by permission of Cell Press (copyright holder) and the authors.

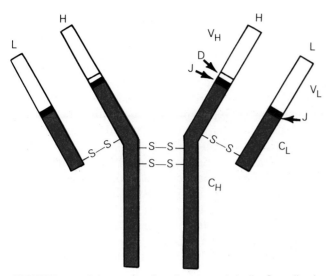

FIGURE 4.5. Schematic of an immunoglobulin G antibody molecule.

tually absent in prokaryotes (Archaebacteria being an exception), rare in yeast, and uncommon in insects. Their function therefore would seem to be selective rather than general.

Ranging from 50 to 20,000 base pairs in length, introns seem to be made by "junk" DNA, since most of the bases within introns do not seem to be relevant to cutting and splicing and can be altered without any apparent effect. The boundary bases at the beginnings and ends of introns cannot be altered without consequences.

Introns may interrupt trinucleotide codons with splicing occurring **in-frame,** or introns may lie between codons. For every intron removed, two cleavages take place. According to the **Chambon rule,** introns are bounded by particular boundary sequences of dinucleotides, a GU dinucleotide on the **5′ donor side** and an AG (CG or UG) dinucleotide on the **3′ acceptor side:**

$$5'\text{ donor side} \quad\quad \text{intron} \quad\quad 3'\text{ acceptor side}$$
$$\text{exon---5'-N} \mid \text{GU} ----- \text{AG} \mid \text{N---exon}$$
$$\text{cut} \quad\quad\quad\quad\quad\quad \text{cut}$$

where N is any nucleotide (Breathnach et al., 1978).

Most splicing is processive or continuous, in effect, binding one exon with the next. The first exon tends to remain in place and join with the second exon. Alternative or jump cut splicing skips exons and turns one or more intervening exons into parts of an enlarged intron:

Processive splicing

$$\text{exon}_1\text{-intron}_1\text{-exon}_2\text{-intron}_2\text{-exon}_3$$
cut and splice $\downarrow$
$$\text{exon}_1\text{-exon}_2\text{-exon}_3$$

Jump cut splicing

$$\text{exon}_1\text{-intron}_1\text{-exon}_2\text{-intron}_2\text{-exon}_3$$
$$\mid \quad\quad \text{enlarged intron} \quad\quad \mid$$
alternative cut and splice$\downarrow$
$$\text{exon}_1\text{-exon}_3$$

Alternative splicing appears to be developmentally important, for example, in the creation of immunoglobulin diversity beyond that already created by gene rearrangement.

Possibly, splicing can even link exons from different initial transcripts (Brody and Abelson, 1985). On the other hand, occasional slippage within introns can also create two mRNA transcripts from the same precursor (King and Piatigorsky, 1983).

Mechanisms of processing seem to be similar for different RNAs and in different phyla. One shared feature of processing is **RNA-mediated catalysis,** in which RNA plays a catalytic role. Another generally shared feature is the utilization of large RNP complexes.

These features are not as unique to processing as was first thought. Instead, they define a class of mechanisms (class I) operating during processing in the nucleus and translation in the cytoplasm. A second class of mechanisms (class II) employs proteins as the primary enzymatic units. This class includes all the well-known enzymatic pathways of cell metabolism, replication, and transcription (see Alberts, 1986).

Much of what is known about RNA processing in eukaryotes comes from work on the ciliate, *Tetrahymena thermophila*. The results of *in vitro* experiments indicate that the removal of introns from the rRNA precursor is an **autocatalytic** or **self-splicing** process. The intron itself seems to play the catalytic role while cleaving itself out of the molecule (see Zaug and Cech, 1986).[4]

In mammalian cells, splicing requires a **small ribonucleoprotein particle** (snRNP, or "snurp," of about 250,000 daltons compared to 4.5 million daltons in ribosomes) containing **small nuclear RNA** (snRNA), specifically, **U1snRNA** (U2snRNA may also be involved), which seems to bind the pre-mRNA near the splice site. Somehow, via its structural organization, or complementary base pairs, U1snRNA selects the sites for cleavage and probably holds the complex together, while the introns are excluded and the exons are connected.

In yeast, pre-mRNA is first bound to a 40S **spliceosome** complex. Introns are excised follow-

[4]T. Cech and S. Altman were awarded the Nobel Prize in Chemistry in 1989 for the discovery of catalytic RNA.

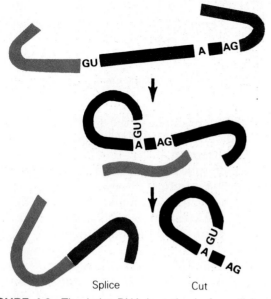

FIGURE 4.6. The lariat RNA hypothesis for splicing two exons together. The intron (black) is bound by the GU of its 5′ donor codon and by the AG of its 3′ receptor codon. While forming a lariat RNA, or branch structure, the hydroxyl end of exon 1 (light gray, left) is freed, and the 5′ terminal phosphate of the intervening sequence is tied in an RNA lariat to the final A of the branching sequence. Exon 1 and exon 2 (dark gray, right) are spliced, and the lariat structure is released. A, adenine; G, guanine; U, uridine. From B. Ruskin, et al., *Cell,* 38:317 (1984) by permission of the authors.

ing formation of an intermediate circular structure, or **lariat RNA** formed within the intron (Fig. 4.6) (Padgett, 1984).

Evolutionary arguments are sometimes made to explain RNA-mediated catalysis. Sometimes thought to lack the elegance expected of cell processes, RNA-mediated catalysis is explained as a relic from an RNP past (see Alberts, 1986). Such thoughts should be dismissed as relics of the argument of biological inertia (i.e., that a molecule exists in a cell because of history rather than function). Alternatively, RNA-mediated catalysis is consistent with polyphyletic concepts of the origin of cells (see Sogin et al., 1989) and with dynamic competition among cell parts.

Theories on the evolution of exons and introns are generally retrospective, linking translated proteins to processed transcripts. The existence of exons and introns are thought to provide a source of variation for cells (e.g., different immunoglobulins) and a source of plasticity during evolution.

Correlations thought to exist between exons and various aspects of protein structure are sometimes inflated with evolutionary significance. For example, intron deletion often occurs at points where the secondary structure of proteins takes sharp bends or forms edges. But facets of proteins and exons may also be independent entities, and splicing *per se* may not have a causal role in shaping proteins.

Exons are also correlated with repeated structures and functional subdivisions of proteins. Single exons may encode individual **domains** (Fig. 4.7, exon 1) or contribute to two domains (e.g., exon 17), **modules** (repeated parts of a domain, e.g., exons 2 and 3, 5 and 6) or **sequence families** (repeated parts of a module, e.g., exon 4). Modules may be as large as 40–50 amino acids and are often repeated as many as 10 times within a polypeptide. One

[5]Joseph L. Goldstein and Michael S. Brown received the Nobel Prize in Physiology or Medicine in 1985 for their studies of cholesterol, including their part in the discovery of the LDL receptor.

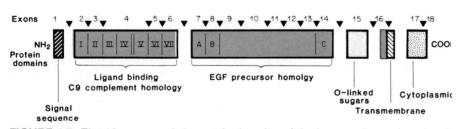

FIGURE 4.7. The 18 exons and six protein domains of the human plasma low density lipoprotein (LDL)[5] receptor (a cholesterol transport protein present in the plasma membrane and associated with coated pits). Exons are numbered 1–17, and the positions of introns are indicated with arrowheads. Seven repeated sequences representing 40 amino acids in the "ligand binding" domain are numbered I–VII, and three other repeated sequences in the "EGF (ectodermal growth factor) precursor homology" domain are lettered A–C. From T. C. Südhof et al., *Science,* 228:815 (1985) by permission of the American Association for the Advancement of Science and the authors.

module in the alpha chain of collagen, for example, is repeated more than 40 times.

The similarity of domains or their subdivisions in different proteins suggests that similar exons have been **recruited** into proteins during the course of evolution. For example, the **low density lipoprotein** (LDL) receptor gene is a mosaic of exons shared with the genes for complement, some clotting factors, and epidermal growth factor. Similar exons in different genes constitute **supergene families**, related by sequence but not necessarily by function.

Some proteins may even have been derived from diverse supergene families through **exon shuffling**, the process of putting split genes "together as mosaics of simpler structures, combinatorial assemblies of a smaller number of minigenes" (Gilbert, 1985, p. 824). On the other hand, split genes may have been created in the course of evolution as a solution to the problem of recombinational deletion (i.e., the loss of small repeated sequences during chromosomal pairing) in proteins with repeated domains.

Gene Rearrangement

The immune system of eukaryotes generates proteins in great variety. Vertebrates, for example, respond to the introduction of virtually any foreign substance, known as an **antigen** or **immunogen,** by producing **antibodies** or **immunoglobulins** with a high affinity for binding with the specific substance.

Antibody molecules are structured stereotypically (Fig. 4.5). Immunoglobulin G, for example, has both heavy (H) and light (L) chains. C portions (C_H and C_L) (see Porter,[6] 1973) have a constant primary structure, and other portions consist of large V regions (V_H and V_L) and small joining (J) regions having a variable primary structure. H chains also have diversity (D) regions with variable primary structures.

The possibility that each complete antibody is coded initially in the genome is eliminated by the sheer number of different proteins potentially synthesized. Not only does this number exceed the magnitude of the available germ-line genome, but an immune cell can generate antibodies in response to substances it has never experienced before, not in the life of the individual or even in the evolution of the species. The creation of antibodies is too innovative to be prescribed.

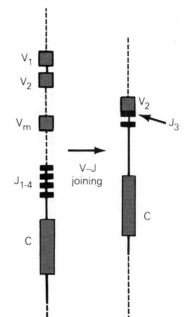

FIGURE 4.8. Scheme for gene rearrangement in the light chain of an antibody molecule. (Left) Different variable (V) and joining (J) coding regions in an embryonic cell's DNA are separated from each other and from the coding region for the constant (C) portion of the antibody. (Right) In the course of lymphocyte maturation, the variable portions of the molecule are assembled by combining particular V and J coding regions of DNA.

Several mechanisms combine to invest antibody-producing cells with creativity. **Somatic mutations,** or base substitutions within genes of somatic cells, introduce some of the variability of immunoglobulins. **Gene rearrangement** introduces more variability and **RNA processing** still more.

The seminal discovery leading to the present understanding of creativity in the immune system was that parts of the genome coding for different portions of antibody molecules (the variable, V_L, and constant, C_L, regions, Fig. 4.5) were farther apart in embryonic cells than in adult lymphocytes (Hozumi and Tonegawa,[7] 1976). Similar discrepancies occur in parts of the genome coding for other portions of the molecule (V_H and C_H). The genome must have been rearranged to bring these portions together—hence, "gene rearrangement."

In the course of **lymphocyte maturation,** the portions of the genes coding for V and J regions or V, J, and D regions are broken and recombined to form complete coding sequences for the variable portions of the respective antibody chains (Fig. 4.8).

[6]Gerald M. Edelman and Rodney R. Porter were awarded the Nobel Prize in Physiology or Medicine in 1972 for their contribution to working out the structure of antibodies.

[7]Susumu Tonegawa was awarded the Nobel Prize in Physiology or Medicine in 1987 for his contribution to the discovery of gene rearrangement.

The process is not without hazard. The regions of chromosomes involved in antibody production (e.g., the heavy-chain region of human chromosome 14) are also prone to errors in rearrangement including translocations of portions to other chromosomes (e.g., human chromosome 8) with subsequent deleterious effects (e.g., the production of lymphomas).

The rearranged gene still does not correspond to a whole heavy or light chain. A small segment of noncoding DNA, known as an **intron** and containing gene control elements, remains between **exons** or coding regions for the potential variable and constant regions. The actual selection of coding regions that will determine the polypeptide chain occurs later through the process of RNA processing.

Rearranging genes leaves introns between exons and permits the cell considerable latitude in choosing which regions will ultimately be joined in fashioning final mRNAs. As it turns out, hundreds of variant V regions for both heavy and light chains and four to five J regions and D regions can combine in tens of thousands of combinations. Moreover, because any light chain can potentially associate with any heavy chain, hundreds of millions of different antibody molecules (i.e., combinatorial variants) may be formed (see Lewis et al., 1985).

These processes plus somatic mutation provide the bulk of variation found in immunoglobulins and possibly operate in the differentiation of other tissues. For example, nervous tissue exhibits structural variability comparable to the biochemical variability of antibodies, but what roles, if any, somatic mutation, gene rearrangement, and processing play in nervous development are unknown.

Translation: RNA-Mediated Polypeptide Synthesis

Each RNA takes part in a different aspect of translation. Each RNA also begins its part in translation at a different time in the process.

For tRNA, translation begins with **charging** by an amino acid. The two-step process is catalyzed by one enzyme: a specific **amino acid activating enzyme,** or **aminoacyl-tRNA synthetase.** The enzyme activates the amino acid through a reaction with adenosine triphosphate (ATP). With the release of a diphosphate group, a high-energy bond of ATP is transferred to a bond between the amino acid and adenosine monophosphate (AMP). The high-energy bond is then transferred to a bond between the amino acid and the tRNA, releasing AMP. This same high-energy bond will later fuel the formation of a peptide bond in the growing polypeptide.

Each of the 20 amino acids has its own aminoacyl-tRNA synthetase, but each enzyme recognizes all the tRNAs capable of transporting that amino acid to a polypeptide chain. Arginine, for example, is encoded by six codons. The enzymes make mistakes only once in a thousand to once in ten thousand reactions between amino acids and tRNAs.

Translation begins for mRNA when its **start codon,** or first 5'-AUG-3' after the cap (left, Fig. 4.9 or the comparable prokaryotic initiation site), is exposed in a site on the small ribosomal subunit. Aided by three **initiation factors** and a molecule of guanosine triphosphate (GTP), an **initiator tRNA** charged with methionine (or formyl-methionine in prokaryotes) adheres by its anticodon to the complementary 5'-AUG start codon to form a composite **initiation complex.** In eukaryotes, flanking se-

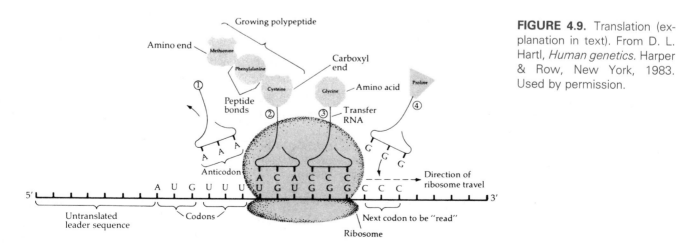

FIGURE 4.9. Translation (explanation in text). From D. L. Hartl, *Human genetics.* Harper & Row, New York, 1983. Used by permission.

quences surrounding the methionine codon (the Kozak consensus sequences) may also support the initiation of translation (Kozak, 1987).

Translation then begins for the ribosome as a whole. The large ribosomal subunit joins the initiation complex by binding with the initiator tRNA; the GTP is hydrolyzed; the initiation factors fall off and another tRNA, complementary to the second codon on the mRNA, drops into place. With the help of **elongation factors** from the cytoplasm, polypeptide growth commences.

Filling the first 5'-AUG-3' codon after the cap or initiation site sets the reading frame; the triplets of bases read as codons for the rest of the length of the message. Only these triplets of bases will constitute codons and will be translated into a polypeptide chain.

Only one polypeptide will ordinarily be translated from a single coding sequence. Exceptions are found in some bacteriophage messengers, and the SV40 large T antigen can start at a 5'-AUG-3' beyond the first one. The reading frame thereby **shifts,** and more than one polypeptide is read from the same sequence of bases.

The large ribosomal subunit is equipped with two binding sites, a **peptidyl-** or **P-binding site** (occupied by tRNA 2 in Fig. 4.9) from which the tRNA discharges its amino acid and an **aminoacyl-** or **A-binding site** (occupied by tRNA 3) where the charged tRNA arrives initially. As the peptide bond forms, the tRNA in the A-binding site acquires the growing polypeptide chain and, fueled by the hydrolysis of a bound GTP, moves or **translocates** to the P-binding site.

Peptide bond formation is catalyzed by **amino transferase** associated with the large ribosomal subunit. The bond is between the carboxyl end of the peptide chain and the free amine group of the upcoming amino acid. Since the amino acids are bound at their carboxyl ends to tRNAs, the amino acid at the P-binding site is simultaneously freed from its tRNA as it reacts with the amino acid on the tRNA in the adjacent A-binding slot.

Most amino acids can be brought to an A-binding slot by several tRNAs. Their different anticodons are compatible with different codons specifying the same amino acid. Redundancy (or degeneracy) in the genetic code makes several codons synonyms for the same amino acid.

The first two bases of a codon are generally rigidly determined, but often any base can occupy the third position without influencing the amino acid assigned to the codon (i.e., the code is virtually reduced to two bases). According to the "wobble hypothesis," ambiguity in the third base is an accommodation to enzymatic modifications of tRNA. Looseness of fit or even mismatched pairs are assimilated without undermining the integrity of coded information (Crick, 1966).

Translation is terminated by **stop codons** (5'-UAA-3', 5'-UAG-3', or 5'-UGA-3') which are not read by tRNA. These codons are recognized by a protein **release factor** and are carried into the A-binding site during translocation. As a result, another charged tRNA is not present in the A-binding site, and when the peptidyl transferase cleaves the bond between the last amino acid and its tRNA, a free amine group is not available to receive the polypeptide. A water molecule completes the carboxyl end of the polypeptide; the polypeptide, tRNA, and release factor are released from the ribosome, and, with the aid of separation factors, the ribosomal subunits separate.

The mRNA, however, is not necessarily free. It may already have complexed (more than once) with other small ribosomal subunits at its start codon, and additional rounds of translation may already have begun. Typically, several ribosomes link up with the same mRNA as long as it is being translated. The group, a **polysome** or **polyribosome,** moves along the mRNA, creating polypeptide chains at the rate of about 15 amino acids per second.

Ultimately, no mRNA lives forever. The poly-A tail shortens; the molecule fails to be recognized as a messenger, and it is degraded by endonucleases and disappears. The forces influencing the longevity of mRNA, like the forces influencing transcription, processing, modification, and transport, can play a role in determining what proteins are synthesized.

CONTROL OF CHANGE

It is clear, therefore, that differentiated organisms may be expected to possess certain types of regulatory mechanisms which are not found in unicellular organisms. The question, however, is whether cellular regulation and differentiation in higher organisms use the same basic mechanisms as bacterial systems, employing similar circuit elements geared in a different way to meet the requirements of higher organisms.

F. Jacob and J. Monod (1961, p. 52)

Until recently, studies on the control of protein synthesis centered on bacteria. The appeal of bacteria was not only their accessibility but the ease with which protein synthesis was turned on or off by manipulating the bacteria's medium. Moreover, microbiologists had accumulated a plethora of biochemical mutants capable of operating differently in response to altered environments.

In the early 1960s, the terms and concepts derived from bacterial studies were widely adapted to eukaryotes, and much of what was learned with bacteria was assumed to apply to differentiation. Recently, technical advances in tissue culture made it possible to handle eukaryotic cells in many of the same ways bacteria were handled formerly and to learn first hand the situation in eukaryotes.

Bacterial Paradigm

Bacterial **induction**[8] is the increased synthesis of an enzyme in response to an **inducer** or **controlling substance** added to the bacteria's media. **Repression** is the decreased synthesis of an enzyme in response to a **corepressor**.

Both induction and repression frequently occur in highly adaptive ways. Induction occurs typically in response to substrates for enzyme catabolism (i.e., breakdown), while repression occurs in response to products of metabolism (i.e., buildup) produced along biochemical pathways. Inducible enzymes are synthesized when their substrates for catabolism are available, while the synthesis of repressible enzymes is tuned down when their physiological end products are already abundant. These

adaptive aspects of induction and repression suggest that the **regulation of gene activity,** like the regulation of other features of living things, is under genetic control and subject to evolutionary pressure.

The nomenclature of gene regulation is standardized if not universally applied and several terms can now be used with relative precision. Generally, inducibility and repressibility depend on **regulatory proteins** with high affinity for specific portions of DNA and the ability to influence transcription by binding to DNA. **Controlling substances,** on the other hand, are **ligands** that exercise their power by complexing with regulatory proteins and altering their affinity for DNA.

By analogy to regulated enzymes, regulatory proteins are said to be equipped with one or more **binding sites** capable of binding the ligand and with one or more **active sites** capable of attaching to DNA. Complexed with the regulatory protein, the ligand alters the affinity of the active site for DNA through an **allosteric effect** (Gk. *all-* other, different, + *stereos* solid). Theoretically, two types of control produce both induction and repression (Table 4.1). In **negative controls,** the regulatory protein is a **repressor protein** capable of preventing or **repressing** transcription and genetic expression, while in **positive controls,** the regulatory substance is an **activator protein** capable of increasing the rate of transcription or **derepressing** transcription. Negative controls tend to dampen fluctuations and maintain stable levels of production, while posi-

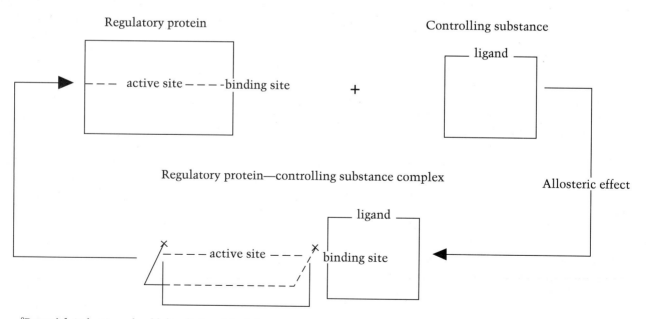

[8]**Bacterial** induction should be distinguished from **embryological** induction, or morphogenesis triggered by tissue interactions.

TABLE 4.1. Induction and repression of bacterial enzymes

Type of regulation			
Positive		Negative	
Control site in DNA			
Promoter region (generally not transcribed)		Operator (overlaps promoter and beginning leader sequence)	
Regulatory substance			
Activator protein (aids RNA polymerase recognition)		Repressor Protein (prevents RNA polymerase binding and transcription)	
Allosteric form of regulatory substance attaching to DNA			
Complex with inducer	Simple	Complex with corepressor	Simple
Consequence when ligand is present			
Induction (derepression)	Repression	Repression (feedback inhibition)	Induction (derepression)
Consequence when ligand is absent			
Repression	Induction (derepression)	Derepression	Repression
Examples			
Catabolite-sensitive operons Lactose operon	Anabolic pathways Arabinose operon	Anabolic pathways Tryptophan operon	Catabolic pathways Lactose operon

tive controls tend to speed up activity. In practice, one or another control may predominate or may be the only active source of regulation.

Some regulatory proteins are attached to DNA when complexed with their controlling substance, while other regulatory proteins are attached to DNA when dissociated from their controlling substance. The transcription of mRNA is inducible or repressible depending on the type of regulatory protein and its relationship to the controlling substances.

Controlling substances or ligands are either **corepressors** or **inducer substances.** A complex of corepressor and repressor protein represses gene activity, while a complex of an inducing substance and an activator protein derepresses or induces gene activity. On the other hand, the dissociation of a repressor–corepressor complex induces, and the dissociation of an inducer–activator complex represses (Table 4.1).

The Jacob–Monod model[9] for negative control of induction, proposed in 1961, is a classic of modern biological theories. It quickly came to dominate the field and the way biologists thought about gene regulation.

The model was based on bacterial mutants. Some mutations turned normally present **constitutive enzymes** into **inducible enzymes** or into **repressible enzymes** whose synthesis in the mutants depended on added controlling substances. Other mutations turned inducible enzymes into **noninducible constitutive enzymes** no longer responsive to controlling substances, and still other mutations blocked the synthesis of enzymes entirely. In addition, the effects of some of the mutations could be undone through bacterial conjugation and the introduction of other genes, while the effects of some other mutations could not be undone.

Since mutations affected enzymes in three different ways, the existence of three sorts of genes was proposed. **Structural genes** were defined as those controlling the amino acid sequence, or primary structure, of polypeptides making up enzymes. **Regulatory elements** or *trans*-acting regulatory elements contained the coded sequences for regulatory proteins capable of moving in the cell.[10] **Controlling elements** or *cis*-acting (Lat. on this side + acting) controlling elements were genes that influenced transcription in adjacent structural or regulatory elements.

Operationally, mutants affecting structural

[9]Francois Jacob and Jacques Monod shared the Nobel Prize in Physiology or Medicine with Andre Lwoff in 1965 for their studies of gene regulation.

[10]Structural and regulatory genes are now considered variations of the same type of gene, differing in the proteins ultimately synthesized rather than in their DNA.

genes altered enzymes. Mutants affecting *cis-*acting controlling elements acted only in the *cis-*arrangement (i.e., when present on the same DNA duplex as the structural genes). Mutants affecting *trans-*acting regulatory elements acted either in the *cis-*arrangement or in the *trans-*arrangement (i.e., present on a different DNA duplex).

The system inspiring the model, and the first transcriptional control system to be thoroughly analyzed, was the system regulating the catabolism of the disaccharide lactose (lactose = milk sugar) in the human colon bacillus, *Escherichia coli* (Fig. 4.10). Genetic mapping experiments showed that three *lac* structural genes were adjacent to each other in the *E. coli* genome. The *lacZ*

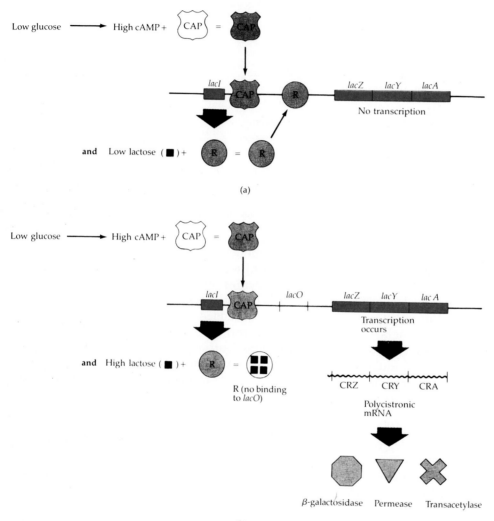

FIGURE 4.10. Regulation at the lactose operon in *E. coli*. When glucose levels are low, cyclic adenosine monophosphate (cAMP) levels are high. The CAP (catabolic activator protein) enters its active form (shaded) and is bound to the CAP site in the DNA promoter region. Transcription is then possible but not automatic. (a) When lactose concentration is low, the lac repressor protein (r, closed circle) binds to the operator, and the lac operon is turned down to minimum levels of transcription. (b) In the presence of lactose, the lac repressor protein does not bind to the operator (*lacO*), and the lac operon is turned up. The structural genes (*lacZ, lacY, lacA*) undergo transcription at the maximal rate into a polycistronic mRNA (CRZ, CRY, CRA); the CR designation means "cross-reacting," referring to immunologically detectable mutant forms of the enzymes), which in turn is translated into the respective enzymes. From D. L. Hartl, *Human Genetics*. Harper & Row, New York, 1983. Used by permission.

gene encoded the β-**galactosidase** enzyme which catalyzed primarily the hydrolysis of the β-galactoside bond in lactose. The *lacY* gene encoded **galactoside permease** which facilitated the passage of lactose into the bacterium. The *lacA* gene encoded **thiogalactoside transacetylase** of unknown function in the bacterium.

The structural enzymes were controlled jointly. Mutations in controlling elements and regulatory genes of the *lac* system simultaneously altered the amounts of all three enzymes. The three genes, it turned out, were transcribed together in a single **polycistronic** mRNA.

The mutations that led to the discovery of the *lac* regulatory gene rendered the lac enzymes constitutive in colonies, while the mutations that led to the discovery of the *lac* controlling elements rendered the lac enzymes **noninducible.** All these mutants appeared on the *E. coli* genetic map in the vicinity of, but not within, the *Z* gene.

The lac repressor protein is made up of four identical subunits that form two symmetrical

grooves enclosing the operator DNA like a "hot dog" in a bun (Steitz et al., 1974). Because the DNA is resistant to DNase digestion when complexed with the repressor protein, operator DNA has been isolated and analyzed (Fig. 4.11). It extends over 24 base sequences between bases −3 and +21, including the first base transcribed (base +1) and the **leader sequence** (Fig. 4.11, compare to Fig. 4.1).

The controlling element responsible for negative control was called the **lac operator** or simply the **operator** and was assigned the symbol *O.* The regulatory gene involved with the operator was called the inducible gene and assigned the symbol *I* (*lacI*, Fig. 4.10). The name was subsequently changed to the *lac repressor* gene or simply the *repressor* gene, although the *I* symbol was retained. The *I* gene was normally expressed continuously, rendering the repressor protein constitutive.

Like other controlling elements, the operator contains **modular elements** (or boxes) with repeated nucleotide **motifs** (or sequence patterns). Specifically, the operator has a high adenine and

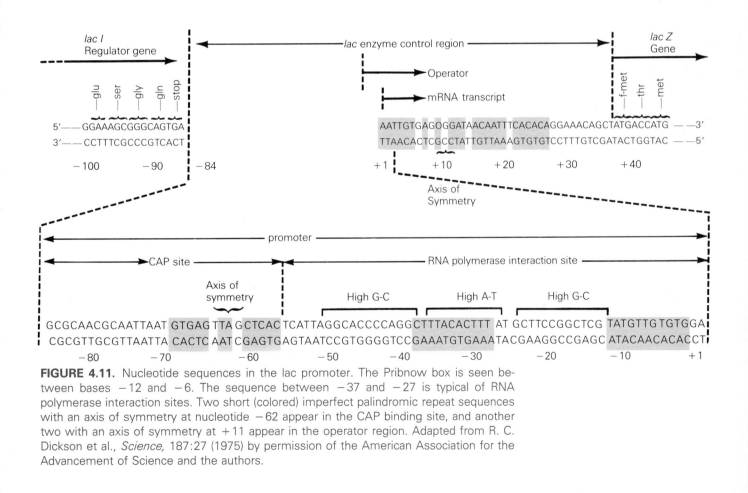

FIGURE 4.11. Nucleotide sequences in the lac promoter. The Pribnow box is seen between bases −12 and −6. The sequence between −37 and −27 is typical of RNA polymerase interaction sites. Two short (colored) imperfect palindromic repeat sequences with an axis of symmetry at nucleotide −62 appear in the CAP binding site, and another two with an axis of symmetry at +11 appear in the operator region. Adapted from R. C. Dickson et al., *Science,* 187:27 (1975) by permission of the American Association for the Advancement of Science and the authors.

thymine content (AT-rich sequences) and includes the greater part of two **imperfectly inverted tandem sequences** (i.e., sequences that are nearly the same on the two strands when examined in opposite directions) or **repeat palindromic sequences** of 34 bases with an axis of symmetry at base $+11$ (e.g., beginning with the top line at base -8 and the bottom line at base $+28$, compare the intervening bases on the two lines, Fig. 4.11). Their twofold symmetry in bases provides the paired recognition sites for the symmetrical grooves of the repressor protein.

The **coding sequence** follows the operator, beginning with the start codon and extending through the polycistronic unit of structural genes. The whole **transcriptional unit** of operator (i.e., virtually the leader sequence) and coding sequence constitutes the **operon.**

How does the operon work? Normally, in the absence of lactose, the constitutive repressor protein is fastened to DNA in the operator region, and the *lac* transcriptional unit is repressed (i.e., not transcribed at a high rate, Fig. 4.10a).

Transcription at a high rate is induced when the repressor protein binds its controlling substance and dissociates from the DNA (Fig. 4.10b). The controlling substance is not actually lactose. Allolactose (differing from lactose at the bond between the monosaccharides and produced from lactose as a minor effect of β-galactosidase) is the ligand that binds with the repressor protein and alters its affinity for DNA. In the absence of the repressor protein, positive control elements take over and transcription is activated.

Positive control of bacterial induction is exercised when activators bind to promoter sites and turn up the rate of transcription. The best known example, and probably the most general, is the lac activator.

This activator is known as **CAP** for **catabolic activator protein.** The controlling factor or ligand is **cyclic adenosine-3′, 5′-phosphate,** or **cyclic AMP** (cAMP). CAP is therefore also known as **CRP** for **cyclic AMP receptor protein.**

CAP consists of two subunits that in the presence of cAMP acquire a symmetry capable of engaging some 30 base pairs of DNA in the promoter site—specifically, the region known as the **CAP site** (Figs. 4.10 and 4.11). An imperfect palindrome with an axis of symmetry at base -62 may serve as a recognition site for the CAP protein. Similar palindromes are found at CAP sites in other promoters.

Together, the *lac* promoter site and operator are controlling elements constituting the **lac enzyme control region.** The bases at the CAP site are not transcribed and are therefore not part of the **lac operon,** but, by binding CAP, they provide a signal that helps RNA polymerase recognize its **binding** or **interaction site.**

The **RNA polymerase interaction site** is downstream to the CAP site (compare to Figs. 4.11 and 4.1). The Pribnow box (covering -12 to -6) and other sequences involved with binding the polymerase (e.g., -37 to -27) are upstream to the operator.

Unlike the lac repressor protein which only regulates activity at the *lac* operator, CAP is part of a system of **catabolite-sensitive** operons governing the breakdown of arabinose, maltose, and other sugars as well as lactose (see de Crombrugghe et al., 1984). CAP sites for these operons frequently employ the 5′-TGTGA-3′ sequence (e.g., between bases at -69 and -65 in Fig. 4.11) or a variation thereof, not at a constant distance from the first nucleotide transcribed.

The positive control of the lac operon, like the negative control, seems remarkably efficient and adaptive. The relationship between lactose and cAMP is obscure, but glucose, the normal carbohydrate source for *E. coli*, has a direct relationship to cAMP production. When glucose levels are low, the cyclase responsible for cAMP production is activated, and the amount of cAMP in the cell increases (Fig. 4.10). If lactose is present as well, the lac operon is transcribed; the lactose enzymes are synthesized, and lactose, as an alternative nutrient, is utilized. On the other hand, in the absence of free cAMP, the complex of CAP and cAMP dissociates; the activator leaves the DNA, and transcription of the operon ceases.

What must be explained are the systems preventing any cell from expressing all its genetic potentialities and the stable transmission of the signals involved in the sorting out of different functions.

F. Jacob and J. Monod (1963, p. 31)

Differential Gene Expression in Eukaryotes

The enormous progress of tissue culture since the advent of antibiotics has made it possible to use eukaryotic cells *in vitro* in some of the same ways bacteria were used in the past. The addition of molecular genetic approaches, cloning of eukaryotic genes, DNA-recombinant technology, and the transfection of cells with artificial plasmids has brought studies of control in eukaryotes abreast of studies in bacteria.

In many ways the bacterial paradigm has been vindicated for eukaryotes. Structural genes and regulatory elements in eukaryotes, like their counterparts in prokaryotes, encode proteins. The activation of transcription by ligand-bound activators has been reproduced *in vitro*, and controlling elements have been identified in eukaryotes for heat shock genes, simian virus 40 (SV40), and adenovirus genes, viral- and cyto-oncogenes (v- and c-oncogenes related to tumor production), and cellular ontogenes (also known as ontogenic genes [development-generating genes]).

Moreover, **transfected** cells (tissue culture cells that have taken up foreign DNA) express their new genes and alter their differentiation (e.g., Davis et al., 1987), and **germ-line transformants** of *Drosophila melanogaster* and **transgenic** mice (organisms receiving foreign DNA through sex cells) show acquired transcriptional controls (see Maniatis et al., 1987). Still, the study of differential gene expression in eukaryotes has its own peculiarities.

Structural genes and regulatory elements in eukaryotes, like their bacterial counterparts, provide templates for protein synthesis. These genes are generally *unique* or relatively unique (also called nonrepeated sequences and single-copy nuclear sequences [scn]) genes. In chromosomes stained with the Giemsa bacterial stain, centers of unique sequences are visualized as light-staining **euchromatic bands** (Fig. 4.12). Beyond this, similarities with prokaryotic genes are strained.

Unlike prokaryotic genes, some eukaryotic structural genes, such as the genes for histones, are present in many copies in the genome. Redundant forms of the same gene may differ in the sizes of intervening sequences if not their number or location.

Ribosomal RNA templates (rDNA) are also present in many copies. For example, the major oocyte 5S rRNA gene (*Xlo*) is present in over 1000 repeats at the ends of most *Xenopus laevis* chromosomes, and even the somatic 5S rRNA gene (*Xls*) is present in about 400 copies on one chromosome.

In contrast to bacterial structural genes which are "leaky" (i.e., they always undergo some transcription), eukaryotic structural genes can be repressed totally. Furthermore, while the amounts of bacterial enzymes change a thousand-fold between uninduced and fully induced states, the amount of specific eukaryotic proteins can change a million-fold between early and late stages of induction.

In prokaryotes, the promoter of one gene can follow directly after the previous gene (e.g., the *lacI* regulatory gene and the *lac* operon, Fig. 4.11), while in eukaryotes, **spacer** sequences (such as the nontranscribed spacer [NTS] of the rRNA gene of mice, Fig. 4.3) typically separate the end of one gene from the initiation sequence of the next. These "nongenic" spacer sequences have variable lengths.

The size of a transcription unit is also different in prokaryotes and eukaryotes. While many bacterial operons are polycistronic (i.e., a single transcript containing sequences for several structural genes, e.g., Fig. 4.10), eukaryotic operons are typically **monocistronic,** encoding only one polypeptide. This does not necessarily mean that eukaryotic transcripts are shorter than their prokaryotic counterparts. In contrast to prokaryotes, eukaryotes make long primary transcripts that are later cut down to size by post-transcriptional processing.

Transcriptional controlling elements or cis-acting regulatory elements in eukaryotic DNA, like those in prokaryotic DNA, contain modular elements with repeated nucleotide motifs. For example, regions with high adenine and thymine

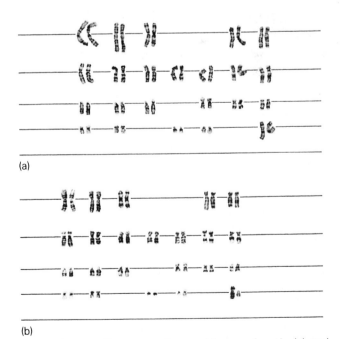

(a)

(b)

FIGURE 4.12. Karyotype of normal human female (a) and male (b) showing Glemsa dark-staining G bands. Courtesy of Silvia F. Pan, M.D., Cytogenetics Laboratory, University of Pittsburgh, Graduate School of Public Health.

content (AT-rich sequences) appear in controlling elements in both eukaryotic and prokaryotic DNA (Fig 4.11) (Wasylyk et al., 1981).

Eukaryotic Goldberg–Hogness boxes (i.e., TATA boxes) are so similar to prokaryotic Pribnow boxes that they are widely thought to have evolutionary as well as physiological parallels. The third T in these boxes may be the same in all promoters, and the first TA is almost always the same. Furthermore, like prokaryotes (Fig. 4.11), the control regions of eukaryotic genes frequently have imperfectly repeated palindromic sequences of 14 and 15 nucleotides.

Promoters stretch over larger regions in eukaryotes than in prokaryotes and are found farther upstream. **Upstream promoter elements** (UPEs) exist in the range of 100 base pairs beyond the gene's TATA box, but they are not unlike prokaryotic promoter elements containing high guanine and cytosine content (GC-rich).

Other features of eukaryotic controlling elements are difficult to relate directly to prokaryotic controlling elements. Above all, eukaryotes are endowed with repetitive DNA, sometimes exhibiting regulatory properties, and with enhancer, activator, and silencer elements controlling gene activity at remote sites.

Repetitive DNA is characteristic of eukaryotes. It occurs in hundreds of thousands to millions of copies per eukaryotic genome. The repeated nucleotide sequences, or reiterated sequences, are short (e.g., in satellite DNA) or long (about 200 base pairs) and may be mixed, in tandem, or dispersed (e.g, in interspersed repeated DNAs).

The case for repeated sequences playing a controlling role is best made for **long terminal repeats** (LTRs). Even at distances of several kilobases, LTRs greatly enhance transcription (e.g., genes for early proteins in SV40).

Possibly, repetitive DNA represents a hierarchical array of transcriptional controlling elements (Britten and Davidson, 1969; Davidson and Britten, 1979). The most likely candidates for *cis*-acting controlling elements are **moderately repetitive** (MR) **DNA** sequences. Their qualifications are that they are transcribed but not translated, and they are interspersed with potentially transcribed structural genes.

MR sequences are also candidates for **transpositional elements** or **mobile elements** capable of moving and inserting themselves anywhere in the genome (McClintock, 1956, 1984), and they may be involved in **genic rearrangement.** MR sequences

may also play a role in activating normal cellular **proto-oncogenes** or **c-oncogenes,** which thereupon behave like cancer-causing **transforming oncogenes** or **v-oncogenes** carried into cells by retroviruses (see Duesberg, 1985).

Repetitive DNA may also play a quantitative role in regulation. As a rule, the percentage of repetitive DNA in various eukaryotes is correlated with the overall DNA content of cells (i.e., the **genomic mass** as estimated from gametes). Fungi, whose genomic mass is low, have only negligible amounts of repetitive DNA. Insects have a greater DNA content per nucleus and a mass of repeated DNA sequences of about 20%. Repeated sequences represent 30–40% of the genomic mass in mammals and as much as 80% in amphibians and higher plants whose genomic mass is among the highest. In the onion, *Allium cepa,* with a spectacular 16.7 picograms (pg) of DNA in its genomic mass, 95% consists of repeated sequences.

Other physiological roles are also played by repetitive DNA. The highly repetitive satellite DNA found in centromeres functions in chromosomal movement, and telomeric repeats function in protecting genes at the ends of chromosomes. The sheer volume of repetitive DNA sequences suggests that they play additional roles in cell behavior (see Brodsky and Uryvaeva, 1985).

Enhancer (also activator, and silencer) elements are loosely defined sequences of nucleotides with broad powers to influence regulated promoters. Occurring either upstream or downstream from their promoters, enhancers are **orientation independent;** they are found at various distances from their promoters, from within the transcription unit to as far as several kilobases away and even on the complementary strand of DNA; their activity is generally quantitative (i.e., "more or less" rather than "all or none").

Different elements can have opposite consequences for transcription. **Silencers** act negatively or subtractively (their absence leading to increased transcription), while **activators** act positively or additively. The nucleotide sequences involved may be overlapping.

Like eukaryotic promoters, enhancer elements are modular, but the sequence units tend to be more numerous than those of promoters. In the SV40 enhancer, for example, **binding elements** (numbered boxes, Fig. 4.13) in two functional domains (A and B) complex with specific and independent protein factors. In different types of cell, different protein factors must recognize one or another of the **sequence motifs** (brackets) and specific

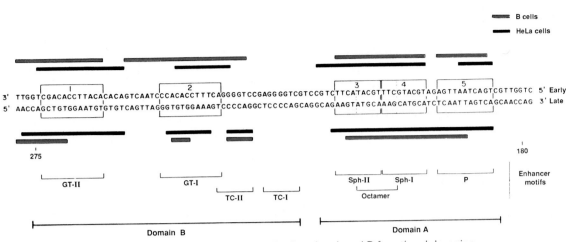

FIGURE 4.13. The SV40 enhancer in B and HeLa cells showing A and B functional domains, recognition sequences required for maximum levels of enhancer activity (boxes 1–5), and repeated sequence motifs (brackets below sequences) including the octamer that is interchangeable with sequences found in several promoters. From T. Maniatis et al., *Science,* 236:1237 (1987) by permission of the American Association for the Advancement of Science and the authors.

base **contact sites** to activate the regulated promoter. The absence of one or another regulatory protein does not necessarily turn off transcription entirely, but full activity may depend on particular sites.

Enhancers are sometimes classified as belonging to one of two categories, although the line between them often blurs (see Maniatis et al., 1987). Inducible enhancers are responsive to changes in their environment. Temporal-, tissue-, and cell type-specific enhancers are active only during development at specific times and in specific tissues.

In general, **inducible enhancers** act positively as shown by their ability to confer inducibility upon unrelated genes tied to them artificially (e.g., the human metallothionein enhancer) (Searle et al., 1985), and the tandem reiterated regulatory elements have additive effects, providing quantitative controls over gene expression. But different inducible enhancers may regulate their promoters in different ways and with different degrees of specificity.

Cellular proto-oncogenes, such as *c-fos*, for example, have inducible enhancers containing several *cis*-acting sequences. A variety of serum factors (platelet-derived growth factor [PDGF], nerve growth factor [NGF], colony stimulating growth factor I [CSF-I], interleukin-3 [IL-3]) activate the enhancer indirectly. The growth factors, as **agonists,** complex with cell-surface receptors, and initiate a cascade of protein phosphorylation leading to elevated Ca^{2+} concentrations, increased pH, and the

production of proteins capable of entering the nucleus and combining with the enhancer sequences. Within minutes of induction the protein encoded by *c-fos* is produced.

Steroid-dependent gene regulation also depends on inducible enhancers containing short *cis*-acting enhancer sequences (called steroid response elements or SREs), but these are regulated by specific steroid receptors in the presence of the hormone. The enhancers for some steroid-dependent genes contain identical sequences (e.g., TGTACA [upstream] and TGTTCT [downstream] half sites in the glucocorticoid and progesterone response elements), while the enhancers for other genes contain unique sequences (e.g., estrogen response elements). The steroid receptors (e.g., progesterone and glucocorticoid receptors) have hormone-binding domains capable of complexing with the specific steroid, and DNA-binding domains capable of recognizing specific enhancer sequences after the hormone-binding domains are occupied.

Some inducible enhancers are close to their regulated promoters or even in the transcription units (e.g., some SREs), while others are remote. For example, the metallothionein enhancer may sit on its promoter, while steroid-responsive enhancers (e.g., for mouse mammary tumor virus and Molony mouse sarcoma virus) are found in remote long terminal repeat (LTR) sequences.

The **temporal-, tissue-,** and **cell type-specific enhancers** are control elements operating in particular cell types and utilized at particular times during the differentiation of the cell. Specificity is established mainly by negative evidence (e.g., failures to detect synthesis when cloned genes are transfected to particular tissue culture cells), but, after applying a variety of sensitive methods for detecting products coupled to DNA sequencing, the possibility of specificity becomes virtually inescapable.

The heavy-chain immunoglobulin enhancer, for example, activates immunoglobulin synthesis when cloned copies of the gene are transfected to cells of the B cell lineage but hardly anywhere else. The specific requirements of other time- and cell-type specific enhancers may likewise only be met in specific cells during the temporal course of differentiation.

The enhancer elements are generally modular, operating in groups of tandem repeat recognition sequences. The Ig heavy-chain enhancer, for example, has four recognition sites containing octameric recognition sequences related to 5'-CAGGTGGC-3' and a fifth site containing a 5'-ATTTGCAT-3' sequence, identical to the "O" Ig octamer sequence of the upstream promoter (Ephrussi et al., 1985).

Tissue-specific enhancers tend to be close to the regulated promoter but not necessarily upstream. The Ig heavy-chain enhancer, for example, is found within the first intron between the variable (V_H) region's joining (J) domain and the constant (C_H) region (see Fig. 4.8 for relationship of regions).

Unlike inducing enhancers with a broad ability to activate different promoters when brought into contact, tissue-specific enhancers are associated with tissue-specific promoters. For example, the Ig heavy-chain enhancer is able to enhance transcription of β-globin and SV40 T antigen in myeloma cells, but Ig enhancer elements fail to stimulate the metallothionein promoter as well as the inducible metallothionein enhancer stimulates it.

What Are Regulatory Proteins?

A great many proteins recognize and contact specific nucleotide sequences in DNA. While generally called regulatory proteins in prokaryotes, these proteins are increasingly called **transcriptional factors** (TFs) in eukaryotes and include factors binding to enhancer sites (in all their variety) as well as promoter sites. The term "regulatory protein" will be used when referring to proteins in both prokaryotes and eukaryotes, while "transcription factors" will be used for specific eukaryotic and animal virus proteins.

Transcriptional factors are sometimes widespread and sometimes tissue specific. The factor called Oct-1, for example, is ubiquitous and binds octameric nucleotide sequences in a variety of cells. Oct-2, on the other hand, binds the _O_ site of B lymphocytes exclusively.

Regulatory proteins and DNA sequences presumably coevolved to achieve the functional intimacy characteristic of their interactions. The structure of typical cellular DNA (i.e., the B form with alternating major and minor grooves) offers proteins a chance to recognize specific nucleotide sequences in the major groove without disrupting the double helix.

The major groove accommodates an alpha helix of a protein and the opportunity for hydrogen bonding between residues on the exterior face of the helix and nucleotide sequences. Unambiguous recognition requires at least two hydrogen bonds per base pair. In addition, proteins are likely to utilize methyl groups and overall DNA structure (e.g., bends in the DNA) to "read" nucleotide sequences.

The symmetry of nucleotide sequences found at many recognition and binding sites reduces the length of protein required for specificity. Moreover, contact with a repeat aids correct positioning and orientation while providing tighter DNA binding than available with a single site.

The DNA-contacting portions of several DNA-binding proteins have been identified by the biochemical analysis of mutant forms in both prokaryotes and eukaryotes. In the most thoroughly studied cases, the structure of these proteins and their relationship to their specific DNA recognition site have been determined through high-resolution electron density mapping and nuclear magnetic resonance.

The proteins binding DNA generally possess two or four identical subunits with two active sites. DNA-binding structures currently identified contain one of two motifs (helix-turn-helix and zinc fingers), but more motifs will undoubtedly be discovered.

The **helix-turn-helix** motif occurs in protrusions on DNA-contacting proteins oriented at about the angle of the major groove in DNA. A short alpha helix within the protrusion lies across the major groove, while an inward pointing second helix, joined to the first by a turn, lies partly within the major groove. The second helix, called the **recognition helix,** is largely responsible for the specific residue-base contacts. The rest of the protein may provide a cushion, nestling the recognition helix and holding it in position for contacting DNA (see Schleif, 1988).

Structures of this sort are found in DNA-binding proteins of prokaryotes and bacteriophage (e.g., the *lac* repressor, the repressor of the tryptophan operon, Trp R, the cyclic adenosine monophosphate receptor protein [CRP] of *Escherichia coli*, the CRO regulatory protein of bacteriophage lambda, and lambda repressor) and in homeodomains (e.g., portions of proteins encoded by the homeotic genes) of eukaryotes. Similar structures are rarely found in proteins that do not bind DNA.

The **zinc finger** motif in DNA-contacting domains is widespread among eukaryotic transcription factors. Diagnosed mainly from similarities in the base sequences of cDNA clones, zinc finger proteins have pairs of cysteine (doublets) separated by a small group of amino acids from pairs of histidine (Berg, 1986). Structurally, the proteins may contain one or more fingerlike loops anchored (i.e., chelated) by zinc to the pairs of cysteine and histidine. Alternatively, the zinc may anchor an alpha helix to a planar portion of the molecule (two antiparallel beta sheets) (see Evans and Hollenberg, 1988).

The zinc finger proteins fall into two major classes. The C_x class has nonrepetitive fingers and a variable number of cysteines available for zinc anchoring. For example, a cell cycle protein of yeast (GAL4) has a single finger that binds to enhancer-like sequences of about 17 base pairs, while mammalian steroid receptor proteins have two finger domains and bind enhancer elements.

In the C_2H_2 class, a finger domain within the protein has a tandem repeated sequences of 12 amino acids with a motif of cysteine, phenylalanine, leucine, and histidine separated by seven to eight additional amino acids. The zinc ions are bound to pairs of histidines and cysteines (hence the 2s in C_2H_2).

A variation on this class of zinc finger proteins contains two or more finger domains. For example, the *Evi-1* gene, which operates in the transformation of mouse myeloid cell lines to myelogenous leukemias, encodes a zinc finger protein having an amino-terminal domain with seven finger repeats and a carboxy-terminal domain with three finger repeats (Morishita et al., 1988).

First described for the *Xenopus laevis* transcription factor TFIIIA (Miller et al., 1985) that binds to the specific internal control region (ICR) of the 5S rRNA gene, TFIIIA has nine zinc fingers. The protein's carboxyl terminus is oriented toward the gene's 5' end, and the amino terminus is oriented toward the 3' end. RNA polymerase III recognizes the TFIIIA molecule and transcribes 5S rRNA (Wolffe and Brown, 1988).

Other C_2H_2 zinc finger proteins are represented by regulatory proteins that bind RNA polymerase II and hence transcribe mRNA. For example, a four-fingered protein is encoded by a gap segmentation gene in *Drosophila melanogaster* (e.g., *Krüppel*), and a 13-fingered protein is encoded by the **testis-determining factor gene,** *TDF,* operating in sex determination in human beings. The human SP1 protein, a three-fingered protein, acts upstream to a variety of genes and binds to a 10 base pair sequence lacking palindromes, while yeast ADR1, a two-fingered protein, recognizes a 22 base pair palindrome.

Another motif is found at the DNA-contact site of Eco RI endonuclease. This enzyme is not, of course, a transcription factor, but its DNA-contact site may illuminate what can be expected in additional DNA-binding sites. The dimeric enzyme contacts the sequence-recognition site (a hexanucleotide, GAATTC) in an atypical form of DNA. Four parallel alpha helices are inserted into the DNA's major groove at their amine terminal ends in what is called a **parallel helix bundle** motif. In addition, extensions called "arms" from each subunit of the enzyme reach around the DNA and strengthen binding (McClarin et al., 1986).

Trans-Acting Controlling Factors in Eukaryotes

The eukaryotic **inducing** or ***trans*-acting controlling factors** include a variety of macromolecules (e.g., immune complexes, transferrin), hormones (both polypeptide, e.g., insulin, and steroid), and serum factors (e.g., epidermal and platelet-derived growth factors). As ligands, the *trans*-acting factors form complexes with **receptors** or **regulatory proteins,** and these attach to single controlling elements or multiple controlling elements of DNA. Hydrocortisone, for example, induces liver enzymes (e.g., tyrosine transaminase), and estradiol and progesterone induce hen oviduct conalbumin (also called ovotransferrin) through the mediation of receptors. The estradiol and progesterone induction of hen oviduct ovalbumin expression requires simultaneous binding in DNA by two receptors.

Some hormones affect DNA after binding to receptors in the cytoplasm. The glucocorticoid receptor, for example, contains several functional domains that allow it to move through nuclear pores and, within the nucleus, attach to the **glucocorticoid response element** (GRE) of DNA (see Beato, 1989). The receptor's **DNA-binding domain** is a cysteine-rich finger region, while its **transcrip-**

tional response domain (or activation site) stimulates specific transcription (e.g., RNA synthesis associated with mouse mammary tumor virus). The thyroxine receptor, on the other hand, is already in the nucleus.

In **cyclic AMP-mediated induction,** cyclic adenosine monophosphate (cAMP) plays the role of a **secondary intracellular message** leading only indirectly to interactions with nuclear receptors. Beginning at the cell surface, ligand-bound receptor proteins in the plasma membrane activate adenylate cyclase, thereby causing an increase in cAMP. In theory, the cAMP condenses with and dissociates a regulatory (R) subunit of a multimeric enzyme (i.e., a holoenzyme), thereby activating a catalytic (C) subunit. Acting as a kinase, the C subunit phosphorylates specific protein substrates, usually enzymes, which alter a number of cellular properties (including the cessation of growth and cell death). The phosphorylated protein may also combine with DNA, either directly or after additional steps, and induce particular enzymes (e.g., phosphodiesterase, Fig. 4.14).

Beyond the greater variety of eukaryotic cell reactions, the eukaryotic cAMP-mediated system

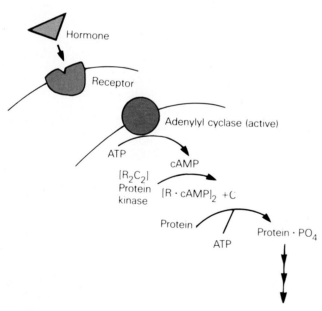

FIGURE 4.14. Sequential steps (top to bottom) in a receptor-mediated hormone response. Activated by the hormone, the receptor is coupled to the enzyme, adenylate cyclase, which mediates the conversion of ATP to cAMP. In turn, cAMP activates cAMP-dependent protein kinase by releasing a catalytic subunit (*C*) capable of activating specific proteins by transferring a phosphate from ATP. Adapted from M. Hunziker-Dunn et al., in P. M. Motta and E. S. E. Havez, eds., *Biology of the ovary.* The Hague, Netherlands (1980). Used by permission of Martinus Nijhoff and the authors.

is strikingly similar to the prokaryotic catabolite-sensitive system (e.g., Coffino, 1981). In the case of bacteria, the catabolite-sensitive control system falls in when the cell is faced with an alternative nutrient source (e.g., lactose in the absence of glucose), while in the case of eukaryotes, the cAMP-mediated system is activated when the cell confronts a hormone arriving from elsewhere in the organism.

> May the concepts derived from the study of regulation in bacteria be of some value in interpretation and analysis of cellular differentiation.
>
> F. Jacob and J. Monod (1963, p. 30)

The Unique Side of Eukaryotic Differentiation

Molecular biologists have covered an enormous distance on the road to understanding selective gene expression and specific protein synthesis, but a huge gap remains. Difficulties in closing this gap arise from differences between prokaryotic and eukaryotic cells and from the reductionist approach to complex problems.

First of all, eukaryotic DNA is not naked like its prokaryotic counterpart but part of the complex structure of chromatin with its own dynamics and controls. Compartmentalization of the eukaryotic cell, moreover, provides opportunities for **posttranscriptional** controls (i.e., controls over processing, translocation of messages out of the nucleus, and translation) not available in prokaryotes.

As important as transcriptional controls are in differentiation, in early embryos, posttranscriptional controls, especially translational controls, predominate (see Chapter 15). Furthermore, the division of labor between organs and the specializations of function among tissues in multicellular organisms offer many more points for control than present in the prokaryotic cell.

While reductionism works well when a eukaryotic cell is studied *in vitro* as if it were a bacterium, problems arise when the eukaryotic cell is studied *in situ* as part of an organism. Differentiation, for example, is vulnerable to reductionist approaches when viewed as the synthesis of specific products, but differentiation is not quite as accessible when viewed from the broader prospective of tissue dynamics. A comprehensive view of multicellular organisms, as opposed to a reductionist view, places protein synthesis in the context of cell populations, tissues, organs, organisms, and the environment (see Slack, 1983).

Chromatin in eukaryotes is much more complicated than the naked DNA of prokaryotes. Chromatin contains a variety of proteins either not found in prokaryotes or not bound to DNA in prokaryotes.

Part of the complexity of chromatin is inherent in eukaryotes' requirement for three RNA polymerases as opposed to prokaryotes' one. Unlike bacterial RNA polymerase, the large and complex eukaryotic polymerases recognize different protein transcription complexes bound to DNA rather than DNA directly. Eukaryotes are required to build these complexes, while prokaryotes get by without them.

The nonhistone chromosomal proteins of the small **high-mobility group** (HMG) are sometimes nominated as candidates for gene regulatory proteins in eukaryotes. Specific HMGs are present in active chromatin, and these can become linked to polypeptides or phosphorylated, presumably with consequences for binding DNA. These proteins lack the essential quality of dissociatability needed for induction and repression, however, and the permanence of the bonds to DNA is incompatible with regulation. The role of HMGs has yet to be ascertained.

Histones are the packaging for the eukaryotic genome. DNA is wrapped around them in interphase chromatin and coiled along with them into mitotic chromosomes (Fig. 2.21). It seems unlikely that DNA-binding proteins could interact with DNA without encountering histones, but the interactions seem more general and managerial than specific and generative.

Histones seem too uniform to operate as specific regulatory factors. Protein (ubiquitin) may be added to the H2A core histone in regions of gene activity, and variants of the H1 histone are found in specific tissue, but unique histones do not seem to be restricted to subregions of the genome.

In eukaryotic cells, the rate of transcription is inversely proportional to the degree of histone phosphorylation and directly proportional to the degree of histone acetylation. Possibly, phosphorylated histones adhere more tightly to DNA, and acetylated histones adhere less tightly. Active chromatin is also less methylated than inactive chromatin. Possibly, the removal of methyl groups prepares chromatin for transcription (Groudine and Conkin, 1985).

At a higher level of organization, nucleosomes of core histones and gyres of DNA (see Fig. 3.7) might operate indirectly to identify DNA protein-binding sites. Not only do bends and kinks in the DNA on nucleosomes broaden the exposure of nucleotides (e.g., base pair opening angles), but they alter other features of conformation (e.g., helical twist) and configuration (e.g., long-range bending). Nucleosomes may thereby extend the definition of a protein-binding site from a sequence of bases to a unique and stable formation that transcriptional factors can recognize (Klug and Travers, 1989; Morse and Simpson, 1988; White and Bauer, 1989). At the same time, sequestering DNA behind core nucleosomal histones and H1 would seem to suppress the potential of DNA for nonspecific binding. The signal-to-noise ratio for RNA polymerases searching for protein-recognition sites may be greatly enhanced by nucleosomes.

Tissue dynamics in eukaryotes, as opposed to cell division in prokaryotes, present additional features of eukaryotic transcriptional controls. Division tends to be part of the development of all tissues and only tends to cease as differentiation reaches its climax. Differentiating cells withdraw from the cell cycle in static, steady-state, and intermediate populations, and cells suspend the cycle in differentiated expanding populations (see Fig. 2.14).

Moreover, unlike bacterial cells, eukaryotic cells undergo a mitotic cycle with conspicuous changes between periods of synthesis and division. Cell division in eukaryotes would even seem incompatible with transcription (prophase of meiosis being the exception). Unlike the naked DNA of prokaryotes that can be transcribed and replicated at the same time, eukaryotic chromosomes are tightly coiled, and their DNA is largely inaccessible.

Inducible and tissue-specific eukaryotic genes become functionally competent during the G-1 period of interphase and possibly the early portion of the S period. The removal of H1 histone and exposure of linker DNA may be part of the mechanism for making eukaryotic DNA available for transcription. For example, in *Xenopus laevis*, removal of H1 from chromatin *in vitro* allows RNA polymerase III to transcribe the *Xlo* gene and produce 5S rRNA. Moreover, the addition of H1 to previously "stripped" chromatin restores the inhibition of 5S rRNA transcription (Wolffe and Brown, 1988).

The possibility that transcription takes place on stretched or loosened portions of G-1 chromatin is also consistent with the results of enzymatic digestion. In the stretched condition (see Fig. 2.21), eukaryotic DNA is sensitive to deoxyribonuclease I digestion and **intermediate deoxyribonuclease I-sensitive domains** of 30–330 kilobases are found in differentiated cells.

The typical **replication-dependent** synthesis of histones during the S period of interphase occurs too late to play a role in differential gene activation. The sequential replacement of nucleosomes (see Fig. 3.7), nevertheless, may be involved in the general inactivation of chromatin in the late S period. The specificity and logistics of specific eukaryotic gene expression may be as dependent on the inactivation of genes as on the activation of transcription.

The failure to replace nucleosomes on freshly replicated DNA may affect transcription in the next generation of sibling cells. The total amount of RNA and the rate of mRNA synthesis increase during the early S period (S-E) but remain constant during G-1, late S (S-L), and G-2 periods. Replicons, or units of replication, initiated after S-E would seem to be transcriptionally incompetent (Flickinger, 1976).

Transcriptionally inert genes appear as late-replicating portions of DNA corresponding to adenine- and thymine-rich (AT-rich) Giemsa dark-staining bands on mammalian chromosomes (Fig. 4.12). A variety of other genes appear as early-replicating, guanine- and cytosine-rich (GC-rich) Giemsa light-staining bands. The latter genes may include MR sequences, inducible and tissue-specific controlling elements, and potentially or actively transcribed structural genes that supply the cell with its requisite **housekeeping enzymes** (Goldman et al., 1984).

Genes that change from a transcriptionally inert status to a transcriptionally active status presumably change from late replicating to early replicating. Giemsa staining patterns are not different in various tissues, however, and staining patterns are not seen to change during differentiation. Possibly, differences are not observed because there are so few working genes among so many unemployed.

The eukaryote's genome is flagrantly lavish compared to the prokaryote's. The difference may be explained by the requirement of eukaryotes to generate great variety during development. While almost all the prokaryotic genome is transcribed, of approximately 124,000 tissue-specific genes in the mammalian genome, about 123,600 are thought to be inactive in any one cell, and 93% of the vertebrate genome is completely suppressed at all times (Goldman et al., 1984).

As evolutionary adventurers, eukaryotes may have traded off the transcriptional economy of the prokaryotic genome for tissue diversity and differentiation. Genic superfluity may have evolved as the *quid pro quo* for variation. In effect, the accumulation of a largely idle genome may be the material price eukaryotes pay for progress.

EMBRYOLOGY AND GENETICS

Embryology, the science which deals directly with the phenomena of heredity . . . is, therefore, the touchstone of every theory of inheritance.

Charles Otis Whitman (1899, reproduced in Maienschein, 1986, p. 246)

Embryology did not spawn the science of genetics, but it assisted handily in the birth. In its classical period, genetics was concerned with how the genotype moved to offspring. Physiological genetics, concerned with how genotypes were expressed, was not easily established, but its heir and beneficiary, developmental biology, has become a dominant branch of biology.

The 19th Century Setting: Mendel Before Mendelism

Gregor Mendel (1822–1884), the parent of genetics, was primarily interested in the effects of outbreeding plants and hybridization, but he undertook a study of differentiated traits in order to understand how heredity operated. He chose the garden pea, *Pisum sativum*, for many of his studies, since it possessed qualitative traits that he could identify unambiguously. This was an extremely fortuitous choice.

The garden pea is normally self-fertilizing, and its varieties therefore tend to be "pure" (i.e., homozygous) for particular traits. Nevertheless, the pistil (ovule-bearing portion of the flower; see below and also Fig. 5.5) of one flower can be artificially inseminated with pollen from another flower. The consistency of traits in garden peas and the ability to artificially inseminate them made Mendel's experiments possible.

His efforts spanned 7 years. By exchanging the sources of pollen and pistil in reciprocal experiments, Mendel showed that the heredity of the traits of interest was not influenced by the source of gametes. He collected the seeds from the first generation (i.e., the **first filial** or **f-1** generation) and raised the seeds into plants. Generally, he let these plants undergo self-fertilization to produce a second generation (i.e., **second filial** or **f-2** generation). His first discovery was that the f-1 generation no longer bred true.

One of the first traits Mendel looked at was height (Fig. 4.15). When a variety of tall plants and of short plants were cross-fertilized, the resulting seeds all grew into tall plants, but the next generation of self-fertilized seeds gave rise to short plants as well as tall ones.

The Parents

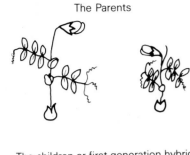

The children or first-generation hybrids

Some of the grandchildren or second-generation hybrids

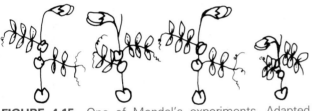

FIGURE 4.15. One of Mendel's experiments. Adapted from C. Auerbach, *The science of genetics.* Harper & Row, New York, 1961. Used by permission.

Hereditary traits had not been blended by cross-fertilization, and they had not disintegrated in a generation in which they had no effects. Instead, traits were passed on in a patent if invisible form. Those passing unnoticed were called **recessive.** Those with effects in every generation were called **dominant.** Mendel counted the plants exhibiting traits and discovered that one-quarter of the f-2 generation exhibited the recessive trait.

Adopting a preformationist convention, Mendel identified the hereditary factors in his experiment as *Anlagen* (pl., *Anlage* sing.), or rudiments. As a shorthand, he represented the dominant *Anlage* as *A* and the recessive one as *a*. His statistical data suggested that pairs of *Anlagen* were present in plants, and fertilization resulted in the redistribution of parental *Anlagen* to offspring.

According to Mendel's explanation, true-breeding dominants possessed a pair of *A*'s (*AA*), while true-breeding recessives possessed a pair of *a*'s (*aa*). After receiving one *A* and one *a* from parents, f-1 hybrids possessed a mixed pair (*Aa*). On passing these *Anlagen* along, one-quarter of the f-2 generation came up as true-breeding dominants (*AA*),

one-quarter as true-breeding recessives (*aa*), and one-half as hybrids (*Aa*). To this must be added that when Mendel looked at additional traits and compared their transmission, he discovered that traits were inherited independently of each other.

In the fashion of his time, he drafted his conclusions into laws. The law of dominance asserted that some traits took precedence over other traits during hereditary passage. The law of segregation stated that factors passed intact between generations. The law of independent assortment declared that factors recombined at fertilization without respect to one another.

Mendel published the results of his now famous experiments in 1866 and 1869. In addition, he corresponded with leading figures in botany, and his reports were cited in important publications available in most important libraries (see Gardner, 1965; Hugo, 1966). But, in general, his experiments were overlooked, unappreciated, or disputed.

Why was Mendel's work so poorly received during his lifetime? One reason, no doubt, was that Charles Darwin's monumental work, *The Origin of Species by Means of Natural Selection or the Preservation of Favoured Races in the Struggle for Life*, published in 1859 monopolized the attention of the scientific world, and Mendel's work did not sit well with Darwinians.

The constancy of factors, which was the essential new element in Mendel's conception of heredity, seemed to be contrary to Darwin's insistence on inheritable variation among members of a species. Had Darwin appreciated the relevance of Mendel's varieties to plant breeders' ''sports'' (now called mutants) and had the parallel between Mendel's results and the emerging cytological discoveries on chromosomes been drawn, the history of development, genetics, and evolution would have been advanced by 30 years. Unfortunately, these connections were not made until the turn of the century, and Mendel's work descended into ill-deserved oblivion until then.

The Darwinian Years

Darwin's work, on the other hand, received well-deserved attention from those interested in heredity and development. Although his evanescent **pangenes** and his uncertain concept of blending inheritance were incorrect, Darwin was the first biologist to propose a single cellular mechanism for heredity and development.

Unlike his predecessor in evolution, Jean Baptiste Lamarck (1744–1829), Darwin discounted the ''inheritance of acquired characteristics.'' Instead, he insisted that heredity could not take direction from development. Darwin argued that descent with modification required changes in reproductive or hereditary materials. All traits had to be passed through the egg, and all changes had to occur during development.

Although skeptical and confused by some aspects of embryology, Darwin concluded in the last edition of *Origin* (1872, Mentor edition, 1958, p. 418):

> Thus, as it seems to me, the leading facts in embryology, which are second to none in importance in natural history, are explained on the principle of variations in the many descendants from some one ancient progenitor, having appeared at a not very early period of life, and having been inherited at a corresponding period. Embryology rises greatly in interest, when we look at the embryo as a picture more or less obscured, of the progenitor, either in its adult or larval state, of all the members of the same great class.

He also undertook research in embryology to test his tentative hypothesis. As he did so frequently in supporting his theory of evolution, Darwin turned to domesticated animals (e.g., Fig. 4.16). The question was whether the anatomical and behavioral deviations that allowed him to draw reasonable "evolutionary" pedigrees for domesticated breeds appeared in a corresponding order in the developmental histories of the breeds.

Adopting von Baer's epigenesis as his model, Darwin made a variety of measurements on freshly hatched and newborn domesticated animals. The results were striking. Varieties considered to be closely related on the basis of anatomy and behavior were more nearly alike at the point of hatching or birth than varieties considered more distantly related. Exceptions were found, but in general the more remote the relationship of adult organisms the earlier differences appeared in development.

Darwin concluded that the magnitude of differences between adult organisms was a function of when changes occurred in the course of development. A small change occurring earlier in development had more impact on how an adult appeared than an equal change occurring later in development. Darwin's theory thus anticipated contemporary concepts of sequential gene action or genetic programs of development.

The "Rediscovery" and Establishment of Classical Genetics

In the early years of the 20th century, botanists monopolized the study of sexual reproduction.

FIGURE 4.16. Various breeds of domesticated pigeons produced by breeders and believed to be descended from the wild rock pigeon seen in the center. From G. J. Romanes, *Darwin and after Darwin*, 4th ed. Open Court Publishing Company, Chicago, 1910.

Plant breeding had become important as a commercial enterprise, and botanists, perusing the literature for insights to help them ply their craft, uncovered and finally appreciated the results of Mendel's earlier experiments.

William Bateson (1861–1926) was the first influential animal biologist to take note of Mendel's work. He had already designed Mendel-like experiments with mice and might have discovered Mendel's laws on his own, but after reading Hugo de Vries' account of Mendel's experiments, Bateson heralded the "rediscovery" of Mendel's laws as the dawn of a new age. He became the first professor of genetics at Cambridge University and proceeded to popularize **Mendelism** in the English-speaking world as well as advance genetic research.

Bateson is credited with introducing the hypothesis that everything about an organism is determined by its genes. A great coiner of terms,[11] Bateson developed a language to allow him to speak of organisms as essentially products of their genes. For example, he extended the definition of "zygote" from the fertilized egg to the entire organism, since, from his point of view, as the first cell containing the entire genome, the zygote was equivalent to the organism. Development resulted merely from the shuffling of genes to somatic cells.

Bateson's view is echoed today when geneticists use his words **homozygote** for an individual with a like pair of genes and **heterozygote** for an individual with an unlike pair of genes, but few now speak of the mature individual as an unmodified zygote. Much more happens in development than a shuffling of genes, or so it seems to those studying development.

In the first decades of the present century, embryologists fought with geneticists over whether development was controlled by the nucleus outward or by the cytoplasm inward (see Gilbert, 1988). Geneticists lined up chiefly with the nucleus, while embryologists lined up with the cytoplasm. Ironically, the strongest argument for nuclear control came from the embryologist–cytologist Theodor Boveri (1862–1915) in the course of studies on chromosomes.

Boveri's **chromosomal theory** proposed that chromosomes were complex structures differing from each other even within the same nucleus and capable of producing qualitatively different effects in cells (Boveri, 1902). His experiment for testing his hypothesis is now a classic and still an excellent example of the power of the experimental method. Moreover, his analysis of data introduced statistical inference to experimental embryology and consequently revolutionized biology (see Davidson, 1986).

Statistics were not what they are today, and evaluating "error" was more a question of reason than rigor, but his results certainly did not eliminate his hypothesis and he concluded:

> Thus, only one possibility remains, namely that not a definite number, but a *definite combination of chromosomes is essential for normal development*, and this means nothing else than that *the individual chromosomes must possess different qualities* (Boveri 1902, translated by S. Glucksohn-Waelsch, reprinted in Willier and Oppenheimer, 1964, p. 84).

Through his analysis of experimental data and the understanding he had already achieved of chro-

FIGURE 4.17. Photograph of Thomas Hunt Morgan as an embryologist (1901). Courtesy of Bryn Mawr College Archives. Used by permission.

mosomal dynamics, Boveri anticipated one of genetics' most central concepts—that uniquely different chromosomes are the vehicles of the genome. W. S. Sutton (1877–1916) took the kernel of Boveri's idea to his mentor, E. B. Wilson (1856–1939) in 1903, and, although it did not germinate, it did not lie totally dormant. Wilson brought Sutton's speculations to the attention of his friend, neighbor, and colleague, Thomas Hunt Morgan (1866–1945), and some 10 years later Morgan and his students wedded the chromosomal theory to genes (Allen, 1983).

When the embryologist Thomas Hunt Morgan (Fig. 4.17) set out to overturn the gene, which he considered absurdly simplistic, he hardly expected he was about to become America's foremost geneticist.[12] His plan was straightforward. He would

---

[11]Bateson coined "genetics," but it was Wilhelm L. Johannsen (1857–1927) who coined the word "gene" in 1903 for a determiner of heredity.

[12]Morgan was awarded the Nobel Prize in Physiology or Medicine in 1933 for his part in establishing the function of chromosomes in the transmission of heredity.

adopt the methods of plant breeders to animals, searching for variants and detecting their passage through generations via controlled breeding regimes. For his animal, he wanted one large enough to be easily observed yet small enough to be raised conveniently in the laboratory. Moreover, he wanted an animal that matured rapidly and bore many young. His inspired choice was the fruit fly, *Drosophila melanogaster.*

Morgan assembled a talented group of young investigators and furnished the "fly room" at Columbia University. There, the breeding and examination of fruit flies went on in an atmosphere of intellectual ferment. The combination was unbeatable and Morgan's laboratory soon became the world's leading center for genetic research (Moore, 1983). Alfred H. Sturtevant, who drew the first chromosomal maps of *Drosophila*, described the "fly room" in 1911 as

> a rather small room (16 by 23 feet), with eight desks crowded into it. Besides the three of us [C. B. Bridges, Morgan and himself] others were always working there—a steady stream of American and foreign students, doctoral and postdoctoral .... There was an atmosphere of excitement in the laboratory and a great deal of discussion and argument about each new result as the work rapidly developed. (Sturtevant, 1965, pp. 46–47).

Following William Bateson's dictum to "treasure your exceptions," Morgan and his associates sought **mutants** (Lat. *mutare* to change), organisms with altered appearances, with which to study heredity. In a few years, they succeeded in isolating more mutants of *Drosophila* than the breeders of domesticated animals had accumulated over centuries (compare Figs. 4.18 and 4.16).

Bridges added cytological studies, and soon chromosomes were seen to be the vehicles for Mendelian genes in much the way they are seen today (Moore, 1986). Many genes were **linked** on the same chromosome and, contrary to Mendel's law of recombination, these genes were passed through heredity more nearly as a unit than as separate units. **Crossing over** between homologous chromosomes occasionally broke the pattern of strict chromosomal inheritance and allowed even closely linked genes to **recombine** on chromosomes.

Unfortunately, in the days of its youthful enthusiasm, the new science ignored complexity and undermined its usefulness to the study of development. What were called mutants were, after all, hereditary differences, but organisms developed from a vast, complicated heredity sameness. The laws of classical genetics said nothing about that

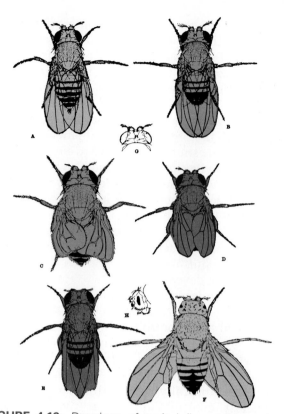

FIGURE 4.18. Drawings of typical (i.e., wild type) and some mutant forms of *Drosophila melanogaster* isolated in Morgan's "fly room." (a) Typical female. (b) Typical male. (c–h) Mutants of chromosome I at loci indicated by numbers in parentheses: (c) rudimentary wing (54.5); (d) miniature wing (36.0); (e) cut wing (20.0); (f) forked bristles (56.5) and bar eye (57.0); (g) white eye (1.5); (h) bar eye (57.0) seen from left side. From E. G. Conklin, *Heredity and environment.* Princeton University Press, Princeton, 1922. Used by permission.

sameness or about all the things mutant and normal organisms had in common (Hubbard, 1982).

Physiological Genetics

Meanwhile, the second branch of genetics was developing slowly. It had gotten off to a good start with Boveri's experimental introduction of polyploidy through polyspermy. The same experiment that demonstrated the requirement of normal development for nuclei containing at least one complete set of chromosomes also established the pathological consequences of abnormal chromosomal combinations. Boveri's concerns extended from cell physiology and the integration of chromosomes to cancer and diseases of chromosomal replication (Koller, 1964), but physiological genetics was not ready to pursue chromosomes. It had first to tackle mutations.

The first reported efforts to track down mutations were made by Archibald E. Garrod (1857–1936), a physician. They appeared in 1902 and were expanded in his book, *Inborn Errors of Metabolism*, in 1909, but the times were not propitious and the work was largely ignored.

Garrod studied metabolic disorders, especially alkaptonuria, a disease characterized by the excretion of alkapton (homogentisic acid) that, on standing, turns urine black. Garrod fed experimental diets to patients with alkaptonuria and to normal individuals. He deduced from the results that homogentisic acid, along with phenylalanine and tyrosine, were normally intermediates in a metabolic pathway leading to the formation of urea (see Fig. 4.19 for a modern version of the pathway).

His case histories indicated that alkaptonuria was inherited and, as suggested by Bateson, was due to recessive genes acquired from both parents. Gerrod's **gene–ferment** (or gene–enzyme) hypothesis proposed that alkaptonuria genes eliminated a ferment (or enzyme) and thereby blocked the breakdown of homogentisic acid. Others at Cambridge extended his idea to mechanisms of pigment formation in flowers and fur, but little headway was actually made.

In Germany, Richard B. Goldschmidt (1878–1958) grasped the importance of physiological approaches to heredity and played with the idea of a dynamic gene. The Great War interrupted his studies, and few others chose to pursue the route he had taken. Even the publication of his book, *Physiological Genetics*, in 1938 failed to turn many heads, although the technique he pioneered, namely, heat shock during development, is used routinely today to study the course of gene action. The idea that the time of gene action plays a role in development would not emerge until Walter Landauer advanced the concept of **phenocopy** (a mutantlike phenotype in a genetically wild-type organism caused by some experimental treatment in the course of development) in his study of chick development.

Meanwhile, the first great step toward a biochemistry of genetics was taken by Boris Ephrussi (1901–1979) with his study of eye color mutants in *Drosophila*. He and his student, George Beadle (born 1903), wondered if mutations affected individual enzymatic reactions or whole biochemical pathways. Employing transplantation in insect larvas, a technique pioneered by Ernst W. Caspari (born 1909), Ephrussi and Beadle discovered that exposure of mutant tissue to diffusible biochemical products could restore normal development (summarized by Ephrussi, 1942). Mutants seemed to short-circuit individual steps in a metabolic

FIGURE 4.19. The metabolic pathway affected by the alkaptonuria gene. The gene causing the metabolic disorder known as alkaptonuria affects the enzyme, homogentisic acid oxidase (HAO), and is part of the normal biochemical pathway involving the metabolism of the amino acid, phenylalanine and tyrosine. From H. Eldon Sutton, ed., *An introduction to human genetics*, 4th ed. Harcourt Brace Jovanovich, New York, 1988. Used by permission.

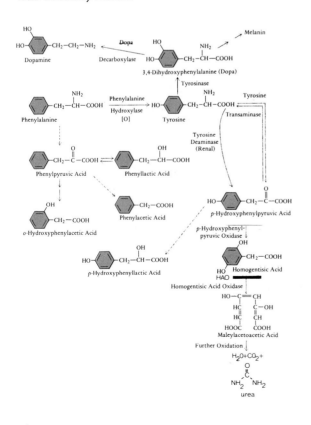

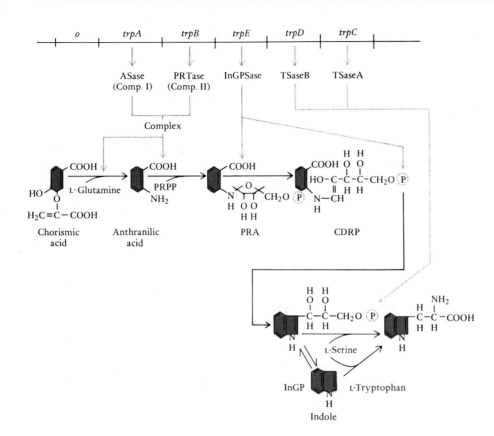

FIGURE 4.20. Skeletal formulas of intermediates in the tryptophan biosynthetic pathway and the position of genes in the DNA of the bacterium, *Salmonella typhimurium. trpA–trpE* and *o*, genes that determine the amino acid sequence in the various enzymes. ASase, PRTase, InGPSasa, TSaseB, and TSaseA, enzymes or subunits of enzymes effecting the pathway. From G. Wuesthoff and R. H. Baurle, *J. Mol. Biol.,* 49:171 (1970) by permission of Academic Press and the authors.

pathway while leaving intact the bulk of the pathway.

Because the actual biochemical steps in the development of eye pigment were technically inaccessible at the time, Ephrussi and Beadle sought more tractable systems. Ephrussi went to yeast and to metabolic disorders. Beadle went to the mold, *Neurospora,* and to mutants affecting biochemical pathways. He chose the mold for several reasons. Its saprophytic life-style meant it could absorb nutrients from its environment. It responded to X-rays by producing varieties, and breeding experiments could be performed to ascertain if the varieties arose from mutations.

Normally, with biochemical pathways intact, the mold synthesized everything it required for growth from a minimal medium. Mutants with defects in biochemical pathways, on the other hand, might be incapable of growth on minimal medium yet capable of growth on media supplemented with metabolites.

The first report on **biochemical mutants** (Beadle and Tatum, 1941) described three demonstrable mutants that grew only on medium supplemented with particular metabolites. Modest but rigorous, the report opened the floodgates, and within a decade a plethora of biochemical pathways and mutants affecting their individual steps were identified.[13] Despite their complexity, the data could be interpreted simply: **one gene–one enzyme** or, more precisely, one mutation–one defective enzyme. Since individual enzymatic steps were disrupted by mutations, normal genes, it seemed, ordinarily produced functioning enzymes.

Only occasionally, results were not quite clean-cut. Some mutations in different genes affected the same enzyme (e.g., Fig. 4.20, *trpA* and *trpB* genes influenced the enzyme tryptophan synthetase), but further research soon clarified the situation. Enzymes consisting of more than one polypeptide chain were affected by more than one mutation (e.g., tryptophan synthetase consisted of ASase and PBTase subunits). The one gene–one enzyme concept therefore was amended to the **one gene–one polypeptide** concept, and the question for researchers moved from what mutations influenced to how they exerted their influence.

Part of the answer came from studies of **thymonucleic acid.** An obscure molecule discovered

---

[13]George W. Beadle, Edward L. Tatum, and Joshua Lederberg shared the Nobel Prize in Physiology or Medicine in 1958 for their studies on biochemical mutants and microbial genetics.

by Friedrich Miescher (1844–1895), it achieved celebrity status in the mid-20th century as DNA, the undisputed genetic material.

Part of the answer was already available but, once again, had to be "rediscovered." In 1933, Jean Brachet hypothesized that **yeast nucleic acid** was present in the cytoplasmic constituents known as basophilic materials (named because of their ability to stain with basic histological dyes). At the same time, Torbjörn Caspersson (born 1910) demonstrated that the basophilic contents in parts of *Drosophila* larvas were proportional to the potential of those parts for growth. Moreover, changes

in basophilia during the course of development were different from one tissue to another and vacillated with the functional state of tissues.

Finally, the correct inference was drawn: the basophilic cytoplasmic counterpart of yeast nucleic acid "must take part in the *synthesis of proteins*" (Brachet, 1945, p. 229, italics original). Chemically purified, yeast nuclei acid was given a new name and, as **RNA,** was soon hailed as the missing link connecting DNA to polypeptides. Within a decade, the central dogma was conceived. Physiological genetics was molecularized, and developmental biology was born.

PART 2 SUMMARY AND HIGHLIGHTS

Periods of mitotic division and interphase rotate through the cell or mitotic cycle. Nuclear DNA replicates and histones are synthesized in eukaryotic cells during the S period of interphase. A G-1 period follows mitosis (M) and precedes S, and a G-2 period follows S and precedes M. These periods may be under the control of chromosomal cycles differing among tissues through which cells move on a stochastic rather than a programmed basis. Cells that tend to remain outside the cycle may have entered a G-0 or mitotically quiescent period.

Mitosis is divided into five phases. In prophase, thin dyad chromosomes consisting of two chromatids appear in place of chromatin, and an amphiaster, or achromatic apparatus, forms in the cytoplasm. Microtubular organizing centers (MTOCs), containing self-replicating centrioles in typical animal cells, form aster fibers directed toward the cytoplasm and polar fibers directed toward the nucleus.

In prometaphase, the nuclear envelope breaks down. Kinetochores develop at chromosomal centromeres, and kinetochore fibers of microtubules rise like trunks while polar fiber branches descend from MTOCs. At metaphase, highly condensed dyad chromosomes line up on the equatorial metaphase plate. At anaphase, the centromeres linking the chromatids divide, and the monad chromosomes, or single chromatids, move rapidly toward the spindle poles. The force for the movement is possibly generated by the disorientation of fibers at the poles and the release of tension among parallel fibers toward the equator. At telophase, the mitotic apparatus disintegrates. Nuclear envelopes reform. Ribosomal synthesis resumes, and nucleoli appear.

Cytokinesis, especially in embryonic cleavage, involves the addition of large amounts of new plasma membrane. In typical animal cells, a contractile ring of actin draws the surface inward; membrane is added, and a furrow in the plane of the former metaphase plate separates the cells. A midbody may continue to connect the dividing cells after the greater part of cytokinesis is ostensibly complete, and a transient syncytium rather than completely independent cells may form briefly following "division."

In the 20th century, interest shifted from the mitotic period toward the interphase period of the cell cycle. Technology, keeping abreast of theory, showed that DNA synthesis typically interrupted eukaryotic interphase. Watson and Crick showed that DNA consisted of complementary chains of nucleotides with internally directed nitrogenous bases. Complementariness suggested how the molecule could be synthesized via semiconservative replication against a template strand of DNA.

The formation of a replication fork depends on topoisomerases and multimeric holoenzymes. Helicases destabilize the double-stranded DNA, and primase or RNA polymerase prepares an RNA primer on the single DNA strands. Gyrases and single-stranded binding proteins relax and stabilize the DNA, while DNA polymerases form a phosphodiester bond between 3′ and 5′ carbons of adjacent deoxyribose sugars, thereby elongating the new DNA chain.

In prokaryotes, multi-subunit enzyme systems identified as prepriming, primosomes, and replisomes are assembled at the beginning of replication on DNA. Moving locomotive fashion along the DNA double helix, the complete replisome opens up a bidirectional replication fork and synthesizes DNA along both strands. The eukaryotic chromatid (or monad chromosome) contains an extraordinarily long loop of DNA, which is replicated at bidirectional forks in several hundred replication units at a time.

Polymerases also have proofreading and correcting functions. While acting as exonucleases, the polymerases cleave out incorrect nitrogenous bases from the new DNA strand, including the RNA primer, and, acting as polymerases again, insert the correct bases.

Polymerase operates processively and forms DNA continuously in the direction of the replicative fork on a leading strand of DNA. Synthesis away from the fork on the lagging strand of DNA is nonprocessive, or distributive, and forms smaller pieces of DNA, called Okazaki fragments.

Ligases join the DNA of the Okazaki fragments and, in eukaryotes, the larger segments of DNA formed on replicating units. Chromatin fibers form when the newly replicated DNA complexes with nucleosomes and coils. The repetitive satellite DNA of the centromere is part of the last DNA to be replicated. Enzymatic elongation of repetitive telomeric DNA may protect gene sequences near chromosomal ends from attrition as RNA primers.

The two divisions of meiosis (meiosis I and II) are linked by interkinesis. Prophase I differs from mitotic prophase by supporting RNA synthesis and the synthesis of zygotene DNA. Interkinesis differs from ordinary interphase by the absence of replication.

Prophase I is divided into five subphases. Leptotene is marked by the appearance of thin dyad chromosomes composed of two chromatids or axial elements. At zygotene, homologous chromosomes synapse to form bivalent or tetrad chromosomes of four chromatids. Synaptonemal complexes zipper chromatids together, and the synthesis of zygotene DNA may be instrumental in recombination and pairing DNA.

In pachytene, chromosomes condense, and chiasmata become obvious between chromatids of paired chromosomes. The chromosome's orientation on the metaphase plate with kinetochores pointing toward poles and the subsequent disjunction of homologous dyad chromosomes to opposite poles may be aided by the ability of chiasmata to stabilize the chromosomes' two-dimensional profile.

Chiasmata may also represent points of crossing over and gene recombination. Resolvase and topoisomerase accomplish recombination, and polymerases repair the damage to DNA. Recombinant DNA formation in prokaryotes via heteroduplexes and chi formations suggests a model for recombination in eukaryotes.

Broadly defined, mutants are phenotypic changes attributable to altered DNA or chromosomes. Different mechanisms of chromosomal alteration may be responsible for spontaneous mutations, but the altered chromosomes and their DNA are treated as normal during replication. Other mutations are attributable to errors in replication, but these are ordinarily corrected by polymerases acting as proofreading enzymes. Only double-stranded errors are likely to escape proofreading, but these errors may be corrected through recombinational repair. By providing new templates as a basis of comparison, meiosis offers this unique opportunity.

At diakinesis, the chiasmata terminalize, and dyad chromosomes separate, while chromosomes move unsteadily to the metaphase plate. Following anaphase I and telophase I, cells enter interkinesis, sometimes without chromosomal dissolution, and move into the second meiotic division.

The sources of variation among eukaryotic cells in an organism are vast. Genes may be more dynamic than formerly thought, and somatic mutation and gene rearrangement occasionally produce totally new genes. Gene amplification may also lead to the production of large amounts of particular genes, but differential transcription or expression of select genes is probably the most general source of differences among cells.

Transcription is the creation of an RNA transcript complementary to a strand of DNA. In prokaryotes one RNA polymerase does the job, but in eukaryotes three enzymes are required: RNA polymerase I for most of the rRNA, II for primary transcripts of mRNA, and III for 5S rRNA and tRNAs.

Recognition and binding sites in promoter regions of DNA include sequences such as TATA boxes and palindromic repeats directing the RNA polymerase "downstream" toward the first nucleotide transcribed and the leader sequence. RNA polymerases imprint coding sequences into complementary trinucleotide codons of mRNA.

Eukaryotic RNAs are processed and modified, especially at their ends. The rRNAs for the ribosome's small subunit and most of those for its large subunit are produced from a single precursor transcript that is cut and modified. Precursors of tRNAs are reduced in length, while additional CCA sequences provide the 3' OH acceptors for amino acids. Through modifications of bases, the tRNAs are able to form characteristic duplex portions and loops. Transfer RNA is charged with specific amino acids and a high-energy bond through the action of specific aminoacyl-tRNA synthetases and ATP.

Messenger RNA from split or interrupted genes is prepared from large pre-mRNA transcripts by removing introns (intervening sequences) and splicing exons (expressed coding sequences) together. Small nuclear RNPs may be involved, and spliceosomes with enzymatic activity, which form lariat RNAs from introns, may be responsible for the precision of splices. Eukaryotic mRNA is also modified by the addition of a poly-A chain at the 3' end and a cap initiation signal at the 5' end.

Translation is the manufacture of polypeptides whose amino acid sequences are determined by the order of codons in mRNA. The start codon is the first AUG sequence behind the cap site. It establishes the reading frame for all subsequent codons. A methionine-charged tRNA attaches to this AUG and, with a small ribosomal subunit, forms an initiation complex. The large ribosomal subunit joins the complex and provides the amino transferase which catalyzes the formation of peptide bonds. A polyribosome forms when several ribosomes read the same mRNA simultaneously.

Additional charged tRNAs match their anticodons with codons on mRNA in the ribosome's aminoacyl- or A-binding site. The growing polypeptide is transferred from the tRNA in the ribosome's peptidyl- or P-binding site to the amino acid complexed with the newest tRNA, and as the tRNA at the P-binding site is released, the tRNA in the A-binding site is translocated to the vacant P-binding site. "Wobble" in the binding sites may account for ambiguity in the third base of the three

base codes and redundancy in the codons specifying most amino acids.

When a stop codon is reached, the polypeptide is released. The ribosomal subunits become available for recruitment into new initiation complexes and complete ribosomes.

The rates of transcription may change in bacteria in response to changes in their medium. Mutations that affect enzymes and rates of transcription have been employed for studying the control of transcription and for devising models of gene regulation.

Jacob and Monod proposed that structural genes determine the primary structure of polypeptides; regulatory genes (e.g., the *lac* inducer) encode *trans*-acting regulatory proteins, either repressor proteins (e.g., the lac repressor protein) or activator proteins (e.g., catabolic activator protein, CAP), while *cis*-acting controlling elements (e.g., the *lac* operator) control the operation of nearby structural genes. Operators are sites in DNA where regulatory proteins bind.

Promoters correspond to sites of RNA polymerase recognition and binding and are either aided (positive control) or hindered (negative control) by regulatory proteins. Induction and derepression occur when the introduction of a ligand changes a regulatory protein's binding to DNA and accelerates transcription. Repression occurs when a similar process has the opposite effect.

Eukaryotic genes contain coding sequences for single polypeptides rather than for successive independent polypeptides (polycistrons) coded in prokaryotic genes. Eukaryotic DNA also has long spacer sequences in contrast to prokaryotes. The controlling elements in eukaryotic and prokaryotic genomes may be similar (e.g., TATA), but, in contrast to prokaryotic DNA, vast amounts of eukaryotic DNA are not ordinarily transcribed.

Enhancers and silencers present in eukaryotic DNA sequences may have additive or subtractive effects, be directionally independent of their promoters, and operate over long distances. The targets for some steroid receptors are enhancers. Polypeptide hormones and thyroxine may also operate through receptor proteins and enhancers. A eukaryotic system involving cAMP as a secondary messenger and cytoplasmic receptor proteins is similar to CAP in bacteria and may operate through activated proteins attaching directly to DNA.

Regulatory proteins or transcriptional factors exhibit different motifs at their DNA-contacting sites: helix-turn-helix and zinc finger. The amine ends of helices identify nitrogenous bases in the major groove of DNA through hydrogen bonds and other cues.

Highly repetitive eukaryotic DNA has no counterpart in prokaryotes. Moderately repetitive (MR) sequences may operate as *cis*-acting control elements or transpositional elements capable of activating genes in early-replicating Giemsa light-staining bands on chromosomes.

Histones and nucleosomes may play roles in the management of gene expression. The stripping of histones may be a condition for transcription, and nucleosomes may aid in DNA recognition.

The cell cycle is uniquely eukaryotic, and the dynamics of tissue growth interacts with transcriptional activity. Cell division in eukaryotes is largely incompatible with differentiation.

Late-replicating, transcriptionally inert tissue-specific genes may become early-replicating upon achieving transcriptional competence. Most eukaryotic genes are not active most of the time, however, and turning them off completely may be an essential part of gene regulation in eukaryotes.

Gregor Mendel drew many substantially correct inferences about the mechanism of inheritance from his qualitative and quantitative analysis of inheritance in garden peas. He described his results in terms of dominance and his laws of segregation and independent assortment.

Theodor Boveri employed statistical inference to test his chromosomal hypothesis about the source of abnormalities in dispermic sea urchin eggs. He concluded that chromosomes were qualitatively different and, through development, influenced normal physiology and pathology.

Darwin attempted to unify the mechanisms of development and heredity through embryological research. He proposed a rudimentary version of the concept of sequential gene action.

William Bateson drew the attention of animal biologists to Mendel's laws and occupied the first professorship of genetics at Cambridge. Thomas Hunt Morgan, an embryologist turned geneticist, launched the classical period of genetics with the flood of mutants rising from his famous "fly room." The gene that emerged was a band on a chromosome, not a biochemical manipulator of cellular activities.

Physiological genetics and the biochemical concept of gene action emerged from the gene–ferment (gene–enzyme) hypothesis of Garrod and the one gene–one enzyme concept of Beadle and Tatum. Brachet linked DNA to protein synthesis through RNA, leading to Crick's formulation of the central dogma: "DNA makes RNA; RNA makes protein."

PART 2 QUESTIONS FOR DISCUSSION

1. How is development related to heredity, and how is a genotype related to a phenotype? What are the dual roles of DNA and how does DNA perform them? Compare a template strand of DNA to a sense strand.

2. Compare and contrast prophase in meiosis and mitosis, interphase and interkinesis. Identify and describe the subphases of meiotic prophase I. What are chromosomal synapses, synaptonemal complexes, and chiasmata? What are recombinant DNA, heteroduplexes, and chi formations?

3. Describe the cell cycle, its periods, and the stages of mitosis. What are the parts of the mitotic apparatus, and how are they thought to function? What drives chromatids apart at anaphase? What is the chromosomal cycle, and how is it related to the mitotic cycle?

4. Describe the complementary pairing of purines and pyrimidines in DNA, and discuss its implications for molecular "reproduction." What is complementarity? Colinearity?

5. Discuss the characteristics of DNA and DNA polymerases that result in the formation of Okazaki fragments. What are replication forks, replicative bubbles, and replicons (RUs)? Discuss the adaptive advantages of both eukaryotic DNA replicating simultaneously at many sites and prokaryotic DNA replicating as a single unit.

6. What is the proofreading function of polymerases? How are pieces of DNA linked together in chromatids, and how are chromatids linked together at centromeres? Define mutant, mutation, segregation, recombination, and recombinational repair. Describe several possible mechanisms of mutation. How can double-stranded errors in base pairs be corrected during meiosis?

7. Describe transcription. What are the similarities and the salient differences between transcription in eukaryotes and prokaryotes? How is the sense strand of DNA chosen by the RNA polymerase? Describe a promoter and what it does. What is the relationship of controlling elements in the Jacob–Monod model to specific nucleotide sequences in DNA?

8. Describe the processing of RNA. What are interrupted genes, introns, and exons? Describe the action of a spliceosome and lariat RNA in splicing.

9. Describe translation. What is an open reading frame, and what are codons and anticodons? How are tRNAs charged? Describe the formation of an initiation complex, a ribosome, and a polyribosome.

10. What is the Jacob–Monod model for the control of gene expression? Describe how the model accounts for what happens in *Escherichia coli* in a low-glucose and high-lactose medium? What are activator and repressor proteins, and what do they do in the presence and absence of ligands? How are allosteric effects thought to influence positive and negative controls of induction and repression?

11. What are the similarities between cortical steroid induction in vertebrates and lactose induction in bacteria? What are the differences between cyclic AMP-mediated induction in eukaryotes and prokaryotes?

12. What is the relationship of moderately repetitive sequences and tissue-specific genes to Giemsa staining regions on chromosomes? How are replication and transcription related in eukaryotic genes?

13. Describe Mendel's theory of *Anlage* and Boveri's chromosomal theory. Why, in your opinion, did it take so long to accept the idea of twin *Anlagen* determining heredity when everyone knew that sexually reproduced organisms had two parents?

14. In what respect did Darwin's embryological research prefigure the present concept of sequential gene action? How did his cellular concept of heredity anticipate the central dogma?

15. Describe the sequence of discoveries leading from alkaptonuria to biochemical genetics and from the gene–enzyme hypothesis to the one gene–one polypeptide concept. In what respects was RNA "the missing link"?

PART 2 RECOMMENDED READING

Allen G. E., 1983. T. H. Morgan and the influence of mechanistic materialism on the development of the gene concept, 1910–1940. *Am. Zool.*, 23:829–843.

Crick, F., 1981. *Life itself.* Simon and Schuster, New York.

Gilbert, W., 1985. Genes-in-pieces revisited. *Science*, 228:823–824.

Jacob, R. and J. Monod, 1963. Genetic repression, allosteric inhibition, and cellular differentiation. In M. Locke, ed., *Cytodifferentiation and macromolecular synthesis.* Academic Press, New York, pp. 30–64.

Judson, H. F., 1978. *The eighth day of creation: Makers of the revolution in biology.* Simon and Schuster, New York.

Jurnak, F. A. and A. McPherson, eds., 1985. *Biological macromolecules and assemblies*, Vol. 2, *Nucleic acids and interactive proteins.* Wiley-Interscience, New York.

Kornberg, A., 1980. *DNA replication and the 1982 Supplement to DNA replication.* Freeman, San Francisco.

Lloyd, D., R. K. Poole, and S. W. Edwards, 1982. *The cell division cycle: Temporal organization and control of cellular growth and reproduction.* Academic Press, New York.

Sang, J. H., 1984. *Genetics and development.* Longman, New York.

Watson, J. D. 1968. *The double helix.* Atheneum, New York.

Watson, J. D., 1987. *Molecular biology of the gene*, Vols. I and II, 4th ed. Benjamin-Cummings, Menlo Park.

REPRODUCTIVE BIOLOGY: WHERE EMBRYOS COME FROM

Sexual reproduction does not make sense (Williams, 1975; Maynard Smith, 1978). Individuals must find mates, usually at some personal risk, and great numbers of gametes are wasted. Still, some form of sex occurs throughout the living world, and judging from its many variations, sexual reproduction would seem to offer many adaptive advantages.

Part 3 attempts to make sense out of sex by placing reproduction in the framework of adaptation. Sexual reproduction is dissected into its components and compared to parthenogenesis, gynogenesis, and asexual reproduction. Gametes are examined, and modes of gametogenesis are scrutinized. The riddle of sexual reproduction is then traced through changing concepts of egg and sperm.

TYPES OF
REPRODUCTION

Sexual reproduction facilitates evolution indirectly by making extinction less likely.

GEORGE C. WILLIAMS (1975, p. 154)

*E*mbryos are produced by zygotes, but reproduction is not limited to zygotes. Organisms also develop from other types of cells, and different kinds of reproduction are defined by the production of these cells.

The classification of reproduction requires no fewer than three categories. In addition to sexual or biparental reproduction, uniparental reproduction and asexual reproduction are independent and mutually exclusive.

Sexual or **biparental reproduction** is reproduction by two parents via a zygote. In contrast, **uniparental reproduction** is reproduction by one parent via a sex cell. It is broadly the province of female members of a species. **Parthenogenesis** (Gk. *parthenos* maiden + *gignesthai* to be born, hence "virgin birth"), or development from an egg without insemination, is found in most animal groups (see Suomalainen et al., 1987), while the similar phenomenon of **agamospermy,** or seed formation without fertilization, is common in plants. **Gynogenesis** (Gk. *gyne* woman + *gignesthai*, hence female birth), or **pseudogamy** (Gk. *pseudein-* to falsify + *gamos* marriage) is development from an egg

following activation by sperm but without sperm contributing to the heredity of subsequent generations.

Reproduction by male members of a species, or **male parthenogenesis,** is rare if it happens at all. The best possibility is *Polypodium*, a hydrozoan, which seems to develop from a spermatozoon that has parasitized a sturgeon egg (Raikova, 1973).

Asexual reproduction is reproduction by cells that are neither egg nor sperm. It occurs within a single phase of the life cycle (see Fig. 1.9) and, in the case of diploid multicellular organisms, involves groups of parental cells rather than isolated cells.

What accounts for different types of reproduction? Inevitably, the answer to questions of this sort is evolution. Like any other part or process in a life cycle, reproduction evolves. But what kind of evolution shapes reproduction?

In general, evolution is attributed to natural selection or to genetic drift. Natural selection proceeds from differential reproductive success within a population, while genetic drift proceeds from the growth of a group after its isolation from an original population. The former is theoretically correlated with some feature of the environment and is equated to adaptation, while the latter is independent of the environment except as it results in a group's isolation.

Research on evolution frequently begins with examining adaptation, since reproductive success can be broken down to measurable components and put in terms of testable hypotheses. The components are survival to breeding age, reproductive life span, fecundity or mating success, and offspring survival. Different hypotheses attribute one or another component to environmental, phenotypic, developmental, or genetic factors (Clutton-Brock, 1988). Genetic drift, on the other hand, is the fallback position adopted when all efforts fail to validate adaptation.

The underlying adaptive advantage of both biparental and uniparental sex is probably their ability to rescue the genome from double-stranded errors through recombinational repair. Asexual reproduction provides no such protection (Bernstein et al., 1985).

Arguments about the adaptive advantage of sexual reproduction frequently turn on the greater number of genes available to the zygote as opposed to the gamete or spore. Over a range of **disomic** conditions, in which one or more chromosomes are paired, to the full-blown diploid condition, in which all the chromosomes are paired, genes double in number. What is more, dominant genes provide an umbrella under which recessive genes can accumulate.

Large numbers of genes, like large data banks in computers, may offer advantages, but a developing diploid organism would seem to have the same advantage whether it were produced biparentally, uniparentally, or asexually. Were developmental mechanisms operating during sexual reproduction to provide greater access to genes, they might offer an advantage, but no such mechanisms are known.

Meiosis and fertilization are often identified as devices for generating variation in a species. Sexual reproduction is portrayed as most likely to introduce variation and asexual reproduction as least likely. But sexual reproduction can also reduce variation, and mutations introduce variation in asexually reproducing organisms just as certainly as in sexually reproducing organisms.

Variation would be favored in a species occupying an unstable environment, while stability would be favored in a species occupying a stable environment. The former species would be expected to evolve a form of reproduction that promoted variation, while the latter would be expected to approach its optimal phenotype through the restriction of variation. Variability would be adaptive when it creates deviant organisms that survive environmental crises, but variability would be maladaptive when it prevents a species from rising to optimal adaptations (Williams, 1975).

Reproductive strategies that promote variation are identified with recombinational sex. Strategies that promote stability are identified with stabilizing sex.

Recombinational sex in diploid multicellular organisms enhances heterozygosity and the accumulation of multiple alleles through the recombination of parts of homologous chromosomes and the complete mixing of sets of chromosomes. The foremost strategy for making new combinations of genes and introducing them fastest is **outbreeding** (also called cross-fertilization or allogamy), or fertilization between gametes originating from different parents.

Actually, no form of reproduction that involves the replication of DNA is totally without potential for introducing new gene combinations. Even some prokaryotes (bacteria) employ strategies that promote genetic recombination.

In **bacterial conjugation,** a part of a DNA molecule from one bacterium is transferred to another bacterium through an **F pillus.** Bacteria may also undergo a form of **transformation** by taking up DNA or **plasmids** from their environments. This DNA may be self-replicating and passively introduced to the bacterium's progeny, or, after incorporation into the bacterium's own DNA as an **episome,** the DNA may be passed along to progeny as part of the bacterium's own genome. In addition, bacteriophage introduce foreign bacterial genes into a bacterium's genome through **transduction.**

Eukaryotes also have the ability to take up DNA from their environment (including the _in vitro_ environment of cultured cells) and undergo **eukaryote transformation.** Moreover, the genome of eukaryotic cells can be altered permanently by viral infection or direct injection of DNA.

The mere generation of recombinants does not move a species closer to an optimal phenotype, however. Recombinational sex is just as likely to break up an adaptive combination as it is to produce one (see Bell, 1982). The "on again off again" type of evolutionary vacillations found among some large breeding populations challenged by interacting species may even be explained by breaking and remaking particular genetic combinations.

Stabilizing sex reduces the likelihood of breaking up adaptive combinations of genes. Several possibilities are available for slowing down the rate at which new combinations of genes are made.

In **inbreeding**, fertilization occurs between the gametes of parents and their offspring or between offspring of the same parents. **Self-fertilization** (selfing, autogamy, or endogamy) employs gametes of the same organism (e.g., the garden pea). Parthenogenic species spurn sperm and accomplish reproduction through the egg alone, while gynogenic species utilize eggs for development but are activated by spermatozoa (usually from related species).

All these forms of stabilizing sex tend to create a high degree of similarity or homozygosity among pairs of homologous chromosomes and preserve homozygosity from one generation to the next.

Parthenogenesis and gynogenesis do this so well that they have even been confused with asexual reproduction. But while stabilizing sex sacrifices the potential for generating varieties through recombination, it preserves the possibility of recombinational repair and continues to remove double-stranded errors.

SEXUAL REPRODUCTION: TURNING THE CYCLE

Sexual reproduction is reproduction requiring the development and fusion of two sex cells (gametes). It might better be called **biparental** or **bisexual reproduction** to distinguish it from uniparental or unisexual reproduction.

Three methods of sexual reproduction seem to have evolved: mating types, oogamy, and sex organs. Each represents an evolutionary channel followed by organisms in different lines of descent.

Mating Types

When sexual reproduction occurs with gametes resembling each other, the process is called **isogamy** or **hologamy,** and the gametes are identified by **mating types.** Found in algae, fungi, and protozoans, **isogametes** (or hologametes) are typically indistinguishable morphologically from ordinary cells. Their fusion is called **conjugation** rather than fertilization.

Unlike the familiar two sexes identified with sperm and egg, isogametes occur in many **mating types.** Cells of one mating type conjugate with cells of other mating types but not with each other. While + and − are occasionally used to identify

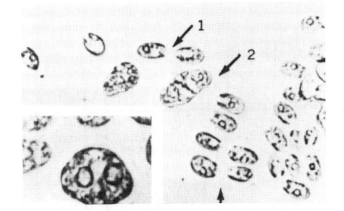

FIGURE 5.1 In *Chlamydomonas eugametos,* a unicellular green alga, morphologically identical isogametes conjugate (arrows 1) and fuse (arrow 2) to form a dormant zygospore (insert). ×500. insert ×1200. From H. C. Bold et al., *Morphology of plants and fungi,* 5th ed. Harper & Row, New York, 1987. Used by permission.

mating types, the large numbers of types found in some species require coded numbers. The nomenclature is engineered so that members of odd-numbered groups conjugate with members of even-numbered groups (i.e., opposites attract).

In the unicellular algae and oomycetes, the gametes are members of the haploid generation, and the cell formed by conjugation is a diploid **zygospore** (Fig. 5.1). Thick-walled, dormant zygospores are representative of a class of zygotes that are extremely resistant to adverse physical environments and frequently serve as a sheltered phase in the life cycle.

In filamentous algae, such as *Spirogyra,* familiarly known as pond scum (Fig. 5.2), isogametes

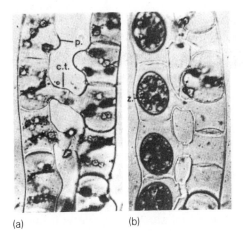

(a)   (b)

FIGURE 5.2 Photomicrographs of *Spirogyra* sp, a filamentous green alga. Early (a) and late (b) stages of conjugation. ×300. From H. C. Bold et al., *Morphology of plants and fungi,* 5th ed. Harper & Row, New York, 1987. Used by permission.

form tubular extensions toward one another. When these extensions meet and fuse, the content of one cell moves into the other cell, forming one zygospore. In filamentous fungi, such as _Rhizopus_, the black bread mold, tubular extensions meet to form a deeply pigmented, resistant zygospore between the conjugating hyphae.

Conjugation in ciliates is also mutual, but it occurs between initially diploid isogametes. Although a ciliate can live and divide for a relatively long time on the activity of its physiologically active, polyploid **macronucleus,** ultimately, the cell undergoes conjugation, and the stock is rejuvenated through the activity of the diploid **micronucleus.**

During conjugation, or sometimes without conjugation (i.e., autoconjugation, analogous to parthenogenesis), the cell's micronucleus undergoes meiosis. Depending on species, a variety of nuclear events then follow. For example, of the four haploid nuclei produced by meiosis in _Paramecium aurelia_, three degenerate, and the survivor undergoes a mitotic division. The presence of two haploid nuclei at this time qualifies the ciliate as an animal gametophyte equivalent. After one of the two haploid nuclei migrates, the new pair of nuclei in each cell perform the role of pronuclei and combine to form a diploid nucleus. The cells then separate into ex-conjugates, and the next diploid generation is started (Sonneborn, 1954).

In unicellular algae and oomycetes, descendants of a single cell (i.e., members of a **synclone**) tend to be members of the same mating type. In the ciliated protozoa and fungi, different mating types may appear among the descendants of the ex-conjugate.

Oogamy

Algae and oomycetes are sometimes ranked in evolutionary series based on their gametes' morphologies and behaviors. The first rank is occupied by species having gametes differing from isogametes only in size (anisogametes). Species with gametes differing in behavior as well as size (heterogametes) come next, and species with massive, stationary eggs and small motile **antherozoids** or sperm top off the series (Fig. 5.3).[1]

In filamentous, coenocytic (i.e., plasmodial), and multicellular algae, the gametophyte's cells-

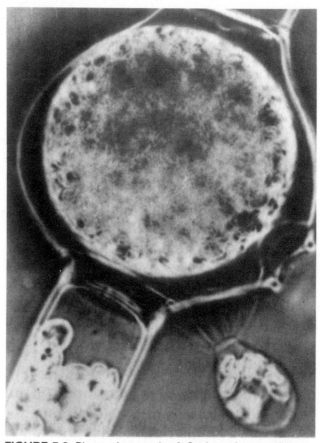

FIGURE 5.3 Photomicrograph of _Oedogonium cardiacum,_ a filamentous green alga. A motile sperm is approaching an encapsulated oogonium. ×150. From H. C. Bold et al., _Morphology of plants and fungi,_ 5th ed. Harper & Row, New York, 1987. Courtesy of Dr. L. R. Hoffman.

turned-gametes are in **gametangia** or, especially in advanced oogamous species, **oogonia** (pl., **oogonium** sing.) when they contain the egg and **antheridia** (pl., **antheridium** sing.) when they release the sperm. Differentiated pores in antheridia allow sperm egress, and pores in oogonia allow sperm entrance (Fig. 5.3).

Every cell in the extremely reduced gametophytes of red and brown algae differentiates in a gametangium. Red algae produce definitive eggs and nonflagellated sperm in male and female gametangia. The massive multicellular brown algae produce morphological isogametes or anisogametes but with distinctively different behaviors. One type of gamete settles on a substratum and releases a diffusible attractant, or sexual pheromone, to which the other gamete is attracted. The type of gamete produced by a gametophyte is determined by a chromosomal mechanism, and different types originate from different gametophytes.

---

[1]Oogamy offers a biological solution to the Plutarchan riddle of the chicken and the egg. Since _Oedogonium cardiacum_ represents a far older group than _Gallus gallus,_ the egg came first.

Sex Organs

Germ cells are harbored in sex organs in most multicellular organisms. In animals, the organs are made of somatic tissue, and the germ cells are derived independently from germ-line cells. In plants, sex organs are made of sterile gametophytic cells and, in monobiontic species, of sporophytic cells as well.

An **archegonium** (sing., *archegonia* pl.) is the characteristic female sexual organ developed by the gametophyte of seedless terrestrial plants (Fig. 5.4). In monobiontic species, sporophytes produce **gametophores** (Gk. *gametes* husband, *gamete* wife + *phoros* carry) which support gametophytes. These may produce sperm (antheridiophores) or archegonia (archegoniophores).

A **reproductive organ** or **gonad** (Gk. *gone*, that which generates + *-ad* suffix denoting connection with) is the sex organ produced by the diploid tissue of animals. A gonad containing sperm is a **testis** (sing., **testes** pl., Lat. bearing witness [i.e., of maleness]). A gonad containing eggs is an **ovary** (sing., **ovaries** pl., Lat. *ovum* egg + *-arium* suffix signifying belonging to thing or place). A gonad

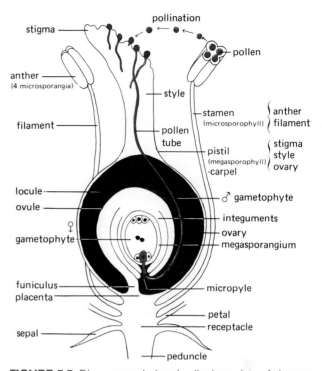

FIGURE 5.5 Diagrammatic longitudinal section of the perfect flower of a hermaphroditic plant. The female gametophyte is contained within the megasporangium within the bulbous ovule. Pollen, the male sex cell, is produced in anthers or microsporangia and gives rise to the male gametophyte as the pollen tube moves through the pistil. From H. C. Bold et al., *Morphology of plants and fungi,* 5th ed. Harper & Row, New York, 1987. Used by permission.

containing both eggs and sperm is an **ovotestis.** In plants, diploid sex organs that enclose male gametophytes are **microsporangia,** and those that enclose female gametophytes are **megasporangia** (Fig. 5.5).

The differentiation of reproductive organs increases following maturity of the animal or plant and during periods of reproductive activity. In animals, the production of different gonads may be influenced by the chromosomal and genetic makeup of the gametes, and gametes may likewise be influenced by the parent.

The type of gamete produced by the organism is reflected in morphology, behavior, and other aspects of **gender** or sexual state. **Primary sexual structures** are those that produce and store gametes, deliver them to the outside, and operate during gestation in viviparous and ovoviviparous species (see below Chapter 6). **Secondary sexual structures** generally develop under the influence of primary sexual structures and are associated with **precopulatory sexual behavior, brooding,** and postgestational or postbrooding **parental care.**

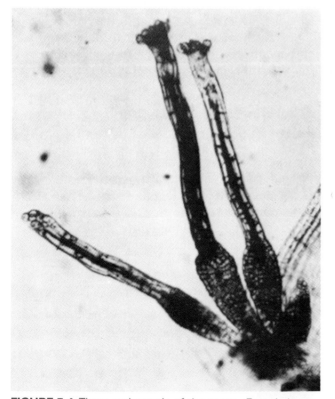

FIGURE 5.4 Three archegonia of the moss, *Funaria hygrometrica.* ×120. From H. C. Bold et al., *Morphology of plants and fungi,* 5th ed. Harper & Row, New York, 1987. Used by permission.

Sex Allocation

The distribution of resources between male and female reproductive functions in a species is called a **sex allocation.** Theories of sex allocation encompass gender (gonochorism and hermaphroditism), sexual behavior (mating systems), mechanisms of sex determination, and numbers of organisms representing each gender (sex ratio). In general, each of these aspects of sex is thought to be vulnerable to environmental pressures (e.g., ranging from the availability of resources to weather and seasonal changes) and is likely to have its own evolutionary history (see Policansky, 1987).

In **gonochors** (Gk. *gone* + *choros* place), including **gonochoristic animals** and **dioecious plants** (Table 5.1), reproductive organs producing sperm and those producing eggs are found in different organisms. **Unisexual** plants and animals are generally identified with one of two **sex morphs. Female** gonochors produce only eggs, and **male** gonochors produce only sperm.

Plants and animals are **cosexual** when only one sexual phenotype or class occurs and organisms function as both female and male. In **hermaphroditic animals** and **monoecious plants,** both female and male types of reproductive organ and system are produced in the same organism. In these plants, the sex organs occur in **imperfect flowers** on different stems or parts of the stem, while in animals, the organs can be different (testis and ovary) or combined (ovotestis). In **hermaphroditic plants,** both female and male types of reproductive organ and system are produced within the same **perfect flower** (Fig. 5.5).

Unisexual and cosexual animal species employ different **mating systems,** or characteristic strategies for achieving fertilization. The cooperation of two organisms is generally required. Joint fertilization is seen in **simultaneous hermaphrodites,** producing both sperm and eggs at the same time (e.g., annelids and gastropods). **Egg trading** is seen in **serial hermaphrodites** where members of a pair alternate between releasing eggs and releasing sperm (e.g., the sea bass family such as the monogamous black hamlet).

In **monogamy** (Gk. *monos-* single + *gamos* marriage), whether permanent or sequential, individuals that otherwise live independently pair while engaged in sexual behavior. Monogamy occurs among females and males and among hermaphrodites (e.g., chalk bass).

Animals living in relatively permanent or stable social groups employ yet other mating systems. In **polygyny** (Gk. *polys* many + *gamos,* sometimes called harem polygyny), the group usually consists of one male (or a small number of males) and several females (e.g., the wrasse) or hermaphrodites (e.g., serranine sea bass). The females and hermaphrodites mate with the male and take few opportunities to mate with others (e.g., competitive pair spawnings or sneak spawnings with opportunistic "streakers"). **Polyandry** occurs within a social group of one female and several males (e.g., bees).

Female and male phenotypes are generally stable in animals, but **labile sex** occurs in some species. **Sex change,** also known as gender change or sequential cosexuality (sequential hermaphroditism, sex reversal, or sex inversion) is **protandrous** when males change to females (e.g., shrimp) and **protogynous** when females or hermaphrodites change to males (e.g., reef fish). Reversion to an earlier sex morph may also occur (e.g., female members of the porgy family becoming males after an earlier sex change of male to female) (Shapiro, 1987).

Although age and size may play a role, social relations signaled through the environment, or **social sex determination,** are frequently the trigger for sex change. In protogynous coral reef fish, death or removal of the male from a polygynous group triggers a female to change into a male. Similarly, death or removal of females in protandrous species

TABLE 5.1. Classification of flowering plants and multicellular animals by reproductive organs

Type	Definition	Examples
Dioecious plant	Separate plants with male and female flowers (i.e., imperfect flower)	Ginko
Gonochoristic animal	Animals with separate male and female organisms	Human beings
Hermaphrodite	Animal producing both egg and sperm simultaneously or in sequence	Some nematodes, gastropods, rare in vertebrates
Hermaphroditic plant	Same flowers produce ovule and pollen (i.e., perfect flowers)	Lily
Monoecious plant	Different flowers on same plant produce ovule and pollen	Maize

having a monogamous mating system triggers males to change into females (Fisher and Petersen, 1987).

In **chromosomal sex determination,** gender is pushed in one direction at fertilization by particular chromosomes. In vertebrates, a **sex-determining chromosome** directs development toward one sex (e.g., Y in mammals directs development toward the male), while in insects, the ratio of **sex chromosomes** (e.g., X in *Drosophila*) to **autosomes** (all other chromosomes) directs development toward the female. A **sex ratio,** or ratio of males to females, is commonly 1:1, when chromosomal sex determination is operating and sex morphs differentiate during embryonic development.

As first described by Nettie Stevens, the sex-determining chromosomes or sex chromosomes may differ morphologically. One sex is identifiable as **heterogametic,** having two types of chromosome (e.g., XY or WZ), while the other sex is identified as **homogametic,** having the same type of chromosome (e.g., XX or ZZ). For example, in mammals and *Drosophila,* males are heterogametic (XY) and females are homogametic (XX), while in birds and reptiles, females are heterogametic (WZ) and males are homogametic (ZZ).

The sex-determining chromosome does not control sex so much as it controls the expression of sex-determining genes. Gene expression is also influenced by environmental signals. In species exhibiting sex change, hormonal and tissue-interactive controls may limit gene expression in the heterogametic sex.

In the **haplodiploidy** system of sex determination found among hymenopterans, diploid eggs are relegated to the female sex morph and unfertilized (initially haploid) eggs to the male sex morph. A reproductive female may adjust the release of sperm stored in her spermathecal organ, thereby determining the sex ratio of her offspring. In wasps and bees, the female's control of the sex ratio (called facultative or labile sex-ratio control) generally has a female bias.

Environmental circumstances also influence sex ratios. The size of a host, for example, seems to trigger the release or withholding of sperm by female parasitic wasps laying their eggs on the host. Diploid eggs tend to be laid on large hosts. The eggs develop into large females capable of producing more eggs than small females. Haploid eggs tend to be laid on small hosts where they develop as males whose reproductive success is independent of size (see Werren, 1987).

Cytoplasmic and other factors also influence

or distort sex ratios. In mites, insects, and plants, maternally transmitted, cytoplasmically inherited microorganisms (rickettsia, spiroplasms, microsporidia, and viruses) change the sex ratio, alter sex determination, or kill males. In addition, in parasitic wasps, a paternally transmitted cytoplasmic factor in sperm, known as paternal sex ratio, destroys paternal chromosomes in fertilized eggs. Haploidy is thereby restored to the egg, and the resulting brood is all male.

The Problem of Bringing Gametes Together

The role of parents in matching egg and sperm is usually performed in acts of **spawning** (shedding of eggs), **milting** or **pollinating** (shedding of sperm or pollen) or, among animals, **mating** (intimate sexual contact). Mating comes in many varieties. Coupled to spawning, mating is **amplexus** (clasping) or **pseudocopulation.** Linked to the introduction of sperm into the female genital tract, or **insemination,** mating is **copulation** (Lat. *copulare* to couple, bind, or join)[2] or **coitus** (Lat. *coire* to come together, hence the sexual joining of individuals). Accompanying the injection of sperm into the skin, mating is **dermal copulation.**

Shedding gametes into the external environment is hardly guaranteed to bring about fertilization. Sperm released in marine, brackish, or freshwater environments may have great difficulty reaching conspecifics (i.e., members of the same species), whether females or hermaphrodites harboring eggs or the eggs themselves. Eggs also have difficulties encountering sperm, since eggs may be shed into a dilute "soup" of sperm, most of which are of the wrong species. Occasionally, cross-fertilizations occur (e.g., sea urchin eggs and mussel or oyster sperm), but these do not produce successful zygotes.

Numerous devices have evolved to promote appropriate fertilization. Massive spawnings at particular seasons and times are performed by many marine annelids. Many fish confine their spawning activities to particular sites such as "nests" (e.g., sticklebacks) or particular freshwater streams (e.g., salmon).

Anatomical structures such as funnel-shaped **gonangia** (the reproductive members of hydroid

---

[2]Copulation is also used to refer to the union of egg and sperm as a parallel to conjugation referring to the union of isogametes.

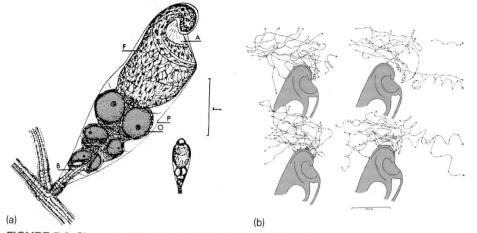

(a)　　　　　　　　　　　　　(b)

FIGURE 5.6 Chemotaxis in the hydroid, *Campanularia calceolifera*. The sperm paths tend to turn toward the aperture to the female gonangium. (a) Drawing of female gonangium: A, aperture; B, blastostyl (supports ova); F, funnel; O, ovum; P, perisarc or gonotheca (acellular covering). Scale = 0.5 mm. (b) Outline of female gonangium and paths of sperm plotted from black field cinematographic records at 10-frame intervals (i.e., 0.45 second). From R. E. Miller, *J. Exp. Zool.,* 162:23 (1966) by permission of Alan R. Liss and the author.

colonies) may concentrate sperm in the vicinity of eggs (Fig. 5.6). Other structures provide sperm access to the eggs. The **micropyle,** or hole, in the **chorions** of numerous marine invertebrates, such as squid, and fish, such as sturgeon, offers the only route to the egg.

Sperm may also be attracted to eggs by **chemotaxis** (i.e., the phenomenon of cells moving toward or away from diffusible substances). Diffusible **sperm attractants** (typical of brown algae, ferns, and mosses) are detected when they make animal sperm turn toward eggs and swim up a concentration gradient (Fig. 5.6). Attractants may be species specific (tunicates and coelenterates) or not (e.g., chitons) (see Miller, 1980). They may be produced by the egg and diffuse through the egg coats; they may impregnate egg coats without the participation of the egg; or they may diffuse out of the micropyle. The same substance that attracts sperm may also bind them to eggs.

Mechanisms barring interspecific fertilization among closely related species may be geared to the likelihood of interspecific encounters. The concept of **character displacement** suggests that **sympatric** species (i.e., having overlapping geographical ranges) would evolve stronger barriers to hybridization than **allopatric** species (with different ranges). The concept is borne out, for example, in sand dollars. Members of closely related species living in a shared portion of their ranges do not hybridize due to a failure in sperm–egg adhesion, while members of the same species living in non-

overlapping portions of their ranges can form hybrids (see Epel and Vacquier, 1978).

Mating overcomes some of the barriers to fertilization, but species that engage in mating confront their own sets of problems. Species specificity in the egg–sperm match is almost always provided gratis, since generally only members of the same species mate, but first they must recognize each other as members of the same species, and male and female or hermaphroditic adults must be able to meet.

One solution to the problem of finding mates is the maintenance of permanent physical contact. For example, males are attached to and maintained by the females in the tongue worm *Bonellia* (phylum Echiurida) and in the parasitic trematodes *Schistosoma mansoni.*

When members of a species tend to live in physical and social isolation, they may be prepared to mate during chance encounters. Alternatively, animals may **advertise** their sexual readiness or employ **sexual signaling** to attract potential mates at times they are prepared to mate.

One possibility for attracting compatible conspecifics utilizes **pheromones** (also pheromene), diffusible or volatile chemical signals released into the environment by members of a species and conveying information to other members of the same species. A variety of female insects release **sexual pheromones** that attract potential sexual partners. Bombykol, for example, is the highly volatile mod-

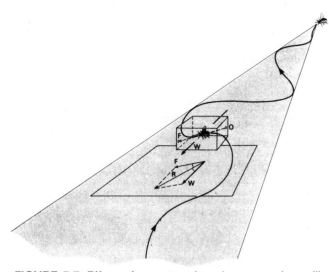

FIGURE 5.7 Effect of sex-attracting pheromone in a silk moth. Bombykol released by the female stimulates a male to fly upwind (serpentine line with arrowheads). The male's flight path (F) is dictated by the wind direction (W, arrow) and a tendency to turn in the opposite direction from the previous turn when the amount of pheromone falls below a threshold level. The resultant track (R, arrow) brings the male to the female. From D. McFarland, *Animal behavior.* Benjamin-Cummings, Menlo Park, CA, 1985. Used by permission.

erately long-chain alcoholic sex pheromone released by female silk moths (Fig. 5.7) (see Schneider, 1969).

Behavioral displays and morphological aspects of sexual dimorphism may also be used in sexual signaling. Male fruit flies, for example, perform wing vibrations and genital licking during courtship dances in the presence of gravid females. The ruffling of feathers by male birds and the ritual combat of rams in rut are also behavioral displays of courtship, while the gorgeous plumage of some male birds and the impressive antlers of bucks are morphological advertisements. Males are generally more striking and perform the more glaring courtship displays, while females seem merely to choose among their potential mates. In species with sex-role reversal, such as wading shore birds known as phalaropes, the female's plumage is conspicuous, and males choose mates.

Male sexual displays often attract females to particular areas. Males may establish **territories** which are used exclusively as mating sites (e.g., monagamous gulls) or as mating and brooding sites (polygamous red-winged blackbirds). Male gulls, for example, attempt to attract (i.e., call down) flying females and induce them to mate and subsequently nest.

Groups of tom turkeys display in a common area, and massed male grouse display at traditional **mating arenas,** called **leks,** each male confined to its territory within the lek (Fig. 5.8) (see Emlen and Oring, 1977). Male mammals, such as kob antelope, may also display at leks visited by females selecting mates, and male fish, such as the bluehead wrasse, stake out territories in leklike behavior while females choose among males.

Aside from predators of other species to which animals are exposed while displaying, animals belonging to carnivorous species may also have to deflect the cannibalistic tendencies of members of their own species during courtship and mating. Some spider males perform courtship rituals which seem to signal friendly intentions and allow the male to approach the female without being consumed. Male empid flies, such as *Empis barba-*

FIGURE 5.8 Prairie chicken (*Tympanuchus cupido*) males displaying at their lek in North Dakota. Courtesy Ed Bry, North Dakota Game and Fish Department.

FIGURE 5.9 Photograph of the fly, *Empis barbatoides,* during mating. The female (below) is devouring prey which the male has given her. From J. Alcock, *Animal behavior,* 2nd ed. Used by permission of Sinauer Associates, Sunderland, MA, 1979, the author, and Carlton Brose.

FIGURE 5.10 Photograph of female grass mantis devouring male (winged individual) during copulation. From J. Alcock, *Animal behavior,* 2nd. ed., Sinauer Associates, Sunderland, MA, 1979. Courtesy of E. S. Ross, California Academy of Sciences.

toides (Fig. 5.9), bring their intended mates a food item which they can devour during mating. The small male fly, *Hilara sartor,* brings its female a hollow silk balloon which occupies her attention during mating (see Kessel, 1955).

The female grass praying mantis, *Heirodula tenuidentata,* will sometimes devour the male's head and thorax during copulation (Fig. 5.10) (see Roeder, 1963). Ironically, because the male's subesophageal ganglion exerts an inhibitory influence, decapacitation improves his sexual performance. Males must develop considerable dexterity to avoid the female's lightning lunge if they are to mate successfully more than once.

The internal environment may also represent a hazardous place for gametes. The female genital tract is not necessarily a hospitable environment for sperm, and the female tract's own defenses against sources of infection can throw up serious barriers to sperm movement. The acidity of the mammalian vagina, for example, which is eminently functional as a bacteriostatic device, is also a formidable spermicidal device. As a coping mechanism, sperm may be ejaculated in a buffered solution, and adaptive morphologies and behaviors may introduce sperm above the vagina into the more hospitable cervix of the uterus (e.g., rats).

Females frequently withhold acidic solutions or secrete neutralizing ones at the time of copulation, and peristaltic movement in the female reproductive tract speeds sperm to the egg. These peristaltic movements are more important in propelling sperm toward the egg than the sperm's own feeble tail-lashings, but, of the tens of billions (i.e., 10^{11}) of spermatozoa in an ejaculum, only a few thousand reach the oviduct, and only a few hundred traverse the remaining distance to the egg. In the mouse, the number of eggs reaching the oviduct may even exceed the number of available sperm (Zamboni, 1970).

Fertilization in mammals usually takes place

in the **ampulla** of the oviduct or uterine tube (see below, Fig. 11.24), the expanded area close to the fimbria. In **spontaneous ovulators,** whose eggs are released periodically without direct environmental stimulation (most mammals), eggs released prior to mating wait in the ampulla for sperm. In **induced ovulators,** whose eggs are released in response to mating, and in spontaneous ovulators mated before ovulation, sperm wait at the **isthmus** (a short, thick-walled, constricted portion of the oviduct adjoining the uterus) for ovulation. Subsequent swimming movements of sperm and the peristaltic movement of egg within the oviduct tend to bring sperm and egg to the ampulla simultaneously, and fertilization is virtually immediate.

Although sperm may not ordinarily swim the greater part of the way to the site of fertilization, and, in some species (e.g., marine shrimp) sperm are immotile, typical sperm's swimming movements may still be important. In mammals, sperm acquire motility during **capacitation** in the female genital tract or as a result of the experimental elution of surface glycoproteins (see Bedford and Cooper, 1978).

Motility may not be sufficient for fertilization, since sperm capacitated *in vitro,* while fully motile, may yet be incompetent *in vivo.* Moreover, sperm cultured in a variety of ectopic *in vivo* sites, such as the colon, bladder, and anterior chamber of the eye, while motile, are incompetent at fertilization (see Yanagimachi, 1981). The secret of intrauterine fertilization may not have been solved, but today's culture methods permit *in vitro* fertilization with a high likelihood of success even in human beings (see McLaren, 1981).

PARTHENOGENESIS AND GYNOGENESIS: SPURNING THE CYCLE

Parthenogenesis is development in an egg containing only an active female genome. **Androgenesis** is development in an egg containing only an active male genome (Beatty, 1957). In **gynogenesis,** the sperm nucleus is not ordinarily incorporated into embryonic nuclei, but if it is, it is not passed on through subsequent generations. In **experimental gynogenesis,** one or another pronucleus is removed, and the active genome is either the male or female.

The differentiation of gametes and passage through a haploid phase suggest that both parthenogenesis and gynogenesis evolved from sexual reproduction rather than from asexual reproduction, and parthenogenesis accompanied by meiosis is sometimes called **unisexuality** or **unisexual reproduction.** Moreover, parthenogenesis in animals is typically part of a reproductive strategy that involves ordinary sexual reproduction as well, and gynogenesis occurs in species or subspecies closely related to others having typical sexual reproduction. In no case, however, has sexual reproduction seemed to evolve from parthenogenesis.

The embryo developing parthenogenically is called a **parthenote** (preferred in the United States) or **parthenogenone** (preferred in Europe, also parthenogen and parthenogon). The embryo developing from gynogenesis is called a **gynogenote** or **gynogenone,** and an embryo developing from androgenesis is an **androgenote** or **androgenone.** In cases of experimental gynogenesis where the source of a nucleus can be identified, a gynogenote derives its nucleus from the egg, and an androgenote derives its nucleus from the sperm. Development employing the egg's nucleus is gynogenetic development, and development employing the sperm's nucleus is androgenetic development without regard to the sex of the developing organism.

Natural Parthenogenesis and Gynogenesis

Natural parthenogenesis and **gynogenesis** are subdivided into categories by functional and cytological criteria. Functionally, normally occurring in many plants and animals (Table 5.2), parthenogenesis may be an occasional (accidental) form of reproduction, as in hens, or an exclusive (obligatory) form, as in some gastrotrichs. It may rotate seasonally with sexual reproduction (cyclical parthenogenesis), as in aphids, or be facultative, as in rotifers, bees, and wasps.

In **cyclical parthenogenesis,** males and sexual females are produced by **sexuparae** females prior to a bisexual reproductive phase, while in **facultative parthenogenesis,** eggs develop with or without fertilization as a function of the availability of males or of sperm. The offspring of facultative parthenogenesis can be of one or both sexes. All females are produced in **thelytoky** (Gk. *thely-* female + *tokein* to bear, hence to give birth to female offspring). In **deuterotoky** or amphitoky (Gk. *deutero-* second or *amphi-* both + *tokein,* hence to give birth to offspring of two or opposite sexes), the offspring include both males and females.

In **arrhenotoky** or strictly speaking **haploid arrhenotoky** (Gk. *arrhen-* male + *tokein* to bear,

TABLE 5.2. Some modes of parthenogenesis

Ploidy of Parthenote	Nuclear Event in Egg Production[a]	Occurs In
Zygoid	Somatic parthenogenesis [g] Meiotic parthenogenesis [b–f]	Artemia (brine shrimp)
Diploid and tetraploid	Premeiotic replication (endomitosis or endoreduplication) [a]	Planarians and earthworms
Usually polyploid (zygoid), embryo remains zygoid	Single meiotic division [d, e]	Angiosperms, most common form in animals (e.g., rotifers, isopods, nematodes, mollusks)
Egg restores diploidy following parthenogenetic meiosis	Fusion haploid nuclei during cleavage	*Drosophila*, moths, stick insects
Haploid (hemizygoid) (produces male; female produced by fertilization)	Meiosis [b] (haploid parthenogenesis)	Insects (chromosomes polytene), rotifers

[a]Letters in square brackets refer to designations in Fig. 5.11.

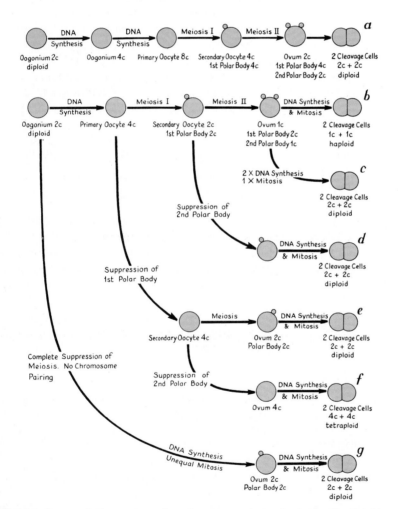

FIGURE 5.11 Some of the routes of parthenogenesis beginning with diploid oogonium and ending with ovum: (a) endomitosis (premeiotic endoreduplication); (b) haploid or hemizygoid development; (c) postmeiotic endoreduplication; (d) suppression second polar body; (e) suppression first polar body; (f) suppression both polar bodies; (g) ameiosis. From U. Mittwoch, *J. Med. Genetics,* 15:165 (1978) by permission of the British Medical Association and the author.

hence to give birth to male offspring), only males are produced. This occurs in honeybees, among others, where females are produced by fertilized eggs and are normal diploid organisms. Ironically, in this **haplodiploidy system** of sex determination, parthenogenesis is utilized to make males.

Among the vertebrates, parthenogenesis is best represented by reptiles. In the order Squamata, 26 lizards and one snake are exclusively (or overwhelmingly) female (Darewskii and Kulikowa, 1961; see Cole, 1975). Spontaneous parthenogenesis also occurs among inbred strains of the domestic turkey (Olsen, 1960a,b) and hen (Sarvella, 1973).

Naturally occurring gynogenesis involving sperm and egg of different species takes place in planarians and nematodes, and a comparable process occurs in the plant *Atamosco mexicana*. Meiosis may or may not occur in these forms, but normal ploidy in body cells is maintained in the gynogenote.

Gynogenesis is represented among vertebrates by fish. Eggs of the silver fish, *Carassius auratus giblio*, for example, are activated by sperm from the related species *Carassius carassius*.

The eggs are probably diploid to begin with, and the embryos remain diploid throughout development. In rare instances when the male pronucleus takes part in the formation of a synkaryon, the resulting organism is triploid, having three times the haploid chromosome number (Schultz and Kallman, 1968). While the fact of gynogenesis is indicative of a high tolerance for foreign material in the cytoplasm, the development of triploid hybrid organisms is indicative of a high, if rare, tolerance for foreign material in the nucleus as well.

In freshwater fish of the Poeciliidae family (including the first recognized vertebrate parthenogenic species, the Amazon molly, *Poecilia formosa*), the formation of hybrid nuclei is a normal part of gynogenesis. Eggs are fertilized by sperm from males of related species (*Poecilia latipinna* or *Poecilia sphenops*), and the male pronucleus actually takes part in development. But the integrity of the diploid maternal nucleus is not lost, and the female nucleus alone is transmitted to future eggs (see Bell, 1982).

Cytological Classification

Cytologically, parthenotes are classified as **meiotic** when produced following meiosis, and **ameiotic** when produced without meiosis (Table 5.3). The **haplontic** (also haploid) or generative form of parthenogenesis (Table 5.2, Fig. 5.11b) found in rotifers and insects is exclusively meiotic. A haploid egg becomes activated and produces a haploid (or azygoid) organism.

The **zygoid** (also diploid) or somatic[3] form of parthenogenesis can be either meiotic or ameiotic and involve either the restoration or preservation of the egg's nucleus. The occurrence of meiosis is sometimes discernible only through chromosomal pairing during division, and the ploidy of the egg may be either retained or increased.

In meiotic forms of zygoid parthenogenesis, equivalents of pronuclei produced following meiosis fuse (**automixis**). Telophase nuclei may fuse into a single **restitution nucleus** or a polar body may be incorporated into the egg. Fusion following meiosis I results in a tetraploid nucleus, while fusion following meiosis II results in a diploid nucleus.

In an incomplete form of meiotic parthenogenesis, a **polyploidizing division**, involving repli-

---

[3]Somatic parthenogenesis is actually a misnomer, since the parthenote is formed from an egg, not from the soma or body of the organism.

TABLE 5.3. Modes of reproduction

	Sexual (Zygogenesis)	Parthenogenic		Gynogenetic (Pseudogamy)	Asexual
		Mictic (Meiotic)			
			Amictic (Ameiotic)		
	Amphimixis	Automyxis	Apomyxis		
Meiosis	+	+	−	−	−
Egg required	+	+	+	+	−
Sperm required	+	−	−	+	−
Male pronucleus required	+	−	−	−	−

cation and separation of chromatids without an intervening cytokinesis, doubles the euploid chromosome number and the cell's DNA content (Brodsky and Uryvaeva, 1985). Similar divisions in normal body cells result in **endopolyploidy** (also called autopolyploidy) and, if repeated, bring about a geometric increase in chromosome number or, at least, in chromatid number within the cell's nucleus (Partanen, 1965). In parthenotes, doubling of chromosome number is usually offset by subsequent division, while recombination, recombinational repair, and even segregation may occur depending on the timing of the polyploidizing division.

Many variations on the theme of polyploidizing division occur in parthenogenesis (Table 5.2; Fig. 5.11). These divisions can occur in the oogonium (a), in the oocyte (e–f), or in the ovum following the second meiotic division (c). This degree of variation is not surprising on theoretical grounds, since no functional cell is actually lost in the process. Rather, a polar body's nucleus is salvaged.

Alternatively, through a process called **endomitosis**, replicated chromosomes separate into chromatids within the nuclear envelope. Karyokinesis does not take place and the new nucleus is **tetraploid** (i.e., having four sets of chromosomes).

Ploidy is also increased by a process called **endoreduplication.** Chromosomes that have already replicated once within a cell replicate again prior to division. The pairs of chromatids thus formed separate early in prophase or condense into **diplochromosomes** with twice the usual number of chromatids (e.g., four instead of two). When these chromosomes separate at anaphase, two tetraploid cells are produced.

The purest form of somatic parthenogenesis (Fig. 5.11g, or eusomatic parthenogenesis) is ameiotic. Known as **apomixis** or **amictogammetic reproduction,** replication occurs without subsequent meiosis, producing a single nucleus capable of supporting development (e.g., weevils). Commonly (Fig. 5.11d, e), the nuclear DNA content of the ovum is doubled, but the meiotic process is incomplete or thwarted during polar body formation. Sometimes called a **restitution mitosis,** the cytoplasm fails to divide (called acytokinesis) and a single cell is formed with two copies of its genome.

In angiosperms (e.g., dandelions), one excess nucleus and one nucleus capable of supporting the formation of the parthenote are produced. Both these nuclei have the zygoid (i.e., diploid) number of chromosomes and the unreplicated body cell content of DNA (see Nagl, 1978).

The "somatic" ovum develops into a multi-cellular organism without the intervention of meiosis, sperm, and amphimixis. Neither meiotic recombination nor segregation therefore occur, and the putative adaptive advantages of sexual reproduction would seem to be lost. On the other hand, the reduced recombinational requirements of somatic parthenogenesis may represent a reproductive compromise, the only escape from sterility or a hybrid's "last resort" (Mittwoch, 1978). Somatic parthenogenesis may have evolved in hybrids whose chromosomes are incompatible with the pairing requirement of meiosis.

Hybridogenic Crosses

Hybridogenic crosses (fertilization producing viable and reproductive hybrids) and **polyploidization** (processes other than fertilization that increase the ploidy of cells) are frequently correlated with parthenogenesis and gynogenesis. Gynogenic strains of goldfish (*Carassius*), for example, are triploid or tetraploid as opposed to normal bisexual diploid goldfish.

Polyploid hybrid species of *Ambystoma* are gynogenic or parthenogenic. For example, the two gynogenic species *A. platineum* and *A. tremblayi* are reciprocal triploid hybrids: 2 × *A. jeffersonianum* + *A. laterale* and 2 × *A. laterale* + *A. jeffersonianum*, respectively. These, hybrid *Ambystoma* species are preponderantly gynogenic.

Likewise, polyploid reptiles originating from the hybridization of individuals from divergent bisexual populations reproduce parthenogenically. For example, triploid representatives of the Australian gekkonid, *Heteronotia binoei*, are parthenogenic while diploid representatives are bisexual.

Still, parthenogenic or gynogenic species are not necessarily hybridized or polyploidized. Many parthenogenic species of reptiles, such as the parthenogenic species of the *Lacerta saxicola* group, for example, are diploid. Furthermore, hybridized or polyploidized species are not necessarily parthenogenic or gynogenic. All polyploid frogs, whether tetraploid, hexaploid, or even octaploid, are bisexual. Even hybridogenic triploid *Rana exculenta* frogs, produced from the cross of *R. ridibunda* and *R. lessonae*, are bisexual (Suomalainen et al., 1987). Similarly, triploid frogs produced experimentally by crossing Japanese *R. brevidopa* and *R. lessonae* revert back to diploidy. In salmonids, **diploidization,** or restoration of diploid inheritance, seems to be occurring following an ancient polyploidizing event, and reproduction is bisexual.

Artificial Gynogenesis

Artificial gynogenesis was pioneered by Oscar and Gunther Hertwig in amphibians and first achieved with sperm incapacitated by heavy irradiation. The eggs gave rise to normal appearing haploid organisms.

The production of live-born mammals by experimental gynogenesis and androgenesis is, at best, rare (see Tarkowski, 1975). Androgenotes and gynogenotes achieve better development than their parthenote counterparts, although deaths occur at the implantation stage. Occasionally, advanced postimplantation development is reached before an embryolethal stage.

One technique for inducing gynogenesis in mammals involves removing or destroying a pronucleus from an artificially fertilized egg via micromanipulation (Modliński, 1975). Another technique employs microsurgery to bisect fertilized eggs whose pronuclei are still separate (i.e., prior to amphimixis at first cleavage) (Tarkowski, 1977). The early mammalian embryos derived by either of these techniques can be transferred to suitably prepared (i.e., pseudopregnant) hosts, but, for the most part, the haploid mammalian gynogenotes abort development at early stages.

Better survival of artificial gynogenotes is achieved with eggs that are "diploidized" (i.e., made diploid). The diploidization is achieved by suppressing a division by treatment with cytochalasin B or D (Markert and Petters, 1977) either before or after removing a pronucleus. Cytochalasin may have other salubrious effects on embryos, however, since treatment enhances survival even without diploidization.

Cytochalasin applied prior to the completion of the second meiotic division results in a triploid state that is reduced to a diploid state by removing the male pronucleus. Live-born mouse gynogenotes (diploidized after removing the male pronucleus) and androgenotes (diploidized after removing the female pronucleus) are all female (Hoppe and Illmensee, 1977).

Why fertilized mammalian eggs develop while gynogenotes generally fail even after diploidization remains a mystery. Since death frequently occurs at implantation, when the paternal genome is active especially in extraembryonic tissues, the participation of paternal chromosomes may be required for normal development. Were this the case, paternal DNA would have to be **imprinted** during spermatogenesis, possibly in its methylation, to distinguish it from the female genome.

Support for this possibility comes from experiments with mice demonstrating the ability of a haploid male nucleus to promote survival of a haploid female parthenote following fusion at the two-cell stage. A haploid female cell is obtained from one of the two cells in a two-cell induced parthenote. The other blastomere is enucleated and provides the cytoplasm for a haploid male nucleus obtained from an androgenote at the two-cell stage. After fusing the two blastomeres, diploid parthenoandrogenotes are able to develop into morphologically normal mice (Barra and Renard, 1988).

Diploid mouse gynogenotes achieved microsurgically by replacing the male pronucleus with a second female pronucleus fail to develop much beyond the blastocyst stage. The development of normally fertilized embryos may depend on the presence of paternal autosomes, since XO gynogenotes show the same inviability as XX gynogenotes (Mann and Lovell-Badge, 1987).

Artificial Parthenogenesis

Artificial parthenogenesis, androgenesis, and **teratomas** in mammals have an intriguing if abnormal relationship. Whether spontaneous or induced (e.g., by treatment with ethanol), early mammalian parthenotes and androgenotes die except when transplanted to ectopic sites where they survive as teratomas, encapsulated tumors containing a variety of differentiated tissues (see Chapter 15, Figs. 15.11 and 15.12). On the other side of the coin, cells derived from teratomas can be brought into line and differentiate into normal tissue, including germ tissue, when introduced into normal blastocysts.

Typically, artificially achieved parthenotes die at implantation stages. Although the development of viable transnuclear parthenotes (i.e., enucleated zygotes receiving a nuclear transplant from a parthenote) has been reported, efforts to repeat the procedure have been uniformly unsuccessful (see Dyban and Baranov, 1987).

Teratomas appearing spontaneously in the ovary and testis of germ-cell origins may represent examples of pathological parthenogenesis and androgenesis (Graham, 1977). Young mice of the LT/Sv strain have both high frequencies of spontaneous ovarian teratomas and parthenogenically activated ovarian follicles. Intrafollicular parthenotes may develop to the blastocyst or even the egg cylinder stages (see Fig. 14.16) before becoming disorganized and transforming to teratomas.

Some ovulated eggs of LT/Sv mice also undergo spontaneous activation and begin parthenogenic development in the oviducts. *In vitro,* these parthenotes may develop to the egg cylinder or early somite stage before dying, and although the blastocysts can induce changes in the uterus associated with pregnancy, parthenotes are resorbed in the uterus.

Spontaneous mammalian parthenotes are generally not haploid, although early haploid parthenotes are occasionally obtained in mice through *in vitro* fertilization. Haploid parthenotes may be eliminated during the initial few days of development.

Haploid parthenotes or some of their blastomeres frequently undergo spontaneous polyploidization to a balanced diploid genome or even a tetraploid genome. The polyploidization may be produced through a blockage of the second meiotic division or may develop through endoreduplication.

Polyploidization improves the development of the embryo, and diploid embryos readily reach the egg cylinder or even neurula stage *in utero* (see Kaufman, 1983). The blastomeres of diploid parthenotes appear more nearly normal than the misshapen blastomeres of haploid parthenotes, and the inner cell mass (see Chapter 14) that gives rise to the embryo proper is of a normal size and larger than that of haploid parthenotes.

Polyploidization parallels the process of endoreduplication normally accompanying the formation of the polyploid trophoblast at the 25–30-cell stage (see Chapter 14, the formation of the early extra-embryonic membrane, Fig. 14.16). Possibly, polyploidization allows the trophoblast to achieve a degree of polyploidization required for implantation. The subsequent failure of these parthenotes to survive cannot be attributed to homozygosity or cell lethals, since the parthenotes transplanted to ectopic sites survive in the form of teratomas.

Naturally occurring, pathological androgenesis occurs in human beings in the common syndrome known as the true or complete **hydatidiform mole.** Characterized by swelling and degeneration in the embryo's portion of the early placenta (i.e., dydropic and cystic degeneration of chorionic villi) and excessive or abnormal increases in the embryo's extraembryonic surface (i.e., hyperplasia or dysplasia of the trophoblast), moles are predisposed to malignant transformation (i.e., they become invasive, destructive tumors), and their growth accompanies the early death of the embryo or fetus. Chromosomal analysis shows true moles to be diploid, while cytological and biochemical analysis shows that both chromosomal sets are of exclu-

sively paternal origin. The egg's nucleus may be expelled following fertilization or an "empty egg" may be fertilized. The chromosomes of a haploid spermatozoon nucleus may then duplicate, or two sperm may fertilize an egg followed by the fusion of nuclei (see Dyban and Baranov, 1987).

Partial hydatidiform moles, on the other hand, or moles with some areas of normally appearing placental tissue, are not prone to malignant transformation and occasionally support the development of malformed fetuses to term. These moles are triploid.

ASEXUAL REPRODUCTION: CHURNING THE LIFE CYCLE

Confined to a phase of the life cycle, asexual reproduction employs neither meiosis nor gametes. Furthermore, asexual reproduction is **amictic,** without the pairing of haploid pronuclei (i.e., without mixis or amphimixis, Table 5.3) (Bell, 1982). While sexual development, or **ontogenesis,** begins with a unicellular zygote, asexual development in multicellular organisms, or **asexual normogenesis,** begins with a multicellular **propagule** (blastema or gemmule). The propagule does not generally resemble a zygote nor does it tend to resemble the embryo during early stages of ontogenesis.

The exceptions are the asexual **gemmules** of sponges (see Hartman and Reiswig, 1971) and the **podocysts** of coelenterates (see Chapman, 1966) in which asexual normogenesis closely resembles ontogenesis. Where asexual normogenesis produces a part roughly comparable to one also produced during ontogenesis (e.g., embryonic-like microvilli on a propagule's surface), the normogenic structure is likely to occur out of order. Such asynchronisms make it difficult to attribute a common developmental process to both ontogenesis and asexual normogenesis (see Brien, 1968). In general, the products of asexual normogenesis only resemble the products of ontogenesis at the end of development (see Berrill, 1961).

Products of Asexual Reproduction

The products of asexual reproduction are of one of two types. (1) Organisms (sometimes called stocks, cormi, or cormidia) break off or **autotomize** from their parents and become members of a **syn-**

FIGURE 5.12 Proliferative asexual reproduction by budding in *Hydra viridis.* Buds develop in the budding region of well-fed hydras in about 2 days. ×15.

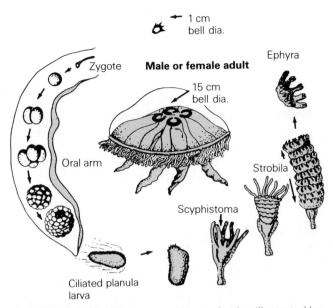

FIGURE 5.13 Modifying asexual reproduction illustrated by *Aurelia aurita.* A polyp (scyphistoma) becomes a segmented strobila and releases ephyras, or immature jelly fish, that grow into the bell-shaped sexual adult medusas. Adapted from W. M. Hamner and R. M. Jenssen, *Am. Zool.,* 14:833 (1974) by permission of the American Society of Zoologists.

clone (also clone).[4] (2) **Members** or **persons** of a **colony** remain connected even after development. Furthermore, asexual offspring can either resemble parents or be quite different. **Proliferative** (facsimile or vegetative) **asexual reproduction** creates morphological replicas of the asexual parent (e.g., Fig. 5.12), while **modifying** (or alternating) **asexual reproduction** creates offspring morphologically different from parents (e.g., Fig. 5.13).

For unicellular organisms, asexual reproduction is generally proliferative and involves cell division whether the cells are haploid spores or diploid products of a zygote. Similarly, algae and fungi living predominantly as multicellular haploid or-

[4]The word "clone" has been corrupted by tissue culturists, who use it to refer to all the products of cell division derived from a single tissue culture cell, and by nuclear-transplanters, who use "clone" to refer to all the products of eggs receiving transplanted nuclei. Although tissue-culture cells are not unicellular organisms nor are the organisms developing from eggs with transplanted nuclei truly asexual, these uses of the word have taken root and are not likely to be uprooted. Clone, in its original meaning, must therefore be sacrificed and "synclone" used to refer to the products of asexual reproduction.

ganisms undergo asexual reproduction by releasing spores capable of germinating into new multicellular haploid organisms.

In diploid multicellular organisms, asexual reproduction is generally confined to characteristic stages of the lifetime and parts of the organism (e.g., a stem in plants, a stolon in animals). Asexual reproduction in an embryonic stage (e.g., **identical** or **monozygotic twinning**) is called **polyembryony.**

Both proliferative and modifying asexual reproduction may occur in a single life history. For example, the polyp, or scyphistoma, of *Aurelia aurita* is capable of both proliferative reproduction by tubelike outgrowths, or stolons, as well as modifying reproduction by ephyra (Fig. 5.13). In the case of trematodes, such as the human blood fluke, *Schistosoma mansoni,* proliferative reproduction allows the organism (as sporocysts) to spread through the intermediate snail host, while modifying asexual reproduction allows the organism (as cercaria) to leave the snail and invade the human host. More often, however, proliferative and modifying asexual reproduction do not occur in a single life history.

Proliferative asexual reproduction is often found in organisms capable of **regeneration,** while modification takes place where regeneration is ab-

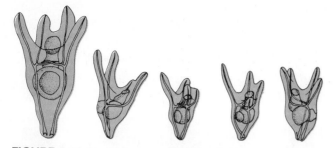

FIGURE 5.14 Normal sea urchin larva (left) and four small larvas obtained from isolated blastomeres of a 4-cell embryo.

sent (Schmahl, 1980). Regulation,[5] or the ability of an embryo to restore a whole or partial structure from a part (Fig. 5.14; see also Figs. 12.4 and 12.12), is sometimes said to occur in propagules, especially proliferative propagules, but modifying propagules may be incapable of regulation.

Asexual Normogenesis

Asexual normogenesis comes in more varieties than ontogenesis. The mechanisms of asexual normogenesis are **morphallaxis** (Gk. *morph-* form +

---

[5]The term regulation should probably be restricted to early embryos which are said to regulate when they give rise to whole organisms following the removal or introduction of blastomeres. Regeneration should be used in other circumstances.

allaxis exchange), the remodeling of existing parental tissue, and **epimorphosis** (Gk. *epi* over + *morph* form + *osis* condition, hence in addition to existing form), outgrowth from parental tissue. Cell death, clearing away existing parental structures, is frequently a part of morphallaxis, while growth is required for epimorphosis. When the destruction of parental parts is obligatory and asexual reproduction is of the modifying type, morphallaxis is analogous to ontogenic metamorphosis (e.g, the formation of ephyra from a strobila, Fig. 5.13).

In colonial animals, such as colonial coelenterates and ascidians (e.g., Fig. 5.15), the asexual production of colony members is commonly epimorphic. **Blastogenic** or **gemmation budding** takes place frequently on relatively undifferentiated tubelike **stolons.** A small group of cells composing the propagule, either aided by the parent or physiologically independent from the start, develops into the complete organism (see Nakauchi, 1982).

Alternatively, multicellular animals may proliferate asexually by morphallactic **regenerative fragmentation** frequently following growth. In sea anemones, the foot of the parent typically undergoes fragmentation (i.e., pedal laceration), and the fragments both remodel internal parental structures and regenerate missing structures. The growth of some colonial corals is similar except that the formation of new structures occurs in the absence of fragmentation. Other coelenterates may

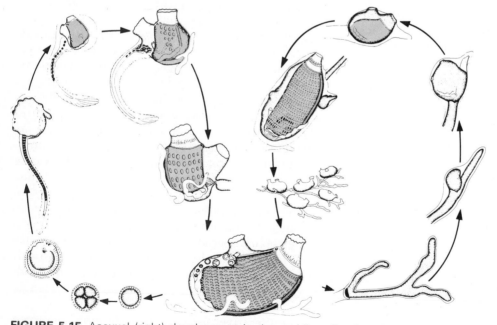

FIGURE 5.15 Asexual (right) development in the ascidian, *Ecteinascidia tortugensis,* by stolonal budding and sexual reproduction (left). From H. H. Plough, *Sea squirts of the Atlantic continental shelf from Maine to Texas,* The Johns Hopkins University Press, Baltimore, 1977. Used by permission.

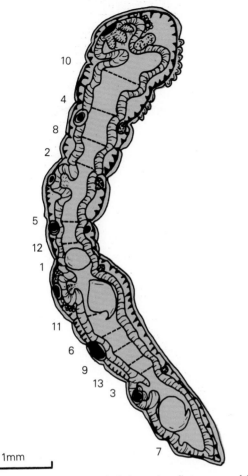

FIGURE 5.16 Strobilation in rhabdocoelan flat worm, *Microstomum lineare.* The numbers refer to the order in which constrictions occur and indicate the sequence of asexual normogenesis.

"split" or undergo fission (longitudinal or, rarely, transverse) as new parts are formed.

In rhabdocoelic flatworms (Fig. 5.16), a form of fragmentation occurs, but regenerative and morphallactic processes precede fragmentation. **Strobilation,** sometimes called **segmentation,** or the isolation of segments, seems to be determined by distance from an existing head or from one just forming.

Asexual normogenesis frequently seems less rigid than ontogenesis. Organisms formed by modifying asexual reproduction, for example, may also be formed by proliferative asexual reproduction. The alternative routes of asexual reproduction, moreover, may proceed through totally different mechanisms. For example, several species of jel-

lyfish formed from two layers during modifying asexual reproduction are formed from only one layer during proliferative asexual reproduction (Bouillon and Werner, 1965). In the latter case, **transdifferentiation** takes place in which tissue left in one state of differentiation by ontogenesis changes to tissue in another state during asexual normogenesis.

Different Functions

Different functions are served by asexual reproduction in the life cycle of multicellular animals. Asexual reproduction is frequently of a **propagative type** called **vegetative reproduction** and based on growth. Alternatively, asexual reproduction may be of a **survival type** elicited in response to adversity, neither requiring nor necessarily involving growth. Survival asexual reproduction may include a dormant stage or one capable of resisting particular conditions and yet germinating new individuals when favorable conditions return.

Morphallaxis is generally associated with survival asexual reproduction, while epimorphosis is associated with propagative asexual reproduction. The distinction blurs in many cases of well-fed organisms where epimorphosis and morphallaxis are interchangeable.

Growth, or increase in mass, underlying the vegetative reproduction of a synclone is theoretically not unlike growth underlying development in individual multicellular organisms. In both cases, the mechanism of growth is synthesis accompanied by cell division, and, since cells produced by ordinary cell division represent the same genome, the members of a synclone are comparable to cells in a multicellular organism.

Vegetative reproduction seems to benefit sexual reproduction in cases of organisms capable of reproducing both ways. Since a synclone can be traced back to a single cell (i.e., the zygote first produced by sexual reproduction), members of a synclone represent extensions of the same parent rather than competitive individuals. Theoretically, members of a synclone are no more in competition with each other than cells in a single organism. The payoff of vegetative reproduction is therefore an increase in the same parental gametic tissue and an increase in the likelihood of producing the next sexual generation (Edwards, 1973).

Chapter

6

THE SPERMATOZOON
AND THE OVUM

Were it not for their ability to fuse during fertilization, spermatozoa and eggs would not seem to have much in common. The sole differentiated representatives of the haploid phase of the animal life cycle, the spermatozoon and egg are about as different as two cells can be.

THE SPERMATOZOON

Types of Sperm

Spermatozoa (or sperm pl., spermatozoon sing.) occur in a variety of forms. More or less amoeboid or T shaped, sperm are produced by nematodes, and a variety of star shapes occur in arthropods (see Jamieson, 1987). Decapod sperm are especially bizarre (Fig. 6.1). With one (in shrimp and prawns) or more (in lobster, crabs, and crayfish) actin-containing spikes, these sperm lack tubulin-containing flagella, mitochondria, and the centrioles normally found at the bases of flagella. The immotile sperm-like gametes of higher plants and fungi also lack flagella (e.g., the pollen of tracheophytes, the ascogonium of ascomycetes).

Other spermatozoa bear **flagella** (pl., **flagellum** sing., Lat. whip; e.g., see Fig. 6.2). The sperm of most green algae or their isogametes or anisogametes generally have two equal flagella at their leading ends, although some have many more flagella (e.g., see Fig. 5.3). Moss sperm have two flagella, while the sperm of ferns and cycads have a helical band of flagella. Stonewort sperm also have two flagella, but these do not originate at the tip of the cell, and brown algae sperm have two lateral and unequal flagella pointed in opposite directions (a forward-directed short tinsel flagellum and a trailing long whiplash flagellum). Dinoflagellate sperm (i.e., the smaller of the anisogametes) have two flagella, one trailing, and most chrysophyte sperm have two unequal apical flagella, but diatom sperm have only one flagellum. Similarly, the sperm of flagellated water molds have a single whiplash flagellum as do most animals.

Flagellated animal sperm consist of three major parts: a **head,** a **midpiece** (or middle piece), and a **tail.** The tail, with the classical "9 + 2" arrangement of microtubules, is the "propeller" that drives the spermatozoon's head and possibly signals the egg that a spermatozoon has arrived. The midpiece, containing mitochondria, is the "powerhouse" that fuels the tail. The head is the "business end" of the spermatozoon that delivers the paternal DNA and converts the egg into a zygote.

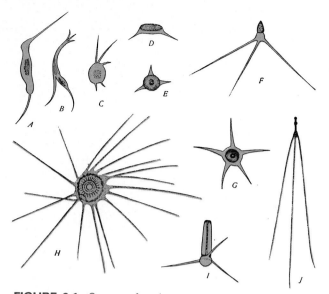

FIGURE 6.1. Sperm of various crustaceans. (a–c) Amoeboid sperm of *Daphnia;* (d–j) star-shaped sperm of crabs, *Dromia* (d, e) and *Porcellana* (j); the sperm of decapods, *Ethusa* (f), *Maja* (g), *Inacnus* (h), and the lobster, *Homarus* (i). From E. B. Wilson, *The cell in development and inheritance.* Macmillan, London, 1896, based on various sources.

The head contains the nucleus and, frequently, a membrane-bound **acrosome** or **acrosomal cap.** The acrosomal contents include hydrolytic enzymes, such as hyaluronidase, and proteolytic enzymes (such as acrosin, resembling trypsin), acid phosphatase, and neuraminidase (capable of removing sialic acid from glycoproteins), all of which aid fertilization.

The acrosomal membrane is morphologically divided into inner and outer portions. The **outer acrosomal membrane** lies below the spermatozoon's plasma membrane, and the **inner acrosomal membrane** faces the nuclear envelope. Precursors of an **acrosomal filament,** especially granular actin (G-actin), may be stored in a **subacrosomal space** between the nucleus and the inner acrosomal membrane, or a preformed filament may be wrapped around the nucleus. With the exception of mammalian sperm, the inner acrosomal membrane provides the spermatozoon's **fusiogenic membrane** and actually fuses with the egg's plasma membrane during fertilization.

The number of midpiece mitochondria is small compared to the number of mitochondria in eggs and somatic cells, and mitochondrial longevity seems to be restricted to the lifetime of the spermatozoon, since sperm mitochondria do not seem to contribute to the zygote although they may en-

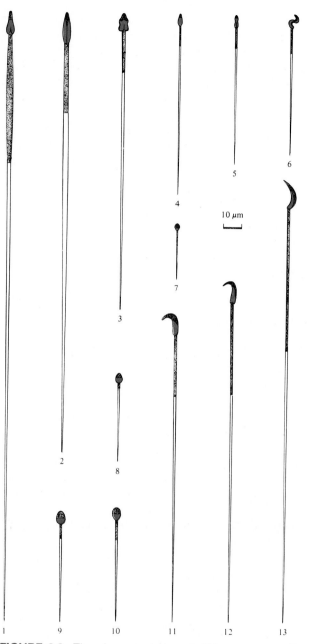

FIGURE 6.2. The shape and size of different mammalian sperm drawn to scale. Acrosome and midpiece shaded. (1–6) Australian marsupials): 1, honey possum; 2, marsupial rat; 3, short-nosed brindled bandicoot; 4, tammar wallaby; 5, brush-tailed possum; 6, koala. (7–13) Eutherian mammals: 7, hippopotamus; 8, human; 9, rabbit; 10, ram; 11, golden hamster; 12, laboratory rat; 13, Chinese hamster. From B. P. Setchell, in C. R. Austin and R. V. Short, eds., *Reproduction in mammals,* Book 1, *Germ cells and fertilization,* 2nd ed. Cambridge University Press, Cambridge, 1982. Used by permission of the author.

ter the egg at fertilization. In *Xenopus laevis*, for example, each spermatozoon contains only about 100 molecules of mitochondrial DNA (mtDNA) in contrast to 10^8 molecules of mtDNA in eggs. None of the spermatozoon's mtDNA subsequently contributes to the zygote's mtDNA pool (Dawid and Blackler, 1972).

Beyond these general properties, differences among flagellated sperm may be considerable. The sperm seem to have diverged several times and independently in the course of evolution due to pressures inherent in different modes of fertilization (Fig. 6.3) (see Afzelius, 1972; Fawcett, 1970). Sperm classified as **primitive animal sperm** are typically discharged into water and usually fertilize externally. Other sperm classified as **modified animal sperm** are ejected with direction and typically are transferred to a female or released in her immediate vicinity. The "modifications" seem to adapt the sperm for directional release.

Primitive Animal Sperm Anatomy

All animal phyla include some species with primitive sperm. They are usually small cells with a rounded or bullet-shaped head and a short flagellum of less than 30 μm. They are not especially energetic, tend to be short-lived (i.e., seconds to hours), and are propelled by tail movements for small distances (i.e., on the order of centimeters). Great numbers of these sperm are usually shed, and success in fertilization seems to depend largely on chance encounters with eggs, although chemicals diffusing from eggs or their receptacles may attract sperm in the vicinity (e.g., see Fig. 5.6).

The presence of an acrosome is correlated with the presence of a primary membrane or envelope surrounding eggs (a vitelline membrane or zona pellucida). No acrosome is present in sperm of species with naked eggs (i.e., coelenterates and sponges) and those with a micropyle penetrating an otherwise impenetrable chorion (see Austin, 1978). As well as functioning in fertilization, the acrosome may function in osmoregulation, since it tends to be larger in marine species than related freshwater species.

A small membrane-bound acrosome may sit atop the nucleus, or separate acrosomal vacuoles may be perched on the nucleus and along its sides. Two centrioles face each other at right angles behind the nucleus. Both contain **microtubular triplets** arranged in a pinwheel fashion. The **proximal centriole** is located just posterior to the sperm head, frequently in a depression at the base of the

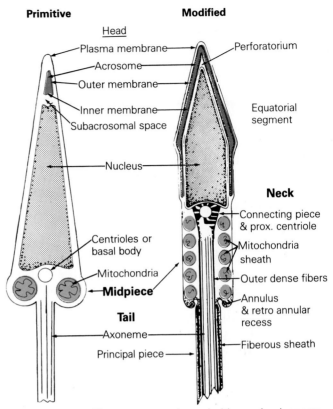

FIGURE 6.3. The anatomy of a primitive animal sperm (*Arbacia*) and a modified animal sperm (human). The head (acrosome and nucleus), midpiece, and tail all show modifications, but many variations are also found among sperm of both kinds.

nucleus called the **centriolar fossa**. The **distal centriole**, or **basal body** of the flagellum, is present at the boundary with the tail.

Nine outer **axonemal filaments** or **microtubular doublets** arise from the basal body and run throughout the length of the tail. Two central doublets complete the 9 + 2 configuration of the **axonemal complex** or **axoneme** (also called the axial filament complex) like that of other flagella and cilia (see Fig. 2.19).

Additional structures are few and simple. Microtubules and dense layers of material may cap and line the nucleus, and a **pericentriolar complex** of fingerlike projections may arise from the distal centriole. In addition, carbohydrate storage products may accumulate alongside the axoneme in the midpiece.

The midpiece generally contains one ring-shaped mitochondrion or no more than four or five simple mitochondria. In ascidians, the midpiece is absent. A single mitochondrion lies alongside the nucleus in the head.

Modified Animal Sperm Anatomy

Modified sperm tend to live longer than primitive sperm, frequently surviving for hours and days after ejaculation or discharge from the male. Human sperm, for example, live 3 days, while bat sperm survive months during the interval between copulation and fertilization. The sperm of honeybees and lizards may even survive 4–5 years within the female body.

Functionally and behaviorally, modified sperm are called on to play a variety of roles in fertilization not played by primitive sperm. Modified sperm are introduced directly into the female by the male during copulation, directed toward the eggs at the time of laying during pseudocopulation, or invade through the skin in dermal copulation or hypodermal impregnation. Devices such as packaging sperm in packets (called spermatophores) or providing sperm with extra microtubular sheaths or even two tails (e.g., turbellarians having dermal copulation) presumably strengthen the sperm and aid their movements through a resistant substratum.

As a variation on directing sperm found in annelids, mollusks, arthropods, and salamanders, males deposit packets of sperm (known as a spermatophores) which are later picked up by the female. In salamanders, the sperm left in the packets are highly modified. A large cytoplasmic fold or **undulating membrane** stretches from the head over the rest of the spermatozoon.

The increase in size of modified sperm compared to primitive sperm parallels an increase in egg size, and the increased complexity of modified sperm is correlated with diminished numbers of sperm released by the male. Heterogeneity is greater among the modified sperm compared to primitive sperm and the proportion of healthy sperm is smaller. Individual selection and competitive strategies would seem to have entered into the evolution of modified sperm, and specialization rather than mere random access may operate in determining a spermatozoon's success.

The anatomy of modified sperm differs from that of primitive sperm throughout the length of the cell. The acrosome of modified sperm is relatively large and prominent, frequently extending back past the equator of the head. The inner acrosomal membrane may adhere to the nuclear envelope, forming a dense **perforatorium** which, in birds, indents the nucleus. Beyond the acrosome, the nucleus is largely bereft of cytoplasmic company but may contain **nuclear vacuoles** (Figs. 6.4 and 6.5). Caudally, some cytoplasm may be present in a narrow **postacrosmal region.**

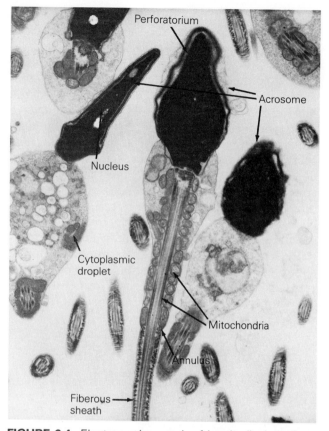

FIGURE 6.4. Electron micrograph of longitudinal sections through human spermatozoon rotated approximately 90° from the spermatozoon seen in Fig. 6.5. × 12,000. From D. M. Phillips, *Spermatogenesis.* Academic Press, Orlando, 1977. Used by permission of Dr. David M. Phillips, The Population Council.

Unlike the situation in primitive sperm, the head and tail of modified sperm are joined by a **neck.** Containing dense **segmental columns** or **striated bands** of a **connecting piece,** the neck links the proximal and the distal centrioles (if present). The connecting piece may surround the proximal centriole (e.g., in insects), or the two may remain distinguishable (e.g., mammals). The material of the connecting piece may penetrate the centrioles to the degree of obliterating them (Fawcett, 1972).

Together or separately, the connecting piece and proximal centriole spread out proximally below a thickened cytoplasmic **capitulum,** fitting snugly into the **implantation fossa** at the base of the nucleus. The thickened outer surface of the nuclear base is the **basal plate,** and the junction of capitulum and basal plate is stiff in the living spermatozoon, indicating a firm bond. Remarkably, this same boundary serves as the plane of **autotomy** or separation in cases where the spermatozoon's tail breaks off during fertilization.

The midpiece of modified sperm is generally

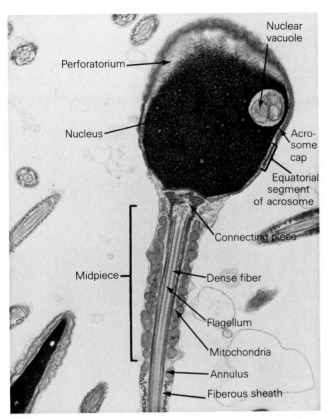

FIGURE 6.5. Electron micrograph of longitudinal sections through human spermatozoon rotated approximately 90° from the spermatozoon seen in Fig. 6.4. × 13,800. From D. M. Phillips, *Spermatogenesis.* Academic Press, Orlando, 1977. Used by permission of Dr. David M. Phillips, The Population Council.

fused with the tail rather than with the head (exceptions include ground squirrel sperm). The midpiece of modified insect sperm typically contains large, longitudinally oriented mitochondria, while modified vertebrate sperm generally have a **mitochondrial sheath** made of a tandem array of mitochondria wrapped around the core of the flagellum.

Throughout, the tail contains the central axoneme. The 9 + 2 arrangement is typical (but not universal, the sperm of the mosquito, *Culex,* for example, have a 9 + 1 arrangement of axonemal filaments). The axoneme of the end piece is naked beneath the plasma membrane, but proximally, additional rods and sheaths accompany the axoneme.

In the beginning of the midpiece of mammalian sperm, segmental columns of the connecting piece give way to **outer dense fibers.** These fibers run peripherally to the axonemal filaments throughout the midpiece and principal pieces. Together, the axoneme and outer dense fibers constitute the **core of the flagellum.** Varying considerably among species in thickness and shape, the outer dense fibers also vary in a single spermatozoon.

They may remain prominent throughout the principal piece (e.g., in hamster sperm, Fig. 6.6) or become reduced to small rods (e.g., in human sperm).

A similar situation is found in insects. In *Drosophila melanogaster,* an accessory microtubule, instead of an outer dense fiber, lies outside each outer doublet of the axoneme. Mutations such as *whirligig* specifically disturb just these microtubules, indicating that their components are different from those of other microtubules.

A dense ring known as the **annulus** (also called a "ring centriole" in insects although it is not a centriole at all) marks the junction of midpiece and principal piece of the tail. A motelike **retroannular recess,** or remnant of the flagellar canal, is tucked in beneath the annulus.

Beyond the annulus, the mitochondrial sheath of the midpiece is replaced by a **fibrous** or **dense sheath** in the principal piece. The fibrous sheath consists of **circumferential ribs** and two **longitudinal columns** which replace two of the outer dense fibers (Fig. 6.6). These columns and the outer dense fibers between them are asymmetrically arranged (Fig. 6.7), but the tail tends toward bilateral symmetry.

The tail beat of modified sperm is stronger than that of primitive sperm and more consistently oriented. In the case of bull sperm, for example, the two-dimensional wave generated by synchronized beating of tails propels sperm at the extraordinary rate of 0.1 mm per second. Human sperm move at a maximal rate of about 0.05 mm per second, or about 7 inches per hour. As a result, modified sperm may travel considerable distances on their own.

Mammalian Sperm Anatomy

While easily accommodated in the category of modified flagellated sperm, mammalian sperm have some unique characteristics (see Bellvé and O'Brien, 1983). The spermatozoon's head is divided by an **equatorial segment** or **zone** (or subacrosomal collar) into an anterior and a postacrosomal (or posterior) region. The morphology of the anterior region is dominated by the **acrosomal cap,** while the postacrosomal region is shaped by the dense **postacrosomal sheath** or **lamella** lying between the plasma membrane and nuclear envelope.

No precursors of an acrosomal filament are stored, and no filament is formed. The fusiogenic membrane is not present in the perforatorium nor any part of the sperm's tip. Instead, the fusiogenic membrane is present in the equatorial segment near the posterior margin of the acrosome.

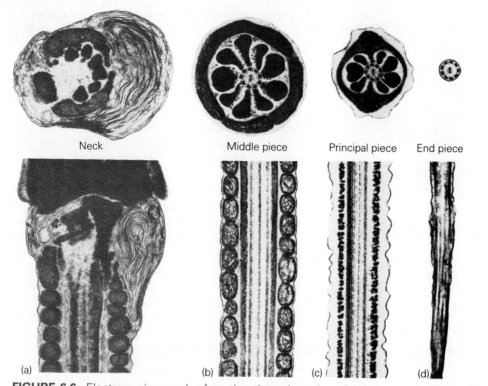

Neck   Middle piece   Principal piece   End piece

(a)   (b)   (c)   (d)

FIGURE 6.6. Electron micrograph of section through neck and tail pieces of mammalian spermatozoa. (Top row) Cross sections. (Bottom row) Longitudinal sections. (a) Neck of Russian hamster spermatozoon. The constriction of the base of the head marks the posterior ring where the plasma membrane fuses with the outer and inner nuclear envelope. The implantation fossa of the nucleus arches above the capitulum. The proximal centriole (right) and the striated connecting piece (left) weld the tail to the head. Redundant nuclear envelope (arrow) may be continuous with an elaborately convoluted membrane system. (b–d) Tail pieces of Chinese hamster spermatozoa. The axonemal core consists of the 9 + 2 axoneme surrounded by outer dense fibers. In the middle piece (b), a mitochondrial sheath surrounds the outer dense fibers. The fibrous sheath replaces the mitochondrial sheath in the principal piece (c), and two longitudinal fibers replace two outer dense fibers. In the end piece (d), the fibrous sheath and outer dense fibers have dropped out, leaving only the axoneme within the plasma membrane. Electron micrographs by Dr. D. Phillips. From D. Fawcett, _Biol. Reprod._ (_Suppl._), 2:90 (1970) by permission of the Academic Press, New York, and D. Fawcett.

The lengths of mammalian sperm vary conspicuously (Fig. 6.2). Human, rabbit, and common domestic mammals have sperm of about 50–60 μm long, while rodent sperm are 150–250 μm long. The longest sperm are those of the tiny marsupial, the honey possum, _Tarsipes rostratus,_ which are about 350 μm long.

The head of a human spermatozoon is more paddle shaped than round (Fig. 6.8). About 4–5 μm long, the head is 2.5–3.5 μm wide (Fig. 6.5) and 1–2.5 μm thick (Fig. 6.4). The heads of rodent sperm are more asymmetric, being hook shaped and flattened (see cover illustration). In the rat, _Rattus rattus,_ the head is as much as 18 μm long. Differences in shape are largely attributed to differences in the acrosomal cap.

Flattening of the acrosome seems to allow sperm to get closer together for packaging into a compact transportable mass. Stacked up in **sperm rouleaux**, sperm are presumably protected from shearing forces during ejaculation.

The mammalian spermatozoon's head ends with a constriction known as the **posterior ring.** The nuclear envelope may be fused with the plasma membrane in this ring.

The short necks of mammalian sperm (i.e., 0.5-μm necks in human sperm) may give rise to a **cytoplasmic droplet** (Fig. 6.4) that extends into the middle piece and contains one or two longitudinally oriented mitochondria. In addition, the nuclear envelope may extend caudally as **redundant nuclear envelope,** and an elaborate membranous

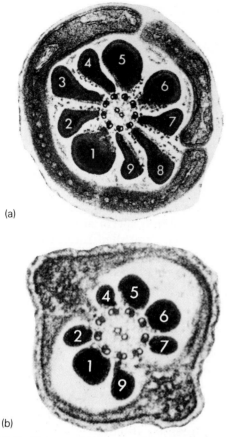

(a)

(b)

FIGURE 6.7. Electro micrograph of cross section through the midpiece (a) and principal piece (b) of the tail of a Chinese hamster spermatozoon. A relatively flat plane passing through both microtubules of the central pair and two outer doublets cuts the tail into one part with three outer doublets (called the minor compartment) and another part with four doublets (called the major compartment). By assigning 1 to the middle of the three doublets in the minor compartment and proceeding to number the doublets in a clockwise direction, each of the outer doublets acquires a unique number (see Fig. 2.19). The third and eighth outer fibers are replaced by the longitudinal columns of the fibrous sheath. The longitudinal column corresponding to the third outer fiber is considered the dorsal while that above the eighth is considered ventral. The strong, well-oriented beat of mammalian sperm is due to this arrangement, of the axonemal filaments, outer dense fibers, and longitudinal columns bound to the fibrous sheath. From D. M. Phillips, *Spermatogenesis.* Academic Press, Orlando, 1977. Used by permission.

FIGURE 6.8. Scanning electron micrograph of human spermatozoa. A distinct postacrosomal region and cytoplasmic droplet are visible in the caudal part of the paddle-shaped area. ×2400. Courtesy Dr. D. M. Phillips, The Population Council.

system of unknown function may also be present (Fig. 6.6). Nuclear pores are only obvious in the redundant nuclear envelope.

The midpiece (4–9 μm in length and 0.05–1 μm in diameter in human sperm) contains the mitochondrial sheath consisting of one or more encircling or helically wound mitochondria. As many

as 300 gyres (i.e., rotations) of mitochondria are found in rodent sperm. Human sperm contain only about 15 gyres.

The axonemal complex runs the entire length of the midpiece and tail (about 45 μm in human sperm). The principal piece is the longest part of the tail (40–45 μm in human sperm). The end piece, beyond the fibrous sheath, is of variable length (5–10 μm in human sperm).

Closer Look at the Nucleus of Primitive and Modified Sperm

No generalities seem to apply equally to all the nuclei of primitive or modified sperm, nor are any characteristics limited to sperm of one or the other type. Still, in general, the nuclei of sperm are smaller than those of somatic cells. For example, the mouse spermatozoon's nucleus is 300 times smaller than that of an average mouse somatic cell. Nuclei may be vacuolated, but they are relatively rigid.

Mammalian and avian sperm DNA is highly

TABLE 6.1. Molecular correlates with nuclear condensation

Degree of condensation	Little or none	Intermediate	Extreme
Example	Crustaceans	Echinoderms	Rodents and insects
Histone (nucleosomal)	−	+	−
Protamine	−	−	+

−, Absent; +, present.

methylated compared to oocyte, fetal germ cell, and blastocyst DNA but undermethylated compared to adult somatic tissue DNA. Methylation is high in the region of unique sequences, but, unlike the DNA of somatic tissue, satellite sequences are undermethylated (Monk et al., 1987).

Sperm chromatin is generally condensed and contains associated nucleoproteins differing from their somatic counterparts. The degree of chromatin condensation ranges from little or none in crustaceans, through intermediate in echinoderms, to extreme in mice where chromatin is even more condensed than in the chromosomes of the metaphase plate (Table 6.1). In the extremely condensed chromatin of some sperm, especially mammals and some insects, dense nuclear packing is correlated with nuclear chemistry. The chromatin lacks histones, is **nonnucleosomal** (i.e., lacks nucleosomes), and contains the relatively small, crystalline arrays of highly basic polypeptides known as **protamines.**

These correlations between nuclear chemistry and density cannot be drawn for less condensed sperm where a variety of additional nucleoproteins are present, ranging from histones to high molecular weight basic proteins. Uncondensed crustacean chromatin lacks protamine, but it also lacks histone and is nonnucleosomal. The partially condensed chromatin of sea cucumbers and more fully condensed chromatin of sea urchins contain histones and are **nucleosomal.** The highly condensed chromatin of chiton sperm also has histone along with nonhistone, so-called intermediate proteins (see Poccia, 1986).

When present, **sperm histones** span a range from relatively somatic-like to decidedly nonsomatic-like tissue-specific nucleoproteins (Table 6.2). **Rana-type** sperm nucleoproteins, also found in goldfish, are most somatic-like but are still on a continuum. **Intermediate** or **Mytilus-type** nucleoproteins have properties of histone mixed with those of more strongly basic **protamines.** In sea urchins and starfish, **sperm-specific histone variants** (i.e., especially variants of H1) are present, but protamines are absent.

Histones may be accompanied by additional protamines or other basic proteins, but in the oyster, *Crassostrea gigas,* histones are the only basic proteins present. The lysine-rich core histones and two H1 linker histones, moreover, are the same as those of somatic tissues (Sellos, 1985). In addition, large protamine-like components of some molluscan sperm may contain lysine.

Originally described in fish, **salmon-type** protamines are widely distributed in sperm from liverworts and mosses, to earthworms, whelks and land

TABLE 6.2. Nuclear proteins of sperm heads

Histone	Intermediate proteins		Protamines Low Molecular Weight Basic Protein	
Lysine-rich			Arginine-rich	
Rana-type	*Mytilus*-type		Salmon type	Mammalian or mouse/grasshopper type
Somatic-like and sperm-specific histone variants especially in H1	Between nonsomatic histones and high molecular weight basic protein	Cysteine:	−	+
		Lysine:	− (except in mollusks)	+
		Amino acids:	30–32	47–50
		Arginine cluster		
		Size:	4–6	4–7
		Location:	Dispersed	Central

−, Absent; +, present.

snails, squid and octopus, centipedes and spiders, and reptiles and marsupials. Salmon-type protamines are small polypeptides (30–32 amino acids, molecular weight about 3000) and contain clusters of polyarginine (4–6 residues) for as much as 90% of their amino acids (i.e., they are arginine rich). Protamines of this type generally lack lysine (unlike lysine-rich histones) and always lack cysteine (Table 6.2). In salmon sperm, only one protamine is present (i.e., the sperm are of a monoprotamine type), although a variety of **intermediary proteins** appear transiently during nuclear condensation.

The **mammalian** or **mouse/grasshopper-type** protamines (sometimes called protamine-like nucleoproteins) are larger than the classic salmon-type protamines and contain about 47–50 amino acids generally including cysteine residues (exceptions occurring among insects). Polyarginine clusters (4–7 residues) are concentrated in a **protamine-like** central domain of about 25 residues. These protamines are present in insects (although they may not contain sulfur) and sharks as well as in eutherian mammals.

A hen is only an egg's way of making another egg.

Samuel Butler, *Life and habit.*

THE EGG OR OVUM

Eggs are adapted for fertilization and for development in particular, and not necessarily the same, environments. Whether fertilized internally or externally, retained, shed, or passed to new sites, eggs must be able to perform their function and deal with their surroundings.

Ovuliparity and Oviparity

Ovuliparous (Lat. *ovum* + *-ulum* diminutive ending, small egg + *parare* to bear, hence bearer of small eggs) species shed unfertilized eggs. **Oviparous** (Lat. *ovum* egg + *parare*) species lay fertilized eggs or early embryos. The former tend to be smaller and have soft or penetrable membranes and frequently jelly coats. The latter tend to be larger and have harder protective coats (see Bertin, 1952).

Most eggs of both types are dependent on their environments for water and swell upon laying. Even some eggs laid in or on the ground absorb external moisture. The eggs of terrestrial amphibians, such as the Apoda, absorb water from the damp ground, and reptilian eggs, such as those of turtles, may increase in weight by almost one-third with water absorbed from damp sand. Eggs laid in

the marine environment acquire salts as well as water from their environment.

The egg's environment is not always accommodating. Eggs laid in fresh water cannot usually obtain salt from the water, and those deposited in terrestrial environments may be threatened with dehydration.

Members of the same species may produce different kinds of eggs during the laying period. In freshwater gastrotrichs, for example, thin-shelled eggs (trachyblastic eggs [Gk. *trachy-* rough + *blastikos* sprouting]) are laid in the spring and in rapidly growing cultures. The eggs start cleaving immediately and cannot survive freeze–thawing or drying. Thick-shelled eggs (opsiblastic eggs [Gk. *opses* appearance + *blastikos*]) are laid at the end of the growing season or if a growing culture is allowed to stand following the depletion of food. A period of dormancy is required for the initiation of development. The eggs can be dried or frozen for several months or years but hatch when favorable conditions and media are restored (see Brunson, 1963).

Self-contained **cleidoic eggs** (Gk. *kleistos* closed) that do not absorb water and produce embryos at the expense of internal reserves are well adapted to life in the terrestrial environment. Protected by thick shells, the cleidoic eggs of birds and insects (with the exception of some grasshoppers, e.g., *Melanoplus* and *Locusta*) require nothing more than warmth, an atmospheric source of oxygen, and a sink for carbon dioxide.

Viviparity and Ovoviviparity

Eggs that are retained by the female parent, whether within the genital tract or following transfer elsewhere in a parent's body, are dependent on a living environment for their nurture. In **viviparous** (Lat. *vivus* alive + *parare*, hence live born) species, the egg's external membranes are shed, and, remaining in the female parent's genital tract, the egg achieves **implantation**, secure physical contact between maternal and embryonic tissue. In **ovoviviparous** species (Lat. *ovum* + *vivus* + *parare*, hence the combination of eggs with live birth), the egg's external membranes are retained during passage through the female parent's genital tract, and implantation is not achieved.

The eggs of the viviparous **therian** or placental mammals (i.e., the metatherian marsupials and the familiar eutherian mammals) arrive in the uterus enclosed within their egg membrane, known as the zona pellucida, and sometimes within a thick outer **mucin coat** (e.g., rabbits), all of which is shed at the time of implantation. In contrast, the eggs of the ovoviviparous **protatherian** mammals (e.g., echidna, the spiny anteater), while thin shelled or even shell-less, retain their zona pellucida in the uterus. The eggs nevertheless absorb maternally derived materials.

In vertebrates, ovoviviparity does not occur among birds, but it occurs among elasmobranchs, teleost fish, amphibians, snakes, and mammals. Many species retain the eggs within the female genital tract, but some species transfer the eggs to other specialized sites in one or the other parent. In the South American marsupial frog, *Gastrotheca ovifera*, eggs are transferred to a dorsal pouch on the mother. *Rhinoderma darwinii* mothers transfer eggs to the father's vocal sac. Male ovoviviparity also occurs in Syngnathidae teleosts, pipefish, and sea horses (see Daly and Wilson, 1979).

Generally, ovoviviparous eggs contain relatively large amounts of nutrients (e.g., Fig. 6.9a).

Viviparous eggs contain relatively small amounts, the embryos relying instead on a **placenta** of maternal and embryonic tissue through which nutrients are passed from the mother.

In some cases, assigning a species to the ovoviviparous or viviparous category is imprecise. For example, embryos of rays usually classified as ovoviviparous are conspicuously nourished by secretions from oviductal villi, while species of galeoid sharks classified as viviparous produce large nutrient-laden eggs that develop first at the expense of reserves. When the reserves are exhausted, the shark forms a **yolk sac placenta** not unlike the earliest placenta formed in pigs (see Chapter 24). Throughout the remainder of development, the shark acquires nutrient material transferred from the lining of the oviducts (see Blüm, 1986).

Viviparity represents the pinnacle of an evolutionary trend paralleling a tendency toward producing fewer young. Among elasmobranchs, for example, oviparous sharks and skates lay barrels-full

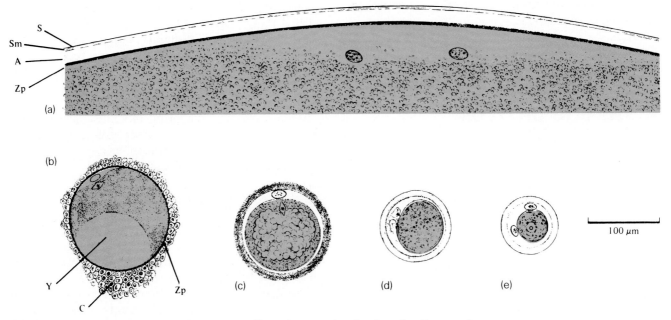

FIGURE 6.9. Mammalian eggs drawn approximately to scale. A, albumin; C, cumulus cells; S, shell; Sm, shell membrane; Y, yolk mass; Zp, zona pellucida. (a) *Tachyglossus*, the spiny anteater (Echidna), a protatherian. Two nuclei indicate that fertilization has already taken place. Granular yolk platelets are present beneath clear cytoplasm and nuclei. (b) *Dasyurus*, the native cat, an Australian marsupial. A large yolk body (Y) is partially surrounded by cytoplasm. The first polar body has formed. The spindle of the second meiotic division lies suspended in metaphase until fertilization. (c) Noctule bat. The bat's egg is surprisingly large, and a newborn can equal one-third the mass of the mother. The zona pellucida is shaded. (d) Guinea pig egg covered by zona pellucida. Both polar bodies have formed, indicating that the egg has been fertilized. (e) Field vole, *Microtus agrestis*. The fertilized egg is covered by zona pellucida. Both polar bodies are present. Adapted from C. R. Austin, in C. R. Austin and R. V. Short, eds., *Reproduction in mammals*, Book 1, *Germ cells and fertilization*, 2nd edition. Reprinted with the permission of Cambridge University Press, Cambridge, 1982. © Cambridge University Press 1972, 1982.

of eggs. In contrast, ovoviviparous sharks produce one to several dozen eggs, and viviparous sharks produce only one to a dozen offspring at a time.

The reduction in egg number accompanying ovoviviparity may be aided by other devices. For example, in the case of the ovoviviparous sand shark, *Carcharias taurus*, only one embryo hatches from its leathery egg capsule originally enclosing 15–20 eggs. While remaining in the oviduct, the newly hatched shark, about 9 inches long, proceeds to eat other capsules until the supply of eggs is exhausted. At birth, the single shark is as much as 40 inches long.

In general, the evolution of viviparity also parallels a tendency toward prolonged development. Unusually long periods of gestation are found in viviparous arthropods. For example, the onycophoran, *Peripatus*, nurtures its developing young for as long as 13 months via a placenta. Some scorpions, although sometimes classified as ovoviviparous, retain their young in the ovary connected by a tube supplying maternal secretions for 6–16 months.

Among prototherian mammals, incubation in the oviparous platypus, *Ornithorhynchus paradoxus*, is 12 days, and gestation in the ovoviviparous spiny anteater, *Tachyglossus aculeatus*, is 12–28 days. Gestation in metatherians, such as kangaroos and wallabies (Macropodidae), lasts 42 days, while pregnancy in the eutherian elephant is 21 months (average 624 days, range 510–730, in *Elephas maximus* and 630–660 days in *Loxodont africana*).

External Membranes

In all animals with the exception of sponges and some coelenterates, the oolemma, or plasma membrane of the egg, is covered with membranes that account for some of the physiological properties of the egg. Membranes offer sites for sperm adhesion, blocks to the penetration of excess sperm (i.e., polyspermy), and protective coats.

Primary membranes or envelopes are closest to the oolemma and are formed while the egg is in the ovary. When thin and penetrable by sperm, the membrane is called the **vitelline envelope** or, in mammals, the **zona pellucida.** When thick and impenetrable by sperm, the membrane is generally called the **chorion.** A **micropyle** (i.e., a small hole) through the chorion provides sperm with access to the egg (see Regier et al., 1980).

Primary membranes are sometimes defined as those produced by the egg itself; secondary membranes are defined as layers added by follicle cells before the egg leaves the ovary (Williams, 1965). Conclusions about the origins of membranes are frequently changed as additional information becomes available, however, and the same membrane may actually originate from both the egg and follicle cells. For example, the zona pellucida of mammalian eggs, with a history of being considered exclusively oocytic in origin, may actually have dual origins (see Guraya, 1985).

The thin vitelline envelope present around the eggs of many marine invertebrates and amphibians provides sites of sperm attachment (Runnström, 1966). Sea urchin eggs also have a primary **jelly coat.** Sperm move through the jelly coat, sometimes within a funnel-shaped hole, but they must penetrate the vitelline envelope to achieve fertilization.

The inner fibrous layer of the vitelline envelope of avian eggs and the zona pellucida of mammalian eggs are also primary membranes. The zona pellucida is thicker in eutherian mammals, where it provides the main barrier between the egg and its environment, than in prototherians and metatherians (compare Figs. 6.9c–e to b) where additional membranes may also be present.

Chorions (Gk. *chori* apart, hence separation) are sometimes defined as primary membranes made exclusively by ovarian follicle cells surrounding the egg. This definition is difficult to apply with confidence.

Thick chorions surround the eggs of insects, tunicates, and teleosts. The eggs lack an additional vitelline envelope and may be fertilized directly by the first spermatozoon passing through the micropyle. In tunicates, **test cells** of maternal origin are incorporated between the chorion and the egg, and the well-developed, thick, and resistant chorion of teleosts, sometimes called the **zona radiata,** contains radiating striations of follicular and oocytic origins.

Layers of follicular cells continue to surround the eggs of most therian mammals (the opossum, Didelphis virginiana, and the phalanger, Trichosurus vulpecula, being the exceptions) during their transport from the ovary to the oviduct. The follicular cells are remnants of the **oocyte–cumulus complex** (OCC) present in developing follicles (Fig. 6.10). The cells add bulk to eggs and seem to allow currents created by oviductal cilia

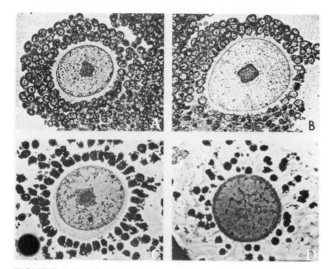

FIGURE 6.10. Sections of rat ovary showing changes in the oocyte–cumulus complex (OCC). (a) Preovulatory OCC in follicle. (b) Nonpreovulatory OCC in follicle. (c) Expansion of OCC and elongation of corona radiata cells prior to ovulation. (d) Denudation of postovulatory oocyte except for some corona radiata cells. ×250. From N. B. Gilula et al., *J. Cell Biol.,* 78:58 (1978) by permission of the Rockefeller University Press and the authors.

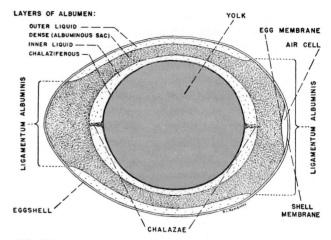

FIGURE 6.11. Diagrammatic cross section of the albumin (white) of a hen egg. From A. L. Romanoff, *Food Res.,* 8:286 (1943). Used by permission of the Institute of Food Technologists.

to sweep eggs into the oviduct (i.e., the cumulus facilitates **egg pick-up**) (see Harper, 1982).

Once in the ampulla, most mammalian eggs quickly shed the cumulus cells wholly or partially (the musk shrew, *Suncus murinus*, is the exception) (see Bedford, 1982). The **denudation** of eggs continues the **uncoupling** and **expansion** of cumulus cells begun in the ovarian follicle prior to ovulation and goes to completion whether sperm are present or not (see Eppig, 1985).

In metatherians and ungulates, denudation is completed quickly and sperm in the ampulla encounter eggs wrapped only in a bare, thin zona pellucida. In most eutherians, denudation is not completed prior to fertilization, and sperm encounter a layer of columnar cumulus cells called the **corona radiata** immediately surrounding and radiating from the zona pellucida. The orientation of corona cells suggests that they guide sperm to the egg.

Sperm may aid the breakup of the cumulus remnant and the corona radiata. In rabbits and rhesus monkeys, the cumulus breaks up faster in the presence of sperm than in the absence of sperm (6 hours compared to 8–10 hours in the rabbit, and 24 compared to 48 hours in the monkey), but in dogs and cats the last cumulus cell remains intact until early cleavage (48 hours or more) (see Austin, 1982).

Secondary membranes or envelopes[1] ***are additional egg coverings acquired after the egg has left the ovary.*** For example, the hard shells of oviparous sharks and rays and the jelly coat of amphibian eggs are secondary envelopes secreted in the oviducts.

The egg white, membranes, and shell of reptile, bird, and monotremes (Fig. 6.9a) are also secondary envelopes. The innermost of these envelopes is a matted, fine fibrous capsule, the **chalaziferous layer,** which spins off at each end of the egg into the **chalazae** (pl., **chalaza** sing., Gk. hail stone or lump, Fig. 6.11). The chalaza at the pointed end of the egg (the cloacal end facing the outside) has two heavy strands with a counterclockwise twist, and the chalaza at the blunt end (the infundibular end) has one strand with a clockwise twist. Anchored peripherally, the chalazal strands function in keeping the egg cell suspended in the center of the egg shell.

The **albumin,** arranged in three concentric layers, while mostly water (85–94%), contains protein woven into fibers and membranes. The **inner liquid layer** is a fluid, viscous albumin in which the egg cell is suspended by the chalazae. The **middle** or **dense layer,** also known as the **albuminous sac,** constitutes more than half the albumin and consists of semisolid fibers dispersed in liquid albumin. The chalazae are anchored to this layer,

---

[1]Membranes added to eggs after leaving the ovary are considered tertiary membranes or envelopes when secondary membranes are defined as those added by follicle cells to ovarian eggs.

which is also attached to the inner shell membrane by the **ligamentum albuminis** at each end of the egg. Finally, an **outer liquid layer** consisting of a viscous liquid with few fibers completes the albumin.

An **inner shell membrane** surrounds the albumin and is cemented to the **outer shell membrane** except where the **air space** or **air cell** develops between the membranes at the blunt end of the egg (Fig. 6.11). Forming only after the egg is laid and has cooled, the air cell functions more or less like an expansion (or contraction) valve in the sense of accommodating to changes in egg volume brought about by changes in temperature.

Unbranched glycoprotein fibers of the inner and outer shell membranes contain carbohydrates, hexosamines, and sialic acid residues. The membranes are sometimes said to be keratin (see Bellairs, 1971), but they probably consist of a collagenous core surrounded by a glycosaminoglycan matrix (see Dumont and Brummett, 1985). Thickening toward the outside, the largest fibers are embedded in **mammillary knobs** on the inner surface of the shell and probably serve as **core proteins** or nucleation sites for mineralization (Fig. 6.12).

The **calcareous shell** begins with basal caps of amorphous calcium carbonate and other minerals, magnesium salts, phosphates, and high concentrations of matrix protein. A lining layer of basal caps gives rise to crystalline cones and the thick, **spongy**

layer (which is actually very compact). The **palisade layer** of calcite crystals, similar in composition to limestone, contains less organic material and magnesium salts than the lining layer, and the outer **crystalline layer** is virtually pure calcite crystals with no organic matrix.

Beyond these layers is the **cuticle,** a layer of protein permeable to air and possibly required for development outside the hen's body. As many as 7000 pores, 40–50 µm in diameter penetrate the shell, permitting gaseous exchange with the air.

The thickness of bird egg shells is a species characteristic. Shells are thinnest in the hummingbird and thickest (as much as $\frac{1}{6}$ inch thick) in the 3-gallon egg of the extinct flightless *Aepyornis*. The shells of hen eggs vary between 0.26 and 0.38 mm, but larger variations are not uncommon. Hens laying eggs daily tend to maintain a constant shell thickness per egg, but the first and last egg in a cycle of three or more eggs will have thicker shells than the others.

The calcareous shell is translucent when laid but quickly becomes opaque as it dries. The calcium of the shell will supply the developing chick with as much as two-thirds of its calcium, and thinning of the shell due to the mobilization of calcium presumably helps the chick to pip the shell at hatching.

The Size and Shape of Eggs

Eggs tend to maximize volume, minimize surface area, and accommodate to movement through narrow passages. Animal eggs vary between spherical and spindle shapes. If the eggs of the primitive agnathous fish, such as the hagfish, are an indication, the primitive vertebrate egg is oval shaped, but typical vertebrate eggs are spherical, while shells are usually oval (i.e., egg shaped) with a pointed and a blunt end.

Compared to the 20–100-µm eggs of angiosperms or even the largest 140-µm eggs of gymnosperms, the eggs of animals vary from large to enormous. Among vertebrates, the smallest eggs of teleosts begin at nearly 400 µm in diameter and exceed 1 mm in prized caviar. Eggs of some species of *Bufo* and *Rana* are as small as 700 µm in diameter, but *Gastrotheca* species have eggs with an overall diameter of 10 mm. Typical reptilian eggs are of modest size, ranging down to 5 mm, but some are very big, such as the 120 x 60 mm python egg with a volume of 226 cm³.

The hummingbird has the smallest avian egg, but it is a strikingly large 6 mm in diameter en-

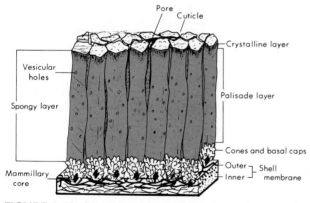

FIGURE 6.12. Diagrammatic section of an avian egg shell and shell membrane. The glycoprotein fibers of the inner and outer shell membrane fuse with the mammillary cores and crystallized basal caps and cones of calcite. The spongy layer consists of calcite crystals comprising a palisade layer in which the crystals have vesicular holes and a crystalline layer of virtually pure calcite crystals. Pores lie between the calcite crystals. From J. N. Dumont and A. R. Brummett, in L. W. Browder, ed., *Developmental biology*, vol. 1, *Oogenesis*. Plenum Press, New York, 1985. Used by permission.

closed in an ellipsoidal shell 14 x 9.5 mm. The ostrich has the largest avian egg, the size of a grapefruit, measuring 80 mm in diameter, with a shell 155 x 130 mm and a volume of 268 cm^3. The ostrich egg is also the largest egg produced by a terrestrial vertebrate. The prize for size, however, goes to the shark, *Chlamydoselachus auguineus*, whose eggs, with a diameter of 150 mm and a volume of 1.77 liters, are probably the largest cells ever produced in the history of life.

The eggs of mammals are disappointing in comparison. The oviparous duck-billed platypus, *Ornithorhynchus paradoxus*, produces an egg 2.5–4.5 mm in diameter enclosed in a dime-sized shell 16 mm in diameter. The ovoviviparous prototherian, *Tachyglossus aculeatus*, has a small hen-like egg about 1 cm in diameter with a thin layer of white (albumin), a shell membrane, and soft shell (Fig. 6.9a).

Metatherian mammals (Fig. 6.9b–c) have small eggs only slightly larger than those of eutherian mammals (Fig. 6.9d–e). Even the largest metatherian egg, that of the Australian native cat, *Dasyurus*, has a diameter of only about 250 μm. The American marsupial, the opossum, *Didelphis virginiana*, has a modest egg, 130–160 μm in diameter.

Most eutherian eggs are in the range of 70–140 μm in diameter, the mouse measuring 70–87 μm, the human 89–91 μm. The range of egg sizes is remarkably narrow from the largest, that of the sheep, *Ovis aries*, measuring 120–180 μm in diameter, to the smallest, that of the field vole, *Microtus agrestis*, measuring about 50 μm in diameter. While small, these eggs are larger by far than

body cells. Mouse eggs, for example, have as much as 1400 times the volume of an average body cell.

The Contents of Eggs

Eggs are frequently said to consist of living parts and **yolk**, or **deutoplasm** (Gk. *deuter-* secondary + *-plasm* formed or molded material) or storage products. The living parts are the cytoplasm, including the egg's **cortex** and plasma membrane or **oolemma**, a spindle, and, when present, a nucleus, usually called the **germinal vesicle.**

Egg chromatin is generally in some state of meiotic arrest and condensed into chromosomes. Oocyte DNA is undermethylated compared to adult somatic cell DNA (e.g., in the mouse) but similarly methylated compared to blastocyst cell DNA (Monk et al., 1987).

The large size of many eggs does not necessarily mean that the egg is distorted. In terms of the ratio of cytoplasmic and germinal vesicle volumes, the egg is comparable to ordinary cells. Furthermore, the concentrations of different cytoplasmic materials are generally no greater in eggs than in body cells and may even be less. For example, amphibian eggs approximately 100,000 times the volume of typical body cells contain 1000–10,000 times more histone, RNA polymerases, DNA polymerase, and ribosomes than body cells.

Mitochondria likewise are present in great numbers in eggs but not disproportionately great. A *Xenopus laevis* egg contains more than 10^8 molecules of mitochondrial DNA (mtDNA), while somatic cells may contain 1000–10,000 mtDNA molecules (Dawid and Blackler, 1972). Likewise,

TABLE 6.3. Partial composition of hen and frog eggs

Material	Hen Egg[a] (Percentage of Total Yolk)	Frog Egg[b] (Percentage of Dry Mass)
"Living" part	20%	
Deutoplasm	80%	
Solid yolk lipovitellin-bound phosvitin (insoluble)	23% (present in yolk spheres)	
Liquid yolk lipovitellin = livetin (soluble)	77%	
Water	48.7%	
Protein	16.6%	45%
Phospholipids and fats	32.6%	25%
Neutral lipids	25.6%[c]	
Phosphatides and cholesterol	7.0%[c]	
Carbohydrates (glycogen)	1%	8.1%

[a]Data from Romanoff and Romanoff (1949) and Balinsky (1981).
[b]Data from Barth and Barth (1954).
[c]Recalculated from Balinsky (1981).

the smaller mature mouse oocytes contain 100–1000 times more mtDNA than somatic cells, and mature bovine oocytes have a 100-fold excess of mtDNA (Hauswirth and Laipis, 1982).

The situation is totally different with respect to deutoplasm. Most of the excess size of eggs is due to the presence of deutoplasm not found in typical body cells at all.

Deutoplasm is different things in different eggs. Some eggs have only energy sources as storage products (if that), while other eggs are "supermarkets" filled with a variety of resources for the future embryo. Even energy reserves vary considerably. For example, the ratio of protein to phospholipids and fats in hen eggs (approximately 1:2; see Table 6.3) is the reciprocal of that in frog eggs (approximately 2:1). The concentration of carbohydrate in the form of its storage product, gly-cogen, varies by nearly an order of magnitude between hen and frog eggs (Table 6.3).

Yolk, frequently the most conspicuous storage product, is a complex mixture of lipid and protein, often in different forms. Neutral lipid, for example, is present as conspicuous fat globules in fish eggs and confined in minute protein-coated **lipochondria** in frog eggs.

Largely in solution, yolk is also present as particles, either crystalline (Fig. 6.13) or, more commonly, granular, and sometimes within vesicles. In hen eggs, 23% of the yolk is particulate (Table 6.3). In reptiles, birds, and prototherian mammals, solid yolk is concentrated in large (as much as 150 μm in diameter) **yolk spheres,** and in cyclostomes, fish, and amphibians, yolk is in membrane-bound **platelets.**

Yolk platelets consist of a core or main portion

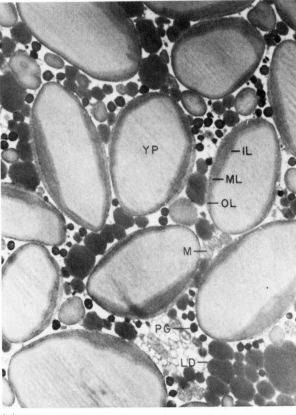

(a)

(b)

FIGURE 6.13. Electron micrographs of yolk platelets in the frog, *Rana pipiens.* (a) Large yolk platelets (YP) from 1-mm egg contain a crystalline main body, roughly hexamonal in outline, covered by inner (IL), middle (ML), and outer (OL) layers. ×4800. (b) The crystalline lattice of a yolk platelet (PY) formed within a mitochondrion in an oocyte. C, apparent limiting membrane formed by enlarged crista, LD, lipid droplet; MC, mitochondrial cortex surrounding yolk; PG, pigment granule. ×60,000. Reproduced from R. T. Ward, *J. Cell Biol.,* 14:309 (1962) by copyright permission of the Rockefeller University Press.

and a superficial layer beneath a bounding membrane. The core portion contains a phosphoprotein, **phosphovitin** (with as much as 8.4% phosphorus), and lipoproteins, **lipovitellins** (with as much as 17.5% lipid) (see Wallace, 1963). How these molecules are associated in yolk crystals is presently uncertain (Lange, 1985).

Classification of Eggs by Yolk

Information about an egg's contents of yolk, or lecithin (Gk. *lekithos* yolk), provides a useful background for understanding some features of cleavage. Caution must be exercised, however, in extrapolating from one egg to another (Table 6.4). In general, criteria for classifying eggs by yolk contents consider the egg's size, amount of yolk, and the yolk's distribution relative to cytoplasm.

Small **alecithal** eggs are completely or almost completely yolkless. The best-known examples of these eggs are the slowly developing eggs of eutherian mammals (Fig. 6.9) and the rapidly developing eggs of many marine invertebrates.

The content of alecithal eggs is called **ooplasm** and may contain lipid storage products. Typically, mammalian eggs contain relatively large amounts of lipid in fat globules or granules. In the case of the guinea pig, *Cavia porcellus*, fat globules are concentrated at a pole dubbed the vegetal pole even though "vegetal" is usually associated with yolk (Fig. 6.9d).

Some eutherian eggs are more properly classified as yolk-poor or **microlecithal.** For example, the eggs of the rhesus monkey, *Macaca mulatta*, contain fine yolk granules; those of sheep, *Ovis aries*, contain nonfat yolk globules; those of the armadillo, *Dasypus novemcinctus*, contain granular yolk; and eggs of the bat, *Rhinolophus ferrumequinum*, have yolk in vesicles.

Metatherian eggs can also be placed in the microlecithal class. The Australian marsupial known as the native cat, *Dasyurus viverrinus*, for example, accumulates yolk in a mass that tends to rest eccentrically (i.e., at a vegetal pole, Fig. 6.9b).

Small **medialecithal** or **oligolecithal** eggs have relatively little yolk, sometimes, as in sea urchins, present in granules. When the yolk is evenly distributed, the eggs are classified as **isolecithal** or **homolecithal** (protochordates). An underlying graded distribution of yolk is sometimes inferred from unequal divisions during cleavage, and the eggs, such as those of sea urchins, are then referred to as having a **tendency toward a telolecithal** distribution of yolk.

Larger **mesolecithal** eggs contain relatively more yolk. In the frog egg, for example, yolk con-

TABLE 6.4. Classification of eggs on basis of amount and distribution of yolk

Amount of Yolk	Distribution of Yolk	Examples
None, *alecithal*		Eutherian mammals and marine invertebrates
Yolk-poor, *microlecithal*		Eutherian and metatherian mammals
Small, *medialecithal* or *oligolecithal*	Even, *isolecithal* or *homolecithal* or Tendency toward telolecithal	Marine invertebrates protochordates (Amphioxus, tunicates, ascidians) Many echinoderms
Modest to heavy, *mesolecithal*	Largely in vegetal pole, *telolecithal*	Primitive fish (ganoids, dipnoans, cyclostomes), amphibians
Lecithotrophic species Great, *megalecithal*	Interior until concentrated vegetally after fertilization, *telolecithal* (cytoplasm in blastodisk)	Teleost fish
	Distinctly concentrated toward center of egg, *Centrolecithal* (cytoplasm in rim and center with nucleus)	Arthropods, especially insects
	Distinctly concentrated in vegetal pole, *telolecithal*	Cephalopods, some gastropods
	In "yolk" or vitellus (cytoplasm and nucleus in blastodisk), *telolecithal*	Some elasmobranchs, reptiles, birds, prototherian mammals

stitutes 80% of the dry weight. This yolk tends to be distributed unevenly. In **telolecithal eggs,** the yolk is concentrated in a **vegetal** or **yolky hemisphere** centered on a **vegetal pole,** while the cytoplasm (and the nucleus) is relegated to the opposite **animal** or **cytoplasmic hemisphere** centered at an **animal pole.** Yolk platelets in the vegetal hemisphere of frog eggs are larger and more closely packed than yolk platelets in the animal hemisphere.

Since the polar bodies form at the animal pole of telolecithal eggs, traditionally, if not by convention, the position of polar bodies is also said to identify the animal pole of isolecithal eggs. Even though yolk may be evenly distributed in an egg, the pole opposite the polar bodies is designated the vegetal pole.

Species producing eggs with conspicuous "yolks" or exclusively yolky areas are considered **lecithotrophic** or yolk-eating species. The actual sizes of the eggs and the distribution of the yolk may differ considerably among these eggs.

The mesolecithal eggs of teleost fish are placed in this class (though not especially large or packed with yolk) primarily because the yolk is conspicuous as a result of its being concentrated interiorly. Cytoplasm and the nucleus are relegated to the periphery, sometimes in a small cap, thereby turning the egg into a scaled down telolecithal egg.

At the time of shedding, some teleost eggs, such as those of the zebra fish, *Brachydanio rerio*, have a relatively uniform content, but following fertilization, the cytoplasm and deutoplasm separate through dramatic movements, known as **cytoplasmic streaming** or **ooplasmic segregation.** The telolecithal condition results from the cytoplasm and dividing nuclei moving into a **germinal disk** or **blastodisk** (Gk. *blastos* sprout + *diskos* platter)[2] and yolk concentrating in a vegetal mass (see below and Fig. 11.15).

Large amounts of yolk are present in **megalecithal** eggs, but the yolk can be distributed in different ways. Eggs of the squid, *Loligo*, the gastropod, *Busicon*, the crab, *Libinia*, and the starfish, *Henricia*, are megalecithal with polarized **telolecithal** distributions of yolk. Insect eggs, on the other hand, have a thick rim of cytoplasm sur-

rounding a mass of yolk. These eggs are megalecithal with a **centrolecithal** distribution of yolk. Typically, an island of cytoplasm contains the nucleus at the very center of a centrolecithal egg, hence the alternate name, **intralecithal** (also ectolecithal).

Insect eggs are frequently bilaterally symmetrical from the start (i.e., slightly flattened regular ovoids or slipper shaped without conspicuous markings, see Fig. 11.7). Presently, points on the egg are identified by reference to **percentage of egg length** (% EL, with 0% at the posterior pole) and **percentage of ventrodorsal perimeter** (% VD, with 0% at the ventral midline) (see Campos-Ortega and Hartenstein, 1985).

The megalecithal eggs of some sharks, reptiles, birds, and prototherian mammals represent extreme forms of lecithotrophic eggs with telolecithal distributions of yolk. The nucleus lies in a cytoplasmic blastodisk or **cicatrix** (also cicatrice [Lat. scar], mistakenly thought to be a mark left by ovulation) lying on the surface of an immense mass of yolk. Surrounded by a thin rim of cytoplasm and the egg's plasma membrane, the "yolk" or **vitellus** is by far the dominant part of the egg cell (Fig. 6.14).

The yolk of reptiles, birds, and monotremes is not perfectly homogeneous. A **latebra** (Lat. *latere* to lie hidden or concealed), or core, approximately 6 mm in diameter, at the center of the vitellus, contains a particularly fluid yolk known as **white yolk.** This type of yolk also covers the vitellus and fills a flask-shaped **neck of the latebra** extending to the **nucleus of Pander** or **nucleus cicatriculae** beneath the blastodisk. About six concentric layers of white yolk branch off this neck. These layers of

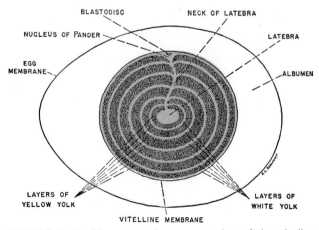

FIGURE 6.14. Diagrammatic cross section of the vitellus of a hen egg. Yellow yolk alternates with white yolk. From A. L. Romanoff, *Anat. Rec.,* 85:261 (1943) by permission of the American Association of Anatomists and Alan R. Liss, Inc.

---

[2]Balinsky (1981, Preface to 2nd edition) notes that "blastodisk" or "blastodisc," as he prefers, "in the past . . . has been used either to denote the concentration of cytoplasm on the animal pole of the uncleaved hen's egg or to denote the mass of cells into which this cytoplasm becomes subdivided during cleavage, or for both." Here the term is used for the uncleaved cytoplasm and nucleus. The spelling with a "k" is that of Patten (1958).

white yolk are less fatty and pigmented than the thicker intervening layers of **yellow yolk.** The differences in yolk may be due merely to diet or access to food inasmuch as the white yolk is laid down during the dark hours of the early morning, while the yellow yolk is laid down during the rest of the day when the hen is actively eating.

Distinctive Areas of Cytoplasm

Inert pigments create some colored areas of cytoplasm. Molluscan eggs, for example, have areas associated with pigment patterns. The scaphopod, *Dentalium*, the elephant tusk mollusk, has pigment over most of its cortex except at two opposite poles (Fig. 6.15). One of these corresponds to the original point of attachment to the ovarian wall. The other represents the animal pole where the egg forms its polar bodies.

Pigments and colored granules are frequently concentrated near the surface of eggs and may be embedded in a cortical layer, free in a subcortical layer, or both embedded and free. The yellow gran-

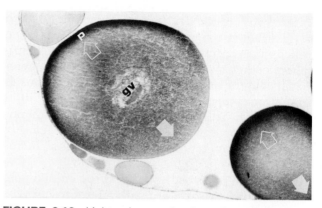

FIGURE 6.16. Light micrograph of section through frog ovary showing eggs at various stages of development. The largest of the eggs has accumulated yolk, giving an opaque, granular appearance, and pigment, darkening the surface. Pigment is embedded in the cortical layer and suspended near the surface throughout the animal hemisphere (p arrow), while yolk is present in platelets that are both larger and more densely packed in the vegetal hemisphere (white arrow). ×36 gv: germinal vesicle.

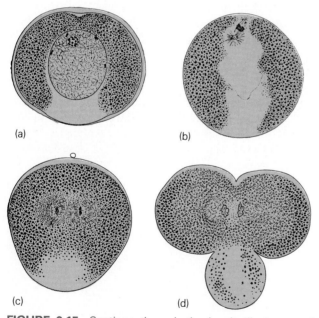

FIGURE 6.15. Sections through the longitudinal axes of eggs of the scaphopod, *Dentalium.* (a) At 5 minutes after release, oocyte has large germinal vesicle, a pigmented girdle, and clear areas at poles. (b) At 30 minutes after fertilization, a polar body is forming at the "animal pole." (c) At telophase of first cleavage 60 minutes after fertilization, egg begins to pulse at "vegetal pole" opposite position of polar bodies (one shown). (d) At 68 minutes, first cleavage proceeds, and a large polar lobe continues to be pushed out of vegetal pole. From E. B. Wilson, *J. Exp. Zool.,* 1:1 (1904) by permission of Alan R. Liss, Inc.

ules and gray yolk of the ascidians (see below and Fig. 9.4) and the purple pigment in eggs of the sea urchin, *Strongylocentrotus purpuratus* (see Fig. 9.5) are largely free and change their distribution dramatically upon fertilization.

In most amphibian eggs, a dark melanotic pigment has a graded distribution opposite that of the yolk (Fig. 6.16). A superficial layer of bound pigment extends over the **animal hemisphere.** As the pigment diminishes, it identifies the **marginal zone** around the egg's equator and the vegetal hemisphere where pigment is sparse. Loose pigment granules are also present interiorly, especially above the equator.

Pigment in amphibian eggs is developmentally inert, and all the developmental events occurring in densely pigmented eggs of some species also occur in sparsely pigmented eggs of other species (such as those of *Xenopus laevis,* the African clawed frog, or even albino variants). Nevertheless, pigment identifies areas associated with particular developmental events. The center of the pigmented animal pole, for example, is the only portion of the egg resistant to penetration by sperm (i.e., nonfusiogenic). It is also the site where the amphibian egg's meiotic spindle resides and the point of polar body formation.

The surface of the egg comprises a **cortical layer** or **cortex** a few micrometers thick. This layer consists of the egg's plasma membrane, the oolemma, folded or poked out into numerous **microvilli,** and a subjacent gel sometimes containing **cortical granules** or particles which when wrapped in membrane are called **cortical vesicles or vacuoles** (Fig. 6.17). Mitochondria and other cell organelles are not present in the cortical layer.

Cortical particles are present in the eggs of bivalve mollusks, some annelids, sea urchins, frogs, fish, hamsters, rabbits, and human beings. Similar particles are not found, however, in gastropod mollusks, insects, urodeles, birds, rats, or guinea pig. Where present, the spherical, membrane-bound vesicles vary in diameter from 1 to 2 μm, but they are hardly simple sacs, nor are they morphologically identical even in closely related species. The complex structures present in the particles include concentrically arranged (Fig. 6.17a) or recurring coils of **lamellar filaments** (Fig. 6.17b). In addition, **extralamellar bodies** may be present near the base of the particle, and the basal region may be connected to the particle membrane (Fig. 6.17c).

HISTORY OF THE EGG AND SPERM

Epigenesis and the Aristotelian Egg

Aristotle (384 to 322 B.C.) deserves credit for beginning the history of the egg (Aristotle, 1984). Aristotle's eggs were the instruments of reproduction in **oviparous** species. Birds and reptiles produced eggs with hard shells, and amphibians, fish, crustaceans, octopus, and squid produced eggs lacking hard shells. In addition, grubs, caterpillars, and other sorts of larvae were the premature eggs of some insects which gave rise to pupae, the real eggs of these insects.

Mammals, human beings, and some other animal species, on the other hand, lacked eggs entirely. They were **viviparous** and reproduced by birth. Instead of eggs, females possessed the **female element.** Virtually dead, this element acquired life and developmental activity from the male princi-

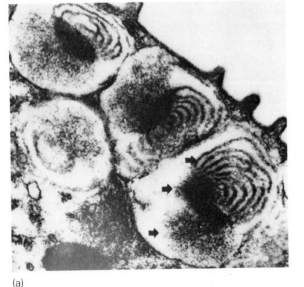

(a)

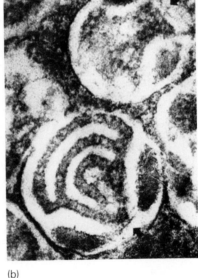

(b)

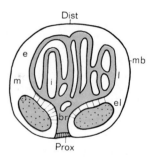

(c)

FIGURE 6.17. Examples of cortical vesicles in sea urchins. (a) Electron micrograph of longitudinal section through particles of *Strongylocentrotus purpuratus.* Lowest arrow, extralamella body; middle arrow, basal region of lamella; upper arrow, lamella ligaments. right horizontal arrow, vitelline membrane. Arrows within particle show its recurring coil. × 29,000. (b) Electron micrograph of longitudinal section through particle of *Paracentrotus lividus.* Left arrow, connections between extralamella bodies and basal region of lamella; right arrow, connection between basal region and vesicle membrane. × 36,000. (c) Diagram of mature *Paracentrotus*-type particle. Abbreviations: l, lamella of coiled filaments and basal region; br, basal region of lamella; el, extralamellar bodies with fibrous connections to lamella; m, matrix; i, interior region of matrix; e, exterior region of matrix; mb, vesicle membrane; Dist, distal pole of particle; Prox, proximal pole of particle connected to basal region and particle membrane. From J. Runnström, *Adv. Morphogenesis,* 5:222 (1966) by permission of Academic Press. Part (a) courtesy of Professor Patricia Harris; part (b) courtesy of Jane Baxandall, Wenner-Gren Institute.

ple present in **semen** (i.e., seed) (see Bodemer, 1971).

The **ovoviviparous** cartilaginous fishes and snakes linked the oviparous and viviparous species in a group that required contact between male and female organisms for reproduction. The female supplied the offspring with the passive vegetative element shared by all living things, embodying nutrition and reproduction. The male supplied the animal *soul* that governed sensitivity and movement.

Beyond the female element was **epigenesis,** incremental development proceeding by stages. In the first stage (prior to birth in viviparous species), the embryo acquired the equivalent of a species soul, and in the second stage (after birth), it added an individual soul, or identity. The human embryo acquired the uniquely human rational soul later still.

And beyond epigenesis was **teleology** (Gk. the doctrine that things are shaped in order to reach a final condition or purpose). An **entelechy** (Gk. to be complete), a final force or cause, pushed things into their future and directed them toward their purpose or effect. In the case of embryos, the force was internal, like a mechanical intelligence (as opposed to a conscious intelligence) whose knowledge of the final state directed the course of development.

In oviparous animals, the souls could be traced to specific elements and a variety of developmental causes. For Aristotle, the yolk and white of the hen's egg were the material cause; the rooster's sperm added the efficient cause; the identity of a bird lent a formal cause; and the identity of a particular hen or cock injected the future or final cause. These causes, like ideas in the mind of an artisan, were not material so much as they were concepts to be worked into material. Like a swordmaker who beats metal into swords, the causes of development extrude organic form (see Needham, 1959).

The elements and causes of development were more difficult to trace in viviparous animals. To explain development in the human female, Aristotle borrowed a theory widespread in India and held by **Hippocrates** (born ca. 460 B.C.) that menstrual flow supplied the soil in which the male planted his seed. Today, when we talk of "blood" relatives, we are taking our words, if not our concept, from this same theory.

The development of the heart and blood was supposed to begin with the coagulation of menses. The coagulum, or clump, was produced in the uterus, and, since one menses had no link to previous menses, the theory maintained that offspring were created *de novo*, that is, beginning afresh. Therefore, neither the coagulum nor the embryo formed around it had continuity with anything that came before, neither a germ plasm nor a heredity material (Oppenheim, 1982).

Aristotle's analysis of sperm, or semen, was likewise flawed and failed to identify semen with living material. He concluded that semen was "a true secretion, and not a homogeneous natural part (a tissue), nor a heterogeneous natural part (an organ), nor an unnatural part such as a growth, nor mere nutriment, nor yet a waste product" (Needham, 1959, p. 39). Semen was the purveyor of form, but semen lacked form itself. It contributed nothing material to the embryo but coagulated the female substance and triggered it to undergo development.

Aristotelian epigenics had several long-lasting ramifications, nevertheless. The idea of incremental development and sequential acquisition of souls gave rise to the views of St. Augustine in the 5th century and to the early canon law that the human embryo had no soul prior to the 40th day of gestation. Moreover, the Aristotelian concept of an entelechy, the actuating or form-giving cause, was adopted by some modern theologians and scientific authorities as a process regulating or guiding an organism to its mature condition (see Needham, 1959).

Hellenistic Egg

Flaws in Aristotle's conceptions of eggs and sperm might have been overcome if he had understood the role of ovaries in viviparous reproduction. But ovaries were only described correctly by **Herophilus of Chalcedon,** the famous Alexandrian anatomist (or infamous "butcher" who performed dissections on slaves) a century after Aristotle's death. Moreover, Herophilus appreciated the parallel roles of the ovary and testis in reproduction.

Galen of Pergamum (131–201), the great Hellenistic physician, while studying in Alexandria, extended Herophilus' view that the ovary and the testis of viviparous animals produced comparable semina. Proposing that the active principle in the male and female mammal was fluid, Galen suggested that the role of the uterus was to attract the female semen down through the oviducts and the male semen up through the cervix. Meeting in the uterus, the two semina were to have comingled and coagulated as the **conceptus** (Lat. *concipere* to take or receive)[3] or unformed stage of prenatal development.

Progress toward understanding the egg and sperm then slowed until the time of **Albertus Magnus** (1206–1280), a thousand years later. More than a popularizer of Aristotle, Albertus probably dissected chick and fish embryos and understood the function of the yolk. He clarified the role of the placenta as a nutritive organ and of fetal vasculature, if not circulation, as the conveyor of nutrition. While deviating from Aristotle to adopt the idea of both male and female semina, Albertus held tenaciously to Aristotelian epigenesis.

Renaissance Egg

Embryology's next spurt of growth came in the 16th century, initiated by **Andreas Vesalius** of Brussels (1514–1564). Working in Padua, Vesalius illustrated structures in the ovaries corresponding to follicles and luteal glands, while his student, **Gabriello Fallopio** (1523–1562), identified, among many other structures, the **Fallopian tubes** (oviducts or uterine tubes) bridging the ovary and uterus.

The mammalian ovaries were considered **female testes** or **testicles** (i.e., diminutive testes) and a source of the female seed (but not, of course, of nonexistent eggs). The idea of the egg arising in the ovary and the embryo arising from the egg was only conceived by **Volcher Coiter** (1534–1576) upon completing his detailed study of chick embryonic development, a study for which he is recognized belatedly as the **parent of embryology** (Short, 1977).

Hieronymus Fabricius ab Aquapendente (1537–1619), the next anatomist in the great succession at Padua, appreciated the role of the ovary in the development of the hen's egg. Indicating that yolk was laid down in the ovary, he proposed that only the white and shell accumulated in the passage down the oviduct.

Moreover, by showing that insect larvae were produced by eggs, Fabricius corrected Aristotle's notion that larvae were premature eggs, although he granted that the pupa was a second egg. Fabricius also described development of the chick and recorded the results of his numerous dissections of mammalian, dogfish, and viper embryos (Adel-

mann, 1942). He did not, however, propose that eggs formed embryos. Unfortunately, despite his excellent illustrations, Fabricius maintained that embryos such as the chick formed from the twisted cords of egg white known as the chalazae (see Fig. 6.11).

Finally, in 1651, the Englishman, **William Harvey** (1578–1657, the same Harvey who theorized that arterial blood must somehow be connected with venous blood), published *Generatione Animalium: Anatomical Exercises on the Generation of Animals* and changed the history of embryology. Above all, Harvey correctly identified the source of the chick embryo within the egg and began the historical process that ultimately led to unifying oviparous and viviparous organisms in the conceptual framework of gametes and fertilization.

Harvey was an admirer of Aristotle ("the philosopher") and managed to defend Aristotelian epigenesis[4] (i.e., the gradual acquisition of form through interactions with the environment), although he attacked spontaneous generation (i.e., the natural production of organisms by the environment). Harvey also refuted the belief held by Aristotle, Hippocrates, and others that the chick was engendered from the vitellus (yolk) and was nourished by the albumin. He also criticized his teacher, Fabricius:

> He likewise saw the Original of the Chicken in the Egge; namely the Macula, or Cicatricula annexed to the membrane of the Yolke, but conceived it to be onely a Relique of the stalk broken off, and an infirmity of blemish onely of the Egge, and not a principle part of it." (From the first English edition [1653], translated from the Latin by Martin Llewellyn, quoted from Needham, 1959, p. 118)

In contrast, Harvey concluded that the cicatrix (Lat. scar) "is, in fact, the most important part of the whole egg, and that for whose sake all the others exist" (Harvey, 1651, p. 359). In 1817, Heinrich Christian Pander confirmed that the cicatrix of a fertilized hen egg was the *Keimhaut* or, in English, the **blastoderm**, from which the chick developed.

In further disagreement with others, Harvey maintained that reproduction was unified by eggs.

> We, however, maintain (and shall take care to show that it is so) that all animals whatsoever, even the viviparous, and man himself not excepted, are produced from ova; that the first conception, from which the foetus proceeds in all, is an ovum of one description or another, as well as the seeds of all kinds of plants." (Harvey, 1651, p. 338)

---

[3]Today, Galen's terms are misrepresented when conception and conceiving are equated with fertilization. To begin with, fertilization takes place in the ampulla of the oviduct and not in the uterus. The embryo does not arrive in the uterus for the better part of a week, and a "conceptus" would therefore be a blastocyst. Since "fertilization" and "blastocyst" are well-defined words, "conception" and "conceiving" should be dropped from scientific discourse.

[4]Harvey probably coined the term epigenesis.

He did not write the succinct phrases attributed to him: "*Ex ovo omnia*" (everything [living] from eggs). These were the words of an unknown artist who engraved the frontispiece of *Generatione Animalium* (Fig. 6.18), but Harvey was responsible for elevating the female element, or female semina of mammals, to the status of an egg (Gray, 1970). He broke down the traditional distinction between nutritive and formative substance and established the viviparous female mammal as contributing more than a mere environment for the male seed.

Harvey's mammalian egg was not at all the same entity we have in mind today, however. While not a coagulum of blood and semen, Harvey sided with Aristotle and concluded that the mammalian egg was formed by the uterus, not the ovary. Like the chick that develops within membranes and a shell, Harvey conceived of the mammalian egg as the chorion-enclosed sac within the uterine shell.

Harvey was a visionary in his espousal of a theory of fertilization even though his idea of fertilization defies the modern imagination. Agreeing with Fabricius that rooster's semen did not enter the hen's uterus, and finding no evidence of semen (only mucous strings) in the uteri of doe he was privileged to dissect as physician to Charles I, Harvey concluded that male semen was somehow absorbed from the cloaca or vagina and activated eggs after moving through the bloodstream.

17th Century Egg Before the Microscope

Meanwhile, the Dutch biologist, **Reijnier de Graaf** (1641–1673), drew attention to the mature **follicles** in the mammalian ovary (follicles that now bear his name although they were described earlier by others including Vesalius). De Graaf incorrectly equated the ovarian follicle with the mammalian egg, but he correctly described the induction of ovulation by coitus in the rabbit and the replacement of follicles by luteal glands after ovulation. Moreover, he showed that the number of luteal glands corresponded approximately to the number of fetuses in the uterus and that these glands disappeared following pregnancy (see Short, 1977).

De Graaf also observed tiny spherical bodies in the oviducts of rabbits 3 days after mating and slightly larger spherical bodies in the uterus a day later. Undoubtedly, these bodies were blastocysts (Fig. 6.19), and their presence in oviducts demonstrated to de Graaf's satisfaction that embryos did not form initially in the uterus.

One of the great paradoxes of embryology's history is that all the major pieces of evidence necessary to solve the puzzle of viviparity were in place by the late 17th century, but the discovery of the viviparous "egg" did not take place for another century and a half. Ovaries were known to discharge follicles, spherical bodies were seen in oviducts (at least in rabbits), and enlarging embryos were found entering the uterus where they presumably developed further. But all this information did not culminate in discovery.

Those who disputed de Graaf's equation of eggs and follicles argued that Graafian follicles were entirely too big to pass down the oviducts to the uterus, and the spherical bodies found in the oviducts and uterus were entirely too small to be follicles. Unfortunately, instead of looking for smaller eggs within follicles, de Graaf's detractors continued to argue that mammalian embryos did not come from the ovary at all.

Moreover, the great Swiss physiologist, **Albrecht von Haller** (1708–1777), who examined the

FIGURE 6.18. Frontispiece of William Harvey's *De generatione animalium* (1651), showing Zeus liberating creatures from the egg inscribed *ex ovo omnia*.

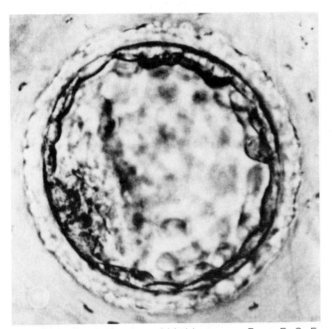

FIGURE 6.19. The 4-day rabbit blastocyst. From E. S. E. Hafez, in R. J. Blandau, ed., *The biology of the blastocyst.* University of Chicago Press, Chicago, 1971. Reprinted with permission of University of Chicago Press. © 1971 by University of Chicago Press.

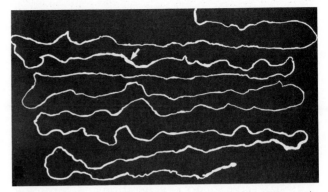

FIGURE 6.20. Pig blastocyst in the uterus, 157 cm in length. The embryonic disk that becomes the fetal pig is a small thickening (arrow). One half actual size. From J. S. Perry and I. W. Rowlands, Early pregnancy in the pig. *J. Reprod. Fertil.,* 4:175 (1962). Used by permission.

uterus and oviducts of sheep, was unable to reproduce de Graaf's observations in rabbits. Instead of spherical bodies, mucous strings, like those in deer, appeared in the uterus, and Haller could not relate the stringlike blastocyst of artiodactyles (e.g., Fig. 6.20) to the spherical blastocysts of rabbits (Fig. 6.19).

Preformation and the Egg and Sperm

At the same time that the role of follicles was disputed in some quarters, a breakthrough had taken place in other quarters which changed the basis for disputation. Aristotelian epigenesis was challenged and with it the related idea of spontaneous generation.

Spontaneous generation and epigenesis were considered intuitively obvious through the Renaissance. They were also generally considered inseparable. According to both concepts, individuals arrived *de novo* and acquired form through sequential interactions with the environment. The major difference between spontaneous generation and epigenesis was the requirement of the latter for a female organism and bodily contact with a male.

The crucial breakthrough came when **Francesco Redi** (1621–1697), a Florentine physician, applied the experimental method to the role of the egg in development. Following Aristotle, Redi believed that new organisms formed by coagulation, the thickening of material into a coherent mass. According to the popular notion of spontaneous generation, the eggs of animals such as flies coagulated in putrefying or decomposing meat, first as maggots (i.e., premature insect eggs) and then as pupae (i.e., mature eggs).

Redi questioned whether reproduction in flies required flies. Did fly seed placed on meat, rather than spontaneous generation, provide the source of maggots?

To answer his question, Redi excluded flies from some pieces of meat by sealing it in flasks, while he allowed flies free access to other pieces of meat in similar but open flasks. The result was unambiguous and the conclusion inescapable: "the flesh of dead animals could not engender worms unless the semina of living ones were deposited therein" (Redi, 1668, p. 78).

Eggs were not formless masses but the components of a generation that moved to the next generation. Both the causes and materials of development were already present in them. Like matter that was neither created nor destroyed in reactions, eggs did not come into existence spontaneously or disappear completely between generations.

Never again would eggs be considered clumps of putrefying flesh, but Redi's interpretation of his experimental results did not stop there. Since the eggs of lowly flies did not come into existence via spontaneous generation, higher organisms could hardly come into existence via epigenesis.

After an agonizing reappraisal, a single alternative to epigenesis and spontaneous generation was found: **preformationism,** the doctrine that the form of animals precedes them, and development is independent of environmental influences. Even in its most naive versions, preformation included the element of continuity so central to modern concepts of heredity and so completely absent in Aristotelian epigenesis. Ultimately, in its more sophisticated versions, preformationism came close to advancing **coded information** as the source of developmentally significant messages. But whether naive or sophisticated, preformation was an all-pervasive theory of generation and, for many embryologists, a passion.

Vitalism, or the doctrine that life depends on uniquely living qualities, took scientific succor from preformationism. Vitalists saw in preformationism a reflection of their belief that a **vital principal** or spirit was the sole conveyer of life. Moreover, Christian fundamentalists found in preformationism the natural explanation for how God, while creating all living things on the Days of Creation, nevertheless allowed living things to produce embryos.

Were preformed parts for every generation stacked within those of the previous generation on the Days of Creation, each generation of embryos could be produced simply by unstacking. The idea of stacking or embedding became the much-abused idea of *emboîtement,* or encapsulation, in which one generation lies within another generation like one Russian doll lies within another.

An especially difficult aspect of *emboîtement* was the size of the embedded structures and the numbers of such structures which would have to have been stacked. Ironically, the preformationists responded to this difficulty by undercutting the very premise of preformationism. Instead of the stacked "dolls" being material, they were conceived of as immaterial.

The preformationists dealt with the problem of size by invoking the ancient doctrine that the law does not apply to the very small (*"De minimis non curat lex"*; Singer, 1959, p. 287). The doctrine made it possible for microbes embedded in microbes to give rise to other microbes without spontaneous generation and for embryos embedded in embryos to give rise to adults without epigenesis.

Because the preformationists argued in terms of invisibility, they were under no constraints to relate the preformed units to anything as concrete as a facsimile of the embryo. Nevertheless, beginning with the classical microscopists of the 17th century, a great deal of effort was aimed at finding such facsimile.

First among the classical microscopists, **Anton van Leeuwenhoek** (1632–1723) of Delft was an indefatigable observer who pushed the lenses he ground to their limits of resolution and scrutinized everything he could hold in suspension before his lenses. In 1677, in the course of investigations on solutions and suspensions, Leeuwenhoek and Johan Hamm discovered "animalcules" or "zoa" within the liquid male seed known as sperm. Later, von Baer renamed the zoa **spermatozoa,** and gradually the idea of a liquid seed was replaced by the idea of **sperm** suspended in a liquid.

Leeuwenhoek and other microscopists saw or conjectured that they would see, if they could, **homunculi** (Fig. 6.21), minute preformed bodies of organisms within the animalcules. Embryos were thought to result from the enlargement of homunculi.

Homunculi and *emboîtement,* which seem so cumbersome today, posed more of a technical challenge than a conceptual problem at the end of the 17th century. At that time, new universes blossomed daily under magnifying lenses. Microscopic parts in eggs and spermatozoa belonged to those universes, and the ability of the microscope to expose them was merely a matter of time, technique, and material.

Ovism and animalism, preformationism's competing schools, were slugging it out by the end of the 17th century. **Ovism** claimed that miniature adult structures were conveyed from

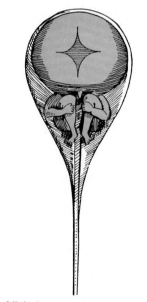

FIGURE 6.21. Nicholas Hartsoeker's hypothetical spermatozoon bearing a homunculus. *Essai de dioptrique.* Paris, 1694.

generation to generation through eggs, and **animalism** contended that miniature adults were conveyed through spermatozoa and, in the case of mammals, developed without eggs.

While Leeuwenhoek was clearly an animalist, other classical microscopists belonged to the ovist camp. **Marcello Malpighi** (1628–1694) of Bologna, for one, is reputed to be (whether rightly or wrongly) (Adelmann, 1965) the foremost architect of preformation in the 17th century and a devout ovist. His work was published by the Royal Society in London and received widespread critical acclaim.

While espousing the idea that embryonic parts were too minute to be detected, Malpighi set about looking for them in hen eggs. He was forced into this contradictory position by the gaps in his knowledge, his purely observational method of deriving knowledge, and his religious–philosophical preferences.

He was thorough and accurate, and his observations on chick development were the most complete of his time (see Fig. 6.22), but he always found miniature chickens growing on the yolk even in unincubated eggs. He could not find any fertile egg which did not have some recognizable parts.

Unfortunately, Malpighi's method was flawed. Judging from his illustrations, embryos he considered at the very beginning of development were considerably older. Not that he misrepresented his data. The eggs he reported as being unincubated were technically not incubated *by a hen*, but, as he carefully noted, he had gathered his eggs during a warm August.

In all likelihood, the heat of the day had triggered development and led to his mistaken conclusion, but Malpighi had confidence in his own

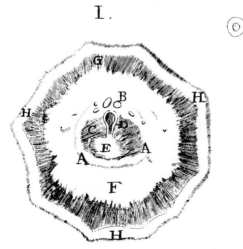

FIGURE 6.22. Unincubated chick embryo as drawn by Malpighi in *De ovo incubato*, 1672. Circle shows blastoderm.

observations. He was not impressed by the arguments of others over temperature, nor was he persuaded by the observations of those who failed to find recognizable parts in fresh eggs. Negative proof proves nothing (even when correct).

Malpighi argued from logic that the lower limits of visibility were not necessarily the lower limits of life. He also argued out of conviction that, since the soul was invisible by definition, preformed units could also be invisible. Religious conviction set preformationism hopelessly adrift in the sea of mysticism. Facts were acceptable if they were consistent with vitalism, but reckoning with contradictory observation and favoring the results of "lifeless" experiments were akin to blasphemy.

Another microscopist and ovist was **Jan Swammerdan** (1637–1680). Although relatively unknown during his lifetime, he had considerable impact through manuscripts published posthumously in 1737. In the course of his study of the developing frog embryo, he became the first to describe blood corpuscles. He also discovered the hermaphroditic gonad of the mud snail, and he vindicated ovism through his detailed study of insect development.

Assuming that the cocoon or chrysalis was the butterfly's equivalent to the egg, Swammerdan dissected cocoons under his microscope. His accurate drawings showed undeniably that the adult butterfly was already formed beneath the cuticular coat of the cocoon. Swammerdan's premise may have been false, since the pupa had already developed from an egg, but his data were unimpeachable.

Ovism's triumph was almost complete in the 18th century. Despite some contradictory evidence, the male seed was relegated to the role of a mere activator. Ironically, the chief evidence supporting this view of sperm was the success of the Italian, **Lazaro Spallanzani** (1729–1799), with **artificial insemination.** Spallanzani reported that frog eggs degenerated in the absence of sperm but developed into normal tadpoles when exposed to sperm. Fish eggs were well known to be fertilized by milt, and Spallanzani succeeded in producing a dog through artificial insemination. Still, he believed that the animalcules swimming in semen were parasites and could not play a role in development.

Other arguments supporting ovism were posited by experimentalists. In an early experiment intended to test the requirement for sperm in mammalian embryos, the oviducts of female dogs were ligated soon after mating. Subsequently, embryos were found above the knot but not in the uterus.

The results were correctly interpreted as demonstrating that embryos originated beyond the uterus; but based on the incorrect premise that sperm had not reached the upper oviducts at the time of ligation, the experimental result was interpreted as demonstrating that sperm were unnecessary for development. Decades would elapse before this error was corrected (see Bodemer, 1971).

The most important arguments favoring ovism were based on empirical observation. The French-Genevan, **Charles Bonnet** (1720–1793), accurately described **parthenogenesis**,[5] virgin birth, in aphids. Throughout the summer of 1742, Bonnet observed female aphids (plant lice) give birth to as many as 10 generations entirely in the absence of males. Only in the fall did males appear. They copulated with females who then laid eggs. Bonnet attributed the change from viviparity to oviparity to lowering temperature (erroneously; as it turns out, shorter day length is the culprit) and argued that in the warm Indies parthenogenesis could continue indefinitely (which it does in some but not all aphid species).

Ironically, the ovist Bonnet had confirmed the earlier discovery of the animalist Leeuwenhoek that summer aphids reproduced without the participation of males. Leeuwenhoek had incorrectly suggested that the aphids were hermaphrodites, while Bonnet correctly identified the aphids as parthenogenetic females (see Suomalainen et al., 1987).

Bonnet's conclusions were carefully reasoned. For Bonnet, an organism was so complex and its parts so interdependent that it simply could not function, much less grow and develop, without having all its parts present and working from the very beginning.

For Bonnet, _emboîtement_ was firmly established by his empirical observations on aphids, a judgment vindicated by modern cytological methods. Embryonic development in parthenogenic aphids begins immediately after oogonia first appear, about the middle of embryonic development. Because embryonic development can go to completion before the mother is mature, three generations of parthenogenetic aphids may lie inside each other.

Extrapolating from his conclusion, Bonnet argued for the preeminence of the female in reproduction. He became the foremost ovist of his time, and his speculations on the mechanism of preformation were widely applauded despite being untestable. At the end, he could declare that the overwhelming endorsement of his views by other scientists represented "one of the greatest triumphs of rational over sensual conviction" in the history of civilization (Needham, 1934, p. 191).

Still, ovism's triumph was not complete. Clues to the role of sperm and egg in sexual reproduction had emerged from studies by botanists on molds and plants. In England, Robert Hooke (1635–1703) described the formation of spores in molds as seen through his microscopes, and Nehemiah Grew (1641–1712) discovered the roles of the parts of hermaphroditic flowers (see Fig. 5.5) as a result of his microscopic investigations.

In Tübingen, in 1694, **Rudolph Camerarius** (1665–1721), Professor of Medicine, showed that removal of anthers from the castor oil plant resulted in empty seeds incapable of germinating into plants, and removal of stigmas prevented seed production entirely. Camerarius also made the first plant hybrids, but a hundred years passed before **Joseph Kölreuter** (1733–1806) showed that the characteristics of seeds were not affected by the direction of **reciprocal** crosses between pollen and pistil. Reciprocity was crucial, since it showed that characteristics were not acquired from one parent but developed from inherent qualities passed equally to the new generation from parents.

Discovery of the Mammalian Egg

At the beginning of the 19th century, viviparity still represented the great weakness in the ovist doctrine. While equipped with ovaries, mammals still lacked a completely satisfactory egg.

Finally, in 1827, **Karl Ernst von Baer** (1792–1876) took a giant step toward discovering the mammalian egg. Seeking the origins of the dog's "vesicles of reproduction" (i.e., blastocysts) that developed in the uterus, von Baer traced similar vesicles to the oviducts and then to vesicles within Graafian follicles (Fig. 6.23). He also established that the same type of vesicles existed within the Graafian follicles of a variety of mammals, from pigs to porpoises, as well as human beings.

Von Baer originally considered the follicle the "maternal egg" (i.e., comparable to the chick's egg). He recognized, however, that the "fetal egg" or "embedded vesicle" within the follicle behaved, "with regard to the coming embryo, as the real egg" (translation quoted by Moore, 1987, p. 451).

The discovery of the tiny "fetal egg" had enormous consequences. Its size alone made a mockery of _emboîtement_, and its formlessness reestablished

---

[5]The term parthenogenesis was coined by Richard Owen in 1849.

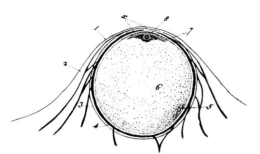

FIGURE 6.23. Graafian vesicle (follicle) of a dog containing an egg (8). 1, Peritoneal epithelium; 2, stroma; 3, theca externa; 4, theca interna; x, stigma; 5, membrana granulosa; 6, follicular fluid; 7, cumulus oophorus; 8, ovum. From C. E. von Baer, *De Ovi Mammalium et Hominis Genesi.* Leopoldi Vossii, 1827; reprinted by Culture et Civilisation, 1966.

epigenesis as a scientific doctrine for the development of embryos.

Coupled with extensive comparative studies on vertebrate embryos, von Baer provided evidence instead of sterile argument to support his ideas, and following his example, embryologists returned to observation as a basis for studying development. Moreover, von Baer's work prepared embryologists to accept the egg as a cell a decade later when the cell theory was promulgated, although von Baer did not appreciate that the egg was a cell when he discovered it.

Impact of the Cell Theory

Martin Barry, an Englishman, was the first embryologist to apply the cell theory to the mammalian embryo even though he had no clear idea of the distinction between nucleus and cytoplasm, nor did he understand cell division via mitosis. Nevertheless, by replacing the globules portrayed in earlier descriptions of embryonic tissue with the concept of cells, Barry clarified the cellular nature of the embryo, and by deriving embryonic cells from the egg, he established the cellular nature of the egg. In his account, the fertilized egg was the original parental cell which was succeeded by two cells, and these by four, each of which gave rise to two more until a mulberry-like object formed of cells too numerous to count (see Oppenheim, 1982).

At about the same time, **Robert Remak** (1815–1865) traced the blastomeres of the frog embryo back to the amphibian ovum and the layers of chick embryos back to the avian ovum. Soon, the cleavage of other eggs was recognized as a form of cell division, and the conclusion that the fertilized egg was a cell became inescapable.

The cellular nature of sperm was also quickly established. In 1841 **Rudolf Albert von Kölliker** (1817–1905) showed that sperm too were produced via cell division and were therefore also cells, but the role of sperm in reproduction was yet to be established.

Fertilization was not easily observed. By 1824, filtered seminal fluid lacking spermatozoa was shown to be sterile, and in 1843, 166 years after Leeuwenhoek's discovery of the animalcules in sperm, Barry saw spermatozoa inside the eggs' envelopes. Finally, a decade later, George Newport (1802–1854) observed the disappearance of spermatozoa into frog eggs and suggested that spermatozoa played a role in fertilization.

Newport immobilized the eggs within a glass cylinder and applied sperm suspensions from the head of a pin. He not only recorded the entry of the sperm but showed that the plane of the first cleavage furrow passed through the point of sperm entry, and the head end of the embryo developed on the opposite surface of the egg. Similar observations on other species of amphibians followed, and the consequences of the egg's fusion with sperm were established.

Still the role of sperm in delivering a pronucleus to an egg was not appreciated. Hints of such a role came first from botanists. A hundred years after Kölreuter studied the effect of reciprocal fertilization, Edward Strasburger (1844–1912) argued that the transmission of characteristics must be a property of the cell nucleus, since the cytoplasm of the ovule was so much greater than that of the pollen tube but had no greater effect on inheritance.

Much of the credit for drawing attention to the nucleus as the seat of heredity goes to the Hertwig brothers, Oscar (1849–1922) and Richard Hertwig (1850–1937), who often worked together especially at their marine station at Roscoff. Richard Hertwig, regarded as the greatest teacher of embryology of his day (Goldschmidt, 1956), discovered the complex nuclear events accompanying conjugation in ciliates. These events seemed to parallel nuclear events during sexual reproduction in multicellular animals but involved the division and transfer of nuclei rather than the fusion of sperm and egg.

In 1875, Oscar Hertwig, among others, showed that one of the two nuclei in fertilized sea urchin eggs was derived from the spermatozoon, the other from the egg. Moreover, the spermatozoon's nucleus, which became the male **pronucleus,** did not disintegrate at fertilization, as was previously thought, but actually fused with the egg's **pronucleus** (at least in sea urchins).

Oscar Hertwig also showed that during the maturation of the sea urchin's egg, unequal divisions gave rise to small polar bodies and a large ovum. From his observation that polar bodies as well as ova contained nuclei, he deduced that nuclear division was important for something other than the mere growth of cells.

The chromosomes also began to make sense as the 19th century moved toward its close.

Cytologists and embryologists, armed with new microscopes and studying a wide variety of invertebrate embryos, discovered regularities in chromosomes that paralleled regularities in development. The thread worm, _Ascaris,_ a nematode and intestinal parasite of humans and domesticated animals, proved especially useful for studies of chromosomes (if unusual in some respects) (see Davidson, 1985).

Nematodes typically have only eight chromosomes, and some species of _Ascaris_ have only four

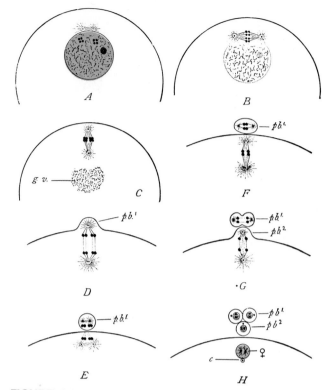

FIGURE 6.25. Diagrammatic summary of chromosomal movements during oogenesis in _Ascaris._ (A) Primary oocyte preparing for first meiotic division. (B) Formation of first metaphase plate. (C) Breakdown of germinal vesicle (g.v.) and movement of spindle perpendicular to oolemma. (D, E) Formation of first polar body (p.b.1) and secondary oocyte. (F, G) Second meiotic division in oocyte and polar body; formation of second polar body (p.b.2) and ootid. (H) Female pronucleus (♀) and centrosome (c). From E. B. Wilson, _The cell in development and inheritance._ Macmillan, London, 1896.

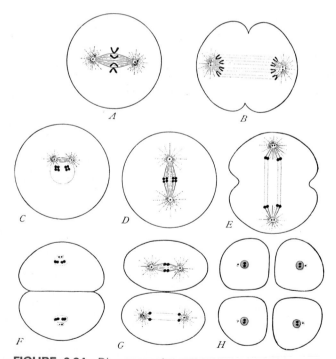

FIGURE 6.24. Diagrammatic summary of chromosomal movements during spermatogenesis in _Ascaris._ (a, b) Spermatogonia divide mitotically. (c–e) Division of primary spermatocyte. (f, g) Division of secondary spermatocytes. (h) Spermatids. From E. B. Wilson, _The cell in development and inheritance._ Macmillan, London, 1896.

or even two chromosomes. The number of chromosomes is the same in germ cells of both sexes. Because fertilization and development occur internally in a narrow duct that does not permit eggs to pass one another, all the events of reproduction and development are followed in sequence down the length of the duct.

Two of the embryologists who capitalized on _Ascaris'_ assets were **Edouard Van Beneden** (1845–1910) and **Theodor Boveri** (1862–1915). They showed that the number of chromosomes decreased by half during the production of sperm (Fig. 6.24) and egg (Fig. 6.25) and doubled during the union of sperm and egg (Fig. 6.26) as each parental gamete contributed equal numbers of chromosomes to the fertilized egg.

Enlarging on earlier suggestions by others (G. Platner and O. Hertwig), Boveri generalized the

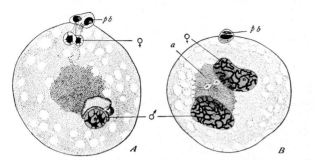

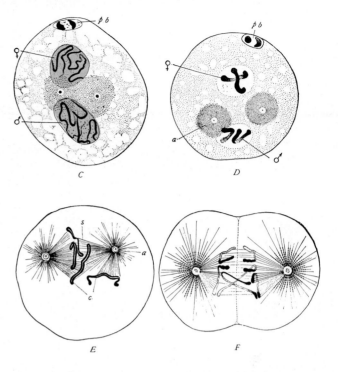

FIGURE 6.26. Boveri's drawings showing syngamy in *Ascaris* and first cleavage division. (A) The egg's nucleus is still small after giving rise to the second polar body (p.b.). (B) Male and female pronuclei swell; centrioles separate in centrosome (a). (C, D) Chromosomes condense during prophase of first cleavage division. Note: Each pronucleus has two chromosomes; the zygote therefore has four. (E) Chromosomes (c) approaching the metaphase plate are double in appearance and attached to centrosomes (a) by spindle fibers (s). (F) Anaphase of first cleavage (only three chromosomes shown). After Th. Boveri from E. B. Wilson, *The cell in development and inheritance.* Macmillan, London, 1896.

chromosomal events of meiosis to include both sperm and egg and thus established the parity of the two gametes in reproduction. Since then, some version of Boveri's illustration (Fig. 6.27) of parallel events in the **gametogenesis** of egg and sperm has been reproduced in virtually every embryology textbook published.

Thus, prior to the 20th century, the concepts of sperm and egg in sexual reproduction had reached their modern maturity. The egg was not a soil waiting for seed, and the spermatozoon did not contain a homunculus. Egg and sperm were not mystical bodies waiting for causes or encapsulating preformed adults in miniature.

Egg and sperm were cells sharing many properties with other cells yet preserving unique properties of their own. As a cell, the egg breached the gulf between viviparous and oviparous species, and as cells, the sperm and egg made their contributions in both heredity and development.

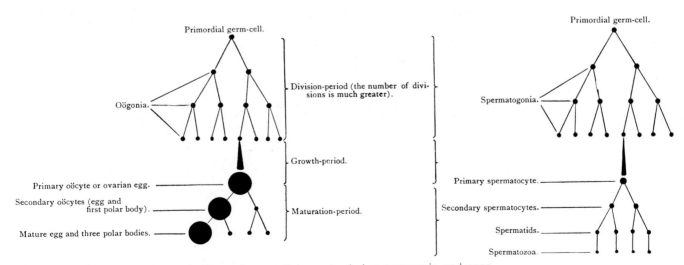

FIGURE 6.27. Boveri's diagrams illustrating parallel events during oogenesis and spermatogenesis. After Th. Boveri from E. B. Wilson, *The cell in development and inheritance.* Macmillan, London, 1896.

GAMETOGENESIS

*A*ccordingly, it is thought that germ cells, in contrast to somatic cells, retain developmental totipotentiality throughout the life of the individual. Nonetheless, it is also true that germ cells become as highly differentiated as somatic cells during ontogenesis.

YOSHIO MASUI AND HUGH J. CLARKE (1979, p. 186)

*G*ametogenesis and differentiation were once thought of as polar opposites. The loss of a cell's genetic potential and plasticity was thought to accompany its differentiation, while the acquisition of a cell's full potential for genetic expression during subsequent development was thought to flow from gametogenesis. **Somatic** or body cells in the adult were differentiated, while gametes were **undifferentiated** and **totipotential.**

These stark contrasts began to fade when the concept of the gene acquired an informational component and when differentiation was linked to differential gene action. Embryologists could then conceive of gamete-producing cells as growing and changing like other differentiating cells. Today, gametes are thought to be specialized for their particular roles like other differentiated cells are specialized for their different roles, and gametogenesis is considered a special example of development and differentiation. The object of research on gametes is now part of the overall enterprise of discovering the controls and mechanisms of development.

Despite their moving into the mainstream, gametes and gametogenesis retain their special significance. In multicellular animals, gametes are the only individual cells that recreate the entire organism, and gametogenesis is uniquely tied to meiosis, restricted to gonads, and controlled largely by humoral agents (i.e., blood-borne messengers) collectively known as hormones. Moreover, gametogenesis in multicellular animals is almost always the province of a discrete line of cellular descent known as the **germ line.** Plants and algae lack a germ line and develop gametophytes not found among multicellular animals. This chapter examines the unique characteristics of gametogenesis in the germ line.

PRINCIPLES OF ANIMAL GAMETOGENESIS

Phases of Gametogenesis

In multicellular animals in general and vertebrates in particular, **primordial germ cells** of the germ line arise outside the gonad and migrate to it early in development. Upon reaching the embryonic gonad the cells are known as **gonocytes** or **primary germ cells.** Gonocytes generally become

TABLE 7.1. The phases of gametogenesis

Phase:	Gonial	Gametogenic		
		-cytes		
Cell type:	-gonia	Primary	Secondary	-tids
Activity:	Mitosis	Meiosis I	Meiosis II	Varied

egg or sperm according to the sex of the gonad in which they develop, although, in some mammals, the presence of the sex-determining chromosome may be incompatible with gonocytes developing into female gametes (see McLaren, 1982).

The "oo-" (Gk. *oion* egg) or, sometimes, "ov-" (Lat. *ovum* egg) identifies germ-line cells in the female gonad or ovary, while the "spermato-" (Gk. *sperma* seed) identifies germ-line cells in the male gonad or testis. Gametogenesis in the ovary, or **oogenesis** (sometimes ovogenesis), leads to the formation of eggs, and gametogenesis in the testis, or **spermatogenesis,** leads to the formation of sperm.

Oogenesis and spermatogenesis involve many activities, from proliferation (i.e., the production of cells) and growth (the accumulation of mass), to differentiation (cytological change) and maturation (the completion of meiosis). Although different for egg and sperm, these activities can be accommodated to general stages of gametogenesis (Table 7.1). Typically, the major stages are a mitotic **gonial phase** and a meiotic **gametogenic phase.**

The gonial phase takes its name from the suffix "-gonia" (Gk. gonos offspring) applied to mitotic cells of both female and male germ lines. Gonocytes of the early embryonic ovary and stem cells of the later ovary are called **oogonia** (sometimes ovogonia pl., **oogonium** or ovogonium sing.), while comparable cells of the testis are called **spermatogonia** (pl., **spermatogonium** sing.).

In the embryonic and differentiating fetal gonad, oogonia or spermatogonia are members of expanding cell populations. Even during early development, however, a vast amount of cell death accompanies growth, and an "expanding" cell population may actually contract.

In the mature gonad, oogonia and spermatogonia are generally proliferative elements in a steady-state population. In adults, especially males, **stem cells** are divisible into two populations. One population consists of **proliferative stem cells** which divide and either reform themselves or give rise to gamete-forming cells. The other population consists of **reserve stem cells** which divide slowly if at all and give rise to pro-

liferative stem cells. Like satellite cells of an intermediate cell population, reserve stem cells may sometimes replace a lost proliferative stem cell population.

Oogonia constitute the proliferative part of a steady-state population in the ovaries of many adult animals, but oogonia are totally absent in the ovaries of adult mammals and birds. The germ-cell population is then static except as germ cells differentiate or degenerate and leave the population. If the reserve of germ-line cells is exhausted prior to the death of the organism, a **postreproductive period** commences during which the organism is sterile. The **postmenopause** in women is such a period.

The retention of gonial cells in adults is correlated with species, sex, and volume of gametes. For example, adult male mammals retain spermatogonia and produce large numbers of sperm, while adult female mammals have no oogonia and produce relatively few eggs. Specifically, a man may produce 50–150 million sperm per day, while a woman may produce only 400–500 eggs in a lifetime.

The gametogenic phase takes over when mitosis is suspended and meiosis has begun. Classically (Fig. 6.27), the onset of cellular growth defines the passage of germ cells from the gonial phase into the gametogenic phase. Today, the transition is widely identified with the initiation of **premeiotic DNA synthesis** (i.e., the S phase prior to meiosis).

In multicellular animals, premeiotic DNA synthesis is a "point of no return," or a **terminal replication.** The major portion of gametic DNA is not synthesized again without fertilization or activation.[1] The G-2 period of interphase prior to meiosis is sometimes known as the **preleptotene phase** and considered part of meiosis.

The suffix "-cyte" is attached to the name of germ cells in every stage of meiosis. **Primary**

---

[1]Claims that spermatids in sharks undergo a mitotic division are unsubstantiated (H. P. Stanley, personal communication).

oocytes (also known as primary ovocytes) and **primary spermatocytes** are engaged in meiosis I. **Secondary oocytes** (or secondary ovocytes) and **secondary spermatocytes** are engaged in meiosis II. The division of a primary spermatocyte yields two secondary spermatocytes, but an unequal division of the primary oocyte yields one secondary oocyte and a diminutive **first polar body** (see Fig. 6.27).

Secondary spermatocytes and oocytes quickly pass through **interkinesis** and enter meiosis II. With the completion of the second meiotic division, the germ cells are identified as "-tids" (Lat. suffix *-ides*, specifying a family). The two cells produced by division of a secondary spermatocyte are called **spermatids.** The unequal division of the secondary oocyte gives rise to a large **ootid** (or ovotid) and a small **second polar body.**

The unequal divisions of oocytes seem well adapted for depositing the bulk of the egg's cytoplasm and deutoplasm in one of the two cells produced by division. Polar bodies may be considered **abortive oocytes,** cells sacrificed to meet the requirements of meiosis with a minimum of loss to the egg.

At the "-tid" stage, meiosis is complete, and the cells are at the haploid or hemizygoid level. Spermatids contain half the amount of prereplication nuclear DNA (i.e., the *C* amount of DNA) and half the number of chromosomes (i.e., the *n* number of chromosomes) compared to somatic cells or spermatogonia. Likewise, the second polar body (and the first, if it undergoes a second meiotic division) is haploid, but the situation in the egg is more complex.

In most species, an egg that reaches the ootid stage has already been fertilized, and the spermatozoon's nucleus is already in residence. While the ootid nucleus can be spoken of as haploid, the fertilized egg cannot. Nor is it diploid. It is a **heterokaryon** (i.e., a cell containing two different nuclei).

The nomenclature of eggs after fertilization has posed a problem ever since meiosis and fertilization were first described. In general, the solution has been to call the female germ cell a **mature ovum** when it is competent for fertilization, a **fertilized egg** after it is fertilized and before it has completed meiosis, and a **zygote** as soon as it contains two haploid nuclei. Mammalian eggs, for example, are mature as secondary oocytes.[2] If fertilized, the second meiotic division is completed as a fertilized egg, and two haploid nuclei appear in the zygote.

[2]Primary oocytes of mammals removed from the ovary are able to fuse with sperm, but they do not undergo cleavage.

Localized Gametogenesis: Cooperation and Integration

In multicellular animals, gametogenesis is probably never **diffuse** in the sense that it can occur anywhere in the organism. Even in sponges, eggs tend to be produced basally, while sperm are produced apically. In coelenterates, such as the familiar *Hydra*, **gonophores** (Gk. *gonos* offspring + *phorein* to bear) occur on the distal half of the animal. Hermaphroditic varieties produce sperm in apical gonophores and eggs in basal ones (Fig. 7.1). In most other multicellular animals, gametogenesis is limited to germ-line cells, which develop exclusively in localized sex organs or gonads, an ovary in the female, a testis in the male.

Germ cells typically exhibit some degree of coordinated development within gonads. Histological sections of testes and ovaries show patches of germ cells in striking if not perfect developmental synchronization. **Intercellular bridges** (Fig. 7.2) connecting germ cells in **syncytial networks** may be responsible for the integration of their activities (see Fawcett, 1972).

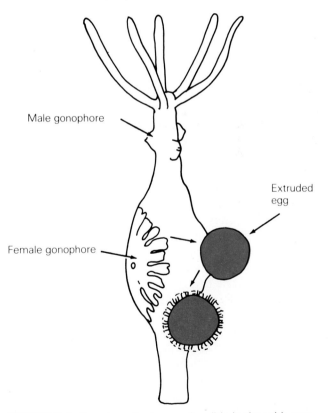

Male gonophore

Extruded egg

Female gonophore

FIGURE 7.1. Drawing of a hermaphroditic hydra with gonophores producing sperm and egg. From R. D. Campbell and H. R. Bode. In H. M. Lenhoff, ed., *Hydra: Research methods,* Plenum Press, New York, 1983. Used by permission.

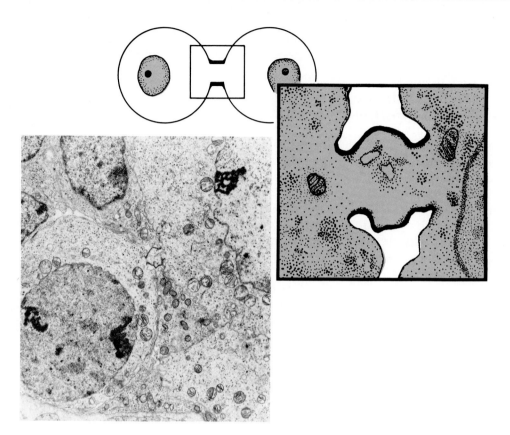

FIGURE 7.2. Diagram of intercellular bridge between germ cells with electron micrograph (×12,000) of section and interpretative drawing. From B. Gondos. In R. E. Jones, ed., *The vertebrate ovary,* Plenum Press, New York, 1978. Used by permission.

Arising during cytokinesis, intercellular bridges are (at least initially) short, cylindrical channels bound by the plasma membrane of the joined cells. The bridges do not appear to be mere passive consequences of incomplete division. On the contrary, they appear to be differentiated structures, since a dense granular material is deposited on the inner bridge membrane, and the channel between cells opens and closes.

Broad enough for the passage of macromolecules and even small cell organelles, the intercellular bridges may provide more central cells with the same sources and sinks available to cells nearer the capillary support network and thus maintain an equality of physiological opportunity among joined cells. The bridges may also provide conduits for the transfer of information coordinating and synchronizing gametogenesis (see Gondos, 1978).

In most vertebrates, the syncytium of connected germ cells forms a compact **nest** (sometimes called a clone). Nests in testes generally have more cells than nests in ovaries. In the female mammal, incomplete gonial cell divisions produce nests not exceeding 8–16 cells, while in the male, nests of 256 or more germ cells may be produced by combinations of six or seven incomplete gonial divisions and two meiotic divisions (see below, Fig. 7.10). In reality, nests of this size are rare, since frequent **degeneration** of cells breaks up nests.

In mammals, nests of male germ cells remain patent until the end of spermatogenesis, and differentiation is relatively synchronized among all the nest cells until the last remnant of cytoplasm linking spermatids is pinched off and finished sperm are released. In contrast, the appearance of primary oocytes in a nest of female germ cells spells the end of their synchronized oogenesis.

Hormones of Sex

Gametogenesis happens in the right place and at the right time. Gametes are produced in the appropriate season and environment, and gamete production in one organism takes place in coordination with gamete production in conspecifics, especially those of the opposite sex. Moreover, in the case of viviparous animals, gametogenesis is linked to the preparation of maternal organs employed in nurturing the developing offspring. Hormones are responsible for the coordination of all these aspects of sexual reproduction.

In many animals, anatomical and behavioral activities advertise stages of gametogenesis and sexual readiness to other animals as part of a strategy that coordinates reproduction. In most female mammals, for example, ovulation is spontaneous, but its occurrence is announced by behavioral changes called **estrus** (or oestrus, Gk. *oistros* gadfly)[3] or **heat,** characterized by receptivity to males and cooperation in mating. Estrus is accompanied by **perineal tumescence** or **sexual swellings** in several Old World primates including macaques, baboons, and chimpanzees, but these swellings are absent in others including gibbons, orangutans, and gorillas, and estrus itself is absent in human beings (see Short, 1984).

The modern study of estrus began with the discovery of C. R. Stockard and G. N. Papanicolaou in 1917 that cyclic changes in the vaginal epithelium of guinea pigs were correlated with cyclical changes in the ovary and uterus. Not only did this discovery make it possible to draw correct inferences about the condition of internal reproductive organs by simple external examination, but the broad correlation of internal and external signs of cellular differentiation suggested that the entire reproductive apparatus operating in female mammals came under the auspices of a single controlling system.

The hormones now known to operate this system and the similar system in males fall into four categories. (1) Lipid hormones, especially sex steroids, are frequently produced by the gonads. (2) Glycoproteins known as **gonadotropins** or **gonadotrophins** are produced in the anterior lobe of the pituitary gland or adenohypophysis (although not exclusively) and activate both ovary and testis. (3) Glycoproteins such as **inhibin** have inhibitory activity. (4) Peptide regulatory hormones such as **gonadotrophin releasing hormone** (GnRH) locally modulate the release of other hormones such as the gonadotropins (see Baird, 1984).

Lipid hormones are widespread among living things and frequently involved in regulating sexual reproduction. Steroids (Gk. *stereos* solid, referring to any solid alcohol related to cholesterol) may be indispensable for gametogenesis. They include **ecdysterones** (or ecdysteroids, Gk. *ekdysis* exit + *stereos*, hence molting steroid) which induce yolk synthesis in insects (see Postlethwait and Giorgi, 1985) and the **sex** or **gonadal steroids** (Lat. *gonad-* genital + *stereos*) of vertebrates. In addition, insect **juvenile hormone,** a nonsteroidal lipid hormone, also plays a role in sexual maturation.

Historically, sex steroids were identified by their activities and presence in sex organs rather

[3]The name refers to swarms of gadflies thought to provoke frenzies of sexual behavior in cows. Not only is the term incorrect in its attribution of cause, but it is erroneous in its equation of frenzy and sexual receptivity. Regrettably, the term is retained, because it is too deeply embedded in common usage to be expunged.

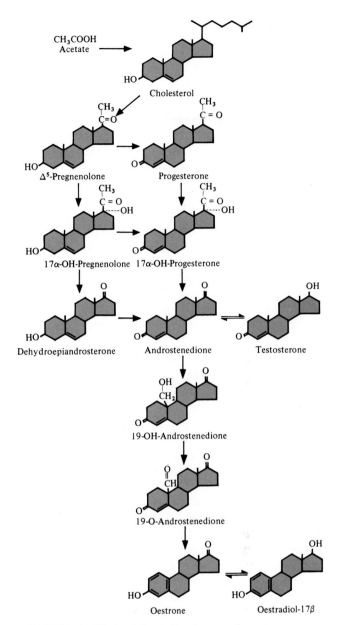

FIGURE 7.3. Skeletal formulas for sex hormones, or gonadal steroids, and other steroids produced in steroidogenesis. From D. T. Baird. In C. R. Austin and R. V. Short, eds., *Reproduction in mammals,* Book 3, *Hormonal control of reproduction,* 2nd ed., Cambridge University Press, Cambridge, 1984. Used by permission.

than by chemical structure. Androgens and estrogens were classes of hormones rather than distinct substances. Today, knowledge of the molecular formulas of specific steroids (Fig. 7.3) with known activities has led to redefining **androgens** as saturated steroids (i.e., having predominantly single carbon to carbon bonds), derived from pregnenolone and progesterone, and **estrogens** as unsaturated steroids (i.e., having double carbon to carbon bonds), derived from androgens.

The **estrogens** (also *oestrogens*, Gk. *oistros* gadfly as in estrus, and *genes*, hence estrus producing), especially **estradiol-17β** (or oestradiol-17β) were named for their ability to provoke estrus and promote the development of female secondary sexual characteristics. Estrogens are present in abundance in female vertebrates, but they are not exclusively female. They are produced by the ovary and placenta of females and, along with other steroids, by other body tissues (i.e., nonglandular tissues, such as bone, brain, fat, and skin, and by the glandular adrenal cortex) of males and females. Furthermore, the testes of the boar and stallion secrete large quantities of estrogen, and, in men, almost half the circulating estrogen is produced by the testes. The rest of the estrogen is derived by the conversion of testosterone to estrogen by body tissues.

Progesterone (Lat. *pro-* for + *gestare* to bear + *-one* signifying a horm*one*), the **hormone of pregnancy,** is essential for the maintenance of pregnancy. It is produced in the ovary, placenta, and adrenal cortex.

The **androgens** (Gk. *andros* man + *genes* producing) were named for their ability to promote the development of male secondary sexual characteristics. **Testosterone** was named in the mistaken belief that it was produced exclusively in the testes. Androgens, including testosterone, are also produced by the ovary, sometimes in greater quantities than estrogens (e.g., in freshwater turtles, domestic chickens, and hamsters). Actually, sex hormones frequently show up in the "wrong" place. For example, cells surrounding a mammalian egg follicle produce androgens!

Many of the physiological activities of androgens depend on their conversion to 5α-dihydrotestosterone (especially in somatic target organs) or to estrogens (especially in the ovary). The integration of hormonally mediated responses is frequently a function of the timely arrival of an androgen and its conversion at the responding tissue to a different steroid with a particular biological activity.

The common biochemical pathways of **steroidogenesis** (Fig. 7.3) facilitates the convertibility of sex hormones in many cells. The pathways begin with cholesterol. Produced from acetate within cells or transported from the blood plasma via low density lipoprotein (LDL), cholesterol is first converted to steroidal intermediates and then to sex hormones. The androgens, androstenedione and testosterone, are produced by the oxidation of pregnenolone to progesterone, hydroxylation, and the removal of the carbon side chain. The most biologically active estrogen, estradiol-17β, is derived from androgens through a series of enzymatic conversions collectively called **aromatization,** which lead to the formation of a benzene ring and hence an unsaturated heterocyclic or **aromatic** compound.

The gonadotropins (GNs) (*gonad-* + Gk. *tropos* turning toward, hence hormones that localize in or primarily affect the gonads, or GTHs, *gonad-* + Gk. *trophikos* nursing, hence hormones that feed the gonads) produced by the adenohypophysis (i.e., **adenohypophyseal hormones**) include **follicle stimulating hormone** (FSH) and **luteinizing hormone** (LH). FSH in females seems to be identical to FSH in males, and LH in females seems to be identical to LH in males, although it is frequently called **interstitial cell stimulating hormone** (ICSH) in males.

FSH and LH are glycoproteins with molecular weights of 28,000 and 35,000. They are heterodimeric (i.e., containing two noncovalently linked different polypeptides): an alpha chain which is identical in both FSH and LH (and in thyrotrophic hormone [TSH], also produced by the adenohypophysis) and a beta chain which is different in the hormones. FSH and LH are thought to be synthesized by the same adenohypophyseal cells known as **gonadotrophs.**

Gonadotropins are similar in a variety of vertebrates, and their alpha and beta chains contain similar repeated subunits. The activities of gonadotropins in different species may not be the same, however. A gonadotropin with a molecular weight resembling LH in one species, for instance, may have physiological activities more nearly resembling those of a gonadotropin with the molecular weight of FSH in another species.

This reversal of activity and molecular size is presumably related to the biological activity of the beta chain. In the course of evolution, recombination of alpha chains with beta chains of different lengths may have changed the molecular weight of one gonadotropin to something more nearly resembling that of the other gonadotropin. Alternatively, the present vertebrate system of two gonadotropins may have resulted from evolutionary convergence and selection for a minimum number of gonadotropins from what was originally a much larger number of gonadotropins.

Another adenohypophyseal hormone, **prolactin** (or lactogenic hormone [formerly called luteotropin]), may also be considered a gonadotropin, since it has activity in both the testis and ovary.

As its name suggests, prolactin also affects the mammary glands. Prolactin consists of a single polypeptide and is produced by **lactotrophs** in the adenohypophysis.

In addition, the placenta and possibly the embryo itself produces other gonadotropins. These include **human chorionic gonadotropin** (hCG), which regulates progesterone production in the corpus luteum and placenta, and **pregnant mare serum gonadotropin** (PMSG), which has actions similar to LH. **Placental lactogen** has activities comparable to those of prolactin and pituitary growth hormone.

Inhibin is a heterodimeric glycoprotein found in both testicular and follicular fluids of several mammals and in the placenta. Circulating inhibin appears to selectively inhibit FSH release from the pituitary and both hCG and GnRH from the placenta. The hCG hormone seems to stimulate inhibin release from the placenta via a cyclic adenosine monophosphate-mediated mechanism (Petraglia et al., 1987).

Gonadotrophin releasing hormone (GnRH) is produced in the hypothalamus (Fig. 7.4) and in the placenta. It is a decapeptide and presumably a single hormone, since the same preparations that cause the release of FSH also cause the release of LH. GnRH in mammals is sometimes known as **luteinizing hormone releasing hormone** (LH-RH).

In the placenta, GnRH may operate as a local modulator of hCG release. Permitting the release of hCG, GnRH seems to operate as an antagonist to the local action of inhibin (i.e., it operates in a paracrine role).

Another hormone, the **prolactin inhibiting factor** (PIF), thought to be the monoamine neurotransmitter **dopamine,** is also produced in the hypothalamus. PIF regulates prolactin release at the same time GnRH stimulates gonadotropin release.

The rate of GnRH production is not uniform. A **pulse generator** in the hypothalamus causes a **pulsatile** or **rhythmic quantal discharge** of GnRH secretion, and thus a rhythmic signal is conveyed to the pituitary. As a result, LH secretion (if not FSH) is also pulsatile (Fig. 7.5) (see Nobil, 1980). The rhythmic secretion of GnRH (like a peristaltic pump) seems essential for its normal ability to control LH release, and changes in the pulse generator mediated by hormones are responsible for puberty and, in different animals, breeding seasons and responses to sexual stimuli.

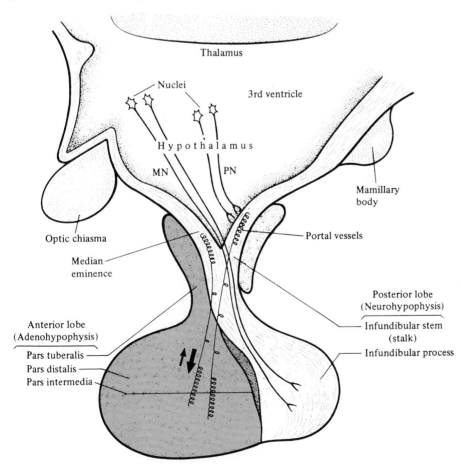

FIGURE 7.4. Hormonal transport between the hypothalamus and the anterior lobe (adenohypophysis) of the pituitary via portal vessels (corkscrews) of the hypothalamic–hypophyseal portal system. While the main route of transport (large arrow) delivers gonadotrophin releasing hormone synthesized by neuroendocrine cells in nuclei in the base of the brain, gonadotropins produced by gonadotrophs in the adenohypophysis also move back through the infundibular stem, or stalk of the pituitary, to the hypothalamus (small arrow). PN, parvicellular endocrine neurons; MN, magnocellular endocrine neurons. From F. J. Karsch. In C. R. Austin and R. V. Short, eds., *Reproduction in mammals,* Book 3, *Hormonal control of reproduction,* 2nd ed., Cambridge University Press, Cambridge, 1984. Used by permission of the author.

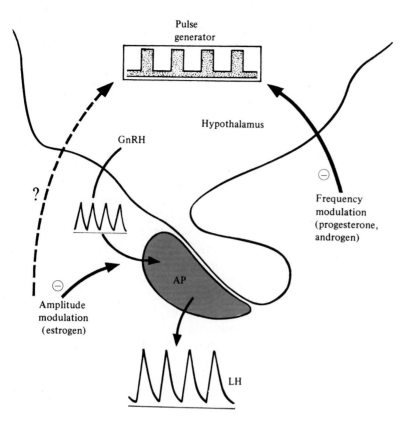

FIGURE 7.5. Interactions influencing the pulsatile outputs of gonadotrophin releasing hormone (GnRH) and luteinizing hormone (LH). Progesterone and androgens operating on nerves in the hypothalamus reduce (−) the frequency of GnRH pulses, while estrogen operating on the adenonypophysis (AP) reduces (−) sensitivity to GnRH and thus the amplitude of LH pulses. The possibility of an estrogen-mediated response in the hypothalamus is uncertain. From F. J. Karsch. In C. R. Austin and R. V. Short, eds., *Reproduction in mammals,* Book 3, *Hormonal control of reproduction,* 2nd ed., Cambridge University Press, Cambridge, 1984. Used by permission of the author.

GnRH is produced by **neuroendocrine cells** (also known as neurosecretory or neural humoral cells) in different parts of the hypothalamus: the medial basal region known as the **arcuate nucleus** above the median eminence, the anterior hypothalamus, and the preoptic area above the optic chiasma. These cells deliver their products to the **third ventricle** (i.e., the cavity of the brain surrounded by the thalamus) and to the **median eminence** via extended **neuroendocrine cell processes** (Fig. 7.4).

GnRH reaches the adenohypophysis through the **hypothalamic–pituitary portal system** (corkscrews in figure), an especially efficient venal system conveying hormones over short distances. Capillaries in the hypothalamus are connected by **hypothalamic–pituitary portal veins** to sinuses in the adenohypophysis. As a result, GnRH is delivered to the adenohypophysis from neuroendocrine cell endings in the median eminence or from the third ventricle. In addition, gonadotropins are also delivered to the hypothalamus by a backwash through the same portal system.

The control of gonadotropins by GnRH is remarkably similar throughout the vertebrates. With the exception of goldfish which have an inhibitor of GnRH and some squamate reptiles (snakes and lizards) which have only one gonadotropin (Licht, 1983), sex hormones are linked to *two* gonadotropins and to *one* GnRH in every major group of vertebrates. In fact, GnRH is found to be very nearly the same molecule wherever it has been isolated. Differences, where they occur, are in the seventh and eighth amino acids of the ten present (Peter, 1983).

Regulation of Hormones

In general, hormones control hormonal secretion. Enormous strides in understanding the relationship of the hormones to each other were made during the 1930s and 1940s (sometimes known as the "age of hormones"), but much of the present understanding of hormonal interactions comes from the 1970s when new methods, especially radioimmunoassays (RIAS), made it possible to measure hormone levels in circulation and detect changes in hormone levels over short intervals.

Today, it seems that two types of control influence hormone levels. First, each hormone regulates its own secretion through **homeostatic feedback loops.** Most endocrine systems are kept at a **tonic** or average **basal** level by **negative** or **inhibi-**

tory feedback loops through which the elevation in the amount of a hormone causes a decrease in its secretion. Second, massive cyclic **surges** or bursts (see below, Fig. 7.34) result from **positive** feedback loops through which a rise in the amount of a hormone (e.g., estrogen) causes a further increase in secretion (the LH surge).

Different lengths of feedback loops operate in hormonal regulation (see Karsch, 1984). GnRH affects the neuroendocrine cells that produce it through an **ultrashort** negative feedback loop localized in the hypothalamus. In addition, pituitary gonadotropins influence their own secretion by **short loop** negative feedback or backwash loops through the pituitary stalk to the neuroendocrine cells of the hypothalamus (Fig. 7.5).

Finally, **long loop** feedback from the gonad through the general circulation influences the level of gonadotropin secretion. Gonadal steroids may affect adenohypophyseal cells, the pulse generator, or neuroendocrine secretory cells. Low levels of estrogen, for example, reduce sensitivity of gonadotrophs to GnRH, thereby reducing LH secretion. Testosterone and progesterone inhibit the frequency of GnRH pulses by acting at neural sites in the hypothalamus. High and prolonged levels of estrogen, on the other hand, cause GnRH release and the LH surge.

Gonadal Activity of Hormones

In general, gonadotropins induce or prepare the gonad for the action of sex steroids and sex steroids prepare the gonad for the action of gonadotropins. Sex steroids also activate or trigger maturation, however. Accordingly, the direct action of steroidal hormones in gonads is faster than the action of gonadotropins. Injected progesterone, for example, can induce ovulation in the frog within hours, while injected gonadotropins take days to have the same effect (see Smith, 1975).

FSH and LH (but not prolactin) affect transcription through fluctuations in the cellular content of cyclic adenosine monophosphate (cAMP, see Fig. 4.14). Reacting with specific receptors in the cell membrane, the gonadotropins activate adenylate cyclase, mediating the conversion of adenosine triphosphate (ATP) to cAMP. In turn, cAMP activates protein kinases, and information cascades through the cell to regulatory proteins capable of interacting with specific gene promoters. FSH, for example, induces the synthesis of LH receptors and enzymes mediating steroidogenesis.

The cellular action of steroids also leads to the induction of specific cellular receptors. For example, estrogen too induces the synthesis of LH receptors. The prerequisite for enzyme induction by steroidal hormones is the presence of specific steroidal receptors. A steroidal–receptor complex in the nucleus, it would seem, binds with the cis-regulatory elements of a steroid-responsive gene and thereby promotes specific gene expression.

SPERMATOGENESIS

The picture of gamete development which the general biologist carries in his mind is the result of accidents of his own education superposed on the vagaries, rather than the systematic progress, of the history of research, and characteristically depends largely on the selection of prototypes of which the best known is that of the mammal.

E. C. Roosen-Runge (1977, p. 1)

Interest in spermatogenesis in mammals is understandable. Not only are human beings most likely to learn about themselves from mammalian studies, but our most prized domesticated animals are also mammals, and research on storage of mammalian sperm has practical implication. The ability to freeze and store mammalian sperm is responsible for the expansion of the artificial-insemination industry, and *in vitro* fertilization has become practical, in part, as a result of improved techniques for culturing sperm. Still, the modified sperm of mammals should not be used as a pad for launching speculations about the sperm of other animals.

Structure of the Vertebrate Testis

Vertebrate testes contain **lobular compartments** where spermatogenesis takes place. In fish and amphibians, the lobules contain **cysts** (or spermatocysts, sometimes called follicles) where germ cells undergo spermatogenesis in synchronized **nests** (sometimes called clones).[4] The lobules of reptiles, birds, and mammals contain **seminiferous tubules** (Fig. 7.6) rather than cysts, and spermatogenesis is synchronized in contiguous nests (see Roosen-Runge, 1977).

---

[4]Nests should not be called clones, since nests include the products of meiosis as well as mitosis.

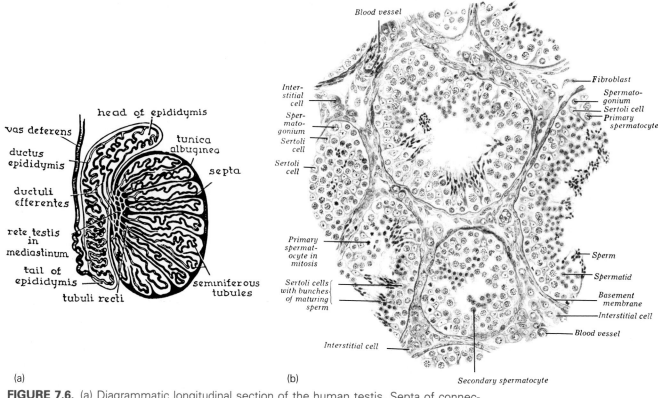

(a)   (b)

FIGURE 7.6. (a) Diagrammatic longitudinal section of the human testis. Septa of connective tissue actually separate 200–300 lobules containing one to four convoluted seminiferous tubules. (b) Semidiagrammatic cross sections of several human seminifierous tubules with an average diameter of 150–200 μm. ×170. Diagram *a* from W. J. Hamilton, *Textbook of human anatomy,* Macmillan, New York, 1957. Used by permission. Diagram *b* from D. W. Fawcett, *Textbook of histology,* W. B. Saunders, New York, 1975. Used by permission.

The walls of testicular cysts and tubules are composed of large supporting **sustentacular cells,** often called Sertoli cells (after Enrico Sertoli, 1842–1910, who first described them), arranged as a simple **seminiferous epithelium** extending from a **basal lamina** to a central **lumen.**[5] In addition germ cells in all stages of spermatogenesis are found among the sustentacular cells (Fig. 7.6b).

The sustentacular cells are tightly bound to each other by **ectoplasmic specializations** or **junctional complexes** containing as many as 50 rows of **tight junctions** accompanied by staggered bundles of microfilaments. Flattened cisternae of the en-

doplasmic reticulum with ribosomes on their cytoplasmic sides lie adjacent to the junctional complexes.

Additional junctional complexes develop between spermatids and sustentacular cells and existing complexes expand. The complexes lack bands of microfilaments on the spermatid side, but otherwise they resemble junctions binding sustentacular cells to each other.

The junctional complexes between processes of sustentacular cells create a **blood–testis barrier** and divide the seminiferous tubule into two major **compartments** (Fig. 7.7; double lines, Fig. 7.8). A **basal compartment** at equilibrium with blood serum lies between the basal lamina and the complexes. An **adluminal compartment** (i.e., toward the lumen, also called an intercellular compartment), at equilibrium with **seminiferous fluid** but not with blood, lies apical to the junctional complexes.

---

[5]Histologists have traditionally classified the wall of the seminiferous tubule as a **stratified epithelium** consisting of sustentacular cells and germ cells (see Ross and Reith, 1985). While the presence of junctional specializations between sustentacular and germ-line cells justifies this tradition, the different origins of the cell types argue against it.

lumen

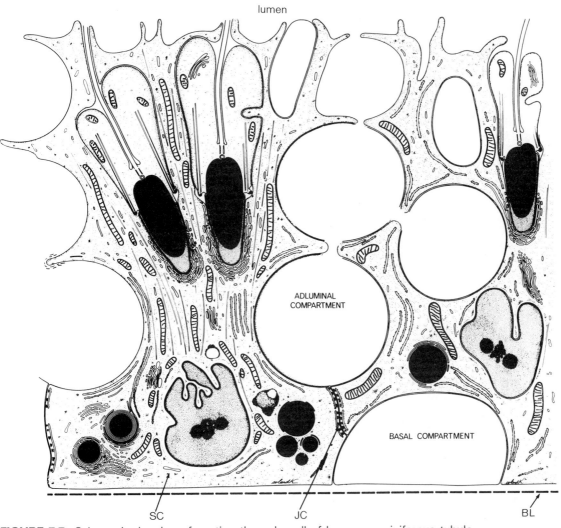

ADLUMINAL
COMPARTMENT

BASAL COMPARTMENT

SC   JC   BL

FIGURE 7.7. Schematic drawing of section through wall of human seminiferous tubule. Sustentacular cells (SC) with their large irregularly shaped nuclei connected to each other through junctional complexes (JC) form a simple epithelium stretching from the basal lamina (BL) to the lumen of the tubule. From D. W. Fawcett, *Handbook of physiology,* Williams and Wilkins, 1975. Used by permission of the American Physiological Society.

The blood–testis barrier appears at about the time of puberty and creates an **immunologically privileged site,** where spermatocytes are shielded from antibodies and other blood-borne proteins, and where spermatogenesis can occur in an environment regulated by the seminiferous fluid. Functioning as a secretory gland, the tubule produces a variety of unique proteins that accumulate in seminiferous fluid and control sperm functions (Setchell, 1982).

Cells in the spermatogonial and spermatogenic phases of spermatogenesis (Table 7.2) are restricted to one or the other compartment. The spermatogonial phase of spermatogenesis occurs in the basal compartment, and the spermatogenic phase occurs in the adluminal compartment. Spermatogonia are cradled among the sustentacular cells, while spermatocytes and spermatids are buried in indentations in the sustentacular cells or in apical **crypts** at the luminal surface.

The transition between the two phases and movement between the two compartments occur as spermatogonia connected by intercellular bridges undergo synchronous premeiotic DNA

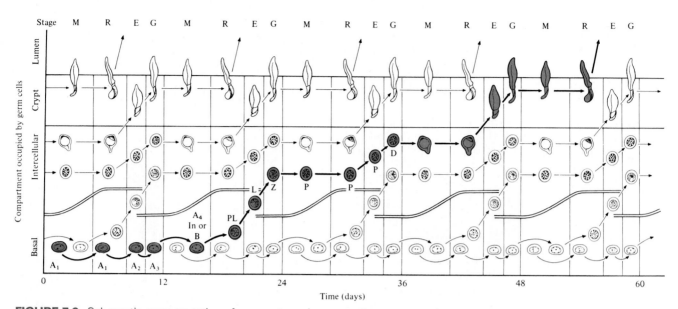

FIGURE 7.8. Schematic representation of spermatogenic waves. Arrows trace the course of spermatogonia and spermatogenic cells in different waves. While spermatogonia (A_1–A_4, In, or B) remain in the basal compartment, spermatogenic cells move through a theoretical intermediate compartment (double lines) into the adluminal or intercellular compartment and apical crypts of the sustentacular cells. Sperm are finally released into the lumen of the seminiferous tubule. Stages or cell associations (correspond to distinctly different groups of cells juxtaposed in cross sections of rodent seminiferous tubules). *E (elongation):* The acrosomes and flagella of spermatids undergo their most conspicuous elongation, while pachytene (P) spermatocytes continue to grow, and the emergence of chromosomes signals the entry of preleptotene spermatocytes into the leptotene (L) subphase. *G (grouping):* As advanced spermatids group at the luminal surface, zygotene (Z) chromosomes appear in some spermatocytes, while pachytene chromosomes in other spermatocytes move up to the diplotene (D) subphase. *M (maturation):* Late spermatids undergo their final maturation, while early spermatids begin forming acrosomes and flagella, and the zygotene chromosomes of some spermatocytes condense into pachytene (P) chromosomes. *R (release):* At the same time sperm are released at the luminal surface, preleptotene spermatocytes (PL) move out of the basal layer of spermatogonia; pachytene chromosomes in primary spermatocytes (P) continue to grow, and the acrosomes and flagella of spermatids develop. From B. P. Setchell. In C. R. Austin and R. V. Short, eds., *Reproduction in mammals,* Book 1, *Germ cells and fertilization,* 2nd ed., Cambridge University Press, Cambridge, 1982. Used by permission.

synthesis. These preleptotene primary spermatocytes move through a theoretical **intermediate compartment** between the basal and adluminal compartments. Within the adluminal compartment, crowded spermatocytes give the appearance

of a continuous layer sometimes called the **germinal epithelium** (see Roosen-Runge, 1977).

Adjacent nests of germ cells frequently initiate spermatogenesis in progressive **spermatogenic waves** (Fig. 7.8). The interval between waves (12

TABLE 7.2. The phases of spermatogenesis

Phase:	Spermatogonial	Spermatogenic			
Stage:		Spermatocyte		Spermiogenic	
Substage:				Spermatid	Spermiation
Cell type:	Spermatogonia	Spermatocytes		Early	Late
		Primary	Secondary	Spermatid	Spermatid
Activity:	Mitosis	Meiosis I	Meiosis II	Differentiation	Separation

days in rats) is proportional to the duration of the **spermatogenic cycle,** the interval between cells leaving the basal compartment and maturing (48 days).

Coupled to relative constancy in the duration of spermatogenesis, spermatogenic waves create a pattern of overlapping stages. Cells in particular stages of spermatogenesis tend to appear together as **cell associations** (also known as stages of maturation) in histological cross sections of tubules. Each cell association is a synchronic slice of history. Put together with other sections, the diachronic flow of history is reconstructed (i.e., E–R, Fig. 7.8) (Setchell, 1982).

In human beings, differentiation in one nest is not readily correlated with differentiation in nearby nests. Cell associations are either too small, their boundaries too interdigitated, or spermatogenic waves do not occur.

Spermatogonial Phase

Mammalian gonocytes or primary germ cells in the fetal male gonad do not divide as prodigiously in the fetal female gonad. Instead, isolated or in small nests (Fig. 7.9), spermatogonia and early spermatocytes enter a **resting stage** from which they do not emerge until puberty. Then, spermatogonia form a large population of **proliferative stem cells** that generate a continuous supply of spermatogenic cells. Spermatogonia also become isolated **reserve stem cells,** sometimes called type A_0 spermatogonia (Clermont, 1962), since they do not move through the cell cycle or they do so only randomly.

Conjoined in **spermatogonial nests** by intercellular bridges (Fig. 7.9), vertebrate spermatogonia divide synchronously, generally through five to six divisions but varying between three divisions in men to 12–13 divisions in some teleosts and sharks. The duration of each cell cycle is about the same, but the S period of DNA synthesis becomes shorter, while the G-2 period, or premitotic gap, becomes longer.

In *Xenopus laevis,* reserve and proliferative spermatogonia are dubbed primary and secondary, respectively, and may be distinguished by the presence of multiple nucleoli in the former (Kalt, 1976). Six spermatogonial divisions usually take place.

In men, reserve spermatogonia (about 12 μm in diameter) called **Ad cells** (d for their dark appearance in typical histological preparations) have dense nuclei and conspicuous nuclear vacuoles. Proliferating spermatogonia called **Ap cells** (p for pale) with lightly staining nuclei ultimately give

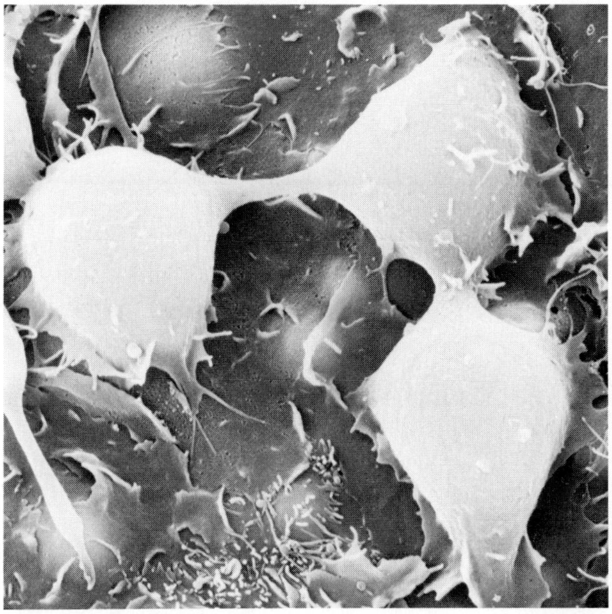

FIGURE 7.9. Scanning electron micrograph of small nest of stellate spermatogonia from explant of 10-day-old rat (i.e., about 1 week before puberty) after 6 days in culture. The spermatogonia are connected by intercellular bridges and attached to a substratum of epithelium by broad, flat pseudopods called lamellapodia. ×2700. From E. M. Eddy and A. I. Kahri, *Anat. Rec,.* 185:333 (1976), by permission of the American Association of Anatomists and Alan R. Liss, Inc.

rise to primary spermatocytes. Spermatogonial proliferation leading from individual Ap cells to spermatocytes takes 58–64 days.

The population of proliferating spermatogonia in mammals consists of several subpopulations. In rodents, such as the rat, **A-type** spermatogonia undergo an average of four successive divisions (A_1–A_4, Figs. 7.8 and 7.10) to become **intermediate** spermatogonia (In). They divide again to become B-type spermatogonia identified by clumped chromatin at the periphery of their nuclei and by large central nucleoli. Division of B-type cells produces two preleptotene primary spermatocytes (PL). The transformation from spermatogonium to spermatocyte takes about 35 days.

The decision to move up the path of spermatozoon differentiation rather than remain in the proliferative stem cell population is made when spermatogonia retain an intercellular bridge after an incomplete division. These A_1 cells are attached by intercellular bridges, whereas stem cells are single.

The last mitotic divisions of spermatogonia leave the cells attached, and at the end of spermatogonial divisions, thick-walled intercellular

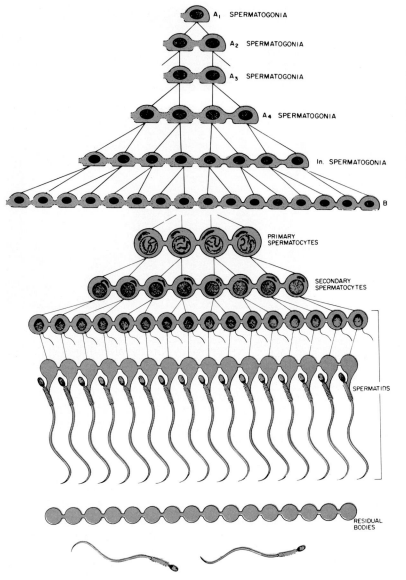

FIGURE 7.10. Schema for proliferation of spermatogonia and association of mammalian spermatogenic cells in nests. A_1 may be compared to dark spermatogonia in human beings and A_2–A_4 to pale spermatogonia. The division of intermediate (In) spermatogonia yields B-type spermatogonia which divide one last time to produce primary spermatocytes. From M. Dym and D. W. Fawcett. In W. Bloom and D. W. Fawcett, eds., _A textbook of histology,_ 10th ed., Saunders, Philadelphia, 1975. Used by permission.

bridges choked with midbodies and remnants of mitotic spindles connect incipient spermatocytes. **Membranous septa** replace the midbodies and temporarily seal off one cell from another. When the bridges are again open, they are intercellular conduits or channels containing typical cytoplasmic elements (arrows in Fig. 7.11) (see Fawcett, 1972).

Spermatogenic Phase

The remainder of spermatogenesis is divided into two stages by the completion of meiosis. In the spermatocyte stage, cells undergo meiosis, while in the spermiogenic stage, spermatids differentiate and are released as spermatozoa from their attachment to sustentacular cells.

**The spermatocyte stage begins when the series of mitotic divisions ends and preleptotene spermatocytes enter premeiotic interphase accompanied by premeiotic DNA synthesis.** Although this S period is longer than previous ones, DNA synthesis is incomplete. The synthesis of DNA in the sex-determining chromosomes (i.e., the X and Y chromosomes in mammals) lags behind that in the autosomes (i.e., all the other chromosomes). Furthermore, a small amount of additional synthesis does not take place until zygotene and pachytene, and, even then, the centromeric DNA remains unreplicated (see Chapter 3).

With the help of sustentacular cells to which they are bound by **desmosomes,** nests of early spermatocytes journey from the basal to the adluminal

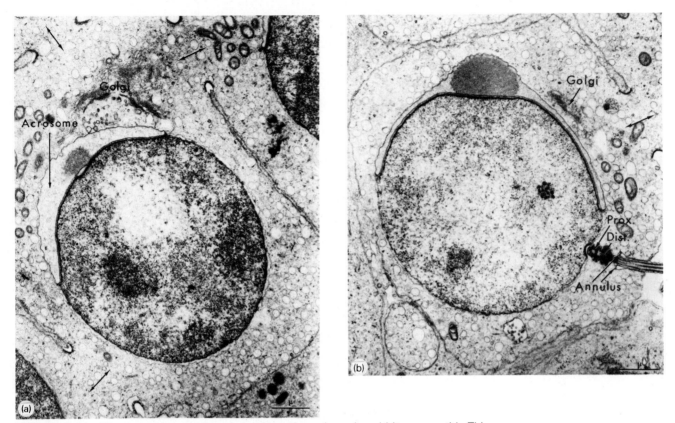

FIGURE 7.11. Electron micrographs of sections through early rabbit spermatid. Thin extensions of sustentacular cells separate spermatids except where they are joined by intercellular bridges (double arrows). (a) The acrosome develops in association with the Golgi apparatus. The nuclear envelop in the vicinity of the acrosome has thickened. (b) Thickened precursor of perforatorium spreads as acrosome covers more of nucleus. The plasma membrane covering the tail doubles back at the annulus. The distal (Dist.) centriole is the basal body for the axonemal fibers emerging in the tail, and the capitulum of the proximal centriole (Prox.) settles into the indented implantation fossa of the nucleus, thickened by a basal plate. From D. M. Phillips, *Spermiogenesis,* Academic Press, Orlando, 1974. Used by permission.

compartment. **Preleptotene** spermatocytes (PL, Fig. 7.8) resemble B-type spermatogonia cytologically even after premeiotic DNA synthesis, although they occupy adluminal sites in the tubule. Soon, with the condensation of chromosomes and an increase in the amount of cytoplasm, spermatocytes are distinguishable from their predecessors.

The chromosomes of primary spermatocytes move rapidly through the gymnastics of meiotic prophase I (see Fig. 3.9). Pachytene is the longest of the subphases (lasting 16 days in men) and the subphase in which spermatocytes reach their greatest size (16 μm in diameter in men). RNA and protein synthesis also peak, but, oddly, many spermatocytes degenerate at the same time.

At diplotene, **lampbrush chromosomes** can sometimes be distinguished (e.g., the Y chromo-

some of *Drosophila hydei* has lampbrush loops), but these do not obtain the dimensions found in oocytes (see Fig. 3.13). The nucleus swells and reaches its maximal size. Sex-determining chromosomes form unusual spherical **heterochromatic** bodies (i.e., bodies that stain darkly with basic nuclear stains) that fail to perform RNA synthesis, while autosomes synthesize both ribosomal and stable heterogeneous RNA transcripts, most of which remain in the nucleus associated with chromosomes.

Protein synthesis during prophase is responsible for the production of a variety of proteins such as those of the acrosome, mitochondria, and the plasma membrane (but not nucleoproteins associated with nuclear condensation). Among the many enzymes that appear for the first time in primary spermatocytes are hexokinases and lactate dehydrogenase which will later be utilized by the spermatozoa.

As spermatocytes continue to move toward the lumen of the seminiferous tubule, desmosomes connecting them to sustentacular cells become more conspicuous, and gap junctions may appear if only temporarily. Soon (only 24 days after leaving the proliferative population in men), the chromosomes undergo diakinesis, and the cell enters the first meiotic metaphase. RNA and protein synthesis cease and will not resume again until the spermatid stage when the cell is haploid. The first meiotic division follows, and two small secondary spermatocytes (about 9 μm in diameter in men) are formed. The second meiotic division follows soon after (only 8 hours in men), giving rise to spermatids.

The spermiogenic stage begins with the end of meiosis. **Spermiogenesis,** or spermateliosis, is broken into two substages (Table 7.2): a **spermatid**

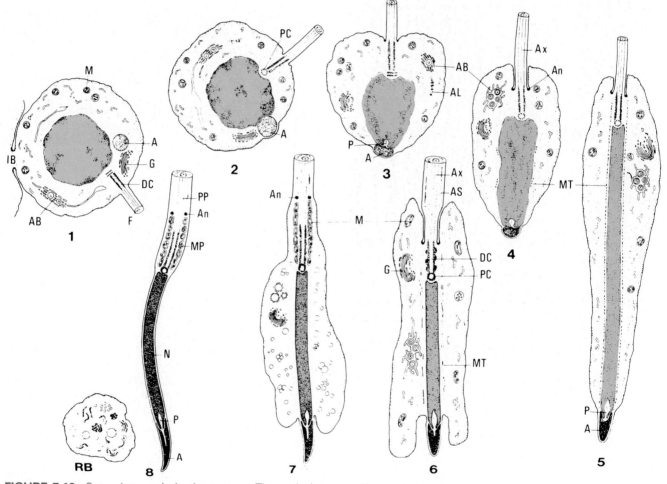

FIGURE 7.12. Spermiogenesis in the rooster. The early (steps 1–3) rounded nucleus is 4.5 μm in diameter. The elongating (step 4) nucleus is 7.5 μm in length, and the maximally elongated cylindrical nucleus (step 5) is 21 μm long and 0.8 μm in diameter. As chromatin condenses, the nucleus shortens to its definitive dimensions of 10 μm long by 0.6 μm wide. A, acrosomic granule; AB, alveolar body; AL, annulate lamellae; An, annulus; Ax, axoneme; AS, amorphous cytoplasmic sheath along axoneme; DC, PC, distal and proximal centrioles; ER, endoplasmic reticulum; F, flagellum; G, Golgi apparatus and residual Golgi apparatus; IB, intercellular bridge; M, spherical and elongating mitochondria; MP, middle piece (midpiece) with mitochondrial sheath (4 μm long); MT, microtubules and manchette; N, nucleus; P, perforatorium; PP, principal piece of tail; RB, residual body. From L. Xia, Y. Clermont, M. Lalli, and R. B. Buckland, *Am. J. Anat.,* 177:301 (1986), by permission of Alan R. Liss, Inc., and the authors.

substage broadly associated with differentiation and a **spermiation** substage associated with the release of sperm.

Differentiation during the **spermatid substage** invites classification into subphases and steps (e.g., as many as 19 steps in the rat but more generally 8, e.g., the rooster, Fig. 7.12). In the simplest classification, based on morphology, **round** or **early spermatids** (steps 1–3) are separated from **elongating** or **midstage** (steps 4–5) and **condensing** or **late spermatids** (steps 6–8). Round and early elongating spermatids move toward the lumen. Midstage spermatids reach the luminal surface while buried in crypts of sustentacular cells. Late spermatids remain in the crypts until completely condensed.

The earliest round spermatid is a small cell (6 μm in diameter in men) with a relatively large and diffuse nucleus. Its cytoplasm contains all the usual intracellular organelles including tubular elements of rough surface endoplasmic reticulum.

The spermatid's cytoplasm also contains, or will soon contain, a variety of smooth surface membranous bodies more or less unique to germline cytoplasm and early blastomeres. The most conspicuous bodies are the **annulate lamellae** (described below in oocytes, Fig. 7.26). In addition, spermatids form **alveolar bodies,** tubular networks among parallel stacks of endoplasmic reticulum cisternae, and **multivesicular bodies,** resembling vesiculated mitochondria. All these bodies disappear during spermatid elongation or prior to spermiation (see Fig. 7.12 and 7.13b).

The histochemical test for polysaccharide, known as the periodic acid Schiff (PAS) test, reveals the presence of **PAS-positive material** in Golgi apparatus granules. Membrane-bound vesicles containing the PAS-positive granules pinch off the Golgi apparatus and coalesce near the nucleus to form an **acrosomal vesicle** containing a conspicuous **acrosomal granule** (Fig. 7.11a). Applied to one end of the nucleus, the vesicle enlarges and stretches to form the double-membrane **acrosome** or acrosomal sheath (Fig. 7.11b) (Phillips, 1974).

The inner acrosomal membrane forms a tight association with the nuclear envelope. In mammals, a proteinaceous material begins to accumulate at this time between the two membranes, fusing with them to form the **perforatorium.** In birds, the anlage of the perforatorium appears in a cavity at the top of the nucleus (Fig. 7.12).

The microtubular components of the axonemal complex are soon synthesized for the first time, and the spermatozoon's **flagellum** begins to form at the periphery of the round spermatid. The **distal centriole** (Dist. in Fig. 7.11b) serves, at least

initially, as the **basal body** or center of microtubular condensation. Microtubular (and other) components are added; the complex of microtubular doublets or axonemal fibers takes shape, and the tail's axoneme emerges.

As the axoneme elongates posteriorly, the centrioles migrate inwardly. The **proximal centriole** (Prox. in Fig. 7–11b), capped by the fibrillar **capitulum,** fits into the indented subnuclear **implantation fossa** lined by a **basal plate.**

The striated **connecting piece** then forms in association with the distal centriole (Fig. 7.13a), infiltrating it, in mammals, to the point of obliterating it. In insect sperm, both centrioles may be disrupted and overtaken by the enlarged connecting piece (Phillips, 1974).

Additional fibrillar material may give rise to a transient extension of the proximal centriole called the **centriolar adjunct.** Together with the connecting piece, the centriolar adjunct seems to cement the tail to the nucleus. Later, the plasma membrane at the proximal end of the neck fuses with the nuclear envelope of the head along the indented **posterior ring** (Fawcett, 1972).

Cytoplasm moves tailward, and the outer layer of the acrosomal sheath juxtaposes the spermatid's plasma membrane. A thick cuff of cytoplasm hangs over the developing tail from a ring-shaped thickening known as the **annulus** (Fig. 7.13a). Later, as excess cytoplasm is reduced to a **cytoplasmic droplet** on the midpiece, the cytoplasmic cuff is scaled down to the annulus alone lying above a shallow **retroannular recess.**

The mammalian midpiece is fashioned by the caudal movement of the annulus. Beginning behind the distal centriole, the annulus moves away from the nucleus as mitochondria converge on the mitochondrial sheath (Fig. 7.14).

In early spermatids of insects, some snails, and ostracods, mitochondria condense to form a **nebenkern** (Fig. 7.15). This knotted ball of mitochondria eventually unravels into two large mitochondria that elongate along the tail behind the annulus or ring centriole.

Following the appearance of the axoneme, the dense outer fibers of the mammalian axonemal core emerge as ridges running along each of the axonemal fibers. As the ridges thicken, they lose their attachments to the axonemal fibers. Later, the fibrous sheath develops from a **tubular complex** pushed ahead of the migrating annulus.

Nuclear elongation brings the nucleus to its maximal size before nuclear condensation reduces the nucleus to its final state. During the condensation of nuclei in mammalian spermatids, part of

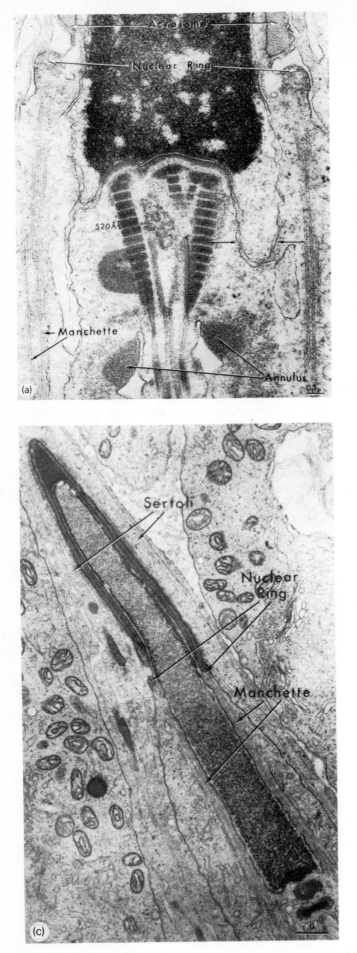

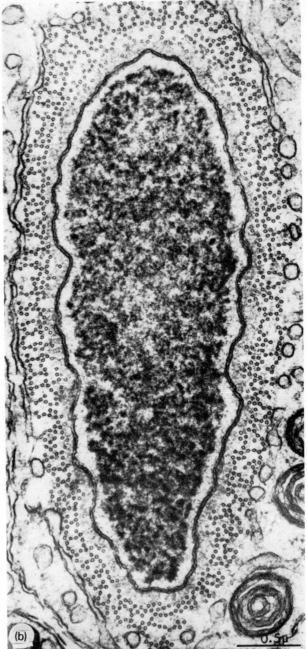

FIGURE 7.13. Electron micrographs of sections through Chinese hamster spermatids. (a) Longitudinal section: The thickened nuclear ring at the posterior margin of the acrosome provides a site for microtubule elongation and manchette formation. Redundant nuclear envelope (between arrows) and the connecting piece with dense striated columns with a periodicity of 52 nm form in the neck. The annulus is surrounded by chromatoid material (at heads of arrows) possibly originating from the remains of nuage. (b) Cross section: Microtubules (seen as "pin heads" in cross section) comprise the manchette surrounding the base of the nucleus. Circular profiles of vesicular endoplasmic reticulum associated with annulate lamellae and mitochondria or multivesicular bodies contain relatively dense material (lower right). (c) Longitudinal section: The *manchette* attached to the *nuclear ring* seems to draw the spermatid's cytoplasm and *acrosome* tightly over the tip of the nucleus. From D. M. Phillips, *Spermiogenesis,* Academic Press, Orlando, 1974. Used by permission.

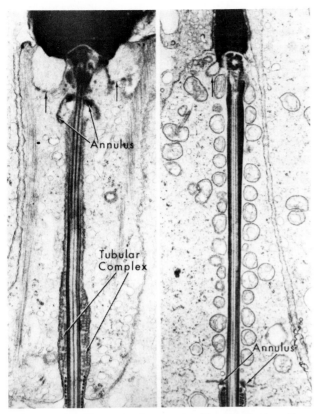

FIGURE 7.14. Electron micrograph of longitudinal sections through rabbit spermatids showing caudal movement of the annulus. The mitochondrial sheath forms behind the annulus, and the fibrous sheath forms from a tubular complex ahead of the annulus. From D. M. Phillips, *Spermiogenesis,* Academic Press, Orlando, 1974. Used by permission.

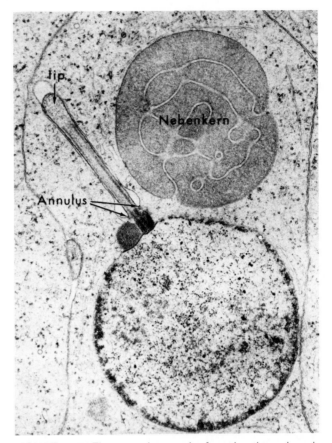

FIGURE 7.15. Electron micrograph of section through early round spermatid of the stink bug, *Euchistus,* showing the nebenkern formed by the condensation of mitochondria. The annulus or ring centriole lies above the true centriole and a mass known as the centriolar adjunct adjacent to the nucleus. The nuclear envelope is thickened in the vicinity and appears to be tightly connected to the centriole and centriolar adjunct. Eventually, the centriolar adjunct surrounds the centriole and binds the tail to the nucleus. From D. M. Phillips, *Spermiogenesis,* Academic Press, Orlando, 1974. Used by permission.

the posterior nuclear envelope is drawn into folds of **redundant nuclear envelope** (arrows, Fig. 7.13a).

A **manchette** (Fr. *manche* sleeve) of bundles of microtubules extends backward from a ring at the posterior edge of the acrosome (called the **nuclear ring** although it is peripheral to the nucleus) over the nucleus and beyond to the developing midpiece (Figs. 7.13a–c). Posterior movement of the manchette seems to draw the apical cytoplasm and acrosome tightly over the apical end of the nucleus to the level of the equatorial segment. Later, the manchette is replaced by a postacrosomal sheath or lamella in the mature sperm.

The story of the sheath's formation begins before the spermatid substage. At the electron microscopic level of observation, a fibrogranular material, called **nuagelike** material (Fr. *nuages* cloud) in vertebrates, appears at the spermatocyte stage in the midst of clumped mitochondria. When the mitochondrial clumps break up in spermatids, the material migrates and adheres to the outside of the nuclear envelope in **dense bodies.** These migrate

posteriorly and seem to accumulate additional material streaming through nuclear pores (Fig. 7.16).

After becoming free of the nucleus, the dense bodies are called **chromatoid bodies** (meaning colored or chromosomal-like bodies) which associate with the annulus (tips of arrows marked Annulus in Fig. 7.13a). The chromatoid bodies and annulus move posteriorly to the junction of the tail's middle and principal pieces as mitochondria concentrate and coil around the core of the flagellum to form the mitochondrial sheath (Fig. 7.14). The chromatoid body may then disintegrate although it remains in rodents as a thickening overhanging an annular recess.

The function of the fibrogranular material in spermatogenic cells is uncertain. Similarly appear-

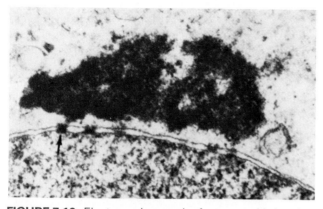

FIGURE 7.16. Electron micrograph of a spermatid showing dense body adhering to outside of nuclear envelope in apparent association with nuclear pores (arrow). From D. W. Fawcett. In R. A. Beatty and S. Glfrom D. W. Fawcett. In R. A. Beatty and S. Gluecksohn-Waelsch, eds., *Proceedings of the International Symposium on the Genetics of the Spermatozoon*, 1972, by permission of D. W. Fawcett.

ing materials are ubiquitous in vertebrate germ-line cells and are morphologically similar to polar granules in insect and nematode eggs. The perinuclear location of dense bodies and the streams of nuclear material passing through pores suggest that the fibrogranular material is a storage depot for RNA.

Several observations are consistent with this suggestion. RNA has been identified in nuagelike material of spermatocyte in the cranefly and ciprinoid teleosts. In *Drosophila* spermatocytes, the appearance of fibrogranular material is correlated with a considerable amount of transcription. Moreover, while fibrogranular material disappears in spermatids, a high level of protein synthesis occurs in the absence of detectable transcription (Brink, 1968). On the other hand, the close association of mitochondria with fibrogranular material in spermatocytes and with the chromatoid body in spermatids suggests that the fibrogranular material provides a pool of raw materials utilized in mitochondrial growth.

Nuclear condensation lags behind cytoplasmic differentiation. Breaking the rule in somatic cells that histone synthesis parallels DNA synthesis, nucleoproteins are synthesized in nonreplicating spermatocytes, and protamine is synethized in spermatids from long-lived messenger RNA.

Nuclear events vary considerably with the type of change occurring in nucleoproteins. In some molluscan sperm, for example, the core histones and nucleosomes are retained, while a second H1 linker histone or a lysine–arginine-rich protamine-like protein replaces the original H1 linker (Sollos, 1985). In other animals, histone is replaced by protamine via somewhat different mechanisms (see Poccia, 1986).

The replacement of nucleoproteins is delayed in the house cricket, *Acheta domestica.* Somatic histones present in spermatocytes and early spermatids are replaced by a series of tissue-specific histones in the late spermatid. At this time chromatin fibers lose their beading, although they still contain histones (Fig. 7.17). Very late spermatids and sperm contain protamine-sized molecules that condense through side-to-side alignment and regular folding into packaging units.

In trout, histones are phosphorylated during DNA synthesis in spermatogonia, but little histone phosphorylation is seen in spermatids. Acetylation of preformed histones, especially H4, continues during the spermatid phase, and triacetylated and tetraacetylated forms of histone become characteristic.

Spermatids are inactive in RNA synthesis presumably due to a decrease in RNA polymerase. Residual protein synthesis in spermatids is dependent on long-lived mRNA and subject to translational controls. Only protamines are synthesized at the

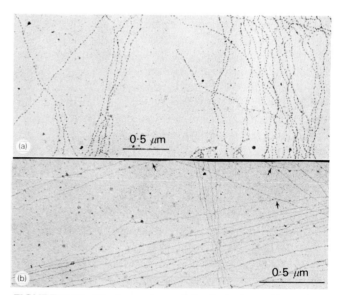

FIGURE 7.17. Electron micrograph of shadowed chromatin fibers from the cricket, *Acheta domestica,* showing beaded appearance in early spermatids (a) and smooth fibers in later spermatids (b). From A. L. Kierszenbaum and L. L. Tres, *J. Cell Sci.,* 33:265–283 (1978), by permission of the Company of Biologists and courtesy of the authors.

mid-spermatid phase, and all nucleoprotein synthesis ceases in late spermiogenesis.

The acetylated and phosphorylated histones of the earlier germ-line cells are replaced by phosphorylated protamines that become dephosphorylated as nuclei condense. Histones are removed starting with H4 and ending with H1.

In rodents, intermediary basic proteins appear before the final protamines, and the transition of nucleoproteins takes place in two stages. Nucleoproteins involved in the first stage may participate in extending chromatin during recombination, while those involved in the second stage may participate more directly in nuclear condensation.

In the first stage, **testis-specific histones** (e.g., H.S, mouse) or **testis-enriched histone variants** (e.g., H1t, THs, rat) appear as early as spermatogonia (e.g., H1t and TH3) and preleptotene spermatocytes (e.g., TH2A and TH2B). Characteristic of spermatocytes, some of these nucleoproteins are synthesized during early meiotic prophase (H2A.S and H3.S), while others appear in late prophase (H1.S and H2B.S).

These histones and others, including some normally found in somatic cells, participate in nucleosome formation in the spermatocyte. They become acetylated throughout but especially in mid-spermatids when they are being replaced in the chromatin and nucleosomes are disappearing.

In the second stage, beginning with spermatids, **transitional proteins** (TPs) are synthesized along with protamines. Later, TPs may help protamines bind to DNA.

In the mouse, early spermatids continue to transcribe mRNA for nucleosomal histones, while protamine transcripts accumulate. As RNA synthesis ceases in early elongating spermatids, an arginine-rich protein fraction accumulates, and both beaded nucleosomal and smooth nonnucleosomal chromatin fibers appear.

The nonhistone TPs replace the histones and in turn are replaced by two (three in humans) forms of arginine- and cysteine-rich protamines. The final cross-linking of disulfide bonds is not achieved until the end of spermiogenesis.

In rats, TPs appear in sequence paralleling nuclear condensation. The first TPs appear in mid-elongating spermatids (TP, TP2, and TP4), and the last (TP3) is added in late-elongated spermatids. The protamine of mature sperm (S) is only added in the final stages.

Release during the* spermiation substage *occurs from crypts at the luminal surface of sustentacular cells. The junctional complexes that have

FIGURE 7.18. The tubulo–bulbar complex of a rat spermatid. The extensions from the head fit into pits in sustentacular cells. From L. Russell and Y. Clermont, *Anat. Rec.,* 185:259 (1976), by permission of the American Association of Anatomists, Alan R. Liss, Inc., and the authors.

bound the spermatids have increasingly become restricted to their apical ends. In mammals at spermiation, junctional complexes disappear completely, and the spermatid head is anchored to the sustentacular cell only by thin bulbous extensions (Fig. 7.18) dipping into **bristle-coated pits.**

The combination of bulbous extensions and pits is known as the **tubulo–bulbar complex.** Emerging from the head in waves, the spermatid extensions are repeatedly **phagocytized** (i.e., ingested) by the sustentacular cells only to be regenerated by the spermatid. As well as anchoring the spermatid, the tubulo–bulbar complexes reduce the amount of cytoplasm around the sperm head (e.g., by up to 70% in rats).

In the rat, a last wave of bulbous cytoplasmic extensions is sent out from the spermatid's sickle-shaped head without making contact with pits. As these extensions are retracted, the sperm head is finally freed.

At the same time, a **cytoplasmic bead** (also known as the kinetoplasmic droplet) moves from the neck of the sperm to the junction of the middle and principal tail pieces. The granular **Golgi rest,** or portion of the Golgi apparatus not incorporated into the acrosome, is conspicuous within the bead.

As the spermatid moves into the lumen of the seminiferous tubule, the cytoplasmic extension between the spermatid and the bead lengthens and finally breaks. The abandoned beads or **residual bodies** (Fig. 7.12) retain the intercellular bridges that once connected all the cells of the spermatogenic nest (Fig. 7.10). Lying within their crypts in the sustentacular cells, the residual bodies, like the spermatid's bulbous extensions, are disposed of by phagocytosis.

Hormonal Control of Spermatogenesis

Germ cells ordinarily lack androgen receptors, but the initiation of spermatogenesis at puberty requires both the pituitary gonadotropins (GNs), follicle stimulating hormone (FSH), and luteinizing hormone (LH, in males sometimes called ICSH, interstitial cell stimulating hormone). LH promotes the growth of testicular **interstitial endocrinocytes** (often called Leydig cells after their discoverer, Franz von Leydig [1821–1908], or interstitial cells of Leydig) and thereby enhances steroidogenesis.

FSH in conjunction with LH promotes growth of seminiferous tubules and overall testicular growth (Fig. 7.19). Testosterone, in turn, works sy-

nergistically with FSH to promote the formation of junctional complexes in sustentacular cells and thus the adluminal compartment where spermatogenic cells undergo meiosis and differentiation.

Once underway, spermatogenesis is independent of further FSH stimulation except in seasonally reproducing animals. In rams and roe bucks, for example, annual changes in the size of testes and the amount of spermatogenic activity in the course of the year reflect changes in both GNs. Changes in day length play the dominant role in regulating spermatogenesis, but temperature is an important variable in amphibians, reptiles, and mammals, if not birds. Working through the central nervous system, environmental variables may also alter the testicular response to GNs (see van Tienhoven, 1983).

Induced by shortened day length, changes in rams' testes parallel increases in the amplitude and frequency of the pulsatile LH secretion and FSH production (Fig. 7.19) (see Lincoln and Short, 1980). The diameter of seminiferous tubules increases as spermatogonia proliferate, and spermatocytes form the germinal epithelium. Interstitial endocrinocytes in intertubular spaces enlarge, and, as additional LH receptors appear, steroidogenesis leads to testosterone secretion.

Destroying or damaging the hypothalamus interferes with the production of gonadotrophin releasing hormone (GnRH) by the hypothalamus and results in testicular atrophy. Normally, the regulation of hypothalamic and pituitary hormonal function in the male depends in part on the steroids secreted by the testis (Fig. 7.5). Testosterone suppresses the amplitude of rhythmic peaks of GnRH secretion and, consequently, the pulsatile release of LH. Estrogens, especially estradiol-17β, secreted by sustentacular cells in the testis retard the release of FSH by suppressing sensitivity to GnRH.

The removal of the pituitary gland (i.e., hypophysectomy) or the administration of specific antisera to GNs leads to the suppression of spermatogenesis and steroidogenesis, the involution of interstitial endocrinocytes, and removal of LH receptors. The number of spermatids and spermatocytes is reduced, and the diameter of seminiferous tubules decreases.

GNs (with the help of prolactin) may prevent testicular atrophy and even lead to the restoration of spermatogenesis in hypophysectomized males. In rams and rats, various androgens can also slow or prevent the loss of spermatogenesis if administered soon after hypophysectomy, but, with few

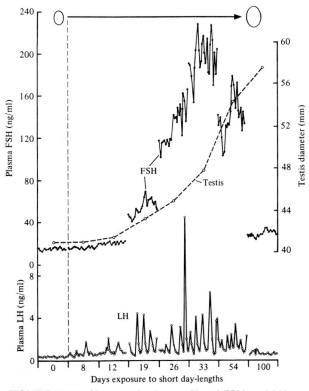

FIGURE 7.19. Changes in the profile of FSH and LH concentrations in blood and in the diameter of ram testes following experimental shortening of day length. At day 0, the length of exposure to artificial light was reduced from 16 to 8 hours per day. Changes began immediately and became dramatic after about 2 weeks. From G. A. Lincoln and M. J. Peet, *J. Endocrinol.*, 74:355 (1977). Used by permission.

exceptions (pigeons, monkeys, and squirrels), testosterone does not restore spermatogenesis once it has disappeared.

This difference in hypophysectomized animals receiving GNs as opposed to gonadal steroids suggests that GNs offer something that is not supplied by testosterone alone. The unknown element may be the **androgen binding protein** (ABP) produced by sustentacular cells in response to FSH (Fig. 7.20). ABP is found in seminiferous fluid and, although it does not seem to concentrate androgens, may

function to move androgens into the adluminal compartment.

Possibly, androgens, both testosterone and its more potent derivative dihydrotestosterone, must bind with ABP in order to reach the lumen of the seminiferous tubule and hence the male genital ducts. There, surface cells pick up the bound complex, degrade the protein, and release the androgen.

Other effects of FSH and LH may also be mediated through sustentacular cells and interstitial endocrinocytes. While FSH can cause sustentacular cells to swell and release sperm, the hormone also causes fluctuations in sustentacular cells' secretion of estrogens. Similarly, fluctuations in gonadal androgens may be due to effects of LH on interstitial endocrinocytes. Both GNs operate through specific receptors and the adenylate cyclase system, but, while LH leads to increases in androgen-synthesizing enzymes, FSH leads to increases in aromatizing enzymes.

The hormone **inhibin** also enters the calculus of spermatogenesis. Produced by sustentacular cells in response to FSH and testosterone stimulation (Fig. 7.20), inhibin enters the circulation and the seminiferous fluid on route to the rete testis where it is concentrated. Upon inhibin injection into males, GnRH production in the hypothalamus and LH release by the pituitary is inhibited. While inhibin has not been detected in the circulation of untreated males (although it has been detected in normal females), it may suppress spermatogenesis in seasonally breeding males by acting on the hypothalamus and pituitary (see Steinberger and Steinberger, 1973). Inhibin also prevents DNA synthesis in spermatogonia in rams and other mammals.

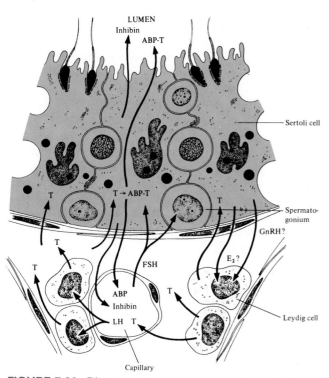

FIGURE 7.20. Diagrammatic cross section of testis showing portion of a seminiferous tubule and surrounding stroma and illustrating hormonal interactions, steroidogenesis, and the production and movement of *androgen binding protein* (ABP) and *inhibin*. Interstitial endocrinocytes produce testosterone (T) in response to LH reaching them through the circulation. Sustentacular cells produce *inhibin* and ABP in response to FSH. Originally promoted by FSH reaching them from the circulation, *spermatogonia* pass into the spermatogenic phase. The effects of estrogen (E) and GnRH on interstitial endocrinocytes are uncertain, although these cells may have receptors for these hormones. Inhibin and testosterone move into the circulation and into the luminal space where they become part of seminiferous fluid. From D. M. de Krester. In C. R. Austin and R. V. Short, eds., *Reproduction in mammals,* Book 3, *Hormonal control of reproduction,* 2nd ed., Cambridge University Press, Cambridge, 1984. Used by permission.

Final Preparation and Capacitation

The final preparations of sperm for fertilization may take place before or after leaving the male's genital tract. In the cases of fish through amphibians, the sperm are ready to fertilize as soon as they are deposited or taken up in the female reproductive tract. In the laboratory, however, when sperm are liberated directly from a testis, an incubation period of several minutes may be required before the sperm begin swimming movements and become capable of fertilizing eggs.

Sperm from birds and prototherian mammals are also ready to fertilize eggs as soon as they leave the male's genital tract. But in marsupials and eutherian mammals, sperm leaving the seminiferous tubule are not motile or competent for fertilization. Their deficiencies may represent a compromise among the requirements of storing sperm, moving them through the male genital tract, and directing them into or toward the female genital tract. While passing through the female genital tract, the mammalian sperm undergo **capacitation** and achieve competence for their role in fertilization.

Upon leaving the seminiferous tubules, mammalian sperm enter the male **genital duct system.** Sperm first enter a labyrinth of epithelial lined cavities known as the **tubuli recti.** From here, they move through the **rete testis** into the **ductuli efferentes,** to the head (capit), body (corpus), and tail (caudad) of the **epididymis,** and finally into the **ductus deferens,** the ejaculatory duct, for passage to the urethra and out of the body (Fig. 7.6a).

Passage through the genital duct system may be interrupted by a period of storage. The results of experiments with radioactive precursors of DNA in rams indicate that spermatogenesis takes about 30 days, but sperm remain in the epididymis for an additional 2 weeks. In men, the comparable figures are 16 days for spermatogenesis and 12 days in the epididymis. In other mammals, sperm may be retained in the epididymis for months without losing their ability to perform in fertilization.

Sperm in the epididymis, known as **epididymal sperm,** may acquire a surface coat, possibly a peptide **decapacitator,** and an **acrosome inhibiting protein** that stabilizes the acrosome and **decapacitates** the sperm during prolonged storage (Oliphant et al., 1985). The sperm may also acquire a surface coating of materials required for subsequent binding to the egg. In addition, the sperm are toughened or **tanned** by the formation of stabilizing disulfide bridges in many of their proteins including nuclear proteins.

Taking place in the female genital tract or under appropriate culture conditions, capacitation **destabilizes** the acrosome and elevates the sperm's functional state. In effect, the **decapacitation** achieved in the epididymis is reversed in the female genital tract, presumably by the removal or modification of stabilizing surface material. The spermatozoon's tail then generally acquires a vigorous deeply undulating beat and the capacity for continuous motility during the spermatozoon's short life.

Finally, ejaculated sperm lose the ability to fertilize. Unless artificially frozen at −79°C, mammalian sperm lose the ability to fertilize eggs a few hours after ejaculation. At the most, the sperm of men and domestic mammals last only a few days in the female genital tract.

OOGENESIS

The maturation of the oocyte always requires the synthetic products of cells other than itself.

Eric H. Davidson (1968, p. 200, emphasis original)

Types of Oogenesis

Multicellular animals form eggs in three different but related ways: cellular cannibalism, panoistic oogenesis, and meroistic oogenesis. In each case, incomplete cell divisions leave oogonia attached to each other by intercellular bridges in **oogenic nests.** Some differences in the types of egg formation grow out of the relationship of nest cells to each other. Other differences arise from the relationship of eggs to nearby maternal somatic cells.

At one extreme, **cellular cannibalism,** found in teleost fish ovaries (see Raven, 1961), results in the formation of one egg from an entire oogenic nest. A single amoeboid oogonium devours the other oogonia and, greatly enlarged, becomes an oocyte when finally surrounded by maternal **follicle cells.**

Similarly, in some sponges and coelenterates, such as the familiar _Hydra_ (Fig. 7.1), only one oocyte emerges from a gonophore originally of several thousand cells. Possibly as a result of fusion, cells provision the successful oocyte with storage products while losing their own nuclei and ultimately disappearing (Brien and Reniers-Decoen, 1950).

At the other extreme, all cells in the nest are potentially released as oocytes. This form of oogenesis is characteristic of vertebrates and primitive insects (Fig. 7.21a). It is called **panoistic** (Gk. _pantos_ all + _oion_ egg) or **follicular egg formation,** since each oocyte is surrounded by maternal **follicular** or **accessory cells.**

The type of egg formation in advanced insects (Fig. 7.21b, c) and many other invertebrates (see Raven, 1961) lies between these extremes. One oogonium in a nest routinely becomes an oocyte, while the others become **nurse cells** or **trophocytes** (nutritive cells) which contribute materially to the egg. Oogenesis is then called **meroistic** (Gk. _meros_ part + _oion_, since only a part of the nest gives rise to an oocyte) or **nutrimentary egg formation.**

Panoistic and meroistic oogenesis are similar in several ways, and differences between them are quantitative rather than qualitative. In insects, both forms of oogenesis take place in **ovarioles** or **egg strings.** Proximal to a terminal filament, oogonia proliferate in a **germarium** or **germinative zone.** Oocytes mature in a **vitallium** or **growth**

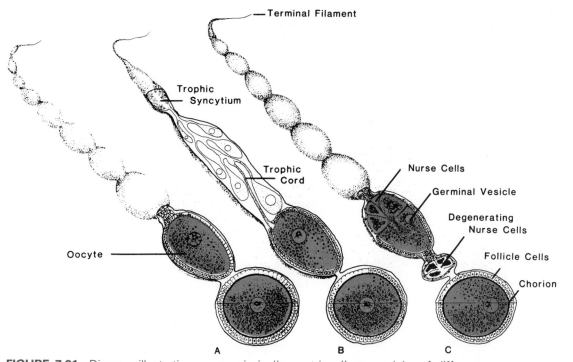

FIGURE 7.21. Diagram illustrating oogenesis in "egg strings" or *ovarioles* of different insects. (a) In *panoistic oogenesis,* the germarium produces independent oocytes. (b, c) In *meroistic* oogenesis, the germarium produces dependent oocytes that receive RNA and other materials from attached *nurse cells.* (b) The nurse cells form a *trophic syncytium* connected to the oocyte by a *nutritive* or *trophic cord* in the *telotrophic* form of meroistic oogenesis. (c) The oocyte is connected to several nurse cells via *ring canals* (shown as gaps) in the *polytrophic* form. From S. J. Berry. In L. W. Browder, ed., *Developmental biology,* Vol. 1, *Oogenesis,* Plenum Press, New York, 1985. Used by permission.

zone, and a layer of accessory, or follicular, cells forms a **cyst** or **follicle** around oocytes in panoistic as well as meroistic oogenesis.

In panoistic oogenesis, no nurse cells are produced, while in the simplest type of meroistic oogenesis one nurse cell is produced. In primitive dipterans (such as chironomids), for example, the oogenic nest consists of two cells. One becomes a nurse cell, while the other becomes the oocyte. Five nurse cells are produced in lice oogenic nests, and seven nurse cells are produced in mosquito (*Anopheles*) and silkworm (*Cecropia*) nests. Meroistic oogenesis reaches its fullest expression in *Drosophila* and other advanced dipterans, which produce 15 nurse cells for each oocyte.

In *Drosophila melanogaster,* the division of a **germ stem cell** (Fig. 7.22) provides a replacement germ stem cell and an oogonium known as a **cystoblast** (Cb; i.e., a cyst stem cell). Four perpendicular and incomplete divisions without growth (M$_1$–M$_4$) result in 16 small **cystocytes** welded into a nest by intercellular bridges known as **ring canals** or **fusomes.**

All the cells in the nest soon grow, but synaptonemal complexes, indicative of meiotic chromosomal pairing, appear only in two cells. These cells, called **pro-oocytes** (i.e., 1^4 and 2^4), are joined to each other by the initial intercellular bridge (single line in figure) and to three cystocytes by additional intercellular bridges. One pro-oocyte abandons meiosis and, along with the remaining cystocytes, becomes a nurse cell, while the other pro-oocyte continues meiosis to become the definitive oocyte.

The type of meroistic oogenesis taking place in *D. melanogaster* is called the **polytrophic form** (meaning fed from multiple sources, Fig. 7.21c). Alternatively, in the **telotrophic form** (meaning fed from afar, e.g., the milkweed bud, *Oncopeltus,* Fig. 7.21b), oocytes are connected to a **trophic syncytium** (i.e., fused nurse cells) through a **nutritive** or **trophic chord.** Although the structures of telotrophic and polytrophic ovarioles are different, the consequences of connecting nurse cells to the oocyte are the same, namely, feeding the oocyte (see Berry, 1985).

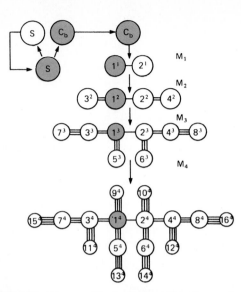

FIGURE 7.22. Diagram illustrating proliferative pattern and changing cell volume during oogenesis in the fruit fly, *Drosophila melanogaster*. Superscripts correspond to the number of the mitotic divisions following the origin of the cystoblast (C_b), and the number of lines joining cells corresponds to the division at which the ring canal formed. Division of a germ-stem cell (S) gives rise to another germ-stem cell and an original cyst-producing cell or cystoblast (C_b). As a result of four incomplete mitotic divisions (M_1–M_4), intercellular bridges connect 16 cystocytes in an oocyte–nurse cell (or O/NC) syncytium. In the intercellular bridges, arrested contractile rings are converted into ring canals and a gelatinous fusome replaces the spindle. Because consecutive mitotic spindles are anchored at one end by the previous fusome, the second and third generations of ring canals are distributed to only one of the two cells formed. Adjacent fusomes subsequently fuse into a single polyfusome which dissolves once the cystocyte divisions are completed. Two cystocytes (1^4 and 2^4), known as pro-oocytes, are distinguished by being linked through the very first intercellular bridge and three additional intercellular bridges. These cystocytes alone begin meiosis, but only one becomes the definitive oocyte and completes meiosis. The other pro-oocyte and the other cystocytes become trophocytes or nurse cells. From R. C. King and P. D. Storto, *BioEssays*, 8:18 (1988), by permission of Cambridge University Press and courtesy of R. C. King.

Large amounts of materials, ranging from mitochondria and ribosomes to proteins, carbohydrates, and DNA from nurse cell nuclei are funneled into the oocytes via the nutrient chords or ring canals (see below, Fig. 7.28). In insects and annelids, oocyte feeding occurs until the nurse cells are exhausted, and in mollusks, such as the snail, *Helix*, the nurse cells are finally engulfed by the developing oocyte.

In the end, the three types of oogenesis produce oocytes, but each type seems to have evolved in response to very different evolutionary pressures. Cellular cannibalism seems to be geared to meet the requirements of rapid oogenesis in **opportunistic breeders** (i.e., those taking advantage of rare and unpredictable opportunities to breed). Although most of the potential oocytes are sacrificed, a few "good ones" are made available quickly.

Panoistic oogenesis, on the other hand, occurs in relatively long-lived animals with regular and reliable breeding seasons. This form of oogenesis tends to be prolonged with every oocyte retaining at least the potential for passing through the entire process.

Meroistic oogenesis represents an intermediate and, in some ways, a compromise between cellular cannibalism and panoistic oogenesis. Occurring primarily in shorter-lived animals or those that breed during a brief period, meroistic oogenesis overcomes the constraints of time by capitalizing on the synthetic capacities of nurse cells even though some potential oocytes are sacrificed.

Stages of Oogenesis

Egg development is divided into a mitotic oogonial stage followed by a meiotic oocytic stage. Rates of growth and subphases of meiotic prophase provide the criteria for further classifying primary oocytes into substages (Table 7.3). A slowly growing **previtelline** or **bouquet substage** leads to a moderately growing **germinal vesicle** or **dictyate substage** (Fig. 7.23) and, finally, a rapidly growing **vitelline** or **vitellogenic substage** capped by cytokinesis of the first meiotic division.

These substages of oocyte growth are readily appreciated in the ovaries of seasonally breeding anurans such as *Rana temporaria* (Fig. 7.24) where discrete generations of oocytes begin growth once every year. Small oogonia and premeiotic and previtelline oocytes are buried in the thin wall of the hollow ovary. The germinal vesicle substage is represented by larger translucent oocytes and the vitelline substage by enormous oocytes with cytoplasm rendered opaque by yolk and pigment (see Fig. 6.16).

The ovaries of continuously breeding species, such as *X. laevis*, contain a nearly continuous distribution of oocytes. In contrast to ranid females which can only be induced to release eggs seasonally and after a period of cooling, *Xenopus* females can be induced to ovulate at any time of year. The year-round capacity to ovulate greatly adds to the attractiveness of *Xenopus* for research.

TABLE 7.3. Increase in size of typical amphibian oocyte[a]

Oocyte substage:	Premeiotic	Previtelline	Germinal vesicle	Vitellogenic
Meiotic subphase:	Preleptotene	Leptopachytene	Diplotene	Late diplotene and diakinesis
Diameter (in μm):	50	80	350	1500[b]
Gross characteristics		Transparent	First appearance of yolk platelets	
Ranges (in μm)		15–225	225–425	
Duration of interval			2 years	28 days
Increase in volume		4×	84×	79×
Overall increase				27,000×
Cubic centimeters of oxygen per cm³ of oocytes[c]	0.69		1.5	1.2 (preovulatory)

[a]Data represent averages from various sources, especially Grant (1953).
[b]*Xenopus laevis* oocytes average 1430 μm.
[c]Data from Brachet (1950).

Oogonia

Primordial germ cells reach the embryonic gonad following a hazardous trek from elsewhere in the embryo. They have relatively large size, abundant juxtanuclear mitochondria, and, in the case of amphibians, persistent yolk platelets. Cytoplasm stains deeply with dyes (such as pyronine) having high affinities for RNA and carries telltale traces of germ plasm or nuage.

Upon entering the gonad, the primordial germ cells are transformed into primary germ cells or gonocytes. Their cytoplasm no longer stains deeply for RNA, and germ plasm has disappeared (see Eddy, 1975).

As the gonad acquires the identity of an ovary, gonocytes become proliferative **primary oogonia.** The cell population expands rapidly into the **reserve** or **stem cell population** of the female germ line. Between the 9th and 17th days of incubation in the chick embryo, for example, mitosis in primary oogonia accounts for a 25-fold increase in the number of primary oogonia (Table 7.4).

In the human embryo, waves of cell division, rather than steady growth, drive the 700–1700 primary germ cells to a population of 600,000 primary oogonia by the beginning of the fetal period at 8 weeks and to 6,800,000 oogonia, the maximum number ever reached, during the 5th month (Baker, 1963). Waves of division also increase the size of oogonial populations in the rat, rabbit, guinea pig, and monkey fetuses.

While complete cell division in primary oogonia gives rise to additional primary oogonia, incomplete cytokinesis gives rise to nests of **secondary oogonia** attached to each other through intercellular bridges. In *Xenopus laevis*, for example,

average nests consist of 16 secondary oogonia (Coggins, 1973).

The nuclei of nest cells are irregularly shaped and contain irregularly shaped nucleoli. The cytoplasm contains newly formed (or reformed) membraneless particles resembling germ plasm and called **nuage particles** (arrows in Fig. 7.25a). Discrete but irregular in outline and closely associated with mitochondria, the particles have a fibrous to granular consistency at the electron microscopic level. They may contain germ-line determinants (see Hardisty, 1978), but, in oocytes, they seem to act as an intermitochondrial cement or mitochondrial resource (see Norrevang, 1968).

In secondary oogonia and early oocytes of many vertebrates, from megalecithal birds to alecithal mammals, an immense number of mitochondria gather temporarily around the centrosome and a mass of Golgi material, smooth and rough surface endoplasmic reticulum, and vacuoles. Sometimes described as a cloud and known as the **Balbiani body** (sometimes, Balbiani's vitelline body or the yolk nucleus of Balbiani), the conglomerate of cell organelles disappears before yolk deposition. In birds, elements of the body (especially the Golgi apparatus and mitochondria) disperse to the periphery rather than disintegrate (Bellairs, 1971).

Depending on species, oogonia undergo a transformation to oocytes approximately simultaneously or only gradually over a lifetime. Primary oogonia either disappear completely and the germ line becomes a static tissue, or they continue to constitute a proliferative subpopulation and the germ line becomes a renewing or steady-state tissue.

FIRST MEIOTIC DIVISION | SECOND MEIOTIC DIVISION

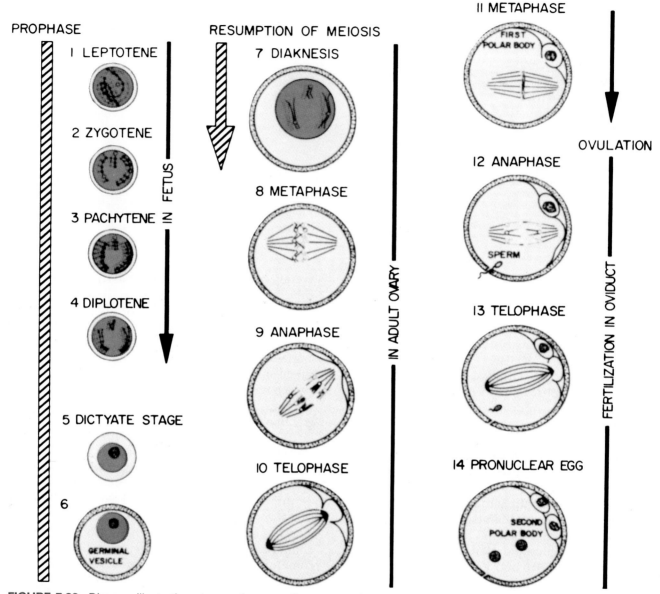

FIGURE 7.23. Diagram illustrating stages of mammalian oogenesis as they are typically found in the fetal and adult ovaries and in the oviduct following ovulation and fertilization. For simplicity, only three pairs of chromosomes are shown. From A. Tsafrir, In R. E. Jones, ed., *The vertebrate ovary,* Plenum Press, New York, 1978. Used by permission of the author.

In the case of adult vertebrates, oogonia are present in most teleosts, all amphibians, the majority of reptiles, and, as an exception to the rule for mammals, in some prosimian primates. Oogonia are absent in cyclostomes, elasmobranchs, all birds, prototherian, and almost all therian mammals.

Premeiotic Oocyte

The transformation of oogonia to oocytes occurs with the cell's entry into **premeiotic interphase** and the synthesis of DNA prior to meiosis. In *X. laevis*, the premeiotic S period, lasts about a week and occurs simultaneously in all the cells of a nest (Coggins, 1973). The premeiotic nucleus

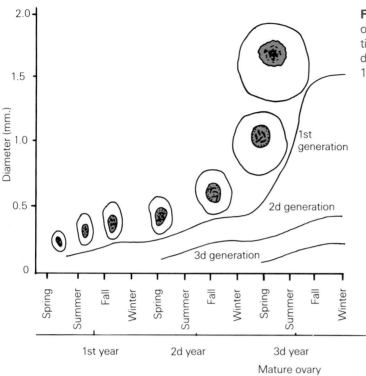

FIGURE 7.24. Seasonal growth of *Rana temporaria* oocytes. The mature ovary contains three "generations" of oocytes, each of which began growing in a different year. Adapted from P. Grant, *J. Exp. Zool.,* 124:513 (1953), by permission of Alan R. Liss, Inc.

(Fig. 7.25b) is rounder compared to the earlier oogonial nucleus (Fig. 7.25a) and has moved to the cell border farthest from the intercellular bridges at the center of the nest. The nucleolus is large, round, and centrally located in the nucleus.

Premeiotic DNA synthesis takes place in the oocytes of most mammals prior to or shortly after birth. All the oocytes of a mouse, for example, are labeled when the mother is injected with tritiated thymidine, between the 12th and 15th days of gestation. No labeling occurs when the injection is postponed until after birth (Borum, 1966). In the ferret, hamster, mink, rabbit, and vole, some mitotically active oogonia remain at birth, and, in

prosimians, oogonia are present throughout the adult lifetime (see Gondos, 1978).

The premeiotic oocyte is not marked by conspicuous growth, but, in vertebrates, **follicular cells** of maternal origin surround the oocyte at this time and form a **unilaminar** (i.e., single layered) or **primordial ovarian follicle.** Otherwise, the premeiotic S period oocyte passes without morphological change into the G-2 **preleptotene oocytes.**

Previtelline or Bouquet Substage

The oogonial stage lapses unceremoniously into the previtelline substage with the appearance

TABLE 7.4. Mitosis and degeneration of oogonia[a]

Species	Period(s) of Mitosis	Amount of Growth	Period(s) of Degeneration	Extent of Degeneration
Chick	9–17 days (of incubation)	25×	17 days incubation to 2 days posthatching	40%
Rat	$14\frac{1}{2}$–$17\frac{1}{2}$ p.c.	6×	$18\frac{1}{2}$–2 p.p.	33%
Human	2nd month gestation	600×	Prior to birth (several waves)	3.5×
	5th month gestation	12×	From birth to 7 yr	7×

[a]Data from Nieuwkoop and Sutasurya (1981). p.c., days postcoitum; p.p., days postpartum.

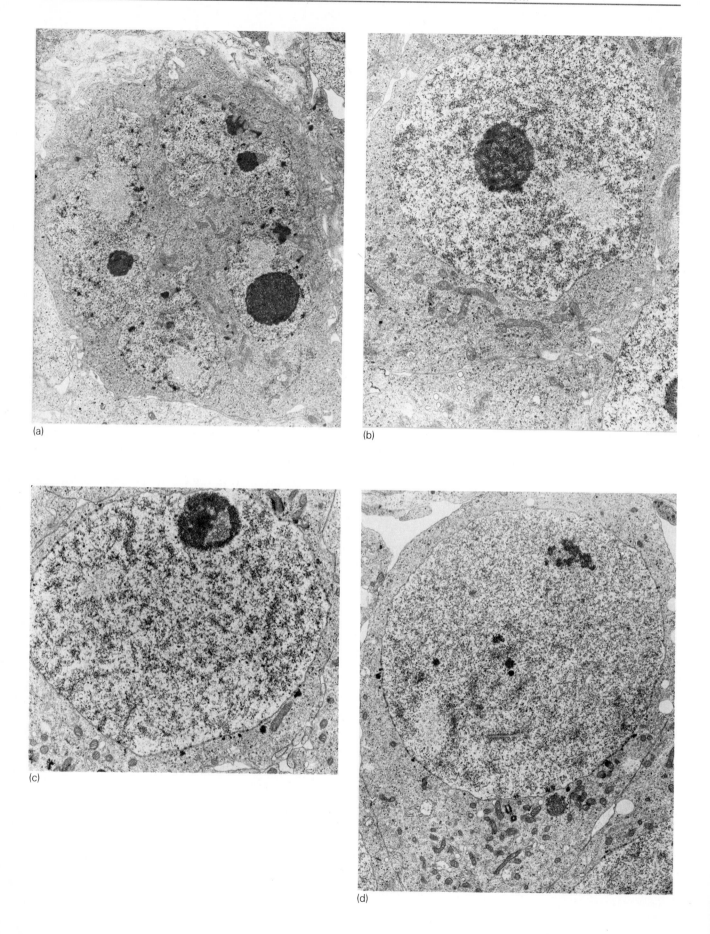

(a)

(b)

(c)

(d)

of leptotene chromosomes at the light microscopic level (oocyte stage 1; Fig. 7.23). The chromosomes synapse, or conjugate, at zygotene (oocyte stage 2) and condense into bivalent chromosomes at pachytene (oocyte stage 3). Attached to a confined region of the nuclear envelope and bulging toward the center, the chromosomes compose a fanciful bouquet (see Fig. 3.10), hence the **bouquet substage.**

Differentiation, in the sense of cell change, first becomes apparent at the electron microscope level. In the cytoplasm, the centrosome, surrounded by numerous mitochondria and Golgi vesicles, has moved to the vicinity of intercellular bridges. In the nucleus, electron-light material from the leptotene nucleus (e.g., Fig. 7.25b) is redistributed as a halo (Fig. 7.25c, arrows) around the nucleolus (n). Unpaired portions of chromosomes, or **chromosomal axes** (also called axial cores, a) are welded together by **synaptonemal complexes** (also see Fig. 3.11).

Intercellular bridges at the center of the nest seem to polarize several features of the oocytes and anticipate the polarity of the mature ovum. The chromosomal axes and synaptonemal complexes are implanted in nuclear plaques at the flattened central edge of the nucleus closest to the intercellular bridges, and the bouquet opens in the opposite direction. The nucleus and its nucleolus lie farthest from the intercellular bridges, and the nuclear envelope has fewer pores in the vicinity of the nucleolus than elsewhere. While the intercellular bridges will disappear, the nucleus will reside in the animal hemisphere of the future ovum, and

mitochondria associated with nuage will cluster in the vegetal hemisphere (see Gerhart, 1987).

Intercellular bridges break down during pachytene, and microvilli appear at the bridge's broken ends as cytoplasmic remnants are drawn back into the cell's surface. No trace of bridges is found as the oocyte surface smooths, and the first layer of extracellular membrane then appears. From this time onward, the progress of former nest cells is no longer synchronous.

Distinctive **annulate lamellae** (AL, also porous cytomembranes, fenestrated lamellae, etc., Fig. 7.26), consisting of double-membrane sacs, or cisternae, laced with abundant and well-aligned pores, appear in the cytoplasm of oocytes of animals from mollusks, crustaceans, and insects, to echinoderms, ascidians, amphibians, and mammals including chimpanzees and human beings. AL resemble cisternae of the endoplasmic reticulum lacking ribosomes, but AL pores resemble nuclear pores and appear to be derived from detached folds of the nuclear envelope. In the case of meroistic insects, the breakdown of nurse cell nuclei may supply oocytic AL (see Williams, 1965).

AL are not confined to oocytes or even to germ cells. They are a common feature of spermatogenic cells, embryonic cells, and transformed tumor cells where their presence is correlated with protein synthesis following a posttranscriptional delay. In the rabbit, where AL are absent until after fertilization, they appear in early blastomeres. While their function is unknown, it is thought to be involved in delivering nuclear material to the cytoplasm and in regulating posttranscriptional gene

---

FIGURE 7.25. Electron micrograph of thin section of *Xenopus laevis* ovary. (a) Secondary oogonium. The nucleus (nu) appears several times in the section due to sectioning through an irregular shape. Several dense nucleoli are associated with electron-light areas within the nucleus. Nuagelike material is occasionally associated with the outside of the nuclear envelope along with mitochondria. ×5400. (b) Premeiotic interphase-leptotene oocyte. The nearly spherical nucleus is at the side of the egg farthest from the central intercellular bridges (lower left) connecting the oocyte to sibling oocytes in the nest. The bulk of the oocyte's cytoplasm lies between the intercellular bridge and the nucleus. The nucleolus has rounded, and the electron-light material within the nucleus is coalesced. ×7200. (c) Zygotene oocyte. The nucleolus now appears at the edge of the nucleus farthest from the oocyte's intercellular bridges, while axial cores and synaptonemal complexes appear on the flattened edge of the nucleus closest to the intercellular bridges. A halo of electron-light material surrounds the nucleolus, and nuclear pores appear in the nuclear envelope. ×8700. (d) Pachytene oocyte in the full-blown "bouquet substage." Synaptonemal complexes are seen at the edge of the nucleus closest to the intercellular bridges and centrioles. The relatively light nuclear cap closer to the outside of the nest contains several irregularly shaped nucleoli and is devoid of synaptonemal complexes. ×6600. From L. W. Coggins, *J. Cell Sci.* 12:71 (1973), by permission of Company of Biologists, Ltd. and courtesy of the author.

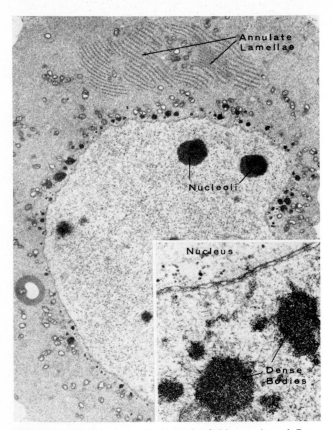

FIGURE 7.26. Electron micrograph of thin section of _Rana clamitans_ early pachytene oocyte. Dense bodies (irregularly shaped black granules) are accumulating in the perinuclear area. Annulate lamellae are seen near top of figure, and multiple nucleoli appear within nucleus. ×4500. (Insert) Granules about 35 nm in diameter appear in the nucleus adjacent to nuclear pores from which "streamers" of material appear to move toward dense bodies. ×17,000. Reproduced from E. M. Eddy and S. Ito, _Journal of Cell Biology,_ 49:90 (1971), by copyright permission of the American Society for Cell Biology, Rockefeller University Press, and the authors.

expression through the release of stored developmental information (see Kessel, 1985).

The **nuagelike material** that first appeared among mitochondria in amphibian oogonia (arrows, Fig. 7.25a) is increasingly associated with the nucleus during the previtelline substage. During pachytene, the material seems to be transformed into perinuclear **dense bodies** (or fibrogranular bodies) resembling similar bodies in spermatids (see Fig. 7.16).

Granules of approximately 35 nm in diameter appear within the nucleus in the vicinity of dense bodies, and **streamers** of granules spring from nuclear pores. Although transcription is rapid during pachytene, RNA does not seem to be present in

the streamers or in the dense bodies (Eddy and Ito, 1971). Whether the nuclear granules give rise to the streamers and add materially to the dense bodies is uncertain, but the amount of nuagelike material increases in the cytoplasm.

**Gene amplification, the selective DNA replication of particular genes without the replication of the remainder of the genome, is the most unusual feature of the previtelline substage.** Specifically, gene amplification results in the temporary appearance of **extrachromosomal bodies** (sometimes called extrachromosomal nucleoli and micronucleoli) which, in the case of oocytes, contain the **nucleolar organizer region** or large ribosomal RNA gene (rDNA, see Fig. 4.3). Already containing as many as 450 iterations of the rDNA sequence in each homologous set of _X. laevis_ chromosomes (see Scheer and Dabauvalle, 1985), amplification accounts for the addition of a million or so excess ribosomal genes and about a thousand extrachromosomal bodies (see Brown and Dawid, 1968; Gall, 1969).

The entire nucleolar organizer region is not equally represented among extrachromosomal bodies, but all the tandem repeats along the length of an amplified rDNA molecule are the same (Wellauer et al., 1976). Selection and different rates of replication among rDNA genes would be explained if amplification employed the **rolling circle** mechanism of replication characteristic of many bacteriophage.

Amplification would begin with the separation from the chromosome of a selected piece of double-stranded DNA containing a ribosomal RNA transcription unit. The DNA would be ligated into a circle, and one strand (the so-called plus or initial strand) would be nicked by an endonuclease-like enzyme (Fig. 7.27).

Then, under the influence of DNA polymerase, a new (so-called minus or complementary) strand would be produced by discontinuous synthesis beginning at the free 5′ end of the nicked plus strand. The amplifying gene would have a "tail" emerging from the end of a "lariat" (Rochaix et al., 1974).

Meanwhile, continuous synthesis from the free 3′ end of the nicked plus strand utilizes the original minus strand as a template. Another plus strand is thereby synthesized, and as the completed double strand is peeled off, the new plus strand becomes available as a template for yet another minus strand.

Revolving continuously, in this way, the minus strand would act as an endless template ex-

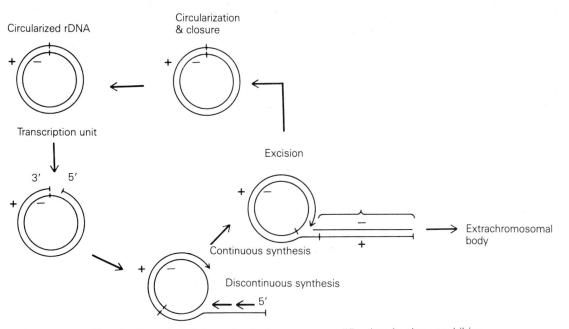

FIGURE 7.27. The "rolling circle" hypothesis for gene amplification in the amphibian oocyte.

tending the plus strand, and double-stranded molecules would be completed by the synthesis of the complementary minus strands. The double-stranded DNA with both plus and minus strands would be excised from the lariat's tail upon the completion of several revolutions and would be incorporated into an extrachromosomal body, or the free gene might be circularized and enter the cycle of replication itself.

The process would seem to have an indefinite capacity for repetition, but (fortunately for the oocyte) it stops before pachytene. Still the amount of amplification is prodigious. In prepachytene *Xenopus laevis* oocytes, each rRNA gene is cranked out an average of 1500 times.

The cytological picture of gene amplification in oocytes begins to take shape in the early pachytene nucleus (Fig. 7.25d). The nucleolus breaks up into numerous nucleoli, and a nuclear **cap** forms containing a high concentration of DNA (McGregor, 1968). Inoculation of female frogs with tritiated thymidine and subsequent autoradiography of oocytes shows that replication begins in the halo around the zygotene nucleolus and expands to the nuclear cap during pachytene. The absence of axial cores and synaptonemal complexes suggests that **cap DNA** is extrachromosomal. The cap acquires a fibrogranular appearance and numerous nucleoli (Coggins, 1973).

Gene amplification is not a universal part of

oogenesis. It does not occur in meroistic oogenesis in insects or even in the panoistic sea urchin. Amplification can occur, however, even when nucleoli are not produced, as in oocytes of some annelids and mollusks. Moreover, in some insects (e.g., in tipulid [crane] flies, gyrinid [whirligigs], dytiscid [diving] beetles, and some orthopterans), gene amplification associated with ribosomal-like DNA occurs in the **chromatin-** or **DNA-body** of oocytes and not in nucleoli (Lima-de-Faria et al., 1972).

Structural and regulatory genes are not generally amplified during the course of development. Exceptions occur in some ciliates where the macronuclei exhibit gene amplification as well as polyploidy. In *Drosophila*, genes for the synthesis of egg membrane or **chorion** proteins are amplified more than 10-fold in follicular cells (Spradling and Mahowald, 1980). In vertebrates, the phenomenon of gene amplification may be of clinical importance, since amplification affects genes conferring methotrexate resistance to tumors (see Schimke, 1982). These genes transcribe mRNA for the enzyme, dihydrofolate reductase (hence they are called *dhfr* genes).

Like the amplified nucleolar organizer gene of the oocyte, the amplified *dhfr* genes are capable of both replication and transcription, and, like amplified extrachromosomal nucleoli, the amplified *dhfr* genes occur as extrachromosomal elements, in this case, known as **double minute** (DM) chromosomes. Unlike amplified nucleoli, however, amplified *dhfr* genes, like bacterial episomes, may become permanently incorporated into a chromosome as conspicuous **homogeneous staining regions** (HSRs).

Extrachromosomal nucleoli and most DM chromosomes tend to be transient structures. The oocyte's amplified nucleoli are shed from the nucleus at the first meiotic division and later disappear from the cytoplasm, while DM chromosomes are frequently lost during cell division.

Germinal Vesicle or Dictyate Substage

The second meiotic substage is a period of moderate growth corresponding to the **diplotene** subphase or, in mammals, a modified **diffuse diplotene** subphase of meiotic prophase. The chromosomes form either typical chiasmata (see Fig. 3.13) or networks of **dictytene** or **dictyate chromosomes** (Fig. 7.23). The nucleus swells into a full-blown germinal vesicle.

In general, the germinal vesicle substage tends to be the longest in oocyte development with a duration roughly proportional to a species' life expectancy. Dictyate oocytes last 3 months in crickets, 4–8 months in *X. laevis*, and 7 months in newts, although, in the chick, the earlier substage is prolonged, and the germinal vesicle substage lasts only 3 weeks.

In mammals, the dictyate substage is sometimes called the **quiescent stage** and equated to the **first meiotic arrest** (as distinct from the second meiotic arrest that occurs in secondary oocytes), and it can be extremely protracted. For example, the oocytes in newborn mice are in leptotene and early zygotene. By 4 days after birth, they have entered the pachytene or early diplotene subphase where they will reside for no fewer than 12 days when the mouse reaches puberty. Thereafter, some oocytes continue meiosis, but others will remain in the dictyate substage for as long as 2 years.

In cats, oocytes present at birth are predictyate, and some will continue there until puberty, but thereafter they will all be in the dictyate substage. In rabbits and several small rodents where oogonia are present at birth, diplotene oocytes appear early in the neonatal period. In most other eutherian mammals (e.g., cows, guinea pigs, humans, monkeys, rats, sheep), oocytes are diplotene (oocyte stage 4, Fig. 7.23) at birth and rapidly move on to the dictyate substage (oocyte stage 5). The dictyate substage can therefore last from soon after birth to

the end of the organism's reproductive life (e.g., 40–50 years in women).

The accumulation of RNA is characteristic of germinal vesicle oocytes, even those of mammals where RNA synthesis occurs at reduced rates. While in panoistic eggs, RNA is manufactured by the oocyte itself, in meroistic eggs, the RNA is "foreign."

Transcription in nurse cells in meroistic insects supplies oocytes with RNA. The oocyte's own chromosomes **spiralize** or form a **karyosome** or **nuclear body** which has an extremely low rate of transcription (see Bier et al., 1972).

Nurse cells, on the other hand, undergo repeated **endoreduplications** (i.e., replication without mitosis) and develop **polytene** chromosomes containing many conjugated chromatin strands capable of transcription at high rates. The genome of *Drosophila* increases 1024-fold, while that of the moth *Antheraea polyphemus* increases 30,000-fold.

Large amounts of RNA accumulate in nucleolus-like material associated with the excess DNA and in giant ring-shaped nucleoli (e.g., moth nurse cells), presumably containing ribosomal RNA. The material is then transported to the oocyte through the connecting nutrient chord or ring canals (Fig. 7.28) (Bier, 1963). Meroistic oocytes utilizing foreign rRNA support protein synthesis at a higher rate than panoistic oocytes utilizing their own rRNA.

Lampbrush chromosomes are distinctive features of germinal vesicle oocytes in panoistic species. Although mammalian dictyate chromosomes are generally too dispersed to give anything more than the impression of a network, they too are thickened by **lateral loops** arising from a central axis. In some panoistic insects, lateral loops give the diplotene chromosomes a hairy appearance, but, in amphibians, loops give the chromosomes a wiry appearance, hence the title **lampbrush chromosomes** (see Fig. 3.13).

The germinal vesicle chromosomes contain a thin **chromosomal filament** or **central axis** of two DNA duplexes and many thick **chromomeres** of condensed chromatin. The lateral loops of the chromosomes are pairs of single DNA duplexes swinging out of chromomeres (Fig. 7.29). Loops range from 6 μm in the oocytes of panoistic insects to 200 μm in amphibian oocytes. As many as 5000 loops per chromosome are present in *Notophthalmus* (= *Triturus*) *viridescens* (see Callan, 1982).

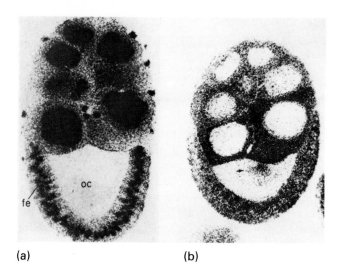

(a)   (b)

FIGURE 7.28. Autoradiographs of sections of the fly, *Musca domestica,* ovary. (a) Fly fixed 30 minutes after incubation with tritiated cytidine. (b) Fly fixed 5 hours later. Nurse cell nuclei, cytoplasm, and the follicular epithelium (fe) are densely labeled after exposure to the nucleic acid precursor, but the oocyte is virtually unlabeled. Later, labeled material has shifted from nurse cell nuclei to cytoplasm, and labeled material (arrow), presumably RNA, has moved from nurse cells to the oocyte through intercellular bridges (ring canal). Reproduced from K. Bier, *Journal of Cell Biology,* 16:436 (1963), by copyright permission of the Rockefeller University Press.

One, but usually two or more, and sometimes as many as 10 **transcription units** (TUs, also known as **thin–thick** regions or "Christmas trees," Figs. 7.29 and 7.30) are present per loop but not necessarily facing the same direction. Each TU contains a 5′ **promoter site** where closely packed transcripts, like branches, are initiated (Miller and Hamkalo, 1972). Not only are the "Christmas trees" tall (averaging 6.3 μm in *X. laevis* and up to 50 μm in *N. viridescens*), but they have extraordinarily long branches of nascent transcripts (e.g., 100 kilobases in the newt).

RNA synthesis on the lampbrush chromosomes is awesome. At rates as much as 100 times faster than those in body cells, germinal vesicle oocytes even transcribe the antisense strand of DNA. In *N. viridescens* oocytes, nascent transcripts of the histone gene cluster include RNA coded by repetitive and spacer sequences of DNA and even highly repetitive satellite sequences not normally transcribed. Possibly the termination signal operating in ordinary body cells does not work in oocytes, and transcripts read through satellite regions and other genes of the same cluster that happen to be on the 3′ side of promoters (Gall et al., 1983).

A loose correlation between TU length, loop length, and the amount of nuclear DNA (i.e., the

Lampbrush chromosome (represented by newt)

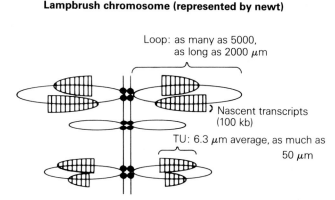

Loop: as many as 5000, as long as 2000 μm

Nascent transcripts (100 kb)

TU: 6.3 μm average, as much as 50 μm

FIGURE 7.29. Diagrammatic view of lampbrush chromosome. Chromomeres along the length of two DNA duplexes (filaments) give rise to lateral loops of single DNA duplexes. Transcription in domains or transcription units (TUs) produces nascent transcripts.

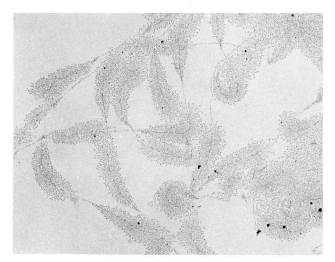

FIGURE 7.30. Electron micrograph of spread preparation showing a transcription unit or "Christmas tree" on part of a lampbrush chromosomal loop from the newt, *Notophthalmus* (= *Triturus*) *viridescens*. The "trunk" consists of a double helix of DNA from which transcripts of RNA grow as "branches." ×11,400. From O. L. Miller, Jr. and B. R. Beatty, *Science,* 164:955 (1969), by kind permission of the author.

_C_ value) in an oocyte suggests a relationship between TU length and the distance between structural genes if not a correspondence between TUs and structural genes. But nascent transcripts of heterogeneous nuclear RNA (HnRNA) in oocytes do not correspond to normal products of genes.

RNA synthesized during the germinal vesicle substage is largely of unknown function.

Only a small portion of oocytic sea urchin and amphibian RNA is even potentially mRNA. As much as 90% is large transcripts containing nontranslatable interspersed repetitive sequences (see Davidson, 1986), and large amounts of oocytic HnRNA are degraded in the nucleus. Moreover, the amount of stored polyadenylated (poly(A)$^+$) mRNA in the cytoplasm decreases during the germinal vesicle substage (Dworkin and Dworkin-Rastl, 1985).

In the sea urchin, the complexity of RNA sequences (i.e., the range of unique gene products) increases during the lampbrush phase (Hough-Evans et al., 1979). In amphibians, on the other hand, RNA synthesis, while continuing throughout the germinal vesicle substage, has produced no new species of RNA. In _X. laevis_, oocytic poly(A)$^+$ oocytic RNA processed before the appearance of full-blown lampbrush chromosomes produces the same profile of proteins _in vitro_ as oocytic poly(A)$^+$ oocytic RNA processed after the appearance of lampbrush chromosomes (Golden et al., 1980).

Still, some mRNA transcripts may be favored and accumulate in the cytoplasm. In the newt, for example, histone mRNA of the variety containing repeated sequences known as satellite 2 accumulates in the cytoplasm during the lampbrush phase, while histone mRNA containing repeated satellite 1 sequences is degraded in the nucleus (Mahon and Gall, 1984).

Some transcripts are peculiarly localized in the animal or vegetal caps of eggs. The **vegetal pole 1** (Vg1) mRNA in _Xenopus_ eggs, for example, is increasingly localized in the vegetal pole during oocyte growth (Melton, 1987). The protein coded by Vg1 is similar to **transforming growth factor β2** (TGF-β2) and may play a role in inducing dorsal mesoderm later in development (Weeks and Melton, 1987).

Oocytic mRNA is sometimes called maternal RNA because it is synthesized while the egg is within the ovary. The RNA is also called masked when not utilized in the oocyte. Bound in RNA-containing particles (sometimes called informosomes), mRNA is transferred to the zygote and utilized at some later time. Neither "maternal" nor "masked" is a completely appropriate adjective for the oocytic mRNA, since it is not synthesized by the maternal (i.e., somatic) tissue, and it may not be masked in the sense of hidden. Better terms are **oocytic** or **stored mRNA**.

Protein synthesis in the early embryo is largely at the behest of translational controls over mRNAs originating in the oocyte.

Several mechanisms seem to be involved in suppressing protein synthesis in the oocyte and several more in restoring protein synthesis in the embryo. The failure of oocytes to translate the largest part of their mRNA does not seem to be due to a deficiency in the mRNA. It is generally capped (an exception being the tobacco hornworm moth oocyte), polyadenylated (an exception being the surf clam, _Spisula_), and at least partially available for translation. In the sea urchin, for example, oocytic RNA recovered under conditions of high sodium concentration is translated in _in vitro_ systems (Moon et al., 1982). Possibly, the rapid loss of hydrogen ions and rise in pH and sodium accompanying fertilization expose mRNA for translation or activate initiation factors.

The difficulty unfertilized eggs have synthesizing secretory proteins may be due to a downstream "log jam." The translocation of secretory proteins into the intracisternal space of the endoplasmic reticulum depends on a multimeric **signal receptor complex** (SRC) that recognizes sequences of amino acids called **secretory signal sequences** on nascent polypeptides. Attached to the amine terminus of the polypeptides, the complex temporarily arrests translation.

Ordinarily, the complex is removed by docking protein on the endoplasmic reticulum, and translation resumes as the secretory protein enters the intracisternal space. In oocytes, however, docking protein is not available, hence the "log jam." Possibly, the synthesis of secretory proteins resumes in the zygote when docking protein becomes available (e.g., when new compartments of the endoplasmic reticulum are opened) (Walter and Blobel, 1983).

The oocyte's ribosomes have also been considered "masked," since they are largely inactive. Ribosomes in the cytoplasm are competent for protein synthesis, however, at least in the presence of exogenous mRNA. Operating with greater efficiency than _in vitro_ systems, _Xenopus laevis_ oocytes translate injected rabbit hemoglobin mRNA from injected reticulocyte 9 S RNA (see Lane and Knowland, 1975).

Control over the production of inert ribosomal subunits takes place within the nucleus. Ribo-

TABLE 7.5. Type and time of RNA synthesis in *Xenopus laevis*[a]

Class of RNA	Numbers of Genes	Percentage of Stored RNA	
		Early Oocyte (50–150 μm)	Late Oocyte (400–1200 μm)
Ribosomal			
28 S	2,000,000 (amplified)	15%	63%
18 S	2,000,000 (amplified)	15%	32%
5 S	24,000–100,000	40%	2%
Transfer	40,000	35%	2%

[a]Data mainly from H. Denis. In R. Harris et al., eds., *Cell differentiation*, 1st International Conference on Cell Differentiation. Scandinavian University Books, Munksgaard, Copenhagen, 1972.

somal RNA is transcribed from two different sources. RNA polymerase I transcribes the amplified nucleolar organizer region as a 40 S rRNA precursor (Table 7.5). During subsequent RNA processing, transcripts are whittled down to 18 S, 5.8 S, and 28 S rRNAs (see Fig. 4.3).

In *Xenopus laevis*, the small **oocyte-type 5 S rRNA gene** (only 120 bases) is highly recurrent (representing a family with as many as 20,000 members) in the amphibian genome and is not amplified. RNA polymerase III (RNA pol III) is active throughout the germinal vesicle substage, and great amounts of the precursors of 5 S rRNA (and of transfer RNA, also transcribed by RNA pol III) are transcribed (Table 7.5).

Transcription of the 5 S rRNA gene is aided by an accessory **transcription factor** called **TFIIIA** (i.e., the first, or A, transcription factor discovered for RNA polymerase III), a protein of 40 kd that binds to a 50-bp control region within the gene. Excessive amounts of the protein (on the order of 3×10^9 molecules per cell) present during early oogenesis may differentially direct RNA pol III toward the 5 S rRNA gene's initiation sites, if only because the promoter regions of tRNA genes do not bind the transcription factor (Hofstetter et al., 1981).

The 5 S rRNA genes come in two varieties differing by as few as three bases in the internal control region of the coding sequence, an oocytic variety transcribed exclusively in the oocyte and a small somatic or adult family transcribed elsewhere. The somatic 5 S rRNA genes have greater affinity for TFIIIA than the oocytic genes, but the large amount of TFIIIA in the oocyte allows transcription of both varieties of rRNA.

After transcription, TFIIIA remains bound to the transcribed rRNA as a stable 7 S RNP storage particle that moves into and accumulates in the cytoplasm. The production of oocytic 5 S rRNA may be self-limiting, since as TFIIIA becomes in-

creasingly tied up in particles, it is proportionally less available for transcription. The greater affinity of the somatic variety of the gene, moreover, precludes synthesis of the oocytic variety when nuclear TFIIIA concentration is reduced (Honda and Roeder, 1980).

Stockpiles of 5 S rRNA and the other rRNAs are balanced by the transcription of 5 S rRNA throughout early oogenesis as opposed to transcription of the nucleolar organizer region primarily in late oocytes (see Krämer, 1985). When finally assembled in extrachromosomal nucleoli, *X. laevis* oocytes will have manufactured approximately 1.1 $\times 10^{12}$ ribosomes and supplied the cytoplasm with 1000–10,000 more ribosomes than a typical somatic cell.

Vitelline Substage

Primary oocytes precede diakinesis and the resumption of meiosis with a period of rapid growth. In mesolecithal and megalecithal eggs, the growth is accompanied with the accumulation of large amounts of yolk, or deutoplasm (Fig. 7.24), hence the name **vitelline** or **vitellogenic substage**.

Even alecithal eggs grow rapidly prior to ovulation. Generally, nucleolar material enters the cytoplasm, presumably passed through nuclear pores. In the sea cucumber, *Thyone*, ribosomes are transferred into the cytoplasm when the whole nucleolus is moved directly out of the nucleus. In alecithal mammalian eggs (Fig. 7.23, oocyte stage 6), a dramatic increase in transcription occurs as the oocyte moves past the germinal vesicle substage (i.e., escapes from its first meiotic arrest) and progresses to diakinesis (corresponding to the onset of folliculogenesis, see below).

The substage may be short-lived in alecithal eggs, however. As the mammalian oocyte and follicle grow, the rate of ribosome synthesis slows

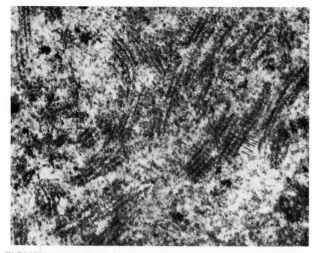

FIGURE 7.31. Electron micrograph of mouse oocyte cytoplasm showing bundles of fibrous material that have replaced ribosomes. Bars (right center) illustrate the periodicity of the fibrous material. ×36,000. From R. B. García, S. Pereyra-Alfonso, and J. Roberto Sotelo, *Differentiation,* 14:101 (1979), by permission of the International Society of Differentiation, Springer-Verlag, and the authors.

again, and monosomes (single ribosomes or subunits) are replaced by polysomes. Finally, as the mammalian oocyte approaches its first meiotic division (also known as maturation), transcription ceases, and ribosomes are tied up in bundles of fibrous material possibly as a protection against adverse effects of fertilization (Fig. 7.31) (García et al., 1979).

In mesolecithal and megalecithal eggs, a generalized growth accompanies vitellogenesis. The germinal vesicle swells, and the amount of cytoplasm increases.

The number and bulk of mitochondria increase enormously. For example, in the marine pseudocoelomate, *Priapulus,* the number of mitochondria increases from 5–8 in oogonia to 40,000 (Nørrevang, 1968). *Xenopus laevis* oocytes end up with 200 times more mitochondrial DNA than chromosomal DNA (Brown and Dawid, 1968). Accordingly, the rates of oxygen uptake (Table 7.3) and of oxidative phosphorylation increase, albeit not proportionally.

In frogs, precursors of cortical particles arise free in the cytoplasm, possibly from the Golgi apparatus. The particles grow and move toward the outer layer of cytoplasm.

The greatest part of the growth of mesolecithal and megalecithal eggs is due to the accumulation of deutoplasm. In amphibians, the oocyte's increase in volume during the month of vitellogenesis rivals the increase registered over the previous 2 years (Fig. 7.24, Table 7.3). The volume of terminal oocytes in the roach, *Blatella germanica,* increases 180-fold during the final 10 days, and the volume of hen eggs increases 200-fold during the last 1–2 weeks of oogenesis.

Yolk production or vitellogenesis in a variety of invertebrates, including primitive insects and crustaceans, is autosynthetic. The yolk protein is synthesized directly by oocytes. In advanced insects and vertebrates, vitellogenesis is **heterosynthetic** and the bulk of yolk is actually "foreign" to the oocyte. From the oligolecithal eggs of echinoderms to the cleidoic eggs of birds, yolk originates in the mother and is supplied to the egg through the mother's circulatory system or body cavity.

Typical of many **precursor–product** relationships, in heterosynthetic vitellogenesis, yolk first appears in a **native form** or **precursor,** produced somewhere in the parent other than where the mature and smaller product is ultimately deposited. This phenomenon was first discovered in the pupa of the silkworm, *Platysamia cecropia,* where a protein produced by the fat body and found in hemolymph was antigenically identical to yolk (Telfer, 1954, 1965).

The yolk precursor–polypeptide in many insects has an initial molecular weight of about 220,000. Before leaving the fat body, the precursor is reduced to a large and small polypeptide with molecular weights of 180,000 and 47,000 known collectively as **vitellogenin** (Vg). In bees, wasps, and primitive dipterans, such as mosquitoes, only a large Vg is produced, while in advanced dipterans, such as the fruit fly, only a small Vg is produced. When taken up by oocytes, Vg is modified to **vitellin** (Vn) by the removal of lipid and deposited as yolk (Postlethwait and Giorgi, 1985).

In the sea urchin, a glycoprotein yolk precursor with a molecular weight of about 220,000 is first found in maternal coelomic fluid. When finally stored in yolk granules, the protein is only 200,000 daltons (Harrington and Easton, 1982). Presumably, the portion of the molecule cleaved off in the process contains signals for transport and absorption without which the cells might not recognize the material or treat it appropriately.

The major vitellogenins (VTLs) in oviparous vertebrates are synthesized in the liver by hepatocytes (i.e., liver parenchymal cells) (see Wahli et al., 1981; Wily and Wallace, 1981). A minor avian yolk lipoprotein known as **very low density lipoprotein** (VLDL II) is also synthesized in the liver and follows the same route to oocytes. Both yolk precursors are normally **sex-limited,** since they are produced in the female under the influence of estrogen. However, in response to estrogen treatment, VTL and VLDL II synthesis can take place in the male's liver as well as in the female's.

VTL synthesis lends itself very neatly to experimentation especially in male birds and in *in vitro* frog liver systems. By cloning the genes for VTL from *X. laevis* in *Escherichia coli*, moreover, probes have become available for studying the hormonal control of transcription in vitellogenesis.

An initial injection of estradiol-17β in roosters increases the number of estrogen receptors in the nucleus of male hepatocytes from 100 to about 1000 and exposes three specific DNase I-hypersensitive sites (i.e., transcription sites) and a demethylated site in the upstream 5' region (Burch and Weintraub, 1983). Detectable amounts of VTL mRNA appear 4.5 hours after the injection, and VTL itself appears within 12 hours. The amount of VTL mRNA continues to increase until it reaches a concentration of 30,000 messengers per cell at 10 days. At that time, VTL accounts for an incredible 70% of the hepatocyte's protein.

At about 8 days after a single injection of estradiol, the number of estrogen receptors in the hepatocyte nucleus begins to decline, but the number levels off at about three times its original level. By 30 days, VTL synthesis has stopped, and by 40 days, VTL is no longer detectable.

One of the DNase I-hypersensitive sites exposed by estradiol appears to be unstable, but the other two remain even following the disappearance of estrogen. Presumably, because only the unstable site has to be reactivated, a second injection of estradiol-17β restores VTL mRNA production at an accelerated rate (within an hour), and after only 1 week, the hepatocyte produces 100,000 VTL mRNAs per day!

The *X. laevis* VTL gene is actually a small class or family of genes (A1 and A2, B1 and B2) lying within a stretch of 6.3 kb of DNA and differing by only 5% of their nucleotides in their coding regions (Wahli and Dawid, 1979). Divergence of about 20% occurs in the noncoding regions, and the 33 introns contained in the genes differ considerably in sequence and length. Nevertheless, the introns interrupt the coding sequences of the different genes at very nearly the same points. Processing seems to be very complex, since it takes different routes in different hepatocytes but ends with the same series of exons in the final mRNA.

Following translation in the hepatocyte, VTL polypeptide chains are **lipidated, phosphorylated, and glycosylated.** The molecules leaving the liver are enormous **phospholipoglycoprotein** dimers of about 460,000 daltons, having identical polypeptide monomers, 12% lipid, 1.5% phosphate with attached calcium, and 1% carbohydrate.

The VTL dimers in vertebrate blood are very stable despite their size, having a half-life of 40 days in ovariectomized females and in males. Vertebrate VTL reaches the oocyte unaltered and moves through intercellular spaces in the surrounding follicular cells without modification.

The packaging of yolk begins when it is taken up by endocytosis in large coated vesicles (Fig. 7.32). The clathrin-stabilized material of the coated vesicles is probably recycled back to the cell surface, while membrane-bound yolk is carried to platelets or spherules. Yolk and other nutritional materials may also be taken up passively by pinocytosis (i.e., cellular "drinking").

The difference between the relatively large yolk platelets of the amphibian vegetal hemisphere and small platelets in the animal hemisphere is a consequence of **directional transport** following endocytosis. Vesicles formed in the animal hemisphere tend to move to the vegetal hemisphere, while vesicles formed in the vegetal hemisphere tend to fuse with nearby yolk platelets (see Gerhart, 1987).

With the exception of reptiles and birds, cell organelles are frequently associated with yolk. The

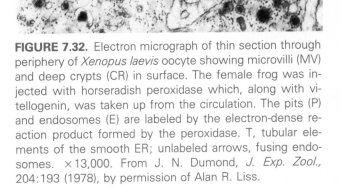

FIGURE 7.32. Electron micrograph of thin section through periphery of *Xenopus laevis* oocyte showing microvilli (MV) and deep crypts (CR) in surface. The female frog was injected with horseradish peroxidase which, along with vitellogenin, was taken up from the circulation. The pits (P) and endosomes (E) are labeled by the electron-dense reaction product formed by the peroxidase. T, tubular elements of the smooth ER; unlabeled arrows, fusing endosomes. ×13,000. From J. N. Dumond, *J. Exp. Zool.,* 204:193 (1978), by permission of Alan R. Liss.

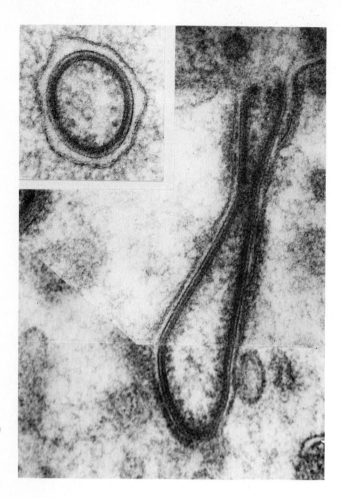

FIGURE 7.33. Electron micrograph through periphery of hen oocyte showing longitudinal section of a *lining body* indenting the oocyte's cell membrane (o.c.m.). (Insert) Same in cross section. × 78,000. From R. Bellairs, *Developmental processes in higher vertebrates,* University of Miami Press, Coral Gables, 1971. Used by permission.

organelles are cisternae of the Golgi apparatus in coelenterates, intramembranous spaces of the endoplasmic reticulum in crustaceans, mitochondria or spaces among mitochondria in gastropods, and the matrix of modified mitochondria in fish and, if rarely, ranid frogs (Fig. 6.13) (see Wallace, 1978; Williams, 1965).

In birds, bulbous extensions from follicle cells dent the oocyte's plasma membrane. Called **lining bodies** (Bellairs, 1971), the extensions may aid the passage of material into oocytes (Fig. 7.33). Alternatively, lining bodies, which have peripheral thickenings, may hold the oocyte in place while its border is unstable due to endocytosis. Were this the case, lining bodies would function as inverted versions of the tubulo–bulbar complexes that attach spermatids to sustentacular cells during cytoplasmic shedding. In amphibians, similar follicular cell processes (sometimes called macrovilli) are attached to oocytes through desmosome-like junctions until ovulation.

In the avian oocyte of about 1 mm in diameter, long before the onset of vitellogenesis, the germinal vesicle is already in position below the blastodisk, and **white yolk** occupies the position of the future latebra. A peripheral layer of white yolk is added, and, when the egg has reached 5 mm in diameter, the first layer of **yellow yolk** is added. Fat and large amounts of carotenoid pigments are responsible for the color of this yolk.

Vitellogenesis begins in earnest when the oocyte is about 6 mm in diameter and continues thereafter for 8–10 days. The **vitellus** (i.e., the "yolk") grows by about 2.5 mm per day to a final diameter of about 32 mm. As existing layers of yolk are thickened, additional layers are added. Thick layers of yellow yolk appear from daybreak to midnight, and thinner layers of white yolk accumulate between midnight and daybreak (Fig. 6.14) (see Bellairs, 1971; Romanoff and Romanoff, 1949).

The formation of granular yolk takes place in-

TABLE 7.6. *Xenopus laevis* yolk polypeptides: their molecular size in daltons and some distinguishing characteristics[a]

	Precursors					
	Vitellogenin (dimer)		460,000			
Monomer peptides	1.5% protein phosphate					
	200,000 (usual estimate)					
	197,000					
	188,000					
	182,000					

Proteins in yolk platelets

Lipovitellin 1 115,000 (average)			Lipovitellin 2 31,000 (average)		
alpha	beta	gamma	alpha	beta	gamma
121,000	116,000	111,000	34,000	31,500	30,500

Phosvitin

40% serine; 7.4% protein phosphate		
(dephosphorylated)	37,500	39,000
(sedimentation)	33,000	34,000

Phosvette 1		Phosvette 2	
4.8% protein phosphate	19,000	10.7% protein phosphate	13,000

[a]Data mainly from H. S. Wily and R. A. Wallace, *J. Biol. Chem.*, 256:8626 (1981).

side the oocyte. The dimer of VTL is cleaved to monomers (Table 7.6), and, in turn, the monomers are cleaved to different yolk proteins: the phosphoprotein, **phosvitin** (PV), and two lipoproteins, **lipovitellin** (LV) **I** and **II** (Wahli et al., 1981; Wily and Wallace, 1981). In addition, two yolk phosphopolypeptides known as **phosvettes** along with lipovitellins seem to be cleaved from still a VTL monomer not containing PV.

Each of the lipovitellins are resolvable into three polypeptides (alpha, beta, and gamma), and phosvitin is resolvable into two polypeptides. These are not only unique in their relatively high phosphate content (7.4%) but in their high serine content (40%). The molecular weights of the phosvitins are uncertain, because different estimates are derived from the results of polyacrylamide gel electrophoresis (requiring prior dephosphorylation) and from equilibrium sedimentation.

The formation of yolk crystals begins with the action of phosphatases. Partially dephosphorylated, phosvitin becomes insoluble. Beyond this, the mechanism for forming granular yolk material is uncertain.

The crystal skeleton of amphibian platelets is probably formed by LV dimers with important inter-LV contacts in various directions and domains for LV–PV and LV–lipid interactions. Phosphory-

lated moieties, especially PV, probably lie next to the water space in the crystal and make contact with LV dimers in a ratio of 1:4 (Lange, 1985).

The vitelline substage of eutherian mammals can only be considered a disappointment in comparison to birds and amphibians. Lipovitellin–phosvitin complexes have not been described in eutherian eggs, and in the mouse, protein levels in blood do not even increase in response to estrogen. Protein is taken up by mouse oocytes, and pinocytosis is observed in guinea pig oocytes, but the materials taken in are probably digested. Possibly, the vitellogenin genes were lost in the evolution of eutherian mammals (see Wallace, 1978).

OOTELEOSIS AND LUTEINIZATION

The secrets of development lie in the egg. In *Drosophila* it takes approximately 10 days to produce a mature egg, whereas embryogenesis only lasts for 22 hours.

W. J. Gehring (1984, p. 4)

The oocyte becomes competent for productive fertilization through **ooteleosis** at the end of oogenesis. Changes center on **maturation,** the resumption

TABLE 7.7. Stage of egg at fertilization and examples[a]

Fertilization during first meiotic arrest	
Class I[b]: Fertilization before meiosis begins or before first prophase (prometaphase) is completed. Activation triggers maturation.	
Immature oocyte still at beginning of meiosis	Trematode platyhelminthes, some polychaete annelids, echiuroids
Meiosis begun but oocyte nucleus intact (i.e., before beginning of meiotic division)	Mesozoa, sponge, *Ascaris* and other nematodes, chaetognaths, other polychaete annelids, some mollusks (e.g. Dentalium), canids (e.g., dog and fox) among mammals
Germinal vesicle broken down (but metaphase plate not established)	Starfish
Fertilization during second meiotic arrest	
Class II: Maturation proceeds to first metaphase before fertilization; activation triggers resumption of meiosis.	
Arrested at first meiotic metaphase	Nemertines, other mollusks, oligochaete and polychaete annelids, many insects, ascidians
Class III: Fertilization at second metaphase; activation triggers resumption of meiosis.	
First meiotic division completed; arrested at second meiotic metaphase	*Amphioxus*, most mammals, and other vertebrates
No second meiotic arrest	
Class IV: Maturation completed before fertilization. Activation triggers initiation of mitosis.	
Both polar bodies extruded	Coelenterates, sea urchins, some mammals (insectivores)

[a]Data from C. R. Austin, *Fertilization*, Prentice-Hall, Englewood Cliffs, NJ, 1962.
[b]Classes based on L. Rothschild, *Fertilization*, Methuen, London, 1946.

of meiosis, and **ovulation,** discharge from the ovary, but the occurrence and order as well as the names attached to changes differ among species (Table 7.7).

Maturation (or its first stage sometimes called preovulatory maturation) resembles the transition from the G-2 period to the M period in the ordinary cell cycles. While in mammals, the prolonged germinal vesicle substage, or the first meiotic arrest, moves imperceptibly into diakinesis, in most animals, maturation begins dramatically with the rupture of the oocyte's nuclear envelope. This event, corresponding to a meiotic prometaphase, is called **germinal vesicle breakdown** (GVBD), especially where it is conspicuous (e.g., echinoderms and amphibians as opposed to mammals).

Controls over maturation and ovulation differ widely among species, but fall into two broad categories: hormonal and genetic control components. The problem is to ascertain how these components operate in the context of the organism's reproductive cycles and within the confines of ovarian follicles.

Hormonal Control Components

Maturation seems to be controlled by hormones in all species of multicellular animals except those in which fertilization precedes maturation (Class I, Table 7.7) (see Wasserman et al., 1984). In *Aplasia*, so-called bag cells release neurotransmitters with maturation-inducing properties, and maturation taking place in the body cavity of many marine invertebrates requires substances released by the brain and other tissues. In the starfish, for example, a polypeptide **gamete-shedding substance** (GSS) or **radial nerve factor** (RNF) promotes follicular synthesis of a **maturation-inducing hormone** (MIH), now known to be **1-methyladenine** (1-MA) (see Schuetz, 1985).

In vertebrates, gonadotropins (FSH and LH) induce maturation and ovulation primarily by promoting the release of steroid hormones from follicular cells. In fish, gonadotropin stimulation of the ovary promotes steroidogenesis and the release of different progesterone-like products, and in frogs, progesterone production by follicular cells is crucial for maturation. Progesterone alone can replace follicular cells and gonadotropins for the induction of maturation *in vitro* (see Maller, 1985).

Ironically, rather than promoting maturation, the vertebrate follicle inhibits it. Transplanted to hormone-free medium, eggs within follicles (i.e., follicle-enclosed oocytes) fail to initiate maturation, while eggs removed from their follicles resume meiosis and undergo GVBD.

Since the addition of LH to medium allows even follicle-enclosed oocytes to initiate maturation *in vitro*, LH seems to overcome the follicular inhibition of maturation (see Tsafriri, 1978). Likewise, the LH surge that precedes ovulation (Fig. 7.34) seems to override one or more inhibitory influences localized in the follicle and possibly residing in follicular fluid. The presence of large amounts of estrogen within the follicle at ovulation suggests that estrogen may also be involved in removing the block to meiosis.

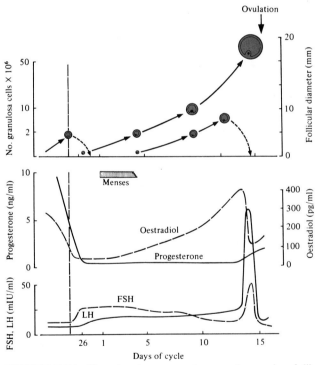

FIGURE 7.34. The simultaneous changes in ovarian follicles and in plasma levels of gonadal steroids and gonadotropins for human beings. (Upper panel) Growth in two follicles that undergo atresia (dashed line) as opposed to a dominant follicle that reaches ovulation (arrow). (Middle panel) Changes in plasma levels of gonadal steroids (oestradiol and progesterone). Menses, the sloughing of the uterine lining, provides a convenient point for counting the days of the cycle. (Lower panel) Plasma levels of gonadotropins (LH and FSH) showing the surge which is followed by ovulation 28–36 hours later. From D. T. Baird, *J. Reprod. Fertil.*, 69:343 (1983). Used by permission.

Genetic Control Components

Mechanisms for switching off the machinery of transcription that operate during the early dictyate stage and for turning on the machinery of meiotic chromosomal condensation that operates during meiosis I incorporate genetic control components. Much of the research on these components has been performed with *Xenopus laevis* oocytes and has depended on the injection of material into the cytoplasm or germinal vesicle at different times prior to, during, and after maturation (see Gurdon, 1982). An *in vitro* cell-free system employing extracts from *Xenopus* eggs at different stages has also proved efficacious (Lohka and Masui, 1984).

The germinal vesicle seems to contain material capable of promoting chromosome formation, since injected DNA is converted to stable **minichromosomes** with nucleosomes protecting the DNA from degradation (Wyllie et al., 1978). An oocyte component, it would seem, can also open the otherwise closed chromatin for transcription. Some of the transcribed genes of injected DNA including those for histones and SV40 proteins are subsequently translated in the oocyte cytoplasm, if only at low level. The oocyte variety of the 5 S RNA gene in injected somatic cell DNA is even recognized and selectively turned on (see Gurdon and Melton, 1981).

Foreign somatic cell nuclei, even those of mammalian cells (such as human HeLa cells), swell conspicuously after transplantation to *X. laevis* oocyte cytoplasm. The swollen nuclei resemble germinal vesicles (Fig. 7.35), and their chromosomes seem to enter a dictyate state.

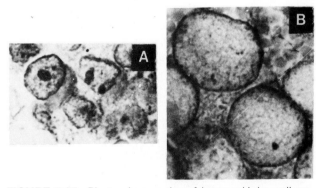

FIGURE 7.35. Photomicrographs of human HeLa cell nuclei 5 minutes after injection into *Xenopus laevis* oocyte cytoplasm (A) and 3 days later (B). ×850. From J. B. Gurdon, E. M. De Robertis, and G. Partington, *Nature* (London), 260:116 (1976), by permission of Macmillan Journals, Ltd. and the authors.

Some of the genes in mammalian nuclei transplanted to *X. laevis* cytoplasm are transcribed as mRNAs and subsequently translated. Other genes appear not to be transcribed despite their previous transcription in intact mammalian cells (Gurdon et al., 1976). The mechanism of both turning on and turning off genes in these mammalian nuclei may involve entry into the nuclei of **gene control proteins** from the oocyte cytoplasm (see Bonner, 1975).

Nuclei from adult *X. laevis* brain cells transplanted into oocytes at various stages of maturation (Table 7.8) perform the same types of activity performed by corresponding germinal vesicles. Actively engaged in RNA synthesis but not DNA synthesis, the previtelline oocyte induces RNA synthesis in brain nuclei but not DNA synthesis. At the end of its first meiotic arrest, at the point when its own chromosomes are condensing, the germinal vesicle oocyte induces chromosomal condensation in transplanted brain nuclei even though brain cells do not normally divide. Appropriately, a protein named **nucleoplasmin** is abundant in the germinal vesicle where it is thought to aid in the assembly of histones. Furthermore, a fertilized or activated egg, which has suspended transcription in favor of replication, induces replication in injected brain nuclei despite their normally residing in an indefinite interphase.

The putative gene control protein, known as **meiosis inhibitory factor** (MIF) or **meiosis-preventing substance** (MPS) in amphibians, is linked to the prolongation of meiotic prophase I and is thought to be eliminated prior to maturation (see Byskov, 1982). A similar substance may be present in mammalian follicular fluid and active in the first meiotic arrest. Able to retard or prevent the maturation of eggs *in vitro* even across species lines, this putative meiotic inhibitor may be a simple molecule (see Masui and Clarke, 1979).

One candidate for MIF is a polypeptide identified in bovine granulosa cells (the **oocyte matur-ation-preventing factor,** OMPF) and in pig granulosa cells and follicular fluid (the **oocyte maturation inhibitor,** OMI). In addition, the nucleotide hypoxanthine and cyclic adenosine 3',5'-pyrophosphate (cAPP) in follicular fluid inhibit maturation.

All these inhibitors may prevent meiosis through a common mechanism involving cAMP. The polypeptide inhibitors may induce cAMP by promoting adenylate cyclase activity, while the nucleotides may raise intraoocytic levels of cAMP by inhibiting phosphodiesterase (Sato et al., 1985).

Maturation, in turn, is induced when cAMP levels fall. The drop is not due to leakage, since cAMP does not accumulate in the medium. A variety of **mitogenic agents** and inducers of maturation in amphibians, from progesterone to insulin and insulin-like growth factor, may lower cAMP levels by inhibiting adenylate cyclase or by promoting phosphodiesterase activity. Diminution of cAMP presumably results in loss of cAMP-dependent protein kinases, which in turn may result in lower concentrations of the active phosphorylated forms of specific cell cycle inhibitors.

The induction of maturation also has a positive control side. A *Xenopus* oocyte protein once known as the **maturation-promoting factor** but now more generally known as **M-phase-promoting factor** (MPF) can overcome the cycloheximide-induced block to GVBD when injected into cycloheximide-treated oocytes. MPF promotes the breakdown of the nuclear envelope, whether the germinal vesicle envelope of oocytes, the prometaphase nuclear envelope of cleaving blastomeres and tissue culture cells, or the envelopes of nuclei reconstituted in a cell-free system by sperm, chromatin, or naked DNA (Dunphy et al., 1988).

In oocytes of the starfish, *Marthasterias glacialis,* **growth-associated** or **M-phase-specific histone H1 kinase** (H1K) plays the role of MPF. Triggered by the hormone 1-methyladenine (1-MA also abbreviated 1-MeAde), H1K kinase activity in-

TABLE 7.8. Changes in nuclear activity or chromosome configuration induced by the injection of adult brain nuclei into oocytes or eggs of *Xenopus*[a]

Type of Recipient Cell	Normal Activity of Recipient Cell	Activity Induced in Injected Nuclei	Time Within Which Response is Seen
Growing oocytes	RNA + DNA −	RNA + DNA −	1 day
Oocytes undergoing maturation	Condensed meiotic chromosomes	Condensed chromosomes	1 day
Activated eggs	DNA + RNA −	DNA + RNA −	1 hour

+, Active synthesis; −, no synthesis.
[a]Based on Gurdon (1982).

creases 30-fold and induces meiotic division. *In vitro*, the process depends on the removal of calcium and cAMP which inhibit it presumably through their activation of calcium- and cAMP-dependent kinases (Arion et al., 1988).

Similarly, the kinase activity of an MPF produced by human HeLa cells oscillates through the cell cycle. Coming on dramatically at the transition between G-2 and mitosis (known as the G-2/M transition), the kinase reaches its peak activity in metaphase and loses activity rapidly before cytokinesis (Draetta and Beach, 1988).

Active forms of *Xenopus* MPF and its starfish and human cognates are protein complexes ranging between 90 and 150 kd. The complexes consist of two or more major components, one between 32 and 34 kd, another between 45 and 62 kd, and, at least in the HeLa MPF complex, an additional 13-kd component.

Increased MPF kinase activity in HeLa cells at the entry into mitosis accompanies four changes in the complex: increased phosphorylation of the MPF's 34-kd component, association of the 13- and 34-kd complex with the 62-kd component, phosphorylation of the 62-kd component, and an unknown activating step of the complete p13–p34–p62 complex (Fig. 7.36). Similar changes may regulate the entry of all eukaryotic cells into mitosis as well as promote the maturation of oocytes.

A network of interacting cell cycle proteins, which has become available for study in molecular detail as a result of cloning yeast genes, has significant implication for MPF and MPF-like complexes (Gautier et al., 1988). In fission yeast, *Schizosaccharomyces pombe*, a key component of the network is the 34-kd p34^{cdc2} kinase that regulates the entry into DNA synthesis and independently the initiation of mitosis. In budding yeast, *Saccharomyces cerevisiae*, the similar and functionally interchangeable component is the p34^{CDC28} kinase that regulates the cell cycle "start" event following mitosis. In addition, fission yeast have several proteins that interact with p34^{cdc2}, one of which, the 13-kd p13^{suc1}, binds to p34^{cdc2}. The association is required for progression through the cell cycle.

The 32–34-kd component of MPF and its cognates resembles fission yeast's p34^{cdc2} kinase and budding yeast's p36^{CDC28} kinase immunologically and, in the case of HeLa MPF, by peptide mapping. The 45–62-kd component of MPF, like a 40-kd protein of budding yeast, may be the natural substrate for the complex's kinase activity, but yeast do not have a H1 histone to act as a substrate for phosphorylation. The 13-kd component of HeLa MPF resembles the 13-kd p13^{suc1} of fission yeast, and both *Xenopus* MPF and starfish H1K bind tenaciously to yeast protein p13^{suc1}. Moreover, bound *Xenopus* MPF loses the ability to induce nuclear-envelope breakdown (Dunphy et al., 1988).

The amount of yeast kinase activity does not oscillate through the cell cycle, but the resemblance of the 32–34-kd component of MPF and MPF-like complexes and the 34–36-kd cell cycle kinases of yeast is remarkable. Indeed, the 32–34-kd component is considered a homolog of the p34^{cdc2} kinase, sharing structural and functional properties, and a *cdc2* family of cell cycle genes is thought to form part of the regulatory gene network controlling the initiation of mitosis from yeast to human beings (Draetta and Beach, 1988).

Alternatively, MPF may operate through the phosphorylation of H1 histone, required for chromatin condensation into chromosomes. MPF may also promote mRNA recruitment by indirectly phosphorylating the 40 S ribosomal subunit (see Maller, 1985). In frogs, RNA and protein synthesis (including mitochondrial protein) is promoted at maturation in conjunction with cAMP-independent protein kinase activity and phosphorylation. The latter is not dependent on the former, since proteins are synthesized on preexisting mRNA (see Smith, 1975).

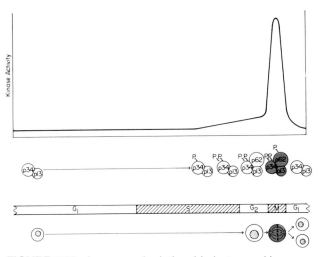

FIGURE 7.36. A proposed relationship between kinase activity, the HeLa MPF complex, and the cell cycle. The interphase complex of components p34 and p13 has no kinase activity until its p34 component is phosphorylated. Kinase activity increases accompanying increasing phosphorylation and association with the p62 component. An unknown activation step at mitosis brings about maximal kinase activity, and reversion to the original components at cytokinesis brings about kinase inactivation. From G. Draetta and D. Beach, *Cell,* 54:17 (1988), by permission of the Cell Press.

While total protein synthesis in frog oocytes increases during maturation two- to threefold to a maximum at the first meiotic metaphase, specific proteins are synthesized at much higher rates. The rate of core nucleosomal histone synthesis, for example, increases from about 50 to about 2500 picograms per hour, 400 times greater than the rate at which DNA will be synthesized during the first cleavage cycles. Surprisingly, the rate of histone H1 synthesis remains relatively low until cleavage (see Benbow, 1985).

Ovarian Cycle

When eggs are discharged from the ovary prior to maturation (i.e., preovulatory maturation, Table 7.9), the process is likely to be mediated by hormonal controls. In mammals, these controls operate an **ovarian cycle** in which the fate of the egg is intimately linked to the fate of a somatic **follicle** surrounding the egg.

The mammalian ovary oscillates between **follicular** and **luteal phases** (Fig. 7.37). During the follicular or folliculogenic phase, the follicle grows and differentiates.[6] Ovulation takes place at the climax of folliculogenesis. In the luteal phase, differentiation of the postovulatory follicle results in **luteogenesis,** the development of the **luteal gland** or **corpus luteum** (Lat. yellow body named by Marcello Malpighi for the structure in the cow which accumulates a yellow pigment, Fig. 7.37) (see Hutchinson and Sharp, 1977).

In the event of pregnancy, retention of the corpus luteum delays the return of folliculogenesis. The involution of the corpus luteum, or **luteolysis,** after abortion or parturition initiates a new ovarian cycle (see Weir and Rowlands, 1977). In the absence of pregnancy, a regular ovarian cycle may

---

[6]Folliculogenesis is also defined as the transformation of nests of oocytes to clusters of small follicles (Peters, 1978).

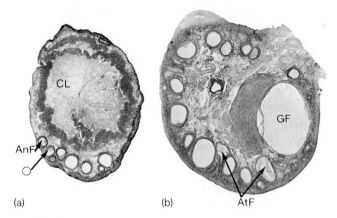

(a) (b)

FIGURE 7.37. Light micrographs of sections through ovaries of the crab eating monkey, *Macacca fascicularis.* (a) In the early follicular–late luteal stage. Section passes through oocytes (O) with their cumulus oophorus within an early antral follicle (AnF). The regressed corpus luteum (CL) of the last estrus fills a large part of the ovary. (b) In the late follicular stage. Large tertiary follicle (Graafian follicles, GF.) dominates the ovary. One (right) is conspicuously close to the edge. AnF: antral follicle, early. AtF: atretic follicle. CL: corpus luteum, regressed. GF: Graafian follicle (mature tertiary follicle). O: oocyte. Courtesy Dr. A. J. Zeleznik, University of Pittsburgh, School of Medicine.

continue indefinitely or throughout a breeding season, depending on species. The pattern of cycles is a species-specific characteristic influenced by nutrition and frequently by seasonal variables such as the duration of daylight (see Herbert, 1977).

Most female mammals are **spontaneous ovulators** with a regular ovarian cycle, although cycles differ greatly with species. Ovulation occurs once a year in **seasonally monestrus** mammals (such as foxes and roe deer), several times per breeding season in **seasonally polyestrus** mammals (e.g., horses, sheep, red deer, and many primates), and perenni-

TABLE 7.9. Ovarian phases of oogenesis in most adult mammals

Phase:	Quiescent			Preovulatory maturation and ooteleosis		
Periods of meiotic arrest:	First					Second
Stage of meiosis:	Dictyate			Diakinesis and GVBD	First division	Second metaphase
Type of egg:	Primary oocyte				Secondary oocyte	
Follicle stage:	Primordial	Primary	Secondary	Tertiary	Ovulation	

ally in **polyestrus** mammals (e.g., cows, pigs, rats, mice, chimpanzees, and human beings).

Some female mammals (rabbit, ferret, mink, field vole, cat, and camel) are **induced ovulators** who ovulate in response to coitus. The same Nembutal-sensitive, central nervous system mechanism for GnRH release operating in spontaneous ovulators operates in induced ovulators, but the crucial hormonal discharge is triggered by a nervous reflex arc rather than by the hypothalamic pulse generator (Fig. 7.5).

In both spontaneous and induced ovulators, ovulation operates as a physiological "circuit breaker" at the crest of a hormonal surge (Fig. 7.34). The profile of hormones secreted by the ovary changes from predominantly estrogen to progesterone, and the buildup of LH promoted by estrogen comes to a grinding halt. The sequence of events, rather than the events themselves, resets hormone secretion to tonic level.

The **estrus cycle** is measured by periodic changes in the sexual behavior of some female mammals and the coordination of ovulation with mating (see Short, 1984). Sexual solicitation or receptive behaviors by females accompanies ovulation at the **estrus** or **heat** phase of this cycle. A transitional period known as **metestrus** follows estrus and merges with **diestrus** (corresponding to the luteal phase of the ovarian cycle) and a period of female unresponsiveness to males. **Proestrus** (corresponding to the follicular phase of the ovarian cycle) reprograms the female's behavior for the next estrus.

The **menstrual cycle** (Lat. *mensis* month, referring to monthly cycles in human beings) seen in many Old World primates (the Cercopithecoidea and Hominoidea) is marked by a vaginal discharge or **menses** over the course of several days known as the **menstrual period** (also period of menses, menstruation, or simply the period). The discharge is debris from the uterine lining sloughed at the end of the luteal phase of the ovarian cycle. In nonmenstruating primates and in most mammals, the uterine lining does not slough or a discharge is not conspicuous.

Menses are provoked by the same internal conditions that regulate the ovarian cycle and therefore provide a convenient marker for identifying the phases of the ovarian cycle (see Speroff and VandeWiele, 1971). In the average 28-day human menstrual cycle, for example, folliculogenesis begins slightly before menses appear. The follicular phase lasts about 2 weeks and culminates in ovulation. The luteal phase follows and, 2 weeks later, another period of menses (Fig. 7.34).

Ovarian Follicle

Containing a single oocyte, an **ovarian follicle** (Lat. *folliculus* small bag or pod) is an epithelial bead[7] surrounded by a basal lamina or, as seen in the light microscope, a basement membrane. The epithelial cells are called **follicular cells** and, in the case of vertebrates, are comparable to sustentacular (Sertoli) cells of the testis by way of supporting and sustaining the germ cell and secreting estrogen.

Transient follicles consisting of one wholly or partially enveloping epithelial cell layer are seen in some echinoderms (asteroidea, holothuroidea, and possibly echinoids). A single layer of epithelial cells surrounds the oocytes of fish and amphibians. These cells are much taller on the animal pole side of fish oocytes than on the vegetal pole side, but the cells surrounding the amphibian oocyte, known as the *tunica granulosa*, are uniformly flattened.

In birds and reptiles, large and multilayered follicles form around oocytes and bulge from the ovary. In mammals, multilayered follicles also form, but these become swollen with a fluid-filled chamber and, rather than bulging, migrate into the ovary, returning to the surface only when they reach maturity.

The primordial follicle[8] (or follicular stage 1; Figs. 7.38 and 7.39) *of mammals consists of a small oocyte (about 30 μm in diameter in the human fetal ovary), generally in the dictyate stage, surrounded by a simple squamous epithelium of follicular cells.* Primordial follicles are concentrated in the **ovarian cortex,** the region just below the surface epithelium and dense connective tissue known as the **tunica albuginea.** The source of follicular cells is uncertain, and they are variously ascribed to the epithelial cells of the **rete ovarii** (remnants of the mesonephric kidney), the surface epithelium, or the stromal cells of the ovary.

Except in primates where a few follicles begin to grow prior to birth, all follicles in the fetus and the newborn mammal are primordial follicles. A few follicles may also begin growing while the mammal is still a juvenile, but, even after **puberty** (i.e., the onset of sexual maturity) and the initiation of the ovarian cycle, 90% or more of the fol-

---

[7]The follicle is sometimes considered the surrounding stroma as well as the epithelium. All the follicle's epithelial cells are then equated with the granulosa.

[8]The designations used here are consistent with the widely used nomenclature of Franchi and Baker (1973). See Dvorak and Tesarik (1980) for alternative nomenclature.

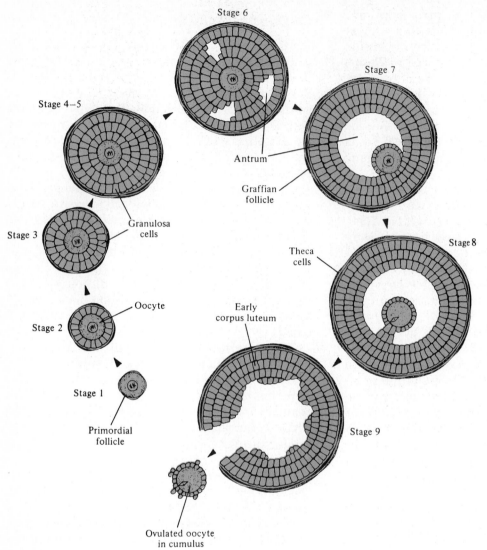

Stage 6

Stage 7

Stage 4–5

Antrum

Graffian
follicle

Granulosa
cells

Stage 3

Theca
cells

Stage 8

Oocyte

Early
corpus luteum

Stage 2

Stage 9

Stage 1

Primordial
follicle

Ovulated oocyte
in cumulus

FIGURE 7.38. Stages of folliculogenesis. The *primordial follicle* represents stage 1. The *primary follicle* includes the slowly growing stages 2–5. Rapid growth commences at stage 6 and *secondary follicles* acquire *antral cavities* which coalesce into an *antrum. Tertiary follicles,* also known as Graafian or mature follicles, correspond to stages 7 and 8 during which the oocyte is progressively elevated on the cumulus oophorus. At stage 9, the follicle ruptures and releases the oocyte with a loose covering of cumulus cells known as the *corona radiata.* The follicle becomes transformed to the corpus luteum. From T. G. Baker. In C. R. Austin and R. V. Short, eds., *Reproduction in mammals,* Book 1, *Germ cells and fertilization,* 2nd ed., Cambridge University Press, Cambridge 1982. © Cambridge University Press 1972, 1982. Reprinted with the permission of Cambridge University Press.

licles remain primordial. These follicles serve as the source of growing follicles throughout life or, in what may be a uniquely human situation, until menopause and the cessation of ovarian cycling (e.g., at 40–50 years).

Primary follicles represent an incipient growing stage in mammalian folliculogenesis. When the squamous cells surrounding the primordial follicle have thickened to **cuboidal** cells (follicular stage 2; Figs. 7.38 and 7.39), and when additional follicular cells produced by mitosis have accumulated (follicular stage 3), the follicle is known as an **early primary follicle.** As still more follicular cells accumulate, and as the original simple epithelium is transformed into a **stratified** (multilayered) **cuboidal epithelium,** the follicle is known as a **late primary follicle** (follicular stages 4–5).

A loose connective tissue layer known as the **theca** (Lat. a case enclosing something) surrounds

the primary follicle and invests it in maternal circulatory vessels. The population of thecal cells closest to the follicle is especially dense and well aligned with the follicle's basement membrane. These thecal cells are called **epithelioid** even though they are derived from the connective tissue. After a peripheral, coarse fibrous connective tissue layer is added to the theca as the **theca externa,** the highly cellular epithelioid layer and richly vascularized thecal layer close to the follicle are known as the **theca interna.**

In some mammals (e.g., rabbits and hares), theca interna cells, known as **interstitial cells** (possibly left over from atretic follicles; see below), are concentrated near the surface of the ovary in **interstitial glands.** The interstitial cells resemble the interstitial endocrinocytes (Leydig cells) of the testis and secrete large amounts of steroid hormones. In humans and some other mammals, similar concentrations of interstitial cells are found at the

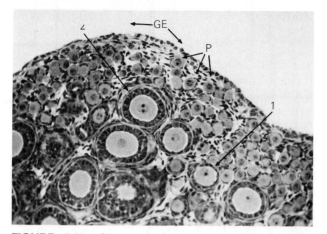

FIGURE 7.39. Phase contrast micrograph of section through cortex of an eight-day-old mouse ovary. GE, germinal epithelium (mesothelium of ovary). P, primordial follicles. 1 and 2, early growing follicles (note deeply staining nucleoli in germinal vesicles). 1, with one layer of cuboidal follicular cells. 2, with two layers of cuboidal follicular cells. ×260. From R. Bachvarova. In L. W. Browder, ed., *Developmental biology,* Vol. 1, *Oogenesis,* Plenum Press, New York, 1985. Used by permission.

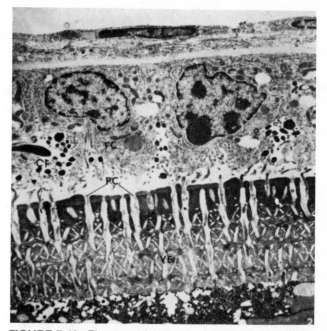

FIGURE 7.40. Electron micrograph of section through mature chorion and follicle cells (FC) of *Fundulus,* a minnow. Chorionic fibrils (CF) appear between follicle cells and pore canals (PC) penetrate the membrane. ×4200. From J. N. Dumont and A. R. Brummett. In L. W. Browder, ed. *Developmental biology,* Vol. 1, *Oogenesis,* Plenum Press 1985, New York, 1985. Used by permission.

hilus of the ovary (Fig. 7.37) along with major blood vessels. These interstitial cells secrete androgens.

As the primary follicle thickens, trypsin-digestible and periodic acid Schiff staining (i.e., PAS-positive) material appears in the spaces between follicular cells near the oocyte. This glycoprotein coalesces into the primary or vitelline envelope known as the mammalian **zona pellucida.**

In most vertebrates, both the oocyte and the follicular cells seem to contribute material to the envelope. In salamanders, two distinct layers form (see Wischnitzer, 1966), but in teleosts, the PAS-positive material becoming the **chorion** seems to be entirely a product of the oocyte (see Dumont and Brummett, 1985).

The formation of the vitelline envelope is influenced by the presence and withdrawal of cellular projections. The surface of the oocyte erupts with **microvilli.** Possibly serving as a passive barrier to laying down too tight an envelope early in development (Guraya, 1978), the microvilli allow the active surface of the cell to increase enormously during the subsequent growth phase.

Projections, sometimes called **macrovilli,** from the follicular cells generally develop later than the oocyte's microvilli. The macrovilli give the membrane a radially striated appearance, hence the name **zona radiata.** In teleosts, **radial** or **pore canals** in the **chorion** are filled with oocytic microvilli and follicular cell processes also known as **chorionic fibrils** (Fig. 7.40). Microvilli not withdrawn at ovu-

lation shape the **micropyle** through which sperm later penetrate the teleostean chorion.

Growth of follicles in mammals is only regular and sustained after puberty. The few follicles beginning growth prior to puberty in primates are presumably stimulated by an early spurt of gonadotropins, but in the absence of a sustained hormonal supply, these follicles degenerate. Only follicles that begin growing after puberty have a chance of sustaining maturation.

Neither the initiation of growth nor the early maintenance of growth in follicles is dependent on gonadotropins, since both occur even in hypophysectomized animals. The number of early growing follicles per ovary, moreover, is relatively constant throughout the normal ovarian cycle, while gonadotropin levels normally change radically (Fig. 7.34).

Much as primordial follicles provide a pool for growing follicles, follicles in early stages of growth provide a steady-state **precursor** population for more advanced growing follicles. During every ovarian cycle, some of the follicles in this precursor population enter a period of rapid growth.

Even the rapidly growing follicles do not develop at the same rate and gradually one or more

dominant follicles emerge, while other follicles degenerate. The phenomenon of selecting particular follicles for growth from the population of quiescent follicles and of choosing dominant follicles from growing ones is called **germinal selection.** Although the number of rapidly growing follicles is correlated statistically with gonadotropin level, the mechanism of germinal selection as a whole is presently unknown (see Baker, 1982).

In human beings, between menarche (the first menstruation) and menopause (the last menstruation), about 20 follicles per month enter advanced stages of growth. Usually only one dominant follicle continues to grow and reach maturity. The frequency with which two dominant follicles mature per month, as indicated by the frequency of **dizygotic twins** (i.e., twinning resulting from the fertilization of two eggs), seems to be greatly influenced by regional and genetic factors (Table 7.10). Multiple ovulations (or superovulation) are promoted by **priming** doses of gonadotropin administered several days prior to the induction of ovulation.

Spurred by increasing amounts of LH present during the cycle, the typical human dominant follicle grows from about 2 mm in diameter on day 1 of the cycle to 23 mm in diameter at the time of ovulation about 2 weeks later. Palpable and visible via ultrasound, the follicle's cell number has increased 100-fold during this period, and the quantity of estrogen released by the ovary into circulation has expanded exponentially (Fig. 7.34, middle panel).

The last stages of follicular growth are stimulated by elevated amounts of FSH. Injections of FSH with LH or with pregnant mare serum gonadotropin (PMSG) are superior to injections of FSH alone in promoting growth, but pretreatment with estrogens is also necessary to promote the final stage of folliculogenesis in hypophysectomized rats (see Baird, 1984).

Ironically, at the same time that the dominant follicle is reaching the zenith of its growth, other follicles that began growing at about the same time suffer the opposite fate. Unable to sustain growth during the long period of diminishing FSH secretion (Fig. 7.34, bottom panel), these follicles degenerate.

Degeneration occurs in imperforate or atretic follicles (Gk. a without + tetrainein to pierce). Ovulation fails and the trapped oocyte degenerates within the follicle.

Massive amounts of degeneration also occur among nests of oogonia in fetal gonads before follicles are formed and in primordial and early growing follicles during prepubescence (Table 7.4) (see Gondos, 1978). In adult mammals, atresia eliminates most of the growing follicles during each month's period of folliculogenesis, and in the case of human beings, atresia ultimately eliminates the population of primordial follicles remaining at menopause (see Weir and Rowlands, 1977).

Atretic follicles (Fig. 7.37) form a heterogeneous population of small scars initially containing the remnants of the follicle's basement membrane and the oocyte's zona pellucida. The first sign of degeneration seen at the electron microscopic level is the retraction of the oocyte's microvilli and the follicular cells' surface projections. At the light microscopic level, the oocyte's nuclear membrane wrinkles, and chromosomes condense. The oocyte may then complete its meiotic divisions (called pseudomaturation) and even appear to undergo cleavage, but the atretic follicle is soon invaded by macrophages, and the egg is destroyed.

The follicular cells of primordial follicles may survive atresia, but those of growing follicles do not. The doomed follicular cells detach from the oocyte. Their nuclei become pycnotic, and, lying in the follicle, the cells degenerate.

Unlike follicular cells, thecal cells survive atresia. The stroma surrounding the atretic follicle thickens, invades, and ultimately takes over the area formerly occupied by the follicle. Because the cytological differences between epithelioid thecal cells and follicular cells is obliterated in the process, follicular cells are sometimes thought to survive atresia by transforming into stromal cells.

Atresia represents a negative form of germinal selection in which the growth of one follicle rather than another is turned off, and degeneration re-

TABLE 7.10. Incidence of dizygotic (two egg) twins

Country (and Racial Identification)	Incidence of Dizygotic Twins Per 1000 Births
Japan	2.7
Spain	5.9
France	7.1
India	6.8
USA (white identified)	7.1
West Germany	8.2
Sweden	8.6
England and Wales	8.8
Greece	10.9
USA (black identified)	11.1
Congo	19.0
Nigeria	42.0

From I. MacGillivray, P. P. S. Hylander, and G. Corney, *Human multiple reproduction.* Saunders, Philadelphia, 1975.

sults. Statistically, atresia seems to be controlled by gonadotropins, since hypophysectomy reduces atresia, and the injection of exogenous gonadotropins promotes atresia; but atresia involves all ovarian compartments, and the mix of interactions is unclear.

Possibly, only the follicular cells of the dominant follicle present enough LH receptors at the time of the LH surge to bring about ovulation. In the absence of sufficient quantities of receptors on the follicular cells, LH may act primarily on stromal cells, causing the secretion of progesterone. The same hormone that stimulates ovulation in the dominant follicle may be the culprit that damages oocytes in less advanced follicles.

Another possibility is suggested by the ability of exogenous estrogen to retard the rate of atresia in hypophysectomized mice. If estrogen is necessary to preserve follicles, flux in estrogen production following the pulsatile secretion of LH by the pituitary (Fig. 7.5) may jeopardize small follicles, while dominant follicles with large reserves of estrogen are unaffected.

Genetic damage and the absence of an X or Y chromosome in human oocytes (e.g., as in the X0 condition known as Turner's syndrome) also cause increased loss of potential germ cells through atresia. Oocytes themselves may therefore give or fail to give signals that influence their survival.

Secondary follicles, also known as vesicular or antral follicles (follicular stage 6; Fig. 7.38) ***contain cavities or antra*** (Lat. caves) ***filled with a PAS-positive material*** (sometimes called Call-Exner bodies). FSH is normally required for the formation of these cavities.

Originally, the oocyte and follicular cells' plasma membranes are simply apposed to each other, but during the antral stage, intercellular contacts develop into intimate nutritive and presumably communicative connections. Points, where desmosomes once fastened follicular cells to oocytes, lengthen into thick projections from the surface of follicular cells, and abundant microvilli from the oocyte surface radiate into the zona pellucida. Gap junctions (Fig. 7.41) also form, presumably providing important highways for the transport of selected molecules (Anderson and Beams, 1960).

As the secondary follicle grows, the antral cavities fuse into a single **antrum** (or antrum folliculi) filled with **follicular fluid** (or liquor folliculi) less PAS positive than the fluid of earlier cavities. Except for high levels of estrogen, follicular fluid resembles blood serum in composition and is pre-

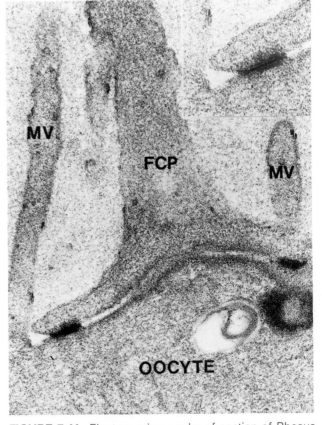

FIGURE 7.41. Electron micrographs of section of Rhesus monkey, *Macacca mulatta,* ovary showing gap junction (GP) between foot of bifurcated follicular cell process (FCP) and oocyte. MV, microvillus arising from oocyte. Lanthanum-impregnated to show gap junctions. ×70,700. Insert (upper right): ×140,000. Reproduced from E. Anderson and D. F. Albertini, *J. Cell Biol.,* 71:690 (1976) by copyright permission of the Rockefeller University Press and the authors.

sumably derived from it with little alteration by the follicular cells.

The tertiary, mature, or Graafian follicle (follicular stages 7–8; Fig. 7.38 right edge, and 7.42) ***consists of a layer of follicular cells surrounding a central antrum and supporting the oocyte near the top of a small cellular hill.*** The wall of the follicle has been known as the **stratum** or **membrana granulosa** (MG) ever since cells (or more likely their nuclei) were first seen as granules (see Fig. 6.23). The cells of the membrana are now called **granulosa cells** and the hill of follicular cells supporting the oocyte is known as the **cumulus oophorus**. The integrated unit of oocyte and cumulus cells is the oocyte–cumulus complex (OCC; see Fig. 6.10).

During the last phases of folliculogenesis, the cumulus narrows to a stalk (Fig. 7.37). Cumulus

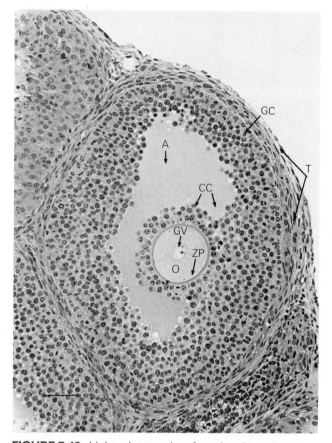

FIGURE 7.42. Light micrographs of section through mouse ovary showing antral follicle. Notice the abundance of mitotic figures among the follicular cells. A, antrum. CC, cumulus cells. GC, granulosa cells. GV, germinal vesicle. O, oocyte. T, theca. ZP, zona pellucida. Bar = 100 μm. Courtesy Dr. J. J. Eppig, Senior Staff Scientist, The Jackson Laboratory, Bar Harbor.

rated cyclic fatty acids that, among other things, induce the rupture of the follicle at ovulation. In addition, inhibin seems to be a granulosa cell product, and the antrum, which contains a large amount of the glycoprotein, may serve as a reservoir for inhibin entering the circulation (Rivier et al., 1986).

Granulosa cells secrete large amounts of estrogen, especially estradiol-17β, which accumulates in the follicular fluid at 1000 times its concentration in blood (see Eckstein, 1977). The presence of high concentrations of estrogen seems to further promote follicular growth. Once in circulation, the estrogen also causes the LH surge that triggers ovulation.

The synthesis of estrogen by granulosa cells takes a roundabout route requiring the presence of

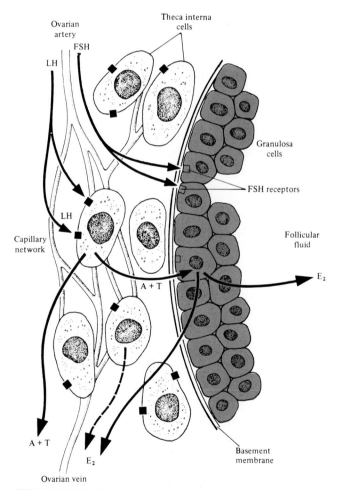

FIGURE 7.43. Diagram showing relationship of theca interna cells and granulosa cells of follicle and relationship of FSH and LH to androsteneodione (A) and testosterone (T) production by thecal cells as opposed to estrogen (E$_2$) production by granulosa cells. From D. T. Baird. In C. R. Austin and R. V. Short, eds., *Reproduction in mammals,* Book 3, *Hormonal control of reproduction,* 2nd ed., Cambridge University Press, Cambridge, 1984. Used by permission.

cells linked through the zona pellucida to the oocyte by projections and gap junctions become a distinct subpopulation of **columnar** cells radiating from the zona pellucida as the **corona radiata** (see Fig. 6.10). Detachment from the cumulus stalk of the oocyte with its corona and additional cumulus cells prior to ovulation is a good indication that **preovulatory maturation** is almost complete.

The fate of granulosa cells is intimately tied to LH levels in circulation. By promoting increased circulation in the ovary, LH makes it possible to meet the metabolic demands of the growing granulosa cells. At the same time, FSH and estrogen stimulate the production of LH receptors on the granulosa cells, thereby preparing them for the final LH surge and ovulation.

Granulosa cells secrete a variety of products. **Prostaglandins,** both prostaglandin E$_2$ (PGE$_2$) and F$_2$ (PGF$_{2\alpha}$), are lipid-soluble, oxygenated, unsatu-

circulating gonadotropins and active thecal cells (Fig. 7.43). FSH receptors on granulosa cells promote the production of aromatase enzymes via cAMP (i.e., FSH couples the FSH receptor to adenylate cyclase activity; Fig. 4.14), but androgens are not synthesized. Instead, they arrive ready-made in the granulosa cells, thanks to thecal cells.

Thecal cells containing LH receptors produce androgen-synthesizing enzymes in the presence of LH via their own cAMP system. Cholesterol is converted to pregnenolone in the thecal cell's mitochondria (see Dimino and Campbell, 1980), and pregnenolone in the cytoplasm is converted to androstenedione (A) and testosterone (T, Fig. 7.43). These hormones either enter the circulation or transverse the basement membrane of the follicle directly. Granulosa cells then convert (i.e., aromatize) the androgens (A and T) to estrogens (Fig. 7.3) (see Heap and Illingworth, 1977).

Final Preparations of the Egg

The plethora of nearly simultaneous events which fertilization may soon trigger suggests that overriding controls coordinate the conclusion of ooteleosis. But coordination does not necessarily mean that events are interdependent, mutually causative, or even caused by the same agents. The egg may achieve its final competence through the maturation and refinement of independent events, especially species-specific events.

Oolemmal specialization allowing fusion with sperm (i.e., the establishment of fusiogenic regions in the oolemma) develops early in oogenesis, but blocks to polyspermy capable of preventing excess fertilization may be installed toward the end of ooteleosis. While lecithotrophic eggs are ordinarily fertilized by many sperm (i.e., polyspermic), small eggs are ordinarily fertilized by one spermatozoon (monospermic) and may be damaged by multiple fertilizations. At the end of ooteleosis, many mammalian and marine invertebrate eggs deploy one or more electrophysiological and biochemical traps to be sprung by the first spermatozoon fusing with the egg.

The sow egg's block to polyspermy is not in place on the 17th day of the sow's 21-day estrus cycle, since 92% of the eggs induced to ovulate in response to human chorionic gonadotropin (hCG) injection at that time become polyspermic during artificial insemination. In contrast, only about 3% of the eggs induced to ovulate on the 20th day become polyspermic (Hunter et al., 1976).

In echinoderms, the ability to activate the so-

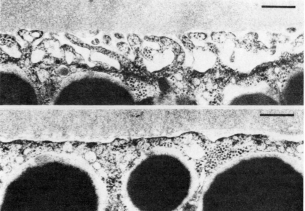

FIGURE 7.44. Electron micrographs through cortex of starfish oocyte before and 1 hour after the induction of maturation by exposure to 1-methyladenine. The copious microvilli present prior to induction have disappeared after induction. The gray area at top is the vitelline membrane. The large black circles are yolk droplets. Bar = 0.05 μm. From W. J. Moody and M. M. Bosma, *Dev. Biol.,* 112:396 (1985), by permission of Academic Press, Orlando, and the authors.

called fast block to polyspermy is acquired late in ooteleosis. This ability parallels the disappearance of microvilli (Fig. 7.44) and a loss of about 50% of the oocyte's surface area. These changes remove potassium channels while leaving calcium channels in place. The oolemma thereby becomes competent for elevating its action potential to levels consistent with blocking fusion with excess sperm (Moody and Bosma, 1985). Additional blocks to polyspermy are deployed as cortical granules move to the cortex during late ooteleosis.

Diakinesis and germinal vesicle breakdown in vertebrates are induced by the LH surge. Oocyte chromosomes condense, and RNA transcription ceases. The bivalent chromosomes separate (Fig. 7.23, oocyte stage 7), and the nuclear envelope ruptures. A stockpile of DNA polymerase, histones, and nuclear precursors are released from the germinal vesicle to the cytoplasm, turning the mature oocyte into a "giant nucleus" (see Benbow, 1985).

In most vertebrates, the oocyte's chromosomes become arranged on the metaphase plate of meiosis I (oocyte stage 8) prior to ovulation. The spindle is displaced toward the egg's surface, and, as the first meiotic anaphase progresses, the spindle assumes a radial orientation (oocyte stage 9).

At telophase (oocyte stage 10), a cytokinetic furrow pinches off the diminutive **first polar body** from the **secondary oocyte.** In mammals, the oocyte is sometimes said to be in the **chromatin mass stage** since the chromosomes form a crescent-shaped mass beneath the oolemma.

Interkinesis and prophase of the second meiotic division are generally brief in vertebrates, and the chromosomes move to the plate of the second meiotic metaphase (oocyte stage 11). In vertebrate oocytes, the second meiotic arrest generally occurs at this time, and most mammalian eggs proceed through ovulation as secondary oocytes.

The second meiotic arrest is conveniently named in most vertebrates, since it occurs during meiosis II. The term refers to the maintenance of the second meiotic metaphase, but in canids (e.g., dogs and foxes), the second arrest occurs during meiosis I, and in insectivores, it occurs after meiosis II. Other phyla have their corresponding meiotic arrests at different stages (Table 7.7) and, like vertebrates, over a range of meiotic stages (e.g., compare the starfish to sea urchin, Table 7.7).

In *Rana pipiens*, the second meiotic arrest seems to depend on a **cytostatic factor** dispensed to the oocyte's cytoplasm from the nucleus by GVBD and the mixing of nuclear sap with cytoplasm. Earlier replacement of the germinal vesicle with a blastula nucleus prevents normal development except if some germinal vesicle material is introduced in the egg (see Smith, 1975).

In mammals, egg size and the follicle's state of differentiation figure in determining the stage of meiotic arrest. In the mouse, for example, preantral follicles respond to LH treatment only by advancing to metaphase I, whereas Graafian follicles advance to metaphase II.

Upon entering the second meiotic arrest and ovulating, eggs become **mature ova** able to participate in productive fertilization. The second meiotic arrest may therefore be more appropriately called the "oogenic climax."

Eggs may be thought of as "arrested" in cases where their competence for fertilization is prolonged, but in most animals, this "prolongation" is brief. Eggs remain fertilizable for not more than 24 hours in humans, about as long in other primates and ungulates, longer by half in the ferret, and shorter by half in rodents and rabbits (see Austin, 1982).

Differences in the timing of the second meiotic arrest among species suggest that nuclear and cytoplasmic events proceed out of synchrony. The cytoplasm in general may be ready for fertilization before the nucleus. Removal of the germinal vesicle in starfish and frogs, for example, does not prevent subsequent parthenogenic activation, and fish

eggs whose germinal vesicle is kept from breaking down develop as haploids following fertilization (see Masui and Clarke, 1979).

Individual cytoplasmic events may also proceed at different paces. For example, sows injected with hCG on the 17th day of their estrus cycle release eggs that can fuse with sperm but fail to resume meiosis following fertilization. The immature sow eggs also have greatly diminished abilities for converting the spermatozoon's head to a male pronucleus (see Baker, 1982). Sows injected on the 20th day yield normal eggs that resume meiosis and undergo further development.

The second meiotic arrest represents a finely balanced moment when nuclear and cytoplasmic events in the egg are finally coordinated. In addition, the arrest or climax is coordinated with ovulation and thereby the opportunity for fertilization. With the exception of animals whose eggs are fertilized while in the ovary (e.g., some fish), vertebrate eggs are ovulated prior to maturity. In the case of human beings, an LH surge early in the morning is followed 28–36 hours later by arrest and ovulation late in the afternoon.

Ovulation

Mammals ovulate at relatively precise times after the surge of LH that accompanies estrus or follows coitus (Table 7.11). While LH or hCG injection alone can stimulate ovulation, the normal mechanism may involve FSH as well as LH or the ratio of the two gonadotropins.

In mammals, the gonadotropin surge (Fig. 7.34) stimulates a last wave of mitosis in follicular cells and increases the permeability of the follicle's basement lamella. The antrum swells rapidly, and the follicle reaches its maximal size (Table 7.11). The follicle is not under greater internal pressure than capillaries, and when its wall finally ruptures at ovulation, the secondary oocyte is not ejected violently.

Mammalian eggs are ovulated at a translucent **stigma** or area of ischemia (i.e., local anemia or deficiency of blood) appearing between the follicle and the ovarian surface. Resulting from the closing down of blood vessels, the stigma is an area of death in both the granulosa and the adjacent ovarian cortex. While some follicular cells are shed into the antrum, others become pycnotic and are either found in the antrum or are destroyed by macrophages.

Follicle rupture occurs under circumstances resembling inflammation. Under the influence of FSH, vascular permeability increases, and the fol-

TABLE 7.11. The follicle at ovulation[a]

Mammal (by Orders)	Diameter (mm)	Hours between LH Surge, Onset of Estrus, or Coitus and Ovulation	Number of Eggs Ovulated and Ovaries Involved	Presence of Corona Radiata
Carnivora				
Canis familiaris (dog)	10	24–48	8–10, both	+
Felis catus (cat)		24–30	4–6, both	+
Mustela furo (mink)		30	8–9, both	+
Lagomorpha				
Oryctolagus cuniculus (rabbit)	2	9–11	10, both	+
Primates				
Homo sapiens (human being)	23	28–36	1, LR	+
Macaca mulatta (rhesus monkey)	10	24–36	1, LR	+
Rodentia				
Cavia porcellus (Guinea pig)	0.8	10	2–4, both	+
Cricetus auratus (hamster)	0.62	9–11	6, both	+
Mus musculus (mouse)	0.22–0.5	2	6, both	+
Rattus rattus (rat)	0.9	7–12	10, both	+
Ungulates				
Artiodactyla *Bos taurus* (cow)	10–20	40	1, R	–
Ovis aries (sheep)	15–19	22	1–2, R	–
Sus scrofa (sow)	7–10	40	6–12, both	–
Perissodactyla *Equus caballus* (horse)	32–55	24–48	1, LR	–

[a]Data from H. Peter and K. P. McNatty, *The ovary*, Paul Elek, London, 1980; and P. L. Altman and D. S. Dittmer, eds. *Growth.* Federation of American Societies for Experimental Biology, Washington, DC, 1962.

+, Present; –, Absent.

R, right ovary predominates; LR, left and right ovaries alternate; both, both ovaries release oocytes.

licular granulosa cells release tissue-type **plasminogen activator** (tPA) and the proteolytic enzyme **collagenase.** Prostaglandin PGF$_{2\alpha}$ would seem to play a mediating role, since inhibitors of its synthesis (i.e., inhibitors of cycloxygenase and lipoxygenase pathways or arachidonic acid metabolism) prevent tPA and collagenase release.

Platelet activating factor (PAF) may also play a mediating role, since the PAF inhibitor BN52021 blocks the increase in vascular permeability, the increase in collagenolysis, and ultimately gonadotropin-induced ovulation. Possibly released from blood cells (neutrophils), PAF directs platelet aggregation and adherence to endothelial cells above the follicle (Abisogun et al., 1989).

Strained by accumulated fluid and weakened by the action of collagenase, the stigma becomes elevated like a blister above the follicle. The flow of a viscid medium from the follicle into the peritoneal cavity announces the rupture of the ovarian wall. The oocyte follows, surrounded by cumulus cells (i.e., the remains of the OCC, Fig. 6.10). Coated by exudate, the egg may stick to the surface of the ovary in the vicinity of the stigma until cilia on the oviductal fringe or **fimbria** sweep the egg into the mouth of the oviduct (or Fallopian tube, Fig. 7.45).

The egg is now moved to the swollen **ampulla** at the upper end of the oviduct distal to a constriction called the **isthmus** (see Fig. 11.24) The egg may meet awaiting sperm or await the arrival of sperm. After about 2 days, unfertilized eggs generally move down the oviduct, but in the mare, unfertilized eggs may remain in the ampulla through an entire estrus cycle and even after an egg from the next cycle is fertilized.

In the event of fertilization, the typical mammalian egg moves to anaphase II (oocyte stage 12,

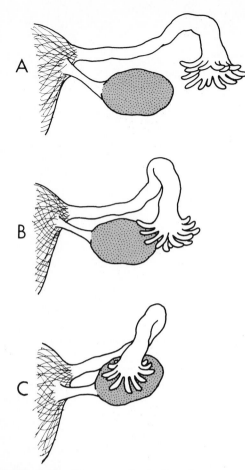

FIGURE 7.45. Diagrams illustrating the movements of ovary and oviduct bringing the fimbria (fringe) into close contact with surface of ovary at time of ovulation. From E. S. E. Hafez. In E. S. E. Hafez and T. N. Evans, eds., _Human reproduction. Conception and contraception,_ Harper & Row, New York, 1973.

Fig. 7.23), telophase II (ootid stage 13), and the formation of the second polar body. The formation of the male and female pronuclei (ootid stage 14) marks the successful completion of maturation. Meanwhile, the ovulated follicle undergoes luteinization.

Luteinization and Maintenance of the Corpus Luteum

The transformation of a follicle and its associated stroma to a corpus luteum, or luteinization, is a process of **modulation** or change from one functional state to another. The original thecal cells become **theca luteum cells** or **small luteal cells,** which collectively form thick walls known as **trabeculae** and **luteal folds.** The granulosa cells become **granulosa luteum cells** or **large luteal cells.**

Following the LH surge, they cease dividing, but their growth accounts for most of the increase in size of the gland (a process called hypertrophy or excess growth).

Just prior to ovulation and continuing dramatically immediately afterward, the membrana granulosa and the attached theca interna become folded (i.e., plicated). The follicle takes up blood from broken vessels in the vicinity (fibrin clot) and is known as the **corpus hemorrhagicum.** When the shrunken antrum is invaded by blood vessels, the luteal folds become infiltrated with vessels from within as well as from without and the former follicle becomes a definitive **corpus luteum** (Fig. 7.37).

Since each ovulated mammalian egg leaves a luteal gland behind, luteal glands provide a record of ovulations. In different species, the number of eggs ovulated can be equal in both ovaries, one ovary can predominate, or ovulation can alternate from one ovary to the other (Table 7.11).

Luteal maintenance, like luteal development, is the consequence of a delicate hormonal balance. LH receptors are essential. Stromal and thecal cells have LH receptors during the follicular phase of the ovarian cycle, but receptors develop on granulosa cells just prior to ovulation. Only follicles that have sufficiently high concentrations of LH receptors on their granulosa cells seem capable of being converted to luteal glands by the action of LH.

Late follicular cells and large luteal cells also contain receptors for prolactin. Unlike LH and FSH receptors, prolactin receptors do not operate through cAMP. Prolactin reverses the temporary diminution or desensitization of LH receptors on granulosa luteum cells which follows the LH surge and, along with LH, is part of the **luteotropic complex** of hormones which maintains the luteal phase of the ovarian cycle.

In the rabbit, estrogen is also required to maintain the corpus luteum. Prolactin, which inhibits estrogen production in follicles when present in high concentrations, also inhibits progesterone production in luteal glands.

In the presence of LH, luteal cell mitochondria convert cholesterol to pregnenolone, and a cytoplasmic enzyme, 3β-hydroxysteroid dehydrogenase, oxidizes pregnenolone to progesterone (Fig. 7.3), the major hormonal product of the gland. In primates, large amounts of estrogen are also secreted, and most luteal glands also secrete two polypeptide hormones: **relaxin,** which relaxes the uterus during pregnancy, and **oxytocin,** which may operate as a local regulator of blood pressure.

The lifetime of the corpus luteum depends on whether pregnancy (or pseudopregnancy) has occurred. In the event of a pregnancy, the gland (sometimes called the corpus luteum of pregnancy) is retained and grows dramatically under the influence of chorionic gonadotropin, but when there is no pregnancy, the gland (sometimes called the corpus luteum of diestrus) involutes. Similarly, at the end of pregnancy, the gland ceases to secrete progesterone and, following an invasion of fibroblasts, is converted to a whitish scar known as the **corpus albicans** (Lat. white body).

Luteal involution may involve $PGF_{2\alpha}$. The ability of LH to couple its receptor to the adenylate cyclase system and generate cAMP is inhibited by this prostaglandin. In ungulates and other mammals but not primates, $PGF_{2\alpha}$ released by the uterus is concentrated in the ovary as a result of cross currents of circulation (i.e., a countercurrent multiplier system). In primates, the estrogen or $PGF_{2\alpha}$ responsible for the gland's involution is produced in the gland itself.

PART 3 *SUMMARY AND HIGHLIGHTS*

Sexual reproduction is almost universal, but accounting for it in terms of adaptive advantage is problematic. Sexual reproduction may offer a way of correcting double-stranded errors through recombinational repair. Variation promoted by meiosis and fertilization may represent an acceptable compromise between the risk of extinction and optimal adaptation.

Many forms of sexual reproduction seem to promote out-breeding or cross-fertilization. These may involve circular DNA (plasmids) that are taken up and sometimes incorporated into chromosomes (episome), the incorporation of a viral (bacteriophage) genome (transduction), and any of the three forms of sexual reproduction found in eukaryotes. Mating types are found in isogamous or hologamous species where isogametes undergo conjugation and nuclei move between cells. Oogamy is the specialization of parts of gametophytes as gametangia and the differentiation of morphologically and behaviorally distinct gametes (in anisogamous, heterogamous, and oogamous species) leading to the formation of distinct eggs and sperm.

Sex organs imply adaptation in the sporophyte or diploid generation and the accommodation, transport, and nurture of gametes. In monoecious plants, male and female sex organs are found in separate places on the same individual, while gonochoristic animals and dioecious plants house the different sex organs in separate male and female organisms. In gonochoristic animals, male sex organs called testes and female sex organs called ovaries are correlated with conspicuous secondary sexual characteristics functioning in sexual behavior and the care of offspring. Individuals are assigned to female and male genders generally according to these characteristics. In addition, some species include (or may be exclusively composed of) hermaphrodites having female and male sex organs either serially or simultaneously and either occupying the same organ (ovotestis) or different organs.

Gender is often correlated with sex-determining chromosomes or ratios of sex chromosomes and autosomes, and one gender frequently has different sex chromosomes (is heterogametic), while the other has morphologically identical sex chromosomes (is homogametic). In bees and wasps, fertile females can withhold sperm and lay unfertilized eggs that develop into males, while all fertilized eggs develop into females. Different mating systems involving social groups or isolated individuals that pair while mating seem to be adapted for meeting different environmental pressures.

The difficulties of bringing gametes together after spawning in an aquatic environment may be overcome through anatomical adaptations and chemotaxis. Mating or intimate sexual contact generally overcomes the problem of species specificity but poses other problems and perils. The problem of finding mates in suitable condition is frequently solved by morphological and behavioral displays advertizing readiness and by pheromones to attract sexual partners.

Parthenogenesis and gynogenesis (pseudogamy), development from an egg without the participation of two parental genomes, seem to reduce whatever advantage sexual reproduction may provide in terms of promoting variation but may still provide the advantages accruing from recombinational repair. These forms of reproduction may or may not involve meiosis (meiotic versus ameiotic parthenogenesis) and the formation of a synkaryon (automixis) as opposed to the retention of the original nucleus (apomixis). The formation of a polyploid nucleus through polyploidizing division (sometimes called endoreduplication) frequently accompanies meiosis. In meiotic parthenogenesis a restitution mitosis without cell division (acytokinesis) restores the diploid condition of the egg. In haplontic parthenogenesis a haploid organism develops. Organisms with polyploid genomes may utilize somatic parthenogenesis and completely suppress meiosis.

Asexual reproduction does not seem to be an alternative to sexual reproduction so much as an adjunct. Asexual reproduction occurs entirely within a phase of the life cycle without involving gametes, meiosis, or pronuclei and seems to be a functional equivalent of growth. The asexual progeny of a unicellular or multicellular organism are members of a synclone. All the cells produced during asexual reproduction are assumed to share identical genomes.

Asexual normogenesis starts with a group of cells (propagule) rather than a single cell (zygote). Sexual reproduction may have evolved from asexual reproduction were cellular cannibalism to have reduced a propagule to a zygote. Asexual normogenesis may resemble regeneration and the products of asexual reproduction may resemble the parent (facsimile or proliferative reproduction), or they may not resemble the parent (modifying reproduction). Even when the products of asexual and sexual reproduction are morphologically identical, the processes giving rise to each may be different and cell types and layers may have different origins (transdifferentiation).

In epimorphosis, new tissue is formed by outgrowth, while in morphallaxis, new tissue is formed by rearrangement of existing parental tissue. Metamorphosis occurs when cell death accompanies morphallaxis, and regulation refers to the development of missing parts. The asexual propagule may be physiologically isolated before developing (blastogenic or gemmation budding), may break off the parent (regenerative fragmentation), or may form from a part of the parent by fission or strobilation.

Sperm come in a variety of forms but flagellated sperm are the most common among animals. Flagellated sperm shed into water are loosely classified as primitive. Their acrosome is small to absent; mitochondria in the midpiece do not form a sheath, and the tail is simply an axonemal complex covered by a plasma membrane.

Modified sperm are specialized for internal fertilization, pseudocopulation, or dermal copulation. The acrosome of these sperm is prominent and influences the size and shape of the head. The nuclear envelope lacks pores except posteriorly where it is folded back as redundant nuclear folds. A neck region contains a connecting piece of segmental columns which may obscure the centrioles. The midpiece contains mitochondria linked or fused in mammalian sperm into a helical mitochondrial sheath around a core of outer dense fibers and an axonemal complex. An annulus connects the midpiece to the tail's principal piece. In mammals, seven outer dense fibers continue into the principal piece of the tail, while two fuse with longitudinal columns of the tail's dense sheath and connecting ribs. The terminal piece of the tail lacks a sheath.

Classified according to the amount of yolk, eggs can be alecithal or virtually yolkless, microlecithal with little yolk or lecithotrophic (yolk eating), mesolecithal or oligolecithal with modest amounts of yolk, or megalecithal with conspicuously great amounts of yolk filling a vitellus. According to the distribution of yolk, eggs can be isolecithal with an even distribution of yolk, telolecithal with a concentration at one end (a vegetal pole or vitellus), or intralecithal (centrolecithal) with yolk concentrated in the egg's center.

A primary membrane is produced while the egg is in the ovary. The membrane may be a vitelline membrane (envelope), zona pellucida, or a chorion (zona radiata) with a micropyle.

The eggs of therian or placenta-forming viviparous mammals are alecithal to microlecithal and isolecithal. The eggs of prototherian mammals resemble those of reptiles and birds in being megalecithal and telolecithal. Neither eutherian nor metatherian mammals produce egg shells, although eggs may accumulate mucous layers during their passage to the uterus. Bird eggs acquire various secondary coverings, a thick albuminous sheath, chalazae, liquid albumin, shell membranes with an air cell at the blunt end, and a porous calcareous shell covered by a thin cuticle of protein.

Ideas about the egg and sperm (i.e., seed) have changed since Aristotle described the egg as a coagulum and development as epigenic, *de novo* in origin, and dependent on environmental influences. William Harvey focused attention on the role of eggs in the reproduction of all animals, and Redi established the principle of reproductive continuity through eggs. Swammerdam, Malpighi, and other ovists established the scientific basis for preformationism and the idea of emboîtement, or encapsulation, of preformed parts within the egg. Leeuwenhoek and other animalists discovered and described homunculi within spermatozoa. De Graaf discovered follicles and probably blastocysts but overlooked the mammalian egg.

The early 19th century saw the reemergence of epigenesis following the discovery by Karl Ernst von Baer of the mammalian egg. By the mid-19th century, the cell theory cast the egg in the role of a cell. Emulating the achievements of the botanists, embryologists like Oscar and Richard Hertwig, van Beneden, and Boveri, studying *Ascaris* and sea urchin gametes and embryos, generalized the processes of meiosis in sperm and egg and clarified the chromosomal features of meiosis and fertilization.

The initiation and maintenance of gamete production in vertebrates are a function of hormones. Lipid hormones include the sex or gonadal steroids, androgens, especially testosterone and androstenedione, progesterone, and aromatized estrogens. The androgens are produced chiefly by interstitial endocrinocytes (Leydig cells) in the testes and stromal cells in the ovary, while sustentacular and follicle cells aromatize androgens to estrogens. Testosterone accumulates in seminiferous fluid, while estrogen accumulates in follicles. Gonadal steroids operate on promoter regions of genes through receptors.

Gonadotrophs in the anterior hypophysis or adenohypophysis produce the glycoprotein gonadotropins, follicle stimulating hormone (FSH), and luteinizing hormone (LH) consisting of identical alpha polypeptides and different beta polypeptides. In both male and female vertebrates, a glycoprotein, inhibin, produced in the gonads inhibits gonadotropin release. Gonadotropins operate chiefly after complexing with surface receptors coupled to the adenylate cyclase system. The accumulation of LH receptors on follicular cells is a prerequisite to ovulation and luteogenesis. In addition, the adenohypophysis produces prolactin which inhibits progesterone synthesis in the ovary.

The hypothalamus produces a decapeptide gonadotrophin releasing hormone (GnRH) which is transported directly to the adenohypophysis via the hypothalamic–pituitary portal system. Pulsatile release of GnRH causes the secretion of gonadotropins, in the case of LH, also in pulses. Gonadotropins promote the production of the sex hormones; sex hormones either promote or retard the release of GnRH. As a result of negative feedback control loops, basal levels of gonadotropins and gonadal steroids are maintained, but positive feedback control loops in female vertebrates cause surges of hormonal release that lead to ovulation, luteogenesis, and the return to the tonic mode. Without follicular cells secreting estrogen the positive feedback loop is broken. Instead, luteal cells secrete progesterone, the hormone of pregnancy, whose activities on folliculogenesis are antagonistic to those of estrogens.

The rotation between periods of folliculogenesis followed by luteogenesis in the mammalian ovary is known as the ovarian cycle. Ovulation, whether seasonal or year-round, a unique event (monestrus) or a repeated event (polyestrus), may be spontaneous, regular, and cyclic or induced by coitus. Increases in female sexual behavior known as estrus or heat may coordinate ovulation with mating. In some Old World primates, a cycle of uterine sloughing known as the menstrual cycle or period follows unsuccessful ovulation.

The ability of androgens to promote ovulation involves conversion to estrogen by follicular cells, but maturation can be induced by removal of the oocyte from the follicle, suggesting that the prolonged dictyate stage, or first meiotic arrest, is due to the presence of a maturation inhibitor. Oocytes themselves appear to produce and contain gene control molecules or components that regulate the synthetic events in the egg. As demonstrated by interspecific nuclear transplantation to oocytes and mature ova, eggs have the means to switch off the machinery for transcription and turn on the machinery for condensing chromosomes.

The gametogenesis of sperm (spermatogenesis) and eggs (oogenesis) begins with the narrowing of gonocyte or primary germ cell populations into reserve and proliferative stem cell populations (spermatogonia and oogonia) of steady-state germ tissue. The elimination of oogonia may further narrow female germ cells to a static tissue of oocytes.

Incomplete cell divisions result in the formation of nests or syncytia of committed cells connected by intercellular bridges. Spermatocytes tend to remain in nest until late in differentiation. Oogonia may be linked in huge nests and differentiate via cannibalism. Smaller nests may undergo meroistic oogenesis (telotrophic or polytrophic), or nests may break up completely, and their oocytes undergo panoistic oogenesis.

After completing meiosis, sperm differentiate and shed excess cytoplasm. Eggs store materials, grow, and differentiate while holding meiosis in abeyance, and they are generally discharged (i.e., ovulated) from the ovary while still in meiosis. Eggs synthesize or acquire vast RNA reserves sometimes involving the amplification of the ribosomal organizer region of DNA. While large numbers of oocytes reside in the mature ovary in a state of dormancy or first meiotic arrest of mammals, spermatocytes in the mature testis are whisked through meiosis without interruption. While great numbers of sperm mature at the same time, in some species such as *Homo sapiens*, as few as one egg in one of the two ovaries present matures at a time.

Following the premeiotic DNA synthesis and throughout the period of meiosis, gametogenic cells are in the "-cyte" phase. Primary spermatocytes and oocytes are in meiosis I; secondary spermatocytes and oocytes are in meiosis II. While the primary spermatocyte forms two secondary spermatocytes, the primary oocyte forms only one secondary oocyte and a diminutive first polar body (an abortive oocyte). Secondary spermatocytes divide equally into spermatids, but the secondary oocyte divides unequally into a large ootid and a diminutive second polar body. Eggs that have acquired competence for productive fertilization are called mature ova.

In the seminiferous tubules of vertebrate testes, spermatogonia are confined to a basal compartment by the presence of extensive junctional complexes between extensions of sustentacular (Sertoli) cells. At the beginning of meiosis, nests of spermatocytes penetrate the blood–testis barrier by moving into the adluminal compartment of the tubule. Junctional complexes become restricted to the apical end of spermatid heads and to luminal crypts on sustentacular cells. Synchronous development within nests and a tendency to initiate meiosis in contiguous regions result in regular cell associations in cross sections of seminiferous tubules and the appearance of spermatogenic waves.

The spermiogenesis of haploid spermatids is divided into a differentiating spermatid phase and a releasing spermiation phase. Glycoprotein-containing granules in the Golgi apparatus coalesce into a membrane-bound acrosomal granule which spreads over the apical end of the nucleus. Nuagelike material initially associated with mitochondria becomes perinuclear and may acquire nuclear materials as it forms dense bodies and moves to the annulus as a chromatoid body. A nuclear ring at the posterior edge of the acrosome attached to a manchette seems to drive cytoplasm into the midpiece where mitochondria become organized behind the chromatoid body. Histones in the nucleus may be replaced with tissue-specific histones, intermediate protein, or protamines, and redundant nuclear folds appear as the nucleus condenses. Microtubular precursors of the axonemal complex are synthesized and assembled while dense outer fibers appear and a sheath encloses the principal piece of the tail.

Late spermatids shed cytoplasm in tubulo–bulbar complexes with sustentacular cells and in residual bodies containing the remnants of intercellular bridges. Following release from seminiferous tubules and storage in the epididymis, sperm are decapacitated and packaged for transport. Modified spermatozoa ejaculated from the male acquire competence for fertilization (are capacitated) in the female genital tract and moved to the point of fertilization (the ampulla of the oviducts in mammals).

The oogenic phase of oogenesis is divided into a nongrowing previtelline stage, a slowly growing germinal vesicle or dictyate stage, and a rapidly growing vitelline or vitellogenic stage. Gene amplification of the nucleolar organizer gene in amphibians occurs during zygotene and pachytene in the previtelline stage. During diplotene, lampbrush chromosomes engage in extraordinary amounts of RNA synthesis. Much of the RNA produced is broken down before leaving the nucleus or germinal vesicle, but some seems to be stored as a stable variety of RNA. At a comparable stage in meroistic insects, the oocyte nucleus is not transcriptionally active, but RNA is transported from nurse cells to oocytes.

Yolk precursors are lipoglycophosphoproteins generally synthesized outside the oocyte (e.g., in the liver of vertebrates, the fat glands of insects) under the influence of steroidal hormones. Transported to the oocyte in serum, the precursors are taken up by oocytes via endocytosis, processed, and stored in yolk platelets, spherules, and granules.

Vertebrate oocytes typically move through the first meiotic division slowly but mammalian oocytes are said to reside in a first meiotic arrest. At the time mature ova are ovulated, they come under a second meiotic arrest, in most mammals, at the second metaphase.

During ooteleosis, eggs pass through diakinesis, and the germinal vesicle breaks down (GVBD). cAMP and calcium inhibit maturation, but a cascade of cAMP- and calcium-independent phosphorylations may be involved in ending meiotic arrest in oocytes and in the transition to mitosis in ordinary cells. The activity of cAMP-independent kinases may depend on their own phosphorylation, and the transition from G-2 to mitosis or escape from meiotic arrest may depend on the phosphorylation of other proteins, such as H1 histone, by active kinases.

One part of this cascade is played by a polypeptide in the M-phase promoting factor or MPF of *Xenopus* oocytes and by cognate polypeptides in starfish oocytes and mammalian cells. These polypeptides are homologs of yeast p34^{cdc2} kinase and are now thought to belong to a *cdc2* gene family.

Oocytes develop within an epithelium of follicular cells. Mammalian oocytes are surrounded by an epithelial bead known as an ovarian follicle. Stromal cells nearest the follicle become epithelioid and form a highly vascular theca interna, while stromal cells farther from the follicle form a fibrous theca externa.

Folliculogenesis and the ovarian cycle begin with the formation of a primordial follicle having a simple flattened epithelial layer around a dormant oocyte. Primary follicles have cuboidal follicular cells and form a stratified layer during a hormone-independent early growth period. Subsequently, secondary (growing) follicles are characterized by hormone-dependent rapid growth accompanied by the formation of the fluid-filled antral cavities. Tertiary (mature or Graafian) follicles contain a central antrum that expands rapidly.

The follicular cells forming the granulosa membrane surrounding the antrum are called granulosa cells, while a pedestal of cumulus oophorus cells supports the oocyte. An interacting oocyte–cumulus complex (OCC) begins breaking down prior to ovulation but generally accompanies the egg during the pick-up stage of transport into the oviduct and may linger as a corona radiata.

Most mammalian follicles become atretic (imperforate) and degenerate as atretic follicles. A follicle selected to develop fully (germinal selection) moves to the surface epithelium of the ovary and releases its mature ovum at a stigma under the influence of prostaglandins.

Relieved of its oocyte, the follicle and surrounding theca cells enter luteogenesis of the ovarian cycle, folding inward and becoming the progesterone-secreting corpus luteum. Modulating follicular cells enlarge (hypertrophy) as nondividing large luteal cells, while epithelioid and theca interna cells become small luteal cells. If pregnancy occurs, one corpus luteum of pregnancy will remain patent for each developing embryo under the influence of placental gonadotrophins (hCG), but if pregnancy fails to occur or when it is terminated, the corpus luteum degenerates into a corpus luteum of diestrus and finally into a scar (corpus albicans). The ovarian cycle is then reset at folliculogenesis.

PART 3   QUESTIONS FOR DISCUSSION

1. What are the theoretical difficulties posed by sexual reproduction? Describe some strategies that overcome these difficulties. What are the differences between reproduction through mating types, oogamy, and sex organs? Between gametangia and gonads? Between archegonia and gametophores?

2. What are gender differences? Define female, male, and hermaphrodite. What is sex determination? What roles do chromosomes play in it? How are sex ratios established? What is a mating system? Monogamy? Polygamy?

3. Suggest some reasons for considering asexual reproduction a form of growth rather than reproduction. Describe various forms of asexual reproduction, and contrast epimorphosis and morphallaxis, modifying and proliferative reproduction, germination budding and regenerative fragmentation. What is transdifferentiation?

4. Critique the arguments for deriving parthenogenetic and gynogenic reproduction from sexual reproduction as opposed to asexual reproduction. What are meiotic parthenogenesis, haplontic parthenogenesis, and somatic parthenogenesis?

5. Describe the main differences between the classes of primitive and modified sperm. Spec-

ulate on how these differences may represent adaptations for fertilization in different environments.

6. Compare typical eggs of oviparous and ovuliparous species to purely cleidoic eggs and to the eggs of viviparous species. What requirements of eggs are met through environmental interactions? Classify eggs according to their amount and distribution of yolk.

7. How does your concept of the egg differ from Aristotle's? From Bonnet's? From von Baer's? How does your concept of fertilization differ from Galen's? From Harvey's? How does your concept of sperm differ from Leeuwenhoek's? From Spallanzani's? How does your concept of the Graafian follicle differ from de Graaf's? How has the concept of the egg changed since von Baer's discovery?

8. How are intercellular bridges involved in oogonial renewal in panoistic species? In meroistic species? What is the role of intercellular bridges in spermatogonial renewal? In cell associations? Describe the release of sperm.

9. What are the common features of spermatogenesis and oogenesis? Describe the wanderings and changes in nuagelike material during spermatogenesis and oogenesis.

10. What are the major differences between spermatogenesis and oogenesis? In "-gonial," "-cyte," and "-tid" stages? What is a mature ovum? Should the second meiotic arrest be renamed the oogenic climax? Why? What are egg envelopes? Where are they made, and how should they be identified? Describe the manufacture and storage of yolk.

11. Name the main gonadal steroids, and describe their relationships in steroidogenesis. What are the roles of cellular receptors in the hormonal controls of gametogenesis in vertebrates? Describe the hypothalamic–pituitary–gonadal axis and feedback controls of hormonal secretion in both the tonic and surge modes.

12. What are the compartments of the seminiferous tubule and how are they separated? Describe the formation of the acrosome, manchette, mitochondrial sheath, and axonemal complex of sperm.

13. Compare and contrast selective gene amplification in panoistic species to endomitosis in meroistic species. Describe transcription in oocytes, and discuss the possibility of "masked" messengers, ribosomes, and proteins.

14. Describe and interpret the consequences of transplanting nuclei and DNA to eggs at different stages of development. What substances are thought to arrest meiosis, and how is the oocyte thought to escape from arrest?

15. Compare and contrast the ovarian cycle, the estrus cycle, and the menstrual cycle. Describe atresia, and speculate on its cause(s). Describe folliculogenesis and luteogenesis as parts of the ovarian cycle and as cellular events. Distinguish between primordial, primary, secondary, and tertiary follicles and between the oocytes occupying them. Describe ovulation and the role of the oocyte–cumulus complex.

PART 3 RECOMMENDED READING

Austin, C. R. and R. V. Short, eds., 1982. *Reproduction in mammals*, book 1, *Germ cells and fertilization*, 2nd ed. Cambridge University Press, Cambridge.

Austin, C. R. and R. V. Short, eds., 1984. *Reproduction in mammals*, book 3, *Hormonal control of reproduction*, 2nd ed. Cambridge University Press, Cambridge.

BioScience, Vol. 37 (No. 7), July/August 1987. Articles by D. Policansky and others on sex change in plants and animals.

Bloom, W. and D. W. Fawcett, 1986. *A textbook of histology*, 11th ed. Saunders, Philadelphia.

Browder, L. W., ed., 1985. *Developmental biology: A comprehensive synthesis*, Vol. 1, *Oogenesis*. Plenum Press, New York.

Guraya, S. S., 1985. *Biology of ovarian follicles in mammals*. Springer-Verlag, Berlin.

Jones, R. E., ed., 1978. *The vertebrate ovary: Comparative biology and evolution*. Plenum Press, New York.

Kaufman, M. H., 1983. *Early mammalian development: Parthenogenetic studies.* Cambridge University Press, Cambridge.

Metz, C. B. and A. Monroy, eds., 1985. *Biology of fertilization,* Vol. 1, *Model systems and oogenesis;* Vol. 2, *Biology of the sperm,* 2nd ed. Academic Press, Orlando.

Mittwoch, U., 1978. Parthenogenesis: Review article. *J. Med. Genet.,* 15:165–181.

Needham, J., 1959. *A history of embryology,* 2nd ed. Abelard-Schuman, New York.

Phillips, D. M., 1974. *Spermiogenesis.* Academic Press, Orlando.

Roosen-Runge, E. C., 1977. *The process of spermatogenesis in animals.* Cambridge University Press, London.

van Tienhoven, A., 1983. *Reproductive physiology of vertebrates,* 2nd ed. Cornell University Press, Ithaca.

Zuckerman S. and B. J. Weir, eds., 1977. *The ovary,* 2nd ed., Vols. 1–3. Academic Press, Orlando.

Part 4

FERTILIZATION AND THE ZYGOTE

Fertilization is a uniquely living process and even novel among living processes. Two gametes, on the brink of death, meet, fuse, and are rejuvenated in the form of a zygote, sparkling with vigor. Many of the structures so carefully built within the sperm are destroyed, and the egg reverses its metabolism from storage to utilization.

In Part 4, fertilization and the zygote are examined with an eye toward unraveling the mechanisms of change. Fusion is traced to the properties and activities of the spermatozoon and mature ovum, and the consequences of fusion are followed through subsequent cytoplasmic and nuclear events. As it turns out, individual aspects of the overall process are not as unusual as their cumulative effects.

Chapter

8

THE EMBRYO'S BEGINNINGS

Life is a continuous stream. The death of the individual involves no breach of continuity in the series of cell-divisions by which the life of the race flows onwards.

E. B. WILSON (1896, p. 9)

*B*eginnings are always difficult to pinpoint, but if one maintains that a process, such as life, is continuous, identifying its beginning becomes impossible. Whether a discrete thing, such as an organism, can be said to have a beginning is also problematic. Establishing the beginning of the embryo is certainly fraught with difficulty. Does it begin with bringing gametes together, with their fusion, or with subsequent internal events?

SYNGAMY

Originally, **syngamy** (Gk. *syn-* by means of + *-gamia* marriage) meant reproduction via egg and spermatozoon. The contrasting term, isogamy or hologamy, meant reproduction via isogametes or hologametes (see Chapter 5).

Today, syngamy has come to mean the fusion of two gametes or the formation of the zygote, whether by sperm and egg or by isogametes. In this sense, syngamy is a synonym for **fertilization,** the act of making fertile—specifically, making a ga-

mete fertile by fusion with another gamete—and both syngamy and fertilization are usually broken down into **early events** immediately preceding fusion and **late events** following fusion.

The distinction between syngamy and fertilization in modern usage is one of emphasis and breadth. Syngamy refers broadly to the concatenation of processes, events, and interactions that produce the first embryonic cells, while fertilization narrows in on the consequence of gametic fusion, namely, the initiation of embryonic development.

Sperm Motility

Flagellar motility is fueled by ATP and results from the action of flagellar **dynein ATPase** activated by intracellular alkalinization. The energy from mitochondria may reach the flagellum through a **creatine phosphate** (CP) shuttle system mediated by **creatine kinase** (CK) isozymes. CP produced in the spermatozoon's midpiece or head by **mitochondrial CK isozyme** seems to diffuse to the tail where it is consumed by **flagellar CK iso-zyme.**

Primitive sperm display some odd changes possibly related to the energy requirements of motility. For example, 2 minutes after sperm attachment to the egg, the lone mitochondrion in the ascidian spermatozoon's head swells and moves down the length of the tail, presumably dispensing ATP, until it is shed at the tip.

In mammals, flagellar movements may aid sperm to move through the cervical mucus at the opening of the uterus and maintain their position at the site of fertilization. Tail-lashing may also be important for moving the sperm once they have reached the vicinity of the egg or, more precisely, the remains of the oocyte–cumulus complex (OCC: oocyte, corona radiata, and any additional cumulus cells still attached).

During its passage through the female genital tract, the spermatozoon's metabolism and oxygen consumption increase. *In vitro* guinea pig and hamster sperm become hyperactive, and they traverse erratic three-dimensional paths with vigorous whiplash-like beatings of their flagella. Hyperactivity is sensitive to temperature, inhibited by zinc ions in the medium, and dependent on an exogenous energy source such as pyruvate (but not glucose or lactate).

The alkaline environment of the ampulla relaxes extracellular bonds in the OCC, and hyaluronidase (the "spreading" enzyme that digests the neutral mucopolysaccharides, hyaluronic acid) may aid sperm's penetration of the **extracellular matrix.** Hyaluronidase is present in follicular fluid and released from sperm's plasma membrane even prior to the acrosomal reaction (Talbot, 1985).

Later, sperm motility will also aid mammalian spermatozoa to penetrate the zona pellucida, and, since immotile mammalian sperm fail to fuse with the egg (Barg et al., 1986), flagellar movements would seem necessary for membrane fusion. Possibly, these movements act as a source of kinetic energy, allowing the spermatozoon to butt against the egg, or the movements may provide a signal inducing the egg's response to the sperm.

Acrosomal Reaction

Before the acrosomal reaction was appreciated by embryologists, the role of egg jelly in fertilization was linked to sperm binding. In echinoderms, **fertilizin** or **sperm-isoagglutinin** from eggs was thought to bind with **antifertilizin** on the sperm's surface in a "lock and key" (i.e., a prototype of antibody–antigen reactions).

Jacques Loeb (see Loeb, 1913) originally showed that **egg water** (i.e., seawater in which eggs have been kept) from the sea urchin *Arbacia*, the marine annelid *Neanthes* (formerly *Nereis*), and from other species stimulates heightened activity in sperm and then causes them to form large (e.g., 2–4 mm in diameter) aggregates called **swarms** with sperm of the same species. Because sperm disaggregated from the swarms in 30 seconds to 10 minutes (the exception being swarms of starfish sperm), Loeb argued that the sperm were not bound but only mutually attracted to each other in the swarm. When the attraction lapsed, the swarm broke up and the sperm soon died.

Frank R. Lillie (see Lillie, 1913) also investigated the effects of the egg water from different species but came to a different conclusion. He proposed that egg water contained a **sperm-isoagglutinin,** a **polyvalent** glue capable of adhering to sperm and cohering to itself, thereby agglutinating sperm in the swarm. Furthermore, since repeated washing eliminated eggs' fertilizability, Lillie suggested that the isoagglutinin was also necessary for fertilization. His **fertilizin theory** proposed that **fertilizin** located at the surface or in the ectoplasm of eggs normally operated to bind sperm to egg and when present in egg water operated to bind sperm to each other (see Just, 1939).

The disagreement between Loeb and Lillie was settled when the importance of one additional observation was appreciated: sea urchin sperm were sterile after disaggregating from swarms. While adherent to each other in the swarm, the sperm had changed. They had lost the ability to adhere to eggs and fertilize them. As it turned out, fertilizin-reacted sperm were **acrosome-reacted** sperm. Their acrosomal reaction had been induced by fertilizin even in the absence of eggs (see Dan, 1967).

Induction of the acrosomal reaction may be requisite in some sperm despite the strikingly swift character of the reaction. In sea urchins, the acrosomal reaction takes 5 seconds or, at most, 30 seconds, but interaction with **fertilizin** released by eggs promotes fusion of the spermatozoon's plasma membrane and acrosomal membrane.

"Jelly" or surface coat material from the egg induces the acrosomal reaction in many primitive sperm (e.g., the hemichordate acorn worm, enteropneusta) (see Colwin and Colwin, 1963). Similarly, in the toad, *Bufo bufo,* the jelly coat (i.e., added to eggs during passage through the midoviduct) induces the acrosomal reaction (Katagiri, 1974).

In the acidified (but not acidic) environment

created by the metabolically active sperm, the jelly coat inducer of *Strongylocentrotus* is leeched into the medium. Originally 30–50 μm thick, the jelly coat ultimately dissolves completely. Solubilized egg jelly consists of a fucose sulfate-rich polysaccharide component (80% of the total mass), with a molecular weight of about 300,000, and a sialoprotein (20%).

Inducing the acrosomal reaction seems to have a different material basis than binding sperm. Induction and binding are even naturally separated in some species, such as the sea urchin, *Lytechinus pictus.* Their sperm do not swarm, but their eggs release fertilizin capable of swarming the sperm of other species (e.g., *S. purpuratus*). The fucose sulfate-rich polysaccharide from sea urchin jelly, it seems, has the ability to induce the acrosomal reaction but not cause aggregation, while the sialoprotein from the jelly can cause aggregation but not induce the acrosomal reaction.

Antifertilizin may not be one material either. Antibodies raised in rabbits against a sperm surface glycoprotein with a molecular weight of 84,000 bind to sperm and prevent the acrosomal reaction, but another surface glycoprotein with a molecular weight of 64,000 is also active in binding sperm (see Lopo, 1983).

Wheat germ agglutinin (WGA), which binds to *N*-acetylglucosamine and sialic acid derivatives, agglutinates the sperm of some species of sea urchins and blocks their acrosomal reaction. In *S. purpuratus*, WGA binds with a 210-kd membrane glycoprotein and prevents the acrosome reaction induced by egg jelly or by the ionophore A23187. The possibility that the target of WGA is the sialic acid of N-linked glycoconjugates is supported by the ability of neuraminidase and endo-β-*N*-acetylglucosaminidase F (an enzyme that hydrolyzes type N-linked oligosaccharides in glycoproteins) to abolish WGA binding (Podell and Vacquier, 1984).

The fertilizin–antifertilizin bond is not especially species specific. The complex may be most efficacious when both components are from the same species, but in 11 crosses among the components from different species of sea urchins, an acrosomal reaction was induced in nine without binding sperm. The failure of specificity may be due to the antifertilizin surface component, since antibodies to a surface antigen from *S. purpuratus* cross-react with sperm from 28 species in seven different phyla (see Lopo and Vacquier, 1981).

Capacitation and acrosomal induction have some parallel features and may be interrelated, ***consisting of interchangeable steps (some of which may be bypassed*** in vitro). In *in vitro* experiments, hyaluronic acid and glycosaminoglycans from the extracellular material of the OCC are capable of inducing the acrosomal reaction. In addition, one member of a family of acidic, highly glycosylated zona pellucida proteins from mouse eggs, specifically, **ZP3,** seems to stimulate the acrosomal reaction *in vitro* in mouse sperm and may do so normally as well (Wassarman, 1987).

Whether the acrosomal reaction is actually induced in capacitated vertebrate sperm is uncertain, however. In mammals, cumulus cells of the OCC may release chemotactic attractants for sperm, but capacitated mammalian sperm undergo their acrosomal reaction in the absence of the OCC (i.e., in the presence of zona-free eggs), and metatherian and some eutherian mammalian eggs (notably those of ungulates; see Table 7.11) lack the corona radiata and present sperm with naked eggs.

The female genital tract may still **prime** sperm or increase the probability that they will undergo the acrosomal reaction. For example, the alkaline environment of the ampulla may break alkali-labile bonds during capacitation, thereby predisposing sperm for the acrosomal reaction. This possibility is supported by the ability of the surface-reactant **hyamine** (a detergent that disrupts ionic bonds) to cause the acrosomal reaction in bovine and guinea pig sperm prior to capacitation.

The requirement to remove material from sperm seems to be central to capacitation. For example, the ability of seminal plasm to reversibly **decapacitate** already capacitated sperm is attributable to a **decapacitation factor** (DF), a peptide with a molecular weight on the order of 500–1000. DF seems to dissociate from sperm during capacitation and lose its decapacitating ability through dilution. In addition, an **acrosome-stabilizing factor** (ASF, a glycoprotein similar to fibronectin) in the outer coat of freshly ejaculated sperm seems to be removed during capacitation, since it is not present in capacitated sperm (see Oliphant and Eng, 1981).

Priming may involve the unmasking or modification of surface properties. Regional differences in charge densities are created (i.e., detected as binding of colloidal iron and ferritin), and changes in the distribution of glycoproteins and terminal oligosaccharides take place.

Vesiculation or the fenestration of the acrosome is the first morphological step in the acrosomal reaction. In the sand dollar, *Echinarachnius parma,* vesiculation occurs along a **rim of dehis-**

cence, and a double-membrane cap flips off the head of the sperm exposing the acrosomal granule (Fig. 8.1). More often, in marine invertebrate sperm and in vertebrate sperm having acrosomes, the **outer acrosomal membrane** and the spermatozoon's **plasma membrane** fuse at several points, forming windows (i.e., fenestrae, Fig. 8.2) between the acrosome and the external environment. The **fenestrated membrane** usually breaks up into vesicles or cisternae but does not disintegrate and may remain as an abandoned **acrosomal cap** or **ghost** (see below, Fig. 8.6).

The reacted spermatozoon's nucleus remains covered by the inner acrosomal membrane and subacrosomal material and is not exposed to the extracellular environment (Huang and Yanagimachi, 1985). In avian and mammalian sperm, the composite structure covering the apex of the spermatozoon's nucleus is the tough, highly nonfluid perforatorium.

Increased permeability to calcium accompanies the induction of vesiculation, and ionophores, such as A23187, that specifically allow divalent cations to enter sperm also induce the acrosomal reaction. On the other hand, lanthanum and the anesthetic procaine, which inhibit calcium transport, also inhibit the acrosomal reaction. In sea urchins, at least, fertilizin–antifertilizin binding in the spermatozoon's plasma membrane may alter ion transport channels and render the membrane more **fluid** (i.e., noncrystalline) by removing or increasing turnover of cholesterol.

An outflow (efflux) of protons (i.e., hydrogen ions) through the plasma membrane accompanies the influx of calcium ions from the extracellular environment (see Yanagimachi, 1981), thereby elevating the spermatozoon's internal pH. Wheat germ agglutinin, while blocking the acrosomal reaction, also blocks the efflux of hydrogen ions and the influx of calcium ions.

Within the sperm, calcium ions may be bound to **calmodulin** present in the sperm heads' postacrosomal region. The abundant, heat-stable, low molecular weight protein (17,000 daltons), which

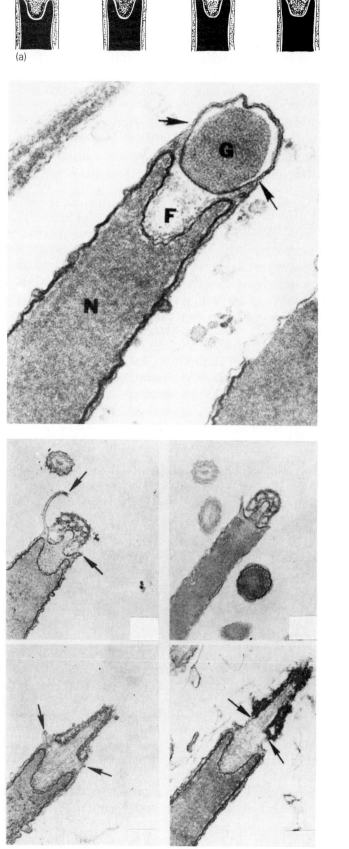

FIGURE 8.1. Schematic drawings (a) and electron micrographs (b) of sagittal sections through rostral sperm head during the acrosomal reaction in the sand dollar, *Echinarachnius parma*. An acrosomal cap opens behind a rim of dehiscence, and an acrosomal filament emerges from the cup-shaped subacrosomal space or nuclear fossa (i.e., indentation in the nuclear tip). Arrows: rim of dehiscence. From R. G. Summers and B. L. Hylander, *Cell Tiss. Res.*, 150:343 (1974), by permission of Springer-Verlag and the authors.

232

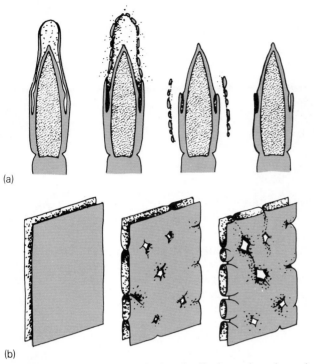

(a)

(b)

FIGURE 8.2. Diagrammatic longitudinal section through typical mammalian sperm (upper) and portions of the plasma and outer acrosomal membranes (lower) illustrating fenestration of the acrosomal cap. Upper diagram from J. M. Bedford and G. W. Cooper in G. Poste and G. L. Nicolson (eds), *Cell Surface Reviews,* vol. 5: Membrane Fusion. By permission of Elsevier/North-Holland Biomedical Press (1978) and the authors. Lower diagram adapted from R. Yanagimachi. In L. Mastroianni and J. D. Biggers, eds., *Fertilization and embryonic development in vitro,* Plenum Press, New York, 1981. Used by permission.

accounts for 12% of the total soluble protein in rabbit sperm, has a high affinity for calcium (see Klee et al., 1980). Bound to calcium, calmodulin activates target enzymes, such as protein kinases, and alters structural proteins, such as polymerized tubulin (see Epel, 1980).

Not every effect of calcium is salubrious. Free calcium uncouples oxidative phosphorylation in mitochondria and starves the sperm of ATP. The early death of sperm following incorporation or release of calcium may represent a first-line **block to polyspermy,** since only fertilizing sperm survive (see Turner, 1987).

Calcium-activated **phospholipase** would create further perturbation in the membrane, and **acrosin,** a proteolytic acrosomal enzyme, could remove protein blocks to membrane fusion (among other effects, see below) (Yanagimachi, 1981). Since surface proteins presumably play a major role in stabilizing the spermatozoon's plasma membrane,

changing these proteins would presumably play a major role in destabilizing the membrane.

Proteins implicated in membrane fusion include the major tissue histocompatibility antigens (e.g., the H-2 family of mouse tissue-recognition and graft-rejection proteins) present in the surface of the outer acrosomal membrane and some **autoantigenic** proteins (i.e., antigens capable of raising antibodies in the same animal in which they are produced) present in the **periacrosomal region** (behind the acrosome) of guinea pig, mouse, and rabbit sperm. In the presence of complement (a cytolytic protein in serum) and serum containing antibodies to the autoantigens, the acrosome breaks down as if undergoing an acrosomal reaction.

Acrosomal processes (also called filaments or tubules) of different kinds are discharged at the apices of many primitive sperm, often with remarkable speed (Fig. 8.3). Short sea urchin processes are about 1 μm, while long sea cucumber processes reach 60–90 μm.

The process consists of a core of **filamentous-** or **F-actin** fibers covered by the former inner acrosomal membrane and material derived from the outer nuclear envelope. The actin fibers may be preformed, as in the case of the horseshoe crab, *Limulus* (Fig. 8.4), or assembled from **profilactin, globular-** or **G-actin** stores concentrated in the cup-shaped **subacrosomal space** or **nuclear fossa.**

In the sea cucumber, *Thyone,* actin filaments emerge from a group of 20–25 filaments anchored to the nuclear envelope in the center of the fossa. Called the **actomere,** these filaments serve as nucleating bodies for actin polymerization and direct the polarity of the self-assembly process outward at amazing rates of 5–10 μm or more per second (Fig. 8.5)!

Sperm Binding Sites

The idea of **binding sites** on sperm and complementary **receptors** on the envelopes of eggs arises from observations on fertilization. In the laboratory, suspensions of sperm in high concentration quickly saturate vitelline envelopes with bound sperm. Extraordinarily large numbers of sperm are bound during an initial **phase of firm attachment** or **binding phase.** For example, a million sperm may attach to the egg of the horseshoe crab, *Limulus polyphemus,* and several thousand to sea urchin eggs.

Still, the egg surface is not entirely covered with sperm, and spaces remain between adhering sperm. Possibly, vitelline envelopes support only a

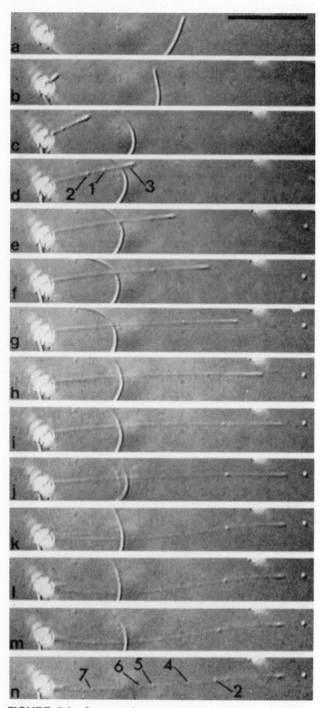

FIGURE 8.3. Sperm of sea cucumber, *Thyone briareus,* during elongation of acrosomal filament. Single frames photographed from TV monitor at 0.75-second intervals beginning 1.95 seconds after initiation of acrosomal reaction. Numbers identify cytoplasmic blebs. Spermatozoon's tail is seen as arc toward left. Bar (in frame a) = 20 μm. ×1100. Reproduced from L. G. Tilney and S. Inoué, *J. Cell Biol.,* 93:820 (1982), by copyright permission of the Rockefeller University Press and courtesy of the authors.

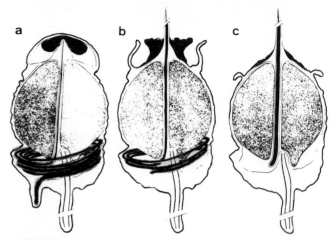

FIGURE 8.4. Elongation of the acrosomal filament in the horseshoe crab, *Limulus polyphemus,* from a preexisting actin filament bundle. The membrane of the acrosomal filament acquires a coat of acrosomal material as it passes through the remains of the acrosomal granule. Reproduced from L. G. Tilney, J. G. Clain, and M. S. Tilney, *J. Cell Biol.,* 81:229 (1979), by copyright permission of the Rockefeller University Press and courtesy of the authors.

limited number of receptors capable of adhering to binding sites on the sperm.

Further observations on fertilization suggest that binding sites and receptors play a dynamic role in fertilization and are physiologically active rather than passive glues. A **phase of sperm detachment** generally follows the binding phase, sometimes very quickly (e.g., 0–25 seconds in the sea urchin). The vitelline envelope's receptor sites seem to be altered at this time and no longer fasten sperm by their binding sites. Moreover, receptor binding requires free calcium ions.

The times at which sperm's hypothetical binding sites appear during fertilization differ with species. In most mammals, the acrosome must be present and intact for the initial adhesion of sperm with the zona pellucida. Mouse sperm, for example, will not bind to the zona if the acrosomal cap is lost. Possibly an initial **zona-adhesion site** in the spermatozoon's plasma membrane is removed by the acrosomal reaction (Wasserman, 1983). On the other hand, guinea pig sperm adhere to the zona pellucida only after the acrosomal reaction. Hamster sperm can bind both before and after their acrosomal reaction, although acrosome-reacted sperm rapidly lose their ability to fertilize.

Conspecific sperm may bind to egg membranes by carbohydrate sequences, since some plant lectins can block fertilization. For example, concanavilin A (Con A), which recognizes α-D-glucose and α-D-mannose, blocks fertilization in the sea ur-

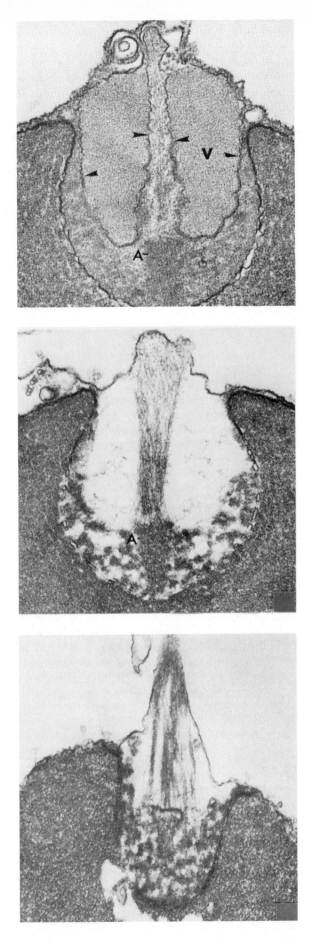

chin, *Anthocidaris crassispina*, by binding to the surface of the spermatozoon's apical end (and to the midpiece). Con A does not bind to the sperm of most echinoderms and binds to the midpiece of *S. purpuratus* without preventing fertilization.

Lectin-labeling experiments with mammalian sperm demonstrate the presence of α-D-mannose, *N*-acetylglucosamine, and D-galactose-like terminal saccharides on the inner acrosomal membrane (see Huang and Yanagimachi, 1985). Rabbit sperm have a glycoprotein possessing oligosaccharide side chains with terminal sialic acid residues, and guinea pig sperm have a demonstrable glycocalyx. The lectin wheat germ agglutinin (WGA), which has its highest affinity for *N*-acetylglucosamine and hence glycoprotein or glycolipid, binds to the periacrosomal surface (i.e., around the acrosome) of mature hyrax sperm. Moreover, the enzyme neuraminidase, which cleaves sialic acid from the carbohydrate portions of glycoproteins and glycolipids, alters the electrophoretic motility of rat sperm (see Bedford and Cooper, 1978).

Polypeptides may play a binding role, and whole *S. purpuratus* sperm yield a variety of polypeptides when extracted with detergents such as sodium dodecylsulfate (SDS) and Triton-X100 (see Lopo and Vacquier, 1981). In several echinoderms, a binding material, called **bindin,** is released from the acrosomal granule as the acrosomal process elongates and precipitates on the portion of the acrosomal membrane remaining with the sperm (between arrows, Fig. 8.1). Isolated from sea urchin acrosomal vesicles, bindin lacks carbohydrate and migrates as a single major component with a molecular weight of 30,500.

The amino acid sequences of bindin from both *S. purpuratus* and *S. franciscous* are similar. The first 45 amino acid residues are identical. The re-

FIGURE 8.5. Electron micrographs of thin sections through apical end of sperm of the sea cucumber, *Thyone briareus,* with no treatment (upper) and after treatment with the ionophore x537A, revealing the fine structure of the actomere (center and lower). (Upper) A bundle of filaments extends from the actomere (A) through the acrosomal granule (V). Dense material is already attached to the inner surface of the acrosomal membrane (arrowheads). ×90,000. (Middle) Exocytosis follows the initial acrosomal reaction. The acrosomal membrane fuses with the cell surface and most of the contents of the granule are evacuated, leaving a clear zone. Filaments begin to extend from the actomere. ×90,000. (Lower) Long filaments now extend from the actomere and from the profilactin lateral to the actomere. ×95,000. Reproduced from L. G. Tilney, *J. Cell Biol.,* 77:551 (1978), by copyright permission of the Rockefeller University Press and courtesy of L. G. Tilney.

mainder differ largely in the number of aspartic acid and proline residues. This degree of similarity would seem to extend to bindin throughout the class Echinoidea, since an antibody to *S. purpuratus* bindin cross-reacts with acrosomal granules from all members of the class tested but not with the granules of members of other echinoderm classes (see Lopo and Vacquier, 1981).

Another carbohydrate-free bindinlike protein with low molecular weight comes from the wormlike echiuroid, *Urechis caupo*. Bindinlike material from the oyster, *Crassostrea gigas*, on the other hand, consists of two glycoprotein components with molecular weights of 65 and 53 kd.

Bindinlike protein is probably not present in the plasma membrane of eutherian sperm, but circumstantial evidence suggests the existence of bindinlike material on the mammalian inner acrosomal membrane. For example, antibodies raised in guinea pigs against the inner acrosomal membranes of rabbit sperm prevent fertilization by acrosome-reacted rabbit sperm (Srivastava et al., 1986). Furthermore, fucoidin, a fucose heteropolysaccharide with an affinity for sea urchin bindin, conjugates with the inner acrosomal membrane of guinea pig sperm and competitively inhibits sperm–zona binding in guinea pigs and hamsters (see Huang and Yanagimachi, 1981).

In the guinea pig, a sperm-surface antigen present over the posterior head of intact sperm is redistributed to the inner acrosomal membrane of acrosomal-reacted sperm. Similarly, in the mouse, initial binding of spermatozoa to the zona pellucida depends on a sperm-surface galactosyltransferase (GalTase) present on the dorsal side of the anterior sperm head overlying the acrosome. GalTase has a high affinity for *N*-acetylglucosamine residues on zona glycoproteins and is repositioned coincident with the acrosome reaction to the lateral surface of the sperm (Lopez and Shur, 1987; Shur, 1989).

In the pig, a surface-bound GalTase mechanism may establish contact between intact sperm and the zona pellucida, but following the acrosome reaction a different carbohydrate recognition mechanism takes over. Boar sperm contain at least one polypeptide soluble in sodium deoxycholate (DOC) with a high affinity for heat-solubilized pig zona pellucida proteins. This 53-kd polypeptide is localized within the acrosome of capacitated, noncapacitated, epididymal, and testicular sperm (Brown and Jones, 1987). It is **proacrosin,** the zymogen (i.e., precursor) form of the proteolytic enzyme acrosin.

Like bindin, proacrosin has proteolytic prop-

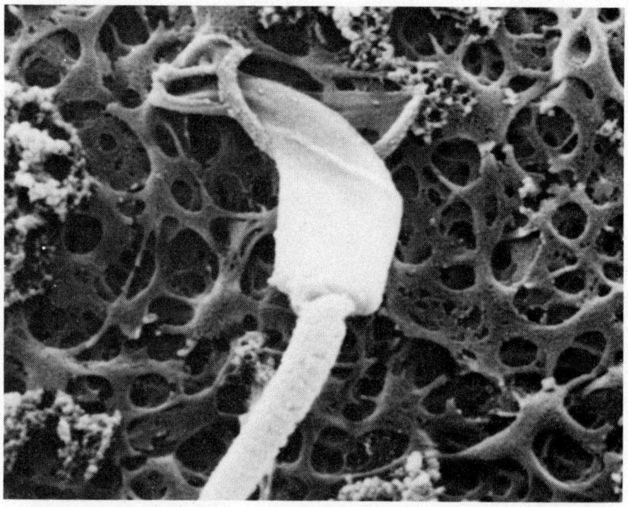

FIGURE 8.6. Scanning electron micrograph of a golden hamster spermatozoon and egg from which cumulus cells were previously removed (i.e., cumulus-free oocyte). The spermatozoon reveals its inner acrosomal membrane (IA) connected to its fenestrated acrosomal ghost (AG). Portions of the split ghost (arrows) are attached to the zona pellucida. Fixed at 15 minutes postinsemination. ×8400. From G. N. Cherr, H. Lambert, S. Meizel, and D. F. Katz, *Dev. Biol.,* 114:119 (1986), by permission of Academic Press and courtesy of the authors.

erties, a hydrophobic N-terminal sequence, and a tendency to adhere to surface structures with multivalent polysaccharide chains. Released proacrosin translocated to the external surface of the acrosomal cap may bind zona carbohydrates, while the localized proteolytic effect of active acrosin may make membrane carbohydrates more accessible for binding (Jones et al., 1988). Possibly, in mammals generally, residual bindinlike acrosomal material sticks to the acrosomal ghost and both keeps the fenestrated membranes intact and glues the ghost to the zona pellucida while allowing the spermatozoon to enter the zona (Figs. 8.6, 8.7).

Receptors on the Egg's Envelope

The hypothetical receptors would have to be distributed on the outer surface of the egg's envelope to bind sperm. In order to play a role in fertilization, sea urchin receptors should lie in the vicinity of the egg's microvilli (Fig. 8.8) that dot the thin (10–30 nm thick) vitelline membrane. In ascidians, fucosyl-containing proteins found in the outer surface of the vitelline coat could play the role of receptor.

Isolated sea urchin vitelline membranes are 90–95% protein with about 3.5% carbohydrate.

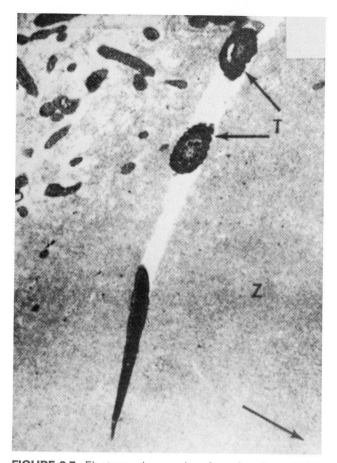

FIGURE 8.7. Electron micrographs of section through the slit made by a rabbit spermatozoon as it penetrates the zona pellucida (Z). Egg lies in direction of arrow. Cross sections of the undulating spermatozoon's tail are seen in the slit (T). ×12,800. From J. M. Bedford, *Biol. Reprod. Suppl.,* 2:128 (1970), by permission of the Society for the Study of Reproduction and courtesy of J. M. Bedford.

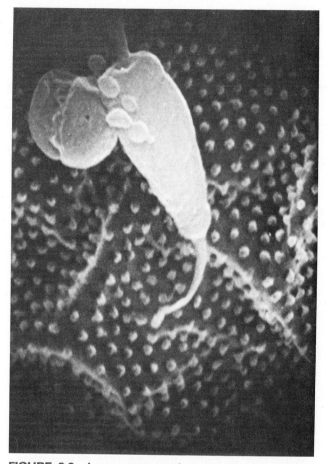

FIGURE 8.8. Acrosome-reacted spermatozoon of *Strongylocentrotus purpuratus* with elongated acrosomal process. ×20,000. From G. Schatten and D. Mazia, *Exp. Cell Res.,* 98:325 (1976), by permission of Academic Press and courtesy of G. Schatten.

Of the numerous proteins dissolved by detergents, such as sodium dodecylsulfate (SDS), and separated by polyacrylamide gel electrophoresis (SDS–PAGE), few stain for carbohydrate. Still, galactosamine is present in glycoconjugates derived from sea urchin vitelline envelopes, and a bindin receptor fraction from *S. purpuratus* with a molecular weight of at least 5 million contains mannose and galactose, if not sialic acid, and 4% sulfate (see Lopo and Vacquier, 1981).

Vitelline membranes from mollusks, however, are about one-third threonine-rich protein and nearly two-thirds carbohydrate. Since proteases released from the egg's cortical granules eliminate the eggs' fertilizability, receptors would seem to be proteins or glycoproteins. Monosaccharides and polysaccharides also block sperm–egg interactions in several invertebrates and plants.

The concentration of sperm receptors on amphibian eggs may be increased by transcription, but the receptors may represent the physical condition of materials rather than the materials themselves. For example, **coelomic eggs,** or eggs in the coelomic cavity which have not passed through the oviducts or acquired their final egg coats, are not fertilized by normal sperm. A rough fibrous coat over the vitelline membrane is reprocessed into a finer fibrous coat while eggs pass through the upper part of the oviducts. Only this coat normally permits sperm adhesion and penetration.

Various lectins (e.g., Con A), monosaccharides, and glycoconjugates bind to the mammalian zona pellucida and block fertilization, presumably by occupying the sperm receptor. Moreover, hydrolysis of glycoproteins prevents fertilization in hamsters, humans, and mice.

In the mouse egg, the putative receptor is ZP3, an 83-kd glycoprotein consisting of a 44-kd polypeptide backbone, 3–4 N-linked oligosaccharides, and O-linked oligosaccharides. ZP3 is similar to a 58-kd glycoprotein of pig zonas, and to a glycoprotein in sea urchin eggs which may play a receptor role (Bleil and Wassarman, 1980).

In mouse ZP3, O-linked glycoconjugates (i.e., oligosaccharides conjugated to the β-hydroxyl groups of serine and threonine) containing N-acetyl-D-galactosamine may provide the molecular receptor sites. Deglycosylation (by trifluoromethanesulfonic acid) but not the removal of N-linked glycoconjugates (i.e., oligosaccharides conjugated to asparagine and removed by endo-B-N-acetyl-D-glucosaminidase F) destroys ZP3's sperm receptor activity. Furthermore, small glycopeptides obtained by extensive pronase digestion and O-linked oligosaccharides obtained by mild hydrolysis and reduction from ZP3 can bind to sperm (i.e., possess *in vitro* sperm receptor activity) (Florman and Wassarman, 1985).

The egg's envelopes frequently provide the main barrier to cross-fertilization. In the presence of the envelope, heterologous (i.e., foreign) sperm may fail to undergo the acrosome reaction, or they may fail to adhere and penetrate the envelope. In teleosts, the chorion alone seems to prevent cross-fertilization, and hybridization is common when the micropyle is enlarged to allow the simultaneous passage of more than one type of spermatozoon.

The match between binding sites and receptors is often species specific, if only weakly. Human sperm, which readily attach to the human zona, fail to attach to baboon, guinea pig, hamster, mouse, pig, rabbit, rhesus monkey, and squirrel monkey zonas but will attach and penetrate the gibbon zona. Guinea pig sperm show a high degree of specificity in their attachment to zonas, but hamster sperm attach to guinea pig, mouse, and rat zonas (albeit without subsequently penetrating the zonas).

Eliminating the egg envelope often reduces the species specificity of fertilization and improves the frequency of cross-fertilization. Removal of the sea urchin's entire vitelline layer by trypsin digestion or dithiothreitol (a reducer of disulfide bonds), for example, allows sperm of different species to extend their acrosomal processes and fuse with the eggs.

Similarly, the species specificity of fertilization of mammalian eggs depends on the zona pellucida, and **zona-free** eggs may be fertilized by sperm of different species. Human sperm, which do not penetrate intact hamster zonas, can fuse with zona-free hamster eggs, and mouse sperm,

which rarely penetrate rat zonas, can fuse with zona-free rat eggs.

Perforating the Egg's Envelopes

Different types of envelope present different challenges for fertilizing sperm. Movement through the micropyle of an otherwise impenetrable chorion and perforation of an egg's primary membrane also offer additional opportunities for species specificity.

Often lacking acrosomes, teleost sperm may adhere transiently to micropyle receptors at lateral binding sites. Present in the spermatozoon's plasma membrane, the binding sites may not only allow the first spermatozoon a species-specific "grip," but while occupied by one spermatozoon, the receptors are unavailable to other spermatozoa. The interaction of micropyle receptors and binding sites may thereby effectively block other sperm's access to the egg. Similarly, in mammals, the glycoprotein ZP2 may operate as a secondary receptor or grip, attaching to binding sites on the freshly exposed inner acrosomal membrane and orienting sperm toward the zona (Fig. 8.9) (Wassarman, 1987).

In sea urchins, the thin vitelline membrane is penetrated almost simultaneously with sperm binding to the membrane (Fig. 8.10), but in animals with thicker primary membranes, such as the zona pellucida of mammals, penetration is gradual. For example, hamster sperm penetrate the zona in 10 minutes (i.e., estimates range from about 7 to 20 minutes), while mouse sperm take about 20 minutes.

The ability of sea urchin sperm to perforate eggs' vitelline envelopes depends on the **exocytosis** of acrosomal enzymes through the acrosomal fenestrations and on the adhesion of enzymes to the exposed inner acrosomal membrane at the apex of the spermatozoon. Ascidian sperm release a trypsinlike serine protease, **acrosin** (27 kd), **spermosin** (32–35 kd), and a chymotrypsin-like enzyme that digests the ascidian vitelline coat (Sawada et al., 1986). These enzymes, along with bindin, seem to adhere to the inner acrosomal membrane at the tip of the spermatozoon.

The vesiculation of the mammalian acrosomal cap is also a requirement for subsequent penetration of the zona pellucida. Immotile and degenerating sperm that shed their caps (in a process called the **false** or **degenerative acrosomal reaction**) without fenestrating the membrane and releasing the acrosomal contents, merely break down without fertilizing the egg.

While mouse sperm bind to eggs before completing their acrosomal reaction, they must complete their acrosomal reaction if they are to remain adherent. The fenestrated acrosome anchors the spermatozoon and orients it for a proper entry into the zona (Fig. 8.6), but the spermatozoon only enters the zona when it splits the acrosomal membrane (about 15 minutes after binding in mice) and leaves the ghost behind (Fig. 8.9) (see Talbot, 1985).

Enzymes known as **zona lysins** released from the acrosome through fenestrae (Fig. 8.2) adhere to the inner acrosomal membrane of the perforatorium and digest a slit through the zona. Mammalian lysins include acrosin (Enzyme Commission number EC 3.4.21.10), which hydrolyzes bonds between arginine and other amino acids (also lysine to lysine bonds and, less vigorously, lysine to other amino acids), β-N-acetylhexosaminidase, acid

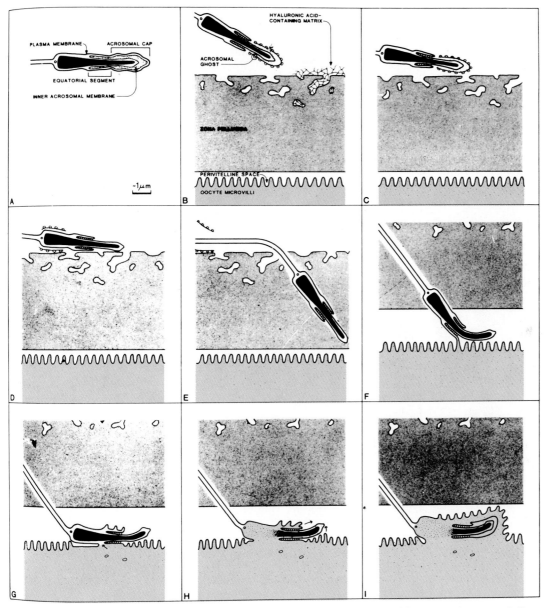

FIGURE 8.9. Diagrams illustrating fertilization in a generalized mammal. (a) Nomenclature and dimensions. (b) The acrosomal membrane is fenestrated but reaction is not completed. (c) The acrosomal ghost adheres to the surface of the zona pellucida and orients the sperm. (d, e) Ghost splits and zona-lysing enzymes digest a slit into which the spermatozoon moves, propelled by tail movements. (f) The equatorial segment of the spermatozoon fuses with oocyte microvilli. The perivitelline space then expands as a result of the cortical response. (g) Microvilli begin to incorporate the spermatozoon's head. (h, i) The spermatozoon's nucleus decondenses posteriorly, while the anterior perforatorium encapsulated in a vesicle is incorporated into the egg. From P. Talbot, *Am. J. Anat.,* 174:331 (1985), by permission Alan R. Liss, Inc., and the author.

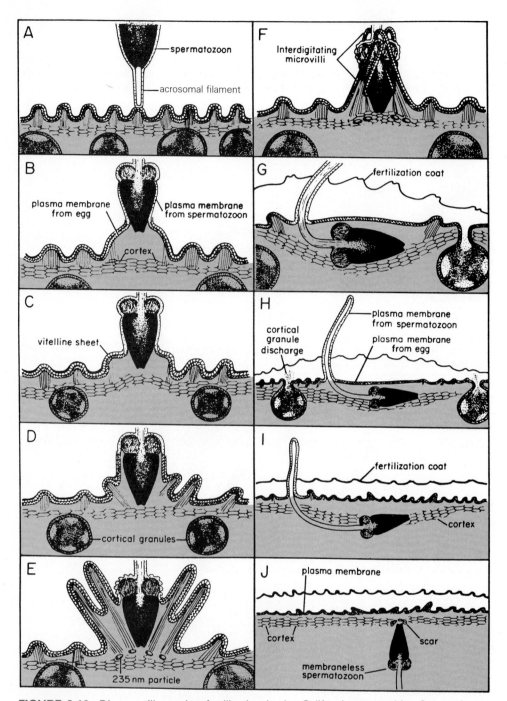

FIGURE 8.10. Diagram illustrating fertilization in the California sea urchin, _Strongylocentrotus purpuratus._ (a) Acrosomal filament (acrosome) reaches the vitelline envelope. (b) Surface of egg bulges outward. (c, d) The spermatozoon's head is covered by the oolemma and a thin layer of cytoplasm as long microvillar processes erupt from the egg's surface. (e, f) The processes ensnare the head and form a fertilization cone, while small particles appear below. (g) As cortical vesicles are discharged (cortical exocytosis), the fertilization envelope replaces the vitelline envelope and lifts off the egg's plasma membrane. The spermatozoon, less its plasma membrane, is drawn into the egg. (h–j) Rotation of the spermatozoon's head brings it into position to penetrate the cortical layer (cortex). Leaving only a small scar to testify to its path, the membraneless spermatozoon escapes into the cytoplasm. Note lower magnification in (h–j). From G. Schatten and D. Mazia, _J. Supramol. Struct.,_ 5:343 (1976), by permission of Alan R. Liss, Inc., and courtesy of G. Schatten.

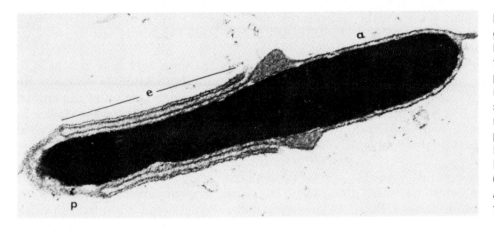

FIGURE 8.11. Electron micrograph of sagittal section through hamster spermatozoon in perivitelline space.[1] Note continuity of plasma membrane with the acrosomal membrane in equatorial segment. (a), Acrosomal segment; e, equatorial segment; p, postacrosomal region. ×48,800. From H. D. M. Moore and J. M. Bedford, *J. Ultrastruct Res.,* 62:110 (1978), by permission of Academic Press, Inc. and courtesy of the authors.

phosphatase, arylsulfatase, collagenase, nonspecific esterase, hyaluronidase, neuraminidase, and phospholipase (see Bellvé and O'Brien, 1983).

Of three zona pellucida glycoprotein families isolated from pig oocytes by SDS–PAGE, only glycoproteins with molecular weights of 90,000 and 65,000 are sensitive to boar sperm acrosin. Their digestion must be sufficient to create the sperm's slit through the zona, since glycoprotein with a molecular weight of 55,000 resists digestion (Hedrich et al., 1986).

Rodent sperm bore slits at an angle between oblique and perpendicular and traverse paths between straight and curved. Wiggling back and forth, the spermatozoon swims through the slit on its way to the egg surface (Fig. 8.7). While passing through the zona, a spermatozoon's head becomes more pointed. The flagellar beat of hamster sperm, if not mouse sperm, is slower than the hyperactive beat of a capacitated spermatozoon, but the bend of the tail may be more profound during passage (Suarez et al., 1984).

After penetrating the zona, the spermatozoon's head moves into the **perivitelline space** prepared for it during the egg's cortical response. The tail may remain in the zona (e.g., mouse and rat) or quickly (e.g., within 10–15 seconds in hamsters) follow the head into the perivitelline space as a result of rotation of the egg or of several vigorous beats of the flagellum.

The sperm may make contact with the egg immediately or move freely within the space for several minutes (about 20 in the rat). In pigs and opossum (but not rodents), hyaluronidase brought into the perivitelline space on the inner acrosomal membrane hydrolyzes hyaluronic acid adjacent to

[1]Immature eggs are employed, because the zona's block to polyspermy is not yet in place, and many sperm may reach the perivitelline space, thus increasing the likelihood of finding a spermatozoon in a section.

the plasma membrane of the oocyte, increasing fluidity in the perivitelline space.

Fusion

"Fusion" rather than "penetration" describes the interaction of sperm and egg membranes during syngamy. Sperm do not pierce the egg's oolemma the way they may bore through vitelline envelopes. Upon meeting, possibly at random, **fusiogenic** (also fusogenic and fusigenic) portions of the spermatozoon's membrane and the oolemma fuse. The subsequent incorporation of sperm into the egg is at least as much the business of the egg as of the sperm.

With the exception of eutherian mammals, the fusiogenic portion of the spermatozoon's membrane lies at the apical tip, often over the apical acrosomal process. In sperm lacking acrosomes (e.g., sponge, coelenterate, and teleost sperm), the plasma membrane at the apex is fusiogenic, while in sperm with acrosomes, even those with minute acrosomes (e.g., tunicate sperm), the apical inner acrosomal membrane is fusiogenic (Fig. 8.10) (see Yanagimachi, 1988).

The fusiogenic membrane of the acrosome-reacted sperm of most vertebrates, from lampreys to salamanders, reptiles, birds, prototherian, and metatherian mammals (but not eutherians), is also at the apex. In the primitive sperm of the lamprey, an acrosomal tubule even leads the spermatozoon's fusiogenic membrane to the egg.

Eutherian mammalian sperm not only lack acrosomal processes but the leading membrane is not fusiogenic. The fusiogenic membrane lies far from the tip of a eutherian spermatozoon in the plasma membrane of the **equatorial segment** circling the spermatozoon head (Fig. 8.11).

The equatorial segment of a eutherian spermatozoon forms during the acrosomal reaction when a rim of dehiscence develops far down the spermatozoon's head and separates the acrosomal ghost from the equatorial zone (Fig. 8.2a). Portions of the inner and outer acrosomal membranes are sandwiched between the spermatozoon's plasma membrane and its nuclear membrane. Ladderlike **septa** between the central layers (Fig. 8.12) seem to stabilize the equatorial segment against fenestration and render it rigid during passage through the zona pellucida.

The mechanism rendering sperm fusiogenic is unknown. Possibly, the inner acrosomal membrane of nonmammalian sperm is always fusiogenic and merely has to be exposed via the acrosomal reaction. Alternatively, bindin released by the acrosomal reaction may actuate the inner acrosomal membrane's fusiogenic potential. The mammalian spermatozoon's plasma membrane in the equatorial zone, it would seem, changes dramatically at the time of or as a result of the acrosome reaction.

Fusiogenic sites can be labile or stable in different sperm. Within 1–4 minutes after completing the acrosomal reaction, sea urchin sperm fail to fuse with the egg's plasma membrane (i.e., are incompetent for fertilization), while acrosome-reacted rabbit sperm removed from one egg's perivitelline space are capable of fertilizing other eggs.

The surfaces of some eggs may not be uniformly competent for fertilization. In hydrozoa, fusion is limited to the small portion of the animal pole where the second polar body was extruded. In tunicates, fusion tends to occur in the vegetal pole, and in frogs, fusion occurs preferentially in a band 60° from the animal pole to the margin of the vegetal hemisphere. In the painted frog, *Discoglossus pictus*, whose sperm are an incredible 2.3 mm long, the region of fusion on the egg is limited to a small recess, known as the "dimple," at the animal pole (see Elinson, 1986). Fusion occurs in the animal pole of bird eggs.

More often, fusion can occur anywhere that microvilli are present on eggs. Sea urchin sperm, for example, associate at random with the surfaces of eggs, and mammalian sperm associate with every part of the mammalian egg except at the "bald" or smooth point lacking microvilli above the meiotic spindle and the site of polar body formation (Fig. 8.13). In the rat egg, however, the nonfusiogenic surface may extend over half the egg's surface.

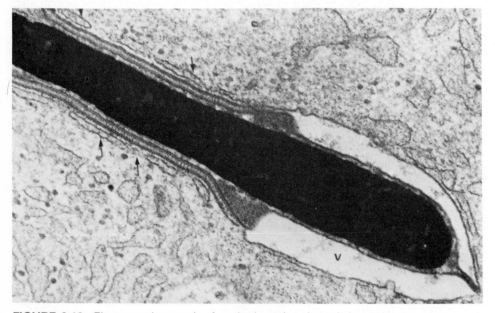

FIGURE 8.12. Electron micrograph of sagittal section through hamster spermatozoon in immature (i.e., diplotene) oocyte removed from ovarian follicle and fertilized *in vitro*. Had the egg been ripe, the nuclear contents would be dispersed. Electron-opaque material between inner and outer acrosomal membranes of the equatorial segment forms dense septae. Vesicles form around nucleus (arrows), and a large phagocytic-like vesicle (v) of oolemma externally and inner acrosomal membrane internally forms around the perforatorium. ×50,000. Arrows, flattened vesicles in egg's cytoplasm parallel and lie adjacent to equatorial segment. From H. D. M. Moore and J. M. Bedford, *J. Ultrastruct. Res.*, 62:110 (1978), by permission of Academic Press and courtesy of the authors.

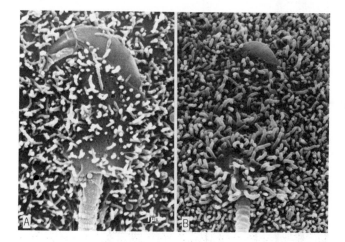

FIGURE 8.13. Scanning electron micrograph of acrosome-reacted hamster spermatozoon fusing with zona-free egg *in vitro*. From R. Yanagimachi. In L. Mastroianni and J. D. Biggers, eds., *Fertilization and embryonic development in vitro*, Plenum Press, New York, 1981. Electron micrographs by Dr. D. M. Phillips. Used by permission of Plenum Press and the author.

In general, the entire oolemma is fusiogenic. The lipid components of the oolemma are relatively fluid and lectins bound to the oolemma show a high degree of clustering.

Even eggs enclosed within impenetrable chorions may be fusiogenic over their entire surface. For example, the chorionated eggs of fish that are normally fertilized only at the micropyle are capable of being fertilized at any point when the chorion is removed (i.e., dechorionated eggs).

For many species, neither the egg nor the spermatozoon's fusiogenic membranes are especially discerning by way of species specificity. If the egg's protective membranes are removed or if the egg is from a gynogenic species, syngamy may be achieved by sperm and egg of different species. Trypsin-treated sea urchin eggs, for example, are readily hybridized, and hamster eggs, freed of their zonas, can be fertilized by a wide variety of mammalian sperm.

In other species, the eggs and sperm are more discerning. Zona-free rabbit eggs, for example, are fertilizable with mouse and rat sperm but not hamster and guinea pig sperm. Zona-free rat eggs are fertilizable by mouse sperm, but mouse eggs are rarely fertilizable by foreign sperm.

Furthermore, "quantitative" if not qualitative species specificity of fusion is detectable even in hamster eggs. Zona-free hamster eggs show much greater affinities for hamster sperm than for guinea pig and human sperm, even though hamster eggs are capable of being fertilized by sperm of other species.

Mechanisms of Fusion

Microcinematography and scanning electron micrography of eggs exposed to sperm show that the successful sperm are generally among the first to arrive at the surface of the egg. In the sea urchin, acrosome-reacted sperm penetrating the jelly coat and adhering to the vitelline envelope rotate counterclockwise as a result of tail beating. After several seconds (e.g., between 10 and 20 seconds in sea urchins, 4 seconds in annelids and enteropneusta), the fertilizing spermatozoon stands up perpendicularly to the egg surface (Fig. 8.14a) and its tail movements stop (Tegner and Epel, 1976).

Flagella motility in marine invertebrate sperm may not be involved in the final incorporation of the sperm. Motionless and even tailless sea urchin sperm can affect fertilization (Epel et al., 1977). Even in mammals where strong tail movements are required for passage of sperm through the zona pellucida and in fish where tail movements seem to aid passage through the micropyle, motionless sperm are effective agents of fertilization once within the perivitelline space.

Like the successful sea urchin sperm, the successful mammalian sperm are seen to become immobilized following contact with the oolemma. Possibly, an explosive release of Ca^{2+} by the egg paralyzes the spermatozoon tail (see Yanagimachi, 1988).

Membrane fusion depends on bringing membranes to within a **primary minimum** or **molecular distance** on the order of 1.5 nm, consistent with establishing short-range attractive forces (e.g., London–van der Waals forces) (Poste and Allison, 1973). Electrostatic repulsion must be neutralized to bring membranes this close together.

Divalent cations may aid in establishing close approximation between membranes by neutralizing negative charges and dispelling water (the so-called hydration barrier) between phospholipid head groups. Ca^{2+} is essential for spermatozoon–egg fusion in guinea pigs, hamsters, and human beings, and Mg^{2+} may be beneficial. Mg^{2+},

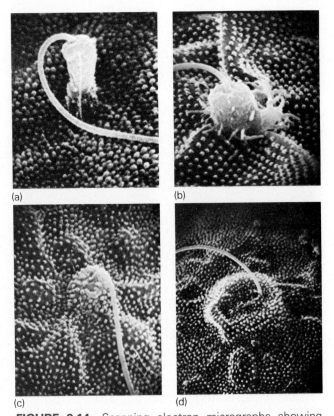

FIGURE 8.14. Scanning electron micrographs showing outer surface of the California sea urchin, *Strongylocentrotus purpuratus*, eggs during fertilization. Knoblike microvilli dot the egg surface. (a) Spermatozoon standing on its head fuses with egg as the surface bulges. Some vitelline papillas throw out fine fibers, about 50 nm in diameter, which may anchor the fertilizing spermatozoon to egg during early incorporation (×3350). (b) About 20 microvilli, some 1 μm in length, erupt from the egg surface and ensnare the spermatozoon (×4500). (c) A fertilization cone engulfs the spermatozoon's head (×3350). (d) The spermatozoon's head rotates from the perpendicular to the plane of the egg surface and sinks into the fertilization cone (×3000). From G. Schatten and D. Mazia, *J. Supramol. Struct.*, 5:343 (1976), by permission of Alan R. Liss and courtesy of G. Schatten.

Ba^{2+}, and Sr^{2+} can also trigger fusion if less effectively than Ca^{2+}. K^+, on the other hand, is not essential for spermatozoon–egg fusion in mouse although guinea pigs require K^+ while undergoing their acrosomal reaction if they are later to prove capable of fusion.

Charge is also minimized where membranes bend with a radius of curvature less than 0.1 μm (Pethica, 1961). This is approximately the radius at tips of acrosomal processes (Fig. 8.10), folds in the equatorial segment of mammalian sperm (Fig. 8.9f), and apices of microvilli on the egg's surface. The acrosomal processes in the mussel, *Mytilus*,

and the surf clam, *Spisula*, are seen to fuse with microvilli (see Longo, 1973). Similarly, while the acrosomal processes of sea urchin sperm make contact at random loci on the egg's surface, fusion seems to occur with the egg's microvilli. Still, intermicrovillous plasma membrane may also be fusiogenic.

The importance of electrostatic repulsion as an energy barrier to fusion is illustrated by the action of erythrosin B. Erythrosin B increases potassium permeability and prevents the appearance of the fertilization potential (see below). Sperm binding is not prevented, but gamete fusion fails to take place.

When the membrane potential is controlled by electrodes in voltage clamp studies, only potentials on the order of the original resting potential are found to be compatible with fertilization. Sea urchin eggs clamped at a positive potential do not fuse with sperm, but a 30-millisecond (msec) pulse at -60 mV allows fertilization in 12% of the trials, and a 70-msec pulse allows fertilization in 90% of the trials. A 30-msec pulse at -90 mV, moreover, allows fertilization in 60% of the trials (see Jaffe and Gould, 1985).

An initial stepwise depolarization (increase in positive charge within the egg) may either be symptomatic of spermatozoon–egg contact or establish a **permissive value** for the membrane potential (E_m) required for fusion. The step (or steps in the case of polyspermy) is small (on the order of 1–2 mV), and the membrane potential remains stable at the new value for a time (about 11 seconds, see below, Fig. 8.22) (Dale and De Santis, 1981b).

The acidity of the medium affects membrane potential. In hamsters, acidic pH prevents fusion. For example, changing the culture medium from 7.1 to 6.1 (a 10-fold change in H^+ concentration) prior to introducing sperm reduces the percentage of fertilized hamster eggs from 100 to 0%, but returning the pH to the normal alkaline range (7.3) allows fusion. Possibly, specific fluid domains are sensitive to changes in pH.

Membrane proteins would seem to be important to fusion in the case of mouse eggs where treatment with trypsin and pronase renders eggs incapable of fusion. Hamster eggs, on the other hand, are unchanged by the same treatment and become fertilized upon exposure to sperm.

Membrane fusion depends on the fluidity of membrane lipids (see Poste and Allison, 1973). In mammals, fusion is retarded below 25°C, at which temperature membrane lipids are frozen. In hamsters, spermatozoon–egg fusion does not occur at

10°C despite prolonged waiting. The response is reversible and spermatozoon–egg pairs fuse following warming.

Fusion ultimately requires the creation of perturbations in membrane lipids. Bindin and lysin may owe their fusiogenic properties to their insertion into the oolemma and consequent ability to perturb membrane lipids (e.g., like the F [fusion] protein of paramyxoviruses [such as Sendai virus]) (see Hoekstra et al., 1987).

Gangliosides and the major sphingophospholipids in the outer leaflet of the cell membrane might be targeted, but exogenous phosphatidylserine and lysophosphatidylserine, associated with the inner leaflet, prevent fusion of guinea pig eggs with acrosome-reacted sperm. Phospholipase C, acting on phospholipids in the inner leaflet of the membrane, could produce disjunctive lysophospholipids and fatty acids. Possibly, phosphatidylinositol turnover and inositol-1-4-5-triphosphate synthesis initiated during fertilization provide phospholipid substrates for membrane fusion (see Epel, 1989).

Incorporation

Following membrane contact and fusion, a fertilizing spermatozoon is incorporated and its contents are wholly or partially drawn into the egg. In the sea urchin, 1–2 seconds after initial fusion, the egg's surface bulges outward toward the spermatozoon, and long microvillar processes erupt from the egg's surface. Like talons, the processes stretch out and grasp the spermatozoon's head. By 3 minutes, these processes, still bearing traces of original knoblike microvilli on their surfaces, fuse to form a **fertilization cone** (or sperm-induced protrusion), engulfing the spermatozoon's head (Figs. 8.14a–c and 8.15).

In the next 30–40 seconds, the sea urchin spermatozoon begins to disappear into the egg. The rate of its incorporation, 5 μm per minute, is constant from beginning to end, some 6 minutes later (Epel et al., 1977).

The entire spermatozoon's head appears to be drawn into the cone by the contraction of the egg's cortex, and shortening of microvilli (Fig. 8.14c–d). Since inhibitors of actin polymerization (cytochalasin B and D) prevent sperm incorporation in the sea urchin (Dale and De Santis, 1981a), microfilament contraction in the egg cortex seems to be the primary force of incorporation. However, fertilization in the clam, *Spisula solidissima,* is insensitive to cytochalasin, indicating that different mecha-

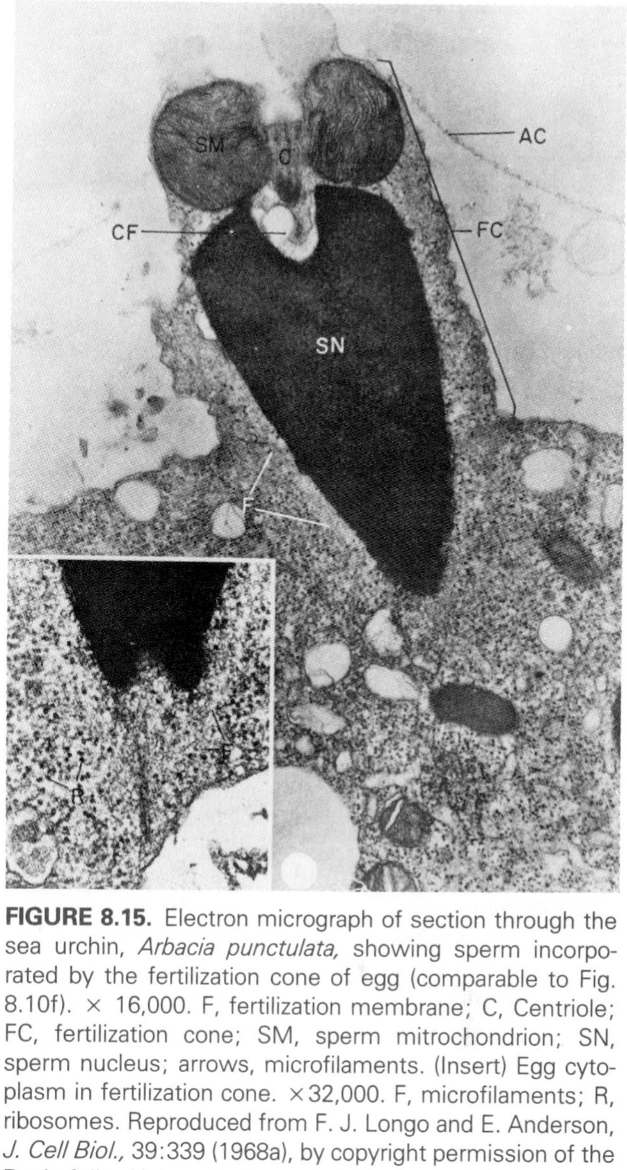

FIGURE 8.15. Electron micrograph of section through the sea urchin, *Arbacia punctulata,* showing sperm incorporated by the fertilization cone of egg (comparable to Fig. 8.10f). × 16,000. F, fertilization membrane; C, Centriole; FC, fertilization cone; SM, sperm mitochondrion; SN, sperm nucleus; arrows, microfilaments. (Insert) Egg cytoplasm in fertilization cone. ×32,000. F, microfilaments; R, ribosomes. Reproduced from F. J. Longo and E. Anderson, *J. Cell Biol.,* 39:339 (1968a), by copyright permission of the Rockefeller University Press and courtesy of the authors.

nisms of sperm incorporation may operate in different species (see Epel, 1978).

Unlike fertilization in most other animals, eutherian mammalian sperm generally lie down on their side upon reaching the oolemma (Figs. 8.9 and 8.13) and, 10–30 minutes later, fuse with the oolemma by the equatorial segment, possibly at the bend of the segment (Fig. 8.9f, g). A girdle of microvilli spreads over the successful spermatozoon's head and, forming an **incorporation cone** comparable to the fertilization cone (see Gaddum-Rosse, 1985), draws the spermatozoon's central portion beneath the egg's surface. The egg's microvilli proceed posteriorly, engulfing the postacrosomal region, while the apical perforatorium remains external (Fig. 8.13).

Fertilization and incorporation cones are not permanent structures, but their duration is variable. In the sea urchin, the cone is usually gone by the time the pronuclei are completely formed, while in the mouse, the incorporation cone may persist through the pronucleate stage and disappear only at cleavage.

The mammalian perforatorium is the last part of the spermatozoon's head taken into the egg (5–10 minutes after fusion begins in the mouse). Fusing with the deflected portion of the outer acrosomal membrane at the anterior fold of the equatorial segment (Fig. 8.9h), the oolemma forms a **phagocytic-like vesicle** around the perforatorium (Figs. 8.9i and 8.12).

The zygote's plasma membrane always incorporates at least part of the spermatozoon's plasma membrane. Membrane components quickly intermingle, and surface labels and antigens present in sperm plasma membranes become dispersed throughout the zygote's surface (Gaunt, 1983).

Parts of the spermatozoon may be unincorporated. In annelids, the tail and middle piece remain on the surface, while in echinoderms, only part of the tail is left outside. *Arbacia* eggs incorporate most of the spermatozoon's tail, but only a small portion of the tail is incorporated in *Mytilus* and *Spisula* (see Longo, 1973).

Tail motility slows conspicuously in mammals immediately after membrane fusion (e.g., within 10–15 seconds in hamsters) but continues intermittently as long as the tail is still at the egg's surface. The disappearance of the tail *in vivo* takes about 23 minutes in the rat, 45 minutes to 1 hour in the mouse, and 2 hours 45 minutes in the Syrian hamster. *In vitro*, incorporation takes much longer (e.g., 4–7 hours in the hamster).

In most mammals (the Chinese hamster and field vole being exceptions), the nucleus and all internal parts of the spermatozoon's head are taken up by the egg (the inner acrosomal membrane, perinuclear material, basal body and centrioles [if present], connecting piece, mitochondria, and internal parts of the tail). The spermatozoon's components largely disappear, but microtubules and possibly centrioles survive and take part in early development. In the exceptions where the tail is autotomized, separation occurs between the capitulum and the basal plate of the nucleus.

The fate of the spermatozoon's mitochondria varies among species, but the sheer preponderance of egg mitochondria dictates that zygotic mitochondria are derived overwhelmingly from eggs.

The conventional view is that the spermatozoon's mitochondria do not contribute to the zygote's pool at all even in species where the contents of the spermatozoon's midpiece enter the egg.

The evidence for this view comes from the analysis of lengths and restriction fragment patterns for mitochondrial DNA (mtDNA). Only maternal mtDNA seems to appear in viable interspecific hybrids (e.g., mules versus hinnies), subspecific hybrids (*Xenopus*), and the progeny of conspecific deer mice, *Peromyscus*, rats, humans, and others. Moreover, extensive backcrossing to the paternal line of the tobacco budworm, *Heliothis* (really a moth), and mice (derived from a cross between *Mus domesticus* and *M. spretus*) fails to concentrate paternal mtDNA in offspring (see Avise and Lansman, 1983).

The **maternal inheritance** of mtDNA (also known as uniparental inheritance and female-mediated or clonal transmission) provides a convenient explanation for the uniformity of mtDNA generally found throughout an organism. Virtually all individuals, whether vertebrates or *Drosophila*, are **homoplasmic** (i.e., carrying only one type of mtDNA). Furthermore, the transmission of the **heteroplasmic** state (carrying different types of mtDNA) by some female crickets (*Gryllus*) to their offspring suggests that recombination and segregation of mtDNA, were they to occur at all, do not occur rapidly (Harrison et al., 1985).

The possibilities of sperm-mediated transfer of mtDNA across female lineages (sometimes called "leakage") and of some paternally derived mitochondria contributing to the zygote's mitochondrial pool are not eliminated by these data, however. First, present techniques for identifying paternal mtDNA may be insufficiently sensitive to detect small amounts in the zygote's pool. Second, while mtDNA and nuclear DNA appear to be transmitted to offspring independently, many mitochondrial proteins (e.g., most ribosomal proteins) are encoded by nuclear genes. The mitochondrion is, after all, more than its mtDNA.

The idea of mtDNA's maternal inheritance is cautiously assumed by population geneticists interested in matriarchal phylogenies (e.g., Moritz et al., 1987). Given the additional assumption that mtDNA production is based on simple replication, differences in mtDNA among organisms show both the direction and rate of evolutionary change. The validity of these assumptions will continue to be tested as more sensitive techniques become available and more examples of heteroplasmicity and heterogeneity in mtDNA sequences are found (see Wilson et al., 1985).

ACTIVATION RESPONSE

Rapid changes in the interval between syngamy and the integration of sperm and egg launch the zygote on its new course. The changes include alterations in cortical morphology, a release of free calcium, fluctuations in electrical excitability, and the formation of prophylactic blocks to polyspermy. Known as the **activation response,** some of the changes seem explosive. Together, they "blast loose the block(s) to development in unfertilized eggs" (Jaffe, 1985, p. 153).

Cortical Morphology

The egg's cortex is more like the shell of a tennis ball than the skin of a balloon (Mitchison and Swann, 1955). The cortex of sea urchin eggs consists of bundles of contractile proteins, such as actin and myosin, elastic proteins, ATPase, and protein kinase (Fig. 8.10) (see Epel and Vacquier, 1978). Prior to fertilization, the cortex is responsible for the egg's elasticity and its slightly elevated internal hydrostatic pressure.

Many egg cortices contain a variety of vesicles, particles, and pigment granules. The most conspicuous of these are the **cortical vesicles** (CVs, i.e., membrane-bound cortical granules) present in eggs that cleave totally (e.g., coelenterates, echinoderms, annelids, mollusks, crustaceans, frogs, and mammals) with few exceptions (e.g., ascidians, urodeles, and primitive insects) and **vacuoles** present in the meroblastic eggs of fish. Cortical vesicles lie below the bundles of cortical proteins (Fig. 8.10) or are fixed to the surface (Fig. 8.16). Differences in estimates of the thickness of the cortex (e.g., 0.2–0.5 to 1.3–1.5 μm in the sea urchin) are attributable to whether the cortical vesicles are included in the estimates.

In *Strongylocentrotus purpuratus* and *S. franciscanus,* the vesicles average 1.3 μm in diameter and appear to be tethered to the oolemma by thin struts that may normally be integral parts of the cortex (see below, Fig. 8.21a). The density of vesicles indicates that an average egg with a diameter of 80 μm and a surface area of 20,000 μm² contains 15,000 cortical vesicles.

Cortical vesicles are probably not comparable in all species where they appear. They differ greatly in size, from sea urchin vesicles of 1.3 μm in diameter to freshwater teleost vacuoles 10–40 μm in diameter (see Fig. 9.6). The vesicles may be inseparable from the oolemma or separable, for example, by centrifugation. Their contents may include a lamella coil and matrix, as in the sea urchin (Figs.

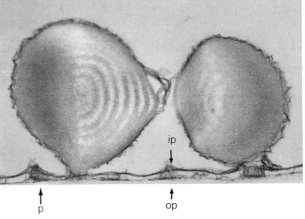

FIGURE 8.16. Electron micrograph of section through cortical vesicles cut perpendicularly to supporting plastic dish (below). The vesicle membrane appears to be in intimate contact with the oolemma. Vitalline papillas dotting the surface appear as thickenings of the vitelline envelope. ×37,200. From V. D. Vacquier, *Dev. Biol.,* 43:62 (1975), by permission of Academic Press and courtesy of the author.

6.17 and 8.17), or they may appear granular, as in frogs and mammals.

Different types of cortical vesicle even appear within one species. For example, along with its large cortical vacuoles, the medaka, *Oryzias latipes,* contains small subcortical vesicles (1.5–5 μm in diameter), and so-called a-granules (0.1–0.3 μm in diameter) are present. In the frog, *Xenopus laevis,* a monolayer of uniformly dense cortical vesicles is distributed beneath the oolemma in the animal hemisphere, while one or more layers of irregularly shaped and less dense vesicles populate the vegetal hemisphere (Grey et al., 1974). Collectively, these may perform the functions otherwise exercised by a single type of cortical granule (see Gilkey et al., 1978).

Cortical Response

Originally related to changes specifically in the egg's cortex following fertilization (see Just, 1939), the concept of a cortical response was expanded to include the formation of the perivitelline space above the cortex and the spreading of a **fertilization wave** around the egg. The event linking these processes and unifying the concept of a cortical response is the **regulated** (or triggered) **exocytosis** of cortical granules, the calcium-dependent **vectorial transport** of material across the oolemma, and the introduction of vesicle membrane into the oolemma.

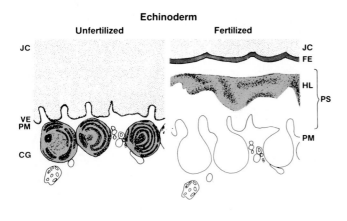

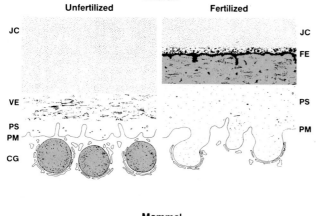

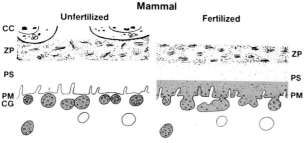

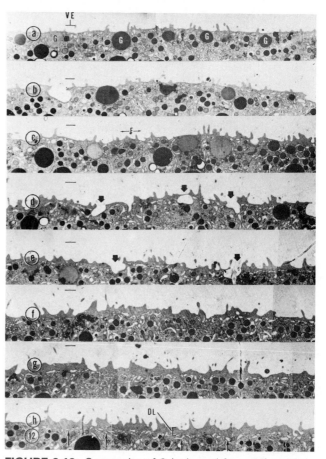

FIGURE 8.17. Drawings based on electron micrographs of sections through periphery of various eggs before and after fertilization. CC, cumulus cell; CG, cortical vesicles; FE, fertilization envelope; HL, hyaline layer; JC, jelly coat; PM, plasma membrane (oolemma); PS perivitelline space; VE, vitelline envelope; ZP, zone pellucida. (Top) Echinoderm (*Strongylocentrotus pupuratus*). Explosive exocytosis creates loose, irregular folds in PM and turns VE into FE. A new hyaline layer appears in the PS. (Middle) Anuran (*Xenopus laevis*). Subdued exocytosis results in expansion of PS, while material formerly in vesicles diffuses into vitelline envelope, hardening it, and forming a new FE on the outside. Cisternae gather below the PM and a dense layer of cytoplasm (not shown). (Bottom) Mammal (mouse). Exocytosis is restrained and the surface configuration not especially altered. Accumulation of material in PS and ZP is responsible for the plasma and zona blocks to polyspermy. From E. D. Schmell, B. J. Gulyas, and J. L. Hedrick. In J. F. Hartmann, ed., *Mechanism and control of animal fertilization,* Academic Press, Orlando, 1983. Used by permission.

FIGURE 8.18. Composite of 8 (selected from 20) overlapping electron micrographs of a section showing the progressive exocytosis of cortical vesicles in *Xenopus laevis.* Although the wave of exocytosis is seen here from left to right (as well as top to bottom), it would actually have taken place in the opposite direction. The contents of the large gray cortical vesicles (G) near the surface (not to be confused with the darker yolk platelets and lipid vesicles) are exteriorized following the fusion of membranes (large arrows). The bar on the left of the micrographs (VE in a) shows the position of the vitelline or fertilization envelope. It is elevated until it is out of the field (in h) as the perivitelline space forms below. Large interconnecting cisternae (arrows in h), formerly beneath the cortical vesicles, and a dense layer (DL) of cytoplasm are seen in the cortex following exocytosis. ×2600. From R. D. Grey, D. P. Wolf, and J. L. Hedrick, *Dev. Biol.,* 36:4–61 (1974), by permission of Academic Press and the authors.

Beginning at the point of spermatozoon–egg fusion, the membranes of cortical vesicles fuse with the zygote's plasma membrane (PM, or oolemma). A wave of exocytosis proceeds to deposit the vesicles' contents on the egg's surface (Fig. 8.18).

Cortical vesicles adhere closely to the plasma membrane in dome-shaped areas free of intramembranous particles. Prior to fusion, the plasma membrane may pucker inward. Fusion of the cortical vesicle's membrane and the plasma membrane gives rise to a small exocytotic pore (10–50 nm in diameter) which lengthens into an aqueous channel. The cortical vesicle then swells to about twice its original diameter, while its laminae dissociate and its core structure breaks down. Small vesicles bud off the lips of the cortical vesicle, and hydration and dissolution of granular structures aid in widening the channel into a broad mouth. The cortical vesicle has now become an **exocytotic pocket,** expelling its disrupted contents (Fig. 8.19) (see Chandler, 1988).

Exocytosis results in a diminution of the egg's volume. For example, in the prawn, *Penaeus*, as much as 30% of the egg's volume is lost. At the same time, exocytosis results in an increase in the egg's surface area. In the sea urchin, surface area more than doubles, and the capacitance of the membrane rises. Some of the additional membrane is taken up 3–5 minutes after fertilization in microvilli, but some membrane disappears.

Fusion between a cortical vesicle and the plasma membrane, like most other types of fusion, is vesicular (i.e., forms vesicles), and portions of the

originally juxtaposed membranes are detached during exocytosis as small vesicles (Fig. 8.19). Moreover, **endocytotic vesicles** forming concomitantly with microvilli further reduce the egg's surface area. The reduction may be as much as 46%, but whether the endocytosis affects original plasma membrane or cortical vesicle membrane is presently unknown (see Longo, 1987).

The plasma membrane is biochemically altered in the wake of cortical vesicle exocytosis. The amounts of triglycerides and phosphatidylserine in the membrane increase, while phosphatidylcholine decreases and chains of fatty acids become shorter and less unsaturated.

Surface proteins are partially hydrolyzed by the trypsinlike enzymes released from cortical vesicles, and polypeptide components of 20,000–30,000 daltons from isolated sea urchin egg surfaces appear for the first time in SDS–PAGE gels. In addition, a Na^+-dependent amino acid transport system is inserted into the membrane.

The exocytosis of cortical vesicles is rapid and comparable to similar processes in other cells (exocytosis at nerve endings takes less than 5 milliseconds). Typically, the wave of exocytosis in eggs is so rapid that it seems to be self-propagating or **autocatalytic** rather than due to diffusion of a substance through the dense egg cytoplasm. Like an explosion, the rate of propagation is consistent, although it is sensitive to temperature. In species from marine brown algae and echinoderms to freshwater fish and frogs, the wave travels at 5–15 μm per second (see Gilkey et al., 1978).

In sea urchin eggs (75–120 μm in diameter), cortical exocytosis begins 10–25 seconds after sperm–egg fusion (top of Fig. 8.17), and about 5 seconds (10–20 seconds maximally) later the final exocytotic pockets open up at the rear of the wave. In the larger medaka egg (1.1 mm in diameter), the wave begins 5–10 seconds after the first spermatozoon darts through the funnel-shaped micropyle and is completed about 2 minutes later. In the frog, *Xenopus laevis* (1.3 mm in diameter), the exocytosis of cortical vesicles (Figs. 8.17, middle, and 8.18) begins at 3 minutes after fertilization and is complete 1–3 minutes later, or almost 4–6 minutes after fertilization. The waves can take even longer (e.g., 7–8 minutes) at lower temperatures (see Elinson, 1980).

In *Xenopus laevis*, cortical vesicles of both animal and vegetal hemisphere are evacuated as the wave of exocytosis moves around the egg, but exocytosis is not completely linear (i.e., some vesicles erupt out of turn), and some vesicles in the vegetal hemisphere may persist even into the cleaving embryo. A flocculent material (arrows in

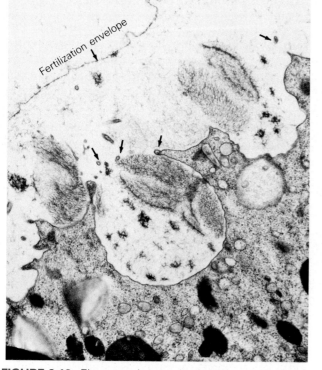

FIGURE 8.19. Electron micrograph of section through surface of quick-frozen and freeze-substituted sea urchin showing small vesicles (arrows) pinched off from lips of exocytotic vesicle. ×30,000. From D. E. Chandler, *J. Cell Sci.,* 72:23 (1984), by permission of Company of Biologists, Ltd. and courtesy of the author.

Fig. 8.18c), apparently evacuated from the vesicles, appears, and the surface becomes deeply crevassed. The plasma membrane between the craters left by ruptured cortical vesicles forms loose and irregular folds, but the microvilli formerly scattered on the surface largely disappear. Cisternae (arrows in Fig. 8.18h), which formerly lay around cortical vesicles, widen and appear below a dense layer (DL) at the egg's surface.

Cortical exocytosis in mammals is relatively slow (about 15 minutes in hamsters). The granular material of the vesicles is not expelled as in other eggs but merely spills over into the perivitelline space (bottom of Fig. 8.17).

Forming the Fertilization Envelope and Perivitelline Space

The vitelline envelope (VE) is converted to the **fertilization envelope** (FE) as a perivitelline space (PS) forms above the plasma membrane. A variety of functions are attributed to the FE, but the PS is generally considered a shearing zone, permitting the egg to rotate, and a reservoir or bladder. If the space is not allowed to expand or the egg is kept from rotating within it, embryos (e.g., frogs) develop abnormally.

In the sea urchin, the blister that forms between the VE and the plasma membrane at the point of fertilization (see below, Fig. 10.1) spreads rapidly into the superficial PS. The rupture of bonds between the VE and oolemma is presumably due to the action of a **vitelline delaminase** or protease released from the cortical vesicles. The best candidate for the job is β-1,3-glucanohydrolase (i.e., β-glucanase, E.C.3.2.1.-39), known to be present in cortical vesicles (Vacquier et al., 1973) but incapable of penetrating the VE.

Materials from the cortical vesicles accumulate in the perivitelline space forming the glassy **hyaline layer** (HL) below the FE. **Hyaline,** the major component of the layer, is a high molecular weight glycoprotein. It may also be replaced by cleaving eggs after experimentally removing it. Hyaline seems to function in pressing blastomeres against each other during cleavage, and its removal results in twinning (see Schuel, 1985).

The FE thickens (up to 0.90 μm compared to the 0.10 μm of the VE) with the accumulation of cortical materials on both sides. Proteoliaisin, a cortical vesicle protein, forms a major part of the FE. Other materials **harden** (or tan) the FE, it would seem, and the fertilization membrane loses its original solubility in mercaptoethanol. In addition,

an **ovoperoxidase,** a major cortical vesicle enzyme, catalyzes the formation of tyrosine dimers and trimers, further cross-linking FE proteins (Epel, 1978).

In fish, the chorion hardens without forming an independent FE as the contents of cortical vesicles are deposited in the perivitelline space. After losing contact with the micropyle, the egg lies totally within the perivitelline space.

In the frog, cortical granular material rapidly permeates the filamentous vitelline membrane despite its thickness (about 1 μm) and reduces its permeability. The membrane's amino acid composition is changed, and its solubility in mercaptoethanol is lost (Grey et al., 1976). In addition, a cortical vesicle lectin diffuses all the way through the VE and is polymerized by galactose residues in the innermost jelly layer (Wolf, 1974). The frog's FE (Fig. 8.17) therefore consists of a new peripheral **fertilization layer** (F layer) just below the jelly coat (JC) as well as an altered VE.

Because F-layer impermeability prevents the diffusion of cortical material and provides an osmotic barrier, water accumulates between the plasma membrane and the FE. As the originally narrow space widens into the definitive perivitelline space, the hydration of mucopolysaccharides gives it a semifluid consistency. A loose floccular material accumulates, but small vesicles originally present in the space are not conspicuously altered.

Some mammalian zona pellucidas harden after fertilization (e.g., 1–2 hours in mouse), paralleling the cortical response. Rabbit and mouse zonas, for example, become resistant to dissolution in mercaptoethanol and to digestion by trypsin and pronase. The zonas of hamster eggs, on the other hand, are not hardened following fertilization.

The mammalian perivitelline space forms following the release of the cortical vesicles. The zona pellucida (about 8 μm thick in the mouse) does not thicken and does not form an additional FE (Fig. 8.17), but it is altered.

In some mammals (e.g., dog, hamster, human, and sheep), additional sperm are not seen in the perivitelline space after *in vivo* fertilization, even though the space may be large and continue to increase in size due to secretory activity by the embryo's first blastomeres. In other mammals (e.g., pocket gopher and rabbit), sperm that have penetrated the zona pellucida accumulate in the perivitelline space as if it were a reservoir, and the width of the space may increase slightly following fertilization due to a slight withdrawal of water from the egg. In a third group of mammals (e.g., cat, ferret, guinea pig, mouse, and rat), sperm accumulate occasionally in the perivitelline space.

Mechanisms of Exocytosis

A few generalizations about regulated exocytosis seem to hold in systems as different as mammalian platelets and marine invertebrate eggs, but other aspects of exocytosis are different (see Baker, 1988). Generally, regulated exocytosis is Ca^{2+} dependent. Exocytotic vesicles and plasma membranes become fusiogenic, and a fusiogenic force drives the membranes together. But while regulated exocytosis in mammalian tissues has a requirement for adenosine triphosphate (ATP), cortical exocytosis in eggs (e.g., sea urchin eggs) does not.

In general, the mechanism of exocytosis in eggs is triggered by contact with sperm. Incorporation of sperm is not required, however, at least in hamster eggs (see Gwatkin, 1977). Furthermore, eggs of the marine polychaete *Sabellaria* undergo exocytosis upon shedding, and the bivalve mollusks *Barnea* and *Spisula*, as well as the echiuroid *Urechis*, do not undergo exocytosis until cleavage (i.e., independently of fusion with sperm).

The fusion of cortical vesicles and plasma membrane depends on overcoming electrostatic repulsion between the plasma membrane and the vesicles' membrane. A correlation in time between exocytosis and the change in membrane potential, known as the fertilization potential (see below), is suggestive of a causal relationship, but the depolarization of mature eggs seems to register rather than cause the cortical response.

Unfertilized sea urchin eggs brought to their fertilization potential of +10 mV by a current (i.e., held in a voltage clamp) do not initiate the discharge of cortical vesicles. On the other hand, the fertilization potential may depend on cortical exocytosis, since the fertilization of immature sea urchin oocytes lacking cortical vesicles does not produce a fertilization potential (Dale and DeSantis, 1981b).

Free calcium ions (Ca^{2+}) are likely to trigger cortical exocytosis by operating directly on membranes. The **calcium signal** for cortical vesicle exocytosis seems to be given by a dramatic increase in subcortical Ca^{2+} concentration (or Ca^{2+} transient) beneath the surface of eggs.

A wave of free calcium may be observed washing over eggs microinjected with the Ca^{2+}-sensitive photoprotein **aequorin** (a calcium-binding protein that scintillates while binding free calcium, Fig. 8.20). Medaka (*Oryzias latipes*) eggs begin to glow with luminescence immediately after fertilization and proceed to beam for several minutes.

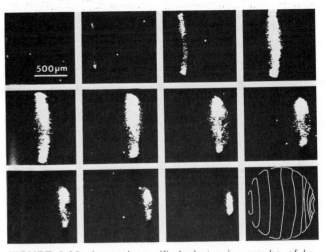

FIGURE 8.20. Image-intensified photomicrographs of luminous wave fronts in aequorin-injected medaka eggs (left to right, top to bottom) and summary tracings showing leading edges of wave fronts (lower right). Photomicrographs were taken at 10-second intervals beginning with fertilization. Luminescence in the presence of free calcium travels across the egg from the point of fertilization at the animal pole beneath the micropyle to the vegetal pole where it weakens and disappears. Reproduced from J. C. Gilkey, L. F. Jaffe, E. B. Ridgway, and G. R. Reynolds, *J. Cell Biol.,* 76:448 (1978), by copyright permission of the Rockefeller University Press.

The 10,000-fold increase in luminescence is indicative of a 300-fold increase in the egg's free calcium and corresponds to a peak Ca^{2+} concentration of about 5 μM (Gilkey et al., 1978; Jaffe, 1985).

Similarly, aequorin-microinjected sea urchin eggs show amounts of free calcium beginning to rise at about 10 seconds after fertilization and increasing 100- and 300-fold to a peak level of 1–5 μM at 1 minute. Other Ca^{2+}-sensitive photoproteins (quin2 and Fura2) indicate a peak of 1.95 μM Ca^{2+} in the sea urchin, *Lytechinus pictus*, and microelectrodes indicate a peak subcortical Ca^{2+} concentration of 1.2 μM within 2 minutes in *Xenopus laevis*, a threefold rise from prefertilization levels. In all cases, the rise in free calcium is quickly dissipated as calcium is dispersed into the environment.

Less direct lines of evidence are also consistent with a calcium signal. Procaine and lanthanum ions, which have high affinities for calcium ion binding sites on membranes, inhibit the release of cortical vesicles. Injection of calcium into eggs frequently induces the cortical response. Parthenogenic activation by "pin-pricking" of a variety of eggs (e.g., hamsters, frogs, sea urchins) requires calcium in the medium or, at least, transfer to calcium-containing medium following pricking.

Divalent cation-transporting ionophores (e.g., A23187 and X537A) are efficient and widely effective activators of cortical exocytosis and, in Ca^{2+}-containing medium, of parthenogenic development (e.g., sea urchin, tunicates, mollusks, amphibians, and mammals) (see Jaffe, 1985).

Results with an _in vitro_ system support the notion of a calcium signal for cortical exocytosis. Fragments of the cortex prepared by the homogenization or entire isolated egg cortices (called hulls) are prepared in artificial calcium-free seawater (CaFSW) at pH 7–8 (Vacquier, 1975) in the presence of so-called chelating agents (actually sequestering agents that tie up divalent cations but keep them in solution), such as ethylene glycol bis(2-aminoethyl ether)-_N,N,N',N'_-tetraacetic acid (EGTA) and ethylenediaminetetraacetate (EDTA). Fixed to slides, the cortices are found to be carpeted by cortical vesicles (hence they are called lawns, Fig. 8.21a), presumably because the cortical vesicles are tethered to the plasma membrane by elements of the cytoskeleton.

When exposed to low concentrations of calcium (i.e., 0.2 mM), the cortical vesicles fuse with each other and with the plasma membrane, their contents disappear, and the plasma membrane separates from the FE (Fig. 8.21b). Like other instances of membrane fusion, strontium and barium ions can replace calcium ions and induce the breakdown of cortical vesicles, but magnesium cannot. Moreover, cortical vesicles isolated from the hulls of eggs by mild trypsinization explode upon the addition of calcium to the suspending medium.

The source of the calcium signal in eggs of marine invertebrates may not normally be the medium, since exocytosis can generally be induced by ionophores in the absence of external calcium (exceptions being the surf clam, _Spisula,_ and the tube-worm, _Chaetopterus_). Furthermore, injection with EGTA blocks the cortical response.

One source of calcium for the egg may be sperm. With their high calcium content, sperm may normally represent a "detonator" for the egg's "calcium bomb" (Jaffe, 1980).

**Internal sources of Ca^{2+} are provided by the egg's own endoplasmic reticulum or, more precisely, subcortical vesicles belonging to a cortical reticulum beneath the plasma membrane.** At least in echinoderms and vertebrates, Ca^{2+} is released from these stores under the influence of **inositol 1,4,5-trisphosphate** (IP$_3$ or InsP$_3$) (Epel, 1989). What is more, experimentally, microinjection of IP$_3$ into

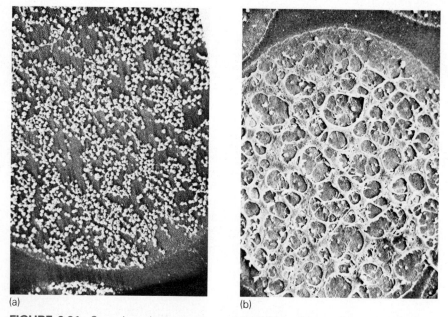

(a)   (b)

FIGURE 8.21. Scanning electron micrograph of hulls from a _Strongylocentrotus franciscanus_ egg before (A) and after (B) exposure to dilute calcium solution. (a) Seen from the inside of the egg, cortical vesicles (white spheres) lying above the vitelline envelope (dotted by microvilli) appear to be tethered to the oolemma by thin struts. (b) Network of fused cortical vesicles after exposure to 0.34 M CaCl$_2$ in artificial seawater. The material released by the vesicles contributes enzymes and proteins to the hyaline layer of the perivitelline space. From V. D. Vacquier, _Dev. Biol.,_ 43:62 (1975), by permission of Academic Press and courtesy of the author.

unfertilized *Lytechinus pictus* eggs causes calcium release and induces cortical exocytosis (Whitaker and Irvine, 1984).

The source of IP$_3$ following fertilization is uncertain. One possibility is that the spermatozoon introduces IP$_3$ directly. Other possibilities involve phosphoinositides (PI) metabolism or **PI cycling** in the zygote's membrane (Table 8.1, lower half).

Under the influence of active phospholipase C, oolemma-bound **phosphatidylinositol 4,5-bisphosphate** (PIP$_2$ or PtdIns[4,5]P$_2$) undergoes rapid turnover immediately after fertilization, and IP$_3$ surges fivefold by 10 minutes after fertilization. During the first 30 seconds, PI concentration declines, but it is restored to its original concentration within 2–5 minutes. Possibly, free calcium introduced by

the spermatozoon promotes the breakdown of PIP$_2$ to IP$_3$. The spermatozoon may also inject its own phospholipase C into the plasma membrane which then hydrolyzes PIP$_2$.

Alternatively, the spermatozoon occupies a surface receptor and sets in motion a receptor-mediated system:

1. Occupied, the sperm receptor activates G protein (i.e., guanyl nucleotide- [or GTP-] binding protein). Its regulatory beta subunit dissociates, and guanosine diphosphate (GDP) on the alpha subunit is exchanged for guanosine triphosphate (GTP) from the cytosol.
2. The alpha subunit in turn activates PI-specific phospholipase C.

TABLE 8.1. Membrane lipid breakdown products as regulators and secondary messengers of fertilization

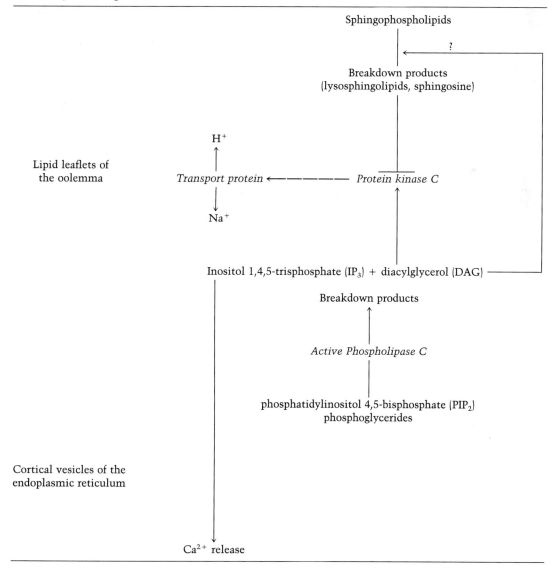

3. The enzyme hydrolyzes membrane-bound PIP_2 into water-soluble IP_3 and diacylglycerol (DAG).

In any case, IP_3 enters the cytosol with explosive results. According to a concept of autocatalytic **calcium-stimulated calcium release,** an initial pulse of IP_3 induces the release of Ca^{2+} into the cortical cytosol, which in turn promotes the further conversion of PIP_2 to IP_3:

localized rise in $IP_3 \rightarrow Ca^{2+}$ release into cytosol $\rightarrow$ PIP_2 hydrolysis $\rightarrow$ progressive rising in IP_3 around the egg's surface

The **traveling calcium explosion** and consequent wave of exocytosis continue until internal stores of calcium are exhausted. Gradually, the progress of the wave slows, and free calcium is pumped out of the egg into the medium (Jaffe, 1985).

Alternatively, calmodulin could provide the final destination for free Ca^{2+} (Trimmer and Vacquier, 1986). The protein comprises as much as 0.5% of the total protein in eggs of *Xenopus laevis* and 0.1% in eggs of sea urchins. A number of cellular proteins are regulated by calmodulin in a Ca^{2+}-dependent mode. Bound to calcium, calmodulin may be a permanent regulatory subunit of an enzyme (e.g., phosphorylase kinase) or an activator added to an enzyme's regulatory subunit (e.g., adenylate cyclase). In either case, the zygote reaches a new equilibrium, if cleavage can be called an equilibrium.

Changes in Electrical Excitability

Like nerves and muscles, unfertilized eggs are electrically active, having the ability to undergo rapid, transient changes in **membrane potential.** Defined as the difference in voltage across the membrane and measured by impaling eggs with intracellular microelectrodes, membrane potential is attributed to differences in charged ion concentration on either side of the cell membrane. Selectively permeable, the egg's membrane passes inorganic ions (especially Na^+, K^+, Cl^-, and Ca^{2+}) only through ion channels in transport proteins and regulates permeability by transiently opening **voltage-gated channels.** The channels open rapidly in response to a change in membrane potential, while they close slowly as the former balance is restored.

The highly negative resting potentials recorded for marine annelid, mollusk, and coelenterate eggs are attributable to their potassium-selective membranes. The neutral to slightly negative resting potentials recorded for tunicate eggs may be due to relatively nonspecific permeabilities bringing the egg into equilibrium with the environment.

The unfertilized sea urchin egg's resting potential is typically recorded between -60 and -90 mV. Actually, its resting potential may be -5 to -20 mV, much closer to neutrality (Dale and DeSantis, 1981b). Low resting potentials may be artifactual and due to the appearance of nonspecific permeabilities in aging eggs. On the other hand, the more neutral measurements can be due to **leakage current** accompanying the impalement of the egg or to the opening of nonspecific channels (as opposed to potassium channels) (see Shen, 1983).

Sea urchin eggs with both high and low resting potentials are electrically excitable, but aging in eggs has consequences for subsequent development. Older sea urchin eggs (e.g., eggs ovulated more than 2 hours before fertilization) undergo changes in membrane potential, but they generally abort development before reaching larval stages.

An **action potential** is a transient depolarization (i.e., addition to a positive internal charge). It is usually brought about by the massive opening of voltage-gated Na^+ channels in response to an impulse or stimulus reaching the cell and ends with the restoration of the resting potential within milliseconds. In the case of eggs, a current applied to an impaling microelectrode can produce action potentials.

The amplitude of action potentials registered by sea urchin eggs in response to a charged stimulus appears to be Ca^{2+} dependent and Na^+ independent. Similar Ca^{2+}-dependent voltage-gated action potentials are also present in unfertilized annelidan, coelenterate, echiuroid, and molluscan eggs. In tunicates and some starfish, however, the egg's action potential has a major Na^+-dependent component.

The changes in membrane potential accompanying the fertilization of eggs is quite different from an action potential. It is stimulated by fertilization, occurs in stages, and lasts several seconds, and at the restoration of the resting potential (or something near it), the membrane has lost its electric activity (Fig. 8.22).

The first stage is a small steplike depolarization recorded at 2–4 seconds after pipetting sperm into the dish containing the impaled egg. This depolarization may correspond to the initial contact between egg and sperm, since it does not occur in parthenogenically activated eggs. Additional steplike depolarizations may also occur later, presumably corresponding to contacts between the egg and late-arriving sperm (Dale and DeSantis, 1981b).

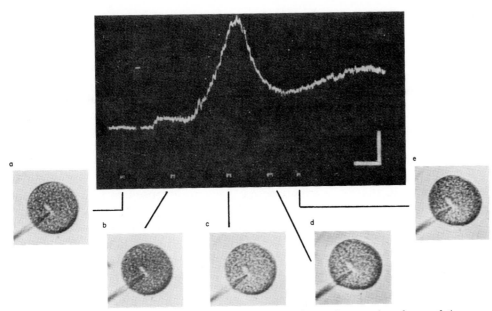

FIGURE 8.22. Oscillograph record and simultaneous photomicrographs of egg of the sea urchin, *Paracentrotus lividus,* during fertilization. Oscillograph record. The resting potential is superseded by a steplike depolarization, indicative of contact with sperm. In this record, two nearly close steplike depolarizations indicate that two sperm made contact with the egg. The more dramatic increase to the fertilization potential follows. The resting potential is slowly restored. The dash on the upper left of the oscillograph record represents the zero level. Horizontal bar = 5 seconds; vertical bar = 10 mV. Photomicrographs. (a) Impaled egg before insemination. (b) Spermatozoa (not shown) have attached. (c) The cortical response (flattened areas of cortex) begins at two points of sperm entry. (d) Fertilization envelope partially elevated. (e) Fertilization envelope completely elevated. The cortical response and the duration of the fertilization potential are especially brief in this record, presumably because of the dispermic fertilization. From B. Dale and A. DeSantis, *Dev. Biol.,* 85:474 (1981), by permission of Academic Press and courtesy of the authors.

Second, an overshoot of potential known as the **fertilization potential** takes over. This new potential may develop slowly (1.8 mV/sec, Fig. 8.22) (Dale and DeSantis, 1981b) or rapidly (Jaffe, 1976). In *Rana pipiens* eggs, a resting potential of -28 mV jumps to a fertilization potential of $+6$ mV within 1 second of fertilization (see Elinson, 1980). An outflow of chloride ions seems to be responsible for the relatively prolonged hyperpolarization of the membrane in the freshwater environment of the frog.

The fertilization potential is resolvable into two phases: an initial rising Ca^{2+}-dependent phase and a slower developing and longer lasting Na^+-dependent phase. In some echinoderms, the calcium channels operating in the initial phase are inactivated before the sodium channels are completely open, and the fertilization potential is biphasic, but, generally, the transition is relatively smooth.

At about the same time the cortical response is nearing completion (beginning 1 minute after fertilization in the sea urchin), the intracellular pH rises and the potential begins its return to something near the original resting potential. This alkalinization is sodium dependent and produced by the activation of a Na^+–H^+ exchange mechanism.

The initiating signal for alkalinization seems to be an increase in diacylglycerol (DAG) which accompanies the hydrolysis of polyphosphoinositides (PIP_2). DAG-mimics (i.e., phorbol esters also known to be potent protein kinase C activators) added to the medium initiate alkalinization. Here then is another example of the enormous role played by breakdown products of plasma membrane phospholipids in regulating the events of fertilization (Table 8.1, upper half).

Like the cascade of early events leading to Ca^{2+} release (see above), **phosphoinositide- (PI-) mediated signaling** causes a second or late cascade:

1. DAG activates membrane-bound protein kinase C and possibly enzymes catalyzing the breakdown of sphingophospholipids.
2. Once active, protein kinase C activates a variety of phosphate-regulated enzymes and transport

proteins including Na^+/H^+ channels (see below).

3. The breakdown products of sphingophospholipids (lysosphingolipids and sphingosine) inhibit protein kinase C, thereby modulating the effects of DAG (Hannun and Bell, 1989).

A gradual elevation of pH (e.g., 45–60 seconds in sea urchin eggs or several minutes in *Urechis* eggs) accompanies the final restoration of the membrane potential to something approximating the original resting potential. The repolarization (i.e., decrease in positive charge within the cell) accompanies inactivation of the sodium channels.

The new resting potential, like the original one, is generally K^+ dependent (starfish being an exception), the repolarization accompanying an increase in potassium permeability. In any case, the membrane's voltage-gated channels are no longer responsive to stimulation, and the membrane is no longer electrically active.

Prophylactic Blocks to Polyspermy

Many eggs are normally fertilized by only one spermatozoon. These **monospermic eggs** (including those of coelenterates, annelids, echinoderms, teleosts, anurans, and mammals) employ prophylactic blocks to polyspermy that prevent the entry of additional sperm.

Block failures or experimental procedures that override the blocks and allow more than one spermatozoon into the zygote may lead to abnormal development, a condition diagnosed as **pathological polyspermy.** Sea urchin eggs, for example, are vulnerable to pathological polyspermy when as few as two sperm are present in the zygote (see Chapter 4).

Many types of prophylactic block have little more in common than their operation at the egg's surface. For example, sturgeon eggs withdraw their microvilli from unused micropyles following fertilization, thereby foreclosing the possibility of additional sperm fusing with the egg. Alternatively, subtle blocks may be thrown at the level of the plasma membrane or the egg's envelopes. Generally identified by the rate at which they appear, some blocks are only inferred from theory, while others are demonstrated by experiment.

A fast or early block to polyspermy in sea urchins was first proposed on the basis of the frequencies with which polyspermy appeared as a function of sperm concentration in fertilizing media (Rothschild and Swann, 1952). Earlier, sea urchins were only thought to throw up a block to polyspermy along with their cortical response beginning 10–25 seconds after fertilization. The fast block was supposed to be thrown up earlier or thrown up in the absence of a cortical response.

For the sea urchin, only monospermy occurs when mature ova are bathed in media containing light concentrations of sperm. Polyspermy occurs under laboratory conditions only when the medium contains heavy concentrations of sperm.

The potential of eggs to become polyspermic following fertilization is tested in double-exposure experiments. Eggs are exposed to a light concentration of sperm followed by exposure to a heavy concentration. The frequency of polyspermy is greatest soon after the first exposure and declines rapidly thereafter, but whether it declines more rapidly than predicted by the onset of the cortical response is uncertain.

When sea urchin eggs are exposed to high concentrations of sperm, 100% of the eggs are fertilized by 3–5 seconds, but 4–5 additional sperm enter the egg in the next 20 seconds (i.e., the period prior to the cortical response). The rate of entry, which is proportional to sperm concentration, is linear and the same before and after the first fertilization. Sperm receptivity therefore may not change during the period of the "fast block" (Byrd and Collins, 1975).

If the fast block exists at all in sea urchins, it seems to be overridden at high sperm concentrations (but not necessarily in other marine invertebrates) (see Epel and Vacquier, 1978). The block is therefore sometimes called the **partial block** (also incomplete or labile block).

Possibly, the fast block is a statistical phenomenon representing differences from egg to egg. Some sperm may continue to fuse with eggs despite the block, or different eggs may produce the block at different rates.

Another concept of a fast block to polyspermy features the egg's electrophysiological properties. In *Strongylocentrotus purpuratus*, the magnitude of the fertilization depolarization within 3–30 seconds of fertilization in individual eggs is inversely correlated with polyspermy. Eggs whose fertilization potentials exceed +5 mV are always monospermic, while those whose fertilization potentials are less than −10 mV are frequently polyspermic (Jaffe, 1976).

The ability of a positive potential immediately following fertilization to inhibit the fusion of supernumerary sperm and eggs is widely identified with the "long-sought-after 'fast block to polyspermy'" (Trimmer and Vacquier, 1986, p. 15). While sperm can adhere to the vitelline membrane of eggs whose membrane potential is maintained in a voltage clamp above $+5$ mV, sperm do not fuse with these eggs, and repeated additions of new suspensions of sperm continue to be ineffective at fertilization. When the current is turned off, however, the eggs are quickly fertilized (see Jaffe and Gould, 1985; for contrary evidence see Dale, 1987).

The notion of a charge-mediated block to polyspermy is consistent with other circumstantial and experimental evidence. In the sea urchin, *Lytechinus pictus*, for example, the rise in membrane potential at fertilization is not as rapid as in *S. purpuratus*, and the change in receptivity to sperm is likewise not as rapid. Moreover, lowering the concentration of extracellular sodium induces polyspermy. Reciprocally, raising internal sodium in toad eggs causes polyspermy.

How changes in electric charge could operate in blocking polyspermy is unknown. One hypothetical mechanism (Fig. 8.23) employs membrane-bound sperm receptors whose orientation and efficacy in binding sperm are altered as a function of membrane charge. At the same time that the change in potential accompanying contact with sperm establishes the conditions required for the fusion of sperm and egg plasma membranes, the configuration of receptors on the surface not already attached to sperm is altered, rendering them

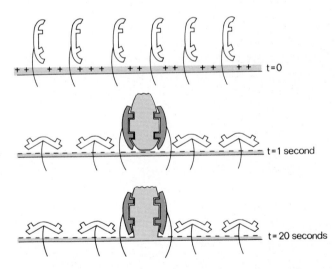

FIGURE 8.23. Scheme for rapid change in receptivity to sperm accompanying contact with sperm and depolarization of plasma membrane. Initially ($t = 0$) the membrane potential is highly positive (+ signs), and oolemmal receptors are available to attach to binding sites on sperm. Upon contact with sperm ($t = 1$ second), the plasma membrane depolarizes (− signs), and the orientation of oolemmal receptors is altered. The new potential allows the bound sperm to fuse with the egg ($t = 20$ seconds) while blocking other sperm from attaching and fusing with the egg. From D. Epel and V. D. Vacquier, In G. Poste and G. L. Nicolson, eds., *Membrane fusion,* Elsevier/North-Holland Biomedical Press, Amsterdam, 1978. Courtesy of the authors.

unavailable for further attachment. Other possibilities, such as charge affecting specific **insertion proteins** required for fusion, are equally plausible.

The sea urchin's fast block (if it exists) is insensitive to cytochalasin B, since the drug fails to alter the frequency of polyspermy, but similar blocks (sometimes called rapid blocks) in other invertebrates may be sensitive. In *Spisula*, for example, 50% of the eggs block polyspermy within 5 seconds of introducing sperm, and virtually all eggs block polyspermy within 15 seconds, but sperm continue to enter the surf clam's eggs in the presence of cytochalasin B. Possibly, the initial entry of calcium ions into *Spisula* eggs activates an actin-polymerizing mechanism which in turn blocks the entry of additional sperm.

No mammalian block to polyspermy is equivalent to the fast blocks to polyspermy. Mouse and hamster eggs show brief, recurring hyperpolarizations following fertilization, but these represent only small shifts in electrical potential. In the mouse, reducing sodium or calcium concentrations in the medium does not induce polyspermy, and in the hamster, the hyperpolarizations occur even when the potential is maintained at a positive voltage. Rabbit eggs show modest fertilization potentials (i.e., changes of only about 8 mV), but these may not be sufficient to represent a barrier to polyspermic fusion.

The ability of mammalian eggs to exclude multiple sperm penetrations is age dependent. Preovulatory eggs with fewer than the mature complement of cortical vesicles and aging ovulated eggs are prone to polyspermy (see Gwatkin, 1977). Freshly ovulated mammalian eggs exhibit unique surface blocks to polyspermy not comparable to blocks found elsewhere.

Surface blocks to polyspermy can occur at the level of the plasma membrane or egg envelope. One mammalian surface block, developing at about 40 minutes after the first fertilization, is known as the **plasma membrane block** (see Wolf, 1981). Its existence is inferred from the presence of sperm in the perivitelline space. These sperm do not fuse with the egg, and their number in mouse eggs is simply a function of sperm concentration. Mouse eggs (if not hamster eggs) artificially reinseminated with additional acrosome-reacted sperm also fail to become polyspermic.

In general, mammalian eggs accumulating sperm in the perivitelline space (e.g., rabbit) have a plasma membrane block. The release of cortical vesicles may be irrelevant to this block, since ionophore-activated mouse eggs release at least half their cortical vesicles but fail to acquire the block.

The plasma membrane block may therefore be electrically mediated, but it is unlike the fast block inasmuch as it is permanent.

A **vitelline reaction** may also operate at the surface of the egg to exclude sperm, but this reaction is relatively slow to develop (e.g., 2–3.5 hours in hamster eggs). Involving an increase in lectin-binding proteins and in negatively charged groups, the vitelline reaction may cause permanent, if not rapid, resistance to sperm entry.

In addition, some mammalian eggs throw a **zona block** in the zona pellucida. Mammalian eggs that exclude excess sperm from the perivitelline space (e.g., dog, hamster, human, and sheep) presumably have this type of block. Mouse and rat eggs seem to represent an intermediate condition with partial plasma and partial zona blocks.

Late blocks to polyspermy were implicit in the first descriptions of the cortical reaction and lifting of the FE (see Just, 1939). This "late," "slow," "complete," or "permanent" block, like the zona response, was associated with the exocytosis of cortical vesicles.

In the sea urchin and other marine invertebrates, the contents of the cortical vesicles may aid in blocking polyspermy in a variety of ways. For example, a trypsinlike proteinase or **sperm receptor hydrolase** from cortical vesicles in sea urchins removes sperm receptors from the VE. During a **phase of detachment** lasting about 30 seconds and corresponding to the period of cortical vesicle exocytosis, all sperm, except the fertilizing one, detach from the VE.

Since no further sperm can bind themselves to the fertilization membrane after sperm receptor hydrolases have done their work, this block is normally irreversible. The plasma membrane does not lose its fusiogenic capacity, however. Envelope-free fertilized eggs whose hyaline layer is dissolved in Ca^{2+}- and Mg^{2+}-free seawater are capable of fusing with sperm even at the two-cell stage (Sugiyama, 1951).

Similarly, hamster eggs remain capable of fusing with sperm in the four-cell stage, but sperm are usually prevented from penetrating the zona and reaching the egg. The hamster's zona block develops about 16 minutes after spermatozoon-egg fusion and 8 minutes after the exocytosis of cortical vesicles. The block may originate in material released from cortical vesicles, since the supernatant of cortical-reacted, zona-free eggs transferred to other eggs promotes resistance to zona penetration by sperm (see Yanagimachi, 1988).

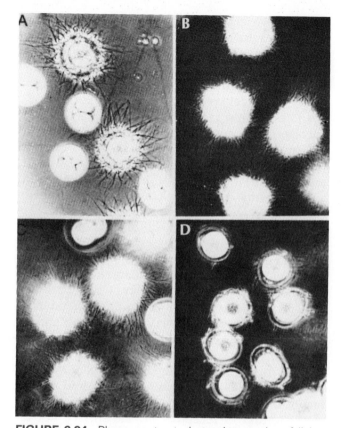

FIGURE 8.24. Phase contrast photomicrographs of living mouse oocytes and early embryos with and without halo of bound sperm. The halos around eggs in (a) and (b) show the range of binding by untreated control sperm. Part (d) shows the nearly complete absence of binding when sperm are exposed to ZP3 extracted from unfertilized eggs. Part (c) shows the high degree of binding when sperm are exposed to $ZP3_f$ extracted from 2-cell mouse embryos. ×133. From J. D. Bleil and P. M. Wassarman, *Cell,* 20:873 (1980), by permission of Cell Press.

Excess mouse sperm fail to fuse with the plasma membrane in mice following release of cortical granules. In addition, the mouse egg has a **zona response** which beyond providing a zona block to polyspermy starts the relatively rapid hardening of the zona and relatively slow (about 4 hours *in vivo* in the mouse) loss of previously bound sperm (note sperm-free zonas of two-cell embryos in Fig. 8.24a, c). Thereafter, sperm may attach loosely to the egg, but they are no longer rigidly bound.

The shedding of bound sperm may depend on the digestion of proteinaceous receptors. A trypsin-like enzyme(s) released by cortical vesicles of sea urchins may be involved, and a serine hydrolase collected from electric-shocked hamster eggs prevents other eggs from binding sperm. Moreover, mild trypsin treatment of eggs, intended to remove receptors, prevents subsequent fertilization, and new protein bands of zona polypeptides are detected by SDS–PAGE after the zona response.

The zona response may also depend on the modification of carbohydrate portions of glycoprotein receptors. Hamster zonas, for example, exhibit diminished binding to potato agglutinin specific for *N*-acetyl-D-glucosamine following fertilization (see Jaffe and Gould, 1985).

The mouse ZP3 (molecular weight 83,000) is also modified during the zona response. While purified ZP3 from unfertilized eggs condenses with sperm and prevents sperm binding to eggs (Fig. 8.24, compare a and b to d), $ZP3_f$ from fertilized eggs (i.e., prepared from the zona pellucidas of early embryos) fails to coalesce with sperm and does not prevent binding to eggs (Fig. 8.24c). Since the two ZP3s are electrophoretically identical, structural changes in the carbohydrate moiety rather than polypeptide backbone may have been altered following fertilization (Bleil and Wassarman, 1980).

Hardening the vitelline coat refers to cortical materials rendering fertilization membranes impenetrable by sperm lysins. Hardened vitelline coats block polysperm in sea urchins and frogs. In *Spisula*, however, removing the surface coat has no effect on blocks to polyspermy.

In mammals, "hardening" refers to the physical properties of the zona and dehydration, not necessarily to the blockage of sperm's passage. For example, the postfertilization rabbit zona, which is hardened, fails to exhibit a zona block to polyspermy and sperm continue to penetrate to the perivitelline space until a mucin coat is deposited around the zona. In the hamster, on the other hand, the zona is not hardened but is highly resistant to penetration by multiple sperm.

The formation of the perivitelline space may also play a role in blocking polyspermy. The separation of the fertilization membrane from the plasma membrane seems to block sperm in the sand dollar, *Echinarachnius*, and in many fish. In sturgeons and trout, a perivitelline jellylike layer secreted during the cortical response seems to be impenetrable to sperm, while in salmon, a perivitelline fluid agglutinates sperm (see Austin, 1978). In the marine annelid, *Neanthes* (formerly *Nereis*), a jellylike layer forced through the FE agglutinates sperm.

Eggs lacking cortical vesicles, such as those of ascidians, may also form hardened envelopes which provide a block to polyspermy. The postfertilization ascidian chorion, for example, provides a block to polyspermy. Tunicates, however, contain subcortical vesicles or enlarged cisternae of the endoplasmic reticulum which may shed hardening substances during fertilization.

Multiple blocks to polyspermy may be required in eggs relying on depolarization for an early block. Without a second block, the return to a relatively low resting potential following the fertilization potential would invite polyspermy.

In the sea urchin, **inducers of polyspermy** block a block to polyspermy by causing a premature return to a low resting potential. The inducers nicotine and ammonia prevent the depolarization of the egg membrane and depress the fast block by causing a rise in pH above 7 (Epel and Vacquier, 1978). The amine anesthetic procaine also causes a premature rise in pH but, by occupying calcium ion-binding sites, also prevents the release of cortical vesicles. Both the early and late blocks to polyspermy are thus depressed, and the fusiogenic plasma membrane of the egg continues to fuse with sperm.

Normally, with the advent of the late block, the possibility of additional fertilizations is eliminated. High pH and low membrane potential, which seem to be required for cleavage, can then return without danger of polyspermy. The early block to polyspermy may reflect an evolutionary compromise that allows the egg to undertake some of its preparations for cleavage speedily even if it must then reverse itself to continue with cleavage.

Chapter

9

REORGANIZING THE ZYGOTE

Preformation Versus Epigenesis

But, we do not know ... what a germinal localization really is. Is it an accumulation of preexisting, maternal messages? Does it contain substances, presumably nonhistone proteins, that will specifically derepress given genes in the nuclei that have colonized a particular cytoplasm? Answers to these questions should provide an answer, in molecular terms, to the great problem raised by the "naturalists" of the 18th century (Spallanzani, Buffon, Haller, Bonnet, Needham, etc.): preformation or epigenesis?

J. BRACHET (1977, p. 182)

The ordinary process of turning the zygote into an ordinary cell is frequently portrayed in extraordinary terms: a "chain-reaction," an "explosion," or a "cascade." The plasma membrane is described as "discharging" the fertilization potential; free calcium is said to "skyrocket"; hydrogen is pictured as "rushing" out of the zygote; and respiration is described as "erupting."

These terms presuppose preformed elements within the egg ready to unleash their powers of development as soon as the egg is fertilized. Are there such elements or do the explosive metaphors exaggerate and distort the events following fertilization?

Not only do events in the zygote often resemble events in ordinary body cells, but some features of the activation response are restrained and gradual. An epigenic alternative of change through process and interaction emerges around every experimental corner in the study of fertilization, yet the issue is not settled: "preformation or epigenesis?"

PHYSIOLOGY AND MORPHOLOGY OF ACTIVATION

"Explosive" View of Activation

Derived largely from research on sea urchins, the "explosive" view has been in the forefront of studies of fertilization for a century. While many aspects of sea urchin fertilization have been confirmed in other species, many other aspects have not, and caution must be exercised in making generalizations.

The respiratory or oxidative burst was first described by Otto Warburg (1908) in sea urchin eggs as a dramatic six- to seven-fold increase in oxygen (O₂) consumption immediately following fertilization (e.g., Psammechinus in Fig. 9.1). The increase was transient and O_2 consumption quickly returned to a level more nearly that of the oocyte. Later, during cleavage, respiration again increased, this time at a more or less uniform rate.

Not coupled to changes in ADP or ATP levels, the postfertilization respiratory burst appears to be nonmitochondrial and occurs in the presence of cyanide. In *Strongylocentrotus purpuratus*, the burst is Ca^{2+} dependent and can be induced in whole eggs with the calcium-transporting ionophore A23187 or in homogenates by the addition of calcium.

Unsaturated fatty acid oxidation may account for part of the burst (see Epel, 1980). A calcium-sensitive, lipid-oxidizing enzyme, lipoxygenase, is

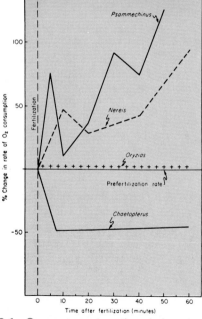

FIGURE 9.1. Oxygen consumption during the first hour after fertilization in several eggs. The percentage change in rate of O_2 consumption is plotted against time after fertilization in minutes. *Psammechinus,* a sea urchin; *Neanthes* (formerly *Nereis*) and *Chaetopterus,* marine annelids; *Oryzias,* a freshwater teleost. From A. Monroy, *Chemistry and physiology of fertilization,* copyright © 1965 by Holt, Rinehart and Winston, Inc., reprinted by permission of the publisher.

activated when free calcium becomes available following fertilization (Steinhardt and Epel, 1974). On the other hand, at least two-thirds of the increased O_2 uptake is due to a burst of hydrogen peroxide (H_2O_2) production from molecular oxygen.

Once produced, H_2O_2 provides an extracellular oxidant. It oxidizes tyrosine residues to dityrosyl cross-linkages, thereby **hardening** the fertilization envelope. The reaction is catalyzed by **ovoperoxidase** secreted along with cortical granules and incorporated into the fertilization membrane in the process of hardening (see Shapiro, 1981).

Embryos (but not sperm in the vicinity) are protected from the potentially cytotoxic hazards of H_2O_2 in two ways. Catalase reduces the peroxide enzymatically, and **ovothiol,** a substituted 4-mercaptohistidine, reduces it nonenzymatically. Ovothiol plays the role of a H_2O_2 scavenger without being depleted in the process. While ovothiol is rapidly oxidized to a disulfide in the presence of H_2O_2, reduced ovothiol is regenerated by the oxidation of glutathione to its disulfide (Turner et al., 1988).

The respiratory burst occurs at fertilization in many sea urchins (a 6–10-fold increase), in *Urechis caupo* (a 1.2-fold increase), and in some annelids (e.g., *Neanthes* and *Chaetopterus*). In starfish (*Asterias forbesii* and *Patiria miniata*), similar respiratory changes occur at the completion of meiosis. But the respiratory burst is not a universal feature of animal fertilization (Fig. 9.1) (see Monroy and Moscona, 1979), and several mollusk eggs actually undergo a decrease in oxygen uptake upon fertilization. Furthermore, teleosts (e.g., the Japanese medaka, *Oryzias*) and amphibians register no change in respiration following fertilization or throughout cleavage.

Protein synthesis following fertilization increases in sea urchins. The rate of incorporation of amino acids into proteins jumps 5- to 15-fold beginning about 5 minutes after fertilization (Fig. 9.2). Transcription does not drive the reaction, since actinomycin D, while inhibiting RNA synthesis, does not prevent the rise in protein synthesis (Fig. 9.3). Even enucleated sea urchin eggs (i.e., merogons) exhibit the rise in protein synthesis following fertilization (Monroy and Tyler, 1963). Protein synthesis in the sea urchin zygote therefore depends on the translation of stable or **stored RNA** (often called maternal RNA and sometimes known as masked RNA) previously stockpiled in the oocyte (see below Chapter 15) (Bachvarova, 1985).

The mechanism for turning on protein synthesis is uncertain. Efforts to locate defects in the oo-

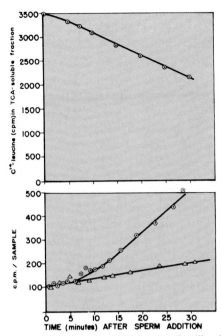

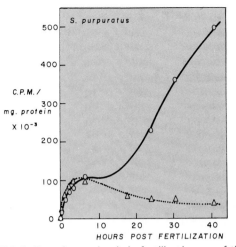

FIGURE 9.2. Protein synthesis following fertilization in the sea urchin, *Lytechinus pictus*. Eggs were preloaded with radioactive ^{14}C-leucine. (Top) Counts per minute (c.p.m.) in the hot trichloracetic acid- (TCA) soluble fraction of fertilized eggs. The decrease of free leucine is at least partially due to its incorporation into protein. (Bottom) Counts per minute (c.p.m.) from leucine incorporated into protein. Triangles, without addition of sperm; circles, after addition of sperm at time 0. From D. Epel, *Proc. Natl. Acad. Sci. USA,* 57:899 (1967), by permission of D. Epel.

FIGURE 9.3. Protein synthesis in fertilized eggs of the sea urchin, *Strongylocentrotus purpuratus,* in the presence (dotted line with triangles) and absence (solid line with circles) of actinomycin D. ^{14}C-L-valine incorporated by fertilized eggs from the medium is detected as counts per minute (C.P.M.) per milligram protein. Radioactive proteins accumulate in all these eggs from fertilization until a plateau is reached at the 64-cell stage. A second wave of protein synthesis beginning at hatching does not take place in the presence of actinomycin D. From P. R. Gross, *J. Exp. Zool.,* 157:21 (1964), by permission of Alan R. Liss, Inc.

cyte's protein synthetic machinery that might be repaired in the zygote have not been fruitful. Oocyte ribosomes do not seem defective, since, in *Xenopus laevis*, they are capable of translating injected hemoglobin mRNA (see Gurdon and Melton, 1981). Similarly, oocyte mRNA cannot be uniformly defective, since it can direct the synthesis of histones in a cell-free system (i.e., *in vitro*) (Gross et al., 1973).

According to the **ionic** or **ionic signaling hypothesis,** the combined effects of calcium and protons activate protein synthesis, at least in sea urchins, in *Urechis caupo*, and in *Spisula solidissima* (Epel, 1980, 1988; Whitaker and Steinhardt, 1985). In tunicates and amphibians, the early rise in free calcium alone may trigger the zygote's increase in protein synthesis.

The range of proteins synthesized in the zygote differ by species. Almost all the proteins synthesized by the sea urchin zygote are identical to those found in unfertilized eggs. On the other hand, *Drosophila melanogaster* eggs synthesize new proteins soon after fertilization, even though new RNA is yet to be transcribed.

While the quantity of newly synthesized proteins can be minute, their qualitative contribution can be enormous. In the clam, *Spisula*, for example, one protein switched on by fertilization is a small subunit of ribonucleotide reductase.

Proteins synthesized following fertilization may release the zygote from meiotic arrest and catalyze the entry into cleavage. For example, two proteins synthesized by sea urchins and clams are mitosis-inducing proteins, called cell-cycle-related proteins or **cyclins.** These proteins are translated from stored mRNA following fertilization and during the S periods of the succeeding mitotic division. Clam cyclin A mRNA injected into *Xenopus laevis* oocytes is not only translated, but the injected cells enter the M phase of their cycle (Swenson et al., 1986).

The gross rate of protein synthesis in mammalian zygotes fails to show a significant increase compared to oocytes, but particular proteins are synthesized following fertilization. Like sea urchins and clams, mouse zygotes produce a set of proteins required for the completion of meiosis and cleavage divisions. Unlike the cyclins that are destroyed after each division only to be synthesized again during the succeeding S period, the mouse protein components, known as **M phase compo-**

nents, are synthesized only up to the first interphase. A role for M phase components in the control of cleavage is predicated on their periodic and reversible modification by phosphorylation during meiosis and cleavage divisions (Howlett, 1986).

Total RNA synthesis in mouse zygotes fails to show a significant increase following fertilization, but particular genes transcribed by RNA polymerase type II show dramatic increases. Herpes simplex thymidine kinase gene injected into zygote nuclei is transcribed and subsequently translated in vastly greater amounts (25- to 45-fold) than the same gene injected into oocyte nuclei. Genes without introns are transcribed at especially high rates in zygotes compared to oocytes, and mRNA injected into zygotes shows decreased stability, indicating greater flux in the machinery for protein synthesis (see Chen et al., 1986).

Processive View of Activation

Many features of zygote physiology exhibit restraint and processive development. Intracellular pH rises slowly; cytoplasmic turbidity (indicative of contraction in the cortex), as measured by light scattering, increases gradually, and normal meta-bolic enzymes, such as nicotinamide–adenine dinucleotide (NAD) kinase, are activated incrementally. Furthermore, amounts of glycolytic intermediates and hexose phosphates rise, probably through ordinary mechanism of glycogen-phosphorylase activation and glycogen mobilization (Table 9.1).

In some cases, processive activation occurs without fertilization. For example, as the marine annelid *Chaetopterus* approaches the time for spawning, its eggs ripen or pass through a prematuration phase that activates the cortical response upon shedding even in the absence of sperm. Other eggs normally shed in the germinal vesicle stage (Table 7.7, class I), such as those of the clams *Barnea* and *Spisula*, and the scallop *Pectinaria* may be activated by elevated levels of potassium in seawater without fertilization.

Contrary to the idea of an explosion that cannot be called back once started, processive activation is reversible or at least suspendible. In *Barnea* and *Spisula* and in the echiuroid *Urechis*, for example, activation is blocked by treatment with acidified seawater (e.g., in the vicinity of pH 7 as opposed to the normal pH 8–9) or by reduced amounts of calcium. Additional sperm in normal

TABLE 9.1. A program of postfertilization changes in *Strongylocentrotus purpuratus* at 17°C[a]

Primary Effector	Seconds	Responses
Unknown events		
Sperm–egg contact	0–0	Opening Na^+ channels
Membrane depolarization	3–3	Early block to polyspermy
PIP/PIP_2 turnover	10–20	Ca^{2+} release from endoplasmic vesicles
Early or phospholipase C related events		
Free Ca^{2+}	20–60	Cortical exocytosis begins; changes in surface (light scattering), actin polymerization in cortex, microvilli, and fertilization cone; late block to polyspermy; protein kinase C activation
Late or protein kinase C related (or pH-dependent) events		
Rise in intracellular pH	30–60	Na^+–H^+ exchange, NAD kinase activation, lipoxygenase action
	35	Burst of O_2 uptake, Ca^{2+} decrease, endocytosis
	60	Hyperpolarization
	300	Resumption of K^+ conductance, increased amino acid and nucleotide transport, nucleoside phosphorylation, increased protein synthesis
	600	Tyrosine protein kinase, pronuclear movements
	1200	Chromosomal merging, DNA synthesis

[a]Based on data in Epel (1978, 1989, and personal communication).

seawater can then "reactivate" the eggs and lead to polyspermic development (see Jaffe, 1985).

In *Urechis*, fertilized eggs inactivated by acidified seawater can be reactivated by ammonia or dilute seawater. Many reactivated eggs initiate cleavage and proceed to develop normally. On the other hand, eggs parthenogenically activated in ammonia (or dilute seawater), which become arrested due to prolonged treatment, develop normally in normal seawater when exposed to sperm and presumably fertilized (Tyler and Bauer, 1937).

Even in sea urchins, late activation responses can be dissociated from early ones. Intracellular pH rises in unfertilized sea urchin eggs placed in dilute solutions (i.e., 1–30 mM) of ammonium hydroxide or ammonium chloride in seawater at pH 9, and protein synthesis increases; mRNA becomes polyadenylated, and chromosomes condense. But the eggs fail to undergo cortical exocytosis, raise their membrane potential to the level of a fertilization potential, or show a burst of oxidation.

Exocytosis and the fertilization potential therefore do not seem to be necessary for the synthetic changes occurring later in the activation reaction (see Epel, 1978). This conclusion is not entirely surprising, since some eggs, such as those of *Urechis*, do not perform the cortical response upon fertilization. The independence of later changes from exocytosis does not necessarily mean that they are independent of free-calcium levels, however, since the calcium pulse or "explosion" normally accompanying exocytosis may occur if only with a lower than normal peak in the presence of ammonia (see Jaffe, 1985).

Possibly, activation consists of stages with their own causes, or primary effectors, and their own responses. In the sea urchin, for example, during the first 60–90 seconds beyond the point of membrane depolarization (i.e., the fertilization potential), free calcium may trigger a set of responses, including the release of acid, which by bringing about an increase in intracellular pH triggers another set of responses (Table 9.1).

Fertilization (the "unknown" period) (Epel, 1978) and the rising intracellular pH in the first 1–4 minutes are accompanied by calcium-dependent **early events**. These may initiate a second causal web of **late events**. In addition to a lowered membrane potential and an increased conductance of potassium by the zygote's plasma membrane, late events include the insertion or activation of systems for nucleoside, phosphate, and amino acid transport and a great rise in protein synthesis.

The early events have a more explosive character than the late events, but they too are interconnected. Free intracellular calcium seems to "ig-

nite" calcium- or magnesium-dependent NAD kinase (or a calmodulin–calcium activator of NAD kinase), but levels of nicotinamide–adenine dinucleotide phosphate (NADP) rise gradually at the expense of NAD.

In *Strongylocentrotus purpuratus* eggs, the amount of $NADPH^+$ rises some threefold, and most of the two- to threefold increase in the amount of NADP occurs within 1 minute, but in *Arbacia* and *Spisula* eggs, a three- to sevenfold increase occurs in the course of 1–3 hours. The reducing power of NADP increases the redox potential of the egg and may play a role in stimulating cellular biosynthesis, promoting DNA replication, and assembling microtubules in the mitotic apparatus.

The change in intracellular pH, triggering the late events, is tied to the activation of sodium and proton channels in the zygote's plasma membrane. As if it is suddenly "uncorked," a Na^+–H^+ **counterexchange system** utilizing separate channels drives intracellular pH upward. As intracellular sodium and hydroxyl ion concentrations increase, protons depart (known as the acid release of fertilization), and "fertilization acid" accumulates in the zygote's microenvironment. In the sea urchin, this proton efflux causes an intracellular pH rise from 6.84 one minute after fertilization to a new equilibrium of 7.27 at about 5 minutes. Similarly, homogenates of unfertilized and fertilized eggs register an increase from pH 6.5 to pH 6.8 (Epel, 1980; Shen and Steinhardt, 1978).

Morphological Aspects of the Zygote's Reorganization

Physiological change is ordinarily assumed to precede morphological change. In the zygote, the two may be hand in hand, and, occasionally, morphology gets the upper hand.

Cytoplasmic streaming and ooplasmic segregation, or directed movements in the zygote's cortex and subcortical cytoplasm leading to zones of cytoplasmic concentration, completely alter the appearance of many small and medium-sized eggs following fertilization. Even ascidian eggs that have no cortical granules and do not form vitelline membranes within their thick chorions exhibit a profound morphological reaction to fertilization (Fig. 9.4). Yellow granules and clear cytoplasm stream downward toward a point farthest from the germinal vesicle where the spermatozoon has fused with the oolemma. The movement extends to the maternal follicle cells (known as **test cells**) sandwiched between the chorion and the zygote.

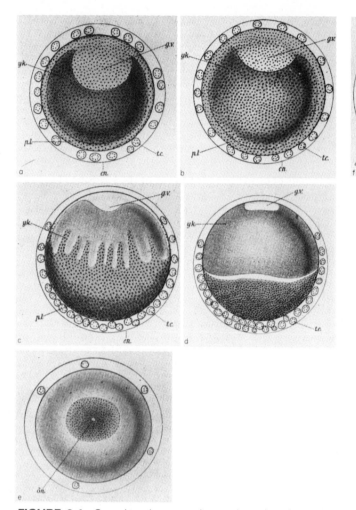

FIGURE 9.4. Cytoplasmic streaming and ooplasmic segregation in the tunicate, *Styela* (= *Cynthia*) *partita*. (a) Unfertilized egg before breakdown of the germinal vesicle. (b) Similar egg after fertilization. (c) Five minutes after fertilization. Peripheral protoplasm and yellow granules (dots) spread to lower (vegetal) pole of egg where the spermatozoon has entered. The test cells (uniquely ascidian maternal cells found between egg and chorion) are also carried to the lower pole, and gray yolk in the upper pole is exposed. (d) Later. Yellow granules and clear cytoplasm have collected in the vegetal pole. (e–g) Views of the vegetal pole at successive 5-minute intervals. (e) The yellow granules are maximally concentrated in zone of clear cytoplasm around sperm nucleus. (f) Yellow granules spread upward. (g) Yellow granules nearly cover lower hemisphere, called yolk. *ch.*, chorion of jelly coat; *g.v.*, germinal vesicle; *p.l.*, peripheral layer of yellow cytoplasm; *t.c.*, test cells; *yk.*, gray yolk; *y.h.*, yolk hemisphere; *O.*, point of sperm nucleus. From E. G Conklin, *J. Acad. Natl. Sci. (Philadelphia),* 13:1 (1905).

Only the **gray yolk** in the egg's interior is left behind as a **yolk body.** As the germinal vesicle's breakdown is completed, the yellow granules become even more concentrated around the spermatozoon's nucleus. Then, as the spermatozoon's nucleus approaches the egg's nucleus, the yellow granules spread upward and cover nearly half the egg. Later, during cleavage, the granules and colored cytoplasm enter particular blastomeres in an extraordinary display of precise ooplasmic segregation.

In many cases, ooplasmic segregation is associated with **germinal localization** or the localization of hypothetical determinants of development in particular regions of the zygote. In *Styela* (= *Cynthia*) *partita*, for example, the **mesodermal crescent** of yellow cytoplasm takes up a position just below the equator where blastomeres will later form mesoderm. The **light gray cytoplasm** appears as a crescent, known as the **notochordal crescent.** Blastomeres containing this cytoplasm are destined to form notochord (Conklin, 1905). The remainder of the vegetal hemisphere is dominated by a **slaty gray crescent** consisting of mitochondria in a subcortical layer and yolk granules. Endoderm will form here. Little yolk and few mitochondria are left in the transparent animal hemisphere. Blastomeres carved out of this region will form ectoderm (see below, Fig. 11.33).

Morphologically significant ooplasmic segregation seems to arise in many molluscan eggs via a phenomenon resembling cytoplasmic streaming. Instead of superficial streams, transient bulges or **polar lobes** form containing interior material (see below, Fig. 11.47). The first polar lobe forms before the first cleavage, and the second polar lobe forms between the first and second cleavages. These polar lobes generally funnel their cytoplasmic contents entirely into one of the two blastomeres arising from the upcoming division. The material funneled into the blastomeres drives their development toward mesoderm and primordial germ-cell differentiation. Annelids may form only small polar lobes or none at all, but cleavage still results in the formation of blastomeres of unequal size and developmental fates.

Cytoplasmic streaming is not necessarily associated with developmental determinants. In sea urchins, such as *Paracentrotus* (= *Strongylocentrotus*) *lividus*, for example, pigment granules leave the animal hemisphere clear while becoming concentrated in a subequatorial band (Fig. 9.5). The pigment granules are not associated with any morphological activity and can be displaced in a centrifugal field without altering the development of the embryo.

Cytoplasmic streaming is also exhibited by many teleost eggs (Fig. 9.6). Large amounts of cytoplasm, rather than particular kinds of granules or yolk, stream into the animal pole, thereby forming the **germinal disk** above the yolky vitellus. Streaming is extremely energetic in some fish, and in

time-lapse films, cytoplasm pulsates as it is pumped upward.

In amphibians, **cytoplasmic determinants** or morphogenic inducing substances may be distributed in the outer layers of yolk or in the cortex. The positions of these determinants may change in the course of development, most conspicuously during the formation of the gray crescent (see Brachet, 1977).

The amphibian gray crescent (in most anurans) or clear crescent (in newts) forms as a consequence of cytoplasmic movement (if not streaming) and gives the appearance (if not the reality) of ooplasmic segregation. Ideally, the gray crescent appears prior to cleavage and disappears prior to the

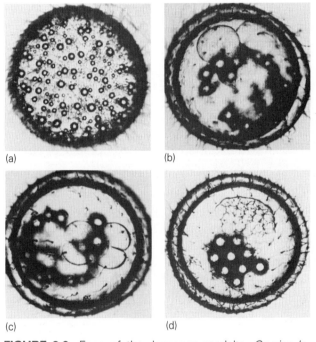

(a)   (b)

(c)   (d)

FIGURE 9.6. Eggs of the Japanese medaka, *Oryzias latipes.* (a) Prior to fertilization. Chorionic filaments decorate outside of chorion. Cortical vacuoles (abundant small circular profiles) line cortex, and oil droplets (larger circular profiles) float in egg. Following fertilization, cytoplasm flows to pole and small oil droplets coalesce into a few large droplets. Cortical vacuoles are eliminated by exocytosis, and the perivitelline space separates the egg from the chorion. (b) 2-cell stage. (c) 4-cell stage. (d) Late cleavage germinal disk. Part (a) From A. Monroy, *Chemistry and physiology of fertilization,* Holt, Rinehart and Winston, New York, 1965. Parts (b)–(d) from A. Monroy and A. A. Moscona, *Introductory concepts in developmental biology,* University of Chicago Press, Chicago, 1979. © 1979 by The University of Chicago. Reprinted with the permission of the publisher and A. Monroy. Photographs courtesy of Dr. E. Nakano.

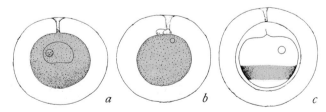

a   *b*   *c*

FIGURE 9.5. Drawing showing distribution and movement off red pigment (stippling) in *Paracentrotus* (= *Strongylocentrotus*) *lividus.* (a) ovarian oocyte. (b) mature ovum after giving off its polar bodies. (c) after fertilization. From T. H. Morgan, *Experimental embryology,* Columbia University Press, New York, 1927. Attributed to Th. Boveri.

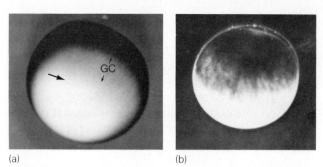

(a)   (b)

FIGURE 9.7. The gray crescents at the beginning of cleavage in *Xenopus laevis* (a, arrow) and a ranid frog (b). Part (a) from R. P. Elinson and B. Rowning, *Dev. Biol.,* 128:185(1988), by permission of Academic Press and courtesy of the authors. Part (b) from G. M. Malacinsky. In C. B. Metz and A. Monroy, eds., *Biology of fertilization,* Vol. 3, *The fertilization response of the egg,* Academic Press, Orlando, 1985. Courtesy of the authors.

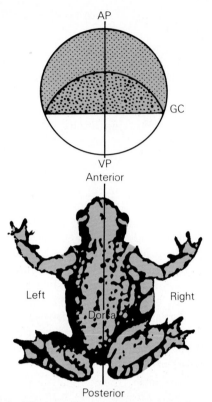

FIGURE 9.8. Diagram illustrating relationship of plane cutting gray crescent (GC) into two equal halves and bilateral symmetry of animal.

blastula stage. It has a lenticular outline (not crescent-shaped) and stretches halfway around the egg's equator between the relatively dark animal hemisphere and the creamy vegetal hemisphere (Fig. 9.7).

In reality, gray crescents are only seen in favorable material and are not identical in every species. Gray crescents emerge over a range of times, usually halfway between fertilization and cleavage-furrow formation (50 minutes in *X. laevis,* 70–130 minutes in *R. pipiens* and *R. temporaria* [= *fusca*], 4–6 hours in *A. mexicanum*). In lightly pigmented species, such as the Italian or crested newt, *Triturus cristatus,* and in darkly pigmented species, such as the European toad, *Bufo bufo,* a crescent is not seen at all, and in moderately pigmented eggs, such as those of *X. laevis,* the "gray" crescent is a diffuse yellow-brown to rust.

Despite these difficulties in harmonizing gray crescents, interest in gray crescents is high. At the very least, the gray crescent's location inspires curiosity. In 60–70% of amphibian eggs, the gray crescent is more or less bisected by the first cleavage furrow and the midsagittal plane of the future animal. Gray crescents always disappear during cleavage, but the future dorsal lip of the blastopore roughly replaces the gray crescent, and cells that will later form dorsal structures, such as the neural tube, notochord, and somites, originate in the vicinity of the former gray crescent (Fig. 9.8) (see Kirschner et al., 1980).

Possibly the correlations of gray crescent and dorsal differentiation are merely fortuitous. In *X. laevis,* the plane of the first cleavage furrow bisects the gray crescent 70% of the time but corresponds

to the plane of bilateral symmetry 100% of the time (Klein, 1987). Possibly, the gray crescent is a consequence rather than a cause, and other processes and events establish dorsal differentiation. But the formation of the gray crescent is still the first sign of incipient bilateral symmetry in the embryo and at the very least an intriguing clue to dorsalization (the initiation of or commitment to form dorsal structures).

Establishing Bilateral Symmetry in the Amphibian Embryo

Over the years, embryologists have tried to test the role of the gray crescent in development. Is it a center of **dorsalizing localization** where determinants of dorsal development are concentrated, or is it a developmentally irrelevant consequence of shifting pigment in the vicinity of developmentally significant events taking place elsewhere? Experiments have centered on nailing down the mechanism of gray crescent formation and on ascertaining the developmental power of the gray crescent.

The formation of the gray crescent involves the timely occurrence of several events. As illustrated by eggs of the common leopard frog, *Rana pipiens*, maintained in the laboratory at 20–23°C (Figs. 9.9 and 9.10), the events and their chronology figure in various concepts of gray crescent formation.

At 5 minutes after fertilization, exocytosis of cortical vesicles begins at the sperm entry point (SEP), and the perivitelline space begins to separate the zygote from the fertilization envelope. At the same time (actually beginning at 3–4 minutes and reaching a maximum at 5 minutes after fertilization), the cortical layer of the egg contracts (I in Fig. 9.10) (see Nieuwkoop, 1977).

Known as **the activation contraction,** the contraction of the cortex seems to involve the spaces between microvilli which become tightly packed in the animal half and dispersed in the vegetal half. Contraction is Ca^{2+} dependent and can be induced by exposure to the calcium-transporting ionophore A23187 or by injection of calcium. Cytochalasin B does not disturb the contraction, indicating that it is not dependent on microfilament assembly (see Elinson, 1980).

The margin of the pigment cap is drawn sharply upward, by as much as 50% (i.e., from line -1—1- to line -2—2- in Fig. 9.10) to within 60° of the animal hemisphere. In deeply pigmented ranid eggs, the gray crescent occupies the remaining 30° to the embryo's frontal surface.

The white **maturation spot** at the animal pole marking the position of the first polar body's extrusion disappears. The contraction also elevates the male pronucleus and brings it closer to the ultimate point of pronuclear meeting.

Cortical rotation follows. It is now equated to the **symmetrization rotation** (Ancel and Vintemberger, 1948, 1949) or **cortical reaction of symmetrization** (Pasteels, 1964) once thought to signal the change from radial to bilateral symmetry in the egg (see Gerhart, 1980). The rotation is strictly cortical. The inner contents of the egg do not follow the cortex. In lightly pigmented species, such as *X. laevis*, eggs spotted with fluorescent marks in the subcortical cytoplasm show the entire cytoplasmic core of the egg turning within the cortex (see Gerhart, 1987). Rotation parallels bands of microtubules arrayed beneath the vegetal cortex (Elinson, 1988).

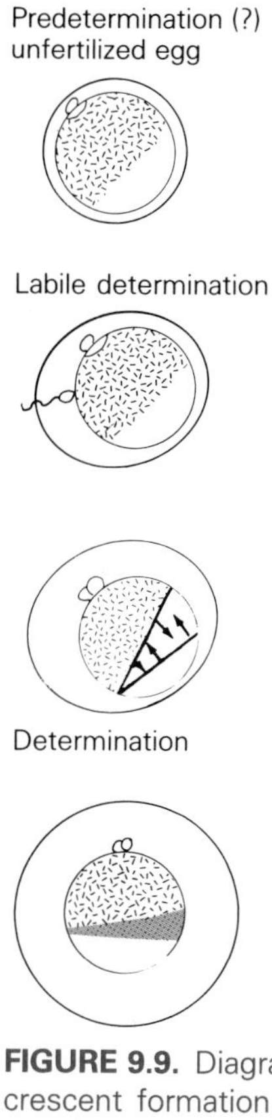

Predetermination (?)
unfertilized egg

Labile determination

	Time (in minutes)
Fertilization	0
Exocytosis and contraction	5
Relaxation	15
Extrusion 2° polar body	35
Disappearance maturation spot Cortical rotation = symmetrization rotation Asymmetric movement of pigmented cap	
Amphimixis	45

Determination

Perivitelline space maximum Gravational rotation	60–120
Gray crescent appears	70–130
First cleavage	150

FIGURE 9.9. Diagrammatic drawing of the phases of gray crescent formation. Times taken from R. P. Elinson, *Int. Rev. Cytol.,* 101:59 (1980).

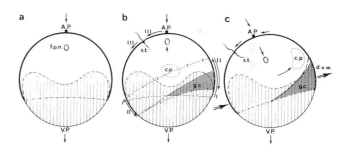

FIGURE 9.10. Detailed scheme for the formation of gray crescent in anurans. (a) Initial configuration. (b) Contraction around point of sperm entry draws pigment cap unsymmetrically upward (I) from its initial position (-1–1-) to a new, lopsided position (-2–2-). Subsequent relaxation (II) expands pigment cap to new (-3–3-) position. (c) Subsequent rotation of cortex (III) rearranges the animal pole and juxtaposes yolk platelets and cortex at the dorsal vitelline wall (DVW) beneath the gray crescent. AP, animal pole; CP, clear cytoplasm; FP, female pronucleus; GC, gray crescent; VP, vegetal pole; I, II, III, arrows indicating directions and extent of sequential contractions and expansions of egg cortex; 1, 2, 3, original position and positions following contraction and expansion of pigmented cap margin. Double stemmed arrow indicates plane of bilateral symmetry. From P. D. Nieuwkoop. In A. A. Moscona and A. Monroy, eds., *Current topics in developmental biology,* Vol. 11 p. 115, Academic Press, Orlando, 1977. Used by permission.

Along with contraction, the second polar body is extruded (20–35 minutes), and a subsequent relaxation of the cortex results in the downward movement of the pigmented margin (II in Fig. 9.10) to a reexpanded position (line -3—3-) not quite corresponding to the original position. The new position of the margin is lopsided, skewed toward the sperm entry point with the animal pole (AP) displaced ventrally (III). Internally, large yolk platelets are drawn close to the surface in a region opposite the sperm entry point called the **dorsal vitelline wall** (DVW).

When the perivitelline space has swelled sufficiently and the zygote is completely suspended, it may be seen to rotate, depending on its initial orientation. Eggs that happen to have been laid with their vegetal hemisphere upward rotate most conspicuously as their center of gravity amidst the large yolk platelets is drawn downward, while the less dense animal hemisphere moves upward. When eggs of deeply pigmented frogs are fertilized artificially, their nearly simultaneous rotation is like a flashing signal informing the eager observer that fertilization has been successful.

In *Xenopus laevis*, one or two **postfertilization waves** now slowly ripple across the surface. Their appearance is correlated with the enlargement of the spermatozoon's aster in the animal hemisphere, and the waves do not appear in eggs treated with vinblastine or colchicine, two antimicrotubule drugs (see Gerhart, 1980). The waves move from the sperm entry point to the opposite pole where the gray crescent appears.

The gray crescent occupies the egg's equator above the thinly pigmented dorsal vitelline wall. The gray crescent's color is attributable in part to thinning in the cortex opposite the sperm entry point (see Nieuwkoop, 1977) and in part to drawing unpigmented vegetal cortex over the lightly pigmented dorsal vitelline wall (see Elinson, 1980).

The position of the gray crescent seems to be determined epigenically by a succession of events (see Brachet, 1977). A weak gray crescent **predetermination** seems to occur within the ovary. Stronger but still **labile determination** is influenced by the sperm entry point, and irreversible **determination** follows rotation.

The possibility of gray crescent predetermination prior to fertilization is inferred from the results of heating axolotl eggs for 10 minutes at 35–65°C. These **heat-shocked** eggs develop their gray crescents immediately rather than after the normal wait of 4–6 hours. Had the gray crescent depended on sperm entry and rotation alone, heat shock would not have induced the premature appearance of the gray crescent.

The power of sperm to dictate the position of the gray crescent was first discovered in the 19th century by Wilhelm Roux, building on George Newport's earlier experiments with localized fertilization (see Brachet, 1977; Moore, 1987). The sperm entry point in anurans is able to override predetermination, since eggs can be fertilized almost anywhere on their animal hemisphere and form their gray crescent on the opposite meridian. But sperm's role in gray crescent determination is not universal. Urodele eggs are normally fertilized by several sperm at different points on the surface (i.e., urodeles are normally polyspermic) and form their clear crescents without regard to points of sperm entry.

In sections of anuran eggs, a pigmented **sperm path** or **tract** (TS, Fig. 9.11a) leading away from the sperm entry point is directed toward the site of the future gray crescent. This movement of pigment does not cause gray crescent formation, since eggs activated by pricking form the crescent without forming pigmented sperm trails. The initiation of exocytosis at the point of spermatozoon–egg fusion does not seem to determine the position of the gray crescent either, since pricking also initiates exocytosis. Furthermore, an anuran egg activated by the divalent cationic ionophore A23187 forms a gray crescent, although exocytosis is initiated simultaneously all around the egg, and a urodele egg forms its clear crescent despite its lack of cortical vesicles and the absence of exocytosis.

The spermatozoon's influence on the gray crescent's position may actually be exerted by the

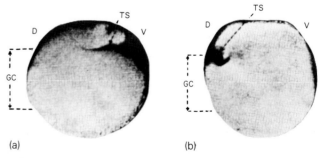

(a)   (b)

FIGURE 9.11. Photomicrographs of sections through *Rana temporaria* (= *fusca*) zygote. (a) Normal zygote showing the sperm path (TS) pointing at the gray crescent (GC) opposite the sperm entry point. (b) Zygote after forced rotation through 360°. The gray crescent (GC) has formed in the vicinity of the sperm entry point and the sperm track (TS). D, dorsal; V, ventral. From P. Ancel and P. Vintemberger, *Arch. Anat. Microsc. Morphol. Exp.*, 31:1 (1948). Used by permission of the authors.

centriole and aster fibers arising from it, since colchicine and vinblastine, which interfere with microtubule assembly, prevent gray crescent formation. Cytochalasin B, which interferes with actin assembly, does not affect gray crescent formation. Furthermore, in the toad, *Bufo arenarum*, sperm homogenates injected into eggs determine the position of the gray crescent but only in the presence of asters (see Elinson, 1980).

In any case, the fertilizing spermatozoon's ability to determine the position of the gray crescent, like predetermination that came before, is labile and can be overridden. For example, an egg whose animal–vegetal axis rotates through more than 135° in the perivitelline space forms its gray crescent at the leading edge of the pigmented cap independently of the point of spermatozoon entry. The more the egg rotates, the greater the likelihood that rotation will override the spermatozoon's influence.

The power of rotation to dictate the final position of the gray crescent is demonstrable in experiments with enforced rotation, either the inversion of eggs immobilized by removing perivitelline fluid or the centrifuging of embedded eggs. For example, the enforced rotation of eggs by 360° at any time within the first 45 minutes after fertilization overrides the effect of spermatozoon entry and places the gray crescent in the vicinity of the pigment path (TS, Fig. 9.11b).

Eggs held upside down form two gray crescents, one in the expected position and a second, known as Born's crescent, on the opposite side of the vegetal hemisphere (see Pasteels, 1964). Even eggs rotated ventral-side up or centrifuged in this position before the first cleavage form multiple planes of bilateral symmetry and caudally attached twins (Fig. 9.12).

The experimental analysis of the role of the gray crescent has yielded mixed results. Regions such as the clear cytoplasm (CP, Fig. 9.10) and the dorsal vitelline wall beneath the gray crescent may be of greater morphogenic significance than the gray crescent. Possibly, these regions override the effects of the gray crescent, or they and the gray crescent may be seats of different determinants.

On one hand, removal of the gray crescent from *X. laevis* zygotes results in the arrest of the embryo as a biradially symmetrical ball. Reciprocally, when gray crescents are removed from embryos and grafted to other embryos, secondary embryos, or partial twins, appear at the site of grafting. All this is consistent with the possibility that gray crescents are **dorsalizing loci** capable of inducing

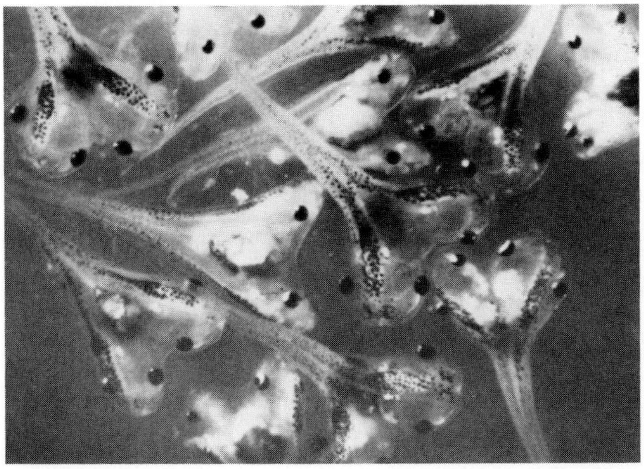

FIGURE 9.12. Double-headed *Xenopus laevis* tadpoles developed from eggs centrifuged at 30 *g* for 4 minutes directed 90° to animal–vegetal axis. The greatest number of double-headed animals is obtained with eggs centrifuged at about the time of the cortical symmetrization rotation. From J. Gerhart et al. In W. R. Jeffery and R. A. Raff, eds., *Time, space, and pattern in embryonic development,* Alan R. Liss, Inc. New York, 1983. Used by permission.

dorsal formations (Curtis, 1962), but the transplantation experiments required immobilizing the eggs, and immobilization, rather than transplantation, might have induced the secondary embryos.

But if rotation rather than the position of the gray crescent establishes the axes of bilateral symmetry in the embryo, what does rotation do? One possibility is that normally large yolk platelets dorsalize the cortical cytoplasm of the dorsal vitelline wall opposite the sperm entry point as they come into contact with it following the asymmetric contraction of the cortex and rotation. In immobilized eggs artificially rotated or centrifuged, the same large yolk platelets induce the development of a secondary embryo at a point where they are forced into contact with cortical cytoplasm (Gerhart et al., 1981).

Distinguishing between effects of rotation as opposed to the gray crescent is complicated, since a variety of treatments that interfere with establishing the gray crescent also inhibit rotation. For example, when colchicine or vinblastine is injected into eggs before or during rotation, rotation and gray crescent formation fail, accompanied by the appearance of abnormalities in gastrulation and the failure of neurulation. The degree of rotation rather than the all-or-none formation of the gray crescent seems to be involved, since the embryos develop a spectrum of defects ranging from the absence of dorsal structures to the production of tails, trunks, and all but complete heads as a function of how much rotation took place (see Gerhart, 1987).

The dorsalizing effects of the gray crescent may simply be coincidental to the presence of nearby large vegetal yolk platelets. The possibility of these yolk platelets having dorsalizing activity during their movement or when closely applied to the dorsal vitelline wall is consistent with the induction of secondary embryos by flattening eggs and by centrifuging them. Moreover, the transplantation of vegetal cells inheriting these platelets results in the induction of dorsal formations (Gimlich, 1985).

Superficial cells formed in the region of the former gray crescent therefore appear to be dorsalized rather than dorsalizing. Underlying vegetal yolk platelets or later vegetal cells and not the descendants of cells bearing traces of the gray crescent inherit whatever dorsal determinants there may be and the ability to dorsalize other cells (Nieuwkoop, 1973).

FORMING THE EMBRYONIC NUCLEUS

As the egg and spermatozoon nuclei move toward their "rendezvous with destiny," they change dramatically. The spermatozoon nucleus metamorphoses into a male pronucleus, and the egg nucleus typically completes maturation and becomes the female pronucleus. Then, for the first time since the preleptotene stage of meiosis, DNA is replicated. The first nuclear division is not far behind.

Despite the speed with which some of these acts take place and the uniqueness of the setting (i.e., the launching of a new diploid generation), nuclear reorganization in the zygote may be accomplished by the same mechanisms and controls operating in more ordinary somatic cells. The zygote is, after all, the first cell of the future organism, and at least some regulators that develop and operate in it may be expected to operate in later cells.

Trail of the Spermatozoon Nucleus

The movements of nuclei within the zygote that bring them together take as little as 8 minutes in *Arbacia* and 25 minutes in chicks or as much as 3–4 hours in mammals. Among the variables affecting the interval are the meiotic state of the egg's nucleus at the time of fertilization (Table 7.7), the size of the egg, and the distance between the

point of sperm entry and the point of polar body formation.

Pigment granules tracing the fertilizing spermatozoon's path in the egg sometimes leave a **sperm tract** (TS, Figs. 9.10 and 9.11a). In frogs, the sperm tract leads directly toward the central cytoplasm where the male pronucleus is joined by the female pronucleus.

Sperm tracts are not always straightforward. While still within the fertilization cone, the apex of a sea urchin's spermatozoon is directed toward the center of the zygote (Fig. 8.10). Once free of the cone, but still confined by the cortex, the spermatozoon's path arches (Figs. 8.14 and 9.13), and the spermatozoon's head completes a rotation of 180°. Breaking through the cortex and entering the zygote's cytoplasm, the sperm's **centrosome** (or centrotubule), containing its centrioles, is surrounded by microtubular **astral rays,** and a **monaster** (or monopolar aster) leads the sperm's nucleus along the sperm tract.

Spermatozoan rotation is variable. In many marine invertebrates, such as the mussel, *Mytilus,* spermatozoan rotation may not be as common or proceed quite as far as it does in sea urchins, and female centrioles, left over from meiosis and associated with their own asterlike structure (see below), may provide the zygotic centrosome.

In frogs, spermatozoan rotation may not take place at all. **Aster fibers** originating from the spermatozoon's centriole trail the nucleus. They penetrate to the cortex where they stiffen the surface of the egg.

In mammals, the degree of spermatozoan rotation is variable. The origin of the zygotic centrosome is uncertain (Fawcett, 1975).

Actin associated with the spermatozoon may play a role in drawing the nucleus into the egg, but nuclear movement within the fertilized egg is retarded by puromycin (an inhibitor of protein synthesis). It therefore seems that *de novo* protein synthesis, not merely preexisting proteins, is required for the pronuclei's conjugal journey (Yanagimachi, 1978).

The breakdown of the spermatozoon's nuclear envelope is the first of a series of events in the metamorphosis of the spermatozoon's nucleus into the male pronucleus. Disintegration may require phosphorylation via maturation (or M-phase) promoting factor (MPF, see above), a cAMP-independent kinase, and it may be aided by swelling due to the movement of nuclear proteins into the nucleus. The process of nuclear envelope disintegration is reminiscent of the acrosomal reaction during which membrane-fusion results in **vesiculation** followed by the enlargement of holes.

In sea urchins, nuclear metamorphosis begins when the poreless nuclear envelope vesiculates and acquires pores. The pores enlarge, and the envelope disintegrates into vesicles or cisternae which disperse into the cytoplasm. The only survivors of the spermatozoon's nuclear envelope are the thickened indented portions lining the **postacrosomal space** at the apex and the **centriolar fossa** at the base (Fig. 9.14).

The mammalian spermatozoon's nuclear envelope begins disintegrating near its equator (Fig. 9.15). The postnuclear cap region also disintegrates, but the perforatorium is preserved temporarily beneath the phagocytic-like vesicle at the spermatozoon's apex. Actinomycin D prevents the disintegration of the nuclear envelope, indicating that the translation of newly transcribed mRNA is involved in the event.

The envelope's breakdown is usually the prelude to chromatin decondensation (Fig. 9.16), but exceptions can always be found. For example, in some nematodes and coccid insects, the spermatozoon's chromatin decondenses without changes in the nuclear envelope. Still, the remaining

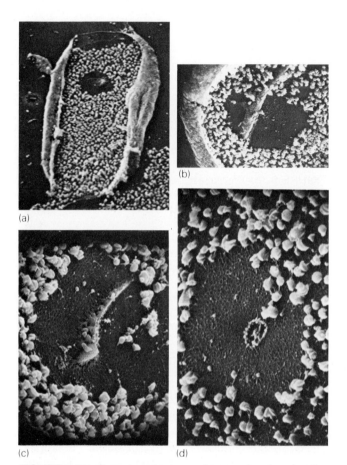

(a) (b)

(c) (d)

FIGURE 9.13. Scanning electron micrographs showing the inner surface of *Strongylocentrotus purpuratus* hulls isolated during sperm incorporation. (a) First sign of fertilization on the inside of the egg. Microvilli are already elongating over the spermatozoon's head (compare to Figs. 8.10 and 8.14c). Cortical vesicles (white "beads") disappear from an area, approximately 2 μm in diameter. ×600. (b) Small particles, about 235 nm in diameter, are seen beneath the fertilization cone. A membraneless spermatozoon with most of its tail already incorporated lies below the oolemma, but the cortical layer blocks the spermatozoon's entry to the egg's cytoplasm. ×1800. (c) Beginning at the apex, the internalized spermatozoon rotates out of the plane of the egg surface. ×4500. (d) A "scar" of small particles persists on the inside of the cortex following rotation and the spermatozoon's entry into the egg's cytoplasm. ×4500. From G. Schatten and D. Mazia, *Exp. Cell Res.,* 98:325 (1976), by permission of Academic Press and courtesy of G. Schatten.

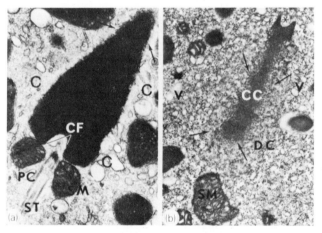

(a) (b)

FIGURE 9.14. Electron micrographs of thin sections through fertilized *Arbacia punctulata* egg showing fertilizing spermatozoon. (a) The nuclear envelope is almost completely vesiculated, except at the centriolar fossa (CF), and cisternae (C) or vesicles appear in the surrounding cytoplasm. Compare to (a) in Fig. 9.16. ST, sperm tail axonemal complex; PC, proximal centriole; SM, sperm mitochondrion; arrows, chromatin dispersing. (b) The thickened portion of nuclear envelope in the postacrosomal region (*) of the nucleus does not vesiculate or disappear. Coarse aggregates of chromatin (CC) give rise (arrows) to dispersed chromatin (DC). Vesicles (V) appear at the periphery of the dispersed chromatin. Compare to (b) in Fig. 9.16. ×11,000. SM, sperm mitochondrion. Reproduced from F. J. Longo and E. Anderson, *J. Cell Biol.,* 39:339 (1968), by copyright permission of the Rockefeller University Press and the authors.

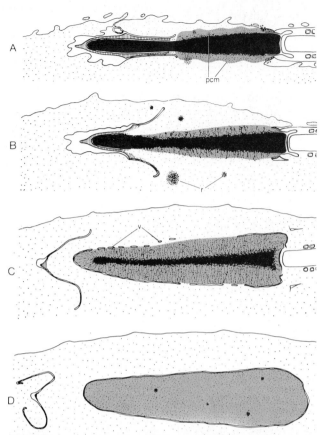

FIGURE 9.15. Metamorphosis of golden hamster sperm nucleus into male pronucleus upon incorporation into a zona-free egg. (A) The perforatorium is brought into the egg within a phagocytic-like vesicle, and post nuclear cap material (pcm) swells. The nuclear envelope disintegrates, and chromatin disperses, while aggregates of ribosome-like particles (r) appear. (B–D) As chromatin dispersal continues, vesicles (v) appear peripherally. They fuse into cisternae and merge into the pronuclear envelope. The remains of the perforatorium do not participate in nuclear envelope formation. From R. Yanagimachi. In S. J. Segal, R. Crozier, P. A Corfman, and P. G. Condliffe, eds., *The regulation of mammalian reproduction,* Charles C. Thomas, Publisher, Springfield, IL, 1973. Used by permission of publisher.

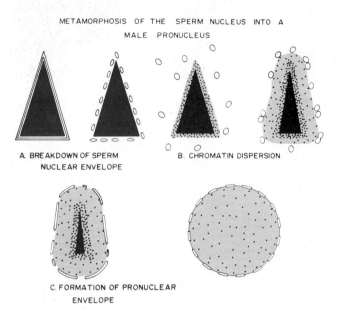

METAMORPHOSIS OF THE SPERM NUCLEUS INTO A MALE PRONUCLEUS

A. BREAKDOWN OF SPERM NUCLEAR ENVELOPE

B. CHROMATIN DISPERSION

C. FORMATION OF PRONUCLEAR ENVELOPE

FIGURE 9.16. Pronuclear development in the *Arbacia* zygote. (a) The nuclear envelope (double line) vesiculates to form cisternae (ellipses) which scatter in the cytoplasm. (b) The nucleus decondenses, and cisternae aggregate at the periphery of dispersed chromatin. (c) The cisternae fuse and coalesce into the nuclear envelope. The spheroidal male pronucleus contains dispersed chromatin and an envelope with pores. From F. J. Longo and M. Kunkle. In A. A. Moscona and A. Monroy, eds., *Current topics in developmental biology,* Vol. 12, *Fertilization, Academic Press,* Orlando, 1978. Courtesy of the authors.

processes leading to male pronuclear development may be similar.

Chromatin decondensation involves the dissociation of the sperm's basic and nonbasic nuclear proteins from DNA. The process may display no regional pattern; decondensation may proceed outward from the periphery to the center (Fig. 9.15) or from a region between the apex and base (arrows in Fig. 9.14).

Decondensation of chromatin, in mammals, begins even before the entire spermatozoon's head

is incorporated and is well advanced by 30–40 minutes, even before the nuclear envelope is entirely lost. The process seems to have an energy requirement, since, in the rabbit, it is inhibited by the metabolic inhibitor sodium azide.

Arginine-labeled sperm proteins disappear from the spermatozoon's nucleus as the female nucleus completes meiotic anaphase II. Antibodies to the basic proteins of sperm no longer detect the antigens after meiotic telophase II.

How the nucleoproteins are dissociated is uncertain. The mammalian sperm's basic, arginine-rich protamine peptides and sperm-specific proteins are heavily cross-linked by cysteine residues into a tough, keratin-like mass requiring equally tough reagents *in vitro* for penetration and dissolution. Disulfide reducing agents, such as dithiothreitol or 2-mercaptoethanol in combination with urea, detergents, or trypsin, decondense sperm heads (see Wolgemuth, 1983).

The naturally occurring disulfide reducing agent, reduced glutathione, may play the role of a decondensing factor *in vivo* (see Matsui and Clarke, 1979). The presence of glutathione reductase in eggs and fluctuations in the concentration of reduced glutathione during the cell cycle support this contention. Proteolysis does not seem to operate, however, and phosphorylation of sperm protamine–DNA complexes may also be involved in dispersing mammalian sperm chromatin (see Longo, 1985).

Chromatin activation occurs in mammals simultaneously with chromatin decondensation. Like DNA in other dividing cells, a variety of factors may promote transcription by interacting with DNA, especially during the replacement of nuclear proteins and the deposition of histone.

Basic and nonbasic proteins resembling those found in the female pronucleus replace those of the spermatozoon. Testis-specific H1 histones are replaced immediately as if their loss is a prerequisite for chromatin dispersion, while sperm histone H2B is replaced in a more or less parallel fashion with dispersion. The histones deposited on paternal (and maternal) DNA are synthesized both during meiotic maturation and after fertilization.

Protein synthesis occurs immediately following fertilization in all the mammalian species studied, and transcription proceeds simultaneously, at least in mouse pronuclei (see Wolgemuth, 1983). The role of pronuclear transcripts in concomitant translation is uncertain, although a microglobulin of specifically paternal origin occurs by the two-cell stage.

Mouse male pronuclei incorporate tritiated uridine into RNA, it would seem, under the direction of maternal cytoplasm. Even during the interval before pronuclear formation, decondensed mammalian paternal chromatin, containing neither basic chromosomal proteins nor nucleosomes, transcribes RNA, and aggregates of ribosome-like particles appear in the vicinity of the decondensing sperm nucleus (r in Fig. 9.15).

In the horse ascarid, *Parascaris equorum* (= *Ascaris megalocephala*), and the pig ascarid, *Ascaris lumbroicoides*, the female genome is tied up in meiosis at the time of fertilization and is unavailable for transcription (cf. class I, Table 7.7). The recently arrived male genome undergoes a massive burst of rRNA synthesis. As a result, a large fraction of the zygote's ribosomes is probably paternally derived rather than maternally derived (Kaulenas and Fairbairn, 1968).

In sea urchins and amphibians, histones and nuclear proteins are replaced in the absence of histone synthesis, indicating that maternal stores of basic proteins are responsible for the change. In sea urchins, actinomycin D (an inhibitor of transcription) reduces tritiated uridine incorporation (see

Longo and Kunkle, 1978), but the embryonic amphibian genome does not synthesize RNA. The zygote is even able to stop transcription of a transplanted nucleus (Gurdon and Brown, 1965).

Nuclear envelope formation for the male pronucleus resembles postmitotic nuclear envelope formation in other cells, except the process within the egg is not preceded by a telophase. A spermatozoon's nucleus intervenes between the last telophase and pronuclear formation.

The male pronucleus is filled with finely dispersed chromatin and is much less dense than its predecessor. In *Arbacia* (Fig. 9.17), the male pronucleus swells to a volume 20-fold larger than that of the spermatozoon's nucleus, while in the mussel, *Mytilus*, the increase in volume is as much as 115-fold. This swelling seems to be controlled by the cytoplasm, and, even after disruption, the cytosol from *Strongylocentrotus* eggs can cause swelling of isolated sperm nuclei *in vitro* (see Longo and Kunkle, 1978).

Similarly, as much as a 60-fold swelling occurs when somatic cell nuclei are injected into *Xenopus laevis* oocytes (Fig. 7.35) (Gurdon et al., 1976). This swelling seems to reflect an uptake of cytoplasmic proteins by the nuclei and underlies the interplay of nucleus and cytoplasm, or **nuclear–cytoplasmic interactions.**

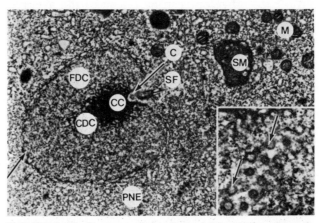

FIGURE 9.17. Electron micrograph of thin section through fertilized *Arbacia punctulata* egg and heart-shaped fertilizing spermatozoon during formation of pronuclear envelope (PNE). Nuclear pores (arrows in insert) appear for the first time. Compare to (c) in Fig. 9.16. × 9300. Insert, × 14,000. C, centriole; CC, condensed chromatin; CDC, coarsely dispersed chromatin; FDC, finely dispersed chromatin; SF, sperm flagellum; SM, sperm mitochondrion; M, mitochondria; white arrow, centriolar fossa; black arrow, apex sperm nucleus. Reproduced from F. J. Longo and E. Anderson, *J. Cell Biol.,* 39:339 (1968), by copyright permission of the Rockefeller University Press.

The process of male pronuclear formation seems to be controlled by egg cytoplasm, since even injected frozen-thawed and freeze-dried human sperm can develop into structures resembling male pronuclei. Some processes can be influenced by intracellular pH and inhibited or reversed by changing ions in the zygote's medium. Other processes seem to be finely controlled by cytoplasmic factors.

The pronuclear envelope begins to form when vesicles appear at the periphery of the finely dispersed sperm chromatin. The vesicles coalesce into cisternae and merge into the final pronuclear envelope. Decondensation may be a necessary condition for pronuclear envelope formation, since nuclear lamina proteins and membrane vesicles require chromatin for binding (Newport, 1987).

The timing of envelope formation differs among species. In sea urchins, vesicles appear even before all the chromatin is dispersed (Fig. 9.17). In *Spisula*, on the other hand, vesicles appear around the paternal chromatin only after it is totally dispersed. When complete, the pronuclear envelope contains an ample supply of pores (insert), and it swells profoundly.

The source of most of the material incorporated into the pronuclear envelope is uncertain. In the clam, *Barnea*, the pronuclear envelope seems to arise from the egg's endoplasmic reticulum (see Longo and Kunkle, 1978). In mammals, none of the pronuclear envelope can be traced to the spermatozoon's nuclear envelope or even to the abundant annulate lamellae usually present in the vicinity (Fig. 9.18). The materials for assembling the pro-

nuclear envelope seem to preexist, since inhibition of protein synthesis in the mammalian zygote to 80% of control levels does not prevent normal pronuclear envelope formation.

In the sea urchin, the pattern of chromatin dispersal draws the pronuclear envelope into a heart-shape, incorporating the nonvesiculated portions of the original spermatozoon's nucleus (Fig. 9.14). Approximately 20% of the pronuclear envelope can be traced morphologically to the spermatozoon's postacrosomal region and centriolar fossa (Fig. 9.17). These portions of envelope remain intact during nuclear decondensation and are incorporated intact into the male pronucleus. They are even incorporated into the zygote's nuclear envelope.

The vesicles that coalesce into the remaining 80% of the sea urchin's male pronuclear envelope are not as easily traced and may well originate in the egg's endoplasmic reticulum. Suggestive evidence favoring this hypothesis comes from an experiment employing centrifuged *Arbacia* eggs. Based on the premise that pronuclear envelopes would form faster where more endoplasmic reticulum were available and slower where less were available, the centrifugation technique devised by Ethel Browne Harvey (1940) was employed to concentrate endoplasmic reticulum in some regions of the egg and dilute it in others (Longo, 1976). The development of pronuclear envelopes was then monitored in the different regions.

Eggs centrifuged in a sucrose solution stretch and even break into halves or **merogons** (Gk. *meros* part + *gonos* offspring, hence, part of the egg, Fig. 9.19). The contents of the egg are redistributed as a function of their densities. Oil droplets and the nucleus move centripetally; yolk and pigment move centrifugally, while the endoplasmic reticulum and mitochondria are concentrated in the clear region between. Upon separation, the nucleated centripetal half contains abundant endoplasmic reticulum and mitochondria. The yolky enucleated centrifugal half contains a sparse population of mitochondria and negligible amounts of endoplasmic reticulum.

Eggs and nucleated halves stratified before germinal vesicle breakdown (GVBD) are capable of incorporating sperm and undergoing cleavage. The centripetal regions of stratified eggs and nucleated halves form complete male pronuclei in 4–6 minutes after fertilization. In contrast, the centrifugal region of stratified eggs forms complete male pronuclei at 12 minutes at the earliest, while nonnucleated halves fail to form male pronuclei even 20 minutes after fertilization.

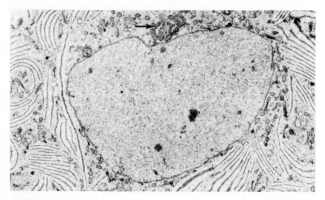

FIGURE 9.18. Electron micrograph of a thin section of a fertilized hamster oviductal egg showing male pronucleus among a plethora of annulate lamellae. Arrow shows the spermatozoon's tail. From R. Yanagimachi. In A. A. Moscona and A. Monroy, eds., *Current topics in developmental biology,* Vol. 12, *Fertilization,* Academic Press, Orlando, 1978. Used by permission.

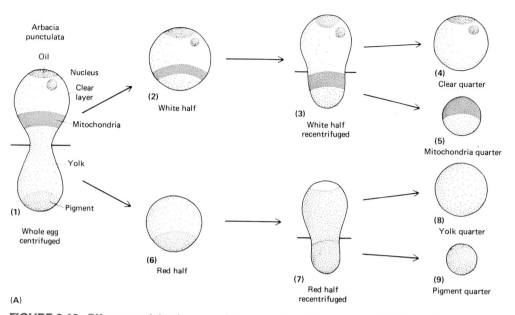

FIGURE 9.19. Effects on *Arbacia punctulata* eggs of centrifugation at 10,000 *g*. Gradually, the contents of the eggs stratify, and the egg elongates. By 3 minutes, some of the eggs have become dumbbell shaped, and some have broken into merogons. Further centrifugation breaks halves into smaller parts. From E. B. Harvey, *Biol. Bull.*, 79:160 (1940), by permission of the Marine Biological Laboratory, Woods Hole.

The time required to develop a male pronucleus is therefore inversely proportional to the abundance of endoplasmic reticulum in the vicinity of the fertilizing spermatozoon. Of course, other elements in eggs are also differentially distributed by centrifugation and may influence pronuclear envelope formation, either positively or negatively, but the results of this experiment raise the odds that the maternal endoplasmic reticulum provides the progenitor vesicles of the pronuclear envelope.

The control of pronuclear formation seems to involve maternal and ovarian factors. Oocytic organization does not seem to play a role, at least in sea urchins and amphibians, since decondensation occurs *in vitro* in homogenates or cytosols prepared from fertilized and unfertilized eggs.

Some control factors may accumulate in the course of the egg's differentiation, since immature eggs, though capable of being fertilized, are not necessarily capable of inducing changes in the spermatozoon's nucleus. Sperm nuclei incorporated into progressively more mature eggs exhibit greater degrees of pronuclear metamorphosis.

Immature rabbit and cow oocytes cultured *in vitro*, for example, are incapable of decondensing fertilizing sperm. Mammalian eggs become fully capable of decondensing sperm only after GVBD. A conspicuous exception occurs among canines (e.g., dogs and foxes) which ovulate primary oocytes before GVBD (Table 7.7). These oocytes are not only capable of being fertilized, but they remove the spermatozoon's nuclear envelope and decondense the nucleus as well.

Generally, vertebrate sperm nuclei in previtellogenic and vitellogenic oocytes become surrounded by cisternae and do not decondense. Meiotic sea urchin oocytes support the metamorphosis of sperm nuclei through the stage of nuclear envelope disappearance, and chromatin decondenses to a limited extent, but cytoplasmic vesicles do not accumulate or form a pronuclear envelope. Only mature **pronuclear** eggs support the full spectrum of events in male pronuclear metamorphosis (Longo and Anderson, 1968).

Early oocyte's cytoplasm may lack a **male pronucleus growth factor** (MPGF) required for nuclear swelling. In the rabbit (an induced ovulator), MPGF activity only appears in mature ovarian oocytes 7 hours after administration of human chorionic gonadotropin or coitus. Since the oocyte has been arrested in metaphase II for 2 hours at this time and has not been transcriptionally active for slightly longer, increasing amounts of MPGF activity do not depend on synthesis in oocytes. Moreover, MPGF activity does not appear in *in vitro* matured eggs, suggesting that MPGF originates in the follicle rather than in the oocyte itself (see Thibault, 1973).

Hamster oocytes removed from the ovary in the dictyate stage and rendered zona-free can fuse with sperm and incorporate them but do not vesiculate their nuclear envelopes or decondense their chromatin (Fig. 8.12) (Moore and Bedford, 1978). Later germinal vesicle stage oocytes can remove a spermatozoon's nuclear envelope but cannot decondense its nuclear chromatin. After prometaphase I, oocytes can partially decondense sperm, and after metaphase I, oocytes completely decondense sperm chromatin. A hypothetical factor dubbed **sperm chromatin-decondensing factor** (SCDF) (Yanagimachi, 1978) or **sperm nucleus-decondensing factor** (SNDF) (Yanagimachi, 1981) may change hamster oocytes' ability to decondense sperm in the course of nuclear maturation.

Fertilization itself does not seem to alter the amount of SNDF available, since sperm nuclei injected into unfertilized eggs decondense. On the other hand, decondensing sperm nuclei may degrade or use up SNDF, since the ability of eggs to decondense sperm nuclei disappears soon after fertilization. The decondensing activity of SNDF reappears in *Arbacia* and hamster zygotes "refertilized" before the first cleavage, but it disappears again after the prometaphase of the first cleavage division.

A hypothetical **sperm pronucleus development factor** (SPDF) (Yanagimachi, 1981) may also be involved in the development of the male pronuclear envelope. Activity attributed to this factor becomes available only upon fertilization, since injected hamster sperm form pronuclei only when introduced into previously fertilized eggs. Moreover, because zona-free hamster eggs that become heavily polyspermic form a maximum of only five male pronuclei, the amount of SPDF per egg seems to be finite. Possibly, sperm nuclei compete for the factor, or, possibly, sperm nuclei degrade the factor as they metamorphose into pronuclei.

Reactivating the Egg Nucleus

Some marine annelid eggs are naturally activated upon shedding into seawater and proceed with nuclear maturation before encountering sperm. Development, rather than activation, requires sperm in these cases, and preparations for cleavage are abandoned if the eggs are not fertilized.

Starfish eggs, for example, are ovulated and shed when their germinal vesicles are still intact. The eggs' maturation proceeds spontaneously, and the optimal time for fertilization is after GVBD but

before the formation of the first polar body (class I, Table 7.7). Eggs that are fertilized after having formed their first polar body frequently become polyspermic and develop pathologically (i.e., the eggs are "overripe").

More often, fertilization plays a role in activating the egg. For example, the eggs of echiuroids and ascarid (both also class I) are activated by fertilization and move rapidly and smoothly through both meiotic divisions while beginning their development.

Eggs activated by fertilization are generally suspended in maturation or caught in a meiotic arrest prior to fertilization. The eggs of many insects, ascidians, and annelids, among others, are arrested at metaphase I (class II) and are triggered by fertilization to extrude their first and second polar bodies, while most vertebrate eggs are arrested at metaphase II (class III, called the second meiotic arrest) and extrude their second polar bodies following fertilization.

The eggs of still other animals, notably sea urchins, are arrested at a **pronuclear** stage (class IV) rather than at a meiotic stage. Maturation has already occurred; both polar bodies have been expelled, and the female pronucleus along with a centrosome is present before ovulation and fertilization.

In each case, fertilization allows the egg to escape from its arrested state, but different sorts of nuclear processes and division are triggered. Possibly, a single factor released at fertilization operates differently in each situation. Free calcium, for example, which shoots up at fertilization, may inactivate a **cytostatic factor(s)** (see Matsui and Clarke, 1979), thereby freeing the oocyte's nucleus from a premeiotic, meiotic, or pronuclear arrest and allowing maturation or cleavage to proceed depending on the particular species.

The egg's second meiotic division giving rise to the egg nucleus is unusual in being grossly unequal. The nucleus retained by the egg becomes the female pronucleus directly upon completing meiosis, much the same way a typical nucleus forms after cell division (i.e., without the metamorphic changes experienced by the spermatozoon's nucleus forming the male pronucleus).

The position of the egg's centrosome determines the point of polar body extrusion. The centrosomes of starfish oocytes, for example, are plastered against the cortex (Fig. 9.20) and may even create a distinctive protrusion at the animal pole.

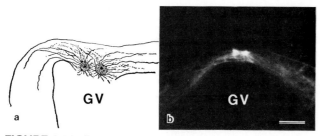

FIGURE 9.20. The premeiotic starfish oocyte centrosome and aster at the animal pole between the cortex and germinal vesicle (GV). (a) Drawing of centrosome and aster in *Asteria forbesii* based on E. B. Wilson and A. P. Mathews, *J. Morphol.,* 10:319(1895). (b) Dark-field immunofluorescent micrograph of *Piaster ochraceus* oocyte stained with fluorescein-labeled antitubulin antibody. Bar = 10 μm. From T. E. Schroeder, *Dev. Growth Differ.,* 27:311 (1985), by permission of the Japanese Society of Developmental Biologists and courtesy of T. E. Schroeder.

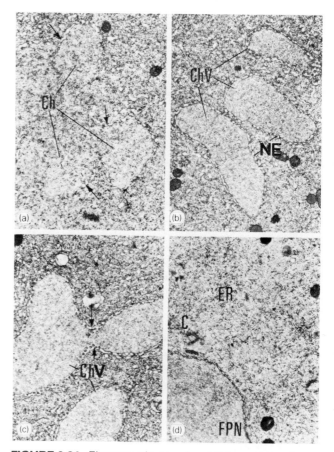

FIGURE 9.21. Electron micrographs of thin sections showing formation of female pronucleus (FPN) in fertilized eggs of *Spisula*. Vesicles (arrows) coalesce into cisternae around diffuse female chromosomes (Ch) forming chromosome-containing vesicles (ChV) and merging into a nuclear envelope (NE). C, centriole; ER, endoplasmic reticulum. From F. J. Longo. In C. B. Metz and A. Monroy, eds., *Biology of fertilization*, Vol. 3, *The fertilization response of the egg*, Academic Press, Orlando, 1985. Used by permission.

Meiotic spindles in this position yield greatly unequal divisions.

The position of the spindle can be altered experimentally and the distinction between the sizes of polar bodies and later blastomeres thereby erased. Eggs of the California echiuroid, *Urechis caupo*, for example, can produce blastomere-sized polar bodies when parthenogenically activated by prolonged treatment with ammonia.

The female pronucleus begins to form immediately once both polar bodies are extruded even when fertilization has yet to occur, as in the sea urchin. The chromosomes remaining in the egg after meiosis disperse, and vesicles appearing at their periphery fuse into cisternae and **chromosome-containing vesicles** or **karyomeres** (Fig. 9.21). The bilaminar structure of these vesicles already resembles a perforated nuclear envelope, and, as they fuse, an irregularly shaped female pronucleus takes form.

Subsequently, the structure swells into a sphere which acquires granule-containing, nucleolus-like bodies plastered randomly against the nuclear envelope (Fig. 9.22). These bodies, which stain for RNA, frequently deform the pronuclear envelope, swell the perinuclear space in their vicinity, and widen nuclear pores. Intranuclear annulate lamellae and crystalline structures may also appear.

The female pronucleus in *Arbacia* is larger than the male pronucleus (see below, Fig. 9.27). The difference may be due to the duration of pronuclear development, since, in the sea urchin, the female pronucleus forms before spawning when it completes meiosis, while the male pronucleus is only formed after fertilization. In other animals, oocytes are fertilized prior to the completion of polar body formation (cf. classes I–III, Table 7.7). The male and female pronuclei have approximately equal periods of time to develop, and they turn out approximately the same size (see Longo and Kunkle, 1978).

In mammals, the cytoplasmic factors controlling female pronuclear development do not seem to be the same as those controlling male pronuclear development. Blocking male pronuclear development, for example, via polyspermy in hamster eggs, fails to prevent female pronuclear development (see Yanagimachi, 1981).

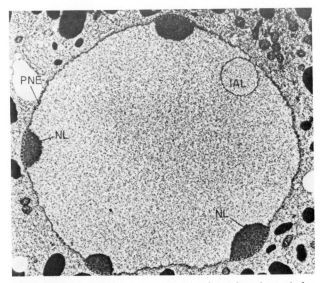

FIGURE 9.22. Electron micrograph of section through fertilized *Arbacia punctulata* egg showing female pronucleus. ×6900. IAL, intranuclear annulate lamella; NL, nucleolus-like structures; PNE, pronuclear envelope. Reproduced from F. J. Longo and E. Anderson, *J. Cell Biol.,* 39:339 (1968), by copyright permission of the Rockefeller University Press and courtesy of the authors.

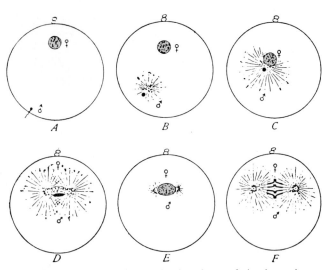

FIGURE 9.23. E. B. Wilson's drawings of the interphase behavior of asters and the formation of the first mitotic spindle in the sea urchin. Soon after the spermatozoon enters (a), an aster (monaster) forms (b) and leads the male pronucleus to the center of the zygote and the female pronucleus. Two centrosomes (c) appear and move to opposite sides of the zygote as the pronuclei merge and fuse (d). Microtubules between the centrosomes establish two asters with a streak between them (disaster). Aster rays extend far into the cytoplasm, but, just prior to the breakdown of the nuclear envelop, they disappear (e). The mitotic spindle (amphiaster) that succeeds the disaster conducts the chromosomes through the first cleavage division. From E. B. Wilson, *The cell in heredity and development,* 3rd ed., Macmillan, New York, 1928.

Aster Cycle

The **aster cycle** is an unusual interphase phenomenon accompanying pronuclear movement in the sea urchin (Fig. 9.23). Once at or near the center of the egg and lying between the male and female pronuclei, the spermatozoon's **monaster** either splits into two centrosomes or disintegrates and is replaced by two centrosomes. The astral rays surrounding the original centrosome separate between the centrosomes, forming an astral **streak** that develops into a **diaster** resembling a mitotic spindle (see Schroeder, 1985).

At the height of diaster development, the edge of the egg is crenated (i.e., wrinkled) as if pulled from within by astral rays. Then, just minutes before the first cleavage division commences with the breakdown of the zygotic nuclear envelope, the astral rays of the diaster disappear (Fig. 9.23e). The astral cycle comes to an end, and a mitotic cycle commences with a genuine mitotic spindle replacing the diaster. Reaching far into the yolky cytoplasm, the new astral rays make contact with the polar cortex and anchor the spindle.

First Round of Replication

The suppression of replication that characterizes the oocyte also characterizes the mature ovum. Suppression is not based on lack of substrates or enzymes. The unfertilized eggs of *Strongylocentrotus purpuratus*, for example, contain both the replicative enzymes for DNA synthesis and an adequate pool of deoxynucleotides to support replication in pronuclei, yet replication fails to occur.

In *X. laevis*, embryonic forms of DNA polymerase₁ and DNA polymerase₂ accumulate more than 100-fold during oocyte growth and double at maturation and ovulation to reach an all-time high five orders of magnitude above the amount found in single somatic cells (Benbow, 1985). Although the DNA polymerase may occur independently from its primase, in *X. laevis*, sea urchins, and chicks, the enzyme also occurs in association with a small polypeptide primase, and in *Drosophila melanogaster* only the associated form is found. Still, DNA synthesis is suppressed prior to fertilization.

DNA replication in fertilized eggs occurs in one of two ways. In **pronuclear replication,** the typical case, a unified zygotic nucleus is never formed, and replication occurs within the pronuclei. In **zygotic replication,** a unified zygotic nucleus is formed. Replication follows.

Sea urchins exemplify zygotic replication. Commencing 10–20 minutes after fertilization, zygotic replication occurs on the heels of pronuclear fusion. The two events nevertheless seem to be under separate controls, since unincorporated pronuclei are capable of DNA synthesis and perform it synchronously in polyspermic eggs (see Longo and Kunkle, 1978).

Pronuclear envelope formation and the anchorage of DNA loops are probably necessary conditions for replication as the corresponding processes are in ordinary mitosis (Newport, 1987). Pronuclear replication in *X. laevis* zygotes commences in both the male and female pronuclei about 5 minutes after the second polar body is extruded. The S period takes about 20 minutes (Graham, 1966). In mammals, pronuclear replication does not begin until some 4–10 hours after fertilization, and S is estimated at 3.5–5.5 hours (see Wolgemuth, 1983).

This rate of replication is remarkably fast compared to rates of replication in spermatogonia, a variety of somatic cells, or cells *in vitro* and seems to depend on a greater number of **initiation sites** or **origins** in fertilized eggs. In sea urchins, amphibians, and birds (but not in dipterans such as *D. melanogaster*), the rate of bidirectional replication along strands of DNA is not more rapid in cleaving embryos, but replicons are smaller and clustered closer together in cleaving cells compared to other cells (Callan, 1972).

Several hypotheses intended to account for the frequency of DNA initiation sites invoke a combination of positive control molecules that recognize the tertiary structure of initiation sites and negative control factors that inhibit replication. For example, **I-factor(s),** putative inducers of DNA synthesis, may exist in unfertilized *X. laevis* eggs, blastulas, and gastrulas, but not in oocytes, late embryos, or adult somatic cells (see Benbow, 1985).

Since unfertilized eggs can take up thymidine but not phosphorylate it, thymidine kinase seems to be blocked. The removal of this block *in vivo* (but not *in vitro*) 10 minutes after fertilization may unplug the entire replicative system. On the other hand, the formation of pronuclear envelopes may introduce initiation sites for DNA synthesis otherwise absent from the gametes' nuclear envelopes.

Other hypotheses account for replication by a change in the location or availability of previously **compartmentalized** material. Traffic in enzymes from the cytoplasm to nucleus is reduced in un-

fertilized eggs but flourishes after fertilization (see Monroy and Moscona, 1979). Pronuclear swelling, for example, may reflect the acquisition by the pronuclei of essential components for the alteration of chromatin and the synthesis of DNA.

Polyacrylamide gel profiles of sodium chloride extracts of combined male and female pronuclei do not differ especially from similar extracts of pure female pronuclei, indicating that the pronuclear protein components of chromatin are largely the same. These extracts differ greatly from those of sperm nuclei, however, indicating that a significant exchange of protein has occurred during the metamorphosis of the spermatozoon's nucleus into the male pronucleus (see Longo and Kunkle, 1978).

Proteins taken up by the pronuclear chromatin probably preexist in the egg cytoplasm, since, in the sea urchin, the rate of protein synthesis is still low in the freshly fertilized zygote. Moreover, sperm nuclei incubated in the cytosol removed from eggs take up large amounts of proteins, increasing their protein/DNA ratio from 0.11 to 3.7.

Among the oocytic proteins taken up by the nuclei are histones. Unlike typical somatic tissue, histone synthesis and DNA synthesis are not linearly coordinated in the zygote or in the cleaving embryo (Adamson and Woodland, 1974). The high rate of histone synthesis found, for example, in *X. laevis* oocytes during maturation, increases after fertilization but still does not keep up with the demand for histone posed by the exponential rate of DNA synthesis in the cleaving embryo. The deficit is made up by histones stored during oogenesis.

DNA polymerase α is also among the proteins taken up from the ovum's cytoplasm. The amount of enzyme decreases in the cytoplasm and increases in the nucleus. The activity of an 8.2 S DNA ligase also increases in the nucleus of *Ambystoma mexicanum* between 3 and 8 hours after fertilization, while activity of another 6 S DNA ligase decreases (see Benbow, 1985).

Amphimixis (Pronuclear Copulation or Karyogamy)

Prior to the first cleavage, pronuclei come together. The pronuclei become intimately associated but do not necessarily fuse. In any event, homologous chromosomes come to share a single mitotic spindle.

Microtubule assembly and disassembly are involved in bringing the male and female pronuclei together. Colcemid, which prevents microtubule assembly, and taxol, which inhibits microtubule depolymerization, arrest pronuclear movements when eggs are treated with either alone. While colcemid prevents aster formation, eggs fertilized in the presence of taxol form especially large monasters that keep the pronuclei at arm's length. The pronuclei in the taxol-treated eggs undergo a cycle of nuclear breakdown and reformation within the spermatozoon's monaster, but fusion and cleavage fail (Schatten et al., 1982). The force drawing nuclei together might be provided by the disassembly of microtubules in the aster, since colcemid reverses and overcomes the effects of taxol.

Two patterns or types of nuclear intimacy are distinguishable. In the **pronuclear fusion pattern** or **sea urchin type** of amphimixis (Fig. 9.24), fusiogenic pronuclear envelopes meet and fuse into a unified nuclear envelope around a single zygote nucleus. In the **chromosomal mingling pattern** or **Ascaris type** of amphimixis (Fig. 9.25), pronuclei do not fuse, although they may become extensively interdigitated (Fig. 9.26). Instead, as chromosomes condense, the pronuclear envelopes vesiculate and become penetrated by microtubules of the mitotic apparatus. Homologous chromosomes mix on the metaphase plate but are only incorporated into uni-

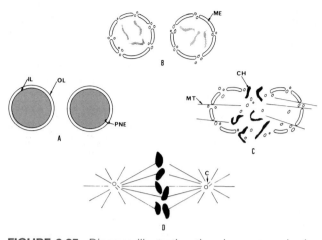

FIGURE 9.25. Diagram illustrating the chromosomal mingling pattern of amphimixis in the surf clam, *Spisula*. C, centriole; CH, condensed chromatin; IL, inner lamina of pronuclear envelope; ME, membranous structures formed during vesiculation; MT, microtubules; NE, nuclear envelope; OL, outer lamina of pronuclear envelope; PNE, pronuclear envelope. From F. J. Longo, *Biol. Reprod.,* 9:149 (1973), by permission of the Society for the Study of Reproduction and courtesy of the author.

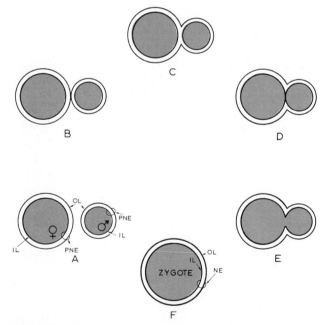

FIGURE 9.24. Diagram illustrating the pronuclear pattern of nuclear fusion in *Arbacia*. Fusion of outer and inner layers of the pronuclear envelopes creates internuclear bridges and a single zygotic nuclear envelope. IL, inner lamina of pronuclear envelope; NE, nuclear envelope; OL, outer lamina of pronuclear envelope; PNE, pronuclear envelope. From F. J. Longo, *Biol. Reprod.,* 9:149 (1973), by permission of the Society for the Study of Reproduction and courtesy of the author.

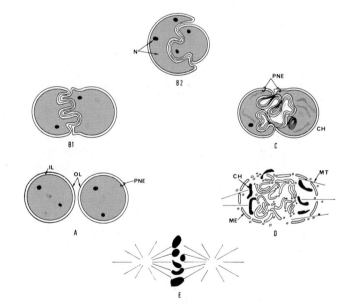

FIGURE 9.26. Diagram illustrating the interdigitation of nuclear processes as a variation of the chromosomal mingling pattern of amphimixis in the rabbit. C, centriole; CH, condensed chromatin; IL, inner lamina of pronuclear envelope; ME, membranous structures formed during vesiculation; MT, microtubules; NE, nuclear envelope; OL, outer lamina of pronuclear envelope; PNE, pronuclear envelope. From F. J. Longo, *Biol. Reprod.,* 9:149 (1973), by permission of the Society for the Study of Reproduction and courtesy of the author.

fied bodies in telophase nuclei after the first cleavage.

The unbridgeable difference between the two patterns is the formation of a zygotic nucleus in the pronuclear fusion pattern and the absence of such a nucleus in the chromosomal mingling pattern. Otherwise, the differences may not be especially great. Taking into account the timing of nuclear events in fertilized eggs, the two types of amphimixis appear to represent extremes in a spectrum of possibilities rather than antitheses (see Longo, 1973, 1987). As a rule, the later in nuclear maturation eggs are fertilized, and the later the male pronucleus forms, the more profoundly male and female pronuclei become associated prior to the first cleavage.

The pronuclear fusion pattern (Fig. 9.24) occurs principally in sea urchins where female pronuclear development precedes male pronuclear development (cf. class IV, Table 7.7), and male pronuclear life tends to be brief (e.g., 10–15 minutes in Arbacia). The typically larger female pronucleus flattens on the side facing the male pronucleus, and **nuclear projections** or **blebs** resembling microvilli emerge (Fig. 9.27). Upon making contact, the inner and outer membranes of one projection fuse with the corresponding membranes of another projection to form **internuclear bridges** (Fig. 9.28).

As bridges broaden, the pronuclei coalesce into a single **zygote nucleus** (Fig. 9.29). The chromatin within the pronuclei is not exposed to cytoplasm during fusion, and the zygote nucleus persists for the duration of an **interphase pause** prior to prophase of the first cleavage division.

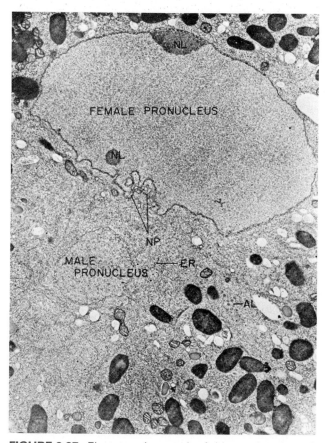

FIGURE 9.27. Electron micrograph of thin section through fertilized *Arbacia punctulata* egg showing female and male pronuclei. The female pronucleus has flattened on the side facing the male pronucleus and formed microvilli-like nuclear projections (NP). Endoplasmic reticulum (ER) surrounding the male pronucleus merges with cytoplasm surrounding the female pronucleus. ×6900. AL, annulate lamellae; NL, nucleolus-like body. Reproduced from F. J. Longo and E. Anderson, *J. Cell Biol.,* 39:339(1968), by copyright permission of the Rockefeller University Press and courtesy of the authors.

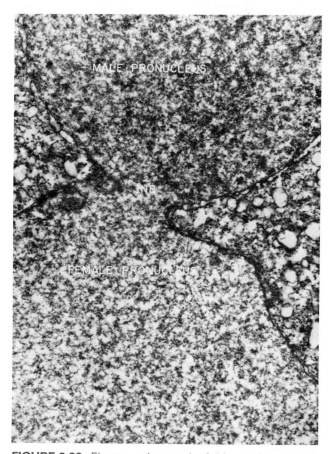

FIGURE 9.28. Electron micrograph of thin section through fertilized *Arbacia punctulata* egg showing fusing male and female pronuclei. The inner and outer membranes of the pronuclear envelopes have fused, forming an internuclear bridge (INB). ×14,000. Reproduced from F. J. Longo and E. Anderson, *J. Cell Biol.,* 39:339 (1968), by copyright permission of the Rockefeller University Press and courtesy of the authors.

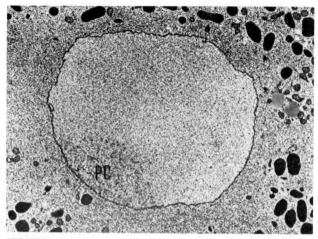

FIGURE 9.29. Electron micrograph of thin sections through fertilized *Arbacia punctulata* egg showing final fusion of male and female pronuclei into a zygotic nucleus. Paternal chromatin (PC) is initially distinguished by its density. ×5700. Reproduced from F. J. Longo and E. Anderson, *J. Cell Biol.,* 39:339(1968), by copyright permission of the Rockefeller University Press and courtesy of the authors.

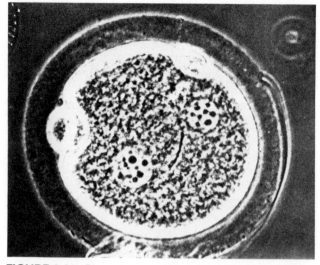

FIGURE 9.30. Phase contrast light micrograph of fertilized hamster egg with two pronuclei. The two polar bodies lie beneath the zona pellucida, and the tail of the fertilizing spermatozoon lies between the pronuclei. ×600. From B. D. Bavister. In L. Mastrioianni, Jr. and F. D. Biggers, eds., *Fertilization and embryonic development in vitro,* Plenum Press, New York, 1981. Used by permission.

The chromosomal mingling pattern of amphimixis occurs in the parasitic round worms, Ascaris (for which E. B. Wilson originally gave the pattern the name, Ascaris type), and, as it turns out, in most other animals. Species exhibiting the chromosomal mingling pattern typically have eggs that are fertilized before completing meiosis (classes I–III), and they frequently produce male and female pronuclei of nearly the same size (e.g., Fig. 9.30).

The surf clam, *Spisula,* represents the extreme version of chromosomal mingling. Its eggs are fertilized while still in the germinal vesicle stage (class I), and produce equal pronuclei simultaneously. These pronuclei hardly have any interactions before vesiculation leads to the disintegration of their envelopes (Fig. 9.25).

The mussel, *Mytilus,* is less extreme. Its eggs are fertilized while arrested in the first meiotic metaphase (class II). They form pronuclei that meet in the cytoplasm beneath the polar bodies and form intimate if not highly convoluted associations.

Closest to the pronuclear fusion pattern, typical vertebrate eggs, such as the rabbit, are fertilized while arrested in metaphase of the second meiotic metaphase (class III). Pronuclei bleb extensively and become highly convoluted (Fig. 9.31), indented, and interdigitated throughout the region of pronuclear association. "Spot welds" may occur between the pronuclear envelopes (Longo, 1987), but fusion does not occur, and, when each pronuclear envelope vesiculates and breaks down (Fig.

9.32), paired cisternae of pronuclear ends (PE) are found amidst the condensed chromosomes (CH). Microtubules soon attach themselves to the kinetochores of the dense mitotic chromosomes, and cleavage commences without a zygotic nucleus having ever truly formed.

Species Specificity and Intracellular Blocks to Polyspermy

Eggs fertilized by sperm of a different species or by more than one spermatozoon may fall back on amphimixis to rescue development from the deleterious effects of hybridization and pathological polyspermy. Amphimixis is generally only weakly species specific, however, and its ability to provide an intracellular block to polyspermy depends on the pattern of amphimixis.

Species specificity is conferred on gynogenic species by amphimixis. Sperm from the fertilizing species are excluded from a future role in development by the complete failure of amphimixis. Otherwise, amphimixis is liable to result in hybrids.

While most hybrids fail at some point in development for a variety of reasons, interspecific pronuclei from related species are often able to complete amphimixis. Associations that produce viable hybrids (e.g., the mule) may not run into major chromosomal problems until meiosis. The

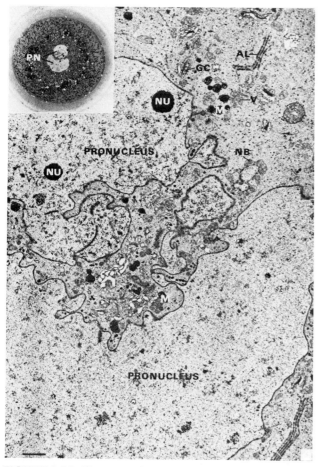

FIGURE 9.31. Electron micrograph of section through fertilized rabbit egg showing pronuclei with nuclear blebs. ×9700. AL, annulate lamellae (cytoplasmic); GD, Golgi complex; M, mitochondria; NB, nuclear blebs; NU, nucleolus-like body; V, vesicles. Bar = 1 μm. ×5800. (Insert) Light micrograph of section showing juxtaposition of pronuclei (PN). ×240 From F. J Longo and E. Anderson, *J. Ultrastruct. Res.,* 29:86 (1969), by permission of Academic Press and courtesy of the authors.

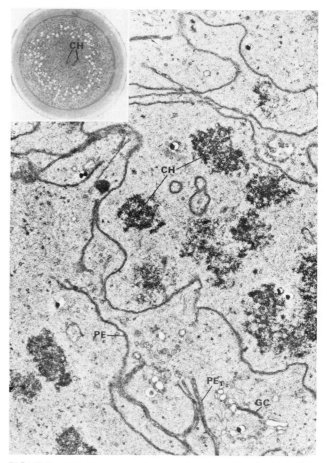

FIGURE 9.32. Electron micrograph of section through fertilized rabbit egg showing breakdown of pronuclear envelopes with emergence of condensed chromosomes. ×15,250. CH, condensing chromatin; GC, Golgi complex; PE and PE$_1$, paired ends of pronuclear envelope. ×15,250. (Insert) Light micrograph of similar section through zygote at prometaphase showing condensing chromatin (CH). ×240. From F. J. Longo and E Anderson, *J. Ultrastruct. Res.,* 29:86 (1969), by permission of Academic Press and courtesy of the authors.

hybrids may be sterile, but their development is otherwise fruitful.

Protection against pathological polyspermy is not offered by the pronuclear fusion pattern of amphimixis seen in sea urchins. When more than one spermatozoon has overridden the prophylactic blocks to polyspermy, fusiogenic pronuclear envelopes are likely to merge into a single polyploid zygotic nucleus. Sea urchin embryos are incapable of handling the excess chromosomes physiologically, and the production of aneuploid blastomeres results in severely abnormal embryos.

Amphimixis of the chromosomal mingling pattern potentially provides an **intracellular block to polyspermy** and a measure of protection against the effects of aneuploidy after polyspermic fertil-

ization. In the rare instances in which a normally monospermic species undergoes a polyspermic fertilization, chromosomal mingling may limit the chromosomes occupying the metaphase plate to those of one male pronucleus and the resident female pronucleus (see Austin, 1978). Only chromosomes on this metaphase plate are then incorporated into the nuclei of the first cleavage blastomeres, effectively eliminating the chromosomes of the excess male pronuclei from a future role in development.

The chromosomal mingling pattern of amphimixis also provides the key to normal development in species exhibiting **physiological polyspermy.** Polyspermy normally occurs in megalecithal cephalopods and gastropods, insects, elasmobranchs, reptiles, birds, and even most mesolecithal urodeles. The eggs of polyspermic species store more materials than eggs of their monospermic relatives, and the number of sperm normally fusing with eggs is correlated with egg size. Fertilization tends to be internal, and sperm are of the modified type. These relationships are not absolute, however. For example, monospermic anuran eggs store material and are about the same size as polyspermic urodele eggs.

Polyspermy Versus Monospermy in Amphibians

Urodeles are typically physiologically polyspermic. Syngamy takes place in both animal and vegetal hemispheres. The eggs do not have a fast block to polyspermy and, in the absence of cortical vesicles, do not produce a slow block. A **capsular chamber** is formed outside the egg's vitelline envelope and within the jelly coat. The egg rotates within this chamber and brings the pigmented animal cap uppermost, but the chamber does not block entry of supernumerary sperm.

Some constraints still operate, and even physiological polyspermy has limits. Only 3–15 sperm enter urodele eggs, and they do so within 5 minutes of insemination. As in anurans, swelling of the jelly coat inhibits sperm penetration.

Pathological development results when the number of supernumerary sperm exceeds the normal upper limit. For example, 10 sperm cause abnormal cleavage and development in *Triturus palmatus*, but small numbers of excess sperm are handled easily.

While each male pronucleus forms a monaster in the company of its own centriole and aster, the female pronucleus is drawn into the orbit of the monaster of only one male pronucleus. This pronucleus thereby becomes the **principal**, while the others become the **accessory** male pronuclei. Presumably, because asters are relatively solid, the accessory male pronuclei are incapable of approaching either each other or the principal pronucleus. Only the paired female and principal male pronuclei go on to form a mitotic spindle (Fig. 9.33), while the unsuccessful accessory pronuclei with their monasters disintegrate without leaving heirs.

Unlike urodeles, frogs develop abnormally in the event of polyspermy in any degree. A single principal male pronucleus associates with the resident female pronucleus, but accessory male pronuclei form spindles rather than mere monasters and divide along with the paired pronuclei. Division of the accessory male pronuclei (up to 10) results in **multipolar cleavage** and gives rise to bi-

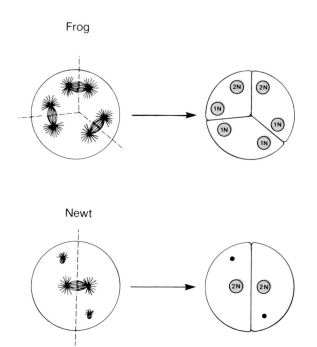

FIGURE 9.33. Drawing illustrates the difference between the reaction to multiple fertilizations of normally monospermic frog egg and normally polyspermic newt egg. In the frog, each spermatozoon directs the formation of a spindle. Cleavage is multipolar and results in binucleate cells with abnormal numbers of chromosomes. In the newt, a complete spindle is established only by the zygotic nucleus combining male and female pronuclei. The other sperm nuclei set up abortive monasters and disintegrate. Only a single cleavage furrow forms and the blastomeres have normal diploid nuclei. From R. P. Elinson, *Int. Rev. Cytol.,* 101:59 (1986), by permission of Academic Press, Orlando and the author.

nuclear cells. In the case of fertilization by three sperm, for example, mitotic division results in the formation of three binucleate cells—one with two haploid nuclei and two with a haploid and a diploid nucleus (see Elinson, 1986).

In subsequent cleavages, the haploid and diploid nuclei continue to form spindles and consequently wind up in blastomeres with single nuclei. The blastula becomes a mosaic of haploid and diploid nuclei with an undetermined number of aneuploid nuclei. Many embryos survive until gastrulation and some longer, but development is ultimately arrested.

Possibly, urodelan polyspermy represents a primitive condition adapted to providing the developing egg with additional sources of material. **Spermic provisioning** provides some eggs with mitochondria, for example. Sperm may also add centrioles or centriolar material, and their tails may supply microtubular proteins. Moreover, in polyspermic species such as birds, excess sperm may provide transient centers of cellularization in peripheral cytoplasm.

Alternatively, polyspermy, especially in large eggs, may result from difficulty in sending a signal throughout the cortex and closing the doors to additional sperm over a storehouse of yolk. The intra-cellular block to polyspermy based on the suppression of amphiaster formation on top of the chromosomal mingling pattern of amphimixis may principally represent an accommodation to polyspermy rather than an adaptation to spermic provisioning.

PART 4 SUMMARY AND HIGHLIGHTS

Fertilization represents the successful conclusion to the life of the sperm and egg. As extraordinary as the events of fertilization are and enormous in their consequences, many of the processes are like those in ordinary cells.

The movement of sperm to the egg is aided in mammals by the loosening of bonds in the oocyte–cumulus complex and corona radiata as a function of pH and possibly of hyaluronidase. In different species, only sperm with a relatively intact acrosome or only acrosome-reacted sperm are capable of binding to eggs. Sperm motility presumably aids sperm moving through the zona pellucida or the micropyle of a chorion but motility is not required for fusion, and tails may cease to gyrate during incorporation.

Materials diffusing from egg coats may induce the acrosomal reaction of the sperm. The reaction generally involves the vesiculation of the acrosomal cap, or random fusion of the spermatozoon's plasma membrane with the outer acrosomal membrane and exocytosis of the acrosomal contents. Fusion may be concentrated at a rim of dehiscence and an acrosomal ghost may be left behind as the spermatozoon makes its way to the egg. The release of acrosomal filaments may involve the polymerization of actin precursors, an actomere anchored to the spermatozoon's nucleus, or the ejection of a preformed coil.

Glycoprotein receptor sites in egg coats such as ZP3 of the mammalian zona pellucida and lectin-like binding protein on the spermatozoon's surface such as sea urchin bindin allow sperm and egg membranes to adhere with species specificity. Specific galactose transferase on the acrosomal surface and proacrosin, a precursor of acrosin, released by the acrosome reaction and transferred to the spermatozoon's tip may play a role in attachment by recognizing carbohydrate substrates in egg glycoproteins.

The apex of the spermatozoon or perforatorium spearheads the penetration of the egg's protective envelopes, generally after being coated with proteolytic acrosin or lysin released from the acrosome. The mammalian zona pellucida is pierced by the action of sperm-bound enzymes. Where the egg is equipped with an impermeable chorion, sperm reach the egg through one or more micropyles.

The egg is instrumental in taking up the sperm. The egg sends out microvilli and a fertilization cone from the egg's surface engulfs the spermatozoon's head and most of the spermatozoon including part or all of its tail.

Syngamy, defined as the permanent fusion of gametes, requires overcoming electrostatic repulsion between membranes. Like fusiogenic portions of the sperm, the egg provides microvilli with a radius of curvature commensurate with reduced electrostatic repulsion.

The fusiogenic portion at the apex of most sperm may be covered with bindin, possibly with the capacity to insert itself into the phospholipid leaflets of the oolemma. In eutherian mammals the fusiogenic portion is in the equatorial segment girdling the head, and the spermatozoon is incorporated sideways. The perforatorium at the apex of the mammalian spermatozoon is the last part of the head to enter the egg.

Activation of the egg has explosive qualities, especially the exocytosis of cortical vesicles and an autocatalytically driven massive release of calcium. The respiratory burst accompanying fertilization in some invertebrates may be due to the uptake of oxygen in hydrogen peroxide production.

Other features of activation are processive or interactive. According to the ionic hypothesis, free calcium is responsible for most of the early events in the activation reaction including the oxygen burst (in echinoderms) and an increase in NAD kinase. Sodium–proton exchange or the release of "fertilization acid" raises the zygote's pH and may cause the rise in protein synthesis accompanying fertilization. Ionophores, pH, and ammonia may activate eggs and mimic fertilization.

Through unknown mechanisms, oolemma-bound G protein is activated upon fertilization, which in turn activates phosphoinositide-specific phospholipase C. Consequently, inositol triphosphate enters the cytosol and, reaching the cortical elements of the endoplasmic reticulum, causes Ca^{2+} release. A perivitelline space forms as a fertilization envelope is raised and hyaline and other products of exocytosis accumulate. Hydrogen peroxide hardens the fertilization membrane by oxidizing sulfhydryl groups.

Upon encountering sperm, permeability changes on the surface allow an influx of Na^+, which produces a small stepwise depolarization in membrane potential followed by a relatively prolonged depolarization or fertilization potential. The stepwise depolarizations may represent permissive voltages required for fusion of sperm and egg, but the positive potential of the fertilization potential prevents the entry of sperm and is sometimes equated with a fast block to polyspermy.

The positive membrane potential of several invertebrate zygotes are incompatible with sperm fusion. Following exocytosis, excess sperm adhering to the primary egg membranes of many animals, including mammals, detach, and a slow or permanent prophylactic block to polyspermy is established.

Phosphoinositide-mediated signaling utilizes the degradation products of phospholipids released in membranes as intracellular (secondary) messengers and regulators of proteins. Lipid-regulated protein kinase C transduces diacylglycerol secondary messengers to diverse activities including the activation of Na^+–H^+ exchange. As a result, internal pH rises and the membrane potential is restored to about its prefertilization resting potential. Sphingosine and lysosphingolipids are potent reversible inhibitors of protein kinase C.

Cytoplasmic streaming results in ooplasmic segregation in marine invertebrates and in the formation of the germinal disk in fish. Similarly, cortical contraction in amphibians is involved in determining the position of the first cleavage furrow and establishing the midsagittal plane. The position of the gray crescent in amphibians is pro-

gressively determined in the oocyte, by the point of sperm entry, and by regional thinning of the cortex. The large monaster formed by the male pronucleus during its migration to the female pronucleus may also influence the formation of the gray crescent.

Large yolk platelets brought into contact with the cortex as a result of cortical contraction and rotation may be instrumental in dorsalizing the portion of the egg normally in the vicinity of the gray crescent. The experimental displacement of these platelets results in caudally fused double embryos.

Calcium may neutralize inhibitory factors and set the egg on a course to postovulatory maturation with the extrusion of polar bodies, but calcium-independent and cAMP-independent mechanisms promote maturation and operate during cleavage. An M-phase or maturation promoting factor is able to overcome cAMP inhibition of maturation and initiate cleavage. One and possibly more components of this factor seem to be widespread among eukaryotes and associated in different ways with the control of the cell cycle.

After incorporation, the sperm may rotate and follow a path leading to the female pronucleus. Sea urchins exhibit an aster cycle, forming a monaster and diaster followed by the cleavage amphiaster. The presence of a centrosome is required for division and its position rather than the state of the nucleus may determine the size of cells produced by division.

The spermatozoon's nuclear envelope vesiculates; basic protamines complexed with the spermatozoon's DNA disappear, and its chromatin decondenses and disperses. The male pronucleus forms when vesicles and cisternae originating in the maternal endoplasmic reticulum surround the decondensed chromatin and fuse into a pronuclear membrane. The ability of egg cytoplasm to induce and support these changes is acquired during germinal vesicle breakdown and activation. The female pronucleus forms after completing its maturation by the merging of vesicles surrounding its chromosomes. Replication begins in the separate pronuclei.

The eggs of most animals are arrested at the time of fertilization, either prior to meiosis, during meiosis, or in a postmeiotic pronuclear phase. Eggs fertilized early in meiosis tend to have less pronuclear interactions exemplified by the chromosomal mingling pattern of starfish. Eggs fertilized later tend to have more intimate pronuclear interactions culminating in pronuclear fusion in sea urchins.

In the chromosomal mingling pattern, a membrane-bound zygote nucleus does not form. The pronuclei may meet and form interdigitating processes, but the pronuclear chromatin condenses into chromosomes, and the pronuclear envelopes vesiculate and disappear without fusing. The chromosomes from both pronuclei mingle on the shared metaphase plate before being incorporated into the nuclei of blastomeres. In the pronuclear fusion pattern, the male and female pronuclei meet and fuse to form a zygote nucleus with common nuclear envelope.

The chromosomal mingling pattern operates in gynogenic species and may provide a degree of protection against the pathological effects of polyspermy and hybridization in normally polyspermic species. Urodeles accommodate to polyspermy by suppressing the formation of amphiasters in supernumerary male pronuclei. Anurans lack this adaptation and develop pathologically when fertilized by excess sperm.

Fertilization is a process of membrane fusion: from the acrosomal reaction of sperm, the exocytosis of cortical granules in eggs, the meeting of fusiogenic membranes between sperm and egg, and coalescence of vesicles in the process of forming nuclear envelopes. All these reactions seem to involve the same mechanisms during fertilization as they do elsewhere, including calcium influx and sodium–hydrogen exchange.

A close look at early development does not identify a single point when an embryo begins. In animals, the embryo's cytoplasm and nucleus have different beginnings, since in most species the spermatozoon and egg's plasma membranes fuse while the egg's nucleus is still in meiosis. With few exceptions (e.g., sea urchins) the embryo's nucleus does not begin in the zygote, since, even after meiosis is complete, the zygote's haploid pronuclei do not merge into a single nucleus. The embryo's nuclei do not form until cleavage.

The best an embryologist can do when asked when an embryo begins is to acknowledge the failure of embryology to provide a discrete answer. Characterizing the mechanisms and controls of fertilization, not belaboring beginnings, is the job of embryologists. Laboring in their own vineyards, embryologists contribute to growing knowledge, not to confounding controversy.

PART 4 QUESTIONS FOR DISCUSSION

1. What is the role of sperm motility in syngamy? Compare and contrast membrane fusion during the acrosomal response, the cortical response, and amphimixis. Contrast exocytosis in the acrosome and in cortical vesicles.

2. Compare and contrast receptors and binding sites involved in the attachment of sperm to egg envelopes. Describe fertilizin and antifertilizin. Distinguish between acrosomal induction, priming, and capacitation.

3. What is mouse ZP3 and how does it work? Compare and contrast the condition of the acrosome and the induction of the acrosomal reaction in sea urchins and mammals.

4. Describe the acrosomal reaction, vesiculation, and the ejection of the acrosomal filament. What is the rim of dehiscence and how does it contrast with vesiculation of the acrosome?

   Compare and contrast the fusiogenic portions of most sperm with eutherian sperm.

5. What do bindin, acrosine, and sperm lysins have in common? Distinguish between sperm binding sites and bindin. How does the egg contribute to its own fertilization? What is the fertilization cone?

6. What is the role of calcium in the acrosomal response? In the cortical response? In other features of the activation reaction? What role is calmodulin thought to play in membrane fusion?

7. What effects does cAMP have on maturation? On the initiation of cleavage? What is the M-phase (maturation) promoting factor, and what evidence implicates it in the control of the eukaryotic cell cycle?

8. What changes occur in oxygen consumption following fertilization? In protein synthesis? In transcription? In intracellular pH? What is the acid of fertilization and how is pH raised in the zygote?

9. What is a resting potential? An action potential? A fertilization potential? Discuss explosive models and processive models for activation apropos of early or phospholipase C related events and late or protein kinase C related events.

10. What are cytoplasmic streaming and ooplasmic segregation? Describe these processes in ascidians and compare them to polar lobe formation in mollusks. How is the position of the gray crescent determined in amphibians? Compare and contrast the crescents of ascidians to the gray crescent of amphibians.

11. Describe and distinguish between prophylactic fast, slow, surface, and intracellular blocks to polyspermy. How might the removal or alteration of receptors provide an irreversible block to polyspermy? What is spermic provisioning?

12. Describe pronuclear fusion and chromosomal mingling patterns of amphimixis. How might chromosomal mingling preadapt a zygote for an intracellular block to polyspermy? In your opinion, is the urodelan or the anuran block to the pathological effects of polyspermy the more primitive? Why?

PART 4 *RECOMMENDED READING*

Bedford, J. M. 1982. Fertilization. In C. R. Austin and R. V. Short, eds., *Reproduction in mammals:* Book 1, *Germ cells and fertilization,* 2nd ed. Cambridge University Press, Cambridge, pp. 128–163.

Chia, Fu-S. and A. H. Whiteley, eds., 1975. *Developmental biology of the echinoderms, Am. Zool.,* 15: Sec. I, 493–565.

Greenberg, A. H., ed., 1987. *Invertebrate models: Cell receptors and cell communication.* Karger, Basel.

Gwatkin, R. B. L., 1977. *Fertilization mechanisms in man and mammals.* Plenum Press, New York.

Hartmann, J. F., ed., 1983. *Mechanism and control of animal fertilization.* Academic Press, New York.

Longo, F. J., 1987. *Fertilization.* Chapman and Hall, New York.

Mastroianni, L. and J. D. Biggers, eds., 1981. *Fertilization and embryonic development in vitro.* Plenum Press, New York.

Metz, C. B. and A. Monroy, eds., 1985. *Biology of fertilization,* Vol. 3, *The fertilization response of the egg.* Academic Press, Orlando.

Monroy, A. and A. A. Moscona, eds., 1978. *Current topics in developmental biology,* Vol. 12, *Fertilization.* Academic Press, Orlando.

Schatten, G., ed., 1988. *Cell biology of fertilization,* 2nd ed. Academic Press, Orlando.

CLEAVAGE AND BLASTOMERES

*W*atching the cleavage of a marine invertebrate egg through the eye pieces of a stereoscope is a breathtaking experience. Although cleavage furrows slice into the egg at a rate just below the limits of visual cognition (about 6° per minute, the same rate as the minute hand on a watch or the sun at sunset), the effect of the movement is unmistakable.

On a scale of developmental potential, blastomeres lie closer to the zygote than to ordinary body cells, and, like the zygote, blastomeres come and go rapidly in the life of the individual, but, with the formation of blastomeres, the embryo begins its trek toward multicellularity. Cells are produced and remodeled into entities capable of constructing a multicellular organism.

Part 5 traces the early development of blastomeres or nuclear–cytoplasmic units representing cells in syncytial embryos. The chapters lay out the rules whereby cleavage creates patterns of blastomeres and explore the consequences of cleavage for cell lineage and for embryonic determination.

10

GENERAL AND UNIQUE FEATURES OF CLEAVAGE

$\mathcal{C}$leavage is different things to different embryologists. Classically, cleavage is defined morphologically, referring to the division of the zygote and early blastomeres. Cleavage is also defined cytologically by the division of nuclei in the embryo, and molecularly by differential translation without concurrent transcription. The cleaving embryo is the beneficiary of prior synthesis of both RNA and protein and many cell functions proceed with the support of stored products often identified as maternal products.

In addition, **cellularization,** the production of cells around preformed units, may be included as a type of cleavage when it occurs in early embryos. In embryos from birds to insects, karyokinesis occurs initially without cytokinesis or after only partial cytokinesis, and a **syncytial blastoderm** is formed with nuclei suspended in a common cytoplasm. Each nucleus and its halo of associated cytoplasm is called an **energid.** Later, plasma membrane completely surrounds and isolates energids. The blastoderm is then **cellularized.**

MORPHOLOGICAL AND PHYSIOLOGICAL ASPECTS OF CLEAVAGE

During the course of cleavage there is a stepwise change in the ratio between active nuclear material and the cytoplasm, and the change proceeds in favor of the chromosomal material.

Alfred Kühn (1971, p. 155)

Producing cells is only the most obvious morphological consequence of cleavage. The cells are also increasingly different from the zygote and more nearly like adult cells in size and physiological qualities. Furthermore, the existence of boundaries within the embryo provides new opportunities for segregating materials and promoting differentiation.

Stages of Cleavage

Cleavage, in the classical sense, is frequently subdivided into "early" and "late" stages. Early cleavage gives rise to a **morula** (Lat. *morum* a mulberry), while late cleavage gives rise to a blastula. As the terms imply, cleavage is not the same from beginning to end, and cleavage as a process seems to change in the course of time. Blastomeres probably do not have the same physiology in late cleavage as in early cleavage.

The line separating early and late stages cannot be drawn sharply. The transition from early to late cleavage is defined in different species by the loss of synchrony in division, the onset of morphogenesis, or the origin of local as opposed to global patterns of cleavage furrows.

Division may be synchronous throughout an embryo during early cleavage and asynchronous in late cleavage. Alternatively, cleavage divisions may be synchronous in some blastomeres while asynchronous in others from the beginning.

A shift away from proliferation toward morphogenesis sometimes occurs at convenient times for separating early and late cleavage. On the other hand, morphogenesis may start earlier in one part of the embryo than in another, and **cryptic** or **hidden morphogenesis** may occur during early cleavage and only become apparent during late cleavage.

In the sea urchin (Fig. 10.1) and many other marine invertebrate embryos, for example, fluid-filled **segmentation cavities** or swellings between blastomeres are present from the first division. Hydraulic pressure developing in segmentation cavities pushes the blastomeres against the fertilization envelope, thereby maintaining the embryo's egg shape, but gradually the cavities become apparent as a clear space, ultimately identified as the blastocoel within the embryo.

A similar space is not visible from the outside in amphibians (Figs. 10.2 and 10.3) but surface events mark the transition from early to late cleavage at some time after the seventh division. The loosely cohering or merely aggregated blastomeres of early cleavage (Fig. 10.4) gradually become stably attached to each other, and junctional complexes appear between cells at the embryo's periphery. The blastomeres flatten against each other and extend their tight borders, while the segmentation cavities coalesce and form a central blastocoel. Welded together at their edges, late-cleavage blastomeres retain the shape of the egg even if the fertilization membrane is removed.

Changes in the **cleavage pattern** inscribed on the surface of the embryo by the order and position of cleavage furrows also mark the transition from early to late cleavage. During early cleavage, successive divisions are assigned Roman numerals.

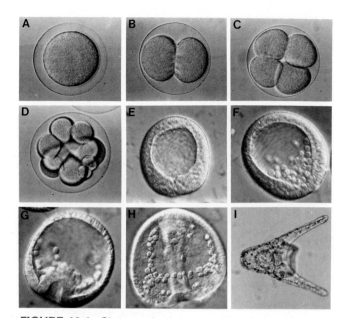

FIGURE 10.1. Cleavage in the sea urchin, *Lytechinus variegatus* (stages and times after fertilization for embryos cultured at 22° C). (a) Fertilized egg. Fertilization membrane has lifted off embryo and the hyaline layer is visible as faint line above plasma membrane. The female pronucleus is visible near the center. (b) Two-cell stage, 1½ hour. Hyaline layer and fertilization membrane visible. (c) Four-cell stage, 2 hours. (d) Sixteen-cell stage, 3 hours. Mesomeres are seen in animal half (four visible upper left), macromeres beneath (two visible), and micromeres at vegetal pole (two visible lower right). (e) Blastula just after hatching, 9 hours. Fertilization membrane is gone. The blastocoel is conspicuous in center. Thickened region (toward bottom) is the vegetal plate. (f) Mesenchyme blastula, 12 hours. Sixty-four primary mesenchyme cells have migrated inward (ingressed) and piled up on top of the vegetal plate. (g) Early gastrula, 14 hours. The first phase of invagination, or primary invagination, produces a blastopore and short archenteron. Primary mesenchyme cells are migrating. (h) Late gastrula, 17 hours. Invagination is complete. The tip of the archenteron, or gut, has established an oral contact with the ectoderm near the animal pole (top). The primary mesenchyme ring is visible at the widest level of the larva and two tri-radiate spicule rudiments, visible within the ring, show the beginning of skeleton formation. (i) A pluteus larva at 36 hours. The large arms in the plane of focus are the anal arms. The oral hood is beneath. Courtesy of Charles A. Ettensohn, Department of Biological Sciences, Carnegie Mellon University. Photos by Seth Ruffins.

The same numbers are used to identify cleavage furrows. For example, cleavage I, or the first cleavage, produces two blastomeres separate by furrow I. Cleavage II in both blastomeres produces two blastomeres separated by furrow II in each of the

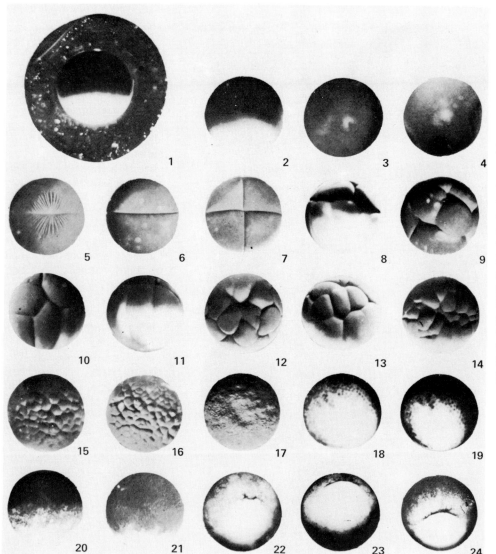

FIGURE 10.2. Cleavage in the frog. (1) Unfertilized egg. (2) A gray crescent forms opposite the point of sperm entry. (3) Dimple appears where the second polar body is forming. (4) The polar body is extruded. (5) Tension lines radiate on either side of the first cleavage furrow. (6) Two-cell stage. Tension in the furrow has relaxed. (7) Four-cell stage. (8–11) Eight- to 16-cell stage. (12, 13) Sixty-four-cell stage. (14) Early blastula. (15, 16) Late blastula. (17–21) Epiboly toward vegetal pole. (22–24) Invagination and formation of dorsal lip of blastopore. Reprinted with permission of Macmillan Publishing Company from R. Rugh, *A guide to vertebrate development,* 7th ed., 1977. Copyright © 1977 by Macmillan Publishing Company.

first two blastomeres (i.e., for a total of four blastomeres). The number of blastomeres or cleavage nuclei (n) present during synchronous division is ascertainable from the cleavage number (c) by the formula $n = 2^c$.

In late cleavage, patterns of new furrows tend to lie within portions of the older pattern. Synchrony largely breaks down, and it is no longer possible to talk about cleavage numbers in the embryo as a whole (although one may still refer to the cleavage number for individual blastomeres).

In practice, at the junction of early and late cleavage, embryologists introduce a new scheme of nomenclature. Instead of identifying cleavages and furrows by numbers, blastomeres are labeled by name. Interest in the division of particular blastomeres replaces interest in divisions in the embryo as a whole.

Cleavage Divisions

Several features of cleavage division distinguish it from ordinary cell division, most conspicuously, the failure of blastomeres to grow in the interval between divisions. While generally the trigger for cell division seems to be pulled by growth, the large size of eggs seems to preset the zygote's trigger for multiple divisions without additional growth.

Blastomeres get smaller during cleavage until they come to resemble adult cells in size. In addition, blastomeres lose water. The generally hydrated condition of the egg is reduced, and blastomeres begin to resemble ordinary body cells in hydration.

Cleavage also restores the cell's **nuclear/cytoplasmic volumetric ratio** to something like that of

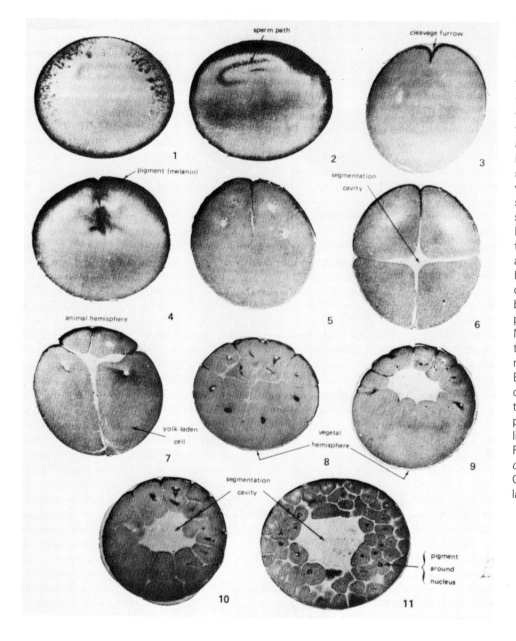

FIGURE 10.3. Cleavage in the frog seen in sections of the egg and embryo. (1) Unfertilized egg. (2) After fertilization. The sperm penetration track is the pigmented curve. (3) The first cleavage furrow begins at the animal pole. (4) The cleavage furrow moves more rapidly through the animal hemisphere than through the vegetal hemisphere. (5) The second cleavage furrow also starts at the animal pole. (6) Four-cell embryo in cross section. The segmentation cavity arises between cells. (7) Eight-cell embryo. (8) Section of approximately 32-cell embryo slightly off center. (9) Approximately 32-cell embryo. Note enlargement of segmentation cavity. (10) Approximately 64-cell embryo. (11) Early blastula. Segmentation cavities have coalesced into the blastocoel. Reprinted with permission of Macmillan Publishing Company from R. Rugh, *A guide to vertebrate development,* 7th ed., 1977. Copyright © 1977 by Macmillan Publishing Company.

the adult cell (Table 10.1). Following the breakdown of the oocyte's germinal vesicle, the egg's nucleus and even the zygote's pronuclei are disproportionately small compared to the volume of zygotic cytoplasm. Early-cleavage nuclei seem out of touch with large amounts of cytoplasm and may be unable to function efficiently in nuclear–cytoplasmic exchanges of information.

In subsequent divisions, the blastomeres' nuclei more or less retain their sizes, while cytokinesis reduces the volume of blastomeres. As each cleavage division proceeds, each nucleus occupies more and more of the volume of the blastomere. The result is a rapid increase in the nuclear/cytoplasmic volumetric ratio and presumably in

opportunities for intimate nuclear–cytoplasmic interactions.

Even in mammals where transcription and translation occur in the early embryo, real growth does not occur. Cleavage reduces the volume of cytoplasm compared to the volume of nuclei, and an adultlike nuclear/cytoplasmic volumetric ratio is reached at about the 120-cell stage.

The ratio of amounts of DNA to RNA in blastomeres also changes compared to the zygote. Because DNA is replicated during the S period prior to every cleavage, the amount of DNA in the embryo as a whole increases exponentially, while the amount of RNA increases only linearly following the resumption of synthesis (Brachet, 1950). On a

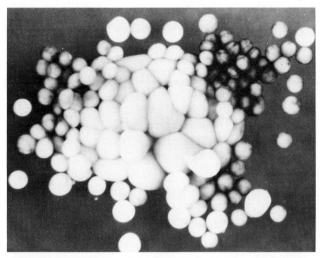

FIGURE 10.4. Photomicrographs of cleaving *Ambystoma* (= *Siredon*) *mexicanum* and dissociated blastomeres. Embryos are treated with $\frac{1}{15}$ M phosphate buffer in sterile tap water, and, after cleaving, cell dissociation is achieved by mechanically removing the vitelline envelope. Dissociated cells after seventh cleavage. Total cell number = 128. ×15.4. From K. Hara, *Wilhelm Roux' Arch. Dev. Biol.,* 181:73 (1977), by permission of Springer-Verlag and courtesy of K. Hara.

per cell basis, the absolute DNA content remains constant, while RNA content diminishes.

Changing ratios of various sorts clearly change the vital statistics of blastomeres, but whether these changes actually influence cleavage is uncertain. In the case of mouse embryos, changing the nuclear/cytoplasmic volumetric ratio fails to alter cleavage or protein synthesis. Cytoplasm can be withdrawn by micropipette or injected, or zygotes bisected, leaving the pronuclei in one half or dividing them among the halves. The stage-specific surface antigen SSEA-1 still appears before the 6–8-cell stage as it does normally (Petzoldt and Muggleton-Harris, 1987).

Cytokinesis as such does not seem to determine the production of enzymes such as SSEA-1

TABLE 10.1. Changes in nuclear/cytoplasmic volumetric ratio accompanying cleavage in the sea urchin, *Echinus*

Stage	Nuclear/cytoplasmic ratio
Oocyte (with germinal vesicle)	1:7
Uncleaved egg	1:400–550
4-Cell embryo	1:18
56-Cell embryo	1:12
Blastula (about 1000 cells)	1:6–7

since cytokinesis can be eliminated by cytochalasin B without interfering with synthesis. Cytochalasin B-treated ascidian embryos produce enzymes such as acetylcholinesterase at the correct time while arrested in cleavage (see Fig. 11.36). Nuclear division continues in treated embryos on a normal schedule, however, suggesting that the **developmental clock** timing and coordinating events during cleavage is run by DNA replication (Satoh, 1982).

Differentiation Without Cleavage

Cell division during cleavage ordinarily provides the boundaries wherein differences can develop. It is hard to imagine how embryonic parts could undergo differentiation without being delimited from each other (Lindenmayer, 1982). Nevertheless, some differentiation, while conspicuously defective and incomplete, can occur in the absence of cleavage furrows or boundaries.

Discovered by Frank R. Lillie at the turn of the century (Lillie, 1902, 1907), oocytes of the New England polychaete, *Chaetopterus pergamentaceus*, exposed briefly to excess potassium chloride, **differentiate without cleavage** in 15–20 hours. The Mediterranean *Chaetopterus variopedatus* reacts the same way (Brachet and Donini-Denis, 1978). Furthermore, brief exposure to the calcium ionophore A23187 (but not excess potassium and A23187) and, if only occasionally, the sulfhydryl reagent dithiothreitol (DTT) in artificial seawater also triggers differentiation without cleavage.

Normally arrested at metaphase I, treated eggs complete the expulsion of the two polar bodies and lift their fertilization membrane in the absence of sperm (i.e., differentiation is parthenogenic). Calcium is necessary for the activation reaction but not for further stages of differentiation.

Wherever some degree of differentiation without cleavage is induced, replication, or at least the formation of monasters, and the movement of cytoplasm is observed (Brachet and Donini-Denis, 1978). The treated eggs proceed to form a polar lobelike protrusion and a cleavage furrow, but the furrow is aborted and regresses. The yolky parts of the egg and the more homogeneous cytoplasm separate at opposite poles, and the concentrated cytoplasm flows over the yolk mass creating a superficial clear ectoplasm, or hyaloplasm, surrounding a yolk-rich endoplasm.

The treated eggs go on to hatch from the fertilization membrane and form cilia. The resulting **unicellular larvae** swim (Fig. 10.5), but, lacking specialized cilia, the unicellular larvae merely

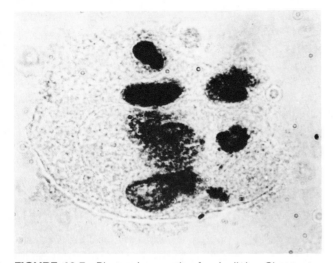

FIGURE 10.5. Photomicrograph of unicellular *Chaetopterus variopedatus* larva slightly flattened under cover glass and stained by the Feulgen reaction to show DNA. Several nuclei are seen in different activities. At the top is a pycnotic nucleus which is presumably physiologically dead. Two metaphase nuclei are seen at the lower right. Their chromosome numbers are different. From J. Brachet and S. Donini-Denis, *Differentiation,* 11:19 (1978), by permission of the International Society of Differentiation, Springer-Verlag, and S. Donini-Denis.

whirl. Even the best cases fail to display the unidirectional movement characteristic of normal larvae.

The unicellular larvae gradually outlive their developmental program. They use up their yolk and acquire vacuoles. A23187 ionophore-treated eggs may form multipolar mitotic figures and occasionally undergo cellularization to form parthenogenic larvae, but most of the larvae undergo cytolysis by 2 days after activation.

Differentiation without cleavage shows how far development can go without manufacturing cellular building blocks. The failure of the unicellular larvae to develop further, on the other hand, implies the limits of development without cells. Cleavage may involve more than the production of cells, but development is incomplete without them.

Mechanics of Cleavage

Cleavage begins with the nucleus (i.e., karyokinesis) and ends with the cytoplasm (i.e., cytokinesis). The furrow separating blastomeres is ultimately complete, if not immediately, then at some later time, and blastomeres are, or become, separate sibling cells (Fig. 10.4). While mitosis may not be immediately manifest as cell formation, nuclei divide during early cleavage, and cells are ultimately formed in powers of 2.

In the axolotl, *Ambystoma mexicanum* (Table 10.2), for example, through the eighth and ninth cleavages, precisely 128 and 256 nuclei are produced corresponding to 2^8 and 2^9 cells. Furthermore, almost all cells undergo their 10th cleavage synchronously, forming 510 of the expected 512 cells (Hara, 1977). Similarly, in the zebrafish, *Brachidanio rerio*, divisions in the blastodisk remain synchronous, and the number of cells present is a product of a strict doubling series through the 512-cell stage (Kimmel and Law, 1985a).

During this period of synchrony, the **law of the rectangular intersection of successive division**

TABLE 10.2. Cleavage in *Ambystoma mexicanum*[a]

								Cleavage Number								
	I	II	III	IV	V	VI	VII	VIII	IX	X	XI	XII	XIII	XIV	XV	XVI
				Period of synchronous divisions												
Rate of furrow elongation animal hemisphere				1 mm/hr												
vegetal hemisphere				0.02–0.03 mm/hr												
Interval between (min)	196	73	65	59					61	61	64	68	74	83	101	150
			Minutes behind animal hemisphere							Cycles behind animal hemisphere						
Equatorial zone					5											
Vegetal hemisphere					15								1		2	
Number of cells expected	2	4	8	16	32	64	128	256	512	1024	2048	4096	8192	16384	32768	65536
Number of cells counted (median)	2	4	8	16	32	64	128	256	510	986	1878	3390	4869	6484	8308	11347

[a]Data from K. Hara, 1977. The cleavage pattern of the axolotl egg studied by cinematography and cell counting. *Wilhelm Roux' Arch. Dev. Biol.,* 181:73–87.

planes tends to be enforced. According to the law, as first proposed by the botanist Julius Sachs (1832–1897), mitotic spindles tend to swing at right angles from one division to the next.

Oscar Hertwig explained these gyrations with two **mechanical principles of cell division:** (1) a nucleus tends to take up a position at the center of the available cytoplasm, and (2) the long axis of a mitotic spindle tends to parallel the long axis of the available cytoplasm.

Since the plane of cytokinesis forms in the plane of the metaphase plate perpendicularly to the long axis of the mitotic spindle, cells tend to divide midway along their length and perpendicularly to their longest cytoplasmic axis. Division reduces the long axis of the cell by half, and, because growth does not follow division, the long axis in a cleaving cell becomes the short axis in the resulting sibling cells. Successive spindles therefore lie at right angles.

Alternately, centriolar reproduction may dictate the positioning of spindles. In the acoelous turbellarian, *Polychoerus*, the axes of the two centrioles at each spindle pole are at right angles to each other and to the spindle. Upon duplicating, the sibling centrioles separate at right angles (see Fig. 3.3). In the succeeding generation, the cleavage spindles lie perpendicularly to each other, whereas in alternating generations, the cleavage spindles lie parallel to each other (see Costello and Henley, 1976).

Still, the impression that cleavage is controlled mechanically and passively may be misleading. For example, rather than mechanically drawing in the cleavage furrow, a **contractile ring** in the egg cortex seems to be induced by asters. The results of displacing asters through various means and at different times during cleavage in sea urchins indicate that the location of the contractile ring is determined by metaphase, and furrowing is not dependent on the spindle thereafter (see Giudice, 1986).

One may grant the possibility that forces positioning the first astral rays in the zygote set into motion the chain of events that position succeeding spindles, but what positions the first spindle? This spindle may be oriented in a highly stereotyped fashion with respect to the point of sperm entry and the embryo's planes of symmetry (e.g., frog eggs).

Rather than being passively positioned by the cytoplasm, the first spindle may be stabilized in a particular position by active interactions with the cell's cytoplasm and cortex. Astral rays, which reach all the way to the zygote's cortex and even push against the surface (e.g., crenation at the surface of sand dollar eggs), may draw free asters toward a pole or a particular portion of the cortex. The rays may search out receptive sites on the egg's inner surface and position the mitotic spindle accordingly (see Schroeder, 1985).

Cleavage Cycle

Periodic cell or nuclear divisions are the most conspicuous events associated with cleavage, but a cleavage cycle includes other periodic events that may occur in the absence of cell division. For example, activated or fertilized *Xenopus laevis* eggs from which nuclei are removed do not divide, but they undergo about five periodic cortical or surface contraction waves (SCWs) at the same rate as normal cleavage (Hara et al., 1980). In addition, kinase activity and the concentration of **maturation promoting factor** (MPF, the same soluble factor that made its initial appearance as the oocyte overcame its first meiotic arrest) oscillate in the absence of a nucleus.

Surface contraction waves seem to be controlled by nuclear material released at the breakdown of the germinal vesicle (GVBD), since the waves do not occur in activated eggs previously deprived of their germinal vesicle. The regulation of MPF and overall kinase activity, on the other hand, is entirely cytoplasmic, since these components oscillate in the absence of any nuclear material and surface contraction waves (Dabauvalle et al., 1988). MPF is widely present in dividing cells (it might better be known as a mitotic factor) and may activate kinases, thereby acting as the essential cell cycle oscillator in the cytoplasm.

In some embryos, the idea of a cleavage cycle is not easily applied. In gastropods, for example, an interruption of several hours (3 in the pond snail, *Lymnaea stagnalis*) occurs at the 24-cell stage. Other embryos seem to lack a global cleavage cycle, and individual blastomeres may exhibit different cleavage rates. In *Caenorhabditis elegans* the first two blastomeres divide asynchronously, for example, and in the rabbit, a range of difference between synchrony and asynchrony begins with the first two blastomeres.

Elsewhere, especially where cleavage is relatively rapid, the interval between successive cleavages is nearly the same throughout the embryo. Nearly simultaneous divisions occur at intervals of 5–15 minutes in dipteran insects (10 minutes in *Drosophila melanogaster*), 20 minutes in the gold fish, $\frac{1}{2}$ hour in the sea urchin (Fig. 10.1), and 1 hour in the frog (Figs. 10.2 and 10.3). At the fifth division in *A. mexicanum*, cells near the animal pole start dividing only about 5 minutes before cells in the equatorial region and only 15 minutes before those near the vegetal pole.

Even in the mouse, the generation time for blastomeres is relatively rapid at 10–12 hours compared to the 15–20 hours of rapidly dividing adult cells. While not simultaneous, the first 4–5 divisions tend to overlap.

Throughout the egg, blastomeres undergoing mitosis round up, divide, and relax. As cleavage proceeds, small phase shifts in division from the animal to the vegetal pole and changes in cell shape during division result in a **cleavage wave** rolling over the embryo. In *A. mexicanum*, nine cleavage waves roll down the embryo, the first seven initiated within 7 minutes of each other from top to bottom.

Despite this high degree of regularity, synchrony gives the appearance of breaking down before it actually does. The discrepancy is especially great in telolecithal eggs, where differences in the rate of furrow elongation in different parts of the embryo exaggerate differences in cleavage rates. Between the second and fifth cleavages, in *A. mexicanum*, for example, furrows in the animal hemisphere elongate at the rate of 1 mm per hour, while those in the vegetal hemisphere elongate at only 0.02–0.03 mm per hour (i.e., a 33–50-fold difference, Table 10.2).

The rate of cleavage is not usually constant throughout the entire cleavage period. In mammals, for example, the onset of cleavage is delayed for 24 hours, and the rate of division increases toward the end of cleavage. Similarly, in *A. mexicanum* at 20°C, the first cleavage does not take place until about 3.25 hours after the extrusion of the second polar body. Cleavage then becomes more rapid. The intervals between the first and second and between the second and third cleavages are 73 and 65 minutes. The interval between the next eight consecutive cleavages is remarkably constant at about 1 hour, but then the pace of cleavage slows. At gastrulation, beginning with the 15th division, the interval between divisions is about 2.5 hours (Table 10.2).

Cleavage rates also tend to become nonuniform throughout the embryo. By the 11th cleavage in *A. mexicanum*, for example, the animal-pole cells are roughly one cleavage ahead of vegetal-pole cells, and, when blastomeres in the animal hemisphere reach the 15th cleavage, those in the vegetal hemisphere are about two cleavages behind (Hara, 1977). In sea urchins, the rates of cleavage develop a gradient toward the vegetal pole with the highest rate in the small cells (i.e., micromeres) (Parisi et al., 1978).

Logistical Problems of Cleavage

Yolk is the chief resource for lecithotrophic eggs, but yolk also poses mechanical problems for cleavage and metabolic problems for blastomeres. Yolk is not nearly as biochemically inert as was once thought. Amphibian platelets, for example, cannot be handled in the cytoplasm as if they were simply freight cars of reserves. Platelets contain a variety of enzymes, including proteases, phosphoprotein phosphatases, and acid phosphatases, and their outer layers also contain inhibitors of these proteases, some DNA, RNA, and glycoproteins.

Yolk also brings with it a legacy of physical problems. Mechanically, yolk is a heavy burden to bear. Crystalline yolk platelets (see Fig. 6.13) and granular forms of yolk have a greater density than most cytoplasmic constituents, and difficulty maneuvering them tends to block the even distribution of cytoplasm and suspended constituents, including organelles, within blastomeres.

In the course of evolution, various strategies have been adopted to counter the obstacles offered by yolk to normal cell physiology. Eutherian mammals, for example, have opted to give up the large amounts of yolk accumulated in the eggs of their reptilian ancestors.

Many marine animals adopted a strategy of keeping yolk to a minimum. This minimum, however, covers a large range. Even among related species, those with large yolky eggs have slower development, while those with smaller less yolky eggs and concentrated ooplasm have rapid development to swimming larvae that can feed themselves.

For example, *Ascidia nigra* (Fig. 10.6a) is a large solitary ascidian having small eggs, while *Clavelina picta* (Fig. 10.6b) is a colonial (or composite) ascidian with large eggs. *Ascidia nigra* eggs, with large amounts of ooplasmic RNA, develop within 11–12 hours into swimming larvae. *Clavelina picta*'s large yolky eggs develop to the swimming larva stage in the course of several days.

The amount of yolk is also correlated with the number of eggs released and the amount of protection offered by the parent. For example, *C. picta* releases only a few eggs at a time and stores them in a protected brood-pouch during development. In contrast, *A. nigra* releases large numbers of eggs at a time directly into the environment (see Cowden, 1976).

Organisms with megalecithal eggs have evolved compromises permitting yolk and blastomeres to coexist. Yolk may be allowed to accumulate in some portions of the egg while it is eliminated from other portions, or, as in the case

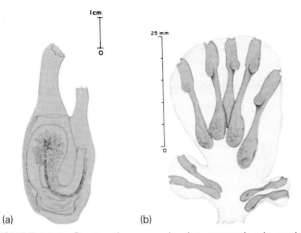

FIGURE 10.6. Contrasting reproductive strategies in ascidians. (a) The large, isolated *Ascidia nigra* lacking a brood pouch but capable of producing large numbers of sparsely yolked, rapidly developing eggs. (b) The colonial (e.g., composite) *Clavelina picta* with brood pouch containing small numbers of highly yolked eggs that develop slowly. From H. H. Plough, *Sea squirts of the Atlantic continental shelf from Maine to Texas*, The Johns Hopkins University Press, Baltimore, 1978. Used by permission.

of teleosts, yolk remains in one region as cytoplasmic streaming shunts cytoplasm to another region.

By influencing the distribution of cytoplasm, yolk also affects cleavage and subsequent development. According to **Balfour's law:**

> The rapidity with which any part of an ovum segments [i.e., cleaves] varies *ceteris paribus* with the relative amount of protoplasm it contains; and the size of the segments formed varies inversely to the relative amount of protoplasm. When the proportion of protoplasm in any part of an ovum becomes extremely small, segmentation does not occur in that part" (Balfour, 1880, p. 99).

In other words, Balfour's law predicts that over a range of yolkiness, divisions will be displaced toward the more cytoplasmic-rich areas, while at the extreme of yolkiness, divisions will simply not occur.

Balfour's law seems to be observed by many cleaving embryos. For example, in medialecithal eggs with a telolecithal tendency (e.g., sea urchins) and in mesolecithal eggs with a telolecithal distribution of yolk (e.g., frogs), the mitotic spindles of the third cleavage division are displaced toward the more cytoplasmic-rich animal hemisphere (e.g., Fig. 10.3). Nevertheless, the role of the asters surrounding the centrioles of animal cells is probably more important than yolk in determining the position and orientation of the mitotic spindle and cleavage planes (Rappaport, 1961; Schroeder, 1985).

Mobilizing Resources

In the course of development, virtually the entire content of the yolk is mobilized and absorbed. In the chick, absorption is active via endocytosis, and an endocytic transport mechanism passes solute transcellularly (see Dunn and Fitzharris, 1987).

Carbohydrate and protein metabolism precede lipid mobilization, but more than 90% of the total energy requirement for development is derived from the oxidation of yolk fatty acid (Romanoff, 1967). Despite the abundance of lipid, especially cholesterol, chick yolk is deficient in polyunsaturated fatty acids and phospholipids required for membrane assembly. These are synthesized in the yolk sac from yolk precursors and stored in the liver. Not surprisingly, the placenta of mammals likewise synthesizes polyunsaturated fatty acids (Noble, 1987).

The amount of phospholipid required for membrane synthesis in an embryo is enormous. Phospholipid for the plasma membrane of a cleaving spherical cell, for example, is equal to half again as much as that originally present around the cell. When cytokinesis is suppressed at the beginning of cleavage, as in insects, the amount of cell membrane required for subsequently dressing cells in membranes is staggering. The amount of plasma membrane utilized for the cellularization of a *Drosophila* embryo (2.3×10^6 μm^2), for example, is equal to 10 times the original surface area membrane (Fullilove and Jacobson, 1971).

Some of the logistic problems of cleavage seem to work themselves out passively. The frog's egg, for example, has all the ribosomes required to bring it to the tadpole stage of development and all the messenger RNA required to get it to the blastula stage. It also has enough histone to get started (although more will be made along the way). As a result of prior preparations, the frog's blastomeres have sufficient amounts of these materials to meet the fundamental needs of rapidly dividing cells.

Similarly, the 5–10% of the egg's mass consisting of **tubulin** (one of the major protein components of the mitotic apparatus) is enough to produce mitotic apparatuses for about 10,000 cells, the number of cells present in the blastula. But by the blastula stage, tubulin synthesis is adequate to maintain required concentrations. Structural proteins in general may normally autoregulate their own accumulation via posttranscriptional control (Brandhorst et al., 1986).

Cleaving embryos decrease in mass (e.g., by as much as 20% in the cow and 40% in sheep). Likewise, the protein content of the embryo falls (e.g., in mouse by 25% in the first 3 days of cleavage).

These losses may be accounted for by metabolism, but they hardly deplete the embryo of reserves, and embryonic cells are certainly not starving. They either tap internal resources, or they are supplied with uterine fluids.

Some hypotheses about how stores are activated center on mechanical features of cleavage. The **compartmentalization hypothesis** (Runnström, 1933) suggests that the cutting up of the egg into small cellular units, brings once separate materials together and bridges gaps in metabolic pathways. Alternatively, cleavage and cellularization may dilute or segregate inhibitors of biochemical pathways. Probably, the events of cleavage are not controlled by one common mechanism (Epel et al., 1974).

MOLECULAR ASPECTS OF CLEAVAGE

An early, important achievement in the application of molecular biology to developmental problems was the formulation of, and acquisition of experimental evidence for, the hypothesis that the sea urchin egg contains a store of mRNA that becomes translationally active only after fertilization .... The stored maternal mRNA hypothesis was rapidly and enthusiastically accepted ... undoubtedly [as] the result of earlier experimental investigations indicating that early embryonic development is largely independent of the zygotic genome but quite dependent on information stored in the maternal cytoplasm.

B. P. Brandhorst et al. (1986, p. 283)

While cleaving cells may appear normal, cell division during cleavage has unique features, and cleaving cells are not typical dividing cells. Most unusually, in a few species, chromosomal material is lost during cleavage, a phenomenon known as **chromosomal diminution.** In *Parascaris equorum* (= *Ascaris megalocephala*), for example, chromosomal diminution is preceded by **chromosomal fracture** in all but cells leading to the germ line (see Fig. 11.55). Most cells in the somatic line lose at least 80% of their DNA (see Davidson, 1986). In the copepod, *Cyclops furcifer,* chromosomal diminution may be selective in the sense of resulting in the loss of the paternal chromosomal set alone (Cowden, 1976).

The chromosomes of early molluscan and annelidan embryos are sometimes **vesiculated.** Persisting until the 16-cell stage, for example, in *Ophryotrocha labronica,* vesicles or **karyomeres** may be empty or contain a dense core of a **nucleo-**lus-like body (NLB). The NLBs do not incorporate much tritiated uridine. Rather, they contain material resembling the nuclear envelope and, like the envelope, incorporate tritiated myoinositol and leucine (Emanuelsson, 1973).

Endopolyploidy is also a feature of cleavage in some organisms. In molluscan development, endopolyploidy occurs in nuclei of the apical plate, head vesicle, prototroch (see below), larval liver, and protonephridia but not elsewhere (van den Biggelaar, 1971b).

Even more ordinary cleavage has its unique side. Cleavage is a **dissociated** form of cell division in which several synthetic acts normally associated fail to be synchronized in the usual fashion.

Replication and the Cell Cycle

The resumption of DNA synthesis by the zygote is an extraordinary event. Not only was replication blocked at the spermatocyte and oocyte stages, sometimes years earlier, but, when it commences soon after fertilization, it does so at a remarkably fast pace.

The period of DNA synthesis in *Strongylocentrotus purpuratus* takes only 10 minutes for the first cleavage and 13 minutes thereafter. In *A. mexicanum*, both male and female pronuclei begin to incorporate tritiated thymidine at 20 minutes after fertilization, and both are labeled by 30 minutes. The incorporation of precursor ceases, and no detectable DNA synthesis occurs at 40 minutes (see Benbow, 1985). The duration of the S period is therefore less than 20 minutes. In the mouse, S is estimated at 4 hours with a range from 3.5 to 5.5 hours, approximately half the time of S in rapidly dividing tissue culture cells (see Wolgemuth, 1983).

In part, the great speed of cleavage is due to the absence of the premitotic and postmitotic gaps. In *Lymnaea stagnalis* zygotes, G-1 cannot be resolved until after the first three divisions (van den Biggelaar, 1971a). Similarly, the G-1 period is not detected in the mouse before the 8-cell stage (see McLaren, 1982). In *Xenopus laevis*, the onset of DNA synthesis follows directly after mitosis and mitosis follows directly after DNA synthesis without conventional G-1 and G-2 periods for the first 12 divisions (Newport and Kirschner, 1982).

Like pronuclear replication, blastomeric replication employs more **initiation sites** or **origins** rather than faster polymerases (see Callan, 1972). As the rate of replication increases during early cleavage, initiation sites occur at extraordinarily small intervals. In dipteran embryos, such as

D. melanogaster, **macrobubbles** appear with a mean length of 5.6 kb and a mean center-to-center distance of 9.7 kb. In cleaving *Paracentrotus lividus*, clusters of small 300-kb **microbubbles** compose 80% of the replication bubbles.

Similarly, in *X. laevis* embryos, very small replication bubbles dominate replication. These microbubbles are evidently under the control of cytoplasmic materials capable of entering the nucleus, since very small microbubbles can be induced in somatic chromatin exposed to the cytosol of eggs.

Transcription

Ordinarily, one assumes that gene activity coordinates cell behavior, and so it comes as a surprise that during cleavage transcription does not run the show. Transcription is not even required for early cleavage.

Nucleoli and ribosomal RNA synthesis are, in general, conspicuously lacking in nuclei of most cleaving embryos, and messenger RNA synthesis frequently proceeds only at low levels. The ability of embryos to undergo cleavage while actinomycin D inhibits RNA synthesis (see Fig. 9.3) indicates that the zygote is already prepared with sufficient RNA to sustain cleavage. In general, protein synthesis during cleavage depends on preexisting RNA.

The transcription of RNA is nevertheless subject to the same adaptive pressures that mold the rest of the organism's biochemistry, and RNA synthesis occurs during the cleavage of some species and in different parts of the same embryo. For example, in the colonial ascidian *Clavelina picta*, with small amounts of ooplasmic RNA, nucleoli appear in the late blastula, and ribosome synthesis occurs prior to cytodifferentiation of larval tissue. In the rapidly developing *Ascidia nigra*, on the other hand, with relatively large amounts of ooplasmic RNA, nucleoli are not formed until the metamorphosis of the larva (see Cowden, 1976) (Fig. 10.6).

In echinoderms, transcription proceeds at a relatively high rate in only one type of cleaving blastomere, the **micromere** (Fig. 10.7). While prior to the 16-cell stage, newly synthesized micromeric RNA resembles that synthesized by other blastomeres and present in the egg itself, at the 16-cell stage, new varieties of RNA are synthesized by the micromeres. Much like typical differentiating cells, sea urchin micromeres have a narrower range of transcripts (i.e., complexity) and less poly(A)+ mRNA compared to other blastomeres (see Giudice, 1986).

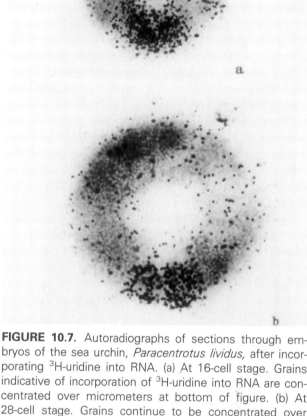

FIGURE 10.7. Autoradiographs of sections through embryos of the sea urchin, *Paracentrotus lividus,* after incorporating ³H-uridine into RNA. (a) At 16-cell stage. Grains indicative of incorporation of ³H-uridine into RNA are concentrated over micrometers at bottom of figure. (b) At 28-cell stage. Grains continue to be concentrated over micromeres. From G. Czinak, H. G. Wittmann, and I. Hindennach, *Z. Naturforsch.,* 22b:1176–1182 (1967), by permission of Verlag der Zeitschrift für Naturforschung and courtesy of the authors.

Despite the general indifference of cleaving embryos to actinomycin D, some features of early cleavage are disrupted by the drug (although not necessarily by its antitranscription activity). Treated sea urchin embryos fail to develop the mitotic gradient normally centering on the micromeres (Parisi et al., 1979) or show the characteristic increase in histone synthesis (Senger and Gross, 1978).

In terms of physiological activities therefore, transcription in the early embryo may have only subtle effects, but these are not inconsequential for later development. The failure of embryos to progress beyond the blastula or early gastrula stage following the suppression of transcription suggests that new transcripts are required for advancing development through gastrulation.

Mammalian embryos represent exceptions to the rule that transcription is suppressed during cleavage. Although sperm and egg are transcriptionally quiescent, fertilization turns on the transcriptional motor. In the mouse, unambiguous incorporation of the RNA precursor tritiated uridine into RNAase-sensitive material occurs by the 2-cell stage (Mintz, 1964), and, at least, tritiated adenosine is incorporated at the pronuclear stage (Clegg and Pikò, 1982).

The synthesis of tRNA and high molecular weight HnRNA precedes the synthesis of ribosomal RNA precursors in the mouse. Nascent (i.e., freshly synthesized) HnRNA and polyadenylated RNA transcripts of the embryonic genome are detected in the 2-cell mouse embryo, but rRNA precursor molecules are not detectable until the 4–8-cell stage. Similarly, in the rabbit, nascent RNA transcripts found at the 2–4-cell stage do not include rRNA.

The function of the transcripts is ambiguous. Showing that transcripts are synthesized in the embryo is not the same as showing that they function as messengers. The HnRNA transcripts may play a role in nuclear regulation without ever entering the cytoplasm and serving as mRNA in protein synthesis.

Possibly, oocytic or stored mRNA alone operates in translation during cleavage. This possibility is consistent with the results of actinomycin D experiments. Moreover, no differences appear in the unique proteins synthesized following fertilization, or **fertilization-dependent proteins,** whether these are from 2-cell embryos, parthenogenically activated eggs, or even enucleated eggs (see Wolgemuth, 1983).

On the other hand, the appearance in the 2-cell embryo of β_2-microglobulin of paternal origin suggests that some transcripts made following fertilization function as messengers for cytoplasmic translation (Sawicki et al., 1981). The resolution of the conflict over the role of nascent transcripts in the mammalian embryo may have to wait for even more sensitive techniques for detecting gene products than those currently available.

Translation

Protein synthesis in mature ova is so slow that less than 0.5% of the zygote's protein is synthesized by the egg itself. Whatever the "bottleneck" in the egg's protein synthesis, cleavage breaks it and sets protein synthesis flowing. In the sea urchin, fertilization is accompanied by a sevenfold increase in protein synthesis, as stored messenger

is recruited and ribosomes are mobilized in polysomes (Fig. 10.8) (Kedes and Gross, 1969; Rinaldi and Monroy, 1969).

The proteins synthesized during early cleavage in sea urchins are not a random sample of the messengers available. The proteins are the ordinary type expected in dividing cells, namely, tubulin and histone, but these are selectively synthesized. For example, stored mRNA for histones located in the cytoplasm surrounding the nucleus supports a high rate of synthesis especially in the rapidly dividing micromeres (see Giudice, 1986).

Few, if any, differences in newly synthesized proteins appear among sea urchin blastomeres. Even micromeres tend to produce the same proteins produced by other blastomeres. Actinomycin D, however, prevents the increase in micromeric histone production, suggesting that some differential transcription is required for protein synthesis even when preexisting stored mRNA acts as a template (Senger and Gross, 1978).

In mollusks, differences in protein synthesis among blastomeres accompanies the distribution of **polar lobe cytoplasm.** During maturation and the initial cleavage divisions, **polar lobes** swell and protrude recurrently from the vegetal cortex. Upon contraction, the polar cytoplasm is funneled into given blastomeres (see below, Fig. 11.47). Protein synthesis in polar lobes seems autonomous and continues even in isolated lobes, but protein synthesis within the embryo is dependent on the polar lobes.

Different patterns of protein synthesis occur in normal embryos and **lobeless** embryos from which

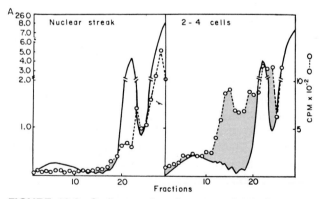

FIGURE 10.8. Sedimentation diagrams of polyribosomes (dashed line) and monoribosomes (solid line) at first cleavage (nuclear streak) and between first and second cleavage (2–4 cells). The two diagrams are similar except for the increase in small polyribosomes at the 2–4-cell stage. From A. M. Rinaldi and A. Monroy, *Dev. Biol.,* 19:73 (1969), by permission of Academic Press.

lobes are experimentally removed. These differences persist even seven cleavages later. The polar lobes therefore seem to contain specific mRNAs or possibly specific regulators of translation that eventually affect transcription (see Cather et al., 1976).

Protein synthesis by cleaving frog embryos fluctuates. There are no unique cleavage proteins synthesized in *X. laevis*, although at least one uniquely blastula protein is synthesized in the blastula stage. But electrophoretically identifiable proteins synthesized in the cleaving embryo can be partitioned into three classes on the basis of synthesis at other stages.

A small class of proteins (class 1) made during cleavage is also made in mature oocytes and in differentiated tissues. This class of **ubiquitous intracellular proteins** includes actin and tubulin.

A large class of **embryonic proteins** (class 2) is distinctly activated during cleavage but not found in differentiated tissues (with one exception). Some of the proteins in this class were not previously synthesized in either unfertilized eggs or in mature oocytes, although the proteins might have been synthesized earlier in oogenesis. The proteins in this class and similar proteins in rabbits are presumably synthesized on preexisting mRNA (van Blerkom and McGaughey, 1978).

Another large class of proteins (class 3) is accumulated during oogenesis and synthesized in oocytes as well as in cleaving embryos and in at least two differentiated tissues. The synthesis of different members of this class is transiently suppressed following cleavage (i.e., in blastulas or gastrulas) only to be reactivated later, while the synthesis of other members of this class is continuous (Bravo and Knowland, 1979).

In addition, a few proteins are made in mature oocytes but not during cleavage. Proteins in this class are synthesized again during embryogenesis and are found in adult tissues. Like the proteins in class 3, the additional proteins synthesized in the oocyte are also synthesized in adult tissues. A similar pattern of selectively turning off the synthesis of particular proteins during cleavage is also found in rabbit embryos where some oocyte and blastocyst proteins are not synthesized during cleavage.

Chromatin Formation

The chromosomes of cleaving cells are distinguishable from those of ordinary dividing cells. Unique **cleavage histones** staining with bromphenol blue at pH 2.3 may appear only in cleaving cells.

When do typical histones appear? Murine **somatic histones** staining with alkaline fast green appear first on condensed anaphase and telophase chromosomes of 4-day blastomeres (see Cowden, 1976). Trivalent iron at low pH, which recognizes somatic histones, stains interphase nuclei more intensely in embryos at later stages of development than at earlier stages. Somatic histones do not appear in nuclei until the gastrula stage of the snail, *Helix aspersa*, and the swimming blastula stage of the sea urchin, *Lytechinus variegatus*.

In *Xenopus laevis*, synthesis of the four core histones (H2A1 [oocyte type], H2A2, H2B, and H3 [adult type]) increases dramatically in cleaving embryos, although the synthesis of histone H1 is not detected until the late blastula stage (Fig. 10.9). The selective increase in synthesis of core histones begins three times above the rate of synthesis in oocytes and becomes 50–80 times the rate in oo-

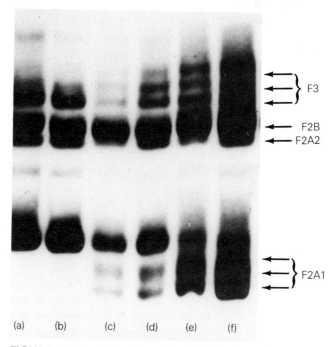

FIGURE 10.9. Autoradiograph of tritiated lysine-labeled histones extracted from *Xenopus laevis* eggs and embryos and analyzed by electrophoresis on acid/urea acrylamide slab gels. (a) From unfertilized eggs incubated in radioactive substratum for 4 hours. (c, d) Incubation began with late cleaving embryos and continued for 4 hours until the pregastrula stage. (e, f) Incubation began with early pregastrula embryos and continued 12 hours until neurula stage. The direction of histone migration is from the top to the bottom. The faster moving part of fraction F2A1 is the adult type. The slower moving part is the oocyte type. Arrows show the position of marker histone fractions. From E. D. Adamson and H. R. Woodland, *J. Mol. Biol.*, 88:263 (1974), by permission of Academic Press.

cytes during early cleavage (Adamson and Woodland, 1974).

Along with the vast stores of histone accumulated in the oocyte (e.g., about 0.3% of its total proteins), synthesis during cleavage provides more than enough core histone to meet the requirements of the cleaving embryo's nucleosomes. Actually, synthesis during cleavage alone occurs about 500 times faster than required to meet the blastomeres' needs. It would seem that the synthesis of core histones in the oocyte and during early cleavage anticipates requirements of the late blastula and early gastrula for core histones rather than needs of the cleaving embryo.

The histones undergo a qualitative shift during cleavage in their load of substituents. The oocyte variety of H2A1 histone (upper portion of band in Fig. 10.9) appears to be deacetylated to monoacetylated and nonacetylated adult varieties of histones (lower double band) presumably as it is incorporated into the chromatin of cleaving nuclei. The H3 histone, on the other hand, moves predominantly in the other direction, acquiring more acetyl groups as it accumulates in more nuclei.

Chapter

11

TYPES OF CLEAVAGE:

Descriptions and Comparisons

Specializations in the configuration of the internal egg structure which restrict or enhance the freedom of movement of the embryonic material in the dynamic egg system are the factors that decide and distinguish the determinate from the indeterminate type (of development).

DIETRICH BODENSTEIN (1955, p. 338)

Yolk is the chief variable in identifying the major types of cleavage (Table 11.1), but variations on these types are provided in different phyla. In descending order of diversity, intralecithal cleavage is simplest, if only because it has evolved exclusively in one phylum (Arthropods), while total cleavage has the most variations, being found in diverse phyla, from marine invertebrates to mammals.

TABLE 11.1. Eggs classified by amount of yolk and type of cleavage

Amount of Yolk and Distribution	Type of Cleavage	Examples
Alecithal, microlecithal, isolecithal	Total/holoblastic (complete)	Mammals, marine invertebrates
Medialecithal (oligolecithal), telolecithal	Total/holoblastic (polar)	Amphibians, primitive fish
Megalecithal, mesolecithal, telolecithal	Meroblastic/partial, meroblastic/discoidal (cleavage only at animal pole)	Cephalopods, teleosts, reptiles, birds
Megalecithal, centrolecithal	Intralecithal/superficial (karyokinesis within yolky portion followed by peripheral cellularization)	Pterygote insects, primitive onycophorans, chelicipods, myriapods, apterygotes

INTRALECITHAL CLEAVAGE

Lecithotrophic arthropods generally exhibit intralecithal cleavage. The conspicuous exceptions occur among crustaceans and some viviparous and parasitic arthropods, but even their cleavage may be derived from the intralecithal type. Moreover, aside from variations in arachnids and primitive apterygotes, cleavage in primitive onychophorans, arachnids, myriapods, apterygotes, and pterygote insects is remarkably uniform (Anderson, 1973).

Basic Pattern

Centrolecithal eggs have cytoplasm concentrated around a zygotic nucleus at the center of the yolk. Cytoplasm may also be distributed in a rim, known as the **periplasm,** beneath the egg's cortex. Cleavage is called intralecithal because the first few mitotic spindles occur within the yolk. Karyokinesis proceeds without cytokinesis to produce **intralecithal nuclei** within an early **syncytial** embryo.

An intralecithal energid, or physiologically active unit of nucleus and cytoplasm, moves from the yolk to the periplasm. The transient syncytial blastoderm is sometimes creased by incipient furrows or studded with **buds.** The embryo is then called a **periblastula** and is said to be at the **syncytial blastoderm stage** or **periblastula stage.**

The cellularization of the syncytial blastoderm transforms it into a cellular blastoderm. Cleavage furrows cut into the periplasm and partially separate nuclei as the embryo enters the **cellular blastoderm stage.** Because blastomeres are only formed superficially, intralecithal cleavage is also called superficial cleavage. The yolk at the center of the embryo is not cellularized (or not permanently), although it may be invaded by a few **yolk energids** or even infiltrated by cells known as **vitellophages** (Lat. *vitellus* yolk + Gk. *phagein* to eat).

As the blastoderm forms, mitosis tends to be **radial** or **tangential.** In radial mitosis, the long axis of spindles lie along radii, and furrows form parallel to the surface. Internal and external blastomeres are separated, and, if coordinated among many blastomeres, internal and external layers of blastomeres arise. In tangential mitosis, spindles lie parallel to the surface, and furrows form perpendicularly to the surface. As a result, columnar cells are added to extant cell layers.

Cleavage According to Classes

The end result of intralecithal cleavage is most often a yolky mass (with or without nuclei or cells) surrounded by a cellular blastoderm. Variations in reaching this end are chiefly related to the size of eggs and the mode of reproduction (see Anderson, 1973).

Onychophorans (Fig. 11.1) with primitive yolky eggs (viviparous and ovoviviparous species) illustrate the basic form of intralecithal cleavage. Ovid eggs more than 1 millimeter in diameter have a centrolecithal distribution of yolk but lack a conspicuous periplasm. Karyokinesis produces a series of intralecithal nuclei which associate with surrounding cytoplasm as energids. The yolk mass divides into **yolk spheres,** but these may not be nucleated.

Energids migrate to the periphery where they become blastomeres enclosed in plasma membrane and gather in a small **disk** of blastomeres. Further migration and proliferation provide a complete **blastoderm** covering the central yolk mass. Blastoderm cells also penetrate the yolk mass and become **vitellophagic** or the yolk-digesting cells (see Manton, 1949).

Chelicerates (Fig. 11.2) exhibit several modifications of superficial cleavage. The typical ovid to spherical arachnid egg has a central nucleus surrounded by a yolk-free cytoplasm connected to a thin periplasm by a reticulum of radial cytoplasmic threads. These threads may separate **yolk spheres** into stacks or **yolk pyramids.**

In the spherical eggs of spiders (range 0.4–1.9 mm in diameter), the first three nuclear divisions

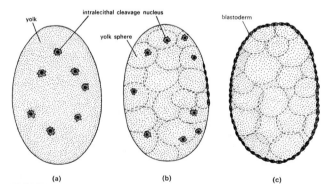

FIGURE 11.1. Intralecithal cleavage in an idealized yolky primitive onychophoran egg of the oviparous (e.g., *Ooperipatus*) or ovoviviparous (e.g., *Peripatoides*) type. (a) Intralecithal cleavage nuclei surrounded by halos of cytoplasm are suspended in yolk. (b) The yolk mass divides into a number of yolk spheres as nuclei move to the periphery and form blastomeres. (c) Further migration of nuclei to periphery and division of blastomeres completes the blastoderm. From D. T. Anderson, *Embryology and phylogeny in annelids and arthropods,* Pergamon Press, Oxford, 1973. Courtesy of D. T. Anderson.

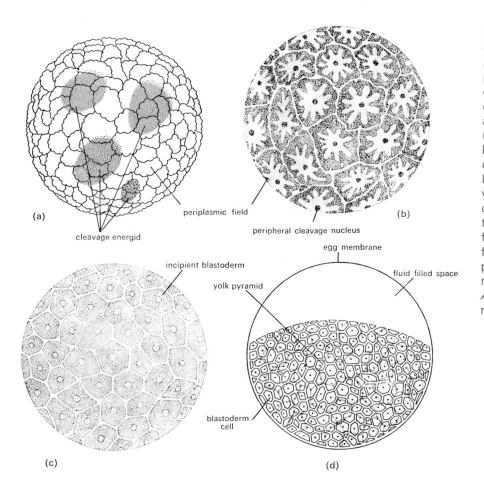

FIGURE 11.2. Modified intralecithal cleavage in the spider, *Cupiennius salei*, showing the relationship of periplasmic fields and superficial cleavage. (a) At the 4-nuclear or energid stage, periplasmic fields lie at the periphery over yolk pyramids. (b) Periplasmic fields coalesce and become occupied by cleavage nuclei. (c) Superficial cleavage establishes an incipient blastoderm as yolk pyramids begin to disappear centrally. (d) The blastoderm contracts leaving a flattened ventral surface of the embryo adjacent to a fluid-filled space. The remaining yolk pyramids merge into a central yolky mass. From K.-A. Seitz, *Zool. Jb. Anat.,* 83:327–447 (1966), by permission of Gustav Fischer Verlag.

are intralecithal and occur without furrowing. **Yolk pyramids,** radiating from a central cavity, match polygonal **periplasmic fields** on the surface. The pyramids proliferate, but the polygonal fields coalesce and ultimately disappear as intralecithal nuclei reach the periplasm, and a cellular blastoderm emerges. The yolky portions of the pyramids then fuse into a yolk mass. Later, cells from the blastoderm invade the yolk mass and become vitellophages.

Intralecithal nuclei and energids of arachnid eggs do not seem to be cast in a pattern. The even distribution of the first eight nuclei in the yolky cytoplasm seems to result from the elongation of mitotic spindles.

In small spider eggs, the yolk pyramids may be of unequal size and the central cavity eccentric, suggestive of an underlying pattern in the yolk. Further evidence of pattern or polarity does not appear until the blastoderm flattens against the central yolk mass. Then, the flattened part of the blastoderm corresponds to the ventral part of the embryo.

Despite their large size (2.5–3.5 mm in diameter) and the polar, rather than central, position of the nucleus, the eggs of xiphosurans (i.e., horse-shoe crabs) retain a form of superficial cleavage. The first few divisions are intralecithal, but the egg cortex adjacent to the nuclei quickly begins to furrow. From the animal pole downward, numerous large blastomeres appear, all equally yolky and containing central nuclei. Divisions soon separate a blastoderm of columnar blastomeres from an undivided yolky central cell. The mechanism may resemble total cleavage, but the end result is characteristic of superficial cleavage.

Ovoviviparous buthid scorpions exhibit a truncated form of intralecithal cleavage (or a polarized form of total cleavage) beginning in a yolk-free polar cap of cytoplasm. The first two divisions are equal and perpendicular, but no cleavage furrow passes through the yolk. When furrows are completed at the 64-cell stage, blastomeres quickly surround the yolk in a form of precocious gastrulation.

Myriapod eggs tend to isolate syncytial yolk pyramids. The surface of a centipede egg is marked by fields overlying yolk pyramids. While some intralecithal nuclei remain within the yolk in association with cytoplasm and become vitellophages, most energids gather at the periphery and concentrate in a germ disk.

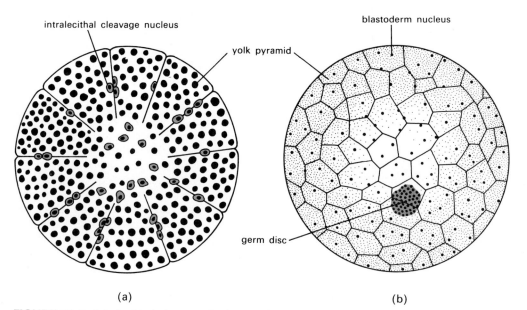

(a)   (b)

FIGURE 11.3. Intralecithal cleavage in the centipede, *Scolopendra cingulata*. (a) Section of embryo at yolk-pyramid stage. (b) Superficial view of blastoderm. From D. T. Anderson, *Embryology and phylogeny in annelids and arthropods,* Pergamon Press, Oxford, 1973. Courtesy of D. T. Anderson.

Blastomeres form when plasma membrane surrounds peripheral energids. Radial mitosis then gives rise to superficial blastomeres, and spreading of blastomeres over the surface produces a complete blastoderm. The furrows initially present in the central yolk mass between yolk pyramids (Fig. 11.3) break down, and a unitary nucleated yolk mass appears in the center of the egg.

Crustaceans generally exhibit radial cleavage derived from spiral cleavage (see below), but cleavage in many crustacea grades into the intralecithal type. Some members of the order Malacostraca exhibit unambiguous intralecithal cleavage.

Copepods produce small eggs which begin cleavage with a truncated intralecithal cleavage. In *Holopedium,* with an egg of 120 μm in diameter, no cytokinesis occurs during the first three synchronous nuclear divisions. Then furrows simultaneously separate eight superficial **octants,** leaving the yolky center intact. Mitosis accompanied by superficial furrowing proceeds through a 16- and 31-nuclear stage when the yolk is finally cleaved into complete, equally yolky, pyramidal blastomeres.

Cleavage in the larger (250 × 190 μm) eggs of *Daphnia* is similar, but penetration of furrows into the yolky center is never more than superficial. A blastoderm ultimately forms around a yolky center.

Among the malacostracans, the amphipods, such as *Gammarus,* with a large 500 × 350 μm egg, and parasitic isopods, with small 150 × 170 μm eggs, exhibit total and approximately equal radial cleavage. However, pyramidal blastomeres merge into a blastoderm surrounding a unitary yolk mass. Similar yolky pyramidal blastomeres are formed by other malacostracan eggs, and *Anaspides,* with an egg 1 mm in diameter, forms a hollow, spherical blastoderm despite its large size.

Decapods with eggs in the vicinity of 400–950 μm in length and 300–600 μm in diameter exhibit total cleavage at first. Yolky pyramidal cells form, and, as their number increases, nuclei with associated cytoplasm migrate peripherally where they are eventually cut off as a superficial blastoderm. The borders between the yolky pyramids break down, and a single yolk mass forms within the cellular blastoderm. In other decapods, such as *Astacus,* with an egg 2.8 × 2.4 mm, the formation of yolky pyramidal cells seems to have been skipped, and blastoderm formation occurs directly by the intralecithal method (Fig. 11.4).

The apterygote insects exhibit a range of cleavage types from the purely intralecithal and superficial to something more nearly total (Jura, 1972). Cleavage is intralecithal in the large (approximately 1.0 mm) ovid apterygote eggs, such as those of *Thysanura* (i.e., bristle-tails), and in the spherical eggs of *Diplura* (i.e., japygids). In thysan-

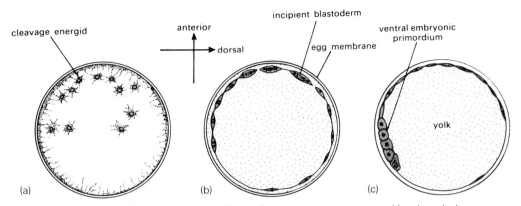

FIGURE 11.4. Intralecithal cleavage in the malacostracan crustacean, *Hemimysis lamornae.* (a) The 12-nuclear stage. (b) The 64-cell stage. Early blastoderm. (c) Germ disk forming on ventral surface of blastoderm. From D. T. Anderson, *Embryology and phylogeny in annelids and arthropods,* Pergamon Press, Oxford, 1973. Courtesy of D. T. Anderson.

urans (Fig. 11.5), the first three to five intralecithal mitoses are synchronous, but many more asynchronous divisions follow. Most of the cleavage energids migrate to the surface where they produce a uniform blastoderm, while a few vitellophagic energids remain in the yolk.

The thysanuran embryo develops from an embryonic disk formed at the posterior end, while the remainder of the blastoderm becomes extraembryonic ectoderm. The differentiation of the blas-

toderm and development of an embryonic primordium are characteristics shared by apterygote and pterygote embryos.

Polarity emerges in thysanuran eggs only when a differentially high rate of mitosis in the emerging posterior blastoderm causes a thickening (Fig. 11.5). A posterior to ventroposterior disk forms which is the compact **embryonic primordium.**

Smaller collembolan apterygote eggs (i.e., snow fleas) exhibit a truncated form of intralecithal

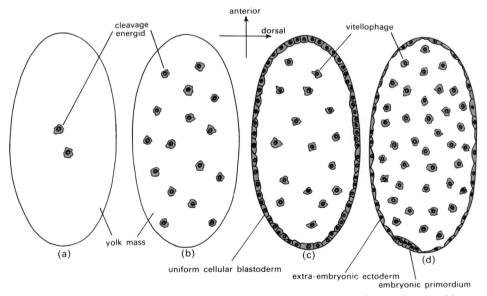

FIGURE 11.5. Intralecithal cleavage in a thysanuran egg. (a–b) Cleavage energids are suspended in the yolk mass. (c) Energids that migrate to the surface form a uniformly cellular blastoderm, while those remaining within the yolk become vitellophages. (d) The embryonic primordium forms as a thickening of the posterior (sometimes ventroposterior) blastoderm, leaving the remainder of the blastoderm as extraembryonic ectoderm. From D. T. Anderson, *Embryology and phylogeny in annelids and arthropods,* Pergamon Press, Oxford, 1973. Courtesy of D. T. Anderson.

cleavage and, if only initially, total (possibly spiral) rather than superficial cleavage of the egg mass. Late cleavage is superficial and limited to a cellular blastoderm. Ultimately, a typical superficial cleavage pattern emerges.

In *Tetrodontophora bielanensis,* two intralecithal nuclear divisions are followed by the total cleavage of the embryo (Fig. 11.6). While other collembolan eggs exhibit several intralecithal divisions, *T. bielanensis* definitely forms a 4-cell stage. Cleavage synchrony then breaks down, and blastomeres are formed with irregular and unequal shapes.

Total cleavage continues to the 8-cell stage when the collembolan egg is a sphere of pyramidal blastomeres, but, at this point, superficial cleavage commences. Tangential divisions produce an outer layer of pyramidal, yolk-free blastomeres, while

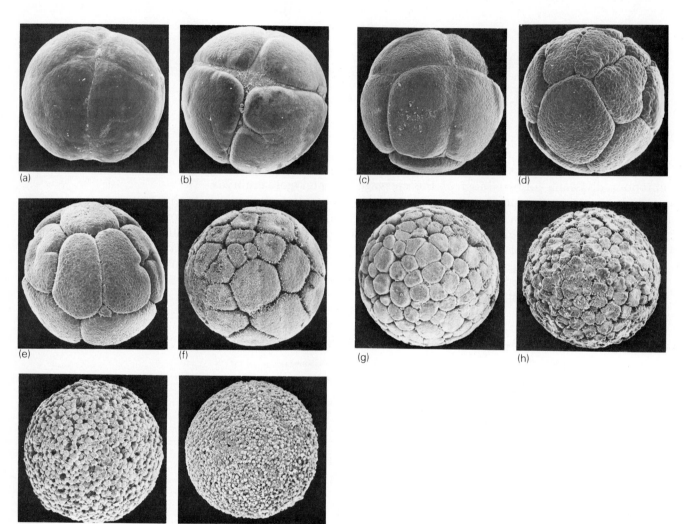

FIGURE 11.6. Scanning electron micrographs of truncated intralecithal cleavage and direct development of the primitive collembolan apterygote *Tetrodontophora bielanensis.* All ×205. (a) Animal pole view about 8 days since fertilization (d.s.f.). Total, equal cleavage beginning. (b) Animal pole view about 10 d.s.f. Divisions are synchronous. 4- to 8-cell stage. (c) Side view about 10 d.s.f. 8-cell stage. (d) About 12 d.s.f. 8- to 16-cell stage. (e) About 14 d.s.f. 16- to 32-cell stage. Cleavage becomes asynchronous and unequal, although furrows (right and left) continue to be perpendicular to old ones. (f) About 16 d.s.f. Advanced embryo 32- to 64-cell stage. Animal pole blastomeres smaller than vegetal pole blastomeres. (g) About 18 d.s.f. "Resting" 64-cell stage prior to next cleavage. (h) About 20 d.s.f. 128-cell stage. (i) About 22 d.s.f. 300-cell stage. (j) About 35 d.s.f. 3,000-cell stage. Premature blastoderm stage. From Cz. Jura, A. Krzyszotofowicz and E. Kisiel. In H. Ando and Cz. Jura *Recent advances in insect embryology in Japan and Poland* (1987), by permission of Arthropod Embryology Society of Japan and courtesy of the authors.

the central yolk mass fails to cleave. The superficial blastomeres shed their yolky portions, and nuclear-free yolk accumulates in a central cavity.

As blastomeres continue to accumulate in the blastoderm, additional yolk concentrates centrally. At about the 64-cell stage (Fig. 11.6f), occasional radial mitosis gives rise to inner **yolky cells.** Initially polygonal in shape, the population of small or irregularly shaped yolky cells moves internally. Shedding yolk in the form of globules, the cells become dispersed in the yolk mass. The cells gathering as a tight central mass are **primordial germ cells,** while those more nearly separated and intermediate in position are vitellophages.

Meanwhile, the outer blastomeres also release yolk globules. Bound by a reticulum of cytoplasm, the nonnucleated globules give the blastoderm a bubbly appearance. Once free of yolk, the outer blastomeres change from pyramidal to more nearly spherical, and nuclei move from the center of cells to their periphery. As the blastoderm becomes definitive, the blastomeres change to columnar and cuboidal shapes. At the blastoderm stage, the collembolan embryo resembles the pterygote embryo although arriving at this condition through a rather different route.

The pterygote insects show less overt variation in cleavage than the apterygotes. From primitive Exopterygota (= hemimetabolous species)

with large yolky eggs to the advanced Endopterygota (= holometabolous species) with relatively smaller eggs, classic intranuclear and superficial cleavage is the rule. Exceptions occur in species with specialized polyembryony (i.e., formation of more than one embryo from a single zygote), in viviparous species, and in parasitic species (Ivanova-Kasas, 1972).

The slipper-shaped eggs of exopterygote insects are about 1 mm in length, with a distinctive anterior–posterior axis, generally, a concave dorsal surface, and a convex ventral surface. The nucleus is central. The periplasm is thin in the eggs of primitive exopterygotes but thick and linked to a **cytoplasmic reticulum** separating **yolk spheres** in the eggs of more advanced species (see Anderson, 1973).

The elongated spheroidal (regular ovoid slightly flattened dorsoventrally) eggs of endopterygote insects have a cytoplasmic reticulum and a distinctive periplasm similar to that of advanced exopterygote insects. Less than 1 mm in length (e.g., 560 µm in *Drosophila*), the eggs of advanced endopterygotes have less yolk than those of other insects.

Synchronous divisions producing intralecithal nuclei may occur throughout cleavage. In the endopterygote insects, between 7 and 10 divisions precede the nuclear invasion of the periplasm (Table 11.2). Another two to five approximately syn-

TABLE 11.2. Synchronous divisions preceding cellularization of the blastoderm in Endopterygota (Holometabola)

Order Species	Hours per Mitosis	Number of Energids Moving into Periplasm	Number of Divisions at Cellularization
Coleoptera	0.5–1	256–512	11–13
Megaloptera	—	128	9
Neuroptera	—		10
Mecoptera	1	1024	—
Lepidoptera	1 (approx.)	512	11
	0.5 (approx.)	512	11
Diptera			
Culex fatigans	0.25	128	12
Dacus tryoni	0.25	128	13
Drosophila melanogaster	0.17	512	12
Lucilia sericata	0.17	512	13
Cochliomyia hominivorax	0.08	256	12
Siphonaptera		128	10
Hymenoptera			
Habrobracon juglandis	0.25	1024	13
Apis mellifera	0.5	1024	12

Source: D. T. Anderson. The development of holometabolous insects. In S. J. Counce and C. H. Waddington, eds., *Developmental systems: Insects,* vol. 1. Academic Press, London, 1972.

chronous divisions occur in the syncytial blasto-derm before cellularization.

The distribution of cytoplasm in the cellular-ized blastoderm is conspicuously nonhomoge-neous. Whether a blastodisk, a disk of blastomeres, a germ disk, saddle of blastoderm, or pole cells, specialized areas of the egg cortex are readily iden-tifiable. The presence of these areas suggests the presence of a preexisting **morphogenic pattern** de-termining the movement of energids to and within the periphery.

The freshly laid egg of *Drosophila melanogas-ter* completely fills the vitelline envelope. The ho-mogeneous yolk is distributed throughout the cytoplasm except for the rim of periplasm (Fig. 11.7). The egg's nucleus with its associated cyto-plasm is situated dorsally, approximately three-quarters of the total distance from the posterior pole (i.e., 75% EL [egg length] measured as a per-centage of distance from the posterior pole).

As the spermatozoon's nucleus becomes a male pronucleus, the oocyte nucleus undergoes its first meiotic division without ejecting a polar body. The male pronucleus approaches the egg nuclei as they undergo the second meiotic division to pro-duce four haploid nuclei (if both nuclei divide, or one diploid and two haploid nuclei if only one oocyte nucleus divides). The innermost of these ootid nuclei becomes the female pronucleus (Cam-pos-Ortega and Hartenstein, 1985). The polar bod-ies are then thrown off anterodorsally (Fig. 11.8a, arrow).

The first two nuclear divisions of cleavage take place within 25 minutes of the egg's being laid (stage 1). The four intralecithal nuclei and halos of cytoplasm (i.e., energids) lie in the anterior third of the egg (Fig. 11.8a).

During the next 40 minutes, the egg forms clefts at the anterior and posterior poles, and the intralecithal nuclei pass through divisions 3 to 8 (stage 2). The rate of division in cyclorrhaphan em-bryos (1 per 5 minutes in screwworm larvas, Table 11.2), botflies, and common houseflies is among the fastest for animal mitosis.

The intralecithal energids move posteriorly until they are distributed over an ellipsoidal field in the center of the egg approximately equidistant from both poles (i.e., between 20 and 80% EL, Fig. 11.8b). In *D. melanogaster* at the seventh cycle of nuclear division (i.e., the 128-nuclear stage), about 26 nuclei and associated cytoplasm remain within the yolk and become vitellophages or yolk cells. Their nuclei divide three more times in synchrony with other nuclei, but, afterward, replication with-out division leads to polyploid nuclei. The other

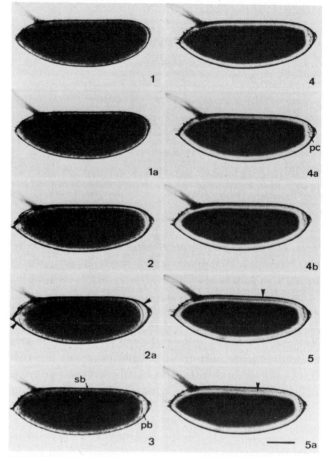

FIGURE 11.7. Photomicrographs of living *Drosophila me-lanogaster* eggs. Eggs were rendered transparent by im-mersion in Voltalef oil. Embryos at stages 1 and 1a are the same embryo; those at stages 2–5a are another embryo. Stages are given in lower right hand corner. Bar = 100 μm. Stages 1–1a: about 25 minutes. Egg maturation and first two nuclear divisions of cleavage. Stages 2–2a: Nu-clear divisions 3–9. Yolk withdraws from periphery, and clefts (arrowheads) appear at ends of egg. Stage 3: Three pole buds (pb) appear at posterior pole. These will divide twice before becoming complete pole cells. Somatic cell buds (sb) appear in periplasm. Stage 4–4b: Pole cells (pc) completed posteriorly. Stage 5–5a: Cellularization begins. Plasma membranes divide periplasm (arrowheads). From J. A. Campos-Ortega and V. Hartenstein, *Embryonic de-velopment of Drosophila melanogaster,* Springer-Verlag, Berlin, 1985. Courtesy of the authors. Used by permission.

102 nuclei at the seventh cycle turn peripherally and migrate to the surface (Fig. 11.8c).

The first nuclei reaching the periphery enter the posterior cleft where they induce the formation of **pole buds** (pb, Fig. 11.7). While dividing in syn-chrony with other nuclei, pole bud nuclei and cyto-plasm form 12–18 **pole cells** (pc, Figs. 11.7). These are the first complete cells in the embryo. For the most part, they are **primordial germ cells,** but a few

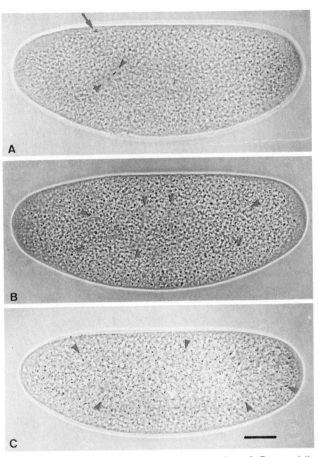

FIGURE 11.8. Phase contrast micrographs of *Drosophila melanogaster* embryos fixed and stained with fuchsin to aid visualizing intralecithal nuclei (arrowheads). Bar = 50 μm. (a) Stage 1 embryo during second intralecithal division. Arrowheads point to chromosomes of two nuclei in the anterior third of the egg. The other two lie in approximately the same position but outside the plane of focus. A polar body is visible at the arrow. (b) Stage 2 embryo. Intralecithal energids have moved posteriorly and now occupy an ellipsoidal field in center of the egg. (c) Later stage 2 embryo. Intralecithal energids move peripherally. From J. A. Campos-Ortega and V. Hartenstein, *Embryonic development of Drosophila melanogaster,* Springer-Verlag, Berlin, 1985. Courtesy of the authors. Used by permission.

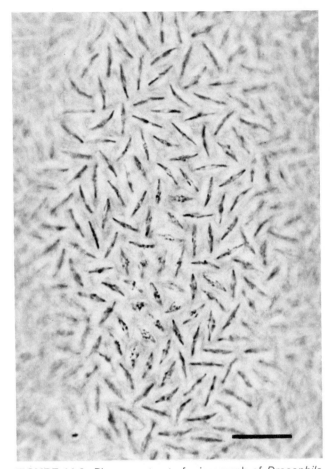

FIGURE 11.9. Phase contrast of micrograph of *Drosophila melanogaster* egg showing synchronous anaphase figures of the 13th division in a portion of syncytial blastoderm. Embryos were fixed and stained with fuchsin to aid visualizing chromosomes. Bar = 10 μm. From J. A. Campos-Ortega and V. Hartenstein, *Embryonic development of Drosophila melanogaster,* Springer-Verlag, Berlin, 1985. Courtesy of the authors. Used by permission.

pole cells interdigitate with posterior blastoderm cells and take part in forming the midgut. After one or two more synchronous divisions, 34–37 pole cells begin dividing asynchronously with the **somatic** nuclei comprising the remainder of the blastoderm.

By the ninth division, which lasts 15 minutes in *D. melanogaster*, **peripheral nuclei** are evenly distributed throughout the periplasm (stage 3). During this division, the surface bubbles with **somatic buds** indicative of incipient cellularization.

For the next 50 minutes, the embryo is in the **syncytial blastoderm stage** (stage 4, Figs. 11.9 and 11.10a). The rim of cytoplasm thickens for four more divisions. These are **metasynchronous** divisions, proceeding in waves. The wave fronts originate at both the anterior and posterior poles and meet at the egg's equator. The size of the nuclei decreases, while the length of the cleavage cycle increases to approximately 20 minutes.

In *Drosophila montana* (Fig. 11.11), cellularization is initiated after approximately 12 cleavage divisions (slightly earlier than in *D. melanogaster*) when the population of nuclei in the syncytial blastoderm is about 3500 nuclei. As tangentially oriented peripheral mitotic spindles break down, nuclei elongate radially. Centrioles and asters appear peripherally amid a ground substance. Microtubules form a sheath around the nuclei, and nu-

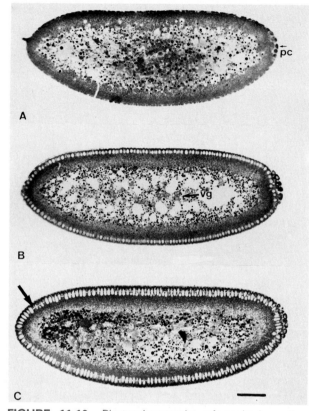

FIGURE 11.10. Photomicrographs of sagittal sections through _Drosophila melanogaster_ embryos during cellularization. Bar = 50 μm. (a) Early stage 4, the syncytial blastoderm stage. Bulges at surface are the buds of incipient cells. The first buds to form and the first cells to be completed are the pole cells (pc) at the posterior end of the embryo. (b) End stage 4. Rounded interphase nuclei are evenly distributed around periphery. Yolk grains and vitellophages (vg) are distributed centrally. (c) Stage 5. Cellularization has begun. Folds of plasma membranes have penetrated periplasm about midway through the peripheral nuclei (arrow). From J. A. Campos-Ortega and V. Hartenstein, _Embryonic development of Drosophila melanogaster,_ Springer-Verlag, Berlin, 1985. Courtesy of the authors. Used by permission.

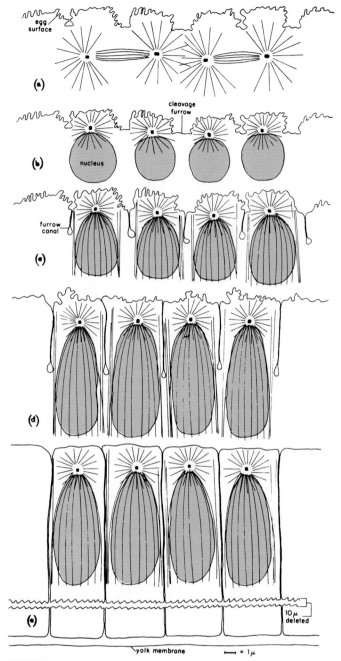

FIGURE 11.11. Diagrammatic illustration of cellularization at the periphery of a _Drosophila montana_ egg. (a) Twelfth synchronous division (third in syncytial blastoderm) supplies about 3500 nuclei. (b) The egg's plasma membrane becomes corrugated in anticipation of cleavage furrow formation. Asters with centrioles appear peripherally. (c, d) Surrounded by a sheath of microtubules, the nuclei elongate and move away from the surface. Furrow canals lead the plasma membrane downward, and the surface begins to flatten. (e) With the completion of cellularization, the blastomeres are surrounded by their own plasma membranes, and the yolk is surrounded by a yolk membrane. From S. L. Fullilove and A. G. Jacobson, _Dev. Biol.,_ 26:560 (1971), by permission of Academic Press and the authors.

cleoli appear at the apices of the nuclei. The presence of these nucleoli signals the transition of energids to cells and the transformation of the syncytial blastoderm to the cellular blastoderm (Fullilove and Jacobson, 1971). Cellularization begins when **furrow canals** crease the plasma membrane and **internuclear folds** jut between peripheral nuclei. Elongating rapidly, the folds take only 30 minutes at 25°C to extend beyond the nuclei.

Desmosomes forming at the apices of the incipient cells seem to preclude the possibility of drawing the surface membrane inward. Instead, the membrane utilized in cellularization seems to be added from **multilamellar bodies** already in continuity with the surface (Fig. 11.12). For 90 minutes, the rolled up and stored multilamellar bodies spin out plasma membrane at an overall rate of about 600 μm^2 per second (stage 5, Figs. 11.7 and 11.10b).

Bubbling at the embryo's surface ceases, and the membrane flattens out peripherally as if stretched. Beneath the surface, furrows on the ventral side of the embryo reach the yolk, their point of deepest penetration, and spread laterally. Soon furrows on the dorsal and lateral sides also spread laterally above the yolk. The wide cytoplasmic bridges that connect the incipient cells of the blastoderm to the central yolk finally narrow, and adjacent furrows fuse, turning the upper membrane (lamella) into the plasma membrane of the blastodermal cells, and the lower membrane into the **yolk membrane** or **yolk sac** (Fig. 11.11).

By about 3 hours after *Drosophila* eggs are laid (stage 6), plasma membrane has clothed approximately 5000 cells and formed the yolk membrane

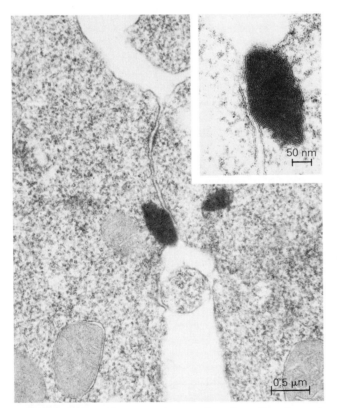

FIGURE 11.12. Electron micrographs of section through *Drosophila* blastoderm at time of cellularization showing superficial multilamellar body (dark body near mitochondria) in continuity with ingrowing furrow (insert). From J. G. Bluemink, W. J. Hage, and S. Bilinski, in H. C. Slavkin, ed., *Progress in developmental biology,* Part A, Alan R. Liss Inc., New York, 1986. Courtesy of the authors. Used by permission.

over the yolk mass. Individualized but closely linked, the blastoderm cells are now ready to commence morphogenic movements and gastrulation.

PARTIAL CLEAVAGE

Cleavage is partial when furrows do not penetrate the yolk. Initially, karyokinesis leads to the formation of a **yolk syncytium,** and furrows forming vertically to the surface leave **open blastomeres** rather than closed or complete blastomeres. Later, undercutting furrows and tangential divisions create complete blastomeres.

Two forms of partial cleavage are distinguishable: (1) the partial cleavage of cephalopods and some other invertebrates, such as ovoviviparous scorpions, and (2) the **discoidal** (also called polar discoidal) cleavage of megalecithal vertebrates (i.e., reptiles, birds, and oviparous prototherian mammals) and some mesolecithal teleosts. In the cephalopod type, division spreads from a cytoplasmic area more or less continuous with the yolk. In vertebrates, division is limited to an insular cytoplasmic **blastodisk** (sometimes blastodisc) or **germinal disk** (sometimes germinal disc) floating on a sea of yolk.

Partial Cleavage in Cephalopods

In the large (1.6 × 1 mm) eggs of the squid *Loligo pealei,* the first two furrows are straight and perpendicular to each other, but the third furrow is curved (Fig. 11.13). While parallel to the first posteriorly, the third diverges at about 30° anteriorly, delineating blastomeres of unequal sizes. The larger blastomeres are called **vegetal** (= macromeres); the smaller ones are called **animal** (= micromeres). This pattern is reminiscent of the pattern of spiral cleavage in other mollusks (see below, Arnold and Williams-Arnold, 1976) and in apparent violation of Balfour's law which requires blastomeres cleaved from equally yolky areas of cytoplasm to be of the same size.

The fourth furrow creates two populations of cells: those connected to the yolk laterally as well as centrally and those connected only centrally. Similarly, the fifth furrow adds blastomeres without lateral connections to yolk.

At the same time, the continuous downward movement and convergence of the first and second furrows separate the central blastomeres from the yolk below and create a **cellular blastoderm** of closed blastomeres. Below, a **yolk syncytium** of nu-

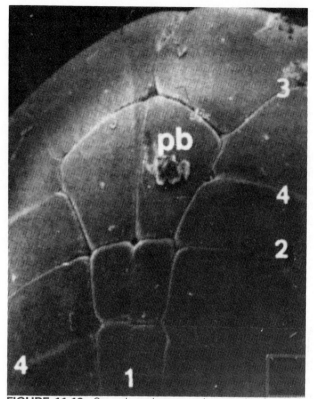

FIGURE 11.13. Scanning electron micrograph of a *Loligo pealei* embryo. The 16-cell stage illustrating partial cleavage. The first cleavage furrow (1) defines the median or sagittal plane of the embryo, and the second (2) separates the embryonic anterior and posterior halves. The third cleavage furrows (3) on either side of furrow 1 diverge at about a 30° angle anteriorly, thus skewing the pattern of blastomeres away from orthogonality. The two fourth (4) cleavage furrows on either side of the second separate a central group of 4 blastomeres from the peripheral yolk, thereby creating two classes of cells: those continuous with yolk peripherally and those separated from yolk peripherally. ×110. pb, Polar bodies; 1–4, cleavage furrows. From J. M. Arnold and L. D. Williams-Arnold, *Am. Zool.,* 16:421 (1976), by permission of the American Society of Zoologists.

clei embedded in cytoplasm continues to lie on the yolk.

Peripherally, the blastomeres acquire triangular outlines and are sometimes called **blastocones.** They divide and may ultimately form complete cells, but, while characteristically syncytial and broadly connected to the yolk, they expand laterally beyond the cellular blastoderm.

Gradually, the cellular blastoderm overgrows neighboring cells at its edge and forms an **outer cellular layer.** Mitotic activity beneath the outer cellular layer then thickens the blastoderm, and a morphogenic cellular layer forms between the outer cellular layer and the yolk syncytium.

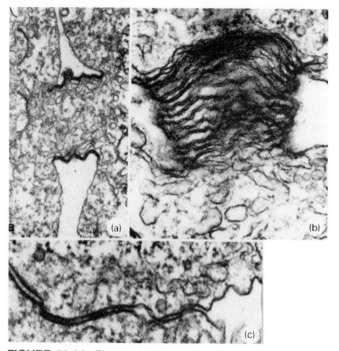

FIGURE 11.14. Electron micrographs of sections through the squid, *Loligo pealei,* blastoderm illustrating different types of intracellular connections. (a) Intercellular cytoplasmic bridge. The thickened walls resemble those of other cytoplasmic bridges. ×19,700. (b) Occluded intercellular bridge. The lamellae filling the bridge presumably block the passage of material. ×53,000. (c) Septate desmosomes between cells in the outer cell layer. ×56,000. From J. M. Arnold and L. D. Williams-Arnold, *Am. Zool.,* 16:421 (1976), by permission of the American Society of Zoologists.

As blastomeres accumulate in the cellular blastoderm, their connections and relationships become complex. Cytoplasmic bridges between sibling cells become occluded during interphase (Fig. 11.14). Gap junctions and **septate desmosomes,** as well as "ball joint" complexes associated with vesicles, and connecting processes associated with gap junctions, appear among the blastomeres. All these structures tend to appear at the end of cleavage as morphogenesis and cell movement become the dominant activities of the blastoderm.

Discoidal Cleavage in Teleosts

Teleosts, such as the zebrafish *Brachidanio rerio* and the Japanese medaka *Oryzias latipes,* undergo discoidal cleavage following a dramatic reshuffling of cytoplasm and yolk (Fig. 9.6). The blastodisk (Fig. 11.15a) formed by cytoplasmic streaming is elliptical, rather than circular, and lies on a side when coming to rest under the microscope. The first cleavage furrow cuts across the short axis

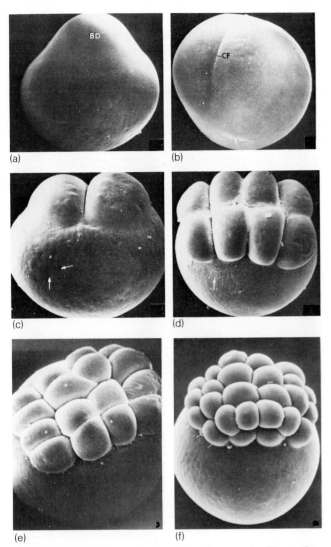

(a)   (b)

(c)   (d)

(e)   (f)

FIGURE 11.15. Scanning electron micrographs illustrating discoidal cleavage in the zebrafish, *Brachidanio rerio.* (a) Cytoplasmic streaming produces the blastodisk, BD. ×150. (b) First cleavage furrow, CF. ×160. (c) The 4-cell embryo. ×155. (d) The 8-cell embryo. ×155. (e) The 16-cell embryo. ×204. (f) The 32-cell embryo. ×180. From H. W. Beams and R. G. Kessel, *Am. Sci.,* 64:279 (1976), by permission of the Society of Sigma Xi and courtesy of the authors.

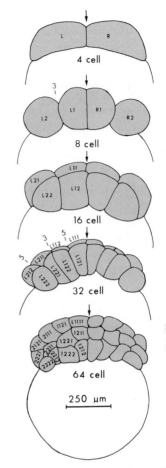

FIGURE 11.16. Camera lucida drawings of live zebrafish, *Brachidanio rerio,* embryos illustrating nomenclature for blastomeres. From C. B. Kimmel and R. D. Law, *Dev. Biol.,* 108:78(1985), by permission of Academic Press and the authors.

of the ellipse (Fig. 11.15b, i.e., perpendicular to the page in Fig. 11.16), while the second cleavage furrow formed 35–40 minutes after fertilization cuts across the long axis (Fig. 11.15c, i.e., in the plane of the page). Identical left (L) and right (R) cells lie in parallel planes (in front of and behind the page in Fig. 11.16).

The third furrow (formed about 70 minutes after fertilization) is parallel to the first, and the fourth furrow (at about 90 minutes) is parallel to the second. Blastomeres are thereby separated into subpopulations of **central** and **marginal** cells (from the center of the blastodisk to the edge). All divisions are restricted to the blastodisk in agreement with Balfour's law, and furrows are **vertical** (i.e., perpendicular to the blastodisk's surface) and alternate at right angles to one another in agreement with Hertwig's mechanical principles.

The fifth furrow (at about 105 minutes) may not appear quite synchronously in all blastomeres, and it may not be quite parallel or perpendicular to earlier furrows (Beams and Kessel, 1976). On occasions when the fifth cleavage is not vertical in one or more blastomeres, **compensatory** cleavage furrows at the sixth division restore the appearance of a blastodisk with five vertical furrows. As a result, an **orthogonal pattern** of rows and columns of blastomeres is present roughly through the 32-cell stage (Fig. 11.15).

During early interphase, transient **cytoplasmic bridges** connect the sibling blastomeres. Intracellular tracer molecules, such as fluorescein isothiocyanate-dextran (FD), which move across cytoplasmic bridges, quickly appear in both sibling blastomeres after injection into only one. As the next mitosis approaches, however, injected FD no longer moves to adjacent cells, indicating that cytoplasmic bridges close prior to mitosis.

Enclosing blastomeres totally within their plasma membrane and separating them from the yolk begins at the fifth cleavage when some furrows undercut the blastomeres. The sixth cleavage is generally parallel to the surface of the blastodisk, or **horizontal,** and completely cuts some cells off from the yolk and others off from the surface (Fig. 11.16, 64-cell stage). Blastomeres are thereby separated into subpopulations of **deep** or **internal cells** and **peripheral cells.**

The nomenclature employed to identify the blastomeres consists of L (left) and R (right) designations and numerical suffixes accumulating after the letters (Fig. 11.16). Every division after the second adds another numerical suffix. The blastomere lying nearest the midsagittal plane or the anterior end of the embryo (depending on the angle of furrowing) is numbered 1. The other blastomere is numbered 2.

Open blastomeres are detectable with the aid of FD. Injected into open blastomeres, FD diffuses into yolk and, as a backwash, into other nearby open blastomeres. Blastomeres separated from the yolk do not permit FD to enter the yolk. Confining FD begins with the sixth cleavage in some central cells (e.g., L1121, Fig. 11.16), but some marginal blastomeres remain broadly open to the yolk until the 10th cleavage (Kimmel and Law, 1985).

Gap junctions appear among the blastomeres after they are entirely separated from the yolk. Lucifer yellow (LY), a small fluorescent dye capable of moving through gap junctions, is seen in neighboring blastomeres soon after injection into a single blastomere. Unlike cytoplasmic bridges formed by incomplete mitosis and generally broken during subsequent development, gap junctions are sites of dynamic differentiation in adjacent cellular envelopes. The accumulation of gap junctions signals the beginning of morphogenic movement by blastomeres and the end of cleavage.

Discoidal Cleavage in Birds

In the hen, *Gallus domesticus*, discoidal cleavage begins in the **blastodisk** or **cytoplasmic germ plasm** with the formation of the first cleavage furrow about 4.8–6.7 hours after laying the previous egg or 0.5–2.4 hours after ovulation (hours uterine age, hr.u.a). The second cleavage furrow appears

1.0–2.7 hr.u.a, and the third appears 1.2–4.2 hr.u.a (Fig. 11.17). Throughout this time, the eggs are in the oviduct, at the edge of the isthmus, surrounded by a layer of albumin.

While furrow formation is visible upon dissecting out the egg, the nuclear events of early cleavage in avian embryos are cloaked in mystery. The ordinary rules of cell division requiring a close

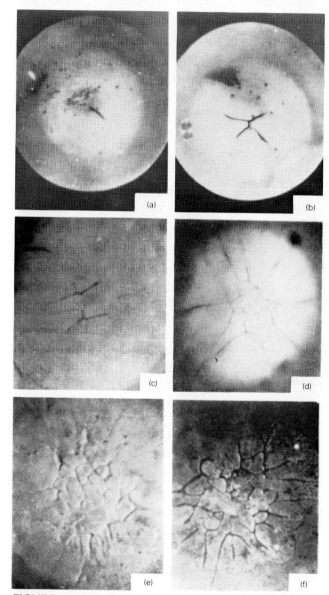

FIGURE 11.17. Photomicrographs of hen's blastodisk. (a) First cleavage furrow. Egg at isthmus of oviduct. ×12. (b) Second cleavage furrow. Egg at isthmus of oviduct. ×18. (c) The 8-cell stage. Egg at isthmus of oviduct. ×18. (d) About 16-cell stage. Egg at uterus. ×12. (e) Discoblastula of about 64 cells. Egg at uterus. ×18. (f) Discoblastula of about 100 cells (about 74 cells are central and 24 marginal). Egg at uterus. ×18. From M. W. Olsen, *J. Morphol.,* 70:513 (1942), by permission of Alan R. Liss, Inc.

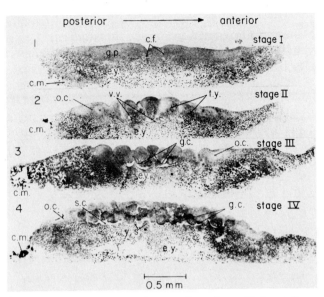

FIGURE 11.18. Photomicrographs of sagittal sections through the blastodisk of cleaving hen's eggs illustrating the Eyal-Giladi and Kochav (E.G&K) stages of the cleavage period at given hours of uterine age (hr.u.a). The blastodisk (or germ plasm, g.p.) was marked by carbon particles (c.m.) and fixed in Sunfelice's fluid which has the effect of precipitating glycogen as a "cap" (g.c.) toward the inner surface of cells. (1) Stage I (0–1 hr.u.a.). The blastodisk sitting upon the ball of yolk is poor in both yolk and glycogen. Vertical cleavage furrows (c.f.) penetrate the cytoplasm toward the egg yolk (e.y.). (2) Stage II (2 hr.u.a.). Giant open cells (o.c.) are formed centrally by vertical cleavage furrows that expand horizontally into ventral spaces (v.s.). Tongues of yolk (t.y.) penetrate the incipient cells as the blastodisk thickens medially and contracts laterally. (3) Stage III (3–4 hr.u.a.). Horizontal cleavage furrows appear separating closed epithelial and subepithelial cells at the thick center of the blastodisk. (4) Stage IV (5 hr.u.a.). Vertical cleavage furrows now reach the periphery of the blastodisk and form a single layer of open cells. The number of subepithelial cells increases centrally, thickening the blastoderm to some 3–4 cells. Ventral spaces merge into a subblastodermic cavity (s.c.). Yolk stumps (y.s.) protrude into the subblastodermic cavity from the underlying subgerminal periblast. c.f., Cleavage furrow; c.m., carbon mark (used to identify the cytoplasmic germ while removing it from the yolk for fixation); e.y., egg yolk; g.c., glycogen caps; g.p., (cytoplasmic) germ plasm; o.c., open cell; s.c., subblastodermic cavity; t.y., tongue of egg yolk; v.s., ventral space; y.s., yolk stumps. From S. Kochav, M. Ginsburg, and H. Eyal-Giladi, *Dev. Biol.,* 79:296 (1980) by permission of Academic Press and courtesy of the authors.

association between cytoplasmic furrows and mitosis do not seem to operate, and eggs may have external manifestations of cleavage without attendant nuclear divisions (Bellairs et al., 1978).

The difficulty in tracking nuclear events is possibly attributable to the large size of the avian egg. Beyond the ball of yolk, the blastodisk (g.p., Fig.

11.18, 1) is 2 mm in diameter at the time of oviposition. Nuclei may simply be indiscernible in this vast amount of cytoplasm.

Alternatively, the cleavage nuclei may not be in the area where investigators have looked for them. Instead of lying in the middle of the cytoplasmic portion of the blastodisk (as required by Balfour's law), the nuclei may lie below the cytoplasm in yolkier regions of the periblast. Possibly, cleavage nuclei within the yolk are overlooked because of their similarity to yolk spheres (Bellairs et al., 1978).

When later cleavage nuclei are finally seen (at about the 12-cell stage), they are not especially close to cleavage furrows. Instead of asters at the poles of mitotic spindles inducing the cortex to furrow, asters originating from some of the excess sperm acquired during polyspermic fertilization may operate during early cleavage.

Furrowing begins with a shallow V-shaped or U-shaped depression at the surface of the cytoplasmic germ plasm. Surrounded by a dense cytoplasm containing microfilaments and membrane-bound bodies, early furrows elongate into narrow pendulum-shaped passages with complex bulbous ends (Fig. 11.19 insert).

Initially, the furrows seem to be centers of pinocytotic activity with surface folds abounding in their walls. As the furrows deepen, the folds become restricted to elaborate furrow wall protrusions emerging from a **furrow base body** or cytoplasmic knot (Fig. 11.19). Tight junctions bind parallel portions of the furrow walls, and glycogen and yolk accumulate at the tips of the protrusions.

The areas demarcated by cleavage furrows should probably be considered syncytia even though they are called **open cells** (sometimes body cells). Communicating broadly with the yolk below (Fig. 11.20), open cells are giants, 500–700 μm in diameter.

Tongues of yolk (t.y., Fig. 11.18, 2) move into open cells in the intervals between divisions and introduce large quantities of yolk into the incipient blastomeres (Kochav et al., 1980). Glycogen also accumulates in the cells and may precipitate as an inverted "cap" (g.c.) during fixation.

The initial furrows do not appear in or reach out to the **marginal periblast** at the edge of the cytoplasmic germ. Centrally, the furrows (cf. Fig. 11.18, 1) burrow down and spread out laterally into **ventral spaces** (Figs. 11.18, 11.20) that merge into a **subblastodermic cavity** (s.c.). By 10–11 hr.u.a (corresponding to Eyal-Giladi and Kochav [E-G&K] stage VI), the subblastodermic cavity extends to the periphery (see Fig. 13.5).

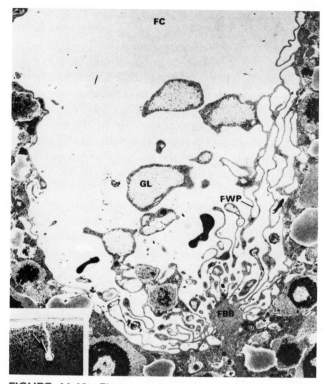

FIGURE 11.19. Electron micrograph of section through bulbous end of pendulum-shaped cleavage furrow of 4-cell chick embryo. ×3600. Section taken adjacent to region indicated by arrows in inserted photomicrograph. ×180. At the base of the furrow cavity (FC), the furrow base body (FBB) fans out into furrow wall protrusions (FWP) that dilate into vesicles containing glycogen particles (gl) and yolk. From I. Gibson, *J. Ultrastruct Res.*, 49:331 (1974), by permission of Academic Press and courtesy of I. Gibson.

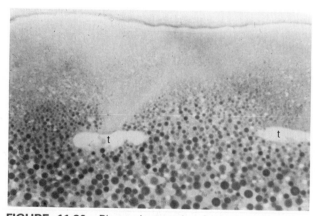

FIGURE 11.20. Photomicrograph of section through a hen's 12-cell embryo showing an open cell between transverse furrows (t). From R. Bellairs, F. W. Lorenz, and T. Dunlap, *J. Embryol. Exp. Morphol.*, 43:55 (1978), by permission of the Company of Biologists and courtesy of the authors.

The first closed blastomeres lie above the subblastodermic cavity forming a **blastoderm** (i.e., a skin of blastomeres). The **subgerminal periblast** lies beneath the subblastodermic cavity. Nonnucleated **yolk stumps** appear in the uneven surface of the subgerminal periblast. Divisions in the blastoderm produces a **superficial epithelium** and a deep layer of **subepithelial cells.**[1] The number of cells in the blastoderm increases as their sizes decrease (Fig. 11.18). At the end of cleavage, the central part of the blastoderm is 4 or 5 cells thick, while the peripheral parts are 1–2 cells thick.

TOTAL CLEAVAGE

Cleavage furrows cut through the entire zygote (cytoplasmic portion and yolk) and through blastomeres during total cleavage. The blastomeres formed are complete (or almost complete) membrane-bound cells from the start.

When blastomeres (especially early ones) are of relatively equal size, cleavage is called **equal.** When the blastomeres are of conspicuously different sizes, cleavage is called **unequal.** "Equal" actually means more equal than not, rather than identical.

Whether the division is equal or not, furrows generally form at a constant rate as they pass through isolecithal eggs. In contrast, eggs having a graded distribution of yolk exhibit **polarized total cleavage** in which furrows tend to slow down in the yolky vegetal hemisphere.

In accordance with Balfour's law, mitotic figures and horizontal furrows are displaced from the vegetal hemisphere and toward the animal hemisphere in eggs with graded distributions of yolk and hence with polarized total cleavage. A modest degree of polarization occurs in echinoderm eggs with a shallow gradation of yolk (i.e., a tending toward the telolecithal). An extreme degree of polarized total cleavage (sometimes identified with partial cleavage) occurs in amphibians and primitive bony fish. Later furrows begin forming in the animal hemisphere before earlier furrows are completed in the vegetal hemispheres. The incomplete furrows produce a transient syncytium. The early 2- and 4-cell stages of a frog embryo, for example, are actually, if only temporarily, 2- and 4-nuclear syncytia.

---

[1]'Epithelium' is sometimes reserved for specialized tissue in adults, and epiblast (or epiblasteme) is used for similar tissue in embryos.

Angles of cleavage in totally cleaving eggs are usually measured with regard to the animal–vegetal axis. This is not difficult in telolecithal eggs, but defining axes in nontelolecithal eggs can be arbitrary. Since polar bodies occur at the animal pole of many telolecithal eggs, the site of polar body emergence in nontelolecithal eggs is often considered equivalent to the animal pole. This site may not correspond to the animal pole by any other criterion.

The egg's pole opposite the site of polar body formation is often considered the vegetal pole. Some types of yolk may be concentrated at this pole. Instead of retarding cytoplasmic activities, the yolk of some totally cleaving eggs induces heightened activity (e.g., polar lobe formation in molluscan eggs).

The idea that animal and vegetal poles are homologous (i.e., have similarities based on common descent) in different species should not be accepted uncritically. Where species can be related by other criteria, the poles of their eggs can also be expected to exhibit homologies. Otherwise, the animal and vegetal poles and the animal–vegetal axis should merely be taken as a convenient starting point for describing the angles scored by cleavage furrows on the egg. Furrows cutting through or parallel to the animal–vegetal axis are called **vertical** (also meridional or longitudinal). These furrows produce blastomeres lying side by side around the axis. Vertical furrows that cut the zygote or blastomeres into equal right and left halves are called **bilateral** cleavages.

Furrows perpendicular to the animal–vegetal axis are called **horizontal** (also transverse or latitudinal) and produce blastomeres lying above and below one another along the axis. Horizontal furrows may also be called **equatorial** when they divide blastomeres equally.

When furrows are more nearly diagonal than parallel or perpendicular to the animal–vegetal axis, they are called **oblique** and cleavage that produces them is termed **oblique** or **spiral cleavage**. The blastomeres produced do not lie directly over one another but lie with their centers more or less over cleavage furrows. The first two cleavages, which come before overtly oblique divisions but cannot be considered oblique for want of morphologically distinct criteria, are sometimes considered **prospectively spiral.**

In general, blastomeres formed by the same horizontal or oblique cleavages and lying about the same distance from the animal or vegetal pole constitute a **tier** or **story.** Vertical divisions increase the number of cells in tiers, while horizontal or oblique divisions create new tiers or heighten existing ones. Early horizontal, vertical, and oblique divisions frequently occur in unison (or nearly in unison) among all the cells of a tier.

Following Hertwig's mechanical principles, horizontal divisions tend to alternate with vertical divisions, and clockwise divisions tend to alternate with counterclockwise divisions within a tier. Still, there are exceptions. In trombidiform and sarcoptiform mites (but not the mesostigmatids), for example, the mitotic spindles in sibling blastomeres are perpendicular to each other.

In general, as cleavage progresses, surface blastomeres bud off internal cells through **periclinal** furrows sloping parallel to the surface. Early divisions of internal cells tend to be bilateral (i.e., producing right and left siblings). The change from total cleavage marked on the surface to internal cleavages not marked on the surface frequently accompanies an increase in morphogenic activity and the end of cleavage.

The complexity of total cleavage is reflected in the number of **patterns** formed by blastomeres. Equal and unequal, uniform and polarized, vertical, horizontal, and oblique furrows all contribute to inscribing totally cleaving eggs with a specific pattern and often with an indelible morphogenic message.

Asymmetric Pattern

The asymmetric pattern (also called anaxonic, aplanar, and chaotic) is the default class for embryos whose cleavage furrows cannot be correlated with planes of symmetry. Two subclasses fall within this class.

Irregular cleavage occurs in some coelenterates. Although the pattern is sometimes considered a category of radial cleavage, cleavage furrows cut the egg along unspecified meridians. The pattern is also called **anarchic** in some hydroids where some blastomeres (excess blastomeres?) fail to be incorporated into the future embryo (see Berrill, 1961). The phenomenon of naturally discarding blastomeres, known as **blastomeric autotomy,** is not easily rationalized in terms of adaptation.

Occasionally, if rarely, when blastomeric autotomy affects the first blastomeres, embryos are produced by the separate blastomeres (Hargitt, 1911). Hydroid **polyembryony** (i.e., the creation of more than one embryo from the same egg) is an example of **natural regulation,** the ability of isolated blastomeres to restore whole embryos.

Spherical or rotational cleavage is the form of cleavage found in the alecithal and microlecithal eggs of eutherian and metatherian mammals (Gulyas, 1971). The early blastomeres are more or less equal, but the relationship of furrows to planes of symmetry is quickly obliterated by **blastomeric rotation.**

In the rabbit, the first cleavage plane (CP-1, Fig 11.21) passes through the axis marked by the polar bodies and more or less cuts the egg into halves. The blastomeres then flatten against each other (Fig. 11.22a).

In about half of the observed eggs, the first two blastomeres divide synchronously or division in one of the blastomeres follows quickly after division in the other. The two cleavage furrows (CP-2A and CP-2B, Fig. 11.21a) fall anywhere over a range from paralleling each other and the polar axis to deviating in a counterclockwise direction by as much as 35–50°.

The first two blastomeres divide asynchronously in the other half of the observed eggs. The first blastomere to divide (2–3 hours after first cleavage) is called the **first blastomere** (Fig.

11.22b–f). The cleavage plane cutting it (CP-2A) separates blastomeres A and B (i.e., right and left, respectively, without regard to planes of symmetry in the future embryo).

The second blastomere to divide (after a 15–20-minute delay) is called the **second blastomere.** This blastomere elongates prior to division (Fig. 11.23) and separates into blastomeres C and D. The cleavage plane (CP-2B, Fig. 11.21b) falls perpendicularly to both the previous planes (CP-1 and CP-2A),

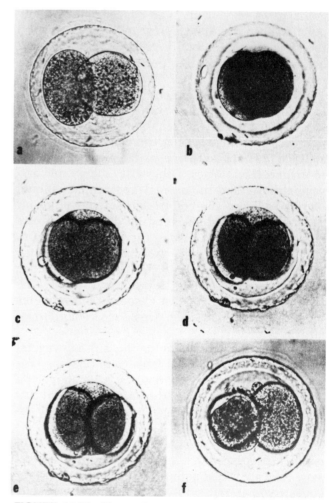

FIGURE 11.22. Photomicrographs of living rabbit embryo during cleavage of "first blastomere. All figures ×180. (a) The 2-cell stage viewed from animal pole. The "first blastomere" on left is elongating in the direction of the long axis of the spindle prior to second cleavage. (b–e) Cleavage of the "first blastomere" viewed from side (animal pole up, "first blastomere" forward). Cleavage plane (CP-2A) separates blastomere A (on right) from B (on left). (f) The 3-cell stage viewed from front (i.e., prior to second division of "second blastomere"). Animal pole up; A blastomere to left; "second blastomere" to right. From B. J. Gulyas, *J. Exp. Zool.,* 193:235 (1971), by permission of Alan R. Liss, Inc.

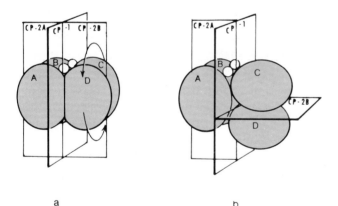

a   b

FIGURE 11.21. Diagram illustrating relationship of cleavage planes (CP) to rotation in rabbit embryo. CP-1 separates the first pair of blastomeres. CP-2A and CP-2B separate the second pairs. (a) When the second divisions are synchronous or nearly synchronous, CP-2A and CP-2B tend to lie along the same plane and cleavage is followed by counterclockwise rotation. (b) When one of the first two blastomeres (designated the "first blastomere") divides before the other (i.e., CP-2A precedes CP-2B), then CP-2B cuts the "second blastomere" perpendicularly to both CP-1 and CP-2A. The blastomeres come to rest in the same relative positions whether brought there by rotation or produced there by the cleavage planes. From B. J. Gulyas. *J. Exp. Zool.,* 193:235 (1971) by permission of Alan R. Liss, Inc.

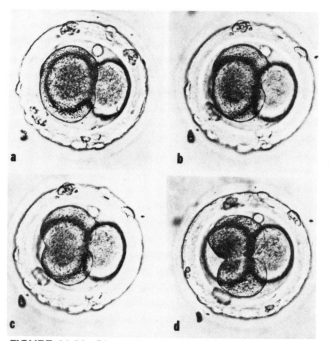

FIGURE 11.23. Photomicrographs of living rabbit embryo during cleavage of "second blastomere." All figures × 180. (a–d) (viewed from side) "Second blastomere" forward, A and B balstomeres back, polar bodies up. (a) The 3-cell stage. "Second blastomere" elongates. (b–d) Cleavage plane CP-2B creases second blastomere and separates it into blastomere C, above, and D, below. Note that CP-2B is perpendicular to CP-1 (in plane or page) and CP-2A (in long axis of page). From B. J. Gulyas, *J. Exp. Zool.,* 193:235 (1971), by permission of Alan R. Liss, Inc.

creating a **crosswise** pattern of furrows and blastomeres.

The difference between embryos with synchronous and asynchronous second cleavages does not last long. Counterclockwise **rotation** (Fig. 11.21a) of the C and D blastomeric pair of synchronous or nearly synchronous embryos rapidly produces the same crosswise arrangement of furrows and blastomeres seen initially in asynchronous embryos.

Similar crosswise arrangements of blastomeres are seen in mouse, rhesus monkey, mink, possibly baboon, and human embryos, but not in rat embryos. Rotation as such has been observed in the mouse as well as the rabbit and is considered likely to be present elsewhere (Gulyas, 1971).

Further divisions in each blastomere appear to be perpendicular to the previous division, but, because of rotation, divisions in the A and B blastomeres are perpendicular to divisions in the C and D blastomeres. Actually, "perpendicular" is not quite accurate, since additional rotations of blastomeres continue to throw the angles of cleavage planes askew.

Closer Look at the Early Eutherian Mammalian Embryo

The possibility of obtaining and fertilizing mammalian gametes *in vitro* was once a remote dream. Today, advances in hormonal physiology and culturing techniques have made *in vitro* fertilization a routine practice.

Obtaining eggs became feasible when the roles of gonadotropins and sex hormones in ovulation began to be clarified in the second quarter of this century. In the normal mammalian ovarian cycle, eggs develop under the influence of the pituitary gonadotropin, follicle-stimulating hormone (FSH), and the ovarian sex hormone, estradiol. At maturity, the ovarian follicle (or Graafian follicle) secretes large amounts of estradiol, triggering a massive release of a second gonadotropin, luteinizing hormone (LH), from the pituitary and triggering ovulation.

When these hormones or their synthetic cognates became commercially available, regimens of hormone administration were devised for controlling ovulation and even promoting it. By mimicking or exceeding the normal hormonal events, especially a large dose of gonadotropins, normal ovulation and **superovulation** (the shedding of unusually large numbers of mature oocytes) were induced.

In the case of women experiencing infertility due to ovulatory failures, hormonal therapy may be employed to induce superovulation. The subsequent high frequency of multiple births of nonidentical twins in these women is testimony to the efficacy of the therapy.

The same hormonal treatment that makes it possible to control ovulation makes it possible to accurately predict the time of ovulation and thus collect mature eggs from the ovary just prior to ovulation and at the moment of their greatest competence for productive fertilization. These are the eggs used for *in vitro* fertilization. In humans, several eggs can be aspirated from ripe follicles, while in mice and rabbits, several hundred eggs can be collected at a time.

Obtaining sperm might not seem to be as difficult as obtaining eggs, since sperm are normally ejaculated (i.e., forcefully expelled) from the male body in an ejaculum (also called ejaculate). But without capacitation, sperm taken from the epididymis, where they are stored prior to ejaculation, or even ejaculated sperm are incapable of fertilizing eggs *in vitro* (see Chapter 7). The passage of sperm through the uterus was once thought to be necessary for capacitation, but with improved culture techniques, epididymal and ejaculated sperm are now capacitated *in vitro*, and mammalian eggs are fertilized *in vitro* with sperm that have never been in a uterus.

Embryo culture has been most actively pursued in convenient laboratory mammals. With today's techniques, fertilized eggs of certain strains of mice and rabbits are cultured from 1-cell zygotes to blastocysts *in vitro*, and early sheep, cow, and human embryos can develop *in vitro* through cleavage into morulas and blastocysts. These are the **preimplantation** stages occurring prior to uterine implantation (Fig. 11.24).

Culturing embryos beyond the 2-cell stage was once considered enormously difficult. The preimplantation embryo was thought to require complex materials or substrates in the culturing medium, but, as it turned out, the requirements for embryos were not as demanding as those for adult cells *in vitro*. The mammalian embryo was not, however, found to be nearly as self-sufficient as the chicken or frog embryo.

The *in vitro* conditions for mammalian eggs were designed to resemble the *in situ* conditions of the oviduct through which the mammalian egg and early embryo normally pass on their way to the uterus. Discovering what these conditions were depended on good guesses and experimental

TABLE 11.3. Ability of different carbon sources to sustain mouse eggs and embryos *in vitro*

Substrate	Stage of Development[a]			
	Oocyte	1-Cell	2-Cell	8-Cell
Pyruvate	+	+	+	+
Oxaloacetate	+	+	+	+
Phosphoenolpyruvate	−	−	+	+
Lactate	−	−	+	+
Glucose	−	−	−	+

[a] −, Does not support development; +, supports development.
Source: Data from J. D. Biggers and R. M. Borland, 1976. *Annu. Rev. Physiol.*, 38:95.

testing. The mouse embryo, for example, was found not to utilize exogenous glucose prior to the 8-cell stage and therefore to require some other carbon source if it were to be cultured at early stages. Experiments demonstrated that the mouse oocyte and the 1- and 2-cell embryo could utilize pyruvate or oxaloacetate (Table 11.3), and adding either of these intermediates of the tricarboxylic acid cycle to the medium for culturing early mouse embryos yielded a successful outcome.

The culture medium for preimplantation embryos also required some protein, such as albumin. The protein was thought to function as a stabilizer of the egg membrane rather than as a nitrogen source, since it could be replaced with the metabolically inert macromolecule polyvinylpyrrolidone after the 2-cell stage. The medium also re-

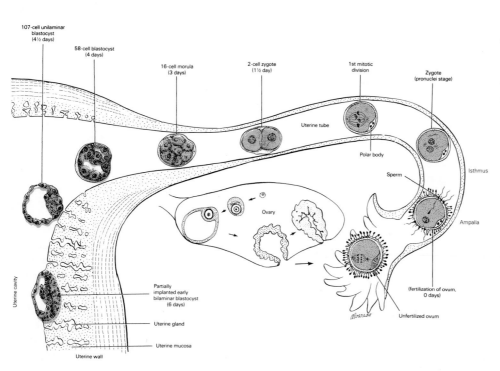

FIGURE 11.24. Schematic drawing of ovary, oviduct (uterine tube), and uterus (uterine cavity and wall) during fertilization, normal passage of embryo to uterus, and early implantation in human being. From R. F. Gasser, *Atlas of human embryos,* Harper & Row, New York, 1975. Used by permission.

quired some buffering salts, such as the bicarbonate buffer of Krebs–Ringer's solution, to maintain an optimal pH and to be osmotically compatible with the embryo.

Beyond the 4-cell stage, blastomeres become increasingly bound to each other. Originally confined by the zona pellucida, the loosely bound blastomeres of the early 8-cell embryo have lobular contours. Profound changes in surface glycoproteins, surface antigens, and microvilli are soon initiated. Changes in the membrane transport system result in fluid entering the spaces between blastomeres, while changes in connecting junctions are responsible for the accumulation of fluid and for intercellular communication (see Ducibella, 1977).

At the 8-cell stage, juxtaposed plasma membranes of mouse blastomeres become active and areas of contact increase. Membranes appear to zipper the blastomeres together in a process called **compaction** (Fig. 11.25).

At first, the blastomere's juxtaposed plasma membranes are **nonjunctional** (i.e., junctional specializations are absent). In the subsequent **morula stage** of mammalian embryos, the blastomeres move into even closer and stronger contact, and spaces disappear between nonjunctional membrane surfaces at **incipient junctional sites** (Figs. 11.26 and 11.27).

Segmentation cavities appear between membranes at other nonjunctional sites, and the spaces coalesce to form the **blastocyst cavity** or blastocoel. The process is called **cavitation,** but the term is a loose generic term applied to many situations involving cavities within a previously solid mass of cells.

Two types of specialized junction form between blastomeres: tight junctions and gap junctions. In the mouse, rat, and rabbit, these junctions first appear at the 8-cell stage. In monkeys, baboons, and human beings, the junctions only appear at the 16-cell stage.

Tight junctions (Fig. 11.27a) consist of cohesive bands attaching the outer leaflets of the plasma membranes of opposing cells. In addition, desmosomes (Fig. 11.26) or macular (spot) junctions accompany tight junctions and increase the strength of cell-to-cell bonds. Together, the tight junctions and desmosomes form a **junctional complex** that, with attached radiating filaments, conspicuously thickens, strengthens, and presumably stabilizes the cells' membranes.

Soon after first appearing, the tight junctions spread out around the blastomeres as **zonula tight junctions** or **zonula occludens** (also zona occlu-

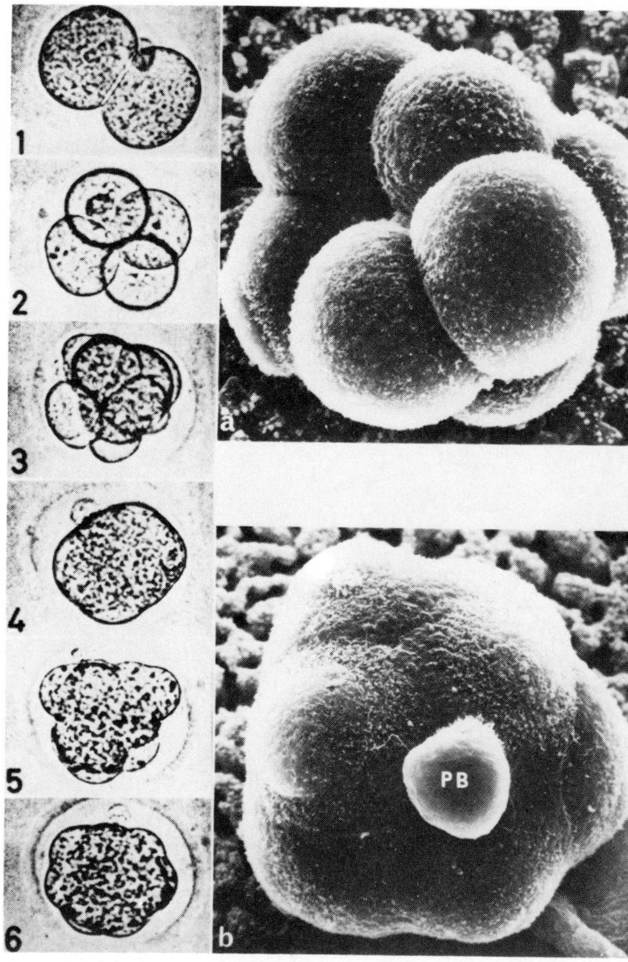

FIGURE 11.25. Single frames from film of mouse embryo cleaving *in vitro* (1–6) and scanning electron micrographs of a zona-free 8-cell embryo (a–b). (1) The 2-cell embryo; (2) 4-cell embryo; (3) early 8-cell embryo during compaction; (4) later 8-cell embryo after compaction; (5) early morula; (6) late morula. The morula's dividing blastomeres ripple the surface, but separate cells do not reappear. (a) Early 8-cell embryo before compaction. (b) Late 8-cell embryo after compaction. PB, polar body. Photomicrographs from J. G. Mulnard, *Arch. Biol.,* 78:107(1967), by courtesy of Professor Mulnard. SCMs from T. Ducibella, in M. H. Johnson, ed., *Development in mammals,* Vol. 1, by permission of North-Holland Publishing, Amsterdam, 1977.

dens). When fully formed, these junctions are leakproof barriers sealing off the embryo from the outside. As early as the 16–32-cell-stage morula in the mouse, tight junctions exclude molecular tracers, such as lanthanum, from the space between cells (Fig. 11.27a, b).

Gap junctions (see Fig. 2.9) are differentiated patches of adjacent cellular membranes mediating electrochemical cellular communication. The patches are made of densely packed **connexons,** complexes of intramembranous proteins arranged around a channel and penetrating the cell's plasma

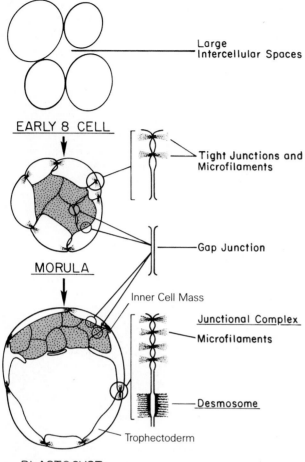

BLASTOCYST

FIGURE 11.26. Summary of steps leading to compaction of mouse embryo. Formation of tight junctions and junctional complexes among peripheral cells leads to differentiation of trophectoderm, while the formation of gap junctions leads to differentiation of the inner cell mass. From T. Ducibella, in M. H. Johnson, ed., *Development in mammals,* Vol. 1, by permission of North-Holland Publishing, Amsterdam, 1977, and the author.

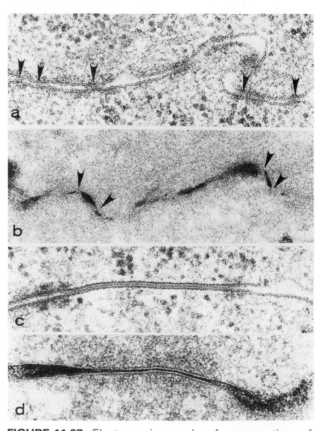

FIGURE 11.27. Electron micrographs of cross sections of tight (a, b) and gap (c, d) junctions between mouse embryo cells. Since the morula stage, tight junctions exclude lanthanum used as a molecular tracer, while gap junctions allow it to pass. (a) Apparent points of fusion of outer leaflets of plasma membrane (arrows). (b) Lanthanum tracer is excluded from points of fusion. (c) Pentalamina structure of gap junction. (d) Narrow channels of intercellular communication infiltrated by lanthanum tracer. ×72,000 approximately. From T. Ducibella, in M. H. Johnson, ed., *Development in mammals,* Vol. 1, by permission of North-Holland Publishing, Amsterdam, 1977, and the author.

membrane. Where the connexons of adjacent cells abut each other, they narrow the gap between the cells (hence the name of the junction), and where the connexons' channels are in register, they form a continuous channel between cells. The junctions do not exclude small molecular tracers (less than 1500 daltons) from intercellular passage (Fig. 11.27c, d), and their presence among blastomeres indicates that the embryo has constructed a "communication network."

The formation of tight and gap junctions is the first act of differentiation in the embryo, since nothing like them occurs previously, and they occur differentially among the blastomeres. Zona occludens form between blastomeres at the periphery of the embryo, while gap junctions form both be-

tween and among peripheral and interior blastomeres (Fig. 11.26). After the tight and gap junctions are in place, the layer of peripheral cells is known as **trophectoderm** (also trophoblast), and the collection of interior cells is known as the **inner cell mass.**

Although not apparent morphologically, the difference between trophectoderm and inner cell mass seems to be anticipated earlier in cleavage. In the early embryo, cells dividing faster than other cells get the jump on inner positions and contribute disproportionately to the inner cell mass. Possibly, differences in the rates of mitosis mean that cells are already differentiated even before their junctions provide morphological evidence of differentiation.

Vertical Pattern of Total Cleavage

Vertical cleavage (as opposed to spiral cleavage) is judged to occur when early cleavage furrows passing through the animal–vegetal axis tend to identify planes of symmetry in the future organism. When cleavage furrows anticipate planes in radially, biradially, and bilaterally symmetrical larvae or adults, the cleavages are called, respectively, **radial, biradial,** and **bilateral.**

In species with **direct development** (i.e., hatching or being born with adultlike features), dorsal–ventral and right–left symmetries are often manifest during cleavage (e.g., *Branchiostoma* [*Amphioxus*] and primitive fish among the protochordates and vertebrates). In species with **indirect development,**[2] the larval planes of symmetry are manifest in the embryo and transmitted to the adult despite an intervening **metamorphosis** (Gk. transformation) during which larval parts are remodeled or, more often, destroyed and replaced with new adult parts. Some species (e.g., sea urchins) delay the onset of their adult symmetry until late cleavage stages. Others (e.g., spirally cleaving embryos) initiate their adult symmetry soon after cleavage is under way. Still other species (e.g., ascidians and amphibians among the protochordates and vertebrates) initiate adult symmetry in the embryo as early as the first cleavage, if not at the point of fertilization.

Radial cleavage (also called monaxonic, polyplanar, and cylindrical) occurs when the animal–vegetal axis is preserved in the planes of the first cleavage furrows, but neither dorsal-ventral nor right-left planes are revealed during cleavage. Radial cleavage is probably not homologous in all species where it occurs. Rather, radial cleavage reflects the suppression of or relatively slow onset of adult types of symmetry especially in organisms having indirect development.

Pseudoscorpions exhibit a form of radial cleavage. Nourished by maternal secretions in a brood pouch, the pseudoscorpion embryo seems to have abandoned the yolk and intralecithal cleavage characteristic of other arachnids. The relatively small (60–130 μm), spherical eggs of pseudoscorpions have large, central nuclei and little yolk in

their cytoplasm. In different species, equal total cleavage creates a 4- or 8-cell stage of similar **macromeres.** The next division is unequal. Radially oriented mitotic spindles form, and each macromere cuts off a small, external, yolk-free **micromere** (Fig. 11.28).

The blastomeres produced by subsequent divisions of one or all the micromeres do one of two things. Some micromeres form a syncytial **trophic membrane** through which the embryo is nourished after transfer to the mother's brood pouch. Alternatively, most micromeres spread over the macromeres and form a blastoderm. The boundaries between macromeres break down, and nuclei remaining in the unified yolk mass become vitellophagic centers.

At the end of cleavage, the embryo's situation resembles that of a typical intralecithal egg after superficial cleavage. Presumably, pseudoscorpion eggs are derived from intralecithal eggs via secondary yolk reduction, and the pseudoscorpion type of radial cleavage is derived from superficial cleavage via precocious cellularization (see Anderson, 1973).

Crustaceans generally exhibit radial cleavage (some malacostracans, parasitic copepods, and Cirripedea being exceptions). Among the branchiopods, the cladoceran, *Polyphemus,* and the free living copepod, *Cyclops,* having relatively small eggs (120–140 μm in diameter) with little yolk, cleave with a simple radial pattern. The result is a hollow, spherical embryo (i.e., a blastula) of a few radially disposed wedge-shaped blastomeres having equal shares of yolk.

The anostracan, *Artemia salina,* also cleaves totally, equally, and radially without morphological evidence of axes. The yolky eggs become balls of 512 narrow cells surrounding a hollow center (Fig. 11.29).

The superficial distribution of nuclei in radially cleaving crustaceans is reminiscent of intralecithal cleavage. Moreover, blastomeres frequently revert to the condition of a blastoderm surrounding a central yolk mass, and some malacostracans seem to have transformed radial cleavage entirely into a form of intralecithal cleavage.

Differences between cleavage forms in crustaceans are correlated with the amount of yolk. Smaller, less yolky eggs tend to total radial cleavage, while larger, yolkier eggs tend toward intralecithal cleavage (see Anderson, 1973).

An unusual feature of radial cleavage in crustaceans is the segregation of presumptive larval or **nauplius areas** at a time when relatively few cells are present. In this respect, crustacean cleavage resembles spiral cleavage rather than other forms of

---

[2]Since larvae are frequently considerably modified before taking on adult form, indirect development is sometimes called **morphallactic development** or **morphallaxis** (Gk. *morphe* form + *allos* other or -*allag*- exchange). Direct development, on the other hand, is sometimes called **epimorphic development** or **epimorphosis** (Gk. *epi* on or upon + *morphe*), implying that the adult grows out of the embryo.

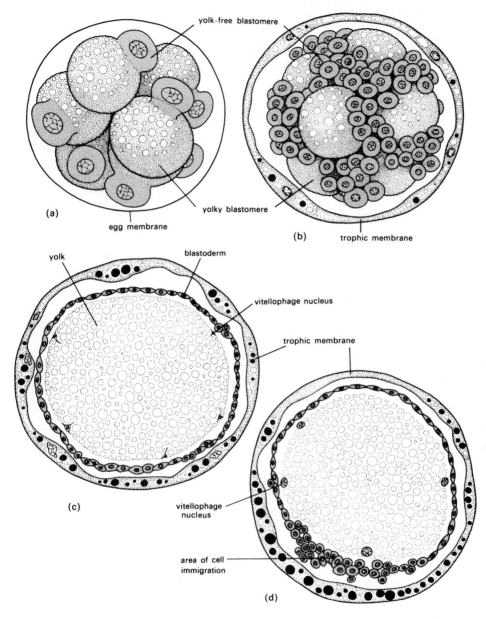

FIGURE 11.28. Schematic drawings of whole embryos during early cleavage, and sections of embryos at later cleavage illustrating radial cleavage and subsequent restoration of the superficial pattern in the pseudoscorpion, *Pselophochernes.* (a) At the radial fourth cleavage, 8 yolk-laden blastomeres (macromeres) spin off 8 superficial, yolk-free blastomeres (micromeres). (b) Micromeres divide and produce (1) a blastoderm that gradually surrounds the macromeres and (2) a trophic membrane through which the embryo is nourished when subsequently passed to the brood pouch. (c) Section through embryo at last cleavage shows completion of blastoderm. Breakdown of boundaries between macromeres leaves vitellophage nuclei within unified yolk mass. (d) Section through later stage shows piling up of micromeres at onset of gastrulation. From D. T. Anderson, *Embryology and phylogeny in annelids and arthropods*, Pergamon Press, Oxford, 1973. Courtesy of D. T. Anderson. Used by permission.

radial cleavage (e.g., echinoderms). Possibly, a primitive spiral pattern of cleavage, such as that present in barnacles (see below), became secondarily radial in crustaceans and finally intralecithal (see Anderson, 1973).

Echinoderm embryos exhibit another type of radial cleavage. Typically, the embryos have an animal–vegetal axis but lack a longitudinal plane between left and right sides and a frontal plane distinguishing dorsal and ventral aspects throughout cleavage (Fig. 11.30).

The sea urchin egg's animal–vegetal axis (defined by the location of polar bodies and the opposite pole of the egg) corresponds to the **anterior–posterior** axis of the future larva. This axis seems to be established before fertilization when the egg is still in the ovary. The vegetal pole is the point of oocyte attachment to the ovarian wall. The animal pole is opposite (see Giudice, 1973, 1986).

Animal–vegetal polarity is visible in some *Paracentrotus lividus* eggs in the form of a graded band of granules known as the Boveri band. Distributed throughout the animal hemisphere, the granules are concentrated in the subequatorial zone, but their abundance falls off rapidly below that zone, and they are absent at the vegetal pole (see Fig. 9.5).

The relationship between the point of sperm entry into the egg and the position of first cleavage furrow in the sea urchin is presently controversial (see Giudice, 1986). On one hand, a relationship is suggested in *Lytechinus variegatus* where the

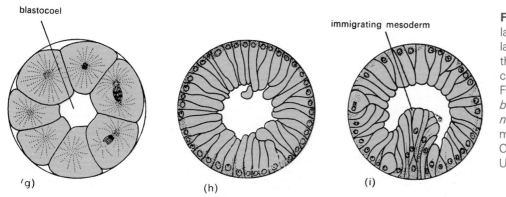

FIGURE 11.29. Diagram of late cleavage and early gastrulation in *Artemia salina.* (a) At the 5th cleavage. (b) The 512-cell stage. (c) early gastrula. From D. T. Anderson, *Embryology and phylogeny in annelids and arthropods,* Pergamon Press, Oxford, 1973. Courtesy of D. T. Anderson. Used by permission.

plane of the first cleavage furrow is seen to form usually within 8° of the direction of pronuclear movement (Schatten, 1981).

Contrary evidence comes from *Strongylocentrotus droebachiensis.* In this sea urchin and others, a **jelly canal** (sometimes incorrectly called a micropyle although it does not occur in a chorion) at the animal pole of the egg allows sperm access to the egg. Since the first cleavage furrow passes through the animal pole, the fertilization cone marking the point of sperm entry tends to be correlated with the plane of this furrow. The jelly canal is not, however, the exclusive point of sperm entry to the egg (Schroeder, 1980a), and when fertilization takes place elsewhere the furrow does

not necessarily pass through the fertilization cone (Schroeder, 1980b).

Less controversy surrounds the question of whether the point of sperm entry or particular early cleavage furrows influence the future dorsal–ventral axis of the larva (see Okazaki, 1975). The answer is clearly no. Marking the surface of blastomeres with vital dyes fails to reveal any dorsal–ventral or right–left differences until just prior to gastrulation. Neither the point of mouth formation (i.e., stomodeal invagination) on the ventral side of the prismatic larva nor the point of fusion of skeletal elements on the dorsal side of the future pluteus (Fig. 11.30) is anticipated by planes in the cleaving embryo.

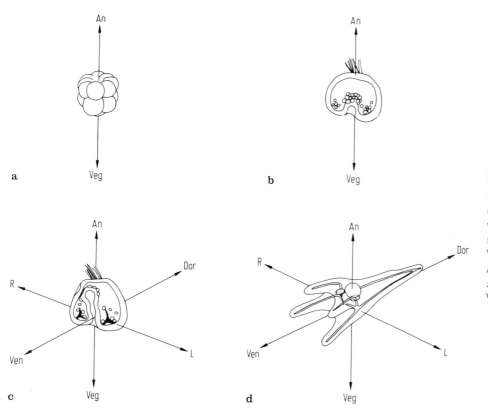

FIGURE 11.30. Diagram showing the positions and emergence of axes of symmetry in the sea urchin. (a) The 16-cell embryo with animal–vegetal axis in place. (b) Early gastrula with midvegetal invagination. (c) Early prism-stage larva. Bending of midgut toward wall with stomodeal invagination signals establishment of ventral aspect, while elongation of triradiate spicules toward point of fusion signals establishment of dorsal aspect. (d) Pluteus larva with bilateral symmetry. An, animal; Dor, dorsal; L, left; R, right; Veg, vegetal; Ven, ventral. Adapted from G. Giudice, *The sea urchin embryo.* Springer-Verlag, Berlin, 1986.

The second cleavage is vertical and perpendicular to the first, and the third cleavage is horizontal and perpendicular to the first and second (Fig. 11.31). As one of the first overt signs of animal–vegetal polarity, the third cleavage spindles are slightly displaced toward the animal hemisphere, and the furrows are not quite equatorial. Four slightly smaller animal-pole cells form above four slightly larger vegetal-pole cells.

The fourth cleavage reveals the inherent polarization of the embryo even more dramatically. The animal-pole cells divide vertically to form a tier of 8 more or less intermediate-sized cells called **mesomeres.** At the same time, the spindles in the vegetal-pole cells are strongly displaced vegetally. The cells divide horizontally and unequally to form an upper tier of four large **macromeres** and a lower tier of four small **micromeres** (Fig 11.31).

The unequal division of vegetal pole cells at the fourth cleavage seems to be determined independently of cleavage number, since suppressing earlier cleavages with hypotonic seawater, vigorous shaking (Hörstadius, 1939), or brief ultraviolet irradiation does not alter the timing of micromere production (see Hörstadius, 1973). The timing for cleaving the micromeres seems to be set by a **cleavage clock** that ticks away in the form of cyclic changes in the concentration of sulfhydryl-rich proteins in the egg cortex (see Sakai, 1968). Suppression of these changes by heat shock also suppresses cleavage (Dan and Ikeda, 1971), but the mechanism for driving mitotic spindles deeply into the vegetal hemisphere for the polarized fourth cleavage remains unknown.

The angles of cleavage continue to differ in the tiers during the fifth cleavage. The mesomeres are divided horizontally and equally into two 8-cell tiers known in descending order as **animal 1** (an$_1$) and **animal 2** (an$_2$), while the macromeres divide vertically, expanding the tier of macromeres to 8 still relatively large cells.

At the next cleavage, an$_1$ and an$_2$ cells divide horizontally, and the macromeres divide horizontally. The an$_1$ and an$_2$ tiers now consist of two 8-cell layers, while the macromeres give rise to two 8-cell tiers of more or less equal size. The upper tier is known as **vegetal 1** (veg$_1$) and a lower tier as **vegetal 2** (veg$_2$, Fig. 11.31).

The micromeres fail to divide synchronously with the tiers at the sixth cleavage, and a 56-cell stage (rather than the expected 64-cell stage) is present briefly. When the micromeres resume division, they do so at an accelerated rate and produce more micromeres which concentrate in a mass at the vegetal pole of the embryo.

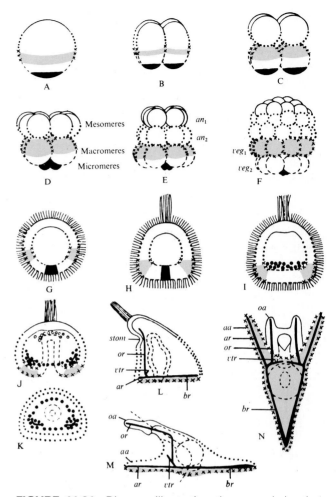

FIGURE 11.31. Diagram illustrating the normal development of the pluteus larva in the sea urchin, _Paracentrotus lividus_ (compare to Fig. 10.1). Solid lines, dots, crosses, dashes, and black show the portions of the fertilized egg involved in the formation of the an$_1$, an$_2$, veg$_1$, veg$_2$ tiers, and micromeres, respectively. (a) uncleaved egg. (b) The 4-cell stage. (c) The 8-cell stage. (d) The 16-cell stage. (e) The 32-cell stage. (f) The 64-cell stage. (g) Early blastula at time of hatching. (h) Late blastula with differentiation among cilia (note apical organ above). (i) Mesenchymal blastula (i.e., blastula after micromeres have migrated inwardly and formed the primary (1°) mesenchyme). (j) Gastrula after invagination of veg$_2$ and movement of secondary (2°) mesenchyme from tip of gut (note formation of two triradiate spicules among 1° mesenchyme cells). (k) Same as (j) seen in transverse (cross) section in vicinity of 1° micromeres. (l) Prism-stage larva showing ventral invagination of stomodeum (stom) by ectoderm derived from an$_1$. The spicules have elongated into an anal rod (ar), a body rod (br), an oral rod (or), and a ventral rod (vtr). (m, n) Pluteus larva from left side and anal side. The larva is bent at its oral arm (oa) and elongated dorsally and at its anal arm (aa). [Other abbreviations same as in (l)]. Note the bent broken line showing the position of the original animal–vegetal axis in the egg. From S. Hörstadius, _Pubbl. Staz. Zool. Napoli,_ 14:251 (1935). Used by permission of the Stazione Zoologie, Naples, Italy.

Biradial cleavage (also diaxonic and biplanar) is found in the biradial ctenophores (i.e., comb-jelly) characterized by direct development (i.e., hatching as a juvenile that develops into an adult without a conspicuous metamorphosis). Cteno-phores have two intersecting planes of symmetry (like an American football): a sagittal plane divid-ing the animal into identical sides, and a tentacular plane (i.e., equivalent to a frontal plane) dividing the animal into identical ends. Differences in the animal appear between the oral and aboral surfaces (i.e., where the football is tied).

Cleavage is highly stereotypic. The first cleav-age is accomplished by an unusual **unipolar infold-ing** (Fig. 11.32 left). In the ctenophore, *Pleurobra-chia bachei*, the first furrow is always initiated where the polar bodies are given off. Since the same site corresponds to the oral surface of the future animal, cleavage would seem to establish the oral–aboral axis (see Freeman, 1979).

The embryos of ctenophore exhibit biradial symmetry from the inception of cleavage. The first cleavage furrow corresponds to the sagittal plane of the future adult. The second cleavage furrow corresponds to the tentacular plane. Subsequently, as each tier of blastomeres forms (Fig. 11.32 right), identical blastomeres are seen on either side of each plane.

The third cleavage establishes two symmetri-cal groups of blastomeres which will continue to characterize the animal through its adult lifetime. The spindle angles and zones of cortical contrac-tion responsible for the critical third cleavage seem to be determined independently of the actual num-ber of cleavage divisions experienced by the em-bryo. In the lobate ctenophore, Mnemiopsis, re-versible inhibition of cleavage by cytochalasin B or 2,4-dinitrophenol at the 2-cell stage results in

"skipping" a division. When the inhibition is re-laxed, most embryos proceed with the symmetrical division and form only 4 cells, while the untreated controls form 8 cells.

The similar phenomenon in echinoderms at-tributed to a "cleavage clock" has implications for ctenophore development. Possibly, the first cleav-age establishes an independent rhythm for subse-quent cleavages. The suppression of cleavage does not block that rhythm. It does not alter the mech-anisms directing spindles to fall at particular an-gles or cortical constrictions to reside at specific levels (see Schroeder, 1985). When permitted to resume, cleavage picks up the cell's rhythm, the cleavage beat, so to speak, not the cleavage number.

Unlike their presumed relatives, the coelenter-ates, which may discard blastomeres, ctenophores seem to produce a blastomere for each place in the future animal. Moreover, the removal of any blas-tomere after the third cleavage results in a specific deficit in the future animal (see Freeman, 1979). Mnemiopsis bisected as an embryo, for example, remains a "1/2 animal" even after growing to re-productive size.

Still, ctenophores may not have specific re-quirements for particular blastomeres (i.e., like those of *Caenorhabditis elegans*, see below). In adults, the removal of structures is followed by the regeneration of those structures, and "1/2 animals" developing from bisected embryos can become whole animals when wounded as adults. Removing blastomeres from embryos may result in deficient development due to difficulty in feeding and lack of adequate nutrition. When hand-fed to overcome nutritional problems, even deficient juveniles grow to normal whole adults (Martindale, 1986).

Bilateral cleavage (also triaxonic and mono-planar) occurs when an early cleavage furrow separates complementary right and left sides (i.e., mirror images). Embryos exhibiting bilateral cleav-age also exhibit precocious localization of the anterior–posterior and dorsal–ventral axes.

Protochordate eggs are generally fertilized in the vegetal hemisphere at the point farthest from the germinal vesicle. Cytoplasmic constituents are then seen to be actively rearranged via **ooplasmic segregation** (see Figs. 9.4 and 11.33a). Cytoplasmic granules and yolk cascade downward at fertiliza-tion and then, as cleavage commences, move up-ward to form **crescents** (Fig. 11.33b).

Cleavage begins with a bilateral division split-ting the cytoplasmic constituents equally between the first two blastomeres (Fig. 11.34a). The blas-

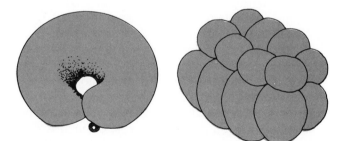

FIGURE 11.32. Cleavage in a ctenophore. (Left) The first cleavage begins near the site of polar body discharge as a unipolar infolding. (Right) At the 16-cell stage, a tier of smaller micromeres is cut off from a tier of larger macro-meres. The biradial symmetry of the embryo is evident at this stage. Two planes divide two groups of four micro-meres and four macromeres.

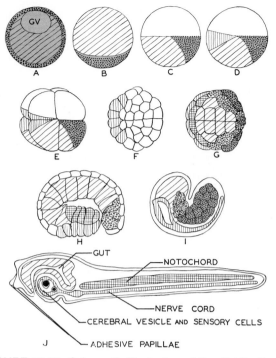

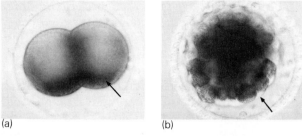

FIGURE 11.34. Nomarski interference phase micrographs of _Styela plicata_ embryos at 2-cell stage (a) and 64-cell stage (b) showing distribution of yellow crescent material (arrow). ×300. From J. R. Whittaker, in S. Subtelny, ed., _Determinants of spatial organization,_ Academic Press, Orlando, 1979. Used by permission.

FIGURE 11.33. Schematic illustration of the distribution of cytoplasmic constituents to particular blastomeres in _Styela_ (= _Cynthia_) _partita_ as described originally by Conklin. Anterior to left, posterior to right. (a–c) Uncleaved stage. The yellow cytoplasm (dots) forms the yellow crescent posteriorly, and the gray yolk (diagonal hatching) becomes concentrated vegetally. (d) The 2-cell stage (AB cell viewed from side). A new light gray crescent (vertical hatching) appears anteriorly and opposite the yellow crescent. (e) The 8-cell stage. The yellow crescent is concentrated in the posterior vegetal cells (B4.1), while the light gray crescent is divided between the anterior vegetal (A4.1) and anterior animal (a4.2) cells. The gray yolk is divided between the anterior (A4.1) and posterior vegetal cells (B4.1), and the clear cytoplasm dominates the anterior (a4.2) and posterior (b4.2) animal cells. (f) The 64-cell stage viewed from animal pole. Light gray cytoplasm is localized bilaterally in prospective cerebral vesicle cells (a7.9, a7.10) and prospective sense organ (sensory) cells (a7.13). Clear cytoplasm is present throughout prospective epidermis (a7.11–a7.16, b7.9–b7.16) and prospective palps or adhesive papillae (a7.9–a7.10). (g) The 64-cell stage viewed from vegetal pole. Light gray cytoplasm is present in prospective brain stem (A7.4) and prospective spinal cord (A7.8). Gray yolk moves to gut (endoderm) cells (A7.1, A7.2, A7.5, B7.1, B7.2), Notochordal cells (cross hatched, A7.3, A7.7) separate from the vegetal blastomeres of the anterior light gray crescent. Prospective mesenchyme cells (small circles, B7.3, B7.7) separate from prospective tail muscle (dots, B7.4–B7.5, B7.8) of the posterior yellow crescent. (h) Longitudinal section of neurula. Presumptive gut and notochord complete invagination. (i) Parasagittal section of early tadpole. Neural tube has formed and tail elongates. (j) Longitudinal section of tadpole. From R. L. Watterson, in B. H. Willier, P. A. Weiss, and V. Hamburger, eds., _Analysis of development,_ Saunders, Philadelphia, 1955.

tomeres (each identified as AB2, Fig. 11.35) form A blastomeres which spawn the **anterior cell lineage** and B blastomeres which give rise to the **posterior cell lineage.** A horizontal, third cleavage then isolates blastomeres in the animal and vegetal hemispheres.

A dual-purpose nomenclature identifies the blastomeres by cleavage number and position (Fig. 11.35). Capital and lowercase letters are used to identify the source of a cell back to the 4-cell stage. A and a signify anterior; B and b signify posterior. Capital letters identify the descendants of blastomeres formed first in the vegetal hemisphere; lowercase letters identify the descendants of blastomeres formed first in the animal hemisphere. In addition, a whole number after each letter specifies the cleavage which that blastomere performs next, and another number following a period identifies the blastomere's position (i.e., lower numbers being closer to the vegetal pole).

The originator of this nomenclature, Edwin Grant Conklin, is also credited with drawing attention to the movement of granules and suggesting a mechanical role of cleavage in segregating the granules. Although the cytoplasms may not be as colorful in other protochordates and orange granules may replace yellow ones, similar cytoplasmic constituents seem to have similar distributions in many representatives of the protochordates (see Berrill, 1961). The ascidian, _Styela_ (= _Cynthia_) _partita_ (Conklin, 1905), represents the stereotypic extreme, but even the Amphioxus, _Branchiostoma lanceolatum_ (Conklin, 1932), which is far less developmentally rigid, shows approximately the same localization of the major cytoplasmic constituents.

The yellow granules, which appear to be pigmented yolk granules adherent to mitochondria (Berg and Humphreys, 1960), form the **yellow pigmented crescent** (Fig. 11.33, dots and circles, Fig.

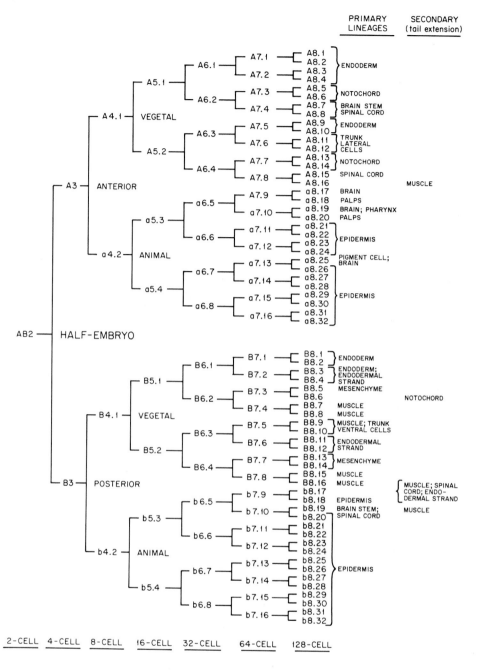

FIGURE 11.35. Stereotypic cell lineage of ascidian embryo. Because the embryo is overtly bilaterally symmetrical after the first cleavage, the fate of only half the embryo is shown. The other half is a mirror image of the half shown. Compare with Figure 11.33. From J. R. Whittaker, in S. Subtelny, ed., *Determinants of spatial organization,* Academic Press, Orlando, 1979. Used by permission.

11.34) and become localized in posterior blastomeres at the 4-cell stage. The granules are transmitted largely (but not exclusively) to the vegetal blastomeres of the 8-cell stage (Fig. 11.33e) and finally to the presumptive mesenchyme and tail muscle cells of the developing larva (Fig. 11.33f–h).

Similarly, an anterior **light gray crescent** (vertical hatching in Fig. 11.33) is shunted to blastomeres destined to give rise to nervous tissue (i.e., the prospective cerebral vesicle, sensory cells, brain stem, and spinal or nerve cord) and the notochord. The **transparent** or **clear cytoplasm** (also called hyaloplasm) of the animal hemisphere ends

up in the epidermis, and the **gray yolk** of the vegetal hemisphere (diagonal hatching) ends in the gut or endoderm.

The specific localization of materials in blastomeres is not limited to granules. Enzymatic activity which appears exclusively in certain larval tissues (i.e., **histospecific enzymes**) also follows precise paths from specific blastomeres. Because the histospecific enzymes might not appear until the specific tissues are formed, however, the distribution of these enzymes is more difficult to follow than the distribution of colored granules.

The trail of enzymes is followed by inhibiting

cytokinesis (but not karyokinesis) with cytochalasin B at various times after fertilization. In a form of **differentiation without cleavage,** enzymatic activity appears in blastomeres at the same time it would normally have appeared in the larval tissue (see Whittaker, 1979).

In *Ciona intestinalis*, acetylcholinesterase activity is ordinarily undetectable in the zygote but is detected at about 8 hours after fertilization in the presumptive tail muscle region of the embryo (Fig. 11.36). At a comparable time, acetylcholinesterase is detected throughout the cytoplasm of cytochalasin-treated, arrested zygotes (Fig. 11.36a) and in both AB2 blastomeres of arrested 2-cell embryos (Fig. 11.36b). Embryos arrested at later stages have a differential distribution of the enzyme. It

appears in the B3 cells of embryos arrested at the 4-cell stage and in the B4.1 cells of embryos arrested at the 8-cell stage (Fig. 11.36c). Like the yellow granules, the enzyme follows the route to the tail musculature (Fig. 11.36d–g).

Equally restricted, but taking a different path, alkaline phosphatase activity passes from each of the blastomeres of the 4-cell stage to only the 4 vegetal blastomeres (A4.1 and B4.1) of the 8-cell stage and finally, largely, to the gut endoderm. Succinic dehydrogenase activity, associated with mitochondria, winds up preferentially in the tail musculature. Tyrosinase (dopa oxidase) activity, associated with pigment formation, is restricted to the pigmented sensory cells (melanocytes) of the brain (Whittaker, 1979). What exactly is being dis-

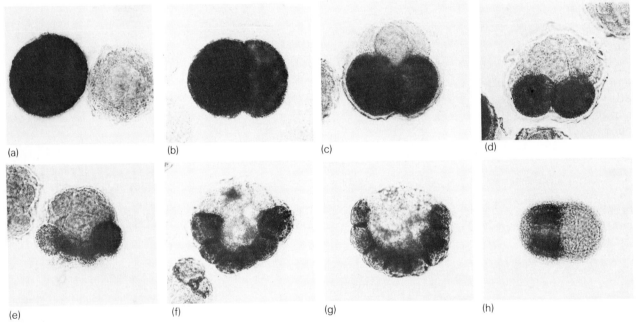

FIGURE 11.36. Photomicrographs of embryos of the tunicate *Ciona intestinalis* showing sites of acetylcholinesterase activity as dark staining. Embryos cultured at 18°C through various cleavage stages were arrested [in each case except (h), the normal control] and tested for acetylcholinesterase at 16 hours after fertilization. The deep staining caused by enzymatic activity becomes concentrated in blastomeres that later give rise to the muscle cells of the tadpole's tail. × 140. (a) Arrested as zygote at 1-cell stage. Staining detected throughout. (b) Arrested at 2-cell stage about 1 hour after fertilization. Staining detected throughout AB2 blastomere. (c) Arrested at 4-cell stage about $1\frac{1}{2}$ hours after fertilization. Staining limited to two B3 blastomeres in posterior cell lineage. (d) Arrested at 8-cell stage 2 hours after fertilization. Staining limited to two B4.1 blastomeres. (e) Arrested at 16-cell stage $2\frac{3}{4}$ hours after fertilization. Staining in four vegetal B5.1 and B5.2 blastomeres. (f) Arrested at 32-cell stage $3\frac{1}{2}$ hours after fertilization. Staining in six B6.2–B6.4 blastomeres. (g) Arrested at 64-cell stage (but before the A6.7 blastomeres have actually completed the sixth cleavage) $4\frac{1}{4}$ hours after fertilization. Staining usually in eight B7.3–B7.8 blastomeres. (h) Untreated 9-hour tailbud embryo. Staining in prospective tail musculature. From J. R. Whittaker, *Proc. Nat. Acad. Sci.,* 70:2096, by permission of the National Academy of Sciences and the author.

tributed (i.e., whether mRNA, precursors of messages, or nuclear-reactive substances such as secondary messengers) remains a mystery, but the prior localization of materials and cleavage's stereotypic angles seem to provide the mechanical cause for distributing something with morphogenic consequences, whatever it is.

Amphibian bilateral cleavage provides the classic examples of Balfour's law and Hertwig's mechanical principles of mitosis (see Fig. 10.2). The spindles lie in the long axis of the available cytoplasm, and the vertical spindles are displaced toward the animal pole. The first and second cleavages are vertical, the second running perpendicular to the first. The third cleavage is horizontal and perpendicular to the first and second, but it is displaced toward the cytoplasm of the animal pole. Further divisions of the amphibian egg do not produce tiers of blastomeres (as in echinoderms) but a range of smaller animal-pole cells (sometimes called micromeres) and larger vegetal-pole cells (sometimes called macromeres).

In *Xenopus laevis* embryos obtained by mating, the first furrow separates right and left sides whether or not it bisects the pale gray crescent. The crescent marks the dorsal side of the embryo whenever the furrow bisects it but not when the first furrow falls elsewhere (Fig. 11.37) (Klein, 1987).

In ranids, the first furrow generally cuts the gray crescent and lies on the midsagittal plane, but if the first furrow misses, the second furrow will probably hit the gray crescent and may even take over as the midsagittal plane. In that event, the second furrow **compensates** for the first furrow's bad "aim."

Even greater latitude in the relationship of furrows to planes of symmetry is found in other amphibians. In *Bombinator*, 40% of the embryos form the midsagittal plane some 30–90° from the plane of the first cleavage division, and in urodeles the furrows and planes are unrelated. Even the linkage between the first cleavage plane and the embryonic axis in *Xenopus* eggs is readily uncoupled when eggs are fertilized artificially (Danilchik and Black, 1988).

In the nomenclature devised for *Xenopus laevis* (Fig. 11.38), the bilateral first cleavage produces a left (L) and a right (R) blastomere. The second cleavage being in the midfrontal plane produces dorsal (D) and ventral (V) blastomeres (Fig. 11.39). The horizontal third cleavage separates blastomeres closer to the animal pole (RV1, RD1, LD1, and LV1) from those closer to the vegetal pole

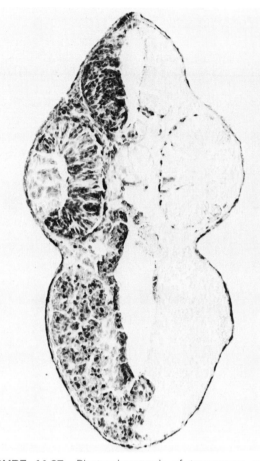

FIGURE 11.37. Photomicrograph of transverse section through the head region of a *Xenopus laevis* tadpole showing sites of horseradish peroxidase activity as dark staining. The horseradish peroxidase was injected into the right blastomere at the 2-cell stage. With the exception of some cells (brackets) that crossed the midline, presumably during gastrulation, all stained cells appear on one side of the midsagittal plane. (Note: The epidermis and pigmented retina on the right side of the photomicrograph are dark due to the normal presence of pigment not to horseradish peroxidase staining.) N, notochord. From S. L. Klein, *Dev. Biol.,* 120:299 (1987), by permission of Academic Press and courtesy of S. L. Klein.

(RV2, RD2, LD2, and LV2). With each subsequent division, a cell closer to the animal pole or the sagittal plane is assigned the number 1, and the sibling cell is assigned the number 2.

Deviations from the strictly horizontal third cleavage also occur in *Xenopus,* and exceptions to the *Xenopus* pattern are sometimes conspicuous. In *Ambystoma mexicanum* and *Bufo vulgaris,* for example, the third cleavage is generally horizontal in two blastomeres and vertical in two.

Differences among embryos do not persist too long. **Compensatory cleavages** or **alternative cleav-**

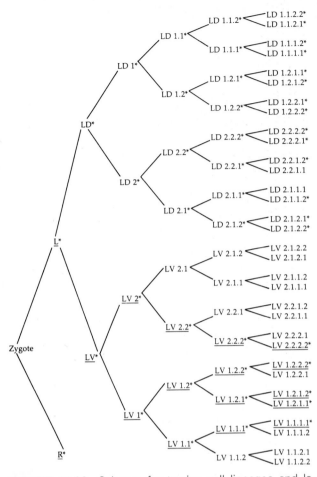

FIGURE 11.38. Scheme for tracing cell lineages and labeling blastomeres on the left side in the frog. L, left; R, right (note that the right side is identical to the left and therefore not shown); D, dorsal; V, ventral. Cells designated 1 are closer to the sagittal plane or animal pole than cells designated 2. From M. Jacobson, in N. C. Spitzer, ed., _Neuronal development_. Used by permission of Plenum Press, New York, 1982, and the author.

ages intervene and **restore** or **regulate** the cleavage pattern. Blastomeres which fail to divide perpendicularly at the third cleavage divide perpendicularly at the fourth cleavage, and vice versa. Thus, differences resulting from the angle of the third cleavage are short lived, and by the fifth cleavage, differences created by variation in earlier cleavages are generally offset (Hara, 1977).

The orientation of furrows in successive divisions therefore does not seem to be independently controlled. Rather, cleavages seem to achieve a pattern, if not one way, then another. Cleavage in amphibians is nowhere near as stereotyped as in some other totally cleaving organisms, but the result tends to be highly stereotyped. Some blastomeres may have different histories than their place in the

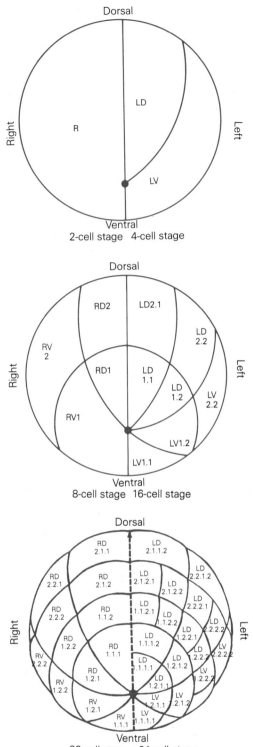

FIGURE 11.39. The animal-pole view of cleavage patterns in the frog at the 2–64-cell stage. The point of convergence corresponds to the animal pole. The vegetal pole is not visible. Adapted from M. Jacobson and G. Hirose, _J. Neurosci.,_ 1:271 (1981), by permission of the Society for Neuroscience and the authors.

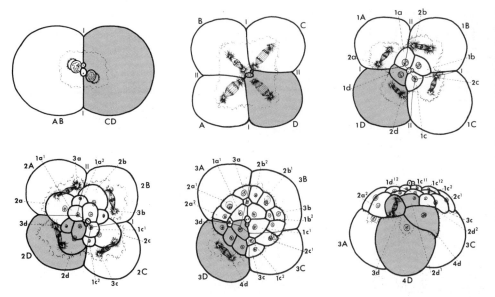

FIGURE 11.40. Cleavage in the slipper limpid, the uncoiled gastropod, *Crepidula fornicata.* Lower right viewed from side. Others viewed from animal pole. Polar bodies are shown as two small cells at center or top. The first two furrows are identified by Roman numerals. Blastomeres are labeled according to the scheme in Fig. 11.44. From D. P. Costello and C. Henley, *Am. Zool.,* 16:277 (1976), by permission of the American Society of Zoologists.

early embryo suggests, and estimates of cell lineages based on position alone may not be accurate, but, at some point, each potential "slot" in the embryo's plan is filled.

SPIRAL OR OBLIQUE CLEAVAGE

Spiral cleavage is the more popular name for oblique total cleavage. The angle of furrowing during the first few divisions is more nearly diagonal than either vertical or horizontal (at least in theory).

In the course of spiral cleavage, consecutive furrows swing back and forth like a pendulum. Divisions are called **clockwise** or right-handed (also called dexiotropic) when the newly formed blas-

tomere closest to the animal pole lies to the right of its sibling (i.e., with animal pole upward). Divisions in the opposite direction are called **counterclockwise** (anticlockwise) or left-handed (also called laeotropic, leiotropic, and laevotropic).

The furrows do not actually spiral or rotate like a corkscrew. Rather, oblique furrowing, blastomeric movement, and readjustment place the blastomeres of one tier over the furrows of another tier, giving an impression of torsion, at least in the best cases (e.g., the limpet, *Crepidula*, Figs. 11.40, 11.41, *Neanthes*, Fig. 11.42).

At some point in early cleavage, furrows cut off tiers of smaller, less yolky **micromeres** from larger, more yolky **macromeres.** The size difference between macromeres and micromeres may be large (e.g., in the leech, *Helobdella*, Fig. 11.43), small, or reversed (i.e., "micromeres" may actually be larger

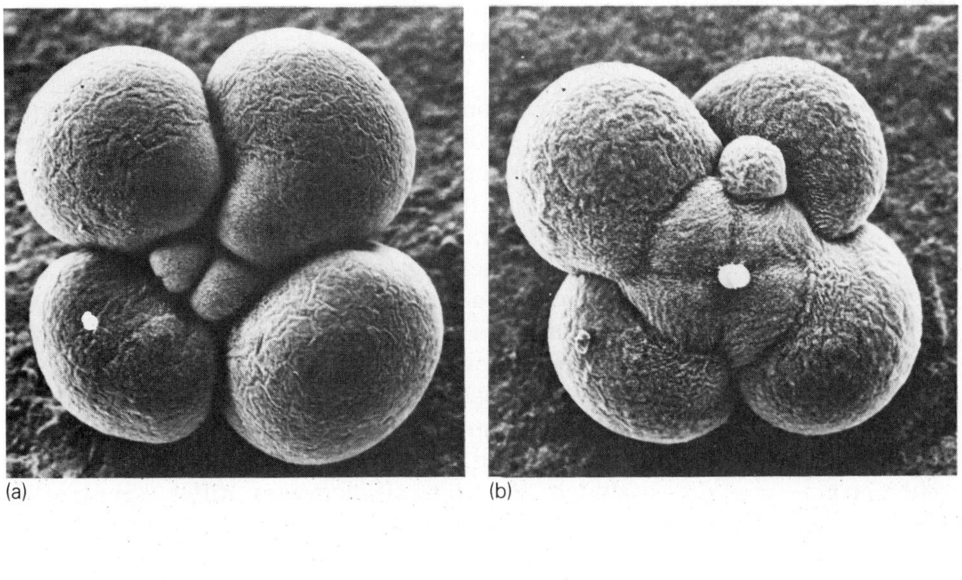

FIGURE 11.41. Scanning electron micrograph of *Crepidula fornicata* embryo during formation of first (a) and second (b) tiers of micromeres. The divisions are asynchronous. The 1d and 1c micromeres form first during the third cleavage, and the 2d micromere forms first during the fourth cleavage. ×250. From M. R. Donmen, in W. R. Jeffery and R. A. Raff, eds., *Time, space, and pattern in embryonic development,* Alan R. Liss, Inc., New York, 1983. Used by permission.

(a)

(b)

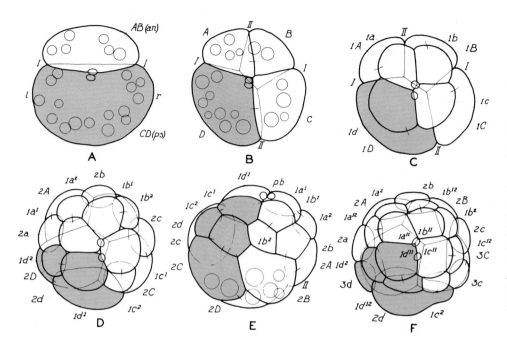

FIGURE 11.42. Cleavage in the clam worm, *Neanthes succinea* (formerly *Nereis limbata*). Parts (a–d) and (f) viewed from animal pole. Part (e) viewed from side. The large circles are oil droplets, the two ovals [p.b. in (e)] are polar bodies. The cross piece formed by bends in the first two furrows is evident in (b–d) and (f). The first two furrows are identified with Roman numerals. The blastomeres are labeled according to the scheme in Fig. 11.44. From D. P. Costello and C. Henley, *Am. Zool.,* 16:277 (1976), by permission of the American Society of Zoologists.

than "macromeres" as in the nemertine, *Cerebratulus,* and the nematode, *Caenorhabditis*).

The particular cleavage division at which the first micromeres and macromeres separate ranges from the first to the third in different phyla. Hence,

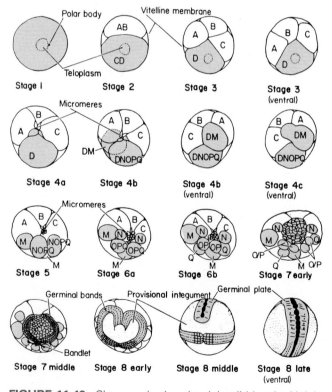

FIGURE 11.43. Cleavage in the glossiphoniid leech, *Helobdella triserialis,* viewed from animal pole. From S. Astrow, B. Holton, and D. Weisblat, *Dev. Biol.,* 120:270 (1987), by permission of Academic Press and the author.

the first tier of micromeres may be constituted by a **monet** (also monad or unit) of one cell (in cirripedes, see below, Fig. 11.53, and aschelminthes including *Caenorhabditis elegans,* Fig. 11.54, and other nematodes, Fig. 11.55), a **duet** of two cells (acoel turbellarian, Fig. 11.52), or a **quartet** (or quadrant) of four cells (polyclad turbellarians and mollusks, Figs. 11.40 and 11.41, annelids, Fig. 11.42, and leeches, Fig. 11.43) (see Costello, 1945). Subsequent unequal divisions of macromeres may then add one to three more tiers of micromeres containing the same numbers of cells.

In general, the blastomeres which will give rise to the ectodermal, mesodermal, and midgut rudiments of the future embryo are segregated from one another before or by the fifth cleavage division. Afterward, blastomeres stop spiraling and switch to bilateral cleavage. The timing of the transition to bilaterality is highly stereotyped and presumably finely tuned to the adaptive advantage of the embryo.

Phyla containing species with spirally cleaving eggs are sometimes lumped together as **spiralians** in the convenient but not officially recognized superphyletic taxon **Spiralia.** In addition to sharing spiral cleavage, spiralian eggs are enclosed in tough, unyielding chorions, and the embryos are schizocoelous protostomes (see Chapter 17). Development is frequently indirect with larvae hatching soon after gastrulation (the delayed hatching of parasitic species and nematodes is exceptional). The number of micromeres and macromeres in spirally cleaved tiers and the site and timing of the switch to bilateral cleavages differ among the phyla, but spiralians are widely considered related through an evolutionary network.

Nomenclature

Methods for identifying the blastomeres of spirally cleaving eggs are dominated by the scheme Edmund B. Wilson devised for annelids and Edwin G. Conklin extended to mollusks. In its present form (as revised chiefly by R. Woltereck for *Polygordius* and K. Okada for *Nereis japonica*), the

scheme easily allows investigators to retrace **cell lineages** from blastomeres' names and name blastomeres from their lineages (see Kumé and Dan, 1968).

For annelids and mollusks, macromeres are identified by capital letters (A–D) and micromeres by lowercase letters (a–d, Fig. 11.44). The letter assignments made to a cell, be it macromere or micromere, never change, and the same letter is also assigned to the descendants of that cell.

Macromeres are also identified by a numerical

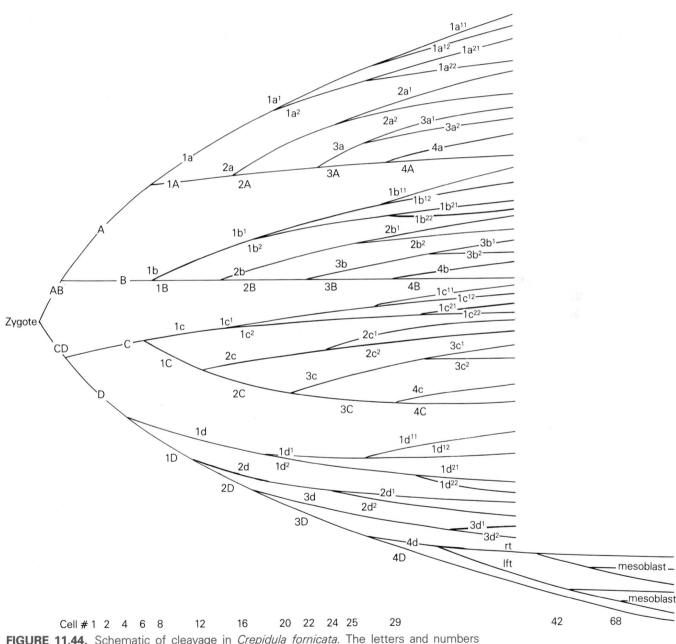

FIGURE 11.44. Schematic of cleavage in *Crepidula fornicata*. The letters and numbers identifying blastomeres, and the cells listed below correspond to those in Figs. 11.40 and 11.42, but the timing and synchrony of events differ in representatives of various phyla.

prefix corresponding to the number of times the macromere has divided. Unlike its letter, the macromere's numerical prefix changes each time the cell divides (e.g., the 1A macromere becomes the 2A macromere upon dividing).

By Wilson's definition, micromeres are originally cleaved from macromeres. The same numerical prefix assigned to the macromere at the time it gives rise to the micromere is therefore assigned to the micromere (e.g., the same division that gives rise to the 2A macromere produces the 2a micromere). Moreover, the micromere's numerical prefix, like its letter, never changes (e.g., the 1a micromere gives rise to two 1a micromeres upon dividing [distinguished as $1a^1$ and $1a^2$]).

Numerical exponents (sometimes suffixes) give the relative position of micromeres produced at each division. Number 1 identifies the micromere closest to the animal pole, and 2 identifies the sibling micromere farthest from the animal pole. These exponents accumulate as divisions proceed, and letters designating micromeres gather

strings of 1s and 2s (e.g., the $1a^{11}$ cell produces the $1a^{111}$ and $1a^{112}$ cells).

This system of nomenclature is so widely used that without having some appreciation of it, most reports on spiral cleavage will make little sense. Its popularity is understandable, since molluscan and annelidan eggs have been the premier subjects for studies on spiral cleavage for a century.

Still, the scheme presents problems. First, devised for the quartets of micromeres formed by annelids and mollusks, the scheme is not easily extended to the monets and duets of micromeres formed by other spiralians (see below). Second, the implication of the scheme that micromeres are initially products of macromeres is difficult to escape but not necessarily correct in all spiralians (Anderson, 1973). Finally, the Wilson–Conklin nomenclature is heavily weighted toward the annelidan and molluscan **trochophore larva** (Fig. 11.45) and not easily translated to larvae formed by other spiralians.

Spiral Cleavage by Quartets

The first blastomeres are usually identified as AB and CD, and cleavage can be equal, as in uncoiled gastropods (*Crepidula*, Fig. 11.40) and coiled gastropods (e.g., *Lymnaea* and *Bithynia*, Fig. 11.46), or unequal, as in polychaete annelids (e.g., *Neanthes* = *Nereis*, Fig. 11.42) and the glossiphoniid leeches (e.g., *Helobdella*, Fig. 11.43). In a phenomenon related to unequal division, some eggs are equipped with **polar lobes** through which as

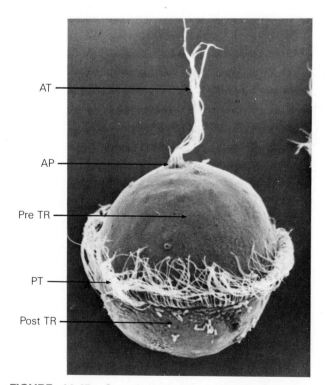

FIGURE 11.45. Scanning electron micrograph of the trochophore of the marine polychaete, *Sabellaria alveolata.* The larva has a characteristic apical tuft (AT) rising from an apical plate (AP), a pretrochal region (PreTR), a ciliated band or prototroch (PT), and a posttrochal region (PostTR). From M. R. Dohmen, in W. R. Jeffery and R. A. Raff, eds., *Time, space, and pattern in embryonic development,* Alan R. Liss, Inc., New York, 1983. Used by permission.

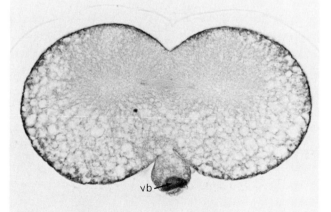

FIGURE 11.46. Light micrograph of section through the freshwater snail, *Bithynia tentaculata,* at the first cleavage showing a small polar lobe. A vegetal body (vb) within the polar lobe stains with the methyl-green pyronine stain for RNA. ×350. From J. N. Cather, N. H. Verdonk, and M. R. Dohmen, *Am. Zool.,* 16:455 (1976), by permission of the American Society of Zoologists.

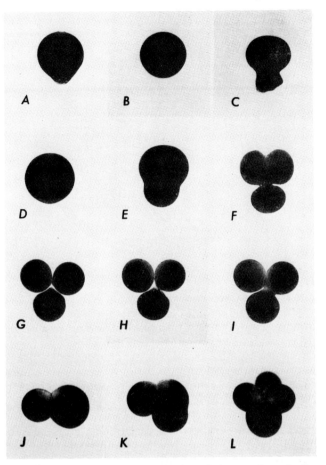

FIGURE 11.47. Photomicrographs of the mud snail, *Ilyanassa obsoleta,* showing formation and withdrawal of polar lobes during maturation and early cleavage. (a) Polar body (top) forms at the animal pole and relatively small first polar lobe at the vegetal pole (bottom). (b) Zygote rounds up during interval between eruption of polar bodies. (c) Irregular second polar lobe forms during second meiotic division. (d) Mature ovum. (e) Rounded polar lobe appears prior to first cleavage. (f) The first cleavage furrow appears as the connection narrows between polar lobe and zygote. (g) In the trefoil stage, the polar lobe is connected to the CD blastomere by a narrow stalk. (h, i) Absorption of polar lobe into CD blastomere. (j) Two-cell stage. The CD blastomere on right is larger than the AB blastomere on left. (k) The fourth polar lobe appears on the CD blastomere prior to the second cleavage division. (l) Four-cell stage. The polar lobe has been absorbed by the D blastomere which is larger than the others. In parts (a–k) animal pole is upward; in part (l) view is from animal pole. From A. C. Clement, *Am. Zool.,* 16:447 (1976), by permission of the American Society of Zoologists.

much as one-third of the cytoplasm is shunted to particular blastomeres rendering them larger than others (Fig. 11.47).

Polar lobes form at the vegetal pole of an egg or even in the vegetal fragment of an artificially divided egg (i.e., a merogon). When the polar lobes are large, as in the marine mud snail, *Ilyanassa obsoleta,* the 2-cell embryo with a narrowly connected lobe is known as a **trefoil.** Small lobes may be present in some annelids (e.g., the polychaetes *Chaetopterus*), and still smaller lobes, sometimes containing discrete **vegetal bodies,** are present in freshwater gastropods such as *Bithynia tentaculata* (Fig. 11.46). Usually, more than one polar lobe is formed preceding and accompanying cleavage, and, in the periwinkle (*Littorina*), lobes may persist to the fifth division.

Polar lobes are absent in other annelids, although multiple protuberances, knobs, or areas of **pole plasm** may be found (e.g., in the oligochaete, *Tubifex*). In leeches, a yolk-deficient region of cytoplasm called animal-pole **teloplasm** follows the path of polar lobe cytoplasm largely (if not exclusively) to a single blastomere (Fig. 11.43).

This is not to say that the blastomeres are always distinguishable. The first two blastomeres of the North American pulmonate snail, *Lymnaea palustris,* for example, seem to be developmentally identical.

In annelids and mollusks generally, the first two divisions are nearly vertical **prospectively spiral cleavages.** Divergence from the vertical is noticeable mainly toward the vegetal pole where portions of the first and second furrows overlap as a **cross piece** or **cross furrow.** The furrows generally are distinguishable, since the cross piece bends the first furrow toward the right and the second cleavage furrow toward the left (when looked at from the animal pole, Figs. 11.40 and 11.42).

Whenever the blastomeres formed by the first division are conspicuously unequal, the larger of the two blastomeres is designated CD, the smaller AB. In mollusks and annelids with polar lobes (e.g., Fig. 11.47), the CD blastomere receives all the contents of the first lobe.

Similarly, whenever the blastomeres formed by the second division are conspicuously unequal, the largest sibling of the CD blastomere is designated D, and it receives all the contents of the second polar lobe when present. The D blastomere typically lies on the **dorsal** aspect of the embryo. D's sibling blastomere, designated C, comes to lie on the right side of the embryo (Fig. 11.48 left).

Division of the AB blastomere produces a B blastomere that occupies the ventral aspect of the embryo, and an A blastomere that occupies the embryo's left side. The A blastomere is frequently the smallest of the four blastomeres.

When the first two cleavages are equal, the polar bodies and the cross piece are used to identify

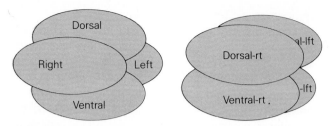

FIGURE 11.48. Orientation of first four blastomeres in quartet-type spiral cleavage (left) and bilateral cleavage (right).

the blastomeres. The positions of equal-sized blastomeres correspond to those of unequal-sized blastomeres: D dorsal, B ventral, C right, and A left (Fig. 11.48 left). In many cases, the developmental fates of blastomeres in each position are remarkably consistent in eggs with equal and unequal cleavage.

The first two furrows lie along the animal–vegetal axis and bisect the angles of the future sagittal and frontal planes of the larva (Fig. 11.48 left) (Anderson, 1973). This relationship of furrows to planes of symmetry in quartet-forming eggs contrasts with the relationship found in bilaterally cleaving eggs (Fig. 11.48 right) where furrows correspond to midsagittal and midfrontal planes, and where pairs of mirror-image blastomeres lie on either side of the midsagittal plane. One is hard pressed to imagine a common ancestor bridging this difference. Rather, it would seem that quartet-forming spirally cleaving embryos and bilaterally cleaving embryos arose from unrelated ancestral populations.

Divisions of micromeres and macromeres are not especially synchronous in the spirally cleaving annelids and mollusks. The D and C macromeres lead in the first round of micromere production, and the 1D macromere leads in the second round (Fig. 11.41). Greater asynchrony follows.

The third division is the first overtly oblique division in annelids and mollusks. It is usually unequal in all blastomeres and gives rise to a first quartet of micromeres identified as 1a–1d and a quartet of macromeres identified as 1A–1D.

In mollusks, the 1a–1d micromeres become the larval **apical plate** and paired adult **cephalic plates.** These micromeres divide unequally and obliquely into **central cells** ($1a^1$–$1d^1$ cells) and peripheral **trochoblasts** ($1a^2$–$1d^2$) that produce the ciliated waistband known as the **prototroch** of the trochophore (Figs. 11.45 and 11.49). In gastropods, lineages arising from the a and b micromeres produce the ciliated **velum** of the **veliger larva** (see Fig. 12.3a) (Morrill, 1982).

Along with cells of the second tier of micromeres, the central cells and trochoblasts constitute

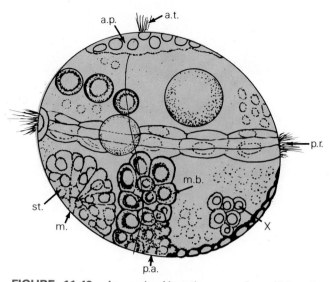

FIGURE 11.49. An early *Neanthes succinea* (formerly *Nereis limbata*) trochophore larva about 20 hours after fertilization viewed from left side. About ×400. a.p., Apical plate; a.t., apical tuft; m., mouth; m.p., mesodermal band, derivatives of second somatoblast (mesoblast); pa., anal pigmented area; pr., prototroch; st., stomodeum; x, derivatives of first somatoblast (ectoblast). Curved line shows apical-basal axis. From E. B. Wilson, *J. Morphol.,* 6:361 (1892).

the **cap** of the embryo. In gastropods, consecutive divisions of the central cells produce a nearly perfect radial **molluscan cross** (Fig. 11.50). Similar crosses also appear in chitons and an **annelian cross** appears, for example, in *Neanthes* (Conklin, 1896).

The macromeres produce micromeres 2a–2d while becoming 2A–2D (Fig. 11.44). In annelids, the 2d cell is larger than the other micromeres and is known as the **first somatoblast** (once designated X). The 3D cell (called the mesendomere or mesentomere) divides precociously to produce the 4d micromere, or **second somatoblast,** and the 4D macromere. A 25-cell stage results, and the macromeres generally become the **endomeres** (E, also entomeres) that give rise to the future endoderm.

Bilateral divisions increasingly replace oblique ones after this stage. For example, the 4d micromere divides bilaterally at the 38-cell stage in annelids (e.g., *Neanthes*) or the 42-cell stage in mollusks (e.g., *Crepidula*, Fig. 11.44).

In annelids, the 4d cell produces left and right **primary mesoblasts** (M_1 and M_r or myoblasts, My*l* and My*r*), although they still lie on the right of the second cleavage furrow. These cells subsequently give rise to **mesodermal stem cells** (msc), arranged in bilateral **mesodermal germ bands** (MGB), and to primordial germ cells in the adult.

In gastropods such as *Crepidula*, the situation is similar except for the addition of one more division. The 4d cell divides bilaterally into two **mes-**

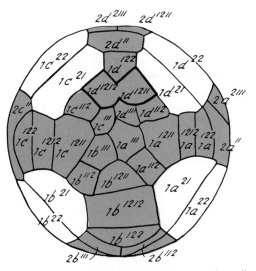

FIGURE 11.50. Cellular outlines of the pond snail, *Lymnaea stagnalis,* embryo viewed from the animal pole. Drawing shows the molluscan cross formed by the descendants of the central cells of first tier of micromeres (i.e., descendants of la¹–ld¹). The beginning of bilateral cleavage is seen in the ''d'' arm of the cross (bold outline). From Chr. P. Raven, *Am. Zool.,* 16:395 (1976), by permission of the American Society of Zoologists.

endoblasts (M*l* and M*r*, also mesentoblasts). These give rise to posterior **primary endoblasts** (or entoblasts), the antecedents of gut endoderm, and (at the 68-cell stage, Fig. 11.44) to anterior **mesoblasts** (sometimes designated M_l and M_r), the antecedents of mesoderm (including mesenchyme and protonephridia in the larva) and primordial germ cells.

The first surface sign of bilaterality in mollusks occurs in the dorsal (i.e., d) arm of the radial molluscan cross (Fig. 11.50) when the $1d^{121}$ cell divides bilaterally. Soon, bilateral cleavages occur in the $1a^{121}$ and $1c^{121}$ cells of the lateral arms of the cross and finally in the $1b^{121}$ cell of the ventral arm.

In oligochaetes, such as *Tubifex*, half the **pole plasm** is shunted at the fourth cleavage into the 2d cell. The second half is subsequently passed on to the **second somatoblast** (i.e., the 4d cell). The 2d cell subsequently gives rise to the $2d^{111}$ **ectodermal proteloblast** which divides bilaterally into the right and left **ectoteloblasts** (designated NOPQ*l* and NOPQ*r* also called the ectodermal teloblast, T*l* and T*r*) (Shimizu, 1982).

Secondary reduction in the course of evolution seems to have simplified mesodermal and ectodermal lineage in leeches. In *H. triserialis*, most of the teloplasm is funneled into the large equivalent of the 2d cell (the DNOPQ cell, Fig. 11.43 or to $2d^{222}$ cell designated SNOPQ by Fernández and Olea, 1982). The remainder of the teloplasm goes to the small equivalent of the 2D cell (the DM cell) (Astrow et al., 1987). The A–C blastomeres produce only micromeres, while the D blastomere alone produces the ectoderm and mesoderm.

The DNOPQ cell is the ectoteloblast precursor (Astrow et al., 1987), and the DM cell is the mesoteloblast precursor. Dividing bilaterally and equally, DNOPQ produces NOPQ$_r$ and NOPQ$_l$ ectoteloblasts, while DM produces M$_r$ and M$_l$ mesoteloblasts.

NOPQ ectoteloblasts give rise to four bilateral pairs of ectodermal **stem cells** or **teloblasts** (N, OP, OP, and Q cells). Their lineages form germ bands and **ventral plate** ectoderm including the precursor of the ventral nerve cord. If one of the NOPQ cells is injected with the intracellular tracer horseradish peroxidase (HRP), only the ectoderm of the corresponding half of the larva subsequently contains the tracer (Fig. 11.51).

In polyclad turbellarians, the first quartet of micromeres produces anterior dorsal ectodermal structures including the nervous system, eye pigmentation, and ganglia. The second and third quartets contribute dorsal epidermis and the **mesectodermal** muscular and mesenchymal parts of the pharynx. The 4d cell alone goes on to form mesoderm and endoderm, like the mesendoblast of mollusks, while the remaining cells of the fourth quartet of micromeres (4a–4c) fail to divide and serve a purely nutritive function.

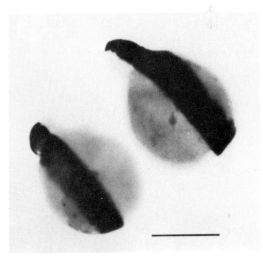

FIGURE 11.51. Photomicrographs of embryos of the leech, *Helobdella triserialis,* following histochemical test for horseradish peroxidase (HRP). The NOPQ$_r$ or NOPQ$_l$ cells (see Fig. 11.43) were injected with HRP, and the embryos were fixed 5 days later. In each embryo, opposite halves and only these halves reveal the presence of HRP. From D. A. Weisblat, R. T. Sawyer, and G. S. Stent, *Science,* 202:1295 (1978), by permission of the American Association for the Advancement of Science and courtesy of the authors. Copyright 1978 by AAAS.

Spiral Cleavage by Duets

In acoel turbellarians, such as *Polychoerus carmelensis* (Fig. 11.52), the first division is equal, but the second divisions are unequal and oblique. As a result, a duet (or dyad) of micromeres is formed near the animal pole, while a pair of macromeres is formed vegetally.

The micromeres divide first, but all blastomeres ultimately divide. The cleavage angle is clockwise. The micromeres expand their tier, while the macromeres produce another duet of micromeres. The next cleavage of macromeres is still relatively oblique and produces yet another duet of micromeres, but the remaining divisions are more bilateral than oblique.

The various fates of the blastomeres resemble those of blastomeres in the polyclad turbellarians with the exception that the acoel produces only duets as opposed to quartets. The acoel also lacks a gut, and because it develops directly to a juvenile stage it has no larval form.

The first duet of micromeres in the acoel turbellarian embryo gives rise to ectodermal structures including the nervous system. The second and probably the third duets give rise to epidermis and peripheral parenchyma or soft tissue. The fourth duet micromeres and macromeres give rise to internal parenchyma (i.e., soft tissue).

The fourth duet of the acoel embryo therefore corresponds developmentally to the entire fourth quartet in the polyclad embryo rather than to any two cells in that quartet. No two letters in the a–d

nomenclature adequately represent the duet of blastomeres.

In the evolution of spiral cleavage in the flatworms, developmental functions seem to have been redistributed among the blastomeres rather than handed down through particular lines of blastomeres. Homologies between the blastomeres of acoelous turbellarian embryos and polyclad turbellarian embryos are obvious but not directly translatable in terms of a single system of nomenclature.

Spiral Cleavage by Monets

Monets are probably not homologous in all the phyla in which they appear. The single-celled equivalents of micromeres and macromeres do not necessarily occupy the same positions or follow the same developmental pathways.

In crustaceans, cleavage spans a large number of possibilities from intralecithal to total and from radial to spiral. Unambiguous spiral cleavage is found in the eggs of thoracican barnacles where a conspicuously oblique division occurs at the very first cleavage (Fig. 11.53). This pattern may be primitive, and all the permutations of crustacean cleavage may be derived from spiral cleavage modified by the accumulation of yolk (Anderson, 1973).

In the small ovid eggs of barnacles (100–300 μm in length), the very first cleavage furrow separates a large, yolky, dorsal macromere from a small, yolk-free ventral micromere. Since it lies dorsally and is the only cell with the potential to produce a variety of adult tissues, it seems to be the crustacean equivalent to the D macromere of mollusks, annelids, and polyclads. The small cell is therefore considered the sole member of the first monet (Costello and Henley, 1976).[3] The micromere is identified as d or 1d and the macromere is labeled 1D. The micromere divides meridionally or obliquely but equally into 1d1 and 1d2 cells and the macromere into 2d and 2D (Fig. 11.53).

After the first two divisions, furrows are more vertical than oblique, and a more or less bilateral symmetry emerges. The progeny of the first three micromeres (i.e., 1d, 2d, and 3d) constitute a blastoderm or **ectoblast** of the future larva and surround the yolky descendants of the endoblast. The

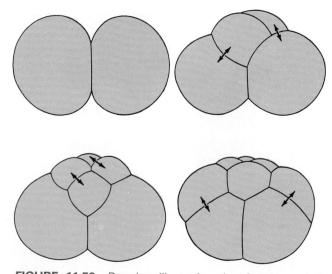

FIGURE 11.52. Drawing illustration duet-forming spiral cleavage pattern in the acoel turbellarian, *Polychoerus carmelensis.* Adapted from D. P. Costello and C. Henley, *Am. Zool.,* 16:277 (1976), by permission of the American Society of Zoologists.

[3]A broad transverse ventral area of attachment established at the second division resembles the cross piece found in quartet-producing spirally cleaving eggs. The smaller cell at the barnacle's 2-cell stage may therefore not be a micromere but the equivalent to an AB cell that has lost its yolk. Were this the case, the larger of the barnacle's cells would be the equivalent to a CD cell (Anderson, 1973).

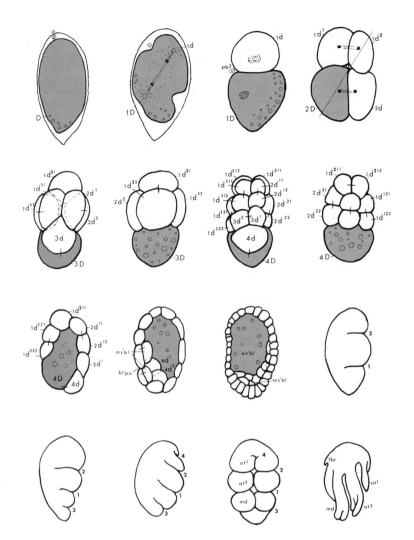

FIGURE 11.53. Drawing illustrating monet-forming spiral cleavage pattern, gastrulation, and organogenesis in small-egged species of the goose barnacles (*Lepas*) and acorn barnacles (*Balanus*). D. P. Costello and C. Henley, *Am. Zool.*, 16:277 (1976), by permission of the American Society of Zoologists. [Anderson (1973), among others, rejects the monet nomenclature adopted here, preferring a quartet-type system of nomenclature.]

superficial ectoblast cells divide synchronously and equally at a faster rate than the macromeres and are presumptive ectoderm.

The next cell cleaved from the macromere, the 4d cell, divides bilaterally into the **primary mesoblast** cells, the $4d^1$ and $4d^2$ cells. The 4D macromere is the **endoblast** (or entoblast) that subsequently forms the midgut.

In Aschelminthes, furrows are more nearly vertical than oblique, but the developmental pattern is spiralian. In the rotifer, *Asplanchna*, and the gastrotrich, *Lepidodermella*, a single macromere is present at the 4-cell stage. After six more cleavages, small micromeres constitute a population of **ectoblasts** and **mesoblasts.** The macromere gives rise to endoderm. In the nematode *Caenorhabditis elegans*, cleavage is also characteristically unequal and rapidly becomes asynchronous among the progeny of different early blastomeres (Fig. 11.54).

In many nematodes, such as *Parascaris equorum*, following the first division, chromosomes in the anterior cell (AB) undergo chromosomal fracturing, while chromosomes in the posterior (P_1) cell retain their integrity (Fig. 11.55). The **acentric** pieces (i.e., lacking centromeres) of the fractured chromosomes fail to be incorporated in the reforming nuclei and are lost from future nuclei. The loss is called chromosomal diminution or elimination (see Chapter 10).

The posterior cell (formerly S or stem cell of the germ line, now, more generally, P_1; Fig. 11.56) proceeds to divide unequally three more times. Each of the new anterior cells, EMS (formerly EMSt), C, and D, undergoes chromosomal fracture and diminution, while each of the new posterior cells (P_2, P_3, P_4) maintains patent chromosomes. The descendants of cells undergoing chromosomal diminution comprise the **somatic line** and give rise to body tissues, while the descendants of cells retaining the full complement of chromosomal material become the **germ line.** By the 32-cell stage, the fracturing of the chromosomes ceases.

About one-quarter (i.e., 22–33%) of the DNA inherited from the gametes (i.e., germ-line DNA)

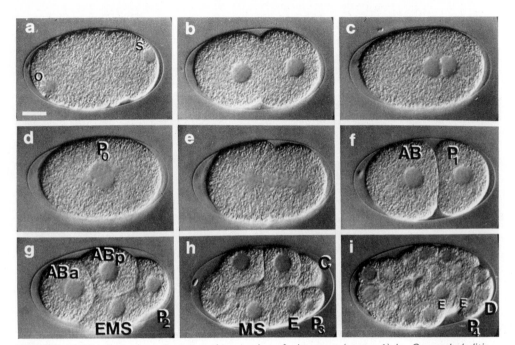

FIGURE 11.54. Scanning electron micrographs of cleavage (stage 1) in *Caenorhabditis elegans*. Anterior toward left. Dorsal toward top. (a) About 30 minutes after fertilization. The oocyte (o) and sperm (s) nuclei are at opposite poles. (b, c) Nuclei migrate (primarily oocyte nucleus) while rippling of cytoplasm gives impression of pseudocleavage. (d) Nuclei migrate toward center and form *Ascaris*-type zygotic nucleus (P_0). (e, f) Cleavage follows immediately forming the larger AB somatic founder cell and the P_1 cell. (g) AB divides to form the anterior AB_a and posterior AB_b cells, while P_1 divides unequally to form the larger EMS cell and the posterior P_2 cell. (h) All cells divide. EMS forms the E endodermal founder cell and the MS cell. P_2 forms the C somatic founder cell and the P_3 cell. (i) The AB cells divide synchronously, but otherwise divisions are increasingly asynchronous, proceeding in the order AB, MS, E, and C. Division of P_3 gives rise to the D somatic founder cell and the primordial germ cell, P_4. The two gut precursor cells (E, E) begin their inward migration, thereby initiating gastrulation. From E. Schierenberg, *J. Embryol. Exp. Morphol.*, 97(Suppl.):31 (1986), by permission of the Company of Biologists and courtesy of the author.

is eliminated by chromosomal fracture and diminution. The bulk of this DNA consists of repeated DNA sequences comprising as much as 99% of the cell's satellite DNA. This DNA is not transcribed and, while qualitatively different from the retained-satellite DNA, is organized around a similar 120-bp repeating unit.

The proportion of ribosomal genes in the genome is unchanged following diminution, suggesting that fracturing does not discriminate for or against particular genes. A small part of the eliminated DNA may also contain unique sequences (Müller et al., 1982; Tobler et al., 1972).

Like the paternal genes for rRNA that were active soon after fertilization, the unique sequences slated for elimination may have been active earlier, possibly during gametic differentiation. The eliminated sequences may therefore be dispensable, but the adaptive value of elimination is unclear (see Davidson, 1986).

In the nematode, *Caenorhabditis elegans*, cell lineage has been followed from the zygote to a larva of 959 somatic cells (Sulston et al., 1983). This lineage, by far the most accurate and complete ever obtained for any organism, has clarified many of the issues surrounding spiral cleavage in nematodes and has required only minor revisions in the nomenclature employed for other nematodes. Little has been done, however, to homologize the nomenclature for quartets, duets, and the monets.

The zygote (sometimes called the primordial or P_0 cell) cleaves unequally to form a relatively small posterior (P_1) "macromere" and a larger anterior "micromere" (AB, Figs. 11.54 and 11.56). The AB cell line (which does not separate into recognizably different A and B lines) forms progeny that spread over the surface of the embryo except in the dorsocaudal region. The majority of the AB cell derivatives are presumptive ectoderm (giving rise primarily to the cutaneous hypodermis and neurons), while some of the cells are presumptive mesoderm

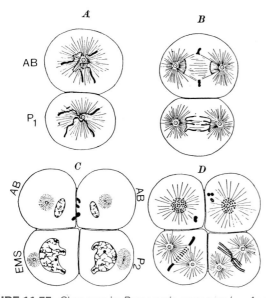

FIGURE 11.55. Cleavage in *Parascaris equorum* (= *Ascaris megalocephala*) illustrating chromosomal fracturing and diminution in somatic cell line and retention of complete chromosomes in primordial germ cell line. (a) The 2-cell stage at prophase. P_1, posterior cell, "macromere," AB, anterior cell, "micromere." (b) Anaphase of second division. Chromosomal fracturing results in chromosomal material failing to move to spindle poles in AB cell. (c) The 4-cell stage. P_2 and EMS are descendants of posterior cell. Chromosomal material not incorporated into nuclei remains in cytoplasm. (d) Metaphase of third cleavage in P_2 and EMS cells. Chromosomal fracture seen in EMS cell. From E. B. Wilson, *The cell in heredity and development,* Macmillan, London, 1896.

(forming four muscles on the ventral surface of the tail and so-called rectal cells).

Division of the P_1 cell produces an anterior EMS cell (named retrospectively for endoderm and mesoderm-stomodeum) and a posterior P_2 cell. Anteriorly, the EMS cell divides unequally to produce

E and MS cells. P_2 produces P_3 and C cells, and P_3 produces P_4 and D cells.

By the 24-cell stage, four anterior **founder cells** or **embryonic blast cells** (MS, E, C, and D) and a primordial germ cell (P_4) are separated. The founder cells will divide equally and relatively synchronously to produce different lines of somatic cell differentiation, while the P_4 cell will cease dividing temporarily and then give rise to the germ line. The founder cell lineages of *C. elegans* and other nematodes are remarkable for their early origins and narrow options for differentiation.

EVOLUTION OF CLEAVAGE TYPES AND PATTERNS

Different types of cleavage presumably evolved in response to specific and local requirements and to the balance between adaptive advantages and disadvantages inherent in competing developmental strategies. In addition, initial conditions such as the sizes of eggs and their yolk contents presumably influenced the evolution of cleavage. But, what came first?

Answers to evolutionary questions are inevitably based on guesswork, especially where no fossil record is available to evaluate alternative possibilities. Still, some statements about the antiquity of one form of cleavage compared to another are justified by comparative evidence outside embryology, and assertions about adaptive advantages are sometimes testable.

Currently, **primitive shared conditions** are defined as those conditions in a present group of organisms which would have been present in the ancestors of the group, while **derived shared conditions** are those conditions in a branch of the

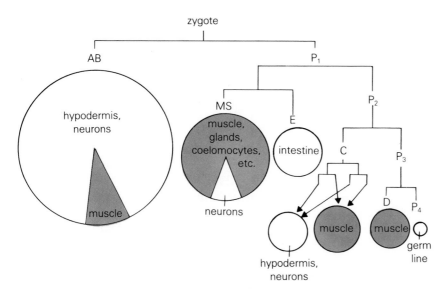

FIGURE 11.56. Scheme for early separation of founder cells in the nematode, *Caenorhabditis elegans.* Areas of circles and of segments within circles are proportional to number of cells of particular types in juveniles. Stippling represents ectodermal derivatives. Color represents mesodermal derivatives. From J. E. Sulston, E. Schierenberg, J. G. White, and J. N. Thomson, *Dev. Biol.,* 100:64 (1983), by permission of Academic Press and the authors.

group but not in other branches. Both primitive and derived conditions can be highly evolved or only modestly evolved, and their evolution may have proceeded linearly or tortuously. **Primary conditions** evolve gradually, whereas **secondary conditions** evolve after a change in evolutionary direction, especially a loss from a previously highly evolved condition.

Traditionally, embryologists emphasize yolk as the supreme determinant of cleavage type and the evolution of yolk as the force shaping the evolution of cleavage. Embryologists generally assume that yolk-poor eggs are primitive and yolk-rich eggs derived and that ovuliparity is primitive and both oviparity and viviparity derived. But did eggs actually evolve in these directions, and did the evolution of yolk dictate the evolution of cleavage?

Which Came First: Total Cleavage or Partial Cleavage?

Answers to this question are closely linked to whether yolky eggs are primary and yolk-poor eggs secondary, or yolk-poor eggs are primary and yolky eggs secondary. In specific cases, both possibilities are plausible.

In the case of mammalian eggs, yolk conceivably evolved in the direction of secondary loss. The small, yolk-poor eggs of therian mammals (i.e., the placental mammals including metatherians, or marsupials, and eutherian mammals) may have begun their descent as a moderately large, yolky egg, like that of the prototherian duck-billed platypus (Fig. 6.9). Gestation, or internal brooding linked to placental nourishment, would have made yolk redundant, and its negative aspects (its high metabolic cost for the female parent and its liabilities for the egg) would have led to its elimination.

Given this scenario, and the further assumption that the moderately large, yolky ancestral egg performed partial cleavage, the total cleavage of eutherian eggs would represent a derived secondary condition. The therian mammals would have "reevolved" total cleavage from partial cleavage at the same time as they evolved viviparity.

On the other hand, the evolution of the cleidoic eggs of birds presumably began with small, less yolky reptilian-like eggs. The avian form of partial cleavage would have taken off from the partial cleavage already installed in reptilian eggs.

Similarly, but independently, the form of partial cleavage found in teleosts would have evolved as yolk accumulated in the course of teleostomic evolution. In primitive ganoids with polarized total cleavage, furrows take the "path of least resistance," while in advanced teleosts with discoidal

cleavage (Fig. 11.15), yolk is shoved aside and cleavage takes place in the relatively unencumbered cytoplasmic-rich blastodisk.

Mollusks also provide examples of an evolutionary trend toward partial cleavage accompanying the accumulation of yolk. The cephalopods seem to have evolved their megalecithal eggs from primitive microlecithal molluscan eggs and their partial cleavage (Fig. 11.13) from total spiral cleavage (Figs. 11.40–11.41).

Which Came First: Total Cleavage or Intralecithal Cleavage?

The answer to this question is also closely linked to whether yolk-rich or yolk-poor eggs are primary or secondary in particular cases. The arthropods provide examples of the parallels between yolkiness and intralecithal cleavage evolving in both directions.

Beginning with microlecithal eggs and radial cleavage (e.g., cladocerans), a few wedge-shaped blastomeres seem to have evolved into a blastoderm of many narrow cells (e.g., *Artemia*, Fig. 11.29) and, as the amount of yolk increases, especially in malacostracans, a highly cellular blastoderm. At the same time, cellular borders seem to have disappeared. The yolky parts of pyramidal cells fuse centrally, and, in megalecithal decapods, a superficial blastoderm appears around a unified yolk mass as a result of unambiguous intralecithal cleavage (Fig. 11.4).

On the other hand, a change from intralecithal to total cleavage appears to have accompanied the secondary reduction of yolk and production of smaller eggs in the evolution of viviparity among terrestrial arthropods. The major features of this trend are traced in schemes for the evolution of scorpions, pseudoscorpions, mites, and onychophorans (see Anderson, 1973).

For example, instead of the intralecithal type of cleavage seen in the yolky eggs of primitive onychophorans, (Fig. 11.1) the spherical, less-yolky eggs of nonplacental viviparous onychophorans exhibit total cleavage. Rather than forming intralecithal nuclei, the egg breaks up into spheres. Most of these are anucleate **pseudoblastomeres,** but some are nucleated spheres and at least one is a blastomere which migrates to the surface and produces the organism.

The small microlecithal eggs of placental viviparous onychophorans also undergo total cleavage. The morula-like early embryo does not resemble a product of intralecithal cleavage. Some blastomeres form a stalk that reaches the maternal tissue, or placenta, while other blastomeres form the

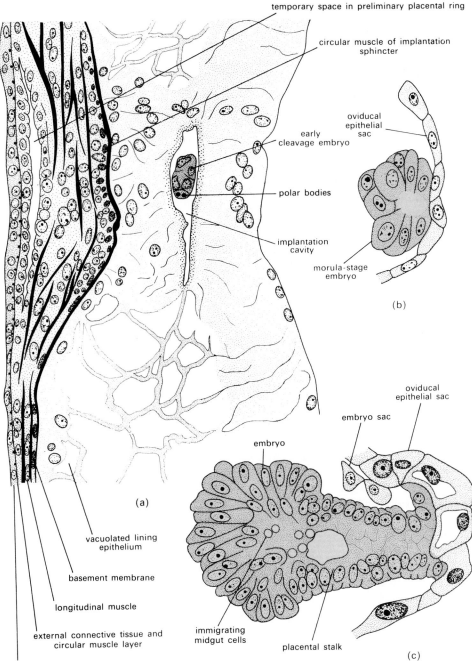

temporary space in preliminary placental ring

circular muscle of implantation sphincter

oviducal epithelial sac

early cleavage embryo

polar bodies

implantation cavity

morula-stage embryo

(b)

vacuolated lining epithelium

basement membrane

longitudinal muscle

external connective tissue and circular muscle layer

external epithelium

(a)

oviducal epithelial sac

embryo sac

embryo

immigrating midgut cells

placental stalk

(c)

FIGURE 11.57. Implantation in viviparous placental onychophorans of Central and South America (e.g., *Epiperipatus trinidadensis* and *Macroperipatus torquatus*). (a) The small naked egg undergoes total, equal cleavage to produce a small embryo that initially lies free in an oviductal crypt. (b) Further division results in a hollow embryo (a morula) that contacts the oviductal wall. (c) Proliferation of blastomeres and growth produces a columnar embryonic blastoderm with migrating midgut cells set upon a placental stalk. From D. T. Anderson, *Embryology and phylogeny in annelids and arthropods,* Pergamon Press, Oxford, 1973. Courtesy of D. T. Anderson. Used by permission.

definitive embryo (Fig. 11.57). The resemblance to an early cleaving mammalian embryo (Fig. 11.24) provides a striking example of parallel evolution.

Which Came First: Blastomeres or Blastoderm? Cleavage or Cellularization?

Because one ordinarily thinks of living things as cells or consisting of cells, blastomeres are thought to have evolved before the cellular blastoderm. If, however, living things as such preceded

the evolution of cells, syncytia might be the primitive condition rather than cells (see Hadzi, 1963). Of course, both paths of evolution might have been traveled independently.

In the case of insects, the total cleavage of primitive collembolan eggs (at least until the 32-cell stage, Fig. 11.6) may represent the primitive condition. Blastomeres would have preceded the syncytial blastoderm of thysanuran (Fig. 11.5) and pterygote insects (Fig. 11.10).

On the other hand, priority for syncytia is suggested by numerous examples of intralecithal cleavage in the eggs of other primitive arthropods.

Moreover, preexisting cytoplasmic networks surrounding the yolk spheres of yolky onychophoran eggs (Fig. 11.1) and laced through the periplasmic fields of spiders (Fig. 11.2) and yolk pyramids of myriapods (Fig. 11.3) provide focal points for later cellularization.

Possibly, yolk wrapped in a cytoplasmic reticulum and occupied by energids served as a stepping stone for more than one evolutionary pathway. Competition among energids could have provided the impetus for secondary reduction in energid numbers leading to the evolution of cells (i.e., single nuclei occupying single cytoplasmic units). Alternatively, increased size and the accumulation of inert material could have buffered nuclei against competition and led to the expansion of energid numbers within a syncytium. Cleavage and cellularization in different phyla might represent unrelated evolutionary pathways.

Which Came First: Radial, Biradial, Bilateral, or Spiral Cleavage?

These different forms of total cleavage occur among a host of planktonic marine invertebrates. The question of which form of cleavage came first must be asked for organisms belonging to the same phylum or superphyletic group, whereas parallels may be sought among groups.

The plankton is a very rich but hostile environment. Morphological adaptations to it range from radial symmetry in freely floating organisms to bilateral symmetry, in swimming animals. Behavioral adaptations include the capacity for directional movements and increased sensory acuity. Developmental adaptations may be geared to a smooth transition between adaptive but not necessarily optimal developmental stages (sometimes called regulative development) or to an abrupt transition through a highly vulnerable metamorphic stage between more nearly optimal developmental stages (sometimes called determinate development).

Organisms producing radially, biradially, and spirally cleaving eggs tend to release them in large numbers. The eggs are generally small with little yolk representing a minimal parental investment, and, with few cleavages, they undergo rapid indirect development. The hatched larvae have a superficial radial symmetry and enter the world of the plankton with the barest of equipment for making a living.

Echinoderms, on one hand, provide themselves with a radially symmetrical larva in the blastula stage (Fig. 11.31i) before paying too much attention to restricting the developmental potential of cells. Development to a bilaterally symmetrical prism and pluteus (Fig. 11.30) is gradual, and metamorphosis to an adultlike condition is undramatic.

Spiralians (e.g., annelids and mollusks), on the other hand, sort out developmental potentials to a few teloblasts (Fig. 11.47) and tuck these away by the gastrula stage (Fig. 11.49). Developmentally, the toplike trochophore (Fig. 11.45) is a loaded spring. At metamorphosis, larval tissues are eclipsed as teloblasts produce the adult.

Protochordates fit neither the echinoderm nor the spiralian pattern. The first three protochordan cleavages bear traces of a prior echinoderm-type radial cleavage (compare Figs. 11.31c and 11.33e), but embryos do not hatch as radially symmetrical larvae. Bilaterality may have been pushed back from the larval stage of an echinoderm-like ancestor to the embryonic stage of protochordates by adaptive pressure to hatch a larva capable of functioning in habitat selection. This possibility may be reflected in the granules and cytoplasms which are sorted out and localized in the protochordate egg prior to cleavage.

The stereotypic pattern of blastomeric development in ascidians (if not other protochordates, e.g., Amphioxus) and the profound metamorphosis occurring at the end of the larval stage (see Fig. 5.15) superficially parallel events in spiralians. Actually, the two situations are quite different. While latent determinants of spiralian development are sequestered in teloblasts and reserved for the adult, the determined blastomeres of ascidians are utilized to produce a complex larva.

Similarly, ctenophore eggs produce biradially symmetrical cydippid larvae (actually juveniles) while sorting out developmental determinants in an uncompromising series of divisions. If ctenophore cleavage evolved from a primitive coelenterate-type of irregular cleavage, it would seem that just about any degree of early developmental determinacy can be superimposed on embryos exhibiting just about any type of total cleavage.

In general, the cleaving embryo seems to have been an evolutionarily plastic stage in the development of the organism. Various aspects of cleavage seem to be quite fluid in moving from stage to stage and superimposing their effects in different cleavage patterns.

Chapter

12

CONSEQUENCES OF CLEAVAGE:

Analytic Studies

*T*he consequences of cleavage are not amorphous masses of cells but organized multicellular masses. Organization is exhibited originally by cleavage patterns and the distribution of cytoplasmic components and later by differentiation. More pointedly than any other stage of development, cleaving embryos raise embryology's central question: Where does organization come from?

Embryos at other stages and even adult tissues raise the same question, but the stakes are greatest during cleavage when most cell lineages are just getting started and morphogenesis has hardly begun. One may wonder about the degree to which blastomeres are representative of other cells and about cleavage's ability to influence determination far down the road of development, but the great issue of embryology is never clearer than it is during cleavage: Do cells differentiate because of what they are or where they are? Analytic studies of cleavage attempt to answer this question.

DETERMINATE AND REGULATIVE DEVELOPMENT

The oocytes and early embryos of many types of animals contain a nonuniform distribution of cytoplasmic components, although the developmental significance of these components is not known. . . . These observations suggest that the oocytes of frogs and the early eggs of nematodes contain systems for cytoplasmic localization, possibly by either unequal segregation or position-dependent stabilization of macromolecules.

K. J. Kemphues et al. (1988, p. 311)

The analysis of development frequently begins with the premise that a course of development is either fixed (i.e., determinate) or flexible (i.e., regulative). For example, spiral and biradial cleavage, bilateral cleavage in ascidians, radial cleavage in the nemertines and crustaceans, partial cleavage in cephalopods, and the postcellularization insect embryo are classified as determinate, while spherical, bilateral, and partial cleavage in vertebrates, radial cleavage in echinoderms, and the intralecithal phase of cleavage in insects are lumped together as regulative.

A presumption of antagonism between determination and regulation is sometimes implied in definitions and an incompatibility is often inferred by usage. When concepts are clarified, however, determination and regulation are seen to refer to different phenomena and to take their direction from different controls. They are not mutually exclusive.

Operational Definitions

Hans Driesch (1867–1941) left his stamp on embryology in many of the terms routinely used to theorize about development (although Paul Weiss [1939] should be credited with codifying and clarifying these terms). Several of the terms have **operational definitions** limiting them to special circumstances or conditions. For example, some definitions apply *in situ*, that is, when cells develop in their normal positions. Other definitions apply *in vitro*, when cells are isolated from the organism (e.g., in tissue culture), or *in vivo*, when cells are **transplanted** from a **donor** to a **host** organism or from a place of origin to an abnormal or **ectopic** site in an organism.

A blastomere's **fate** or **prospective significance** is the normal differentiated state of its cellular lineage developing *in situ*. Fate is not necessarily equal to the full range of developmental possibilities.

Determination is the fixation of a cell's fate. Theoretically, it is the irreversible activation of genes by cytoplasmic and nuclear controls leading cells toward particular types of differentiation.

A blastomere or tissue's **competence** is its ability to reach its fate *in vivo* or to differentiate into other tissues beyond its fate in response to conditions in a host or at an ectopic site. **Regulation** is said to occur when competence permits a transplanted blastomere or tissue to harmonize with new surroundings, and when developmentally significant local control, or **induction,** integrates the transplant into a morphogenic pattern with its surroundings. Competence is sometimes redefined therefore as the ability of tissues to regulate or to react to inductive influences.

An isolated blastomere (i.e., a blastomere *in vitro* or in a potentially neutral *in vivo* environment) is also said to regulate when it gives rise to the complete morphogenic pattern of the embryo

or larva. A physical **field,** such as a magnetic field which instantly fills in missing parts, provides a superior model compared to induction for this sort of regulation.[1]

A transplanted blastomere or tissue may also **self-differentiate** and form a **secondary structure** (also called an ectopic or additional structure) not normally present in the region. Self-differentiation by individual blastomeres is called **autonomous cellular differentiation.** The secondary structure does not become integrated into other structures in the vicinity, but host tissue may be integrated into the secondary structure by inductive influences arising from the transplant.

A blastomere's **potency** (or prospective potency, also possible fate, and sometimes potential) encompasses all the tissues that the blastomere's lineage gives rise to *in situ, in vivo,* and *in vitro* with or without interaction with other cells. Ideally, the full range of a blastomere's potency for self-differentiation is revealed when cells are isolated, but the requirements of development and differentiation are often met only in an *in vivo* environment, and potency is equated with self-differentiation in such a potentially neutral environment.

Determination generally accompanies the imposition of limitations or **restrictions** on potency and competence (Fig. 12.1). A blastomere is determined when its fate corresponds to its potency and competence. Blastomeres determined early are **precociously determined** or self-differentiating, and their differentiation is called **determinate development.** When an early blastomere's fate does not correspond to its potency and competence, its subsequent differentiation is called **indeterminate** or **regulative development.**

Determinate Development

Probably no species is purely determinate in its development. Spiralians provide the classic examples of determinate development, but they also have regulative features and may even acquire regulatability through development. Insects, on the other hand, may be regulative in early development but exhibit unique features of determination in specific embryonic tissues. Ascidians are justifiably classified as determinate, but their determination is not absolute and may represent a uniquely derived condition. Certainly, determi-

---

[1]These definitions of regulation are not incompatible. Both imply that global forces shape blastomeres into whole embryos and differ only insofar as these forces arise from a host or from isolated blastomeres.

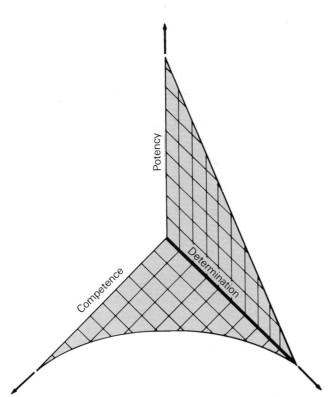

FIGURE 12.1. Restriction: the theoretical relationship of potency, determination, and competence in blastomeres moving toward their fate.

nation is not universal among protochordates. The lancet *Branchiostoma lanceolatum* (= Amphioxus), for example, is considered quintessentially regulative.

Nematodes, in general, and* Caenorhabditis elegans, *in particular, exhibit many features of determinate development (see Schierenberg, 1982). Founder cells are quickly established and give rise to lineages with limited prospective significance. Blastomeric lineages are generally autonomous, and the experimental destruction of blastomeres leads to defects corresponding to a blastomere's prospective significance (Sulston et al., 1983).

Only about 5000 genes are present in *C. elegans* (about three times the amount of DNA present in *Escherichia coli*), running the organism in all its aspects, and about half of these genes are presumably of the "housekeeping" variety not involved in determination. Studies on the remaining **developmental genes** employing mutation and transformation by microinjection of exogenous DNA indicate that almost everything affecting development, including programmed cell death, is controlled by genes. From the timing and orientation of divisions to the death of certain cells, determination in *C. elegans* seems to be "hardwired" in genes.

Various types of developmental genes seem to take part. **Lineage genes** operate throughout cellular lineages, and mutations of these genes affect cells lying in several body regions with multiple pleiotropic effects. Other developmental genes may operate in only one body region, and their mutations may transform one region to another (e.g., the *map-5* mutation anteriorizes the posterior body region and prevents the development of male copulatory structures). Finally, other genes (e.g., the *glp-1*$^+$) operate in cells of specific lineages differentiating in specific pathways (e.g., muscle differentiation among pharyngeal cells of the AB lineage). Mutations prevent only one path of differentiation in only one lineage (see Emmons, 1987).

Determination is nevertheless incomplete during cleavage. First, potency and competence are not entirely restricted in early blastomeres. Specifically, only intestinal endoderm and germ cells arise from single founder cells (E and P_4), while muscle is produced by each of the other founder cells (AB, MS C, and D, see Fig. 11.56).

Second, inductive influences seem to deliver graded **positional cues** for differentiation. Positional signaling on a global scale is suggested by the reversal of fates in AB blastomeres when their positions are exchanged. Moreover, local inductive interactions are indicated by the failure of AB lineages to differentiate pharyngeal muscle cells when the P_1 cell or its anterior progeny is removed.

Genes may control inductive influences, but they do so at a distance. For example, the eggs of mothers homozygous for some **maternal effect mutants** fail to support induction. In embryos of mothers homozygous for *glp-1*, induced pharyngeal muscle cells of the AB lineage fail to differentiate, while similar autonomous cells of the P_1 lineage succeed. The mutant genes acting in the P_1 lineage seem to have undermined its ability to induce, leaving the AB lineage in much the same position it occupies when the P_1 is absent.

Finally, some developmental pathways may not be entirely separate by cellular lineage. **Linearly equivalent progenies** normally give rise to cells of more than one type along the length of the embryo. On occasion, cells from contralateral branches of linearly equivalent progenies may replace destroyed cells. The same lineage that gives rise to hypodermis, for example, may compensate for defects in motor nerves caused by laser-microbeam cautery during embryonic and postembryonic periods. In these instances, cell fate may not be fixed until differentiation (Sulston and White, 1980).

Molluscan embryos are often portrayed as mosaics of self-differentiating blastomeres with very limited and highly stereotypic fates. For example, in eggs of the tusk shelled mollusk, *Dentalium entalis,* the AB blastomere normally gives rise to much of the trochophore's pretrochal area (except the apical tuft formed by 1d). Isolated, the AB blastomere gives rise to a larva having a defective pretrochal region and conspicuously lacking posttrochal tissue. The CD blastomere normally gives rise to much of the posttrochal tissue. Isolated, the CD blastomere forms a defective posttrochal larva largely lacking the pretrochal region (Fig. 12.2) (Wilson, 1904).

Determination is also apparent in micromeres. Each of the first 32 blastomeres of *Neanthes,* for example, develops according to its fate upon isolation (Costello, 1945), and many defects resulting from the elimination of these blastomeres are consistent with the missing cell's prospective significance (see Clement, 1976).

In the mud snail, *Ilyanassa obsoleta,* elimination of first quartet micromeres causes specific but not widespread deficiencies (Table 12.1). Removing 1a results in the loss of the left eye, while removal of 1c results in the loss of the right eye

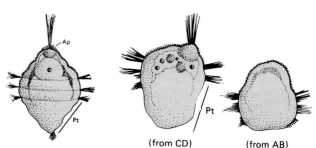

FIGURE 12.2. Trochophore larva of the scaphopod (tusk shell), *Dentalium entalis,* and abnormal larvas formed from isolated blastomeres (see Fig. 11.44 for names of blastomeres). CD, larva developing from CD macromere. AB, larva developing from AB macromere. PT, post-trochal region. From E. B. Wilson, *J. Exp. Zool.,* 1:1 (1904), by permission of Alan R. Liss, Inc.

(Fig. 12.3). Elimination of 1b creates a deficiency in the ectoderm between the two eyes, but eliminating 1d has no apparent effect.

Elimination of second quartet micromeres of *Ilyanassa* results in much more severe defects, although elimination of 2b has no effect at all. Normally, the 2c and 2d micromeres are fated to take part in the development of the shell. Removal of 2c results in reduction of the shell by half. In ad-

TABLE 12.1. Defective development of *Ilyanassa obsoleta* due to removal or killing of polar lobe or micromere

		Removed or Killed												
		Blastomeres of[a]												
		1st Tier				2nd Tier				3rd Tier				4th Tier
Larval Part	First Cleavage Polar Lobe[a]	1a	1b	1c	1d	2a	2b	2c	2d	3a	3b	3c	3d	4d
Eye														
Left	−	−				−								
Right	−			−										
Ectoderm between eyes	?		−											
Pulsating heart	−							−			def			−
Shell	−						dim	−						
Velum lobes														
Left	−									dim				
Right	−										dim			
Foot (halves)														
Left	−											−		
Right	−											−		
Stomodeum							(everted)							
Intestine	−													−
Statocyst														
Left												−		
Right												−		
Any apparent effect		−	−											

[a]−, Absent; ?, uncertain; def, defective; dim, diminished.
Source: Data from Clement (1967, 1976).

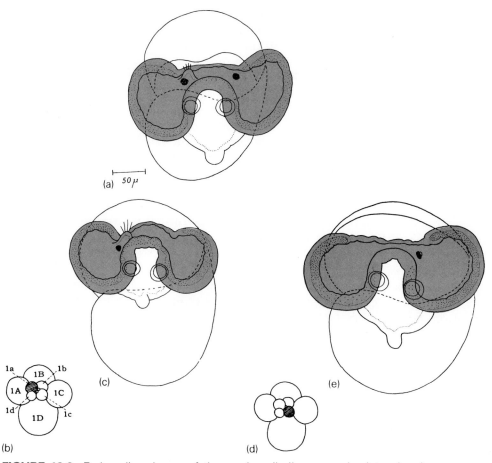

FIGURE 12.3. Early veliger larvas of the mud snail, *Ilyanassa obsoleta,* showing consequence of removing specific first tier micromeres. (a) Normal veliger larva. The right tentacle (hill with hairs over right eye) appears precociously. (b) Diagram of operation. The 1a micromere (diagonal hatching) is deleted from the first tier of micromeres at the 8-cell stage. (c) Result of deleting 1a. The left eye fails to appear, but the embryo seems otherwise normal. (d) Diagram of operation. The 1c micromere (diagonal hatching) is deleted from the first tier of micromeres at the 8-cell stage. (e) Result of deleting 1c. The right eye and right tentacle fail to appear, but the embryo seems otherwise normal. Stippling, pigmentation on velar lobes (velar cilia not shown); blackened spots, eyes; double circles, statocysts. From A. C. Clement, *J. Exp. Zool.,* 166:77 (1967), by permission of Alan R. Liss, Inc.

dition, the removal of 2c results in the absence of a pulsating larval heart. Removing 2d generally results in the absence of the shell. In the freshwater mussel, *Unio,* and the slipper limpet, *Crepidula,* removing 2d results in the absence or impaired development of foot as well as shell (i.e., posttrochal ectodermal derivatives).

The fourth quadrant of micromeres is dominated by 4d. Elimination of 4d alone results in the absence of a pulsating larval heart and intestine.

Despite all these instances of determinate development, some features of regulation are also found or suggested by the results of experiments on cleaving spiralians. Even in some of the classic examples of species with early determination (e.g.,

Chaetopterus, Sabellaria, and *Neanthes*) a latent ability to regulate can be demonstrated by conditions that extend the connection of the first polar body from the CD blastomere to the AB blastomere. The conditions vary enormously, ranging from exposure to heat, cold, high concentrations of potassium chloride, deprivation of oxygen, compression, and centrifugation, but the resulting embryos form duplicate parts in much the same way (Fig. 12.4). If the treated blastomeres are totally isolated, more or less complete larvae develop (see Watterson, 1955).

Early determination is to some degree a species-specific rather than a phylum-specific property, and some annelids and mollusks do not seem

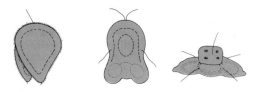

FIGURE 12.4. Partial twins induced in *Neanthes succinea* (= *Nereis limblata*) by broadening the base of the polar lobe and allowing polar lobe cytoplasm to enter the AB blastomere as well as the CD blastomere. From A. Tyler, *J. Exp. Zool.*, 57:347 (1930), by permission of Alan R. Liss, Inc.

especially determined at the time of the first cleavages. In the oligochaete pipe worm, *Tubifex*, for example, isolated CD blastomeres and even D blastomeres produce complete small embryos, and in the pond snail, *Lymnaea palustris*, eliminating one or two (or even three, in one case, Morrill et al., 1973) first quartet micromeres from an 8-cell embryo has no effect on subsequent development (Morrill, 1985).

Normally, the 3c and 3d micromeres provide the principal primordia of the foot in *Ilyanassa*. The muscle of the larval heart comes in part from 4d, while its ectodermal vesicle comes from 2c and possibly in part from 3c. Nevertheless, elimination of third quartet micromeres has smaller effects on the embryo in general than eliminating earlier quartets. Removing 3a or 3b results in a diminution in the size of the ciliated left and right velar lobes (locomotory organs), respectively, and removing 3c or 3d results in defects in the right and left statocyst (balancing organs) and sides of the foot.

The ability to regulate and produce parts corresponding to missing blastomeres is even greater when macromeres are involved. Elimination of C from a 4-cell *Ilyanassa* embryo, for example, should result in the production of a larva lacking a heart and a right statocyst inasmuch as removal of 2c results in the loss of the larval heart, and removal of 3c results in the loss of the right statocyst.

In fact, elimination of C does not prevent heart or right statocyst development in a substantial number of larvas. It is hard to imagine how these structures could have developed without induction and cell interactions playing a role (see Clement, 1976).

Insects exhibiting the internal development of adult body parts, imaginal disks (i.e., endopterygote or holometabolous insects, see below, Fig. 14.5), provide dramatic examples of determination.

These organized groups of larval cells normally develop into specific complex adult structures at the time of metamorphosis into the adult. The determination of imaginal disks is demonstrated by fragments of disks transplanted repeatedly through larvae (i.e., via serial transplantation). When permitted to differentiate by allowing the larval host to undergo metamorphosis, the imaginal disk fragment develops into the same structures that would have been formed by the original imaginal disk.

Even deviations from strict determination do not produce chaotic or random changes, but whole parts normally formed by other imaginal disks. Known as **transdetermination**, imaginal disk fragments follow a limited number of options (see Hadorn, 1966).

The extreme degree of determination represented by imaginal disks may not be representative of the precursors of embryonic and larval tissues. In advanced dipterans, such as *Drosophila melanogaster*, the determination of germ-line cells, at least, may occur quite early, but nuclei in the syncytial blastoderm of other insects are still undetermined, and the blastoderm is capable of regulation (see below).

In ascidians, the prospective fate of blastomeres is extremely limited (Fig. 11.35). Killing blastomeres generally results in uncompensated deficiencies in the embryo corresponding to the dead blastomeres' fates, and isolated blastomeres generally differentiate according to their prospective significance.

Most deficiencies in embryos caused by killing blastomeres parallel deficiencies in the distribution of cytoplasmic granules. Eliminating blastomeres containing yellow granules produces defects in presumptive tail muscle and mesenchyme, while eliminating blastomeres in the gray yolk crescents disturbs presumptive endoderm.

Cells of the tail-muscle lineage (B4.1, Fig. 12.5) synthesize their characteristic acetylcholinesterase enzyme even in isolation, but the embryo lacking these blastomeres fails to synthesize the enzyme. Similarly, muscle-specific calcium currents developing synchronously with acetylcholinesterase synthesis appear at approximately the normal time in presumptive tail-muscle cells isolated from gastrulas (Simoncini et al., 1988).

Determination is not an absolute quality or regulation entirely absent in ascidians. Regulation is observed when embryos are grafted together with parallel animal–vegetal axes. The half embryos do not self-differentiate but combine into a giant larva. Even if these larvae are constructed by

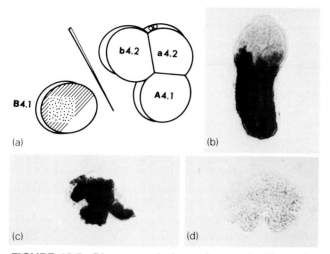

FIGURE 12.5. Diagram and photomicrographs illustrating experimental isolation of the B4.1 blastomeres and the development of acetylcholinesterase activity. (a) Diagram of blastomere isolation. (b) Normal tailbud control larva showing enzyme localization in developing tail. × 140. (c) Enzyme activity in isolated B4.1 cell lineage. × 140. (d) Absence of reaction in remainder of embryo. × 140. From J. R. Whittaker, G. Ortolani, and N. Farinella-Ferruzza, *Dev. Biol.,* 55:196 (1977), by permission of Academic Press and courtesy of the authors.

point-for-point matching of cells in right and left halves, some degree of interaction between embryonic parts would seem to have taken place.

Isolation experiments also fail to show self-differentiation in every instance. For example, isolated blastomeres containing clear cytoplasm (presumptive ectoderm, see Fig. 11.33) fail to produce a recognizable epidermis, and isolated blastomeres containing the light gray crescent (presumptive neural ectoderm, cerebral vesicle, sensory cells, and nerve cord) fail to produce any nerve tissue.

To a degree, these failures can be reversed by combining the presumptive ectodermal and neural ectoderm blastomeres with other blastomeres (Reverberi, 1972). Brain formation, for example, is restored when the presumptive neural ectodermal blastomeres are combined with presumptive notochord blastomeres. It would seem that intercellular interactions are involved in nerve cord differentiation.

Mechanisms of Determinate Development

Determination has both positive and negative features. Determined blastomeres may possess the means for their own differentiation, but they are also unresponsive to intercellular influences. At the same time determined cells have the virtue of

differentiating in abnormal circumstances, their resistance to intercellular influences keeps them from filling in missing parts and cooperating in differentiation when disturbed (see Freeman, 1979).

Determination does not imply any lack of influences on development. On the contrary, mechanisms of self-differentiation are often thought to involve cytoplasmic determinants of development. These determinants are not necessarily granular, although species with colored granules or zones of distinguishable cytoplasm have been widely exploited in research. Why then should determination be linked to a lack of intercellular and global influences?

Possibly, determination is like the setting of a developmental clock. The clocks in the various presumptive tissues have different settings, but, overall, differentiation will be synchronized at some later time. Once determined, a presumptive tissue would not be well served were it subject to extracellular influences that might reset its clock.

Were this the case, resistance to outside influences would represent an adaptation of determined cells rather than an intrinsic part of determination. Mechanisms that reduce communication among cells, including mechanisms for the suppression of intercellular signals, might exist independently of determination.

In nematodes, cytoplasmic determinants seem to exercise control over gene activation. Determinants such as the P granules segregating to primordial germ cells (Fig. 12.6) (Strome and Wood, 1983) are visualizable as granules present in the oocyte.

Other putative determinants are not present at fertilization. So-called rhabditin granules, for example, associated with the gut, appear after the E cell has divided three times. Of course, invisible cytoplasmic determinants may exist prior to the appearance of visible ones.

In mollusks, some of the determinants of development may be transmitted through the polar lobes, since removal of the lobes at the first cleavage alters development profoundly (see Dohmen, 1983). The cleavage pattern and anterior–posterior axis of **lobeless** embryos fail to develop normally (Fig. 12.7). Since removal of the second polar lobe (L-2) results in less severe defects than removal of the first polar lobe (L-1), different determinants would seem to be sequestered in blastomeres during succeeding cleavages.

In *Dentalium*, defects in L-1 lobeless embryos resemble defects caused by elimination of the CD

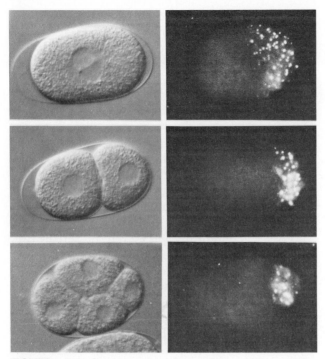

FIGURE 12.6. Segregation of P granules to the posterior P cell of *Caenorhabditis elegans* eggs. Each row shows the same embryo in right and left panels. (Left panel) Nomarski images. (Right panel) P granules stained by indirect immunofluorescence with monoclonal antibodies. (Top) Zygote at pronuclear meeting. Earlier evenly distributed P granules now move to posterior pole. (Middle) The 2-cell embryo. P granules exclusively in P_1 cell and concentrated where P_2 cell will form. (Bottom) The 4-cell embryo. P granules in P_2 cell. From S. Strome and W. B. Wood, *Cell,* 35:15 (1983), by permission of M.I.T. Press and courtesy of the authors.

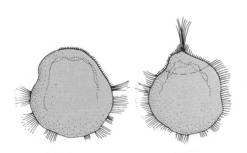

(a) (without L_1)   (b) (without L_2)

FIGURE 12.7. Trochophore larvas of the scaphopod (tusk shell), *Dentalium entalis,* formed from lobeless blastomeres (i.e., blastomeres from which a polar lobe was removed. (a) Polar lobe removed at first cleavage. (b) Polar lobe removed at second cleavage. From E. B. Wilson, *J. Exp. Zool.,* 1:1 (1904).

blastomere (compare Fig. 12.7 to Fig. 12.2), the normal recipient of the polar lobe. L-2 lobeless embryos develop further, but they have defective post-trochal development and other defects associated with the D blastomere, the normal recipient of the second polar lobe (Fig. 12.7) (Wilson, 1904).

In *Ilyanassa*, L-2 lobeless embryos develop much like micromere-deficient embryos. For example, lobeless larvae typically fail to form eyes. The digestive track lacks an intestine. The foot, heart, and external shell also fail to appear, and velar tissue, while abundant, is not organized in the typical bilobed fashion (Table 12.1).

Although polar lobe determinants are widely thought to operate within cells, the possibility that they operate across cell boundaries has not been excluded. For example, L-2 lobeless *Ilyanassa* embryos lack a properly organized velum, whereas embryos developing in the absence of the D macromere have a reasonably complete velum. Since polar lobe determinants of the velum do not operate on the D cell lineage, they would seem to operate through intercellular interactions.

Similarly, deletion of D quadrant micromeres does not prevent the formation of the eyes and right half of the foot. It would seem therefore that polar lobe material operates across micromere boundaries and directs cells belonging to the a–c lineages to develop eye and foot structures (see Clement, 1976).

The biochemical form of polar lobe determinants is unknown or disputed. The polar lobes of *Ilyanassa* contain yolk, while those of *Bithynia* are nearly free of yolk. *Bithynia*'s lobe contains a **vegetal body** (Fig. 11.46) which *Ilyanassa*'s lacks. This body is first seen as a lenticular mass lying against the cortex of the egg at the vegetal pole. After germinal vesicle breakdown, the vegetal body acquires the shape of an inverted cup. Along with the other contents of the polar lobe, the vegetal body is funneled into one of the blastomeres.

Bithynia's vegetal body stains for RNA, suggesting that RNA may be the polar lobe determinant or associated with it. Moreover, ultraviolet (UV) irradiation of polar lobes results in abnormalities. Effective UV irradiation is in the range of 260 nm (corresponding to the peak for nucleic acid absorption as opposed to the 280-nm peak for protein absorption), and the results are dose dependent.

Determinants in the cortex of pond snail eggs also have links to RNA. Freshly laid *Lymnaea stagnalis* eggs contain six lenticular cytoplasmic bodies called **subcortical accumulations** (SCA, Fig. 12.8). The positions of the SCAs correspond to sites that were formerly occupied by follicle cells.

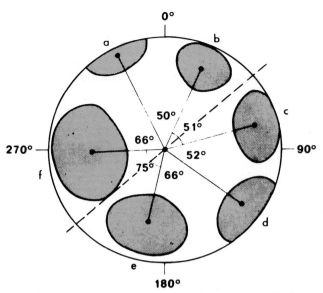

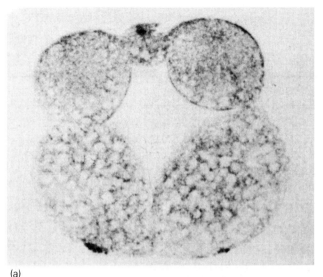

(a)

FIGURE 12.8. Ventral side of egg of the pond snail, *Lymnea stagnalis,* showing the locations of subcortical accumulations (SAC, stippled). The dashed line indicates the approximate median plane of the future embryo. From Chr. P. Raven, *Am. Zool.,* 26:395 (1976), by permission of the American Society of Zoologists.

Cleavage divides the SCAs among the blastomeres, and, at the 8-cell stage, the substance of the SCAs forms a subcortical layer in the vegetative pole of the macromeres. RNA-rich granules soon appear in the same locations (Fig. 12.9a), and, as cleavage progresses, the granules form **ectosomes,** irregularly shaped RNA-rich bodies at the inner ends of the macromeres.

After a rearrangement of blastomeres in the 24-cell embryo, the large 3D blastomere comes into contact with the micromeres and the 3D's ectosome disappears, possibly by transferring RNA to the micromeres (Fig. 12.9b). While SCAs and ectosomes have some of the earmarks of intracellular cytoplasmic determinants, they are particularly intriguing because of their potential role as **intercellular cytoplasmic determinants** (see Raven, 1976).

The results of centrifugation experiments support the possibility that small RNA-rich granules tethered to the cortex play the role of cytoplasmic determinants. Centrifugation of sufficient force to displace granules and disrupt the pattern of ooplasmic substances in early cleaving gastropods frequently results in abnormalities of head development. When performed early enough in the cleavage cycle to allow the displaced ooplasmic materials time to return to their original positions, however, centrifuged eggs develop normally.

UV irradiation of developing cephalopod embryos also suggests that RNA is present in deter-

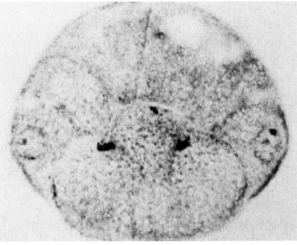

(b)

FIGURE 12.9. Photomicrographs of longitudinal section through *Lymnaea stagnalis* embryos. RNA shows up as dense precipitate (gallocyanin staining). (a) The 8-cell embryo. Densely staining RNA-rich granules appear at the vegetative surface of the relatively large macromeres (toward bottom). A large segmentation cavity occupies the space between the blastomeres. (b) The 24-cell embryo. The large cells with densely stained, compact RNA-rich ectosomes are the 3A and 3C macromeres. Cellular rearrangement has occluded the segmentation cavity and brought the apex of the 3D cell into contact with the micromeres. The light-staining ectosome appears in 3D near its apex. From Chr. P. Raven, *Am. Zool.,* 26:395 (1976), by permission of the American Society of Zoologists and the author.

minants. Irradiation of the presumptive equatorial region of the surface during the first cleavage of a *Loligo pealei* embryo, for example, results in the absence of arms and the reduction of an eye. Cells from the unirradiated portion of the embryo ultimately take over the irradiated area, but development is halted (Arnold and Williams-Arnold, 1976).

The effects of irradiation are **positionally specific** rather than specific to nuclei. Similar effects caused by the experimental ligation of cytoplasmic regions (i.e., separation by a nooselike ligature) also suggest that development is controlled by determinants in the embryo's surface.

The determinants of cephalopod development do not seem to be suspended in the cytoplasm, since gentle centrifugation to the degree of displacing cytoplasm does not cause a displacement of organs in the resultant cephalopod embryo. Determinants may be associated with surface microfilaments, since cytochalasin B, the inhibitor of microfilament assembly, applied to the surface of the early blastoderm, produces regionally specific abnormalities in later development.

In ascidians, such as Styela, *centrifugation for various lengths of time and at various velocities displaces cytoplasmic constituents and redistributes them.* When cleavage follows soon after centrifugation, structures develop in the abnormal positions corresponding to the positions of the cytoplasmic constituents (Fig. 12.10).

When centrifuged earlier with respect to cleavage, however, granules return to their original position and normal development follows. An intriguing, and as yet unaddressed, question is what causes the felicitous redistribution of cytoplasmic constituents.

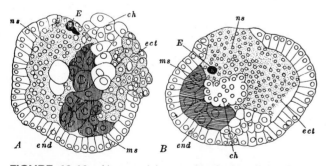

FIGURE 12.10. Abnormal larvas developing from *Styela* (= *Cynthia*) *partita* eggs centrifuged shortly before cleavage. Displaced cytoplasmic factors did not return to normal positions prior to being isolated in blastomeres. The larvas are largely turned inside out. Endoderm (*end*) faces the outside, and ectoderm (*ect*) is turned in. *ch,* Notochord (chorda); *E,* eye; *ns,* nervous system. From E. G. Conklin, *Heredity and environment in the development of men,* 5th ed. rev., Princeton University Press, Princeton, 1922. Used by permission.

Of course, the cytoplasmic constituents that are displaced and then redistributed may not be primary components of determinants. The cytoplasmic granules and yolk that create the colored cytoplasms and crescents of the ascidian egg may represent prosthetic groups or companions of determinants. A deeper structure in the egg cortex or nondisplaceable granules may contain the actual determinants. The appearance of directed movement may be only the consequence of random movement and the trapping of granules and yolk by stable determinants.

The ascidian determinants are not likely to be proenzyme forms of enzymes, since the appearance of the enzymes is prevented by treatment with puromycin, an inhibitor of protein synthesis. Preformed messenger RNA may represent the determinant of alkaline phosphatase in the ascidian gut, since the formation of the enzyme in *Ciona intestinalis* is insensitive to actinomycin D, an inhibitor of transcription. The synthesis of acetylcholinesterase and tyrosinase, on the other hand, is sensitive to actinomycin D, suggesting that the determinants operate by controlling transcription (Whittaker, 1979).

Regulative Development

Regulative embryos tend to restore the whole embryo from a part following injury to a part. In terms designed to characterize regulation, a blastomere is regulative when potency and competence exceed prospective significance or fate. The capacity to interact with inductive signals and receive positional cues is either maintained or regenerated in embryos exhibiting regulation.

In echinoderms, the development of small but otherwise normal larvae from blastomeres isolated at the 2- and 4-cell stage (see Fig. 5.14) *provides the classic example of regulation in the sense of compensation for experimentally induced defects.* The isolated blastomeres continue to cleave as if the missing parts were still present (Fig. 12.11). Half and quarter embryos do not divide like whole embryos, and regulation occurs without the regeneration of missing blastomeres (see Hörstadius, 1973).

Blastomeres soon become restricted, and the ability of echinoderm embryos to regulate is reduced. Blastomeres isolated at the 8-cell stage fail to regulate completely (Fig. 12.12), and isolated tiers of blastomeres at the 64-cell stage no longer regulate.

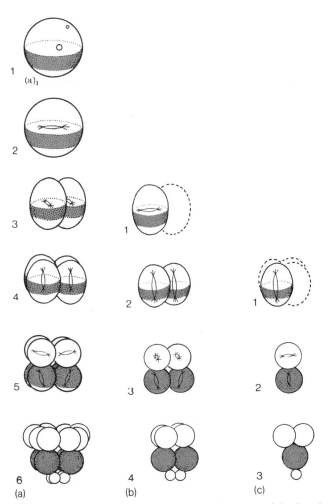

FIGURE 12.11. Diagram illustrating cleavage of isolated blastomeres of the sea urchin, *Paracentrotus lividus.* (a) Complete embryo. (b) Blastomere isolated at 2-cell stage. (c) Blastomere isolated at the 4-cell stage. From S. Hörstadius, *Experimental embryology of echinoderms,* Clarendon Press, Oxford, 1973.

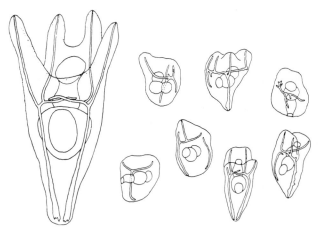

FIGURE 12.12. Drawings of control pluteus and abnormal plutei developing from seven blastomeres isolated from 8-cell embryo. From S. Hörstadius and A. Wolsky, *W. Roux' Arch. Entwicklungsmech. Organ.,* 135:69(1936), by permission of Springer-Verlag.

The potencies of tiers are different and limited (Fig. 12.13). Isolated animal tiers develop into balls with exaggerated cilia. Isolated vegetal tiers develop into slightly more complex balls with some internal structure and reduced cilia.

These restrictions are not the same as determination, since they are reversible. The quantitative addition of portions of vegetal tiers or micromeres to isolated animal tiers promotes regulation in a dose-dependent fashion (Fig. 12.13) (see Hörstadius, 1973).

Embryos at various stages can be dissociated into individual cells to test their potencies or competencies for regulative development. Allowed to reaggregate *in vitro*, the cells reform masses and go

on to differentiate into relatively normal plutei (Fig. 12.14). Competence and the ability to differentiate, it would seem, do not depend on the maintenance of embryonic structure. Once isolated cells have reaggregated, they seem to sort themselves out and resume their interrupted functions (see Giudice, 1986).

Amphibian blastomeres may be isolated by gentle ligation during early cleavage (Fig. 12.15a). Hans Spemann (1869–1941), one of the first embryologists to perform the experiment, used eggs of the newt, *Triturus vulgaris* (= *taeniatus*), and employed a fine human hair for his ligature. Tightening the knot with watchmaker's forceps, he constricted the embryo at intercellular borders until he separated blastomeres or small groups of blastomeres. In some cases, both isolated halves regulated completely and **identical twins** developed (i.e., two isolated larvae, Fig. 12.15b), but, even more often, one half regulated while the other half did not, and **unequal twins** developed (Fig. 12.16).

Alerted by variation in the coloring of eggs, Spemann guessed that cleavage furrows cut eggs at different angles. To test the possibility that furrows separated qualitatively different blastomeres, he ligated eggs around the first cleavage furrow. By the gastrula stage, he could see that some ligatures bisected the embryos longitudinally (Fig. 12.17a) while others bisected them frontally (Fig. 12.17b).

Identical twins were produced when the first cleavage furrow lay along the median longitudinal plane and separated right and left blastomeres. Un-

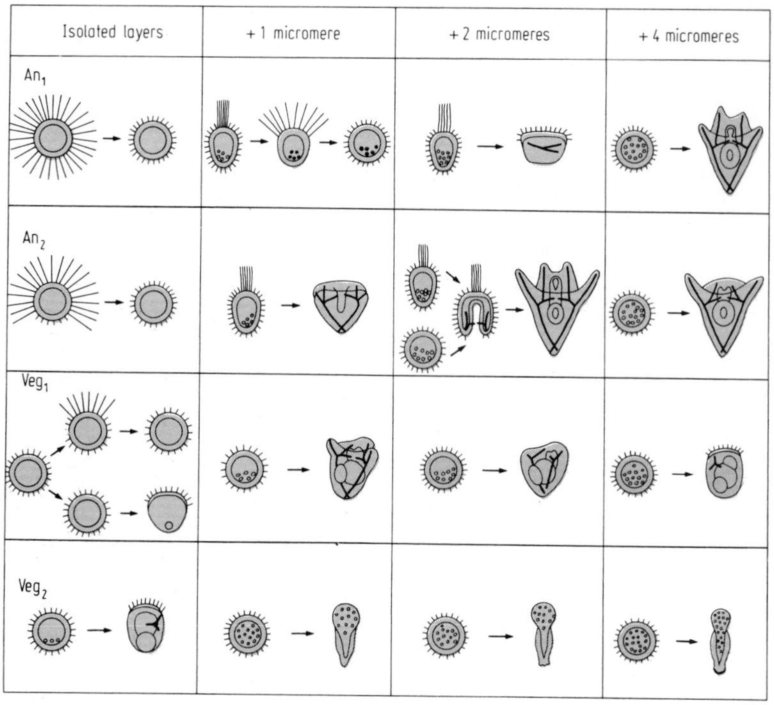

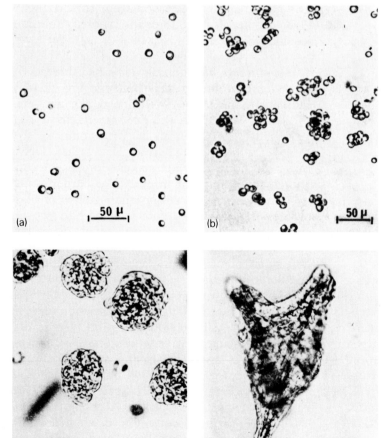

FIGURE 12.13. Prospective potencies of animal and vegetal tiers of the early embryo and the alteration of differentiation by the addition of micromeres. From S. Hörstadius, Pubbl. Staz. Zool. Napoli, 14:251 (1935). Used by permission of the Stazione Zoologie, Naples, Italy.

FIGURE 12.14. Photomicrographs of isolated mesenchyme blastula cells and reaggregates giving rise to relatively normally appearing pluteus larvas. (a) Freshly associated cells. (b) After 1 hour of aggregation in rotatory flask. (c) After 9 hours of aggregation. (d) Pluteus-like aggregates 3 days after aggregation. From G. Giudice and V. Mutolo, in M. Abercrombie, J. Brachet, and T. J. King, eds., _Advances in morphogenesis_ Vol. VIII, p. 115, Academic Press, Orlando, 1970. Used by permission.

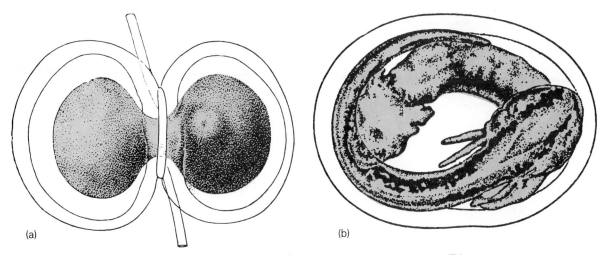

(a)   (b)

FIGURE 12.15. The results of "hair-loop" experiments with the common newt, *Triturus vulgaris* (= *Triton taeniatus*). Equal twins formed following tight ligation. From H. Spemann, *Z. Wissenschaft. Zool.,* 123:105(1928).

equal twins were produced when the cleavage furrow lay along the frontal plane and divided the embryo into dorsal and ventral halves. Only the dorsal half regulated to form an embryo.

The ability to regulate is therefore limited to right, left, and dorsal blastomeres. Development of the ventral blastomere's lineage depends on continuity and presumably interactions with cells of the dorsal lineage. The ventral blastomere seems to be caught in a preregulative condition in which it can respond to regulating influences while part of a left or right half but is incapable of either regulating or self-differentiating on its own.

In insects, intralecithal cleavage is incompatible with separating early blastomeres, but the results of ligating across the entire blastoderm prior to cellularization demonstrate an ability to regulate (Fig. 12.18). Inspired by Spemann, Friedrich Seidel (born 1897) ligated freshly laid eggs of the dragonfly, *Platycnemis pennipes*. Intralecithal division had started, but the nuclei had not yet reached the cortex (Seidel, 1929). When tight ligatures were placed in the middle of the presumptive germ band (i.e., the embryo-forming portion of the blastoderm) whole embryos developed behind the ligature (Fig. 12.18a). When similarly tight liga-

tures were placed extremely posteriorly, embryos developed in front of the ligature (Fig. 12.18c). But ligatures placed about one-sixth the overall distance from the posterior pole completely suppressed development of embryos in both portions, although extraembryonic membranes developed in front of the ligature (Fig. 12.18b).

It seems that embryos only develop where they have enough material to form and where they are continuous with a portion of posterior cytoplasm lying about one-eighth the overall distance from the posterior pole. Ligation that deprives the blastoderm of continuity with this posterior cytoplasm stops development or limits it to extraembryonic membranes.

Gross regulation ceases after cellularization, but damage to presumptive larval tissues may still induce compensatory changes. In *Drosophila melanogaster*, for example, the early neural ectoderm compensates for cells destroyed in laser-microbeam experiments.

Even following cellularization, interactions with the local environment may yet play a role in directing the differentiation of individual cells. When presumptive neuroblasts begin to enlarge, in *Drosophila*, their determination becomes irreversible, but the cells acquire their identities according

a   b

FIGURE 12.16. Unequal twins of the common newt, *Triturus vulgaris* (= *Triton taeniatus*), produced by constriction. The ball of cells is a *Bauchstück* or belly piece. From H. Spemann, *Embryonic development and induction,* Yale University Press, New Haven, 1938. Used by permission of Yale University Press.

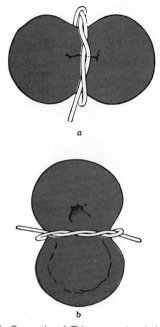

FIGURE 12.17. Gastrula of *Triturus vulgaris* (= *Triton taeniatus*) loosely constricted between the blastomeres at the 2-cell stage to mark the angle of cleavage. (a) The first cleavage furrow was in the median plane. (b) The first cleavage furrow was not in frontal plane. From H. Spemann, *Embryonic development and induction,* Yale University Press, New Haven, 1938. Used by permission of Yale University Press.

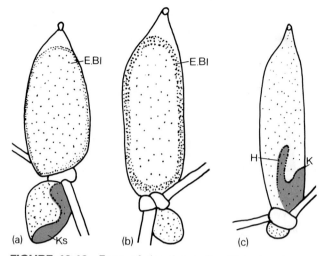

FIGURE 12.18. Eggs of the dragonfly, *Platychemis pennipes,* about a day after having been tightly ligated. (a) A blastoderm (ks) forms behind a ligature placed in what would have been the middle of the embryo. Only an empty extraembryonic membrane (E.Bl) appears ahead of the ligature. (b) A ligature placed about one-sixth the overall distance from the posterior pole suppresses development except of the extraembryonic membrane (E.Bl). (c) A blastoderm with a tail (H) and a head (K) develops ahead of a ligature placed as far posteriorly as possible. From F. Seidel, *W. Roux' Arch. Entwicklungsmech. Organ.,* 119:323 (1929). By permission of Springler-Verlag, Inc.

to the positions occupied at the time of enlarging and then give rise to characteristic families of ganglion mother cells by invariant cell lineages. Moreover, while enlarging, neuroblasts may inhibit neighboring neural ectodermal cells from enlarging, thereby preventing these cells' differentiation into neuroblasts and inducing their differentiation into nonneuronal cells (see Doe et al., 1988).

Finally, in endopterygote insects, the development of imaginal disks, even while determinate, exhibits regulation within specific organs. When a disk fragment undergoes differentiation, a whole structure is formed. This **organotypic regulation,** or the formation of whole parts, would seem to depend on global influences capable of promoting regulation (see Hadorn, 1966).

Are Mammalian Eggs Determinate or Regulative?

Testing the potency of mammalian blastomeres was first accomplished via the destruction of cells and suppression of cell division (see Snow et al., 1981). When one of two blastomeres of a mouse embryo was destroyed, the cultured embryo gave rise to a blastocyst with fewer by about half the normal number of cells. When three out of four mouse blastomeres were killed, an abnormal blastocyst formed, and when seven out of eight blastomeres were killed, the embryo died. In the case of the rabbit, although the number of survivors diminished drastically, a few (11%) normal blastocysts were produced even when seven out of eight blastomeres of an 8-cell embryo were destroyed.

Sheep provided the material for actually isolating blastomeres. When the blastomeres of 2- and 4-cell embryos were isolated and placed separately into zona pellucidas, subsequent transfer to prenatal foster mothers produced sets of identical twins (Fig. 12.19) and quadruplets (see Willardsen, 1979).

Aggregating blastomeres from different embryos was accomplished independently by Beatrice Mintz (1962) and Andrzei Tarkowski (1961; Tarkowski and Wroblewska, 1967). The method (Fig. 12.20) took advantage of the "stickiness" of blastomeres at about the time of compaction. The **reaggregation** of blastomeres was also enhanced by the addition of plant lectins.

After removing blastomeres from their zona pellucidas, little more than pushing them together was necessary to get the blastomeres to form a sin-

FIGURE 12.19. Photograph of three sets of identical lamb twins produced from sets of isolated blastomeres. From S. M. Willadsen, *Nature (London),* 277:298 (1979), by permission of Macmillan Journals.

gle mass and round up into a single embryo. The aggregates formed a blastocyst *in vitro* and, upon transfer to prenatal foster mothers, developed to term.

The first mice produced by combining blastomeres from two embryos were called **tetraparental** (four parented), since each blastomere represented two parents. Since mice were soon created with blastomeres from six parents (i.e., with blasto-

meres from three embryos) (Markert and Petters, 1978) or more, a more general term was coined. Animals developed from aggregates of blastomeres originating in different embryos were called **allophenic** (Gk. *allos-* different + *phainein* to show referring to phenotype) or **chimeras** (from the mythological Greek monster in which parts of different origins were combined) (McLaren, 1976). When the blastomeres were chosen from stocks of parents with different genetic markers (e.g., particular fur colors), the allophenic offspring were also **genetic chimeras** containing phenotypic contributions from different parents (Fig. 12.21) (see review by Herbert and Graham, 1974).

Regulation in mammals seems to be even more complete than in sea urchins, since mammals regulate to normal size and are born at the normal time, while sea urchins produce larger or smaller embryos depending on whether blastomeres are combined or separated. Initially, blastocysts derived from single mouse blastomeres have fewer than the normal number of cells, and most embryos remain about half the normal size between days 6 and 10 postcoitum (p.c.) of development. But by day 12, or about halfway through mouse gestation, the embryos are indistinguishable from normal (Tarkowski, 1959).

Embryos developing from larger than normal numbers of blastomeres are also born at the normal

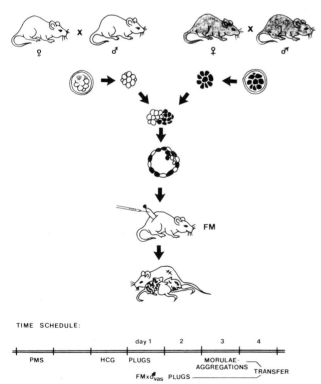

TIME SCHEDULE:

FIGURE 12.20. Diagrammatic illustration of method for producing chimeric mice by fusion of two morulas from genetically different parents. FM: prenatal foster mother. From K. Bürki, *Experimental embryology of the mouse,* Karger, Basel, 1986. Used by permission of S. Karger AG.

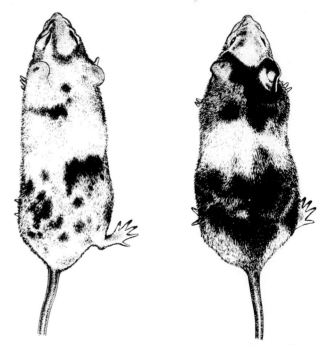

FIGURE 12.21. Chimeras obtained by combining blastomeres. From R. L. Gardner, in C. R. Austin and R. V. Short, eds., *Embryonic and fetal development,* Cambridge University Press, Cambridge, 1982. Used by permission.

time and of the normal size. The large embryos formed from aggregates of blastomeres remain large for up to 5.5 days, or approximately 1.5 days after implantation. At that time they begin to resemble normal embryos, and from 6 days onward, they are the same size as normal (Buehr and McLaren, 1974).

Both upward and downward regulation may depend on nutrition supplied by the mother (Snow et al., 1981). Downward regulation may involve the nutritional constraints put upon growth that normally takes place before uterine blood vessels break down and supply the implanting embryo with a circulating source of energy and raw materials. Upward regulation may have to wait until the embryo's placenta is working adequately, at 10–10.5 days, to supply the additional nutrition required for **catch-up** growth. Upward regulation may also involve additional internal controls, since a normal embryo does not experience the same spurt of growth seen in reduced embryos following the nutritional hookup of maternal and embryonic tissue.

GERM CELL ORIGINS AND CELL-LINE SEGREGATION

We do not know what is special about germ cells. We do not know whether nuage is a germ-cell determinant, and whether it will be found in pregastrulation embryos, and if so, whether it will be present in all cells or only a subset of cells.

Ann McLaren (1981, p. 94)

Determinate and regulative development differ in the timing of cell-line segregation and possibly in the mechanism of segregation. Cells in embryos exhibiting both types of development will ultimately move down separate pathways of differentiation, but the early entrance into some pathways by cells in determinate embryos may require mechanisms not employed by cells in regulative embryos.

The segregation of the germ line has been central to the study of cell-line segregation. Theories that played a pivotal role in elevating the nucleus to its present position of preeminence in cellular controls placed a premium on the early segregation of the germ line. Efforts to test or vindicate these theories catapulted the germ line into prominence and made it the most thoroughly studied cell lineage in the annals of embryology.

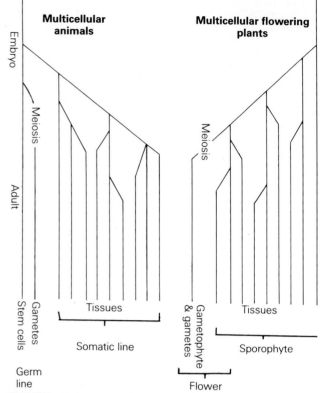

FIGURE 12.22. Diagram comparing germ and somatic lines in an animal with localized gonads to gametophyte and sporophyte in a flowering plant.

Narrowly defined, the germ line is unique to animals (Fig. 12.22). It consists of gametes and cells giving rise to gametes as distinguished from all somatic cells. Data accumulated on the origins of the germ line provide a vast resource of information on cell-line segregation and cellular determination.

Theories of Germ and Somatic Separation

Originally, the **germ-plasm theory** (Weismann, 1885, 1892) proposed a complete and absolute separation of somatic- and germ-cell lines during cleavage. The withdrawal of germ plasm from cleaving cells into specific blastomeres was also supposed to be a once-in-a-lifetime phenomenon, irreversible and irrevocable.

More recent versions of the germ-plasm theory allow for more latitude in timing the separation of the germ and somatic lines and in the degree of separation. Three basic patterns of germ-line formation are identified: a more or less predetermined germ-plasm pattern of early and complete segregation resembling Weismann's version, an epigenic pattern with sensitivity to environmental influences and parallels to the formation of somatic lines, and an intermediate pattern combining features of the other patterns.

The germ-plasm pattern of germ-cell origins occurs when precursors of gametes are segregated early in development, especially during the first few cleavage divisions. The progenitor germ cell is called a *germ-stem cell* or, more commonly, a **primordial germ cell** (PGC or P). Additional PGCs derived from an original PGC are frequently identifiable by cytological and cytochemical criteria or by chromosomal characteristics.

In mollusks and annelids, large PGCs are identifiable quite early in development by their size and position, and their pedigree can be traced retrospectively to the 4d cell in the 64-cell embryo (Fig. 12.23, D-cell line), but no chromosomal events occur as a guide for following the germ line. In the nematode, *Caenorhabditis elegans*, the P-cell lineage (Fig. 11.56) is the sole source of germ cells and

the sole host to **P granules** segregated by cleavage divisions (Fig. 12.6).

In some other nematodes (e.g., *Parascaris*), chromosomes in the germ-line cells remain patent, while chromosomes in somatic-line cells undergo fracture and diminution (see Fig. 11.55). Despite the appearance of nuclear determination in these nematodes, responsibility for preventing chromosomal fracturing in the germ line rests with the cytoplasm of the cell cortex.

The first experimental evidence for a cytoplasmic role in germ-line determination (Hogue, 1910) involved centrifuged *Parascaris equorum* embryos. Presumptive somatic-cell nuclei displaced by centrifugation to the pole of the egg retained their chromosomes like PGCs, while other nuclei remaining in somatic portions of the embryo underwent chromosomal diminution (Fig. 12.24).

The behavior of **B chromosomes** in various grasses and **accessory chromosomes** in some insects, birds (e.g., chicks), and mammals (harvest mouse and fox) resembles that of germ-line chromosomes during chromosomal diminution. The small, heterochromatic chromosomes do not occur in homologous pairs but as variable elements with weak if any effects on the phenotype. Accessory chromosomes may increase the rate of crossing over, thereby diminishing fertility and vigor. They are concentrated in the germ line and are somehow eliminated during the segregation of the somatic line.

Cleavage nuclei taking part in forming the somatic line exhibit chromosomal diminution without chromosomal fracture in mite and some insect

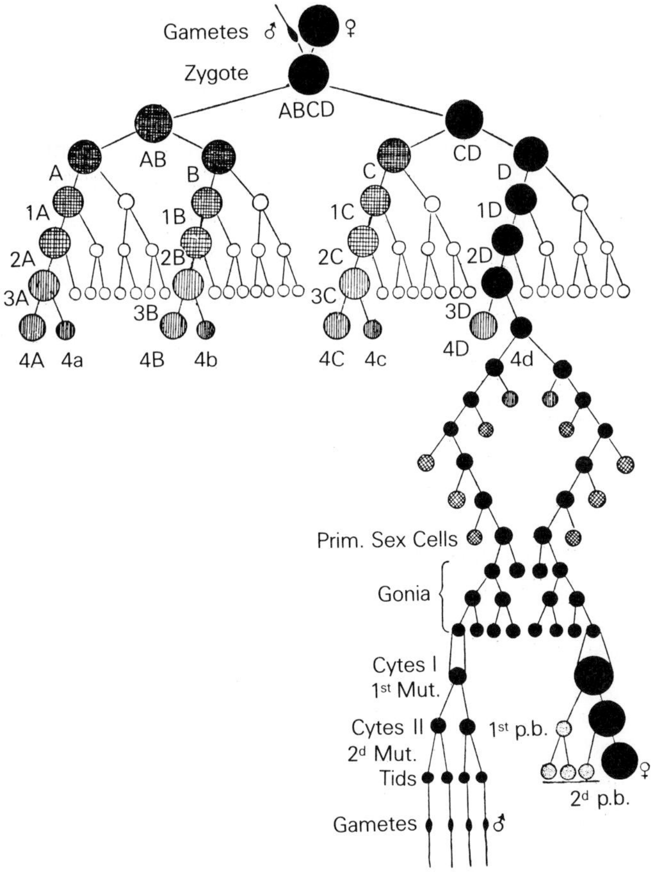

FIGURE 12.23. Diagram illustrating origin of primordial germ cells (filled circles) and various lines of somatic cells (shaded and clear circles) in a mollusk or annelid. Numbers and letters refer to cell lineages. From E. G. Conklin, *Heredity and environment in the development of men,* 5th ed. rev., Princeton University Press, Princeton, 1922. Used by permission.

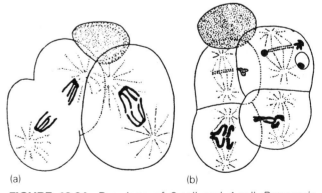

(a)   (b)

FIGURE 12.24. Drawings of 2-cell and 4-cell *Parascaris equorum* (= *Ascaris megalocephala*) embryos following centrifugation. Instead of only one primordial germ cell retaining all its chromosomal material, both cells in the 2-cell embryo (a), and two of the four cells in the 4-cell embryo (b) have retained complete chromosomes. From M. Hogue, *Arch. Entwicklungsmech. Organ.,* 29:109 (1910).

embryos (beetles, flies, butterflies, grasshoppers, gall midges). After the first three divisions of the zygotic nucleus have taken place in a central core of cytoplasm, one nucleus moves to the posterior pole of the egg where a granular cytoplasm, called **polar cytoplasm,** awaits. A cell membrane encloses this nucleus and much of the polar cytoplasm creating the first PGC. Through subsequent divisions, this nucleus and its descendants retain all chromosomes, while other nuclei undergo chromosomal diminution, eliminating as much as three-quarters of their chromosomes. At the end of cleavage, the newly formed blastoderm consists of somatic cells with relatively small nuclei and small numbers of chromosomes (Fig. 12.25).

Higher dipterans, such as *Drosophila*, do not undergo chromosomal diminution, but eggs contain a polar cytoplasm characterized by RNA-staining **polar granules** (Fig. 12.26). When cytomembranes form around nuclei reaching the polar cytoplasm (Fig. 11.7 stage 4), distinctive **pole cells** form. The granules fragment within the pole cells and lose their RNA staining, and polysomal-like clusters reassociate in ill-defined masses sometimes called **nuagelike** for their irregular, fanciful "cloudy" appearance and resemblance to the nuage of vertebrate germ cells (Kalt, 1973).

At the same time, the pole cells develop hollow nuclear organelles called **nuclear bodies.** The developmental significance of these bodies is unknown (see Mahowald et al., 1979), but their presence allows the pole cells to be traced as they migrate from the pole to the midgut of the embryo

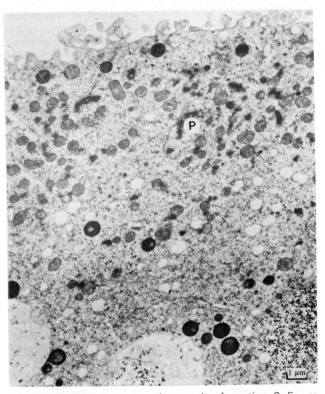

FIGURE 12.26. Electron micrograph of section 3–5 μm into the posterior polar cytoplasm of a *Drosophila melanogaster* embryo. Fine polar granules (P) in the vicinity of mitochondria are surrounded by clusters of polyribosomes. From A. W. Mahowald and R. E. Boswell, in A. McLaren and C. C. Wylie, *Current problems in germ cell differentiation,* Cambridge University Press, Cambridge, 1983. Used by permission.

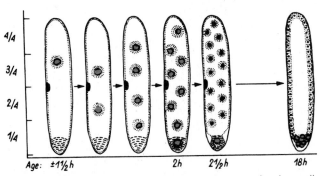

FIGURE 12.25. Diagram illustrating cleavage in the gall midge, *Wachtiella persicariae.* Fractions on ordinate give relative distances from posterior (germinal) pole. Nuclei undergoing mitosis without chromosomal reduction are indicated by open circles; those undergoing chromosomal reduction are shown by filled circles. After the third division, one nucleus moves into the polar cytoplasm (lower tip of egg) and becomes the discrete pole cell. All the other nuclei become somatic-cell nuclei of the blastoderm. Adapted from I. Geyer-Duszynska, *J. Exp. Zool.,* 141:391 (1959). By permission of Alan R. Liss, Inc.

(see Fig. 18.8). There, the nuclear bodies fragment and disappear but not before half of the pole cells die or become incorporated into the anterior portion of the hind gut. The other half of the pole cells are PGCs.

Polar granules are implicated in germ-line determination by both negative and positive evidence. On one hand, polar granules are largely absent or abnormal in sterile individuals bearing *grandchildless* mutations. On the other hand, following the microinjection of polar cytoplasm in the apical end of embryos, some somatic cells came to resemble PGCs. When these cells are transplanted to posterior sites in other embryos, from where the cells can enter the gonad, the cells develop into gametes as testified by the appearance of their genetic markers in offspring (Fig. 12.27) (Illmensee and Mahowald, 1974).

The present version of the germ-plasm theory differs from Weismann's. Nuclear differentiation is now seen as a consequence of cytoplasmic influence rather than the initial cause of germ-line determination. Furthermore, since in insects, some

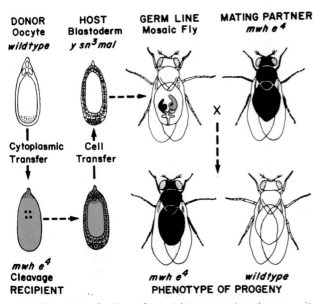

DONOR Oocyte *wildtype*

HOST Blastoderm *y sn³ mal*

GERM LINE Mosaic Fly

MATING PARTNER *mwh e⁴*

Cytoplasmic Transfer

Cell Transfer

mwh e⁴ Cleavage **RECIPIENT**

X

mwh e⁴   *wildtype*
PHENOTYPE OF PROGENY

FIGURE 12.27. Outline of experiment testing the capacity of polar cytoplasm to induce germ-line cell differentiation in the fruit fly, *Drosophila melanogaster.* Polar cytoplasm (stippled) removed from an immature oocyte of a wild-type fly is injected into the anterior tip of an early embryo having multiple wing hair (*mwh*) and ebony (*e⁴*) genetic markers. At the blastoderm stage of development, ectopic pole cells found at the anterior tip are transferred to the posterior pole region of a UV-irradiated host embryo bearing yellow (*y*), singed-3 (*sn³*), and maroonlike (*mal*) genetic markers. The irradiation has the effect of reducing the number of potential gametes formed with the host's genome. The host flies are allowed to develop and subsequently mate with genetically marked flies. The appearance of the multiple wing hair and ebony markers in progeny demonstrates that germ cells originated from the ectopic pole cells under the influence of the injected polar cytoplasm. From A. P. Manowald, *Am. Zool.,* 17:551 (1977), by permission of the American Society of Zoologists.

pole cells seem to take part in forming the larva's gut, the presence of germ plasm is no longer considered incompatible with all somatic-cell differentiation (but see Mahowald and Boswell, 1983).

The epigenic pattern of germ-cell origins implies no germ-line determination in the embryo prior to the appearance of germ cells. Rather, gametogenesis takes place in growing somatic tissues.

The flowering plants were always problematic for the Weismann germ-plasm theory, since gametes differentiated from somatic tissue (Fig. 12.22 right). Moreover, the sporophyte of plants is exclusively diploid, and the gametophyte is exclusively haploid, while the animal germ line includes the diploid progenitor or stem cells of the gametes as well as the haploid gametes.

The germ line of flowering plants exhibit epigenic origins. **Meristematic tissue** consisting of dividing cells at growing points can form ordinary plant tissue or a flower depending on environmental cues (Evans, 1972). The ordinary plant tissues are exclusively somatic, but the flower includes the gametophyte with its germ-line gametes.

Similarly, in sponges, germ cells are derived from amoeboid cells or even the more typical flagellated cells called choanocytes. In flatworms, **neoblasts** are capable of forming germ cells or a variety of differentiated somatic cells. In coelenterates, **interstitial cells** give rise to nerves, stinging cells (cnidoblasts), gland cells, mucous cells, and germ cells (but see Littlefield, 1985). Hydra's germ cells can also be formed from other body cells (Davis, 1973). Similarly, in colonial ascidians, circulating blood cells (hemoblasts) give rise to germ cells and to somatic cells (see Raven, 1961).

In all these animals, including the hydroids studied by Weismann, germ tissue is not localized in gonads. The germ and somatic lines are not separated until gamete-forming cells make their epigenic appearance under environmental influences (see Nieuwkoop and Sutasurya, 1981).

The intermediate pattern of germ-cell origins occurs in most vertebrates. A germ plasm is segregated early in development but not by the first cleavage divisions. Epigenic influence comes to bear on germ-line determination, but localized cytoplasmic determinants may also play a role. Early germ-line cells are still called PGCs for convenience, but they arise conspicuously after cleavage.

In amphibians, birds, and mammals, PGCs are generally identified by large size, spherical or ellipsoidal shape, a relatively high nucleocytoplasmic ratio, and perinuclear concentrations of mitochondria. In addition, PGCs have a high content of the enzyme alkaline phosphatase and unique surface antigens detectable with monoclonal antibodies (Eddy and Hahnel, 1983).

Presumptive PGCs are detectable in the vegetative endoderm of anuran blastulas (Fig. 12.28) or in the marginal zone of presumptive lateral plate mesoderm of early urodelan gastrulas. Removal of endoderm from anuran neurulas or the presumptive lateral plate mesoderm of urodeles results in the production of sterile adults (Fig. 12.29 left). It would seem that presumptive PGCs are removed with the endoderm and that the remaining tissue does not regenerate PGCs.

Transplantation or exchange of genetically marked marginal mesoderm in axolotls fosters germ-cell production of the transplant-type (Smith,

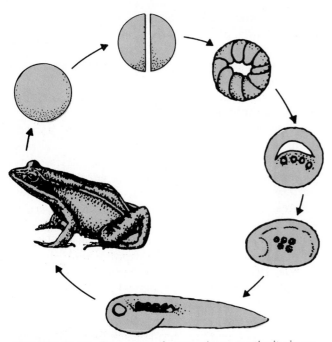

FIGURE 12.28. The route of germ-plasm continuity in anurans. The putative germ plasm (stippled) in the vegetal pole of eggs moves anteriorly during early cleavage. It becomes localized in large, yolky cells of the presumptive endoderm of the blastula and in primordial germ cells of the neurula. After moving into the genital ridge of the tadpole, the primordial germ cells colonize the gonad (ovary or testis) and give rise to the next generation of gametes.

1964). Similarly, transplantation of blocks of anuran endoderm from two strains of *Xenopus laevis* (Blackler and Fischberg, 1961), from subspecies of *X. laevis* (Blackler, 1962), and even from the species of *X. laevis* and *X. mülleri* (Blackler and Gecking, 1972) introduces foreign germ cells into host gonads (Fig. 12.29).

In urodeles, the germ line is experimentally inducible in animal-pole cells by contact with the vegetal hemisphere (Tiedemann, 1975; Nieuwkoop and Sutasurya, 1979; but see Mahowald, 1977). The production of ectopic PGCs suggests that microenvironmental or epigenic influences play a role in normal germ-cell determination (Smith, 1964).

The dark-staining cytoplasm of most anuran PGCs (the exception being *Rana esculenta*) has been equated with **vegetal cytoplasm** (also called germinal plasm or sex plasm) found in **islands of vegetal cytoplasm** among the yolk platelets in the vegetal pole of cleaving embryos (Fig. 12.30). In the course of cleavage, these islands move preferentially to one spindle pole. After 20–25 divisions, the vegetal cytoplasm is concentrated in only two dozen cells which then give rise to PGCs (Fig. 12.28) (Blackler, 1970).

By pricking the vegetal pole of *X. laevis* embryos at the 2- and 4-cell stage and draining the vegetal cytoplasm and yolk, embryos are produced

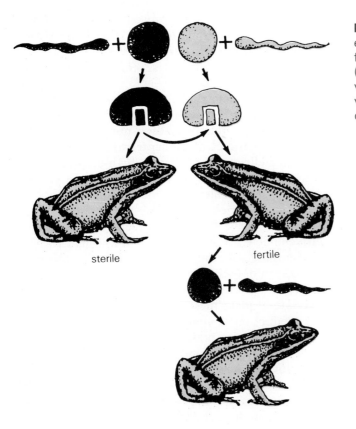

sterile   fertile

FIGURE 12.29. Scheme of experiment testing the presence of primordial germ cells in the vegetal endoderm of frog neurulas. When the vegetal endoderm is removed (left), sterile animals generally develop. When ablated vegetal endoderm is replaced (right), fertile animals develop. The origin of the germ cells in the donated tissue is demonstrated through cytological or genetic markers.

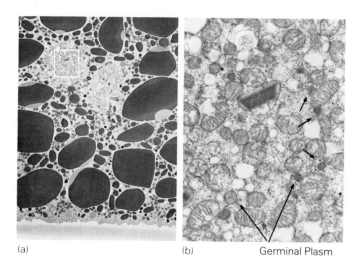

(a)   (b)   Germinal Plasm

FIGURE 12.30. Electron micrograph of section through cortical cytoplasm at the vegetal pole of 2-cell *Rana pipiens* embryo. Islands of germinal plasm (square and b) contain concentrations of mitochondria and small (0.2–0.3 μm in diameter), membraneless, round, dense, fibrogranular masses (tips of arrows in b) called germinal granules. a ×3600. b ×16,000. From E. M. Eddy, *Int. Rev. Cytol.*, 43:229 (1975), by permission of Academic Press and the author.

without PGCs or with severely limited numbers of PGCs (Buehr and Blackler, 1970). Damage to the egg does not seem responsible for the lack of PGCs, since germ cells differentiate when vegetal-pole cells are transplanted to the drained embryos (see Blackler, 1970). Moreover, extra germ cells appear when excess vegetal cytoplasm is injected into eggs (Wakahara, 1978).

The vegetal cytoplasm contains **germinal granules** (Fig. 12.30, arrows) associated with dense clusters of mitochondria. Germinal granules are RNA-staining, electron-dense bodies averaging 0.2 μm in diameter but ranging up to 0.9 μm in diameter (Mahowald and Hennen, 1971). The germinal granules in *Rana pipiens* and *X. laevis* eggs look very much alike and contain electron-dense foci primarily 10–20 nm in diameter (Smith et al., 1983).

As PGCs form in the frog's endoderm and move toward the embryonic gonad (i.e., the genital ridge), the vegetal cytoplasm moves from the periphery closer to the cell's nucleus. When the PGCs enter the rudimentary gonad, a nuage of irregularly shaped fibrous material replaces the fibrogranular material of the germinal granules.

Nuage in various guises accompanies germ cells throughout their differentiation (e.g., see Figs. 7.16 and 7.25), but a hiatus occurs in early immature oocytes when neither islands of vegetal cytoplasm nor germinal granules can be found. The loop back to the embryo begins with a diffuse fibrogranular material (FG in Fig. 12.31a) in fully grown but immature oocytes that condenses into germinal granules prior to maturation and ovulation (GG in Fig. 12.31b). Neither RNA precursors nor amino acids are incorporated into germinal granules in mature eggs (Smith and Williams, 1975).

In avian embryos, PGCs first appear well beyond cleavage at the primitive streak stage of development, at about 16 hours of incubation (Fig. 12.32 left). These large cells are conspicuous because of their round nuclei, great quantities of yolk, and large centrosomes (Fig. 12.33).

First identified in the incompletely cleaved yolky endoderm at the junction of the chick blastoderm and yolk, the PGCs accumulate between endoderm and ectoderm along a **germinal crescent.** They increase in number throughout most of the first day of incubation until mesoderm has entered the germinal crescent, and the PGCs begin the wanderings that will take some of them to the genital ridge and the primitive gonad.

Removal of the germinal crescent by microsurgery (Willier, 1937), cautery, or ultraviolet (UV) irradiation (Fig. 12.34) eliminates PGCs and results in sterility (Reynaud, 1969). Intravenous injection of suspended germinal crescent cells into UV-irradiated hosts restores germ cells to the gonad, and when PGCs from chicks and turkeys are exchanged or when different varieties of chicks are used as donors and hosts, gametes of the donor type mature. Not surprisingly, the turkey sperm and eggs fostered by chicks and hens are not normal, but the fostered sperm fertilize or activate chick eggs with about the frequency expected for turkey sperm. The fostered turkey eggs develop into relatively normal turkeys for as many as 15 days of incubation (Reynaud, 1976).

In mammals, PGCs also fail to appear until the primitive streak stage, well after cleavage. Identified with the help of histochemical staining for alkaline phosphatase, the first PGCs are found in the endoderm or mesoderm in the vicinity of the hind gut invagination (Fig. 12.35). The darkly staining

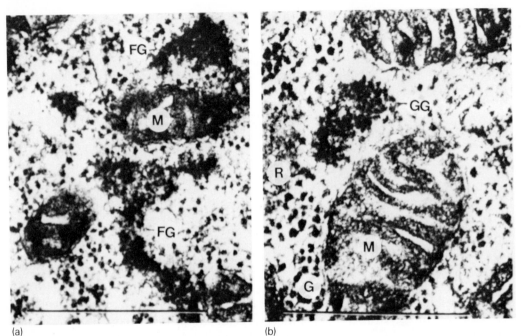

(a)   (b)

FIGURE 12.31. Electron micrographs of section through cytoplasm of fully grown *Xenopus laevis* oocytes. (a) Immature oocyte with fibrogranular material (FG). (b) Later preovulatory oocyte (B) with germinal granules (GG). M, mitochondrion, R, ribosome, G, glycogen. Bars = 0.5 μm. From L. Dennis Smith and M. A. Williams, in C. L. Markert, ed., *The developmental biology of reproduction,* Academic Press, Orlando, 1975. Used by permission.

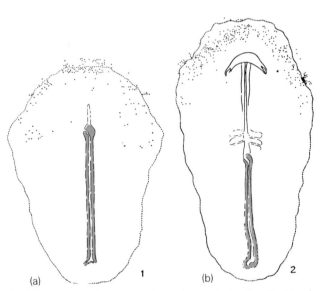

(a)   (b)

FIGURE 12.32. Census of large "wandering" cells (dots) identified as primordial germ cells in germinal crescents of early chick embryos. The germ cells first appear in the primitive streak stage and continue to increase in number through the notochordal-process stage (a) until the embryo has three pairs of somites, at about 22 hours of incubation (b). (a) ×26. (b) ×16. From B. H. Willier, *Anat. Rec.,* 70:89 (1937), by permission of Alan R. Liss, Inc.

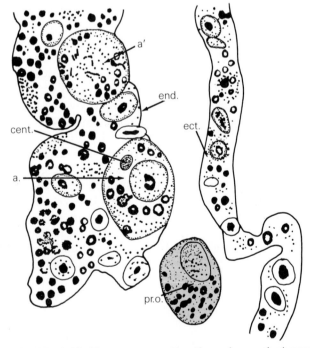

FIGURE 12.33. Transverse section through germinal crescent of chick embryo at primitive streak stage. A primordial germ cell (pr.o.) in space between ectoderm (ect.) and endoderm (end.) resembles large dividing cells (a and a') appearing to emerge from endoderm. The cells are conspicuously large and contain abundant yolk, round nuclei, and a large centrosome (cent.). ×670. From C. H. Swift, *Am. J. Anat.,* 15:483 (1914), by permission of Alan R. Liss, Inc.

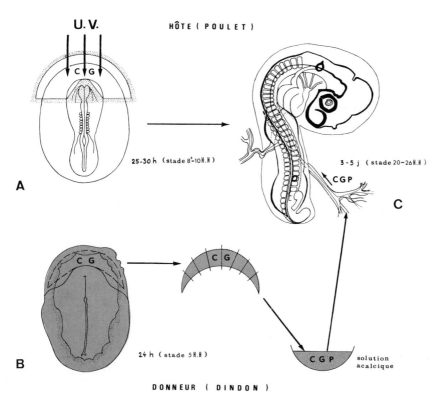

FIGURE 12.34. Outline of experiment testing the origin of primordial germ cells (CGP) in the germinal crescent (CG) of early chick embryos. (a) Chick embryos at 25–30 hours of incubation are sterilized by ultraviolet (UV) radiation over the germinal crescent. (b) Germinal crescents from primitive streak stage turkey embryos (donneur dindon) are isolated and dispersed. (c) The sterilized chick host embryo (hôte poulet) at 3–5 days of incubation receives the dispersed turkey germ cells. After the host chick has developed and matured, artificial insemination demonstrates that the donor cells are the source of germ cells. From G. Reynaud, *W. Roux' Arch. Dev. Biol.,* 179:85 (1976), by permission of Springer-Verlag and the author.

PGCs are transformed into large, smooth surface PGCs in later embryos (Eddy and Hahnel, 1983).

Because the requirements of mammalian embryos for a uterus have not been overcome, experiments to determine the origins of germ cells have not been performed on advanced mammalian embryos. The results of killing blastomeres in early mammalian embryos or adding additional blastomeres to embryos suggest that mammalian germ cells are not determined during cleavage (Gardner, 1978). Even in the late mammalian blastocyst, some cells retain the ability to form somatic-cell or germ-cell lineages (see Eddy and Hahnel, 1983).

Cellular plasticity and germ-line differentiation even extend to malignant mouse **teratocarcinoma** cells (Mintz and Illmensee, 1975). When injected into mouse blastocysts in small numbers, teratocarcinoma cells contributed to various cell lines including the germ-cell line. The teratocarcinoma cells were probably derived originally from primordial germ cells, and malignant cells may operate under their own rules, not necessarily those influencing normal tissue. Still, rather than indicating an early determination of the germ line, the results suggest that in mammals an **undetermined** stem-cell line present in the embryo is capable of giving rise to a variety of tissue lines including the germ line (Eddy et al., 1981).

The possibility that early mammalian germ cells are undetermined is consistent with data on the frequency of 5-methylcytosine, or **DNA methylation,** in cytidine-phosphate-guanosine (CpG) clusters. While the DNA of embryos at 7.5 days postcoitum (p.c.) of separated embryonic germ layers, yolk sac, and sperm is hypermethylated, germ cells obtained from the gonads of male and female mouse fetuses at 12.5 and 14.5 days p.c. are markedly undermethylated (Monk et al., 1987).

Methylation in fetal mouse germ cells' DNA is distinctly oocyte-like and different from somatic cells' DNA. Even the X chromosome, inactivated in other fetal tissues, is active in germ cells of the female gonad. To the degree that methylation determines gene inactivation, fetal germ cells would seem to retain greater indeterminacy than fetal somatic tissue.

Characterizing the Germinal Plasm

Theoretically, in animals with both germ plasm and intermediate patterns of germ-cell origins, germ plasm (whether nuage, germinal granules, or something else) flows from generation to generation exclusively through the germ line. The plasma would have to be augmented in every generation, or it would be effectively diluted out

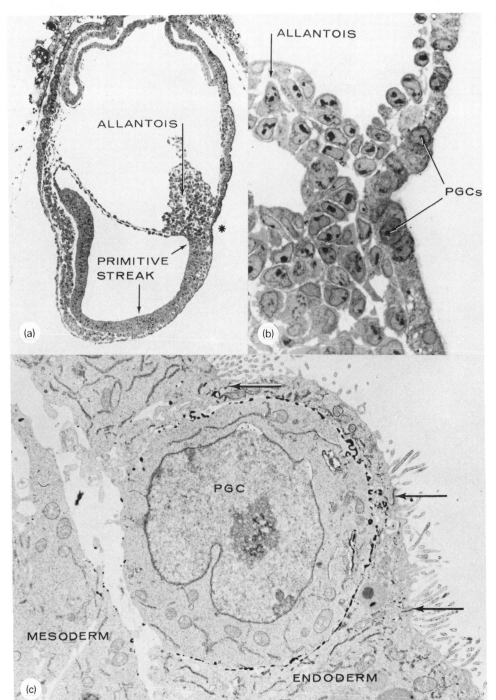

FIGURE 12.35. Sagittal sections through posterior end of mouse embryo at 8 days postcoitus. Presumptive primordial germ cells (PGCs), detected by their relatively intense alkaline phosphatase staining, appear in the hindgut region at the posterior end of primitive streak (star in yolk sac). In electron micrographs, a fine precipitate (black dots) represents the alkaline phosphatase reaction product. It appears mainly at the surface of the PGC. Endodermal cells are linked by tight junctions (arrows). (a) ×110. (b) ×740. (c) ×9300. From J. M. Clark and E. M. Eddy, *Dev. Biol.,* 47:136 (1975), by permission of Academic Press and courtesy of the authors.

of existence by reproduction. Moreover, like mitochondria and chromosomes, germ plasm would have to be self-replicating, since its production and maintenance could not depend on somatic plasm other than for raw materials.

The active ingredients in germ plasm would be **determinants** of the germ line. One determinant could do everything, or different determinants could do different jobs. One could be responsible for preventing chromosomal fracture, one for preventing chromosomal diminution, and others for the differentiation of egg and sperm.

Structurally, the portions of cytoplasm identified as **germinal cytoplasm** are remarkably similar in eggs with the germ plasm pattern and with the intermediate pattern of germ-line segregation, from insects, annelids, and mollusks to sea urchins and vertebrates (see Eddy, 1975). Germinal cytoplasm is conspicuously visible after staining for RNA; it is packed with mitochondria, and it contains small dense fibrogranular masses (e.g., Figs. 12.26, 12.30, and 12.31). The bulk of experimental research on germinal cytoplasm has been directed at insect and amphibian eggs where material with some of the characteristics of germinal cytoplasm is transplantable (see Mahowald and Boswell, 1983).

In *D. melanogaster* the polar granules contain a basic protein of 95 kd capable of binding a messenger RNA. The granules stain for RNA but are not associated with ribosomes (see Mahowald, 1977).

The results of UV-irradiation experiments suggest that the active ingredient in the germinal cytoplasm of both insects and amphibians is RNA or an RNA–protein complex. Exposed to UV irradiation, insect polar cytoplasm loses its ability to influence primordial germ-cell differentiation.

In the fly, *Smittia,* the action spectrum for induced sterility has an absorbance peak at 260 nm (i.e., characteristic of nucleic acids). Moreover, the effect of UV irradiation is photo reversible as it might be were the target a nucleic acid (Brown and Kalthoff, 1983).

Similarly, UV irradiation of the vegetal pole of frog eggs, even *Rana esculenta,* produces sterile frogs, while treatment of the egg's animal pole has no effect on gamete formation (see Smith and Williams, 1975). The response to UV irradiation is dose dependent (Fig. 12.36), reaching total sterility at about 8000 ergs/mm^2, and a wavelength of 254 nm is more effective in reducing numbers of primordial germ cells than 230, 278, and 302 nm (Smith, 1966). Moreover, the absorption spectrum of the vegetal material resembles that of nucleic acids rather than protein.

The UV-sensitive germinal material lies like a cap over the vegetative pole of uncleaved *Xenopus laevis* eggs and moves upward after the first cleavage (see Fig. 12.28). Unlike the results of pricking, irradiated embryos may recover or compensate for the loss of germinal material (Bounoure et al., 1954), although photoreactivation does not seem to occur (Smith, 1966). The injection of vegetal-pole material from incompletely cleaved unirradiated donors into similar UV-irradiated embryos results in the production of some PGCs in 23% of the recipients, while the injection of vegetal-pole material from 4-cell unirradiated donors into 4-cell UV-irradiated embryos results in the production of some PGCs in 47% of the recipients. In no case do PGCs develop in irradiated animals receiving animal-pole material.

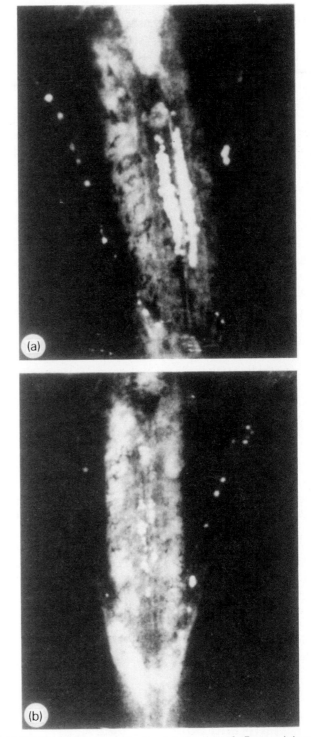

FIGURE 12.36. Anesthetized tadpoles of *Rana pipiens* (Shumway stage 25, tadpoles begin to eat) dissected to show genital ridge areas. (a) Normal unirradiated control has abundant bright primordial germ cells in ridges. (b) Tadpoles developing from eggs at first division irradiated with 5300 ergs per mm^2 of UV. The number of primordial germ cells is reduced. About ×23. From L. Dennis Smith, *Dev. Biol.,* 14:330 (1966), by permission of Academic Press.

Surprisingly, UV irradiation does not affect the appearance of the fibrogranular, germinal granules or the irregular nuage in frog eggs (Smith and Williams, 1975), but UV irradiation does several other things to the zygote. UV irradiation delays cleavage, prevents germ-cell migration, and retards mitosis in PGCs, any of which may be linked to sterility (Zust and Dixon, 1977). A UV-sensitive material may not be the one and only irreplaceable determinant of germ cells (but see Smith et al., 1983).

Germ plasm may not be exclusively germinal either. The granular materials present in developing mammalian eggs and PGCs are not limited to these cells but are also present in other embryonic cells. These materials are not segregated during cleavage in mammals, and their roles in determination may not be limited to germ cells (Eddy et al., 1981).

Thus, the identity of germ-line determinants is still a matter of speculation. Researchers seem to agree that something in the polar and vegetal cytoplasm influences the formation of PGCs, but what it is and how it works are unknown. Determinants may affect the course of future primordial germ-cell differentiation directly or by protecting nuclei from influences that would otherwise determine differentiation in a somatic line.

The continuity of germ plasm from generation to generation is also uncertain. On the one hand, both germinal plasm and PGCs are sensitive to UV irradiation, suggesting continuity between fertilized eggs and germ cells. On the other hand, nothing resembling polar or vegetal cytoplasm is present in insect or frog eggs at the earliest immature stages of development. One possibility suggested to reconcile the difference is that nuage or nuagelike material in germ cells is an empty protein framework for germinal determinants. On the rare occasions when the germ line is actually segregated the framework is filled with RNA (see Eddy, 1975).

How Distinctly Different Are the Germ and Somatic Lines?

The germ line is no longer generally considered a biological expression of eternity, but the germ line is still credited with greater potential for producing a variety of tissues than somatic lines. Even in anurans, where germ-line cells do not differentiate from transplanted somatic-line cells, somatic lines have differentiated from transplanted germ-line cells (Wylie et al., 1985).

Independent of the limitations of the somatic line, hereditary and developmental functions of the germ and somatic lines no longer seem to be quite as separate as once was thought. Since in the epigenic pattern of germ-line origins the somatic tissue conveys hereditary information to the next generation of gametes as much as gametes convey hereditary information to the next generation of somatic tissue, the transmission of hereditary information is not limited to the germ line. Likewise, since gametes undergo a course of differentiation as intricate as the differentiation of cells in the somatic line, the unraveling of developmental information is not the exclusive prerogative of the somatic line.

Moreover, artificial (if not natural) selection can alter inherited characteristics in germ cells at least in rotifers. One line of rotifers selected from older eggs declined markedly and ultimately underwent extinction, while another line developing from younger eggs remained viable (Lansing, 1952).

Comparative evidence for evolutionary changes in germ cells is as convincing as evidence for evolutionary change in somatic cells. Much like the evolution of the somatic line, the evolution of the germ line may be driven by environmental influences selecting individuals against a background of genetic variations.

The expression of genes in haploid gametes, or **haploid expression,** is difficult to document, but spermatids can synthesize both ribosomal and polyadenylated messenger RNA (Monesi et al., 1978). The production of two surface antigens by sperm from heterozygous mice is indicative of haploid expression, since transcription prior to the completion of meiosis would have produced sperm with the same surface antigen or antigens. Furthermore, as one would expect from a chromosomal-linked mechanism of expression, about half the sperm population expresses one surface antigen, while the other half expresses another (Palmiter et al., 1984).

Selection in haploid gametes is inferred from deviations in the expected ratios of some traits. For example, as many as 90% of the offspring express the recessive *t* tailless mutation when heterozygous male mice are mated. *t*-Bearing sperm may survive better in the female genital tract than other sperm and thus live to fertilize more eggs (see Setchell, 1982), or differences in the sperm's surface proteins may allow better binding to the egg's zona (Scully and Shur, 1988).

Germ cells as well as somatic cells, it would seem, face contingencies and hazards. In animals, sex cells are by no means immune to the forces that shape living things in general. Rather than life being the "continuous stream" (Fig. 12.37a) projected by Weismann's germ-plasm theory, life is a "bumpy road" (Fig. 12.37b) of interacting forces and cellular groupings (see McLaren, 1982).

The animal zygote has, after all, an ambiguous status in the classification of somatic and germ cells, since the zygote gives rise to blastomeres of the germ line and the somatic line (Fig. 12.23). Some early blastomeres likewise give rise to both germ-line cells and somatic-line cells until the lines are separated. The germ and somatic lines therefore share a common environment during these early phases of development.

In the case of vertebrates, the embryo forms the blastula or epiblast (that portion of the embryo containing the rudiments of all the germ layers) before the germ and somatic lines separate. All the embryo's cells prior to the epiblast stage are influenced by the same environment, and, in animals with internal fertilization, the environment of the maternal body (dashed line, Fig. 12.37b) continues to influence the germ line.

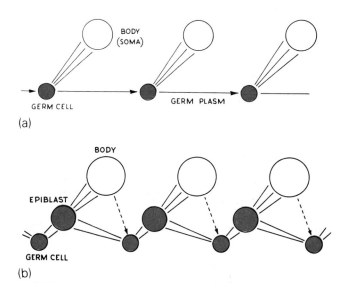

FIGURE 12.37. Contrasting views of the relationship between germ cells and the body (soma). (a) Weismann's concept of the germ plasm stream with somatic plasm branches. (b) Alternative concept of an interacting system in which the germ cell is subject to influences (dashed arrows) from the organism. The epiblast is the embryonic precursor of the fetus before the germ and somatic lines separate. From A. McLaren, *Germ cells and soma: A new look at an old problem,* Yale University Press, New Haven, 1981. Used by permission.

CLEAVAGE IN HISTORICAL PERSPECTIVE

By the end of the 19th century, embryologists had begun to understand cleavage quantitatively and moved on to probe it qualitatively. They wondered how the production of great numbers of new cells was related to morphogenesis and differentiation. Did the change in cell number reflect a change in cell type? Were blastomeres qualitatively different? Efforts to answer these questions left an indelible imprint on 20th century embryology (see Maienschein, 1985).

Segregation of Germ and Somatic Cells During Cleavage

The possibility that cleavage separated different cell lines arose first with regard to germ- and somatic-cell lines. In 1870, the embryologist Wilhelm Waldeyer (1836–1921, who is also credited with naming chromosomes) asked where mammalian oocytes came from, since they were a nonrenewing population in the adult ovary. He suggested that female mammals were born with a finite stock of oocytes, and these were used up in the course of a lifetime. Although his concept

was broadly rejected, most sharply by the great American endocrinologists Herbert M. Evans and Edgar Allen, Waldeyer's view made inroads in embryology.

Within a decade, embryologists looking at frog and trout embryos drew a sharp distinction between germ cells and somatic cells. While sharing a common origin in the zygote with the somatic cells that made the individual, qualitatively different germ and somatic cells were declared to go their separate ways in early development and to share no common source in the adult individual.

In 1882, August Weismann (1834–1914) took another step toward a theory of separate germ and somatic cells. Concerned with the duration of life, Weismann proposed that everything about an organism, including its life expectancy, was predetermined by its somatic cells. The definitive lifetimes of cells composing an organism forecast its development, senescence, and death. Only cells of the germ line potentially escaped this fate.

The germ-plasm theory was Weismann's attempt to explain differences in germ and somatic lines as he perceived them. The germ line of cells was seen as floating in an eternal stream of germ plasm, while the somatic line of cells was seen as sinking in a mortal sea of somatic plasm (Fig. 12.37a).

A devout 19th century preformationist, Weismann believed firmly that form was predetermined at the earliest stages of development. For him, the constancy and repeatability of heredity and development could only be explained by equally constant and repeatable causes. Only the transmission of preformed units from one generation to the next seemed to meet this requirement.

According to Weismann, epigenesis, the theory that form developed through an interplay of microenvironmental and macroenvironmental interactions, was nothing more than an intraorganismic version of Lamarck's already discredited theory for the interorganismic inheritance of acquired characteristics. Epigenesis was an "impossible" *ad hoc* hypothesis pasted together by ignorance.

Attacking epigenesis and the inheritance of acquired characteristics simultaneously, Weismann mobilized experimental evidence for the absence of environmental influences on heredity. He showed that despite repeated mutilation of newborn mice, no consistent change was passed on to the progeny of the unfortunate animals. Somatic tissue could be ravaged by the environment without as much as leaving a scratch on heredity.

Weismann's attention was drawn toward the nucleus as he contemplated a safe haven wherein the germ could reside, while the soma rode out the storm of life. The choice was timely. With the advent of the aniline dye industry, nuclear chemistry and cytochemistry had moved to the forefront of modern biology. Cytologists discovered mitosis and correctly described meiosis and fertilization following Weismann's shrewd deduction that a reduction division would accompany sexual reproduction. Results with all the latest techniques seemed to shore up Weismann's theorizing.

Carl Nägli (1817–1891), whose pioneering chemical analyses of nuclei and cytoplasm provided the scientific basis for cytochemistry, had earlier suggested that part of a cell was self-perpetuating (i.e., replicating). Nägli coined the term **idioplasm** (Gk. *idios* one's own or arising from oneself + *-plasm* form or mold) for substances providing the physical basis of heredity, but other cytologists equated idioplasm with Weismann's germ plasm and both with chromatin or, at least, with its "inherent organization" (Wilson, 1898).

Weismann theorized that a germ line of cells consisting of gametes and their progenitors uniquely transmitted idioplasm from generation to generation. The zygote and blastomeres were originally neutral **trophoplasm,** but they gave rise to all the cells of the somatic line and produced the organism by sorting out hereditary information in the course of development.

Not since Darwin proposed his ill-starred pangene theory had anyone offered a comprehensive cellular solution to the problems of both heredity and development. Unlike his predecessor, however, Weismann was applauded and his theory warmly received. "Weismann's contribution, like that of so many influential figures in the history of science, was the expression of his doctrine to a century philosophically ripe for its acceptance" (Oppenheimer, 1967, p. 161).

As a theory of nuclear activity, Weismann's had the virtue of concreteness. It conformed to or easily accommodated much of the empirical evidence available at the time. For example, Weismann proposed that idioplasm was partitioned and germ and somatic plasmas were segregated during cleavage, while Boveri showed in *Parascaris* that patent chromosomes were found in primordial germ cells, and fractured and diminished chromosomes were found in somatic cells (see Fig. 11.55). According to Boveri:

> **The original nuclear constitution of the fertilized egg is transmitted as if by a law of primogeniture, only to one daughter-cell, and by this again to one, and so on (quoted from Wilson, 1898, p. 111).**

Weismann's theory had great heuristic value, and it undoubtedly helped prepare the next generation of biologists for the acceptance of genes as preformed units. It was so widely adopted that, in effect, it achieved the rank of scientific dogma.

The weaknesses of the theory lay in its concept of how idioplasm operated in development. Weismann did not conceive of separate hereditary and developmental materials (i.e., today's DNA and RNA, respectively). Theoretically, intact idioplasm was passed by cleavage to germ-line cells, and fragments of idioplasm were segregated to somatic-line cells.

Weismann suggested that developmental factors of idioplasm were arranged in hierarchical arrays like stones in a pyramid. While all the factors were present in the zygote, consecutive cleavages left fewer and fewer factors in each of the somatic blastomeres. Ultimately, when the complement of factors present in a given somatic cell was reduced to a certain minimum, the factors became active, and the cell differentiated accordingly.

Weismann argued for the importance of normal chromosomes for normal development by pointing to abnormal chromosomes accompanying pathological development (including dispermic sea urchin eggs and some tumors), but, in general, he could not decipher normal changes in chromosomes accompanying normal differentiation. His theory also failed to tie quantitative differences in chromosomes to qualitative differences in the embryo, but the theory was in no danger of foundering.

Overwhelming support for Weismann's theory came from the laboratory of Wilhelm Roux (1850–1924). In 1888, Roux reported the results of an experiment that, on the surface, vindicated Weismann's concept for the segregation of developmental factors by cleavage.

Roux's Hot Needle Experiment

Roux conceived of his "pricking experiment" out of his own deep-seated belief in preformationism. For Roux, preformation was the only embryological theory with scientific merit, since in his opinion, it alone was accessible to experimental verification (Roux, 1894).

Inasmuch as any preformed unit governing development would have to be material, and material was potentially manipulatable, preformed determinants of development would have to be manipulatable. They were therefore potentially suitable subjects of experimentation.

Epigenics, on the other hand, offered few opportunities for experimentation. Environmental influences such as gravity and the earth's electromagnetic field were unchanging or of no consequences where they changed. Roux concluded that "we therefore have to look for the formative forces in the egg itself" (Roux, 1888, p. 5).

Suggesting that cleavage provided the mechanism for parceling out formative forces, or developmental determinants, Roux hypothesized that, at every cleavage division, half of the determinants were distributed to each blastomere. Since the first cleavage would segregate half of the zygote's determinants to each of the first two blastomeres, by the 4-cell stage, the embryo would already be a **mosaic** with each blastomere carrying only the factors involved in the development of quarters of the embryo.

Roux thought of testing this hypothesis by killing some blastomeres and allowing others to continue developing. If the hypothesis were correct, the surviving blastomeres would undergo autonomous change, or what he called **self-differentiation,** and form parts they would normally form. Parts corresponding to the dead blastomeres would be missing. If the hypothesis were incorrect, the surviving blastomeres would differentiate according to prevailing environmental cues, presumably the same as those normally experienced by the whole egg, and develop into complete embryos.

Initially, Roux attempted to kill frog blastomeres by puncturing them with a room-temperature needle. Even though blastomeres released cytoplasmic material at the point of puncture, they often survived and underwent normal development. Roux interpreted the results as illustrating the nucleus's ability to restore lost cytoplasmic determinants.

In his next experiment, Roux resorted to heating the needle to be sure he killed the punctured blastomeres. The results (Fig. 12.38) were, in a word, spectacular.

> [An] amazing thing happened; the one cell developed in many cases into a half-embryo generally normal in structure, with small variations occurring only in the region of the immediate neighborhood of the treated half of the egg (Roux, 1888, translated by Hans Laufer, p. 12).

The surviving blastomeres gave rise only to the portions of the embryo to which they would have given rise in whole animals.

To make his case airtight, Roux performed state-of-the-art procedures available for microscopic examination of solid tissue. He succeeded in showing that the external appearance of half-embryos frequently continued internally (Fig. 12.39). However, distinct exceptions were also discovered in which the half-embryos seemed to regenerate missing parts.

In one of the great "happy accidents" of science, Roux chose to minimize these instances of regeneration. He did not ignore them, but he labeled them as "postgenerations" and consequences of "reorganization processes" and shunted them aside for follow-up experiments. Had Roux made more of the exceptions, he would not have been able to interpret his results as succinctly as he did, and the fireworks he lit in embryology might never have gone off.

For Roux, the results led to a simple conclusion:

> In general we can infer from these results that each of the two first blastomeres is able to develop independently of the other and therefore does develop independently under normal circumstances. . . . Each of these blastomeres contains therefore not only the formative substance for a corresponding part of the embryo but also the differentiating and formative forces (Roux, 1888, pp. 25–26).

As the 20th century approached, Roux's substances and forces were equated with the factors in Weismann's theory. The idea that developmentally significant substances were the same as hereditary factors funneled into blastomeres during cleavage became known as the **mosaic theory of develop-**

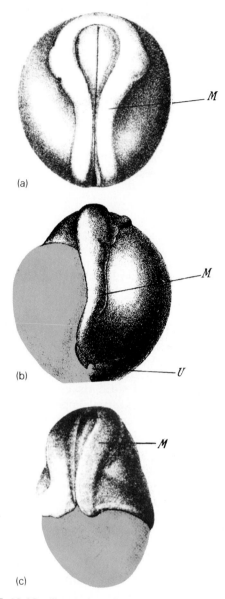

(a)

(b) M U

(c) M

FIGURE 12.38. Illustrations from Roux's 1888 report on his "hot needle" experiment. (a) Dorsal view of normal green frog (*Rana esculenta*) embryo at the neural fold stage. (b) Lateral-embryo half (right) at organogenesis stage developing from one of the first two blastomeres following pricking of other blastomere. The colored area is the remains of the pricked blastomere. (c) Anterior-embryo half at neural fold stage developing from one of the first two blastomeres following pricking of other blastomere. The colored area is the remains of the pricked blastomere. *M,* medullary plate or neural folds; *U,* remnant of yolk plug. From W. Roux, *Virchows Arch. Pathol. Anat. Physiol. Med.,* 114:113 (1888), reproduced in B. H. Willier and J. M. Oppenheimer, *Foundations of experimental embryology,* Prentice-Hall, Englewood Cliffs, NJ, 1964.

ment or the **Roux–Weismann hypothesis,** and embryologists around the world set out to test it.

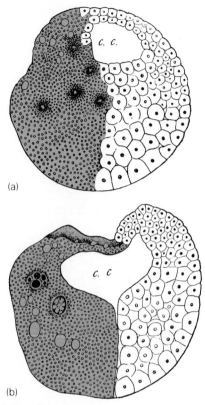

(a)

(b)

FIGURE 12.39. Drawings of sections through lateral-embryo halves developing from one of the first two blastomeres following pricking of other blastomere. The colored areas on the left are remains of the pricked blastomeres. C.C. blastocoel. From W. Roux, *Virchows Arch. Pathol. Anat. Physiol. Med.,* 114:113 (1888), reproduced by B. H. Willier and J. M. Oppenheimer, *Foundations of experimental embryology,* Prentice-Hall, Englewood Cliffs, 1964.

Finding Consistency with the Roux–Weismann Hypothesis

Although "consistency is not proof," a hypothesis should be consistent with what is already known, or it is hardly worth investigating further. From its inception, the mosaic theory was consistent with a great deal of empirical evidence and was the inspiration for efforts to find even more evidence.

The theory predicted that blastomeres would be different from each other as soon as they were cleaved. This meant that differences in later organisms should be traceable to particular blastomeres and to no other blastomeres in early organisms. When the mosaic theory of development was first promulgated, some **cell lineages** (see Fig. 11.44) were already known, but many more would soon be traced.

The place was the new Marine Biological Laboratory (MBL) at Woods Hole, Massachusetts. Viewed by some as a summer resort, it was also a community of scientists (see Pauly, 1988). Jacques Loeb, who had made artificial parthenogenesis practical, worked there, and Charles Otis Whitman (1842–1910), who first described cell lineage in the leech, *Glossiphonia* (= *Clepsine*) *complanata*, was director. Edmund B. Wilson (1856–1939) studied *Neanthes succinea* (= *Nereis limblata*), *Patella*, and *Dentalium*, and Edwin G. Conklin (1863–1952) studied *Ilyanassa*, *Styela* (= *Cynthia*) *partita*, and *Branchiostoma lanceolatum* (= *Amphioxus*). With extreme diligence and infinite patience, they immobilized nearly transparent marine embryos in glass tubes and, turning them over under magnifying glasses, located the embryo's polar bodies, oriented the cleavage furrows (e.g., see Figs. 11.40 and 11.42), and eventually identified every cell in the blastomeric lineages up to stages of early differentiation (Fig. 11.49).

At MBL, the young Americans cast modern embryology in the mold of Roux's mechanism and Weismann's preformationism. Nobly bearing the indecorous badge of "egg rollers," they set records for tracing cleavage patterns that are only now being broken by work on *Caenorhabditis elegans*.

Left *in situ*, each blastomere revealed its fate or destiny (otherwise known as prospective significance). Lineages connected blastomeres to differentiating cells (see Fig. 11.44), and fate maps identified sites on the egg's surface (e.g., Fig. 12.40) and areas on cleaving embryos that would later be incorporated into actual parts of gastrulas, larvas, and adults.

In many species, cleavage patterns were highly stereotypic, and cell lineages and fate maps were extremely rigid. In many spirally cleaving eggs and in the bilaterally cleaving eggs of ascidians (see Fig. 11.33), the progeny of blastomeres took unique pathways and differentiated exceptionally early. Cleavage in these organisms was called **mosaic cleavage** in recognition of its consistency with the mosaic theory of development.

Moreover, cytoplasmic streaming and ooplasmic segregation seen in ascidians and polar body activity seen in mollusks (see Fig. 11.47) seemed consistent with the mosaic theory. The precocious distribution of cytoplasmic substances to certain blastomeres during cleavage preceded differentiation, suggesting that development was predetermined and that cytoplasmic substances were or contained developmental determinants.

Finally, the case of **handedness** in snails seemed to provide convincing empirical evidence

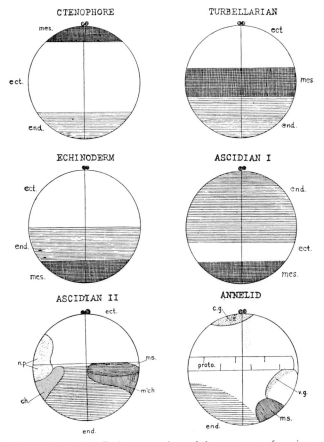

FIGURE 12.40. Early examples of fate maps of various phyla (ca. 1922). The designated areas become specific parts of the organisms. The fate maps of the ctenophore, turbellarian, echinoderm, and ascidian I are drawn for embryos at the end of the first cleavage. The ascidian II fate map is drawn for an embryo at the end of the second cleavage. The annelid fate map is drawn for an advanced embryo. Crosshatch, mesoderm or mesenchyme (mes); horizontal lines, endoderm (end); clear, ectoderm (ect); c.g., cerebral ganglion; ch., notochord; m'ch., mesenchyme; ms., muscle; n.p., neural plate; proto., prototroch (ciliated band in the larva); v.g., ventral ganglion. From E. G. Conklin, *Heredity and environment in the development of men*, 5th ed. rev., Princeton University Press, Princeton, 1922. Used by permission.

for the mosaic theory of development. Here, adult characteristics could be traced back directly to cleavage.

The difference between right-handed and left-handed snails (or univalve mollusks in general) extends to every organ in the animal's body (Fig. 12.41 g and g') but is most easily seen in the shell (Fig. 12.41 h and h'). The handedness of a shell is ascertained by laying the opening toward you and the point upward. If the shell opens on the right (Fig. 12.41 h'), the shell is right-handed; if it opens on the left (Fig. 12.41 h), it is left-handed.

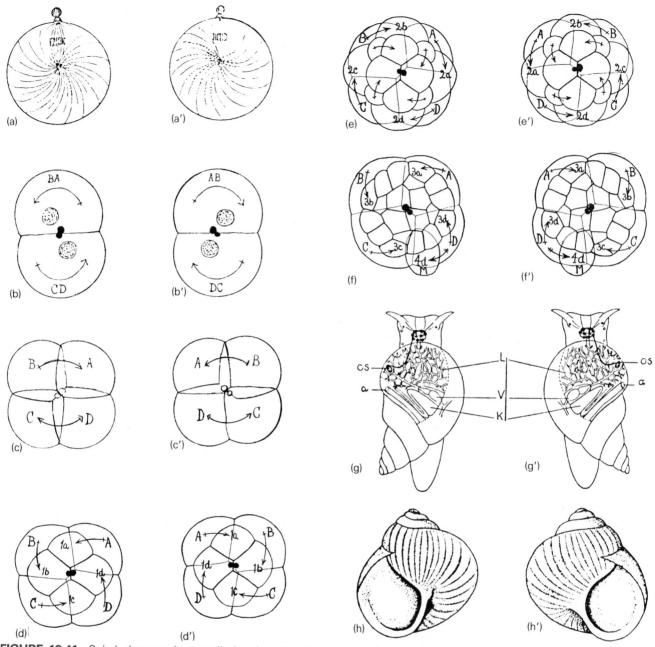

FIGURE 12.41. Spiral cleavage in a snail showing directions of spiraling in left-handed (a through f) and right-handed (a' through f') varieties. In the adults (g' and g), the organization of all organs are inversely symmetrical. a, Anus; K, kidney; L, lung; os, olfactory organ; V, ventricle; h' and h, right-handed and left-handed shells of adults. Adapted from E. G. Conklin, *Heredity and environment in the development of men,* 5th ed. rev., Princeton University Press, Princeton, 1922. Used by permission.

In most univalve species (e.g., *Limnaea* and *Crepidula*), shells are typically right-handed, while in some species (*Physa* and *Planorbis*), shells are typically left-handed. In some right-handed species, left-handed races appear, and in some left-handed species right-handed races appear (see Raven, 1966).

The differences in handedness may be diagnosed as early as the second cleavage division when the cross piece forms at the juncture of B and D blastomeres. The cross piece bends to the left in prospectively left-handed snails and to the right in prospectively right-handed snails. Embryos forming right-handed shells subsequently perform right-handed (clockwise) divisions on odd cleavages and left-handed (counterclockwise) divisions on even cleavages, while embryos forming left-handed shells proceed with left-handed divisions on odd cleavages and right-handed divisions on even cleavages (Fig. 12.41).

Still the question remained whether cleavage controlled handedness or something controlling handedness also controlled cleavage. The first possibility, which complemented the mosaic theory of development, was widely accepted, but many "egg rollers" interpreted the observation more conservatively. These embryologists resisted mosaic determinism as long as other hypotheses were not excluded.

The MBL embryologists lacked Roux's conviction that developmental influences could be localized in material factors. Fearing that the mosaic theory of development would churn up the same sterile debates over preformationism that prevailed in European embryology well into the 19th century, the Americans argued that the antidote for contention was experimentation.

Ultimately, the results of breeding experiments illuminated the basis of handedness (see Sturtevant, 1965). Handedness was inherited maternally (i.e., in a Mendelian fashion but only after the delay of one generation). A single pair of genes with right-handed dominance determined handedness. For example, only right-handed snails were produced by mating a heterozygous right-handed mother to a homozygous left-handed father, but half the progeny carried no genes for right-handedness and produced all left-handed offspring in the next generation. The handedness of the offspring was determined even before polar body formation and fertilization therefore, and genes inherited by the mother from the previous generation controlled cleavage and handedness.

Experimentally Testing the Roux–Weismann Hypothesis

The mosaic theory required that cells undergoing particular developmental events house nuclei with developmentally significant substances not present in other cells. This requirement led to several verifiable predictions.

To test the mosaic theory, embryologists attempted alternatively to exclude or to include hypothetical substances in places where they could prevent or could promote particular pathways of development. Procedures included both deletion or ablation experiments in which parts of the embryo were removed (i.e., *in vitro* isolation) and addition experiments in which parts were added (i.e., *in vivo* transplantation, see Watterson, 1955).

The obvious way to begin was by repeating Roux's hot needle experiment on other species.

Conklin killed blastomeres in the tunicate, *Styela* (called *Cynthia* at the time), and traced the development of surviving blastomeres. He found, as Roux had found for the frog, that even as early as the 2- and 4-cell stages, the blastomeres were self-differentiating. For the most part (the nervous system being the exception), surviving blastomeres formed the same parts they would have formed if nothing had been done to the embryo, while dead blastomeres left holes in the embryo in place of the parts they might have formed.

The results of other experiments left the interpretation of Conklin's hot needle experiments in doubt, however. When centrifugation displaced the colored cytoplasms of freshly fertilized ascidian eggs, embryos sometimes developed with dislocated tissues (Fig. 12.10). Cytoplasmic constituents, rather than developmental factors arising in the nucleus, therefore seemed to play the role of developmental determinants.

Woods Hole embryologists also separated the blastomeres of small molluscan and annelidan embryos by flushing them through the mouths of pipettes. Shearing forces broke the eggs' extracellular envelopes and freed the blastomeres. The treatment was crude and rough on embryos, but surviving blastomeres were often self-differentiating (Fig. 12.2) in agreement with the mosaic theory.

Still, the results of other experiments cast doubt on the role of the nucleus in mosaic cleavage. When polar lobes were removed from molluscan embryos without further disturbing cleavage, lobeless embryos developed severe defects (Fig. 12.7). Polar lobes, rather than cleavage nuclei, seemed to be repositories of developmental factors.

The hypothesized developmental role for substances seemed to be confirmed experimentally for the eggs of a variety of spiralian species and over a range of experimental conditions, but there were also exceptions. Every spiralian species did not exhibit purely mosaic cleavage nor was the development of isolated blastomeres always completely predetermined. Furthermore, despite similarities with ascidians and characteristic protochordate-type localizations of cytoplasmic constituents, blastomeres of the lancet, Amphioxus, failed to exhibit self-differentiation.

Additional contrary evidence came from another quarter. Soon after Roux published the results of his hot needle experiment, Hans Driesch (1867–1941) called attention to sharply contradictory experimental results obtained with sea urchins (Driesch, 1892). Driesch, Oscar Hertwig, and others had obtained a few isolated blastomeres by vigorously shaking eggs in tubes. Occasionally, these blastomeres developed into relatively normal larvas. Later, when gentler methods (such as separation in seawater free of calcium ions) were employed, each of the first four blastomeres was shown to produce small, complete, and normally appearing larvas (see Fig. 5.14).

The blastomeres could not have received only halves and quarters of the developmental factors required by the Roux–Weismann hypothesis. Cleavage, it would seem, did not mechanically parcel out different developmental factors and create unequal blastomeres in echinoderms. The early blastomeres were equal and capable of doing each other's developmental jobs when called upon by the experimental environment.

Driesch interpreted his results to show that the embryo was a **harmonious equipotential system** capable of regulating its form to make up for deficits such as the death or removal of blastomeres. In contrast to isolated blastomeres of mosaic eggs, those of regulative eggs could develop relatively normally.

For Driesch, regulation was a uniquely living phenomenon, inimitable by machines. In his opinion, Roux's reductionism could never explain regulation, and efforts to analyze causal components of developmental processes were useless.

With the idea of uniquely living phenomena, Driesch opened the door to vitalism in the 20th century, and, by promoting model systems of development, he challenged Roux's attempt to convert embryology to a reductionist science. On balance, the effect was positive, since at the time, embryology was in danger of "throwing the baby away with the bath water" (i.e., studying cellular autonomy to the exclusion of global and interactive aspects of development).

The issue of **holism**, or what the organism can accomplish because it is integrated and part of its environment, should never be suppressed merely because it is inconvenient or does not yield to reductionist approaches. Today, many **Drieschians**, including Lewis Wolpert, promote broadening the range of developmental problems to include position and polarity and the scope of approaches to include modeling and simulation (see Wolpert, 1985).

Considering himself more the oppressed than the oppressor, Roux sprang to the defense of reductionism. In 1894, he founded the journal that became *Wilhelm Roux's Archives of Developmental Biology*, to provide an unimpeded forum and safeguard for analytical approaches to development.

Nature was not a grab bag of exceptions, and experimentation was not blind groping. In Roux's opinion, experimentation alone could solve the riddles of development including those presented by contradictory results.

In the last analysis, experimentation did resolve the conflict but not in Roux's favor. One difference in how Roux and Driesch carried out their experiments proved crucial. While Roux had left the dead blastomere inside the egg membranes, Driesch had either removed dead blastomeres or isolated blastomeres entirely from each other.

When blastomeres killed with the hot needle were removed and when blastomeres were isolated completely (even in *Rana esculenta* used by Roux), complete tadpoles often developed (Fig. 12.15). Amphibian embryos, like sea urchin embryos, were regulative and not mosaic. Roux and his followers had considered the results with sea urchins exceptions to the rule of mosaic development, but, in the end, the "exceptions proved the rule."

How had Roux gone wrong? In hindsight, one can see that Roux's experimental procedure was flawed. He failed to anticipate all the possible variables that might have affected his result and had not done sufficient control experiments to eliminate all alternative explanations.

A crucial variable in Roux's experiments that he only considered tangentially was the continued presence of dead blastomeres adjacent to the living blastomere (Figs. 12.38 and 12.39). Possibly, they prevented the normal rotation of the egg or pressure from dead cells obstructed movement by descendants of living blastomeres (Hamburger, 1988).

Alternatively, in the presence of the dead blastomere, the living blastomere may not have recovered from the traumatic effects of separation quickly enough to reestablish normal polarity by the onset of gastrulation. This possibility is suggested by the results of isolating *Xenopus laevis* blastomeres in medium containing dilute calf serum. Tending to prolong the presence of open wounds in the cell surfaces, the calf serum turns blastomeres into half-embryos resembling Roux's (Kageura and Yamana, 1983).

After dominating embryological research for a quarter of a century, the Roux–Weismann hypothesis was vanquished by its own method, experimental negation. Roux was forced to acknowledge the existence of an early period of regulative development, at least in some species. Furthermore, the preformationist underpinnings of the hypothesis were shaken. Even if later development were mosaic, room had been made for epigenesis to play some role in early development (see Maienschein, 1985).

Ultimately, Roux's reputation did not hang on his hot needle experiment, but on experimentation itself which he, possibly more than anyone else, brought to embryology. In that respect he was completely vindicated.

Explaining Regulative Development

Hans Spemann (1869–1941) and other European and American embryologists picked up the challenge of analyzing regulation. Unlike Driesch, they did not appeal to models for explanations but employed experimentation. Unlike Roux, they did not abandon epigenesis but restored it to a place in the pantheon of embryological theories.

The differences between mosaically and regulatively developing embryos seemed to be the range over which their cells were responsive to environmental conditions. While a cell's environment included gravity and the earth's electromagnetic field, it also included other cells, their products, and the organization of the embryo.

Possibly, each blastomere in the intact embryo communicated its presence to its companions and told them they were not alone. A **communal message,** or sum of cellular communication, might instruct blastomeres not to proceed as if they were alone and make a whole organism by themselves. Isolated blastomeres might continue for a while to make blastomeres as if still part of the whole, but gradually memory of the communal message would be lost. Cells would then act as if they were blastomeres arising from whole eggs and make a whole until they received messages, once again, that they were not alone.

If not quite as reductionist in his conception of the embryo as Roux and certainly more tolerant of ideas that touched upon vitalism, Spemann was no less mechanistic in his way of thinking. He had an epigenic rather than preformationist bias, and, for him, epigenic hypotheses were as testable as preformationist ones. Ultimately, Spemann's approach, not Roux's, dominated embryology in the first half of the 20th century, but Roux's mark, like Weismann's before him, was left on the emerging science of genetics (see Moore, 1987).

A New Perspective

The advent of genetics changed the prevailing perspective on development. The zygote continued to impress (if not overwhelm) embryologists with its power to release developmental energy, and blastomeres spawned by cleavage continued to bewilder with their contradictory qualities, but cleavage was driven out of the forefront of embryological concerns. A modified version of the Roux–Weismann hypothesis came to epitomize purely genetic approaches to development, and the concept of induction introduced by Spemann was equated to epigenic approaches.

Recently, efforts to bridge the gap between organisms used in genetic studies and those used in developmental studies produced a rash of work on the development of *Drosophila* and on the genetics of *Xenopus*. In addition, *Caenorhabditis elegans* has recently emerged as the premier subject for both genetic and developmental research. Egg rolling was revived, but the objective shifted from cleavage patterns and mosaicism to cell lineage and the action of determinants.

The use of molecular **intracellular tracers** such as horseradish peroxidase (HRP) makes cell lineage studies much easier to conduct and vastly extends their range. Unlike vital dyes that mark the surfaces of groups of cells, intracellular markers identify the progeny of single cells. In spiralians, results with intracellular tracers validate much of the literature on cell lineages based on direct observation and aid in carrying the analysis of cell origins to advanced stages. In vertebrates, especially anurans, with opaque embryos and multicellular layers, intracellular tracers add an internal dimension to studies of blastomeric descent.

New theories arise from the results, but they sometimes boil down to new words and not new ideas (see Wolpert, 1985). The spontaneous emergence of morphogenic patterns in cleaving embryos is offered as evidence for a **prepattern** or an **organizational field** preexisting in the zygote. The destinies of embryonic cells carved out of the zygote by cleavage or by cellularization are thought to be determined by positions in the field, but the nature of the field remains uncertain.

Alternatively, the embryo as a whole is sometimes portrayed as an ecosystem with cells occupying the position of individual representatives of species. Since two blastomeres cannot occupy precisely the same niche at the same time in the cleaving embryo, a blastomeric **exclusionary** or **competitive principle** must operate during cleavage.

If each aspect of a cell's relationship to its environment is represented by a dimension, a cell's complete identity and place in the organism has n dimensions. Competition throughout cleavage could exclude blastomeres or parts of the embryo from occupied dimensions and allow them to shuffle into unoccupied dimensions. The development of barriers in the form of cleavage furrows (whether complete or incomplete) and the separation of groups of cells would ordinarily land cells irreversibly in particular niches, while the experimental removal of cells from niches would restart the developmental shuffle.

PART 5   SUMMARY AND HIGHLIGHTS

Cleavage transforms the zygote into a multicellular embryo. The blastomeres or cellular blastoderm produced by cleavage begins to resemble adult cells or tissue in several ways. Because growth does not accompany cleavage division, embryonic cells become increasingly small and acquire a nuclear/cytoplasmic ratio characteristic of adult cells. Cleavage distributes materials and reduces their absolute amounts per cell from the levels found in eggs. While an egg contains massive amounts of ribosomes and mRNA, blastomeres have amounts found in typical rapidly dividing cells.

Blastomeres give rise to cell lineages which in turn give rise to differentiating tissue. Different species stick more or less rigidly to cleavage patterns defined by the size and number of blastomeres and by the angles of furrows. Erratic division producing blastomeres of atypical shapes or in abnormal positions may be followed by compensatory divisions restoring the typical pattern.

For the purpose of following lineages, blastomeres are assigned names encoding their position, the number of divisions preceding their production, and sometimes their fate. A variety of methods including "egg rolling" and the injection of intracellular tracer molecules such as horseradish peroxidase and fluorescent dyes are used to follow sibling cells through their lineages. In several spiralians, cell lineages have been traced to larval stages and, in the nematode, *Caenorhabditis elegans*, all the way to the adult.

Cleaving or cellularizing the embryo seems to be a requirement for development, but several aspects of cleavage are unique to this phase of embryonic life and presumably represent adaptations to the egg's circumstance. Mitotic spindles in successive divisions frequently alternate at right angles, and furrows may not pass through the yolky parts of eggs rapidly if at all. In agreement with Balfour's law, spindles are generally excluded from yolk-rich areas, but asters, rather than yolk, seem to determine the position of furrows, and the positioning of centrioles is probably responsible for determining the angle of cleavage. Plasma membrane seems to be added to cleavage furrows near the point of elongation rather than as a result of stretching at the surface.

Cycles of karyokinesis or mitosis and waves of cytokinesis or furrowing occur in synchrony as though under the control of a cleavage clock. Changes in sulfhydryl-containing cortical proteins in the sea urchin egg are correlated with stereotypic changes in the spindle angle. As if following an internal schedule, some differentiation may occur without cytokinesis in treated *Chaetopterus* eggs, and ascidian eggs produce tissue-specific enzymes on schedule even when cytokinesis is inhibited.

The postmitotic gap (G-1) may be suspended during cleavage. Replication (S) is especially rapid. Large numbers of microbubbles (indicative of increased numbers of origins or initiation sites) are detected autoradiographically at the electron microscopic level.

Transcription is not conspicuous in most cleaving embryos, and fertilization-dependent proteins may not be different from egg proteins. In the same embryo, some cells (e.g., echinoderm micromeres) support transcription, while others do not. In mammals, transcription begins during cleavage, but the products may not be required for cleavage, and nascent heterogeneous nuclear RNA and tRNA are synthesized precociously, but rRNA is not.

Protein synthesis may begin quite abruptly with cleavage, but it generally increases gradually. The types of protein selected for synthesis may not be unique to cleavage. Some proteins (e.g., actin, tubulin) synthesized in oocytes and adult tissues are also synthesized during cleavage. Some embryonic proteins not generally found in adult tissues may be synthesized during cleavage, while other proteins synthesized before and after cleavage are present in adult tissues. Different cleavage histones may be synthesized during a short interphase and packaged into the new chromosomes. The substituents of these histones are then gradually altered in conformity with adult types of histone. Polar lobes may be associated with specific proteins.

Different types of cleavage have presumably evolved in response to the different amounts and distributions of yolk. A return to total cleavage without a restoration of polarization seems to accompany secondary yolk reduction in species evolving a viviparous style of reproduction.

Megalecithal eggs thoroughly loaded with yolk and mesolecithal teleostean eggs restrict cleavage to cytoplasmic-rich portions. Intralecithal (or superficial) cleavage occurs in megalecithal eggs with a centrolecithal distribution of yolk, while partial (meroblastic or discoidal) cleavage occurs when the distribution is telolecithal as in cephalopods, fish, and cleidoic vertebrate eggs. Microlecithal to mesolecithal eggs tend to exhibit total (or holoblastic) cleavage in which yolk as well as cytoplasm is divided.

Intralecithal cleavage is characteristic of arthropods with the exception of crustaceans. Karyokinesis precedes cytokinesis, and energids of intralecithal nuclei with surrounding halos of cytoplasm are present in a syncytial embryo. The energids migrate to the surface or to a thickened zone of cytoplasm known as the periplasm where they form a syncytial blastoderm.

The surface of the egg then cellularizes by the elongation of internuclear folds led by furrow canals to form a single-layered blastoderm of more or less evenly distributed cells. Surrounded by a yolk membrane or sac, a unified yolk mass containing vitellophages forms beneath the cellular blastoderm. The yolk mass may be differentiated into yolk spheres or yolk pyramids surrounded by a cytoplasmic reticulum, and the superficial cytoplasm may have periplasmic fields in anticipation of cellular boundaries.

In pterygote insects, such as *Drosophila*, after several rounds of synchronous karyokinesis, intralecithal nuclei enter the pole cytoplasm at the posterior tip of the egg. They form pole buds and, after additional divisions, pole cells which provide the primordial germ cells. Other intralecithal nuclei enter the periplasm where they induce somatic buds and somatic cells. The cellularization of the insect periplasm is an incredible feat of synchronous cytokinesis. In *Drosophila*, for example, 3500–5000 cells form simultaneously.

Cleavage is partial when division occurs exclusively in a circumscribed cytoplasmic-rich portion or blastodisk (periblast, germ, or germinal disk). Partial cleavage in cephalopods seems to have evolved from spiral cleavage, while partial cleavage in vertebrates may have evolved several times from eggs with bilateral total cleavage.

Cleavage inscribes cephalopod eggs with a pattern of diverging furrows. A yolk syncytium of incompletely separated nuclei gives rise to a cellular blastoderm and, in turn, to cellular layers lined by septate desmosomes and gap junctions. In teleosts, consecutive vertical and perpendicular furrows divide the elliptical blastodisk into an orthogonal pattern of central and peripheral blastomeres. After central cells are undercut by the completion of furrows, horizontal divisions separate deep and peripheral cells. Cytoplasmic bridges break down, and gap junctions accumulate among the blastomeres.

Partial cleavage in the avian egg begins during the egg's passage down the oviduct. The relationship of furrows to cleavage nuclei is uncertain, but open syncytial cells broadly in contact with yolk become complete cells when the deep ends of pendulous furrows spread horizontally. Tangential divisions give rise to superficial epithelial cells and deep subepithelial cells, while ventral spaces merge into a subblastodermic cavity above a subgerminal periblast.

Total cleavage frequently produces stereotypic patterns of furrows and blastomeres. Cleavage may be equal and produce blastomeres of equal size, or it may be unequal and produce blastomeres of unequal size. Furrows may be vertical (meridional or longitudinal) paralleling the animal–vegetal axis, horizontal (transverse or latitudinal) cutting across this axis and dividing the egg into tiers (or stories), or equatorial cutting across the egg midway. Oblique or spiral cleavages produce blastomeres lying over the furrows of those below either toward the right (dexiotropic, right-handed, or clockwise) or to the left (laeotropic, left-handed, counterclockwise) when viewed with the animal pole up. Periclinal furrows occur within the blastoderm parallel to the surface.

Irregular or anarchic cleavage in coelenterates and spherical or rotational cleavage in mammals have no relationship to future planes of symmetry in the embryo. In other cases of total cleavage, the planes of furrows have discrete relationships to the future planes of symmetry in larvas and adults.

The study of cleavage in therian (placental) mammals was greatly advanced by techniques for obtaining and culturing gametes and embryos. Gonadotropin regimens induce superovulation, and eggs aspirated from the ovary and fertilized *in vitro* with artificially capacitated sperm develop through cleavage and the preimplantation phase. Transferred to the uterus of a primed or pseudopregnant prenatal foster parent (i.e., several days after induced ovulation), the embryo continues development to normal term.

Mammalian blastomeres may not divide in synchrony but tend to rotate with respect to one another forming a crosswise 4-cell embryo. The blastomeres undergo compaction by flattening against one another and forming a morula with nonjunctional complexes between cells. Cavitation results in the formation of a blastocyst cavity surrounded initially by trophectodermal cells and processes. Junctional complexes seal an inner cell mass within a trophectoderm or trophoblast. Communicating gap junctions later appear at the borders of cells in the inner cell mass.

Radial cleavage is common in crustaceans and some other arthropods and in echinoderms. Yolky crustacean eggs frequently achieve an intralecithal-type distribution with a yolky mass surrounded by a cellular blastoderm.

The microlecithal echinoderm egg divides twice vertically and then horizontally into animal and vegetal tiers with sizes ranging from equal to unequal according to a tendency toward telolecithality. The fourth and fifth cleavages are asymmetric in the tiers, and the sixth cleavage establishes two animal tiers (an_1 and an_2), two vegetal tiers (veg_1 and veg_2), and a group of micromeres.

Biradial cleavage occurs in ctenophores. Unipolar infolding of cleavage furrows parallels the future animal's planes of symmetry.

Bilateral cleavage in protochordates frequently follows the ooplasmic segregation of cytoplasmic material to distinctive crescents. The yellow pigment crescent is sequestered by cells of the tadpole's tail musculature and mesenchyme. A light gray crescent segregates with cells of the nervous system, and a transparent or clear crescent segregates to future epidermal cells. Gray yolk ends up in gut or in endoderm cells.

The protochordate cleavage pattern resembles that of echinoderms with a superimposed bilaterality.

A similar pattern is found in the early stages of amphibian cleavage where the first furrow frequently passes through the gray crescent and lies on the midsagittal plane of the future larva.

Spiral or oblique cleavage occurs in the small eggs of phyla sometimes lumped together as Spiralia. Tiers of one, two, or four micromeres arise from similar numbers of macromeres in different species. In barnacles and aschelminthes, an unequal first cleavage separates a single micromere or monet from a macromere (but not necessarily the larger cell). The first cleavage of acoelous turbellarians is equal, and two macromeres proceed to form duets of micromeres (sometimes called half spiral cleavage). In polyclad turbellarians, annelids, rhynchocoels, mollusks (with the exception of cephalopods), echiuroids, and sipunculids, the third cleavage cuts off a quartet of micromeres.

In quartet-forming annelids and mollusks, furrows tend to be stable and may aid in tracing cell lineages into the trochophore larva. The largest macromere or a developmentally similar macromere occupies the dorsal (D) position in the embryo. The ventral (B) macromere makes contact with the dorsal macromere at a transverse ventral cross furrow, while the left (A) and right (C) macromeres make contact at a sagittal dorsal furrow. Several tiers of micromeres are generally cleaved from macromeres before bilateral cleavage replaces oblique cleavage.

Tiers of micromeres and specific micromeres are frequently associated with specific larval and adult structures. In many annelids, the first three quartets of micromeres form the ectoderm; the 2d cell, or first somatoblast, gives rise to the ectoderm of the trunk, and the 4d cell, or second somatoblast, forms the mesoblast and mesodermal bands.

In the nematode, *Caenorhabditis elegans*, all the major tissue types are sorted out in MS, E, C, and D founder cells and the P_4 primordial germ cell. The analysis of mutants indicates that determination is under tight genetic control. Cells may receive particular granular material and have very little latitude to deviate from their prospective significance. Determination may not be complete, however, since some blastomeres may require induction and positional information, and cells of some lineages may utilize cellular interactions in choosing their pathway of differentiation.

Mosaic or determinant development (epitomized by spiralians and protochordates) and regulative or indeterminant development (epitomized by echinoderms and amphibians) are defined in several ways: by the specificity with which blastomeres *in situ* are traceable to particular larval or adult structures, and by the results of *in vivo* or transplantation experiments and *in vitro* or isolation experiments. A blastomere's prospective significance or fate is the tissue into which its lineage differentiates *in situ*. Self-differentiation occurs when a lineage differentiates *in vitro* according to its fate. Potency measures the range of cell types into which a blastomere's lineage can differentiate *in vitro* and *in vivo*. Competence is the range of cell types into which a blastomere's lineage differentiates under the influence of induction or intercellular interactions *in vivo*.

Determination is the first irreversible step taken toward differentiation. Cellular autonomy and a lack of competence among blastomeres are the chief characteristics of determinate development. Mechanisms operating one characteristic may integrate it with the other, or both could occur simultaneously without integration.

The results of experimentally ablating or adding blastomeres and removing or displacing cytoplasm are generally consistent with early determination in mosaic development and with late determination in regulative development. As determination increases and fate unfolds, potency and competence are restricted.

Regulation in terms of the ability of isolated blastomeres to produce entire embryos shows the absence of determination. Regulation as the integration of a transplanted part into its surroundings is indicative of continued receptivity to local influences.

Echinoderms illustrate both types of regulation. Blastomeres isolated from the 2–4-cell embryo may develop into whole larvas, and deficiencies caused by the isolation of tiers are remedied by the timely addition of other cells presumably capable of transmitting appropriate intercellular messages.

Mammalian embryos are also regulative. Sheep blastomeres isolated at the 2-cell stage are capable of developing into normal organisms. Similarly, individual blastomeres from the 4-cell stage in mice and even the 8-cell stage in rabbits are capable of developing. Mammalian blastomeres from different embryos of the same species may also be combined and form a single integrated chimeric or allophenic organism of normal size expressing genes from four or more original parents.

The prototype for the concepts of determination is the separation of somatic lines of cells from the germ line. In August Weismann's germ-plasm theory, the zygote is portrayed as a branching point in the stream of life where a mortal somatic line diverges from a potentially immortal germ line.

In contemporary theories, the germ line is not quite as isolated. It has heredity, differentiation, and vulnerability to selective pressures (e.g., eggs of rotifers). The absence of determination in the early germ line is consistent with data on the frequency of 5-methylcytosine, or DNA methylation, in early mammalian germ cells.

While in germ-plasm theories of germ-line origins, the germ and somatic lines separate early, in intermediate theories, embryonic stages contain regions (e.g., the epiblast of vertebrates) whose cells combine germ and somatic potential until later into development, and, in epigenic theories, the same cells form germ and somatic tissues under local influences (animals with diffusely distributed potential germ cells and flowering plants).

So-called germinal cytoplasm or polar cytoplasm associated with germ-cell segregation has been identified experimentally in frogs and insects as a cytoplasmic determinant of germ-cell differentiation. A fibrogranular material, sometimes associated in germinal granules and possibly related to the nuage or nuagelike material found in germ cells, may be or contain the active ingredient. The putative determinant in cleaving embryos and primordial germ cells is sensitive to ultraviolet irradiation, and animals irradiated as early embryos are sterile. Germinal granules may not be present continuously from generation to generation, as required by a Weismann-type germ plasm.

Current ideas about the localization of hereditary material in the nucleus are traceable to Weismann's idea that germ plasm resided in the nucleus and to the idea of early cytochemists of a self-perpetuating hereditary material or idioplasm. Studies on chromosomes and the discovery of chromosomal fracture and diminution in the somatic line of some nematode embryos were consistent with the separation of somatic and germ lines and with the identity of germ plasm, idioplasm, and chromatin.

The idea of separate germ and somatic lines served as a model for the Roux–Weismann hypothesis and the mosaic theory of development according to which somatic characteristics were distributed to blastomeres by cleaving nuclei. The results of Roux's hot needle experiment vindicated the major prediction of the hypothesis and suggested that early determination was a consequence of funneling hereditary factors to blastomeres through their nuclei.

Whitman, Wilson, and Conklin working at Woods Hole and others followed the intricacies of cytoplasmic segregation and mosaic cleavage in spiralians and ascidians. Their results generally supported the Roux–Weismann hypothesis but suggested a greater role for cytoplasm compared to the nucleus in sorting out determinants. Handedness in snail development showed that the geometry of the adult could be traced to cleavage angles, but genetic rather than mechanical explanations ultimately explained the control of handedness.

Regulation in echinoderms, demonstrated by Driesch, and in frogs, demonstrated by Spemann among others, showed that cleavage did not always distribute determinants. Explanations for the embryo's ability to fill in missing parts are offered by concepts of harmonious equipotential systems with fieldlike properties, positional information, and communal messages that suppress or induce blastomeric potential.

The roles of cell history and cell position in determination are studied today in the clonal analysis of cell lineage especially in *Caenorhabditis elegans*, *Drosophila melanogaster*, and *Xenopus laevis*. In general, cells seem to take some messages with them from their past while responding to other messages delivered to them from their present surroundings. Both blastomeric descent or history and a range of local and global influences seem to determine development.

PART 5 QUESTIONS FOR DISCUSSION

1. What is the significance of cleavage for embryonic development? Compare and contrast blastomeres with adult somatic and germ cells. How do replication, transcription, and translation differ in species with different forms of cleavage?

2. What is Balfour's law? What are Hertwig's mechanical principles? Describe the relationship of different types of cleavage to the concentration and distribution of yolk in eggs. How is the secondary loss of yolk thought to influence cleavage in viviparous species?

3. Compare and contrast intralecithal and partial cleavage, radial and biradial cleavage, bilateral and spiral cleavage. What is the relationship of cleavage to cellularization? How are blastomeres related to energids? Vitellophages? The yolk mass? Yolk spheres? Yolk pyramids? Pseudoblastomeres? How are furrows produced?

4. How would you name blastomeres in order to describe their nuclear histories and positions in various types of cleavage? Can you devise a system of nomenclature that corrects for compensatory divisions resulting in blastomeres with different histories occupying the same position?

5. What is the relationship between planes of cleavage furrows and planes of symmetry in embryos exhibiting different forms of total cleavage? What are the inferences of blastomeric rotation for the polarity of early mammalian blastomeres? What is embryo transfer? A pseudopregnant animal? Prenatal foster mother? What is a tetraparental animal? An allophenic or chimera animal?

6. Distinguish between macromeres and micromeres in various forms of spiral cleavage. What are the difficulties in extending the quartet system of nomenclature to spiralians with duet and monet forms of oblique cleavage?

7. Evolution seems to shape the cleaving embryo as much as it shapes the organism at every other stage of its life cycle. How has cleavage been adapted to meet the contingencies of terrestrial life in cleidoic and in viviparous vertebrates?

8. Suggest plausible evolutionary links between spiral and partial cleavage in mollusks, radial and intralecithal cleavage in arthropods, radial cleavage in echinoderms, and bilateral cleavage in vertebrates.

9. Give operational definitions of perspective significance, potency, and competence, and illustrate how they differ in so-called determinate and regulative development. Compare and contrast the evidence for determinate development in a gastropod (nematode or ascidian)

to regulative development in a sea urchin (amphibian or mammal). What is the evidence that mammalian embryos are regulative?

10. Compare and contrast Weismann's original conception of the germ plasm and current ideas about the separation of germ and somatic lineages. Describe germ-plasm, intermediate, and epigenic types of germ-cell origins. What are chromosomal fracture and diminution, and what do they imply about a germ plasm?

11. Examine current ideas on determinants. How might determinants be acquired or activated during determinate and regulative development? Describe the content and characteristics of germinal cytoplasm in insects and amphibians. Critique the evidence linking germinal granules to nuage or nuagelike material.

12. Describe the results of Roux's hot needle experiment, and compare them to those of Driesch's egg shaking experiment and Spemann's hair loop experiment. What conclusion do you draw from the differences in these results?

13. What is a founder cell? A teloblast? Do you think determination depends on a cell's history of divisions or its position in an embryo and why? Do you think determination depends on the segregation of nuclear plasma or cytoplasmic determinants?

PART 5 RECOMMENDED READING

Anderson, D. T., 1973. *Embryology and phylogeny in annelids and arthropods.* Pergamon Press, Oxford.

Campos-Ortega, J. A. and V. Hartenstein, 1985. *The embryonic development of Drosophila melanogaster.* Springer-Verlag, Berlin.

Costello, D. P. and others, 1976. Symposium on spiralian development. *Am. Zool.,* 16:277–625.

Czihak, G., 1975. *The sea urchin embryo: Biochemistry and morphogenesis.* Springer-Verlag, Berlin.

Davidson, E. H. and R. A. Firtel, eds., 1984. *Molecular biology of development.* Alan R. Liss, New York.

Giudice, G., 1986. *The sea urchin embryo: A developmental biological system.* Springer-Verlag, Berlin.

Harrison, F. W. and R. R. Cowden, eds., 1982. *Developmental biology of freshwater invertebrates.* Alan R. Liss, New York.

Horder, T. J., J. A. Witkowsky, and C. C. Whylie, 1985. *A history of embryology,* 8th Symposium of the British Society for Developmental Biology. Cambridge University Press, Cambridge.

Kumé, M. and K. Dan, 1957. *Invertebrate embryology* (translated from the Japanese, by J. C. Dan, 1968 reprinted, 1988). Garland Publishing, New York.

Maienschein, J., ed., 1986. *Defining biology: Lectures from the 1890s.* Harvard University Press, Cambridge.

Nieuwkoop, P. D., A. G. Johnen, and B. Albers, 1985. *The epigenetic nature of early chordate development: Inductive interaction and competence.* Cambridge University Press, Cambridge.

Raven, C. P., 1966. *Morphogenesis: The analysis of molluscan development.* Pergamon, Oxford.

Sawyer, R. H. and R. M. Showman, eds., 1985. *The cellular and molecular biology of invertebrate development.* University of South Carolina Press, Columbia.

Weiss, P., 1939. *Principles of development: A text in experimental embryology.* Henry Holt, New York.

BLASTULATION AND BLASTULAS

*T*he **blastula** (Gk. *blastos* sprout + diminutive ending) is the forgotten embryo of embryology (Nieuwkoop, 1973). Although recognized as a stage of ontogenesis in *Nomina embryologica*,[1] blastulas are not accorded subheading status by the Library of Congress or by the Index Medicus. **Blastulation,** the development of a blastula, is also largely ignored (even by *Nomina embryologica*), and many features of blastulation are considered parts of cleavage and gastrulation.

Part 6 rectifies this oversight and elevates the blastula to a subject for contemporary consideration. The blastula is shown to occupy a critical and in some ways a strategic position in the dynamics of development.

Blastulas bridge the gap between embryos making blastomeres and embryos making germ layers. The blastula breaks the developmental tether stretching back to the oocyte and enters the pathway of morphological, physiological, and biochemical independence leading to the gastrula.

[1]Approved by the 11th International Congress of Anatomists.

Chapter

13

THE DESCRIPTIVE ANATOMY
OF BLASTULAS

*B*lastulas typically have more cells than cleaving embryos, exhibit surface modifications (e.g., cilia and junctional complexes), and often conceal internal cavities. Beyond these criteria, blastulas are usually defined on a species-by-species basis. Embryos with a variety of shapes and forms are identified as blastulas, and a series of embryos with blastula-like characteristics but changing morphologies may be lumped in a **blastula stage** without regard to which embryo is the definitive blastula.

Some blastulas have a cavity known as the **blastocoel** (sometimes blastocoele, Gk. *blastos* + *koilos* hollow [as in cave])[1] or cleavage cavity (also von Baer's cavity). Without any prefix, "blastula" ordinarily refers to the embryo with a blastocoel (e.g., the amphibian blastula), but blastocoel-containing blastulas of invertebrates are sometimes called **coeloblastulas** (i.e., hollow blastulas, also blastospheres or true blastulas).

The process leading to the formation of the blastocoel is called **cavitation.** Although in other contexts, cavitation implies the excavation of a

hole in a solid, in the case of the blastocoel, cavitation usually involves the coalescence of separate segmentation cavities originally produced during cleavage and lying between blastomeres. In addition, the formation of the blastocoel involves polarized secretion through intracellular vesicles.

A solid blastula is a **stereoblastula** (Gk. *stereos* solid + blastula). When spherical, stereoblastulas are sometimes called **morulas** (Lat. *morum* mulberry + diminutive ending; e.g., see Fig. 11.25), and when flattened or consisting of plates of blastomeres, they are called **placulas** (Gk. *plakos* plate + diminutive ending; e.g., see Fig. 11.32). Morulas are sometimes considered preblastulas, especially when they later become hollow (e.g., mammals), but the distinction is only loosely drawn and cannot be made in cases where cavities do not appear prior to gastrulation.

BLASTULATION IN LECITHOTROPHIC EGGS

Intralecithal Blastulation

In insects, blastulation completes a **periblastula** or uniform **cellular blastoderm** (e.g., see Fig.

---

[1]The spelling "blastocele" is inappropriate, since "-cele" suggests the Greek *kele* meaning hernia or tumor rather than cavity.

11.5) consisting of a single layer of cells enclosing a central yolk. No blastocoel is found.

In *Drosophila melanogaster,* cellularization is not quickly completed (stage 5, Campos-Ortega and Hartenstein, 1985) (see Fig. 11.11). Especially in the dorsal and lateral blastoderm, cytoplasmic bridges connect blastomeres to the underlying syncytial yolk cytoplasm. These connections will not disappear until gastrulation when the **yolk membrane** or sac covers the yolk and when cell membranes completely surround the blastomeres.

Meanwhile, further adjustments occur in the periblastula. The pole cells formed earlier shift their positions dorsally followed by the dorsal shifting of the posterior blastoderm. At the same time, midventral cells darken at the point where they will soon invaginate.

At the periblastula stage, nuclei are locked into cells via cellularization, and **cellular lineages** are established. Specific cell lineage **compartments** or multicellular domains emerge in anticipation of future segments and imaginal disks (e.g., the anterior and posterior compartments of *Drosophila* wings; see below and Fig. 14.6). No cell will normally move from one compartment to another.

Cells belonging to compartments are linked by gap junctions, but cell-to-cell movement of injected fluorescent molecules is restricted at defined **communication restriction boundaries. Communication compartments** within these boundaries may coincide with cell lineage compartments, suggesting that gap-junctional communication plays a role in compartment organization (Weir and Lo, 1984). Moreover, mutant *Drosophila* with disrupted boundaries between cell-lineage compartments (e.g., *engrail* mutants) exhibit wild-type communication compartments, suggesting that networks of gap junctions may normally specify lineage compartments (Weir and Lo, 1985).

Partial Blastulation

Following partial cleavage, blastulation produces a **discoblastula**[2] consisting of a cap of cells resting on a yolk membrane. This membrane may cover yolk or a **yolk syncytium** containing nuclei and cytoplasm, the outer layer of which is sometimes called a periplasm.

In cephalopods, partial cleavage produces a flat discoblastula cap (or germ disk; also germ-

---

[2]The discoblastula is sometimes called a blastodisk or blastodisc, but these terms should be reserved for the disk of cytoplasm in the mature ovum.

disc, Fig. 13.1) ***conforming to the surface of the egg.*** In *Loligo pealei,* the fusion of furrows beneath and lateral to the blastomeres completes cellularization of the discoblastula and provides the blastocoel or **subcellular space.** Cytochalasin B inhibits the undercutting of the blastomeres as well as the formation of the blastocoel, suggesting that the formation of the blastocoel is a passive consequence of membrane spreading and cleavage (Arnold and Williams-Arnold, 1976).

Small central cells are surrounded by syncytial blastocones oriented radially especially at the posterior edge. At the seventh division (or afterward in yolk-rich eggs), the cap's edges begin to thicken into a multilaminar or multicellular ring which soon covers the blastocones (see Boletzky, 1988). The peripheral migration of ring cells marks the beginning of gastrulation.

In teleosts, discoidal cleavage produces a high discoblastula (or blastodisk). Diffuse, continuous segmentation cavities above the uncleaved yolk are collectively considered a blastocoel (Fig. 13.2). **Deep blastomeres** (DB) lie next to the yolk. They are covered by an **enveloping layer** (EVL, Fig. 13.3) of epithelial cells which is tenaciously attached at its periphery to a **yolk syncytial layer** (YSL, formerly periblast, Fig. 13.4) (Betchaku and Trinkaus, 1978). Beyond the discoblastula, the YSL merges with a nuclear-free **yolk cytoplasmic layer** (YCL, formerly yolk gel layer) (Lentz and Trinkaus, 1967) that covers the remainder of yolk sphere.

The EVL's epithelial cells are linked by tight junctions expanded circumferentially as zonula occludens between the cells' outer edges. The junctions seal off the segmentation cavities and deep cells from the outside (Fig. 13.4). Gap junctions develop beneath the zonula occludens.

Some deep cells are also linked by junctional complexes to EVL cells. These deep cells may later become part of the EVL. Cell division in the EVL is not sufficient to provide the more than 5000 cells of the gastrula's EVL, and some deep cells, it would seem, must enter the EVL (Betchaku and Trinkaus, 1978).

Other deep cells show evidence of differentiation in anticipation of germ-layer formation. The endoplasmic reticulum and Golgi apparatus develop, and nucleoli and polyribosomes appear. Differentiating deep cells form rounded pseudopods, or lobopodia, which may function in subsequent gastrulation.

In birds, partial cleavage produces a multilaminar discoblastula (see Fig. 11.18). Blastulation

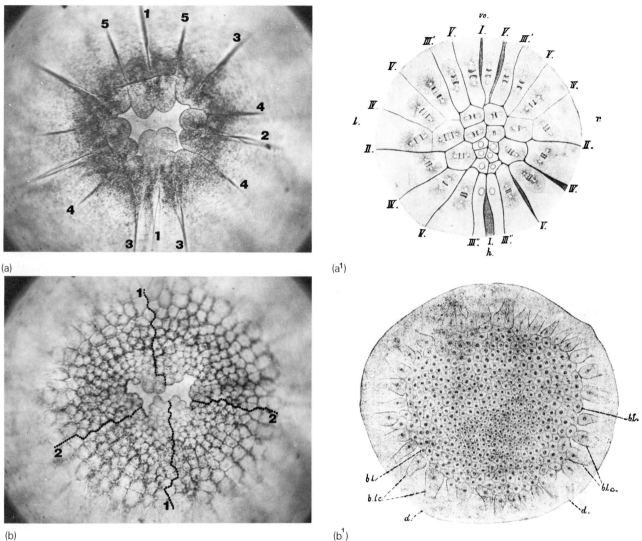

(a)

(a¹)

(b)

(b¹)

FIGURE 13.1. Recent photomicrographs and 19th century drawings of the squid, *Sepia officinalis,* in late cleavage (a, a') and as a discoblastula (b, b'). The central gap (a and b, compare to Figure 11.13) occurs in large squid eggs. Cleavage furrows are identified by numbers (1–5) and dotted lines. bl., blastomere; blc., blastocone; d., deutoplasm (yolk). (a) and (b) from S. v. Boletzky, in M. R. Clarke and E. R. Trueman, eds., *Mollusca,* Vol. 12, *Paleontology and neontology of cephalopods,* Academic Press, Orlando, 1988, p. 185. Used by permission. (a') and (b') from E. Korschelt and K. Heider, *Embryology of invertebrates* (translated from German by E. L. Mark and W. McM. Woodworth), Swan Sonnenschein, 1895.

molds the discoblastula. An outer epithelium of tightly bound cells covers a subepithelial layer of loosely bound deep cells (d.c., Fig. 13.5). The slit-like subblastodermic cavity (s.c., also subgerminal cavity)[3] separates deep cells from the yolk membrane of the subgerminal periblast except at the periphery where open blastomeres meet the peripheral periblast in a **germ wall** (Kochav et al., 1980). Peripherally, cleavage continues for 12–14 hours following ovulation and fertilization (or hours of uterine age, hr.u.a.) as the egg traverses the oviduct and uterus in its **preincubation** or intrauterine phase.

By the end of this period, the discoblastula has acquired its anterior–posterior axis. Rotating, with its pointed end forward, the egg moves obliquely

[3]The subblastodermic cavity is sometimes considered a blastocoel, but the title of blastocoel is more often assigned to cavities between deep cells or to a latent cavity between the epiblast and hypoblast of gastrulating embryos (e.g., see Balinsky, 1981).

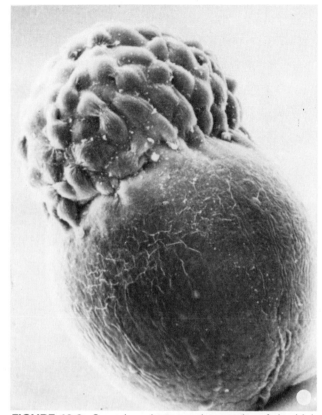

FIGURE 13.2. Scanning electron micrographs of the high discoblastula of the zebrafish, _Brachidanio rerio._ ×196. From H. W. Beams and R. G. Kessel, _Am. Sci.,_ 64:279 (1976), by permission of the Society of Sigma Xi and courtesy of the authors.

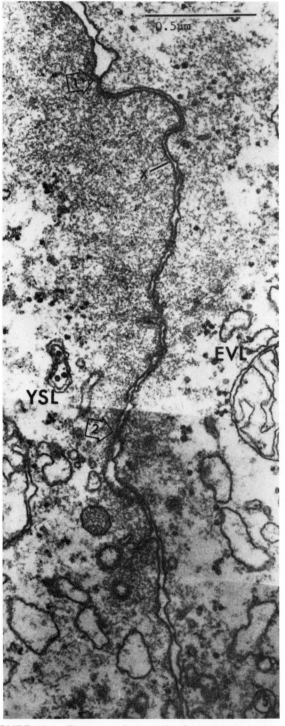

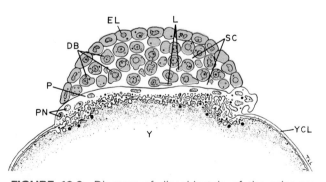

FIGURE 13.3. Diagram of discoblastula of the minnow, _Fundulus heteroclitus,_ lying on yolk (Y). The discoblastula consists of an enveloping layer (EL) enclosing deep blastomeres (DBs) mounted on a yolk syncytial layer (P) containing nuclei. Segmentation cavities (SCs) appear among DBs, and rounded pseudopods called lobopodia (l) project from them. The YSL continues as the yolk cytoplasmic layer (YCL) around the egg. From T. L. Lentz and J. P. Trinkaus, _J. Cell Biol.,_ 32:121 (1967), by permission of Rockefeller University Press and the authors.

FIGURE 13.4. Electron micrograph of section through portions of the enveloping layer (EVL) and the yolk syncytium layer (YSL) of a _Fundulus heteroclitus_ discoblastula. The cell membranes are closely apposed (distal to arrow marked by X) in what seems to be a zonula occludens. From T. Betchaku and J. P. Trinkaus, _J. Exp. Zool.,_ 206:381 (1978), by permission of Alan R. Liss, Inc., and courtesy of the authors.

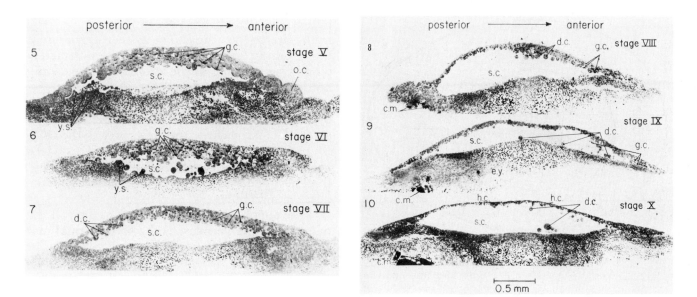

FIGURE 13.5. Photomicrographs of meridional sections through chick discoblastulas at Eyal-Giladi and Kochav (E.G&K) stages V–X, 8–20 hours uterine age (hr.u.a) corresponding to the period of area pellucida formation and enlargement. (5) Stage V (8–9 hr.u.a.). Cells become smaller as cleavage continues, but some giant open cells (oc) are still found peripherally. The subblastodermic cavity (sc) has expanded, presumably as a result of accumulating fluid. Glycogen caps (gc) and yolk platelets are prominent in cells. (6) Stage VI (10–11 hr.u.a.). The embryo is 4–5 cells thick at the center, and open cells are no longer found at its periphery. Yolk stumps (y.s.) protrude into the subblastodermic cavity from the underlying subgerminal periblast. (7) Stage VII (12–14 hr.u.a.). The discoblastula consists of large rounded cells. Glycogen deposits (g.c.) are present within anterior cells, while deep cells (d.c.) at the posterior margin have begun to detach from a subepithelial layer and fall into the subblastodermic cavity (s.c.). (8) Stage VIII (15–17 hr.u.a.). Cell division has continued, and cells in the epithelial layer are now smaller; but anteriorly, the epithelium is thinner due to the dropping away of deep cells. Glycogen deposits are reduced over the posterior half of the discoblastula. (9) Stage IX (17–19 hr.u.a.). The wall of the discoblastula is reduced for the most part to a simple epithelium of polarized cuboidal to columnar cells having nuclei at their peripheral surface. Deep cells (d.c.) continue to be seen in the subblastodermic cavity. (10) Stage X (about 20 hr.u.a. and the time of laying). The wall of the discoblastula is thin throughout its length. Occasional hypoblast cells (h.c.) appear on the inner surface of the epithelium in anticipation of gastrulation. c.m., Carbon marker used to identify the limits of the area pellucida during isolation. e.y., egg yolk. From S. Kochav, M. Ginsburg, and H. Eyal-Giladi, *Dev. Biol.,* 79:296(1980), by permission of Academic Press and courtesy of the authors.

through the oviduct and uterus. The posterior end of the embryo's axis tends to form at the uppermost rim of the discoblastula (Kochav and Eyal-Giladi, 1971).

According to **von Baer's rule,** the embryo's anterior–posterior axis normally lies across the egg with the anterior end directed away from the viewer when the pointed end of the egg is toward the viewer's right (Fig. 13.6). Gravity seems to play a role in the establishment of the axis, since inverted eggs develop a new polarity according to their orientation rather than the direction of

pointed and rounded ends (Eyal-Giladi and Fabian, 1980).

Meanwhile, subepithelial cells in the posterior margin of the discoblastula lose their glycogen deposits (g.c., Fig. 13.5 corresponding to Eyal-Giladi and Kochav [E.G&K] stage VII), and heavily yolked deep cells (d.c.) detach and sink into the subblastodermic cavity. The consequent thinning of the discoblastula renders its posterior margin translucent, and a definitive **area pellucida** (i.e., a shiny area) begins to emerge posteriorly above the subblastodermic cavity.

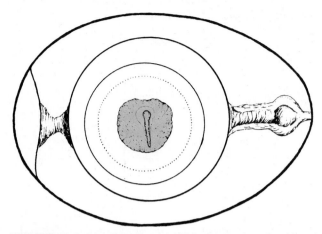

FIGURE 13.6. Von Baer's rule. With the egg's pointed end toward the right and the blunt end (with air space) toward the left, the embryo's anterior end is directed away and its anterior–posterior axis lies across the egg. From K. E. von Baer, _Über Entwickelungsgeschichte der Thiere,_ Erster Theil, Königsberg, 1828.

As cell shedding spreads anteriorly and the area pellucida enlarges, the underlying yolk surface becomes smooth. At 15–17 hr.u.a. (Fig. 13.5, E.G&K stage VIII), the area pellucida reaches the center of the discoblastula. Glycogen deposits continue to disappear, and dividing cells remaining in the roof of the discoblastula become smaller.

By 17–19 hr.u.a. (Fig. 13.5, E.G&K stage IX), deep-cell shedding reaches the anterior margin, and the discoblastula is reduced to a layer 1–3 cells thick. The total number of cells appearing in meridional sections of the discoblastula decreases by one-fifth at this stage due to the loss of the subepithelial cells. In contrast, the epithelium's outer layer actually increases in epithelial cell number.

Finally, at about 20 hr.u.a., shedding is complete (Fig. 13.5, E.G&K stage X), and the area pellucida coincides with the breadth and width of the subblastodermic cavity. Some of the detached deep cells are free in the subblastodermic cavity, but most have disappeared. The outer layer of the discoblastula is now organized as a **polarized** epithelium with nuclei preferentially distributed at the cells' outer edges. This epithelium overlies an incomplete basal lamina and becomes the **epiblast.** Small cells appearing on its undersurface spread out as **hypoblast cells** (h.c., Fig. 13.5) and form the **hypoblast.**

At the same time (although more generally assumed to be 25–26 hr.u.a.), the egg enters the **pre-primitive streak** stage. Gradually, cells pile up in the posterior half, and, as the embryo becomes an **embryonic shield,** gastrulation commences. Eggs may be laid at this time, but the moment of laying

differs considerably from egg to egg. Those arriving at the uterus (i.e., the end of the hen's oviduct) in the morning are likely to be laid immediately, while eggs arriving in the afternoon are generally not laid until the next day.[4]

The large central area pellucida is surrounded by a thick rim known as the **area opaca** (i.e., a dense area). Microscopically, the area opaca is the cellular area at the edge of the embryonic **blastoderm** beyond the subgerminal periblast. With its greater thickness and attachment to yolk, the area appears dense and granular.

The region where cells are open to the yolk and merge with the periblast is the **blastodermal margin** or marginal zone (also junctional zone or germ wall). A layer of the glycoprotein **fibronectin** (FN, also known as large external transformation-sensitive [LETS] protein and, less often, cell surface protein [CSP]) accumulates in this zone at the pre-primitive streak stage (E.G&K stage X) and in the early gastrula (i.e., H&H stage 3, Fig. 13.7).

Fibronectin also appears at the periphery of the area opaca and in particulate material and parallel fibrous bands at the germinal crescent between the anterior boundary of the area pellucida and the area opaca (Sanders, 1982). It is laid down in the embryonic basal lamina of the presumptive ectoderm in the roof over the subblastodermic cavity. Similarly, in amphibians, fibronectin lines the roof of the blastocoel (see below, Figs. 20.11, 20.12).

Fibronectin accumulation at the blastula stage anticipates some morphogenic movements (Critchley et al., 1979). In the chick, accumulations

[4]The times for the final E.G&K stage (Kochav et al., 1980) is as much as 13 hours ahead of the time for the comparable first H&H stage (Hamburger and Hamilton, 1951). The discrepancy is at least partially due to differences in estimates for the duration of the intrauterine phase.

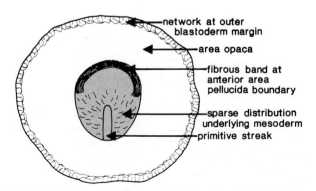

network at outer blastoderm margin
area opaca
fibrous band at anterior area pellucida boundary
sparse distribution underlying mesoderm
primitive streak

FIGURE 13.7. Diagram of early chick gastrula (H&H stage 3) showing distribution of fibronectin. From D. R. Critchley, M. A. England, J. Wakely, and R. O. Hynes, _Nature (London),_ 280:498(1979), by permission of Macmillan Journals.

at both the blastodermal margin and the germinal crescent identify sites of active cell movements. Cells at the blastodermal margin will migrate peripherally to form the vitelline endoblast and spread over the yolk sphere as the endodermal yolk sac. Primordial germ cells concentrated in the epiblast of the germinal crescent (see Fig. 12.32) will emerge and begin their journey to the gonad (Eyal-Giladi et al., 1981), while mesoderm will occupy the vacated region. Still, fibronectin is only sparsely distributed in the main part of the area pellucida where the bulk of gastrulation movements will take place.

BLASTULATION IN TOTALLY CLEAVING EGGS

Blastulation following Spherical Cleavage

Embryos produced by spherical cleavage begin to show signs of polarity as blastulas. Coeloblastulas develop conspicuously different cell types, and stereoblastulas form distinctive inner and outer cells.

Sponge embryos developing internally in hermaphroditic parents become blastulas as early as the 8–16-cell stage. The simplest blastula is a coeloblastula sometimes called an **archiblastula** (Gk. *archein* to begin + blastula, implying a primitive blastula) with a cellular wall of identical columnar cells flagellated toward the cleavage cavity or blastocoel.

Blastulas of calcareous sponges go through several stages and differentiate precociously. They form their first blastulas with few cells but go on to form more highly multicellular blastulas through their blastula stages (Fig. 13.8). For example, in *Sycandra raphanus*, 8 cells in the 16-cell embryo are micromere-like and divide rapidly while producing flagella in the direction of the blastocoel. A second blastula named a **stomoblastula** (Gk. *stoma* mouth + blastula) develops a mouth in a layer of slowly dividing macromere-like cells and ingests parental cells through the mouth (Fig. 13.8e, f; also Fig. 19.10).

A third and definitive coeloblastula, called the **amphiblastula** (Lat. *amphi-* around, signifying both sides of a thing + blastula) is produced when the embryo undergoes an **inversion** of layers by passing through its own mouth and turning itself inside out. The flagella of the micromere-like cells then face outward. Surrounded by a nutritive maternal

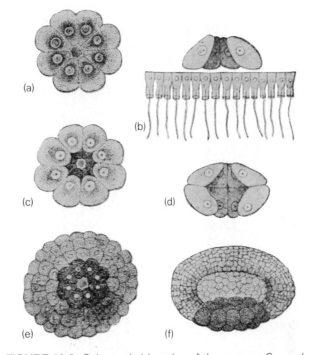

FIGURE 13.8. Schematic blastulas of the sponge *Sycandra raphanus*. Flagella are not shown on the embryo, but are indicated in B on the collared cells of adjacent radial tube in parent. (a) Surface view of 8-cell embryo showing mouth. (b) Optical section in vertical view of 8-cell embryo showing relationship to parent. (c) Surface view of 16-cell embryo. (d) Optical section in vertical view of 16-cell embryo. (e) Eight granular (macromere-like) cells everted at lower surface. (f) Side view of coeloblastula (blastosphere). Granular cells below and an epithelium of columnar cells surround a blastocoel. From E. Korschelt and K. Heider. *Embryology of invertebrates*. Swan Sonnenschein, London, 1895.

layer, the blastula's cells proliferate and elongate around an enlarging blastocoel. Ultimately, the amphiblastula works itself out of the parent and is "born."

Coelenterates also form different blastulas. Among hydrozoans, the amount of yolk accumulated by the egg and the constraints on space seem to be important variables in determining the type of blastula.

The yolk-rich eggs of irregularly cleaving tubularia-like species and some radially cleaving filamentous species form solid stereoblastulas (Fig. 13.9a). Restrained by the wall of the gonophore, the yolk-laden blastomeres seem to take up all available room, leaving none for a blastocoel.

Lacking a fertilization membrane, an intercellular cement seems to be responsible for holding blastomeres together during blastulation. The same cement may prevent blastomeric autotomy.

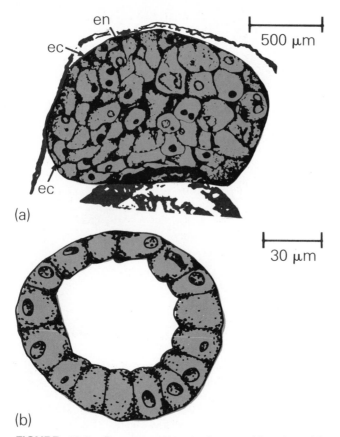

(a)

(b)

FIGURE 13.9. Representative hydrozoan blastulas. (a) Stereoblastula of *Hydractinia echinata.* The blastomeres undergo precocious differentiation as ectoderm (or ectoblast, ec) and endoderm (or endoblast, en). The superficial covering is the wall of the gonophore. Parental hydroid tissue is seen below. (b) Coeloblastula of *Coryne eximia.* The blastomeres show no sign of differentiating, although some are enlarged with yolk. Part (a) from Fennhoff, in P. Tardent and R. Tardent, eds., *Developmental and cellular biology of coelenterates,* Elsevier/North-Holland Biomedical Press, New York, 1980. Part (b) from G. van de Vyver, in P. Tardent and R. Tardent, eds., *Developmental and cellular biology of coelenterates,* Elsevier/North-Holland Biomedical Press, New York, 1980.

In contrast, the less yolky eggs of most radially cleaving species form coeloblastulas (Fig. 13.9b) (van de Vyver, 1980). Similarly, coeloblastulas are produced in species which bear eggs in free-floating medusas unrestrained by the walls of a gonophore.

The blastula of placental mammals (i.e., therian mammals) is identified with the morula following compaction and with the preimplantation blastocyst and the expanding blastocyst prior to implantation. The hollow blastocysts of marsupials and of primitive eutherian insectivores (Fig. 13.10 right) form soon after cleavage, whereas most eutherian blastulas undergo compaction at an early

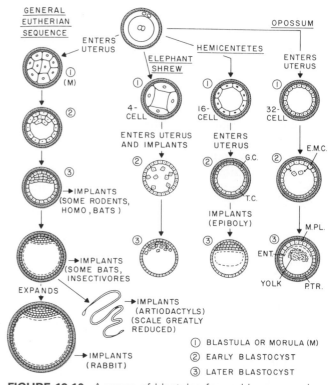

FIGURE 13.10. A range of blastulas formed by mammals from marsupials (represented by the opossum) to various eutherians including insectivores (elephant shrew and hemicentetes [i.e., the tenrec]). From W. A. Wimsatt, *Biol. Reprod.,* 12:1 (1975), by permission of the Society for the Study of Reproduction.

stage and blastomeres become sealed into a solid morula before undergoing cavitation and forming a **blastocyst cavity** (see Figs. 11.26, 13.10 left M, 13.11 stage IV, Table 13.1). The berrylike appearance of the 16–32-cell morula (see Fig. 11.25 5, 6) is caused by dividing cells becoming spherical and bulging at the embryo's surface.

During compaction, the blastomeres become polarized, or they may manifest an inherent polarity. An apical **polar cap** of microvilli overlying a meshwork of microfilaments forms at the blastomeres' outer surface between tight junctions (Figs. 13.11 and 13.12). The microvilli are nonadhesive, and blastomeres do not migrate over the polar cap of other blastomeres.

The distribution of nuclei is also polarized. They become concentrated basally, and endocytotic vesicles accumulate between the nuclei and the apical surface, suggesting that the blastomeres' basal surfaces are actively taking up or transporting intercellular fluid.

Microtubules become aligned along the cells' highly adhesive basolateral surfaces, and gap junctions (see Fig. 11.27) make their initial appearance.

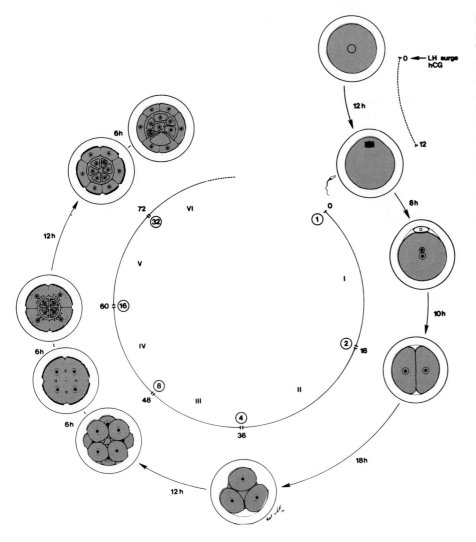

FIGURE 13.11. Scheme for the formation of the mouse blastocyst. Times between stages and elapsed time postcoitum (0–72) are given in hours. Cell numbers are circled, and stages corresponding to the number of cleavage divisions are given in Roman numerals. Thickened edges represent microvilli. From M. H. Johnson, in G. M. Edelman and J.-P. Thiery, eds., *The cell in contact: Adhesions and junctions are morphogenetic determinants,* A Neurosciences Institute Publication, Wiley, New York, 1985. Used by permission.

In addition, sibling blastomeres may be connected by an **intercellular bridge** (Fig. 13.13). The polarization of these blastomeres may lay the foundation for further events (Johnson and Pratt, 1983).

At the fourth cleavage in the mouse, on the average, one cleavage furrow cuts perpendicularly to the surface, while the other seven cut more or less parallel to the surface (i.e., periclinally or para- tangentially, Fig. 13.11 stage V). Whether spindle orientation is achieved randomly or not, it results in the separation of 9 larger **outer** (also known as apical, outside, external, or polar) cells and 7 smaller **inner** (also known as basal, deep, internal, inside, or apolar) cells. **Partially internal** cells that are neither totally inner nor totally outer are produced by diagonal furrows.

TABLE 13.1. Stages, timing, and cell number in the mouse blastula

Stage		Beginning	
Descriptive Name	Number	Days Postcoitum	Cell Number
Compaction	IV	2	8
Morula (inside and outside cells)	V	2.5	16
Late morula or early blastocyst	VI	3	32
Midstage blastocyst		3.5	40
Unilaminar blastocyst	VII	4	64
Late blastocyst (expanding)	VIII	4.5	about 100

Source: Data from Johnson and Pratt (1983) and Johnson (1985).

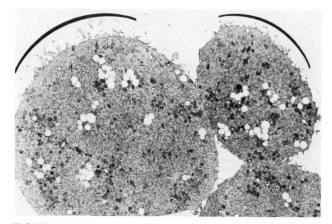

FIGURE 13.12. Electron micrograph of section of early 8-cell mouse morula showing portions of three blastomeres forming microvilli at apical surfaces (beneath brackets). ×6,700. From T. Ducibella, in M. H. Johnson, ed., *Development in mammals,* Vol. 1, Elsevier/North-Holland Publishing, Amsterdam, 1977. Used by permission.

In addition to the orientation of spindles, the choice of cells for inside and outside positions in the morula may be influenced by (or reflect differences in) the rates of division among blastomeres. In the pig embryo, beginning with the 2-cell stage, slowly dividing and larger so-called primitive blastomeres give rise to more actively dividing, smaller cells which soon cover the larger cells (Heuser and Streeter, 1928).

In the pig, the failure of embryos to produce at least one internally located cell at compaction may account for the formation of blastocyst vesicles without embryos. The total cell number of such vesicles is within the normal range, indicating that cell division is not retarded (Papaioannou and Ebert, 1988). The high mortality of pig embryos (22–41%) (Perry and Rowlands, 1962) may, in part, be attributed to the absence of internal cells.

The inner cells of mouse embryos lack polar caps. The cells have highly adhesive surfaces that flatten against each other, forming a compact cell cluster.

Some cell movement over the surface, or **epiboly,** may occur (Soltynska, 1982), but the outer cells, adhesive only on their basolateral surfaces below their polar caps, do not move to the inside. Instead, they flatten over and envelop the inner cells, while focal tight junctions expand circumferentially into zonula occludens and junctional complexes (see Fig. 11.26).

In the mouse, the ratio of approximately 9:7 outer to inner cells first established in the 16-cell morula continues more or less in the 32-cell late morula or early blastocyst (18–20 outer cells to

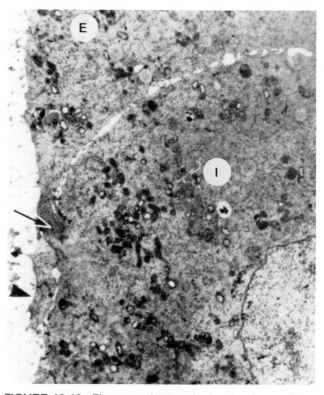

FIGURE 13.13. Electron micrograph showing intercellular bridge (arrow) between sibling external (E) and partially internal (I) cell. A fingerlike projection of another cell is seen at arrowhead. ×3000. From M. S. Soltynska, *J. Embryol. Exp. Morphol.,* 68:137(1982), by permission of the Company of Biologists.

12–14 inner cells) and as late as the **midstage blastocyst** of 39 cells. Sibling cells produced at the fifth division (Table 13.1, stage VI) may also divide between inside and outside positions, but beginning with the sixth division, sibling cells are confined entirely to one or the other position, and the populations of outer and inner cells become **clonally separate** (Cruz and Pedersen, 1985).

The rate of cell division increases and peaks at the early blastocyst stage. In the pig, the total cell number doubles four and five times between days 5 and 8. At the late blastocyst stage, the growth of internal cells slows down compared to the growth of external cells and, in mouse embryos, cell death may be found among internal cells.

Cavitation and the formation of the blastocyst cavity occur at different times among eutherian species. The cavity emerges at the fourth cleavage in the pig and approximately one division later in the mouse. At this time, the mouse embryo is identified as a **late morula** or **early blastocyst** (Table 13.1, stage VI) of 20–32 cells. Human morulas and those of most other eutherians undergo cavitation later still, but, in the rabbit, the blastocyst cavity does not appear until about 200 cells have accumulated.

The secretion of extracellular matrix glycoproteins accompanies cavitation. Laminin, the ubiquitous protein of basal lamina, is the first glycoprotein to appear. It is found at the late 8-cell to early 16-cell stage in the mouse embryo and accumulates thereafter.

The blastocyst cavity is most nearly homologous to the avian subblastodermic cavity, since the inner cells above the blastocyst cavity form the mammalian embryo much like the blastoderm above the subblastodermic cavity forms the avian embryo. Still, the two cavities do not occupy precisely the same positions. The initial subblastodermic cavity (Figs. 13.5 and 13.14) lies between the yolk membrane and the subepithelial deep cells of the discoblastula, while the inner cells of the mammalian embryo are initially separated

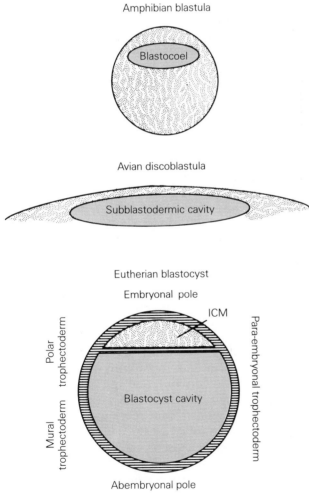

FIGURE 13.14. Schematic sections through amphibian blastula, avian discoblastula, and mammalian blastocyst showing the relationship of cavities (colored) to presumptive germ-layer cells (stippling) and extraembryonic trophectoderm (horizontal hatching).

from the blastocyst cavity by outer cell processes (Johnson and Pratt, 1983).

Alternatively, the blastocyst cavity is considered homologous to the blastocoel even though the walls of the cavity do not seem to be homologous with those of the archetypic vertebrate blastocoel (e.g., of amphibians, Fig. 13.14). Actually, the lining of the blastocyst cavity shifts in the course of development from one germ layer to another. Unlike the walls surrounding an amphibian blastocoel, the blastocyst cavity does not lie above the embryo's presumptive yolky endoderm; it is not walled by the marginal zone, and it is not roofed by presumptive ectoderm (see Chapter 19).

Finally, the cavity is sometimes portrayed as a hole representing the absent yolk, especially since the cavity will be lined by a layer of primitive endoderm during gastrulation and renamed the yolk sac (albeit sans yolk). Initially, however, the walls of the cavity do not resemble anything remotely associated with a vertebrate yolk sac. The early blastocyst cavity is lined by outer blastocyst cells and their cellular processes, not presumptive endoderm.

A **trophectoderm** (TE, referring to a [troph] nutritive [ecto] outer [derm] skin) or **trophoblast** (referring to the blast or "sprout" of the nutritional layer or placenta)[5] of epithelial cells and their processes initially lines the entire blastocyst cavity (Fig. 13.11, stage VI). Many desmosomes, presumably functioning as "spot welds," are added to the terminal junctional complexes between TE cells at the onset of cavitation (see Fig. 11.26). The junctional complexes already present might not be able to withstand the increased stress generated by hydrostatic pressure in the cavity of the blastocyst without the additional desmosomes.

In typical eutherians, an eccentric **inner cell mass** (ICM, also known as the embryoblast and germ disk) is distinguished from the trophectoderm (Fig. 13.10). In metatherians, no ICM forms, and in eutherian insectivores, the formation of the ICM is delayed (Fig. 13.10 right).

The portion of the blastocyst containing the ICM is the **embryonic pole** (Fig. 13.14), while the opposite portion is the **abembryonic pole** (Lat. *ab-* away from). Proliferative activity is concentrated

[5]Currently, "trophectoderm" is preferred by researchers using mouse and primate embryos, while "trophoblast" is more commonly used by descriptive embryologists discussing pig or other ungulate embryos. Historically, the term "trophoblast" referred to a solid mass of cells rather than the outer layer of a hollow blastocyst (Streeter, 1938). The term "trophectoderm" avoids any confusion that may result from earlier interpretations of "trophoblast."

at the embryonic pole. The blastocyst's TE is divisible into the **polar TE** at the embryonic pole and the remaining **mural TE** (Lat. wall).

The ICM does not shift its position in the blastocyst, although the polar TE expands. Gap junctions and small adhering junctions between the ICM and the TE integrate and fix their relationship prior to implantation. Division in the polar TE produces cells that move into the mural TE, but most of the TE's later expansion is due to cell flattening. Division in the ICM produces cells of increasingly homogeneous size and content, and, although they are generally not arranged as a single layer, they are referred to as the **unilaminar embryonic disk** (also blastodisk or blastoderm).[6] The entire blastocyst is called a **unilaminar blastocyst** at this time (about 4 days in the mouse).

At this time, the blastocysts have entered the uterus. The rodent and primate blastocysts then **hatch** from the zona pellucida. In the mouse, a trypsinlike enzyme, strypsin, produced on the outer surface of mural trophectodermal cells digests a hole in the zona pellucida.

At the end of the preimplantation phase in rodents and most primates, TE cells stimulate proliferation and cause a **decidua swelling** in the adjacent uterine wall. Secreting plasminogen activator (another trypsinlike enzyme), the trophectoderm also begins digesting the uterine wall in anticipation of implantation.

The mural TE cells of rodents no longer divide after the **late blastocyst stage** (beginning about 4.5 days in the mouse). Nuclei undergo DNA endoreduplication and become polyploid. If these cells are placed in contact with a 3.5-day ICM, they can still give rise to proliferative polar TE (Gardner et al., 1973). Afterward, the TE cells are irreversibly committed to mural TE differentiation.

Prior to implantation, the blastocyst expands conspicuously. Sodium is pumped into the blastocyst cavity by the action of sodium pumps in trophectodermal cells and water follows. In mice, TE cells flatten around the blastocyst cavity between the sixth and seventh divisions, and the TE cellular processes that isolated the ICM from the blastocyst cavity are withdrawn or disappear at the seventh division cycle.

The swelling of **expanding blastocysts** may be spectacular. In the rabbit, for example, the blastocyst (although still enclosed within a thinned zona) swells from a volume of 2 nl on the third to fourth

days postcoitum (p.c.) to about 2.5 ml on the 10th day p.c. (see Fig. 6.19). The blastocyst later reaches a diameter of 3–4 mm in an embryo containing several thousand cells.

In ungulates, implantation is delayed, and **polarized expansion** may produce an enormous stringlike blastocyst. In sheep and cattle, blastocysts reach a length of 20 cm before attaching to the wall of the uterus sometime between the second and third weeks, but pigs take the prize. Elongating 300-fold between the 9th and 16th days, the blastocysts may exceed 1 meter in length (McLaren, 1982) (see Fig. 6.20).

Expansion depends on the ability of the trophectoderm to move fluid from outside to inside the embryo without producing major leakages. The presence of as many as six parallel zonula tight junctions in the trophectoderm seems to render it a **tight epithelium** (see Fig. 11.27a, b), and the release of cytoplasmic vesicles into the cavity of the blastocyst provides fluid for the cavity, but since both zonula tight junctions and cytoplasmic vesicles are present before cavitation (i.e., in the morula), they alone cannot explain the expansion of the blastocyst cavity. Possibly, the necessary cause of expansion is the opening of channels of ion transport (see Ducibella, 1977).

The embryo may still be called a blastocyst at the end of the preimplantation phase, but the unilaminar embryonic disk and the TE have moved to new plateaus. In rodents and primates, the unilaminar blastodisk will soon become a **bilaminar blastodisk,** and the trophectoderm will differentiate into a syncytial penetrating portion and a cellular proliferating portion.

Blastulas of Radially Cleaving Species

Crustaceans with yolky eggs and radial cleavage produce blastulas resembling those produced following intralecithal cleavage (e.g., see Fig. 11.28), although cell migration rather than cellularization is responsible for covering the yolk mass. Cleavage in less yolky crustacean eggs produces a coeloblastula with a single cell layer (e.g., see Fig. 11.29). Cellular migration in these embryos may not be evident until gastrulation (see Anderson, 1973).

The initial blastula of echinoderms is considered a morula (Fig. 13.15), although a blastocoel or segmentation cavity is present. The definitive blas-

---

[6]The title of inner cell mass or ICM is sometimes retained and applied to any disk lacking separate germ layers.

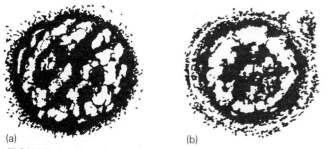

(a)   (b)

FIGURE 13.15. Photomicrograph of morulas of the sea urchin, *Clypeaster japonicus*. (a) During cell division, blastomeres round up and bulge from the surface. (b) Between divisions, the relaxed cells spread out relatively smoothly against the hyaline layer and each other. From K. Dan and T. Ono, *Biol. Bull.*, 102:50 (1952), by permission of the Marine Biological Laboratory and the authors.

tula is reached when cilia appear on its surface (Figs. 11.31g and 13.16a).

The mechanism of blastocoel formation involves the hyaline layer (see Fig. 8.17) and segmentation cavities formed between blastomeres during cleavage. Prior to each cleavage division, protein is secreted into the subhyaline space. Water then moves through the fertilization envelope, swelling the subhyaline space. Following each division, this fluid is transferred to the segmentation cavity (Fig. 13.17). Because the blastomeres are bound peripherally to the hyaline layer via **attachment fibers** of microvilli, the segmentation cavities dilate centrally (Fig. 13.18). As cleavage proceeds, blastomeres continue to be held in place peripherally; fluid continues to move centrally, and the blastocoel emerges from the coalescence of the distended segmentation cavities (Dan and Ono, 1952).

The fluid of the segmentation cavities resembles that in the blastocoel (Dan, 1952), but at about 5 hours after fertilization, an extracellular **fuzzy layer** about 0.1 μm thick lines the blastocoel. The layer, secreted by the blastomeres, is composed of acid mucopolysaccharides with their sulfate groups masked by additional proteins.

Small spaces remain between blastomeres, and filopods, or fine extensions of the cells' surfaces, move over adjacent cell surfaces (Fig. 13.16b). Increasing in their activity after the 32-cell stage, filopods from one cell may attach to another cell and couple electrophysiologically through gap junctions.

Before each division, the blastomeres become spherical, and the surface of the embryo becomes globular (hence the embryos are considered morulas, Fig. 13.15). Early synchronous divisions are re-

(a)

(b)

FIGURE 13.16. Scanning electron micrographs of fractured *Lytechinus variegatus* blastulas at about an hour prior to hatching. (a) Blastula at 7 hours after fertilization (cultured at 25°C). Cilia are present on the outer surface. Rounded cell ends face the blastocoel. (b) Higher magnification of blastocoel in slightly older blastula. ×2500. Spreading inner surfaces and fine pseudopods (i.e., filopodia) smooth out the blastocoel lining. From J. B. Morrill and L. L. Santos, in R. H. Sawyer and R. M. Showman, eds., *Cellular and molecular biology of invertebrate development*, University of South Carolina Press, Columbia, 1985. Used by permission.

placed by waves of cell division spreading upward from the vegetal pole in the early blastula and by a vegetal to animal gradient of mitotic activity in the late blastula (Agrell, 1960). As blastomeres become smaller during cleavage, the wall of the blastula becomes thinner and the blastocoel larger.

The micromeres divide into 4 **larger micromeres** and a ring of **smaller micromeres**. At the 64-cell stage, when each animal and vegetal tier (an_1, an_2, veg_1, and veg_2) is represented, the morula's

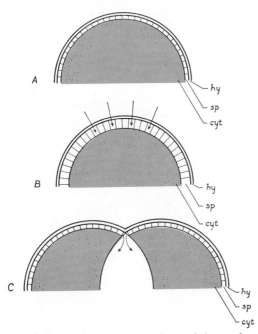

FIGURE 13.17. Schematic illustrations of the surface of a sea urchin embryo showing hypothetical mechanism of segmentation cavity and blastocoel fomation. (a) Following fertilization and during amphimixis only a small fluid-filled space (sp) is present between the cytoplasm (cyt) and the hyaline layer (hy). (b) At 3 minutes prior to cleavage, fluid from the environment swells the space. (c) After cleavage the same fluid moves into the segmentation cavity between the blastomeres. cyt, Cytoplasm; hy, hyaline layer; sp, fluid-filled subhyaline space. From K. Dan and T. Ono, *Biol. Bull.,* 102:50 (1952), by permission of the Marine Biological Laboratory and the authors.

blastomeres become linked and transformed into an embryonic tissue, at least for a brief period prior to hatching (Dan, 1952). Soon (e.g., at about the 10th cleavage in *Mespilia*), septate desmosomes seal the periphery (Wiley, 1979), and each blastomere acquires an external cilium (Fig. 13.16a). Internally, the rounded inner surfaces of the cells stretch and flatten, like shingles on a roof, smoothing the lining of the blastocoel from the animal pole downward (Fig. 13.16b).

The blastula's wall now secretes the **hatching enzyme** apically, and the cilia poke through the hyaline layer into the perivitelline space. With impulses between cells coordinating ciliary beating, the embryo rotates within the fertilization envelope. As tighter junctions form between cells, the blastula's cellular wall becomes impermeable, and the blastocoel's fluid becomes trapped.

Finally (at about 8 hours after fertilization in *Lytechinus variegatus*), the hatching enzyme lyses the fertilization membrane, and the blastula hatches. Known as the **premesenchyme blastula** in anticipation of developments which climax the next series of changes, the larva acquires few cells and undergoes very little growth. With 1000–2000 cells and a concerted ciliary beat, the larva moves unidirectionally with its animal pole forward.

At first, little cellular rearrangement or movement occurs over most of the surface, but the swimming blastula elongates in the animal–vegetal axis, and an apical tuft of elongated cilia is produced by cells descending from the original an_1 tier (Fig. 11.31h).

The remaining vegetal cells are incorporated in a pattern called the **Okazaki pattern** (after K. Okazaki). These cells become increasingly columnar while radiating from a point on the blastula's wall corresponding to the junction of the original an_2 and veg_1 blastomeres. The blastula is pear or ovid shaped, and the blastocoel acquires a funnel-shaped outline (Fig. 13.19a) (see Morrill and Santos, 1985).

At about the same time, a thin basal lamina begins to line the inner wall of the blastula except at the vegetal pole (Fig. 13.19b). In *Lytechinus*, about 30 descendants of the four larger micromeres separate from the hyaline layer and from the blastula wall. In the course of about 1 hour, these cells enter the blastocoel. The process is called **ingression,** and the cells become the **primary mesenchymal cells** (PMCs) of the **mesenchymal blastula.**

The smaller micromeres of the original micromeric group and the tall columnar cells of the veg_2 lineage replace the larger micromeres in the blastula's wall (see Fig. 11.31i). In other sea urchins (e.g., *Cidaris*), the formation of PMCs is delayed until gastrulation.

Typically, in the hour following their appearance, the PMCs migrate toward the animal pole along the wall of the blastocoel. The columnar cells of the blastula's wall shorten and the vegetal

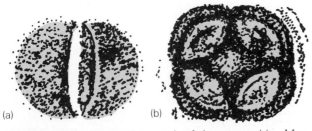

FIGURE 13.18. Photomicrograph of the sea urchin, *Mesphilia globulus,* embryo at the 2-cell stage (a) and 4-cell stage (b) showing swollen segmentation cavities. From K. Dan and T. Ono, *Biol. Bull.,* 102:58 (1952), by permission of the Marine Biological Laboratory and the authors.

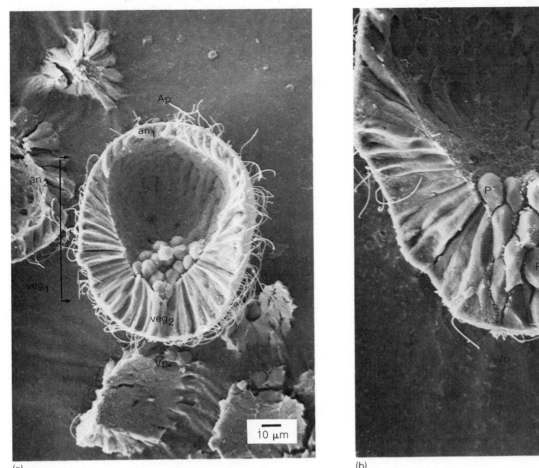

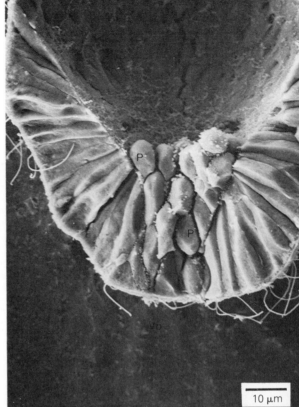

(a)   (b)

FIGURE 13.19. Scanning electron micrographs of fractured *Lytechinus variegatus* blastulas during the separation of primary mesenchyme cells (p) from the blastula wall and their release into the blastocoel. (a) As primary mesenchyme cells (p) enter the blastocoel, they are replaced at the vegetal pole (Vp) by small micromeres and cells of the veg_2 tier. Cells of the veg_1 and an_2 tiers expand the surface. The an_1 tier occupies the apical end (Ap) of the embryo and forms the apical tuft of elongated cilia. (b) Primary mesenchyme cells separating from the vegetal pole of the blastula. From J. B. Morrill and L. L. Santos, in R. H. Sawyer and R. M. Showman, eds., *Cellular and molecular biology of invertebrate development,* University of South Carolina Press, Columbia, 1985. Used by permission.

pole forms the flattened **vegetal plate** (also known as the gastral or endodermal plate). The mesenchymal blastula is now poised for gastrulation.

Blastulas of Bilaterally Cleaving Protochordates and Vertebrates

Bilateral cleavage tends to produce coeloblastulas. The size of the blastocoel is inversely proportional to the amount of yolk, while the number of cells spanning the thickness of the blastula's wall is correlated with the distribution of yolk. The yolk-poor isolecithal eggs of Amphioxus (*Branchiostoma lanceolatum*, Fig. 13.20), for example,

produce a large central blastocoel surrounded by a simple cellular layer, while the yolk-rich telolecithal eggs of anurans produce a relatively smaller acentric blastocoel surrounded by a thickened multicellular layer (see Fig. 10.3). The thickness of the wall reflects the yolk content of the blastomeres. At its thinnest point, in the animal pole, the roof over the blastocoel is two (salamanders) to four (anurans) cells thick, while at its thickest point in the vegetal pole, the floor under the blastocoel is a dozen or more cells thick.

In the ribbed newt, *Pleurodeles waltlii*, fibronectin (FN) is detected in the early blastula. As a component of a dense fibrillar extracellular matrix,

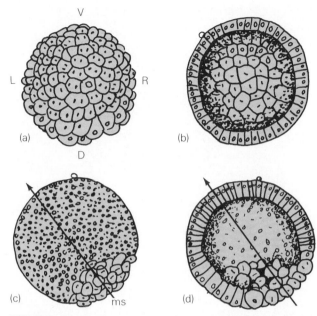

FIGURE 13.20. Drawings of Amphioxus, *Branchiostoma lanceolatum,* blastulas between 4 and 5.5 hours after fertilization. (a) View of anterior surface at 4 hours. About 256 cells are present. Presumptive endoderm cells at the lower margin are in interphase. Presumptive ectodermal cells are dividing. (b) Optical section from animal pole at 5 hours. Presumptive endoderm cells are dividing. (c) View of right side at 5.5 hours. Small cells of the mesodermal crescent (ms) are dividing at the future posterior pole of embryo. Large presumptive endodermal cells are in interphase. (d) Same in optical section (i.e., focused at midsagittal plane). Dividing cells bulge into the blastocoel. v, Ventral; D, dorsal; L, left; R, right; ms, mesodermal crescent; arrow, anterior posterior axis. From E. G. Conklin, *J. Morphol.,* 54:69(1932), by permission of Alan R. Liss, Inc.

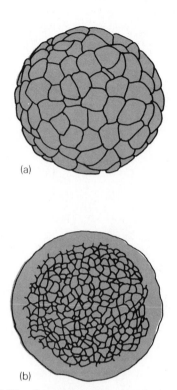

FIGURE 13.21. Camera lucida drawings of the animal hemisphere surface of axolotl, *Ambystoma mexicanum,* blastulas showing reduction in cell size accompanying blastulation. (a) Ten minutes before the onset of the eighth cleavage (refined Harrison stage 7). (b) Ten minutes before the onset of the 12th cleavage (refined Harrison stage 8¼). From K. Hara and E. C. Boterenbrood, *W. Roux' Arch. Dev. Biol.,* 181:89 (1977), by permission of Springer-Verlag and the author.

fibronectin is part of the embryonic basal lamina underlying the roof of the blastocoel (i.e., the inner presumptive ectoderm surface). Absent from the outer surface of the blastula, amphibian fibronectin, like avian fibronectin, seems to be an extracellular material or ECM involved in cell migration during gastrulation (Boucaut et al., 1984).

Divisions continue during the blastula stage, and cells become smaller (Fig. 13.21), but when the blastocoel reaches its maximal size, the blastula enters its **midblastula transition** (MBT, also blastular transition) (Gerhart, 1980; Newport and Kirschner, 1982). Occurring at the 11th division in *Ambystoma mexicanum* (corresponding to Harrison stage 8) and at the 12th division in *Xenopus laevis,* the MBT is marked by the **desynchronization** and slowing of division (see Table 10.2). In *X. laevis,* the duration of the average cell cycle decreases

from 35 minutes between cleavages 2 and 11 to 70 minutes between cleavages 12 and 14, and the G-1 and G-2 periods appear for the first time.

The starkness of the MBT suggests an abrupt transition to the gastrula, but other activities in the late blastula change gradually into activities in the gastrula. For example, prior to gastrulation, occasional vegetal blastomeres contract at their external surface and migrate inward. The contraction is responsible for the deposition of pigmented spots in the otherwise creamy vegetal hemisphere, and the migration is responsible for the convex bulge in the floor of the blastocoel.

In *Ambystoma punctatum,* the pigmented spots are linked in a reticulum, but in the blastula of *Ambystoma mexicanum,* the pigmented spots coalesce into a crescent. The reticulum seems to have no bearing on the site of any future events, but the crescent marks the site of the future blastopore (i.e., the point at which the embryo's surface invaginates at the beginning of gastrulation).

Blastulas of Spirally Cleaving Species

The title of blastula is often conferred on cleaving spiralian embryos whose outer configuration is still that of the egg. Different varieties of blastulas are identified on the basis of internal structure.

Freshwater and brackish-water snails form placulas with animal and vegetal cells in juxtaposition (see Fig. 12.9b, *Lymnaea stagnalis*), and marine gastropods (e.g., *Crepidula* and *Ilyanassa*) form stereoblastulas with yolk-laden macromeres surrounded by yolk-poor micromeres. Stereoblastulas are also formed by annelids and leeches having unequal cleavage and forming large nondividing yolky endodermal precursor cells (e.g., the leech *Helobdella* and the polychaete *Neanthes*, Fig. 13.22b). Small eggs with nearly equal cleavage and an even distribution of yolk among the blastomeres may have segmentation cavities and small

blastocoels (e.g., the tube dwelling polychaete *Eupomatus uncinatus*, Fig. 13.22a), and large eggs with unequal cleavage but with a relatively even distribution of yolk may have conspicuous blastocoels (e.g., *Scoloplos armiger*, Fig. 13.22c).

A blastocoel may form by the coalescence of lenticular cleavage cavities (not unlike those in echinoderms, Fig. 13.18). No significant cellular rearrangement occurs, and the blastocoels may function primarily in embryonic osmoregulation.

In land and freshwater varieties of snails, conspicuous segmentation cavities appear between early blastomeres. Blastomeres take up albumin from the surrounding albuminous capsule and transport vacuoles with capsular fluid to the cavities. As osmotic pressure is raised, the cavities swell, but the animal pole soon ruptures and discharges the blastocoel's contents. These **recurrent cleavage cavities** thereby pump fluid out of the blastula and may do so several times between divisions (see Verdonk and van den Biggelaar, 1983).

Nematodes also form a variety of blastulas. *Parascaris equorum* (= *Ascaris megalocephala*)

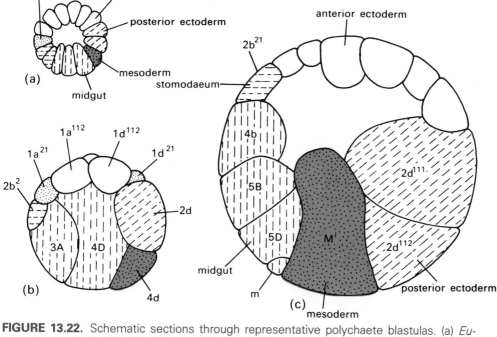

FIGURE 13.22. Schematic sections through representative polychaete blastulas. (a) *Eupomatus uncinatus* blastula. The small blastocoel is surrounded by a single layer of nearly equal blastomeres. (b) *Neanthes succinea* (= *Nereis limbata*) blastula. The interior is totally occupied by the 4D macromere (precursor of the midgut). (c) *Scoloplos armiger* blastula. The conspicuous blastocoel is surrounded by a single layer of blastomeres of different sizes. From D. T. Anderson, *Embryology and phylogeny in annelids and arthropods*, Pergamon Press, Oxford, 1973. Courtesy of D. T. Anderson.

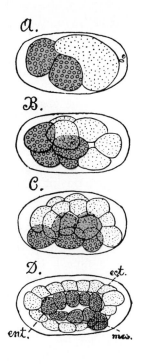

FIGURE 13.23. Schematic drawings of the nematode, _Rhabditis nigrovenosa,_ embryo. (a) The 3-cell cleaving embryo. (b) The 9-cell cleaving embryo. (c) The 16-cell blastula. A layer of presumptive ectodermal cells (stippled) overlies a layer of presumptive endodermal cells (small circles). (d) Gastrula. Cellular rearrangement and shifting have brought the presumptive endoderm (_ent._) inside while leaving the presumptive ectoderm (_ect._) and presumptive mesoderm (_mes._) outside. From E. Korschelt and K. Heider, _Embryology of invertebrates_ (translated from German by E. L. Mark and W. McM. Woodworth), Swan Sonnenschein, 1895.

has a coeloblastula with a small but definitive blastocoel. _Rhabditis nigrovenosa_ has a stereoblastula (Fig. 13.23), while _Cucullanus elegans_ forms a plate of two cell layers (i.e., a bilaminar blastula). In _Caenorhabditis elegans_, a stereoblastula is present briefly during the interval following founder cell formation and the beginning of gastrulation about 100 minutes after fertilization (Fig. 11.54h).

Variety in the morphology of blastulas is not paralleled by variation in the pattern of spiralian determination. The fate of spiralian blastomeres, it seems, is in the hands of factors that work equally rigidly in blastulas of different forms.

Chapter

14

PATTERN FORMATION IN THE BLASTULA

[In] a rich Persian carpet, the beautiful figures and patterns ... can be shown only by spreading and extending it out; when it is contracted and folded up, they are obscure and lost.

PLUTARCH, *Lives:* Themistocles

*I*n the first half of the 20th century, a school of embryologists led by Ross G. Harrison (1870–1959) argued that pattern formation was as fundamental to development as differentiation (see Harrison, 1969). Another school led by Thomas Hunt Morgan (1866–1945) maintained that the idea of pattern formation was a metaphenomenon tacked onto the phenomenon of differentiation and not independently decipherable (see Morgan, 1927).

Recently, the weight of opinion shifted toward the first school (see Malacinski, 1984). The turning point came when Louis Wolpert (1969) rephrased older views of pattern in terms of **positional information.** While patterns were not necessarily consequences of positional information, positional information provided a way of talking about patterns, of simulating or modeling them, and formulating quantitative hypotheses about their development.

A biological **pattern** is the coherent structure or design of an organism or its parts. **Pattern formation** is the development of spatial organization. Embryos develop patterns when structural varia-

bles that can have different values within a set of possible values take on particular values.

The examination of all sorts of patterns has given rise to several concepts. For example, blastomeres sometimes seem to combine labile tendencies with actual and kinetic properties. Lability, called **dynamic determination** or **commitment,** implies weak, reversible, or unstable change consistent with further regulation. In contrast, kinetic activity, or **material determination,** implies strong, irreversible, or stable change leading to self-differentiation.

Dynamic determination does not mean that something real does not happen or that some change is not actually taking place within the cell. Modern techniques, including indirect immunofluorescence and genetic probes, detect minute molecular changes indicative of even more minute changes, any of which may represent dynamic determination. Material determination may represent only the tip of a vast iceberg of dynamic determination awaiting embryological exploration.

415

MAPS AND LINEAGES

Terms and Definitions, Concepts and Conflicts

Some patterns are already apparent in the blastula (e.g., columns and rows of cilia), but the reality of other patterns is largely inferred from later development. The pattern of areas with different fates, for example, is plotted as a **fate map** from the results of *in situ* studies on the movements of markers in embryos.

Fate maps show what will become of embryonic regions in the course of normal development. Historically, **marker experiments** are employed for plotting fate maps. Regions in embryos are marked with vital dyes or other labels and the marked regions are found by dissecting later embryos. The marked germ layers of the gastrula or organ rudiments of the neurula, for example, are projected back onto the original position of the marked regions in the early embryo and identified there as **presumptive germ layers** or **organ rudiments.**

Like similarities among cleavage patterns, similarities among fate maps may be attributed to common ancestry. Possible homologies among fate maps are sometimes striking, but fate maps also show signs of evolutionary divergence. Major differences occur especially in perspective **cenogenic** (Gk. *zan-* recent + root of *gignesthai* to be born, hence of recent origin) structures. These structures perform temporary functions during periods of gestation or incubation and tend to appear early in development.

Fate maps are also drawn for regions in later embryos and even for parts of larvas. Late fate maps are usually fine grained, containing small areas of prospective significance and corresponding to narrow cellular lineages.

Do fate maps, like latent photographic images, represent an invisible **prepattern** or **organizational field**? Does development merely bring out what already preexists? If so, how is the latent pattern embedded in the substance and operation of the blastula, and how does the pattern become manifest in structure and organization? At the cellular level of organization, how are the properties of blastomeres combined and integrated into properties of the blastula?

Other properties of embryonic regions and blastomeres, in addition to fate, may also be plotted. Patterns of self-differentiating areas, or **determination maps,** may be plotted from results with explant *in vitro,* and maps of competencies and potencies may be inferred from the results of **deletion experiments,** in which cells or groups of cells are removed from embryos, and from the ability of transplants to differentiate *in vivo* but at ectopic sites.

Patterns of cellular clones with different histories or **cellular lineages** are traced by a variety of methods. Work on small, transparent embryos may be accomplished with the aid of **Nomarski differential interference contrast** (DIC) microscopy. Lineages in larger or opaque embryos may be traced with the aid of **intracellular markers,** such as horseradish peroxidase (HRP) or fluorescein dextran complexes, and by pleiotropic **genetic markers.** In addition, labeled cells may be grafted **orthotopically** from a labeled donor embryo to an identical place in an unlabeled host embryo and traced through later development.

In general, maps are concerned with regions, while lineages are concerned with cells. The two may correspond in small embryos where cleavage may separate regions into blastomeres, but in larger embryos, regions are generally multicellular and blastomeres contribute to different regions.

In practice, the difference between maps and lineages can be great. Maps are generally made for the late blastula or early gastrula so that markers can be traced into the later gastrula or neurula before they become too diffuse or diluted to be found. Cell lineage is often traced from earlier stages of development when distinct blastomeres are recognizable even in embryos whose cleavage pattern tends to become obscure at later stages. As a consequence, maps are constructed for embryos with multicellular surface and subsurface layers, while lineages are traced for blastomeres that extend from the surface to the blastocoel. Especially in later embryos, cellular layers are not necessarily derived from the blastomeres that once occupied the same site.

Cell lineages derived from unrestricted or **pluripotent** individual blastomeres are classified by their fates and by their ability to **mingle** with the descendants of different blastomeres (Table 14.1). A **clonal domain** is a region of a postgastrula embryo or larva containing cells from an earlier blastomere group. Similarly, a **polyclonal domain** is a region of a postgastrula embryo or larva containing cells from several earlier blastomere groups.

Cell lineage **compartments** or **polyclonal compartments** (also called multicellular domains) containing the descendants of determined **founding cells,** or a **founder cell group,** are regionally restricted in differentiated structures. The **lineally**

TABLE 14.1. Nomenclature devised to describe progressive determination of *Xenopus laevis* blastomeres involved in the formation of the central nervous system[a]

	Stages of Development		
Cleavage	Early Embryo (In *X. Laevis*: Until 8th Division)	Early Blastula (After 9th Division)	Postgastrula or Larva
Fate not restricted (cells pluripotent)[b]			
Blastomere	Blastomere group or clone		Clonal domain
Several blastomeres	Blastomere groups		Polyclonal domain
		Fate regionally restricted[c]	
		Founder cell group or founding cell	Compartment or polyclonal compartment

[a]See M. Jacobson (1983).
[b]Progenies of pluripotential blastomeres differentiate into cells of many types.
[c]Progenies of founder cells differentiate into cells of different histotypes within a localized portion of a predictable structure.

equivalent progenies of such cell groups have a range of probabilities for differentiating into cells of different types, and **lineage-restricted** cells have only one set of probabilities.

In theory, founder cell groups in localized compartments produce cells throughout a region, including cells of the same or of different cell types (A. G. Jacobson, 1985). The variety of cells produced (e.g., neuronal- and sensory-cell types in the case of the vertebrate CNS or hypodermal and neuronal cells in the case of nematodes) is limited by the group rather than any consideration of presumptive germ layers.

Lying within or between different areas of a fate map, founder cell groups and compartments coexist like neighborhoods within a community. They maintain their borders and their local identities while cooperating in the larger enterprises of regional development (see Lawrence, 1981).

Cellular mingling within compartments may be extensive (e.g., germ layers of *Xenopus laevis*) or limited (e.g., *Drosophila melanogaster*), but it is absent or minimal between compartments (e.g., see below Fig. 14.12). The preference of cells to remain in one compartment rather than mingle with cells of other compartments is reproducible *in vitro* where cells that are intentionally mixed tend to reassort and cohere to other cells of the same compartment (see Jacobson and Klein, 1985).

Founder cell groups in the blastula may not be irreversibly determined, and their preferential cohesion may represent a form of positional information that specifies the location at which cells are normally forced to reside. This type of restriction, called **location specification,** can have major consequences for the subsequent determination of cells. Limited to a specific embryonic locale, the progeny of the founder cell groups may come under specific, determining influences that **assign** or fix their fates.

Arthropods with Intralecithal Cleavage

Initially, cleavage energids (i.e., cleavage nuclei with halos of cytoplasm) in the syncytial blastoderm are distributed either evenly as a **uniform blastoderm** or differentially as a **differentiated** or **polarized blastoderm.** Eventually, the syncytial blastoderm is replaced by the cellular blastoderm, and even an initially uniform blastoderm becomes a differentiated blastoderm. In *Drosophila melanogaster*, the right and left halves of each segment of the future insect are founded by discrete groups of cells at this time (Garcia-Bellido and Merriam, 1969).

Cell movement tends to begin immediately upon (if not before) the formation of the cellular blastoderm. Concentrated at the posterior pole, a high density of cells (or even energids) is achieved near the ventral midline (Fig. 14.1). The more densely cellular part of the blastoderm is called the **embryonic primordium,** and it alone gives rise to the embryo proper. The remainder is the embryonic ectoderm, and it forms extraembryonic membranes.

Some of the fate maps drawn for the blastoderms of onycophorans (Fig. 14.2a, compare to Fig. 11.1), myriapods (Fig. 14.2b, compare to Fig. 11.3), and pterygotes (Fig. 14.2c, compare to Fig. 11.10) are remarkably consistent. The ventral embryonic primordium ends in presumptive gut ectoderm in the form of the presumptive stomodeum (i.e., mouth) and proctodeum (i.e., anus).

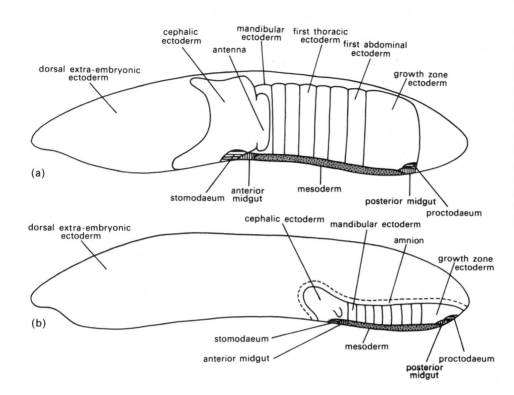

FIGURE 14.1. Proposed fate map for the exopterygote dragonfly, *Platycnemis,* seen in left lateral view, showing the expanded embryonic rudiment at the uniform blastoderm stage (a) and the differentiated blastoderm stage (b). From D. T. Anderson, *Embryology and phylogeny in annelids and arthropods,* Pergamon Press, Oxford, 1973. Courtesy of D. T. Anderson. Used by permission.

In *Drosophila* (Fig. 14.3) a ventral **germ band** of presumptive mesoderm is flanked by presumptive neurogenic ectoderm and capped by presumptive endoderm at both ends. The **dorsal blastoderm** is an area of variable size representing the presumptive extraembryonic ectoderm.

The similarities in fate maps are not unexpected, since all insects develop in segments, and the structures produced by a segment in one species tend to be produced by a comparable segment in another species.

The results of deletion and ligation experiments usually corroborate the fate maps at least as far as the development of larval structures, and the maps are consistent with the mosaic mapping of adult structures (see below). Still, these fate maps must be considered tentative, since they are drawn from observations on cell movement in living specimens without benefit of cell markers (see Anderson, 1973).

In addition to small size and inaccessibility of eggs, the rapid rate of change in blastoderms makes drawing fate maps problematical. Moreover, different fate maps may be required when different structures are targeted. For example, presumptive germ layers and larval structures may be circumscribed in corresponding areas of the early cellular blastoderm, but adult structures may be circumscribed in other areas.

Different types of insect also require different fate maps. In **exopterygote insects** (i.e., with wings developing from external pads and incomplete metamorphosis [= hemimetabolous]), **nymphs** (Gk. *nymphe* bride [as in nuptial] referring to a sexually immature form) are hatched. Terrestrial nymphs resemble the adult, but adaptations to an aquatic environment render aquatic nymphs quite different from the terrestrial adults. In **endopterygote insects** (i.e., with wings developing under the integument and complete metamorphosis [= holometabolous]), wormlike **instars** (Lat. figure or form, e.g., the caterpillars of moths and butterflies and the maggots of flies) are hatched. Nymphs and instars molt several times, and while nymphs become adults following a final molt, instars metamorphose into adults following a pupal stage.

Instars form the segmental integumentary organs of the adults, such as legs, wings, and eyes, gonads, and sexual organs from reserved cells located on each side of the body. When localized in discrete groups the reserves are called **imaginal disks** (also discs, sometimes formative disks) or primitive **anlagen** (pl., *anlage* sing.), but all the disks may not be discrete at the time the larva hatches, and in different species, they become morphologically distinct at different times.

The determination of presumptive larval structures occurs before the determination of imaginal

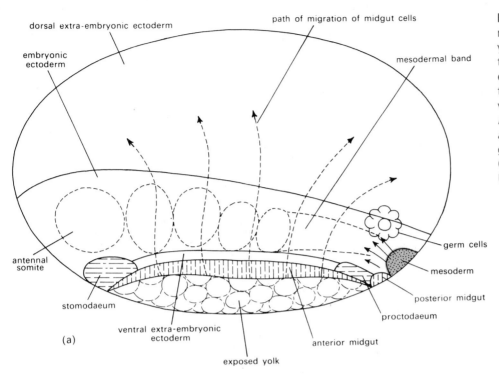

(a)

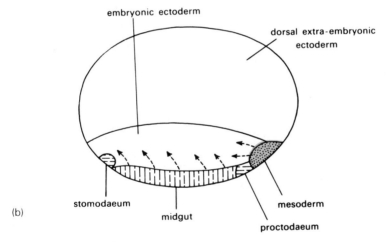

(b)

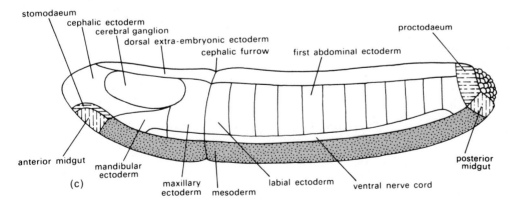

(c)

FIGURE 14.2. Proposed fate maps for the blastoderms of a yolky onycophoran (a), a centipede, Chilopoda (b), and an endopterygote insect, the dipteran, *Dacus,* in left lateral view. From D. T. Anderson, *Embryology and phylogeny in annelids and arthropods,* Pergamon Press, Oxford, 1973. Courtesy of D. T. Anderson. Used by permission.

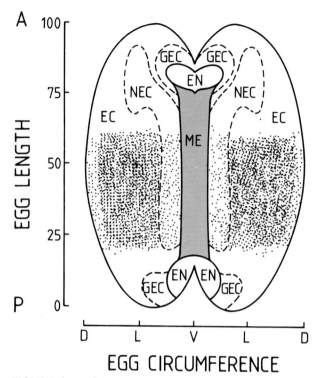

FIGURE 14.3. Fate and determination map for the cellular blastoderm of *Drosophila*. The blastoderm is drawn as if cut open along the dorsal midline and spread out. UV irradiation of the shaded area causes defects in the larval hypoderm (i.e., integument). The axes show percentages of egg length (0 = posterior pole) and distance around circumference. A, anterior; D, dorsal; L, ventral; P, posterior; V, ventral; EC, presumptive ectoderm; EN, presumptive endoderm; GEC, presumptive gut ectoderm (= stomodeum and proctodeum); ME, presumptive mesoderm (= ventral germ band); NEC, presumptive neurogenic ectoderm. From C. Nüsslein-Volhard, in S. Subtelny, ed., *Determinants of spatial organization,* 37th Symposium of the Society for Developmental Biology, Academic Press, Orlando, 1979. Used by permission.

disks. The determination of parts within the disks occurs afterward, turning the disks into "mosaics of separate fields" (Bodenstein, 1955) each capable of regulating within itself but no longer apt to regulate across field lines. Finer determination within the fields may not take place until shortly before pupation, and the venation of lepidopteran wings may only be determined in the pupa.

The existence of imaginal disks has given the embryos of endopterygote insects the reputation of being double embryos, a realized embryo forming the larva and a reserve embryo forming the adult. Actually, the presence of quiescent or presumptive adult anlagen more or less held in morphogenic suspension is widespread in embryos having indirect development. The growth zone of annelidan

larvas, for example, like an imaginal disk, remains relatively inactive until metamorphosis when it begins to spin off the trunk segments of the adult (see below). What is unusual about insects is the number and complexity of the structures formed by imaginal disks (Fig. 14.4).

The fate maps of imaginal disks are plotted from data on somatic mosaic adults having external regions expressing different phenotypes (Sturtevant, 1929) (Fig. 14.5). Somatic mosaics may be spontaneous or induced by X-rays via somatic crossing over. In addition, the mosaics can be produced by variegated or V-type position effects (i.e., effects brought about by translocating genes next to heterochromatin) and by chromosomal elimination. Recessive genes on the remaining chromosome of a homologous pair are then able to alter the phenotype of the affected cells. For example, **gynandromorphs** with male and female tissue develop when an unstable ring-X in a female *Dro-*

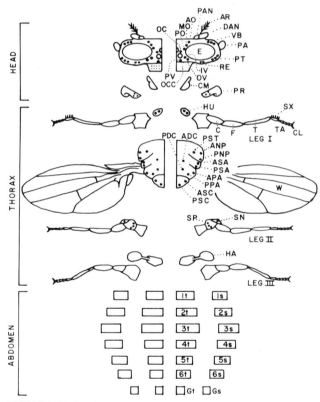

FIGURE 14.4. Exploded view of adult *Drosophila melanogaster* showing areas corresponding to independent imaginal disks or similar primordia present in the larva. Parts and bristles are identified with two- and three-letter codes. Compare to Fig. 14.5. From Y. Hotta and S. Benzer, in R. H. Ruddle, ed., *Genetic mechanisms of development,* 31st Symposium of the Society for Developmental Biology, Academic Press, Orlando, 1973. Used by permission.

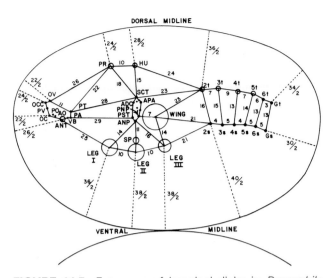

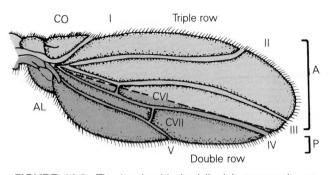

FIGURE 14.5. Fate map of imaginal disks in *Drosophila melanogaster* obtained by mosaic mapping seen from the inside of the right half of the blastoderm. The symmetrical left half is indicated by the curve below. The data used in constructing this map are from the two sides of 703 mosaic flies. Distances between disks (numbers along solid map lines) and the nearest midline (numbers along dashed lines) are given in sturts. Consult Fig. 14.4 for positions of designated parts and bristles in adult. From Y. Hotta and S. Benzer, in R. H. Ruddle, ed., *Genetic mechanisms of development,* 31st Symposium of the Society for Developmental Biology, Academic Press, Orlando, 1973. Used by permission.

FIGURE 14.6. The border (dashed line) between the anterior (A) and posterior (P) compartments of the normal *Drosophila melanogaster* wing does not follow morphological cues such as longitudinal veins (II–VI) or crossveins (CV1–CV2). CO, costa lobe; AL, alar lobe; triple row of bristles on anterior margin; double row of bristles on posterior margin. From P. A. Lawrence and G. Morata, *Dev. Biol.,* 50:321 (1976), by permission of Academic Press and the authors.

sophila melanogaster embryo is lost. Cells with XO male nuclei produce a patch of tissue with a male phenotype (see De Pomerai, 1985; Sang, 1984).

Because the patches with the recognizable phenotypes in somatic mosaics are patent, they seem to represent cellular **clones** and progenies of a single **founding cell** present at earlier times. Little mingling seems to occur among clones, and their sharp borders indicate that clones are **autonomous** (i.e., showing little evidence of morphogenically significant interactions).

Clones producing adult structures recognize borders. The cells of a clone divide until reaching the clone's border but do not cross it. **Polyclonal compartments** are territories occupied by related clones. Increasingly refined polyclonal compartments are established in the course of time.

Soon after cellularization of the blastoderm, segments are determined and anterior and posterior compartments are fixed. Initially, these compartments extend to different structures formed in the same body segments, but soon thereafter the anterior and posterior compartments are limited to individual structures (i.e., the imaginal disks for wings, halteres, legs, and antennae). In the case of

a wing, for example, a straight line separates anterior and posterior compartments without regard to any anatomical structure (Fig. 14.6) (Crick and Lawrence, 1975).

Limits in cellular connectedness identify the borders of a compartment. Cells in one compartment are capable of passing dye among themselves but not to cells in adjoining compartments.

Dorsal versus ventral and proximal versus distal compartments are also present in imaginal disks formed by single segments, but these compartments are established in the first instar larva, long after the establishment of anterior and posterior compartments. Within the large imaginal wing disks, slowly growing cells restrict the dorsal and ventral compartments and form margins between the two surfaces. Proximal and distal compartments define the body portion (notum) and the wing blade extending from the body.

Mosaic mapping is based on the premise that boundaries between regions in a somatic mosaic occur at random (i.e., depend on the angle of the mitotic spindle in the clone's founding cell). The probability of a boundary occurring between any two sites, therefore, is a function of how far apart those sites are in the blastoderm. Much like the frequency of crossing over among genes on a chromosome is used to plot one-dimensional genetic maps (in units now commonly called **morgans** after T. H. Morgan), the probability that a boundary in a somatic mosaic will pass between two blastodermic sites is used to plot two-dimensional fate maps of the blastoderm. In this **mosaic mapping** technique, map units called **sturts** (after A. H.

Sturtevant) equal the probability of 1% that two structures on the same side of the body will be of different phenotypes in a set of adult somatic mosaics (Fig. 14.5).

The determination of larval structures in the blastoderm and of adult structures in imaginal disks are, in large part, entirely separate affairs even from one side of the body to the other. While larval and adult segments are arranged in similar orders in the blastoderm, little overlap is found in the larval and adult integument. Furthermore, little overlap is seen in the fate maps of imaginal disks for integumentary structures and internal tissues and structures (Hotta and Benzer, 1973).

Mosaic mapping has also been applied to behavioral mutants (i.e., sturts are calculated from the frequencies of boundaries occurring between somatic mosaics and flies exhibiting different behaviors). Comparisons between behavioral and morphological maps show that behaviors involving some organs, such as legs, may be affected by mutations that do not alter morphology. This is not surprising, since nerves from thoracic ganglia innervating legs and the legs themselves can be traced by mosaic mapping to different parts of the blastoderm. Genes affecting the morphology of other organs, such as eyes that produce their own nerves, are identical to or closely linked to genes controlling organ-dependent behavior. Moreover, muscle defects responsible for "wings-up" behavior may be coded in the presumptive mesoderm.

Chordates with Total and with Partial Cleavage

From the mosaically developing ascidian to the regulatively developing vertebrate, presumptive germ layers are localized in roughly comparable positions. Whether drawn with the help of naturally occurring cytoplasmic markers, vital stains, carbon particles attached to cells, or grafts labeled with tritiated thymidine, fate maps show an animal–vegetal or anterior–posterior distribution of presumptive areas much like that in echinoderms (see Fig. 11.31).

Protochordate blastulas are frequently marked by naturally colored cytoplasms which are segregated and progressively localized (e.g., see Fig. 11.33). The distribution of these visibly distinct cytoplasms in the gastrula and subsequently in the larva is used retrospectively to plot the fate map of the blastula.

While the sizes of areas on the fate maps differ, their relationships are similar across species lines and independent of determination. Fate maps are relatively the same in the mosaic ascidian and regulative lancelet (see Fig. 13.20).

Amphibian blastulas are the classic material for tracing fate maps. The method employs vital dyes to stain cytoplasm in late blastula cells. Rather than diffusing throughout the egg, the vital dye is confined by cell membranes. Moreover, because division slows in the late blastula, the dye has few opportunities to become diluted. It remains concentrated in the cells to which it is applied and allows the investigator to rediscover the stained patches of cells within the gastrula.

Amphibian blastulas are also well suited for tracing potency, since blastula cells contain their own yolk platelets and require nothing more than buffered salt solutions to divide and survive *in vitro*. Differentiation seems to require something in addition to buffered salts, however, since clumps of cells are more likely to differentiate than isolated cells, and the formation of an epidermal covering seems to aid differentiation. In addition to determination therefore, **inductions** (i.e., local interactions that prescribe or alter fates, see Chapter 21) are reflected in potency maps.

Interest in vital staining and cell movement during the early years of the 20th century provided the background for Walther Vogt's pioneering marker experiments culminating in his seminal 1929 report on the fate maps of frogs. Using small chips of agar colored with vital dyes (e.g., neutral red, Nile blue sulfate), Vogt stained small patches of cells on frog blastulas. After watching the patches move toward the blastopore and disappear inside, he dissected gastrulas and neurulas to rediscover the stained patches in their new positions. His fate maps were drawn retrospectively from the combined results of many embryos and different stained patches. In effect, the amphibian fate map represents the position of embryonic cells at the beginning of their journey into the germ layers of the gastrula and the embryonic rudiments of the neurula.

Jean Pasteels, Osamu Nakamura, and others repeated Vogt's method and expanded it to other amphibians (see Holtfreter and Hamburger, 1955). A general picture of an amphibian fate map emerged (Fig. 14.7). Since then, virtually the same procedure (Fig. 14.8) was used to draw the fate map for *Xenopus laevis* (Keller, 1975, 1976). In addition, spikes of colored agar were inserted into embryos to stain underlying **deep cells.**

While still accepted in principle today, the amphibian fate map is refined in consideration of the

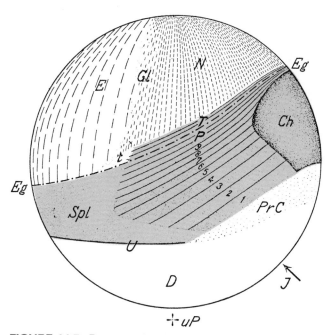

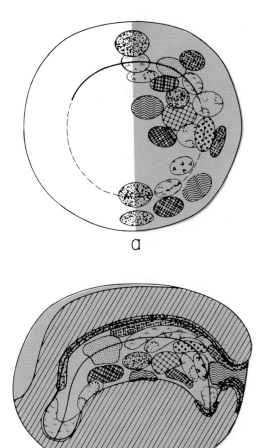

FIGURE 14.7. Fate map for the left side of blastula of the Japanese newt, *Cynops* (= *Triturus*) *pyrrhogaster*. Ch, presumptive notochord; D, presumptive endoderm of archenteron; E, presumptive epidermis; Eg, limit of invagination; GI, presumptive neural crest; J, initial point of invagination; N, presumptive neural plate; P, presumptive posterior trunk somites; PrC, presumptive prechordal mesoderm; Spl, presumptive lateral plate mesoderm of trunk; T, presumptive tail somites; t, presumptive lateral mesoderm of tail; U, boundary between presumptive mesoderm and endoderm (prospective lateral lip of blastopore); 1–9, numbers of presumptive trunk somites. From O. Nakamura, Y. Hayashi, and M. Asashima, in O. Nakamura and S. Toivonen, eds., *Organizer—A milestone of a half-century from Spemann,* Elsevier/North-Holland Biomedical Press, New York, 1978. Used by permission.

FIGURE 14.8. Summary of experiments on movement of vitally stained patches into *Xenopus laevis* embryo during gastrulation. (a) Vegetal view of early gastrula (stage 10⁺) showing initial positions of stained patches on right side (dorsal toward top). Solid line, blastopore; dashed line, margin of future blastopore. (b) Optical section of right side of neurula (stage 18) showing new positions of stained patches. From R. E. Keller, *Dev. Biol.,* 42:222 (1975), by permission of Academic Press.

deep distribution of some presumptive areas in contrast to the **surface distribution** of other areas (Løvtrup, 1975). The movement of dye through the surface of embryos in earlier experiments is thought to have confounded the identification of some areas.

At issue is a large part of the area posterior to the **limit of invagination** (e.g. in Fig. 14.7 = LI in Fig. 19.29) or presumptive blastoporal rim. The surface of this area seems to be occupied only by presumptive endoderm (formerly thought to occupy the vegetal part, D, below the boundary, U, with presumptive mesoderm, Fig. 14.7) and presumptive notochord (Ch).

In urodeles and anurans, including *X. laevis,* most, if not all, of the remaining **marginal zone** or presumptive mesoderm (T, t, P, 9–1, Spl) and PrC lies below the surface. This **deep mesoderm** lies under the superficial presumptive endoderm, espe-

cially in the region of cephalic endoderm, antecedent to the foregut. **Superficial mesoderm,** or surface mesoderm, covers a relatively larger surface area of the blastula in urodeles than anurans, but *X. laevis* has no superficial mesoderm at all. *Xenopus'* entire presumptive mesoderm lies beneath the surface of the blastula.

The presumptive epidermis (or cutaneous ectoderm, E), neural plate (or neural ectoderm, N), and intervening presumptive neural crest (GL) constitute the collective presumptive ectoderm. Originally occupying about one-third of the surface, the presumptive ectoderm expands to cover the entire surface of the embryo (see Chapter 19, Fig. 19.29). Everything else in the embryo is ultimately **inter-**

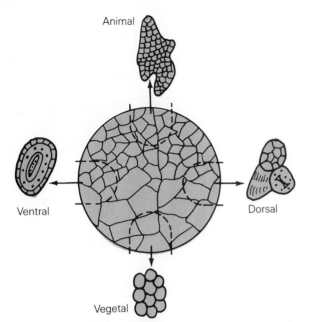

FIGURE 14.9. Diagram illustrating the differentiation of explants taken from different parts of the newt blastula. Adapted from O. Nakamura and T. Matsuzawa, *Embryologia*, 9:223 (1967) and O. Nakamura and H. Takasaki, *Proc. Jpn. Acad.*, 46:546 (1970).

nalized beneath the presumptive ectoderm during gastrulation or shortly thereafter.

Potency maps of amphibian blastulas differ sharply from fate maps especially toward the animal and vegetal poles of the blastula. Explants from the animal hemisphere form only a ciliated epidermis, but explants from the vegetal pole fail to develop. However, explants from the ventral marginal zone including some tissue from each of the presumptive germ layers differentiate an epidermis surrounding some loose connective tissue,

or mesenchyme (i.e., embryonic connective tissue), and blood. Explants from the dorsal marginal zone including presumptive neural ectoderm and mesoderm produce notochord, muscle, and neural tube-like structures (Fig. 14.9) (see Holtfreter and Hamburger, 1955).

The potencies of topographic regions of the blastula (Fig. 14.10), rather than presumptive germ layers, have also been tested *in vitro*. Unlike annuli of anterior tissue (regions I and II) and plugs of vegetal tissue (region IV), central rings (region III), including the marginal zone tissue, form the complete spectrum of embryonic tissue organized into something resembling an embryo. This embryo has more tissues than indicated by the fates of tissues in the marginal zone (Fig. 14.7) (see Nieuwkoop, 1977).

An equally complete spectrum of differentiation and embryogenesis (from epidermis and atypical gut, to blood, muscle, kidney, notochord, and other mesodermal derivatives) can be obtained from the animal hemisphere (regions I and II) when fused to portions of the vegetal hemisphere (region IV) in the absence of the marginal zone (region III). The animal hemisphere provides all the mesoderm and the endoderm of the abnormal head region. Under the influence of the vegetal-pole endoderm, animal hemisphere tissues of the blastula have greater competencies than suggested by their fates or potencies in isolation.

Mesoderm, it would seem, arises by inductive interactions between the prospective endoderm and ectoderm. In contrast to ectoderm and endoderm that acquire their identities in the blastula autonomously, mesoderm acquires its identity through cellular interactions.

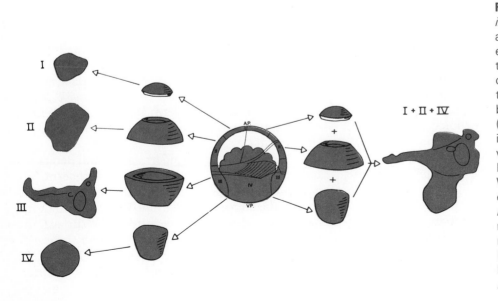

FIGURE 14.10. Differentiation *in vitro* of explants cut from axolotl blastulas. (Left) Differentiation of explants in isolation. (Right) Differentiation of composite containing polar portions. A relatively complete embryo is formed by the marginal (III) region and by the composite. Reprinted (in adapted form) with permission of Macmillan Publishing Company from M. W. Slack, in G. M. Malacinski, ed., *Pattern formation: A primer in developmental biology*, Macmillan, New York, 1984. Copyright © 1984 by Macmillan Publishing Company. P. D. Nieuwkoop, *Adv. Morphogenet.*, 10:1 (1973).

Cell lineage studies in *Xenopus laevis*, employing injected horseradish peroxidase (HRP) (Hirose and Jacobson, 1979; M. Jacobson, 1983) or fluorescein-dextran-amine (FDA) (Dale and Slack, 1987) as intracellular molecular markers also indicate that blastomeres have greater ranges of potency and competence than their fates would indicate. With the exception of cells at the extreme vegetal pole, which invariably differentiate as endoderm cells, blastomeres from the second- to the ninth-cleavage division (i.e., the 512-cell stage) are pluripotent in their development (Table 14.1). Their progenies give rise to cells of different germ layers which differentiate in cells of diverse histotypes in different organs. Moreover, the same blastomere in one embryo may give rise to different cells within the same polyclonal domain of another embryo. Cell mingling and the probability that cells will take an alternative pathway of differentiation account for the variation.

The situation changes dramatically in the animal pole after the ninth division. Founder cell groups consisting of 14–26 blastomeres are established at that time for each of seven compartments in the tadpole's central nervous system (CNS). The founder cells for the anterior–median compart-

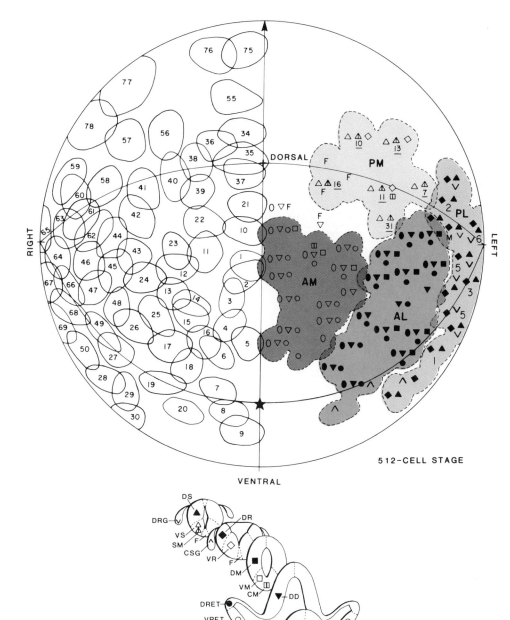

FIGURE 14.11. Compartments of the central nervous system (dashed lines, right) and outline drawings of blastomeres (left) injected at 512-cell stage with horseradish peroxidase. The embryo has the same orientation as embryos in Fig. 11.39. The star is at the animal pole; the arrow, lying in the dorsal midline, points toward the vegetal pole (not shown). The seven compartments identified by the failure of cells to cross borders are identified by boldface capitals. AM, anterior medial (single symmetrical compartment around midline); AL (paired), anterior lateral; PL (paired), posterior lateral; PM (paired), posterior medial. Below: symbols identify locations of injected cells in schematic sections of early larval central nervous system. From M. Jacobson, *J. Neurosci.*, 3:1019 (1983), by permission of the author and the Society for Neuroscience.

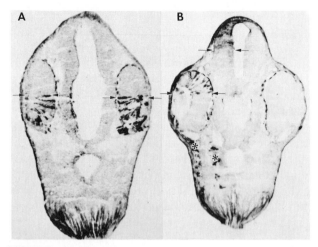

FIGURE 14.12. Photomicrographs of cross sections through mid-tailbud *Xenopus laevis* embryo (Nieuwkoop and Farber stage 36) showing labeled cells derived from blastomeres injected with horseradish peroxidase at 512-cell stage. The dark precipitate formed by the histochemical reaction for peroxidase identifies labeled cells and should not be confused with normal pigment in the epidermis outlining the sections and in the pigmented retina around the eyes. (a) Injected cell in the anterior-medial founder cell group (AM in Fig. 14.11). The labeled cells reside throughout the ventral portions of the eyes on right and left sides but do not cross an anterior-dorsal boundary (arrows). (b) Injected cell in the left anterior-lateral founder cell group (AL in Fig. 14.11). The labeled cells reside throughout the dorsal portion of the eye and brain on the injected side but do not cross an anterior-ventral boundary (arrows) or move from one side of the embryo to the other side. In addition, labeled cephalic neural crest cells appear ventrally after migration (stars). Courtesy of the author and the Society for Neuroscience.

ment (AM in Fig. 14.11) span the dorsal–ventral midline (Fig. 14.12), while those for the anterior–lateral (AL), posterior–medial (PM), and posterior–lateral (PL) compartments are separated on each side of the midline.

Further changes occur at the 12th division. Prior to this division, cells in the various compartments are immobile, and the clones of cells descended from particular blastomeres remain in discrete contiguous patches or **blastomere groups.** Following the 12th division, **migratory cells** mingle at the borders of patches and form mixed groups. Cell **lineage restriction** is not yet complete among these cells, but their **assignment** to germ layers and cell types is increasingly determined as gastrulation commences (M. Jacobson, 1983). The transition from stable to moving clones of cells occurs in the *Xenopus* blastulas at the **midblastula transition** (MBT).

The forces narrowing potential and driving cells to their fates may still be lying in wait, however. Especially in the vicinity of the midline, the convergence of cells, rather than any prior experience in lineage groups, may yet determine cell fate (Danilchik and Black, 1988).

Fish, reptiles, birds, and mammals appear to have similar fate maps or, at least, not irreconcilably different fate maps. Moreover, the fate maps are compatible with the amphibian's (except for the complete absence of presumptive extraembryonic membranes in most amphibians, Fig. 14.13).

Blastular tissues fall into two categories: those

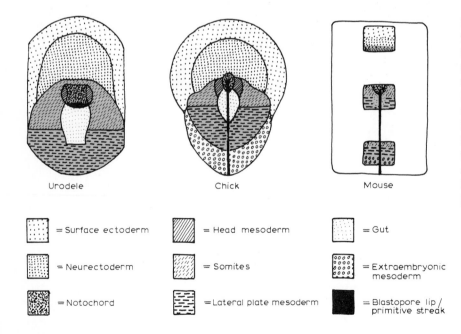

FIGURE 14.13. Diagrammatic fate maps for the surface of an early urodele gastrula, a chick epiblast at the primitive streak stage, and a mouse embryonic cylinder with primitive streak. The large blank areas of the mouse fate map are uncharted. The urodele and mouse embryos are shown as if flattened. Anterior is at top in each case. From R. Beddington, in M. H. Johnson, ed., *Development in mammals,* Vol. 5, Elsevier Science Publishers, Amsterdam, 1983. Used by permission.

Urodele

Chick

Mouse

= Surface ectoderm

= Neurectoderm

= Notochord

= Head mesoderm

= Somites

= Lateral plate mesoderm

= Gut

= Extraembryonic mesoderm

= Blastopore lip / primitive streak

that are totally internalized (i.e., presumptive endoderm and mesoderm), and those that are only partially internalized (i.e., presumptive ectoderm) during gastrulation. In vertebrates generally, at least some internal tissues initially lie on the surface at the blastula stage. The two exceptions (teleosts and eutherian mammals) may not be too difficult to explain. In teleosts, cells forming internal embryonic tissues lie below the ectodermal enveloping layer (EVL in Fig. 13.3) from the beginning, but the EVL may be more of a covering for the discoblastula than its outer surface (Fig. 14.14a).

In eutherian mammals, embryo-forming parts lie beneath the trophectoderm (TE), but this situation seems to have evolved from that in prototherians and metatherians in which all early embryonic cells are part of the blastocyst surface (Fig. 13.10). Moreover, the TE of eutatherians generally opens up (the exception being rodents), thereby exposing the embryo-forming parts to the surface.

Vertebrate embryos also tend to form discrete presumptive areas at the blastula stage (Fig. 14.13).

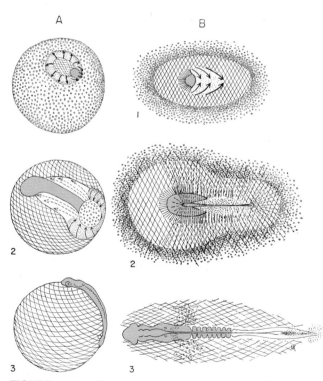

FIGURE 14.14. Schematic drawings of teleost (*Fundulus*, a) and chick (b) embryos showing distribution of presumptive external ectoderm (cross hatched) in discoblastulas (1), gastrulating embryos (2), and tailbud embryos (3). The cells forming the embryo proper (colored) occupy contrasting positions in the two types of embryo (i.e., initially peripheral in fish versus central in chicks). Adapted from D. Rudnick, in B. H. Willier, P. A. Weiss, and V. Hamburger, eds., *Analysis of development,* Saunders, Philadelphia, 1955.

The apparent differences between the chick and mammalian fate maps (beyond those due to incomplete data in the case of mammals) seem more quantitative than qualitative. Overlapping boundaries may provide only small areas in mammalian embryos with unique fates compared to avian embryos.

Possibly differences are consequences of the techniques employed to draw the fate maps. For chicks, maps are based on extensive marking experiments, while for mammals, the maps are largely extrapolated from the results of colonization and transplantation experiments (see Beddington, 1983; Snow, 1981).

The largest discrepancies between the fate maps of vertebrates in general involve the cenogenic extraembryonic membranes (see Amoroso, 1981). Yolk sacs seem to have evolved independently several times (including in amphibians with direct development), and three additional extraembryonic membranes (i.e., the amnion, chorion, and allantois) are produced by **amniotes** (the reptiles, birds, and mammals) but not by **anamniotes** (fish and amphibians).

In fish, the **ectodermal yolk sac** (crosshatched in Fig. 14.14a) develops as the EVL spreads over the yolk syncytial layer (YSL) behind the contracting yolk cytoplasmic layer (YCL). In contrast, the **endodermal yolk sacs** of reptiles and birds are derived from peripheral **presumptive endodermal cells** and a syncytial yolk periplasm. In chicks, the **vitelline endoblast** (i.e., source of the yolk sac endoderm) is localized initially in the germ wall at the blastodermic margin (see Fig. 13.7).

In eutherian mammals, histological observations suggest that the **primitive endoderm** of the yolk sac arises from the inner cell mass (ICM) (see Streeter, 1938). The results of marking experiments employing melanin granules (instead of the vital dyes) in blastocysts *in vitro* indicate that the yolk sac endoderm does not arise from the trophectoderm (TE), and the results of interspecific **rat/mouse chimeras** (see below and Fig. 14.16) show that ICM cells can contribute to the yolk sac endoderm. It would seem that the presumptive endoderm is localized in the floor of the ICM or unilaminar embryonic disk (see Rossant and Papaioannou, 1977).

In reptiles, birds, and prototherian mammals, the source of extraembryonic ectoderm for the amnion and chorion is a relatively large area on the surface (e.g., Fig. 14.14b). In therian mammals, the source of the chorion ectoderm is the trophectoderm (TE), while except in some eutherian mammals (rodents) amniotic ectoderm is formed by the

primary (or primitive) **ectoderm**[1] or unilaminar embryonic disk.

Colonization and transplantation experiments in eutherian mammals provide most of the data incorporated into fate maps. Marked embryonic disk cells in rabbits have been tracked and found to migrate, but the classical amphibian approach to drawing fate maps has not been extensively utilized in mammals (see Snow, 1981). In addition to small size and inaccessibility, the rapid growth of mammalian embryonic disks limits the usefulness of amphibian-type cytoplasmic markers.

Normally, at 4 days postcoitum (p.c.), at the beginning of implantation, the embryonic mouse ICM contains about 15 cells. At about 6½ days p.c., at the beginning of gastrulation, the unilaminar embryonic disk has 600 cells. The duration of the mitotic cycle is then dramatically reduced by about half (from 10–12 to 5–6.5 hours), and the rate of proliferation doubles. At 7 days p.c., the disk has 2000 cells and bulges out in an **egg cylinder.** This enormous growth makes marking experiments all the more difficult, since it tends to spread and dilute markers applied to cells and prevents their relocalization.

In colonization experiments, cells bearing a unique label, such as an electrophoretically distinguishable isoenzyme of glucose phosphate isomerase (GPI) or a gene detectable by molecular hybridization, are introduced into embryos. Ideally, the labeled cells colonize a part of the chimeric embryo or its extraembryonic membrane where their presence can be detected at some later time.

One of the first applications of colonization experiments led to an estimate of just how many ICM cells are actually **embryo formative cells** or **embryo clonal initiator cells** (see Mintz, 1970). Cleaving embryos from genetically different mice were removed from donor females and cultured in a high-serum medium permitting the embryos to reach the blastocyst stage. Then the zona pellucidas were dissolved with pronase and pairs of denuded embryos were brought into contact. At 37°C, the cells adhered, and, after one more day in culture, a unified, double-sized embryo was formed

[1]During the CIBA Symposium on Embryogenesis in Mammals, 1975 (*Ciba Foundation symposium; 40 [new series]*, Elsevier/North-Holland) discussion on nomenclature for the mammalian embryo, "primary" was preferred to "primitive" as a modifier for the late blastocyst's "ectoderm," but "epiblast" and "hypoblast" were preferred to "ectoderm" and "endoderm" for the first two germ layers formed during early gastrulation.

and transferred to a prenatal foster mother (Fig. 12.20). Following implantation, the embryos regulated to normal size, and about one-third survived to birth.

Analyses for genetic markers in a variety of tissues revealed that about 25% of the mice were formed entirely of cells from one blastocyst, while the other chimeras had some tissue with genetic markers originating in both blastocysts. If the embryos were formed from only one clonal initiator

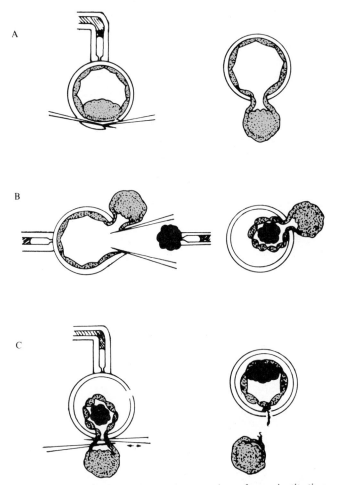

FIGURE 14.15. Experimental procedure for substituting the inner cell mass (ICM) of one embryo for that of another and creating a chimeric blastocyst with ICM and mural trophectoderm (TE) from different embryos. (a) Held by a suction pipette, the zona pellucida of the blastocyst is slit. The ICM protrudes on its own accord. (b) An ICM is injected into the evacuated mural TE sac. (c) The extruded ICM is amputated. If the original embryos are genetically distinguishable, cellular lineages derived from the ICM and TE can be traced in differentiated tissues. From V. E. Papaioannou, in P. F. Baker, ed., *Techniques in cellular physiology: Techniques in the life sciences,* Elsevier/North-Holland Scientific Publishers, Amsterdam, 1981. Used by permission.

cell in the ICM, no mice would have been chimeras, since one cell would not have had both genetic markers. The very existence of chimeric mice meant that embryos could be formed from at least two nonsibling cells, but what is the most likely number (n) of embryo formative cells?

If a and b represent the different genetically labeled cells in the mice, nonchimeric mice have only a or b cells, while chimeric mice have both a and b cells. Assuming that blastocyst cells mix randomly in the aggregate and that cells from neither blastocyst have any advantage in forming the embryo, an n of 3 predicts the observed ratio of 1 nonchimeric to 3 chimeric mice.[2]

A small advantage for cells of either blastocyst would increase the ratio of nonchimeric to chimeric mice and lead to an underestimate of the number of clonal initiator cells. Three should therefore be considered a minimal estimate of the number of embryonic founder cells until confirmed by results with other approaches.

Colonization experiments have also been used to detect the contribution of the ICM and TE to extraembryonic membranes (Fig. 14.15). Chimeric embryos constructed of an ICM with one marker and a blastocyst or isolated mural TE with a different marker were subsequently dissected and the developing embryo's tissues tested for their markers. The TE invariably produced the chorion (i.e.,

ectoplacental cone, extraembryonic ectoderm, and trophoblast giant cells, Fig. 14.16), while the entire embryo and other extraembryonic membranes were derived from cells of the ICM (Papaioannou, 1982).

While the ICM forms the embryo, many ICM cells evidently do not. Shortly before the beginning of gastrulation, isolated unilaminar embryonic disk cells may be classified either *smooth* or *rough* on the basis of appearance. The results of injecting cells bearing a specific glucose phosphate isomerase into blastocysts marked by a different isoenzyme indicate that the smooth cells alone give rise to the embryo. These cells also form the amnion and the extraembryonic mesoderm of the yolk sac and chorioallantoic membrane. The rough cells give rise to the endoderm of the yolk sac (Gardner, 1982; Gardner and Rossant, 1979).

In contrast, $3\frac{1}{2}$-day midblastocyst mouse ICM cells injected into the blastocyst cavity of similar midblastocysts give rise to the full range of embryonic tissue. Likewise, single cells from the primary ectoderm of the $4\frac{1}{2}$-day late mouse blastocysts transplanted to $3\frac{1}{2}$-day midblastocysts also give rise to the full range of embryonic tissue (see Gardner, 1975). Furthermore, pieces of the primary ectoderm of pregastrula mouse and rat blastocysts differentiate into mesodermal and endodermal as well as ectodermal derivatives upon transplantation to sites in adults (see Škreb et al., 1976).

The failure to identify determined areas of the mouse embryo prior to $4\frac{1}{2}$ days p.c. can be interpreted in one of two ways. Pluripotential cells in

---

[2]Expanding the binomial for $n = 3$ gives $a^3 + 3a^2b + 3ab^2 + b^3$, and collecting terms gives $a^3 + b^3$ nonchimeric mice to $3a^2b + 3ab^2$ chimeric mice.

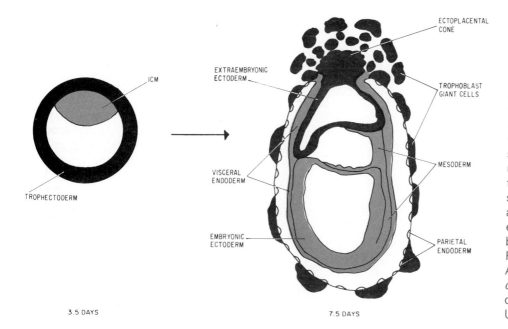

ECTOPLACENTAL CONE

ICM

EXTRAEMBRYONIC ECTODERM

TROPHOBLAST GIANT CELLS

VISCERAL ENDODERM

MESODERM

TROPHECTODERM

EMBRYONIC ECTODERM

PARIETAL ENDODERM

3.5 DAYS

7.5 DAYS

FIGURE 14.16. Diagram of chimeric blastocyst (left) consisting of inner cell mass (ICM) from *Mus caroli* (colored) and trophectoderm from *Mus musculus* (black) and summary of their subsequent development into parts of a 7.5-day embryo (right). *In situ* hybridization with species-specific DNA probes is used to distinguish between the lineages. The TE is the source of all the chorionic derivatives (trophoblast giant cells, ectoplacental cone, extraembryonic ectoderm). From J. Rossant, in N. Le Douarin and A. McLaren, eds., *Chimeras in developmental biology*, Academic Press, Orlando, 1984. Used by permission.

discrete areas may have regulated in response to the experimental technique. Alternatively, determined cells completely mixed in the embryonic disk may sort themselves out only during gastrulation.

The results of transplantation experiments show differences in parts of the mouse egg cylinder beginning 7 days p.c. at the midgastrular stage. Fragments of the embryo transplanted to ectopic sites, such as the adult kidney capsule, differentiate in tissue-specific fashions (Fig. 14.17).

Presumptive extraembryonic endoderm lies ventrally while embryo-forming cells lie dorsally. Presumptive embryonic ectoderm (both cutaneous and neural ectoderm) lies at the anterior dorsal margin of the embryo-forming region; a mixture of presumptive embryonic endoderm and mesoderm (notochordal, somite, and lateral plate mesoderm) lies centrally, and a presumptive extraembryonic mesoderm lies posteriorly. By $7\frac{1}{2}$ days, these regions have moved into the definitive locations at which they construct embryonic rudiments.

Spiralians

The fate maps of highly mosaic spiralian species are closely related to cleavage patterns and cell lineage. From the monet system of spiral cleavage in barnacles (see Fig. 11.53) and nematodes, to the duet system in acoelous flatworms (see Fig. 11.52) and the quartet system in polyclad flatworms and annelids (see Fig. 11.42), and most mollusks (see Fig. 11.40), early micromeres are specialized for sensory, motility, and protective functions, while late micromeres and macromeres tend to operate in digestion and internal movement. Still, the pathways connecting nuclei in cell lineages and the pathways partitioning the blastula's surface in fate maps may differ significantly among species.

Actual fate maps (as opposed to cell lineages) have been drawn from empirical observations mainly for annelids (Fig. 14.18) (see Anderson, 1973), but some generalities may be extended to mollusks and even to arthropods with spirally cleaving eggs. Above all, fate maps of the surface or cortex appear more regular in interspecific comparisons than cleavage patterns and may even illuminate the evolution of differences in cell lineages.

The fate maps for annelid blastulas appear to be composites of sharply zoned presumptive areas or **areas of determination.** These areas occupy different amounts of the blastula's surface in different species. The mechanism for shifting presumptive areas seems to reside in the cell surface, since the nuclear lineages in the different species may be the same. **Cortical determination,** assessed from fate maps, seems to have priority over **nuclear determination,** assessed from cleavage patterns, in establishing the fates of cells.

The evolutionary shifting of areas seems to be fueled by a tendency toward direct development. The reduction or elimination of the larva may be adaptive for minimizing the time spent in the hazardous plankton.

In general, species with more complex trochophore larvas incorporate more primary blastomeric materials into larval structures, while spe-

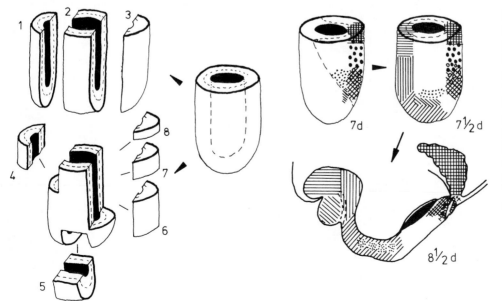

FIGURE 14.17. Diagram illustrating microsurgical fragmentation of a mouse embryo cylinder (left) and determination maps for 7- and $7\frac{1}{2}$-day embryos based on results of transplantation to ectopic sites (right). From M. H. L. Snow, in G. M. Edelman, ed., *Molecular determinants of animal form,* Alan R. Liss, New York, 1985. Used by permission.

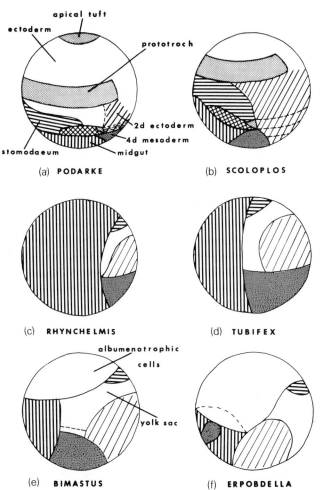

FIGURE 14.18. Fate maps of annelids and a leech. (a) Primitive polychaete with well-developed trochophore. (b) Polychaete with reduced trochophore. (c, d) Oligochaetes with yolky eggs. (e) Advanced oligochaete with specialized albumin-digesting cells (albuminotrophic cells). (f) A leech with specialized areas resembling those of advanced oligochaetes. Adapted from D. T. Anderson, *Embryology and phylogeny in annelids and arthropods,* Pergamon Press 1973. Used by permission.

cies with reduced trochophores utilize more cleavage blastomeres for teloblasts giving rise to presumptive adult structures. In addition, blastomeres no longer utilized for the production of larval structures may be adapted for the production of specialized cenogenic structures which perform temporary functions exclusively in the embryo.

Indirect development, or larval production, is epitomized by primitive polychaetes (Fig. 14.18a), while larval reduction or direct development reaches its climax in advanced freshwater clitellates (oligochaetes and leeches, Fig. 14.18e, f). The shifting of determination areas affects both those

areas remaining on the surface and those internalized during gastrulation.

The area remaining on the surface of primitive polychaetes with well-developed trochophores (e.g., see Fig. 11.45) is divided by the presumptive **prototroch** (Fig. 14.18a) into the presumptive **anterior ectoderm** and presumptive **posterior ectoderm.** The band-shaped prototroch is formed primarily by the peripheral cells of the first quartet of micromeres ($1a^2$–$1d^2$, see Fig. 11.42) with the addition of different numbers of anterior quartet cells depending on species. The presumptive anterior ectoderm corresponds to central cells of the first quartet of micromeres and forms the pretrochal region (or episphere). It produces surface epidermis and cerebral ganglia. In addition, a presumptive **apical tuft** or **cap** may be formed by anterior cells of the first quartet of micromeres ($1a^{111}$–$1d^{111}$).

The presumptive posterior ectoderm corresponds to the second and third quartets of micromeres and forms the posttrochal region (or hyposphere). It produces the trochophore's posterior structures (e.g., ciliated telotroch and neurotroch and the caudal or pygidial plate). The **2d cell** (and occasionally some adjacent cells, especially 3d) supplies a band of presumptive **ectoteloblasts** (see Fig. 11.49) also called the first somatoblast that produces the surface epithelium and ventral nerve cord of the late larva and adult (see Chapter 22, Fig. 22.1).

In polychaetes, the area to be internalized is posterior and ventral, like a dish stretching from the presumptive stomodeum to the edge of the ectoderm, while the area remaining externalized lies anterior and dorsal in the concavity of the dish. The internalized cells form the midgut, stomodeum, and mesoderm, while additional **ectomesodermal** cells (crosshatched in Fig. 14.18a and b) form the mesoderm of specialized head structure (e.g., the proboscis of *Scoloplos*).

Ectomesodermal cells are descendants of the 3a, 3b, $3c^2$, and $3d^2$ cells. The stomodeum forms from the 2b cell and other cells (especially 3b) in its vicinity. Generally, the mesoderm is formed from the descendants of the **4d mesentoblast** or **second somatoblast.** The cell also gives rise to some midgut cells and primordial germ cells. Bilateral mesodermal teloblasts or mesoteloblasts (M_l and M_r) derived from 4d give rise directly to paired ventrolateral mesodermal bands (see Fig. 11.49).

Polychaete species with reduced trochophores (Fig. 14.18b) and oligochaete species with trochophores lacking a prototroch or with no trochophores at all develop a unified ectoderm from descendants of the same cells that produce the anterior and posterior ectoderms of primitive polychaetes. In advanced oligochaetes, a temporary **yolk sac ectoderm** (Fig. 14.18e) derived from similar cells covers the large, yolky presumptive midgut cells prior to their incorporation into the embryo.

Generally, the midgut is formed by descendants of the yolky macromeres remaining after cleavage. In polychaetes and primitive oligochaetes, 3A, 3B, 3C, and 4D constitute the presumptive midgut, but in advanced oligochaetes, only 3A, 3B, and 3C take part in midgut formation. The 3D cell is totally preoccupied with mesoderm formation. Similarly, in leeches (Fig. 14.18f), 3D (also called DM, Fig. 11.43) divides equally into M_l and M_r cells, both of which produce mesoderm.

In marine species of annelids (i.e., polychaetes, Fig. 14.18a, b), mollusks, and primitive freshwater oligochaetes (Fig. 14.18c, d), relatively large amounts of yolk may be accumulated and function to support extensive development prior to hatching. In these species, the morphogenic function of yolk-laden cells is reduced. At the extreme, cephalopods achieve the ultimate in yolk accumulation and the elimination of any morphogenic role for the yolk-laden mass.

Advanced freshwater clitellates lay their eggs in cocoons or **albuminous capsules** that are digested during development. Earthworms (Fig. 14.18e) produce especially large first and second somatoblasts (i.e., 2d and 4d). Macromeres give rise to cenogenic **albuminotrophic cells** rather than micromeres and form an **albuminotrophic organ** (i.e., albumin eating) rather than a larval midgut. After functioning in albumin digestion, the organ is overgrown and internalized during gastrulation whereupon it degenerates. By producing somatoblasts (i.e., an ectoteloblast and mesoteloblast), the D blastomere retains its role as the major source of structures in the adult.

In leeches, a small group of ventral cells derived exclusively or mainly from 2D makes the entire midgut. In advanced leeches, the A and B macromeres give rise to single micromeres and the C macromere forms only two micromeres before becoming a nonproliferative albuminotrophic cell. The descendants of 2c, 3d, and 4d comprise the presumptive endoderm (Fig. 14.18f).

Many polyclad flatworms also digest cocoons during early development. Similarly, freshwater and terrestrial gastropods digest albuminous capsules by pinocytosis at their borders but these cells retain their morphogenic function.

Teloblast production is widespread among spiralians. In some respects, teloblasts parallel imaginal disk in insects, but teloblasts also interact with each other in the development of complex structures, whereas imaginal disks are largely autonomous.

For example, descendants of the 2d ectoteloblast and 4d mesoteloblast in many annelids form a unified annular **growth zone.** Initially, the ectoteloblast and bilateral mesodermal bands formed by the mesoteloblast combine efforts and elongate into larval segments. The growth zone then enters a period of quiescence, but, like an imaginal disk, at metamorphosis the composite growth zone becomes active again and elongates into the ectoderm and mesoderm of the adult trunk.

Similarly, in mollusks, the interaction of the somatoblasts is responsible for the formation of the shell gland in the posttrochophore velliger larva (see Fig. 22.1). Even in flatworms, interactions between comparable D-derived cells seem to be responsible for the coordinated development of adult structure.

Despite the well-known mosaicism of spiralian embryos, isolated embryonic segments containing composite teloblastic structures may be capable of regulating development. For example, normally appearing acoelous flatworms can form from half-embryos as long as sufficient endomesoderm and ectoderm is present, and, in annelids and mollusks, the mesoderm seems to induce differentiation in the ectoderm (see Cather, 1971).

PATTERN FORMATION AND REGULATION

The regional specificity of differentiation that begins during cleavage matures in the gastrula. Understandably, embryologists studying emerging patterns in the blastula use terms that are more appropriate for cleaving eggs (such as animal and vegetal poles) and for gastrulas (such as endoderm, mesoderm, and ectoderm). Moreover, because studies on the origins of symmetry and on the anatomy of germ layers preceded the formal analysis of spatial organization, embryologists borrow the language of regulation and determination to phrase the problems of pattern formation. Where and when is spatial organization regulative, and how is it determined?

Animal–Vegetal Axis

Probably no facet of embryology owes so much to research on one animal as the origin of polarity owes to studies on sea urchins. The very idea of a role for the animal–vegetal axis in pattern formation goes back to the turn of the century when Boveri and others tried to make sense of the contrasting results obtained from isolated blastomeres of 4- and 8-cell sea urchins (compare Fig. 5.14 to Fig 12.12).

Was this difference in the plutei due to the difference in size of the blastomeres or to the bisection of a hidden animal–vegetal axis by the third cleavage furrow? In a long series of experiments begun in the mid-1920s, Sven Hörstadius (born 1898) attempted to answer this question (see Hörstadius, 1973).

To eliminate the possibility that cleavage itself caused the abnormal development of plutei, longitudinal, animal, and vegetal halves of unfertilized eggs were separated, fertilized, and allowed to develop. Longitudinal halves regulated, but each animal and vegetal half produced an abnormal pluteus sometimes containing structures appropriate for the half from which it was derived but not for the other half. Differences along the animal–vegetal axis therefore seemed to precede cleavage and not follow it.

To eliminate the possibility that the deficiencies in plutei were due to the small volume of blastomeres isolated from 8-cell embryos, Hörstadius isolated and combined blastomeres of even smaller volume from 16-cell embryos. Occasionally, a macromere combined with an animal-pole blastomere produced a dwarf pluteus with a total volume even less than that of an average blastomere from an 8-cell embryo. Small volume alone was therefore not the determining factor in abnormal development.

To eliminate the possibility that the normal development of blastomeres isolated from 2-cell embryos was due to their greater volume, Hörstadius isolated animal and vegetal halves from 16–32-cell embryos. After developing for 1–2 days, animal halves produced even more abnormal larvas than the blastomeres of 8-cell embryos (Fig. 14.19). Greater volume was therefore not the limiting factor in normal development. Rather, the results suggested intrinsic animal and vegetal biases in the respective halves.

The structures formed by isolated animal and vegetal halves were similar to those formed normally by the halves *in situ*, if sometimes exagger-

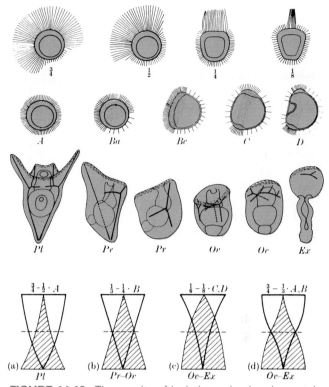

FIGURE 14.19. The results of isolating animal and vegetal halves of sea urchin embryos and gradients of hypothetical animalizing and vegetalizing influences. Top row: Animal halves 24 hours after isolation. The fractions indicate the portions of surface with stereocilia. Second row (a–d): Fully differentiated animal halves. Third row: Fully differentiated vegetal halves. *Pl*, pluteus; *Pr*, retarded larvas resembling the normal prism stage; *Ov*, ovoid larvas; *Ex*, exogastrula. Bottom row: Hypothetical gradients of animalizing (heavy line) and vegetalizing (hatched area) influences in animal and vegetal halves separated along dashed lines. Differences in the distributions of animal and vegetal parts in isolated halves are attributed to normal variation in gradients. From S. Hörstadius. *Pubbl. Stn. zool. Napoli,* 14: 251 (1935). Used by permission.

ated and off-center. The long **stereocilia** (i.e., nonmoving cilia) produced by animal halves on the first day of development resembled those of the normal apical (or animal) tuft, but instead of covering only about one-eighth of the thickened animal plate (Fig. 11.31), stereocilia sometimes covered one-quarter, one-half, or as much as three-quarters of the surface. Moreover, instead of the expanded zones of stereocilia being centered at the animal pole, they often seemed to be centered on the ventral surface in the vicinity of the prospective mouth or stomodeum (Fig. 14.19, top row).

By the second day, the stereocilia had disappeared in isolated animal halves much as apical tufts disappeared in normal plutei, and occasionally stomodeal invaginations appeared (Fig. 14.19, second row from top). In some of the abnormal plutei, remaining cilia were uniformly distributed (a), while in others, densely ciliated areas shared the surface with areas covered predominantly by a thin-walled **pavement epithelium.** Occasionally, a stomodeum dented the surface (d), but in no case was the gut or associated tissues produced.

In contrast, isolated vegetal halves rarely produced an apical tuft. Instead, they tended to form ovoid larvas (*Ov*, Fig. 14.19) tightly packed with skeletal spicules and typical gut components but lacking a mouth. In addition, isolated vegetal halves occasionally developed into diminutive plutei (Pl) and into a variety of intermediate types (Pr) resembling the prism stage between gastrula and pluteus.

Larvas known as **exogastrulas** (Ex) had everted gutlike structures protruding from their vegetal ends. While appearances may be deceiving, exogastrulas gave the impression of a reduced larval body unable to incorporate an excess of gut (see Gustafson, 1965).

Most batches of eggs produced a preponderance of halves which differentiated animal and vegetal structures in **intermediate** degrees (Fig. 14.19, center). Some batches produced halves that differentiated more strongly toward animal structures (left side), and other batches produced halves that differentiated more strongly toward vegetal structures or toward both animal and vegetal structures (right side). In effect, the result with any egg seemed to be driven as much by chance as by determination, but, overall, a pattern of causes was decipherable.

Double-Gradient Hypothesis in Echinoderms

In 1928, John Runnström (1888–1972) advanced a comprehensive hypothesis to explain the origins of polarity in echinoderms (see Runnström, 1975). In order to account for the differences between batches of eggs and the exaggerated differentiation of some halves, Runnström suggested that dynamic aspects of the embryo's physiology rather than static features of the embryo's surface controlled the morphogenesis of animal and vegetal structures. Instead of determinants, he proposed that **gradients** in animal and vegetal morphogenic activities, or **morphogenic gradients,** controlled spatial organization.

Conceivably, animal and vegetal gradients peaked at their respective poles and declined toward the opposite poles while overlapping throughout the embryo in opposite directions (Fig. 14.19, bottom row). According to the double-gradient hypothesis, **morphogenic substances** (subsequently called **morphogens** by Turing [1952]) poured into the cytoplasm at the poles and became distributed in polarized concentration gradients throughout the embryo.

An **animalizing** morphogen pushed the determination of cells toward animal properties or promoted **animalization** (formerly ectodermization, Fig. 14.19d to a). A **vegetalizing** morphogen pushed the determination of cells toward vegetal properties or promoted **vegetalization** (formerly endodermization, a to d).

Gradients of animalizing and vegetalizing morphogens were thought to compete with each other for dominance over morphogenic territory. In the "no-man's land" around the equator where neither gradient prevailed, potentials were expressed that were neither totally animal nor totally vegetal. Feedback controls determined the balance between the differentiation of animal and vegetal structures.

The development of the full range of structures in normal embryos depended on competition between the two gradients. Vegetal influences were indispensable for preventing overanimalization and obtaining normal differentiation of animal structures, and animal influences were indispensable for preventing overvegetalization and obtaining normal differentiation of vegetal structures. Ultimately, a normal developmental pattern emerged from the combination of gradients, competition, and balance.

Attempts to test the double-gradient hypothesis in echinoderms involved efforts to throw normal animal and vegetal gradients off balance or to restore balance to upset gradients. The alteration of gradients was generally accomplished by isolation and transplantation experiments.

Results were generally consistent with the double-gradient hypothesis. For example, in isolated half-embryos, morphogenesis associated with the missing pole was moderated, while that associated with the remaining pole was exaggerated (Fig. 14.19). Likewise, transplantation of vegetal material to animal halves initially reduced the amount of animalization (Fig. 14.20 center column [subscript 2]) and ultimately increased the amount of vegetalization of the animal tissue sufficiently

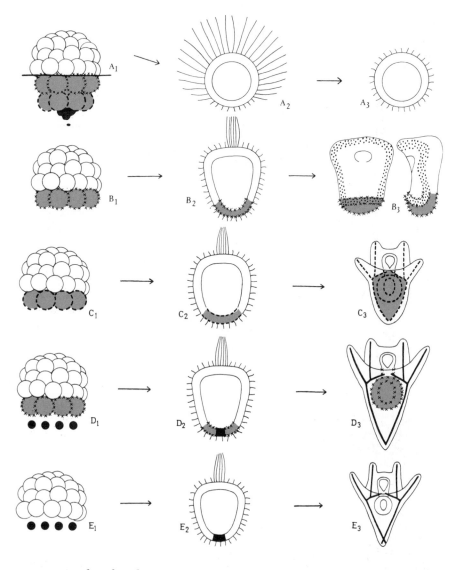

FIGURE 14.20. Vegetalizing influences from different tiers of the vegetal hemisphere. Subscript 1, portion or graft left to differentiate; subscript 2, 1 day later; subscript 3, 2 days later; A, isolated animal half; B, animal half plus veg_1 tier; C, animal half fused with veg_2 tier; D, animal half plus veg_1 tier fused with micromeres; E, animal half fused with micromeres. From S. Hörstadius, *Pubbl. Stn. Zool. Napoli,* 14:251 (1935). Used by permission.

to permit the development of plutei (Fig. 14.20 right column [subscript 3]).

The vegetalizing ability of transplanted vegetal tissues also increased quantitatively with proximity to the vegetal pole. For example, animal halves were vegetalized more by the addition of the veg_2 tier (Fig. 14.20c row) than the veg_1 tier (b row). Blastomeres from the vegetal tiers were also vegetalized in excess of their prospective significance following transplantation of micromeres (d row).

Transplanted micromeres were by far the best **vegetalizers.** Groups of four micromeres even vegetalized isolated animal halves, inducing them to form small but complete guts and to differentiate into relatively normal dwarf plutei (row e compare to row a). Moreover, the vegetalizing activity of micromeres was correlated with the number transplanted (see Fig. 12.13). Still, morphogenic control was more complex than a simple formula for mi-

cromere numbers. Position and time also played controlling roles.

The ability of cells to regulate in response to animalizing and vegetalizing influences was not uniform but changed both spatially and temporally. Cells closer to the animal pole seemed to be more resistant to vegetalization, and cells closer to the vegetal pole seemed to be more resistant to animalization. In addition, resistance increased in the course of time.

For example, isolated an_2 tiers were sufficiently vegetalized by two micromeres to form dwarf plutei, while an_1 tiers required four micromeres to achieve the same degree of vegetalization. Likewise, veg_1 tiers produced their most complete plutei with the addition of only one micromere, while the addition of four micromeres resulted in the development of ovoid larvas resembling those produced by isolated veg_2 tiers. Additional micromeres resulted in veg₂ tiers forming increasingly high proportions of exogastrulas (Fig. 12.13).

Moreover, while exaggerated properties appeared in halves isolated during cleavage and early blastulation, the development of halves isolated from late blastulas increasingly approximated structures formed *in situ*. Isolated animal halves became refractory to vegetalization by micromeres, and, as shown by the results of tracing vitally stained blastomeres, vegetal cells were not generally recruited for animal structures. The double-gradient hypothesis as such provided no explanation for these differences and changes in sensitivity and resistance.

A metabolic parameter of morphogenesis was first suggested by Charles Manning Child (1869–1954) on the basis of parallels between axial metabolic gradients and the developing axes of polarity. Child (1914) showed that vulnerability to lethal (cytotoxic) doses of potassium cyanide and other poisons was greatest in the animal pole of annelids and sea urchin embryos and in the mouth ends of hydroid polyp and medusa buds. While his attributing **dominance** and elevated oxygen utilization to regions of greater vulnerability was an extrapolation (an unjustified one, as it turned out), inferring a role for physiology in morphogenesis was a direct and enormously influential insight.

The hypothetical gradients of animalizing and vegetalizing influences in sea urchin blastulas were soon found to parallel **reduction gradients** shown by oxidation–reduction (redox) indicators such as the vital dye Janus green (Fig. 14.21). Embryonic sea urchins stained blue with Janus green changed to red and finally to clear when placed in an anoxic environment (e.g., provided by crowding). Acting as a terminal electron acceptor in place of oxygen, the dye changed color as a function of dehydrogenase activity and the rate of electron transfer (see Child, 1941; Hörstadius, 1973).

Two gradients of Janus green reduction appeared in intact embryos. The first gradient emerged during cleavage and was centered at the animal pole. The second gradient emerged at the late blastula stage and was centered at the vegetal pole. Prior to gastrulation, when both gradients were present, the second appeared stronger (Fig. 14.21a).

Isolated animal halves showed only the animal-pole gradient (Fig. 14.21b), and isolated vegetal halves usually showed only the vegetal-pole gradient (c). Occasionally, an additional gradient appeared in isolated vegetal halves after a delay.

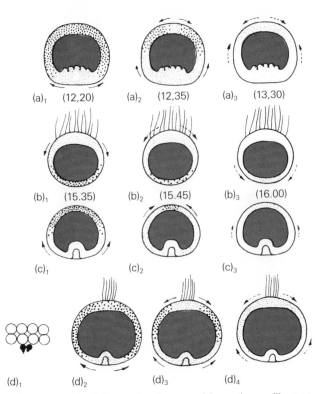

FIGURE 14.21. Schematic of sea urchin embryos illustrating metabolic gradients visualized through Janus green reduction. Color changes from blue to red to clear are indicated by changes from large dots to fine stippling, to open. Arrows show direction of change, and times after fertilization are given in parentheses. (a) Intact late blastulas. The vegetal gradient and animal gradient in intact embryos combine to reduce dye at a fast rate. (b) Isolated animal halves as blastulas. A single gradient is centered at the animal pole. (c) Isolated vegetal halves as early gastrulas. A single gradient (the typical but not universal result) is centered at the vegetal pole. (d) Micromeres (filled circles) fused with isolated animal half. A second gradient is induced at the vegetal pole. From S. Hörstadius. *J. Exp. Zool.*, 120: 421 (1952) by permission of Alan R. Liss.

This gradient was centered at what then became a new animal pole.

Many of the transplantation procedures that induced new or altered old morphogenic gradients also changed the reduction gradients. Micromeres implanted in isolated animal halves, for example, induced vegetally centered gradients in Janus green reduction (Fig. 14.21d). Moreover, inhibitors of cytochrome oxidase, such as 2,4-dinitrophenol, vegetalized isolated animal halves (although not whole embryos), and chloramphenicol, known to inhibit protein synthesis in mitochondria, vegetalized embryos (see Gustafson, 1975).

Rates of protein synthesis also paralleled the rates of metabolism and reduction gradients.

Amino acid incorporation was first higher in the animal region and then in the vegetal region, but the rate of protein synthesis per unit volume rather than per cell was not substantially different in micromeres compared to other cells.

Regional gradients in RNA synthesis also crudely paralleled reduction gradients and reflected animalizing and vegetalizing activities (see Wall, 1973). The rate of radioactive adenine incorporation into nonribosomal RNA was highest in the animal hemisphere of young blastulas and in the vegetal hemisphere of advanced blastulas. It was especially high in micromeres (see Fig. 10.7).

Some inhibitors of RNA synthesis, such as azauridine and azaguanine, tended to animalize if not very intensely, but other inhibitors, such as 5-fluorodeoxyuridine, fluorouracil, and 6-methylpurine, arrested the development of blastulas. Actinomycin D also arrested development at the blastula stage, but it also depressed animalizing effects (Lallier, 1975).

Chemical ways of exaggerating morphogenesis were first discovered by Curt Alfred Herbst (1866–1946) in the last decade of the 19th century. **Animalizers** pushed determination toward structures typically associated with the animal hemisphere, and **vegetalizers** pushed determination toward structures typically associated with the vegetal hemisphere.

Some agents are both animalizing and vegetalizing depending on dose. Trypsin, like other proteolytic enzymes, animalizes at high doses. It causes the reduction (if not elimination) of the vegetal gradient in Janus green reduction (Fig. 14.22a) and enhances the gradient centered at the animal pole. Yet, at low doses, trypsin vegetalizes isolated animal halves (Ortolani et al., 1982).

Some animalizing agents work against vegetalizing agents. Zinc ions, for example, are antagonistic to Li⁺, and vice versa. In general, while vegetalizers slow down the rate of development, animalizers speed it up, and, while some vegetalizers suppress oxygen consumption, animalizers require oxygen.

Cell-surface properties may be modified by animalizers and vegetalizers, and cellular cohesiveness or permeability may be altered. Animalizers may denature protein structures, while vegetalizers stabilize them. Ultimately, these agents must alter the cell's differentiation by either transmitting an effect internally or transferring the target of activity from the surface to the interior.

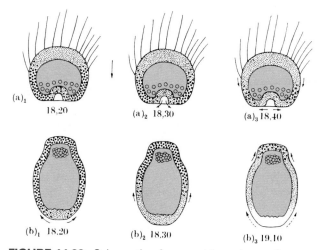

FIGURE 14.22. Schematic of sea urchin embryos illustrating differences in the gradients of Janus green reduction in animalized and vegetalized sea urchin embryos (compare to Fig. 14.21 row A). (a) Whole embryos animalized by trypsin. The vegetal gradient of Janus green reduction is retarded. (b) Whole embryos vegetalized by lithium ions. The animal gradient of Janus green reduction is retarded. The color changes from blue to red to clear are indicated by changes from large dots to fine stippling to open. Arrows show direction of change. Times are hours after fertilization. From S. Hörstadius. *J. Exp. Zool.*, 129: 249 (1955) by permission of Alan R. Liss.

Vegetalizers come in a variety of forms, the best known being the potent teratogen, the lithium ion Li⁺. Herbst demonstrated that subcytolytic (i.e., nonlethal) concentrations of Li⁺ added to seawater caused the production of exogastrulas and abnormal plutei. The amount of gut vastly increased, while the amount of pavement epithelium decreased. Endoderm and mesoderm seemed to form at the expense of ectoderm (see Lallier, 1975).

The embryos' vulnerability to Li⁺ extended from fertilization to the early blastula stage. Lithium also induced gastrulation in animal halves isolated from early blastulas, but halves of later blastulas were progressively refractory to vegetalization by Li⁺.

In these and other respects, Li⁺ mimicked the qualitative and quantitative effects of micromeres (see Hörstadius, 1973). Like transplanted micromeres, Li⁺ reduced the number of mitochondria at the animal pole and eliminated the animal gradient in Janus green reduction (Fig. 14.22b).

The normal increase in oxygen uptake following fertilization and accompanying cleavage was depressed by Li⁺ proportionally to the degree of vegetalization (see Gustafson, 1965). Glycogenolysis and carbohydrate catabolism, which normally increased in sea urchin blastulas, was also depressed by Li$^+$ (see Giudice, 1986). Inhibitors of glycolysis failed to vegetalize, but several enhanced the vegetalizing activity of Li$^+$, and potassium cyanide and sodium azide induced exogastrulation and vegetalization in whole embryos as well as in animal halves. Vegetalization therefore seemed to depend on a reduction of ATP synthesis brought about by the depression of carbohydrate catabolism (see Lallier, 1975).

The vegetalizing effect of Li$^+$ is ameliorated by potassium in an appropriately balanced medium. Potassium may act as a competitor of Li$^+$, possibly substituting for it or acting at the surface to prevent lithium's entry into cells.

The situation with RNA and lithium is complex. Normally, RNA content increases in the blastula stage mainly due to an increase in messenger RNA. The bulk of mRNAs synthesized by exogastrulas is significantly different from mRNAs synthesized by normal embryos (see Giudice, 1986).

In *Strongylocentrotus purpuratus* and *Lytechinus pictus*, Li$^+$ does not alter the rate of protein synthesis during cleavage, but it depresses protein synthesis in the blastula and at later stages. RNA synthesis is not comparably depressed and, it would seem, is not responsible for the depression of protein synthesis (Wolcott, 1982).

Potassium ions prevent the effect of Li$^+$ on protein synthesis if present simultaneously prior to hatching. The addition of K$^+$ to the posthatching medium of animals already affected by Li$^+$ does not restore protein synthesis or reverse lithium's other effects (Wolcott, 1981).

The vegetalizing activity of Li$^+$ in sea urchins has its counterpart in a wide variety of embryos. In mollusks, for example, subcytolytic concentrations of Li$^+$ produce head malformations such as the cyclocephalic or cyclopean malformation which leaves a larva with one eye and a compressed brain.

The induction of the cyclocephalic malformation of *Lymnaea stagnalis* is anticipated by the disruption of the normal sequence of divisions in the quartets of micromeres. The 3D macromere does not achieve contact with the micromeres at the 24-cell stage (see Fig. 12.9b), and the onset of bilateral cleavages is postponed. Instead of the 1d^{12} cell dividing bilaterally (see Fig. 11.50), it divides frontally. The consequences of lithium's suppression of bilaterality is the shifting of the pattern of development toward the midline and the disappearance of the median parts of the larva's head.

The effects of Li$^+$ in mollusks does not require penetration deep into the egg. Lithium probably competes with Ca^{2+} by binding to negative groups on phospholipids in surface membranes. Cell nuclei and spindles are rotated, and the chronological order of cleavages is disturbed (see Raven, 1976).

Animalizers of sea urchin embryos also come in a variety of forms (see Lallier, 1975). They include thiocyanate and iodide (whose animalizing activity was also discovered by Curt Herbst), ammonium, low pH, and heavy metal ions, such as mercury and especially zinc ions.

The timing of animalizing activity varies with the agent. Iodosobenzoic acid animalizes sea urchins up to the early blastula stage; rhodanide ions animalize before fertilization, but thiocyanate only animalizes if applied before fertilization. 8-Chloroxanthine animalizes both before and after fertilization, but sulfate-free seawater only animalizes embryos at the blastula stage at which time it completely inhibits mucopolysaccharide synthesis and suppresses 1° mesenchyme formation.

The effect of animalizers on RNA is variable. Ribosomal RNA synthesis does not normally increase in blastulas despite an increase of 50% in nucleolar activity. Likewise, transfer RNA synthesis increases only modestly in the blastula. Animalization only prevents the increased ribosomal synthesis occurring after gastrulation especially in the gut of the pluteus.

The electrophoretic pattern of total proteins synthesized by normal embryos is different from those synthesized by animalized embryos (see Guidice, 1986). For example, animalized *Arbacia punctulata* embryos have fewer bands in their electrophoretic protein patterns compared to controls. Posttranscriptional controls may be operating, but some specific mRNAs may be either turned on or off by animalization.

Further tests for effects on mRNA were performed in *Strongylocentrotus purpuratus* with the help of a library of cloned cDNA sequences representing a proportion of the most abundantly synthesized polysomal poly(A)$^+$ RNA species in mesenchyme blastulas. Screening RNA from embryos animalized by Zn^{2+} revealed that at least two poly(A)$^+$ RNA species (Fig. 14.23 a5 and b9) were repressed, while another species (g11) increased in concentration (Shepherd et al., 1983). The enhanced RNA species seemed to encode for sea urchin metallothionein (MT), a small, cysteine-rich, Zn^{2+}-inducible protein (Nemer et al., 1984).

Normally, MT is synthesized under intrinsic control (i.e., independently of exogenous heavy metal ions), first on stored (i.e., maternal) mRNA and, after the 8-cell stage (9–12 hours), on newly

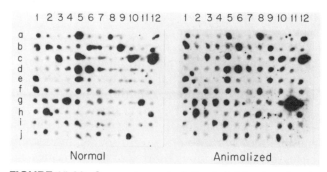

FIGURE 14.23. Comparison of relative hybridization signals from RNA of normal (left) and animalized (right) *Strongylocentrotus purpuratus* embryos. Autoradiographs show dot blots of unique cDNA clones hybridized with ^{32}P-labeled polysomal RNA from normal embryos and those animalized with zinc. The cDNA clones were prepared from some of the most abundant mesenchymal blastula polysomal RNAs. The size of a spot is proportional to the amount of specific poly(A)$^+$ RNA in the samples. A highly zinc-inducible clone (g11) is identified by the exaggerated spot at the intersection of row g and column 11. Zinc-repressible clones (a5 and b9) are identified by their diminished spots. From G. W. Shepherd, E. Rondinelli, and M. Nemer, *Dev. Biol.,* 96:520 (1983), by permission of Academic Press and the authors.

synthesized MT mRNA. MT is low in mesoendodermal derivatives of the pluteus, but it appears to be constitutive in the ectoderm and present at high levels even in the absence of exogenous Zn^{2+}.

Treatment with Li^+ results in a decrease in MT mRNA, but Zn^{2+} causes a 25-fold increase and an accumulation of MT in the animalized blastula and gastrula. MT synthesis is sustained even after removal of the Zn^{2+} inducer.

The increase in MT may be due entirely to animalization and the conversion of presumptive mesoendodermal tissue to ectoderm. Isolated mesoendodermal preparations of pluteus larvas register increases in MT, but no changes occur in isolated ectodermal cells already producing abundant MT.

Double Gradients in Amphibians

A double-gradient hypothesis for amphibians was suggested by Dalcq and Pasteels (1938) on the basis of tissues seen to form at the junction of graft and host tissues in embryos. The hypothetical gradients were placed perpendicular to each other, as opposed to the sea urchin's antiparallel gradients, and constructive interactions were generally attributed to the amphibian gradients, as opposed to more antagonistic activities attributed to sea urchin gradients.

If an amphibian gradient lying along the animal–vegetal (A/V) axis is arbitrarily assigned values from 1 to 4, and another gradient lying along the dorsal–ventral (D/V) axis is assigned values from 1' to 4', then the high points of the two gradients coincide at one point (4, 4'). According to the Dalcq–Pasteels theory (refined by Slack, 1984), portions of the late blastula represented by high numbers can make portions with lower numbers fill in missing parts of the gradients (Fig. 14.24, second and third rows), but portions of the blastula

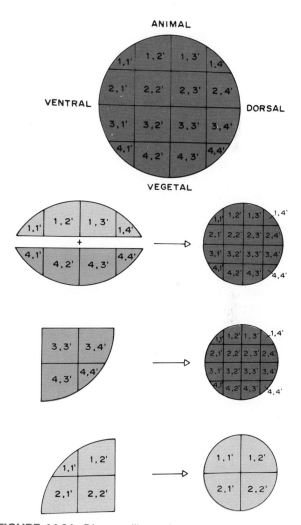

FIGURE 14.24. Diagram illustrating perpendicular axes of polarity in the amphibian blastula and the regulation of parts in combination and isolation. Numbers represent hypothetical animal–vegetal (1–4) and ventral–dorsal (1'–4') gradients. Tissue from higher parts of the gradients can induce tissue from lower parts to fill in the missing numbers, but only tissue from the very highest part can induce the formation of the whole spectrum of graded values. From J. M. W. Slack, in G. M. Malacinski, ed., *Pattern formation: A primer in developmental biology,* Macmillan, New York, 1984. Used by permission.

represented by low numbers cannot make other tissue fill in missing parts (bottom row). The region containing the combined high points corresponds to the **dorsal lip** of the future blastopore, also known as the **organizer.** As predicted, its transplantation anywhere in the embryo induces the formation of a complete set of parts.

In some ways, double gradients in amphibians correspond remarkably to double gradients in echinoderms. For example, amphibian animal-pole material is vegetalized by vegetal-pole material to the degree of forming endodermal structures (Fig. 14.10). Likewise, the effects of ammonia, zinc, and low pH (i.e., acidity) are antagonistic to the action of lithium in both echinoderms and amphibians, but lithium is not a vegetalizing agent in amphibians.

Exposure to Li$^+$ soon after fertilization and during cleavage up to the 32–64-cell stage promotes, rather than reduces, neural differentiation in amphibian embryos (Breckenridge et al., 1987). Furthermore, agents that animalize sea urchins do not convert amphibian presumptive endoderm into mesoderm and ectoderm or have other overtly animalizing effects (see Nieuwkoop et al., 1985).

Lithium has **anteriorizing** activity in amphibians. In *Xenopus laevis*, phenocopies induced by lithium (dissolved in their medium or injected into blastomeres or into the blastocoel) have abundant anterior structures (cement glands, eye), axial neural structures (brain parts), notochord, and a tubular beating heart, but very little muscle (somite) and posterior tissues (tail). The treated "dorsoanterior embryos"[3] (also called "axial" or "ventroposterior-deficient" embryos) may also fail to develop bilateral symmetry (i.e., right and left sides), developing instead circumferentially repeated anterior structures (adhesive glands, mouth parts, eyes) culminating in completely **radialized** (radially symmetrical) **dorsoanterior-enhanced embryos** (Kao and Elinson, 1988).

Alternatively (if not conversely), a **ventralizing** effect is produced by ultraviolet (UV) irradiation (at 254 nm) of the vegetal half of freshly fertilized frog eggs or the application of cold or hydrostatic pressure soon after fertilization. Treated *Rana pipiens* embryos fail to develop a gray crescent (Manes and Elinson, 1980) and have delayed gastrulation. Cell division occurs at a normal rate in UV-treated *Xenopus laevis* embryos, since normal numbers of cells are produced, but the midblastula transition

is either retarded or absent in the animal cap (Cooke and Smith, 1987).

UV-induced defects in anterior–dorsal morphogenesis may extend to the elimination of the entire nervous system and axial mesoderm (i.e., lacking notochord and somites). On the other hand, affected embryos form abundant lateral plate or ventral structures, especially blood islands producing hemoglobin-synthesizing erythrocytes. In extreme cases, differentiation is limited to those tissues and structures that would normally form around the point of sperm entry. The extreme **dorsoanterior-deficient** embryos (also called nonaxial embryos) fail to exhibit bilateral symmetry, forming instead a cylindrical or completely **radialized ventroposterior-enhanced embryo.**

Embryos can be partially rescued from the effects of UV irradiation by treatment with lithium during early cleavage. When UV-treated embryos are subsequently exposed to lithium at the 32–64-cell stage, or when marginal zone tissue from lithium-treated embryos is grafted to UV-treated embryos, some dorsal and anterior structures are produced and the embryos acquire bilateral symmetry (Kao and Elinson, 1988; Kao et al., 1985).

By implication, UV irradiation and lithium affect the same gradient in an amphibian-type morphogenic gradient system. Alternatively, UV irradiation and lithium may affect different, interacting gradients in a sea-urchin-gradient system.

Possibly, lithium and UV both have their effects on phosphosinositol turnover (see Table 8.1). A gradient in phosphoinositol turnover is highest ventrally and lowest dorsally in cleaving *Xenopus* embryos. Since lithium inhibits inositol phosphatase, it could flatten the gradient to uniformly ventral levels, while UV could raise the gradient to uniformly high dorsal levels. Results of injections into specific blastomeres support this possibility. Injection of lithium into presumptive dorsal blastomeres of UV treated embryos results in more nearly normal tadpoles, while the teratogenic effect of lithium injected into ventral blastomeres of unirradiated embryos is blocked by coinjection with inositol (Bursa and Gimlich, 1989).

The problem in interpreting results such as those with UV- and lithium-treated frogs is the classical problem of pattern formation cited by Morgan (1927): researchers assign axes of symmetry to embryonic parts, while embryonic parts only differentiate. More precise definitions of axes of symmetry, or how parts differ along particular axes (e.g., how somites on the right and left sides of an embryo differ), might remove ambiguity and advance research on pattern formation. Otherwise,

_____

[3]The term is regrettable, since somites are also dorsal organs but are missing, and the heart is not a dorsal organ but is present.

the arbitrariness of assigning "dorsal" or "ventral," "anterior" or "posterior," and "right" or "left" will continue to plague research on morphogenic gradients.

Studying the Development of Bilateral Symmetry

Eggs are rarely totally bereft of polarity. Features such as the position of polar bodies or the location of a micropyle suggest an underlying polarity even in the most spherically symmetrical egg. The question here is not so much whether the egg is polarized or not, nor what is the relationship of the egg's polarity to the polarity of the embryo (see Chapters 11 and 12), but when and how are the embryo's planes of symmetry fixed in the course of development.

Sea urchin oocytes and freshly fertilized zygotes lack the dorsal–ventral (D/V) axis visible in some other echinoderm eggs (e.g., sea cucumbers and star fish). Nevertheless, the sea urchin embryo's future ventral, or oral, surface may be latent in the egg. A region especially sensitive to a variety of cytotoxic agents seems to correspond to the prospective ventral surface, and a secondary high point in the animal Janus green reduction gradient may identify the future dorsal surface. The presumptive ventral surface also corresponds to a site of concentrated cytochrome oxidase activity (see Gustafson, 1975).

D/V determination increases through the blastula stage, although it is not firmly installed until gastrulation. When 16-cell embryos are cut meridionally along what is to become the frontal plane of the larva (Fig. 14.25 *a*), the isolated dorsal (b_2–c_2) and ventral (b_1–c_1) halves restore complete D/V axes. Regulation in the original dorsal half proceeds through a reversal of polarity, the cut surface becoming dorsal and a new dorsal surface replacing the original dorsal surface. (*d*)

The differentiation of the ventral surface of sea urchins may be inhibited by early exposure to weak solutions of Li^+ in seawater. Among other defects, the resulting **radial larva** has a mouth at its anterior pole rather than on its ventral surface and a straight digestive tract rather than one bending ventrally. This radializing effect of Li^+ may be initiated at the surface, since the detergent sodium laurylsulfate, which would be expected to denature proteins at the cell surface, radializes *Psammechinus miliaris* embryos exposed through the early blastula stage (see Gustafson, 1965).

Ventral dominance over D/V polarity is observed in half-embryos formed by bisecting blastulas along the future midfrontal plane (Fig. 14.25e_1). The larvas formed from original ventral halves are abnormally broad ventrally (e_2 and e_4), while those formed from original dorsal halves have reversed D/V polarity and are contracted ventrally (e_3 and e_5).

Bilaterality can ordinarily be regulated in composite *Paracentrotus lividus* embryos formed by fusing meridional halves from cleaving embryos of

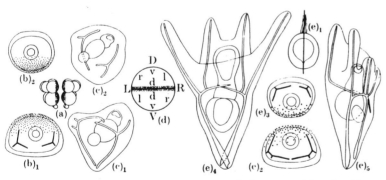

FIGURE 14.25. Fixing the dorsal–ventral axis in sea urchins. Lacking any overt sign of dorsal–ventral polarity at the time of surgery, the embryos are cut along every meridian and stained with a vital dye at their edges. The judgment that an embryo is cut along the dorsal–ventral plane is based on the disposition of the stain in the resulting larvas. (a) A 16-cell embryo cut in half and vitally stained along the cut surface. (b)₁, (c)₁ Gastrula and pluteus from the original ventral half. The embryo maintains polarity while regulating. (b)₂, (c)₂ Gastrula and pluteus from the original dorsal half. Dorsal development is retarded at cut surface (stippled), and dorsal–ventral polarity is reversed. (d) Orientation of cut (capital letters) and polarities during regulation (lowercase letters). (e)₁ Meridional cut through blastula. (e)₂, (e)₄ Gastrula and pluteus from the ventral half maintains polarity while regulating. Note broad ventral surface. (e)₃, (e)₅ Gastrula and pluteus developing from dorsal half shows reversal of polarity. Note narrow ventral surface. From S. Hörstadius, *Experimental embryology of echinoderms*, Clarendon Press, Oxford, 1973.

different ages, but fusing meridional halves from blastulas of different ages results in the development of unintegrated ventral and dorsal structures. By the late blastula stage therefore, the ability to regulate axes of bilateral symmetry seems to have disappeared (Hörstadius, 1973).

Right–left polarity in the sea urchin is also fixed gradually during the blastula stage. Cleaving embryos cut along the midsagittal plane slowly regulate and restore structures on their wounded sides (Fig. 14.26a–b). Cut blastulas form only incomplete skeletons and seem to find it difficult to restore missing right and left structures (d). Meridionally cut gastrulas form near-perfect right and left half-plutei without restoring absent structures at all (e).

Amphibians express dorsal–ventral polarity following fertilization in circumferentially oriented, parallel bundles of microtubules in the vegetal hemisphere (Elinson and Rowning, 1988). The midsagittal plane ordinarily corresponds to the first cleavage furrow (Klein, 1987), but the right–left axis of the embryo is not necessarily fixed at that time and may be reestablished in a variety of ways (see Rudnick, 1955).

In *Xenopus laevis*, symmetry is still sufficiently flexible prior to cleavage that enforced rotation (Cooke, 1987) or centrifugation (Gerhart et al., 1983) produce partial twins (see Fig. 9.12), and, at the 2–8-cell stage, isolated lateral halves develop into small, normal tadpoles. Polarity is also sufficiently determined, however, to prevent the development of normal tadpoles from dorsal or ventral halves (Kageura and Yamana, 1983) (also see Figs. 12.15 and 12.16).

Amphibians establish bilaterality in several **polarizing phases,** only the last of which is irreversible. By the time the embryo reaches the blastula stage, it has already moved up the developmental escalator past a sperm entry phase, a rotation-sensitive phase, and a gray crescent phase (see Chapter 9). As a blastula, the embryo enters a secondary dorsoventral polarization phase which it completes when it enters a final neuralizing phase in the gastrula.

This epigenic concept of polarization or progressive dorsoventral determination evolved from a series of experiments beginning in the late 1960s by Pieter Nieuwkoop (born 1917) and associates (see Nieuwkoop, 1985). While explants of animal blastomeres (Fig. 14.10, regions I and II) and vegetal blastomeres (region IV) from newt blastulas form atypical ectoderm and endoderm when cultured separately, they form a range of dorsal mesodermal and neural structures as well as ectoderm and ventral endodermal structures when combined (Fig. 14.10). In newts and *X. laevis*, vegetal-pole material forms vegetal **flask cells** and invaginates in the fashion of a gastrula, while animal-pole material forms **dorsal structures** (neural and mesodermal) and endodermal structures.

The differentiation of dorsal structures in the original animal-pole material is not determined by the position of the original gray crescent but by the orientation of the vegetal plug (region IV). Dorsal–ventral polarizing activity formerly localized in the vicinity of the gray crescent seems to have moved to the vegetal tissue. Polarizing activity takes the form of **mesoendodermal induction** (Fig. 14.27, solid arrows) passing between the **vegetal yolk mass** (vym) and the **totipotent animal cap**

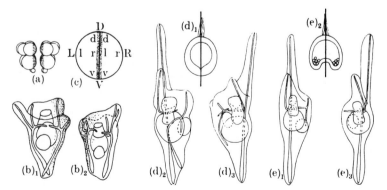

FIGURE 14.26. Fixing the right–left axis in sea urchins. Lacking any overt sign of bilaterality at the time of surgery, the embryos are cut along every meridian and stained with a vital dye at their edges. The judgment that an embryo is cut along the midsagittal plane is based on the disposition of the stain in the resulting larvas. (a) A 16-cell embryo cut in half and vitally stained along the cut surface. (b)$_1$, (b)$_2$ Plutei showing delayed development along left and right sides, respectively. (c) Orientation of cut (capital letters) and polarities during regulation (lowercase letters). (d)$_1$ Meridional cut through blastula. (d)$_2$, (d)$_3$ Resulting plutei (ventral view) showing impaired regulation at left and right sides, respectively. (e)$_2$ Meridional cut through gastrula. (e)$_1$, (e)$_3$ Resulting right and left half plutei showing lack of regulation. From S. Hörstadius, *Experimental embryology of echinoderms,* Clarendon Press, Oxford, 1973.

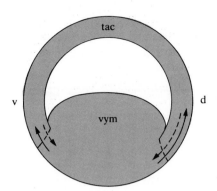

FIGURE 14.27. Diagrammatic midsection of an amphibian blastula illustrating hypothetical interactions of the vegetal yolk mass (vym) and the totipotent animal cap (tac). The extent of the interactions are greatest dorsally (d) where the blastopore will form at the beginning of gastrulation. The blastocoel may act as a barrier to interactions. From P. D. Nieuwkoop, *J. Embryol. Exp. Morphol.,* 89 (Suppl): 333 (1985), by permission of the Company of Biologists.

(tac). In addition, the animal cap induces (dashed arrows) gastrulation-type activities among vegetal cells.

The position of the blastocoel in amphibian eggs between animal and vegetal tissue (i.e., between the vegetal yolk mass and totipotent animal cap) may have strategic significance. Possibly, the spread of animal–vegetal influences is blocked by the blastocoel and thereby restricted to the cellular wall of the marginal zone (region III).

The results of centrifugation experiments with blastulas are consistent with this possibility. When the centrifugal force collapses, the blastocoel and animal-pole material is brought into contact with vegetal-pole material, and the animal pole undergoes morphogenesis toward dorsal structures. When the blastocoel remains patent, morphogenesis is not altered (see Nieuwkoop, 1973). Still, "the ultimate task is to elucidate some molecular transduction machinery" (Cooke, 1987, p. 426) for establishing polarity.

Stored (or maternal) mRNAs transcribed in the oocyte on **maternal-effect genes** may be involved in the induction of presumptive mesoderm in blastulas. The line of evidence leading to this possibility begins with mRNA sequences identified by screening against cDNA libraries for cleaving embryos (3000 transformants) and gastrulas (5000 transformants). A few mRNAs (i.e., 9) seem to be enriched specifically in dorsal areas of the blastula, while other sequences (i.e., 8) seem to be enriched in ventral areas (Wilt and Phillips, 1984).

In particular, transcripts of the *Vg1* gene synthesized during oogenesis are localized in a narrow crescent at the vegetal end of mature *Xenopus* oocytes. Released from bound cortical positions during cleavage, the *Vg1* transcripts are distributed to nearly all presumptive endodermal cells before disappearing in the gastrula. The sequence of nitrogenous bases in *Vg1* indicates that the *Vg1* protein is a member of the **transforming growth factor-β** (TGF-β) family implicated in mesoderm induction (Melton, 1987; Weeks and Melton, 1987; see Chapter 21).

Activation of mRNAs by posttranscriptional processes is also implicated in the acquisition of polarity. While the evidence is circumstantial, **selective polyadenylation** may play a role in polarizing the *Xenopus* blastula. The distribution of sites rich in polyadenylic acid (presumably as poly(A)$^+$ tails on mRNAs) changes conspicuously in *Xenopus laevis* during early development. A column of large yolk platelets containing a high concentration of poly A rises from the vegetal hemisphere to become associated with the dorsal side of the blastula (see Fig. 14.28).

Polyadenylation (i.e., the enzymatic addition of a poly(A)$^+$ tail of 200–300 adenylate residues to the 3′ end of an RNA transcript) prepares RNA for splicing, transport to the cytoplasm, and translation. The presence of a poly(A)$^+$ tail is a requirement for recruitment into polysomes. Cytoplasmic

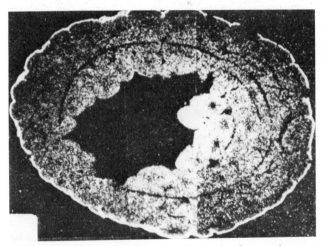

FIGURE 14.28. Dark-field autoradiograph of a section cut perpendicularly to the animal–vegetal axis of a *Xenopus laevis* blastula. Hybridization with ^{3}H-poly U reveals the presence of a high concentration of poly A in an area bordering the blastocoel (bright area right of center). The poly A-rich area is associated with a column rising from the vegetal pole of presumptive endodermal cells containing large yolk platelets. From F. H. Wilt and C. R. Phillips, in E. H. Davidson and R. A. Firtel, eds., *Molecular biology of development,* Alan R. Liss, New York, 1984. Courtesy of the authors. Used by permission.

maintenance of poly(A)$^+$ tails on mRNAs, moreover, is necessary for their continued translation, while the shortening of poly(A)$^+$ tails terminates translation.

Polarizing Centers in Insect Eggs

Experimentally induced phenocopies and mutant animals with **pattern aberrations** or **reversals of polarity** are frequent subjects for the analysis of symmetry. In particular, embryos with **mirror-image duplications,** both those with double heads and thoracic segments without abdomens (the **double cephalon** syndrome, DC, Fig. 14.29c) and those with double abdomens without heads (the **double abdomen** syndrome, DA, Fig. 14.29b), have figured prominently in research on pattern determination.

A common theme emerging from much of this research is that the cytoplasm is the operating base for genetic mechanisms of continuous patterning. The phenotype seems to be under the influence of cytoplasmic determinants that do not determine. Rather, their concentration at any level of the embryo prescribes a degree of differentiation (see Lawrence, 1988).

Hypotheses invoking cytoplasmic determinants to explain anterior–posterior (A/P) symmetry and pattern in insects generally envision positional information mediated by one or two morphogens. According to a dual-control hypothesis, **anterior determinants** are responsible for the differentiation of head and thoracic segments, while **posterior determinants** are responsible for the differentiation of abdominal segments (Fig. 14.30) (Kalthoff, 1983). According to a single-control hypothesis, a concentration gradient of only one determinant dictates both anterior and posterior differentiation (Fig. 14.31) (Meinhardt, 1982).

Dual control has been advanced to explain how the chironomid midge, Smittia sp., forms double-abdomen (DA) embryos under a variety of influences: UV irradiation of the anterior pole, ribonuclease (RNase) exposure, puncture, and centrifugation at an early intravitelline cleavage stage. The hypothesis requires two morphogenic agents capable of interacting.

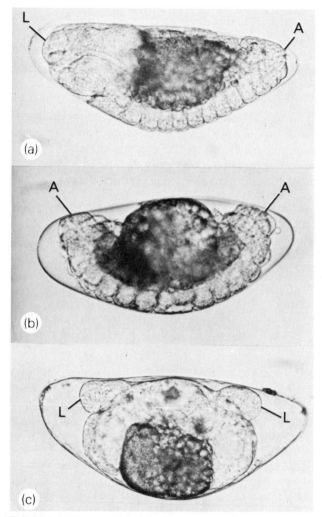

FIGURE 14.29. Embryos of the midge, *Smittia* sp. after centrifugation during the cleavage. (a) Normal (b) Double cephalon embryo (DC). (c) Double-abdomen embryo (DA). A, anal papilla (i.e. end of abdomen); L, labrum (i.e., tip of head [cephalon]). From K. G. Rau and K. Kaltoff, *Nature (London),* 287:635 (1980), by permission of Macmillan Journals and the authors.

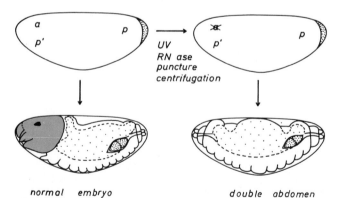

FIGURE 14.30. A hypothetical mechanism for the development of normal embryos of the midge, *Smittia,* and double-abdomen embryos induced by a variety of treatments. An abdomen develops in the absence of an *anterior activator* (*a*) and the presence of a *posterior morphogen* (*p* o *p'*) normally suppressed by the presence of the anterior activator. From K. Kalthoff, in S. Subtelny, ed., *Determinants of spatial organization,* 37th Symposium of the Society for Developmental Biology, Academic Press, Orlando, 1979. Used by permission.

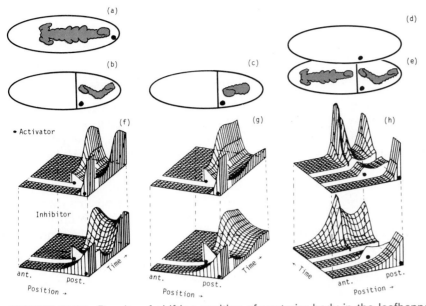

FIGURE 14.31. Results of shifting position of posterior body in the leafhopper, *Euscelis plebejus* (a–e) and interpretations according to the theory of autocatalysis (f–h). Ligation is indicated by a vertical line, the posterior body by a dot. Polarity in the embryo is indicated in the outline drawings of the embryo by the bend in the abdomen (i.e., tail fold). (a) Control. Position of posterior body undisturbed. Normal embryo. (b–c) Posterior body shifted anteriorly and embryo ligated immediately. No embryo anteriorly and abnormal embryos posteriorly. Ligation bars access to polar body material in anterior portion. (b) Embryo with double-abdomen syndrome in posterior half. (c) Dwarf embryo with reversed polarity in posterior half. (d–e) Posterior body shifted anteriorly and ligation delayed. Diminutive embryo with normal polarity anteriorly and abnormal embryo posteriorly. (f–h) Concentration gradients generated by theory of autocatalysis for an activator (top) and inhibitor (bottom) along the length of the egg (position) and in the course of time. The two positions of the posterior body are represented by the dots. The plane of ligation corresponds to the abrupt break along the position axis in (f) and (g) and to the bold line in (h). (f, g) Activator fails to reach or become established anterior to ligation. (f) Two peaks posteriorly (DA syndrome). (g) One peak posteriorly (reversal of polarity). (h) Delay in ligation allows activator to become established anterior to ligation. Three peaks (one anteriorly, two posteriorly). In other individuals, only one peak is established posteriorly (as in g). From H. Meinhardt, *Models of biological pattern formation,* Academic Press, Orlando, 1982. Used by permission.

If the midge's anterior structures are determined by an **anterior** determinant (*a* in Fig. 14.30), it would not seem to be in the nucleus, since UV irradiation of the anterior pole prior to the arrival of energids induces DA formation. The determinant would not seem to be in the egg's cortex either, since centrifugation seems to redistribute it and allow the formation of DA embryos.

The maximum effective wavelength of UV (280–285 nm) indicates a protein target, but wavelengths characteristic of an RNA target (265 nm) are also effective, and UV-induced effects are photorepairable which suggests light-induced, enzymatic repair of pyrimidine dimers in nucleic acids (see Phillips and Kalthoff, 1986). Possibly, the

determinant is a particulate, nucleic acid–protein complex of submitochondrial size, free in the cytoplasm and dense enough to be displaced by moderate centrifugation.

If the midge's posterior structures are determined by a **posterior morphogen,** it would have to resist the experimental treatments that induce DA formation or be quickly regenerated after treatment in order to account for the formation of the abdomen at the anterior end of treated embryos in place of the head and thorax. Possibly, a single morphogen diffuses from the posterior pole, or comparable morphogens (Fig. 14.30, p and p') are produced along the length of the embryo. The normal development of the entire pattern (from head and thorax at one end to abdomen at the other end) could result from the actions and interactions of the anterior determinant and the posterior morphogens (see Kalthoff, 1979).

Dual controls have also been advanced to explain the anteroposterior pattern of *Drosophila melanogaster* (see Nüsslein-Volhard, 1979). One gradient originating at the posterior pole is still entirely hypothetical, but the *bicoid* (*bcd*) gene seems to be responsible for the gradient originating at the anterior pole.

Evidence for *bcd* control of anterior structures is both direct and indirect. Embryos from *bcd⁻* females have terminal posterior structures in place of head and thorax, while the transplantation of cytoplasm from the anterior tip of wild-type embryos to *bcd⁻* embryos restores a nearly normal pattern. The mRNA encoded by the *bcd* gene is strictly localized at the anterior tip of the oocyte and early embryo, but the protein translated from this mRNA in the embryo is distributed in an exponential concentration gradient from an anterior maximum to a minimum about two-thirds down the length of the embryo.

Moreover, the concentration of the *bcd* gene's protein in embryos with different doses of the *bcd* gene is correlated with the position of anterior structures on the embryo's fate map. The *bcd* gene's protein may therefore be a morphogen that autonomously determines position in the anterior half of the *Drosophila* embryo (Driever and Nüsslein-Volhard, 1988a, b).

Single control has been advanced to explain DA formations and other pattern aberrations in the leafhopper, Euscelis plebejus.

Only one morphogen is hypothesized, but its distribution is thought to be determined by more than one factor.

Morphogenesis in the leafhopper depends on the presence of symbiotic bacteria introduced into each egg during oogenesis as a spherical **posterior body** (one of the more unusual examples of coevolution affecting development). Eggs fail to develop in the absence of the bacteria which are thought to release a **posterior-pole material** or morphogen.

Evidence for the posterior-pole material's morphogenic role comes from the results of shifting the posterior body anteriorly and ligating the egg (see Sander, 1984). Separated from the posterior body by a ligature, the anterior portion fails to undergo morphogenesis at all, while a diminutive embryo with the DA syndrome (Fig. 14.31b) or

with reversed polarity (Fig. 14.31c) develops behind the ligature. Possibly, the difference between these two results involves posterior-pole material left behind or lingering effects of the material. When ligation is delayed after shifting the posterior body (Fig. 14.31d), the same sort of defective embryo develops behind the ligature, but a complete embryo develops in the anterior portion (Fig. 14.31e).

One possible interpretation of these results is offered by the theory of **autocatalysis** or self-enhancing activation and regulation (see Meinhardt, 1982). An **activator** generates a strong short-range positive feedback that promotes its own production and the production of an inhibitor. The **inhibitor** is an antagonist to the activator, providing long-range negative feedback and lateral inhibition of activator production. The inhibitor is also more rapidly displaced from its source than the activator, for instance, by diffusion.

The theory requires a **morphogenic center** where the activator and inhibitor are produced. Normally (Fig. 14.31a), the morphogenic center radiates more or less parallel gradients in levels of activator and inhibitor throughout the embryo. Except at the morphogenic center, the concentration of inhibitor exceeds **critical values** or **threshold levels,** preventing the establishment of additional morphogenic centers by the activator.

When experimental manipulation shifts even small amounts of activator to new sites, the threshold of inhibition may be exceeded. The activator would promote its own production and establish a new morphogenic center. An additional gradient of inhibitor would follow.

In *Euscelis*, polarity could be determined by the concentration gradient of inhibitor doubling in the role of morphogenic gradient. Abdomen development would be directed by the inhibitor in its role as morphogen. The experimental manipulation of the posterior body would result in establishing new morphogenic centers (Fig. 14.31c, g) or additional centers (Fig. 14.31b, f, e, h). Reversed polarity could be explained by a new center established at the anterior end of a constricted piece, while DA pieces could be explained by an additional center added at the anterior pole of a piece retaining its original center.

The development of terminal segments (i.e., rather than subterminal segments) at the ends of DA embryos suggests that autocatalytic centers represent the extremes of the blastoderm. Lapses in middle segment development (known as the gap phenomenon) in midge embryos with the DA syndrome may also be explained by overlapping inhibitory gradients at the center of the blastoderm (Sander, 1975).

Cells at any level in the morphogenic gradient would read the concentration of morphogen as a **positional signal** and interpret it against a background **program** stipulated in the genome. If insect blastoderm cells form the most anterior structures first, cells at any level of the blastoderm might read "backward" through more posterior determinations. When the cells reached the particular determination specified by their local positional signal, they would not be permitted to read any further forward. The cells would thereupon be determined, the developmental switch thrown "on" for selective gene activity, and structures produced.

The structures would normally be appropriate for the level, segment, and embryo producing them, but when the source of morphogen is moved and when ligation interferes with the diffusion of the morphogen, incorrect positional signals or no signals at all would be given, and morphogenesis would be upset. Normal morphogenesis, regulation (e.g., Fig. 14.31e, h, in anterior portion), or the production of a developmental aberration (in posterior portion) could all flow from a single mechanism. The difference is not in the genes but in the distribution of morphogen.

Chapter

15

ACTIVATING THE BLASTULA'S GENOME AND INAUGURATING HIGHER LEVELS OF CONTROL

*A*rmed with the instruments and methods of molecular biology, developmental biologists have identified a host of new gene products and provided direct evidence for differential gene action. In amphibian embryos, the profile of molecular activity shows a massive shift from DNA synthesis to RNA synthesis at the blastula stage (Fig. 15.1), after which the embryo's genome takes charge of the embryo's destiny.

Still, in general, the early embryo's genome is either switched off or early genetic expression is irrelevant to early development. Furthermore, turning on the embryo's genome is not a high priority of early embryos. Enucleated eggs, whether fertilized or parthenogenically activated, cleave at the normal rate and develop to a blastula-like stage. Even the complete experimental suppression of RNA synthesis does not stop morphogenic progress before late blastula or early gastrula stages (Gross et al., 1964; Wallace and Elsdale, 1964).

Development of the blastula poses dual challenges: How is development controlled in the absence of gene activity and how is gene activity imposed on development? Answers have not been forthcoming, and a variety of techniques have been brought to bear in arriving at the present understanding of the subject.

ON THE TRAIL OF GENE ACTIVITY

We assume that differentiation proceeds in a stepwise fashion and that . . . although developmental strategies differ and the development of an oak tree, a frog, and a mouse are guided by different control mechanisms, they probably all boil down to essentially the same principle, namely, expressing different gene sets in different cell types (differential gene expression).

D. Solter and B. B. Knowles (1979, p. 139)

Using Hybrids to Detect Gene Activity

Hybrids have been produced ever since the invention of artificial insemination. The premise for

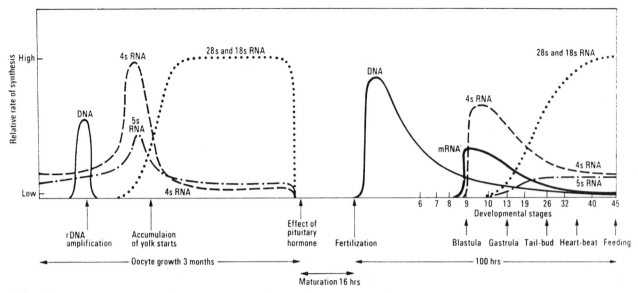

FIGURE 15.1. Profile of DNA and RNA synthesis during amphibian development. mRNA = heterogeneous RNA. From J. B. Gurdon, *The control of gene expression in animal development,* Harvard University Press, Cambridge, 1974. Used by permission.

using hybrids for research on gene activation is that products of paternal genes may be so incompatible with the maternal cytoplasm that synthesis is quickly manifest in abnormal development.

Hybrids of remotely related sea urchins arrest at a blastula-like or early gastrula-like stage, and most teleost hybrids and amphibian hybrids arrest at the gastrula stage, but cleavage proceeds as if normal. In hybrids between closely related sea urchins (e.g., *Echinus* × *Strongylocentrotus*), effects of the paternal genome become evident at the mesenchyme blastula or gastrula stage. Earlier events occur according to the dictate of the maternal species. Early primary (1°) mesenchyme formation and endodermal invagination (i.e., gastrulation), for example, occur by the maternal timetable. The number of 1° mesenchyme cells in the blastula also resembles the number in the maternal species, but the site of 1° mesenchyme cell formation seems to be influenced by the paternal genome and occurs in gastrulas rather than blastulas (Table 15.1) (see Chen, 1967).

The ability of isolated cells from embryos of hybrids to reaggregate with cells from the parental cell lines has also been used to test the state of the embryonic genome. Isolated pregastrula cells from hybrids between *Tripneustes esculentus* and *Lytechinus variegatus* reaggregate with cells of the maternal species but not with those of the paternal species. Possibly, the hybrid's cells have specific maternal-cohesion sites inherited from the egg. Cells isolated from gastrulas of the same hybrids, on the other hand, reaggregate with cells of the paternal species, suggesting that paternal-cohesion sites have been synthesized under the control of an active paternal genome.

The appearance of paternal types of isozyme for lactic dehydrogenase (LDH) in arrested blastulas of *R. pipiens* and *R. sylvatica* hybrids must be due to activity in the parental genome, but in other amphibian hybrids, paternal varieties of isozymes may not appear until the heartbeat stage. The late translation of ubiquitous **housekeeping** enzymes such as LDH is not surprising, since cells presum-

TABLE 15.1. Development of sea urchin hybrids from *Cidaris* eggs and *Lytechinus* sperm

	Archenteron Invagination (Hours)	Mesenchyme Formation (Hours)		Site of Origin of Primary Mesenchyme Cells
Cidaris (♀)	20–33	23–26	(follows invagination)	Archenteron tip
Lytechinus (♂)	9	8	(precedes invagination)	Archenteron base and sides
Hybrid	20	24	(follows invagination)	Archenteron base and sides

Source: E. H. Davidson. *Gene activity in early development,* 2nd ed. Academic Press, New York, 1968. Collated with data from D. H. Tennent. *Carnegie Inst. Wash. Publ.* 182:129 (1914).

ably inherit them ready-made from the maternal cytoplasm among other **maternal proteins.**

Effects of maternal proteins on the nucleus and of paternal nuclear products on the cytoplasm are studied in **nuclear–cytoplasmic hybrids** created by transplanting the nucleus from one species into the enucleated egg of another species. Deleterious effects may not be irremediable. Nuclear–cytoplasmic hybrids of the common carp, *Cyprinus carpio*, and the crucian carp, *Carassius auratus*, for example, are generally arrested in a blastula-like stage, but occasionally one escapes and develops to adulthood. The adult hybrid bears features of both parental species, it would seem, as a result of compatible interactions between the maternal cytoplasm and the paternal nucleus (Yan et al., 1986).

Similarly, nuclear–cytoplasmic hybrids between subspecies of frogs contrasting in color markings may also develop to adulthood. The frogs have the color pattern of the nuclear-donor species, indicative of compatible interactions among the products of the paternal genome and maternal cytoplasm.

Why then do most hybrids fail? Why isn't the maternal genome adequate to supply the embryo with whatever products it requires for development?

Occasionally, even in lethal hybrids (e.g., in *Rana pipiens* × *R. sylvatica* hybrids), some tissue, such as muscle, differentiates internally, suggesting prolonged activity in the maternal genome. Normally, oocytic (maternal) mRNAs stored in the zygote's cytoplasm are sufficiently abundant and diverse to support whatever translation is required for cleavage and early blastula morphogenesis (see below). Possibly, the fault with hybrids lies with the maternal cytoplasm rather than the paternal nucleus. The arrest of development in hybrids may be due to maternal cytoplasm sending inappropriate signals (or ineffective ones) to the paternal genome and the paternal genome responding inappropriately or not at all to the maternal signals (Moore, 1960).

Clumped, nondisjunctive, dicentric, ring-shaped, and fragmented chromosomes in cells from prearrested hybrid blastulas are indicative of early cytoplasmic effects on the nucleus. Hybrid nuclei may even be incapable of mitosis. In nuclear–cytoplasmic hybrids between *Rana catesbeiana* (nucleus) and *R. pipiens* (enucleated egg), the deleterious effects of the foreign cytoplasm on the nucleus are irreversible even when nuclei are transferred back to enucleated eggs of the donor species. The recipient fails to develop beyond cleavage. In some teleost and ascidian hybrids, se-

vere chromosomal abnormalities seem to result in the preferential elimination of the paternal genome.

Immunological Methods for Detecting Gene Activity

Until recently, immunology has provided the premier material for studying gene activity. Embryological interest in immune studies arises from two quarters. First, the legendary specificity of antibodies makes immunological studies a prime tool for studying the synthesis of new **antigens,** especially proteins, in embryos. Second, antibodies may provide direct access to surface **determinants** of cell behavior and differentiation.

Immunological studies of embryos generally employ antibodies to detect antigenic materials on or in cells, tissues, and extracts. **Antibodies** are immunoglobulins. Those reacting with embryonic tissues (and teratocarcinoma stem cells, see below) are usually **macroimmunoglobulins** (IgM) with multiple reactive sites (ten in the secreted form) or **gamma immunoglobulin** (IgG), the major immunoglobulin of blood, with two reactive sites.

Both types of immunoglobulin are produced by lymphocytes known as **immunocytes** or **B cells** following exposure to immunizing **antigens** (immunogens), but IgM tends to be produced during an early or primary immune response, while IgG is the predominant immunoglobulin produced later during a secondary immune response. Both forms of antibody complex with the antigen that elicits their production or **cross-react** with virtually identical **homologous antigens,** but IgM is more efficient than IgG in activating complement (see below) after binding with antigen.

Antibodies are raised or elicited by inoculating animals with immunizing antigens. Antibodies are typically prepared by the **xenoimmunization** of animals (typically rabbits or sheep) unrelated to the source of antigen (e.g., mouse). Blood serum (or peritoneal fluid) obtained from the inoculated animal is used in dilute solutions called **antisera** (pl., antiserum sing.) or, more precisely, **xenogeneic antisera.** Alternatively, **alloimmunization** of members of the same species but of different inbred strains yields **alloantisera,** and **syngeneic immunization** of members of the same inbred strain differing in sex or particular genetic loci yields **syngeneic sera.**

Not all biological materials are "good antigens," capable of eliciting the production of antibodies with a high titer and affinity for the particular antigens. Some antigens can be helped by mixing an **adjuvant** (e.g., Freund's adjuvant of mineral oil and heat-killed tuberculi bacteria) into the inoculum. The route of immunization (i.e., intravenous versus intraperitoneal) and the number of immunizing injections are also important variables in eliciting antibodies.

Separating (if not purifying) a desired antibody from an antiserum may be approached with the technique known as **adsorption.** The desired antibody, referred to as the **adsorbed** antibody (or antiserum), is separated from other antibodies either by binding them to an adsorbing material or by first binding the desired component and then redissolving it. For example, in order to separate an antibody to a **species-specific** antigen in xenogeneic antisera, tissues from the immunized species are extracted and used to adsorb the antisera. When the tissue extracts are bound to Sepharose in an **affinity column,** the adsorbed antibodies pass through the column. In alternate schemes, the columns are used to pick up the desired antibodies. Subsequent elution of the antibodies from the column frees them for use in detecting the antigens of interest.

The properties of antisera (although not the specificity) may also be altered by enzymatic digestion. Unlike native antibodies which have two or more reactive sites (i.e., multivalent antibodies), **Fab** (fragmented antibody or antigen-binding fragments, usually called Fab fragments), produced by incomplete digestion of antisera with the proteolytic enzyme papain, have only one reactive site (i.e., they are monovalent). The absence of additional reactive sites is useful for controlling immunological reactions and limiting the size of antibody–antigen complexes.

Specificity is the ability of an antibody to recognize a unique antigen, while sensitivity is inversely proportional to the amount of antigen required for detection by antisera. Specificity and sensitivity are best served when purified antigens are used to raise antibodies, but this is not always possible, and the two qualities may be antagonistic if not mutually exclusive when unknown antigens are used.

Even purified antigens can have more than one **immunogenic site** (or immunogenic determinant, a reactive site or site of antigen–antibody reactivity). Specific residues in the antigen and the conformation of the antigen (its shape in relationship to receptors on the potential B cell) may provide a variety of immunogenic sites. As a result, the antibodies present in antisera are **polyspecific,** reacting to different sites on complex antigens, and the

serum collected from even one immunized animal is generally **polyclonal,** having been produced by unrelated B cells.

The multiplicity of polyspecific antisera can be problematic, above all because different batches of antisera to the same antigen can be of very uneven quality. Still, by reacting to different parts of the antigen, polyspecific antisera can be extremely sensitive and provide an edge for identifying rare antigens.

Specificity may be enhanced when a small part of a molecule serves as an immunogenic site. Caution must be taken before attributing specificity under these circumstances, however, since the smaller the immunogenic site the more likely it will appear in unrelated molecules.

The same must be said for **plant lectins** which bind to specific carbohydrate antigens (epitopes) in surface glycoproteins. Each lectin has its highest affinity (i.e., binds most tenaciously) for particular carbohydrate moieties (Magnuson and Epstein, 1981):

Concanavilin A (Con A) binds D-mannosyl residues.

Isofucose-binding proteins (FBPs) from lotus binds L-fucose.

Peanut agglutinin (PNA) binds D-galactose.

Wheat germ agglutinin (WGA) binds *N*-acetylglucosamine.

The lectins provide great specificity for these carbohydrates coupled to great generality for carbohydrates containing them.

The **monoclonal antibody technique**[1] represents another approach to specificity. Antibody-producing lymphocytes are obtained from the spleens of mice previously inoculated with a given antigen. By fusing these nondividing spleen cells with dividing **plasmacytoma** cells, dividing antibody-producing cells known as **hybridoma** cells are prepared whose descendants produce **monoclonal antibody** (MAb) *in vitro.*

MAbs are **monospecific** (containing just one antibody), and they react with only one immunogenic site; but if that site is small, it may also be widespread. Consequently, an MAb may actually take part in **heterologous interactions** with otherwise unrelated molecules containing the same immunogenic site as the immunizing antigen.

---

[1]The 1984 Nobel Prize in Medicine or Physiology was awarded to Cesar Milstein and Jeorges J. F. Köhler for developing the monoclonal antibody technique and to Niels K. Jerne for contributions to theories of immunologic development.

As one might have guessed, results with antisera and MAbs are not always identical. For example, syngeneic F9 antisera react weakly with embryos at the 2-cell stage and maximally with morulas at the 8-cell stage (Artz et al., 1973). F9 MAb does not react with embryos at all before the 8-cell stage and reacts only weakly with 8-cell embryos (see Solter and Knowles, 1979).

MAbs have still other problems. While theoretically available in unlimited amounts, hybridoma colonies break down and cease producing MAb. MAbs also have their own type of specificity problems. MAb IgM may bind unspecifically due to high concentration or to the unspecific (F$_c$) portion of the macromolecule. Furthermore, nonspecific MAb is occasionally produced when hybridoma cells producing **autoantibodies** (i.e., antibodies produced by an organism without immunization) or antibodies produced independently of the specific immunization are inadvertently selected for cloning. Nevertheless, the promise of heightened specificity, reliability, uniformity, and commercial supply make MAbs the antibodies of choice today.

Techniques for detecting antigens on, in, or around cells utilize different properties of antibodies (see Tijssen, 1985). **Immune reactivity** is also a function of the antigen's condition.

An *in vivo* test employs the bursting or lysing of immunologically reacted cells. Usually, **complement** (a complex system of about 20 interacting extracellular protein components "complementing" immunological processes) is added to the embryonic or cellular medium to facilitate lysis.

Precipitates known as **immunoprecipitins** or just **precipitin** formed by antigens and antibodies are also useful for detecting antigens. The **precipitin test** depends on the formation of one or more rings of precipitin at the interphase of antisera and antigen-containing extract gently layered in a tube. In **double-diffusion** agar gel plates, bending bands of precipitin running together identify single antigens, while crossing bands identify different antigens (Fig. 15.2). When linked to electrophoresis, immunoprecipitation in agar gels can resolve individual antigens at higher resolution than immunoprecipitation alone.

The ability to detect antigens on cells in microscopic preparations depends on **conjugating** or coupling a visualizable label to antibodies. For example, **fluorescein-labeled antibodies** are made by conjugating antisera with fluorescein isothiocyanate. Bound by the antibody to antigens in histological preparations, fluorescein produces an apple-

FIGURE 15.2. Dark-field photograph showing bright immunoprecipitin bands formed in a double-diffusion agar plate with anti-fly antisera (E) and antigens from *Drosophila melanogaster* at different ages. (6) Extract of 3-hour embryos. (7) Extract of 6-hour embryos. (8) Extract of 9-hour embryos. (9) Extract of 12-hour embryos. (5) Flies. Antigens in the extracts and antibodies in the antisera diffuse through the agar in opposite directions and form precipitin bands where they meet. Differences in the rates of diffusion account for the separation of bands. Bands that bend midway between wells and fuse are formed by the same antigen. Bands that cross are formed by different antigens. The number of common antigens increases with the age of the embryo. From D. B. Roberts, *Nature (London),* 223:394 (1971), by permission of Macmillan Journals and the author.

green fluorescence when irradiated by ultraviolet (UV) light. Since the UV light is invisible to the human observer, the unlabeled part of the microscopic field appears dark, and the technique is called dark-field microscopy.

In **direct methods,** a label is conjugated with the antibody directed at the antigen of interest. In **indirect methods,** an unlabeled **primary antibody** to the antigen of interest is detected by a labeled **secondary antibody.** For example, if a rabbit immunoglobulin is the primary antibody employed to detect a mouse antigen, a conjugated secondary antibody (e.g., made in sheep or goats), capable of reacting specifically with rabbit immunoglobulin, is used to detect the rabbit antibody adhering to the mouse antigen.

The indirect methods offer several advantages, especially for antigen detection in microscopic preparations, and are generally used today. First, avoiding the conjugation step preserves higher titer and specificity in primary antisera. Second, polyspecific secondary antibodies can pile up on an antigen, providing an amplifying effect. But above all, highly reliable and highly active secondary antibodies are obtainable from commercial sources.

Fluorescent dye-conjugated secondary antibodies are visualized in the UV microscope (called **indirect immunofluorescence** or **IF**); radioactive iodine- (^{125}I) conjugated antibodies are detected autoradiographically (called **indirect antibody-binding radioimmunoassay** or **RIA**), and peroxidase-conjugated antibodies are detected histochemically by reacting with hydrogen peroxide and a coloring agent (called the **indirect immunoperoxidase** test or **IP**). Because the peroxidase acts as a catalyst, it amplifies the reaction product and may be used to detect antigen at the electron microscopic level.

These methods can be performed relatively easily with commercially available secondary antibodies saturated with their conjugated label. In addition, ferritin (an iron-containing protein) and colloidal gold-conjugated antibodies are used to detect primary immunoglobulins electron microscopically. Furthermore, new materials, such as the staphylococcus derivative protein A, react quantitatively with portions of IgG immunoglobulins and are used as surrogates for the secondary antibody in quantitative immunohistochemical tests for specific antigens.

The failure of certain antisera (e.g., those against major histocompatibility complex [MHC] antigens) to give a positive indirect immunofluorescence or to cause the lysis of cells in the presence of complement does not necessarily mean that the specific antigen is not on the tissue. The antigens may be present in low concentrations or in **privileged sites** which the antibody cannot penetrate. Despite negative results with other techniques, immune electron microscopy may reveal low levels of antigens.

Likewise, the loss of a detectable antigen does not necessarily mean that the antigen has been eliminated from the surface, since something may be covering it and preventing the attachment of the specific antibody. Since antigenic covers are frequently found to be carbohydrate side chains of proteins, one way of getting around the cover is to gently digest it with an enzyme such as neuraminidase.

Species-specific antigens are useful in viable hybrids for distinguishing between paternal (or embryonic) gene activities and the action of stored

oocytic (maternal) materials. As long as the antibodies are not cross-reacting, the appearance of antigens characteristic of the paternal genome is unmistakable evidence of *de novo* gene action in the hybrid embryo.

In sea urchin hybrids between *Tripneustes esculentus* eggs and *Lytechinus variegatus* sperm, anti-*L. variegatus* antiserum binding to early gastrulas is detected by indirect immunofluorescence. The reciprocal hybrids likewise show the appearance of paternal antigens in early gastrulas. Furthermore, Fab fragments of anti-gastrula antibodies prevent reaggregation of isolated hybrid cells with *L. variegatus* cells, suggesting that paternally derived antigens have appeared at the surfaces of early gastrula cells (Fig. 15.3) (see McClay, 1979).

Tissue-specific embryonic antigens (TSEAs) are detected with antisera raised against tissue extracts and adsorbed with other tissue extracts to remove unspecific antibodies. TSEAs exhibit **differential expression** when they appear in specific tissues and **differential restriction** when they disappear in specific tissues (see Monroy and Rosati, 1979). Frequently, tissue-specific antigens are present in germ layers before differentiated tissues. Antigens present later in distinct tissues may also occur earlier in common precursor germ layers.

In the course of development, some tissue-specific antigens are sorted out to particular tissues rather than appearing *de novo* at the time of differentiation (Fig. 15.4) (Clayton, 1953). In amphibians, several antigens make their debut simultaneously in the blastula. Based on their ultimate distribution, these antigens can be assigned to three classes: common (C) antigens found in both

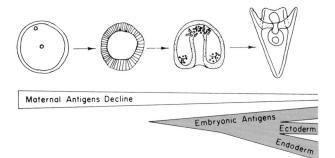

FIGURE 15.3. The decline of maternal antigens already present at fertilization and synthesized on stored oocytic mRNA is accompanied by the appearance of embryonic antigens synthesized *de novo* on both paternal and maternal mRNA. From D. R. McClay and G. M. Wessel, in E. H. Davidson and R. A. Firtel, eds., *Molecular biology of development,* Alan R. Liss, New York, 1984. Used by permission.

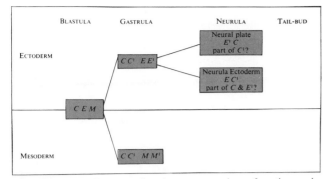

FIGURE 15.4. Scheme for the segregation of antigens in embryos of the alpine newt, *Triturus alpestris*. C E M, classes of antigens present in the blastula; C' E' M', classes of antigens added in the gastrula. C and C', common antigens present in ectoderm and mesoderm, E and E', ectodermal antigens, M and M', mesodermal antigens. In the neurula, different fractions of common and ectodermal antigens become localized or restricted in the neural plate as opposed to the remainder of the neurula's ectoderm. Adapted from R. M. Clayton, *J. Embryol. Exp. Morphol.,* 1:25 (1953), by permission of the Company of Biologists and the author.

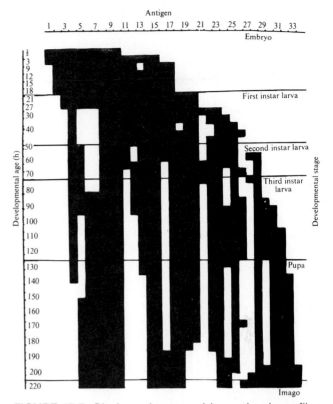

FIGURE 15.5. Block graph summarizing antigenic profiles of *Drosophila melanogaster* at different stages of development. Antisera were prepared in rabbits against extracts of whole organisms at the stages marked by horizontal lines. Extracts of the organisms at more finely divided stages (i.e., according to hours) were tested against the antisera, and individual antigens were identified by immunoelectrophoresis. From D. B. Roberts, *Nature (London),* 233:394 (1971), by permission of Macmillan Journals and courtesy of D. B. Roberts.

ectoderm and mesoderm, ectodermal-specific (E) antigens, and mesoderm-specific (M) antigens. Other antigens (C', E', and M') appear in the germ layers of the gastrula (E', M') and may be further sorted out at a later time (E').

Stage-specific embryonic antigens (SSEAs, also embryo-specific and phase-specific antigens) are transient antigens located in embryos only at particular stages of development. SSEAs are theoretically activated at the particular stage in which they appear, and the control of that activation is presumably related to gene activity.

Ideally, specific anti-SSEA antisera are prepared against antigens peculiar to a stage, but in practice, SSEAs are rarely limited to particular stages. Anti-SSEA antisera are generally prepared against antigens in collections of embryos at loosely defined stages and then adsorbed with extracts from tissues at other stages to remove unspecific antibodies.

In *Drosophila melanogaster*, at a gross level, some SSEAs appear early and disappear later. Antibodies raised in rabbits against whole embryos (Fig. 15.5, top block) precipitate 10 antigens in newly fertilized eggs (i.e., collected in the first $\frac{1}{2}$ hour after laying), 8 of which are also present in unfertilized eggs and probably represent products of oogenesis. The other two antigens may represent products of embryonic synthesis dependent on previously masked or inaccessible mRNA.

Five new antigens are added during the 3 hours prior to cellularization. Two new antigens are added during the period of gastrulation and germ band elongation. At this time, paternal-type isozymes are also introduced, and a variety of mutants exercise their lethal powers.

Antibodies against first instar larvas (second block in figure), second instar larvas (third block), third instar larvas (fourth block), pupas (fifth block), and imagos (i.e., adults, bottom) share about half of the embryonic antigens. Some of these disappear at one or another time only to come back later (Roberts, 1971).

In amphibians, a class of SSEAs has been detected in the late blastula with an *in vivo* assay. An antiserum was prepared in rabbits against gastrulas of the alpine newt, *Triturus alpestris*, and adsorbed with blastulas. This gastrula-specific antisera killed 63% of the gastrulas tested and 78%

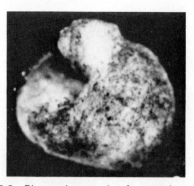

FIGURE 15.6. Photomicrograph of an embryo incubated since the blastula stage in rabbit anti-gastrula antiserum. Gastrulation has been inhibited, and the dorsal lip is either exaggerated or partially extruded. From R. M. Clayton, *J. Embryol. Exp. Morphol.,* 1:25 (1953), by permission of the Company of Biologists and the author.

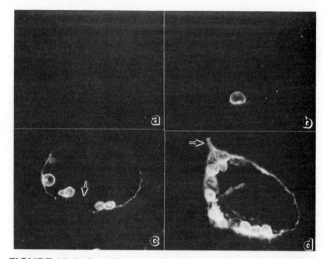

FIGURE 15.7. Dark-field immunofluorescent micrograph of sections of the sea urchin, *Lytechinus variegatus,* stained with monoclonal antibody to the mesodermal antigen AA1g8. Absent in the early blastula (top left [a] frame), the antigen appears on primary mesenchyme cells only after they migrate into the blastocoel (b). The antigen remains in primary mesenchyme cells and also covers the surface lining the blastocoel (c). At the prism stage (d), the antigen penetrates the ectoderm beneath the site of spicule synthesis (arrow). From D. R. McClay and G. M. Wessel, in E. H. Davidson and R. A. Firtel, eds., *Molecular biology of development,* Alan R. Liss, New York, 1984. Used by permission.

of the late blastulas (Fig. 15.6), while 92% of the controls incubated in dilute serum from unimmunized rabbits (i.e., lacking gastrula-specific antibodies) survived (Clayton, 1953).

Most of the tested late blastulas die on the threshold of gastrulation, and half of those surviving show abnormal gastrulation. Possibly, late-blastula SSEAs are required for gastrulation, since antigens may mediate cell-to-cell contacts. In *Rana pipiens,* for example, a rabbit antiserum prepared against gastrulas blocks the adhesion of isolated gastrula cells and prevents their sorting themselves out in reaggregation masses (Spiegel, 1954).

In the sea urchin, some TSEAs are also SSEAs. For example, anti-mesodermal serum (Fig. 15.7) reacts with primary mesenchyme cells as they make their appearance at the mesenchymal blastula stage. Typically, TSEAs are widespread when they first appear in the embryo and become restricted afterward (Fig. 15.4).

An endodermal antigen (Fig. 15.8), absent in the mesenchyme blastula (a), first appears in the vegetal pole just prior to gastrulation (b). It becomes concentrated apically in endoderm cells (c) but disappears from the foregut of the gastrula. In the pluteus larva (d), the antigen is restricted to the midgut and hindgut.

Antigens that seem to be specific to the primary mesenchyme appear for the first time at the very moment primary mesenchyme emerges from the vegetal pole (Fig. 15.7). The cells remain positive for the antigen, and the antigen spreads through the blastocoel's lining as the embryo hatches.

In mammals, early-appearing, stage-specific antigens frequently disappear later. The male H-Y antigen of sperm, for example, is also present on half of the mouse embryos tested (presumably male embryos) beginning with the 6–8-cell stage. The antigen later disappears on all but the male germ-line cells.

Unlike sperm, the cell surface of the unfertilized egg does not express immunogenic antigens (even when an animal is subjected to a prolonged regime of inoculations with adjuvant added to the inoculum). Antisera raised in rabbits against zona-free unfertilized mouse eggs fail to show any specific antibodies after adsorption with adult mouse tissue (kidney, liver, spleen).

Similarly adsorbed antisera raised against zona-free mouse blastocysts and exhibiting an immune response to 8- and 12-cell embryos register only weak binding to unfertilized eggs. In general, antibodies cross-reacting with blastocyst antigens tend not to be concentrated prior to the 8–12-cell stage.

Antisera raised against the portion of the TE known as the ectoplacental cone of 7.5-day mouse embryos (see Fig. 14.16) fail to react with unfertil-

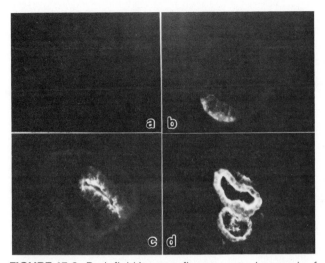

FIGURE 15.8. Dark-field immunofluorescent micrograph of a section through the sea urchin, *Lytechinus variegatus,* stained with fluorescein-conjugated monoclonal antibody against the endodermal antigen, De25c7. Absent in the mesenchymal blastula (a), the antigen appears in the vegetal plate of the embryo just prior to gastrulation (b). The elongating gut of the gastrula loses the antigen at its tip (c), and ultimately the antigen is restricted to the midgut and hindgut of the pluteus (d). From D. R. McClay and G. M. Wessel, in E. H. Davison and R. A. Firtel, eds., *Molecular biology of development,* Alan R. Liss, New York, 1984. Used by permission.

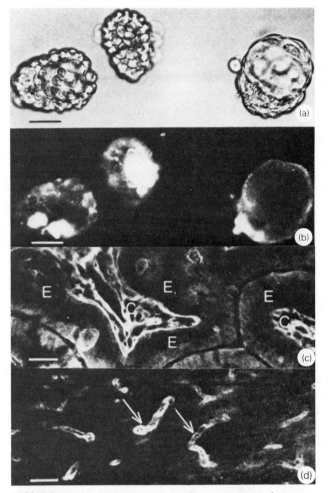

FIGURE 15.9. Indirect immunofluorescence of mouse blastocysts and sections of adult tissue exposed to rabbit anti-mouse L-cell fibroblast antisera adsorbed with mouse liver cells. (L-cells are a mouse fibroblast cell line.) (a) Phase-contrast micrograph of blastocysts after removal of the zona pellucida. (b) Same with dark-field UV illumination. (c) Section of adult mouse small intestine showing positive reaction in the connective tissue cores (C) of villi and negative reaction in the epithelial lining (E). (d) Section of fetal mouse brain showing positive reaction in capillaries (arrows) and negative reaction in neural material. Bar = 25 μm. From L. M. Wiley, in A. A. Moscona and A. Monroy, eds., *Current topics in developmental biology,* Vol. 13, Academic Press, Orlando, 1979, p. 167. Used by permission.

ized mouse eggs and zygotes. Specific anti-ectoplacental cone antigens are detectable at the 8-cell stage, in the inner cell mass (ICM), if only weakly, and strongly and persistently in the TE and extraembryonic ectoderm (see Wiley, 1979).

Antisera raised against adult tissues may also cross-react with gametes and embryos. Antisera to mouse neural tissue cross-react with sperm and preimplantation embryos (Solter and Knowles, 1979). Similarly, antisera raised against mouse L-cell fibroblasts (adsorbed on liver cells) begin to react positively with morulas. Both TE and ICM are positive in blastocysts (Fig. 15.9), but only connective tissue elements remain positive in fetal and adult tissues. Possibly, complete sets of tissue-specific antigens appearing at the morula-to-blastocyst transformation are expressed widely and only later restricted to their adult tissues.

Developmental blocks can be erected by antisera. The same antisera that react with fixed and dead embryos in microscopic preparations may also cause **developmental arrest** when applied to living embryos.

The study of developmentally significant antigens through the effects of antibodies on embryos

was pioneered on amphibians (Spiegel, 1954). Antisera to amphibian surface antigens prevented the reaggregation of isolated embryonic cells without necessarily killing the cells (in the absence of complement).

Since unspecific antisera can also be cytotoxic to the embryo, an essential criterion of antibody specificity is the viability of embryos treated with antisera or, better still, the reversibility of the antisera's effects. Antisera raised against sperm or eggs can kill mouse blastocysts, and antisera raised against blastocysts can inhibit development, but brief exposure to antisera causes only transient developmental arrest.

Fab fragments of rabbit anti-F9 antisera prevent compaction of blastomeres and arrest further development (see Jacob, 1979). The fragments even reverse compaction or **decompact** already compacted mouse morulas. Nothing more complex than steric hindrance by anti-F9 Fab may prevent the normal reaction of the antigen and its receptor thereby blocking compaction, since embryos undergo compaction and continue developing normally when removed from the Fab solution.

The antisera may be reacting with protein "p140," recognized by a monoclonal antibody (ECCD-1). The protein is present on the surface of F9 cells previously treated with trypsin in the presence of calcium ions (Ca^{2+}) but absent from cells treated with trypsin alone. Ca^{2+} seems to protect the protein from degradation by trypsin, thereby preserving it for immunological recognition. Called **cadherin,** the antigen is a Ca^{2+}-dependent cell–cell adhesion molecule and part of a Ca^{2+}-dependent system (CDS) (Yoshida-Noro et al., 1984; see Chapter 20).

When the cavity of the blastocyst begins to appear, the effects of rabbit anti-F9 Fab fragments are no longer reversible. At that time, the antisera become toxic, but isolated ICMs can still be decompacted. The appearance of macula (spot) junctions (desmosomes) during compaction may spell the end of reversibility, suggesting that the F9 antigen is involved in close cell–cell adhesion.

Immunologic techniques as "chemical scalpels" dissect the embryo. In the case of mouse blastocysts, for example, the ICM can be freed of TE by exposing blastocysts to rabbit anti-mouse antisera and lysing the TE by further exposure to guinea pig complement. The tight zonula junctions between TE cells prevent the antisera from reaching the ICM, and, when the guinea pig complement dissolves the TE, the ICM is unveiled, untouched by antibody.

Previously undetected antigens may now be identified on the exposed ICM. For example, laminin, a major component of basal laminae, absent during cleavage and in the trophoectoderm of early (3-day) mouse blastocysts, is found, if only weakly, on immunosurgically exposed ICM cells of late blastocysts and on ectodermal cells of isolated ICMs cultured for 2 days *in vitro.*

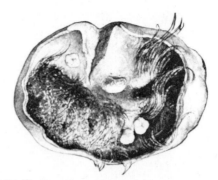

FIGURE 15.10. Drawing of an opened teratoma cyst of the human ovary showing a variety of tissues and structures growing within. Hair, oil glands, teethlike structures, and a tonguelike structure with papillas are distinguishable. From L. B. Arey, *Developmental anatomy,* 7th ed., Saunders, Philadelphia, 1965. Used by permission.

Using Tumors for Studying Gene Activity in Embryos

An important model for early embryonic events has recently been fashioned from studies on tumors known as **teratomas** (Gk. *teratas* monster + *oma* mass, usually specifying a kind of tumor). These tumors are typically found in the ovary (Fig. 15.10), but they also occur at other locations. Nonmalignant teratomas contain disorganized but differentiated tissues of all germ layers. Malignant teratomas, called **embryomas,** are poorly differentiated, highly proliferative, invasive, and destructive tumors.

A variety of teratomas appearing in inbred strains of mice are the material of the mouse **teratoma system** and model for embryonic differentiation (Stevens, 1960). The system includes **testicular teratomas** (Fig. 15.11) arising spontaneously in mouse strain 129 from primitive germ cells, spontaneous **parthenogenic ovarian teratomas** especially those from the LT strain, and teratomas of other strains induced by grafting 4–7-day embryos into the testes of syngeneic mice. These teratomas, among others, are commonly called **teratocarcinomas,** since they are malignant when injected at particular sites and contain epithelial-like cells (i.e., carcinoma).

Teratocarcinomas may be serially transferable from mouse to mouse of the same strain especially in ascites fluid (i.e., intraperitoneally) where the cells proliferate and ultimately kill the host mouse. The same cells are also propagated *in vitro* where they have given rise to various stable cell lines known as **embryonal carcinoma** (EC) cell lines.

These tumor cells frequently have tumor-cell antigens that are particularly "good antigens," and

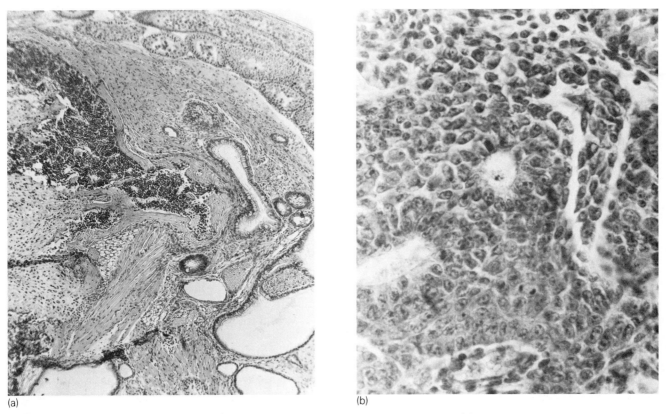

(a)   (b)

FIGURE 15.11. Photomicrographs of histological sections through teratocarcinomas. (a) A testicular teratoma of an adult showing differentiated tissues representing different germ layers. Mesoderm is represented by testicular tubules (top), mesenchyme, cardiac muscle, cartilage and bone with marrow, and skeletal muscle (bottom). Endoderm is represented by mucous epithelium lining cyst (right), and ectoderm is represented by neuroepithelium (lower right). (b) Primitive undifferentiated teratocarcinoma cells have large nuclei and no typical shape. Note mitotic figures. From L. C. Stevens in C. L. Markert and J. Papaconstantinou (eds.), *Developmental biology of reproduction, Academic Press, 1975. Used by permission.*

the antibodies they elicit may cross-react with surface antigens of other tumors and with embryos. Because tumor cells are often relatively uniform and available in virtually unlimited supply through *in vivo* serial transfer and through *in vitro* tissue culture (compared to limited quantities of mammalian embryos), tumor cells have become an important source of embryonic antigens.

Nullipotent teratocarcinomas and EC cell lines, such as F9 (the source of the F9 antigen) derived from the OTT6050 teratoma, typically fail to differentiate and produce tumors of virtually pure EC cells. The failure to differentiate should not be considered absolute, however. In the presence of retinoic acid, for example, "nullipotent" F9 cells differentiate as endodermal cells resembling those lining the yolk sac and even form embryoid bodies (see below) when treated as cell aggregates.

Multipotent or pluripotent teratocarcinomas and EC cell lines are capable of differentiating in vivo and in vitro. Undifferentiated **primitive teratocarcinoma cells** (PTCs) become **differentiated teratocarcinoma cells** (DTCs). For example, PCC4 cells derived from the OTT6050 teratoma may differentiate into tissues representing all germ layers (Fig. 15.11). In the process of differentiating, the cells lose their malignancy and moderate their rate of division.

In addition, some teratocarcinoma cells transferred through ascites fluid form **embryoid bodies** of PTCs wrapped up in a layer of endodermal-like cells (Fig. 15.12). Embryoid bodies retain their malignancy, but their PTCs have the potential for differentiating. Some malignant embryonic teratocarcinoma cells injected into mouse blastocysts even differentiate into normal tissues including fertile gametes (Mintz and Illmensee, 1975).

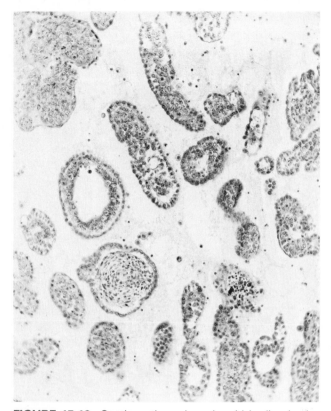

FIGURE 15.12. Sections through embryoid bodies in the ascites fluid of a mouse with a transplantable teratoma. An outer layer of endoderm-like cells surrounds a core of generally undifferentiated embryonal carcinoma cells. From L. C. Stevens in C. L. Markert and J. Papaconstantinou (eds.), *Developmental biology of reproduction,* Academic Press, 1975. Used by permission.

Teratocarcinoma antigens are of special interest because they share some physical characteristics with transplantation antigens present on all adult cells and coded by genes of the major histocompatibility complex (MHC, e.g., the **H-2** *gene complex of mice).* MHC antigens stimulate the adult organism's immune response and govern its ability to accept or reject grafts of tissue and organ transplants. All other antigens normally present in adult tissues, with the exception of **minor histocompatibility** antigens (e.g., the H-Y antigen of male mice), are lumped in the category of **nonhistocompatibility** antigens.

MHC antigens are present in mouse sperm, absent in unfertilized eggs, and present in blastocysts up to 6 days if only at the low concentrations detected by the immunoperoxidase technique at the electron microscopic level. These low concentrations would seem fortunate for implantation inasmuch as the embryo with few MHC antigens is not likely to elicit a strong graft-rejection reaction by the mother (Edidin, 1976).

By 6.5 days, mouse blastocysts express the H-2 gene products in their ICM, and, in the course of implantation, MHC antigens replace teratocarcinoma antigens expressed earlier (e.g., the F9 antigens). The replacement of teratocarcinoma antigens by MHC antigens in the blastocyst parallels the loss of tumor-cell properties and gain of adult cell properties related to cell–cell interactions. Possibly, teratocarcinoma antigens are precursors of MHC antigens (see Wiley, 1979).

Cross-reacting anti-teratoma antisera may be useful for probing embryos for the appearance of significant new antigens. Antibodies raised against EC lines are frequently broadly, but not universally, cross-reactive with embryonic antigens.

For example, anti-402AX antisera (raised in rabbits against the 402AX subline of a 129 strain teratocarcinoma) cross-react with several cell lines (SV40-transformed mouse 3T3 fibroblasts, mouse L-cell fibroblasts, numerous tumor cell lines) and with some cell-surface antigens on normal embryonic cells, although the antisera do not react with any normal adult somatic cells. Similarly, syngeneic antisera to multipotent PCC4 cells and syngeneic antisera to nullipotent F9 cells react with surface antigens present on all multipotent EC cell lines and mouse spermatozoa, but these antisera do not react with any adult somatic tissue or even with the differentiated cells derived from the multipotent PCC4 cells.

Anti-402AX antisera actually contain three antibodies specific for tumor antigens. Two of these antibodies are cross-reactive with antigens on the surface of the embryo. Embryonic antigen I is expressed on unfertilized eggs and the early embryo (Fig. 15.13a). It becomes restricted to the ICM in blastocysts prior to implantation and is absent on the TE of hatched blastocysts. Antigen II, also present on hepatoma cells, is first expressed on the mural TE of blastocysts (Fig 15.13b). Later, the antigen is restricted to the ICM.

Syngeneic F9 antisera and F9 MAb are broadly cross-reactive, although they fail to react with unfertilized eggs and with SV40-transformed mouse 3T3 fibroblasts. Not species specific, the antisera react with sperm and with morulas from every mammal tested, from "kangaroo to man" although not with chicken, axolotl, and frog (see Jacob, 1979). Furthermore, a human male germ-line antigen cross-reacts with F9 antisera.

Still, no stage-specific antigen or definitive pattern for the coming and going of cross-reacting tumor-specific antigens has surfaced in the embryo

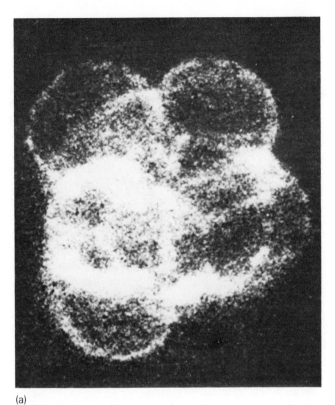
(a)

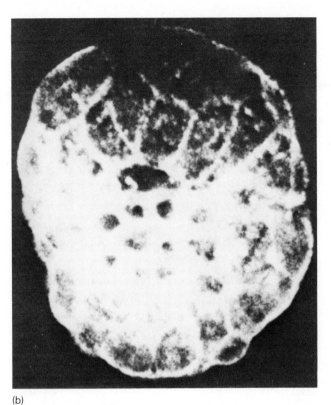
(b)

FIGURE 15.13. Indirect immunofluorescent micrograph of mouse embryo. (a) Morula exposed to rabbit antibodies to antigen-I (from an anti-402AX antiserum eluted from an affinity column) and then to fluorescein-labeled goat anti-rabbit antiserum. The fluorescent outlines of cells indicate the presence of antigen on the cells' surface. About ×600. (b) Blastocysts exposed to whole rabbit anti-402AX antiserum and then to fluorescein-labeled goat anti-rabbit antiserum. The fluorescent outlines of mural trophectodermal cells indicate the presence of antigens on the cells' surfaces, but cells at the embryonal pole and inner cell mass appear only mildly reactive. About ×600. From M. Edidin, *Embryogenesis in mammals,* Ciba Foundation symposia 40 (new series), Elsevier/Excerpta Media/North-Holland, Amsterdam, 1976. Courtesy of M. Edidin. Used by permission.

(see Wiley, 1979). F9 antisera react increasingly with antigens generally distributed over the embryo prior to implantation, and PCC4 antisera are restricted to the ICM even prior to implantation (Gachelin et al., 1977). F9 antigens become restricted to the embryonic ectoderm (i.e., the epiblast) of postimplantation embryos and gradually disappear except in male germ-line cells.

Nevertheless, the possibility continues to exist of using anti-tumor antisera as probes for the analysis of the preimplantation embryo. The cross-reactivity of tumor and embryonic antisera is not totally surprising, since the surface properties of some embryonic cells, such as the ability to migrate and become insinuated between other cells, resemble the malignant properties of tumor cells to metastasize, invade, and destroy normal tissue.

Mimicking Gene Activity: Teratology (Gk. *teratas* monster + *logia* study)

Abnormal organisms resembling mutants are frequently produced following the application of physical or chemical agents to an embryo. The products are called **phenocopies,** and the agents are called **teratogens.**

Since the teratogens are effective at different stages, durations, and concentrations, the possibility arises that the mutant genes whose effects are mimicked likewise act at different times and doses. The study of normal gene activities by treating embryos with heat or chemicals was first suggested by Richard Goldschmidt based on his research on pigment patterns in butterfly wings, but Walter Landauer brought the study of phenocopies to the attention of a wider audience through his vigorous research on chicks (see Oppenheimer, 1981).

Some teratogens seem to produce their effects by causing stable changes in nuclei of blastomeres at the blastula stage. Ether, for example, a teratogen in insects, affects the nuclei of blastoderm cells. Ether-treated *Drosophila melanogaster* resemble **bithorax** mutants, a homeotic mutant affecting differentiation in the metathoracic segment (see Fig. 22.8). Since activated genes are not affected (i.e., only the activation of genes in the **bithorax complex** is affected), ether may operate by imitating the effects of a **super-repressor** regulator mutation (*RB-pbx*, see below).

Teratogens have been used extensively to mimic genetic effects later in development, but their use in early development is largely restricted to the period in which genes are active. Like the activity of zygotic mutants, teratogenic effects emerge at the late blastula stage. The treatments **canalize** development (i.e., narrow its scope and restrict its potential) prematurely in contrast to normal development that exposes the embryo's potential and expands the scope of possibilities.

PROBING GENE ACTIVITY

The question of how the one-dimensional sequence information stored in the DNA is converted into the three-dimensional structure of an embryo, or four-dimensional formation if we also include time, is the fundamental problem of developmental biology.

W. J. Gehring (1987, p. 1245)

Recombinant DNA technology now makes it possible to manufacture cloned DNA from virtually any region of a cell's genome. Excised with restriction nucleases, inserted into plasmids, and grown in bacteria or yeast transformants, an entire genome can be stored in a DNA library, and banks of cloned DNAs can be used to monitor developmental changes in mRNAs. Specific portions of a cell's DNA can be made available in virtually any amount, mutated biochemically, and even reinserted into cells.

Cloned DNA is utilized in different techniques to detect transcripts in cells or locate genes on chromosomes. Most of these techniques employ radioactive, single-stranded nucleic acids called **probes** capable of **hybridizing** with complementary nucleic acids. In addition to cloned DNA, probes may be prepared with mRNA or with **copy DNA**

(cDNA, also called complementary DNA) produced with reverse transcriptase from mRNA.

Probes are used in a variety of ways. In the *in situ* **hybridization** technique, the probe is applied directly to a cytological preparation, a nucleus, a cell, or a tissue section. In **blotting** techniques, nucleic acid on an electrophoretic gel is absorbed and baked into nitrocellulose paper prior to hybridization. DNA:DNA hybridization (known as Southern blots after E. M. Southern who developed the technique) makes it possible to count the number of gene copies in a genome, while similar RNA:DNA hybridization (known as Northern blots) allows one to measure the number of RNA copies made by a gene. Differences between cDNA and cloned nuclear genes also distinguish introns from exons.

Sequencing the nitrogenous bases in the cloned DNA leads to identifying the open reading frame and, in turn, predicting the amino acid sequence of the polypeptide. Cell-free translation of RNA selected by cloned genes (also known as hybrid-selected translation) makes it possible to correlate genes with specific proteins.

Elegant in its simplicity, and more direct for studying gene action than any of its predecessors, the new technology is nevertheless not without problems. The question of specificity, for one, is dogging. Repeated portions of mRNAs interspersed with unique portions may not only be complementary with the DNA that encodes them but with DNA that encodes related transcripts and with DNA that does not encode any transcript. Several techniques have now been used to improve specificity, and more will undoubtedly follow.

A larger problem is the biological "uncertainty principle" inherent in molecular biology: one may either identify a gene without knowing its developmental role, or one may know a developmental role without identifying the gene. Attempts to close the window of uncertainty have released an avalanche of research. Some developmental biologists approach the problem by trying to fill in the open space; others search for ideal systems in which the relationship between genes and developmental roles is direct, but the window remains open.

Ultimately, the current genetic framework for viewing development may prove too restrictive. Gene action in terms of the transcription of structural and regulatory genes may not encompass the whole picture of development. Alternative forms of gene activity have hardly been considered, much less eliminated (Bell et al., 1972). The presence of reverse transcriptase in rabbit trophectodermal cytoplasm (Manes et al., 1981), for example, raises the possibility that RNA also provides genetic information.

For the moment, methods utilizing recombinant technology, cloned DNA, and *in situ* hybridization are enormously popular. Research frequently proceeds by identifying mutations with widespread effects. The mutant genes are then pursued to loci on genetic maps via standard genetic techniques, and hypotheses concerning the genes' activities are tested by manipulating other genes or events governing the timing of gene activity.

Two types of model for the *de novo* production of gene products have emerged from this sort of research. In **transitional models,** the genome is switched on at an abrupt transition. Based chiefly on results with regulative species in which the appearance of new genetic products follows a period of quiescence, genic activity is seen as the antidote to an early indifferent state and a key feature of determination.

Alternatively, in **coordinated models,** synthetic activity undergoes a gradual integration of earlier-acting and later-acting genes. Genes whose products are synthesized prior to fertilization are **maternal-effect genes,** while those whose products are only synthesized following fertilization are **zygotic-effect genes.** The laying down and sequestration of gene products is associated with regulation (e.g., in insects up to the syncytial blastoderm stage; see Fig. 11.10), while the operation of these products is associated with determination (e.g., beginning at the cellular blastoderm stage). This distinction may not be especially neat, however. Gene products may be synthesized both before and after fertilization, and different segmentation genes exhibit features of both maternal and zygotic effects.

Transitional Models

New transcripts begin to come into play in the late-blastula stage. Their appearance may be sudden or their effects suddenly apparent. Materials synthesized from the new transcripts may or may not be utilized at the time, and much may be additions to pools of old products.

In sea urchins, translation during cleavage and blastulation is specific despite transcriptional quiescence. About half the tubulin in the cilia of sea urchin blastulas and the hatching enzyme that facilitates the release of mesenchyme blastulas from the fertilization membrane are synthesized even if transcription is totally arrested by actinomycin D. Tubulin and hatching enzyme even appear in enucleated merogons. Still, pools of rapidly turning over materials, such as histones and tubulin, are supplemented by new transcripts in the blastula (see Davidson, 1986).

The initiation of transcription has been followed closely in *Strongylocentrotus purpuratus* for two mRNAs with the aid of cDNA clones known as pSpec 1 and pSpec 2 differing at their 3' ends (Brushkin et al., 1982). The respective hybridizing transcripts, 1.5 kilobases (kb) and 2.2 kb long, follow different courses of activation. The 1.5-kb transcript begins to increase in prevalence at hatching of the mesenchyme blastula, and the 2.2-kb transcript begins to accumulate 10 hours later. The amount of both transcripts increases greatly and peaks during the gastrula stage, but pSpec 1-hybridizing transcripts are 10 times more abundant than pSpec 2-hybridizing transcripts. The transcripts therefore seem to be independently regulated.

Instead of translating single proteins, about 10 acidic, low molecular weight proteins are translated by the rabbit reticulocyte lysate cell-free system from the sea urchin pSpec 1- and pSpec 2-hybridizing mRNAs. Several larval ectodermal proteins comigrate in two-dimensional electrophoresis with the Spec 1- and Spec 2-proteins. Moreover, the prevalence of three proteins in particular in larval ectoderm follows the mRNA pattern. These proteins appear in the blastula prior to hatching, accumulate preferentially in presumptive ectoderm, are enriched by the early-gastrula stage in ectoderm, and reach their peaks in ectoderm at the late-gastrula–early-prism stage.

The proteins are probably troponin C–related, calcium-binding proteins, but neither the cloned DNAs nor their predominant mRNAs can be assigned to specific proteins. Possibly, closely related *Spec 1* and *Spec 2* genes encode a small family of proteins functionally related to contractile proteins and capable of activating actin in the presence of free calcium. Intriguingly, the expression pattern of *Spec 1* and *2* genes resembles the expression pattern of at least one actin gene.

Actin is encoded by a complex, nonallelic multigene family. Five of the genes (*CyI, CyIIa, CyIIb, CyIIIa,* and *CyIIIb*), differing in their untranslated 3' terminal regions, encode cytoskeletal (Cy) actin proteins, while a sixth unlinked gene (*M*) encodes actin in adult muscle tissues (Table 15.2) (Shott et al., 1984). In addition, the family includes *CyIIc* and *CyIIIc* pseudogenes lacking initiation sequences at the 5' end.

The Cy actin transcripts are reduced from a primary transcript of about 5 kb by the excision of

TABLE 15.2. Expression of the actin gene family in *Strongylocentrotus purpuratus* larvas and adult tissues

Gene	Transcript Length (kb)	First Appearance in Embryogenesis	Expression in Adult Tissues Relative to Pluteus						
			Testis	Ovary	Coelomocytes	Intestine	Tubefoot	Lantern Muscle	Pluteus
CyI	2.2	Maternal mRNA	+ + + + +	+ + + + +	+ + + + +	+ + + + +	+ + + + +	+ + + + +	+ + + + +
CyIIa	2.2	Gastrula (40 hr)	−	−	+/−	+	+/−	+/−	+
CyIIb	2.1	Early blastula (14 hr)	+/−	+/−	+	+	+	+	+
CyIIIa	1.8	Maternal mRNA	−	+	−	−	−	−	+ + + + +
CyIIIB	2.1	Early blastula (14 hr)	−	+	−	−	−	−	+
M	2.2	Early pluteus (62 hr)	+/−	+/−	−	+	+ + +	+ + + + +	+

Source: R. J. Shott, J. J. Lee, R. J. Britten, and E. H. Davidson, *Dev. Biol.* 101:295(1984), by permission of Academic Press.

introns at amino acid positions 121/122 and 203/204. M actin primary transcripts contain additional introns at amino acid positions 41/42 and 267/268. While sharing none of these introns, vertebrate actin genes have introns at all four corresponding positions.

While three actin genes (*CyI, CyIIa,* and *CyIIb*) are linked 5′ → 3′ with the same transcription orientation, they are not coordinately expressed. CyI actin transcripts are ubiquitous. Although they are present only in the class of rare stored (maternal) mRNAs in eggs, they appear throughout embryos at the beginning of development and continue to appear thereafter.

The CyIIb actin gene is also expressed in all tissues after becoming activated for the first time in the blastula. CyIIa is activated in the gastrula. It is expressed predominantly in the vegetal plate cells, and transcripts accumulate in the larval gut.

The CyIII subfamily is also linked 5′ → 3′. CyIIIa and b are separated by a 6-kb spacer and transcribed in the same direction. Transcripts are produced in the ovary, presumably as a result of expression in developing oocytes, and throughout development, if only at low levels prior to the blastular stage. Peak transcription occurs in the gastrula where it is limited to the dorsal (aboral) ectoderm.

The 1.8-kb CyIIIa actin mRNA is the major actin gene product of the embryo. The 2.1-kb CyIIIb actin mRNA is the minor product. CyIIIa transcripts account for over 50% of the total actin messenger and are presumably responsible for most of the actin microfilaments present in the blastula's presumptive dorsal ectoderm. The contraction of these microfilaments, triggered by increased free intracellular calcium ions, flattens the cuboidal presumptive ectodermal cells of the dorsal blastula into the squamous pavement epithelial cells of the gastrula (see Fig. 14.19).

DNA sequences in close proximity or within the actin genes are probably responsible for the qualitative and quantitative modulation of the genes' expression. Linkage of actin genes is more likely a consequence of the evolution of the multigene family than a significant factor in developmental regulation of expression (Shott-Akhurst et al., 1984).

In amphibians, transcription begins in the blastula at the midblastula transition (MBT).
New species of mRNA are transcribed, followed by transfer RNA (4S RNA) and ribosomal RNA (28S, 18S, and 5S rRNA, Fig. 15.1). The processing and translation of the new transcripts follow in rapid-fire succession.

The absence of transcription in earlier embryos is not due to deficiencies in RNA polymerases, since these enzymes are present in abundance (i.e., 10^4-fold per cell), and extracts of early embryos support transcription *in vitro* (see Roeder et al., 1976). Transcriptional cofactors are also present, and fertilized eggs are able to transcribe foreign DNA (e.g., yeast leucine tRNA) during a brief period between microinjection and incorporation into chromatin.

The fault may lie in part with **DNA-binding transcriptional inhibitors,** since permanent transcriptional activity can be induced prematurely by injecting fertilized eggs with excess DNA (Fig. 15.14). The amount of injected DNA required to induce transcription equals the DNA of a 4000–5000-cell blastula (i.e., a blastula at the 11th to 12th cleavage division), remarkably close to the amount of DNA in embryos at the MBT. Possibly, inhibitors are depleted by binding progressively to the exponentially expanding population of DNA in the cleaving embryo. Alternatively, the MBT may be reached when new or endogenous **gene-activat-**

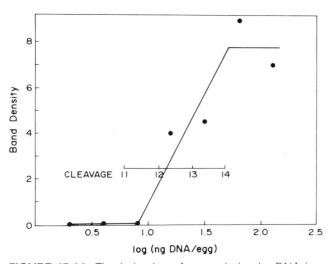

FIGURE 15.14. The induction of transcription by DNA injected into fertilized and centrifuged *Xenopus laevis* eggs. [The eggs were centrifuged in order to prevent cytokinesis (but not replication) and provide a uniform field for the pulse-label experiment.] Eggs injected with a DNA clone containing the pYLT20 (yeast leucine tRNA) gene were subsequently injected with known amounts of cloned pBR322 DNA. The graph shows transcription measured as the density of a 120 nucleotide pYLT20-dependent band in 5% polyacrylamide-urea gels (band density) as a function of the total DNA injected per egg (log ng DNA/egg). Brackets show the amount of DNA contained in embryos after 11–13 cleavages. From J. Newport and M. Kirschner, *Cell,* 30:687(1982), by permission of M.I.T. Press and the authors.

ing substances (specific or nonspecific) switch on the embryo's genes.

Activation does not require the compartmentalization of the zygote by cleavage furrows, since it occurs in cytochalasin B–arrested and in centrifuged eggs that undergo replication in the absence of cytokinesis. Furthermore, the process operates independently of the nucleus, since the MBT occurs on schedule for the cytoplasm when late-cleavage nuclei are introduced into uncleaved cytoplasm (see below, Figs. 16.4 and 16.6) (see Gerhart, 1980; Newport and Kirschner, 1982a).

The activation of ribosomal genes, which form the 7.5-kb precursor of the 18S, 5.8S, and 28S RNAs of ribosomes in *Xenopus* (see Fig. 4.3), is especially intriguing, since the new products are ultimately required by all cells. Ribosomal genes are repeated in tandem about 500 times at a single chromosomal site per haploid genome and have a large (about 140 bp) **gene promoter** at their 5' end. Similar **spacer promoters** (or Bam islands), differing by 14 or 15 nucleotides, appear at intervals farther out in the spacer between genes.

All these promoters are able to load RNA po-

lymerase I and initiate transcription. In addition, repetitive elements in the spacer act as upstream enhancer sequences. Independent of position and orientation, the repetitive elements seem to be attraction sites for binding transcription factor(s) that ultimately activate the gene promoters.

The size of the genetic unit, the number of spacer promoters and repetitive sequences influence the amount of transcription, and the combination of all may ultimately be responsible for activating the genes following the MBT. The question is: What controls the activity of the control elements? The answer may lie in chromatin itself. Possibly, spacer regions are normally under a degree of superhelical stress that keeps them repressed, thereby dampening gene activity (see Reeder et al., 1984).

Complex controls follow on the heels of the MBT. The controls may be positive, negative, or both. The 5S rRNA genes, for example, belong to a single class of genes (including transfer RNA genes) transcribed by RNA polymerase III, but while the oocyte type of 5S gene is turned off after the later-blastula stage, the somatic type remains turned on.

The subtlety of other controls continues to defy analysis. For example, the commitment to form muscle made by presumptive mesoderm cells takes place 10 hours before the actual transcription of skeletal and cardiac muscle alpha-actin mRNA (see Gurdon et al., 1984).

The mammalian version of the MBT occurs at the morula-to-blastocyst transformation (also MBT). Transcription is not abruptly activated, but translation increasingly utilizes fresh embryonic transcripts; asynchrony increasingly characterizes divisions, and cell surfaces mature (see Fig. 11.26).

Transcription of the paternal genome occurs early in eutherian mammals, but translation seems to be dominated by stored oocytic mRNA up to compaction (i.e., the 8-cell stage in mice, Fig. 11.25). The contribution of the active embryonic genome, in any case, cannot be great, since hybrids (e.g., between the rabbit, *Oryctolagus cuniculus,* and the hare, *Lepus europaeus,* or the cottontail, *Sylvilagu floridanus*) and severely aneuploid mice undergo cleavage. Arrest comes at the blastocyst stage.

Development beyond the blastocyst stage seems to depend directly on the activity of the embryonic genome. The rate of messenger turnover is relatively high in mammalian embryos, and stored oocytic mRNA is quickly degraded. Normally, by the blastocyst stage, the embryonic genome is in complete control of development (see McLaren, 1982).

Transcription in the blastocyst is abundant, but most of it is of unknown function. In late rabbit blastocysts (more than 90% TE), 1.8% of the total RNA seems to be from unique-sequence genomic DNA (i.e., informational or structural genes). Based on this minimal estimate for the content of single-copy DNA and an average mRNA of 2000 nucleotides, more than 6400 diverse proteins could be specified in rabbit TE cells, but only 1000–2000 peptide spots are resolved in two-dimensional electrophoresis on polyacrylamide gels. Most of the peptide spots are synthesized on classes of genes whose sequences are judged abundant and mid-abundant (i.e., regulatory) rather than unique. The peptides produced by genes represented by less than 10 copies (i.e., at least two-thirds of the active rabbit TE genome) remain unresolved (see Manes et al., 1981).

Along with turning on genes, mammalian embryos turn off genes. In the case of the X chromosome, all the genes are turned off on one of the two chromosomes normally found in females, possibly via DNA methylation (Monk et al., 1987).

Known as **heterochromatinisation** (i.e., chromatin condensation during interphase) or **X chromosomal inactivation** (Lyon, 1961), X changes into a late-replicating dense body at the edge of the nucleus called **sex chromatin** (Fig. 15.15) or the **Barr body** after its discoverer, the Canadian anatomist, Murray Llewellyn Barr. Females and males with additional X chromosomes turn off all but one X. The presence of a single Barr body in embryonic cells obtained by **amniocentesis** (the insertion of a needle into the uterus) ordinarily signals the development of a female embryo.

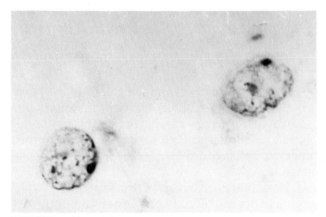

FIGURE 15.15. Photomicrographs of nuclei showing dense heterochromatic sex chromatin or Barr bodies. Courtesy of Sylvia F. Pan, M.D., Cytogenetics Laboratory, University of Pittsburgh, Graduate School of Public Health.

Each of the two X chromosomes normally present in female embryos is active in all cells between the 8-cell embryo and blastocyst. In mice and rats, the paternally derived X undergoes heterochromatinisation in the blastocyst's trophectodermal (TE) cells and, later, in parietal endodermal cells (see Gartler and Riggs, 1983).

Joint activity of both X's stops in the embryonic disk of female embryos at the onset of gastrulation, the primitive streak stage. In general, X inactivation is a random process affecting the paternally derived X or maternally derived X equally, but there are several exceptions. Human embryos tend to inactivate any X that is grossly defective. Kangaroos preferentially inactivate the paternally derived X in most cells of the somatic line, and the creeping vole, *Microtus oregoni*, eliminates the X from the male germ line (see Short, 1982).

Coordinated Models

Maternal effects are defined by the ability of the maternal genome to determine the phenotype of offspring, and zygotic effects are defined by the ability of sperm to rescue an egg from a mutant phenotype. The effects are also distinguished by other criteria.

In **temperature-shift experiments,** putative periods of gene action are equated to periods of temperature sensitivity. The premise is that, since temperature affects synthesis, different temperatures may affect transcription or the functional conformation of translated proteins. Hypomorphic **temperature-sensitive** mutants would suffer from inadequate translation or impaired proteins and produce defective or lethal phenotypes.

A range of **permissive temperatures** is identified by the development of animals with the normal phenotype, and **nonpermissive,** restrictive, or inactivating temperatures are identified by the appearance of animals with abnormal patterns of development. The point of changing temperature is called a shift, either a down-shift to a lower temperature or an up-shift to a higher temperature. A **temperature-sensitive period** (TSP) for lethality is defined as the "execution stage" or the biggest "window of vulnerability" to nonpermissive temperatures affecting development. Operationally, the TSP is an interval between the most extreme times temperature shifts still permit survival and relatively normal phenotypes, typically, the last down-shift to the permissive temperature and the first up-shift from the permissive temperature.

Theoretically, mutants classified as **maternal effects** have their TSP during oogenesis and operate through gene-controlling components synthesized prior to fertilization. Mutants classified as **early zygotic effects** have their TSP during early development and operate through products synthesized after fertilization.

Maternal-Effect Genes

The concept of **maternal-effect genes** goes back to 19th century debates over blending inheritance and whether maternal and paternal effects on development are distinguishable in the offspring. Since the advent of modern genetics, maternal effects have been redefined as traits that are not inherited equally in reciprocal crosses.

Maternal effects are transmitted through the cytoplasm and cortex of the egg rather than through the female pronucleus. They are conveyed to the oocyte before the completion of meiosis and fertilization. Self-replicating bodies in the oocyte's cytoplasm, such as mitochondria and infective agents including viruses, may transmit maternal effects, but the chief operative seems to be **stored** (masked or maternal) mRNA produced in the oocyte or in nurse cells and conveyed to the oocyte.

In one sense, maternal-effect mutants are not "maternal," since they originate in the germ line rather than in somatic maternal tissue. In the sense that these mutants are "maternal," they are active and vulnerable to experimental manipulation of the mother while eggs are developing.

Likewise, temperature-sensitive mutants are as much **temperature-induced phenocopies.** In the sense that the abnormal flies are "mutants," they are affected through specific pathways of development (e.g., gastrulation, segmentation), rather than through generalized debilities, and the mutations are traceable to loci on chromosomes. Still, like phenocopies, expression is variable and some normal flies develop even at optimal nonpermissive temperatures.

In spiralian species, temperature-sensitive periods are associated with determination. The more than 300 strains bearing mutations mapped to the nematode's six linkage groups (five pairs of autosomes and two X chromosomes in hermaphrodites, one X chromosome in males) demonstrate different ways gene action can be linked to development.

Maternal-effect or embryonic-arrest mutations of **embryogenesis genes** (*emb*) may be induced by treatment with ethyl methane sulfonate acting as a mutagen. The mutations are keyed in by a temperature shift during oogenesis (e.g., from the 16°C permissive temperature to the 25°C nonpermissive temperature) and affect the timing of cleavage divisions rather than the pattern of divisions. Breeding experiments are used to home in on the chromosomal location of the genes (Brenner, 1974).

The *emb* mutant phenotypes frequently suggest that the timing of divisions is controlled by stored (maternal) cytoplasmic components, and timing establishes regional differences in the embryo. Whether the cells of mutants divide slower or faster, and even when differences are as small as 10–20%, the mutant embryos fail to regulate toward normal development. Divisions out of sequence may slow migration and allow differentiation in the wrong place. Other mutants may block differentiation. Cultured at the permissive temperature (16°C), normal gene products would program the normal rate and sequence of divisions, cell migration, and differentiation.

Zygotic-effect mutants of postembryonic cell lineages, on the other hand, include general abnormalities in cell division and specific hypodermal or vulval defects in the adult. **Developmental decisions** normally accompanying each division may not be made, and a series of such decisions may be derailed at some early point. The reiterative effects of some mutants suggest that cells are bipotent (i.e., capable of only linear developmental decisions) and limited to entering one or another pathway (see Schierenberg, 1982).

In Drosophila melanogaster, almost all the species of mRNA present in early embryos (Fig. 15.16, lanes D, F, H) are stored or maternal RNA synthesized in the previtellogenic and early vitellogenic stages of oogenesis (i.e., early to middle oogenesis, lane C). The RNA transcribed earlier, in cystocytes (lane E) or in the cystoblast during mitosis and nurse-cell formation (lane G), does not seem to contain all the mRNAs present in the early embryo (Phillips et al., 1984).

Maternal-effect mutants known as female-sterile (fs) mutants produce defects in the germ line (see Gans et al., 1975). In the recessive *grandchildless (gs)* mutant of *Drosophila subobscura,* for example, normally appearing homozygous females produce eggs in which polar granules (i.e., putative germ plasm) disappear at the end of oogenesis. Lacking pole cells, the offspring (i.e., F_1 progeny) are sterile, much as if they were UV-irradiated in their posterior pole (see Mahowald, 1983).

One class of female-sterile mutants affects patterning during early development. Homozygous mutant females produce eggs that begin to develop

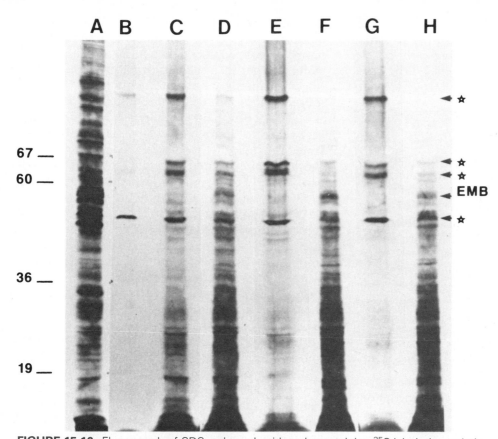

FIGURE 15.16. Fluorograph of SDS–polyacrylamide gels containing ^{35}S-labeled translation products of embryonic *Drosophila melanogaster* RNA. Flies were fed briefly (i.e., pulsed) with an algal paste labeled with heavy ^{13}C, ^{15}N, and ^{2}H nuclides. Thereafter, the flies were fed a diet of unlabeled yeast (i.e., chased). In the course of time, eggs were produced containing different amounts of "dense" RNA synthesized during the pulse and labeled with the heavy nuclides and "light" RNA synthesized mainly during the chase. Subsequently extracted, "dense" and "light" RNAs were separated by centrifugation through a NaI/KI density gradient and translated *in vitro* with rabbit reticulocyte lysate. "Dense" RNA synthesized between early oogenesis and the beginning of vitellogenesis (lane C) translates into the same peptides as "light" RNA synthesized from the remainder of oogenesis to the early embryonic period (lane D). An additional peptide (arrow *emb*) is synthesized only under the direction of the "light" embryonic RNA. "Dense" RNA synthesized during the establishment of nurse cells and the initial period of oogenesis (lane E) or earlier still during mitosis (lane G) does not contain all the stored RNA present in embryos. Fewer peptides seem to have been translated than those translated by the chase "light" RNAs (lanes F and H). Translation products of total *Drosophila* embryo RNA without centrifugation (lane A) and the *in vitro* rabbit system without the addition of *Drosophila* RNA (lane B) are presented for comparison. Starred arrows show endogenous peptides of rabbit translation system. From W. H. Phillips, J. A. Winkles, and R. M. Grainger, in E. H. Davidson and R. A. Firtel, eds., *Molecular biology of development,* Alan R. Liss, Inc., New York, 1984. Courtesy of the authors. Used by permission.

normally only to become abnormal during the blastoderm stage and ultimately die (class 1, amorphs or true female steriles) (Mahowald et al., 1984). A number of these alleles in *D. melanogaster* are found at the *gastrulation-defective* (*gd*) locus on chromosome II.

In another class of female-sterile mutants, even homozygous mothers' eggs are rescued when fertilized by sperm bearing the wild-type allele (class 2, hypomorphs of lethal loci). The *rudimentary* mutant, for example, is lethal due to deficient pyrimidine biosynthesis but is rescued by early production of new bases in the presence of the normal allele introduced by sperm. Both diet and injection of pyrimidines or normal cytoplasm also rescue the embryo.

Zygotic-Effect Genes

Zygotic-effect genes are theoretically turned on following fertilization. In *Drosophila*, early zygotic-effect genes become active during the syncytial and early cellular blastoderm stages. Mutants affected by the male or entire embryonic genome, rather than exclusively by the mother's genome, are **zygotic-effect mutants.**

Theory does not require these genes to be separate from maternal-effect genes, and, frequently, zygotic- and maternal-effect genes cannot be distinguished operationally. For example, the TSP for the female-sterile **gastrulation-defective** mutation is the terminal portion of oogenesis and the first hour after oviposition. Normal expression results in a blastoderm capable of undergoing gastrulation $2\frac{1}{2}$ hours later.

So-called **male-rescuable** mutants are transmitted to offspring from mothers, but the dire phenotype is avoided when eggs are fertilized with normal sperm. The mutants have maternal effects, but they are, by definition, zygotic-effect mutants. The gooseberry-like *fused (fu)* mutant, for example, is male rescuable, and the *fu* phenotype does not appear in the offspring of homozygous mothers and wild-type fathers. Once again, a gene operating in the zygote seems to operate earlier, and transcripts made by nurse cells and transported to oocytes during oogenesis, it would seem, can prevent the dire consequences of homozygosity (i.e., rescued by a maternal effect).

The operations of some maternal-effect, female-sterile mutants also overlap with early zygotic effects. The mutants may be unrescuable or only partially rescuable by normal alleles introduced by the male. For example, sperm with a chromosomal duplication containing the wild-type neurogenic gene, *Notch⁺*, partially rescue *Notch⁻* mutants from the maternal effect (see Mahowald, 1983).

These female-sterile mutants thereby fall into the ambiguous (but growing) category of genes active during oogenesis and in the period prior to cellularization. The phenotype is not necessarily conveyed through the egg nor does it necessarily require zygotic activity, but coordination in expression is required.

Segmentation Mutants

Segmentation mutants are embryonic lethal mutants expressed in homozygotes by a breakdown in the number, symmetry, or polarity of segments. The mutants and their normal counterparts exhibit various degrees of maternal and zygotic effects.

In *Drosophila*, a permanent record of the segmentation genes' activities is inscribed as a **cuticular pattern** of dorsal and ventral **denticle belts** in

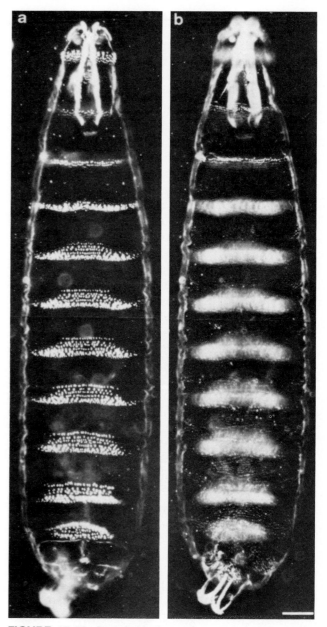

FIGURE 15.17. Dark-field photomicrographs of the fixed and cleared cuticle of first instar larvas of *Drosophila melanogaster* showing the normal ventral (a) and dorsal (b) hypodermal patterns. Anterior at top. Scale = 50 μm. The patterns are formed by belts of *denticle hook rows* marking the borders of segments. On the ventral side, thinner belts identify the anterior borders of three thoracic segments. Thicker belts mark off the eight abdominal segments. A segmentally repeating pattern of fine hairlike structures covers the dorsal side. From C. Nüsslein-Volhard, in S. Subtelny, ed., *Determinants of spatial organization,* 37th Symposium of the Society for Developmental Biology, Academic Press, Orlando, 1979. Courtesy of C. Nüsslein-Volhard. Used by permission.

the cuticle of the near-hatching or freshly hatched first instar larvas. Cuticular preparations are especially useful for studying lethal mutants, since cuticles tend to be produced by the embryonic hypodermis even when very little differentiated tissue develops.

Segmentation mutants arrest development in the blastoderm or during gastrulation but still allow the hypodermis to produce a cuticular segmentation pattern. The pattern in the normal larva (Fig. 15.17) may appear qualitatively different from that in a mutant (e.g., *dorsal,* [*dl,*] Fig. 15.18) or appear quantitatively different (e.g., *bicaudal, bic,* Fig. 15.19). Moreover, mutants may act in separate dorsal/ventral and proximal/distal compartments as well as anterior/posterior compartments (see Chapter 22 for a discussion of compartments).

Early lethal (and some nonlethal) segmentation mutants are grouped into three categories (Nüsslein-Volhard and Wieschaus, 1980). **Pair-rule mutants** (e.g., *fushi tarazu* [*ftz*], named from the Japanese for "too few segments") are defined by their ability to cause deletions in alternating segments. **Gap mutants** (e.g., *Krüppel* [phenotype resembles *bic,* Fig. 15.19]) cause the deletion of **regional domains,** continuous stretches of segments. **Segment polarity mutants** (e.g., *engrailed* [*en*], and *fu*) are pattern duplication mutants in which one part of a segment (e.g., an anterior part) is replaced with its opposite member (e.g., a posterior part) in reversed polarity without altering the number of segments.

Pair-rule and segment-polarity mutants tend to be early zygotic-effect mutants, while gap mutants are predominantly maternal-effect mutants (*Krüppel* and others being exceptions), but many genes are in a dubious class somewhere between. For example, the zygotic-effect *fu* mutant is also a maternal-effect mutant (see above). The normal gene product, it would seem, is produced either or both maternally and zygotically.

Sites of transcription and translation for several segmentation genes have recently been identified in the cleaving and cellularizing embryo. Probes of single-stranded cDNA and products of fusion genes in transformant flies have been used to identify transcripts of the normal gene's product *in situ,* and antibodies have been used to identify the location of the normal gene's protein. The transcripts and proteins of the pair-rule segmentation gene *ftz*+ appear in seven periodic stripes (i.e., the zebra pattern, Fig. 15.20), while products of gap genes, such as *hunchback* and *Krüppel,* are found in broad areas, and those for segment polarity genes, such as *engrailed,* are found in 14 fine bands.

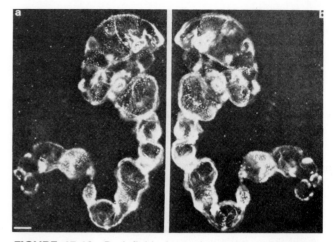

FIGURE 15.18. Dark-field photomicrographs of the right and left sides of a fixed and cleared cuticle of a dominant-type *dorsal* mutant of *Drosophila melanogaster.* Scale = 50 μm. From C. Nüsslein-Volhard, in S. Subtelny, ed., *Determinants of spatial organization,* 37th Symposium of the Society for Developmental Biology, Academic Press, Orlando, 1979. Courtesy of C. Nüsslein-Volhard. Used by permission.

In general, portions of the normal *Drosophila* blastoderm producing transcripts of segmentation genes are the same as the portions affected by the corresponding segmentation mutant. *ftz*+ stripes, for example, coincide with deleted primordia in homozygous *ftz*− mutants. Homozygous *en*− mutants show failures in posterior portions of most segments and widespread anterior duplications (e.g., eye, antenna, wings, legs, halteres, and abdominal segments).

X-ray induced somatic mosaics also implicate products of the *en*+ gene in the determination of the posterior portions of segments. Expression of the mutant in patches of the anterior wing has no effect, but expression in patches of the posterior wing results in anterior-type patterns of veins and bristles. Moreover, these patches may spill over into the anterior wing, indicative of the affected cells' affinity for the anterior compartment.

Putting all this together, it seems that normal gene products direct development in particular **developmental territories** (Martinez-Arias and Lawrence, 1985). The territories identified by bands for the segment polarity *en*+ gene are posterior **compartments;** the territories identified by stripes for the pair-rule *ftz*+ gene are **parasegments** consisting of anterior and posterior compartments of alternating segments; the territories identified by areas for gap genes are **regional domains.** Segment polarity genes affect individual compartments, pair-rule mutants affect parasegments, and gap mutants affect regional domains.

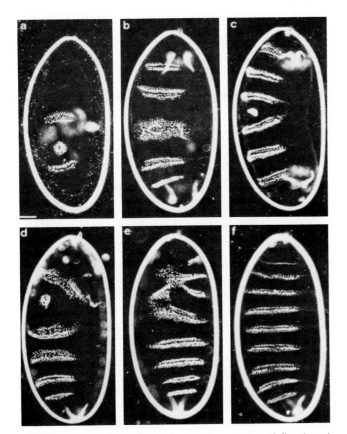

FIGURE 15.19. Dark-field photomicrographs of fixed and cleared cuticles of *bicaudal* mutants. Scale = 50 μm. (A–C) Symmetrical phenotypes with different numbers of duplicated posterior segments: (A) $1\frac{1}{2}$ segments; (B) 3 segments; (C) $3\frac{1}{2}$ segments. (D) Asymmetric phenotype with $4\frac{1}{2}$ posterior segments (bottom) with normal polarity and $1\frac{1}{2}$ duplicated posterior segments (top) with reversed polarity. (E) A unilateral (one side only) asymmetric phenotype. (F) Headless embryo having all three thoracic and eight abdominal segments. From C. Nüsslein-Volhard, in S. Subtelny, ed., *Determinants of spatial organization,* 37th Symposium of the Society for Developmental Biology, Academic Press, Orlando, 1979. Courtesy of C. Nüsslein-Volhard. Used by permission.

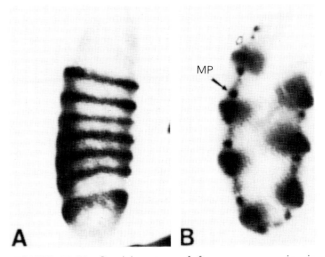

FIGURE 15.20. Spatial pattern of *ftz* gene expression in whole mounts of wild-type *Drosophila melanogaster.* Transformant (i.e., transgenic) embryos received the fusion gene, *ftz/lacZ,* containing the *ftz* promoter linked to the *lac* operon. The *ftz* gene product and the β-galactosidase appear at the same times and with the same pattern, but the β-galactosidase is longer lasting and therefore detectable over a longer period of time. Here, whole mounts are stained with antibody to β-galactosidase expressed with the *ftz* gene product. Anterior is up. (a) Lateral view gastrulating embryo with seven stripes. (b) Lateral view of embryo at extended germ-band stage (about 5 hours). Staining in stripes disappears, but midline neural precursor cells (MP, arrow) have appeared in ventral ectoderm from the third cephalic segment to the eighth abdominal (toward bottom). Note absence of M_P in more cephalad segments (open arrowhead). From C. Q. Doe, Y. Hiromi, W. J. Gehring, and C. S. Goodman, *Science,* 239:170 (1988), by permission of the American Association for the Advancement of Science. Copyright 1988 by the AAAS.

Most of the future larval segments (Fig. 15.17) seem to comprise **double-segment units,** representing adjacent anterior and posterior compartments. In addition, the terminal larval segments consist of **half segments** (Table 15.3).

Various attempts have been made to integrate the differential expression of maternal and segmentation genes into coordinated models of pattern formation (see Meinhardt, 1986). The timing of gene expression and the size of units affected are the variables.

Ideally, maternal-effect genes set up anterior–posterior gradients which gap genes interpret according to nonoverlapping regional domains.

Pair-rule genes then come into play, reading the positional information available and defining parasegmental units, both those expressing the gene and alternate units not expressing it. Finally, segment polarity genes enter the picture, operating in one or another compartment in every segment.

In reality, the difference between maternal and zygotic effects may not be as large as theory would have it. For example, segmentation mutants, which are typically classified as zygotic effect mutants, may not be purely zygotic and may even have some typically maternal effects.

Moreover, genes that are classified one way may turn out to be classified another way when different criteria are applied. For example, because young females cultured at high temperatures produce eggs with especially severe *bicaudal* phenotypes, *bic* is classified as a maternal-effect mutant. On the other hand, treating the early embryo with

TABLE 15.3. Expression of the *ftz*⁺ gene in the cellularizing blastoderm of *Drosophila melanogaster*

Body part:	head			thorax			abdomen								
Larval segments[a]	Ma half segment	Mx	Lb	T1	T2	T3	A1	A2	A3	A4	A5	A6	A7	A8	A9 half segment
								double segment units							
Parasegments: Compartments[b]	1 P|A /	2 P|A /	3 P|A /	4 P|A /	5 P|A /	6 P|A /	7 P|A /	8 P|A /	9 P|A /	10 P|A /	11 P|A /	12 P|A /	13 P|A /	14 P|A	
Parasegments in wild-type expressing *ftz*⁺:	X|X		X|X		X|X		X|X		X|X		X|X		X|X		
Compartments represented in mutant *ftz*⁻:	P|A		P|A		P|A		P|A		P|A		P|A		P|A		

[a]Larval segments: Ma, mandibular; Mx, maxillary; Lb, labial; T1–T3, thoracic; A1–A9, abdominal.
[b]Compartments: A, anterior; P, posterior; X|X, a stripe.
Source: Based on A. Martinez-Arias and P. A. Lawrence, *Nature* (*London*) 313:639(1985).

high temperature also promotes the bicaudal phenotype. *bic* should therefore be classified as a zygotic-effect mutant as well.

The distinction between maternal and zygotic effects, which are blurred when genes have two or more temperature-sensitive periods, would be erased if genes operated more than once (see Doe et al., 1988). In the *ftz*⁺ wild-type phenotype, for example, a gene product that once appeared in stripes in the cellular blastoderm later appears in a subset of segmentally arranged neuronal precursor cells, neurons (*ftz*⁺ neurons), and glial cells (Fig. 15.20a, b) (Hiromi and Gehring, 1987).

Although the operation of different genes is not ruled out, identical oocytic and early embryonic mRNA transcripts and indistinguishable translation products are indicative of the operation of the same gene (see Davidson, 1986). Rather than operating and shutting down, genes contributing to early development may also contribute to late development through multiple periods of transcription (see Mahowald et al., 1984).

Mutually exclusive categories of maternal and zygotic effects apply only at the level of phenotype. At the level of gene action, there is plenty of room for overlap, and both maternal and zygotic effects can operate consecutively. It would therefore seem desirable to identify gene action by stages of development rather than as maternal or zygotic.

SEQUENCES AND PATTERNS OF GENE ACTIVITY: SELECTED EXAMPLES

Instead of searching for an illusory genetic program, developmental biologists must try to fathom the complex network of causal relations to which the sequence of developmental events owes its regularity.

Gunther S. Stent (1985, p. 1)

Many conventional models of sequential and patterned gene activity reflect mechanistic conceptions of life. For example, contemporary tapedeck models of development equate an embryonic cell to a playback, and computer models equate an embryonic nucleus to a disk drive. Genes are considered prerecorded tapes or disks. Some may be replayed or reloaded, while others are virtually erased.

The major questions raised by these models concern the decisions that turn the systems on and keep them running as well as the decisions that turn them off and store them away for later reference. Theoretically, developmental transcription can be turned on by action at promoters and tuned in (i.e., regulated) by action at enhancers. Stage-specific and tissue-specific structural genes can be cued by *cis*-acting regulatory substances, and spent maternal materials can send signals to begin playing new zygotic or embryonic genes. An integrated program with sufficient feedback can coordinate spatial and temporal events at higher levels of organization. But what experiments test these models, and do the results support them?

A variety of approaches are taken to test hypothetical sequences and patterns of genetic controls. Temporal controls by putative **gene-controlling components** are tested by introducing materials into eggs through injection or other means. Temperature-sensitive periods should be consistent with the hypothetical sequence.

Spatial controls are more difficult to test inasmuch as they seem to involve genes performing different roles. Broadly speaking, two classes of genes may operate: **patterning genes** that set the stage and **interpretation genes** that act the parts. Cloned DNA and *in situ* hybridization should make it possible to test these hypothetical controls.

Gene-Controlling Components

The overall control of nuclear activity has been studied experimentally by injecting proteins, RNA, DNA, and the nuclei of adult cells into frog oocytes, unfertilized eggs, and fertilized eggs (e.g., Gurdon, 1982; Lane and Knowland, 1975). Mouse oocytes have been fused with oocytes and embryonic cells (Tarkowski, 1982), and tetraploid embryos have been formed by fusing zygotes (Graham, 1971). The dictatorial powers of the cytoplasm, revealed in the results of all these manipulations, are consistent with the assumption that inspired the experiments, namely, that oocytes, zygotes, and blastomeres contain stable gene-controlling components.

Some of these components may be present in the egg's cortex, implanted there during oogenesis or even deposited by sperm during fertilization. Other components may be introduced into the cytoplasm at the breakdown of the oocyte's germinal vesicle (GVBD) or at the breakdown of male and female pronuclei during amphimixis. Still other gene-controlling components may be released at the breakdown of cleavage nuclei or introduced into the cytoplasm via pores in the nuclear envelopes during interphase.

Gene-controlling components may be inhibitory substances or activating ones; they may be activated or inactivated, covered or uncovered, and they may be either specifically synthesized materials or general nonrenewing materials. Some components may be involved in the assembly of genes into chromatin, some in the mechanism of transcription, and others in repression–derepression. The rapid rate of cell division during cleavage may dilute gene-controlling repressors sufficiently to effectively eliminate their activity on gene expression. Alternatively, constructive interactions among accumulated components may activate derepressors and turn on gene expression.

Mutations in quantitative genes may produce gene-controlling components in abnormal amounts or defective components that elicit abnormal responses. **Hypermorphic** mutants have exaggerated phenotypes, possibly as a consequence of producing excess amounts of a gene-controlling component; **hypomorphic** mutants have reduced phenotypes, possibly because of less than normal amounts of a gene product, and **amorphic** mutants have qualitatively defective phenotypes (i.e., a normal structure or product is missing) possibly due to the absence of a functional gene-controlling component.

In gastropods, clockwise and counterclockwise rotation of shells and internal structures (see Fig. 12.41) *are maternal effects imposed on the oocyte by a single dominant gene inherited by the mother* (see Chapter 12, Sturtevant, 1965). For example, in the freshwater snail, *Lymnaea peregra,* the clockwise gene is dominant. This gene may create a molecular asymmetry in the oocyte's cortex (Meshcheryakov and Beloussov, 1975), or a stable product of the dominant clockwise gene may accumulate in the egg's cytoplasm during ontogenesis.

The latter possibility is supported by the results of cytoplasmic-transfer experiments. Cytoplasm from an *L. peregra* egg injected into an uncleaved counterclockwise egg (e.g., *Laciniaria biplicata* and *Partula suturalis*) causes a clockwise third cleavage. On the other hand, cytoplasm from a counterclockwise egg does not alter cleavage in an uncleaved clockwise egg. Counterclockwise rotation of the snail's shell may represent the default pattern developed in the absence of a signal from a dominant clockwise gene (Freeman and Lundelius, 1982).

In the Mexican axolotl Ambystoma mexicanum, the recessive ova deficient (o) *mutant is a maternal-effect mutant affecting fecundity in homozygous females.* o/o Females produce eggs that are morphologically indistinguishable from normal. Cleaving eggs and early blastulas also appear normal, but division slows in the midblastula to late blastula, and the intense wave of transcription normally commencing in late blastulas does not occur. DNA synthesis is sharply reduced, and while dorsal lips appear, gastrulation is only rarely completed, and neurulation does not even begin.

Eggs bearing the o mutation are not rescuable by fertilization with wild-type (o^+/o^+ or $+/+$) or heterozygous (o^+/o) sperm.[2] Deleterious effects are preventable, however, by injecting eggs prior to the first cleavage with a factor dubbed o^+ substance recovered from normal axolotl germinal vesicles (Fig. 15.21) (see Brothers, 1979).

The o^+ substance is also found in wild-type mature oocyte cytoplasm, in the cytoplasm of fertilized eggs, and in blastomeres in early-cleavage

---

[2]Homozygous males are sterile.

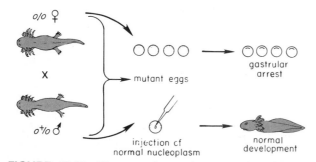

FIGURE 15.21. Diagram of experiment testing ability of normal germinal vesicle nucleoplasm to rescue eggs of *o/o* females from gastrular arrest. From A. J. Brothers, in S. Subtelny, ed., *Determinants of spatial organization,* 37th Symposium of the Society for Developmental Biology, Academic Press, Orlando, 1979. Used by permission.

stages. It would therefore seem that o^+ substance is synthesized or stored in the normal germinal vesicle and enters the cytoplasm at the GVBD. Since the substance disappears in blastomeres of the late blastula and is absent from adult tissues (i.e., testis, liver, and spleen), it would seem to be bound or degraded in the blastula and not synthesized again except in oocytes.

The substance may be (or its action may depend on) a high molecular weight protein, since it is labile to both heat and trypsin. Injections of nuclear sap from the oocytes of several species of amphibians permit *o/o* eggs to develop, showing that o^+ substance is not species specific. Injection into only one blastomere of the 2-cell embryo rescues that blastomere but not the other, indicating that the substance is not transmitted across cell borders.

Nuclei of normal early blastulas transplanted to enucleated eggs of *o/o* females (Fig. 15.22) do not prevent developmental arrest at the early-gastrula stage, but nuclei transplanted from normal mid-blastulas and embryos at later stages occasionally support development completely to sexually mature adults. The o^+ substance may have reentered nuclei in mid-blastula embryos and produced irreversible and stable changes capable of withstanding nuclear transplantation.

In the reciprocal experiment, *o/o* nuclei from the blastomeres of late cleavage embryos and from cells of early blastulas are rescued by the normal cytoplasm of recipient eggs and support the development of complete embryos. The *o/o* nuclei from cells of midblastulas to late blastulas are not res-

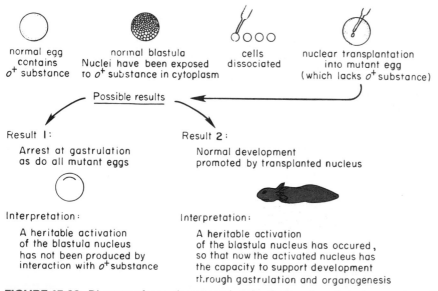

FIGURE 15.22. Diagram of experiment testing the timing of nuclear changes hypothetically triggered by o$^+$ substance in axolotl eggs. Eggs having o$^+$ substance are allowed to develop for different periods. Cells isolated from blastulas of slightly different stages provide nuclei for injection into the enucleated eggs of *o/o* females lacking o$^+$ substance. Result 1: Nuclei from normal early blastulas fail to support development in *o/o* eggs beyond the early gastrula. Result 2: Nuclei from normal midblastulas support development in *o/o* eggs to varying degrees including the development of swimming tadpoles. From A. J. Brothers, in S. Subtelny, ed., *Determinants of spatial organization,* 37th Symposium of the Society for Developmental Biology, Academic Press, Orlando, 1979. Used by permission.

cued by transplantation, however, and fail to support development beyond gastrulation. In the absence of o^+ substance, it would seem, axolotl nuclei are irreversibly damaged in the midblastula to late-blastula stage.

Like the products of enucleated eggs, o/o blastulas seem to lack a functioning genome. While normal late blastulas show a shift in the pattern of isotope incorporation into selected proteins, eggs of o/o females retain the earlier more general pattern of protein synthesis. Conceivably, o^+ substance normally modulates transcription and tunes in late protein synthesis.

Patterning and Interpretation Genes

Theoretically, subsets of patterning and interpretation genes could perform different duties. Patterning genes could include **coordinate genes** that define spatial coordinates laid out in stable gradients, **determination genes** that carry information required for determination, and **selector genes** that trigger differentiation in targeted cells. **Interpretation genes** could interpret messages laid down for them by patterning genes and promote or retard the regional expression of one or more genes involved in differentiation.

Practically, ascertaining which genes operate in which way is a formidable task. The approach generally taken utilizes mutants that disrupt pattern formation.

Different segmentation mutants affect anterior–posterior (A/P), dorsal–ventral (D/V), and proximal–distal (P/D) axes. Mutants such as *almondex* (Fig. 15.23) and *dorsal* (*dl*, Fig. 15.18), for example, derail dorsal/ventral polarity, while *bicaudal* (*bic*) and *dicephalic* (*dic*) alter anterior/posterior polarity.

Ideally, a **cascade of interactions** among gene products should progressively determine all axes, and mutants affecting one axis should not affect another axis. In practice, the axes of the embryo are not independent, and anteroposterior gene expression responds to dorsoventral positional information. For example, the stripes expressing ftz^+ gene product in *Drosophila* blastoderms, which normally lean toward each other dorsally (Fig. 15.20), are parallel and more nearly perpendicular to the long axis in the blastoderms of homozygous dl^- mutants and other dorsalizing maternal-effect

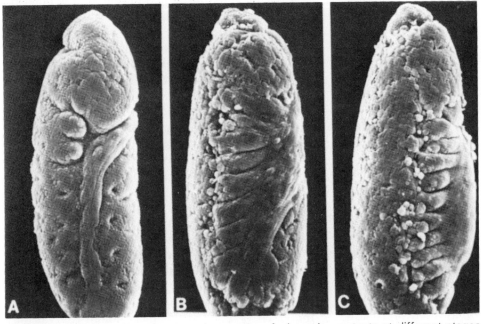

FIGURE 15.23. Scanning electron micrographs of *almondex* mutants at different stages showing the breakdown of pattern formation. (a) At 8 hours segmentation of the blastoderm appears normal. (b) At 9–10 hours, neurulation and dorsal closure commence. (c) Enlargement (hypertrophy) of the ventral nervous system due to the transformation of ventral hypoderm cells reduces the dorsal hypoderm to small patches of cells and results in herniation of the ventral nerve cord. From A. P. Mahowald, in W. R. Jeffery and R. A. Raff, eds., *Time, space, and pattern in embryonic development,* Alan R. Liss, Inc., New York, 1983. Courtesy of A. P. Mahowald. Used by permission.

mutants. It seems as if the stripes are "dorsalized" around the entire circumference of the embryo (Carroll et al., 1987).

About 20 *Drosophila* genes influence D/V polarity, 12 of which are maternal-effect genes expressed during oogenesis. One of these genes, *Toll*$^+$, encodes an integral membrane protein but somehow controls a cytoplasmic substance capable of determining ventral-pole differentiation upon injection into eggs.

Another of these genes, the amorphic *dl* mutant on *Drosophila*'s second chromosome, may act as a selector gene. A concentration gradient of **dorsal protein** (albeit not *dl*$^+$ transcripts) normally found may impose D/V polarity by repressing genes for ventral differentiation. The *dl* mutants dampen this concentration gradient at the same time that they dorsalize the ventral compartment. "Dorsal" seems to represent a "default state" into which embryos slip when other directions are missing (see Carroll et al., 1987).

Similarly, activity at the *almondex* locus is effective during oogenesis and in the brief period prior to blastoderm formation. The altered pattern of the mutant phenotype begins at the late blastoderm stage. In the absence of gene products normally present or produced during the preblastoderm stage, flies expressing the *almondex* phenotype have excess nervous tissue and a ventral rupture of the ectoderm (Fig. 15.23). Cellular determination of specific blastoderm cells fails, and the cells are transformed to new cell types. Only a part of the blastoderm is generally affected, but the pattern of cell determination in the entire blastoderm is disrupted (see Mahowald et al., 1984).

At least eight more genes affecting D/V polarity are expressed only after fertilization (i.e., zygotic-effect genes). These seem to be interpretation genes whose localized expression (e.g., *twist* expressed in the ventral-most region of the gastrula) is influenced by the earlier expression of selector genes (e.g., *dl*) (see Levine, 1988).

The A/P axis is determined by no less than 30 genes. Some genes influence conditions in the egg chamber that cause the deposition of nurse-cell transcripts at the anterior pole of the oocyte. Among these transcripts, some (e.g., encoded by the *bicoid* gene) subsequently play a crucial role in segmentation.

Gap genes and similar maternal-effect genes involved in A/P determination may be coordinated genes encoding DNA-binding proteins. Diffusion through the cytoplasm may account for an initial uniform distribution of transcripts, and some **primary spatial determinants** may establish gradients and belts of gene products.

Mutants derail patterning at the same sites where normal gene products accumulate (see Gehring, 1987). The maternal-effect mutants *bic* (Fig. 15.19) and *caudal* (*cad*), for example, cause the deletion of groups of adjacent segments. The mutants' actions involve both stored transcripts and those synthesized after fertilization, and the distributions of normal transcripts and translation products anticipate the sites of deletions in mutant phenotypes.

The *cad* transcripts are distributed uniformly in the egg, but they acquire an A/P concentration gradient with a caudal peak after fertilization. Following blastoderm formation, a zygotic *cad* transcript appears and accumulates in a single posterior belt.

The *cad* protein is not translated until after fertilization. It accumulates in nuclei and is distributed in a concentration gradient during the migration of energids (about the eighth cleavage). Only a single posterior belt of cells translates the *cad* protein at the cellular blastoderm stage. Homozygous *cad* mutants lacking both the maternal and zygotic expression of normal genes fail to develop caudal segments.

The bicaudal mutant *BicD* causes the deletion of head segments and their replacement with caudal segments leading to a double-abdomen (DA) phenotype. *BicD* is epistatic (i.e., able to suppress the activity of a nonallelic gene) and alters the distribution of the *cad* protein (Fig. 15.24, compare A to B, C to D). Possibly, *BicD* affects a primary spatial determinant to which *cad* responds.

In all *bic* phenotypes, the anterior–posterior pattern of segments, their total number, and their polarity are affected, but *bic* expression is variable and phenotypes differ greatly (Fig. 15.19). Typically, homozygous *bic* produces a preponderance of normal larvas and **symmetrical** larvas (Fig. 15.19a–c) with mirror-image symmetry in abdominal and caudal regions. Less frequently, the mutants have **intermediate** bicaudal phenotypes covering a range of **asymmetrical** larvas. A small number of duplicate segments with reversed polarity may be added to posterior segments with normal polarity (Fig. 15.19d). Occasionally, larvas may show different asymmetries on right and left sides (Fig. 15.19e). The least affected mutant phenotype is the **headless** larva, which retains normal polarity throughout but lacks the anterior-most pattern (Fig. 15.19f) (Nüsslein-Volhard, 1979).

Possibly, functioning as a hypomorphic mu-

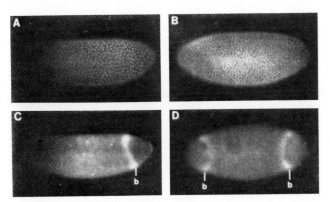

FIGURE 15.24. Immunofluorescent micrographs of the *caudal* protein in normal *Drosophila melanogaster* embryos (A, C) and *Bicaudal* embryos (B, D) at the syncytial blastoderm stage (A, B) and the cellular blastoderm stage (C, D). Normally, the protein, which is concentrated in nuclei and distributed with a concentration gradient peaking posteriorly, is synthesized in a belt (b) corresponding to the most posterior abdominal segments. The *Bicaudal* mutant eliminates the concentration gradient, but symmetrical belts (b, b) announce the development of the symmetrical bicaudal embryo. Adapted from M. Mlodzik and W. J. Gehring, *Cell,* 48:465 (1987), by permission of Cell Press.

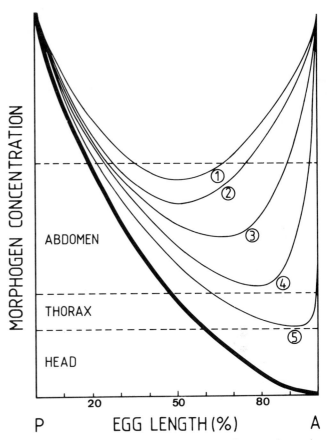

FIGURE 15.25. Hypothetical autocatalytic morphogenic system for the production of different *bicaudal* phenotypes. Under the influence of an autocatalytic activator, an inhibitor morphogen forms a concentration gradient from the posterior (P) to the anterior (A) pole of the egg. Different types of segments (head, thorax, abdomen) differentiate as functions of threshold amounts (dashed lines) of the inhibitor morphogen. Normally, the amount of inhibitor released is sufficient to prevent the formation of an additional center of activator production, and a monotonic gradient results (thick curve). Hypomorphic *bicaudal* phenotypes (1–5) develop when reduced amounts of inhibitor morphogen production permit the formation of an anterior center for activator production. Depending on the amount of inhibitor morphogen released from the new center, *bicaudal* phenotypes appear with normal phenotypes, headless phenotypes (lines 4, 5, corresponding to f in Fig. 15.19), various asymmetric phenotypes (line 3, corresponding to d), and symmetrical phenotypes (lines 1, 2, corresponding to a–c). From C. Nüsslein-Volhard, in S. Subtelny, ed., *Determinants of spatial organization,* 37th Symposium of the Society for Developmental Biology, Academic Press, Orlando, 1979.

tant, *bic* may diminish the amount of a normal inhibitor morphogen in an autocatalytic morphogenic system. Diffusing from a source at the posterior pole under the influence of an autocatalytic activator, the inhibitor morphogen may normally prevent the formation of any additional center of autocatalysis and emerge in a concentration gradient (Fig. 15.25, thick curve).

When less inhibitor is produced in the mutant, inhibitor concentrations would fall below the threshold of inhibition at the anterior pole, and a new center for activator production (and hence inhibitor production) would develop. Depending on local conditions, the inhibitor would be distributed in a variety of reflected or duplicated gradients. Acting in its capacity as morphogen, the abnormal gradients would cause the different bicaudal phenotypes (Fig. 15.19).

According to the positional information concept, specific segment differentiation would "kick in" at given threshold levels of morphogen (Fig. 15.25, dashed lines). While the monotonic gradient (thick curve) would result in a linear arrangement of polarized segments, the asymmetric gradients (thin curves 3–5) would result in the headless phenotypes (Fig. 15.19f) and various asymmetric phenotypes (Fig. 15.19d). The duplicated symmetrical gradient (curves 1 and 2) would result in the symmetrical phenotypes (Fig. 15.19a–c). Small lo-

cal differences in the rates of diffusion through right and left halves of the egg would result in differences in the phenotypes on the two sides of the larva (Fig. 15.19e).

Chapter

16

GENETIC VERSUS EMBRYOLOGICAL CONCEPT OF DEVELOPMENT

Every structure in an organism is determined by two principally different factors: the nuclear genes of the chromosomes and cytoplasmatic factors. The nuclear genes co-operate, so one gene does not give rise to one structure, but the combined action of numerous genes directs a developmental process, giving rise to a certain structure. The cytoplasmatic hereditary mechanism is more conservative than the chromosomatic one.

BENGT KÄLLÉN (1961, p. 138)

In the first quarter of the 20th century, Thomas Hunt Morgan among others cast the gene in a preformationist mold. Reinforced by their success in tracing inheritable variations to units on chromosomes, geneticists became convinced that genes were the only things of importance given to the egg at fertilization and that the products of genes were the causes of development (see Allen, 1983).

Like geneticists, embryologists were firmly committed to analytical approaches and had rallied around the Roux–Weismann hypothesis as an antidote to recapitulationism. They too hoped to guide 20th century biology into the mainstream of hard-nosed science, but embryologists saw the gene as a disguised homunculus and rejected it. They returned to epigenics and promoted it as rival to the theory of the gene.

Both geneticists and embryologists wanted to account for stability during heredity and for change during development. The problem was that the gene accounted for stability, and epigenics accounted for change, but neither accounted satisfactorily for both. With its emphasis on environmental causes, epigenics predicted that formative causes arose in the cytoplasm and passed to the cell nucleus, while with its emphasis on hereditary causes, genetics predicted that formative causes arose in the nucleus and passed to the cell's cytoplasm (see Gilbert, 1988).

Changes in genes (i.e., mutations) were unsatisfactory as models for changes in development, since organisms with altered genes (i.e., mutants) also had altered development. Likewise, stability in epigenesis (i.e., programmed determination) was unsatisfactory as a model for stability in heredity, since heredity in organisms capable of altered development (i.e., those with regulative or indeterminate development) is as stable as heredity in organisms incapable of altered development (i.e., mosaic or determinate development).

Resolving the conflicting concepts depended on discovering relationships between stable genes and unstable epigenesis. Progress was not achieved easily.

GENETICS AND PREFORMATION VERSUS EMBRYOLOGY AND EPIGENESIS

Conflict

Today, one hardly realizes that Morgan began his career as an embryologist (Fig. 4.17) and throughout his career invested part of his summers in research on regeneration of marine organisms at the Marine Biological Station at Woods Hole, Massachusetts. His books, _Regeneration_ (1901) and _Experimental Embryology_ (1927), were embryology classics in their own time, and his numerous articles on development were widely read and admired (Mountain, 1983).

Morgan was relieved that Haeckelian recapitulation was out of vogue at Woods Hole. He appreciated the contribution of embryologists who worked out cleavage patterns (see Oppenheimer, 1983), but he disapproved of their penchant for labeling phenomena instead of analyzing them. He was unabashedly reductionist, at least in his later years, and eschewed complex explanations for development as veils for ignorance.

Because cleaving embryos and blastulas lacked gene products (i.e., variation), Morgan concluded that blastulas were simply the fruits of cell division sometimes containing blastocoels. Gene products began to appear with "embryo formation" at gastrulation.

Originally suggesting that the linear order of genes on chromosomes corresponded to the order with which they became active, he was disappointed to discover no correlation between the sequence of genes and the sequence of developmental stages. Although Morgan conceded a role for epigenics, "if it be admitted that the reaction of the genes is affected by the changes in the cytoplasm" (Morgan, 1927, p. 8), in general, he argued that "genetic production" was the exclusive cause of differentiation.

Embryologists disagreed. "This is not to say that the new genetics was entirely without influence on the then prevailing conceptions (and misconceptions) of ontogeny or that it did not elicit renewed debates over the issues of preformation vs. epigenesis and heredity vs. environment (it did both)" (Oppenheim, 1982, p. 42).

Many embryologists were profoundly anti-preformationist. Even the earlier defection by embryologists to recapitulationism, in part, represented a reaction to preformation. They took their inspiration from von Baer's elucidation of the germ layers, disallowing Weismann's contempt for epigenics and Roux's facile dismissal of environmental forces.

As early as 1896, in the first edition of his classic text, _The Cell in Development and Inheritance_, E. B. Wilson placed the question of gene products in an epigenic framework of gene action (p. 311):

> **If chromatin be the idioplasm in which inheres the sumtotal of hereditary forces, and if it be equally distributed at every cell-division, how can its mode of action so vary in different cells as to cause diversity of structure, _i.e. differentiation?_ [italics original]**

The contrast between Wilson's "mode of action" and Morgan's "genetic production" is stark. Gene actions, which could illuminate differentiation for Wilson, cast shadows over the gene for Morgan, while genetic production, which distilled heredity into differentiation for Morgan, muddied the mechanism of development for Wilson.

Mediation: Neopreformation and Neoepigenics

At the same time that epigenics and preformation seemed most incompatible, a potential resolution to the conflict was already at hand. The scientific diplomat who began negotiating the settlement was Charles Otis Whitman (1843–1910), director of the Marine Biological Laboratory at Woods Hole (see Maienschein, 1986).

Whitman was well known for his study on the leech, _Clepsine_ (= _Glossiphonia_), one of the very first cell lineage studies. Comparing development in the highly mosaic leech to that in other organisms, Whitman observed that the delineation of parts happened either early and rapidly or late and gradually. Ordinarily, preformation seemed to be at work in mosaic eggs where parts were delineated earlier (e.g., during cleavage or in the blastula), and epigenics seemed to be at work in regulative eggs where parts were delineated later (e.g., in the gastrula).

Suggesting that the stumbling block to a unified theory of development in mosaic and regulative eggs was time, Whitman offered a new concept of developmental time that did not pass at the same rate in all species. Developmental time was measured in unseen units of determination which did not necessarily correspond to morphological stages of ontogeny.

By introducing time as a variable, Whitman corrected some of the errors of others. Roux and Weismann had been mistaken in equating divisions with developmental time as the agent of change, and Driesch was guilty of ignoring time as a source of difference.

While preformationists saw development as determinant, Whitman saw determinant development as the precocious acquisition of determination. While epigenecists saw development as indeterminant, Whitman saw indeterminant development as the late acquisition of determination. Cells might be invisibly differentiated from the start, or they might remain in relatively indifferent states for some time. Corrected to include differences in timing, epigenics became **neoepigenics,** and preformation became **neopreformation.**

Instead of looking at development in opposing ways, neoepigenecists and neopreformationists looked at development from opposite poles. Neopreformationists saw development from the gene and cell up, while neoepigenecists saw development from the organism and tissue down.[1]

Still, the merging of neoepigenesis and neopreformation was not complete. Spokespersons for the two schools continued to clash over the appropriateness and consequence of each other's methods. Neoepigenecists tended to be more theoretical and concerned with concepts of development, pattern, symmetry, and, later, induction, while neopreformationists tended to be more resolutely reductionist and concerned with gene products and differentiation.

The two schools also remained divided on the fundamental question of how genes affected the embryos. That breach could not be crossed without

first answering the question: Do genes change in the course of development?

HYPOTHESIS OF NUCLEAR EQUIVALENCE

For most of the 20th century, embryologists interested in genic change had no access to genes. Researchers turned to the nucleus, since it could sometimes be manipulated experimentally, and the number of mitotic divisions in blastomeres provided an objective measure of nuclear (if not developmental) age.

The concept of **nuclear equivalence** is that nuclei, and hence genes, are not altered by ordinary mitotic division. In the course of the century, experimental results have advanced estimates of the number of divisions through which nuclei apparently pass unscathed from cleavage to the blastula, and recently the number has been extended to include terminal divisions prior to cellular differentiation.

Spemann's Ligation Experiment

Evidence for nuclear equivalence was discovered serendipitously by Hans Spemann at the turn of the century when he looked at the control experiments he performed on constricted newt eggs (see Spemann, 1938). Methodically pursuing his plan for testing the Roux–Weismann hypothesis, Spemann performed control experiments to check the viability of the eggs he used and to detect any extraneous influences of constriction. The controls consisted simply of constricting some eggs prior to the first cleavage rather than after and leaving some ligatures loose during cleavage (see Fig. 12.17).

In most cases, when he later tightened the miniature nooses, one portion of the egg underwent development, while the other portion did not. In a few cases, both sides of loosely constricted eggs underwent development (Fig. 16.1). One side always started cleavage later than and lagged behind the other side, but both "twins" developed completely.

Spemann reasoned that typically the portion of the egg originally containing the zygote's nucleus underwent cleavage and developed, while the portion lacking the nucleus did not. In the exceptional

---

[1]The contrasting connotations of **epigenics** and **genetics** were further blurred when C. H. Waddington (1962 and elsewhere) introduced the term **epigenetics** and attempted to harmonize the contrasting views under the title **epigenetic action system.**

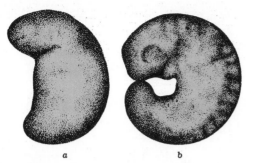

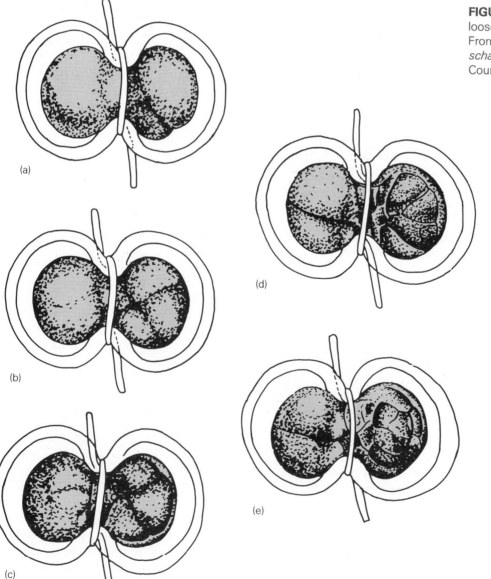

FIGURE 16.1. Twins produced by median construction of an uncleaved egg. Embryo 22a (left) underwent cleavage later than embryo 22b (right), presumably after receiving a nucleus through the bridge that connected the half-eggs at the time. From Hans Spemann, *Embryonic development and induction,* Yale University Press, New Haven, 1938; courtesy of Sylvia F. Pan.

cases, a nucleus moved across the bridge connecting the two sides of the egg (Fig. 16.2) and initiated development. In one case, the itinerant nucleus had even come from an early blastula of 16 cells.

> **According to Weismann this nucleus should contain a fraction only of the germ plasm, and, consequently, allow a part only of the embryo to arise. Instead of this, a whole formation is developed. (Spemann 1938 [1962 reprint, pp. 27–28])**

Spemann's interpretation of his result was narrowly focused but far reaching. First, the exceptions had broken Weismann's rule. Germ plasm could not have been irrevocably partitioned by nuclear divisions during cleavage.

FIGURE 16.2. Development of loosely constricted *Triturus* egg. From H. Spemann, *Z. Wissenschaft. Zool.,* 123:105 (1928). Courtesy of Sylvia F. Pan.

(a)

(b)

(c)

(d)

(e)

Second, the formation of whole embryos by half-eggs with new nuclei added a new twist to the concept of regulation. While Driesch, among others, had demonstrated that sea urchin blastomeres did not regulate completely after three divisions (see Fig. 12.12), in Spemann's hands, a newt nucleus supported complete regulation in uncleaved half-eggs even after passing through four divisions.[2] The cytoplasm of the half-egg had retained its capacity to elicit the full spectrum of developmental activities from the freshly received nucleus despite a delay equal to the duration of four cleavage cycles.

Neither the cytoplasm nor the nucleus had lost the ability to support complete development in the time frame of cleavage. Functional interchangeability with the zygote suggested the absence of change, although the possibility remained that qualities lost during early cleavage could also be regenerated.

Testing Later Blastular Nuclei: Nuclear Transplantation Experiments

The complete equivalence of nuclei in early blastulas and zygotes was consistent with the idea that genes did not change, at least, prior to their activation. The question remained: Did nuclei continue to be equivalent after the initiation of gene action?

The feasibility of determining the equivalence of nuclei by **nuclear transplantation,** the transfer of nuclei from one cell to another or to an enucleated cell, was first demonstrated in protistans (e.g., the protozoans, *Amoeba* and *Stentor,* and the

[2]A nucleus does not regulate in the sense that a blastomere regulates, since a nucleus in isolation does not develop at all.

gigantic alga, *Acetabularia*) (see Brachet, 1957). Nuclei and associated cytoplasm were successfully grafted or pushed with the aid of glass needles into enucleated cell halves. Among the surviving **transnuclear** cells, some cellular activities and aspects of phenotype changed in accordance with the genotype of the nucleus.

The microinjection of a single nucleus and the adherent cytoplasm of a ruptured cell into a mechanically enucleated *Rana pipiens* egg (Fig. 16.3) was first performed by Briggs and King (1952). Since then, nuclear transplantation has been performed in many amphibian species (see Briggs, 1979). In addition, single nuclei have been injected into enucleated ascidians (see Whittaker, 1979) and fish eggs (Yan et al., 1986), and multiple or single nuclei have been injected into nucleated *Drosophila* eggs (see Gehring, 1973).

In *Xenopus laevis* and *Ambystoma mexicanum* activated by electric shock or heat treatment, ultraviolet (UV) irradiation effectively **enucleates** eggs. Embryos fail to develop at all unless they receive an unirradiated nucleus, and the results of reciprocal nuclear transplantations between heterozygous *X. laevis* having only one nucleolus and normal *X. laevis* having two nucleoli show that nuclei exposed to UV irradiation are not represented in the tissues of developing organisms (see Gurdon, 1963).

From the start of the nuclear transplantation experiments in amphibians, the frontier of nuclear equivalence was pushed well beyond Spemann's 16-cell stage. **Transnuclear embryos** developed normally to the swimming tadpole stage (Fig. 16.4) and, following metamorphosis, into frogs with competent germ cells even when the nuclei came from animal- or vegetal-pole cells of blastulas with 1000–3000 cells, or after some 11 and 12 divisions.

FIGURE 16.3. Technique of nuclear transplantation in *Rana pipiens.* The actual maturation spindle and micropipettes are smaller than indicated. Compare to method in *Xenopus laevis,* Fig. 16.8. From T. J. King, in D. J. Prescott, ed., *Methods in cell physiology,* Vol. II, Academic Press, Orlando, 1966, p. 3. Used by permission.

Nuclear Transplantation in *Rana*

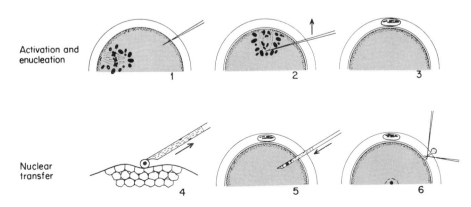

Activation and enucleation

Nuclear transfer

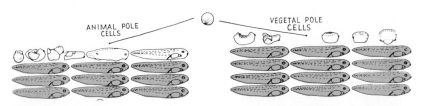

FIGURE 16.4. Results of transplanting nuclei from animal and vegetal poles of a late blastula. The great majority of successfully transplanted nuclei support normal development at least up to the swimming tadpole stage, but some nuclear-transfer embryos develop abnormally (top row of figures). Adapted from J. B. Gurdon, *J. Embryol. Exp. Morphol.,* 8:327 (1960). By permission of Cambridge University Press.

In *X. laevis,* nuclei from gastrulas were also capable of supporting the complete development of transnuclear embryos.

The transnuclear *R. pipiens* that failed to develop completely generally did not become totally disorganized monsters but **arrested** at recognizable stages (Fig. 16.5). Some failed to cleave or under-

went **partial cleavage** and formed blastomeres in only part of the egg. Large numbers of embryos arrested as blastulas and as exogastrulas. Other embryos and larvas suffered from more or less specific endodermal deficiencies (called the endodermal syndrome) or ectodermal deficiencies (called the ectodermal syndrome). Even the occasional normal-appearing adult frog developing from an enucleated egg and a late gastrula's nucleus showed germ-line defects and failed to produce gametes.

The amount of development supported by a nucleus is a measure of **nuclear potency.**[3] The degree to which the nuclear potency of a transferred nucleus resembles that of a zygote's nucleus is a measure of **nuclear equivalence.** Completely equivalent nuclei are **totipotent.** Nuclei supporting development to an advanced point before arrest are **multipotent** (pleuripotent) or **partially equivalent.** Nuclei failing more completely to reproduce the performance of the zygote's nucleus are considered **impotent** or **nonequivalent.**

The arrest of development is related to the source of nuclei in embryos and the age of the embryos. Even nuclei from cells of the different presumptive germ layers of identical blastulas do not yield the same results upon transfer. In *X. laevis* (but not *R. pipiens*), larger proportions of nuclei from presumptive mesodermal cells fail to support development to the stage of normal swimming tadpoles (Fig. 16.6) than nuclei from presumptive ectodermal (animal pole) or presumptive endodermal cells (Fig. 16.4, vegetal pole).

In *R. pipiens,* higher proportions of transnuclear embryos develop abnormally when nuclei are from endodermal cells of gastrulas than from presumptive endodermal cells of blastulas. Similarly, in *X. laevis,* proportionally more nuclei from ectodermal cells of neurulas produce abnormal trans-

TYPES OF EMBRYOS

FIRST

TRANSPLANTATION   GENERATION

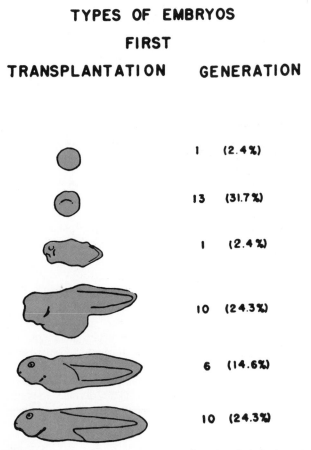

1	(2.4%)
13	(31.7%)
1	(2.4%)
10	(24.3%)
6	(14.6%)
10	(24.3%)

FIGURE 16.5. The percentages and types of defects produced in enucleated *Rana pipiens* eggs injected with nuclei from the endoderm of a late gastrula. From T. J. King and R. Briggs, *Cold Spring Harbor Symp. Quant. Biol.,* 21:271 (1956), by permission of Cold Spring Harbor Laboratory.

[3]**Nuclear potency** is sometimes equated with cellular **potential** in explants or transplants. This usage should be avoided, since a cell's nucleus is not comparable to a whole cell.

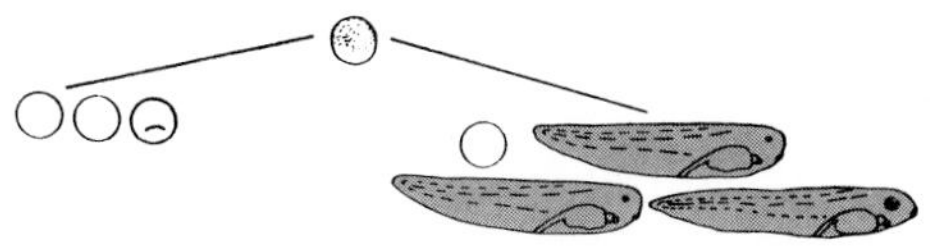

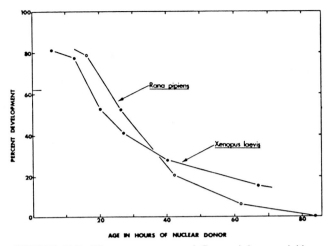

FIGURE 16.6. Results of transplanting nuclei from presumptive mesodermal cells of a late blastula. While the majority of the nuclear-transfer embryos develop abnormally, a few develop to the normal swimming tadpole stage. Adapted from M. Fischberg and A. W. Blackler, in G. E. Fogg, ed., *Cell differentiation,* Symposia of the Society for Experimental Biology, Number XVII, Academic Press, Orlando, 1963, by permission of the Company of Biologists on behalf of the Society for Experimental Biology.

FIGURE 16.7. The percentages of *Rana pipiens* and *Xenopus laevis* nuclear-transfer embryos developing normally to swimming tadpoles as a function of the age in hours of nuclear donors. From R. G. McKinnell, in R. Harris, P. Allin, and D. Viza, eds., *Cell differentiation,* Proceedings of the 1st International Conference on Cell Differentiation, Williams & Wilkins, Baltimore, and Scandinavian University Books, Munksgaard, 1972. Used by permission.

nuclear embryos than nuclei from endodermal cells.

In general, in the course of development, the adequacy of nuclei to support the complete spectrum of developmental events falls off (Fig. 16.7). In both *R. pipiens* and *X. laevis* the fall occurs at about the same time in hours of development for the nuclear donor, but developmental time does not correspond to stage of development. In *R. pipiens* embryos, nuclear potency declines sharply in the late blastula, while in *X. laevis* embryos, nuclear potency does not fall off until the late gastrula (McKinnell, 1972).

The changing proportions of transnuclear embryos developing normally and abnormally leave the impression that an original pool largely of totipotent nuclei gives way to a pool of multipotent nuclei, and then to a pool of impotent nuclei (see Fischberg and Blackler, 1963). These results do not indicate whether homogeneous populations of nuclei produce a statistical spread of different possible results or whether a heterogeneous population of nuclei supports development to different extents.

Serial Nuclear Transfer

In order to measure the homogeneity of nuclei in an embryo, a uniform supply and an adequate population of nuclei were required. Serial nuclear transfer or **nuclear retransfer** was first conceived as a way of supplying both.

The premise was that nuclei from blastulas were largely totipotent and that cleavage through the blastula stage merely proliferated nuclei without changing them. The blastula therefore provided uniform nuclei and a sufficiently large number of them. Nuclear retransfer to enucleated eggs (generally from the same clutch) produced a new round of **nuclear retransfer embryos** (Fig. 16.8).

A **nuclear clone** consists of all the nuclear retransfer embryos derived from a transnuclear blastula. The nuclear retransfer embryos belonging to the same clone generally develop more homogeneously than first-generation transnuclear embryos. At the extremes, it seems that originally totipotent nuclei uniformly produce complete tadpoles, and originally impotent nuclei uniformly produce arrested blastulas. Stable individual differences among the original nuclei therefore seem to account for the preponderance of developmental types produced by clones, and changes in individual nuclei account for the changing proportions of those types in the course of development.

What Makes Nuclei from Postblastular Cells Poor Candidates for Nuclear Transfer?

Several possibilities have been tested. Nuclei may be vulnerable to damage, capable of recovering from damage, or incapable of recovering.

Changes in nuclear equivalence resemble mutations and chromosomal aberrations (e.g., aneuploidy), since defects are inheritable through serial transfer. But unlike mutations and chromosomal aberrations, which are only reversed rarely, defects in nuclei are reversed frequently, especially by serial transfer (Fig. 16.8).

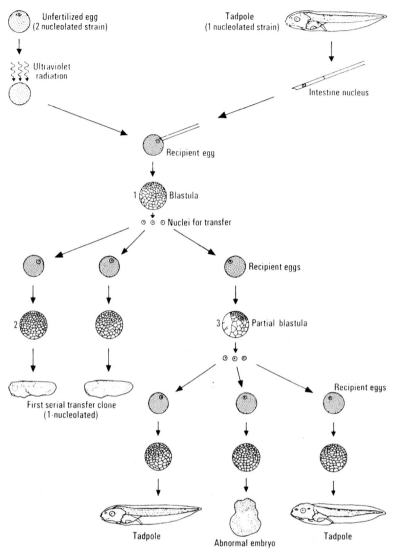

FIGURE 16.8. Serial nuclear transplantation in *Xenopus laevis* and the resulting improvement of embryonic development. From J. B. Gurdon, *Control of gene expression in animal development,* Harvard University Press, Cambridge, 1974.

If nuclei from older embryos are more sensitive to deleterious effects of transfer than nuclei from younger embryos, serial transfer from the blastula stage might not reverse nuclear changes so much as prevent nuclear damage. Possibly, nuclear vulnerability to damage during or after transfer has more to do with arrested development than changes in nuclear equivalence before transfer. Support for this hypothesis comes from cytological evidence of chromosomes replicating normally in embryos prior to transfer but replicating abnormally after transfer to the egg's cytoplasm (Fig. 16.9) (DiBerardino, 1979).

The results of efforts to protect nuclei, stabilize DNA, and maintain chromosomal integrity also suggest that transferred nuclei suffer from deleterious effects of egg cytoplasm. Enucleated eggs receiving transfer nuclei supplied with pro-

tamine or the DNA-binding polycationic amine spermine, at low temperatures, develop into swimming tadpoles in greatly increased percentages compared to eggs receiving unprotected nuclei (see Briggs, 1979).

Possibly, blastular nuclei are less likely than older nuclei to be "shocked" by the conditions they encounter in the egg's cytoplasm. Blastular nuclei may also be more in tune with the rhythms of cleavage and able to get into the "swing" of things. Serial nuclear transfer may "warm up" a nucleus during the first passage through the egg and restore the ability to respond developmentally to the egg in the second round. Acclimatization may also explain how a nucleus from an intestinal epithelial cell of a *Xenopus laevis* swimming tadpole is able to support complete development after serial nuclear transfer (Fig. 16.8).

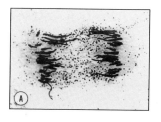
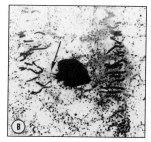

FIGURE 16.9. Photomicrographs of anaphase figures from normal 16-cell stage embryo (a) and from 4-cell nuclear-transfer embryos (b). Frequent and severe chromosomal aberrations and aneuploidy appear in the nuclear-transfer embryos' cells especially those arrested earliest. The nucleus in (b) was derived from an endodermal cell. The abnormal anaphase shows reduced chromosome numbers at one pole (i.e., hypodiploidy) and aberrant chromosomes attached to a clump (arrow). From M. A. DiBerardino, in J. F. Danielli and M. A. DiBerardino, eds., *Nuclear transplantation,* Suppl. 9 to the *International Review of Cytology,* Academic Press, Orlando, 1979. Used by permission.

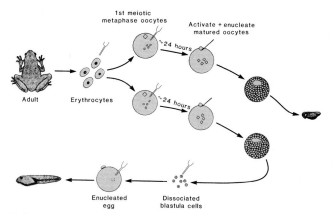

FIGURE 16.10. Testing the potency of *Rana pipiens* erythrocyte nuclei by transfer to oocyte and retransfer to enucleated egg. Erythrocytes are broken by osmotic shock in distilled water and microinjected into oocytes. A day later, the matured oocytes are activated by pricking, and the egg nucleus is removed. Some eggs developed to prehatching tadpoles before arresting (upper). Other eggs serve as donors of blastula cell. Injected into a new generation of activated, enucleated eggs, about one-third of the retransfer embryos become swimming tadpoles. From M. A. DiBerardino, N. J. Hoffner, and L. D. Etkin, *Science,* 224:946(1984), by permission of the American Association for the Advancement of Science and the author. Copyright 1984 by the AAAS.

In order to test the effects of acclimatization, nuclei were injected into oocytes and allowed additional time to adjust to the eggs' cytoplasm before being called upon to divide (Hoffner and DiBerardino, 1980). The nuclei involved were *R. pipiens* erythrocyte nuclei, which hardly support development at all when transferred to enucleated mature ova. But following transfer to immature oocytes, erythrocyte nuclei supported development through the prehatching tadpole stage. Serial transfer through blastulation further improved the nuclei's performance (Fig. 16.10).

Blood from triploid frogs was used in order to provide an unambiguous label of donor nuclei. The nuclear retransfer tadpoles obtained survived for up to a month, at which time they had differentiated hind limb buds, the most extensive development so far obtained with documented differentiated somatic nuclei (DiBerardino et al., 1986).

Possibly, nuclei from the cells of older embryos are unable to meet the demands of the embryo to replicate at the high rate of cleavage. To test this possibility, nuclei were taken from rapidly dividing cells or cells whose rate of division was beefed up experimentally.

The results supported the hypothesis. Some nuclei from rapidly dividing Lucké adenocarcinoma renal tumor cells were effective in promoting relatively normal tadpole development (DiBerardino and King, 1965; McKinnell et al., 1969). Moreover, while nondividing erythrocyte nuclei failed to promote development (Brun, 1978), dividing *X. laevis* erythroblast nuclei occasionally supported development of nuclear retransfer embryos to an early abnormal tadpole stage. Furthermore, when the rate of cell division in larval melanophores, adult keratinizing skin cells, and lymphocytes was experimentally accelerated in tissue culture, the frequencies of successful nuclear serial transfers increased (see DiBerardino, 1988; Du Pasquier and Wabl, 1977; Gurdon et al., 1975).

Like cleavage nuclei, early amphibian blastular nuclei are rushing through mitosis; their chromosomes stay relatively condensed, and their genes remain largely repressed. In older nuclei, chromosomes are stripped of some chromosomal proteins as cells prepare to differentiate. They may have already begun some preliminary stage of differentiation and committed themselves to transcription while shutting down the machinery for rapid chromosomal replication. Possibly, the *Rana* blastula and *Xenopus* gastrula represent the last stages at which the majority of nuclei can still routinely make a successful "handshake" with the enucleated egg's cytoplasm.

Has Nuclear Totipotency Been Demonstrated?

The answer is either a reluctant "no" or a cautious "yes," depending on one's point of view. Perhaps the technique of nuclear transplantation is too coarse and too demanding of the fragile nucleus and egg, but the results have not shown that all nuclei in the cells of differentiated tissues are equivalent.

On the other hand, mitosis and cell division as such do not alter nuclei. The results of serial transplantation experiments indicate that blastula nuclei can reliably transfer the genomic potential for complete development after 100 or more divisions. Differentiation certainly alters nuclei, but some nuclei retain totipotency and, subject to "reprogramming" by serial transfer or incubation in oocytes, support complete development. The restriction of nuclear potency in the other nuclei may flow from changes in chromatin or nucleoplasm, rather than genetic potency.

At present, research on the nuclear equivalence of differentiated cells has moved away from the embryo and development toward the cell and differentiation. The potential of individual genes in differentiated cells is examined via **cell fusion** (i.e., a technique for combining cells in which nuclei fuse after division). The results consistently

indicate that genes which are turned on in fused cells despite having been turned off during prior differentiation seem to have survived dormancy in their native state. "A high degree of genomic information inherited from the zygote must be maintained during cell specialization because the stable phenotypes are reversible under appropriate experimental conditions" (DiBerardino et al., 1984, p. 946).

Ultimately, the question of genomic change is one of DNA physiology, and DNA is far more physiologically active than was once thought. Some changes in DNA, such as its methylation, are potentially reversible, while others, such as gene rearrangement, are not.

The frequency with which genes, like those coding for immunoglobulins (Fig. 4.8), change is small but not insignificant. Other highly complex systems in addition to the immune system (e.g., the nervous system) may also utilize genetic changes to generate variety. Possibly, less malleable and adaptive systems than the immune system get along without gene rearrangement, but possibly they do not.

PART 6   SUMMARY AND HIGHLIGHTS

Blastulas straddle the frontier between zygotic simplicity and multicellular complexity. The role of the blastula in development is presently being reevaluated, and its position as the first distinctively multicellular stage of the embryo is increasingly recognized as a point of departure for the changes that will take place later in development.

The blastula is the last embryonic stage dominated by the egg. Morphologically, the blastula retains the size and shape of the egg, and the outer surface of peripheral blastomeres retains much of the egg's surface, but considerable morphological change also takes place during blastulation.

On the surface, flattened regions anticipate areas of directed cellular movement and the emergence of patterns. Internally, a blastocoel may arise as intercellular cohesion and changes in cell shape accompany membrane specializations and the secretion of materials into and out of segmentation cavities. In terrestrial and freshwater mollusks, the

blastocoel functions in osmoregulation. In amphibians, the blastocoel may present a barrier to animal–vegetal interactions.

Spatial organization or pattern and symmetry are also determined or passed on through the blastula. In symmetrical eggs, axes of symmetry are laid out, and reversible or labile axes are stabilized. Regulative eggs become determined as their developmental pattern is fixed. A hierarchical command structure or cascading of interactions seems to initiate determination, allocating and restricting cells to their fates.

A variety of early embryos are called blastulas. The cellular blastoderm or periblastula of insects, the flattened discoblastula of cephalopods, and the stereoblastulas of many spiralians, typically lack any form of cavity.

Cavities or blastocoels are associated with other blastulas. The teleost's discoblastula, containing a diffuse blastocoel, is a caplike blastoderm attached to a rim of cytoplasm covering the yolk. The avian discoblastula emerges with the thinning of the blastoderm via the massive loss of deep cells and the excavation of the subblastodermic cavity beneath the area pellucida. Totally cleaving eggs also produce a variety of hollow coeloblastulas or true blastulas.

A series of blastulas rather than a definitive blastula may constitute the blastula stage. In sponges, inversion of layers may lead to the differentiation of an amphiblastula. The compact eutherian morula forms a blastocyst cavity and develops into a blastocyst with trophectoderm and eccentric inner cell mass. Echinoderm's coeloblastulas hatch as larval mesenchymal blastulas.

The blastula represents a switching point for many types and levels of genetic control. For the purposes of analysis, the areas depicted on fate maps are sometimes interpreted as organizational fields specified by parameters of positional information, values, and signals. The dimensions and coordinates of morphogenic ecosystems or morphogenic values within sets may also be equated with the areas of fate maps.

Fate maps are plotted from direct observations, especially in small transparent eggs, or with the help of dyes and tracers in opaque and large eggs. The parallels between fate maps and cell lineages in mosaically cleaving eggs suggest that determination may be influenced by specific areas of the egg's surface as well as by cytoplasmic factors.

The results of isolation experiments with blastomeres (or injury-induced isolation of some blastomeres) indicate that most tissues in the blastulas of mosaic eggs are self-differentiating, while the opposite is true in the blastulas of regulative eggs. Potency *in situ* then exceeds potency *in vitro*, and some amount of interaction between the presumptive tissues would seem necessary for normal development.

Maps plotting determination show the ability of explants to differentiate *in vitro* or *in vivo*, and maps charting regulatability illustrate the capacity of embryonic parts to compensate for specific defects. The different maps are similar when the potencies of blastular areas do not exceed their fate and when determination occurs early.

In exopterygote and endopterygote insects, fate mapping of larval structures employs the same techniques used elsewhere, but mapping the imaginal disks of endopterygote insects utilizes mosaic mapping. Depending on the uniformity of nuclei in the syncytial blastoderm and on phenotypic consequences of nuclear alterations in somatic cells, mosaic mapping measures distances in terms of the probability that a clonal boundary will appear between two areas.

All the clones lying within a clonal boundary and nowhere else comprise a compartment or polyclonal compartment. In *Drosophila melanogaster*, separate mutants affect anterior–posterior, dorsal–ventral, and distal–proximal compartments. In *Xenopus laevis*, compartments within the central nervous system are established by founder cell groups or founding cells in the blastula at approximately 512 cells.

Similarities in the fate maps of different vertebrates are suggestive of evolutionary affinities, and the fate maps of arthropods and annelids may be more instructive than cell lineages for clues to evolutionary relationships. The amount of yolk present and the introduction of extraembryonic membranes and cenogenic structures alter fate maps profoundly. The secondary loss of yolk accompanying the evolution of viviparity tends to erase the boundaries of fate maps.

Small differences in amphibian fate maps seem to reflect differences in the thickness of the blastula's wall and in the timing of germ-layer formation. In *X. laevis*, where mesoderm is determined early, presumptive mesoderm lies internally within the wall rather than on the surface of the blastula's wall. The presumptive notochord and presumptive endoderm alone lie on the surface below the prospective limit of invagination. Presumptive ectoderm alone lies above the limit of invagination.

The ability of sea urchin tissue to differentiate into animal-type structures is concentrated in the animal hemisphere, while the ability to differentiate into vegetal-type structures is concentrated in the vegetal hemisphere, especially micromeres and bordering veg_2 cells. According to Runnström's double-gradient theory, as developed and tested by Hörstadius and others, gradients of hypothetical morphogens stream out of the respective poles of eggs and animalize or vegetalize embryonic tissue. Antagonisms between the morphogens promote normal development by broadening the range of possibilities for tissue differentiation, while determination for one type of differentiation results in resistance to determination for the polar-opposite type.

Gradients in the destaining of electron-acceptor dyes such as Janus green under anoxic conditions parallel the animal–vegetal gradients and, as suggested by C. M. Child, presumably reflect graded levels of metabolic activity. In echinoderms, both metabolic and morphogenic effects of the gradients are mimicked by vegetalizing and animalizing agents especially lithium and zinc ions, respectively.

In amphibians, gradients in a double-gradient hypothesis run perpendicularly to each other. The single highest point in both gradients corresponds to the organizer or dorsal lip of the blastopore. Transplanted tissue from the vegetal hemisphere can vegetalize tissue in the animal hemisphere, but the opposite is not observed. Lithium induces neural differentiation or anteriorizes; UV irradiation of the vegetal hemisphere posteriorizes, and both may prevent bilaterality.

In insect eggs, anterior and posterior centers may influence pattern formation. While pole cells appear to be determined by posterior pole plasm, a posterior activity center may also be the source of a posterior determinant. In the absence of an anterior determinant capable of driving tissue toward anterior differentiation, a posterior determinant may cause the formation of larvas with double abdomens resembling the phenotype of *bicaudal* mutants.

A hypothetical autocatalytic mechanism provides models for patterning mutants, producing gradients of morphogens, and triggering thresholds of positional signals. An activation center is the source of both an activator and an inhibitor. The activator operates over a short range to promote the synthesis of itself and the inhibitor. The inhibitor operates at long range to inhibit the production of the activator.

In the guise of a morphogen, the inhibitor may also affect development. Mutant genes that alter the amount or effectiveness of the activator or inhibitor morphogen lead to the formation of additional activation centers and consequently the development of animals with symmetrical or asymmetrical defects, reversals of polarity, or alterations of differentiation within segments.

The emergence of new proteins in embryos and possible surface determinants is studied by a variety of techniques. The development of hybrids produced by artificial insemination may be arrested at the initiation of gene activity. The nuclei of hybrids and nuclear–cytoplasmic hybrids produced by interspecific nuclear transplantation seem to become vulnerable to irreversible damage as they become genetically active.

Changes in proteins are frequently charted with the help of immunoprecipitin reactions and immunohistochemical techniques, especially indirect techniques employing secondary antibodies conjugated with ^{125}I, fluorescent dye, or horseradish peroxidase. Antisera are obtained from animals of different species, different strains of the same species, or different sexes of the same strain. Following adsorption to remove undesired antibodies, the antisera may be directed against tissue-specific and stage-specific embryonic antigens. Monoclonal antibodies may be specific for particular parts of antigens.

Results with adsorbed antisera against tissue-specific antigens generally show that widely distributed antigens in the early embryo later become restricted to particular tissues. Stage-specific embryonic antigens detected by adsorbed antisera vacillate, but antigens shared by mouse blastocysts and several tumor cell lines, especially the teratocarcinoma lines derived from germ-line cells, appear early and then disappear. These antigens contrast with histocompatibility antigens shared by adult cells but lacking in tumor cells. Like ICM cells, teratocarcinoma cells obtained from the central cores of embryoid bodies (a highly malignant tumor when serially propagated through ascites fluid) injected into blastocysts may become integrated into normal differentiated tissues of embryos.

Recombinant DNA technology and hybridization techniques now make it possible to chart gene action directly. Results reveal that the same polypeptides may be coded in more than one part of the genome by genes differing in their untranslated 3' terminal regions. These genes seem to be independently regulated, and closely related genes coding for small families of proteins seem to be differentially activated.

Genes expressed after fertilization are zygotic-effect genes. Gene-controlling compounds produced earlier may influence zygotic-effect genes. Selector genes may operate particular genic networks involved in differentiation.

Genes are also turned off in blastulas. The inactivation of an entire X chromosome in the eutherian blastocyst begins in the TE and primary endoderm where it may be selective for the paternal X. Random X inactivation occurs later in the ICM.

Much of the mRNA undergoing translation in early embryos is transcribed earlier in the oocyte or in nurse cells during oogenesis. Stored (or maternal) mRNA seems to be sufficiently abundant and diverse to support everything required for morphogenesis through blastulation, and posttranscriptional controls direct early development. In amphibians, selective polyadenylation of transcripts associated with large yolk platelets may operate during blastulation to determine dorsal differentiation.

Maternal effects are caused by materials already present in the egg and required for normal embryogenesis. Maternal-effect mutations bring about mutant phenotypes through changes in the amounts or types of stored materials. The expression of maternal-effect mutants such as temperature-sensitive female steriles is manifested during oogenesis in the mother.

The effects of some maternal-effect mutants are reversible and eggs may be rescuable by injection of normal egg cytoplasm following fertilization. In the axolotl blastula, a putative activator alters overall nuclear capacities and makes gastrulation and neurulation possible. The substance is released from the germinal vesicle at GVBD and adsorbed by late-blastular nuclei from the cytoplasm. Later, transcription precedes translation directly and transcriptional controls become dominant.

The midblastula represents the turning point between posttranscriptional and transcriptional controls. In amphibians, the paternal or embryonic genome is switched on at the midblastula transition simultaneously with the loss of synchronized cell division, the beginning of cell movement, and the acquisition of a nuclear/cytoplasmic ratio characteristic of somatic cells. Changes in the blastula's synthetic activities also coincide with the onset of lethality in hybrids, in zygotic-effect mutants, and in embryos whose transcription is experimentally inhibited.

Patterning is sensitive to several zygotic-effect mutations. In *Drosophila*'s cellularizing blastoderm, the normal products of the segmentation genes appear in several patterns. The transcripts and proteins of pair-rule genes appear as a striped zebra pattern across the blastoderm. Alterations in pattern differentiation produced by mutations in seg-

mentation genes are anticipated by changes in the pattern of transcripts and proteins, and the altered segments or portions of segments in mutants correspond to those where the normal gene products appear.

Teratogens are chemicals and physical conditions that mimic zygotic-effect mutations. Like mutations, teratogens may channel development by operating as selectors for secondary genes.

Historically, genetics and embryology divided along lines of preformation and epigenics and of gene production and gene action in the early embryo. Epigenesis foreshadowed gene activation and transcriptional controls; preformation anticipated stored or maternal gene products and posttranscriptional controls.

Charles O. Whitman negotiated a new synthesis by turning time into a developmental variable. Encapsulated in the concepts of neopreformation and neoepigenics, development was looked at from either the gene up or from the organism down. Developmentally significant factors were portrayed as precociously localized in mosaic eggs and gradually localized in regulative eggs.

Hans Spemann interpreted the delayed development of some embryos in ligation experiments as a consequence of the late arrival of a nucleus and suggested that the nucleus was competent to support the full spectrum of developmental events despite previous divisions. The nucleus switched on the cytoplasm of the half-egg at the same time the cytoplasm switched on the nucleus.

Subsequently tested through nuclear transplantation by Briggs, King, Gurdon, and others, Spemann's concept was vindicated up to the blastula stage, and serial nuclear transplantation or retransfer through blastulas showed that mitosis as such did not diminish nuclear potency. The subsequent loss of nuclear potency can be fitted to a stochastic model of decay or a directed model of progressive nuclear differentiation.

Today, the idea of nuclear equivalence, that nuclei, and hence genes, are not altered by cell division, is a widely held conviction among developmental biologists. Inherited on the homologous chromosomes of male and female gametes, replicated unerringly, and passed on unchanged through every round of cell division, very nearly the same genes are thought to be present in virtually every nucleated cell of an organism.

PART 6   QUESTIONS FOR DISCUSSION

1. What aspects of blastulation are anticipated by intralecithal cleavage? By meroblastic cleavage? By total cleavage? Describe a stereoblastula and a coeloblastula (an amphiblastula and morula, a discoblastula and a blastocyst), and discuss why they are all considered blastulas.

2. Compare the formation of the area pellucida in the chick and the inner cell mass of a eutherian mammal, and contrast both to the formation of the blastocoel of a sea urchin or amphibian embryo.

3. What are fate maps, and how are they drawn for amphibians? For endopterygote and exopterygote insects? For spiralians? Compare and contrast the concepts of fate maps and cellular lineages. How does the determination of regions compare to their fates plotted in fate maps? How do cenogenic structures alter fate maps within a group?

4. What is mosaic mapping and how does it compare with fate mapping? What evidence suggests that different genetic controls determine the origins of polarized compartments in imaginal disks? Define a cellular clone or clonal domain, and discuss how it differs from a founding cell or founder cell group and a compartment or polyclonal compartment.

5. Describe the double-gradient hypothesis in sea urchins and explore its extension to amphibians and insects. How might effects of animalizers and vegetalizers on metabolic activities alter morphogenesis? Define positional information and discuss its merits as a theory.

6. How have hybrids been used to study gene action? What are the major objectives of immunological studies on blastulas and how well are these objectives met? Describe some techniques for detecting antigens in extracts and in histological preparations of embryonic tissues. Discuss the advantages and disadvantages of monoclonal antibodies for identifying tissue-specific and stage-specific embryonic antigens.

7. What are teratocarcinomas, and why are some teratocarcinoma cells useful as models for embryonic cells? Describe the coming and going of embryonic antigens and compare them to teratocarcinoma antigens. What are phenocopies, and how are they useful in studies of development?

8. How is gene action studied in embryos with the help of recombinant DNA technology and hybridization? Compare gene activation in sea urchins and amphibians.

9. What is the amphibian midblastula transition? What evidence suggests that the embryonic genome is turned on at the midblastula transition? What is the equivalent of the midblastula transition in other organisms? Define X chromosomal inactivation and discuss its significance for theories of gene control.

10. Define latent pattern and dynamic determination. Contrast maternal-effect mutants to zygotic-effect mutants and describe the phenomenon of rescue. What are gene-controlling components and how might they be instrumental in producing the various phenotypes associated with hypermorphic, hypomorphic, and amorphic mutants?

11. Describe autocatalysis and explain how a system of activators and inhibitors could establish a gradient. How might a morphogen lead to the emergence of a morphogenic pattern in the presence of cells with set morphogenic thresholds? Show how the concept of autocatalysis can be applied to patterning mutants in insects.

12. How do you imagine Morgan and E. B. Wilson would react to the concept of positional information? What position would Whitman take?

13. What is the hypothesis of nuclear equivalence, and what was the contribution of Spemann? Of Briggs and King? Of Gurdon? What is serial nuclear transfer, and how does it affect the performance of transferred nuclei?

PART 6 RECOMMENDED READING

Constantini, F. and R. Jaenisch., eds., 1985. *Genetic manipulation of the early mammalian embryo.* Banbury Report, 20. Cold Spring Harbor Laboratory, Cold Spring Harbor, NY.

Czihak, G., ed., 1975. *The sea urchin embryo: Biochemistry and morphogenesis.* Springer-Verlag, Berlin.

De Pomerai, D., 1985. *From gene to animal, An introduction to the molecular biology of animal development.* Cambridge University Press, Cambridge.

Davidson, E. H. and R. A. Firtel, eds., 1984. *Molecular biology of development.* Alan R. Liss, New York.

DiBerardino, M. A. and J. F. Danielli, eds., 1979. *Nuclear transplantation, International Review of Cytology,* Suppl. 9. Academic Press, Orlando.

Friedlander, M., ed., 1979. *Immunological approaches to embryonic development and differentiation,* Part 1, *Current topics in developmental biology,* Vol. 13. Academic Press, Orlando.

Hörstadius, S., 1973. *Experimental embryology of echinoderms.* Clarendon, Oxford.

Jeffery, W. R., and R. A. Raff, eds., 1983. *Time, space, and pattern in embryonic development.* Alan R. Liss, New York.

Le Douarin, N. and A. McLaren, eds., 1984. *Chimeras in developmental biology.* Academic Press, London.

Malacinski, G. M., ed., 1984. *Pattern formation: A primer in developmental biology.* Macmillan, New York.

Meinhardt, H., 1982. *Models of biological pattern formation.* Academic Press, London.

Sang, J. H., 1984. *Genetics and development.* Longman, London.

Subtelny, S., ed., 1979. *Determinants of spatial organization,* 37th Symposium of the Society for Developmental Biology. Academic Press, Orlando.

Part 7

GASTRULAS AND GASTRULATION

It is not birth, marriage or death, but gastrulation which is truly the most important time in your life.

J. M. W. SLACK (1983, p. 3, remark attributed to Lewis Wolpert)

Gastrulation occurs as the embryo faces its greatest crisis since fertilization. The high frequency of developmental arrests accompanying early gastrulation is mute testimony to the difficulties faced by the embryo. In fact, gastrulas are in a life-threatening situation from which many fail to emerge.

Every thread so carefully spun and woven in the course of gametogenesis has begun to unravel, and some resources are no longer present in adequate amounts to sustain activity. Building blocks, already utilized, are no longer available; waste has accumulated, and the problematic products of some testy genes have cropped up.

Part 7 examines how gastrulation solves the early embryo's pressing problems. Gastrulas tap new resources; they access stores of nutrients and build sinks for wastes. The embryos of yolky eggs harness a food supply and provide a distribution system in the form of a coelom. Embryos of less yolky eggs construct the equipment through which they exploit resources in their environments, and embryos that rely on the transfer of wastes and nutrients to and from maternal circulation establish placental connections.

Chapter

17

DEFINING THE GASTRULA

We do not think we can specify a common starting point for gastrulation in all animals, temporally or spatially, nor can we have a common specific anatomical end-point. The conventional view is that gastrulation is post-blastula and preneurula. There is too much diversity.

I. BRICK AND C. WEINBERGER (1984, p. 537)

Since it has proved impossible to define a gastrula in a way that specifies precisely comparable stages in the development of vertebrates representing the major classes—for instance a shark, a salmon, one or another of the amphibia, a chick, and a mouse—the latest usable definition of gastrulation in this phylum makes no reference to a gastrula at all . . . i.e., it is the morphogenetic movements.

W. W. BALLARD (1984, p. 539)

*E*mbryologists have struggled to define the gastrula as a stage of development ever since Ernst Haeckel coined the term (Gk. *gaster* belly, paunch, or womb + diminutive suffix) more than a hundred years ago. The word is applied to embryos in the process of acquiring two or more layers of cells, but these embryos may not otherwise resemble one another. What is a gastrula anyway?

Occasionally, gastrulas are said to represent the embryonic stage at which determination is first manifested, but, in some species, cells, such as germ cells (e.g., see Fig. 12.23), are determined and segregated long before the gastrular stage. The cytoplasmic distribution of determinants (e.g., see Fig. 12.6) as well as determined cells may also be found among the presumptive germ layers of the blastula.

The gastrula is also portrayed as the embryonic stage at which cell movement is initiated. This claim is contradicted by the earlier commencement of cell movements in some embryos (e.g., the migration of primary mesenchyme cells in sea urchins, see Fig. 13.19, and cellular mixing within compartments in amphibia) and by the withholding of movement until later stages in other embryos (e.g., the migration of primordial germ cells in mammals, Fig. 12.35, and birds, Fig. 12.32).

Gastrulation, the process of making a gastrula, is equally ambiguous when it comes to specifying qualities applicable to all species. Typically, definitions emphasize three facets of the process.

Gastrulation is a process characterized by [1] the onset of an integrated complex of movements and cell shape changes which generate the primary germ layers, [2] the establishment of the immediate basis of axial organization of the embryo, and [3] the placement of the organ system rudiments in their definitive locations. (Brick and Weinberger, 1984, p. 537)

Gastrulation leads directly to the establishment of definitive germ layers in the embryo (e.g., the vertebrate ectoderm, endoderm, and mesoderm); it fixes the axes of symmetry in the organism as a whole; it places embryonic tissues in positions where they normally assume their fates or may undergo the interactions that allow them to develop according to some portion of their competence and potential. Yet embryos generally perform these feats separately, and even embryos called gastrulas doing all these things together may be conspicuously different in other respects.

HOMOLOGY AMONG GASTRULAS

In many gastrulas, future body layers are represented by pairs of cells or even single cells known as **founder cells** in nematodes or **teloblasts** in mollusks and annelids, but, in other gastrulas, groups of cells form layers known as **germ layers** superficially corresponding to adult body layers (e.g., amphibians). Many similarities and differences among gastrulas reflect similarities and differences in the relationship of founder cells and teloblasts to germ layers. The question is: Are the cells and layers of gastrulas homologous? Do they represent similarities due to common descent?

The embryological criterion for homology is similarity in developmental fate, not necessarily similarity in form. Similarity in form may represent analogous adaptations to common problems in development.

Many gastrulas have cavities and depressions continuous with the outside environment. Invaginations, pockets, or folds of the surface also raise the question of homology. A fold or pocket representing the mouth and a pocket representing the anus are not homologous even if their mode of formation is superficially similar. Likewise, extraembryonic membranes may not be homologous despite their function. Dorsal folds of tissue drawn over the yolk in insects, organogenic layers covering the yolk of cephalopods, the ectodermal yolk sacs of teleosts, and the endodermal yolk sacs of mammals are not good candidates for homologous structures.

Confinement raises another issue. The tough coverings of the egg may serve the embryo as protection, but they can also limit growth. The more or less spherical or oval shape of many eggs has a low surface area to volume ratio and severely limits the potential of embryos to expand their developing surfaces. Round worms and insects confront this problem by creating deep folds around which the embryo develops only to stretch out upon hatching. Among the vertebrates, rodent gastrulas form in a closed cylinder, but while the fold in worm and insect embryos expands their integument, the cylinder in rodent embryos expands their guts. Even the characters of their inner and outer surfaces are shaped by adaptation and not by homology.

Fortunately, embryologists do not have to solve the mysteries of homologies with only embryological data. Evolutionary relationships are diagnosed by bringing together criteria from different biological disciplines. Evolutionary histories suggested by the sum of information provide the rationale for identifying homologies.

Classification of Gastrulas, Germ Layers, and Other Structures in the Gastrula

Definitions of gastrulas traditionally identify embryonic **germ layers** with cells that serve as the substratum from which larval or adult tissues and organs arise. To qualify as a germ layer, cells must be broadly attached to each other or even fused rather than present in a loosely woven network. Layers must be **epithelial** or **syncytial** at least partially or for some period, although they can have contractile (i.e., muscular) and excitatory (nervous) features as well.

Pure **mesenchyme** or embryonic connective tissue does not constitute a germ layer by this criterion. Only when mesenchymal cells are reconstituted as an epithelium do they constitute a germ layer.

The dual germ layers of diploblastic embryos and the inner and outer layers of triploblastic animals are the **ectoderm** (sometimes called epiblast or ectoblast) and the **endoderm** (sometimes hypoblast or endoblast, also entoderm), respectively. Triploblastic embryos, generally equated with Hatschek's Bilateria and including all three-layered invertebrates and vertebrates (Table 17.1), have an additional middle layer, or **mesoderm.**

The middle layer emerges in one of two ways: from the coalescence of loose mesenchymal cells or from folding of an epithelium (sometimes called a mesoblast of mesothelium), most often the original endoderm (sometimes called mesendoderm, mesendoblast, mesentoderm, or mesentoblast). Mesoderm is sometimes recognized only when it is derived from endoderm, and the term enterocoel is reserved for animals forming their mesoderm in

TABLE 17.1. Classification based on gastrulas

	Diploblastic	Triploblastic			
		Protostomes			Deuterostomes
		Acoelomates	Pseudocoelomates	Eucoelomates	
				Schizocoels	Enterocoels
Primary germ layers					
Ectoderm	+	+	+	+	+
Endoderm	+	+	+	+	+
Mesenchyme of ectodermal origin	+	+	+	+	+
Mesoderm from particular cells or masses of cells	−	+	+	+	−
or					
Mesoderm from archenteron or prim. streak	−	−	−	−	+
Stomodeum formed first	na	na/+	+	+	−
Proctodeum formed near blastopore or equivalent	na	na	−	−	+

+, Present or yes; −, absent or no; na, not applicable. *Source:* Mainly from Hyman (1940).

this way (see Hyman, 1940). In other embryos, the middle layer is called mesoblast (among other things) and the three-layered embryo is referred to as an acoel, pseudocoel, or schizocoel.

The coelom (also coelome and sometimes celom or celome) is the cavity between the embryo's body wall and gut. Coeloms form in gastrulas, often at the same time that mesoderm in general makes its initial appearance. The mode of origin of the coelom provides the grounds for making further distinctions (Table 2.1). Animals in the **schizocoelous** phyla (including the Mollusca, Echiuroidea, Annelida, and Arthropoda, among others) form their coelom by the splitting or cavitation of a mesodermal or mesenchymal mass. Animals in the **enterocoelous** phyla (including the Chaetognatha, Echinodermata, Protochordata, and Vertebrata) form coeloms from evaginations of the embryonic gut or tissue continuous with (if not derived from) the embryonic gut.

Acoelomates and pseudocoelomates are loosely aligned with the schizocoelous eucoelomates by determinate development, while the enterocoelous eucoelomates generally (but by no means always) have regulative development. This alignment of groups also extends to the typical mode of mesoderm formation. The schizocoelous eucoelomates and their allies typically form their mesoderm from single cells or small groups of **teloblasts** or **mesoteloblasts.** Enterocoelous eucoelomates, on the other hand, typically form a major

portion of their mesoderm from a multicellular layer (i.e., more properly called a **germ layer**).

Another difference among the groups is found in their mode of forming their mouth or **stomodeum** (also stomodaeum, Gk. *stoma* mouth + *-odeum* resembling) at the beginning of the digestive system and their anus or **proctodeum** (Gk. *proktos* buttocks + *-odeum*) at the end of the digestive system (Fig. 17.1). **Protostomes** (Gk. *pro* ahead or first + *stoma*) form their stomodeums around an early pocket and their proctodeums around a late pocket, and **deuterostomes** (Gk. *deuteros* second + *stoma*) form their proctodeums in the vicinity of an early pocket and their stomodeums around a later pocket. The deuterostomes correspond to enterocoelomates. The protostomes include all the other triploblastic invertebrates (Tables 17.1 and 2.1).

The blastopore (a pore in the "blast" or embryo) is the name typically given to any large opening in protostome and deuterostome embryos independently of the homology of the tissue surrounding the pore or the fate of that tissue. At various times, and in various organisms, both stomodeums and proctodeums have been labeled blastopores without any more justification than their being holes.

Today, common practice is to identify the linings of both the stomodeum and proctodeum as ectoderm and to distinguish between these invaginations and the blastopore (Fig. 17.1). The **blas-**

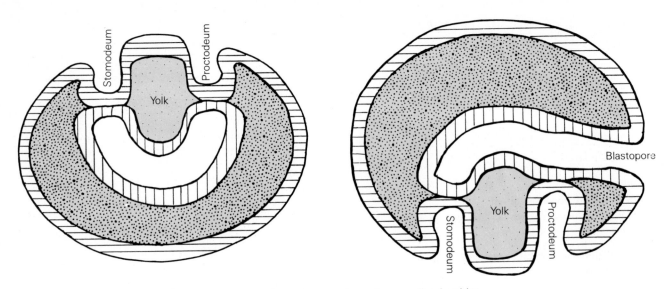

FIGURE 17.1. Schematic longitudinal sections through gastrulas of generalized schizocoelous protostome (left) and enterocoelous deuterostome (right). Light stippling, yolk; heavy stippling, mesoderm; horizontal hatching, ectoderm; vertical hatching, endoderm.

topore (sometimes called gastropore) is defined as a pocket on the embryo's surface where ectoderm flows into endoderm and often, but not necessarily, into mesoderm. In the relatively small-yoked deuterostomes (Fig. 17.2), this pocket is a discrete invagination constituting a true blastopore, while in large-yolked avian deuterostomes and in mammals the developmentally comparable "pocket" is a mere groove called the **primitive streak.**

No such point or groove occurs in protostomes, which therefore have no blastopore or primitive streak. Their stomodeums typically form first and extend to the foregut where they meet endoderm already present within the embryo. The stomodeum does not give rise to endoderm but merely lines the initial section of the digestive tract.

When a blastopore is present on the surface of a gastrula, the tissue surrounding it constitutes the **lips of the blastopore** or blastoporal lips (sometimes further identified topographically as dorsal, lateral, and ventral). Internally, the blastopore is continuous with a cavity called the **gastrocoel** (i.e., cavity of the gastrula), and the lips of the blastopore are continuous with the **archenteron** (Gk. *archein* to begin + *enteron* intestine) or wall surrounding the gastrocoel. Occasionally, the entire archenteron is endodermal (e.g., in *Xenopus laevis*), but more often the wall contains mesoderm and may even contain germ cells.

For example, *Sagitta*, the chaetognath or arrow worm, gastrulates by forming an enormous invagination at the blastopore (Fig. 17.2, bl). The sto-

modeum (st) forms where the cells at the tip of the archenteron make contact with the ectoderm (ec). Folds in the lining of the archenteron proceed to establish the endoderm (en), and two layers of

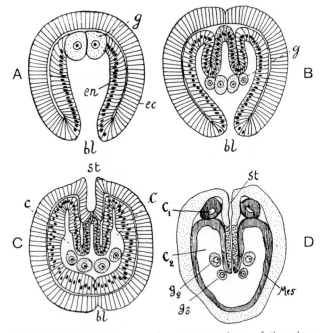

FIGURE 17.2. Diagrammatic cross sections of the chaetognath, *Sagitta,* illustrating three stages of gastrulation. bl, Blastopore; c, coelomic sac; ec, ectoderm; end, endoderm; g, primordial germ cells; st, stomodeum. From E. Korschelt and K. Heider, *Embryology of invertebrates,* Swan Sonnenschein, 1895.

mesoderm, a **visceral layer** or **splanchnic meso-derm** surrounding the endoderm and a **parietal layer** or **somatic mesoderm** lining the ectoderm, surround coelomic sacs (c).

Pitfalls in Classification

As useful as they are for viewing development, germ layers are also problematic. An historical grasp of embryology is required for an understanding of germ layers.

Germ layers are by no means always delineated. Distinctions among the germ layers are most apparent in the highly yolky megalecithal eggs of vertebrates and appear when the yolk is being enclosed. The definitive mesoderm may become obvious only as it surrounds a coelom or provides circulation to and from the yolk (see Hyman, 1940).

Nomenclature concerning the mesoderm requires further qualification. Regrettably, the mesoderm is placed among the primary germ layers in most recent embryology textbooks. Since germ layers are present only in the embryo, attaching "primary" to all of them adds nothing to the designation of germ layers.

In the 19th century, the endoderm and ectoderm were designated **primary germ layers** until the mesoderm emerged (e.g., Korschelt and Heider, 1895). Then, the three germ layers were considered **secondary,** having passed a plateau on their way to greater complexity. Two-layered gastrulas were considered **dipulas** as opposed to three-layered **tripulas.**

A comparable practice today distinguishes between germ layers arising *during* gastrulation and germ layers originating *in* the gastrula (or gastrula-derived germ layers) (Hopper and Hart, 1985). The latter, however, are not limited to mesoderm. For example, the endoderm (as opposed to the hypoblast) of avian embryos is gastrula-derived. Furthermore, identifying the mesoderm as a secondary germ layer by its order of appearance distorts the situation in many embryos where the third germ layer appears simultaneously with the other two.

Characterizing layers by gastrulation and the gastrula places too much burden on morphological appearance and too little on underlying processes. Clarity might better be served by specifying when a layer is determined rather than when it appears.

Designations of primary and secondary might be quite different where determination is concerned. In amphibians, for example, endoderm would be ranked as "primary" inasmuch as it meso-

dermalizes ectoderm in the blastula. Mesoderm might then be considered "secondary" and ectoderm considered tertiary, since it is not fully "ectodermalized" until gastrulation.

What can be done to clarify the "primary germ layers"? Probably, terms such as "primary" and "secondary" should be dropped unless they refer to sequence, and then only if more precise terms are unavailable.

The difficulty in establishing the sequence of events in gastrulation is in part one of technique. Mechanisms involved in forming germ layers are obscure in megalecithal eggs, primarily because yolk is opaque, and the relatively small size of cells hinders direct observation of movements below the surface. A variety of *in vivo* and *in vitro* techniques have been brought to bear on the problem, but, in general, less is known about how gastrulation takes place in yolky eggs where germ layers are distinct than in less yolky ones where germ layers are vague.

For example, while gastrulation in anamniotes via a blastopore seems straightforward enough, mesoderm as well as endoderm may share the lining of the archenteron (e.g., see Fig. 19.25). On the other hand, while the germ layers of amniotes are more completely separated in the gastrula, the presumptive germ-layer cells may be mixed in the primitive streak (see Fig. 14.13).

Embryologists trying to describe and explain gastrulation often resort to extrapolations from what is known about one embryo's gastrulation (e.g., a frog's) to what is unknown about another embryo's germ layers (e.g., a chick's). The notion of homology provides guidance for embryologists when used judiciously, but it is no substitute for data.

MECHANISMS OF GASTRULATION

The formation of germ layers requires the cohesion of cells in a layer and the creation of recognizable boundaries between different layers. Cohesion frequently accompanies the elaboration of junctional complexes at the outer surfaces of cells, but the ability of both indeterminate and determinate embryos to operate as integrated systems follows the formation of gap junctions (see Serras and van den Biggelaar, 1987).

Boundaries between cell layers are formed when originally juxtaposed cell surfaces differentiate or when already differentiated cell surfaces are brought into juxtaposition as a result of cell movement. Mechanisms involving the differentiation of juxtaposed surfaces may be lumped together under the heading delamination (*de-* away from + Lat. *lamina* a thin plate), while mechanisms involving cell movement can be separated into categories of individual cell movement and mass cell movement (see Løvtrup, 1974).

Delamination

Delamination may involve cell divisions with furrows parallel to the surface layer (i.e., tangential or periclinal divisions). This form of delamination (sometimes called primary delamination) is the exclusive mode of endoderm–ectoderm separation in the coeloblastulas of a family of trachyline medusas. More often, delamination by cell division is mixed with delamination without cell division (sometimes called mixed delamination), for example, in the formation of the neuroectoderm of insects.

Delamination not involving cell division or at least not involving it directly (sometimes referred to as secondary delamination) is common in gastrulas having neither a blastopore nor archenteron. In birds and mammals, delamination without cell division provides a portion of the endoderm (i.e., hypoblast or yolk sac endoderm), and it allows mesoderm to separate into the visceral and parietal portions surrounding the coelom.

Whether or not delamination involves cell division, the separation of cellular "lamina" involves the dissolution of adhesive molecules and possibly the insertion of new proteins at the apposing cell surfaces. The affinity of cell membranes to each other or to extracellular materials would have to be reduced, and the introduction of new materials would have to create the potential for swelling the intercellular space. Changes in cell shape, cell movement, secretion, and the configuration of the embryo as a whole would also promote enlarging the new space.

Individual Cell Movement

Individual cell movement or migration is called **ingression** when it takes cells from the surface to the inside of an embryo. When just about any cell on the surface of a blastula is as likely as any other cell to ingress (as in some hydrozoas), the process is called **multipolar ingression.**

In **polar** or **unipolar ingression,** cells from the surface at only one end of the embryo move to the

inside. The movement of yolky vegetal pole cells into the pregastrula amphibian embryo is an example of unipolar ingression from the vegetal pole (see Fig. 19.27). The animal pole, marked by the presence of polar bodies, may also be a site of endodermal ingression (e.g., some other hydrozoans).

Unipolar ingression may also be involved in the formation of mesoderm. In echinoderms, for example, unipolar ingression is responsible for the formation of primary and secondary mesenchyme cells.

Mass Cell Movements

Mass cell movements are responsible for some of the more dramatic forms of gastrulation. **Invagination** (*in-* into + Lat. *vagina* sheath) or **emboly** (Gk. to put into) is recognizable when a discrete pocket or fold of the blastula's surface is seen pushing in, or when negative or positive pressure (pushing or pulling) seems to account for the topographical change. Different invaginations may be found on the surface of an embryo (Fig. 17.1), the most common being the stomodeum, proctodeum, and blastopore.

The defining characteristic of a true blastopore is continuity between the lumen of the archenteron and the outer environment. In microlecithal protochordates, the archenteron may be formed by the invagination of half the egg's surface, but in birds, the shallow primitive groove hardly qualifies as an invagination at all and is not continuous with any internal lumen.

The opposite of invagination, namely **evagination,** is frequently a feature of mesoderm formation, especially the formation of mesoderm lining the coelom. **Enterocoelous** folds of the archenteron wall into *Sagitta*'s body chamber, for example, form the peritoneal mesoderm by evagination (Fig. 17.2).

Cells turning around an edge into the embryo **involute.** In amphibians, the initial portion of the archenteron is formed by invagination, but the remainder requires involution around the lips of the blastopore.

Surface layers frequently reach the site of involution after migration. **Epiboly** (Gk. *epibole* something thrown onto) is the movement of cells or syncytia on the surface of the embryo or over the yolk. Whether epiboly is motivated by individual cell movements (as it probably is in the case of amphibians) or the stretching of cells in a sheet (as in teleosts), it is a collective and coordinated act, which justifies placing it in the category of mass cell movements. Epiboly accounts for the massive shifting of the amphibian embryo's surface during gastrulation (i.e., covering the entire embryo with ectoderm and replacing the cells lost by invagination and involution). Epiboly also characterizes the migration of enveloping cells over the teleost's yolk and the movement of the avian germ wall over the vitellus.

Epiboly may occur without involution and even without cell division (e.g., teleosts). Epibolizing layers may thin and stretch or have cells inserted into them from below (e.g., some coelenterates). But in all cases of epiboly, one or another surface associated with the migrating cells is a **hyaline** (or glassy) **membrane.**

Mass cell migration without such a membrane is **epiauxesis** (Løvtrup, 1974). Thought to depend on the properties of new plasma membrane introduced as a result of cell division, epiauxesis occurs in spiralian eggs, especially those with unequal total cleavage, where small animal-pole micromeres grow over large yolky macromeres (see Figs. 19.33 and 19.34). It may also occur in vertebrates especially in the migration of the neural crest (see Figs. 23.8 and 23.16).

When single germ layers are plastered against one another, they are **laminated.** Lamination frequently accompanies the synthesis by one or both layers of extracellular materials (ECMs) such as *laminin* present in the *basal lamina* of epithelia. Layers sharing a single basal lamina may be welded together (sometimes called fused), for example, the somatic mesoderm and ectoderm, constituting the **somatopleure,** or the splanchnic mesoderm and endoderm, constituting the **splanchnopleure.**

Layers may **segregate** as in the formation of the avian hypoblast from the epiblast or the formation of bands of insect neuroblasts from ventral ectoderm. Mechanisms of segregation may involve the synthesis of new ECM, autonomous cell timing (e.g., the nesting of cells in the epiblast), or a cell's lineage, its individual history of cell divisions.

Invagination or evagination incorporating more than one germ layer is **entypy** (e.g., the evagination of the rodent embryonic disk into an egg cylinder; see Fig. 19.4). A variation of entypy is **body folding.** The folds are identified by the portion of the body overlaying them (e.g., the head fold of birds; see Fig. 18.38) or by a membrane created through the fold (e.g., amniotic fold of insects; see Figs. 18.5 and 18.12).

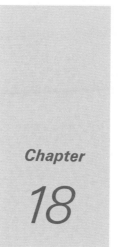

Chapter

18

DEVELOPMENTAL ANATOMY OF LECITHOTROPHIC GASTRULAS

*L*ecithotrophic embryos pay a premium for having so much yolk; namely, they spend energy gaining access to it. Only the surface of the yolk is exposed to cells in the blastoderm or discoblastula. The rest is inaccessible. Inevitably, adaptations for mobilizing yolk and supplying the thickening germ layers of the embryo with yolk products have their impact on gastrulation.

Despite relatively large size, lecithotrophic eggs can sometimes only afford to allocate a portion of their cells to **germinal** functions (i.e., making the embryo). Another portion, and sometimes the bulk of cells, serves the **nongerminal** function of enclosing yolk.

Transient **cenogenic** structures produced by nongerminal cells aid in digesting yolk. In addition, **yolk cells** or **vitellophages** may invade the yolk and, scattering through it, digest and distribute its products.

Gastrulation in the sense of constructing embryonic germ layers is protracted, especially with regard to the endoderm, and a midgut may not form until embryogenesis is well underway. A higher priority is placed on constructing a **coelom** through which the products of yolk breakdown move along to cells.

GASTRULATION FOLLOWING INTRALECITHAL CLEAVAGE

[In yolky, arthropod embryos] gastrulation tends to be prolonged and to be broken up into a number of discrete, though interrelated processes woven around the presence of a large mass of yolk and the requirement of making this yolk available to the developing tissues.

D. T. Anderson (1973, p. 415)

Onycophorans

Onycophorans with yolky eggs form large surface blastoderms around the central yolk mass (see Fig. 11.1). **Presumptive endoderm** invaginates on the ventral surface where a split in the midventral line exposes yolk (Fig. 18.1a). Zones of proliferative activity arise on either side of the split and release **temporary vitellophages** that digest and transport yolk as they migrate through and around the yolk mass. Ultimately, these cells unite at the surface of the yolk to form the endoderm of the anterior midgut epithelium. They continue to function in yolk digestion in their new position.

The posterior midgut is formed later. The con-

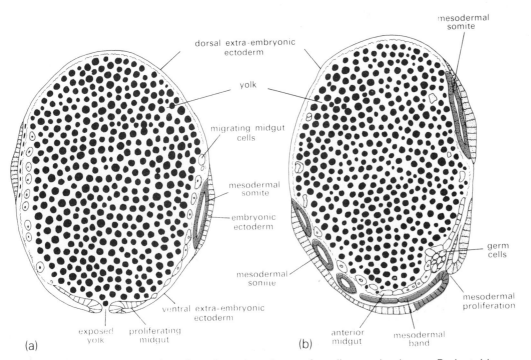

FIGURE 18.1. Diagram of sections through embryos of a yolky onychophoran, *Peripatoides orientalis,* illustrating gastrulation. (a) The midventral sheet of presumptive midgut endoderm has already invaginated and slipped below the surface. Proliferative centers on either side of a midventral slit release cells that digest and distribute yolk as they migrate through the peripheral yolk. (b) The migrating cells form a midgut epithelium around the yolk. Mesodermal cells also invaginate. They too proliferate in zones near the surface, but they remain patent and give rise to mesodermal bands. From D. T. Anderson, *Embryology and phylogeny in annelids and arthropods,* Pergamon Press, Oxford, 1973. Courtesy of D. T. Anderson. Used by permission.

tributing endodermal cells also arise by proliferation in the vicinity of an invagination, but additional temporary vitellophages are not formed. Rather, endodermal cells join the midgut epithelium behind the yolk mass (see Fig. 14.2).

In a similar fashion, two ventrolateral streams of mesodermal cells flow from proliferative zones on either side of a posterior invagination (Fig. 18.1b). The streams reach the anterior parts of the embryo and ultimately provide the mesoderm for all segments.

Presumptive ectoderm is divided into three bands: dorsal and ventral bands of **extraembryonic ectoderm** and a central band of **embryonic ectoderm.** The ventral extraembryonic ectoderm spreads over and heals (i.e., closes) the midventral split, while the dorsal extraembryonic ectoderm remains in place covering the yolk above the embryo. Both extraembryonic ectoderms enclose portions of the yolk mass until, along with the yolk, they disappear. Then the embryonic ectoderm, which hardly moves during gastrulation, covers the embryo (see Anderson, 1973).

Onycophorans with secondarily reduced yolk form their germ layers without very much cell movement, relying instead on **proliferative organogeny.** Dividing cells delaminate to nearby sites where organogenesis takes place.

Crustaceans

Crustaceans also employ proliferative organogeny as the chief means of gastrulation. With few exceptions (e.g., the shrimplike *Anaspides*), only a transient groove marks the depot for cellular immigration (Fig. 18.2).

Cells capable of differentiating into a wide range of tissue types migrate inward, proliferating before, after, or both before and after leaving the surface. Presumptive midgut cells proliferate on the surface and invade the yolk mass without invaginating. Mesoderm forms mainly after the **mesendoblasts** leave the surface and are replaced by ectoderm formed by **ectoteloblasts.**

Unlike terrestrial arthropods, crustacean ectoderm is formed without much ectodermal spread-

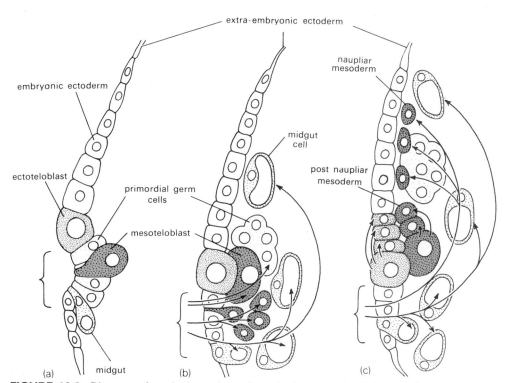

FIGURE 18.2. Diagram of sagittal sections through the gastrula of the small, shrimplike, malacostracan mysid, *Hemimysis lamornae*. Invagination is limited to a shallow, transient groove. Endoderm and mesoderm formation depend on cell division and migration. (a) Primordial germ cells, mesoteloblasts, and presumptive midgut cells still share the surface with embryonic ectoderm. (b) Proliferation of presumptive midgut cells leads to the internal release of midgut cells. The mesoteloblast and primordial germ cells migrate inwardly and are replaced on the surface by ectoteloblasts. (c) Mesoderm and endoderm spread out on the internal surface of the ectoderm. From D. T. Anderson, *Embryology and phylogeny in annelids and anthropods,* Pergamon Press, Oxford, 1973. Courtesy of D. T. Anderson. Used by permission.

ing or epiboly. Except for filling in the areas vacated by the departing mesodermal and endodermal cells, ectoderm in the gastrula remains largely where it was in the crustacean blastoderm (see Anderson, 1973).

Terrestrial Arthropods

In **terrestrial arthropods,** the beginning of gastrulation is hidden from view by yolk, and the end is rendered ambiguous by regional differences. Gastrulation-like morphogenic movements of nuclei take place even before the cellular blastoderm forms and lead to the concentration of nuclei in the **germinal region** or **germ disk** (see Figs. 11.3 and 14.1). Endoderm formation may be anticipated by the segregation of yolk cells or yolk blocks, but the completion of the endoderm may not take place until long after the folding and fusing of mesoderm.

The arthropod germinal region gives rise to the embryo, while the nongerminal region gives rise to

extraembryonic membranes. Germinal regions occupy different amounts of the arthropod surface. In the primitive apterygote, *Collembola,* the germinal region is proportionally much larger than in the advanced bristletail, *Thysanura* (Fig. 18.3), while among pterygotes, advanced dipterans have vastly larger germinal regions than primitive exopterygotes (compare Fig. 14.2c to Fig. 14.1b).

Identified by shape, position, and fate (see Figs. 14.1–14.3), the germinal region increases in size, cellular density, and length to become the **germ band** (also called anlage or embryonic rudiment, primordium, or ventral plate) as gastrulation commences in earnest. In general, the longer or larger the initial germinal region and germ band, the more it is determined, and the faster it develops.

The relative length of the germ band provides a basis for classifying **germ types:** "short-germ" embryos (e.g., *Thysanura,* Fig. 18.3c, also termites and stone flies), "semilong-germ" embryos (e.g., exopterygotes [hemimetabolans], such as the dra-

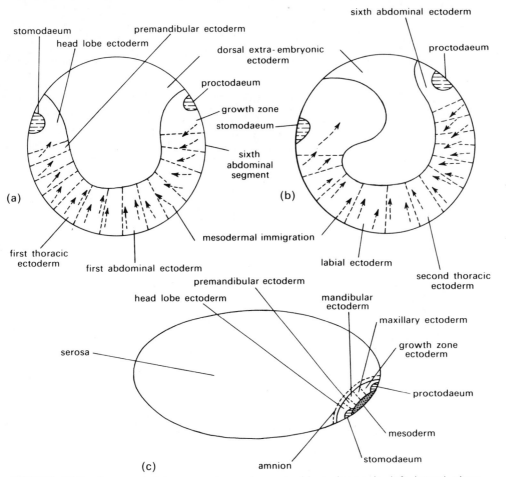

FIGURE 18.3. Proposed fate maps for apterygote blastoderms in left lateral view: (a) *Diplura*; (b) *Collembola*; (c) the bristletail, *Thysanura*. From D. T. Anderson, *Embryology and phylogeny in annelids and arthropods,* Pergamon Press, Oxford, 1973. Courtesy of D. T. Anderson. Used by permission.

gonfly, *Platycnemis*, Fig. 14.1), and "long-germ" embryos (endopterygotes [holometabolans], such as dipterans, Fig. 14.2c). The relatively long-germ embryos of onychophorans (a), myriapods (b), and primitive apterygotes (Fig. 18.3a, b) suggest that the shortened germ bands of *Thysanura* and exopterygotes are derived from long-germ embryos, but the long-germ embryos of endopterygotes are presumably derived secondarily from short-germ embryos (see Anderson, 1972b).

Gastrulation is generally accompanied by **blastokinesis,** or germ band movement, and **germ band lengthening.** The movement tends to bring the germ band into a sheltered environment, either tucked away within the yolk mass (Fig. 18.4) or surrounded by extraembryonic **amnion** (am) and **serosa** (se, sometimes called the chorion, Fig. 18.5). In some exopterygotes, growth draws the embryo conspicuously upward (called anatrepsis, Gk. *ana-* up + *trepein* to turn) and sometimes downward

(called katatrepsis, Gk. *cata-* down + *trepein*) as well. The caudal end of *Platycnemis*, for example, initially moves upward and then turns downward (Fig. 18.6). Other insects take a 180° morphogenic spin as well.

In *Drosophila melanogaster*, the germ band elongates posteriorly, stretches over the posterior tip, and turns anteriorly to produce a U-shaped embryo (Fig. 18.7). After moving up the embryo for three-quarters of its length, the band reaches its maximal length (stage 8 C-O&H, Campos-Ortega and Hartenstein, 1985) and turns inward to produce a G-shaped embryo (Fig. 18.8). The extraembryonic **amnioserosa** covers·the **yolk sac** or yolk membrane over the open part of the G. Mitotic activity, which has been in suspension, soon returns to the germ band, but elongation is not a consequence of pressure from new cells.

Gastrulation in the germ band produces **inner germ layers** (i.e., endoderm and mesoderm). In acar-

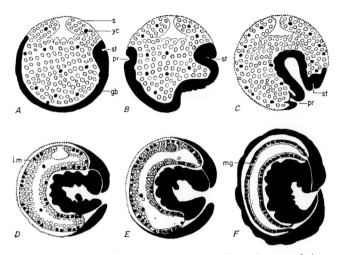

FIGURE 18.4. Diagrammatic sections through egg of the collembolan, *Tetrodontophora bielanensis,* during formation of the serosa and the dorsal organ. The tall columnar epithelium of the embryonic membrane provides dorsal organ material that invaginates to form the wedge-shaped digestive organ. At the same time, yolk-digesting vitellophages and endoderm-forming yolk cells (yc) segregate from the blastoderm. A blastodermic cuticle is secreted. From Cz. Jura, in S. J. Counce and C. H. Waddington, eds., *Developmental systems: Insects,* vol. 1, Academic Press, Orlando, 1972. Courtesy of C. Jura. Used by permission.

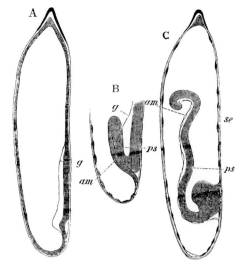

FIGURE 18.5. Schematic longitudinal sections through the embryo of the hemipteran, *Calopteryx,* showing the formation of the amnion (am) and serosa (se) during blastokinesis and involution of the germ band (g, edge of ventral plate, ps). From F. M. Balfour, *Comparative embryology,* Vol. I, Macmillan, London, 1880.

ine embryos (e.g., mites and ticks), proliferation in a ventral groove yields a sequence of cells beginning with yolk cells and vitellophages, followed by posterior midgut epithelial cells, and ending with bilateral groups of small mesodermal cells. In scorpions, the first cells produced in the midline of the germ disk are primordial germ cells. Other inner germ-layer cells follow (see Anderson, 1973).

In the more primitive apterygotes, inner cells are produced diffusely throughout the blastoderm by delamination resulting from tangential divisions. The collembolan midgut epithelium is formed from yolk cells (yc, Fig. 18.4). After the yolk has broken down considerably and has formed a lumen, the yolk cells segregate beneath a limiting membrane and bind themselves together into the midgut epithelium. Similarly, spiders form the major part of their midgut epithelium from temporary yolk cells, originating throughout the germ band but principally in an invagination at the posterior end of a **gastral groove.**

From primitive exopterygotes to advanced endopterygotes, endoderm is formed without the participation of yolk cells (see Anderson, 1973). Typically, endoderm arises from **anterior** and **posterior midgut rudiments** located initially in the germinal

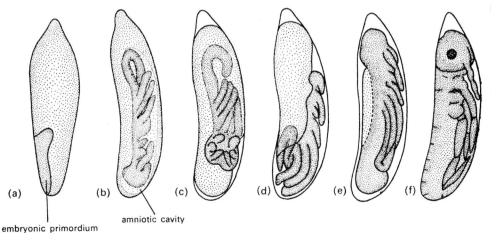

(a) (b) (c) (d) (e) (f)

embryonic primordium

amniotic cavity

FIGURE 18.6. Blastokinesis in *Platycnemis pennipes* during segmentation. From F. Seidel, *W. Roux' Arch. Entwicklungs-mech. Organ,* 119:322(1929), by permission of Springer-Verlag.

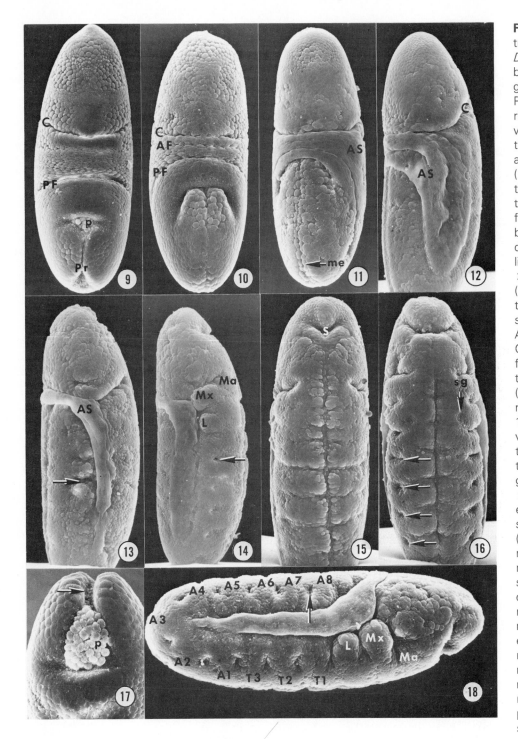

FIGURE 18.7. Scanning electron micrographs of intact *Drosophila melanogaster* embryos during blastokinesis and germ-band segmentation. (9) Posterior midgut pocket carries pole cells (p) into interior while moving anteriorly with the proctodeal inpocketing along the dorsal side. ×200. (10–12) Germ-band elongation is completed, and extraembryonic amnioserosa forms peripherally to the germ band. Mesectoderm (me) is a double row of cells in the midline of the germ band. (10) ×230; (11) ×220; (12) ×240. (13–16) Stages of segmentation. (13) Initial indications of segments (arrow). ×230. (14) At 6½ hours (midstage 11 C-O&H) of cephalic-appendage formation and appearance of tracheal pits (arrows). ×225. (15) ×290. (16) Stage of segmentation (7 hours, late stage 11 C-O&H). ×200. (17) Dorsal view of early gastrula's posterior end showing posterior transverse fold leading to longitudinal proctodeal furrow. ×450. (18) Lateral view of embryo nearing completion of segmentation. Tracheal pits (arrow) are located near anterior edges of the last two thoracic and the eight abdominal segments. ×285. A1–A8, abdominal segments; Af, anterior transverse fold; AS, amnioserosa (extraembryonic ectoderm); C, cephalic furrow; L, labial rudiment; Ma, mandibular rudiment; me, mesectoderm; Mx, maxillary rudiment; P, proctodeum; PF, posterior transverse fold; S, stomodeum; sg, salivary gland primordium; T1–T3, thoracic segments. From F. R. Turner and A. P. Mahowald, *Dev. Biol.,* 57:403 (1977), by permission of Academic Press and courtesy of the authors.

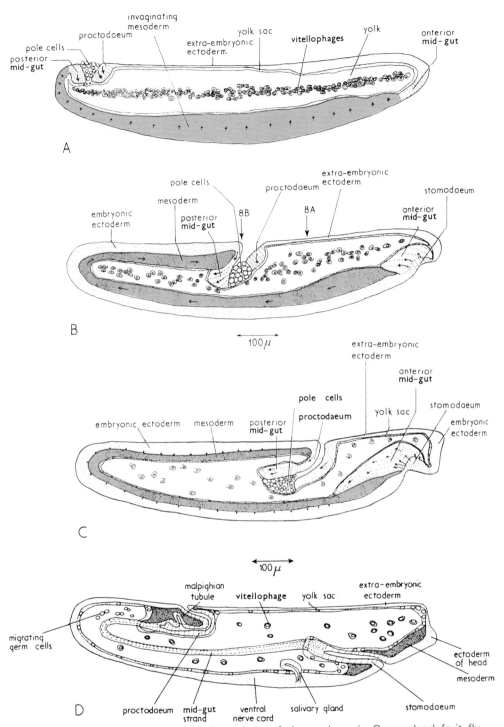

FIGURE 18.8. Schematic longitudinal sections of the embryonic Queensland fruit fly, *Dacus tryoni,* during germ-band elongation and blastokinesis. From D. T. Anderson, *J. Embryol. Exp. Morphol.,* 10:248 (1962), by permission of the Company of Biologists.

epithelium at the ends of the embryo (see fate map, Fig. 14.3).

In fruit flies, the midgut is internalized in two pieces, beginning during gastrulation. The anterior midgut rudiment is internalized in a small **anterior midgut invagination** at the tip of the germ band. Later (stage 10 C-O&H), a superficial **stomodeal invagination** pushes back to the end of the closed midgut rudiment. The second invagination forms the mouth and elongates into several parts (pharynx, esophagus, central portion of proventriculus) sometimes called collectively the foregut (Fig. 18.8).

Posteriorly, a cell plate carrying pole cells migrates dorsally from the tip of the ventral furrow and plunges inward. The plate is the posterior midgut rudiment, and its invagination is the **posterior midgut invagination.** Like its anterior counterpart, the posterior midgut rudiment grows out and, splitting into strands, later moves around the yolk sac. Moving along the lengthening midgut primordium, some pole cells (e.g., 16 in *Dacus tryoni*) begin their migration to the future gonad (i.e., at abdominal segments A5–A6). Other pole cells may contribute to the proctodeum or the middle region of the midgut.

A proctodeal furrow (Fig. 18.7, 17) appears in line with (but not part of) the earlier ventral furrow. Ectoderm heals over the furrow, turning it into the rectum and **proctodeum** (sometimes called the hindgut). Dividing cells in the proctodeal epithelium become organized into primordia of the Malpighian tubules (Fig. 18.8).

Mesoderm also takes different routes into the interior of arthropod embryos, but it is generally of ventral origin. In chelicerates (i.e., Xiphosura, Pycnogonida, and Arachnida), proliferative mesodermal cells delaminate along the length of a shallow groove. In apterygotes, mesoderm is formed by rapid proliferation and migration of cells in the ventral midline (see Jura, 1972).

In pterygotes, mesoderm forms by invagination or by ingression through the **ventral** or **gastral groove** in the midline of the germ band (Fig. 18.9). Except in the holometabolous hymenopterans and dipterans (Fig. 18.10), where the mesodermal band is especially broad, the pterygote mesoderm consists of loosely cohering cells that quickly spread out laterally beneath the ectoderm.

In *Drosophila melanogaster*, the presumptive mesoderm consists of about 1000 cells or one-sixth of the blastoderm, and invagination is confined to a brief 20 minutes (stages 6–7 C-O&H). When all the mesoderm is internalized, the entire surface of the embryo (even presumptive endoderm) is called **embryonic ectoderm.**

Presumptive neural ectoderm becomes distinctive when surface ectoderm heals over the ventral groove as a **median strand.** Cells around the strand segregate as **presumptive neuroblasts** of the **ventral nervous system** (stage 9 C-O&H). The ventromedial segmental **germ-band neuroblasts** will give rise to subesophageal ganglia and thoracic and abdominal ganglia of the ventral nerve cord. **Ventral ectoderm** lies peripherally to the neuroblasts (Fig.

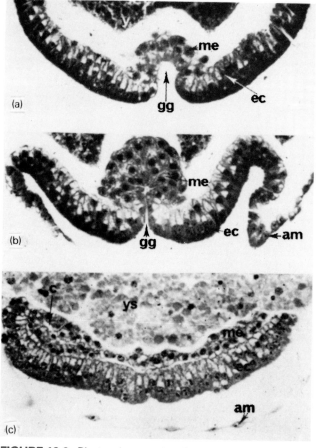

FIGURE 18.9. Photomicrographs of cross sections through the embryonic beetle, *Calomela parilis,* showing the germ band and surrounding extraembryonic membrane. The gastral groove (gg) invaginates and mesodermal cells (me) spread over the segmental ectoderm (ec), while amniotic folds reach beneath the embryo. The inner layer of the folds forms the amnion (am), and the outer layer forms the serosa (not shown). From D. T. Anderson, in S. J. Counce and C. H. Waddington, eds., *Developmental systems: Insects,* Vol. 1, Academic Press, Orlando, 1972. Used by permission.

18.9), and dorsolateral **procephalic lobes** give rise to the subraesophageal ganglion.

Three temporary transverse furrows mark the boundaries of transient **pseudosegments** on the surface in *Drosophila*. The transverse **cephalic furrow** (C, Fig. 18.7), marking the anterior border of the future cephalic appendages, disappears later as the appendages emerge. The **anterior** and **posterior transverse furrows** (AF and PF) crease the ectoderm at the border with the extraembryonic amnioserosa.

Five regular temporary folds appear in the extraembryonic ectoderm behind the cephalic furrow. Like the cephalic fold, which is **intrasegmental** with respect to the permanent segments formed

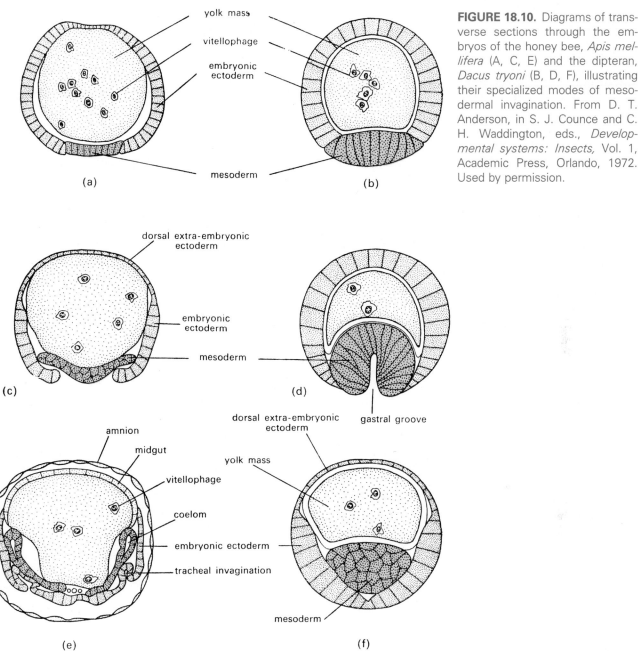

FIGURE 18.10. Diagrams of transverse sections through the embryos of the honey bee, *Apis mellifera* (A, C, E) and the dipteran, *Dacus tryoni* (B, D, F), illustrating their specialized modes of mesodermal invagination. From D. T. Anderson, in S. J. Counce and C. H. Waddington, eds., *Developmental systems: Insects,* Vol. 1, Academic Press, Orlando, 1972. Used by permission.

later on in development, the extraembryonic folds may represent intrasegmental (i.e., compartmental) boundaries in more posterior segments (see Table 15.3).

Cell division, which ceased during cellularization of the blastoderm, resumes following mesoderm formation. Preneuroblasts of the presumptive procephalic lobe (Fig. 18.7) and cells within the anterior lip of the cephalic furrow are the first cells to divide. All cells of the mesodermal tube and the median band (but not the surrounding presumptive neuroblasts) undergo two rounds of mitosis. Divi-

sion in the median band may compensate for the loss of surface cells, but cells produced by division in the mesoderm spread tangentially (Anderson, 1962). The tubal character of the germ band mesoderm disappears as cells migrate into a regular epithelium beneath the ectoderm (Fig. 18.9).

The transverse anterior and posterior furrows and the boundaries of other pseudosegments soon disappear, but ectodermal cells of the cephalic furrow divide and it remains a profound fold, if only ventrally. Ectodermal cells also divide elsewhere, especially in lateral stripes arising from the proctodeal region. The new cells provide a pool of ectoderm that will ultimately move dorsally and cover the embryo (Turner and Mahowald, 1979).

Ventrally, on either side of the median band, some ectoderm destined to give rise to neuroblasts does not exhibit division. Mitosis is seen in part of the neurogenic territory of the dorsolateral procephalic lobe, but mitosis is absent in other parts. It would seem that some subpopulations of neuroblasts do not divide again after the time of cellular-blastoderm formation, while other subpopulations divide at least once more while still at the surface.

Later (stage 10 C-O&H), mitosis is restored in the segmental germ band neuroblasts. **Asymmetric divisions** parallel to the surface of large precursor neuroblasts (i.e., tangential divisions) delaminate small **ganglion precursor cells** (also called ganglion mother cells) interiorly. Dividing procephalic lobe neuroblasts also delaminate ganglion precursor cells by asymmetric divisions. The ganglionic populations grow by subsequent **symmetrical divisions** among the precursor cells.

Extraembryonic membranes in insect eggs are generally simple epithelia, yet they can be highly differentiated locally and sometimes fused in double-membrane systems. In the symphylan myriapods and apterygotes, the **serosa** (Lat. *serum* watery liquid + *-osus* referring to membrane enclosing liquid) stretches over the dorsal side of the embryo, but it fails to enclose the ventral side of the embryo. This extraembryonic membrane is always replaced by embryonic ectoderm arising from the embryo (Fig. 18.4).

A thickened portion (Fig. 18.11, dom) of the nongerminal blastoderm forms the cenogenic **dorsal organ** that penetrates and digests the yolk. External **tendrils** (t), capable of absorbing dissolved products of the yolk's breakdown, may also develop (see Jura, 1972).

In most pterygotes, the serosa represents the outer layer of an extraembryonic evagination, the **amniotic fold** (Fig. 18.9), that spreads over the entire embryo (Fig. 18.12). The inner membrane is the **amnion** (Gk. *amnos* lamb referring to the membrane surrounding the lamb at birth). The formation of the amniotic fold provides the space required for blastokinetic movements (Fig. 18.5).

In the highly advanced dipterans, such as fruit flies, extraembryonic membranes are secondarily reduced. Double membranes do not appear, and the only single extraembryonic membrane is the **amnioserosa** lining the back of the postgastrula embryo (Fig. 18.7, AS). The function of the amnioserosa is uncertain, and, in *Drosophila melanogaster*, it can be removed without adversely affecting development.

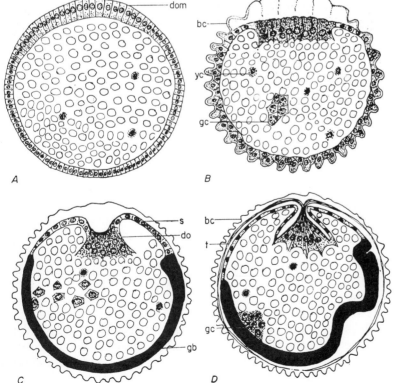

FIGURE 18.11. Diagrammatic sections through eggs of the apterygote, *Tetrodontophora bielanensis,* during formation fo the cenogenic dorsal organ (do). The tall columnar epithelium of the dorsal organ material (dom) invaginates to form the wedge-shaped digestive organ with tendrils (t) emerging from the top. At the same time, yolk cells (yc) segregate from the blastoderm. Some of these cells are yolk-digesting vitellophages, while others form endoderm at the surface of the yolk mass. A blastodermic cuticle (bc) is secreted around the blastoderm. gb, germ band; gc, germ cells. From Cz. Jura, in S. J. Counce and C. H. Waddington, eds., *Developmental systems: Insects,* Vol. 1, Academic Press, Orlando, 1972. Used by permission.

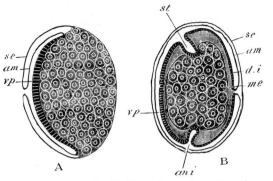

FIGURE 18.12. Longitudinal sections through schematic insect embryo illustrating formation of extraembryonic membranes. The amnion (am) and serosa (se) close over the ventral plate (vp) or germ band and seal the embryo within. From F. M. Balfour, *Comparative embryology,* Vol. I. Macmillan, London, 1880.

GASTRULATION FOLLOWING PARTIAL CLEAVAGE

This [rearrangement of surface cells during epiboly in teleosts] is an absolutely unexplored mode of cell movement which should be on the agenda of future research. But it will not be easy.

J. P. Trinkaus (1984a, p. 394)

The mechanism of the evocation of a primitive streak forming reaction is unknown.

L. Vakaet (1984, p. 560)

Partially cleaving eggs turn a two-dimensional discoblastula into a three-dimensional organism through different modes of gastrulation. The cephalopod discoblastula folds; the teleost discoblastula proliferates, and the bird discoblastula undergoes ingression. Unlike intralecithal eggs, partially cleaving embryos invariably cover the yolk by epiboly.

Cephalopods

The direct development of cephalopods leading to a juvenile stage is conspicuously different from the indirect development of other mollusks leading to a trochophore larva. Beginning with a large, yolky egg, gastrulation in cephalopods is delayed and involves mass cell movements over yolk (see Arnold and Williams-Arnold, 1976).

At about the ninth cleavage, the rim of the squid's simple blastoderm (see Fig. 13.1) stratifies into a cellular **germ ring** 2 or 3 cells thick. Nuclei in the underlying **yolk syncytium** (also called the yolk epithelium) divide once more and cease dividing. The yolk syncytium then advances over the

yolk followed by the margin of the germ ring and cellular blastoderm.

Deep cells, nestled between the outer cellular layer and the inner syncytial layer, proliferate rapidly, with cytokinesis occurring in every direction. The blastoderm thickens, and tongues of cells dart out over **blastocones** (areas between superficial cleavage furrows in the yolk syncytium, Fig. 18.13). The migrating cells divide perpendicularly to the surface of the embryo and leave a new outer layer one cell thick in their wake.

Ruffled membranes ripple at the migrating edges of cells indicative of active cellular movement. Migration is inhibited by cytochalasin B, suggesting a role for microfilaments in the process, and by colchicine, suggesting a role for microtubules, possibly for mitosis in driving the movement (Arnold and Williams-Arnold, 1976). At the same time, germ layers begin to sort themselves out, and gastrulation commences.

The embryo now has three **organogenic layers.** The surface blastoderm cells form a smooth paving layer on the outside and an irregular surface on the inside (Fig. 18.14). The middle layer thickens toward the animal pole to become the organ-forming part of the embryo. Its cells have angular shapes more or less conforming to each other's surfaces and those of outer cells and yolk syncytium. An average of 42 microprojections of different lengths arise from these cells, mainly from the edges. The yolk syncytium gives the impression of a sea in turmoil where wave crests mark the borders of overlying cells.

Following the formation of the organogenic layers, the remaining yolk is covered by cells (Fig. 18.15). The outer layer of cells becomes the flattened covering known as the **outer** (or external) **yolk sac.** The middle-layer cells, which have followed the outer layer over the yolk, form an incomplete (i.e., diffuse) layer. Although gastrulation is still incomplete, embryogenesis begins at this time.

Internal structures arise in the cephalopod embryo from thickenings and surface invaginations that have properties of more than one germ layer. Embryonic parts may also arise from layers atypical of their origins in vertebrate germ layers. For example, parts of the nervous system are derived from the middle organogenic layer rather than from the ectoderm, and thickenings of the outer cellular layer form structures including the funnel folds (Figs. 18.15 and 18.16). The shell gland forms in an invagination of the mantle cap, and the esophagus is formed by an inner cell thickening. Although the source of the midgut is uncertain, the

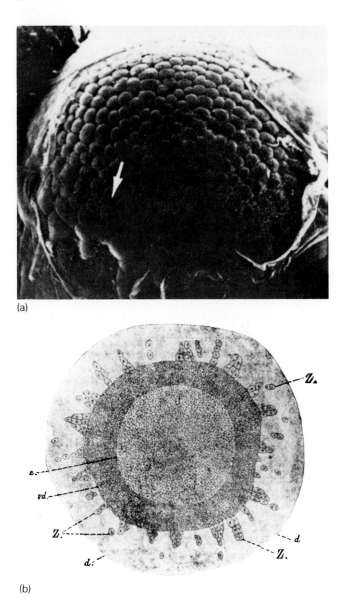

(a)

(b)

FIGURE 18.13. Scanning electron micrograph of the squid, *Loligo pealei* (×110), and drawing of *Sepia officinalis* embryos at the beginning of blastodermal migration and germ-layer formation. bl, Blastomeres; blc, blastocones; d, yolk (deutoplasm); e, simple epithelial (unilaminar) portion of discoblastula; vd, striated epithelial (multilaminar) portion of discoblastula; Z, advancing or separating cells. Part (a) from J. M. Arnold and L. D. Williams-Arnold, *Am. Zool.,* 16:421 (1976), by permission of the American Society of Zoologists. Part (b) from D. Korschelt and K. Heider, *Embryology of invertebrates,* Swan Sonnenschein, 1900.

yolk syncytium forms only part of a yolk sac and does not contribute to an endoderm in the adult organism.

The products of yolk digestion are distributed through the **hemal space** that develops between the outer layer of cells and the yolk syncytium (Fig.

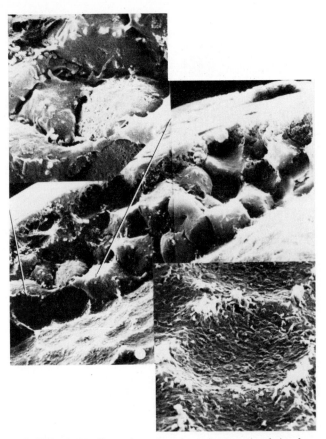

FIGURE 18.14. Scanning electron micrograph of the fractured blastoderm of the squid, *Loligo pealei* (stage 13). ×1150. (Upper insert) Area between arrows showing projections from edges of cells. ×3000. (Lower insert) Depression in surface of yolk syncytium formerly occupied by blastoderm cell. From J. M. Arnold and L. D. Williams-Arnold, *Am. Zool.,* 16:421 (1976), by permission of the American Society of Zoologists.

18.16). Muscle cells which develop across this space contract in synchrony, thereby aiding in the circulation of dissolved materials and incipient blood.

Living Chondrichthyes

Selachians (dogfish, rays, and sharks), like cyclostomes, gastrulate more like tetrapod vertebrates than like teleosts (see Cohen and Massey, 1982; Witschi, 1956). Beginning with birdlike eggs, the selachian discoblastulas of ovoviviparous and viviparous species form **discogastrulas** (Fig. 18.17).

The mechanism of germ-layer formation is not at all teleost-like. Originally, the entire presumptive mesoderm and most of the endoderm lie on the surface (Fig. 18.18). Invagination tucks the first components of endoderm and mesoderm into the

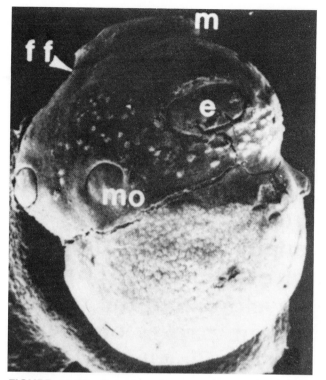

FIGURE 18.15. Scanning electron micrograph of *Loligo pealei* embryo following the formation of the three organogenic or germ layers covering the yolk. Divided at an equatorial region by a waist and arm primordia (unmarked knobs near waist), the embryo consists of an anterior organogenic part and a posterior outer or external yolk sac. The rudiments of eyes (e) and a mouth (mo) are evident. The mantle (m), still just an anterior cap, will later cover the body of the animal. Funnel folds (ff) will converge to form the siphon and slip below the mantle. ×40. From J. M. Arnold and L. D. Williams-Arnold, *Am. Zool.,* 16:421 (1976), by permission of the American Society of Zoologists.

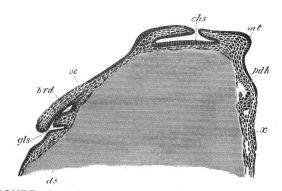

FIGURE 18.16. Vertical section through embryo of squid, *Loligo*. Hemal spaces (pdh) form in the diffuse cellular space between the outer cellular layer and the yolk syncytium or epithelium (ds). Invaginations give rise to a variety of structures. An invagination in the mantle (mt) gives rise to the shell gland (chs), and invaginations elsewhere give rise to the radula sheath (brd) and salivary glands (gls). Cellular thickenings give rise to other structures. The esophagus (oe) and region between funnel folds (x) form as thickenings. From F. M. Balfour, *Comparative embryology,* Macmillan, London, 1880.

roof of the archenteron, and involution brings the remainder inside.

The dorsal part of the embryo develops anterior to a discrete blastopore and above an archenteron (Fig. 18.17b–f). Like the cyclostome gastrula (see Fig. 19.20), the dorsal side of the selachian gastrula is a fold of tissue jutting over the yolk. This resemblance is remarkable, although unlike the blastoporal lip of amphibians (e.g., Fig. 19.28) and reptiles (see below Fig. 18.28), the edge of the selachian fold or flap is convex rather than concave, and the blastopore remains open until late in organogenesis.

The endoderm of the foregut or **embryonic hypoblast** is epithelial, while a **vitelline hypoblast** lining the extraembryonic yolk sac is syncytial except for **deep cells.** Mesoderm accumulates beneath swellings, and mesodermal cells grow out between the ectoderm above and endoderm below (Fig. 18.17f). Behind the cephalic endoderm, the prechordal mesoderm and notochord shift into an intermediate position, allowing endoderm to occupy the entire roof of the archenteron.

As development proceeds, the edge of the blastoderm lying ahead of the blastopore becomes four layered. Mesoderm and endoderm, fusing in a **splanchnopleure,** spread directly over the yolk laterally and anteriorly as a **germ wall.** Ectoderm and endoderm, fusing in a **somatopleure,** are separated from the splanchnopleure by an incipient **coelom.** The layers enclose the yolk, placing it within the endoderm virtually inside the embryonic intestine (e.g., comparable with Fig. 24.2). There, it is digested as if it were an ingested item of food.

Bony Fish

Teleost gastrulation consists of two processes: **epiboly,** which lays down a unique extra embryonic "skin" over the yolk, and **gastrulation proper,** which produces the germ layers of the embryo and yolk sac (Fig. 18.19). Teleost epiboly creates a **periderm** under which germ layers form and the embryo develops.

A cellular blastoderm or **discogastrula** (Figs. 18.20 and 18.21; compare to Fig. 13.3) begins as a mass of loosely bound **deep cells** (D-cells or DCs) covered by an **enveloping layer** (EVL, also called *Deckschicht*) of tightly bound enveloping cells (E-cells). The rim of the EVL adheres to the underlying syncytial **yolk syncytial layer** (YSL, also periblast or vitelline hypoblast).

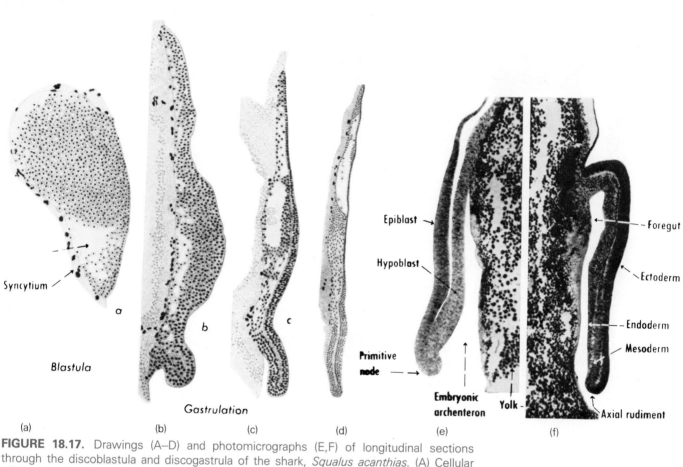

FIGURE 18.17. Drawings (A–D) and photomicrographs (E,F) of longitudinal sections through the discoblastula and discogastrula of the shark, *Squalus acanthias.* (A) Cellular *discoblastula* (stage 8) lies above syncytial yolk layer. (B–F) Gastrulation. Germ layers emerge from thickened discogastrula. (b) Stage 9; (c) stage 10; (d) stage 11; (e) stage 12. (f) Neurula; stage 13. Mesoderm spreads; foregut emerges within the cephalic fold; neural plate thickens. From E. Witschi, *Development of vertebrates,* Saunders, Philadelphia, 1956. Copyright © 1956 by Saunders College Publishing, a division of Holt, Rinehart and Winston, Inc., reproduced by permission of the publisher.

Typical zonula occludens and zonula adherens between E-cells are present as early as cleavage, but desmosomes are only rudimentary until the gastrula stage when abundant and organized tonofibrils are added. Gap junctions may also appear proximally along with additional desmosomes, but communicating gap junctions between E-cells and the YSL disappear during gastrulation (Kimmel and Law, 1985).

The junctional complexes among E-cells and between the EVL and YSL serve dual functions. The high tensile strength of the zonula adherens keeps the EVL intact despite great stretching under tension during epiboly, and the impermeability of the zonula occludens offers a barrier to the outside that protects against adverse conditions. In effect, the teleost embryo developing below the EVL lives in an environment of its own creation (see Trinkaus, 1984a).

Epiboly in teleosts refers to the dramatic expansion of the YSL over the yolk sphere and the EVL over the YSL (e.g., about 250 μm per hour). In the small-yolked killifish, *Fundulus heteroclitus,* the surface of the EVL increases by an order of magnitude, and in large-yolked species, like trout and salmon, the expansion is even greater. The duration of epiboly and the relative timing of embryogenesis are also tied to the dimensions of the yolk sphere (Fig. 18.19).

The YSL contains nuclei derived from invading D-cells. In the zebrafish, *Brachidanio rerio,* marginal blastoderm cells with cytoplasmic bridges to the yolk collapse to form the YSL at about the 10th cleavage division (Kimmel and Law, 1985). The nuclei may divide in the YSL, but in *F. heteroclitus* and other teleosts the number of nuclei in the syncytium remains nearly constant throughout gastrulation.

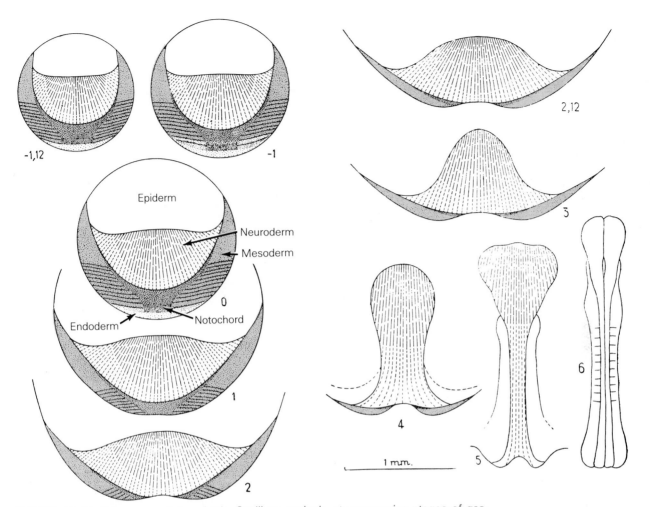

FIGURE 18.18. Fate maps of the shark, *Scyllium canicula,* at progressive stages of gastrulation illustrating the effect of posterior-dorsal involution on the formation of germ layers. From G. Vandebroek, *Cuvier. Arch. Biol.,* 47:499(1936), by permission of the Archives de Biologie.

Peripherally, the YSL is continuous with the thin and nonnucleated rim of cytoplasm covering the yolk (i.e., the **yolk-cytoplasmic layer,** YCL, or yolk cortex). The border between the YSL and the YCL is marked by the presence of a row of nuclei within the YSL and their absence from the YCL. Moreover, the upper and lower plasma membranes of the YSL form microvilli, while the YCL surfaces are flat (Fig. 18.21, stages 16 and 18), and endocytotic vesicles are present only in the YSL at the junction with the YCL.

Even though the YCL exerts contractile tension tangentially and uniformly in all directions and exerts hydrostatic pressure on the yolk mass (hence the spherical shape of the egg), contraction by the YCL does not draw the YSL or EVL over the egg. Instead, the YSL seems to take up the YCL surface in endocytotic vesicles and replace it with YSL surface. Stretching and thinning accompany

YSL epiboly, and the general shortening of microvilli provides additional YSL surface (see Trinkaus, 1984a, b).

In *F. heteroclitus,* epiboly begins while the embryo is still at the blastula stage (Fig. 18.21, stage 12). The outer portion of the YSL spreads far into the YCL beyond the cellular blastoderm. The outer portion of the YSL is the **external yolk syncytial layer** (E-YSL, also marginal periblast). The **internal YSL** (I-YSL) lies below the blastoderm.

The blastoderm soon catches up to the E-YSL (stage 14), and at the beginning of gastrulation proper (stage 16), the gap between the EVL and the outer margin of the E-YSL narrows. For the rest of epiboly, the E-YSL is only slightly ahead of the EVL.

The YSL is the natural substratum for the EVL, and EVL removed from the YSL can adhere to denuded YSL. The YCL, on the other hand, is not an

Epiboly   *Embryogenesis*

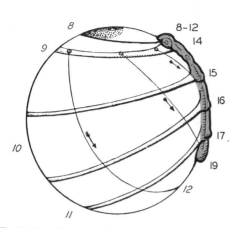

FIGURE 18.19. Diagram illustrating embryogenesis and teleost epiboly of the germ ring over the yolk. Embryogenesis lags behind epiboly, but the relative rate of epiboly is also influenced by the size of the egg. From E. Witschi, *Development of vertebrates,* Saunders, Philadelphia, 1956. Copyright © 1956 by Saunders College Publishing, a division of Holt, Rinehart and Winston, Inc., reproduced by permission of the publisher.

adherent surface, and since a readhering EVL can resume epiboly even after a YSL has engulfed the YCL, the YCL does not seem to be relevant for EVL epiboly.

The normal force for EVL epiboly seems to be contraction of the YSL (not the YCL). The YSL sets the outer limit for the EVL and is never outrun by the EVL. For its part, the EVL offers considerable resistance to YSL spreading. A denuded YSL expands rapidly, and one that has covered about three-quarters of an egg before the EVL is experimentally removed snaps closed over the remaining yolk, even cutting off some yolk in the process.

E-cells stretch and thin as they are pulled by the contracting YSL (Fig. 18.21). In *F. heteroclitus,* the EVL consists of about 5000 cells throughout

epiboly. No cell division is observed, but **cellular rearrangement** occurs despite tight junctional complexes (Fig. 18.22) as cell numbers increase over the egg's equator and decrease in the dash to the embryo's posterior pole (Keller and Trinkaus, 1987).

E-cells are destined to remain at the surface. They do not invaginate, involute at the margin of the blastoderm, or undergo cellular ingression (see Ballard, 1981; Trinkaus, 1984a and b). Neither the expanding EVL nor YSL produces anything comparable to the blastoporal lips of the amphibian gastrula, and neither the hole at the contracting rim of the EVL nor the juncture of YSL and YCL (Fig. 18.19, edge of expanding ring) is a blastopore. The EVL is the thin, transient **periderm** where it lies over the embryo and a simple **extraembryonic ectoderm** where it lies over the YSL in the yolk sac. Strands of YSL penetrate the yolk and become the digestive organ of the yolk without contributing to endoderm.

Gastrulation proper begins with the movement of D-cells in the space between the EVL and I-YSL and ends with the **segregation** of D-cells into germ layers. In *B. rerio,* sibling D-cells (i.e., cell clones) that remain clustered prior to epiboly undergo **clonal dispersion** and mingle freely early in gastrulation (Fig. 18.23). It is unknown whether this movement is active or passive.

By the midgastrula or midepiboly stage (i.e., when the EVL has covered half the yolk sphere, stage 16, Fig. 18.21), some D-cells have accumulated among fibronectin fibrils in a ridge known as the **germ ring** (GR) at the margin of the EVL, and some have converged in a wedge-shaped **embryonic shield** (ES) directed apically from the GR. The GR follows the EVL over the yolk, while the shield thickens with additional cells as if the lagging rate of epiboly in the vicinity of the shield (Fig. 18.19) allows more cells to accumulate in it than elsewhere.

As the germ ring spreads, the diffuse blastocoel

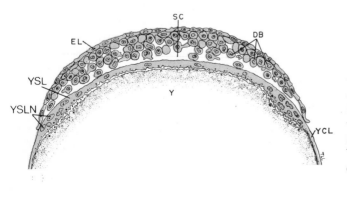

FIGURE 18.20. Diagrammatic section through the early gastrula of the killifish, *Fundulus heteroclitus,* at the beginning of epiboly. The discoblastula consists of an enveloping layer (EL) lying over deep blastomeres (DB) with segmentation cavities (SC) beneath and among cells. The yolk syncytial layer (YSL) containing nuclei (YSLN) is continuous at its edge with the yolk cytoplasmic layer (YCL) covering the vitellus or yolk (Y). Microvilli and endocytotic vesicles beneath the YSL are indicative of digestion. ×140. From T. L. Lentz and J. P. Trinkaus, *J. Cell Biol.,* 32:121 (1967), copyright permission of Rockefeller University Press.

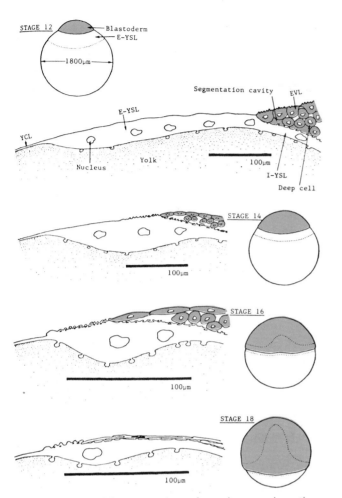

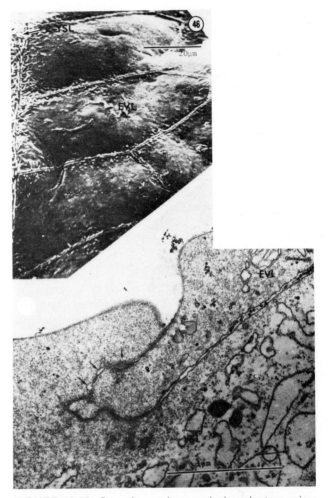

FIGURE 18.21. Diagrammatic surface views and sections of gastrulating killifish, *Fundulus heteroclitus,* showing expanded outer external-yolk syncytial layer (E-YSL) and underlying internal-yolk syncytial layer (I-YSL). The cellular blastoderm of loose deep cells and tightly bound cells of the enveloping layer (EVL) sits above and adheres at its rim to the yolk syncytial layer (YSL). Thickening at the periphery of the blastoderm produces the germ ring (stage 16) which expands asymmetrically over the yolk and grows into the embryonic shield (stage 18). From T. Betchaku and J. P. Trinkaus, *J. Exp. Zool.,* 206:381 (1978), by permission of Alan R. Liss, Inc., and the authors.

FIGURE 18.22. Scanning and transmission electron micrographs of surface of *Fundulus heteroclitus* late gastrulas (stage 18) near attachment between the enveloping layer (EVL) and the yolk syncytial layer (YSL). The leading edges of tightly bound, smooth-surface EVL cells are embedded in the wavy YSL surface. A distal zonula occludens is followed by occasional spaces (arrows) and several areas of zonula adherens where clusters of microfilaments make contact with the closely parallel membranes. (a) Scanning electron micrograph. ×1300. (b) Transmission electron micrograph. ×52,000. From T. Betchaku and J. P. Trinkaus, *J Exp. Zool.,* 206:381 (1978), by permission of Alan R. Liss, Inc., and courtesy of the authors.

of segmentation cavities (SC in Fig. 13.3) coalesce into a **subgerminal cavity** (SC in Fig. 18.20). Lying between deep cells and the YSL, the SC extends between the margins of the germ ring (Fig. 18.24).

Deep cells undergo the same variety of movements associated with gastrulation elsewhere— epiboly and involution. During their vegetal progression, many D-cells at the vegetal margin of the GR sink inward and are replaced by their more central neighbors (Fig. 18.25). The sunken D-cells proceed to migrate back toward the embryonic shield as a loosely adhering **hypoblast** (above the YSL, Fig.

18.26). Later, overlying deep cells, coated by a network of fibronectin fibrils, make contact with the migrating hypoblast, and the SC collapses (see Wood and Timmermans, 1988).

Cells forming the ES are from dispersed clones which have mingled completely, but **clonal restrictions** are imposed during gastrulation in *B. rerio,* perhaps when cells have already become fixed in the ES. Thereafter, many cells act as **founder cells.** Clones arising from the division of these cells and their progeny migrate in regular patterns and tend

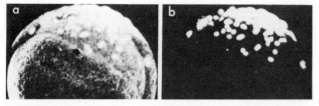

FIGURE 18.23. Micrographs of living *Brachidanio rerio* gastrula illuminated with epifluorescence and bright-field optics (a) and with epifluorescence alone (b). The bright fluorescence is due to a fluorescent intracellular tracer that was injected into a cell at the 64-cell stage. The clone of fluorescing cells is scattered throughout the blastoderm. From C. B. Kimmel and R. M. Warga, *Science,* 231:365(1986), by permission of the American Association for the Advancement of Science and the author. Copyright 1986 by the AAAS.

to populate single tissues (Kimmel and Warga, 1986).

As gastrulation proceeds, the ES elongates into a rodlike **axial rudiment** and gives rise to the embryo's **epiblast.** Segregation soon becomes apparent in the epiblast (Fig. 18.26), and a solid notochord, neural tube, and gut emerge centrally, while an ectoderm forms peripherally under (but not from) the EVL-derived periderm. The neural tube and gut subsequently become hollow, and mesoderm surrounds them and lines the ectoderm. Cavitation within the mesoderm gives rise to the coelom.

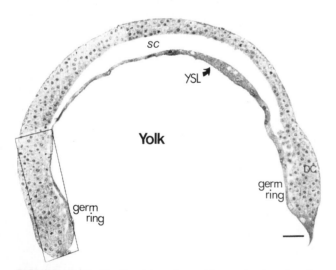

FIGURE 18.24. Section bisecting midgastrula showing the germ ring on opposite sides. D-cells involute at the margin of the germ ring and migrate back over the yolk syncytial layer (YSL) as large, loosely adherent hypoblast cells. A thin layer of enveloping-layer cells lies above the deep cells, and the cavernous subgerminal cavity (SC) lies below. Bar = 50 μm. From A. Wood and L. P. M. Timmermans, *Development,* 102:575 (1988), by permission of the Company of Biologists and the authors.

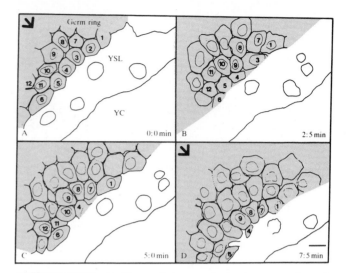

FIGURE 18.25. Outlines of deep cells in the germ ring of the Rosy Barb, *Barbus conchonius,* gastrula viewed with Nomarski optics (differential interference contrast) and drawn from a 16-mm time-lapse sequence. Arrows show the direction of epiboly. From A. Wood and L. P. M. Timmermans, *Development,* 102:575 (1988), by permission of the Company of Biologists Limited and the author.

In the **annual fish,** such as *Austrofundulus,* that emerge briefly in temporary water holes in South America and Africa, the sub-EVL space provides a refuge for D-cells during the dry season. In a unique form of gastrulation, D-cells disperse and then reaggregate into a mass that produces an embryo without first producing an embryonic shield. In the rainy season, the embryo develops rapidly. Larvas hatch, grow, mature, and quickly lay the next round of eggs.

Reptiles

Like birds, reptiles begin gastrulation by delaminating an underlying **hypoblast** (also called a primary hypoblast, primitive endoderm, or anterior endoderm, Fig. 18.27, end) from the discoblastula. The bulk of the discoblastula is thereafter known as the **epiblast** or **embryonic ectoderm.** Both the hypoblast and epiblast lie above the subblastodermic (or subgerminal) cavity (compare to cavity in birds, Fig. 13.5).

In contrast to the single-layered epiblast and hypoblast of birds, the reptilian layers are several cells thick, and the endoderm is corrugated (Fig. 18.28). Furthermore, while both reptilian and avian gastrulas form a posterior embryonic thickening (the **primitive node** in birds, the **primitive, blastoporal,** or **archenteric plaque** in reptiles, bl n in Fig. 18.27), the development of the thickening is different in embryos belonging to the two classes.

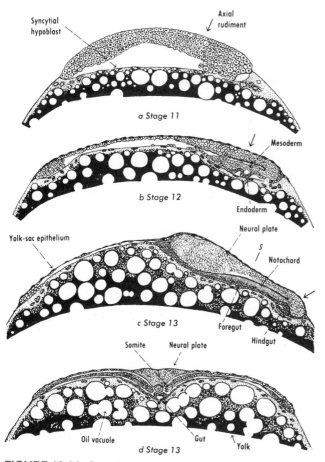

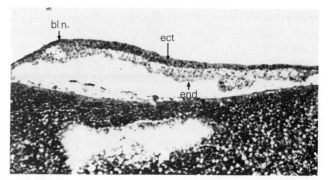

FIGURE 18.27. Photomicrograph of a sagittal section of the early gastrula (stage 7) of the lizard, *Lacerta vivipara jacquin*. An epiblast or embryonic ectoderm (ect) has delaminated a hypoblast or primitive (anterior) endoderm (end) above a subblastodermic (subgerminal) cavity. Posteriorly, the epiblast thickens as the primitive or blastoporal node or plaque (bl n). From J. Hubert, *Arch. Anat. Microsc. Morphol. Exp.*, 51:11 (1962), by permission of Masson et Cie and courtesy of J. Hubert.

FIGURE 18.26. Drawings of sections of gastrulas and early neurula of the trout, *Salmo fario.* (a) Stage 11, sagittal section. An epiblast of deep cells lies under the EVL. (b) Stage 12, sagittal section. Segregation of epiblast cells has given rise to mesoderm and endoderm. (c) Stage 13, early neurula, sagittal section. This neurula has a thickened neural plate above a solid notochord and a gut that is beginning to hollow out as an anterior foregut and posterior hindgut. (d) Same as (c), cross section through S. From E. Witschi, *Development of vertebrates,* Saunders, Philadelphia, 1956. Copyright © 1956 by Saunders College Publishing, a division of Holt, Rinehart and Winston, Inc., reproduced by permission of the publisher.

The archenteron may represent a primitive structure retained from amphibians, but, unlike the situation in amphibians, no yolk plug of endodermal cells protrudes through the reptilian blastopore, and reptilian and amphibian endoderm originate in entirely different ways (Hubert, 1962). The reptilian archenteron is therefore as likely to be secondarily derivative as it is likely to be primitive.

Extension of the archenteron's caudal and lateral edges soon forms a new and continuous en-

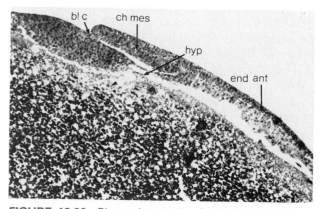

FIGURE 18.28. Photomicrograph of a sagittal section of the midgastrula (stage 7) of the lizard, *Lacerta vivipara jacquin*. The thick chordamesoblast (ch mes) is seen in the roof of the archenteron (or blastoporal canal, bl c), and thin hypoblast (hyp) forms the floor. Anteriorly (left) extraembryonic anterior endoderm (end ant.) is corrugated above the subblastodermic (subgerminal) cavity, while posteriorly a thick ventral blastoporal lip sits on top the yolk. ×140. From J. Hubert, *Arch. Anat. Microsc. Morphol. Exp.*, 51:11 (1962), by permission of Masson et Cie and courtesy of J. Hubert.

The most conspicuous difference between reptilian and avian gastrulas is the presence of an **archenteron** or **blastoporal canal** in reptiles that either opens to the outside by a distinct blastopore (Fig. 18.28) or ends in a **blastoporal plug**. The edge of the thickened reptilian epiblast (sometimes called the blastoporal plaque) invaginates to form an archenteron with a thick roof of mesoderm and thin floor of primitive endoderm. Fused with the hypoblast, the primitive endoderm does not form the definitive endoderm of the embryo but ruptures along with the hypoblast, and the archenteron opens into the subblastodermic cavity.

dodermal sheath. The subsequent covering of yolk with this endoderm provides the embryo with a digestive organ. The ectoderm, in the meantime, forms as a thickening above the archenteron's roof.

The walls of the archenteron are mesodermal (hence the archenteron is sometimes called a **mesodermal archenteron**) (see Nieuwkoop, 1978) and supply the embryo with almost all its mesoderm. Laterally, the archenteron may evaginate double-mesodermal sheets, and the roof gives rise to the prechordal mesoderm and notochord. Only the posterior and lateral extraembryonic mesoderm is supplied by the **ventral lip of the blastopore** (sometimes equated to the primitive streak of birds) (see Bellairs, 1971).

Birds

Represented by the chick, *Gallus gallus*, birds have been popular subjects in the long history of marking experiments and fate mapping (see Waddington, 1952), culminating in autoradiography (Nicolet, 1971), quail-chick xenografting (Le Douarin, 1976), and cinemicrography (see Vakaet, 1984). Complex, three-dimensional morphogenic movements establish the germ layers in embryonic and extraembryonic areas without benefit of a blastopore.

Gastrulation begins with the delamination (frequently called polyinvagination or vertical polyinvagination) of the **primary (1°-) hypoblast** (also called endophyll and primitive or early endoderm, Figs. 18.29 and 18.30). 1°-Hypoblast cells arise in the posterior region of the discoblastula, the **blastoderm** or **epiblast.** Forming simple tight junctions (but no desmosomes), increasing numbers of 1°-hypoblast cells unite and spread centrally. Additional 1°-hypoblast cells delaminate from the **marginal zone** at the edges of epiblast and migrate centrally to unite with the 1°-hypoblast and extend it. The marginal additions seem to have a profound effect on organizing the embryonic axis in the epiblast (Kochav et al., 1980).

Never more than a single cell layer thick, the 1°-hypoblast lacks its own basal lamina. Like a safety net, the 1°-hypoblast catches early **mesoblast** or **middle layer cells** (m.c.) above it.

The entire period of 1°-hypoblast formation takes place while the embryo is still grossly a simple discoblastula (Fig. 18.31a). The subblastodermic cavity (or subgerminal cavity, also called the gastrocoel), delimited by the marginal zone, underlies the circular **area pellucida.** The **germ ring** or **germ wall** surrounds the area pellucida and joins it to the **area opaca** where **extraembryonic ecto-**

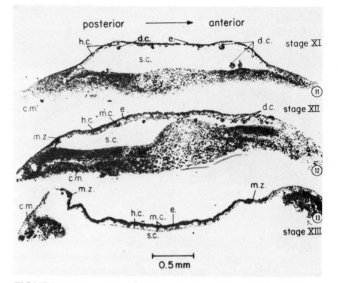

FIGURE 18.29. Photomicrographs of sagittal sections through the blastoderm of the chick at primary hypoblast formation and the beginning of gastrulation. Stage XI E.G&K. (Stages are according to Eyal-Giladi and Kochav, 1976.) Posteriorly, fragmentary sheets of primary hypoblast cells (h.c.) are seen. Only scattered detached cells (d.c.) are present anteriorly. Stage XII E.G&K. The sheet of primary hypoblast covers the posterior half of the blastoderm. Stage XIII E.G&K. The primary hypoblast covers the central region of the blastoderm. The ringed marginal zone (m.z.) is not covered by primary hypoblast. Middle-layer cells appear between the primary hypoblast and epiblast. d.c., detached cell; e., epiblast; c.m., carbon particles used to locate blastoderm; h.c., primary hypoblast cell; m.c., middle layer cell; m.z., marginal zone; s.c., subblastodermic cavity. From S. Kochav, M. Ginsburg, and H. Eyal-Giladi, *Dev. Biol.,* 79:296 (1980), by permission of Academic Press and courtesy of the authors.

derm or **blastoderm** overlies open cells still fused with the yolk (Fig. 18.29).

Pushing against the **periblast,** or peripheral rim of cytoplasm, the extraembryonic ectoderm spreads from a ring about 2 mm in diameter at 5 hours of incubation to one 4.5 mm in diameter at 10 hours. The extraembryonic ectoderm (now also called extraembryonic ectophyll or sometimes trophectoderm) outdistances the underlying hypoblast and forms an **epibolic organ** 1 mm wide that leads the race around the yolk sphere. Hypoblast follows and encloses the yolk (if not completely) in an endodermally lined digestive organ, while extraembryonic mesoderm (or mesoblast) is not far behind.

The extraembryonic mesoderm thickens in a rim (Fig. 18.32) peripheral to the central embryo-forming region (known as the area centralis). The differentiation of blood vessels, known as **angiogenesis** (Gk. *angei* blood vessel + *-genesis*) begins

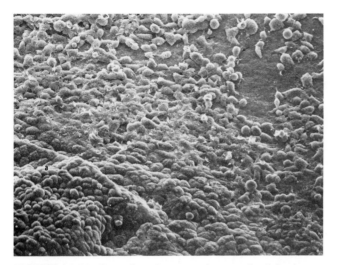

FIGURE 18.30. Scanning electron micrograph of the lower surface of a chick blastoderm (stage XII E.G&K). Delaminating hypoblast cells accumulating below the basal lamina of the epiblast are uniting. The hypoblast at the posterior margin of the embryo (lower left) is conspicuously more advanced. ×100. From C. Weinberger, P. L. Penner, and I. Brick, *Am. Zool.,* 24:545 (1984), by permission of the American Society of Zoologists and courtesy of the authors.

in the rim mesoderm. Flattened mesodermal cells surround larger mesodermal cells in **blood islands** that rapidly unite into a **blood island network** or **angioblastema** (from *angei* + *blastema* offspring) in the more central **area vasculosa** portion of the area opaca (Fig. 18.33). No hemoglobin synthesis is detectable at first, but soon bright red clusters ap-

pear of hemoglobin-containing primary erythroid cells. At 2.5 days of incubation, area vasculosa cells differentiate even in isolated blastoderms *in vitro,* suggesting that the cells are autonomous at this time and synthesis is genetically programmed (Zagris, 1980).

While yolk sac formation and angiogenesis proceed extraembryonically, embryogenesis proceeds in the central area pellucida. The symmetrical discoblastula present prior to laying (Fig. 18.32, stage -1) is asymmetric as incubation begins (stage 0). While still in a **prestreak** stage (designated stage 1 by Hamburger and Hamilton [1951]), the surface layer converges toward the primitive streak like a folding fan (Spratt, 1946).

A superficial ring of **primary endoblast** begins to sink below the surface posteriorly, forming a thickening known as **Koller's crescent** (stage 1, Fig. 18.32). The cells sinking into the blastoderm in the first 8–10 hours of incubation form a thin sheet of **secondary (2°-) hypoblast** or endoderm (Fig. 18.34) which migrates anteriorly and laterally, displacing the 1°-hypoblast peripherally.

Different terms are often employed for the hypoblast by authors using different labeling techniques. In general, the primary endoblast is equivalent to the endophyll and comprises the 1°-hypoblast formed by delamination and the 2°-hypoblast formed by ingression. The **embryonic endoderm** is the **definitive endoblast** or tertiary hypoblast.

The discoblastula, no longer quite a "disk," is called the **embryonic shield.** By 6–7 hours of in-

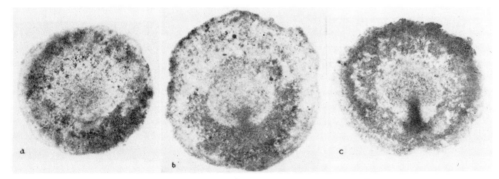

FIGURE 18.31. Photomicrographs of whole mounts of chick blastoderms during early gastrulation (stained with hematoxylin). The area pellucida is the relatively light circular region in the center of the blastoderms. The area opaca is the dark ring surrounding each area pellucida. (a) Stage 1 H&H prestreak (stages according to Hamburger and Hamilton, 1951). The embryonic shield appears at the posterior edge of the area pellucida. ×20. (b) Stage 2 H&H initial streak or short-broad beginning streak. The embryonic shield has extended forward from the posterior border of the area pellucida. ×20. (c) Stage 3 H&H intermediate streak or short streak. Continued lengthening results in a parabolic streak. ×20. From N. T. Spratt, Jr., *J. Exp. Zool.,* 120:109 (1952), by permission of Alan R. Liss, Inc.

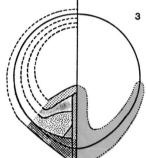

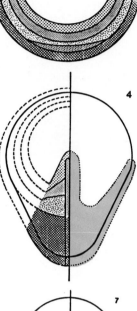

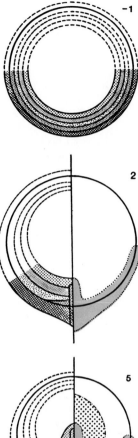

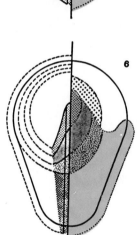

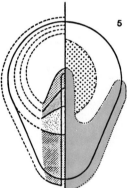

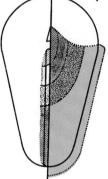

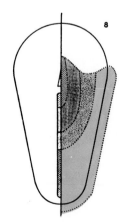

ENDOBLAST

▦ upper layer

▨ deep layer

MESOBLAST

▦ extraembryonic

MESOBLAST

▦ axial

▨ paraxial

▨ lateral plate

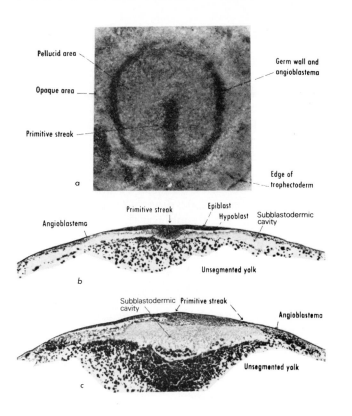

FIGURE 18.33. Photomicrographs of whole mount of chick embryo at 11 hours of incubation (intermediate streak stage, about stage 3 H&H) and sections of sparrow embryo at a comparable stage. The deep staining at the border of the area opaca around the area pellucida in the whole mount (a) is due to the angioblastema present on both sides of the embryo as seen in the cross section (b) and in the posterior part of the sagittal section (c). All ×33. From E. Witschi, *Development of vertebrates,* Saunders, Philadelphia, 1956. Copyright © 1956 by Saunders College Publishing, a division of Holt, Rinehart and Winston, Inc., reproduced by permission of the publisher.

cubation, the posterior thickening extends centrally, and a **short-broad streak** (stage 2) appears. At the same time, the **embryonic ectoderm** (also embryonic ectophyll or ectoblast) thickens in a disk (the *Scheibe*) lying above the embryonic **mesoblast** (Fig. 18.31b) (Spratt, 1952). The ectodermal cells elongate to form a primitive **neural ectoderm** (see Gallera, 1971).

A narrower **medium-broad streak** then appears, and at 12–13 hours, a **short** or **intermediate streak** (stage 3; Fig. 18.31c) is present. By 14–18 hours of incubation, the **definitive primitive streak** has

FIGURE 18.32. Diagrams of the spreading germ layers seen in surface view during progressive stages of the chick embryo. Each diagram is identified by its stage number (Vakaet, 1985) at the upper right. In lower three rows, the converging surface layers are shown on the left, the diverging deeper layers are shown on the right. (−1) About 5 hours before laying, the presumptive areas are still presumably symmetrical as polarity is established. A rim of extraembryonic mesoderm or mesoblast (outside heavy circle) surrounds the central presumptive embryonic area (known as the area centralis). (0) Upon laying, asymmetries appear as the surface blastoderm begins its posterior convergence and concentrates the presumptive areas at the posterior edge. (1) In the prestreak embryonic shield, a crescentic posterior area known as Koller's crescent contains the presumptive 2°-hypoblast. (2) An embryonic shield at 6–7 hours of incubation. Convergence continues to concentrate cells posteriorly, but their piling up now extends centrally into a short-broad primitive streak. (3) Both anterior and posterior elongations produce an intermediate streak (also called the short streak) at 12–13 hours of incubation. (4) The definitive primitive streak almost 2 mm in length at 18–20 hours of incubation. (5) An early notochordal process at 19–22 hours of incubation. (6) A medium notochordal process. (7) A late notochordal process or retreating primitive streak at 23–25 hours of incubation. (8) A later notochordal process just prior to the head fold or neural plate stage (stage 9). The primitive streak continues to regress and shorten. Open arrow, notochord. From L. Vakaet, in G. M. Edelman, ed., *Molecular determinants of animal form,* Alan R. Liss, Inc., New York, 1985. Used by permission.

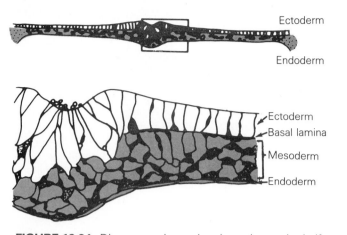

FIGURE 18.34. Diagrammatic section through anterior half of fully developed primitive streak. Thickened ectoderm extends on either side of the primitive streak. Ingressing elongated and flask-shaped cells leave the streak and flatten into endoderm or migrate between endoderm and ectoderm as mesoderm. From R. Bellairs, *Developmental processes in higher vertebrates,* University of Miami Press, Miami, 1971. By permission of John Wiley & Sons, Inc.

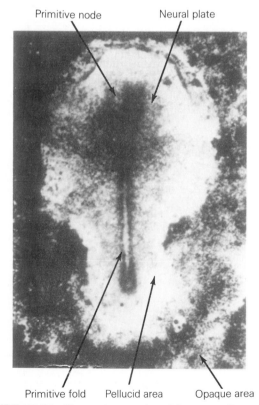

FIGURE 18.35. Photomicrographs of hematoxylin-stained whole mount of chick blastoderms at definitive streak stage (stage 4 H&H, 18–20 hours). ×20. From N. T. Spratt, Jr., *J. Exp. Zool.,* 120:109 (1952), by permission of Alan R. Liss, Inc.

formed, almost 2 mm in length and reaching the center of the original area pellucida (stage 4, Fig. 18.35).

At this time, the endoderm has entirely left the surface of the primitive streak. The first ingressing cells are completely incorporated into a deep layer of 2°-hypoblast, while other endoblast cells have formed the definitive **embryonic endoderm.** This deep epithelial layer beginning with pregut endoderm and extending posteriorly (5 and 6, Fig. 18.32) contrasts with the hypoblast by possessing complex tight junctions, desmosomes, and a basal lamina. Invading the hypoblast, the definitive endoderm expands both anteriorly and posteriorly (see Bellairs and van Peteghem, 1984).

Displaced to the periphery along with the 1°-hypoblast, the 2°-hypoblast does not form part of the embryonic endoderm. The combined hypoblast or primary endoblast is relegated to extraembryonic endoderm. Peripherally, it meets the germ wall still spreading over the yolk, and anteriorly, it is concentrated as a **germinal crescent** along with primordial germ cells (see Fig. 12.32). Fibronectin is concentrated in the same area as a fibrous band of basal lamina between the hypoblast and the ectoderm (Sanders, 1982) (see Fig. 13.7).

Earlier (toward the end of stage 4), a **primitive pit** dents the thickened **primitive node** (or Hensen's node) at the anterior end of the streak. The depression lengthens posteriorly into the **primitive groove** flanked by **primitive folds** (stage 5, Fig. 18.32). Within the primitive groove, surface cells **deepithelialize** and undergo cellular **ingression** (also called polyingression, Fig. 18.34). The ingressing cells become flask-shaped, form blebs and coated pits, and penetrate the basal lamina (Vakaet, 1984) (Fig. 18.36).

After 20 hours of incubation all the cells moving through the streak are **mesodermal** or **mesoblast** cells (Fig. 18.32 5). The ingressing mesoblast migrates in the **intermediate zone** between the ectoderm and endoderm where it forms a mesenchyme several cells thick (Fig. 18.34).

Superficial **convergent movement** brings cells toward the primitive streak where they pile up and thicken the primitive folds. The primitive pit advances anteriorly and lengthens the primitive groove by cutting through the surface layer of presumptive mesoblast (Fig. 18.32 6). The thickened horseshoe-shaped area at the head of the primitive streak is now called the **primitive node** or **Hensen's node.**

The earliest portion of the mesoblast to migrate in the intermediate zone moves farthest anteriorly; mesoblast coming later migrates to the

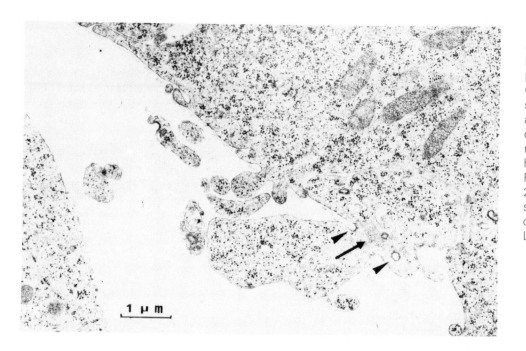

FIGURE 18.36. Electron micrograph of section through blastoderm showing blebs penetrating basal lamina. Coated pits (arrowheads) suggest that basal lamina is actively removed, and bundles of microfilaments (arrow) indicate that blebs are healthy cellular extensions. From L. Vakaet, *Am. Zool.,* 24:555 (1984), by permission of the American Society of Zoologists and courtesy of L. Vakaet.

rear. The most central mesoblast (Fig. 18.32 6–7) stays closest to the axis and forms the **axial mesoderm** or **notochord,** while more posterior mesoblasts partially loop around and form the **paraxial mesoderm** or **somite mesoderm.** As far down as the middle of the primitive streak, mesoblast partially loops around to form the **lateral plate mesoderm,** and, more posteriorly, extraembryonic mesoderm stretches laterally and anteriorly to the area vasculosa.

All these changes are not rigid or entirely unidirectional. The primitive streak is a dynamic structure, and its flux affects everything about it. Cells from one side of the streak can cross to the other side, and even endoderm cells can be recycled through the streak and end up elsewhere in the embryo (e.g., notochord) (Fraser, 1954).

Histologically, axial presumptive notochordal cells condense into a rod, vaguely distinguishable from surrounding mesodermal cells in cross section (Fig. 18.37c) but grossly visible (Fig. 18.37a) as a translucent zone ahead of Hensen's node. This zone has traditionally been called the **head process,**

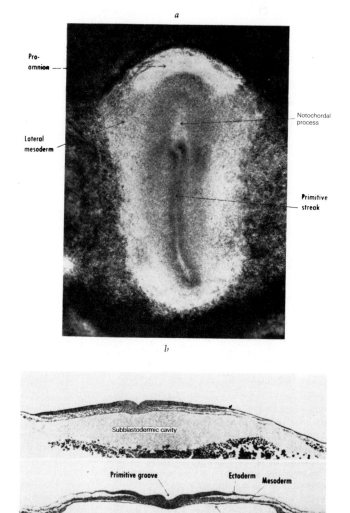

FIGURE 18.37. Photomicrographs of whole mount of chick embryo at notochordal- (= head-) process stage (a, Stage 5 H&H, 19–22 hours, ×25), and cross sections of sparrow embryo (b, ×50) and chick (c, ×50) at comparable stage. From E. Witschi, *Development of vertebrates,* Saunders, Philadelphia, 1956. Copyright © 1956 by Saunders College Publishing, a division of Holt, Rinehart and Winston, Inc., reproduced by permission of the publisher.

but it is really an optical effect due to the translucence of the notochord and should be called the **notochordal process** or **notochordal plate.** The stage identified by the notochord's appearance (7–8), moreover, should be called the **notochordal process stage** rather than the head process stage.

From the time the notochordal plate appears, the streak begins to **regress** and shorten. Soon an **anterior fold** (also called the head fold, Fig. 18.38b–d and b'–d', stage 9, Fig. 18.39) passes under the anterior end of the embryo, and gastrulation passes into organogenesis. Actually, gastrulation continues posteriorly where the primitive streak is still present and mesodermal cells continue to ingress (Fig. 18.32 8).

Further regression of the primitive streak accompanies the elongation of the head process and its undercutting by the head fold. As if the streak's cells are exhausted by ingression, the streak finally disappears, but by this time, the embryo is well advanced (Fig. 18.38d).

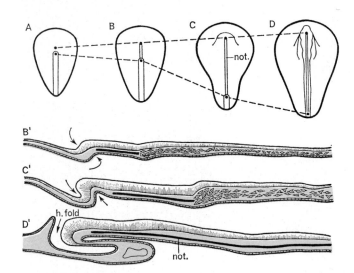

FIGURE 18.38. Schematic drawings showing the extension of the notochordal process, the formation of the head fold, and the regression of the primitive streak. The dots in A–D correspond to markers placed on the embryo. From R. Bellairs, *Developmental processes in higher vertebrates,* University of Miami Press, Miami, 1971. Reprinted by permission of John Wiley & Sons, Inc.

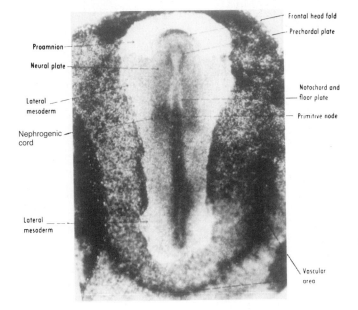

FIGURE 18.39. Photomicrograph of whole mount of chick embryo at head-fold stage. (Stage 6 H&H, 23–25 hours), corresponds to Vakaet stage 9). ×25. From E. Witschi, *Development of vertebrates,* Saunders, Philadelphia, 1956. Copyright © 1956 by Saunders College Publishing, a division of Holt, Rinehart and Winston, Inc., reproduced by permission of the publisher.

DEVELOPMENTAL ANATOMY
OF GASTRULAS FROM
MICROLECITHAL TO
MESOLECITHAL EGGS

Comparative biology is often disparaged by investigators not concerned with evolutionary problems as being merely a poor substitute for imagination in designing experiments. Yet, comparative methods provide (at both anatomical and molecular levels) our basic data on phylogenetic relationships and our major test for ideas of evolutionary mechanisms.

Rudolf A. Raff and Elizabeth C. Raff (1987, pp. xiii–xiv)

GASTRULATION IN
THERIAN MAMMALS:
A COMPARATIVE APPROACH

For centuries the chick blastoderm has been a surrogate for the mammalian embryo. Today, comparisons between chicks and mammals continue to be instructive especially for sorting out the details of mammalian gastrulation.

Of course, gastrulation is not identical in birds and mammals. The evolution of oviparous repro-

duction via cleidoic eggs in birds and of viviparous reproduction via a placenta in the therian or placental mammals (i.e., metatherians and eutherians) inevitably leads to differences. But differences can also be instructive.

Fundamental Similarities
with the Chick

The development of germ layers in birds and mammals begins with the conversion of a single layer of cells into a double layer. In the chick, the single-layered blastoderm in the area pellucida (see Fig. 18.29) takes part in the process, while the peripheral area opaca is still virtually in a state of cleavage.

The metatherian equivalent to the single-layered chick blastoderm is the **germinal region** (i.e., region where the embryo proper forms) or **embryonic epiblast** (also embryoblast) at one pole of the blastocyst. The eutherian embryonic cell line arises from the **inner cell mass** (ICM) or **embryonic**

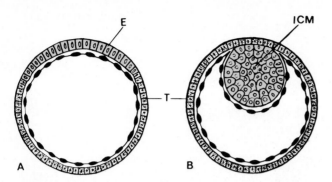

FIGURE 19.1. Schematic diagrams of metatherian (a) and eutherian (b) blastocytes at the bilaminar vesicle stage. The embryonic epiblast (E) occurs at the surface of the metatherian blastocyst, while the comparable ICM of eutherians is beneath a surface layer of trophectoderm (T, stippled). From W. P. Luckett, in M. K. Hecht, P. C. Goody, and B. M. Hecht, eds., *Major patterns in vertebrate evolution,* Plenum Press, New York, 1976. Used by permission.

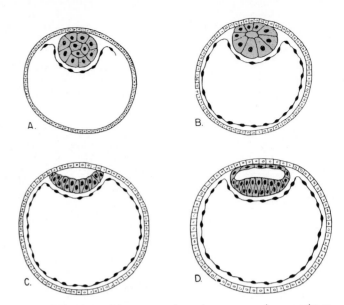

FIGURE 19.2. Diagrams of a rhesus monkey embryo showing the development of the hypoblast. Produced by the epiblast (circle of cells), the hypoblast (beaded line) spreads over the trophectoderm. From W. P. Luckett, in M. K. Hecht, P. C. Goody, and B. M. Hecht, eds., *Major patterns in vertebrate evolution,* Plenum Press, New York, 1976. Used by permission.

ectoderm[1] (sometimes called embryonic epiblast, Fig. 19.1). The remainder of the blastocyst consists of the hypoblast and **trophectoderm** (TE) whose cells are committed to, if not already differentiated as, an extraembryonic epithelium. Extraembryonic development in mammals is in advance of chicks at the beginning of gastrulation, but embryonic development lags.

The germinal regions of therian embryos are sometimes considered **unilaminar** embryos, although the ICM may not be a single cell layer. Like the chick blastoderm (see Fig. 18.29), the eutherian ICM gives rise to an underlying **hypoblast** or early endoderm (also primitive endoderm) and an overlying **epiblast** or early ectoderm (also primitive ectoderm). And like chicks, delamination (or polyinvagination) from the embryonic ectoderm is the mode of hypoblast formation in mammals (Figs. 19.2 and 19.3). Delamination of early endoderm may represent the mode for vertebrates in general, since delamination also produces the hypoblast in reptiles and may even operate in establishing endoderm in the roof of the amphibian archenteron.

In mammals, early endodermal cells migrate both around and away from the epiblast, but they do not arise from the extraembryonic TE (Gardner, 1982). The early endoderm separates into two distinct portions, especially in rodents, with different functions and fates. One portion, known as the **visceral endoderm** (VE), lies under the epiblast; the other portion, known as the **parietal** or **distal en-**

doderm (PE), underlies the trophectoderm (Fig. 19.4b).

Fused to the TE, the parietal endoderm forms the lining of the **bilaminar yolk sac,** generally

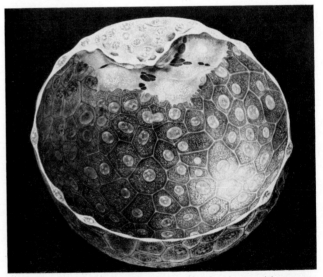

FIGURE 19.3. Reconstructed half of primate blastocyst showing spreading of primary yolk sac endoderm (parietal endoderm, thin, light cells) from vicinity of inner cell mass. From George L. Streeter, "Characteristics of the primate egg immediately preceding its attachment to the uterine wall," in Carnegie Institution of Washington publication 501, *Cooperation in Research,* 1938.

---

[1]The term embryonic ectoderm is unfortunate, since it implies that endoderm and mesoderm arise from ectoderm.

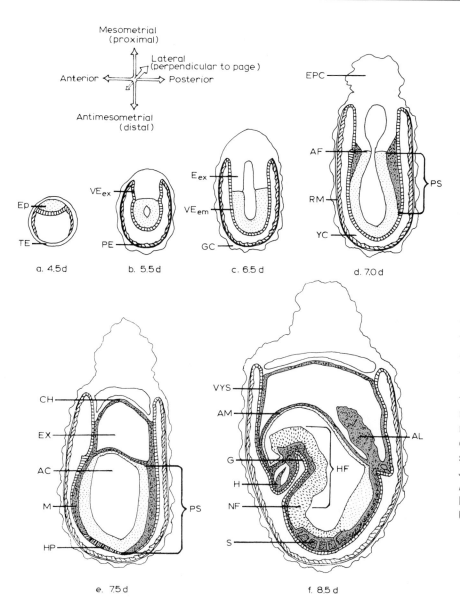

FIGURE 19.4. Diagrammatic longitudinal (a–d) and sagittal (e, f) sections of mouse embryos illustrating germ-layer origins and changes in the morphology of embryonic parts. Arrows show the embryo's polarity and its orientation in the uterus (determined by the position of the mesometrium or mesentery supporting the uterus). The numbers below each figure are days postcoitum. AC, amniotic cavity; AF, amniotic fold; AL, allantois; AM, amnion; CH, chorion; E_{ex}, extraembryonic ectoderm; Ep, epiblast; EX, exocoel (extraembryonic coelom); EPC, ectoplacental cone; G, gut; GC, giant cells; H, heart; HF, head fold; HP, head process; NF, neural fold; PE, parietal endoderm; PS, primitive streak; RM, Reichert's membrane; S, somite; TE, trophectoderm; VE_{em}, visceral embryonic endoderm; VE_{ex}, visceral extraembryonic endoderm; VYS, visceral yolk sac; YC, yolk sac cavity. Crosshatching: horizontal = mainly embryonic, diagonal = extraembryonic endoderm; dashes: neural ectoderm; heavy stippling: mesoderm; light stippling: epiblast and embryonic ectoderm of primitive streak. From R. Beddington, in M. H. Johnson, ed., *Development in mammals,* Vol. 5, Elsevier/North-Holland Biomedical Press, Amsterdam, 1983. Used by permission.

called the primary yolk sac or just yolk sac. In mammals, "yolk sac" is a misnomer, since the fluid-filled cavity is totally devoid of yolk. The name is justified by similarities to the yolk sac of birds, indicative of homology complicated by secondary reduction of yolk.[2]

Despite the absence of yolk, the concentration of amino acids and other nutrient materials is greater in mammalian yolk sac fluid than in uterine fluid (Renfree, 1973). The yolk sac's wall of

[2]The development of the yolk sac in the relatively yolkless metatherian and eutherian mammalian egg is sometimes cited (e.g., Carlson, 1988) as a biological reminiscence or recapitulation of the avian or reptilian condition. This interpretation is misleading, since the yolk sac is a cenogenic structure or adaptation to embryonic life and not an adult structure truncated and inserted into the embryo.

fused extraembryonic ectoderm (i.e., TE) and endoderm constitutes an **omphalopleure** (Gk. *omphalos* navel + *pleura* side or rib) which, in the early gestation of eutherians and throughout the gestation of metatherians, serves as the major organ for absorbing nutrients from uterine fluid.

Above all, as the outer layer of the empty yolk sac, the extraembryonic ectoderm of the original TE serves an active absorbing function. The TE is thus a unique **nutritive ectoderm,** which vindicates its earlier and still common name, **trophoblast,** signifying a primitive placenta or outer embryonic portion of the placenta.

Following the segregation of the hypoblast, or endoderm, the mammalian embryo as a whole (Fig. 19.2b) is properly called either a **bilaminar vesicle** or **bilaminar blastocyst** (see Wimsatt, 1975), but,

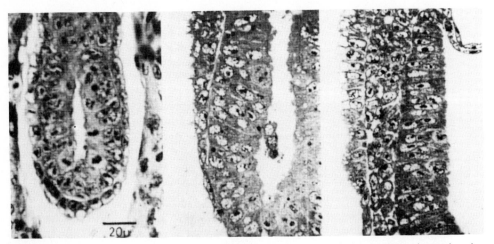

FIGURE 19.5. Photomicrographs of sections through part of the embryonic region in mouse embryos on successive days of development. (Left) At 5.5 days, the cylindrical epiblast is coated with a layer of visceral hypoblast. (Center) At 6.5 days, the epiblast has thickened but remains more or less one layer. (Right) At 7.5 days, a distinct mesodermal layer appears between the thickened ectoderm and hypoblast. From M. H. L. Snow, *Embryogenesis in mammals,* Ciba Foundation Symposium 40 (new series), Elsevier / Excerpta Medica / North-Holland, Amsterdam, 1976. Used by permission.

regrettably, it is most often merely called "blastocyst." This practice is unfortunate even when modifiers referring to time (i.e., early blastocyst versus late blastocyst) or state of implantation (i.e., partially implanted versus totally implanted) are added, since they are imprecise and overly general.

When the mesoderm is added, the embryo as a whole becomes a **trilaminar vesicle** or **trilaminar blastocyst.** Mesoderm first appears in the thickened posterior portions of both mammalian and avian embryos. The embryos develop a **primitive streak** (Fig. 19.4) creased by a **primitive groove** and capped by a **primitive node** or **Hensen's node,** and mesoderm emerges from the streak (Fig. 19.5). Furthermore, as in chicks, the **notochordal** (or head) **process** appears anterior to the primitive node (see Snow, 1976).

Actually, the mammalian primitive streak is more reptilian-like than avian-like. Birds lack an archenteron, but the primitive node of some mammals (including humans) is penetrated by an **archenteric canal** resembling the **blastoporal canal** of reptiles (see Fig. 18.28).

Still, the fate map of mammals is roughly compatible with those of birds and amphibians (see Fig. 14.13) except for the absence of presumptive extraembryonic parts in amphibians. At least, there are no irreconcilable differences.

Results with orthotopic grafts (i.e., transplants to the normal position in a host) of labeled mouse cells (Tam and Beddington, 1987) and with colonization chimeras (Snow, 1985) indicate that cells ingressing through the anterior part of the mammalian streak give rise mostly to gut, paraxial mesoderm, and notochord, while cells adjacent to the anterior part of the streak contribute mainly to paraxial mesoderm and ectoderm (both cutaneous and neural). The middle portion of the streak gives rise to lateral plate mesoderm, and the posterior portion gives rise to extraembryonic mesoderm, if only in younger streaks. Moreover, embryonic stem cell (ES) lines derived from the embryonic parts of mouse egg cylinder-stage embryos (but not from extraembryonic parts) show a high degree of pluripotency *in vitro* and a large degree of developmental potential (Doetschman et al., 1985).

Fundamental Differences with the Chick

Major differences between mammalian and avian embryos are related to the evolution of **viviparity** and placentation in mammals as opposed to **oviparity** and cleidoic eggs in birds. For example, when the therian blastocyst reaches the uterus, its zona pellucida is shed, and the embryo "hatches." In contrast, the chick egg acquires its albuminous coatings, membranes, and shell. The eutherian trophectoderm covers the ICM at least temporarily (Fig. 19.6), while the chick blastoderm resides openly on the surface.

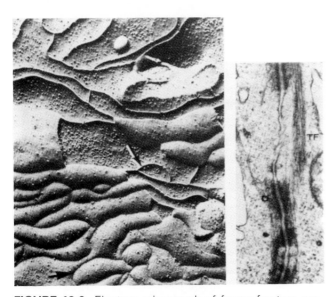

FIGURE 19.6. Electron micrograph of freeze-fracture replica of mouse embryo trophoblast cells after "implantation" *in vitro*. A tight junctional network is seen on and between the faces of the cells. X68,000. From C. W. Lo, in M. H. Johnson, ed., *Development in mammals,* Vol. 4, Elsevier/North-Holland Biomedical Press, Amsterdam, 1980. Courtesy of C. W. Lo. Used by permission.

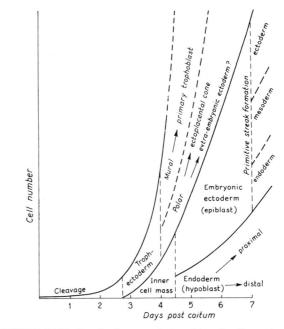

FIGURE 19.7. Graph shows relationship of cell number to days postcoitum for the mouse trophectoderm, inner cell mass, and derivatives. From A. McLaren, *Embryogenesis in mammals,* Ciba Foundation Symposium 40 (new series), Elsevier / Excerpta Medica / North-Holland, Amsterdam, 1976. Used by permission.

Other differences are quantitative but not without qualitative consequences. Unlike the situation in chicks, mitosis in the mammalian trophectoderm tends to be polarized. Centered at the **embryonic pole** (i.e., site of the ICM), **polar trophectodermal cells** divide in the plane of the TE and move into the **mural TE** toward the **abembryonic pole** (see Fig. 13.14) (Cruz and Pedersen, 1985). The mammalian TE population increases linearly, and, with cell stretching and flattening, the blastocyst expands.

At gastrulation, cell division in the epiblast begins to gain momentum (Fig. 19.7). In the mouse, the cell cycle time picks up from one division every 12 hours during cleavage to one division every 5–6 hours (Snow, 1976). Moreover, the divisions are differentially skewed in favor of the ICM and germ-layer formation. Specifically, at 3.5 days postcoitum (p.c.), the mouse has 39–45 TE cells and 15–16 ICM cells (see McLaren, 1976). At 4.5 days p.c., it has about 65 TE cells, 47 ICM cells (Fig. 19.4), 25 epiblast cells, and 22 hypoblast cells.

The artiodactyl ungulates represent exceptions to the general mammalian rule. Their blastocysts expand greatly prior to gastrulation. For example, 570 cells are present in the preimplantation pig blastocyst of nearly 7 days, all but a hundred of which are TE cells (Heuser and Streeter, 1928).

Still, the number of cells present in the ICM or epiblast at the beginning of gastrulation is strikingly less in mammals than birds. In contrast to the paltry 112-cell mouse at 4.5 day, the freshly laid chick blastoderm at about 16 hours postovulation and the beginning of hypoblast formation has about 500 epiblast cells (Kochav et al., 1980).

Gastrulation is also, generally, protracted in mammals compared to birds. The chick reaches the notochordal-process stage at about 20 hours of incubation (or about 1½ days after ovulation) and the head-fold stage (marking the beginning of organogenesis) 6 hours later (still less than 2 days after ovulation). In contrast, mouse embryos are gastrulating at 6.5 days p.c. (Fig. 19.4) and human embryos at 15 days p.c. Neither reaches the notochordal process stage for yet another day.

Unlike the cleidoic eggs of birds which have the leisure to exploit the rich resources made available to them through the yolk, the small yolk-poor eggs of metatherian and eutherian mammals must use their paltry resources judiciously. Instead of investing them in gastrulation, it seems, these mammals devote the bulk of their cells to the production of **extraembryonic membranes** through which they effect nutritional exchange *in utero*.

Not only does the TE represent the greater part of the early blastocyst but the first and most conspicuous products of mouse embryonic stem cell (ES) lines *in vitro* are the alpha-fetoprotein (AFP) and transferrin normally associated with the visceral endoderm. ES lines of blastocyst cells that form cystic embryoid bodies also develop beating heartlike muscle and, when cultured with human umbilical cord serum, blood islands normally associated with splanchnopleure (Doetschman et al., 1985).

After establishing its support system, the embryo's rate of development increases dramatically. The epiblast's axes of symmetry become apparent in the mouse of about 600 cells at $6\frac{1}{2}$ days p.c., and histotypic allocation, if not determination, may begin as early as the primitive streak stage in embryos of 2000 cells at 7 days p.c. Considerable cellular autonomy is apparent in transplants and in ES lines from embryos of fewer than 15,000 cells. Within 12 hours, fragments from 40,000- to 100,000-cell embryos in the neural plate stage express regional restriction (Snow, 1985).

Maintaining the Mammalian Embryo

Maintenance *in vitro* of the mammalian embryo beyond the normal stage of implantation presents difficulties. Successful culturing of **post-implantation** mouse embryos to the midsomite stage with beating heart has so far been achieved only if fetal calf serum is supplied up to the 2-cell embryo for the first 2 days, followed by the addition of human umbilical cord serum to the embryo's medium. Mouse, rat, and hamster embryos dissected out of uteri after normal implantation survive *in vitro* to late somite stages in plasma clots or in circulating medium containing serum.

On the other hand, successful transfer of a preimplantation embryo to the uterus is not as difficult as one might have thought. The first successful transfer of a fertilized mammalian egg to a uterus actually took place nearly a century ago. Progress in making the technique practical, however, was slow.

Embryo transfer is the technique of transferring a preimplantation embryo to the uterus of a **uterine** (or **prenatal**) **foster mother** who has been prepared to receive the embryo. The embryo's development appears to be perfectly normal despite its extraordinary origin. In the case of human beings suffering some forms of infertility and therapeutically receiving an embryo by transfer, no significant birth defect is easily attributed to the procedure (see Edwards, 1974).

Only when the uterus is **primed** by estrogen followed by progesterone will it accept the preim-

plantation embryo. Priming is accomplished in several ways depending on species. In the case of rabbits, the uterus is receptive either when it is pregnant or rendered **pseudopregnant** by mating with a sterile male. In other mammals, hormonal therapy is employed. In the case of human beings, the same conditions that lead to ovulation frequently provide a suitable uterus for embryo transfer a few days after ovulation. The hormonal regime is as convenient for *in vitro* fertilization and embryo culture as it is for implantation, since the embryo normally takes a number of days to reach the implantation stage and the implantation site (see Fig. 11.24).

GASTRULATION FOLLOWING VERTICAL CLEAVAGE

Sponges and Diploblastic Metazoans

Gastrulation provides these morphologically simple organisms with rudiments of the same two epithelial layers and mesenchyme generally found in adults. Gastrulas are loosely identified as any embryo having two layers of cells somewhere between a simpler blastula and a more complex larva. The form achieved by the gastrula may not be especially adultlike. Development may be histologically direct while morphologically indirect.

Sponge gastrulas form from coeloblastulas (Fig. 19.8) and from larval amphiblastulas (Fig 19.9) that have already passed through one or more rounds of inversion. Ectodermal **foundation** cells or **pinacocytes** form the epidermis and provide some mesenchymal cells. Endodermal cells line the spongocoel (i.e., sponge cavity) and form most of the **wandering cells** (amoebocytes and archeocytes) of the mesenchyme.

Coeloblastulas form endoderm by **unipolar ingression** (Fig. 19.10) or invagination (Fig. 19.8) and either contain a spongocoel or become a solid **stereogastrula** or **parenchymula** (Gk. meaning to pour but referring to entrails + the diminutive ending). Polarity seems to be determined at the time of gastrulation, at least, in the case of differentiated amphiblastulas, since the point of invagination is opposite the **osculum** or excurrent pore of the mature sponge.

Coelenterate embryonic ectoderm and endoderm persisting in adults should probably be called **epidermis** and **gastrodermis** (Hyman, 1940), but tradition (and priority) justify calling these layers

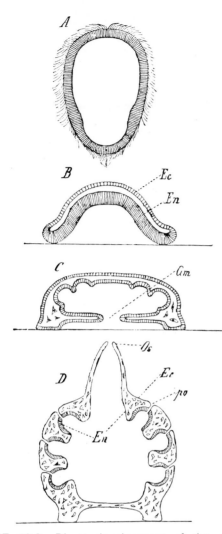

FIGURE 19.8. Direct development of the calcareous sponge, *Oscarella*. (a) Coeloblastula. (b) Attached gastrula. (c) Gastrula mouth (Gm) closes as endodermal sac forms and mesenchyme appears. (d) Young sponge consisting of ectoderm (Ec) and endoderm (En) with mesenchyme between. Inhalent pores (po) open into spongocoel having an exiting osculum (Os). From E. Korschelt and K. Heider, *Embryology of invertebrates,* Swan Sonnenschein, 1895.

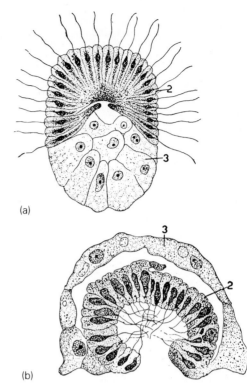

FIGURE 19.9. Sections through (a) amphiblastula and (b) gastrula of the calcareous sponge, *Sycon*. Numbers show changing positions of cells. From L. H. Hyman, *Invertebrates: Protozoa through Ctenophora*, McGraw-Hill, New York, 1940. Used by permission.

ectoderm and **endoderm** (Figs. 19.11 and 19.12). A predominantly **fibrogelatinous** mesenchyme containing loose cells lies between the layers except in larval planulas and hydrozoan polyps where a thin and apparently acellular **mesoglea** glues the layers.

Coelenterates with either a solid stereoblastula or a hollow coeloblastula (see Fig. 13.9) gastrulate to form the two-layered **planula** larva or tentacled **actinula** (Fig. 19.13). Generally, a well-differentiated ectoderm or **ectoblast** is separated from a less well-differentiated endoderm or **endoblast.** A lumen in the endoderm, especially in the case of

stereogastrulas, is formed by cavitation and sloughing of central material (Fig. 19.12).

The mesenchyme typically lying between ectoderm and endoderm seems to be derived from the ectoderm. In contrast to relatively uniform epithelial cells in the ectoderm and endoderm, mesenchyme cells differentiate into a variety of discrete cells of various types including wandering (amoeboid) cells, muscle, and nerve cells.

Gastrulation may begin during cleavage. For example **polarized** gastrulation (i.e., taking place at one end of the embryo), occurs at the animal pole (identified by polar bodies) where cleavage also starts. Polarity therefore is already determined at cleavage, preserved in the gastrula, and passed on to the larva and adult. The site of endoderm formation corresponds to the posterior end of the swimming planula larva (see Freeman, 1980) and, upon metamorphosis, to the oral–anal end of the polyp (i.e., the common point of ingestion and egestion).

The germ layers may form from the inside out (i.e., the ectoderm forming from the peripheral endoderm) or from the outside in (endoderm forming from ectoderm). In the former case, cells seem to

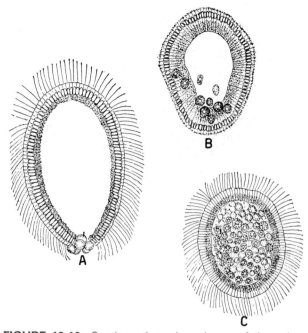

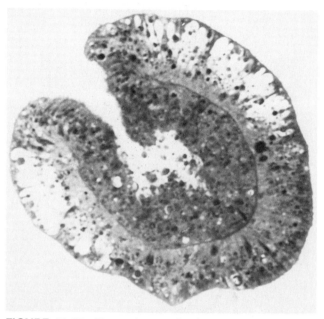

FIGURE 19.10. Sections through embryos of the calcareous sponge, *Leucosolenia*, showing a hollow coeloblastula (a) undergoing unipolar ingression (b) to form a solid stereogastrula or parenchymula (c). From L. H. Hyman, *Invertebrates: Protozoa through Ctenophora*, McGraw-Hill, New York, 1940. Used by permission.

FIGURE 19.12. Photomicrograph of section through gastrula of the anemone, *Actinia equina*. The ectoderm and syncytial endoderm are sharply demarcated. From M. A. Carter and M. E. Funnell, in P. Tardent and R. Tardent, eds., *Developmental and cellular biology of coelenterates*, Elsevier/North-Holland Biomedical Press, Amsterdam, 1980. Used by permission.

become **entrapped** in the surface, and an ectodermal epithelium is constructed of stationary but rapidly dividing cohesive cells. A mass of less compact and occasionally degenerating cells (sometimes in the form of a syncytium) **delaminates** or is left behind in the center of the gastrula to form the endoderm.

Alternatively, endoderm forms from the ectoderm by one of several devices. Cell migration or **ingression** may occur either throughout the length of the embryo (known as **multipolar ingression**) or in only the posterior larval pole (i.e., unipolar ingression). Delamination (or the splitting of layers) may begin as a cellular event accompanying cell division or take place as a mass phenomenon independent of cell division (see van de Vyver, 1980). Unlike the situation in sponges, invagination is not a conspicuous part of gastrulation in coelenterates.[3]

Gastrulation in the biradially cleaving ctenophores also employs delamination. The highly determinate ctenophore embryo has hardly begun cleavage when delamination signals the beginning of gastrulation. The eight macromeres (i.e., endoderm) bud off a layer of micromeres (i.e., ectoderm) through radially directed divisions (Fig. 19.14). The embryo acquires a basket shape and a central cavity running through both micromeres and macromeres. Proliferating micromeres gradually roof over the concavity and migrate laterally and ventrally over the macromeres to form the two-layered gastrula.

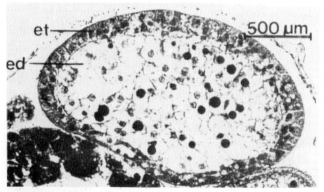

FIGURE 19.11. Photomicrograph of section through a young actinula of the hydroid, *Tubularia crocea*. ed, Endoderm; et, ectoderm. From F. J. Fennhoff, in P. Tardent and R. Tardent, eds., *Developmental and cellular biology of coelenterates*, Elsevier/North-Holland Biomedical Press, Amsterdam, 1980. Used by permission.

[3]The failure of typical coelenterates to invaginate during gastrulation was an embarrassment for the biogenic law and prompted Haeckel to propose the hypothetical gastrea rather than the coelenterates as the ancestral eumetazoan.

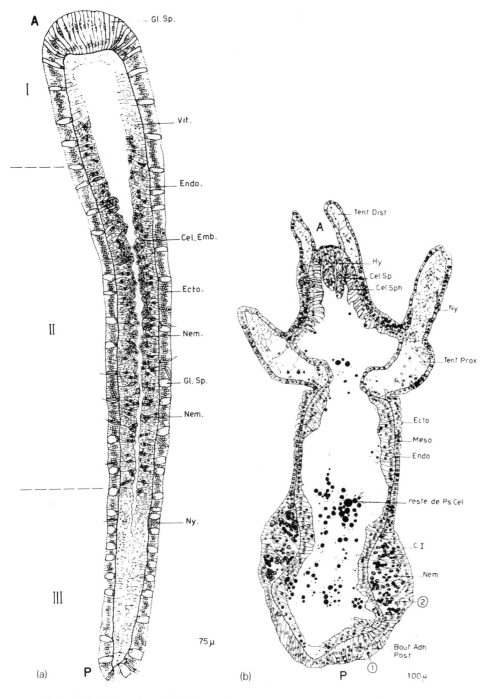

FIGURE 19.13. Longitudinal sections through the planula of the hydroid *Hydractinia echinata* (a) and the actinula of hydroid *Tubularia ceratogyne* (b). From G. Van de Vyver, in P. Tardent and R. Tardent, eds., *Developmental and cellular biology of coelenterates*, Elsevier / North-Holland Biomedical Press, Amsterdam, 1980. Used by permission.

Radial Cleaving Echinoderms

A variety of different processes take part in echinoderm gastrulation. **Primary** (Figs. 19.15d–e and 19.16a) and **secondary invaginations** (Figs. 19.15f–g and 19.16b) especially employ different mechanisms of cell movement.

Primary invagination is **radially** or **rotationally symmetrical** around the animal–vegetal axis (see Fig. 11.30). Tall columnar cells in the **vegetal plate** (see Fig. 13.19, also known as gastral or endodermal plate) of the mesenchymal blastula invaginate or

pocket inward to form an **archenteron**[4] around a **blastopore**. Not too many cells are involved (e.g., about 66 cells in *Lytechinus pictus*), and not more than a ring of one additional cell **involutes** around the **lips of the blastopore.** Even with additions due to cell division, the small, squat archenteron (of 110–120 cells) reaches only between one-fifth to one-third the way up the blastocoel depending on the diameter of different species.

[4]Archenteron is also used to designate the lumen of the pocket or internal continuation of the blastopore.

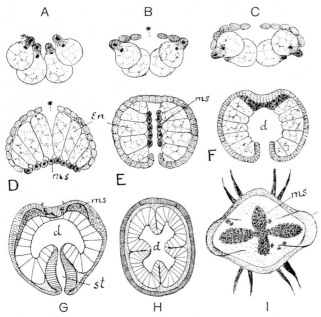

FIGURE 19.14. Diagrammatic optical sections of gastrulating ctenophore embryo and early adult (I). d, intestinal cavity and diverticula; en, endoderm; ms, mesenchyme; st, stomodeurm. From E. Korschelt and K. Heider, _Embryology of invertebrates_, Swan Sonnenschein, 1895.

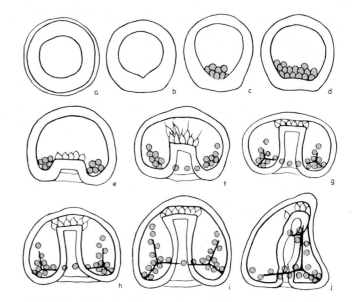

FIGURE 19.15. Outline drawings of the sea urchin, _Pseudocentrotus depressus_. (a–b) Blastula hatching; (c–d) primary mesenchymal cell formation; (e–f) primary invagination; (f–g) secondary invagination; (g–h) secondary mesenchymal cell formation; (j) attachment archenteron to stomedeum. From K. Okazaki, _The sea urchin embryo: Biochemistry and morphogenesis_, Springer-Verlag, New York, 1975. Used by permission.

When isolated surgically from the rest of the larva, vegetal plates of various echinoderms (e.g., the sand dollar, _Dendraster excentricus_, the star fish, _Patiria miniata_, the sea urchin _L. pictus_) buckle inward at about the normal time and to approximately the normal extent. Primary invagination seems to be an autonomous undertaking of the vegetal plate requiring nothing from the remainder of the blastula.

The result of extensive cinematographic studies (Gustafson and Wolpert, 1967) suggests that primary invagination is the result of diminished intercellular adhesion and cell rounding in the vegetal plate. But cell profiles drawn from sectioned material (Fig. 19.17) do not show cells rounding. Rather, cells are roughly **bottle** or **flask shaped,** and, at the end of the primary invagination, they are taller at the tip of the archenteron than in the walls and taller throughout the archenteron than in the noninvaginated surface ectoderm. The invaginated cells seem to have undergone **apical constriction** and active migration but not diminished intercellular adhesion.

Basally (i.e., toward the blastocoel), archenteron cells lack the basal lamina present throughout the rest of the gastrula. Apically, invaginating and surface cells continue to be joined by zonulae adherens (Fig. 19.18) and junctional complexes.

Pulsatory activity in the vegetal pole just prior to invagination and apical constriction accompanying invagination may both result from the contraction of thin bands of actin filaments associated with zonulae adherens.

Following primary invagination, lobous pseudopods on cells at the tip of the archenteron give way to filamentous pseudopods or **filopods** (Fig. 19.15f). Reaching out and moving around the lining of the blastocoel, the filopods seem to explore the basal lamina on the underside of the surface cells (Fig. 19.19). **Secondary invagination** takes place when these filopods shorten and stretch the archenteron across the entire blastocoel (Fig. 19.15f–g) (Gustafson, 1975).

At about this time, the gastrula also shows signs of bilaterality (Figs. 11.30 and 19.15g–j). The apical tuft shifts to an **animal plate** that tilts toward one side, and the tip of the archenteron turns toward the same side and makes contact with the presumptive **stomodeum.** Arching and elongating into a **prism-shaped** larva, an anterior **oral lobe** forms above the ventral mouth, while **anal arms** flare out below it, and a dorsal body arch swells on the opposite surface.

Then, in a process reminiscent of primary mesenchymal cell (PMC) formation (see Fig. 13.19),

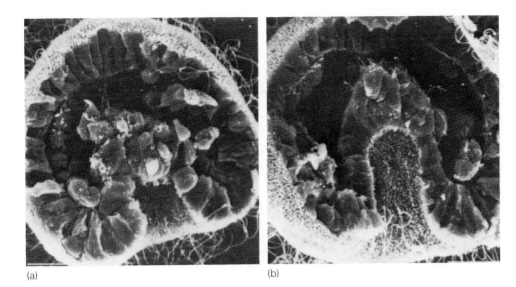

(a)   (b)

FIGURE 19.16. Scanning electron micrographs of the sea urchin, *Authocidaris crassispina*, fractured to reveal the archenteron during primary invagination (a) and secondary invagination (b). From S. Amemiya, K. Akasaki, and H. Terayama, *J. Embryol. Exp. Morphol.*, 67:27 (1982), by permission of the Company of Biologists and courtesy of the authors.

cells at the tip of the archenteron detach or **delaminate** and **ingress** into the blastocoel as **secondary mesenchymal cells** (SMCs, Fig. 19.15h–j). Some SMCs form pigment cells, while others form muscle surrounding the gut and still other cell types. At the same time, the PMCs that entered the blastocoel earlier form the calcified spicules of the larva's skeleton.

Below the tip of the archenteron, the mesodermal lining of the coelom forms by **enterocoelous** evaginations. Descendants of small micromeres (and possibly some SMCs) (Okazaki, 1975) balloon into **coelomic vesicles** that later detach. Following the segregation of all the mesodermal components from the archenteron, the remaining cells constitute the **endoderm,** while the entire surface of the larva is now **ectoderm.**

> The amphibian strategy of development is one of enormous flexibility and power. It is not restricted by lineage-specific cell division programs, requirements for exact cleavage geometry, numbers of cells, . . . or blastomere ancestries. Nor can the initial specification processes depend on sequential cascades of genomic regulatory events.
>
> Eric H. Davidson (1986, p. 257)

Bilateral Cleaving Chordates

The protochordates and anamniote chordates exhibiting bilateral cleavage share several features

A 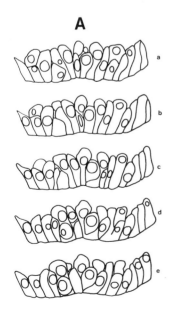 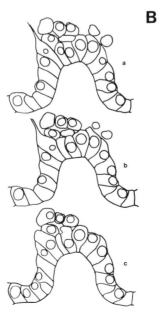 **B**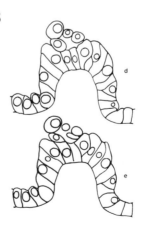

25 μm

FIGURE 19.17. Tracings from section of vegetal plate cells in the sea urchin, *Lytechinus pictus*. The tip of the invaginating archenteron is "hopping" with cell movement: (a) At 22 hours (prior to invagination); (b) at 25.5 hours (following primary invagination). From C. A. Ettensohn, *Am. Zool.*, 24:571(1984), by permission of the American Society of Zoologists.

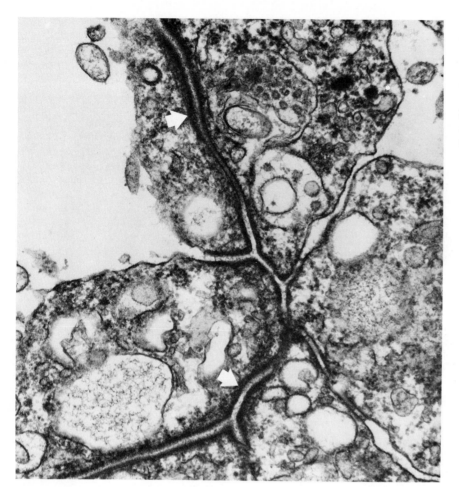

FIGURE 19.18. Electron micrograph through section of archenteron of *Strongylocentrotus purpuratus* showing zonulae adherens (open arrows) encircling the apexes of cells. ×30,000. From C. A. Ettensohn, *Am. Zool.,* 24:571(1984), by permission of the American Society of Zoologists and courtesy of C. A. Ettensohn.

of gastrulation as well. Other features of gastrulation, however, are not shared.

Protochordate gastrulation demonstrates considerable evolutionary plasticity. Modes of gastrulation span the range from invagination in Amphioxus, *Branchiostoma lanceolatum* (Fig. 19.20a, d), to cellular rearrangement in the larvacean, *Oikopleura* (Fig. 19.20c, f), with numerous intermediates represented by ascidians such as *Styela* (Fig. 19.20b, e). Epiboly of ectoderm is common, but involution at the edges of a blastopore is not.

Just prior to gastrulation, the coeloblastula of Amphioxus flattens ventrally (Fig. 19.21 [35]). A roughly triangular **endodermal plate** (pointed to the rear) with lateral and posterior borders of presumptive mesoderm (37) invaginates to form the **archenteron** (55). The triangular **blastopore** (39) has a **dorsal lip** (dl) of larger cells. Two lateral lips conceal cells of the mesodermal crescent just beneath the surface.

The blastopore narrows as the archenteron and the embryo elongate (Fig. 19.21 [63, 46]). The large, slowly dividing endodermal cells elongate as they invaginate, while the small, more rapidly dividing mesodermal cells become round. Ectoderm becomes cuboidal as it stretches and grows over the surface, but no involution takes place at the lips of the blastopore (Conklin, 1932). Even the dorsal lip

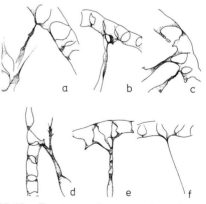

FIGURE 19.19. Tracings of mesenchymal pseudopods from tip of archenteron attached to inner surface of sea urchin ectoderm. From T. Gustafson, *Exp. Cell Res.,* 32: 570(1963), by permission of Academic Press and the author.

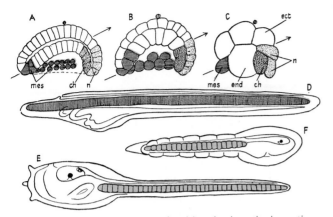

FIGURE 19.20. Diagrams of midsagittal optical sections through the gastrulas and larvas of Amphioxus (a, d), an ascidian, *Styela* (b, e), and a larvacean tunicate, *Oikopleura* (c, f). ms, Tail mesoderm (dense stippling) seen against wall of embryos; n, presumptive neural tissue (lightly stippled); nt, presumptive and differentiated notochord (horizontal hatching). From N. J. Berrill, Size and organization in the development of ascidians, in P. Medawar, ed., *Essays on growth and form*, Oxford University Press, Oxford, 1950. Used by permission.

of the blastopore that has inductive capacity remains external.

The invagination of the endomesodermal wall of the blastula resembles primary invagination in echinoderms although pseudopods may help later. Initially, individual cell movement seems to be restricted to cellular rearrangement associated with masses or blocks of cells. Integrated mass cell movement also characterizes later stages of development.

Presumptive notochordal cells soon line up like a stack of plates and differentiate as the notochordal rudiment (Fig. 19.21 [75, ch]). **Tail bud** cells separate this rudiment from the remains of the blastopore and give rise to the notochord and mesoderm of the tail (see Miyamoto, 1985, for comparable events in ascidians). When the neural ectoderm, in the form of a **neural plate** (Fig. 19.21 [70, np]), dips to form a **neural groove** (ng), gastrulation lapses into neurulation and embryogenesis.

While the blastocoel provides room for the archenteron to invaginate, Amphioxus' relatively large number of smaller cells organized as a monolayer provides flexibility for invagination. In ascidians with small blastocoels and larvaceans with stereoblastulas and no blastocoels, invagination is hardly more than ventral flattening, and the stretching and epiboly of the ectoderm perform a more prominent role in gastrulation (Fig. 19.20).

Evagination is also a feature of mesoderm for-

mation in Amphioxus (Fig. 19.22 183). Arising enterocoelously as periodic right and left diverticula (rd and ld) from the archenteron, segmented mesoderm, like links in a chain (175), surround a coelom extending backward throughout the length of the embryo (179–181). Similarly, movement by small endodermal cells pushes the notochord out of the archenteron.

In contrast, germ layers in the ascidians are relatively solid (Fig. 19.23). The notochordal plate (colored cells, a) above the small archenteron rearranges itself into a notochordal process (b–c). Mesoderm moves laterally as the lumen of the archenteron disappears, but no coelom appears in its midst, although a pericardial cavity that appears later can be considered a coelom.

Since the protochordate eggs are originally about the same diameters (Table 19.1), the differences in the sizes of the blastocoels and the mode of gastrulation do not seem to be due to differences in yolk content. When traced back to cleavage, germ-layer and notochordal origins seem to be very nearly the same (e.g., Fig. 19.20) in the urochordates (e.g., *Styela* and *Oikopleura*) and Amphioxus. Moreover, early gastrula fate maps are broadly comparable not only among protochordates but between protochordates and primitive chordates represented by lampreys (Takata and Hama, 1978).

The most obvious differences in the gastrulation of the protochordates are in the size and number of cells participating in the process. While Amphioxus does not gastrulate until the 9th–10th divisions when it is approaching 1000 cells, ascidians gastrulate two to three divisions earlier when they have about one-tenth as many cells, and the larvacean tunicate gastrulates still earlier with half again fewer cells (Berrill, 1950, 1961).

It would seem that evolutionary novelties or **heterochronies** (i.e., changes in the timing or rates of development) were introduced at gastrulation in protochordates. If the intermediate type of invagination seen in ascidians (e.g., *Styela*) represents an ancestral condition, invagination in Amphioxus may have been derived by the retardation of gastrulation and the production of a coeloblastula, while cell rearrangement in larvaceans may have been derived by the acceleration of gastrulation to an early cleavage stage.

The course of germ-layer determination may have followed the same scenario. Amphioxus, for example, is more regulative than ascidians and can reconstitute a complete dwarf larva from a half-embryo as long as the mesodermal crescent is included. Similarly, neural competence extends throughout the ectoderm of Amphioxus but is re-

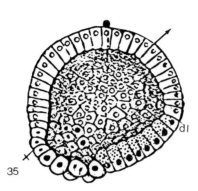

35

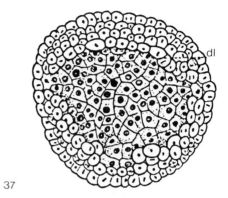

37

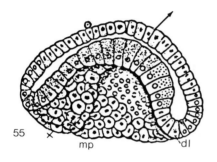

55

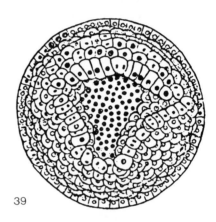

39

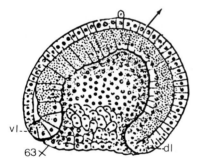

63

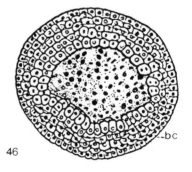

46

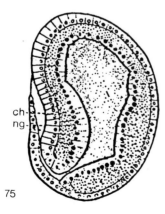

75

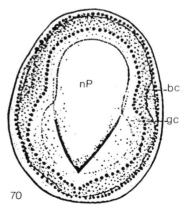

70

FIGURE 19.21. Drawings of optical sections and surface views of Amphioxus illustrating gastrulation movements. Parts 35, 55, 63, and 75 are optical sections showing inside surface beyond midsagittal plane. Parts 37, 39, and 46 are views of ventral surface; 70 is a view of the dorsal surface. ×333.

Part	Hours after fertilization	Description
35	5.5	Prior to gastrulation; endodermal cells become columnar as vegetal plate flattens.
37	6.5	Triangular vegetal plate and lateral–posterior mesoderm invaginate behind dorsal lip (dl).
55	7.5–8.5	The blastocoel collapses as invagination produces an archenteron.
39	7.5	Triangular blastopore is pointed posteriorly; involution does not occur at the blastoporal lips.
63	11	The roof of the archenteron is in contact with underside of ectoderm; the embryo elongates.
46	9.5	The blastopore narrows by convergence of the blastoporal lips.
75	15	The blastopore is covered by ectoderm which also begins to migrate over a shallow neural groove (ng) above the differentiating notochordal process (ch).
70	15–16	Ectoderm converges above the neural plate (np).

bc, Blastocoel; ch, notochord; dl, dorsal lip or presumptive dorsal lip; gc, gastrocoel; mp, mesodermal pouch; ng, neural groove; np, neural plate; vl, ventral lip; arrows, anterior–posterior axis.

Adapted from E. G. Conklin, Embryology of amphioxus. *J. Morphol.,* 54:69(1932), by permission of Alan R. Liss, Inc.

stricted to the presumptive neural ectoderm of ascidians (see Takata and Hama, 1978).

Evidence of evolutionary plasticity in the protochordates is especially significant, because the resulting embryos and larvas are considered our nearest invertebrate relatives (see Berrill, 1961). The primitive, sessile ascidian adult is hardly vertebrate-like (e.g., see Fig. 5.15), however. It only shares the notochord, dorsal hollow nerve cord with enlarged brain, sensory organs, and segmental muscle bands (if not full-fledged somites) of vertebrates as embryos and larvas. Following a brief larval phase, the ascidians lose most of the vertebrate-like features and metamorphose into the "little wine sacs" for which they are named.

Unlike ascidians, larvaceans and Amphioxus

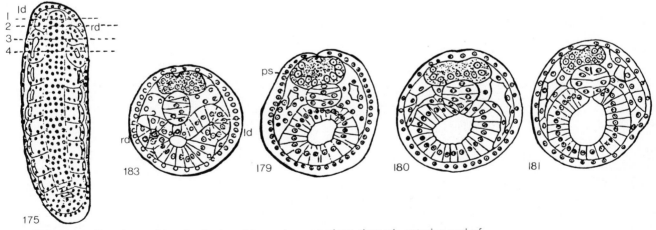

FIGURE 19.22. Drawings of longitudinal and transverse sections through anterior end of elongating 24.5 hour larva. ×667. ld, Left diverticulum; rd, right diverticulum; ps, pigment spot in flor of neural tube. Numbers correspond to levels of sections: 1-183, 2-179, 3-180, 4-181. Adapted from E. G. Conklin, Embryology of amphioxus, *J. Morphol.,* 54:69(1932), by permission of Alan R. Liss, Inc.

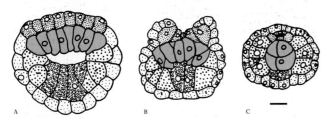

FIGURE 19.23. Diagrammatic transverse sections through embryos of the ascidian, *Styela clava*, illustrating role of cellular rearrangement in formation of notochord, lateral mesodermal cords, and neural tube. Bar = 50 μm. Open cells, notochord; dense small stippling, neural epithelium; sparse small stippling, ectoderm or epidermis; large stippling, mesoderm or muscle; open stippling, endoderm. From D. M. Miyamoto and R. J. Crowther, *J. Embryol. Exp. Morphol.*, 86:1(1985), by permission of the Company of Biologists.

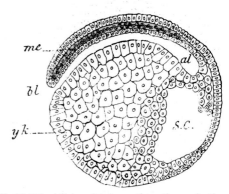

FIGURE 19.24. Midsagittal section through the gastrula (136 hours) of a cyclostome, *Petromyzon planeri*. al, Archenteron (alimentary tract); bl, blastopore; me, mesoderm; S.C., segmentation cavity (blastocoel); yk, yolky cells. From F. M. Balfour, *Comparative embryology*, Vol. II, Macmillan, London, 1881.

do not metamorphose but retain many embryonic and larval features as adults. **Paedomorphosis** (i.e., the retention of ancestral juvenile characters in the later ontogenic stages of descendants) (see Gould, 1977) in these protochordates accompanies **neoteny,** the retardation or elimination of later developmental processes. Possibly, similar evolutionary processes may have been responsible for the evolution of the highly innovative vertebrate forms from embryos or larvas of ascidian ancestors.

Cyclostomes develop from remarkably amphibian-like eggs despite the adults' totally unamphibian-like morphology and lowly position in the scale of vertebrate evolution. The similarity to amphibians reflects the role of egg size (about 1 mm in diameter), yolk content (mesolecithal), and yolk distribution (telolecithal) in early devel-

opment, since the later development of cyclostomes reveals greater affinities with teleosts than with amphibians.

Gastrulation begins with an asymmetric invagination. A pit near the blastocoel deepens, and a dorsal fold of tissue overgrows part of the yolky vegetal pole leaving another part still on the surface (Fig. 19.24). The fold is the roof of the **archenteron** (or mesenteron), and the cavity below (al) is continuous with the blastopore (bl).

The gastrula retains a large blastocoel or segmentation cavity (S.C., Fig. 19.24). It does not disappear until the neurula stage. At the edges of the blastocoel, small surface cells grade into large vegetal cells. The single layer of columnar ectoderm (or epiblast, Fig. 19.25) that eventually covers the embryo is formed by proliferation of these small

TABLE 19.1. Number of cells in some representative chordate embryos at gastrulation

		At Gastrulation	
Genus	Diameter of Egg (mm)	Number of Cleavages	Number of Cells
Oikopleura (a larvacean)	0.09	5–6	38
Styela (an ascidian)	0.13	6–7	76
Amphioxus (lancet)	0.12	9–10	780
Petromyzon (lamprey)	1.00	11	2,200
Trituris (newt)	2.60	14	16,000

Source: Data from N. J. Berrill, *Growth, Development, and Pattern,* Freeman, San Francisco, 1961.

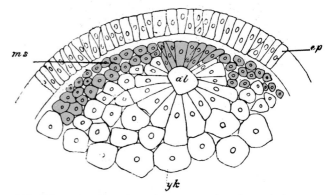

FIGURE 19.25. Drawing of cross section through the gastrula (160 hours) of a cyclostome, *Petromyzon planeri*. al, Gastrocoel (alimentary tract) lined by mesenteron; ep, epiblast (ectoderm); ms, mesoblast (mesoderm); yk, yolky cells. From F. M. Balfour, *Comparative embryology*, Vol. II, Macmillan, London, 1881.

surface cells and by their epiboly over the larger yolky cells.

The **mesoblast** or forerunner of the mesoderm is formed by small cells lying on either side of the archenteron which proliferate and spread between the endoderm and ectoderm. The notochord is later lifted out of the roof of the archenteron by the convergent movement of proliferating endodermal cells.

Primitive osteichthyes are also astonishingly froglike in their early development, but chondrostean and holostean (i.e., ganoid) gastrulas share some affinities with teleosts as well. Like amphibians, the yolk mass of primitive osteichthyes' eggs is carved into blastomeres, but only a dozen or so yolky blastomeres are completely segregated. These are arranged radially, like segments in an orange (Fig. 19.26), and, through unequal proliferation beneath the animal pole, give rise to **hypoblast** or early endodermal cells.

Lacking the teleostean syncytial periblast, gastrulas of primitive osteichthyes enclose their yolk via the epiboly of the small animal-pole cells. As a result, the yolk comes to lie directly in the lumen of the gut where it is broken down and subsequently digested (see Fig. 24.2).

On the other hand, like a teleost, the holostean bowfin, *Amia calva*, forms its **mesoblast** or mesoderm precursor from deep blastodermal epiblast cells without invagination, involution, or ingression from the surface cells. The outer two cell layers of the **enveloping ectoderm** (also called epiblast, outer enveloping epithelial layer, cellular envelope, and *Deckschicht* by analogy with teleosts)

gives rise only to ectoderm. Cells of the central nervous system, eyes, and ear vesicles are derived from the basal layer of the enveloping ectoderm without involution or invagination, while the apical layer forms epidermis exclusively (Ballard, 1984).

Amphibian gastrulation seems to represent an evolutionary compromise. Working against yolk, rather than around it, amphibians substitute one method of moving cells for another (see Elinson, 1987).

Gastrulation begins with cellular ingression. Throughout the yolky vegetal hemisphere, occasional individual endodermal cells disappear from the surface. Undergoing **apical constriction** or narrowing at their bases, the cells acquire the **flask-shaped** appearance of **bottle cells** (Fig. 19.27).

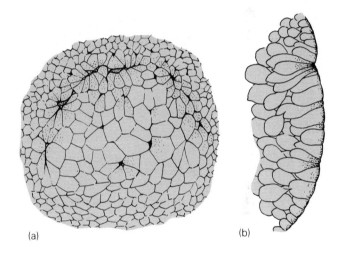

(a)　　　(b)

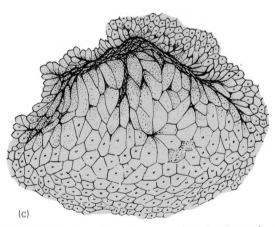

(c)

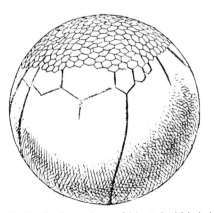

FIGURE 19.26. Surface view of blastula (third day) of the ganoid, *Lepidosteus*, about 3 mm in diameter. The large radially arranged yolky segments proliferate endodermal cells beneath the cap of small animal-pole cells. From F. M. Balfour, *Comparative embryology*, Vol. II, Macmillan, London, 1881.

FIGURE 19.27. Vegetal surface (a, c) and schematic section (b) perpendicular to surface showing apical contraction and formation of flask-shaped vegetal cells. The blastoporal crescent (c) results from collective cellular ingression. From J. Holtfreter, *J. Exp. Zool.,* 94:261(1943), by permission of Alan R. Liss, Inc., and the author.

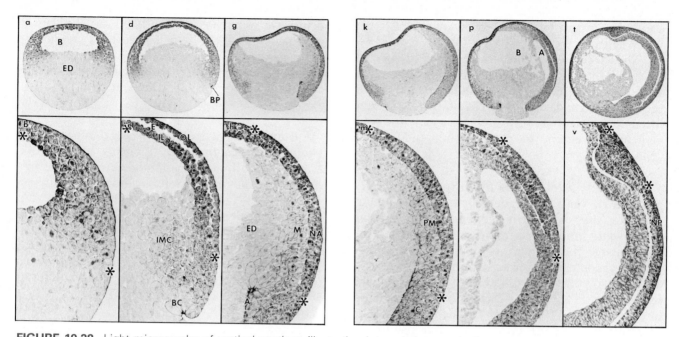

FIGURE 19.28. Light micrographs of vertical sections illustrating internal changes in the frog, *Xenopus laevis*, during gastrulation. (a–b) Stage 9; (d–e) stage 10$^+$; (g–h) stage 11; (k–m) stage 11.5; (p–r) stage 12; (t–v) stage 14; (a, d, g, k, p, t) ×22. (b, e, h, m, r, v ×68. A, archenteron; B, blastocoel; BC, bottle cell; BP, blastopore; ED, endoderm; IL, inner layer of ectoderm; IMC, inner marginal cell; M, mesoderm; NA, neurogenic area; NP, neural plate; OL, outer layer of ectoderm; PM, prechordal mesoderm. Asterisks identify marginal belt of RNA-rich neurogenic cells. From H. Imoh, *Dev. Growth Differ.*, 27:1(1985), by permission of the Japanese Society of Developmental Biologists and the author.

The cell's contracted bases appear as randomly distributed pigmented dots in the pale yolky endoderm at the blastula's surface. The vegetal cells' ingression causes the **vegetal plate** or **base** to contract and flatten on the outside and bulge into the blastocoel on the inside (Fig. 19.28, compare blastocoel in a to d), but not much endoderm actually moves into the embryo.

Invagination comes next in the scheme of gastrulation. As if the ingressing cells decided to "pull" together, new dots marking additional points of ingression congregate in a crescent-shaped **blastoporal pigment line** that invaginates collectively into a **blastopore** or **blastoporal groove** (Figs. 19.27 and 19.28d–e) and elongates toward the center of the embryo.

The force for invagination may be generated by individually ingressing cells, but the coordination of the mass movement depends on the tight junctions reinforced by terminal webs that bind the cells' apical surface. This **surface coat** of the outer part of cells is so tough that it can be removed as a whole and was once thought to be a separate extracellular layer.

Invagination continues at the ends of the crescentic blastopore, extending it laterally and ventrally. The blastopore then circles back around a **yolk plug** of large, yolky, relatively inert endoderm cells. Upon completing this **circular blastopore**, invagination as such comes to an end.

Actually, very little movement of cells has been accomplished by mere invagination. The blastopore (BP, Fig. 19.28d) would hardly be more than a circular dimple in the surface of the gastrula were it not for involution. The great bulk of embryonic tissue taking up residence in the interior of the embryo is not invaginated but involuted around the lip of the blastopore.

Involution, or the movement of cells around an edge, supported by epiboly, or the movement of cells over a surface (toward an edge, the blastoporal lip, in this case), and followed by **extension** (sometimes called **spreading** or respreading), or movements over an internal surface (away from the lip), are the chief mechanisms of amphibian gastrulation. They come into play immediately after invagination provides a lip around which cells involute, and they move the bulk of cells for about a day until gastrulation is virtually complete (see Gerhart and Keller, 1986).

Epiboly accounts for the stretching of ectoderm over the entire surface once shared by cells belonging to other presumptive germ layers (see Fig. 14.7). Notochordal and mesodermal cells as well as foregut and hindgut endodermal cells are brought to the blastoporal lip via epiboly and move into the gastrula via involution. Moreover, bound tenaciously to each other, involuting cells "pull and tug" the vegetal plate anteriorly and lift the yolky endodermal cells into the floor of the archenteron (Fig. 19.29 second column from left).

Involution gets the bulk of cells inside the gastrula, but extension or spreading brings them to

their final positions. After turning the corner of the blastoporal lip, bottle cells (BC in Fig. 19.29) migrate anteriorly along a layer of laminin and fibronectin lining the roof of the blastocoel. The narrow slit initially created by the cells' migration (Fig. 19.28g–h,k–m) swells as it approaches the remnant of the blastocoel into the massive **gastrocoel** within the **archenteron** (A, Fig. 19.28). Ultimately, while confined to its corner of the embryo by the **completion bridge** (i.e., the thin band of cells between the gastrocoel and blastocoel), the blastocoel collapses, and the gastrocoel becomes the major cavity of the gastrula.

The surface coat, drawn in by the involuting cells as they turn around the blastoporal lips, now

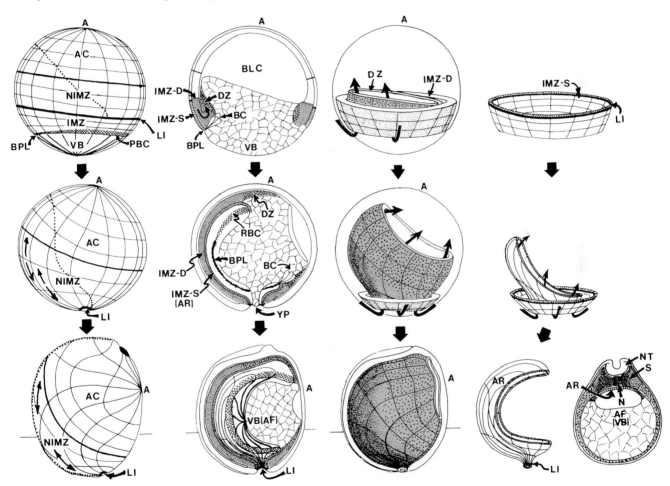

FIGURE 19.29. Diagrams illustrating gastrulation in different layers of *Xenopus laevis*. Top row: early gastrula; middle row: late gastrula; lower row: late neurula. First column (left): ectoderm; second column: midsagittal sections; third column: the involuting marginal zone deep cells (IMZ-D). Lower right insert: cross section of neurula. A, animal pole; AF, archenteron floor; AR, archenteron roof; BLC, blastocoel; IMZ, involuting marginal zone; IMZ-D, deep layer of IMZ; IMZ-S, superficial layer of IMZ; LI, limit of involution; NT, notochord; S, somite mesoderm; YP, yolk plug. Arrows indicate direction of movement. From the *Annual Review of Cell Biology*, Vol. 2 (1986), by permission of Annual Reviews Inc. Courtesy of J. Gerhart and R. Keller.

lines the archenteron (Fig. 19.28g,h,k,m,p,r,t,v). The walls and roof of the archenteron are composed of cells that were formerly parts of the **marginal zone** (MZ) of the blastula and early gastrula, but exactly which presumptive region occupies which part of the archenteron differs according to species, the thickness of the archenteric wall, and time.

The thin marginal zone of urodeles contains cells that are either in or capable of moving into the surface layer bound by the surface coat. Upon involuting, the presumptive notochord and mesoderm regions form the **mesodermal mantle** lining the roof and walls of the archenteron (Fig. 19.30). Endoderm only takes over the lining of the archenteron slowly and does not finish the job until the neural tube stage (Fig. 19.31).

The situation is different in anurans where the marginal zone of the blastula is multilayered. Involuting cells called **outer** or **superficial cells** are attached to the surface coat, while involuting cells called **inner** or **deep cells** are not attached to the surface coat. The **involuting marginal zone** (IMZ) includes **deep** (IMZ-D) cells of the presumptive mesoderm and **superficial** (IMZ-S) cells of the presumptive endoderm (Fig. 19.29).

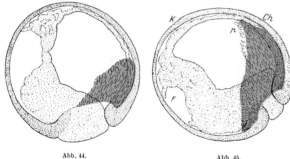

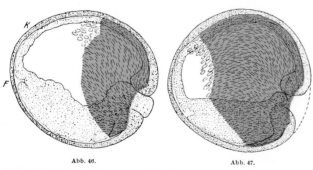

FIGURE 19.30. Diagrams of optical sections through gastrulas of the ribbed newt, *Pleurodeles waltli,* illustrating movement of mesodermal mantle in wall and roof of archenteron. From W. Vogt, *W. Roux' Arch. Dev. Biol.,* 120:385(1929), by permission of Springer-Verlag.

In the ranid anurans, IMZ-S endodermal cells come to line the anterior half or so of the archenteron and have only to replace mesoderm on the archenteron's posterior roof and walls. In *Xenopus laevis*, all the marginal zone's presumptive mesoderm is in IMZ-D cells. No mesodermal cells appear in the lining of the archenteron. Endoderm alone is present throughout the archenteron (AR and AF in Fig. 19–29) (Keller, 1976).

GASTRULATION FOLLOWING SPIRAL CLEAVAGE

Many of the guesses that have been made in the past about the ultimate fates of embryonic cells are correct in general but wrong in detail: the assignment of cell function follows certain broad rules to which there are numerous exceptions.

J. E. Sulston, E. Schierenberg, J. G. White, and J. N. Thomson (1983, p. 110)

Gastrulation tends to take place early in spiralian development, even during cleavage, when presumptive germ layers are represented by single cells or small groups of cells. Endoderm may invaginate and form a blastopore that expands into the lumen of an archenteron (Fig. 19.32), but all that invaginates is not necessarily endoderm. Stomodea and proctodea, pseudoblastopores and pseudoarchenterons lined by ectoderm are also formed by invagination (Fig. 19.33).

In relatively yolky eggs, epiauxesis or epiboly (Løvtrup, 1974) may spread a layer of ectoderm over the gastrula (Fig. 19.34), thereby internalizing cells without their taking an active part in the process. Even apparent inward turning at the edge of a cellular sheet (Fig. 19.35) is probably due to cellular rearrangement rather than involution. In less yolky eggs, individual cell movement or ingression rather than mass cell movement characterizes gastrulation (Fig. 19.36).

Monet Pattern

Gastrulation begins during cleavage and is largely due to cellular rearrangement, ingression, and epiauxesis. As ectoderm spreads over the surface, preendodermal cells are internalized without forming a blastopore or archenteron. Only a transient ventral depression or slit may accompany the internalization of premuscle.

Nematode gastrulation, in general (see Fig. 13.23), ***seems to take place with the fewest num-***

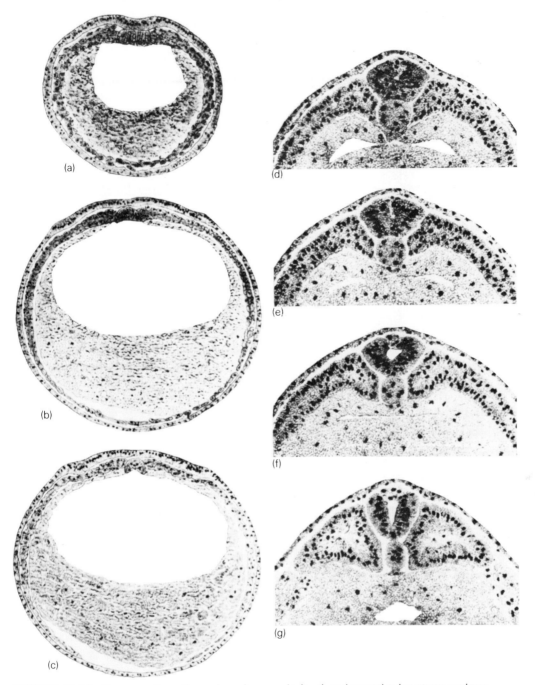

FIGURE 19.31. Cross sections through early neurula (a–c) and neural tube stage embryo (d–g) of the newt, *Plurodeles waltli*, showing the closing of endoderm over the roof of the archenteron. From W. Vogt, *W. Roux' Arch. Dev. Biol.,* 120:385(1929), by permission of Springer-Verlag.

ber of cells possible. In *Caenorhabditis elegans*, gastrulation begins at about 90–110 minutes after the first cleavage when the embryo is at the 28-cell stage and the P_3 cell has just divided into the D and P_4 founder cells (see Fig. 11.56). The presumptive anterior and posterior endodermal cells (Ea and Ep in Fig. 19.36b), near the posterior end of

the embryo, sink inward from the ventral side, followed by the ingression of the primordial germ cell, P_4.

Pharyngeal precursors originating from the MS founder cell ingress en masse and open an **entry zone** (cleft or slit, actually a depression) in the ventral surface (arrow Fig. 19.36). The zone widens and

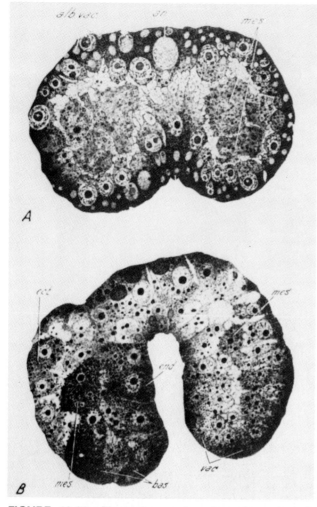

FIGURE 19.32. Photomicrographs of sections through early (a) and late (b) gastrulas of the pond snail, *Limnaea stagnalis*. Invagination depends on pseudopods from cells at the tip of the future archenteron. The gastrula shown in (b) has a normal morphology, although it has a high concentration of basophilic yolk granules (bas.) on the left as a result of centrifugation prior to cleavage. From Chr. P. Raven, *Morphogenesis: The analysis of molluscan development*, Pergamon Press, Oxford, 1966. Used by permission.

of large, granular **hypodermal cells** derived from the AB cell with a contribution from the C founder cell.

Cirripedia gastrulation begins with the fifth cleavage division when the presumptive mesodermal cell migrates inward beneath the ventral ectoderm (Fig. 19.37). The single, large, yolky 4D presumptive midgut cell is still exposed on the posteroventral surface. 4D undergoes its first bilateral division into two yolky cells and then unequal division into two yolk-rich anterior midgut cells and two yolk-poor posterior midgut cells.

The ectoderm is formed by the yolk-poor derivatives of the first three monets (1d–3d). Since little surface is vacated by the ingression of the presumptive endoderm and mesoderm, little epiauxesis takes place.

The barnacle's stereogastrula (Fig. 19.37) is reminiscent of the cellular blastoderm of other crustaceans even if their yolk mass is not divided into cells. Presumptive mesodermal cells are juxtaposed to presumptive midgut endodermal cells while still on the surface, but they ingress independently (see Anderson, 1966).

Duet Pattern

Acoelous flat worms lacking a gut do not form an archenteron of any sort. Progenies of the first, second, and probably third duets of micromeres form ectoderm by spreading over the fourth duet of micromeres and the fourth generation macromeres. Along with some peripheral parenchyma formed by the third and fourth duets of micromeres, the internalized macromeres give rise to the internal mass of the adult (see Costello and Henley, 1976).

Quartet Pattern

The linchpin for understanding gastrulation in annelids, mollusks, and related phyla is the amount of yolk and the degree of unequal cleavage. The host of variations set into motion by these variables is impressive.

Annelids and hirudineans internalize their mesoderm by individual cell migration (e.g., Fig. 19.38a, M-cell, b) and their endoderm by unipolar ingression (19.38c, d sometimes considered delamination) or occasionally by invagination (Fig. 19.38a). The volume of the blastocoel and the size of blastomeres influence the mode of gastrulation.

lengthens posteriorly to incorporate body premyoblasts (i.e., muscle-forming cells) derived from the C and D founder cells and extends anteriorly to include buccal cavity precursors originating from the AB founder cell.

Gastrulation ends with the disappearance of the entry zone in a posterior-to-anterior direction (at 250 minutes), but the last two surface premyoblasts undergo their terminal division and sink inward later (at 290 minutes). Internally, as alimentary track precursor cells become arranged linearly, they are surrounded by body premyoblasts. Externally, the embryo is covered by a layer

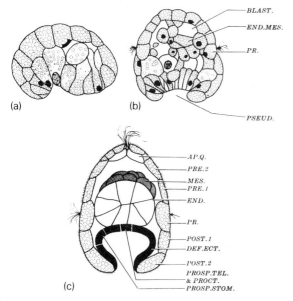

(a)   (b)

BLAST.
END.MES.
PR.

PSEUD.

AP.Q.
PRE.2
MES.
PRE.1
END.

PR.

POST.1
DEF.ECT.
POST.2
PROSP.TEL.
& PROCT.
PROSP.STOM.

(c)

FIGURE 19.33. Schematic sections through early (a) and late (b) gastrulas and hatched trochophore (c) of the wormlike *Neomenia* (Aplacophora). The pseudoblastopore (PSEUD.) of the gastrula is lined by a common pseudoarchenteron (black) containing the definitive ectoderm (DEF.ECT.) of the adult trunk, prospective stomodeum (PROSP. STOM.), telotroch (TEL., caudal organ), and proctodeum (PROCT.). AP.Q., apical quartet; BLAST, blastocoel; END., endoderm; END.MES., endomesoblastic cells; MES., mesoderm; PRE.1,2, pretrochal tiers of test cells; PR., prototroch; POST.1,2, posttrochal tiers of test cells. From T. E. Thompson, The development of *Neomenia carinata* Tullberg (Mollusca, Aplacophora). *Proc. R. Soc. London Ser. B,* 153:263–278(1960). Used by permission of The Royal Society, London, and the author.

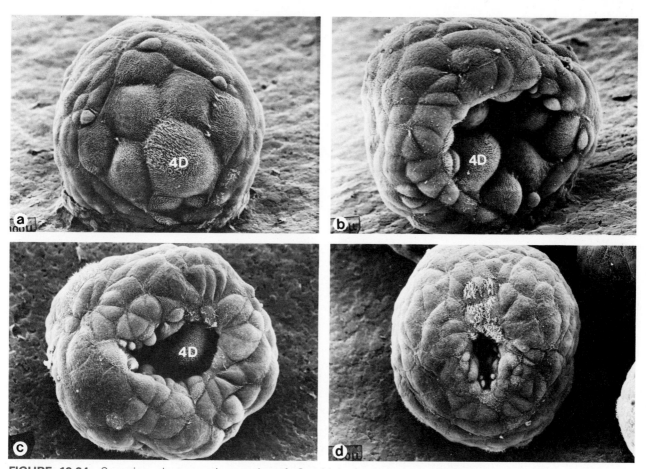

FIGURE 19.34. Scanning electron micrographs of *Crepidula fornicata* gastrula during epiauxesis of ectodermal cells. From N. H. Verdonk and van den J. A. M. van den Biggelaar, in N. H. Verdonk, van den J. A. M. van den Biggelaar, and A. S. Tompa, eds., *The mollusca*, Vol. 3, *Development*, Academic Press, Orlando, 1983. Courtesy of the authors.

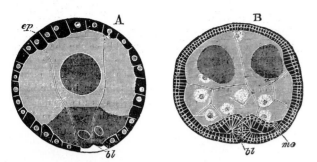

FIGURE 19.35. Diagrams of longitudinal sections through gastrulas of the marine worm, *Bonellia* (Echiurida). bl, blastopore; ep, epiblast; me, mesoblast. From F. M. Balfour, *Comparative embryology*, Vol. I, Macmillan, London, 1880.

In general, the 3A, 3B, 3C, and 4D cells are the stem cells of the presumptive midgut endoderm. In advanced oligochaetes only CD forms the midgut, and in primitive hirudineans only 3A–3C become endoderm. In advanced hirudineans the endoderm forms from 3c, 3d, and 4d. Where presumptive midgut cells represent only a small part of the surface, they may be replaced by small lateral movements of surrounding cells (Fig. 19.38b, d). More extensive epiauxesis takes place where presumptive midgut cells occupy relatively large parts of the blastula's surface (e.g., the oligochaetes, see Fig. 14.18).

Ectoderm forms from micromeres, but in oligochaetes lacking a trochophore entirely (e.g., the lumbricids), AB forms no micromeres and performs merely a nutritive function. Spreading over the surface, the progeny of the 2d cell or **primary somatoblast** gives rise to **ectoteloblasts** and to **ectoblast bands** or **rings** that form the trunk ectoderm and nervous system.

The 4d **second somatoblast,** the **mesoderm stem cell** or **teloblast,** ingresses either directly or in the form of its progeny to establish mesodermal

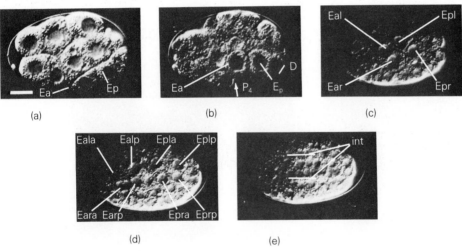

FIGURE 19.36. Differential interference contrast (DIC, also known as Nomarski optics) microflash photomicrographs of living *Caenorhabditis elegans* embryos. The embryos are oriented with their anterior ends toward the left, and their dorsal surfaces toward the top. Embryonic stages 1–5 (a–e) are identified by the numbers (2–16) and positions of E cells (presumptive endoderm or intestinal precursor cells), not all of which may be present in a single optical plane. Times refer to the duration of stages in embryos cultured at 25° C. Bar = 10 μm. (a) Stage 1. Pregastrula (cleavage). 1–28 cells. 110 min. Two E cells (anterior, Ea and posterior, Ep) come to lie on the ventral surface. (b) Stage 2. 2 E. 28–44 cells. 10 min. Gastrulation begins with the dorsal (upward) and inward migration of the anterior Ea and posterior Ep cells. Their disappearance from the surface leaves a ventral depression (arrow). The primordial germ cell, P_4, and premyoblasts derived from the D founder cell follow the E cells into the interior. (c) Stage 3. 4 E. 46–93 cells. 40 min. A bilateral division of E cells produces left and right sibling cells (Eal, Ear, Epl, Epr). (d) Stage 4. 8 E. 94–204 cells. 80 min. A longitudinal division of E cells produces another set of anterior (a) and posterior (p) cells. The MS mesodermal founder cell line has also entered the embryo at this time, leaving a ventral cleft (not in plane of micrograph). (e) Stage 5. 16 E. 205 to about 500 cells. E cells form cylindrical intestinal rudiment (int). Gastrulation virtually ends and organogenesis begins in the embryo at about 250 cells. Unlabeled photomicrographs courtesy of Einhard Schierenberg. Labels based on Sulstrom et al., 1983. Cell numbers, stages, and durations from Schierenberg, 1982.

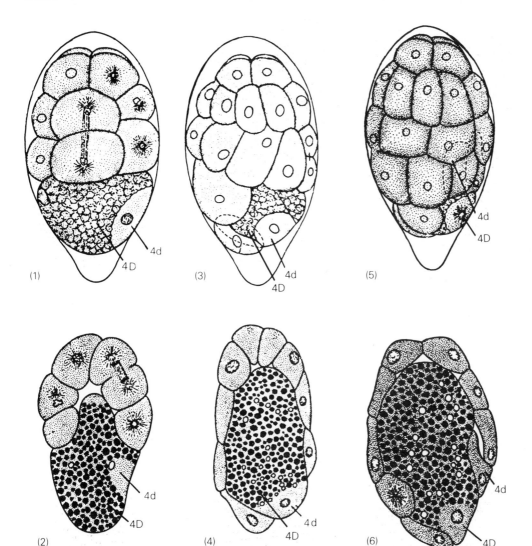

FIGURE 19.37. Schematic surface (1, 2, 3) and sagittal sections (4, 5, 6) of the gastrulating thoracian barnacle, *Tetraclita rosea*. Adapted from D. T. Anderson, *Embryology and phylogeny in annelids and arthropods*. Pergamon Press, Oxford, 1973.

bands (see Fig. 11.49). Similarly, in primitive hirudineans (i.e., leeches), 3D divides equally to give rise to bilateral mesodermal teloblasts which migrate inward and are covered by spreading ectoblast bands derived from 2d.

In advanced oligochaetes and hirudineans (collectively known as the clitellates, see Fig. 14.18e,f), which tend to more direct development, the 3D-equivalent (M) forms equal secondary somatoblasts. They form most of the adult's internal structure after being internalized by spreading germinal bands derived from the 2d-equivalent (DNOPQ, see Fig. 11.43).

Mollusks gastrulate with relatively few cells (Table 19.2). With the exception of involution,

most modes of cellular movement are employed in gastrulation.

Epiauxesis by **ectoblasts** is the predominant mode of ectoderm formation (Fig. 19.34). The 2d

TABLE 19.2. Number of cells present at the beginning of gastrulation in some mollusks

Species	Number of Cells
Paludina	80
Lymnaea	120
Littorina	150
Physa	200

Source: Data from Verdonk and van den Biggelaar (1983).

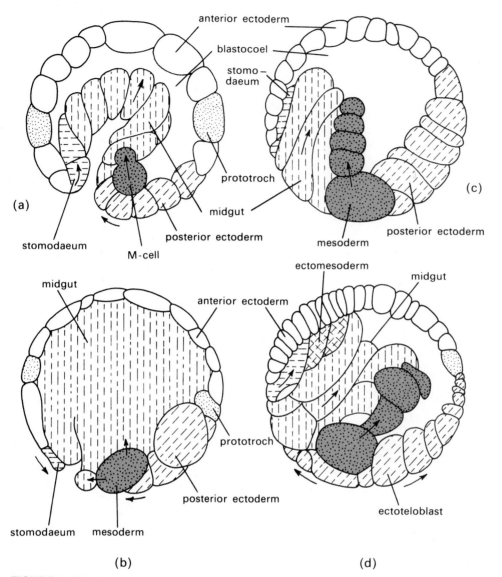

FIGURE 19.38. Diagrams of slightly parasagittal sections of polychaete gastrulas illustrating relationship of cell size and blastocoel size to invagination and cellular rearrangement during gastrulation. (a) *Eupomatus*. The large blastocoel accommodates an invaginated archenteron. (b) *Neanthes*. In yolk-filled eggs, endoderm is internalized by the epiboly of ectoderm, while mesoderm migrates inward independently. (c, d) *Scoloplos*. A modest blastocoel accommodates a solid midgut core formed by ingression and individual cell migration. From D. T. Anderson, *Embryology and phylogeny in annelids and arthropods*, Pergamon Press, Oxford 1973. Used by permission.

micromere, known as the **first somatoblast,** in particular, gives rise to the **somatic plate** which becomes **trunk ectoderm** in the adult (e.g., lamellibranchs). In gastropods, the 2c micromere seems to share responsibility for the somatic plate.

Endoderm is derived from macromeres (4A–4D) and **endomeres** (4a–4c) plus additional cells (also called endomeres) cleaved from 4d, and an archenteron typically forms by invagination. Exceptions are very yolky eggs (e.g., *Teredo*, a no-

torious shipworm) which form a solid archenteron filled with yolky endoderm.

Mollusks with coeloblastulas (i.e., Archaeogastropoda, freshwater and most marine Lamellibranchiata, Polyplacophora, and Scaphopoda) invaginate via the apical constriction of macromeres resembling **primary invagination** in sea urchins. In contrast, endomeres in mollusks with stereoblastulas (i.e., some marine lamellibranchs and gastropods) or with flattened **placula** (i.e., gastropods) invaginate via pseudopods as in **secondary invagination** (Fig. 19.32) (see Raven 1966).

Invaginating macromeres are followed by endomeres. Derivatives of the third and fourth quartets line the rim of the blastopore, and ectodermal **stomatoblasts** of the second and third quartets invaginate to form the mouth. An exception is the prosobranch, *Paludina* (= *Viviparus*) *vivipara*, where the "stomodeal" invagination becomes the anus. Invagination is not accompanied by involution.

In the aplacophoran, *Neomenia* (a wormlike parasite of coelenterates), a **pseudoblastopore** invaginates (Fig. 19.33). Mesoderm and endoderm migrate out of the wall of the **pseudoarchenteron**, leaving a lining of proctodeal and stomodeal ectoderm that everts at metamorphosis to form the adult **trunk ectoderm** (see Cather, 1971).

The primary source of larval mesoderm or **ectomesoderm** is micromeres of the second (i.e., lamellibranchs and scaphopods) and third quartets (i.e., gastropods). Individual cells ingress from the surface and proliferate. Typically (e.g., *Crepidula*, see Fig. 11.40), the 4d cell (known as the **second somatoblast**) gives rise to the **endomesoderm** or **adult mesoderm**. Usually larger than the 4D "macromere," at some time between the 24- and 72-cell stages (e.g., 68-cell stage in *Crepidula*), 4d undergoes an equal bilateral division placing the **primary mesoblasts** $4d^1$ and $4d^2$ or M cells (Mr and Ml, sometimes MEr and MEl cells) on either side of the dorsal median plane (i.e., across the second cleavage furrow). Either the 4d cell itself or its derivatives then ingress individually.

In general, the M cells cleave off some **enteroblasts** or **endomeres** before becoming purely **mesoblasts** or **mesodermal teloblasts** (see Fig. 11.44).

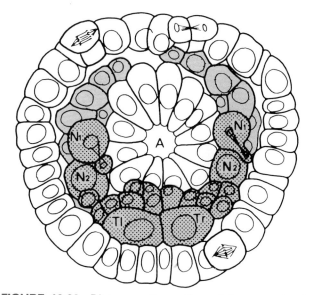

FIGURE 19.39. Diagrammatic section of the gastrula of *Physa fontinalis* showing archenteron (a) lined by endoderm and almost surrounded by mesoderm (colored, endomesoderm stippled, ectomesoderm, above). From N. H. Verdonk and van den J. A. M. van den Biggelaar, in N. H. Verdonk, van den J. A. M. van den Biggelaar, and A. S. Tompa, *The mollusca*, Vol. 3, *Development*, Academic Press, Orlando, 1983. Used by permission.

In species with well-developed mesoderm, proliferation produces enough mesoderm to nearly surround the archenteron by the end of gastrulation (Fig. 19.39) (see Verdonk and van den Biggelaar, 1983).

Numerous exceptions may be cited. In the gastropod, *Paludina* (= *Viviparus*) *vivipara*, for example, the 4d cell is an ordinary endoblast. In *Dentalium* and *Chiton*, $4d^1$ and $4d^2$ invaginate with the endomeres and remain temporarily in the archenteron.

Chapter

20

MECHANISMS AND CONTROLS
OF GASTRULATION

Gastrulation seems to be triggered by an endogenous timing mechanism set into motion at fertilization. Neither cytoplasmically based oscillators timing cleavage nor the titration of substances timing a midblastula transition seems to switch on gastrulation, since both can be perturbed without altering gastrulation's commencement (see Kirschner et al., 1985). The timer of gastrulation would have to be more global than cellular but what is it and how does it work?

The distinguishing features of gastrulas are cell numbers, cell movements, and cellular milieus produced through novel gene expression and the synthesis of new cell-surface and extracellular materials. How are these features related to the initiation of gastrulation?

CELL NUMBER
IN GASTRULATION

A handful of cells is all that is needed in some embryos to begin gastrulation (e.g., microlecithal *Caenorhabditis elegans*, see Fig. 19.36), while thousands of cells are required by other embryos

(e.g., lecithotrophic *Platycnemis pennipes*, see Fig. 18.6). For convenience, gastrulas with small cell numbers (SCN-gastrulas) can be separated from those with large cell numbers (LCN-gastrulas), although many gastrulas are of intermediate size.

Classification of Gastrulas
Based on Cell Number

In both LCN-gastrulas and SCN-gastrulas, the ability of some cells to move is correlated with their being approximately the small size of adult cells. Geometry dictates that these cells are produced after fewer divisions in the SCN-gastrulas of microlecithal eggs than in the LCN-gastrulas of megalecithal eggs. Of course, the cells acquire other properties at the same time, and their extracellular matrix presumably changes as well.

Gastrulation generally depends on the behavior of individual cells in SCN-gastrulas. The formation of mesoderm most frequently involves **individual cell movement** such as ingression, while endoderm and ectoderm are formed by combinations of ingression and cellular epiboly. In contrast, highly choreographed mass cell movements known as **morphogenic movements** (also morphogenetic movements) are characteristic of LCN-gastrulas.

Small numbers of **founder cells** produce clones of cells in SCN-gastrulas, while a variety of blastomeric lineages produce clonal domains or polyclonal domains in LCN-gastrulas (see Table 14.1). Founder cells may be identified by their position and even by cytological characteristics, but they are not generally **specified** by the region they come to occupy. Rather, they tend to be determined for specific histotypes before they arrive at their destination. Their descendants express a restricted range of potentialities, produce a narrow range of differentiated cell types, and have few developmental options.

Large numbers of cells in **founder cell groups** produce compartments or polyclonal compartments. Determination is controlled by intrinsic cellular events rather than extrinsic environmental influences. In LCN-gastrulas, these groups are determined later in development, and different aspects, such as axes of symmetry, may be determined at different times.

Occasionally, in SCN-gastrulas, determination of **lineally equivalent progenies** is stepwise, implying that incompletely determined cells receive some information from their environment. They may not be completely determined until they reach their final destination where they receive their mandate for specific courses of differentiation.

Some founding groups in LCN-gastrulas, on the other hand, are assemblages of indifferent and equivalent **location-specific progenies**. Determination is generally gradual, cumulative, and, to a large degree, **site specific.** Initially, the boundaries between regions are not sharp, but, by gastrulation, clones occupying limited regions are **compartmentally determined** for one or another compartment and **regionally restricted** or **locationally determined** (also called regionally specified) for locomotion to and adherence at a specific site.

Information about the destination of gastrulating cells may be a consequence of statistical variation in cell groups, or cells may be programmed or imprinted with information about their destination through intercellular interactions and influences emerging from elsewhere (see Davidson, 1986). As gastrulation begins, founding groups in LCN-gastrulas have only locational specificity, while founder cells in SCN-gastrulas are already broadly determined. Founding cells are only completely determined when they reach their site of differentiation.

Examples of SCN- and LCN-Gastrulas

SCN-gastrulas tend toward mosaic or determinate development. LCN-gastrulas, on the other hand, are at least initially more regulative or indeterminate, but most gastrulas exhibit features of both tendencies in different regions.

Nematodes, such as *Caenorhabditis elegans*, provide good examples of SCN-gastrulas with founder cells. Gastrulation is reduced, for the most part, to placing already specific founder cells in position where they meet their fates. The zygotic genome of some cell lineages is already differentially active when only a few hundred cells are present. However, linearly equivalent progenies of hypodermal cells are determined, at least in part, by position.

Sea urchin gastrulation is a mixture of characteristics associated with both SCN- and LCN-gastrulas. Like an SCN-gastrula, sea urchin gastrulas have an **autonomously differentiating** somatic-cell lineage in the form of the skeletogenic primary mesenchyme. Specific divisions produce the primary mesenchymal cells (PMCs) both *in situ* and *in vitro*. In addition, the anterior–posterior axis is transferred to the larva directly from the animal–vegetal axis of the egg.

On the other hand, like an LCN-gastrula, sea urchin gastrulas have cell lineages, such as the veg_1 cells, that are variable with respect to cell fate and remain pluripotential far into development. Likewise, secondary mesenchymal cells (SMCs) can become PMCs or cooperate with PMCs to make an integrated skeleton in experimentally PMC-deficient embryos (Ettensohn and McClay, 1988).

Regional specification also occurs in sea urchins in the formation of the stomodeum. Although animal-pole cells have a propensity to form a stomodeum, in general, its appearance is a function of prior contact with the invaginating archenteron.

Eutherian ·mammals segregate the inner cell mass (ICM) and trophectoderm (TE) with earmarks of both locational specification and founder cell determination. To the degree that this segregation is purely a function of the position of cells in the morula (see Fig. 19.4), locational specification would seem responsible, but the early transcriptional activity of cleaving blastomeres smacks of founder cell determination.

Like SCN-gastrulas, mammalian embryos are derived from a few cells (i.e., as few as 3 ICM cells in the 64-cell embryo), but like LCN-gastrulas, most cells in the eutherian ICM do not seem to be precociously determined. The trophectoderm and the primary endoderm delaminating from the ICM prior to implantation (i.e., at about 4 days p.c. in the mouse), however, seem to be determined prior to gastrulation.

Amphibian gastrulas are LCN-gastrulas. Some *Xenopus laevis* cells begin moving when about 10^4 cells are present, and cells begin moving even later in amphibians with larger eggs. Gastrulation follows a few divisions later.

Like other animals with LCN-gastrulas, amphibians have regulative development, but their early embryos also possess a degree of regional specificity, and the fates of cells in different regions are predictable. Still, amphibian structures and tissues are not generally derived by linear descent from individual progenitor cells (see A. G. Jacobson, 1984).

The amphibian strategy for producing a tadpole is one of flexibility and power. Development is not restricted by lineage-specific programs, blastomeric ancestry, or cleavage geometry, and different numbers of cells can contribute to given structures. The same strategy seems to be utilized by amniotes.

CELL MOVEMENT DURING GASTRULATION

In the gastrula, intracellular and extracellular factors take responsibility for directing cells, but cells are also part of their own environment. Differences between the capabilities of individual cells and organized groups are obvious in the different accomplishments of cells *in vitro* as opposed to cells *in situ*. The challenge for the embryologist is turning the obvious into a hypothesis.

Years before Vogt plotted the first amphibian fate map in 1929, his vital-dye experiments showed that cells moved along regular tracks during gastrulation. In SCN-gastrulas, individual cells move, but in LCN-gastrulas, individual cells may march en masse or a mass of cells may move as if an individual. These movements may be active and self-generated as opposed to passive and under the control of extracellular forces.

Active Individual Cell Movement

The possibility that cells actively orient and direct their movement in response to heterogeneities in the environment was suggested by Paul Weiss on the basis of results with cells *in vitro* (i.e.,

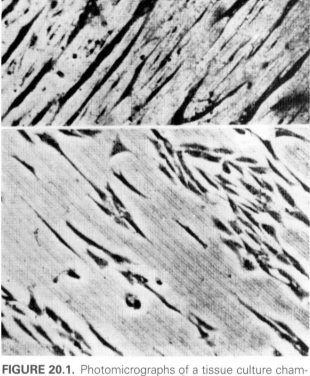

FIGURE 20.1. Photomicrographs of a tissue culture chamber showing 6-day embryonic chick heart cells 4 days after plating on a fish scale or etched glass. Hematoxylin stain. From P. A. Weiss, *Int. Rev. Cytol.,* 7:391(1958), by permission of Academic Press and the author.

in tissue culture) (see Weiss, 1958). He plated embryonic cells on natural (e.g., fish scales) and unnatural (e.g., glass) substrates and observed the cells stretching and moving along edges in both kinds of substrata (Fig. 20.1).

Michael Abercrombie pioneered the study of cell movement *in vitro* and described several relevant cell organs and behaviors. The cells' locomotory organs *in vitro* are their **lamellipodia** or **ruffled** (undulating) **membranes** (RM) and **filopodia** that sometimes jut from the lamellipodia (see below, Fig. 26.24).

Stationary cells seem to be "stuck" to the substratum at **adhesion plaques** (also called focal adhesions and focal plaques) in lamellipodia where the plasma membrane achieves its closest approximation to the substratum (10–15 nm). Microscopic **stress** (or tension) **fibers** containing bundles of actin are anchored at adhesion plaques and stretch from one plaque to another (Fig. 20.6).

Rapidly migrating cells *in vitro* lack adhesion plaques and stress fibers and presumably have less highly ordered arrays of microfilaments (see Burridge et al., 1987). Contacts with the substratum are made continuously under the leading edge of the advancing lamellipodium.

Adhesion plaques are not only sites of mechanical linkage between the cell and its *in vitro* substratum, but regions of communication between the cell and its environment. The cell's *in vitro* extracellular substrate is linked to the cell's cytoskeletal proteins at adhesion plaques in a **transmembrane chain of attachment.** Contact is made at the extracellular domain of integral membrane receptors (or integrins, see below) which bind **talin** (a 215-kd glycoprotein from cardiac and smooth muscle = p-235 from platelets) and, in turn, cytoskeletal associated proteins (e.g., vinculin and α-actinin) mediating attachment to **actin** and other elements of the cytoskeleton.

The contraction of actin provides the force for cell movement; resistance by adhesion plaques may determine direction. The dynamics of cytoskeletal–plaque interactions offer many opportunities for control. Extracellular **mechanoregulatory molecules** may promote plaque formation, and surface materials that "fit like needles in the eye of intracellular proteins" may orient actin bundles (see Buck et al., 1985).

Adhesion plaques may also be sites where regulatory signals are transmitted, especially since plaques contain various tyrosine kinases and calcium-dependent protease. A calcium-dependent talin protease at plaques, for example, may disrupt cytoskeletal attachments and allow cells to redirect their movement (Beckerle et al., 1987).

Transcription may even be influenced by extracellular receptors or **morphoregulatory molecules** in plaques. The evidence is circumstantial, but chick fibroblasts transcribe the vinculin gene at faster rates when forming adhesion plaques than when not (Bendori et al., 1987).

Competition between factors inhibiting and promoting cells' locomotory organs seems to orient cells. Factors that promote RMs tend to attract or guide cells toward them, hence **contact guidance,** while factors that inhibit RMs tend to paralyze cells, hence **contact inhibition of movement** or **contact paralysis** (see Abercrombie, 1961; Harris, 1982).

Contact guidance is exhibited by freshly explanted cells and by some established lines of tissue culture cells. The phenomenon differs among cells and substrates. Some cells exhibiting contact guidance (e.g., human PMN tissue culture cells) show a persistence of locomotion. Other cells (e.g., BHK, CHF, CH), with greater cytoskeletal inflexibility, show different degrees of adhesiveness to a series of mechanical cues.

Adhesive blockage is not all-or-none. It is gradual. As a rule, the larger the environmental feature, the greater the impediment to cell movement.

The alteration of cellular behavior by topographical features of the environment may be due to sites of focal adhesion influencing microfilament orientation within cells. Alternatively, topographical features may alter the probability of a cell's developing a successful protrusion and contacting the substratum in a given direction. Cell orientation and movement would then be determined stochastically (Clark et al., 1987, Dow et al., 1987).

A cell in a "rut" faces two sorts of environments. (1) Radially, large steps and sharp angles may limit cellular protrusion to short processes unable to exert much traction on the cell. (2) Longitudinally, cylinders with small radii of curvature may permit long cellular processes to form parallel to the surface feature. If isolated microfilament bundles in small areas have little influence on cellular orientation, topography may effectively trap a cell and direct its protrusions along long axes of extension.

Contact paralysis occurs when cells encounter each other. Among the heterogeneities in the *in vitro* environment that inhibit RMs are RMs of other cells.

In tissue culture, a normal fibroblast's RM (Fig. 20.2) ceases undulating upon meeting the RM of another fibroblast. The development of new RMs at opposite ends of the cells brings about mutual cellular withdrawal.

Relative degrees of inhibition and promotion represented by cell surfaces and noncellular substrates have the potential to move cells across an entire tissue culture dish. In dense tissue cultures of freshly explanted cells, mutual contacts become persistent zones of paralysis or **adhesiveness,** linking the cells in a confluent **monolayer.** Presumably, mass cell movements during gastrulation and the stasis of cells in new positions can also result from the relative degrees of contact guidance and contact inhibition offered by heterogeneities in cells' environments.

The RMs of some established cell lines and of tumor cells (e.g., sarcoma or connective tissue tu-

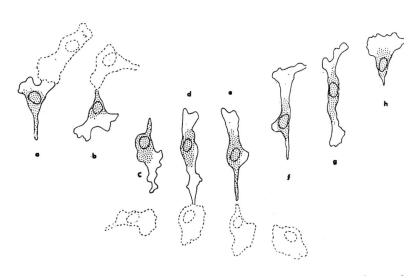

FIGURE 20.2. Tracings from a cinematographic record of chick mesenchyme cells (fibroblasts) in tissue culture. Interval = 4 hours. From P. A. Weiss *Int. Rev. Cytol.,* 7:391(1958), by permission of Academic Press and the author.

mor cells) do not show normal degrees of contact inhibition to each other. Instead, contact guidance allows these cells to pile up or **fasciculate** and produce dense cords of cells. Unlike the monolayer of freshly explanted cells, dense tissue cultures of tumor cells form multilayered nodules and fascicles.

Contact paralysis is linked to **contact inhibition of cell division,** the cessation of division in confluent cultures. Both may be determined by the same cell-surface morphoregulatory substance.

Various treatments and exposure to oncogenic viruses, such as Rous sarcoma virus, may cause the loss of contact inhibition of movement and of cell division in tissue culture. Cells are said to be **transformed.** They are rounder in cross section, lack adhesion plaques and stress fibers, produce less fibronectin (see below) and have less on their surfaces, resume movement and division, even in confluent cultures, and fasciculate.

In addition, the cells acquire the ability to divide *ad infinitum*. Ordinarily, freshly explanted cells divide for a limited number of generations, at which time the cells cease dividing. Like established tissue culture lines, transformed cells divide without constraint. They are said to be **immortalized.**

Freshly explanted fetal fibroblasts resemble normal adult fibroblasts *in vitro* by exhibiting contact paralysis, but the fetal cells show considerable overlapping in confluent cultures, more like virus-transformed cells. Fetal cells also resemble virus-transformed cells and fibroblasts from cancer patients more than freshly explanted normal adult fibroblasts in their abilities to form fascicled colonies when plated on monolayers of contact-inhibited epithelial cells and to migrate into three-dimensional collagen gels at high cell densities (see Schor and Schor, 1987).

In addition, fetal, transformed, and tumor cell-derived fibroblasts synthesize more hyaluronic acid, share increased agglutinability with concanavilin A, and have decreased requirements for serum *in vitro* compared to normal adult fibroblasts. Differences with adult fibroblasts may be accounted for by matrix synthesis in fetal and tumor fibroblasts, and the transition of fetal-type fibroblasts to adult-type fibroblasts may either follow from the expression of genes for new matrix materials or the selection for cells no longer synthesizing fetal-type matrix materials.

Passive Individual Cell Movement

The possibility that individual cell movement is passively controlled by extracellular materials was first suggested by H. V. Wilson (1907). He showed that disassociated sponge cells **reaggregated** and reformed new sponges and suggested that an extracellular "glue" was responsible. While ruffled membranes are now known to appear, and cell movements take place within aggregates, a "glue" or **extracellular material** or **matrix** (ECM) can still account for bringing cells together in the reaggregation masses (Humphreys, 1963).

The ECM may have the quality of **species specificity** (or affinity), since dissociated sponge cells in interspecific mixtures (of *Microciona* and *Haliclona* but not all species) reaggregate with members of their own species rather than with members of the other species (see Misevic and Burger, 1982). Similarly, cells isolated from sea urchin embryos reaggregate with species specificity (see Giudice, 1986).

Other examples of specific adhesiveness are not species specific. Various avian and mammalian fetal tissues reaggregate across species' lines, but these tissues exhibit **tissue-specific** or **histotypic adhesiveness** (see Monroy and Moscona, 1979).

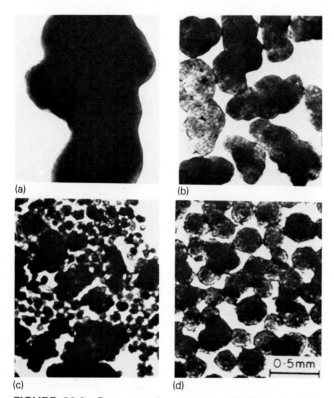

FIGURE 20.3. Reaggregation masses of different chick embryo tissues after 24 hours in gyratory cultures. (a) Liver cells from 7-day embryos; (b) neural retina cells; (c) mesonephros; (d) limb-bud from 4-day embryos. From A. A. Moscona, in E. N. Willmer, ed., *Cells and tissues in cultures,* Academic Press, Orlando, (1965). Used by permission.

Cells isolated from chick and mouse embryos following treatment with trypsin form characteristic **reaggregation masses** (Fig. 20.3) when cultured in conical gyratory flasks and gently rotated.

When mixed, the cells from birds and mice may form common reaggregation masses across species lines and proceed to **reassort** (reassociate or self-isolate) in **tissue-specific** patterns. Some combinations of mouse and chick cells may also form species-specific groups that later **reshuffle** into cross-species aggregates (Burdick, 1970). Like the species specificity of sponge-reaggregation masses, the tissue specificity of embryonic-cell reaggregates may be explained by ECM secreted into the cells' microenvironment.

Alternatively, reassortment may represent an equilibrium configuration of cells with **quantitative differences** in the adhesiveness of cell surfaces. The sizes and shapes of reaggregation masses may also be explained by configurations that minimize the surface free energy of adhesiveness (see Steinberg, 1964).

Tissue specificity and normal cell cohesive-ness may be built into cell-surface interactions through **glycosyltransferase** enzymes which assemble the complex carbohydrate-bound materials present on the surface of plasma membranes. By adding sugars one at a time from specific sugar nucleotides to the nonreducing terminals of polysaccharides, the glycosyltransferases may be instrumental in creating a variety of surface materials (e.g., lightly glycosylated proteoglycans and glycoproteins, heavily glycosylated glycosaminoglycans, glycolipids, gangliosides, blood group substances) (Roth et al., 1971).

Outside the cell, surface glycosyltransferases could be involved in cell-to-cell recognition and in binding with these complex "latchkey" glycoconjugates created by the enzymes (see Roth, 1979). For example, the reaction between β-1,4 galactosyltransferase (GalTase) and its lactosaminoglycan substratum (a complex glycoconjugate of repeating galactose and N-acetyllactosamine residues) is implicated in chick gastrulation, in the compaction of the mouse morula, the formation of the mouse blastocyst, the cohesion of embryonal carcinoma (EC) cells, and sperm–zona adhesion.

The evidence is immunological, morphological, and biochemical. (1) Rabbit anti-GalTase antibody reduces cell-to-cell contact. (2) GalTase is localized on the cell surface. (3) The GalTase regulatory protein α-lactalvumin (α-LA), which lowers the enzyme's affinity for terminal N-acetylglucosamine, and the noncompetitive GalTase inhibitor uridine diphosphodialdehyde inhibit cell contacts (Bayna et al., 1988). GalTase may act as a **surface receptor,** or cell-adhesion molecule, by binding lactosaminoglycans on adjacent cellular surfaces or by binding its substratum in the ECM.

Theoretically, cells would be stuck wherever they happen to be bound to their substrate. The cells would be liberated ("unstuck") by the completion of the enzymatic reaction. The incorporation of exogenous sugar nucleotides into cell surfaces during gastrulation supports this possibility.

Furthermore, the **sugar donor,** uridine diphosphate galactose (UDPGal), required for completing the reaction between GalTase and its lactosaminoglycan substratum, disrupts contacts among EC cells *in vitro*. In the absence of UDPGal, the enzymatic reaction fails to go to completion and cells remain stuck (Shur, 1983).

Active Mass Cell Movements

Movements that are active, in the sense of requiring changes in cells, and massed, in the sense of depending on intercellular relationships, may operate through **selective affinity** (or cohesiveness), linkages that both bind cells to each other and allow cells to exchange partners within a mass. The idea of active mass cell movements was first suggested by Johannes Holtfreter (1944) to explain observations on **bottle** or **flask-shaped cells** in grafted explants and during gastrulation.

After grafting a group of surface cells from the endodermal region near the blastopore of a frog gastrula to the blastocoelic surface of a mass of inner endodermal cells, Holtfreter observed the graft sink below the surface (Fig. 20.4). Bottle cells, like those formed at the blastopore (Fig. 19.27), extended proximal processes and squeezed between the inner endodermal cells, while distal processes remained embedded in the surface and pulled the graft or the lip of the blastopore inward. Since the cells were obviously more tightly bound at their apical surfaces than at their basal surfaces, they must have had different types of cell junctions or adhesiveness. But could differences in adhesiveness direct cell movement?

To answer the question, combinations of tissue fragments from all three germ layers (Holtfreter, 1939) and of aggregates of cells isolated from these layers (Townes and Holtfreter, 1955) were prepared. Not only did all the combinations pro-

duce cell movement, but, when present in particular proportions (i.e., with an amount of ectoderm sufficient to enclose the other cells), aggregates of all three germ layers reconstituted quasiembryos with an outer ectoderm, a partially or totally encapsulated endoderm, and an in-between mesenchymatous mesoderm (Fig. 20.5). Possibly, the germ-layer arrangement normally arrived at through gastrulation is likewise controlled by the selective affinity of cells en masse.

In active mass cell movements, cells are collectively subjugated to the mass and take their direction and rate of movement from the mass. For example, oriented cytoskeletal elements (Fig. 20.6) are restricted to the leading edge of cells of migrating cell sheets. Morphoregulatory molecules (e.g., receptors in the cell surface) would have to convey messages from cell to cell and transduce them to mechanoregulatory molecules (i.e., molecules like actin and myosin with mechanochemical functions) controlling cell shape and movement.

Passive Mass Cell Movements

Finally, cells may also take direction from the **differential adhesiveness** and changing adhesive-

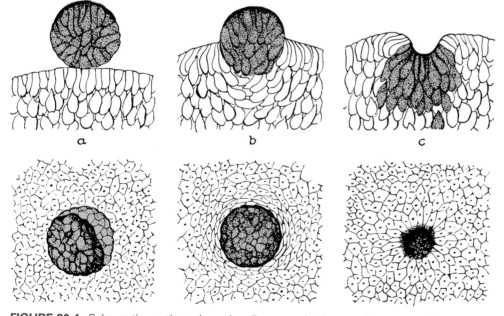

FIGURE 20.4. Schematic sections through cell masses in tissue culture illustrating results of grafting surface endoderm (stippled) from the region of the blastopore to blastocoelic lining of inner endoderm (open outlines). The surface endoderm is "coated" with "surface coat" (indicated by dark band) now known to consist of pigmented apical regions of cells bound together by zona occludens. The surface endoderm forms a "blastopore" as bottle-shaped cells pull the surface coat into the substratum of inner endoderm. From J. Holtfreter, *J. Exp. Zool.,* 95:171(1944), by permission of Alan R. Liss, Inc., and the author.

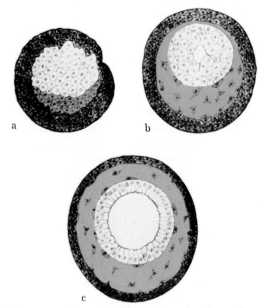

FIGURE 20.5. Diagrams illustrating the formation of a quasiembryo as the result of frog germ-layer reassortment. Cutaneous ectoderm (stippled) and endoderm (large open cells) segregate entirely as mesoderm (small stellate cells) appears as a mesenchyme in the fistula between endoderm and ectoderm. The endoderm forms columnar epithelium surrounding a cavity, while the ectoderm forms a two-layered epidermis. From J. Holtfreter, *Arch. Exp. Zellforschung,* 23:169, translated and reprinted in B. H. Willier and J. M. Openheimer, eds., *Foundations of experimental embryology,* Prentice-Hall, Englewood Cliffs, NJ, 1961.

ness of their substratum (see Johnson, 1970). Substrate adhesive molecules (SAMs, see below) may offer differentially adhesive surfaces by varying in type and amount. Rarely are cells completely disaggregated by one blocking antibody, for example, suggesting that several SAMs are normally present in cell substrates.

Cells, as part of the substratum for other cells, also have a potential for altering adhesion through **cell-surface modulation.** Matrix receptors at cell surfaces, or integrins, are not necessarily tissue specific, and the same molecule may appear in the embryo at different positions and at different developmental stages. The integrins may be expressed simultaneously by cells of one type, and combinations of expression may differ among cell types.

Gastrulation may be affected in several ways by adhesive molecules. Loss of adhesive molecules may lead to the separation of cell layers and delamination. Acquisition of adhesive molecules may permit the recognition and binding of cells from different origins into composite structures. The

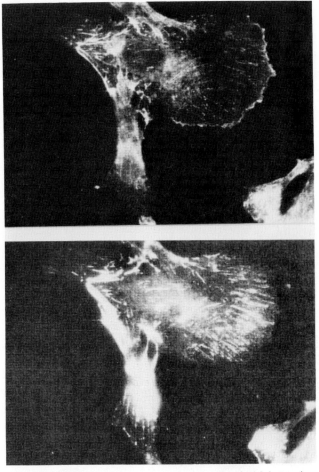

FIGURE 20.6. Epifluorescent micrograph of chick tendon fibroblast stained for the cell-substratum attachment glycoprotein complex antigen (CSATag, top) in the lamellipodia and the surface adhesion molecule fibronectin (bottom) in the *in vitro* substratum. Both antigens are concentrated at focal adhesion plaques at the convergence points of linear stress or tension fibers. Fibronectin is present outside the confines of the cell while CSATag is absent. CSATag is present in the thin ruffled membranes and other parts of the cell's ventral or endofacial surface where fibronectin is absent. From C. A. Buck, K. A. Knudsen, C. H. Damsky, C. L. Decker, R. R. Greggs, K. E. Duggan, D. Bozyczko, and A. F. Horwitz, in G. M. Edelman and J. P. Thiery, eds., *The cell in contact: Adhesions and junctions as morphogenetic determinants,* Wiley, New York, 1985.

timing of changes in adhesive molecules may determine developmental schedules, and the pattern of expression on the surfaces of individual cells may be responsible for arranging cells in particular patterns. Finally, progressive restriction in binding molecules may create discontinuous patterns of binding strength between layers and, in turn, shape germ layers and emerging organ systems.

MOLECULES OF ADHESION

Adhesion molecules are frequently identified by antibodies that block cell adhesion or dissociate adhering cells *in vitro*. Extracted from tissues, putative adhesion molecules neutralize the antibodies and remove the block to cell adhesion and allow cells to associate. The adhesion molecules are thought to play both dynamic morphoregulatory roles and static purely binding roles, but the classification of adhesion molecules has not reached the point of relating structure to function.

Presently, adhesive molecules fall into two broad classification schemes. One scheme is based on the ability of calcium ions (Ca^{2+}) to protect the molecule from enzymatic digestion. The other scheme emphasizes the distribution of molecules.

CIDS and CADS

CIDS are adhesion molecules that belong to a **Ca^{2+}-independent system. CADS** or **cadherins** are adhesion molecules that belong to a **Ca^{2+}-dependent system.**

CIDS are relatively resistant to proteolytic enzymes in the absence of Ca^{2+} and are not further protected by calcium, although LFA-1, a leukocyte-to-macrophage recognition protein, is a CIDS protein requiring Mg^{2+}. In their most conspicuous role, CIDS are the **integrins** that span the plasma membrane and link the cell's internal and external environments (see SAM-receptors below for further discussion of CIDS).

Cadherins are sensitive to proteolytic enzymes but are "protected" (i.e., not broken down) in the presence of calcium (Yoshida-Noro et al., 1984). Cadherins are also temperature dependent, suggesting that they may be more physiologically sensitive than CIDS. Generally, cadherins are similar to cell adhesion molecules or CAMs and junctional adhesion molecules or JAMs (see below).

Some cadherins are active in early embryos. Anticadherin antibodies decompact mouse morulas, perturb morphogenesis of cell layers and embryonic organs, and dissociate cells *in vitro* (see Takeichi et al., 1985). Different cadherins may be transient features of embryonic cells but characteristic of adult tissues. Cadherins also show unique patterns of spatiotemporal expression in embryos, but cadherins are not tissue specific in embryos as they are in adults.

P-cadherin is from the mouse placenta. Found first in extra embryonic layers, P-cadherin is present in the mouse's ectoplacental cone and visceral endoderm at implantation. Maternal decidual cells (but not luminal uterine epithelium) in the vicinity of implanting embryonic tissue also express P-cadherin. While not involved in the initial process of attachment, P-cadherin may connect the embryonic and maternal portions of the placenta.

Epithelial cadherins, or **E-cadherins,** include the 124-kilodalton (kd) glycoprotein of the F9 teratocarcinoma and the preimplantation mouse embryo and the 120-kd **uvomorulin** glycoprotein of blastomeres. In addition, Cell-CAM120/80 from human and mouse inner cell mass (ICM), canine Arc-1 from trophectoderm (TE), L-CAM from chicken liver cells, and the gp140 from *Xenopus laevis* resemble E-cadherin. They are found in embryonic ectoderm, endoderm, and, later, in most epithelial tissues as well as in many epithelial cell lines *in vitro*.

E-cadherin is expressed in the mouse zygote and blastomeres where it plays a role in compaction. At implantation it is expressed throughout the inner cell mass (but not extraembryonically), and it continues to be expressed in endoderm and most of the ectoderm, although it disappears from the neural ectoderm (including the neural crest) and from mesodermal cells migrating through the primitive streak.

N-cadherin from mouse neural tissue is related to A-CAM from chicken cardiac muscle intercalated disks and mesoderm, and N-Cal-CAM from chicken neural retina and notochord and older embryonic nervous, muscle, and epithelial tissues. It makes its initial appearance in the epiblast at the primitive streak stage.

As presumptive neural tube and mesodermal cells lose their E-cadherin, they express N-cadherin. Neural crest cells express N-cadherin transiently before they begin migration and after they reach their destination (e.g., in ganglia), but not while migrating. Presumptive endodermal cells, on the other hand, retain E-cadherin and lose N-cadherin.

cDNAs encoding mouse E-cadherin and uvomorulin, mouse P- and N-cadherin, chicken N-cadherins, and chicken L-CAM have been sequenced and their primary structure of 723–748 amino acids deduced. The molecules appear to be integral membrane proteins with an N-terminal extracellular domain capable of binding cells together by homophilic interactions in the presence of calcium. The molecules also contain a transmembrane hydrophobic domain and a cytoplasmic domain capable of binding actin bundles. The polypeptides are similar (between 50 and 65%), and those of E-cadherin and uvomorulin are identical (i.e., they can be referred to as uvomorulin/E-cadherin). The group seems to be encoded by a cadherin/L-CAM gene family with most heterogeneities occurring in the basal portion of the N-terminal extracellular domain (see Takeichi, 1988).

When L cells are transfected with E-cadherin cDNA joined to a virus promoter, E-cadherin is expressed and colony morphology changes from a dispersed to a compact pattern, suggestive of cadherin's ability to promote cell cohesion. Similar results occur with P- and N-cadherin cDNAs as well.

CAMs, SAMs, SAM-Receptors, and JAMs

Another classification scheme places adhesion molecules in one of three categories (or four when receptors are added). Large glycoprotein **cell adhesion molecules** (CAMs) on the membrane of cells carry out cell-to-cell interactions. **Substrate adhesion molecules** (SAMs, ECM adhesion molecules, or adhesion-promoting matrix proteins) mediate substrate-to-cell binding when joined to a cell by a **SAM-receptor**. **Junctional adhesion molecules** (JAMs) or **cell junctional molecules** (CJMs) appear at specialized cell junctions (see Edelman, 1986).

CAMs (cell adhesion molecules) comprise a broad class of cell-surface integral membrane proteins found in vertebrates (from sharks to humans) and in insects (e.g., grasshoppers and Drosophila). They may be mutually cohesive (i.e., homophilic) or adhesive to other materials (i.e., heterophilic).

CAMs include the mouse **neural cell adhesion molecule** (N-CAM), the chick **liver cell adhesion molecules** (L-CAM), and **neuron-glia cell adhesion molecule** (Ng-CAM). All three contain cytoplasmic regions with serine and threonine residues available for phosphorylation, hydrophobic transmembrane regions, and extracellular regions containing multiple asparagine-linked oligosaccharides (Fig. 20.7) (Edelman, 1986). Other CAMs include vertebrate **myelin-associated glycoprotein** (MAG), glial adhesion molecule, and, in insects, the **fasciclins** and **amalgam.**

Primary or **early (E-) CAMs** occur on mitotic cells, early in development, and in more than one germ layer. Secondary or **adult (A-) CAMs** occur only on postmitotic cells, later in development, and only in specific germ-layer derivatives. CAMs may undergo an **E-A conversion** in the course of development.

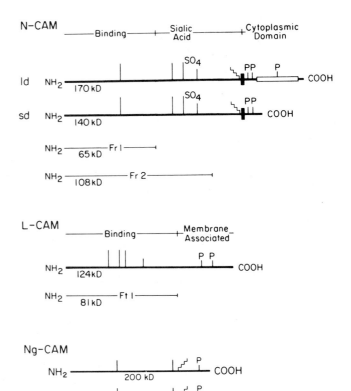

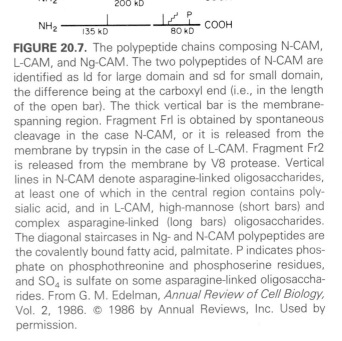

FIGURE 20.7. The polypeptide chains composing N-CAM, L-CAM, and Ng-CAM. The two polypeptides of N-CAM are identified as ld for large domain and sd for small domain, the difference being at the carboxyl end (i.e., in the length of the open bar). The thick vertical bar is the membrane-spanning region. Fragment Frl is obtained by spontaneous cleavage in the case N-CAM, or it is released from the membrane by trypsin in the case of L-CAM. Fragment Fr2 is released from the membrane by V8 protease. Vertical lines in N-CAM denote asparagine-linked oligosaccharides, at least one of which in the central region contains polysialic acid, and in L-CAM, high-mannose (short bars) and complex asparagine-linked (long bars) oligosaccharides. The diagonal staircases in Ng- and N-CAM polypeptides are the covalently bound fatty acid, palmitate. P indicates phosphate on phosphothreonine and phosphoserine residues, and SO_4 is sulfate on some asparagine-linked oligosaccharides. From G. M. Edelman, *Annual Review of Cell Biology,* Vol. 2, 1986. © 1986 by Annual Reviews, Inc. Used by permission.

Cells may modulate their CAMs at times of cell migration. For example, the concentrations of N-CAM and L-CAM in the chick blastoderm diminish greatly in cells after their ingression, but noningressing neural plate cells retain and increase their content of N-CAM while losing L-CAM completely (Fig. 20.8) (see Edelman et al., 1985; Thiery et al., 1985a).

By using standard DNA cloning techniques, cDNA clones for various CAMs have been isolated

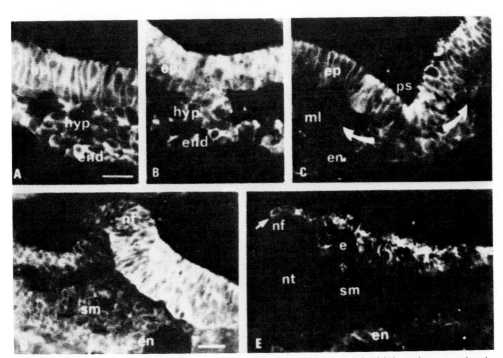

FIGURE 20.8. Epifluorescent micrographs of cross sections through chick embryos stained with antibodies to N-CAM and L-CAM. (a, b) At early notochordal plate stage 7, both N-CAM (A) and L-CAM (B) are present in the epiblast (ep), the endophyll (end), and hypoblast (hyp). (c) At the head fold stage 9, in the region of the primitive streak (ps), L-CAM staining diminishes as cells ingress through streak and disappears entirely as they migrate (arrows) in intermediate layer (ml) and lower definitive endoderm (en). N-CAM staining (not shown) is similar. (d) At the 10 somite stage, N-CAM staining has returned to germ layers including ectoderm (e), somite mesenchyme (sm), and endoderm (en), but it is especially strong in presumptive neural tube (nt; nf, neural fold). (e) L-CAM staining, on the other hand, while present in the ectoderm (e), is absent from the presumptive neural tube (nt, except over the neural folds, nf), somite mesenchyme (sm), and only weakly present in the endoderm (en). Bar = 30 μm. From G. M. Edelman, S. Hoffman, C.-M. Chuong, and B. A. Cunningham, in G. Edelman, ed., *Molecular determinants of animal form,* Alan R. Liss, Inc., New York, 1985. Used by permission.

and sequenced. The deduced amino acid sequences for coding regions frequently predict remarkable similarities in the proteins (e.g., vertebrate N-CAM, Ng-CAM, and insect fasciclins). Similarities in the extracellular region include repeats of immunoglobulin-like domains (Ig-like repeats of the C2 type [containing two cysteine residues separated by approximately 50 amino acids]) near the amino terminus, and of fibronectin-like domains (Fn-like repeats of the Fn type III) near the transmembrane region (see Harrelson and Goodman, 1988).

The Ig-like repeats are also found in vertebrate MAG and in insect amalgam. Possibly, a gene coding for Ig-like domains began as a cell adhesion molecule in primitive eumetazoans and through duplication and divergence expanded into an immunoglobulin superfamily that added functions in cell recognition and immunity.

L-CAM is specified by one mRNA but may be coded in two to three genes (Cunningham, 1985). It is a single polypeptide (110 kd) with a cytoplasmic region and an N-terminal binding domain containing three complex and one mannose-rich asparagine-linked oligosaccharide (for a total glycosylated molecular size of 124 kd).

Mouse N-CAM polypeptides seem to be specified by alternative splicing of mRNAs transcribed by a single gene on chromosome 9. Chick N-CAM polypeptide also seems to be encoded by a single gene.

N-CAMs are calcium independent and temperature independent (i.e., like CIDS). They form homophilic linkages (i.e., N-CAM to N-CAM linkages), although the larger molecules can also form heterophilic linkages (N-CAM to Ng-CAM).

The polypeptides of N-CAM differ near the carboxyl end of the membrane-associated region, and small minor components are reduced to extracellular peripheral proteins bound to membrane phospholipid. The N-terminal binding domain has an unusual middle region coupled to three **polysialic polymers** (with sialic-acid repeats at least 5 residues long) (Fig. 20.7).

The negatively charged 10 carbon sugar, sialic acid, determines N-CAM's conformation. Because electrostatic repulsion inhibits molecular binding in the N-terminal region, cells with less sialic acid in their polymers have greater abilities to adhere (Edelman, 1985).

While E-forms of N-CAM constitute a heterogeneous group with a molecular weight of 200–250 kd, A-forms of N-CAM resolve themselves into predominantly 180- and 140-kd polypeptides. The same polypeptides are 170 and 140 kd following removal of sialic acid by neuraminidase or acid digestion, and 160 and 130 kd following the further removal of asparagine-linked oligosaccharides by endoglycosidase F digestion. The difference between E- and A-forms is mainly in the amount of sialic acid associated with the molecules, the E-form having as much as three times more sialic acid than the A-form (Edelman et al., 1985).

The other CAMs lack polysialic acid, although they may contain some sialic acid residues. L-CAM (a cadherin related to uvomorulin) is capable of binding with concanavilin A (Con A) but not with wheat germ agglutinin (WGA). The molecule would seem to possess glucose and mannose therefore, but not acetylglucosamine residues.

Ng-CAM of chicks is probably identical to **nerve growth factor–inducible large external glycoprotein** (NILE) and to the L1 antigen of mammals. The molecule is calcium independent and heterophilic (i.e., a CIDS). It is probably produced from a precursor polypeptide with a molecular weight on the order of 200 kd. Enzymatic cleavage produces multiple polypeptides of 80 and 115 kd containing four asparagine-linked oligosaccharides having sialic acid residues (total 135 kd).

SAMs (substrate adhesion molecules) are extracellular ligands, high molecular weight glycoproteins such as laminin, hexabrachion, and fibronectin. Like other adhesion molecules, SAMs are identified by antisera capable of blocking the adhesion of cells to ECMs. For example, some **broad-spectrum antisera** (raised against serum-free

LAMININ RECEPTOR MODEL

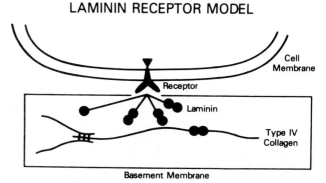

FIGURE 20.9. Scheme for laminin bridging the interface between the cell and basal lamina. From L. A. Liotta, U. M. Wewer, C. Nageswara Rao, and G. Bryant, in G. M. Edelman and J.-P. Thiery, eds., *The cell in contact: Adhesions and junctions as morphogenetic determinants,* Wiley, New York, 1985. Used by permission.

media conditioned by mouse cell lines or by human mammary carcinoma, or against hamster-fibroblast glycoprotein GP120–160) can block the adhesion and spreading of mouse blastocysts on substrates *in vitro* and *in vivo* and even remove adhering blastocysts *in vitro* (see Damsky et al., 1985).

Laminin (LM) seems to provide structural organization for the membrane and for the adhesion of cells (Fig. 20.9) (Chung et al., 1979). It is a cruciate trimer with an A chain (400 kd) backbone for the long arm, and branching B 1 (215 kd) and B 2 (205 kd) chains linked by disulfide bonds. Type IV collagen binding sites are at the ends of the B branches; heparin and axonal-outgrowth sites are at the amine terminus of the A chain and long arm, and a cell binding site with polarizing activity is present toward the carboxy terminus of the A chain near the B branching points (see Klein et al., 1988).

Chick hexabrachion (= cytotactin) is a large, disulfide-bond, extracellular matrix molecule of 220 kd having a six-arm structure (Erickson and Taylor, 1987). It appears in extracellular spaces of the gastrulating chick blastoderm, first anteriorly in the basement membrane of the neural ectoderm and notochord and then progressively posteriorly, but it is not found in lateral plate mesoderm or nonneural ectoderm. Its sequential, site-restricted appearance suggests that hexabrachion may aid in organizing localized regions of extracellular matrix.

Fibronectin (FN), or the family of fibronectins and closely related molecules differing slightly in amino acid composition and glycosylation, is just about ubiquitous in basement membranes except for those of the central nervous system. Secreted by a variety of vertebrate and invertebrate cells and usually present in serum added to tissue culture medium, fibronectin is also present along the migrating pathways of sea urchin primary mesenchyme cells and along the routes of massive cell movements of vertebrate gastrulas.

Disulfide-bonded dimers of two 220-kd polypeptides comprise soluble or **plasma FN** found in blood (also called cold insoluble globulin) and insoluble **cellular** or **matrix FN** found in fibrillar arrays in pericellular spaces, basement membranes, and the ECM. The two forms of fibronectin and the polypeptides making up each appear to be encoded by a single gene. In chicks, an enormous 48-kilobase (kb) gene isolated as a set of five recombinant DNA clones spans the fibronectin gene sequence. The gene contains an astonishing 48 or more exons, with an average length of 150 bp, and longer intervening introns. Selected exons are expressed as several different messenger RNAs as a result of differential splicing (see Yamada et al., 1985).

Multifunctional, fibronectin plays a mediating role for cell migration and structuring the ECM. In addition to its cell-surface binding region, fibronectin is armed with distinct domains capable of forming **multimeric** or **multivalent complexes** with ECM ligands such as collagens I–IV, fibrin and fibrinogen, heparin and heparan sulfate, hyaluronic acid and other glycosaminoglycans, and proteoglycans (Fig. 20.10). Transglutaminase-mediated cross-linking to fibrin or collagen also takes place at the amino-terminal domain.

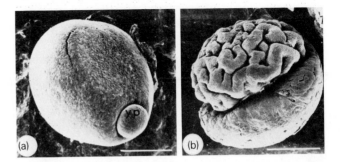

FIGURE 20.11. Scanning electron micrographs illustrating the effect of monovalent fragments (Fabs) of antifibronectin antibodies (anti-FN) injected into the blastocoel of ribbed newt, *Pleurodeles waltlii,* blastulas. (A) Normal late gastrula controls incubated with anti-FN in medium or injected with preimmune serum or bovine serum albumin. (B) The blastocoel roof has become extensively convoluted in embryos 24 hours after injection of monovalent anti-FN into late bastula. Scale bars = 320 μm. From J. C. Boucaut, T. Darribère, H. Boulekbache, and J. P. Thiery, *Nature (London),* 307:364(1984), by permission of Macmillan Journals and the authors.

In amphibians, stored (maternal) fibronectin mRNA is slowly translated during cleavage by all cells. Despite its widespread synthesis, fibronectin is accumulated only in the roof of the blastocoel of the midblastula where it is thought to bind invaginating and involuting cells during gastrulation.

Injecting monovalent anti-newt fibronectin Fab antibodies into the blastocoels of late blastulas or early gastrulas inhibits gastrulation presumably by blocking fibronectin-to-receptor interactions (Fig. 20.11) (Boucaut et al., 1985). Moreover, injec-

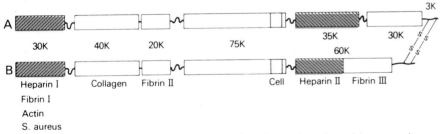

FIGURE 20.10. Map of the structural and functional domains of human plasma fibronectin. The two subunits designated A and B with their amino termini at the left and their carboxy termini at the right differ mainly toward the carboxy end. Protease-susceptible lengths of polypeptides are shown as wavy lines between protease-resistant domains. Different domains capable of binding to the same ligand are designated with Roman numbers. From K. M. Yamada, M. J. Humphries, T. Hasegawa, E. Hasegawa, K. Olden, W.-T. Chen, and S. K. Akiyama, in G. M. Edelman and J. P. Thiery, eds., *The cell in contact: Adhesions and junctions as morphogenetic determinants,* Wiley, New York, 1985. Used by permission.

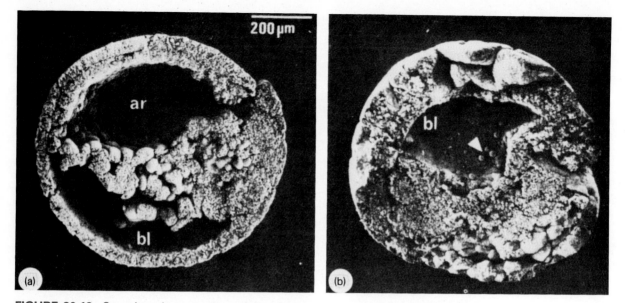

FIGURE 20.12. Scanning electron micrographs of a bisected normal gastrula (a) and an arrested blatula (b) of the ribbed newt, *Pleurodeles wallii*. Both embryos are the same age, but the arrested blastula was injected with the polypeptide (arginine-glycine-aspartic acid-serine-proline-alanine-serine-serine-lysine-proline) from the cell-binding site for FN. As in Fig. 20.9, gastrulation is blocked. The roof of the blastocoel becomes convoluted on the outside while remaining smooth on the inside. Cells reaching the blastocoel detach from the surface (arrowhead) rather than migrate. ar, archenteron cavity; bl, blastocoel. From the *Annual Review of Cell Biology,* Vol. 2, 1986, by permission Annual Reviews Inc. Courtesy of J.-P. Thiery, J.-L. Duband, and G. C. Tucker.

tion of cell binding polypeptides containing the sequence[1]

arginine-glycine-aspartic acid-(serine, sometimes)

into the blastocoels prevents gastrulation (Fig. 20.12) (see Thiery et al., 1985a, b). This remarkably small sequence of amino acids occurs in human, bovine, and rat fibronectin and is part of the binding site for human fibroblasts. In isolation, the peptide presumably occupies enough of the fibronectin receptor for steric hindrance to block fibronectin binding. The peptide may act as a **recognition signal** in fibronectin, since so few amino acids are unlikely to represent a complete binding site.

In chicks, the ECM beneath the early epiblast contains large amounts of fibronectin (Fig. 20.13), glycosaminoglycans (hyaluronic acid and chondroitin sulfate), and possibly some collagens. The basal lamina of the later epiblast contains laminin, collagen IV, entactin, and other glycosaminoglycans. Epiblast cells migrate into the mesoderm and endoderm at points where the basal lamina is disrupted (see Fig. 13.7).

[1] Known as RGDS after the first letter of the amino acids' single-letter abbreviations.

Mouse fibronectin first appears between the early endoderm and epiblast at 4.5 days postcoitum (p.c.) in the embryo, coinciding with the commencement of gastrulation (see Fig. 19.4). Fibronectin forms a layer under the migrating early endoderm of both the visceral and parietal yolk sac by 7.5 days p.c. (Fig. 20.14) and is especially concentrated in **Reichert's membrane** between the parietal endoderm and trophectoderm. Serological activity suggests that this fibronectin is secreted by the trophectoderm. Collagen type IV, laminin, and other basement membrane molecules in Reichert's membrane are probably secreted by the endoderm.

Fibronectin is also present in the amnion and chorion between extraembryonic ectoderm and somatic mesoderm and in the cytoplasm of the trophoblastic giant cells. It is not detected in mouse ectodermal cells (although it is present in chick germinal crescent ectoderm) but occurs in apical granules in endodermal cells and in mesodermal cells.

The mechanisms of anchorage and spreading of epithelial cells by fibronectin are uncertain, but the consequences of anchorage may be profound. Fibronectin may regulate contact paralysis, since exogenous fibronectin inhibits the migration of fibroblasts into three-dimensional gels, and trans-

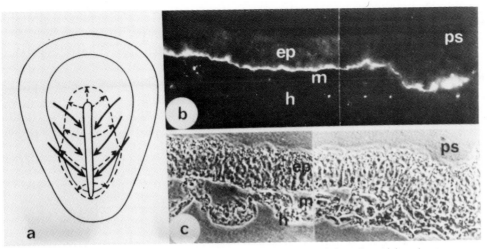

FIGURE 20.13. Diagram of definitive primitive streak stage chick embryo (a) and transverse section through the primitive streak (ps) and adjacent blastoderm viewed by epiflourescence (b) and phase contrast (c). The epiflourescence is due to staining by anti-fibronectin antibodies. Solid arrows show convergent direction of epiblast cells toward primitive streak. Dashed arrows show the subsequent direction of migration under the epiblast. ep, epiblast; h, hypoblast; m, mesoderm; ps, primitive streak. From J. P. Thiery, J.-L. Duband, and A. Devouvée, in G. M. Edelman and J.-P. Thiery, eds., *The cell in contact: Adhesions and junctions as morphogenetic determinants,* Wiley, New York, 1985.

formed fibroblasts synthesize greatly reduced amounts of fibronectin (see Schor and Schor, 1987). Furthermore, fibronectin-poor virus-transformed cells *in vitro* develop adhesion plaques and stress fibers upon the addition of fibronectin to the medium and temporarily reverse the transformed state (see Burridge et al., 1987).

Possibly fibronectin is alternately produced and broken down. Stabilizing matrix fibronectin from soluble precursors could involve disulfide bonding, interactions with collagen, sulfated glycosaminoglycans, or cross-linking by transglutaminases. Alternatively, proteolytic enzymes may break down fibronectin containing matrices; compounds with amino groups may solubilize them, or hyaluronic acid may destabilize them (see Wartiovaara and Vaheri, 1980).

SAM-receptors or integrins (also called matrix receptors) comprise a family of membrane-span-

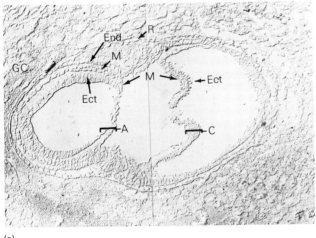

(a)

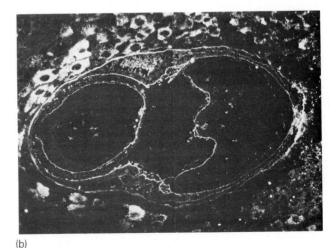

(b)

FIGURE 20.14. Section of a 7.5-day p.c. mouse embryo seen with Nomarski optics (a) and with epifluorescence after indirect staining with anti-fibronectin antibodies (b). Prominent fibronectin staining is seen in Reichert's membrane (R), the amnion (A), chorion (C), and cytoplasm of trophoblastic giant cells (GC). X 270. From J. Wartiovaara, I. Leivo, and A. Vaheri, *Dev. Biol.,* 69:247(1979), by permission of Academic Press and the author.

ning glycoproteins (related to CIDS) bridging the _extracellular and intracellular matrices._ Concentrated in plaques, SAM-receptors may complex (Fig. 20.6) simultaneously with matrix proteins (e.g., fibronectin or laminin) and hence basement membrane collagen type IV on one side of the plasma membrane, and with talin and hence vinculin, α-actinin, and actin on the other side (Fig. 20.9) (Chen et al., 1985; Damsky et al., 1985).

Binding sites are different on either side of the plasma membrane, and different polypeptides compete with binding at the two sites. While binding with fibronectin and laminin is prevented by the cell-binding polypeptide containing the arginine-glycine-aspartic acid sequence (Arg-Gly-Asp or RGD, see above), integrin-talin binding is not disrupted. Binding with talin is blocked by a synthetic decapeptide corresponding to a tyrosine kinase phosphorylation site in integrin's cytoplasmic domain (see Buck and Horwitz, 1987).

Whatever SAMs do, from morphoregulatory (controlling form) to mechanoregulatory (controlling behavior) functions, they must operate through integrins. The far-reaching power of SAMs _in vitro_ may result from their ability to immobilize integrins in the basal portion of cells while recycling integrins through endosomes and internalizing matrix ligands on the free apical surfaces of cells.

Three distinct glycoproteins are resolved from avian integrin by SDS–PAGE under nonreducing conditions. The gene coding for the third band has been isolated and sequenced. The deduced amino acid sequences have multiple glycosylation sites, four cysteine-rich repeats in the extracellular domain, a hydrophobic transmembrane domain, and a cytoplasmic domain of 47 amino acids containing the tyrosine kinase phosphorylation site.

Integrins may not be highly specific for their substrates. The same integrin that binds to fibronectin may bind with laminin. The same monoclonal antibody (e.g., CSAT) that blocks binding with fibronectin blocks binding with laminin, and the cell-binding polypeptide competes with both laminin and fibronectin for binding with integrin.

Furthermore, while integrins are probably heterodimers, they may not be a homogeneous lot. For example, a laminin-integrin protein is about 70 kd (see Liotta et al., 1985), while one fibronectin-integrin is a 47-kd glycoprotein and another is a 140-kd glycoprotein. Antibodies to each block fibronectin adhesion and spreading by fibroblasts (Yamada et al., 1985). In addition, gangliosides with several sialic acid residues may perform the role of fibronectin-integrins. But antibodies to the

CIDS glycoprotein LFA-1, to chick glycoprotein GP140, and to "cell-substratum attachment glycoprotein complex antigen" (CSATag) appear indistinguishable (Damsky et al., 1985).

Some fibronectin-integrins are resistant to trypsinization in the presence of Ca^{2+}, suggesting that they are dependent on a protein component stabilized by divalent cations (i.e., a cadherin). The dependence is not a strict requirement for adhesion, however, since cells bind to fibronectin in the presence of metal ion-sequestering agents (e.g., EGTA).

Fibronectin-integrins have a relatively low affinity for fibronectin, appropriate for a role in modulating cell behavior. In contrast to other bound ECM components, fibronectin may be washed off cells.

The dynamics of fibronectin–cell interactions may reflect the weak affinity of cell receptors to fibronectin. Easily saturated, cells may bind to matrix fibronectin when only low amounts of plasma fibronectin are present in the cellular environment. Cells may then detach from matrix fibronectin when high amounts of plasma fibronectin are available (see Yamada et al., 1985).

JAMs (junctional adhesion molecules or cell _junctional molecules, CJMs) are generally associated with cell junctions, or junctional complexes,_ _differentiated regions of the plasma membrane_ _joining adjacent cells (and sometimes cells to matrix)._ Morphologically diverse and differing in shape, dimensions, and spatial relationships, cell junctions are classified in at least three categories: (1) adhering junctions of a banded type known as **zona adherens** or **adhering zonule**, (2) belted intercellular-space sealing junctions called **zona occludens, occluding zonules,** or **tight junctions,** and (3) adhering junctions of a buttonlike or macular type called **desmosomes** or **adhering maculae.** In addition, gap junctions and other **microdomains** (if not attachment sites) in membranes may be associated with JAMs (see Table 2.2).

Like adhesion plaques, junctional complexes are usually associated with limited cell spreading, blocks to motility, and arrested cell growth. In contrast, cell–substratum contacts and nonjunctional cell–cell contacts are associated with cell spreading, translocation, and normal growth. The types of adhesion molecules present at the different types of contact parallel the behaviors of the contacts, presumably reflecting differences in the molecules' functions.

Zona adherens have been the premier subject for research on cell junctions. The zona are stabilized by three **junctional domains** formed (induced?) during intercellular contact and laid down from the outside-in. (1) A **membrane domain** contains an **adherens-type cell adhesion molecule** (A-CAM or sometimes L-CAM) bridging the **junctional cleft** or intercellular space and spanning the plasma membrane. (2) A membrane-associated **plaque domain** contains **plaque molecules,** plakoglobin, and vinculin. (3) A **cytoskeletal domain,** mediating the attachment of actin to the plaque, contains **actin-associated proteins** (e.g., vinculin, α-actinin, filamin, myosin, tropomyosin) (see Geiger et al., 1985, 1987).

A-CAM, a Ca^{2+}-dependent CAM (i.e., a cadherin), is an integral membrane glycoprotein of 135 kd. It occupies the position in zona adherens that integrin occupies in cell–matrix contacts, but unlike integrin, which is a receptor and a heterophilic adhesive, A-CAM tends to form homophilic bonds (although it can also bind the related L-CAM). A-CAM is most abundant in embryonic chick epithelia, and it is replaced by L-CAM (or uvomorulin) in adult epithelia.

Plaque molecules in zona adherens (and in desmosomes) occupy the position of talin in cell–matrix contacts. **Plakoglobin** or desmoplakin (83 kd) bridges A-CAM and the actin-associated protein vinculin.

Vinculin alone can bind directly to plaque molecules, since talin is generally absent in cell-to-cell adhesion sites (as opposed to cell-to-matrix adhesion plaques). But talin is present alone in the adhesion sites of cytotoxic T lymphocytes and their target cells, suggesting that the two proteins may play different physiological roles.

Two proteins called **desmoplakins,** 215 and 250 kd, make up the dense plates of desmosomes. Other glycoproteins (desmogleins and desmocollins) are present in the desmosomal membrane and the extracellular plates (see Cowin et al., 1985). In addition, half-desmosomes, making contact with the extracellular matrix, resemble adhesion plaques, containing both talin and vinculin in their plaque domains.

Although ordinarily not involved in cell adhesion, coated pits and vesicles are differentiated parts of membranes. **Clathrin,** which coats their internal surface, is a 180-kd protein that seems to stabilize membranes during transport.

Finally, although **gap junctions** are not thought to play a role in cell adhesion, they figure prominently in the consequences of cell adhesion through their role in cell communication (see Gilula, 1985). The major component of rat liver gap junctions is a 28-kd protein. When poly(A)$^+$ mRNA for this component is injected into *Xenopus laevis* oocytes, translation leads to the spontaneous formation of functional gap junctions. The electrical properties of these gap junctions are closer to those for rat liver than for early *X. laevis* embryos (Dahl et al., 1987).

The components of cell-to-cell channels are said to be widely conserved (Warner et al., 1984), since antibodies raised against the major gap junction proteins from rat liver interact with antigens from *X. laevis* oocytes and early cleavage-stage embryos. Moreover, injection of antibodies to rat liver gap junction protein into blastomeres at the 8-cell frog embryo reduces the channel-conducting properties of the 32-cell frog embryo with grave consequences for later development.

Chapter

21

GERM LAYERS
AND GASTRULAS

Phylogeny and Ontogeny

*T*he idea of germ layers predates the idea of gastrulas by a century. Still, the two ideas are not easily separated. Perhaps the clearest distinction today is that germ layers still support the germ-layer theory (in one of its permutations), while gastrulas no longer support the gastrea theory (in any form).

GERM-LAYER AND GASTREA THEORIES

Totally devoid of scientific value, [Haeckel's] gastrea theory was the culmination of the early work on the germ-layers. . . . But the beautiful unity of Haeckel's scheme was too seductive. Huxley and the English embryologists spent their days apotheosizing its author and looking at embryos only for the purpose of fitting the facts of ontogeny into the ideal of phylogeny.

Jane M. Oppenheimer (1940, reprinted, 1967, pp. 269–270)

Haeckel said organisation came only with the germ layers. . . . For the first time with the gastrula,

then, can the organism be considered organised and subject to the adaptive responses through natural selection. . . . Haeckel believed that the earlier stages hold no interest for researchers, serving instead simply to multiply the primary nutritive material. Furthermore, after gastrulation, ontogeny follows phylogenetic patterns of development, or ontogeny recapitulates phylogeny as stated in Haeckel's biogenetic law.

Jane Maienschein (1985, pp. 84–85)

Germ-Layer Theories

The history of the germ-layer theory is complicated by differences between two of its major versions. One of the versions can largely be attributed to Karl Ernst von Baer (1792–1876), for whom germ layers constituted the backbone, so to speak, of comparative embryology. Another germ-layer version can be attributed to Ernst Heinrich Haeckel (1834–1919) and Thomas Henry Huxley (1825–1895) who tried to use germ layers to break the back of resistance to Darwinism (see Oppenheimer, 1940).

The first germ-layer theory began in the 18th century with the discovery by Caspar Friedrich Wolff (1738–1794) that the chick embryo forms layers or "leaves" which later form adult structures. Buried by opposition, the discovery was not validated until 1817, when Christian Pander described the formation of three embryonic layers in the chick, and 1825, when Martin Heinrich Rathke identified outer and inner germ layers in an invertebrate embryo, the crustacean decapod, *Astacus.*

Then, in 1828, Karl Ernst von Baer published *Über Entwickelungsgeschichte der Thiere. Beobachtung und Reflexion* (*On the developmental history of animals. Observation and conjecture*),[1] Part I. This seminal work is largely devoted to an examination of epigenesis and the comparative method, but true to its title, von Baer begins with detailed observations, largely on the development of the chick. Part II of the work, which was not published until 1837, describes development in the embryos of several mammals (e.g., dog, edentate, human, monkeys, narwhal, pig, rabbit, sheep, and sloth), salamanders, frog, shark, and fish and concludes that the development of all vertebrate embryos is fundamentally similar.

An inner mucous layer served a digestive function. An outer serous or skin layer served a protective and sensory function, and two internal layers lining the primitive body cavity produced skeleton and muscle, and blood vessels and blood. The central nervous system formed by tucking in the serous layer, and the foregut formed by folding the mucous layer.

Acknowledging many earlier sources, von Baer showed that the germ layers gave rise to adult structures by thinning and thickening, folding and pocketing, growth and transformation. He claimed, moreover, that the germ layers were present in all vertebrate embryos and that they accounted for the origins of all adult tissues and organs.

The theory was verified for mammals by Theodor Ludwig Wilhelm Bischoff (1807–1882), who named the vesicular mammalian embryo the **blastocyst,** and in the early 1850s, by Robert Remak who traced the germ layers in mammals as well as in birds back to the single ovum and ahead to their histological fates. At about the same time, with the help of improvements in microscopy and inspired by the cell theory, germ layers were shown to consist of cells and extracellular spaces.

Remak distinguished between two **primary germ layers.** An upper layer was subdivided vertically into an epidermal plate and a neural plate, the latter giving rise to the brain and spinal cord. An underlayer was subdivided horizontally into a purely nutritive innermost layer that gave rise to the gut tube and a middle motor-germinative layer that was further split horizontally by the embryonic body cavity into two cellular plates.

With some corrections (such as the sources of peripheral nerves) and for the want of names for the germ layers, the first version of the germ-layer theory was then complete. With a basis in topography, the theory maintained that germ layers composed of cells and their products were universally present among metazoan embryos, had comparable origins in embryos, and gave rise to comparable tissues in adults.

Contemporary versions of von Baer's germ-layer theory cover a range of possibilities. At one extreme, "endoderm," "mesoderm," and "ectoderm" are expunged totally of their former theoretical significance and used merely to identify the topographical lining, middle, and covering layers of embryos. The germ layers are thereby reduced to a shorthand for describing physical relationships. At the other extreme, von Baer's own idea of germ layers crops up in modern research on determination disguised as bifurcations on logic grids and sometimes wrenched into models of linear descent.

Between these extremes, von Baer's germ layers are modified along several lines. Germ layers are no longer simply equated with closely related cells that give rise to cells of the same histotype and form the same structures in embryos of different species. Primordial germ cells, for example, form in different layers even among related groups (e.g., mesoderm in urodeles, endoderm in anurans).

Germ layers are no longer individuated without qualification, since they are not necessarily separated completely. The notochord, for example, is only arbitrarily assigned to the mesoderm, since its cells frequently arise in the endodermal roof of the archenteron (e.g., Fig. 19.25).

Germ layers are also no longer assumed to be homogeneous entities, or, at least, the possibility of homogeneity is raised as a question rather than as a conclusion. The mesoderm, for example, gives rise to cells of so many diverse histotypes (from the epithelial lining of the body cavity and blood vessels, to muscle, all kinds of connective tissue, and blood) and participates in the formation of so many diverse organs (from the dermis of the skin to the muscle of the intestine) that considering it a single embryonic tissue may not be justified.

Possibly, a case can be made for endoderm as

---

[1]The book was dedicated to Christian Pander.

the source of a single histotype, the epithelial cells of the gut and associated glands. To make this case, one must ignore, among other things, the tailbud of *Xenopus laevis* (see Fig. 23.19). The neural ectoderm is connected to postanal endoderm through the tailbud, and the cavities of the neural tube and gut are continuous through a **neurenteric** canal. Without a clear distinction between the germ layers, the tailbud forms the tail's neural tube, notochord, and somites, none of which can be considered epithelial, or at least not for long.

Germ layers are no longer considered universal. The imagination stretches to the breaking point when equating a cell on the surface of one embryo (e.g., an annelidan somatoblast) with an internal layer of cells in another embryo (e.g., the chordamesoderm of a frog). Wrenching facts to fit theory is poor scientific practice and, in this case, fortunately, no longer in vogue.

Histotypes formed by one germ layer are not necessarily formed exclusively by that layer. Not only are salivary glands formed almost interchangeably by endoderm and ectoderm, but, in amphibians, the enamel organs of teeth form from endoderm as well as ectoderm in different parts of the jaw. The difference between endoderm and ectoderm wears thin where the germ layers meet. The germ layers would seem to have more in common than a clean-cut concept of germ layers would permit.

Similarities between tissue types formed by ectoderm and mesoderm are even more striking. Ectodermal **neural crest** or **ectomesenchyme** forms pigment cells, the adrenal medulla, the mesenchyme of amphibian dorsal fin, dermal papilla of hair and feather follicles, the odontoblasts of teeth, part of the walls of large arteries arising from the heart, the ciliary muscles of the eye, cartilage of the visceral skeleton, and part of the neurocranium, all of which have been mistakenly thought of as exclusively mesodermal! Possibly, the historical reluctance of many embryologists to accept the neural crest's multiple roles in development flows from reluctance to sacrifice the last vestige of belief in von Baer's germ layers (see Le Douarin, 1982).

Still, von Baer's version of the germ-layer theory is not entirely abandoned. Like a bulletin board, the idea of germ layers is a useful place for advertising facts and raising questions (Fig. 21.1).

The second germ-layer theory is intimately associated with the history of the germ layers' names. In 1853, George J. Allman coined the terms **ectoderm** and **endoderm,** but he was not naming embryonic cell layers. He was naming the outer cell layer of hydroid polyps the ectoderm and the inner cell layer the endoderm (see Fig. 19.13). How did the names of cell layers in coelenterates come to be applied to the germ layers of vertebrate embryos and to embryos in general?

For almost 20 years, Allman's names were not transferred to embryos even though, in 1849, the English anatomist and naturalist Thomas Henry Huxley had compared the two cell layers of coelenterates to the outer and inner germ layers of vertebrate embryos. Even the publication of Darwin's 1859 *Origin of Species,* which fired debate on evolution and elevated embryos to the stature of evolutionary symbols, failed to link the names of the coelenterate's body layers to germ layers.

That link was forged by Ernst Haeckel as he pondered observations made by his Russian student, Alexander Onufrievitch Kowalevsky (also Kowalewski, 1840–1901). Between 1867 and 1871, Kowalewski documented two fundamental observations. First, he detected affinities among the embryos of ascidians, Amphioxus, and vertebrates (e.g., as embryos, they all possessed a dorsal neural tube, a notochord, and lateral bands of muscles, e.g., see Fig. 19.20) despite their decidedly different appearances as adults. Second, he enumerated similarities among the embryos of vertebrates and the adults of Amphioxus (e.g., the foregut of vertebrate embryos, like the gullet of adult Amphioxus, was penetrated by "gill slits").

Evolution could, of course, account for similarities among embryos as well as it could account for similarities among adults, namely through phylogenic relationships. But why would the embryos of advanced forms (e.g., vertebrates) resemble the adults of primitive forms (e.g., Amphioxus)?

For Haeckel, the answer was **biogenesis** or the **biogenic law** with its parallels between stages of embryonic development and ancestral forms in phylogenic progressions. Ontogeny recapitulated phylogeny.

As a consequence, in 1871, Haeckel applied the names ectoderm and endoderm to the primary germ layers of the gastrula, and a few years later Huxley coined the word **mesoderm** for the middle layer. Soon, a race for naming took place that led to the present state of redundancy. Oscar and Richard Hertwig called the germ layers ectoblast, mesoblast, and entoblast, and the Englishmen Francis M. Balfour and Edwin Ray Lankester used epiblast, mesoblast, and hypoblast for the same germ layers (see Oppenheimer, 1940).

The differences in names reflected differences in the importance attached to invagination versus

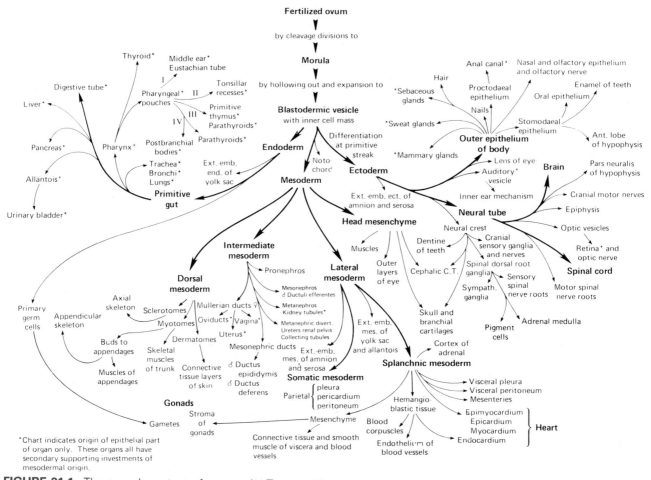

FIGURE 21.1. The germ-layer tree of mammals. Turn upside down for comparison with Fig. 21.2. From B. M. Carlson, _Patten's foundations of embryology_, 5th ed., McGraw-Hill, New York, 1988. Used by permission.

delamination in forming the mesoderm and the embryo's body cavity or **coelom** (also coelome, named by Haeckel). But the meaning and intent of all the authors and all the names were the same: to capture in words the concept of the embryo's germ layers as reminiscences of the ancestor's body layers. This was the essence of the second germ-layer theory.

In the new germ-layer theory, the evolution of metazoans or multicellular animals met the development of metazoan individuals at the level of the germ layers. The same rigid linearity Haeckel imagined taking place in evolution also had to take place in the development of germ layers. Primary germ layers were to form first in embryos as they would have formed in the ancestors of metazoans. The middle germ layer was to form last, as it would have in the course of evolution. Furthermore, organs had to arise from only one germ layer, and organs arising from a specific germ layer in one

species had to arise from the same germ layer in other species. These consequences of biogenesis for Haeckel's germ-layer theory were embodied in his own **gastrea theory.**

Gastrea Theory

In 1874, Haeckel invented **gastreas** (also gastreads) as ancestral metazoans and modeled them after the two-layered saclike embryo of Amphioxus (see Fig. 19.21) described by Kowalewski. Distinguished by their bilaminar structure, the hypothetical gastreas were supposed to have been the first animals to achieve sufficient complexity in the course of evolution to rank as ancestors to living metazoans (Fig. 21.2). Similarly, bilaminar embryos represented the first embryonic stage to have parts sufficiently delineated to develop into the structures of today's metazoan adults.

To cement the comparison between ancestors

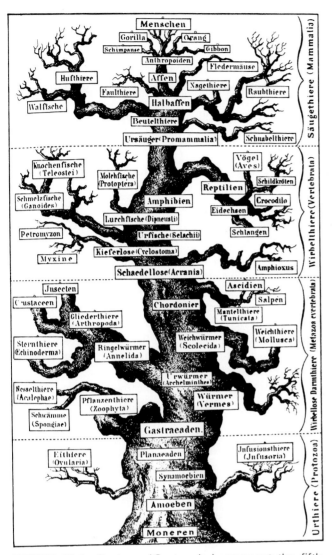

FIGURE 21.2. Gastreas (*Gastraeden*) represent the fifth stage in Haeckel's evolutionary tree and the origin of all living metozoans. From E. Haeckel, *Anthropogenie: Keimes- und Stammes-Geschichte des Menschen*, Engelmann, Leipzig, 1874.

and embryos, Haeckel named embryos with germ layers **gastrulas** after his hypothetical gastreas. Ironically, the name stuck to the embryo long after gastreas were forgotten. But the gastrea and Haeckel's version of the germ-layer theory left their marks.

In Defense of the Gastrea: Gastrulas as Multicellular Organisms

The idea of homology among gastrulas epitomized by the gastrea had largely, but not exclusively, negative reverberations in the history of

embryology. The gastrea provided the rationale for studying gastrulas comparatively, and it served as a corrective for the prevailing impression that embryos were *organisms containing many single cells* rather than *multicellular organisms* (see Russell, 1930).

In the recapitulationist atmosphere of the late 19th century, early embryos were generally considered aggregated clones or colonies of unicellular organisms. The difficulty with this view was that, in unicellular organisms, cells were masters of the organism, while in multicellular organisms, cells were normally servants of the organism. At some point in ontogeny, it would seem, the cells of a multicellular embryo ceased being independent unicellular organisms and submitted to the discipline of differentiation. How did the change happen?

In revising the cell theory, Rudolf Virchow proposed that, ideally, multicellular organisms were societies of cells living under the democratic rule of law. Haeckel and others of his school substituted the dictatorship of the multicellular organism over its cells. Distancing himself from others who derived metazoans from platelike colonies of protozoans, Haeckel derived a unilaminar or single-layered ancestor from a spherical colony of flagellates resembling volvox. Lacking a clear division of labor, cells in the volvox-like colony were subservient to the organism. Haeckel coined the term **blastea** to represent this ancestor and named the **blastula** to represent its present incarnation.

Haeckel evolved the first true metazoan by adding dependency, determination, and ultimately differentiation to the properties of the blastea. Haeckel's concept of **terminal addition** (namely, that new stages are added to the ends of old stages) obliged him to add the additional metazoan characteristics to the end of the blastea's development, while his concept of **condensation** (that phylogenic stages are collapsed in ontogeny) required him to push back the ancestral adult into the advanced embryo. The resultant "ancestor" was the gastrea; the resultant embryo was the gastrula.

The version of a multicellular gastrula proposed by Haeckel is not far removed from many current versions, and Haeckel's contribution to these versions cannot be exaggerated. Above all, he turned embryologists' attention to organismic control of cellular development and differentiation. Despite his commitment to a preformationist form of development based on evolution, he set the stage for the resurgence of interest in epigenesis that has come to dominate embryological research (with certain exceptions).

Distortions Arising from the Germ-Layer Theory

In the decades before the turn of the 20th century, prominent embryologists, like Oscar and Richard Hertwig, argued convincingly that similarities in the **fates** of germ layers were as important in establishing homologies as were similarities in fully formed adult tissues. After all, unlike other fields of biology that could only produce static evidence in support of homologies, embryology provided direct evidence of change from common sources. Regrettably, the difference between evolution as one kind of change and development as another kind was not appreciated (see Gould, 1977).

All too many embryologists, then and now, maintained that embryos were forced to recapitulate their evolutionary past as the price of their own progress. Drawn out of evolutionary retirement, ancestral adults reappeared on the stage of development, recostumed and made up for their new circumstances but still the same old actors.

In effect, embryonic recapitulation resurrected the Lamarckian version of evolution with a new wrinkle. Instead of adults inheriting the acquired characteristics of ancestors, embryos became the beneficiaries of ancestral largess.

Moreover, the preformationist view of development, which embryology had only just abandoned, reemerged with vigor. Ancestral forms, if not homunculei (see Fig. 6.21), were passed along to embryos fully formed. Adaptation in the embryo in the form of cenogenic structures, such as extraembryonic membranes, had to be accepted, but history was the supreme determinant of development.

Embryologists at the time of Haeckel were not ignorant of the facts that argued against his theory. These facts were almost as well documented then as they are now (see de Beer, 1958). Why then did embryologists by and large adopt Haeckel's beliefs?

One possibility is that embryologists were not well-enough grounded in philosophy to recognize the circularity of their arguments. Had they studied syllogisms, they might have realized that germ layers change, and species change, but the changes in germ layers (i.e., development) and in species (i.e., evolution) are not necessarily the same.[2]

---

[2]The same problem is inherent in the dogged insistence of some molecular biologists to identify all similarities among molecules as homologies. Whether or not the attribution of homology is justified by independent evidence, science is not well served by tautology.

Another possibility is that the potential of biogenesis to bolster and advance evolutionary theory lured embryologists away from their usual restraint and good scientific judgment. Possibly, the ease with which biogenesis explained development and aided teaching embryology (e.g., teaching the fate of germ layers) rendered Haeckel's version of the germ-layer theory irresistible.

Evidence Against the Gastrea Theory

The implicit preformationism of biogenesis runs counter to evidence for epigenic interactions and coded developmental determinants (e.g., gene products). Among the many arguments against the biogenic law in general, three stand out as they apply to the gastrula.

First, embryos called gastrulas do not constitute a homogeneous group. They have different numbers of cells (e.g., see Tables 19.1 and 19.2) arranged in different patterns, as masses, layers, and even individually determined cells. Furthermore, cells in some gastrulas may be determined, while similar cells in other gastrulas remain indifferent.

Second, gastrulas do not form via a single mechanism. In addition to differences in the migratory habits of cells, syncytia, and layers, mechanisms of determination in gastrulas differ among species. For example, the syncytial blastoderm of insects is determined as it becomes a cellular blastoderm, while the blastoderm of chicks is determined as cells migrate past each other in different germ layers.

Third, gastrulas do not necessarily form their germ layers in the order prescribed by biogenesis. In _Xenopus laevis_, for example, the three layers are formed simultaneously, while in most amphibians the completion of the endoderm (i.e., one of the primary germ layers) may actually trail behind the formation of the mesoderm (one of the secondary germ layers). Furthermore, determination frequently precedes or follows the gastrula stage and, upon analysis, is found to be a function of interactions among tissues or embryonic parts rather than an inherent property of the specific germ layer.

For the first half of the 20th century, confusion prevailed on precisely where the mechanisms of evolution and development merged, where metaphor left off and reality began. The origin of species and the development of individuals were fused in the biogenic version of the germ-layer theory, and disentangling them occupied a great deal of embryologists' time and effort.

Today, the biogenic chimera of development and evolution is dead (or, at least, asleep). Like many other scientific issues, the problem of the biogenic law has been self-correcting. Random genetic events are now assumed to underlie mechanisms of evolution (e.g., natural selection and genetic drift), while differential gene action is assumed to underlie mechanisms of development (i.e., beginning with selective processing, translational controls, etc.). Development places limits on evolution, and evolution constrains development, but evolution does not mechanically drive development, and development does not slavishly follow evolution.

INDUCTION AND THE "ORGANIZER"

I do not wish to devise hypotheses as long as exact knowledge is attainable by experimental work. Besides, I believe that if the facts are not compiled at random, but gained in a logical proceeding, they will by themselves join together to build up a genuine theory in the original meaning of this word, i.e., a comprehensive view of all facts afforded by experience.

Hans Spemann (1938, p. 367)

When you have eliminated the impossible, whatever remains, however improbable, must be the truth.

Arthur Conan Doyle, *Sherlock Holmes: The Sign of the Four*

At the turn of the 20th century, many embryologists were still trying to strengthen the Haeckelian version of the germ-layer theory. They occupied their time with debates over how gastrulation via multipolar migration and delamination could have evolved from unipolar migration and invagination. Other embryologists were attempting to overthrow the theory completely.

Curt Herbst (1866–1946), for one, showed that the sea urchin's germ layers were not as independent as Haeckel's germ-layer theory required. In the presence of lithium salts, for example, the amount of endoderm increased at the expense of ectoderm (e.g., see Fig. 14.21). Herbst's results received little notice, but after Hans Spemann (Fig. 21.3) described **embryonic induction,** the days of biogenic germ layers were numbered.

Spemann used "induction" in the popular sense of the word, meaning an initiation or an

FIGURE 21.3. Portrait of Hans Spemann (left) and Ross Harrison (right) at the time of the Silliman lecture, 1933. From R. M. Eakin, *Vertebrate embryology, a laboratory manual.* University of California Press, Berkeley, 1971. Courtesy of R. M. Eakin.

introduction to a new state (e.g., as civilians are inducted into the army). Similarly, he used "to induce" in the sense of "to lead" or "to influence."

Epigenic in conception, embryonic induction implies communication. Ideally, an **induction system** is a labile and potentially dynamic network of an **acting system** transmitting signals and a **reacting system** receiving signals. Determination is altered in the process.

In theory, inductive communication networks operate in normal embryos, while in practice, embryologists detect induction under experimental circumstances. Induction is said to have taken place (1) when one of two embryonic tissues brought together experimentally acts like a reacting system by differentiating along lines that it would not otherwise have taken; (2) when experimentally separated embryonic tissues fail to differentiate along lines that they would otherwise have taken. This failure is attributed to the absence of the acting system and signal. Presumably, the same signals that induce the differentiation of tissues in abnormal positions under experimental circumstances operate and induce the differentiation of the same tissues in their normal position under normal circumstances.

Induction implies nothing about the signal, its message, or the medium through which it is transmitted. What is implied is that embryonic tissues or germ layers communicate, and cells within these tissues reach their fates as a result of the information received from their environments. Communication, not ancestors, determines development.

Spemann's Theory

The research that led to the discovery of embryonic induction began at the turn of the century when Spemann was a student of Theodor Boveri in Würzburg and continued in Spemann's own laboratories at Rostock, the Kaiser Wilhelm Institute for Biology in Berlin, and finally at Freiburg, where ironically he occupied the chair once occupied by August Weismann. Spemann was concerned with the development of **composite organs** (like the eye with its retina, lens, and cornea) consisting of two or more parts in intimate association. He wondered if the functional adaptations these organs achieved through morphological interactions were reflected in **coordinated development.**

Hypothetically, developmental integration of the different parts of composite organs may be achieved in either of two ways: epigenically or preformationally. The epigenic alternative posits a schedule of environmental influences dictating the determination of parts, while the preformation alternative posits synchronized acts of self-differentiation.

For Spemann, the environment of a part in a composite organ included all the local conditions now considered the cellular microenvironment. Inclined toward epigenic explanations as a result of his experience with ligation, Spemann imagined that one part of a composite organ influenced another part through their shared environment.

He and his associates and students tested this idea with elegant and incisive experiments, often performed with simple glass needles, micropipettes, and syringes of his own invention. Spemann's (1938) recounting of the history of his discovery in *Embryonic Development and Induction* reads more like a detective story than a scientific treatise, and, even today, it stands as a monument to the scientific method pursued rigorously and critically.

Historically, the type of embryonic induction presently called secondary induction (since it follows an earlier step of primary induction) was discovered first. While interspecies variation and conflicting experimental results did not initially inspire confidence or attract too much attention to secondary induction, its discovery probably made the subsequent discovery of primary induction possible.

Secondary Induction

Spemann began experimenting with the eye. Since the optic vesicle forms before the lens and lies adjacent to the ectodermal cells that form the lens (Fig. 23.10 and 23.12), the epigenic alternative predicted that the optic vesicle influenced the environment of the ectoderm and caused it to form a lens. If this were so, the ectoderm which normally formed the lens would not form it in the absence of the optic vesicle.

Spemann used hot needles or electrocautery to destroy the optic vesicle, and later microsurgery to remove it from embryonic frogs and toads that had not yet formed lenses. His first experiments on the common European grass frog, *Rana fusca* (= *temporaria*), showed that embryonic ectoderm normally covering the optic vesicle failed to form a lens in the absence of the optic vesicle. Other experiments with the toad, *Bombinator pachypus*, gave similar results (Fig. 21.4) as did subsequent experiments with the bullfrog, *Rana catesbiana*.

Still, all the results were not consistent. Under similar circumstances, or circumstances differing in temperature but not preventing development on the unoperated side, embryos of the edible frog, *Rana esculenta*, the marsh frog, *Rana palustris*, some salamanders, and fish formed lenses at the normal site despite removal of the optic vesicle.

Spemann was exceedingly cautious in interpreting the results of his and of others' experiments. He was especially concerned with controls, with environmental factors, possible differences in the embryos' stages at the time the operation was performed, and differences in the sizes of lenses produced.

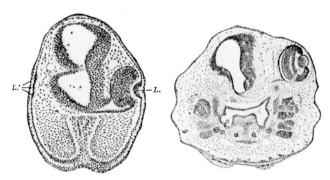

FIGURE 21.4. Sections through the head of a toad embryo from which the left optic vesicle was removed. (Left) Soon after operation. Cells that would otherwise have formed a lens continue to divide but do not form the thickened lens placode. On the unoperated side, the optic vesicle lies just under the lens-forming cells. (Right) Later. No definitive lens is formed on the operated side, while a lens has formed on the unoperated side. L, lens; L', presumptive lens-forming cells. From H. Spemann, *Embryonic development and induction*, Yale University Press, New Haven, 1938; reprinted by Hafner, New York, 1962.

It seemed to him that a system of **double assurance** might be operating in different species to different degrees. In some species, lenses seemed to be determined earlier than in other species. When determination was earliest (i.e., before the optic vesicle was adjacent to the presumptive lens ectoderm), lenses appeared to be self-differentiating or determined independently of the optic vesicle. When determination was later (i.e., after the optic vesicle was present below it), development appeared to be dependent on optic vesicles. But even the most strikingly self-differentiating lens could be influenced by its environment (if not by an adjacent optic vesicle), and even the most overtly enthralled lens could harbor a relic of independence.

Spemann decided to investigate influences on lens development in those species in which determination appeared to take place later. Would a lens be formed by ectoderm that did not ordinarily form a lens if it were experimentally brought in contact with an optic vesicle? He attempted to answer this in two ways.

First, Spemann (and others) severed the connections on one side of embryos between brain and optic vesicles and then pushed the vesicles back toward the trunk where they encountered fresh ectoderm. Lenses frequently developed at the new site. The best results occurred when the severed optic vesicle was not pushed too far back, but occasionally lenses even formed in trunk ectoderm.

The decline with distance in the ectoderm's ability to form lenses in response to optic vesicles was one of the first indications for a role of sensitivity or **competence** in induction. Two possibilities suggested themselves: (1) competence, like a diffusion gradient, diminished as a function of distance from the normal site of lens development; and (2) local conditions at remote sites prevented competent interactions.

In order to test the ability of ectoderm from remote sites to form lenses, Spemann replaced embryonic head ectoderm with trunk ectoderm (Fig. 21.5). The experiment employed the techniques of **grafting** or **transplanting** trunk ectoderm from its original site to the region of the head. The grafted or transplanted part was called the **graft**. The embryo providing the graft was called the **donor,** and the embryo receiving it was called the **host.**

Taking his lead from Ross Harrison of Yale University, Spemann routinely used donors and hosts whose ectoderms had different shades of pigmentation. He then traced the cellular source of the lens via pigment granules.

The results with grafts among members of the

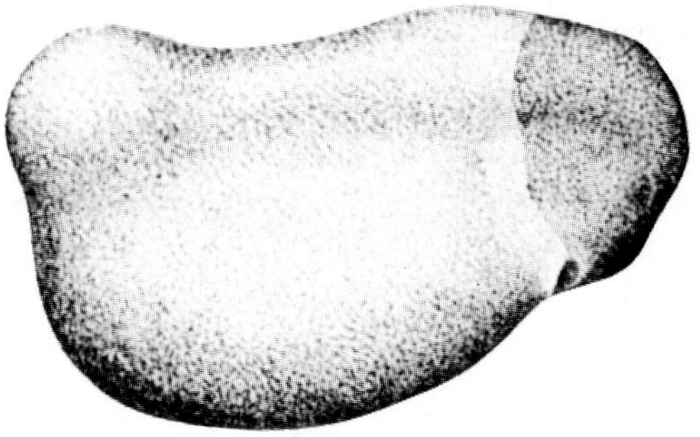

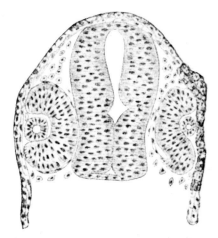

FIGURE 21.5. An operated embryonic frog illustrating a dark piece of ectoderm from the trunk over the right optic vesicle. From H. Spemann, *Embryonic development and induction*, Yale University Press, New Haven, 1938; reprinted by Hafner, New York, 1962.

same species or between members of different species were consistent, but, as in the earlier experiments, different results were obtained with different species. For example, *Rana pipiens*, which seemed to have early lens determination, proved incapable of growing lenses from ectoderm other than that which ordinarily covered the optic vesicle. On the other hand, *Bombinator pachypus* could form lenses from head ectoderm that did not ordinarily form lenses but not from trunk ectoderm, while another toad, *Bufo vulgaris*, could form lenses even from trunk ectoderm (Fig. 21.6).

In interpreting the results, Spemann emphasized cases in which lenses formed when ectoderm was brought into contact with optic vesicles. Others, like Wilhelm Roux, emphasized cases in which lenses failed to develop. Roux went as far as pre-

FIGURE 21.6. Dark trunk ectoderm formed a lens after being grafted over the right optic vesicle. From D. Filatow, Ersatz des linsenbildenden Epithels von *Rana esculenta* durch Bauchepithel von *Bufo vulgaris*. W. Roux' Arch. Entwicklungsmech. Organ., 105:475(1925), by permission of Springer-Verlag.

formationism could go when he allowed the possibility that lens formation could represent "dependent determination" of the sort occurring between a switch and a light bulb. Spemann, on the other hand, followed epigenesis to the logical extreme of cellular interactions when he originated the concept of primary embryonic induction. The problem was to find a model system in which the alternatives could be tested.

Primary Induction

Spemann was punctilious. Beyond paying strict attention to details, both practical and theoretical, he paid attention to the little things that did not fit. Like his contemporaries in genetics, he too "valued exceptions."

One of those exceptions cropped up in his ligation experiments on newt eggs. When the "noose" around an egg was not pulled tight and embryos did not separate, they frequently developed "double formations," that is, animals with two heads or even two heads and trunks. The double formations never had double caudal regions as long as they were joined anywhere (Fig. 21.7).

Spemann pondered how a constriction, which squeezed the embryo equally all around, could cause the doubling of anterior parts but not posterior parts. Following a mechanistic line of reasoning suggested to him by Hans Petersen (see Hor-

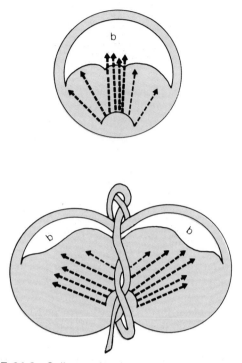

FIGURE 21.8. Cells moving into embryos along paths of least resistance. Dashed lines, hypothetical paths of cell movement; b, blastocoel; (Upper) Normal gastrula. (Lower) Constricted gastrula.

der and Weindling, 1985), Spemann conjectured that if cells in the lips of the amphibian blastopore moved inside the gastrula along paths of least resistance, cells would move toward the blastocoel whether it were single, as in the normal egg, or double, as a result of central constriction. The cells entering the ligated gastrulas would split into two groups and move toward the edges (Fig. 21.8). Spemann's attention was thus drawn from the outside of the embryo to the inside and from his ligature as the cause of double formations to the underlying cell layer.

By 1902, Spemann could formulate the question that was to preoccupy him for the rest of his life: Was there something in the lip of the blastopore which caused the formation of the embryo's specific parts? The trail of experiments he designed to answer the question was picked up by many embryologists and, in 1935, Spemann was awarded the Nobel Prize in Physiology or Medicine, the only biologist ever cited purely for research in embryology.

Two possible answers loomed: the first cells to move inside the embryo caused the formation of the embryo's head by altering the environment of other cells (i.e., the epigenic alternative); having once moved inside, the cells self-differentiated and formed the embryo's head by themselves (i.e., the

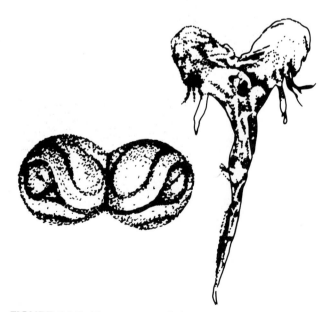

FIGURE 21.7. Two stages of double-formation development from loosely constricted eggs. From H. Spemann, *Embryonic development and induction*, Yale University Press, New Haven, 1938; reprinted by Hafner, New York, 1962.

preformation alternative). Having formulated his question about embryonic head formation in the same terms he had earlier formulated his question about embryonic lens formation, Spemann set out to perform the same sorts of experiments on heads as he had previously performed on lenses.

First, he ascertained that the embryo's head did not form when the dorsal lip was taken away, while head structures formed in tissue that did not normally form them when presented with an ectopic dorsal lip. One crucial question remained: Did the head form from the underlying tissue or did underlying tissue organize its surroundings to form a head?

The experiments designed to answer this question were performed by Hilde Pröscholdt, one of Spemann's students (see Hamburger, 1985). She made a series of grafts employing gastrulas of lightly pigmented *Triturus* (= *Triton*) *cristatus* as donors and gastrulas of various darker species such as *Triturus vulgaris* (= *Triton taeniatus*) as hosts. This particular combination provided the best opportunity to distinguish between host and graft pieces during subsequent differentiation and the chimeric embryos were relatively viable.

Still, only six survived to the point of embryogenesis, but these were quite enough. Grafted to the ventral side of other gastrulas (Fig. 21.9a, b), the pieces of dorsal blastoporal lip did not conform to the new site and develop into ectoderm, like the surrounding tissue. Instead, the grafts sank into and sometimes below the surface as if they were still on the lip of the blastopore.

The transplanted dorsal blastoporal lips did not lose their ability to act like dorsal lips, but ventral tissue in the vicinity of the graft began to act like tissue bordering a dorsal lip. Most conspicuously, the host tissue formed neural folds (Fig. 21.9d) resembling normal structures present on the host (c) but out of place.

When allowed to develop further, a second, small, incomplete, but unmistakable **secondary embryo** formed on one host where the graft had disappeared beneath the surface. At the same time, the host or **primary embryo** continued to develop normally (Fig. 21.10).

In histological sections, the small transplanted piece of dorsal blastoporal lip is seen to have elongated and differentiated into several tissues in the secondary embryo, while surrounding tissue of host origin has formed parts of the secondary embryo. The axis of the secondary embryo contains central nervous system tissue made largely of host tissue (Sek. Med. in Fig. 21.11) and a notochord made entirely of donor tissue. Somites along the

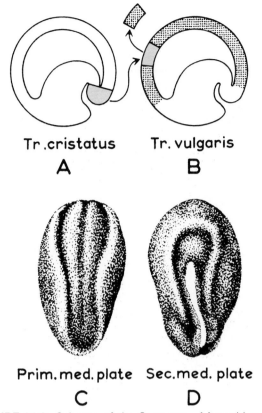

FIGURE 21.9. Scheme of the Spemann—Mangold experiment and results. (a, b) Graft between dorsal blastoporal lip from *Triturus cristatus* gastrula to ventral side of *Triturus vulgaris* gastrula. (c, d) Two views of specimen 1921 Um 8b, *T. vulgaris* host bearing *T. cristatus* graft. In this experiment the graft tissue did not totally pass below the surface of the host. Prim.med.plate, primary embryonic axis; Sec.med.plate, secondary embryonic axis.

sides of the central nervous system and the notochord contain some cells of both host and donor types, but the pronephros and gut lying near the secondary embryonic axis are made solely of host tissue.

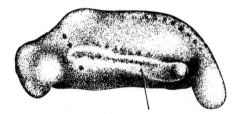

FIGURE 21.10. The specimen 1922 Um 132b at tailbud stage of development. Arrow, secondary embryonic axis. From H. Spemann and H. Mangold, *W. Roux' Arch. Entwicklungsmech. Organ.*, 100:599(1924); reprinted in B. H. Willier and J. M. Oppenheimer, *Foundations of experimental embryology*, Prentice-Hall, Englewood Cliffs, NJ 1964.

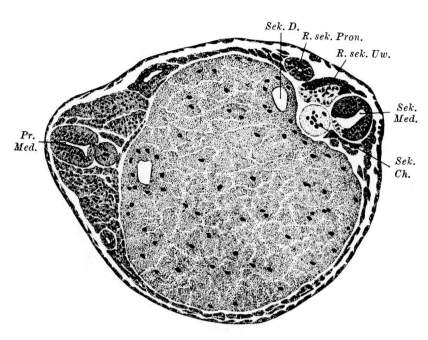

Sek. D.
R. sek. Pron.
R. sek. Uw.
Sek. Med.
Sek. Ch.
Pr. Med.

FIGURE 21.11. Cross section through specimen 1922 Um 132b, shown grossly in Fig. 21.10. The primary embryo is toward the left; the secondary embryo is toward the right. Only some of the parts of the secondary embryo are made up of the lightly pigmented *T. cristalis* cells. The remainder of the embryo is made up of deeply pigmented *T. vulgaris* cells. ×100. Pr. Med., medullary plate (i.e., central nervous system) of primary embryo; Sek. Med., medullary plate of secondary embryo; Sek.D., lumen of secondary intestine; R.Sek.Pron., secondary pronephros; R.Sek.Uw., secondary somite; Sec.Ch., secondary notochord. From H. Spemann and H. Mangold, *W. Roux' Arch. Entwicklungsmech. Organ.,* 100:599(1924); reprinted in B. H. Willier and J. M. Oppenheimer, *Foundations of experimental embryology,* Prentice-Hall, Englewood Cliffs, NJ, 1953.

In the words of Spemann and Hilde Mangold (née Pröscholdt, who died in a household accident before publication of her spectacular results):

> To the extent that they [the secondary embryo's tissues] are not formed by the *cristatus* cells of the implant, they must have originated from the parts of the host that either were already on the spot, or that came there under the influence of the organizer. This is quite evident for the neural tube; it is formed of cells which otherwise would have formed epidermis of the lateral body wall. (Spemann and Mangold, 1924. Translated by Hans Laufer, p. 173 in Willier and Oppenheimer, 1964)

Spemann coined the name **organizer** (i.e., *Organizator* or *Organisationszentrum*) for the dorsal lip of the blastopore and cells underlying the neural plate destined to produce the chordamesoderm, since he believed these cells unified tissues along an axis and integrated them into a coherent structure. He did not attribute static or preformational properties to the organizer but imagined it to be a *"Führer"* for dynamic morphogenic **fields** in the changing embryo. Neither a comprehensive theory nor a purely poetical metaphor, his organizer was to be a psychological tool for thought and inspiration, a provisional term intended to interpret new facts (see Horder and Weindling, 1985).

Spemann further suggested that two sorts of influences might operate in induction. An **organizer field** might move from cell to cell within a layer and integrate cells into tissues and structures. A **determination field** might pass from layer to layer altering determination and causing new tissues and structures to differentiate. Spemann con-

sidered both influences as examples of double assurance and dubbed the sum of these influences **primary embryonic induction.**

In no time, embryologists throughout Europe and in Japan were studying induction and testing hypotheses about the organizer. Then, as a result of the exodus of leading embryologists from Germany and the Third Reich, research on induction moved to America. The "golden age" of embryology arrived, and as far as active research was concerned, the age of Haeckel's germ-layer theory was over.

RESEARCH ON INDUCTION: EMBRYOLOGY'S GOLDEN AGE

Few compounds, other than the philosopher's stone, have been searched for more intensely than the presumed agent of primary induction in the amphibian embryo. This work has been seriously hampered by the fact that the specificity resides in the responding cells, not in the inductor which merely functions as a trigger.

S. Løvtrup, V. Landström, and H. Løvtrup-Rein
(1978, p. 24)

The rash of research on the organizer accompanying and following the publication of Spemann's and Mangold's (1924) seminal paper validated many of their assumptions and generally confirmed their results. The assumption that cells moved around the blastoporal lip of the amphibian gastrula was substantiated by Vogt (1925) and others as they traced the paths of vitally stained cells, and the conclusion that inductive influences moved between germ layers was vindicated when neural ectoderm was induced by chordamesoderm inserted in the blastocoel as a separate tissue.

Research on both the acting and reacting systems of inductive systems was intense and quickly extended the range of developmental events to which induction seemed relevant, but efforts to find inductive substances dominated the field. Spemann had no idea how induction actually worked. He believed induction was mediated through the cellular environment and speculated that inductive influences or **inductors** consisted of chemical or physical entities, but he made few assertions beyond that. In effect, he left the field open for others to move in and explain induction.

Techniques and Methods for Testing Systems

Ideally, the active properties of a system were tested when a reactive system with known properties was exposed to an unknown stimulus. In practice, the transplantation technique of grafting tissues from a donor gastrula to a host was long and laborious, and clear-cut results were rare. Nevertheless, Hermann Bautzmann persevered and tested all early gastrula tissues for their abilities to induce.

All inductively active tissue was found to lie in a **center of organization** or **determination** ranging roughly 90° dorsal and lateral to the blastopore (see Hamburger, 1985). Compared to Vogt's fate map, the center of organization corresponded to the presumptive prechordal mesoderm, notochord, and somite regions (see Fig. 14.7). The dorsal ectoderm that later formed the neural tube and other parts of the early gastrula had no inductive ability.

In time, three techniques originally developed in Spemann's laboratory replaced transplantation for the further study of induction. These techniques employed the ventral ectoderm of the amphibian early gastrula as the reacting system and assumed that it was capable of detecting and responding to any inducing activity present in a test material.

The **implantation** technique simply involved stuffing an **implant** into the blastocoel through a slit in a host embryo's ectoderm (Fig. 21.12). Were the implant to come into contact with the ventral ectoderm and induce it, subsequent differentiation could reveal the quality of the induction. Other events taking place in the embryo as a whole and

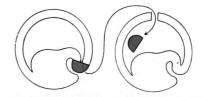

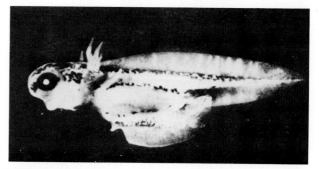

FIGURE 21.12. Drawings illustrating implantation of early and late dorsal blastoporal lips into gastrulas at an early age and photographs of resulting newts with secondary regions. (Upper) Dorsal lip from early gastrula transplanted to blastocoel of gastrula of same age. A head is induced. (Lower) Dorsal lip from older gastrula transplanted to blastocoel of gastrula at the same age as that in upper panel. A portion of the trunk and tail is induced. From L. Saxén and S. Toivonen, *Primary embryonic induction*, Prentice-Hall, Englewood Cliffs, NJ, 1962. Used by permission.

changes in the blastocoel might also influence the results, however.

The other techniques were extensions of Johannes Holtfreter's efforts to provide a **neutral environment** for tissue in which to study induction. The result was *in vitro* methods for culturing amphibian embryonic **explants** in a physiological salt solution (now known as Holtfreter's solution or the standard solution).

Isolated, the ventral ectoderm tends to curl into a mass offering limited access to inducers.

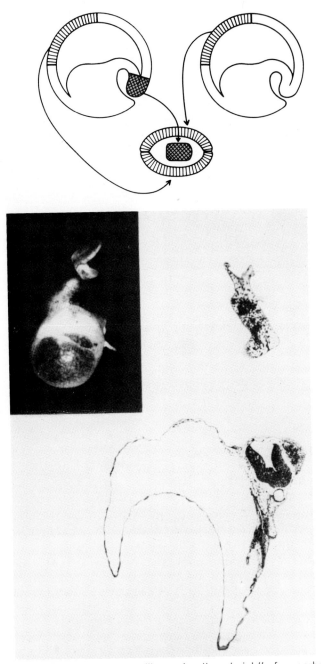

FIGURE 21.13. Drawing illustrating "sandwich" of an early dorsal blastoporal lip between two pieces of ventral ectoderm. Accompanying plate shows resultant "embryo" and photomicrograph of a section through the "embryo." Without the dorsal blastoporal lip, the gastrula ectoderm would have formed a vesicle of abnormally clumped epithelial cells. In the presence of the early dorsal lip, neural structures of the head, trunk, and tail, as well as a conspicuous tail fin within a well-differentiated epidermis, have appeared. From L. Saxén and S. Toivonen, *Primary embryonic induction*, Prentice-Hall, Englewood Cliffs, NJ, 1962. Used by permission.

Holtfreter compensated for this tendency and developed his **sandwich technique** by allowing explants of ventral ectoderm to heal around a **test piece** of tissue or a neutral material such as agar which had absorbed a test solution (Fig. 21.13).

Alternatively, ventral ectoderm was flattened with silk cloth *in vitro* and exposed to test solutions (see Tiedemann, 1978). In the absence of induction, the ventral ectoderm formed an abnormal epithelium consisting of strands of cuboidal cells. Induction, on the other hand, caused the production of a variety of differentiated tissues including well-differentiated epidermis.

While more difficult than implantation, the *in vitro* techniques minimized the pitfalls of implantation. Isolated or in a sandwich, the ectoderm was not influenced by events in the embryo as a whole, and differentiation was attributable directly to action of the test piece or test solution. Furthermore, the sandwich technique eliminated ambiguity over the duration of contact between the test piece and the reacting tissue.

Ideally, reactive properties of a system were tested by introducing a known active system to an ectopic *in vivo* site or novel *in vitro* situation. Of course, the possibility always remained that the site or situation was part of an active system, and, after years of experience with amphibian tissue culture, Holtfreter utterly despaired of devising a completely neutral environment with no inductive effects of its own. Not even the simplest *in vitro* situation could legitimately be considered "neutral" by way of having no influence on tissues.

Reactive capacities were measured in both host and donor tissues at transplantation sites, but primarily in host tissue at implantation sites and in donor tissues in *in vitro* experiments. In addition, the reaction of presumptive ectoderm to different active systems was ascertained by appending folded ectoderm from early gastrulas to the neural ectoderm above the archenteron roof of gastrulas or later stage embryos (Fig. 21.14d) (see Nieuwkoop et al., 1952). Similarly, the reaction of ectoderm in different stages of induction was ascertained by appending folded ectoderm from later gastrulas to the ventrocephalic region of gastrulas (Fig. 21.14, e).

Reacting System: How It Acquires and Loses Reactivity

In the course of development, parts of the embryo gained and lost reactive capacities. Spemann equated the maximum ability to react with an **indifferent state,** but Conrad H. Waddington (1905–1975) equated the ability of tissue to re-

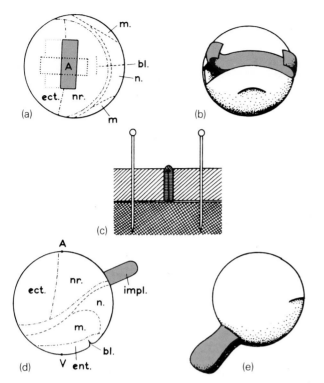

FIGURE 21.14. Scheme for testing reactive systems. A ribbon of reactive tissue cut from the animal pole of a donor gastrula (a, b) is folded between slabs of agar (c) and implanted at the desired position in a host gastrula; for example, at the border of the presumptive neural ectoderm and notochord (d) or in the ventral ectoderm (e). A, animal pole; bl., blastopore; ect., presumptive ectoderm; ent., presumptive endoderm; impl., implant; m., presumptive myotome; n., presumptive notochord; nr., presumptive neural ectoderm; V, vegetative pole. Adapted from P. D. Nieuwkoop, *J. Exp. Zool.,* 120:1(1952), by permission of Alan R. Liss, Inc.

spond to inductive influences with **competence** and considered degrees of competence as functions of a tissue's physiological state. In the context of induction, **potency,** the range of inducible histo-

types into which an embryonic tissue differentiated, merged with competence, and Waddington proposed measuring potency on a physiological scale.

Competence for neural induction was already present in the midblastula presumptive ectoderm of *Cynops* (= *Triturus*) *orientalis,* since it formed neural tissue after exposure to neural-inducing conditions, such as brief treatment with calcium-free medium (see Nieuwkoop et al., 1985). In the axolotl, presumptive ectoderm acquired competence in stages. Competence for mesoendodermal induction (Fig. 21.15, mes.c.) appeared in the morula stage (stage 6), and competence for neural induction (neur. c.) appeared in the blastula stage (7).

In 1926, Fritz Erich Lehmann reported that ectoderm lost its ability to respond to inductive influences as the embryo moved past the gastrula stage. Tested as folds appended to the presumptive hindbrain (i.e., rhombencephalic) region of neurulas, presumptive ectoderm lost its competence for mesodermal and neural induction by the late gastrula stage (Fig. 21.15, stage 12). So-called **transformation competence,** or the ability of a tissue to respond to strong inductors by changing regional determination but not tissue determination, replaced or outlasted competences for mesodermal and neural induction, but transformation competence also disappeared in the neurula (stage 14).

Inductive tissues seemed to alter the environment of competent reacting tissues in such a way as to activate specific latent potencies. At the same time that cells became **committed** to specific lines of differentiation, they were **restricted** or compelled to lose competences and potencies for other lines (see Fig. 12.1).

The loss of competence did not seem to be an entirely passive event, since it was influenced by both RNA and protein synthesis. Actinomycin D, while inhibiting transcription, reduced compe-

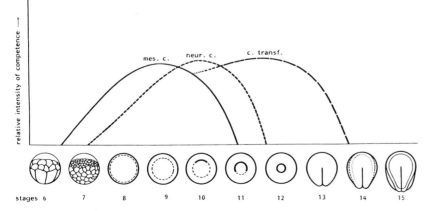

FIGURE 21.15. Relative intensities of competence for induction as a function of stages of development in the axolotl. mes. c., competence for mesodermal induction; neur. c., competence for neural induction; c. transf., competence for transformation from hindbrain structures to forebrain structures and vice versa. From P. D. Nieuwkoop, A. G. Hohnen, and B. Albers, *The epigenetic nature of early chordate development: Inductive interaction and competence,* Cambridge University Press, Cambridge, 1985. Used by permission.

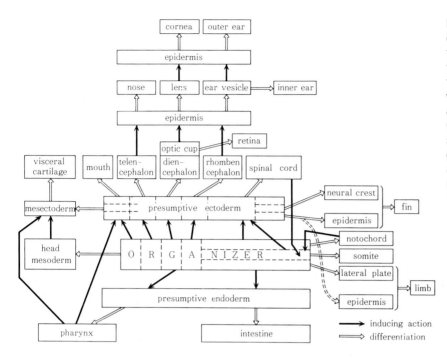

FIGURE 21.16. Hypothetical interactions controlling morphogenesis and differentiation in a variety of organs. Beginning with the organizer, primary, secondary, and tertiary inductions (closed arrows) dovetail with progressive states of determination and differentiation (open arrows). From O. Nakamura, Y. Hayashi, and M. Asashima, in O. Nakamura and S. Toivonen, eds., *Organizer—a milestone of a half-century from Spemann*, Elsevier/North-Holland Biomedical Press, Amsterdam, 1978. Used by permission.

tence in reacting cells, and cycloheximide, while inhibiting translation, also interfered with mesoderm formation (see Tiedemann, 1978).

For Spemann, **determination** was the sum of changes accompanying induction. Induction, not determinants, produced determination. Adapting Roux's term **self-differentiation** to the phenomenon of induction, Spemann redefined it as an acquired quality directing cells toward their ultimate fate. A self-differentiating or determined tissue was one that fulfilled its destiny without further prodding by induction.

Theoretically, before a tissue was self-differentiating, it could have undergone different types of induction and one or more rounds in a series of progressive inductions and determinations. In the test situation, **contact induction,** or the initial interaction of active and reactive systems upon contact, could alter determination, and **assimilative induction,** the integration of tissues from different sources into structures (sometimes called regulation), could drive a tissue to fill in local and ectopic structures.

Normally, secondary, tertiary, and further inductions could follow primary induction. Multiple steps or series of inductions could lead a tissue into comparably advanced stages of determination and direct development toward increasingly precise and narrow channels.

Were this the case, primary embryonic induction would not be a single massive event but only the first event in a cascade or chain reaction of small inductive events (Fig. 21.16). Determination

would not then represent a state so much as a tier in a scale of determination steps. Possibly, at each level, a more determined tissue induced a less determined tissue, until all the steps in the hierarchy were occupied.

Acting System: How It Acquires and Loses Activity

An acting system must resist inductive influences generated by other acting systems as well as generate its own inductive influences. In general, resistance to local inductive influences and the ability of a tissue to act as an inducer are related to the tissue's determination. But are inducing tissues also determined in the sense of differentiating in accordance with developmental fate?

Originally, explants containing presumptive endoderm were reported to differentiate, and endoderm was considered the first of the germ layers to be determined. Had this been the case, the germ layer with the least inducing ability would have been judged the first to be determined. Disputes about the limits of presumptive endoderm in different species of amphibians and consequent errors in identifying the content of explants, however, obscured the results and interpretation of the experiments. It seemed that explants of presumptive endoderm only differentiated when contaminated with presumptive mesoderm.

Then, in 1939, Holtfreter published the results of a large-scale *in vitro* study with explants from salamander and frog embryos. The only explants to

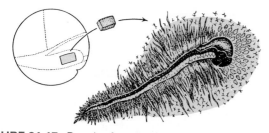

FIGURE 21.17. Result of explanting a relatively large piece of presumptive mesoderm. An embryonic axis with notochord and neural tube expanded at one end and capped with epidermis is surrounded by laterally migrating mesenchyme cells and giant head mesenchyme cells. From J. Holtfreter and V. Hamburger, in B. H. Willier, P. A. Weiss, and V. Hamburger, eds., *Analysis of development*, Saunders, Philadelphia, 1955. Used by permission.

differentiate were from the marginal zone containing portions of Bautzmann's center of organization and corresponding to presumptive mesoderm and chordamesoderm (see Holtfreter and Hamburger, 1955). Explants of ventral and dorsal ectoderm failed to differentiate normally, and explants of endoderm disintegrated (see Nakamura et al., 1978).

The ability of marginal zone explants to differentiate *in vitro* was strongly influenced by the size or mass of the explanted tissue. Relatively large pieces of presumptive mesoderm formed an axis and differentiated an axial neural tube and notochord capped by epidermis and accompanied by laterally migrating mesenchyme cells (Fig. 21.17). Smaller pieces of presumptive mesoderm developed relatively disorganized tissues without regard to their sites of origin (see Fig. 14.9). Differentiation was forthcoming, and organized development followed when the size of an explant or its complexity was sufficient to provide for induction.

Since the center of organization in general and the dorsal blastoporal lip of an early gastrula in particular self-differentiated, while other parts of the gastrula were still indifferent and subject to induction, the dorsal lip seemed to be the first part of the embryo to be determined. Curiously, while induction of other embryonic parts preceded determination, determination in the active organizer preceded induction.

The organizer is **mesodermalized** or determined in a unique way by early endoderm (see Smith et al., 1985). In 64-cell morulas and in blastulas, transplanted presumptive dorsal endoderm induces mesodermal organs of various types, even though the endoderm is not determined at these stages. The influence may move through polarized cellular contacts with the dorsal marginal zone, or

the blastocoel may act as a passive barrier to mesodermal influences (see Fig. 14.27) (Nieuwkoop, 1973).

Other ideas on what determines the organizer are more convoluted. Hypotheses on physiological aspects of competence and potency or metabolic gradients within the embryo, in particular, have frequently paralleled ideas on the organizer's determination.

Induction and Physiological Gradients

Initially, inductive influences were thought to arise from an embryonic center of organization and diffuse through tissues in a single layer. Spemann acknowledged Boveri's priority (see Spemann, 1938) to the idea of morphogenic gradients, but Charles Manning Child (1946) deserves the bulk of credit for drawing attention to parallels between physiological and morphological gradients in vertebrate embryos.

Child's **axial gradients** were polarized gradients of physiological activity that originated in dominant regions and followed the utilization of oxygen. Child's claims were tested in Joseph Needham's and Lester G. Barth's laboratories where susceptibility to metabolic inhibitors (e.g., cyanide) was shown to begin in the animal pole and organizer region. Likewise, the distribution of metabolically relevant substances (e.g., sulfhydryl compounds, alkaline phosphatase, and RNA) and the higher respiratory quotient of the dorsal lip compared to the ventral lip were consistent with Child's theory.

Physiological gradients attracted a great deal of attention and, like any good scientific hypothesis, contained the seeds of their own negation. Ultimately, metabolism failed the test for relevance to morphogenesis (see Brachet, 1950). The sharp animal–vegetal gradient in respiration of the early gastrula was linked to the accumulation of yolk in the vegetal pole, and respiration in the organizer was found to be less than that in the ectoderm.

Metabolic models for induction were not without benefits for embryology, especially for the emergent branch of theoretical embryology. Concepts of metabolic gradients turned embryologists' attention back to Spemann's more general concept of an **organization field** with definite direction and extent.

The field concept implied by Spemann's was an extrapolation of the field concept in physics, namely, a continuously distributed entity that accounts for action at a distance. The dorsal lip had not merely differentiated, after all, but had exerted its influence over the host throughout the area incorporated into the secondary embryo.

Two additional concepts of fields evolved from Spemann's organization field. **Morphogenic fields** theoretically integrated developmental influences over time and space. For example, a morphogenic field might integrate complex signals emitted by the organizer, thereby creating the regional specificity characteristic of the secondary embryo.

Embryonic fields, on the other hand, are the widest areas potentially capable of developing into particular structures or tissues. Such a field is not merely competent as a reactive system but capable of generating its own active system as well.

Complexity of the Organizer

Spatial and temporal complexities of the embryo imply the existence of similar complexity in the organizer. The challenge is to understand how the exercise of primary embryonic induction properties is integrated over space and time (see Nieuwkoop et al., 1985).

Regional specificity was implied in Spemann's explanation for double embryos (Fig. 21.7). He also interpreted the results of his own early transplantation experiments in terms of regional specificity in invaginated tissue, but efforts by Holtfreter to experimentally interrupt gastrulation provided the first direct evidence for progressive induction by invaginated tissues.

Gastrulation was interrupted when membrane-free urodelan blastulas were placed in slightly hypertonic salt solutions. Cells moved outward rather than inward (i.e., the embryos exogastrulated), and **exogastrulas** rather than gastrulas formed. The ectoderm was left a wrinkled vesicle, and the endoderm and mesoderm became an externalized mass (Fig. 21.18). Like ectodermal explants, the ectodermal vesicle failed to differentiate, forming only an abnormal epithelium. The mesoderm, on the other hand, became covered with endoderm, elongated, and differentiated a variety of mesodermal derivatives.

Similarly, **partial exogastrulas** resulted from the treatment of frog gastrulas with basic salt solutions (and a variety of other treatments). Allowed to develop, the malformed embryos showed various degrees of ectodermal organization in proportion to the amount of invaginated tissue underlying the ectoderm (Fig. 21.19). The persistence of caudal abnormalities in embryos with more or less normal cephalic regions suggested that cephalic and caudal organizers were separate entities.

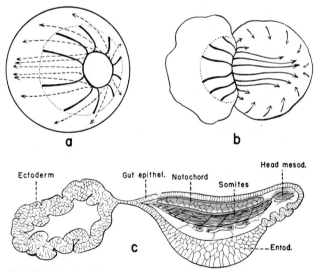

FIGURE 21.18. Drawings illustrating direction of cell movements (arrows) during normal gastrulation (a), exogastrulation (b), and differentiation of mesodermal derivatives in the exogastrulated endomesodermal part of an exogastrula (c). From J. Holtfreter and V. Hamburger, in B. H. Willier, P. A. Weiss, and V. Hamburger, eds., *Analysis of development*, Saunders, Philadelphia, 1955. Used by permission.

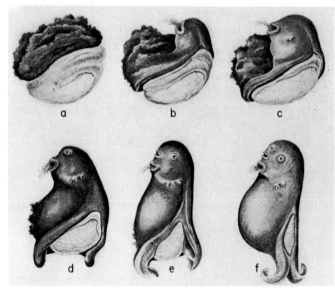

FIGURE 21.19. Drawings of a *Rana pipiens* exogastrula (A) and of malformed embryos developing from partial exogastrulas (B–F) treated briefly in an alkali medium as early gastrulas. All the embryos are of the same age. The degree of ectodermal organization is proportional to the extent of invagination. From J. Holtfreter and V. Hamburger, in B. H. Willier, P. A. Weiss, and V. Hamburger, eds., *Analysis of development*, Saunders, Philadelphia, 1955. Used by permission.

Implantation experiments (Fig. 21.12) revealed that dorsal blastoporal lips of early gastrulas induced head structures, while those of late gastrulas induced trunk and tail structures. The cells that migrated around the dorsal lip first and became prechordal mesoderm and anterior notochord (according to the fate map) induced more anterior structures in the ectoderm. Cells that migrated around the dorsal lip later and became posterior portions of the notochord induced more posterior structures.

The idea of **regional specificity** in the organizer was also consistent with the ability of implants from the roof of the urodelan archenteron to induce different structures in early gastrulas. Sandwiches of ectoderm responded differentially to test pieces of cephalic and caudal mesoderm, and even previously induced neural tissue induced neural ectoderm (so-called homoiogenetic neural induction). Additional efforts to test tissues and materials for specificity turned embryologists' attention away from the organizer and toward specific substances or **inducing substances** whose actions might organize the ectoderm.

Abnormal Inducers

The search for inducing substances began in earnest in 1932 when Alfred Marx induced secondary embryos with organizers of embryos treated with the anesthesia chloreton. If physiologically suppressed or inactive organizers could still induce a secondary embryo, induction could have depended on passive processes such as the diffusion of chemical substances.

In the same year, Bautzmann and others in Spemann's group showed that organizers treated with

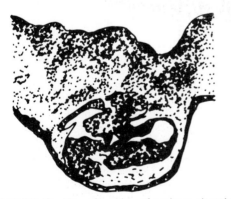

FIGURE 21.20. Section of portion of embryo showing good neural induction under influence of implanted dead organizer from which fat was removed. From J. Holtfreter, *W. Roux' Arch. Entwicklungsmech. Organ.,* 132:225 (1934), by permission of Springer-Verlag.

TABLE 21.1. Summary of Holtfreter's observations on the inducing ability of test pieces

Very powerful inductors:
  Coagulated chick embryo extract.
  Liver, kidney, adrenal, heart, brain of mouse.
  Thyroid, kidney, liver, brain, tooth of man.
  Bottom layer of centrifuged brei of calves' liver.
  Liver of lizard, frog, *Triton.*
  Brain and retina of salamander.
  Heart, ovarian egg, muscle and liver of fish.

Powerful inductors:
  Lens of eye of mouse.
  Centrifuged calves' liver brei (middle and upper layers).
  Thyroid, kidney, liver, testicle, adrenal, fat of bird.
  Kidney and testis of lizard.
  Heart and limb bud of *Triton.*
  Frog muscle.
  Dragonfly ganglion substance.
  Lymph from larva of *Sphinx.*
  Limnaea liver.
  Daphnia extract.

Weak inductors:
  Blood, fatty tissue from mice.
  Liver, heart of salamander.
  Retina of *Triton* larva.
  Fatty tissue from dragonflies.
  Imaginal discs from caterpillar of *Vanessa.*
  Foot muscle of *Planorbis* and *Limnaea.*
  Subcutaneous muscle of enchytrids.

Inactive inductors:
  Ectoderm, endoderm of living gastrula.
  Tadpole gills.
  Starch, agar, egg albumin, pork fat, animal charcoal.

Source: J. Brachet, *Chemical embryology* (translated from the French by L. G. Barth), Interscience, New York, 1950.

alcohol, heat, or frozen also had inducing activity, even though these organizers were dead. Induction, it seemed, was chemical.

Holtfreter and many others proceeded to test a great many materials as implants (Fig. 21.20), as test materials in ectodermal sandwiches, and in solutions bathing explants *in vitro* (Table 21.1). Inducing activity appeared in some very unexpected **heterogenous** (meaning originating outside the body) sources, but perhaps strangest of all was dead amphibian presumptive epidermis and endoderm. These inducers had no inductive activity whatsoever while living.

Structures induced by heterogenous inducers were generally bigger than those induced by implants of the organizer. Beyond their size, the structures never had a completely normal histological appearance, and parts were not generally in their appropriate anatomical relationships. Instead of the normal regional characteristics of brain vesicles and diverticula (such as eye vesicles and pineal rudiment) appearing on specific regions, results with heterogenous inducers were frequently complex jumbles of abnormal neural structures (Fig. 21.20).

Waddington called the ability to induce a regionally organized rudimentary central nervous system **individuation** and the ability to induce nervous tissue in an atypical mass **evocation.** The active substances from heterogenous inducers possessing limited abilities to organize nervous tissue were tentatively classified as **evocators,** while **modulators** controlled the individuation of neural parts. The dorsal lip of the blastopore retained sole right to the title of organizer, since it alone induced all parts of the central nervous system.

Evocators

Evocator activity was ultimately attributed to a large number of compounds. Waddington, Needham, and others working at Cambridge produced weak evocation with sterols that they extracted with ether from amphibian gastrulas, neurula tissues, and even cow liver. Success with synthetic polycyclic hydrocarbons led these authors to the "sterol theory" of induction.

The German-Swiss school found stronger evocator activity in a saponified lipid fraction containing nonsterol fatty acids. Their "acid-stimulus theory" elevated acidic chemicals such as adenylic acid, ATP, and nucleic acids to the post of evocators, while in Belgium, on the basis of his pioneering histochemical studies, Jean Brachet pronounced RNA the active factor in induction.

Basic lipids and several vital dyes also acted as evocators, and Japanese workers induced neural tubes with kaolin and crystals of *p*-quinone. Finally, Barth, working in America, showed that even lipid-free remains of dead organizers were excellent evocators (Fig. 21.20). Ultimately, the range of unrelated chemicals with evocator activity proved so large that an indirect mechanism for evocation seemed more likely than a direct response.

The possibility that evocators were inherently morphologically inert or neutral was tested by Holtfreter in 1944 by briefly exposing explants of ectoderm to a variety of innocuous agents (pH changes produced by CO_2). The **spontaneous neuralization** or **autoneuralization** of these explants and similar explants exposed to ammonia and simple salts such as NaCl and LiCl suggested that induction did not depend on the quality of the inducing agent. Since all the agents tested produced a brief or reversible cytolysis, sublethally damaged cells might have released their own inducers.

Induction may, after all, depend on the release of active ingredients from an inactive complex in damaged cells. The extensive rearrangements of large yolk platelets and actin filaments in early embryos (Wilt and Phillips, 1984) may place associated masked RNA in position where, as first suggested by Barth, evocators or reversible cellular trauma operates to **unmask** them.

"Unnatural" Inducers and Modeling

Although regional specificity is not associated with the idea of evocators, "unnatural" inducers are found with strongly regionally specific inductive activities. Tissues from rats to carp, and metallic ions from lithium to zinc (normally absent in gastrulas) have polarized or regionally specific inductive effects (Table 21.2).

Regionally specific "unnatural" inductors were first described in 1938 when Sulo Toivonen showed that guinea pig kidney induced spinal cord and associated structures in newt gastrulas and when Hsiao-Hui Chuang demonstrated that mouse kidney induced forebrain structures, while newt liver induced trunk and tail structures. Liver lost its trunk- and tail-inducing activity and acquired forebrain-inducing activity upon brief boiling, while both kidney and liver lost forebrain-inducing activity only after prolonged boiling.

Following Lehmann's suggestion, the head or cephalad inductor was called an **archencephalic** (forebrain, nose, eye) inductor (sometimes prosencephalic inductor), and a trunk, tail, or caudal inductor was called a **spinocaudal** inductor. Originally, a distinct **deuterencephalic** (hindbrain and

TABLE 21.2. "Unnatural" inductors

Neural inductors that induce archencephalic structures: Guinea pig liver (alcohol-treated, etc.), boiled *Triturus* liver, heat-treated HeLa cells, Zn^{2+}, NH_4^+, sodium thiocyanate in conjunction with Li^+-free, Ca^{2+}-free, and Mg^{2+}-free solutions.

Mesodermal inducing factors (MIFs) of ectoderm: Conditioned media from *Xenopus laevis* XTC cells and some other cell lines, heparin-binding growth factors, fetal calf serum in conjunction with Ca^{2+}-free and Mg^{2+}-free solutions (sometimes mammalian liver).

Combined neural–mesodermal inductors that elicit rhombencephalic and spinocaudal structures (destroyed by heat treatment): Guinea pig kidney, fresh *Triturus* liver.

Primarily endodermal inductors also known as vegetalizing factors (require prolonged contact compared to neural inductors): Guinea pig bone marrow, carp swim bladder, untreated HeLa cells, 9–12-day-old chick embryos, Li^+.

Source: Data compiled mainly from Nieuwkoop et al. (1985).

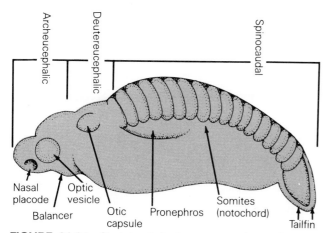

FIGURE 21.21. Sketch of Ambystoma embryo showing structures characteristic of archencephalic, deuterencephalic, and spinocaudal regions.

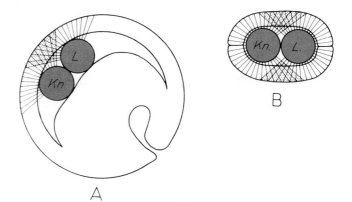

FIGURE 21.22. Double implant and double test material experiments. Guinea pig liver (L) and bone marrow (Kn) employed simultaneously as implants (a) or test pieces within sandwiches of competent ectoderm (b) hypothetically release overlapping active fields of archencephalic and spinocaudal inductive activities. From S. Toivonen, in O. Nakamura and S. Toivonen, eds., *Organizer—a milestone of a half-century from Spemann,* Elsevier/North-Holland Biomedical Press, Amsterdam, 1978. Used by permission.

ear) inductor was also thought to exist (Fig. 21.21) (see Toivonen, 1978).

Instead of inducing specific parts or regions of the central nervous system, "unnatural" inducers may actually induce different tissues. Archencephalic inductors are primarily inducers of nerve tissue (i.e., neuronal inductors), and spinocaudal inductors are inducers of mesodermal derivatives (i.e., mesodermal inductors) or mesoendodermal derivatives (i.e., mesoendodermal inductors earlier called vegetalizing factors).

Inasmuch as the ectoderm and mesoderm are continuous around the blastoporal lips in the posterior region of the gastrula, it is not too surprising that inducers of posterior parts of the central nervous system are also inducers of mesoderm. Presumptive ectoderm retains its competence for mesodermal induction late into the gastrula stage (Fig. 21.15) and its competence for neural ectoderm later still. The ability of the tailbud to give rise to both spinal cord and underlying mesoderm suggests that inductive competence may continue well into organogenesis if only in the tailbud.

Double-gradient systems were first proposed by Holtfreter as the simplest solution to the induction of intermediate structures lying between two distinctly different inductive fields. In 1950, Toivonen and Lehmann simultaneously proposed that two principal inductors were distributed along opposite gradients on the dorsal side of the gastrula. A neuralizing principal was thought to be strongest in the forebrain region and a mesodermalizing principal in the trunk and tail regions.

Impressive evidence in favor of a doublegradient system comes from the results of dual-

implant experiments (Fig. 21.22). Under the influence of liver plus bone marrow tissue, even explanted ectoderm formed "quasiembryos" with taillike structures at one end and eyelike structures at the other end (Fig. 21.23).

In order to test the roles of neural and mesodermal inductors quantitatively, implants were prepared of inductors mixed in various proportions. For example, good mesodermal or spinocaudal inductors (e.g., unheated human tissue culture cells known as HeLa cells) were mixed with good neural or archencephalic inductors lacking spinocaudalinducing activity (e.g., heated HeLa cells or cells exposed to heated human serum or protein-free medium). The mixtures induced more tissues and structures, including deuterencephalic structures, than either inductor alone (Fig. 21.24).

A three-dimensional model extrapolated from these results (Fig. 21.25) accommodates bilateral features of primary embryonic induction as well as axially polarized features. A laterally graded distribution of neural inducer (N, i.e., the neural gradient) is uniform in the anterior–posterior direction, while a longitudinally graded distribution of mesodermal inducer (M, i.e., the mesodermal gradient) is less conspicuously graded laterally.

According to Saxen's and Toivonen's (1962) concept, the neural gradient is a prerequisite for induction of every region. Archencephalic (A) induction only requires the neural inducer. Bilaterally distributed archencephalic structures, such as eyes, nasal placodes, and balancers, appear at ap-

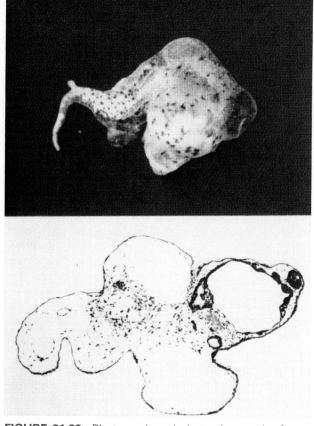

FIGURE 21.23. Photograph and photomicrograph of section through quasiembryo developed from sandwich of competent ectoderm exposed simultaneously to both guinea pig liver and bone marrow. A taillike structure at one end is combined with an eyelike structure at the other end. From L. Saxén and S. Toivonen, *Primary embryonic induction*, Prentice-Hall, Englewood Cliffs, NJ, 1962. Used by permission.

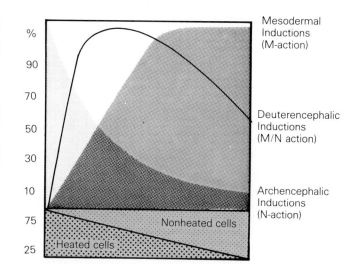

FIGURE 21.24. Double-gradient hypothesis based on results with mixtures of unheated human HeLa cells and heated HeLa cells. The heated cells are strong archencephalic or neural (N-action) inductors, while the unheated cells are strong spinocaudal or mesodermal (M-action) inductors. Intermediate or deuterencephalic induction appears when heated and unheated cells are combined (M/N action). From L. Saxén and S. Toivonen, *Primary embryonic induction*, Prentice-Hall, Englewood Cliffs, NJ, 1962. Used by permission.

propriate levels of the neural gradient on either side of the embryo. Spinocaudal (S) structures also require high levels of mesodermal induction, while deuterencephalic (D) structures are induced at lower levels of the mesodermal gradient.

Time-graded properties of inductors from different sources also appear in models of primary embryonic induction. The role of time has been ascertained by removing portions of the neural ectoderm after exposure to the archenteron roof for various periods (Fig. 21.14). Archencephalic induction proceeds from the briefest period of exposure, whereas deuterencephalic induction and spinocaudal induction require considerably longer periods (Table 21.3).

Nieuwkoop's temporal model for neural induction (Fig. 21.26) provides a sliding posterior-to-anterior scale for regionalization of the central nervous system. As prechordal endomesoderm migrates along the roof of the blastocoel, posterior regions of the presumptive neural ectoderm are exposed to an inductor for longer periods of time and hence undergo stronger transformations (shown as denser hatching).

Molecular Approaches to the Acting System

Interrupted by World War II, interest in induction waned. Frustrated by difficulty in distinguishing between a substance that merely caused sublethal cytolysis and a substance that actually changed a cell's determination, many embryologists abandoned the search for inducing substances. Others resumed work in the postwar years on induction by testing the prevailing molecular prejudices.

Sulo Toivonen and Lauri Saxén and co-workers in Finland tested various centrifugation fractions and extracts of tissues for archencephalic-inducing activity (Table 21.2). They found high activities in a guinea pig kidney fraction containing granules of RNA and protein. Similarly, Tuneo Yamada and others in Japan attributed archencephalic induction to ribonucleoprotein (RNP) isolated from guinea pig liver. But while the Japanese workers also attributed spinocaudal and mesodermal induction to RNP from guinea pig kidney, the Fin-

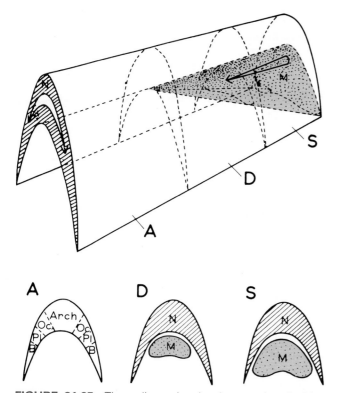

FIGURE 21.25. Three-dimensional scheme of a double-gradient model for primary embryonic induction (above) and cross sections of gradients at different levels (below). A, archencephalic region; D, deuterencephalic region; S, spinocaudal region. N, neural inducer; M, mesodermal inducer. Arch, archencephalon; Oc, eye; Pl, nose; B, balancer. From S. Toivonen, in O. Nakamura and S. Toivonen, eds., *Organizer—a milestone of a half-century from Spemann*, Elsevier/North-Holland Biomedical Press, Amsterdam, 1978. Used by permission.

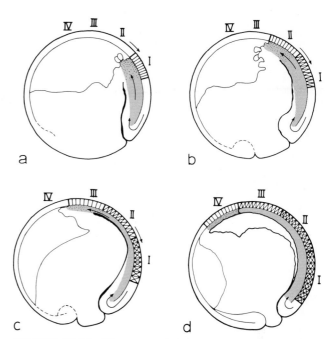

FIGURE 21.26. Diagrammatic illustration of Nieuwkoop's temporal model for neural induction. Prechordal endomesoderm (stippled) initiates neural induction (a) as it migrates along the roof of the blastocoel, but stronger induction (denser hatching) occurs as regions of the presumptive neural ectoderm (I–IV) are exposed to inductive influences for longer periods of time (b–d). From P. D. Nieuwkoop, A. G. Hohnen, and B. Albers, *The epigenetic nature of early chordate development: Inductive interaction and competence*, Cambridge University Press, Cambridge, 1985. Used by permission.

nish workers identified a mesodermalizing factor in a guinea pig bone marrow protein (see Toivonen, 1978).

Membrane-bound ribosomes of guinea pig liver and chick embryo cells have inducing activity, while all other cell fractions lack inducing activity. Curiously, the results of enzymatic digestion of active fractions of liver RNP indicate that most inducing activity is found in protein rather than the RNA.

In Germany, Friedrich Tiedemann and others isolated a **vegetalizing factor** from 9–12-day chick embryos capable of inducing mesodermal structures in explants of animal caps from frog blastulas. The factor is a 28–30 kilodalton (kd) protein consisting of 13-kd subunits. In addition, an inactivator of this protein was found in chick embryos (Tiedemann, 1978).

Recently, interest in induction has shifted to the earlier induction of mesoderm in blastulas (see Chapter 14) where presumptive ectoderm forms mesodermal structures under the influence of presumptive endoderm. Mesodermal inducing activ-

TABLE 21.3. Time-graded induction

| | Duration of Induction (Hours) | | |
Species	Archencephalic	Deuterencephalic	Spinocaudal
Ambystoma (= *Siredon*) *mexicanum*	0.5–1	4–16	12–16
Cynops (= *Triturus*) *pyrrhogaster*	3	15	18

Source: Data summarized from Nieuwkoop et al. (1985).

ity is attributed to **mesoderm-inducing factors** (MIFs) related to growth factors.

MIFs are substances that induce ectodermal explants to form blood, mesenchyme, mesothelium, muscle, and, at high concentrations, notochord. Neural tissue is also induced at high concentrations presumably as a result of progressive induction by notochord or muscle.

MIFs were first found in heat- or acid-activated **conditioned media** (CM) in which *Xenopus laevis* tissue culture cells (XTC) were previously grown (hence, XTC-CM) (Smith, 1987). XTC-CM MIFs were accompanied by MIF-inhibitors, possibly glycoproteins or proteoglycans in the conditioned media.

Several growth factors have been nominated as candidates for MIFs. **Heparin-binding growth factors** (HBGHs) did not fare well, since MIFs do not bind to heparin nor otherwise resemble the HBGHs (Godsave et al., 1988). **Transforming growth factor β2** (TGF-β2), on the other hand, has prospects, since XTC-CM has TGF-β2-like activity, as measured by several biological assays, and the medium's mesodermalizing activity is inhibited by antibodies to mammalian TGF-β2 (Rosa et al., 1988). Moreover, the *Vg1* gene implicated in mesodermal induction in the blastula, encodes a member of the TGF-β2 protein family (Weeks and Melton, 1987).

Purified bovine basic **fibroblast growth factor** (FGF) induces primarily ventral mesodermal products like blood cells in presumptive ectodermal animal caps (Slack et al., 1987), while mammalian TGF-β2 in the presence of FGF induces dorsal mesodermal products like muscle α-actin (Kimelman and Kirschner, 1987). Different growth factor-like proteins working synergistically may be responsible for inducing a spectrum of mesodermal tissues.

In general, MIFs, like Tiedermann's vegetalizing factor from chicks, are acid-stable proteins with 10–13-kd subunits. Although the MIFs seem to fall into a small class of molecules resembling each other, heterogeneity, the nemesis of studies on neural inducing factors, also daunts studies on MIFs.

Molecular Approaches to the Reacting System

Even if the concept of inducing substances has lost its robustness, the idea of primary embryonic induction or an organizer guiding early embryogenesis through tissue interactions is still very much alive. Moreover, recombinant-DNA technology is offering new possibilities for approaching the organizer problem.

The induction of an embryonic tissue that commits it to a specific pathway of differentiation may be mediated by the transcriptional activation of genes. The neural cell adhesion molecule, N-CAM, for example, is translated in chick neural plate, notochord, and myotomes but not in the surrounding tissues, and in *Xenopus laevis*, the presence of N-CAM in the nervous system is dependent on induction.

Used as a probe and in *in situ* hybridization, two clones of *X. laevis* N-CAM cDNA (N1 and N5) show that N-CAM RNA increases during gastrulation (from the midgastrula to the neural plate stage) from uniformly low levels of stored (maternal) mRNA to high levels of fresh transcripts in the neural plate. N-CAM is not differentially expressed in the frog embryonic epidermis or muscle (Fig. 21.27) although it is expressed in avian and mammalian embryonic muscle.

In the frog, detectable amounts of N-CAM RNA are not transcribed in the presumptive ectoderm of the blastula unless brought into contact with inducing tissue. For example, cultured *in vitro* alone, neither presumptive ectoderm nor presumptive endoderm of blastulas forms nervous tissue, and RNA from both fails to hybridize with the N-CAM probe. On the other hand, cultured together, the endoderm induces mesoderm in the ectoderm, and, in turn, the mesoderm induces neural tissue. Under these conditions, RNA extracts hybridize with the N-CAM probe, indicative of N-CAM mRNA transcription (Kintner and Melton, 1987).

If induction can be reduced to gene activation, the organizer can presumably be reduced to strings of gene activation. Hypotheses concerned with genic strings may yet be tested with the aid of techniques made possible by recombinant DNA technology. Existing gene libraries would have to be monitored for potential probes; cloned genes could be injected into developing embryos to heighten gene expression, or antisense strands of DNA could be injected to block gene expression.

The application of molecular approaches to induction have already established that induction consists of several separate steps. Soon after N-CAM transcription begins in *X. laevis*, transcription commences for a spinal cord probe contained within a homeobox-encoding gene (XlHbox6; see Chap. 22) (Gurdon, 1987).

Some of the molecular steps of induction may be independent and additive rather than dependent

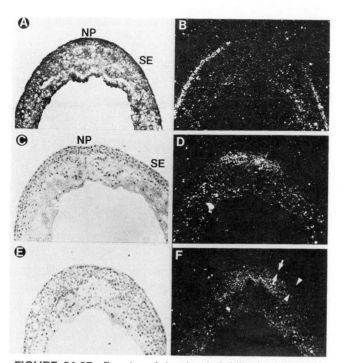

FIGURE 21.27. Results of *in vitro* hybridization showing localization of specific mRNAs. Transverse sections through the neurula of *Xenopus laevis* are seen in phase contrast (a, c, e) and in dark-field optics where autoradiographic grains appear white (b, d, f). (a, b—control) *In situ* hybridization with an epidermal keratin Xek3 RNA probe in section from posterior region of embryo. Keratin mRNA is differentially concentrated in the surrounding ectoderm (se) but not in the neural plate (np). (c–d) *In situ* hybridization with the N5 N-CAM RNA probe in section from posterior region of embryo (i.e., comparable to a and b). N-CAM mRNA is concentrated in neural plate and not detectable above background in surrounding ectoderm. (e, f) *In situ* hybridization with the N5 N-CAM RNA probe in section from anterior region of embryo. N-CAM mRNA is concentrated in the medial, deep region of the neural plate (destined to give rise to the bulk of the central nervous system) in apposition to dorsal mesoderm. The probe is not detectable above background in the surrounding ectoderm, the superficial ectoderm (arrows), destined to give rise to the nonneural lining of the neural tube (i.e., the ependyma), or the lateral portions of the neural plate (arrowheads), destined to give rise to neural crest, and hence the peripheral nervous system. From C. R. Kintner and D. A. Melton, *Development*, 99:311(1987), by permission of the Company of Biologists and the authors.

and interactive. For example, the induction of N-CAM transcription would seem to be uncoupled from later steps of neural induction in the exogastrulas, since ectoderm in exogastrulas of *X. laevis* does not form nervous tissue although it transcribes N-CAM mRNA. Possibly, one inducer has

different effects in different circumstances or regions, or each major induction consists of separate processes.

Current Status of Primary Induction: A Critical Reappraisal

While partially purified "unnatural" inducing substances can cause dramatic inductive effects in various reactive systems, and heparin-binding growth factors with mesodermal inducing activity are available in purified form, no neural inducer has yet been purified to homogeneity nor has one naturally occurring neural inducing substance been identified (see Nakamura et al., 1978; Saxén and Toivonen, 1962, 1985). Has the search for neural inducers been a wild goose chase?

Recently, one review of efforts to find inducing substances associated with the primary organizer (Jacobson, 1982, p. 79) concludes that the Spemann–Mangold observation on the results of transplantation

> ... was the first brick of an edifice that was rapidly built, an edifice in which the concepts of the "organization center" and of "primary embryonic induction" were enshrined. To cap the elaborate construction with a gilded dome, all that remained to be discovered was the inductor itself, or at least, the mechanisms by which the organizer organized. The failure to discover the mechanism of action of the organizer has left the edifice without a roof. ...
>
> More than 50 years of effort have failed to reveal the putative inductor substances, nor has any progress been made in discovering the cellular mechanisms of release, transmission, reception, and interpretation of the developmental signals that are supposed to result in regional differentiation.

Even embryologists who have built their careers and reputations around the search for inductor substances sometimes despair of finding naturally occurring inductors. In their concluding remarks to the celebration of 50 years of induction research, Sáxen, Toivonen, and Nakamura (1978, p. 316) ask:

> Can it be that the pioneering work of one great scientist [Spemann] has led his successors astray or along a pathway that will come to a dead end, despite the superb methods and wealth of information available today? Scientific odysseys are by no means unknown in the history of biology, and this name would perhaps fittingly describe some of the work performed by embryologists in the 1930's. They hunted unremittingly for the magic molecule that would transmit embryonic induction, the "organizin" or "evocator", being convinced that a single active compound was responsible for primary embryonic induction.

The skepticism of these writers does not extend to the concept of interacting tissues epitomized by the concept of the organizer, however. Saxén, Toivonen, and Nakamura (1978, p. 319) end on an upbeat:

> We are convinced that future projects should still be based on the fundamental ideas of Spemann and his school, and

we do not consider that his way of thinking will lead us down the wrong track or into a blind alley.

Jacobson (1984, p. 129) agrees on this point. Writing of his own experiments, he concludes:

> In spite of the remaining uncertainties . . . these findings may be regarded as a vindication of Spemann's theory of the organizer.

PART 7 SUMMARY AND HIGHLIGHTS

A gastrula is an embryo forming its germ layers or an embryo with cells moving into positions where they will produce tissues and organs. Gastrulation shifts parts in the early embryo, bringing them to their positions in the adult. Animals having two and three generalized body layers as adults form the rudiments of these layers during gastrulation and establish the conditions for their further development.

Cell movements include individual multipolar or unipolar ingression, mass invagination, evagination, folding, involution (around an edge), epiboly (over a glassy layer), and epiauxesis (surface movement in absence of a glassy layer). The segregation of juxtaposed cell surfaces is delamination or polyinvagination.

Germ layers are more obvious in the yolk sacs of highly yolky eggs. Many protostomes form a stomodeum at an early point of invagination, while some deuterostomes form a proctodeum in the vicinity of their blastopore. The stomodeum and proctodeum are lined by ectoderm, and the midgut is identified with endoderm. In deuterostomes, an archenteron, where present, is lined by endoderm and may contain dorsolateral mesoderm, chordamesoderm, and ventral germ cells. Schizocoelomates generally form a coelom by cavitation in mesodermal masses, while some enterocoelomates form a coelom between folds of mesoderm originating from endoderm.

Gastrulation in species with intralecithal cleavage frequently provides a dorsal extraembryonic ectodermal membrane or serosa that covers the yolk and an amnion that covers the embryo proper. Folded inward, the insect embryo may sink below the surface or develop within a protected amniotic cavity. Ventrally, a germinal membrane, bands, disks, or plates invaginate to form the mesoblast

followed by the delamination or inward movement of marginal ectodermal bands forming the ventral nervous system. Vitellophages involved with the digestion of yolk may also aid in forming the definitive endoderm, but, in pterygotes, endoderm is formed by anterior and posterior midgut rudiments that ingress from the ends of the midventral band and form the midgut by enclosing the yolk.

Gastrulation in species with meroblastic eggs is dominated by covering the yolk with a digestive yolk sac. Cephalopods form a yolk sac with an outer epithelial layer, a diffuse middle layer containing hemal spaces, and a digestive syncytial inner layer. Three organogenic layers form embryonic structures.

Selachians form their discogastrulas by invagination and involution, bringing a dorsal flap of tissue over an archenteric cavity. Similar flaps appear in cyclostomes, holosteans, and vertebrate tetrapods but not teleosts. The advance of a splanchnopleuric germ wall forms an external yolk sac with a syncytial endoderm and brings the large yolk mass into the gut cavity where it is digested.

In teleosts, epiboly by a germ ring is the chief device for covering the yolk, but the multilayered embryo is formed by the rearrangement of cells in a thickened embryonic shield. The marginal yolk cytoplasmic layer (YCL) of the egg is replaced as a yolk syncytial layer (YSL) followed by an outer enveloping layer (EVL) of tightly bound E-cells cover the yolk sphere. An external YSL (E-YSL) running ahead of the EVL supplies the substratum and force for EVL movement, while the internal YSL (I-YSL) covered by EVL forms the digestive layer of the yolk sac.

The EVL forms an impermeable periderm over the embryo which develops in an environment of its own creation. Clustered deep- (D-) cells turn inward at a germ ring and converge on an axial rudiment. Free mingling and clonal dispersion among D-cells give way to clonal restriction and the segregation of founder cell precursors of germ and tissue layers. An incipient coelom within the mesoderm leaves the yolk mass within the body cavity where it is digested by extensions of the embryonic gut.

Reptiles and birds form a primitive endoderm or primary hypoblast by delamination from an upper epiblast. In reptiles, a mesodermal archenteron surrounding a blastoporal canal is formed by invagination and involution. Following the breakdown of the thin floor of the archenteron and hypoblast, the blastoporal canal opens into the subblastodermic cavity. Caudal and lateral endoderm moves in and covers the yolk.

In the chick, a marginal zone produces additional primary hypoblast cells as the germ ring flows over the yolk. The mesoblast pursues the hypoblast over the yolk. Angiogenesis and embryonic blood cell formation begin precociously in blood islands of the yolk sac.

In the early avian discogastrulas, known as the prestreak and embryonic shield stages, secondary hypoblast displaces the primary hypoblast, and primary endodermal cells sink below the surface, forming a thickened Koller's crescent at the posterior rim of the area pellucida. The thickening builds centrally and lengthens into the primitive streak, and, as mesodermal cells ingress through the streak, an anterior primitive pit lengthens within the streak into a primitive groove.

Cells converging on the streak pile up in primitive folds and ingress in the vicinity of the groove like a "folding fan" of presumptive areas. The primitive pit deepens to the presumptive surface mesoblast, while the primitive node or Hensen's node thickens anteriorly. The notochordal process extends anteriorly from the node as the primitive streak regresses posteriorly. A head fold tucks under the anterior end of the embryo, lifting the head off the yolk sac and forming a subcephalic pocket.

In totally cleaving therian mammals, presumptive embryonic cells separate in a germinal portion of the surface or an inner cell mass beneath the trophectodermal surface. An epiblast differentiates as an epithelial layer as a hypoblast or visceral yolk sac delaminates and migrates peripherally as a parietal yolk sac. The embryo proper and blastocyst thereby become bilaminar. The omphalopleure formed by the fusion of trophectoderm and hypoblast can be a major organ for absorption of uterine contents. With the entry of mesoderm from the embryonic ectoderm of the thickened posterior primitive streak, the embryo proper and blastocyst become trilaminar.

As the meeting place for the ectoderm and endoderm, the primitive streak of microlecithal mammalian eggs resembles the primitive streak of megalecithal avian eggs and the blastoporal lips of amphibian eggs. These streaks and lips are generally considered homologous but considerable variation is also apparent.

Amphibian gastrulation involves the movement of cell sheets, whereas amniote gastrulation involves the movement of independent cells. The primitive streak of birds is marked by a shallow primitive groove and even a depressed primitive pit but not by an invagination. Lacking endoderm, the invagination sometimes found in mammals resembles the reptilian blastoporal canal, but neither matches the endoderm-containing archenteron of amphibians.

Coeloblastulas of totally cleaving sponges form archenterons by invagination, while stereogastrulas may be formed by unipolar ingression. Coelenterates, which may effectively gastrulate during cleavage, tend to form endoderm and ectoderm by delamination or ingression.

In echinoderms, primary invagination via apical constriction of bottle cells in a vegetal plate is independent of the remainder of the embryo. Secondary invagination via filopodial contraction draws the archenteron across the blastocoel with little involution. Ingression or detachment of cells from the tip of the archenteron liberates secondary mesenchymal cells destined to become pigment cells. Enterocoelous folds of mesoderm evaginate from the archenteron, leaving a midgut endoderm connected at one end to the stomodeum and at the other end to the blastopore.

Amphioxus gastrulates by invagination through both primary and secondary mechanisms with little involution. Enterocoelous evaginations form mesoderm. In larvaceans with fewer cells and smaller blastocoels, invagination is restricted and cellular rearrangement is more prominent in gastrulation. Ascidians represent an intermediate type and may have provided an ancestral source for both amphioxus and larvaceans.

Gastrulation in cyclostomes and in primitive teleosts resembles gastrulation in amphibian embryos. Invagination accompanied by involution introduces layers of endoderm and mesoderm beneath the surface, while epiboly covers the gastrula with ectoderm.

In amphibians, ingressing bottle cells swell the floor of the blastocoel and aggregate to form a crescent-shaped blastoporal groove. Deepening and lengthening, the groove becomes the circular blastopore surrounding the yolk plug. The invagination of the amphibian blastopore is a consequence of collective ingression by cells tightly bound to each other at their outer surfaces. Dragging in the surface coat at their tightly adhering apical surfaces, bottle cells of the anterior endoderm lead a spreading sheet of cells along the basal lamina of the blastocoel roof.

In *Xenopus laevis,* the superficial involuting marginal zone (IMZ-S) gives rise to presumptive endoderm and notochord occupying the roof of the archenteron, while the deep involuting marginal zone (IMZ-D) gives rise to all the mesoderm. Surface cells in the pregastrula are relatively small and tightly bound to each other. The mixing of cells from different regions is restricted to members of founder cell groups. Related blastomeres mix during gastrulation, but cells tend to remain within the confines of polyclonal compartments.

Among the spiralians, coeloblastulas are candidates for gastrulation by invagination, while stereoblastulas are more likely to gastrulate by ingression and epiauxesis. Yolk content is a factor, and eggs with more yolk tend to cover their surface with an ectoteloblast and internalize cells by default, while eggs with less yolk are more likely to rely on individual cell ingression to rearrange presumptive regions. The earlier the gastrulation, the more likely primary and secondary somatoblast, primordial germ cells, and mesodermal stem cells or teloblasts will ingress individually.

Macromeres frequently move inside first, either before or after proliferating endomeres or endodermal precursors. In coeloblastulas, pseudopodia at the basal ends of endomeres may connect to the overlying animal hemisphere cells and shorten to invaginate the archenteron. Stomodea and proctodea lined by ectoderm may also invaginate, and pseudoblastopores or pseudoarchenterons, also lined by ectoderm, may later evaginate to supply adult trunk ectoderm.

Gastrulation may be directed by active individual cell movement via contact guidance and contact inhibition or by passive species-specific affinities, histotypic adhesiveness to extracellular materials

or matrix (ECM), or glycosyltransferase-mediated intercellular interactions. Movement of masses of cells may be passively governed by differential or changing adhesiveness or morphoregulatory molecules and mechanochemical functions. Alternatively masses of cells linked by tight junctions may be actively led by bottle cells.

Cadherins, or CADs, of calcium-dependent adhesive molecules and CIDs of calcium-independent adhesive molecules may operate during gastrulation, and integral membrane glycoproteins, such as N-CAM, L-CAM, and Ng-CAM, may be present in changing patterns. Substrate adhesive molecules (SAMs), such as fibronectin and laminin, may bind to other ECM elements and attach to cell-surface receptors. Junctional adhesive molecules (JAMs) specific for intercellular junctions may form stable occluding and adhering attachments.

Gastrulas with small cell numbers (SCN-gastrulas) are less flexible and more likely to exhibit early determination, while gastrulas with large cell numbers (LCN-gastrulas) are more likely to gastrulate via invagination and tend toward regulative or indeterminate development. One or small numbers of founder cells produce sets of cells with given histotypes in SCN-gastrulas, while a variety of cell lineages or founder cell groups produce location-specific progenitors in LCN-gastrulas.

Historically, the study of gastrulation is intimately associated with germ layers and biogenesis. Karl Ernst von Baer characterized germ layers as building blocks of vertebrate embryos, a position they continue to occupy today if less narrowly identified with particular adult tissues. Following the lead of British biologists, especially Thomas Huxley, Ernest Haeckel focused attention on the germ layers as tangible proof of evolutionary recapitulation. To critics, the weak point of the theory was determination or the fixity of germ layers. The lability of germ layers ultimately invalidated Haeckel's version of the germ-layer theory.

Hans Spemann contributed enormously to current epigenic concepts of development through his research on nuclear equivalence and the experimental analysis of development in composite organs, secondary induction, primary induction, and the organizer. Spemann, his students, and others moved portions of germ layers from their sites *in situ* to hypothetically neutral sites *in vivo* and *in vitro* and to potentially inducing sites elsewhere in the embryo. The abilities of the transplants to self-differentiate or enter alternative paths of differentiation were interpreted in terms of determination and the competences of embryonic regions to respond to induction. The dorsal lip of the blastopore was identified as the organizer of the embryo and embryology entered its golden age of research on induction.

Research with unnatural inductors in sandwiches of explanted ectoderm or in the embryonic blastocoel led to the discovery of several types of inductor. Evocators triggered differentiation already prescribed for a tissue, while complex inductors organized the embryo. Anterior structures were induced by head or neural inductors, and posterior inductors were induced by trunk, tail, or mesodermal inductors.

Although all the materials found to have inducing activity shared some properties, such as a degree of cytotoxicity, a common denominator among inducing substances escaped detection. After an initial period of frenzied and enthusiastic research, the 20th century saw many embryologists abandon the search for inducing substances. The greatest benefit derived from research on induction has been the documentation of sequential changes in development. Sequence is consistent with causal roles for cellular history or lineage as well as induction.

PART 7 QUESTIONS FOR DISCUSSION

1. Define the gastrula and gastrulation. Describe the blastopore or primitive streak, archenteron, stomodeum, and proctodeum of some gastrulas. Describe the consequences of yolk, cell size, and cell number for gastrulation. How do triploblastic animals form coeloms? How does gastrulation differ in coeloblastulas and stereoblastulas?

2. Discuss the similarities and differences between blastoporal lips and the primitive streak of birds and mammals, between amniote and anamniote gastrulas, between the mammalian and avian gastrulas, and between reptilian and avian gastrulas.

3. What are the gastrula's greatest problems, and how do gastrulas of different types solve these problems? Describe the mechanisms of endoderm formation in embryos with intralecithal cleavage. What are the functions of vitellophages in various arthropods? How do arthropod embryos enclose their yolk? Cephalopods? Meroblastic vertebrates?

4. Describe the approaches to yolk digestion taken in species with intralecithal cleavage and those with meroblastic cleavage. Distinguish between the yolk sacs and the position occupied by yolk in arthropods, selachians, teleosts, and birds.

5. How are extraembryonic and embryonic ectoderm formed in insects? Describe the serosa, amnion, and amnioserosa and discuss their

functions in insect development. How is the neural ectoderm formed in insects? In spiralians? In vertebrates? What are the functions of the yolk cytoplasmic layer, yolk syncytial layer, and enveloping layers during epiboly and gastrulation proper in the teleost?

6. What are the differences between the archenterons of urodeles, ranid anurans, and *Xenopus laevis*? What are the differences between the deep and superficial involuting marginal zones (IMZ-D and IMZ-S) of amphibians?

7. How does gastrulation differ in species with early founder cells and teloblasts as opposed to those with typical germ layers? What similarities are apparent in cephalopod gastrulation and molluscan gastrulation that were not apparent during cleavage?

8. Distinguish between delamination and fusion, invagination and involution, epiboly (or epiauxesis) and ingression. Distinguish between proliferative organogeny and cell migration in insect gastrulation. What are primary and secondary invagination, and where do they occur?

9. Discuss differences in gastrulas with small cell numbers (SCN-gastrulas) and large cell numbers (LCN-gastrulas). How might contact guidance and contact inhibition work during gastrulation? Describe the role of individual cell movement in gastrulation and evaluate studies of reaggregation and reassortment for concepts

of adhesiveness. Identify and describe CADs, CIDs, CAMs, SAMs, and JAMs.

10. Are gastrulas in general homologous or analogous? Describe different versions of the germ-layer theory. How do embryologists define homology without relying on structures? Make up a plausible scheme for the evolution of protochordate and vertebrate gastrulas.

11. What are primary and secondary induction, and how did Spemann's experience with one influence his ability to study the other? Describe the Spemann–Mangold experiment, and discuss why induction is considered devastating for Haeckel's germ-layer theory. What is the relationship between determination and inductive activity? Distinguish between an evocator and an inducer.

12. How can time-graded properties of natural inducers be tested with the help of unnatural inducers? What do head inducers, trunk or tail inducers, neural inducers, and mesodermal inducers have in common? What are the prospects for using recombinant DNA technology and _in situ_ hybridization for solving the problem of induction?

PART 7 RECOMMENDED READING

American Zoologist, 24:535–688 (1984). Symposium on gastrulation.

Edelman, G. M., 1986. Cell adhesion molecules in the regulation of animal form and tissue pattern. _Annu. Rev. Cell Biol._, 2:81–116.

Edelman, G. M. and J.-P. Thiery, eds., 1985. _The cell in contact: Adhesions and junctions as morphogenetic determinants._ Wiley/Neurosciences Institute Publication, New York.

Hamburger, V., 1988. _The heritage of experimental embryology: Hans Spemann and the organizer._ Oxford University Press, New York.

Hecht, M. K., P. C. Goody, and B. M. Hect, eds., 1976. _Major patterns in vertebrate evolution._ Plenum Press, New York.

Holtfreter, J., 1939. Tissue affinity, a means of embryonic morphogenesis. English translation by Konrad Keck, amended by Professor Holtfreter. Reprinted in B. H. Willier and J. M. Oppenheimer, eds., _Foundations of experimental embryology._ Prentice-Hall, Englewood Cliffs, NJ, pp. 186–225.

Horder, T. J., J. A. Witkowski, and C. C. Wylie, eds. 1985. _A history of embryology_, 8th Symposium of the British Society for Developmental Biology, Cambridge University Press, Cambridge.

Journal of Embryology and Experimental Morphology, 97: Suppl. (1986). Determinative mechanisms in early development.

Johnson, M. H., ed., 1984. _Development in mammals_, Vol. 5. Elsevier Science Publishers, Amsterdam.

Løvtrup, S., U. Landstróm, and H. Løvtrup-Rein, 1978. Polarities, cell differentiation and primary induction in the amphibian embryo. _Biol. Rev._, 53:1–42.

Nakamura, O., and S. Toivonen, eds., 1978. _Organizer—a milestone of a half-century from Spemann._ Elsevier/North-Holland Biomedical Press, Amsterdam.

Nieuwkoop, P. D., 1985. _The epigenetic nature of early chordate development, inductive interaction and competence._ Cambridge University Press, Cambridge.

Spemann, H., 1938. _Embryonic development and induction._ Yale University Press, New Haven. (Reprinted in 1962 by Hafner, New York.)

Spemann, H. and H. Mangold, 1924. Über Induktion von Embryonalanlagen durch Implantation artfremder Organisatoren. _W. Roux' Arch. Entwicklungsmech. Organ._, 100:599–638. (Reprinted in B. H. Willier and J. M. Oppenheimer, eds. _Foundations of experimental embryology._ Prentice-Hall, Englewood Cliffs, NJ, pp. 144–185.)

Thiery, J. P., J. L. Duband, and G. C. Tucker, 1985. Cell migration in the vertebrate embryo: Role of cell adhesion and tissue environment in pattern formation. _Annu. Rev. Cell Biol._, 1:91–113.

Trinkaus, J. P., 1984. _Cells into organs: Forces that shape the embryo_, 2nd ed. Prentice-Hall, Englewood Cliffs, NJ.

Vakaet, L., 1984. Early development of birds. In N. M. Le Douarin and A. McLaren, eds., _Chimeras in developmental biology._ Academic Press, Orlando, pp. 71–88.

Verdonk, N. H., J. A. M. van den Biggelaar, and A. S. Tompa, eds., 1983. _The Mollusca_, Vol. 3, _Development._ Academic Press, Orlando.

Part 8

EMBRYOGENESIS AND MORPHOGENESIS: ACQUIRING SPECIES CHARACTERISTICS

Our enquiry lies, in short, just within the limits which Aristotle himself laid down when, in defining a "genus," he shewed that . . . the essential differences between one "species" and another are merely differences of proportion, of relative magnitude, or (as he phrased it) of "excess and defect."

D'ARCY WENTWORTH THOMPSON (1963 reprint, p. 1034)

Embryogenesis gives the postgastrular embryo the body plan of its species, and morphogenesis furnishes the embryo with the rudiments of larval and adult structures. These rudiments produced in the postgastrula are typically "old" or **palingenic** (Gk. *palin* again + *gignesthai* to give birth), frequently thought of in terms of **homology** (Gk. *hom-* same *-logos* reason, hence a doctrine of likeness) and Haeckelian recapitulation. In addition, "new" or **cenogenic** (Gk. *kainos* new + *gignesthai*) structures are produced, reflecting embryonic adaptation and evolutionary change.

Part 8 raises the question: "Are the controls of development likewise divisible into palingenic and cenogenic categories?" Homology is defined, and determinate development and regulative development in spiralians with small cell numbers and in insects with large cell numbers are reexamined according to their type of homology. The development of the vertebrate body plan is analyzed as an example of palingenesis, and the development of extraembryonic membranes and the implantation of mammalian embryos are scrutinized as examples of cenogenesis.

Some progress has been made toward the identification and analysis of palingenic and cenogenic controls, but more headway might be made if palingenic and cenogenic features of genes were delineated. Like palingenic and cenogenic structures, structural genes, regulatory genes, and other devices governing development have coevolved. Cenogenic controls appear to have evolved from palingenic controls as much by reduction as by innovation.

Chapter

22

HOMOLOGY: THE GENETICS OF CELL LINEAGE, HOMEOTIC MUTANTS, HOMEOBOXES, AND POU

He [Etienne Geoffroy Saint-Hilaire] sought to establish a homology—that is, a true similarity based on common origin, rather than a mere analogy or likeness developed independently by two groups in response to shared functional needs—between the complete segmentation of insects and the partial repetition of vertebrates (seen primarily in the vertebral column and paired ribs).
STEPHEN JAY GOULD (1985, p. 12, writing on the debate between
Etienne Geoffroy Saint-Hilaire and Georges Cuvier in 1830)

*T*he contemporary molecular concept of homology and the application of statistical criteria for assessing homologies in the parts of molecules are relatively recent developments. According to this new concept, similar molecules in different species are **homologous,** and similar molecules within a species are **paralogous.** Molecular sequences tending to correspond or having a high level of **agreement** are dubbed **consensus** sequences, and macromolecules sharing large parts of their primary structure are said to be highly **conserved** (see Reeck et al., 1987).

Legitimized by technology and spruced up with quantitation, the molecular concept of homology has a considerable following among molecular biologists. Advocates extrapolate from the monotonous replication of genes in an orga-nism to unchanged replication through generations and hence through species, phyla, and, indeed, life itself, past and present. But prudence requires demonstrating a plausible evolutionary relationship among structures, including molecules, before proclaiming them homologous and attributing their similarity to evolutionary inertia (i.e., conservation). Anything else is circular and a debased form of scientific reasoning.

The temptation to loosely label structures homologous because they are similar might be more easily resisted if the concept of homology and its history were better understood. Actually, the concept of homology anticipated the concept of the gene by centuries. Homology was even debated before Lamarck gave substance to the idea of evolution in the 18th century (see Källén, 1961), and, a hundred years ago, "homologous relations" were the scientific equivalent to popular "blood relations."

Nature philosophy provided three concepts of homology. **General homologies** were correspondences among general structures present in different organisms. **Serial homologies** were correspondences among structures in different segments of the same animal. A fly's prothoracic legs, for example, were serially homologous with its antennas and generally homologous with a crustacean's legs. **Sexual homologies** (now considered a special case of serial homology) were correspondences among sexual organs in the different genders of the same species.

Evolutionary theory changed all that. **Homogenous** or **homologous** organs were redefined phylogenically as organs that diverged in the course of evolution from their form in an ancestral species. Homology was also redefined ontogenically. In keeping with the **biogenic law,** that "ontogeny epitomizes phylogeny," organs and structures were considered homologous when they arose at comparable stages or from comparable embryonic parts, such as germ layers.

By the advent of the 20th century, comparative embryologists had screened vast numbers of embryos in search of ontogenic homologies. General homologies were thought to reflect the evolution of species, while serial homologies were supposed to illuminate evolution within a species.

At mid-20th century, genetic or hereditary mechanisms were invoked to explain general homologies, but cytoplasmic influences were called upon to explain serial homologies. The possibility of genetic controls over serial homologies gained adherents only with the elucidation of **homeotic mutants.**

Today, the normal development of segmental and partially segmental animals is widely thought to require the superimposition of genetic controls over periodic patterns and over sequential patterns. In *Drosophila*, **segmentation genes** and the mutations that alter the number and polarity of segments seem to be independent of homeotic genes and the mutations that alter segmental identities (see Gehring, 1987). General homology and serial homology therefore seem to be superimposed but independent.

GENERAL HOMOLOGY

These constant relations must be important because, if something always happens in the same way, it is assumed that it is a fixed phenomenon, presumably with cause–effect relationships.

John A. Moore (1987, p. 543)

Annelids, Mollusks, and Cognates

Quartet-forming spiralians epitomize determinate development, and, for the most part, they deserve their reputation. Their cleavage is a paragon of sufficiency in cellular versatility, and their embryogenesis is the acme of efficiency in cellular utilization. Structures and functions that are created and operated with just a few cells in spiralians (e.g., nerve cells) require vast numbers of cells and long periods of development in vertebrates.

The more primitive annelids and mollusks tend to have the more complex plankotrophic trochophores adapted for longer planktonic phases. The gastrulas of these species utilize greater amounts of their embryonic tissue to make the trochophore. Derived species tend to have less complex or reduced trochophores that venture into the plankton for only short (if any) periods and hardly perform any distinctly larval functions (see Rice, 1976). These species utilize less embryonic tissue to make the larva (see Fig. 14.18).

Embryogenesis, whether of larval tissue (i.e., indirect development) or of adult tissue (i.e., direct development), is rapid when **intrinsic** or **autonomous controls** truncate the lines of command between nucleus and cytoplasm. Pattern formation may play an important role in determining cell fate. Cortical or cytoplasmic determinants may operate close to the level of nuclear controls and funnel the tokens of cellular identities into or out of founder cells and teloblasts. Cell lineage may also play an important role through cycles of DNA replication or pathways of chromosomal alteration. **Extrinsic controls,** requiring external factors, tissue recognition, and guidance, play relatively smaller roles.

The indirect development of many annelids and mollusks through a trochophore (Fig. 22.1) seems to be largely at the behest of intrinsic control. Nevertheless, a degree of local regulation may occur, and even the rigidly determined gastrula cells may communicate through gap junctions (Serras and van den Biggelaar, 1987). In some cases, removing early blastomeres only diminishes the size of structures, and removing other blastomeres has no effect at all (e.g., the 2b micromere, Table 12.1). Moreover, manipulation of the polar lobe cytoplasm can sometimes result in severe regula-

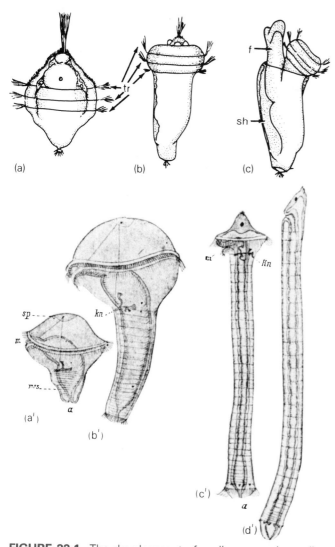

FIGURE 22.1. The development of molluscan and annelidan larvas. (a–c) Elephant tusk mollusk *Dentalium* showing the trochophore of 24 (a) and 30 (b) hours, and larva of 72 hours. (a′–d′) Marine annelid *Neanthes* showing the trochophore, juvenile, and early adult form. a, Anus; d, digestive tract; f, foot; m, mouth; mes, mesoderm present in bilateral bands; kn, nephridium (larval kidney); st, stomodeum; sh, shell; tr, cilia of trochophore; vel, velum; sp, super trochal region. A through C adapted from E. B. Wilson, *J. Exp. Zool.,* 1:1(1904). A′ through D′ adapted from C. Dawydoff, *Embryologie comparée des invertébrés,* Masson et Cie, Paris, 1928.

tion indicative of inductive interaction (see Fig. 12.4).

Indirect development requires placing adult-forming cells in appropriate positions as well as manipulating presumptive germ-layer cells of the larva. Embryonic **stem cell-like** or **asymmetrical divisions** generate cells required for the development of specific structures, while **blast cell-like** or

symmetrical divisions (also called proliferative divisions) generate the kinds and numbers of cells required to form structures. **Teloblasts** are the annelidan and molluscan equivalents of **founder cells** (also called stem cells or embryonic blast cells) in nematodes and **founder cell groups** in endopterygote (holometabolous) insects. After contributing to the larva, the teloblast-cell progenies may operate like imaginal disks by becoming sequestered in a developmentally inert or retarded condition.

The mesoblast (Fig. 11.49) coupled to derivatives of the ectoteloblast (derived from the 2d primary somatoblast) form the larval **growth zone** (see Cather, 1971) and elongate both anteriorly and posteriorly into **primary mesoblast bands.** In polychaete annelids, three pairs of larval segments are derived from the growth zone, and, after metamorphosis (Fig. 22.1b′–d′), mitosis and renewed growth in the growth zone's caudal remnant provide the ectoderm and mesoderm of the adult.

Cavities appear in the mesoderm, and hollow somites differentiate. The enteroblasts enter the formation of the distal end of the intestine and form the endoderm-derived structures of the adult.

Abbreviated development is exhibited in the asymmetry of organs in some univalve mollusks. For example, the left kidney primordium may remain rudimentary (*Paludina, Ampullaria,* and *Pomatias*) or not appear at all (in *Physa* the right kidney does not appear).

The unpaired structures of the posterior part of gastropods are brought to their adult, functional positions during **torsion,** or twisting of organ rudiments (not to be confused with the spiral coiling of the shell which usually occurs later, see Fig. 12.41). The kidney is shifted from the ventral right to the dorsal left. In the mud snail, *Lymnaea* (= *Limnaea*), the single lung primordium is displaced to the right side, and the connection between the lung cavity and the visceral sac (the pleurovisceral connection) becomes oblique.

Torsion also twists other organs. Generally, ventral structures (e.g., hindgut) turn toward the right, while dorsal structures (e.g., shell gland) turn toward the left. The rudimentary straight gut flexes dorsally into a median intestinal loop and then rotates 180° to the right. The anus is brought from the median plane into an anterodorsal position approaching the mouth. The nervous system is also twisted. While torsion depends on the asymmetric development of the snail's retractor muscles, the asymmetry of these muscles "must be due to preformed asymmetry of the early embryo" (Raven, 1966, p. 174).

The direct development of oligochaetes and hirudineans (i.e., clitellates) contrasts quantitatively with the indirect development of more primitive polychaetes. The somatoblasts are larger than in primitive polychaetes and form adult structures immediately rather than after a period of dormancy. Blastomeres assigned to presumptive larval structures elsewhere are reduced or lost.

The D quadrant tends to be the most highly conserved throughout the annelids. Macromeres in other quadrants are lost or modified. In oligochaetes (e.g., lumbricids) lacking a trochophore entirely, A and B form no micromeres and perform merely nutritive functions. Only C and D form the midgut. The 2d cell or primary somatoblast still gives rise to ectoblast bands or rings forming trunk ectoderm and nervous system, and the second somatoblast or 4d cell still produces the mesoderm stem cell or teloblast.

Similarly, in primitive hirudineans (leeches), 2d forms the **ectoblast bands.** Median ventral bands become nerve cord, while lateral rows form circular muscles, somites, and other internal structures. The 3D cell divides equally to give rise to the mesodermal teloblast. Epiboly by micromeres results in covering 3A–3C which become endoderm.

In advanced hirudineans (see Fig. 14.18f), A, B, and C quartets become cenogenic digestive organs (i.e., the albuminotrophic organ). The endoderm forms from 3c, 3d, and 4d. Autonomous controls of determination, if anything, are exaggerated in spiralians exhibiting direct development.

Nematodes

Exemplified by *Caenorhabditis elegans*, nematodes carry the streamlining of development to the ultimate. Developing directly into an adultlike organism, the newly hatched larva provides an elegant example of intrinsic developmental controls (Sulston et al., 1983). Effects of the 300 mutations mapped to the nematode's six linkage groups (five pairs of autosomes and two X-chromosomes in hermaphrodites, one X-chromosome in males) demonstrate abundantly just how tightly gene action can be linked to developmental patterns of cell division, differentiation, and cell death (see Chapter 12).

The developmental anatomy of C. elegans is highly stereotyped. In the first half of embryonic development (about 5 hours at 20°C), a constant amount of cytoplasm is partitioned among cells without the embryo acquiring a wormlike shape (see Fig. 19.36). At the end of gastrulation, at about 4 hours after the first cleavage, the embryo has formed about half (250) its total number of cells. An hour later, it reaches the plateau of 558 cells in the hermaphrodite (560 in males), but it will not hatch for another 5 hours. In the second half of embryonic development, cell elongation, association, and rearrangement shape and stretch the embryo into a nematode virtually without further division.

Throughout the second half of embryonic development, an increasingly profound cleft appears on the underside of the embryo, and the elongating embryo curls upon itself (Fig. 22.2). Like many other invertebrates enclosed in resistant shells, the doubling over of nematodes during development increases the surface area available for morphogenesis (i.e., two for the price of one), while twisting allows elongation within the shell.

The boundaries between clones of cells (i.e., clonal boundaries) in nematodes resemble boundaries between polyclonal compartments in insects more than boundaries between germ layers in vertebrates. Some clones produced by precursor cells generate only cells of a given type but not all the cells of that type, and some produce all the cells of a given type plus cells of other types. But most lineages include cells of different types produced even at terminal divisions, and cells with the same fate may have different ancestries.

The differentiation of founder cell progenies is not limited to specific germ layers except for the intestine (i.e., endoderm) arising from the E lineage, body muscle arising from the D lineage, and germ cells arising from the P_4 lineage. Moreover, only the E and P_4 lineages generate identifiable tissues as single exclusive clones (i.e., lineal boundaries coincide with functional boundaries).

The production of body muscle occurs in several lineages beyond the D lineage. The production of muscle cells by the largely ectodermal AB lineage and the production of neurons by the largely mesodermal MS lineage (see Fig. 11.56) occur at several points in late stages of development. Furthermore, some cells (e.g., G1 and G2) **modulate** rather than differentiate and exchange one function for another (e.g., nervous replacing excretory) in later larvas. The cells may compete for a slot or **primary fate** which, when filled, turns the remaining cells over to an alternative fate.

Hypodermis, the outer cell layer (Fig. 22.3), gives the nematode its shape, but it does not secrete the cuticle until about $1\frac{1}{2}$ hours before hatching (600–750 minutes after the first cleavage). Meanwhile, hypodermal cells are held together by desmosomes or fused together in a syncytium.

LIMA BEAN

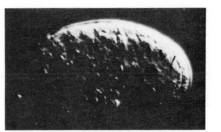

COMMA

TADPOLE

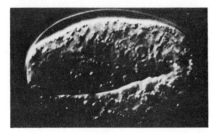

PLUM

LOOP

PRETZEL

HATCH

FIGURE 22.2. Postgastrula stages of living *Caenorhabditis elegans* viewed with differential interference contrast (DIC, also known as Nomarski) optics. The stages are identified by the embryos' contours. Anterior is toward the left. Dorsal is upward. Times refer to the duration of stages at 25° C. (a) Stage 6, lima bean, 45 min. (b) Stage 7, comma, 30 min. (c) Stage 8, tadpole (= 1½ fold), 15 min. (d) Stage 9, plum (= 2 fold), 15 min. (e) Stage 10, loop, 45 min. (f) Stage 11, pretzel, 3 hr. 15 min. (g) freshly hatched worm. Photomicrographs courtesy of Einhard Schierenberg. Durations from Schierenberg, 1982. Reprinted by permission of Alan R. Liss, Inc.

The AB founder cell generates the hypodermis except for the dorsal and some of the tail parts generated by the C founder cell. Beginning anteriorly, hypodermal cells spread out over the surface. At the end of gastrulation, AB pharyngeal precursors move inward from the ventral aspect of the head. Later, four AB muscle derivatives and rectal cells sink inward from the ventral surface.

Dorsally, the hypodermis on each side is "stitched" together by interdigitating cytoplasmic extensions. Nuclei (mainly from the C lineage but including some from the AB lineage) migrate at midembryogenesis across the midline in these extensions (i.e., countermigrate) and establish themselves on the contralateral side (i.e., opposite the side of their origin). Similar extensions and migrations occur ventrally among cells of the C lineage.

Hypodermal cells fuse together into a series of cylindrical hypodermal syncytia (hyp 1–hyp 7, Fig. 22.3) linked together by desmosomes. The last of the anterior syncytia (hyp 7, sometimes called the large hypodermal syncytium) extends dorsally to a cylindrical anal syncytium. The tail is covered by three mononucleate cells (hyp 8, 9, and 11) and one binucleate cell (hyp 10).

Laterally, longitudinal rows of specialized **seam** cells (R and L H0–H2, V1–V6, and T) join the syncytium and, with one exception (H0), become blast cells for the adult. Ventrally, linearly stacked PR and PL cells constitute the ventral cord hypodermal blast cells. Postembryonic division of the seam and PR and PL cells add to hyp 7 until they occupy most of the body surface.

Internally, body muscles are arranged symmet-

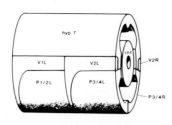

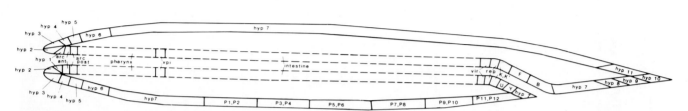

FIGURE 22.3. Stereodiagram of central region and schematic longitudinal section of L₁ larva showing hypodermal syncytial territories and parts of the alimentary track. The corner grooves between the hypodermis and intestine in the stereodiagram are normally occupied by longitudinal body muscles. From J. E. Sulston, E. Schierenberg, J. G. White, and J. N. Thomson, *Dev. Biol.,* 199:64(1983), by permission of Academic Press and the authors.

rically in four longitudinal rows insinuated between the hypodermis and the intestine. Eighty of the individual muscles are generated by the MS, C, and D founder cells. One is generated by the AB founder cell. Additional muscles, generated by AB, are associated with the pharynx, intestine, and rectum.

The entire alimentary track develops from cells of E and MS lineages with AB and C lineages making additions to the mouth, buccal cavity (known as anterior and posterior arcades), and anus. The pharynx, a pumping organ that moves food through the mouth and into the intestine, is derived from MS pharyngeal precursor cells (Figs. 22.3 and 22.4, cells m8–vpi3) that, along with de-

rived buccal cavity precursor cells, sink inward through a ventral slit into the embryo. The progenies of pharyngeal muscle (or myoepithelial) cells m1–m5 fuse into single or paired multinucleate muscle rings before or soon after hatching.

The intestinal endoderm (precursor cells int1–int9) is produced by the E founder cell after dividing into the Ea and Ep (anterior and posterior) cells and migrating inward at the beginning of gastrulation (90 minutes). The valve at the junction of the intestine and rectum (vir) is derived from the AB founder cells as are the next three rectal cells (to B/Y), while a hypodermal syncytium surrounding the anus (hyp 7) is derived from the C and AB founder cells. The alimentary track begins

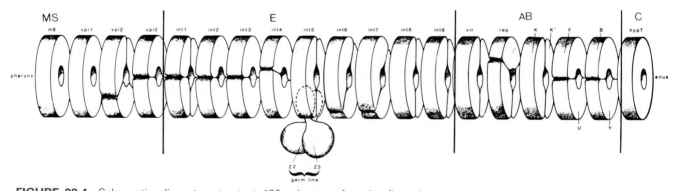

FIGURE 22.4. Schematic alimentary tract at 430 minutes after the first cleavage with germ-line cells, Z2 and Z3, attached to intestinal precursors by cytoplasmic lobes. Twenty intestinal precursor cells are derived from the E founder cell. The pharynx, made of MS founder cell derivatives, extends to the vpi3 cells, and rectal AB founder cell derivatives extend from the vir cells. The hyp7 cell leading to the anus is a C founder cell derivative. From J. E. Sulston, E. Schierenberg, J. G. White, and J. N. Thomson, *Dev. Biol.,* 199: 64(1983), by permission of Academic Press and the authors.

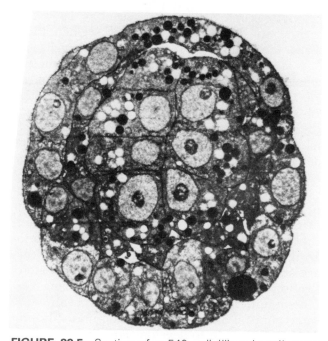

FIGURE 22.5. Section of a 540 cell "lima bean" stage embryo showing simple germ-layer construction at the end of the proliferative phase and beginning of the embryogenic phase of embryonic development. An inner layer of intestinal precursors (endoderm) is surrounded by future body muscle cells (mesoderm). The covering layer consists dorsally of hypodermis (ectoderm) and ventrally of neural precursor cells. C. Krieg, T. Cole, U. Deppe, E. Schierenberg, D. Schmitt, B. Yoder, and G. von Ehrenstein, *Dev. Biol.*, 65:193(1978), by permission of Academic Press and courtesy of the authors.

to emerge at 430 minutes, at about the time the embryo first moves and has folded over about halfway.

At the end of gastrulation (250 minutes after the first cleavage), the ventral surface is occupied by neuroblasts. They undergo their last round of division in the next hour (Fig. 22.5) and are then covered (i.e., internalized) by lateral hypodermal cells. Elsewhere, dorsal and lateral neuroblasts sink inward and are covered by hypodermal cells. The bulk of the nerve cells in the first stage larva (L_1) are derived from the AB lineage (222 neurons compared to 2 from the C lineage and 6 from the MS lineage).

Most of the nematode's nerve cells are in the head where a nerve ring surrounds the pharynx. A ventral nerve cord originating from the ring contains 57 motor neurons and gives rise to circumferential fibers and a dorsal nerve cord.

As the cell number approaches its final number, germ layers become distinguishable (Fig. 22.5); differentiation begins, and the first cell deaths are recorded. Early cell death is followed immediately by phagocytosis, almost always by a sibling cell. Later, hypodermal cells and some others scavenge for the cellular corpses.

In all, one in six cells produced during embryonic development dies (for a loss of 113 cells in the hermaphrodite, and 111 in the male). About 90% of the dead cells belonged to the AB lineage, the remainder to the C and MS lineages.

After four larval stages, separated by molts (i.e., shedding the cuticle), 55 (59 in males) nongonadal **postembryonic** or **larval blast cells** are added to the hermaphrodite's 816 general body nuclei (976 in males). Two somatic precursor cells of the gonad (Z1 and Z4 generated by the MS founder cells) produce 143 somatic gonadal cells in hermaphrodites (55 in males), for a total of 959 nuclei in hermaphrodites (1031 in males). The two primordial germ cells (Z2 and Z3 generated from the P_4 cell; Fig. 22.4) then produce about 2000 germ cells (about 1000 in the male).

Analysis of development in C. elegans *reveals a degree of cellular dependency coupled to broadly autonomous cellular controls.* Removed from their shell and cultured *in vitro*, blastomeres cleave to several hundred cells and form irregular masses capable of twitching (i.e., exhibiting muscle contraction). In contrast to differentiating muscle, however, cell positioning, or **topogenesis**, does not proceed autonomously, and little recognizable morphogenesis is observed. Rather, the potential for topogenesis seems to depend on the microenvironment provided by the intact egg shell.

Maternal-effect *embryogenesis* (*emb*) genes affect the timing of cleavage divisions (rather than the pattern of divisions) as attested by *embryonic-arrest* mutations induced with the mutagen ethyl methane sulfonate (Brenner, 1974). The mutations are keyed in by a temperature shift during oogenesis (e.g., from the 16°C permissive temperature to the 25°C nonpermissive temperature) (see Hirsh et al., 1976).

The cells of *emb* mutants may divide slower or faster, and rates may deviate from normal by only 10–20%. Still, affected embryos fail to regulate toward normal development. The abnormal *emb* mutant phenotypes suggest that deviations in the timing of divisions establish irreconcilable regional differences in the embryo.

The *partitioning defective* (*par-1–par-4*) mutations are maternal-effect mutations causing defective spindle placement, abnormal synchrony in division (but normal overall division rates), and improper localization of germ-line P granules (see Fig. 12.6). Homozygous mothers for "strong" *par* mutations produce embryos that arrest as amorphous masses lacking intestine but otherwise consisting of differentiated cells. Homozygous mothers for "weak" mutations produce some embryos that become infertile adults lacking germ cells (i.e., grandchildless). The *par* genes are possibly required for the proper localization of cytoplasmic determinants in the early embryo (Kemphues et al., 1988).

Zygotic-effect mutants of postembryonic cell lineages include general abnormalities in cell division and specific hypodermal or vulval defects. *Cell death (ced)* genes program cell death into the normal course of development. Cells that autonomously execute themselves in the wild type fail to do so in *ced-3* mutants. Instead of carrying out their programmed cell deaths, the cells express programs of differentiation found in nearby sublineages (e.g., differentiating as a nondopaminergic neuron while a sibling cell becomes a dopaminergic neuron) (Horvitz et al., 1983).

Cuticular mutants (such as *lin-14* mutants) are **heterochronic.** Mutations cause temporal displacements such as the development of larval cuticles in the adult (*lin-14*[d]) or adult cuticles in the larva (*lin-14*[o]). In general, mutations seem to change developmental **decisions** normally accompanying each division. Decisions may not be made, and series of such decisions may be derailed at some early point. The reiterative effects of some mutants suggest that cells are bipotent (i.e., capable of only linear developmental decisions) and limited to entering only one or one other pathway.

Insects

The capacity of the blastoderm to regulate varies within and between different groups of insects. The blastoderm of even the most regulative species gradually and progressively becomes determined, while the blastoderm of the most determinant species shows some degree of regulation (see Bodenstein, 1955).

The dragonfly, *Platycnemis*, for example, remains regulative throughout the syncytial blastoderm stage and possibly into the germ band stage, producing twin dwarf embryos with normal proportions when ligated more or less evenly (Seidel, 1929). Honeybees become determined at an early blastoderm stage, while dipterans may be relatively mosaic or determined as early as fertilization. Even *Drosophila melanogaster,* however, compensates for cells destroyed by laser microcautery, and imaginal disks regenerate when transplanted back to larvas.

Cytoplasmic interactions seem to be fundamental to the determination of segmentation. Thus, the number of segments formed in ligated eggs is frequently less than the number formed in whole eggs, and UV-irradiated eggs of the midge, *Smittia*, have fewer segments than normal (i.e., 14–16 compared to 19) (see Kalthoff, 1979). Since the separation of ligated parts restores a more normal number of segments, some of the interactions determining segment number would seem to be inhibitory.

The results of ligation experiments in *Platycnemis* (Seidel, 1929) also show that development of the blastoderm can be "turned off" by interrupting continuity with tissue in the posterior 1/8th of the embryo (see Fig. 12.18). At early cleavage stages (e.g., the fourth intralecithal division), ligation in front of this point prevents later blastoderm differentiation, but ligation behind the same point allows a complete and normally proportioned embryo to develop. The effect is "all or none" with no intermediates or partially differentiated embryos formed. The ligature seems to operate a decision-making mechanism or determination "switch."

As the fate maps of several arthropods illustrate (see Figs. 14.1 and 14.2), determination lays down a **basic longitudinal body pattern** (Kalthoff, 1979). In pterygotes a **differentiation center** or **zone** in the vicinity of the **cephalic furrow** (Figs. 14.2c and 18.7) may serve as a center for **segmentation** or **metamerization,** the establishment of segments, and the differentiation of appendages in both directions.

As gastrulation ends, the thin rod of internalized mesoderm thickens, and clusters of segmental masses bulge against the yolk sac, while neuroblasts reproduce the segmental or metameric pattern on the surface (e.g., see Fig. 18.7 15) (Turner and Mahowald, 1977). The piles of abdominal mesectoderm on the ventral surface begin their metamorphosis from the previously homogeneous germ band into the **segmented germ band.** Head lobes produce the head segments up to the antennas of the embryo (i.e., procephalic segments), while the postantennal region produces all the postantennal segments (i.e., gnathocephalonic, thoracic, and abdominal segments).

In many insects, both apterygote and pterygote, segments from the head to the thorax are formed directly, but a posterior **growth zone** buds off the abdominal segments. In other insects, the growth zone next to the cephalic furrow produces the entire postantennal region, and, in termites and stone flies, not even head lobes appear initially, but the entire organism is budded off a growth zone.

In endopterygotes, **cephalic lobes** (also called the protocephalon or sometimes procephalon) form the **postantennal** region or protocorm (Fig. 22.6). All together, six segments seem to fuse into the head. The next three segments combine (rather than fuse) into the compact thorax, more or less filled in the adult with the muscles that move wings and legs. Finally, the abdominal segments are hinged together and filled with viscera and reproductive organs.

Fruit Flies

Embryonic development is conspicuously rapid in *Drosophila* and other fruit flies where a first stage larva or **instar** is produced within 24 hours from oviposition. The acquisition of the body plan and embryonic structures is already well underway by gastrulation, and the determination that began before or during cellularization becomes manifest during blastokinesis and segmentation.

At the end of gastrulation, boundaries between segments, or **intersegmental boundaries,** appear as deep folds of the embryonic **epidermis** (stage 11 C-O&H, Fig. 18.7 13–16). Soon, the cephalic furrow disappears, and the cephalic appendages appear. Mandibular rudiments (Ma) lie just posterior to the cephalic furrow; maxillar rudiments (Mx) sit in-between, and the labial rudiments are farther posteriorly and ventrally (L, Fig 18.7 18). Deep dorsocaudal grooves in the procephalic lobe identify the primordia of the optic lobes (O, Fig. 22.6).

Internally, the mesoderm has undergone its third and last round of mitosis and has split into two layers (i.e., the outer somatic layer and the inner splanchnic layer) with the **coelom** between them. Salivary gland rudiments emerge on the ventral surface of the labial appendages, and transient tracheal pits appear on the anterior lateral surfaces of the last 10 segments (see Figs. 18.7 18, and 22.6 arrow).

Germ band shortening is the next major morphogenic event. The separation of the posterior tip of the embryo from the yolk sac (stage 12 C-O&H) signals the beginning of germ band shortening. As the posterior portion of the embryo withdraws toward its original position, the thin extraembryonic amnioserosa (AS, Fig. 22.6 1) stretches and, without any cell division, covers the additional yolk sac on the dorsal surface (Fig. 22.6 2). The shortening of the germ band is registered laterally in deepened

furrows and an exaggerated appearance of segments.

Internally, germ band shortening brings the anterior and posterior midgut strands together as ventrolateral bands on either side of the yolk sac. Once fused longitudinally, the midgut begins the long process of growth over and under the yolk sac (see Fig. 18.8d). Endodermal cells do not encompass the bulk of the yolk for about 4 hours (i.e., until stage 14 C-O&H), and then anterior and posterior dorsal protrusions of yolk are not withdrawn for another half hour (stage 15 C-O&H).

Extensions of the tracheal pits are also brought together by germ band shortening. They soon fuse into rudiments of the branched tracheal tree. The pits close but presumptive posterior spiracles open at the ends of the tree (sp, Fig. 22.6).

The internalization of the paired segmental **germ band neuroblasts** accompanies germ band shortening. The median strand is drawn inward too and becomes incorporated into the ventral nerve cord.

Paralleling the onset of differentiation, the embryo sustains its first wave of cell death. The main site of shrunken dead cells with pycnotic nuclei is the epidermis adjacent to the underlying nervous system. Possibly, the separation of epidermis and nervous system is predicated on the death of these cells. Large **macrophages,** mobilized or induced from nearby mesoderm cells, phagocytized the dead cells.

When germ band shortening is complete (stage 13 C-O&H), the anal plate occupies the posterior pole of the embryo. The yolk sac then protrudes dorsally, behind an **anterior dorsal gap** at the interface of cephalic and truncal regions, and ahead of a **caudal dorsal gap** at the level of the proctodeal opening.

Dorsal closure, or the movement of ectoderm around the protruding yolk, begins with the formation of dorsal ridges or folds of epidermis (dr, Fig. 22.6) *from the procephalic lobe.* The ridges appear on both sides of the anterior dorsal gap and grow dorsally until fusing across the dorsal midline.

The major portion of the protruding yolk is enclosed during **dorsal closure** (except when prevented, e.g., in *almondex* mutants, see Fig. 15.23). The epidermis grows over the exposed amnioserosa, sealing the embryo within a continuous ectoderm (at stage 15, C-O&H, Fig. 22.6 7). No trace of the amnioserosa is found after closure. At the end of the process, cuticular synthesis begins, and the epidermis is transformed into a **hypodermis.**

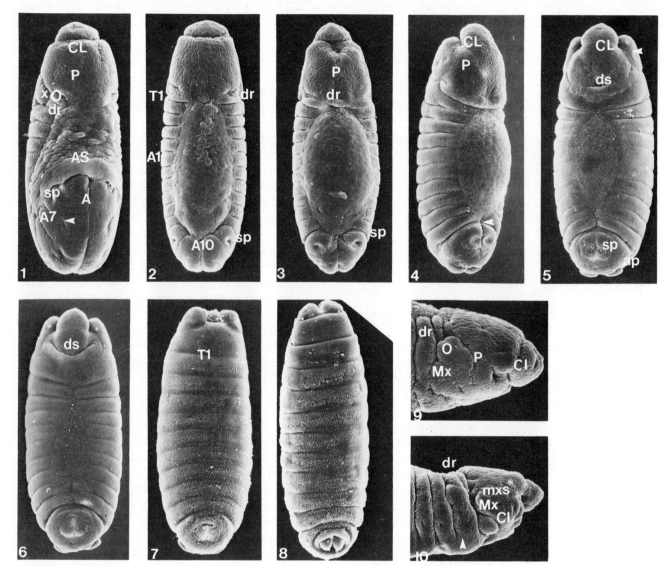

FIGURE 22.6. Scanning electron micrographs of *Drosophila melanogaster* embryos illustrating the climax of germ-band shortening (1, 2, 9), dorsal closure (3–7, 10), head involution (5–7), anal differentiation, and cuticle formation (6–8). (1) At 9 hours after first cleavage (corresponding to stage 12 C-O&H), germ band shortening is halfway. A tracheal pit (arrow) is still present. ×240. (2) At 9.5 hours, germ band shortening complete (corresponding to stage 13 C-O&H). ×290. (3) The dorsal ridge is continuous across posterior border of procephalic lobe (corresponding to early stage 14 C-O&H). Dorsal closure begins. ×260. (4) Dorsal closure continues. Spiracle bearing lobes of anal segment come together on midline (arrow). Procephalic lobe and clypeolabrum fuse, obscuring groove between them. ×280. (5) Dorsal closure continues. A dorsal sac begins to grow over head from the procephalic lobe. A projection on the fused mandibular and maxillary appendages (arrow) is the presumptive mandibular sense organ. ×300. (6) Dorsal closure almost complete. Cuticle formation begins. ×260. (7) At 13 hours (corresponding to stage 15 C-O&H), dorsal closure is complete and the head is almost completely involuted within the dorsal sac. Cuticle differentiation is apparent in numerous small projections. ×250. (8) Nearly complete embryo showing characteristic pattern of denticles cuticle. ×310. (9) Enlarged lateral view of head end of embryo comparable to (1). ×660. (10) Enlarged lateral view of head end of embryo comparable to (4), showing anlage of antennal sense organ (asterisk) and presumptive sites of clusters of sensory hairs (arrows). ×490. A, anal segment; A1, first abdominal segment; A7, seventh abdominal segment; A10, rudimentary 10th abdominal segment; ap, anal pads; AS, amnioseros a; C1, maxillary cirri; CL, clypeolabrum; dr, dorsal ridge; ds, dorsal sac; Hy, hypopharyngeal lobe; L, labial appendage; Mx, maxillary appendage; mxs, anlage of maxillary sense organ; O, optic lobe; P, procephalic lobe; sp, posterior spiracle; T1, first thoracic segment. From F. R. Turner and A. P. Mahowald, *Dev. Biol.,* 68:96(1979), by permission of Academic Press and courtesy of the authors.

Head involution is the withdrawal of the head into the dorsal sac or pouch under the atrium. The process begins with the closure of the dorsal ridges (stage 14 C-O&H; Fig. 22.6 3) and continues until the anterior most tip of the head (the clypeolabrum) is drawn under the atrium (stage 16 C-O&H, Fig. 22.6 8). Ventrally, the head is also displaced into the stomodeum, and laterally, the cephalic appendages become hidden behind lateral borders of the stomodeum and atrium.

The separation and differentiation of organ primordia continues through hatching. The larval heart forms (stage 16 C-O&H), and fat bodies extend laterally and anteriorly from gonads.

The formation of imaginal disks takes place over a protracted period of time. Surface placodes of the ventral thoracic disks are seen during the first instar. Wing disks invaginate and dorsal and ventral mesothoracic disks are contiguous in the blastoderm, possibly to be invaginated during dorsal closure. Antennal and eye disks are attached to the deep lateral portions of the dorsal sac, probably at the site of the optic lobe invagination. The genital disk forms ventral to the anus at the end of the embryonic period.

Separate from the hypodermis, the ventral nerve cord contracts. Ganglia fuse and lose their segmental pattern while concentrating within the first four abdominal segments. The optic lobes become integrated into the supraesophageal ganglion.

Cuticle formation and the differentiation of larval rudiments characterize the remaining stages of embryogenesis. The cuticle-lined portions of the so-called foregut and hindgut are distinguished from the cuticle-free portions of the midgut. Finally, the outer cuticle is embroidered with hairs and bristles (stage 17 C-O&H), and hatching liberates the first larval instar (see Fig. 15.17).

SERIAL HOMOLOGY

Flies almost certainly evolved from insects with four wings instead of two and insects are believed to have come from arthropod forms with many legs instead of six. During the evolution of the fly, two major groups of genes must have evolved: "leg-suppressing" genes which removed legs from abdominal segments of millipede-like ancestors followed by "haltere-promoting" genes which suppressed the second pair of wings of four-winged ancestors. If evolution indeed proceeded in this way, then mutations in the latter group of genes should produce four-winged flies and mutations in the former group, flies with extra legs.

E. B. Lewis (1978, p. 565)

Most of the research on the genetics of serial homologies has been done on the homeotic mutants of *Drosophila melanogaster*. Today, various representatives of the **homeobox** superfamily first identified in *Drosophila*'s homeotic genes offer a good chance for analyzing overriding genetic controls, if not the controls of serial homology (see Mercola and Stiles, 1988).

Homeotic Genes

Homeosis (originally homoeosis, Gk. *homoioun* to make alike + -*sis* suffix of action), a term coined by William Bateson, refers to the change of one member in a series of segments or structures to the form of another member. All mutations that change the identity of segments are **homeotic** (also homoeotic), mutations and their normal genic counterparts are **homeotic genes.**

Homeotic genes are expressed in a segment-specific manner during embryogenesis and larval development (e.g., see Fig. 15.24). They seem to be regulated by positional information and to specify or assign regional identities to parts. The accumulation and diversification of homeotic genes in primitive arthropods with largely identical body segments may have led to the evolution of insects with increasingly complex segments and parasegments (see Chapter 15).

What are homeotic genes? Most homeotic mutants in insects fall into the category of **zygotic mutants** requiring the diploid genome (the maternal-effect *caudal* gene of *Drosophila* is an exception). The mutants are cellular autonomous, and effects of embryonic-lethal homeotic mutations can be studied in patches of cells produced by mitotic recombination in heterozygous embryos. The patches are not affected by normal gene expression around them.

The segment-specific pattern of larval denticles of *Drosophila* (see Fig. 15.17) may be altered by homeotic mutants, but the best known mutants are those affecting the differentiation of imaginal disks and the phenotype of adults. External effects on ectoderm structures are most conspicuous, but homeotic mutations also affect internal structures from the central nervous system to mesodermal derivatives, possibly including muscle. Only the germ line seems to be entirely free of homeotic effects.

In *Drosophila*, homeotic mutants produce deviations in the head and the 11 body segments (e.g., Figs. 22.7 and 22.8) of the adult. Major clusters of closely linked homeotic mutants with similar effects are found in homeotic **gene complexes** (or

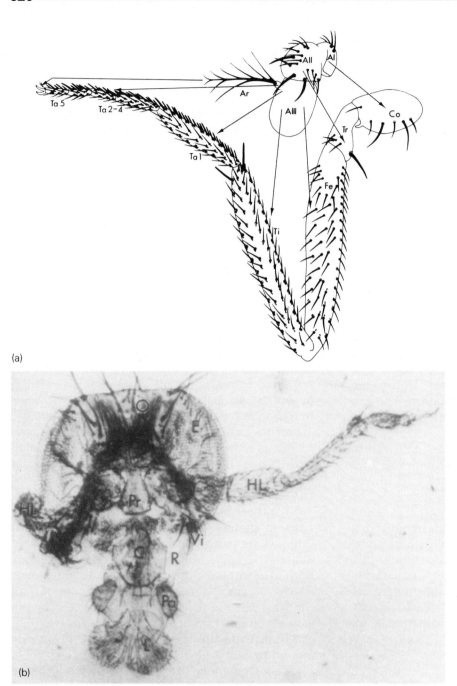

(a)

(b)

FIGURE 22.7. The antenna and homologous parts of the prothoracic leg (a), and an *Antennapedia* mutant (*Antp*[R]) showing large mesothoracic homeotic legs (HL) in place of antennas (b). From J. H. Postlethwaite and H. A. Schneiderman, *Dev. Biol.,* 25: 606(1971), by permission of Academic Press and courtesy of the authors.

pseudoallelic series also known as operons and polycistronic units) located on the right arm of the third chromosome.

Fewer proteins are encoded by genes in the complexes than would be required to specify the diversity of each segment, but nuclear transcripts produced in the complexes are strikingly long, and processing them seems to be intricate. Individual genes (e.g., *Antp*) may contain alternate promoters and termination-processing regions, as well as several exons and introns. The different transcripts produced nevertheless retain the same open-reading frame. Additional complexity may be built into exons in the form of internal microexons or multiple donor splice sites (e.g., the *Ubx* transcription unit). Other genes in the complexes (e.g., *ftz*) may be shorter and relatively simple with two exons joined by one intron, and some may be regulatory, not containing start codons or having only short open-reading frames.

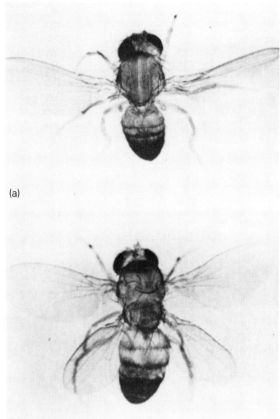

(a)

(b)

FIGURE 22.8. Wild type *Drosophila melanogaster* (a) and an *ultrabithorax* mutant (b, genotype: bx^3 pbx/Ubx^{105}). The mesothoracic wings have been extended to show halters on the metathorax of the wild type and wings on the corresponding segment of the mutant. From E. B. Lewis, in M. Locke, ed., *The role of chromosomes in development*, 23rd Symposium of the Society for Developmental Biology, Academic Press, Orlando, 1964. Used by permission.

The homeotic loci are defined by deletions and inversions. Deletions tend to produce **recessive loss-of-function mutants, hypomorphic mutants** whose phenotypes result from producing less than normal amounts of gene products or **amorphic mutants** whose phenotypes result from producing none of the normal gene products. Inversions tend to produce **dominant gain-of-function mutants,** or **hypermorphic mutants** whose phenotypes result from overproduction of the normal gene products.

The **antennapedia complex** (ANT-C) contains at least eight homeotic genes (Fig. 22.9). Dominant gain-of-function mutants **posteriorize** anterior segments of the head and thorax (Table 22.1). In the most extreme expression of the *Antennapedia* (*Antp*, Fig. 22.7) mutant, an antenna is replaced with a leg (hence the name).

The **bithorax complex** (BX-C, Fig. 22.9) contains at least 12 homeotic genes. Recessive loss-of-function mutants **anteriorize** posterior segments of the thorax and abdomen. The different *bithorax* (*bi*) genes are detected as deletion mutants within the *Ubx* domain. They are expressed narrowly in specific compartments (e.g., anterior or posterior) or parasegments suggestive of specific intracellular or cell-lineage control.

The sequence of BX-C genes or **genetic domains** is remarkably parallel to the proximal-to-distal order of regions over which the genes exercise morphogenic control (see Lewis, 1964). The *Ubx* domain nearest the centromere, for example, specifies the region from the posterior mesothorax (T2p) to the anterior first abdominal segment (A1a), while the abdominal domain (*iab-2–iab-8*), farthest from the centromere, specifies abdominal segments (from posterior A1 to A8). Conceivably, the genes arose by tandem duplications, and functional interactions between the genes prevent their dispersion from clusters.

Mutations affecting different compartments in the same structure may be expressed convergently, and animals bearing combinations of homeotic mutations may possess harmoniously integrated structures affected independently by different mutations. For example, the combination of mutations changing the anterior compartment of the third thoracic segment, or metathorax, into the anterior compartment of the second thoracic segment, or mesothorax, and the posterior compartment of the metathorax into the posterior compartment of the mesothorax change the entire metathorax into a mesothorax when working together and create a fly with two apparent mesothoraxes. These mutations also transform the club-shaped, diminutive wing, known as the **halter** (or haltere sing., halteres pl.), normally on the fly's metathorax, into a full-blown wing resembling that normally found on the fly's mesothorax (Fig. 22.8) (Lewis, 1964).

What do homeotic genes do? Suppose that, in the wild-type *Drosophila melanogaster*, products of normal bi^+ genes promote halter formation not so much by inducing a halter as by limiting the ability of normal $Antp^+$ gene products to induce a wing. The Ant^+ gene, on the other hand, ensures that mesothoracic legs develop on the mesothorax not so much by inducing them but by limiting the ability of the normal bi^+ genes to induce metathoracic legs. A hypomorphic or amorphic bi^- mutation would then cause the mutant phenotype from the unopposed presence of the normal $Antp^+$

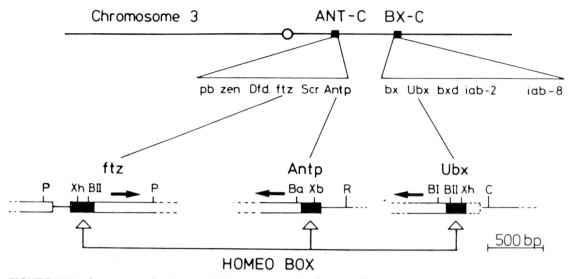

FIGURE 22.9. Gene map of right arm of chromosome 3 in *Drosophila melanogaster* showing the antennapedia complex (ANT-C) and the bithorax complex (BX-C). The homeoboxes (closed bars) in the *ftz, Antp,* and *Ubx* transcription units are present in the 3′ exons. Arrows show the direction of transcription at each locus. Exons are indicated by bars (open and closed), introns by lines, and uncertain areas by dotted lines. From W. J. Gehring, in E. H. Davidson and R. A. Firtel, eds., *Molecular biology of development,* Alan R. Liss, Inc., New York, 1984. Used by permission.

gene product in the metathoracic compartments (Table 22.2).

The possibility that the amount of the *Antp* gene product can alter phenotype was recently tested with the help of recombinant DNA technology and the heat-shock technique for making phenocopies originally explored by Richard Goldschmidt. An *Antp* cDNA under the control of a heat-shock promoter gene was transferred into the germ line of normal flies of different strains. Transformants had two *Antp*⁺ genes, one of which (the fusion gene) was capable of producing additional amounts of gene product under heat-shock control (Gibson and Gehring, 1988).

While at normal temperatures transformants did not show any change in phenotype, some strains of transformants heat-shocked prior to 3 hours resulted in denticle-belt transformation from the prothoracic (T1) to the mesothoracic (T2) patterns. Heat-shock prior to 9 hours produced a range

of developmental failures (e.g., defects in the anterior head anlage and failure of head involution) anterior to the normal domain of *Antp* expression but not posterior (i.e., in abdominal segments). Finally, one strain of transformants heat-shocked during the early third larval stage developed legs in place of antennas. The overexpression of *Antp* protein in the third larval stage would seem responsible for the mutant phenotype in adults.

Homeotic genes do not function in a genetic vacuum within a cell. Rather, operating as **selector genes,** homeotic genes seem to encode DNA-binding proteins which select the **cellular response** or **realizer genes** expressed in developmental pathways (Garcia-Bellido, 1975; Gehring and Hiromi, 1986). Selector genes may be **master control genes** that program cellular determination. They may be switching points in genetic control circuits or guard rails that keep the vast numbers of realizer genes on the path leading to specific structures (see Gehring, 1984).

Segmental differentiation or identity, whether of the normal or mutant type, may represent a tissue's response to a set of competing pressures generated by different homeotic gene products. Downstream gene expression may be regulated by various homeotic selector gene products competing to bind with the same promoters (Gehring, 1987).

TABLE 22.1. Direction of changes initiated by *Antennapedia* (*Antp*) and *Ultrabithorax* (*Ubx*) mutations

... T1	T2	T3	A1	A2 ...

←——— *Antp Ubx* ———→

TABLE 22.2. How homeotic mutants may determine phenotypes through interactive gene expression

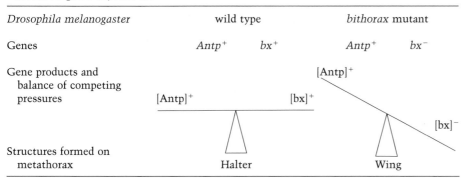

Drosophila melanogaster — wild type — *bithorax* mutant

Genes: $Antp^+$ bx^+ | $Antp^+$ bx^-

Gene products and balance of competing pressures: $[Antp]^+$ $[bx]^+$ | $[Antp]^+$ $[bx]^-$

Structures formed on metathorax: Halter | Wing

Homeobox, Homeobox Superfamily, and POU

The homeobox is the fruit of recombinant DNA technology. Homeoboxes were discovered during efforts to map overlapping segments of cloned homeotic genes by hybridizing cDNA to a gene library, and their analysis is still largely at the molecular level.

Initially, weak hybridization was detected between DNA of the *Antp* locus and a sequence to its left. Soon, poly(A)$^+$ transcripts from this sequence were recognized as originating from the *fushi tarazu* (*ftz*) segmentation gene (Figs. 22.9 and 15.20), and another similar sequence was localized in a transcription unit of the *Ubx* domain (see Gehring, 1984). These "cross-homologous" sequences were named **homeoboxes** because of their presence in both homeotic gene complexes (ANT-C and BX-C) and in homeotic genes (e.g., *Dfd, Scr, Antp*, and the *iab* genes as well as the *bx* genes, Fig. 22.9). The catchy name has stuck despite the appearance of homeoboxes in genes outside the homeotic complexes (e.g., *caudal* and *engrailed*) and in nonhomeotic genes (e.g., *ftz*).

Homeobox sequences were soon discovered in vertebrate genomes probed with heterologous homeobox-containing cDNA (i.e., Southern blotting experiments), and cross-reacting sequences were cloned from *Xenopus*, mice, and humans. The homeoboxes and homeobox-like regions, generally found in the 3' exon (i.e., the last exon, Fig. 22.9), resembled "cassettes," copied very nearly precisely and inserted into several gene complexes. Similarities among the base pairs of homeoboxes and the base pairs of homeotic genes even suggested that the genes were also homologous and evolved from the repetition of a single gene.

The sequence of 180–183 base pairs in *Antennapedia*-like homeoboxes is remarkably similar (i.e., "conserved") to sequences found in different genes and in different species (Fig. 22.10). Most notably, nucleotide similarity between 60 and 80% is found between the *Drosophila* homeobox and sequences in a beetle, *Tenebrio molitor*, earthworms, *Lumbricus* sp., the frog, *Xenopus laevis*, the laboratory mouse, *Mus musculus*, and ourselves, *Homo sapiens*. The homeobox in the *Mo-10* gene on chromosome 6 of the mouse shares two-thirds of its base pairs with the *Antennapedia* homeobox of *Drosophila* and the *antennapedia*-like homeobox of *Xenopus*.

Translated to amino acids, homeobox sequences encode **homeodomains** of 60–61 amino acids. Because of redundancy in the genetic code, the peptides of homeodomains are even more similar than the nucleic acids of homeoboxes (Fig. 22.11) (Ruddle et al., 1985). Nine amino acids are invariant in the 16 homeodomains of *D. melanogaster* and 22 homeodomains of other species. Out of a total of 60 amino acids in one *X. laevis* homeodomain, 58 and 59 are the same as the two major variants of *Drosophila*'s homeodomains. Moreover, the two human homeodomains (Ho-1 and Ho-2) share 90% of their amino acids with homeodomains of the ANT-C and BX-C homeodomains of *Drosophila*. On the other hand, the homeodomains of other proteins share no more than 21 amino acid residues with the prototypic homeodomains of *Drosophila*.

What do homeoboxes and homeodomains do?
The homeobox encodes a DNA-binding domain of the helix-turn-helix motif capable of recognizing sequence-specific cis-acting regulatory elements (Levine and Hoey, 1988). Homeodomains seem to

```
                                                                -21                                  -1
Antp                                                            ATT TAC TTG GAA CCA ACA GAA
Mo -10          -56                                             GGA CCG CCT GGG CAG GCC TCG
Hu-1          GAA TAA GTG TCG TTG CGG CTT TCC TCT ATC TGC       TCC AGA TAT GAC GGG CCG GAC
Hu-2          TGA CCG CAG GCC TCA GCA TCT CCA CTC TGC GTA       ACA GGT TCC TCC TTT GGG CCC
AC1                                                             GGC GTG GGC TAC GGG TCG GAC
MM3                                                             CTT TCT CTT GCA GGG GCG GAC


              1                                       30                                       60
Antp          CGC AAA CGC GGA AGG CAG ACA TAC ACC CGG TAC CAG ACT CTA GAG CTA GAG AAG GAG TTT
Mo-10         TCC AAG CGC GGC CGC ACG GCG AGG CCG CAG CTG GTA GAG CTG GAG GAG GAG TTC
Hu-1          GGG AAA AGG GCC CGG ACC GCG TAT ACC CGC TAC CAG ACC CTG GAG CTG GAA AAG GAG TTC
Hu-2          ACG GCC GGA GGA CGC CAG ACA TAC ACA CGT TAC CAG ACG CTG GAG CTG GAG AAG GAG TTT
AC1           AGG AGG AGA GGA CGC CAG ATC TAT TCC CGT TAC CAA ACT CTG GAG CTC GAG AAA GAA TTT
MM3           AGG AAG AGG GGT CGC CAG ACC TAC ACA AGG TAC CAG ACC CTG GAG CTG GAG AAG GAG TTT


              61                                      90                                       120
Antp          CAC TTC AAT CGC TAC TTG ACC CGT CGG CGA AGG ATC GAG ATC GCC CAC GCC CTG TGC CTC
Mo-10         CAC TTC AAC CGC TAC CTA ATG CGG CCG CGC CGG GTG GAG ATG GCC AAC CTG CTG AAC CTC
Hu-1          CAC TTC AAC CGC TAC CTG ACC CGG CGA CGG CGC ATC GAG ATC GCC CAC GCA CTC TGC CTG
Hu-2          CAC TAC AAT CGC TAC CTG ACG CGG CGG CGG CGC ATC GAG ATC GCG CAC GCC CTG TGC CTG
AC1           CAC TTC AAT CGC TAC CTG ACC CGG CGC AGG AGG ATC GAG ATC GCC AAT GCA CTT TGT CTC
MM3           CAC TTT AAC CGC TAC CTG ACC CGG CGG AGG CGC ATC GAG ATC GCC CAC GTT CTG TGT CTG


              121                                     150                                      180
Antp          ACG GAG CGC CAG ATA AAG ATT TGG TTC CAG AAT CGG CGC ATG AAG TGG AAG AAG GAG AAC
Mo-10         ACC GAG CGC CAG ATC AAG ATC TGG TTT CAG AAC CGG CGC ATG AAG TAC AAG AAA GAC CAG
Hu-1          TCC GAG CGC CAG ATC AAG ATC TGG TTC CAG AAC CGG CGC ATG AAG TGG AAG AAG GAC AAC
Hu-2          ACG GAG AGG CAG ATC AAG ATA TGG TTC CAG AAC CGA CGC ATG AAG TGG AAA AAG GAG AGC
AC1           ACA GAG CGG CAG ATC AAA ATC TGG TTC CAG AAC AGG AGG ATG AAA TGG AAA AAG GAG AGC
MM3           ACC GAG CGA CAA ATC AAA ATC TGG TTC CAG AAC CGC AGG ATG AAA TGG AAA AAG GAA AAC


              181                     201                                                      240
Antp          AAG ACG AAG GGC GAG CCG GAT
Mo-10         AAG GGC AAA GGC ATG CTG ACC
Hu-1          AAA TTG AAA AGT ATG AGC CTG GCT ACA GCT GGC AGC GCT TCC AGC CCT GAG CCC GCC CAG AGG AGC CAA
Hu-2          AAA CTG CTC AGC GCG TCT CAG CTC AGT GCC GAG GAG GAG GAA GAA AAA CAG GCC GAG TGA AGG TGC TGG
AC1           AAC CTC TCA TCT ACC CTT
MM3           AAG GCA TCC AGC CCT TCC TCT AAC AGC CAG GAA AAG CAG GAG ACT GAG GAA GAG GAG GAG GAA TGA AGT
```

FIGURE 22.10. Base sequences in the homeodomain and surrounding regions of the *Antennapedia* (*Antp*) gene of *Drosophila melanogaster*, the Mo-10 gene of the mouse, the Hu-1 and Hu-2 genes of humans, and the CC1 and MM3 genes of *Xenopus laevis*. The homeodomain extends from nucleotides 1 to 180. From R. H. Ruddle, C. P. Hart, and W. McGinnis, in F. Costantini and R. Jaenisch, eds., *Genetic manipulation of the early mammalian embryo*, Banbury Report, Cold Spring Harbor Laboratory, Cold Spring Harbor, NY, 1984. Used by permission.

be present in some, but not all, general and cell-type-specific positive transcription factors or trans-activators.

These tentative conclusions, reached through the comparative method, were not at all obvious when homeoboxes were first discovered. Originally, the correlation of homeoboxes with homeotic genes was thought to tie homeoboxes functionally to serial homology. Like *Drosophila*, many organisms exhibiting homeoboxes also exhibited differential segmentation, while organisms ordinarily thought of as unsegmented seemed not to have homeoboxes in their genomes. The correlation was never quite airtight, however.

Even early on, some *Drosophila* genes containing homeoboxes were known to control specific cell-type differentiation rather than regional differentiation involving multiple cell types. Moreover, the *zerknüllt* homeobox-containing gene controlled the differentiation of embryonic cells along the dorsal–ventral axis independently of segmentation and homeotic genes.

Nor are genes containing homeoboxes tightly correlated with the control of homeotic genes. Mutations in homeoboxes (as opposed to elsewhere in cognate homeotic genes) do not ordinarily induce homeotic mutant phenotypes (see Regulski et al., 1985). What is more, some homeobox-containing genes, like the yeast regulatory gene *PH02*, operate a physiological mechanism having nothing to do directly with determination, cell type, or differentiation (Bürglin, 1988).

Other exceptions may prove even more instructive. Yeast *MAT* mating-type genes, for example, sharing a major portion of their nucleotide sequences (i.e., bp 40–57, Fig. 22.10) with the sequences of animal homeoboxes, may be master selector genes regulating sizable groups of realizer genes through the products of sequence-specific DNA-binding homeodomains.

												-7						-1
Antp												Ile	Tyr	Leu	Glu	Pro	Thr	Glu
Mo-10												Gly	Pro	Pro	Gly	Gln	Ala	Ser
Hu-1	Glu	Trм	Val	Ser	Leu	Arg	Leu	Ser	Ser	Ile	Cys	Ser	Arg	Tyr	Asp	Gly	Pro	Asp
Hu-2	Trм	Pro	Gln	Ala	Ser	Ala	Ser	Pro	Leu	Cys	Val	Thr	Ser	Phe	Gly	Pro	Pro	
AC1												Gly	Val	Gly	Tyr	Gly	Ser	Asp
MM3												Leu	Ser	Leu	Ala	Gly	Ala	Asp

	1									10										20
Antp	Arg	Lys	Arg	Gly	Arg	Gln	Thr	Tyr	Thr	Arg	Tyr	Gln	Thr	Leu	Glu	Leu	Glu	Lys	Glu	Phe
Mo-10	Ser	Lys	Arg	Gly	Arg	Thr	Ala	Tyr	Thr	Arg	Pro	Gln	Leu	Val	Glu	Leu	Glu	Lys	Glu	Phe
Hu-1	Gly	Lys	Arg	Ala	Arg	Thr	Ala	Tyr	Thr	Arg	Tyr	Gln	Thr	Leu	Glu	Leu	Glu	Lys	Glu	Phe
Hu-2	Thr	Ala	Gly	Gly	Arg	Gln	Thr	Tyr	Thr	Arg	Tyr	Gln	Thr	Leu	Glu	Leu	Glu	Lys	Glu	Phe
AC1	Arg	Arg	Arg	Gly	Arg	Gln	Ile	Tyr	Ser	Arg	Tyr	Gln	Thr	Leu	Glu	Leu	Glu	Lys	Glu	Phe
MM3	Arg	Lys	Arg	Gly	Arg	Gln	Thr	Tyr	Thr	Arg	Tyr	Gln	Thr	Leu	Glu	Leu	Glu	Lys	Glu	Phe
													hpo	Ala			hpo		hpo	
													*				*		*	

	21									30				-----	Helix 2	-----				40
Antp	His	Phe	Asn	Arg	Tyr	Leu	Thr	Arg	Arg	Arg	Arg	Ile	Glu	Ile	Ala	His	Ala	Leu	Cys	Leu
Mo-10	His	Phe	Asn	Arg	Tyr	Leu	Met	Arg	Pro	Arg	Arg	Val	Glu	Met	Ala	Asn	Leu	Leu	Asn	Leu
Hu-1	His	Phe	Asn	Arg	Tyr	Leu	Thr	Arg	Arg	Arg	Arg	Ile	Glu	Ile	Ala	His	Ala	Leu	Cys	Leu
Hu-2	His	Tyr	Asn	Arg	Tyr	Leu	Thr	Arg	Arg	Arg	Arg	Ile	Glu	Ile	Ala	His	Ala	Leu	Cys	Leu
AC1	His	Phe	Asn	Arg	Tyr	Leu	Thr	Arg	Arg	Arg	Arg	Ile	Glu	Ile	Ala	Asn	Ala	Leu	Cys	Leu
MM3	His	Phe	Asn	Arg	Tyr	Leu	Thr	Arg	Arg	Arg	Arg	Ile	Glu	Ile	Ala	His	Val	Leu	Cys	Leu
				Ile/Val					hpo											
				*					*											

	41	-----		Helix 3		-----				50										60
Antp	Thr	Glu	Arg	Gln	Ile	Lys	Ile	Trp	Phe	Gln	Asn	Arg	Arg	Met	Lys	Trp	Lys	Lys	Glu	Asn
Mo-10	Thr	Glu	Arg	Gln	Ile	Lys	Ile	Trp	Phe	Gln	Asn	Arg	Arg	Met	Lys	Tyr	Lys	Lys	Asp	Gln
Hu-1	Ser	Glu	Arg	Gln	Ile	Lys	Ile	Trp	Phe	Gln	Asn	Arg	Arg	Met	Lys	Trp	Lys	Lys	Asp	Asn
Hu-2	Thr	Glu	Arg	Gln	Ile	Lys	Ile	Trp	Phe	Gln	Asn	Arg	Arg	Met	Lys	Trp	Lys	Lys	Glu	Ser
AC1	Thr	Glu	Arg	Gln	Ile	Lys	Ile	Trp	Phe	Gln	Asn	Arg	Arg	Met	Lys	Trp	Lys	Lys	Glu	Arg
MM3	Thr	Glu	Arg	Gln	Ile	Lys	Ile	Trp	Phe	Gln	Asn	Arg	Arg	Met	Lys	Trp	Lys	Lys	Glu	Asn

	61						67																
Antp	Lys	Gly	Lys	Gly	Met	Leu	Thr																
Mo-10	Lys	Gly	Lys	Gly	Met	Leu	Thr																
Hu-1	Lys	Leu	Lys	Ser	Met	Ser	Leu	Ala	Thr	Ala	Gly	Ser	Ala	Ser	Ser	Pro	Glu	Pro	Ala	Gln	Arg	Ser	Pro
Hu-2	Lys	Leu	Leu	Ser	Ala	Ser	Gln	Leu	Ser	Ala	Glu	Glu	Glu	Glu	Glu	Lys	Gln	Ala	Glu	Trм	Arg	Cys	Trp
AC1	Asn	Leu	Ser	Ser	Thr	Leu																	
MM3	Lys	Ala	Ser	Ser	Pro	Ser	Ser	Asn	Ser	Gln	Glu	Lys	Gln	Glu	Thr	Glu	Glu	Glu	Glu	Glu	Glu	Stop	

FIGURE 22.11. Amino acid sequences deduced from nucleotide base sequences in the homeodomains of the *Antennapedia* (*Ant*) mutant of *Drosophila melanogaster,* the Mo-10 gene of the mouse, the Hu-1 and Hu-2 genes of humans, and the CC1 and MM3 genes of *Xenopus laevis*. Some sites in the homeodomain would seem to be remarkably conserved. The helical domains (Helix 2 and Helix 3), hydrophobic residues (hpo) and other sites (stars), especially the characteristic residues at 35 and 45, correspond to amino acids in prokaryotic repressors. From R. H. Ruddle, C. P. Hart, and W. McGinnis, in F. Costantini and R. Jaenisch, eds., *Genetic manipulation of the early mammalian embryo,* Bambury Report, Cold Spring Harbor Laboratory, Cold Spring Harbor, NY, 1985. Used by permission.

Similarly, viral and prokaryotic repressor proteins have regions resembling animal homeodomains (e.g., Helix 2 and 3 of lambda Cro, the hydrophobic [hpo], and stared amino acids). These portions of the "homeodomain" specify the DNA-recognition site of the helix-turn-helix motif (see Chapter 4). Identified by tryptophan (Trp or W, in the single letter code) and phenylalanine (Phe or F) toward the carboxyl terminus, the helical regions of the animal homeodomains are indicative of **gene-controlling-gene** functions (see Gehring, 1987).

Ultimately, the exclusive link between homeoboxes and serial homology was broken by the discovery of homeoboxes in echinoderms and in the classic "unsegmented worm," the nematode *Caenorhabditis elegans* (Holland and Hogan, 1986). This is not to say that homeoboxes may not function in specifying regional fate, but their general function may be more sweeping.

At present, at least three selector genes in *C. elegans* are known to contain homeoboxes. The *mab-5* gene specifies epidermal, neuronal, and mesodermal cell division, differentiation, migration, fusion, and programmed cell death in the posterior region (Costa et al., 1988). The *mec-3* gene specifies the differentiation of six mechanosensory neurons, the touch receptors (Way and Chalfie, 1988). The *unc-86* gene controls several neuroblast lineages and the differentiation of specific adult neurons (Finney et al., 1988).

The *mab-5* gene is not only more like typical homeotic genes in coordinating morphogenesis of different cell types in a region, but the *mab-5* homeobox is also more like the *Antennapedia* homeobox (44 of 66 amino acids) than the homeoboxes of the other *C. elegans* genes. Moreover, like segmentation genes carrying homeoboxes (e.g., *ftz*), the region of *mab-5* gene expression corresponds to the region affected by _mab-5_ mutants (see Fig. 15.20).

The _mec-3_ and _unc-86_ homeoboxes are less _antennapedia_-like (sharing only 18 amino acids). The _mec-3_ gene has no similar sequences to genes encoding any known protein except for its homeobox and an acidic region at the C terminus, resembling regions in regulatory proteins exercising positive controls over transcription. On the other hand, the _unc-86_ gene encodes a protein, UNC-86, with sequences similar to three positive mammalian trans-activators: rat pituitary transcription factor Pit-1 (also called growth hormone factor-1 [GHF-1]) and two **nuclear factors** (NFs) or **transcription factors** (TFs), the human ubiquitously expressed constitutive Oct-1 (also called NF-A1 and OTF-1) and the cell-type specifically expressed Oct-2 (the lymphoid transcription factor, member of the NF-A2 or OTF-2 family).

The Pit-1, Oct-1, Oct-2, and UNC-86 proteins share 58 of 170 amino acids (i.e., 34% identity) making up the so-called **POU** (a contraction of Pit, Oct, and UNC, pronounced "pow") **domain.** The degree of sharing is greatest between Oct-1 and Oct-2 (87%) and least between Oct-1 and Pit-1 (52%) and between Oct-2 and UNC-86 (42%).

The **POU domain** consists of a **POU-related homeobox subdomain** toward the carboxyl terminus, a **POU-specific subdomain** or **POU box** toward the amine terminus, and a **linker region** between the subdomains.

and is not a region of high amino acid identity. POU-specific subdomains, on the other hand, share at least 43% of their 75 amino acids (a striking 74 of 75 in Oct-1 and Oct-2 or 99% identity). Two subregions of the POU-specific subdomain (A = acidic and B = basic) share even more amino acids (65% and 53% identity). The POU-specific subdomain may encode an aid to the nearby DNA-recognition helix, since deletions in Oct-1 POU A diminish Oct-1 DNA-binding (Ko et al., 1988).

**What are homeo- and POU-boxes doing in vertebrate genes?** Vertebrate homeobox-containing and POU genes may or may not control cell fates during development. The situation will probably be clarified soon, but, at the moment, while the vertebrate proteins containing POU domains are transcriptional factors they are not yet known to be developmental control proteins (see Marx, 1988).

Pit-1 is an inducible rat transcription factor regulating gene expression in lactotrophic and somatotrophic cells of the pituitary gland. The protein recognizes an ATGNATA(A/T)(A/T)T[1] promoter sequence and activates prolactin and growth hormone genes in the respective cell types. Pit-1 expression in HeLa cells _in vitro_, moreover, results in the selective activation of prolactin and growth hormone expression (Ingraham et al., 1988).

Oct-1 and Oct-2 recognize the octa-decameric sequence ATGCAAAT(NA) (known as OCTA,

POU Domain

N terminus	POU-specific subdomain	—	linker region	—	POU-related homeobox subdomain	C terminus

Twenty-two of the 60 amino acids in the POU-related homeobox subdomain are identical in all four proteins (53 of 60 in Oct-1 and Oct-2), while as few as 14 are shared with the prototypic _Drosophila_ homeodomains. The greatest degree of identity is found among amino acids in the carboxyl terminus third (71% of 17 amino acids). This subregion is called the WFC for the single-letter codes of tryptophan (W) and phenylalanine (F), found at its end in all homeodomains (Fig. 22.11, Phe = F and Trp = W), and the adjacent cysteine (C), found in the POU-related homeobox subdomain as opposed to glutamine (Gln) found in other homeodomains. The WFC subregion ends the putative DNA-recognition helix of a helix-turn-helix motif (see Herr et al., 1988).

The linker region is short (15–27 amino acids)

hence the proteins' names) found in many genes, from household genes, such as those for small nuclear RNA and the core histone H-2B, to the cell-type specific immunoglobulin genes of B-type lymphocytes. Oct-1 is expressed ubiquitously, while Oct-2 seems to be B-cell specific (Sturm et al., 1988) (see Chap. 4).

Downstream from its POU domain, Oct-2 contains a region absent in Oct-1 where a series of four leucine residues are separated by seven amino acids. The leucines would fall out along one side of an amphipathic alpha-helix (i.e., having polarized hydrophobic and hydrophilic regions on the helix), and two such helices (not necessarily of the

[1]A, C, G, and T are adenine, cytosine, guanine, and thymine. N is a nonspecified nitrogenous base.

same monomers) could link up by a **leucine zipper** into a dimeric protein. Interactions between Oct-2 and other proteins may underlie the cell-type specificity of Oct-2 as opposed to the ubiquitous transcriptional activity of Oct-1 (Clerc et al., 1988).

Other circumstantial evidence links vertebrate homeobox-containing genes to cell-type specific transcription. Above all, the expression of murine *homeobox-containing* (*Hox*) genes is often restricted temporally and spatially during embryogenesis.

The mouse *Hox2.1* gene (formerly the *H24.1* and identical to the *Mu-1* gene) is only expressed in adult tissues when it is first expressed in their embryonic precursors. Expression occurs as early as $7\frac{1}{2}$ days postcoitum and is greatest in some fetal tissues (e.g., gut, kidney, liver, lung, spinal cord, visceral yolk sac) while absent in others (amnion, brain, heart, muscle, placenta, and parietal yolk sac) at 12.5 days.

Different homeobox-encoding genes seem to be expressed differently. For example, *in situ* hybridization reveals the presence of *Hox2.1* transcripts in the lungs' enveloping mesenchyme but not in their lining epithelium. Moreover, like *Hox3* transcripts, *Hox2.1* transcripts appear in the fetal spinal cord, but while *Hox2.1* expression extends to the hindbrain, *Hox3* does not (Krumlauf et al., 1987).

Other homeobox-containing genes in vertebrates may be involved in patterning. The *Xenopus* homeobox *Xhox-1A* gene, for example, transcribes its mRNA predominantly in axial structures, especially the segmental somites. What is more, "deregulating" the gene by microinjection of synthetic homeobox mRNA into one blastomere of *Xenopus* embryos at the 2-cell stage disrupts the segmental somite pattern developing on the injected side. Grossly, the embryos "kink" toward the injected side, while microscopically, abnormalities range from uncoordinated somites to fused and totally chaotic arrays of myocytes (Harvey and Melton, 1988) (Fig 22.12).

The expression of vertebrate homeobox-containing genes, like their *Drosophila* prototypes, may occur in complex temporal as well as spatial patterns. The *Xenopus laevis* homeodomain protein XlHbox 1, for example, is restricted to limbs. The protein appears first in the somatopleure corresponding to the presumptive forelimb bud-field. Overlying ectoderm of both the forelimb and hindlimb buds contains the XlHbox 1 protein, but mesodermal expression is absent in the hindlimb. In *Xenopus*, mice, and chicks, the immunologi-

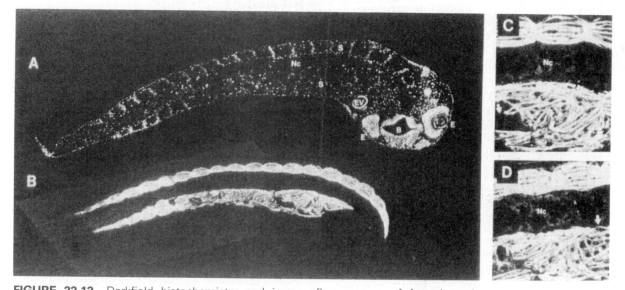

FIGURE 22.12. Darkfield histochemistry and immunofluorescence of frontal section through a Xhox-1A injected embryo. The embryo shows "kinking" toward the lower injected side. (a) A nuclear stain shows nuclei aligned in normal somites (s) on uninjected side of embryo and posterior portion of injected side. (b) Indirect immunofluorescence with muscle-specific monoclonal antibody 12/101 highlighting membranes of myocytes shows disarray in portion of injected side. (c and d) Enlargements (2.5 x) show unstained tissue (arrow in c) and patch of myocytes rotated 90° so as to be seen in cross section (arrow in d). B, brain; E, eye; EV, earl vesicle; Nc, notochord; S, somite or somitic mesoderm. From R. P. Harvey and D. A. Melton, *Cell,* 53:687(1988), by permission of Cell Press and courtesy of the authors.

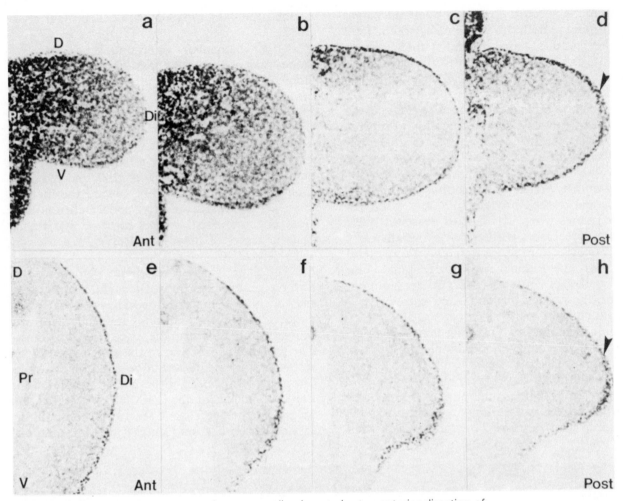

FIGURE 22.13. Serial transverse sections proceeding in anterior to posterior direction of the forelimb (a–d) and hindlimb (e–h) buds of a 10-day mouse embryo stained with antisera against the last 106 amino acids of the human homolog of the XlHbox 1 gene. A gradient of staining (mesodermal gradient) is present in the forelimb but not in the hindlimb. Ant, anterior; D, dorsal; Di, distal; Post, posterior; Pr, proximal; arrowheads, apical ectodermal ridge. From G. Oliver et al., *Cell,* 55:1017(1988), by permission of Cell Press and courtesy of the authors.

cally recognized protein is distributed in early fore-limb nuclei in a concentration gradient declining posteriorly (Oliver et al., 1988) (Fig. 22.13).

Mammalian "homeotic" genes cannot be expected to operate in lineage-restricted compartments like *Drosophila*'s homeotic genes, since early mammalian embryos probably lack these compartments. The results of labeling and grafting experiments in the 9-day mouse embryo *in vitro*, for example, show that, instead of remaining in polyclonal compartments, primitive-streak-derived cells are continuously added to paraxial mesoderm (Tam and Beddington, 1987).

The extent of similarity between mammalian genes and *Drosophila* homeotic genes sometimes goes beyond the homeobox. The mouse *Cdx-1* (Caudal-type homeobox-1) gene, contains several sequences resembling those of *Drosophila*'s *caudal* gene. Like *Drosophila*'s *caudal* gene, *Cdx-1* is differentially expressed in intestinal epithelium, but while *caudal* is a maternal-effect gene also expressed zygotically at the posterior pole and subsequently in posterior midgut, hindgut, and the attached Malpighian tubules, *Cdx* expression is not detectable in the ovary and only appears in the intestinal epithelium at 14 days postcoitum, when tissue differentiation is already advanced (Duprey et al., 1988).

In addition to the homeobox, homeotic genes, like *Antp*, contain multiple independently regulated promoters mediating both positive and negative regulation. Moreover, the genes are capable of generating different RNA transcripts by alternative RNA splicing. Similarly, human *Oct-1* and mouse *Hox2.1* genes contain multiple promoters and participate in complex patterns of expression that produce multiple transcripts. Murine *Hox2.1* is also part of a complex of at least four homeobox-encoding genes on chromosome 11.

These mouse *Hox* genes and the human *Hox2.1*-homolog have extensive similarities, including encoding for high concentrations of serine and proline residues. Moreover, these genes and several (but not all) homeobox-encoding genes from *Drosophila* and *Xenopus* encode a sequence of six amino acids (i.e., a hexapeptide) between 5 and 18 amino acids upstream from the homeobox. Whether the homeodomain and the encoded hexapeptide are functionally related to expression remains to be seen.

Possibly, all these similar sequences represent **molecular tokens,** affecting some aspect of protein binding to DNA. They may identify genes that determine patterns or sequences of events; they may cue related genes and integrate them into functional units; they may specify tissue-specific genes or stage-specific genes.

On the other hand, mammalian homeobox-encoding genes may not operate under any of the rules directing homeotic genes. The look-alikes may not even be limited to transcription factors, and homeotic genes may yet be found lacking homeoboxes.

It should be pointed out that the body plan of protostomes differs considerably from that of deuterostomes, and that the formation of the (sic) their body segments is basically different. However, the presence of the homeobox in vertebrates raises the possibility that homologous genes may be present, and the homeobox provides a tool to isolate such genes. (Gehring and Hiromi, 1986, p. 160)

Chapter

23

ESTABLISHING THE VERTEBRATE BODY PLAN: PALINGENESIS VERSUS PLASTICITY

> The embryonic features that we share with all vertebrates represent no previous adult state, only the unaltered identity of early development. Though they do not allow us to trace the actual course of our descent in any way, they are full of evolutionary significance; for, as Darwin argued, community of embryonic structure reveals community of descent.
>
> Stephen Jay Gould (1977, p. 213)

*I*n contemporary terms, palingenesis is broadly identified with the parallel development of characteristics throughout a relatively high taxonomic category (see Gould, 1977). For example, the development of a dorsal neural tube, notochord, aorta, anterior pharyngeal pouches with aortic arches, a ventral heart, and posterior somites are broadly associated with the development of vertebrates in the immediate postgastrular stage.

As first defined by Haeckel, however, **palingenesis** was the ontogenic repetition of evolutionary stages in an individual's development, or the linear ontogeny of homologous structure. According to Haeckel, embryos were supposed to be subject only to forces from the past and to truncation.

Darwin was one of the rare 19th century biologists who understood that even embryos evolved. He realized that embryos and larvas also adapted to their environments. Embryos developing within egg shells showed adaptations to confined spaces, and fetuses developing within their mother showed adaptations to foreign sources of nutrition.

Today, Darwin's concept is widely accepted (see Gould, 1977). Like cenogenic structures, or specialized embryonic and larval adaptations, embryonic structures that deviate from Haeckelian expectations no longer provoke embarrassed nods from embryologists. The problem is not deciding whether palingenesis is a rule and adaptation an exception but whether palingenic structures and adaptations develop through similar or different mechanisms.

VERTEBRATE POSTGASTRULA

The postgastrula is a nodal stage in vertebrate development. The forms of different vertebrate embryos converge in the postgastrula stage (see Fig.

1.8) for a brief moment before taking off irretrievably in different directions. Actually, the appearance of similarity depends on cropping away large parts of the picture in amniotes, since only the palingenic parts converge. The cenogenic parts do not resemble each other or not nearly as much as palingenic parts.

Compared to their eggs, the convergence of the vertebrate postgastrula is breathtaking. Probably no phylum comes close to vertebrates in sheer variety of eggs and types of early development. Vertebrate eggs span the range from megalecithal to alecithal; fertilization is monospermic and polyspermic; cleavage goes from meroblastic to total; blastulas include nearly solid balls and swollen vesicles, and gastrulation uses virtually all the different mechanisms imaginable for moving cells.

At the end of gastrulation, however, when embryogenesis commences, remarkably similar morphologies and morphogenic processes prevail in the embryo proper. The silhouette of postgastrulas seem almost interchangeable. Many of the same developmental events are taking place in the same linear order and at the same relative pace. Of course, there are exceptions.

Some overtly qualitative deviations from the overall pattern are difficult to explain (e.g., the teleostean neural rod, see below). Other more or less quantitative deviations known as **heterochronies,** or differences in developmental rates among parts, are less difficult to explain and more instructive. After all, the embryo proper, as well as its cenogenic structures, is also adapted to its environment.

Instances of **heterochrony** are recognized when homologous parts in members of different species develop out of synchrony. Such heterochronies are endemic among vertebrates and go both ways. **Promotion** refers to speeding up the rate of a part's development; **retardation** refers to slowing it down (see Chapter 19).

Even within a class, some species will have more rapid development than other species and take short cuts to specificity. For example, large-

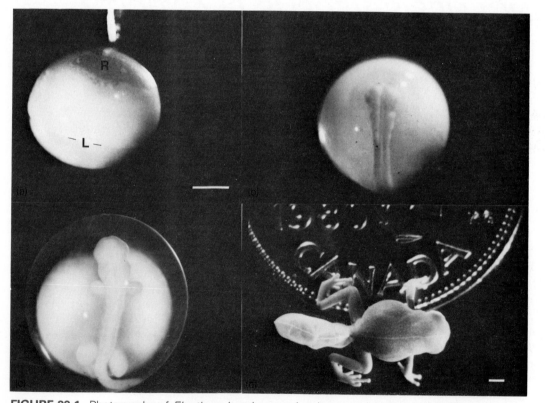

FIGURE 23.1. Photographs of *Eleutherodactylus coqui* embryos and freshly hatched froglet. (a) Gastrula. L, blastoporal lip; R, transparent roof of blastocoel. (b) Neurula. (c) Tailbud (about a day after complete closure of neural tube). (d) Froglet hatches from egg capsule at 3 weeks. Shown next to Canadian penny for comparison in size. The tail, which is a respiratory organ, is swollen as an artifact of fixation. Bar = 1 mm. From R. P. Elinson, in R. A. Raff and E. C. Raff, eds., *Development as an evolutionary process,* Alan R. Liss, Inc., New York, 1987. Used by permission.

yolked eggs of the small hylid anurans (e.g., the Puerto Rican *Eleutherodactylus coqui*) develop directly to froglets while being brooded on the male. The froglets are produced without a tadpole stage (Fig. 23.1) and without the production of the full range of aortic arches.

Other heterochronies involve **neoteny,** accelerated sexual development (or retarded bodily development). Widespread among urodeles, the neotenous larva, or **paedomorph** (Gk. *paid-* child + *morph* form), becomes sexually mature while retaining the larval way of life and larval structures (i.e., metamorphosis to an adult does not occur).

Varying degrees of **morphogenic plasticity,** the capacity to vary in developmental pattern, accompany heterochrony. For example, the neotenous Mexican axolotl, *Ambystoma mexicanum* (Fig. 23.2), can be induced to metamorphose in the laboratory by exposure to thyroxine, the normal hormone of amphibian metamorphosis. On the other hand, paedomorphs found in the wild (e.g., *Necturus*) are permanently committed to the larval state and do not undergo metamorphosis under any circumstance yet imposed in the laboratory (see Gould, 1977).

In deuterostomes, in general, greater developmental plasticity is exhibited where a larval (e.g., pluteus or tadpole) stage is functional or a fetal stage is prolonged. Large mammals, for example, develop slower than other amniotes (birds and reptiles) and are far less rigid in laying down organ rudiments (e.g., the nervous system). Heterochronies are indicative of different, possibly competing, control systems operating in the embryo.

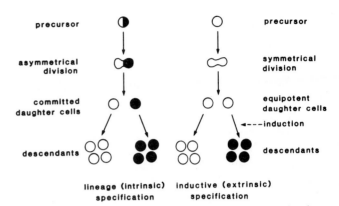

FIGURE 23.3. Two control systems for determination. From W. G. Hopkins and M. C. Brown, *Development of nerve cells and their connections,* Cambridge University Press, Cambridge, 1984. Used by permission.

Autonomous and Interactive Control Systems

Autonomous and interactive control systems rule the postgastrula. Cells and groups of cells may tend to behave autonomously, their determination independent and intrinsic (i.e., determinate development), or they may behave interactively, their determination dependent and extrinsic (Fig. 23.3) (see Davidson, 1986).

In **autonomously differentiating** somatic cell lineages all the cells produced through a set number of cleavage divisions express the same differential functions, *in situ* and *in vitro*. **Interactive lineages** are more variable in their responses to conditions, since determination is delayed.

In the **clonal pattern of differentiation,** lineages retain their plasticity and capacity to **regulate** or express different fates only briefly and do not conform to new surroundings when transplanted. Other cell lineages retain **cellular pluripotentiality,** or the ability to respond to local conditions, until the gastrular stage, and still other cell lineages retain their plasticity until postgastrular stages.

Intercellular interactions play a crucial role in determining the lineages of pluripotential cells. **Regional inductive interactions,** which may be common in all deuterostomes if not in all deuterostome tissues, allow pluripotential cells of different origins to participate in local differentiation. The diverse parts of embryonic systems may thereby be unified.

Theoretically, inductive interactions are **permissive** when a signal or trigger engages an intrinsically controlled developmental mechanism. Inductive interactions are **instructive** when new properties dictated by the source of the induction are introduced into the reactive part of the system.

FIGURE 23.2. The neotenous aquatic axolotl (bottom: *Siredon mexicanum*) and its terrestrial adult counterpart, the spotted salamander (top: *Ambystoma mexicanum*). Note the presence of larval gills and tail fin in the axolotl and their absence in the adult. From A. Duméril, *Ann. Sci. Nat. Zool.,* 7:229–254(1867).

Regulatory cofactors with low inherent informational content may be vehicles for the permissive induction of competent responding tissue. On the other hand, either (or both) a high level of tissue plasticity or the transfer of high information containing macromolecules may operate during instructive induction (see Hardy, 1983).

Evolution has presumably been at work in shaping all these developmental controls. It might confer stability on development by relegating controls to intrinsic genetic mechanisms. Development should then be as resistant to error as the molecular mechanisms of replication and transcription. Spiralians and arthropods seem to have followed this evolutionary path. Vertebrates, on the other hand, seem to have taken a different route to stability.

Vertebrate embryos have a penchant for extrinsic controls and for variety in controls. Different controls may operate on the same developing organs even in closely related species (e.g., large-egg and small-egg frogs).

The logic of the vertebrate model is elusive.

Perhaps vertebrates invest more in extrinsic controls than intrinsic ones as an adaptation to growth. Generally lacking the ability to throw off outgrown equipment via molting or metamorphosis, vertebrates must keep their organ systems growing in synchrony. Where synchrony in growth is based on feedback controls among organ systems, an adaptive advantage may accrue to extrinsic controls.

Alternatively, the preponderance of extrinsic controls in vertebrate development is explained by biological inertia (i.e., "good enough is good enough," Moore, 1987). Development in ancestral vertebrates is assumed to have been under extrinsic controls (i.e., more like the situation in Amphioxus than in ascidians). The development in modern vertebrates of vestigial structures that seem redundant, if not maladapted, is explained by the retention of the instruments of primitive extrinsic controls. For example, the development of the notochord and branchial arches may confer an adaptive advantage in terms of the retention of inductive control systems.

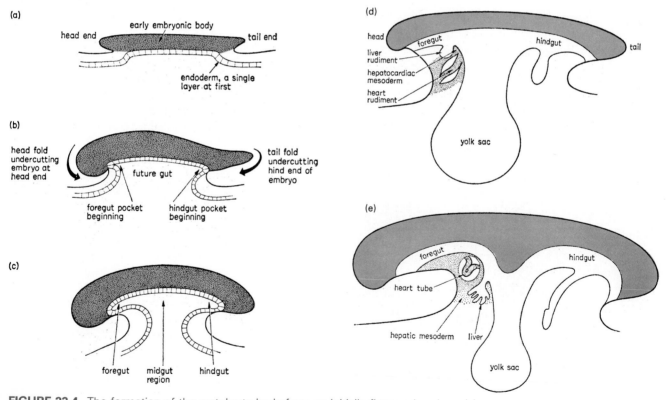

FIGURE 23.4. The formation of the vertebrate body from an initially flattened embryo (a). Body folds lift the embryo off the yolk (b), and the intestine is folded into foregut and hindgut pockets (c). Liver and heart rudiment form in mesenchyme below foregut (d and e) as connections to yolk sac narrow and allantois emerges. From E. M. Deuchar, *Cellular interactions in animal development,* Wiley, New York, 1975. Used by permission.

SHAPING THE VERTEBRATE BODY

Vertebrate embryogenesis occurs primarily in the **neurula** and **tailbud stage** (also called the branchial stage or pharyncula) as the embryo elongates and its egg shape is replaced by new contours. The chief externally visible events delineating the neurular and tailbud stages are the formation of a **neural tube** from a **neural plate** and the segmentation of **somites** from **somitomeres** and a **segmental plate**. The complete closure of the neural tube marks the transition between the neurula and the tailbud stage, and the completion of the organism's complement of somites marks the conclusion of the tailbud stage.

The invagination of **body folds** in fish and amniotes transforms the embryo from a disk or shield to something looking more like a vertebrate (Fig. 23.4). In fish, the tailbud is the first part of the embryo proper freed from the yolk. As the **tail** grows, a **caudal body fold** undermines it. An **anterior body fold** later frees the head, but **lateral body folds** only slowly separate the pharynx, trunk, and abdomen from the yolk (Fig. 23.5).

In amniotes, the anterior body fold forms first (see Figs. 18.38 and 19.4), while the caudal body fold forms last. In chicks, the anterior body fold begins as a dent at the tip of the notochordal process (see Fig. 18.39).[1] By 24 hours of incubation, the dent lengthens into the **subcephalic pocket** (Figs. 23.6 and 23.7) lying between the emerging

---

[1]"Fold" should be limited to bends in flexible membranes. Regrettably, the term "head fold" is not available for the anterior body fold. Priority for "head fold" is held by the enlarged "head process" or cephalic end of the amniote neurula. This regrettable situation has produced a confusing state of nomenclature. The remedy lies in renaming the "head process" the **notochordal process** and expunging "head fold" from current usage.

FIGURE 23.5. Development of a shark with a moderately small yolk mass. The area vasculosa or angioblastema (blood vessel-forming area) is stippled. Arteries are solid; veins are hatched. (a) Early neurula. (b) Neurula with 16 somites (note broadly open posterior neuropore). (c) Neurula with 22 somites (posterior neuropore about to close). (d) Tailbud stage with about 30 somites. Two aortic arches carry blood from primitive heart to dorsal aorta and hence to vascular area of yolk sac via vitelline arteries. (e) Embryo prior to hatching (only vitelline artery and vein shown). Six gill arches carry external gills, and six gill slits allow water to move past the gills. From E. Witschi, *Development of vertebrates,* Saunders, Philadelphia, 1956. Used by permission.

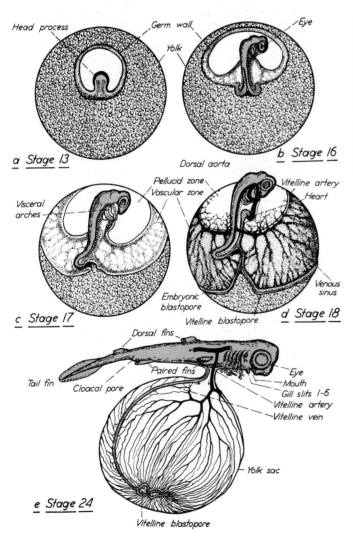

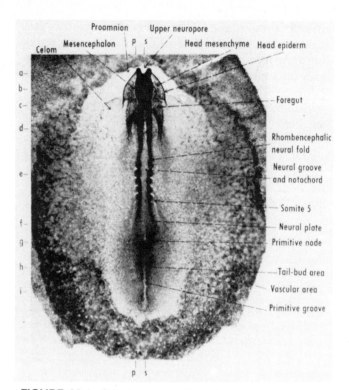

FIGURE 23.6. Whole mount of chick neurula (stage 14c Witschi, stage 8 Hamburger & Hamilton, 26–29 hours of incubation). (H & H stages from V. Hamburger and H. L. Hamilton, _J. Morphol;_ 88:49, 1951.) Neural folds, 4–5 pairs of somites, coelom, and blood islands (vascular area) have formed. Marginal letters correspond to sections in Fig. 23.7. ×16. From E. Witschi, _Development of vertebrates,_ Saunders, Philadelphia, 1956. Used by permission.

head and a layer of fused ectoderm and endoderm called the **proamnion.** Soon thereafter, the mass of yolk dips and accommodates the freed cephalic end of the embryo in a shallow **embryonic trough** (Fig. 23.8).

Internally, endodermal pockets forming the rudimentary gut are the correlatives of the external body folds (Fig. 23.4). A **foregut** pocket tucks into the cranial end of the embryo, and a **hindgut** pocket tucks into caudal end. The foregut lengthens conspicuously as the cranial end lengthens, and the opening of the pocket, or **anterior intestinal portal,** moves halfway down the length of the embryo (compare longitudinal sections in Figs. 23.7 and 23.10). The hindgut opening does not lengthen as much, and the opening of this pocket, or **posterior intestinal portal,** remains close by. The **midgut** develops between the foregut and hindgut pockets in conjunction with the **yolk sac.**

Beginning at the proamnion, the anterior body fold moves caudally, freeing the head end of the embryo from the yolk as far back as the great ves-

sels leading out of the newly formed heart (Fig. 23.9). But caudal progress slows dramatically upon reaching the heart (at about 33 hours in chicks, Figs. 23.9 and 23.10, pp. 639–640).

Posterior to the heart, **lateral body folds** sweep backward and join the **posterior body fold** beneath the tailbud, but the heart is not enclosed within ectoderm until much later. Instead, the heart develops partially outside the embryo proper (i.e., at the juncture of the embryonic and extraembryonic coeloms, Figs. 23.11 and 23.12, pp. 641–642). The midgut will also develop largely outside the embryo proper and will not be totally covered by the body wall until the approach of hatching.

INTERNALIZING THE NERVOUS SYSTEM

During cleavage, cells fated to participate in the formation of the embryo's nervous system are restricted progressively to ectoderm, especially to the presumptive **neural ectoderm** or **neuroepithelium.** In _Xenopus laevis,_ at the 16-cell stage, every blastomere gives rise to some neural cells, while only 24 cells of a 32-cell embryo and 38 cells of a 64-cell embryo give rise to neural cells. At the 512-cell stage, cells at the extreme ventral region give rise to no neural cells (Jacobson and Hirose, 1981).

In the frog blastula, a mass of peripheral RNA-rich cytoplasm called **anterior cytoplasm** or **mesoplasm** is localized in the marginal zone between deep and superficial marginal zone cells (asterisks, Fig. 19.28). During gastrulation, this cytoplasm is seen only in the **neurogenic** (i.e., nerve-producing) **deep cells** of the **neural plate** lying upon the notochord and over the dorsal mesoderm (Imoh, 1985).

In amniotes, the thickened neural ectoderm is confined to a **neural plate** (actually more like a half-moon-shaped area) covering the embryonic rudiment anterior to the primitive node and trailing off posteriorly on either side of the primitive streak (see Figs. 14.13, 18.37, and 18.39). As neurulation commences, the axial notochord or notochordal process underlies the **notoplate** (Fig 23.13, p. 643) in the midline of the neural plate, while paraxial mesoderm or segmental plate underlies lateral portions of the plate.

At its borders, the tall columnar epithelium of the neural plate turns abruptly into the cuboidal epithelium of the cutaneous ectoderm or epidermis. Centrally, the notoplate creases, and the neural plate bends into an inturned longitudinal

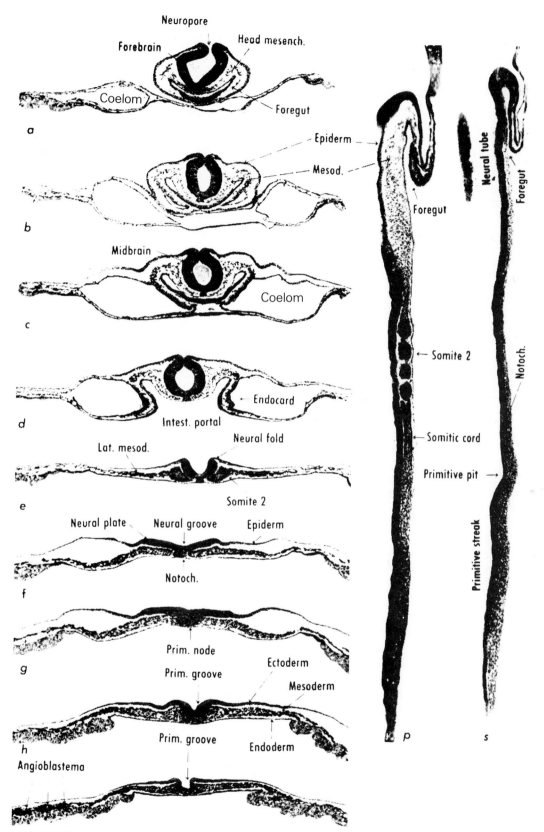

FIGURE 23.7. Cross and longitudinal sections of chick neurula at levels of marginal letters in Fig. 23.6. ×36. p, Parasagittal; s, midsagittal. From E. Witschi, *Development of vertebrates,* Saunders, Philadelphia, 1956. Used by permission.

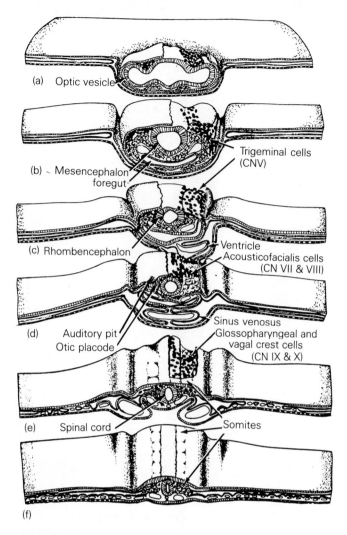

(a) Optic vesicle

(b) Mesencephalon
     foregut

Trigeminal cells
(CNV)

(c) Rhombencephalon

Ventricle
Acousticofacialis cells
(CN VII & VIII)

(d) Auditory pit
     Otic placode

Sinus venosus
Glossopharyngeal and
vagal crest cells
(CN IX & X)

(e) Spinal cord

Somites

(f)

FIGURE 23.8. Stereograms of chick embryo with 7–10 somite (stage 15 a–b Witschi, stage 9–10 Hamburger & Hamilton, 29–33 hours of incubation). The epidermis has been removed on the upper right stereograms (A–E) to show location of neural crest cells (closed circles). From J. P. Thiery and J. C. Boucaut, in M. Fougereau and R. Stora, eds., *Cellular and molecular aspects of developmental biology,* Elsevier Science Publishers, Amsterdam, 1984. Used by permission.

neural groove. Below, the segmental plate converges toward the midline.

At the same time, the embryonic axis elongates. Possibly, like a rubber sheet that rolls up at its edges when stretched across the middle (as once suggested by Wilhelm His), elongation in the notoplate may initiate folding at the neural plate's edges (see Jacobson and Gordon, 1976). **Neural folds** rise at the borders of the plate and turn inward. As epidermal cells migrate over the cusps of the neural folds, **neural crest cells** retreat and migrate outward beneath the epidermis.

In teleosts, the thickened neural ectoderm condenses at the midline into a **neural rod** with a keystone-shaped cross section (Fig. 23.14, p. 643). The rod sinks into the surface as a solid mass that becomes hollow secondarily.

In amphibians and amniotes, the **neural tube** or rudiment of the central nervous system is formed when the lateral edges of the neural folds

roll together and fuse. The epidermis heals over the tube and seals it in (Fig. 23.15, p. 644) (Jacobson et al., 1985).

Neural crest cells are extruded at the junction of neural and cutaneous ectoderm. The ectoderm **deepithelializes** on either side of the midline, releasing the mesenchymal cells.

In the head, large numbers of these cells quickly congregate at sites where their progress is blocked by the emerging brain and eye (Fig. 23.8). Elsewhere in the head and in the trunk, neural crest cells pour out of the dorsal midline and migrate in steady streams.

A **dorsal** or **dorsolateral pathway** takes the cells to sites in the skin where they become the precursors of melanocytes (i.e., integumentary pigment cells). Several **ventral pathways** take the neural crest cells to a variety of locations where they become precursors of neuronal, neuroendocrine, and supportive tissues.

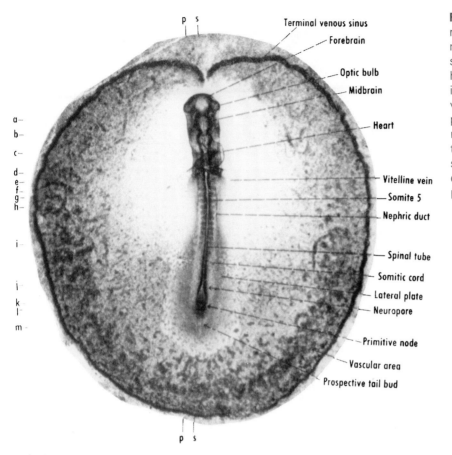

p s

Terminal venous sinus

Forebrain

Optic bulb

Midbrain

Heart

Vitelline vein

Somite 5

Nephric duct

Spinal tube

Somitic cord

Lateral plate

Neuropore

Primitive node

Vascular area

Prospective tail bud

a—
b—
c—
d—
e—
f—
g—
h—

i—

i—
k—
l—
m—

p s

FIGURE 23.9. Whole mount of late neurula (anterior neuropore closed, posterior neuropore open, stage 15c Witschi, stage 10 Hamburger & Hamilton, 33–38 hours of incubation). The brain is divided into forebrain, midbrain, and hindbrain vesicles. Twelve pairs of somites are present, and the heart protrudes toward right. Marginal letters correspond to sections in Fig. 23.10. ×12. From E. Witschi, *Development of vertebrates,* Saunders, Philadelphia, 1956. Used by permission.

Although the pathways may differ by species, the migrating neural crest cells end up in homologous positions. In *Xenopus laevis,* labeled neural crest cells are traced migrating ventrally down the neural tube and notochord and through the dorsal mesentery suspending the embryonic intestine, but few cells migrate within the somite. In contrast, labeled avian neural crest cells take routes in the clefts between somites and, after the deepithelialization of the somite (see below), through the loose cephalic portions of the somites. Avian cells do not appear in the caudal portions of the somites or in the region around the notochord, but cells mechanically obstructed by somites may back up along the neural tube (Fig. 23.16, p. 644). Cells may also migrate rostrocaudally along the aorta (Bonner-Fraser, 1987).

In the head, neural crest derivatives range from neuronal ganglia and dental papilla to aortic arches and portions of the cranial skeleton (see Chapter 25, Table 25.2). In addition, neural crest cells become calcitonin-producing C cells of the avian ultimobranchial body or the parafollicular thyroid cells of mammals, functioning in calcium homeostasis, and chemoreceptor and supporting cells

of the carotid body, functioning in respiratory control.

In the trunk, neural crest cells become sensory and motor ganglion cells of the **peripheral nervous system** (see Chapter 26). In addition, chromaffin cells, including the adrenaline-secreting cells of the mammalian adrenal (or suprarenal) medulla and paraganglia glands, are neural crest derivatives.

To test the ability of the microenvironment in *Ambystoma mexicanum* embryos to influence neural crest cell differentiation, extracellular material was collected on minute nitrocellulose filters (3–5 µm thick, 0.15 by 0.4 mm) inserted either beneath the epidermis (i.e., subepidermal filters for the dorsal pathway) or between the somite and notochord (paraxial filters for ventral pathways 1 and 2, Fig. 23.16). Premigratory axolotl neural crest cells exposed to subepidermal filters *in vitro* dispersed and synthesized pigment, but they did not synthesize the neural cell adhesive molecule N-CAM. Cells exposed to the paraxial filters aggregated, produced N-CAM, and sprouted neurites, but they did not synthesize pigment.

The subepidermal and paraxial filters seem to have exerted regionally specific influences on the

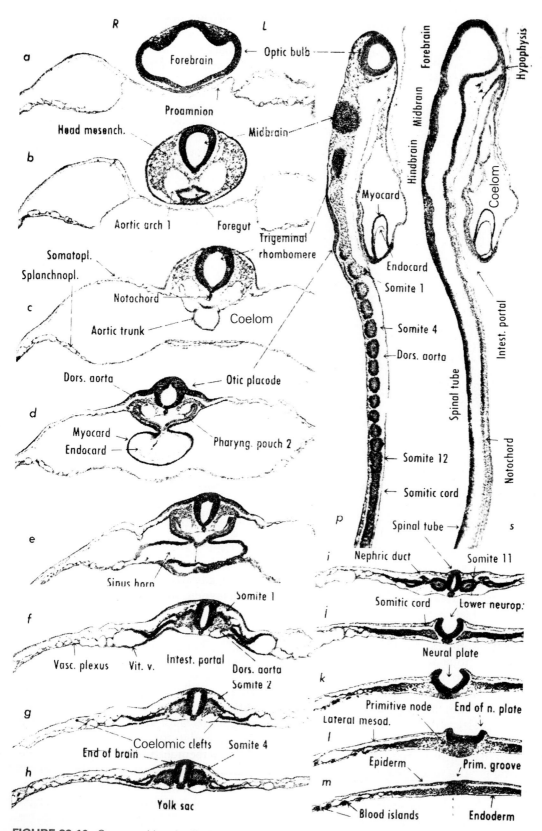

FIGURE 23.10. Cross and longitudinal sections of late neurula at levels of marginal letters in Fig. 23.9. ×32. p, Parasagittal; s, midsagittal. From E. Witschi, *Development of vertebrates,* Saunders, Philadelphia, 1956.

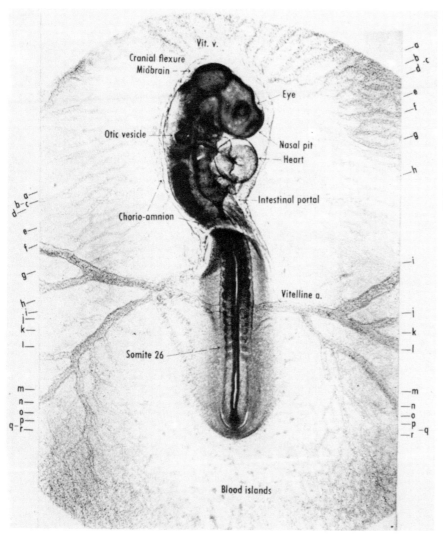

FIGURE 23.11. Whole mount of chick tailbud (pharyncula) with closed posterior neuropore (stage 17d Witschi, stage 15 Hamburger & Hamilton, 50–55 hours of incubation). The head at the midbrain (cranial flexure) and the anterior end of the embryo turns onto its left side (torsion). The fold of chorioamnion covers half the embryo, and 26 pairs of somites are present. Marginal letters correspond to sections in Fig. 23.12. ×12. p, Parasagittal; s, midsagittal. From E. Witschi, *Development of vertebrates,* Saunders, Philadelphia, 1956.

differentiation of neural crest cells *in vitro*, since neural crest cells did not differentiate under control conditions lacking conditioned filters. Possibly, the differentiation of migrating neural crest cells is also directed by specific extracellular substrata *in vivo* (Perris et al., 1988).

Many functional and enzymatic similarities are shared by the crest-derived neuronal and endocrine cells and the peptide-secreting paracrine and endocrine cells of the gastrointestinal mucosa, lung epithelium, and endocrine glands. Functionally, the paracrine and endocrine cells belong to the **diffuse neuroendocrine system** (DNES) which, in conjunction with the peripheral nervous system, is responsible for the homeostatic control of endodermally derived organs. Enzymatically, all these cells belong to the class of **amine precursor uptake and decarboxylation** (APUD) cells (Pearce, 1976).

APUD cells probably do not have a single origin, however, since, in the embryo, the uptake and decarboxylation of amino acid precursors is a property of some mesodermal cells (e.g., in the notochord) as well as neuroectodermal cells. One would have to look to the primary ectoderm of the embryonic disk or a primitive "ectomesoblast" to find a common ancestor for the endoderm-associated APUD and neural crest cells (see Le Douarin, 1988).

The **neural folds** do not close simultaneously throughout their lengths, and they close at different rates in different species. In amphibians and chicks, closure begins midbrain, producing a "keyhole" neurula, while, generally, closure begins more posteriorly. In mammals, closure begins in the presumptive spinal cord at the level of the future neck (i.e., the first pairs of somites, Fig. 23.17, p. 644). In all cases, the tube elongates as it closes.

The last parts of the tube to close are the ends, the **anterior** (upper or rostral) and **posterior** (lower or caudal) **neuropores.** Uncertainty creeps into es-

(text continues on p. 645)

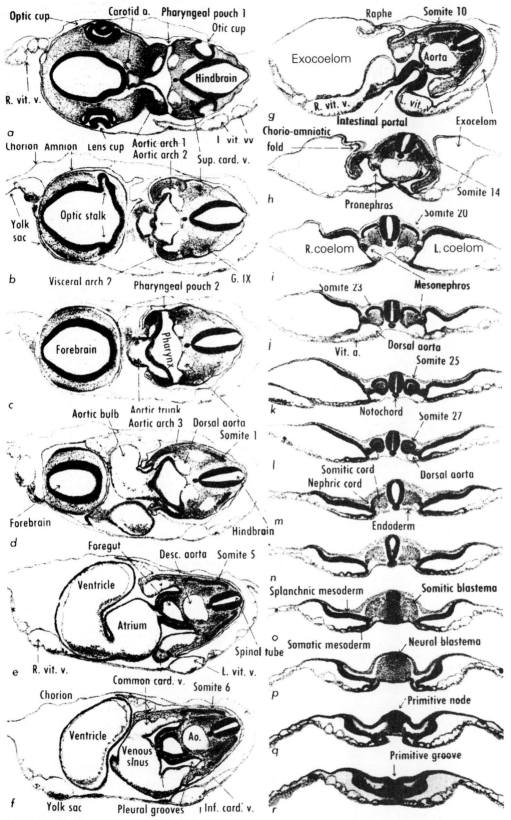

FIGURE 23.12. Cross and longitudinal sections of late tailbud (pharyncula) chick embryo at levels of marginal letters in Fig. 23.12. ×25. p, parasagittal; s, midsagittal. From E. Witschi, *Development of vertebrates,* Saunders, Philadelphia, 1956.

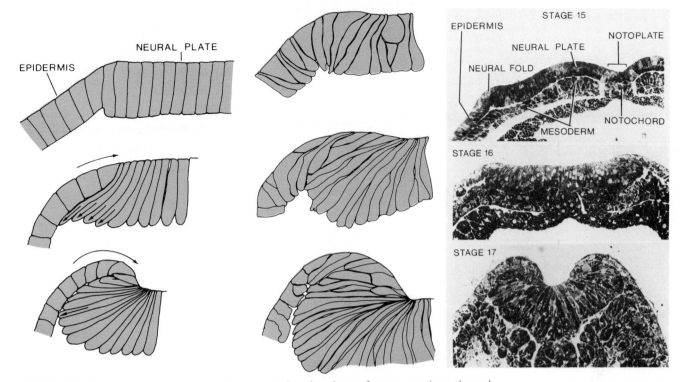

STAGE 15
EPIDERMIS
NEURAL FOLD
NEURAL PLATE
NOTOPLATE
NOTOCHORD
MESODERM
STAGE 16
STAGE 17

FIGURE 23.13. Photomicrographs and interpretative drawings of cross sections through early newt neurulas. Stage 15: early neural plate with incipient folds. Stage 16: neural plate closing. Stage 17: late neural plate. As neurulation progresses, neural plate cells become more columnar, the plate thickens, and contracts in width (from P. L. Anderson, *Anat. Rec.,* 86:58, 1943). From A. G. Jacobson, G. M. Odell, and G. F. Oster, in G. M. Edelman, ed., *Molecular determinants of animal form,* Alan R. Liss, Inc., New York, 1985. Used by permission.

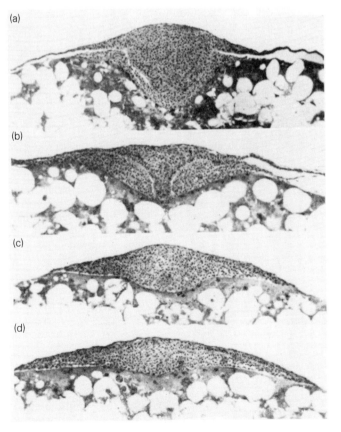

FIGURE 23.14. Photomicrograph of cross sections through neurula and terminal knob at the undifferentiated posterior end of the trout *Salmo furio* (1.7 mm, stage 13 Witschi). From top to bottom sections go through brain, contracting neural plate, margin of terminal knob, terminal knob. ×70. From E. Witschi, *Development of vertebrates,* Saunders, Philadelphia, 1956. Used by permission.

STAGE 18

STAGE 19

STAGE 21

NEURAL
CREST

0.1mm

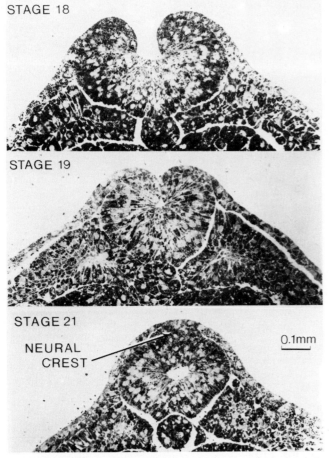

FIGURE 23.15. Photomicrographs of cross sections through late newt neurulas. Stage 18: neural folds fusing. Stage 19: fused neural folds, 5–6 somites present. Stage 21: with olfactory pits and optic cups present. From A. G. Jacobson, G. M. Odell, and G. F. Oster, in G. M. Edelman, ed., *Molecular determinants of animal form,* Alan R. Liss, Inc., New York, 1985. Used by permission.

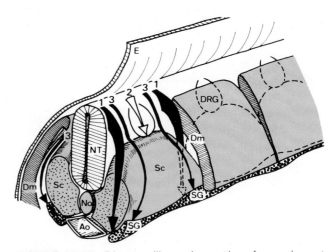

FIGURE 23.16. Diagram illustrating paths of neural crest cell migration in trunk of 3-day chick. The ectoderm (E) has been lifted and the dermatome (DM) of the first somite has been removed to aid visualization. 1, Pathway between somites takes cells to primary sympathetic ganglia (SG) in vicinity of aorta (Ao), 2, pathway blocked by bulk of sclerotome allows cells to accumulate in dorsal root ganglia (DRG), 3, intermediate pathway between dermatome and sclerotome allows some cells to reach orthosympathetic chain. Ao, aorta; DM, dermatome; DRG, dorsal root ganglion; NT, notochord; Sc, sclerotome; SG, sympathetic ganglion and orthosympathetic chain. From N. M. Le Douarin, P. Cochard, M. Vincent, J.-L. Duband, G. C. Tucker, M.-A. Teillet, and J.-P. Thiery, in R. L. Trelstad, ed., *The role of extracellular matrix in development,* Alan R. Liss, Inc., New York, 1984. Used by permission.

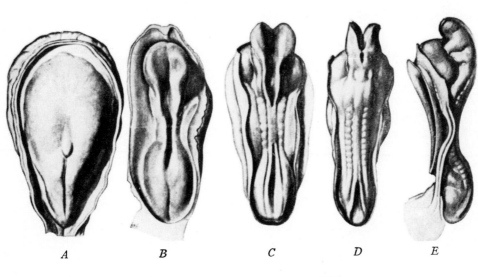

A   B   C   D   E

FIGURE 23.17. Human embryos at the neural stage. (a) Presomite embryo. The primitive streak lies behind the neural plate. Three somites are present. ×40. (b) Neural groove creases embryo. Optic depression is apparent. ×37. (c) Neural fold begins to close at middle of seven cervical somites. ×31. (d) Closure extends into brain. Ten somites are present. ×31. (e) Neuropores remain open, but the neural tube is closed longitudinally. ×20. From L. B. Arey, *Developmental anatomy,* 7th editions Saunders, Philadelphia, 1965. Used by permission.

TABLE 23.1. Consensus statistics on chick and human embryos at closing of neuropores

Event	Embryo	Age	Number of		Stages According to[a]			
			Somites	Branchial Arches	Witschi	H&H	Streeter's Horizons	Carnegie Standard[b]
Closing of anterior neuropore (neurula)	Chick	33–45 h	12	0	15–16	10–11[c]		
	Human	24–25 d	20–24	3	17		XII	11[d]
	Rat	10.5 d	17	2	16			
Closing of posterior neuropore (beginning tailbud stage)	Chick	50–63 h	22	2	17	14[e]		
	Human	25–26 d	21–29	3–4	18		XII	12[f]
	Rat	11.5 d	24	3	18			

Somites: occipital (4), cervical (8), thoracic (12), lumbar (5), sacral (5), caudal (5 plus).
Branchial arches: I (mandibular), II (hyoid), III, etc.
h: hours of incubation.
d: days postcoitum (fertilization).
H&H: Hamburger and Hamilton stages.
[a]Data mainly from P. L. Altman and D. S. Dittmer, *Growth including reproduction and morphological development*, Federation of American Societies for Experimental Biology, Bethesda, 1962.
[b]Carnegie Institutions' Developmental Stages in Human Embryos.
[c]Figure 23.9 and 23.10.
[d]Slightly later than Fig. 23.17e, later than Fig. 25.1c.
[e]Slightly later than Figs. 23.11 and 23.12.
[f]Approximately Fig. 25.1d.

timates of the time of closure based on histological sections and utilizing different schemes for staging embryos (hence the range in Table 23.1). Nevertheless, the anterior neuropore is generally acknowledged to close first. For example, in chicks, the anterior neuropore closes at the 12-somite stage (Figs. 23.9 and 23.10) while the posterior neuropore closes at the 22-somite stage (Figs. 23.11 and 23.12p).

The posterior neuropore of amniotes closes just anterior to the primitive node (Fig. 23.9) in a chevron-shaped area (the rhomboid sinus or sinus rhomboidalis). Failure to close the posterior neuropore results in a common birth defect known as **spina bifida** (cleft spine) (see Gordon, 1985).

In amniotes and anamniotes, a **tailbud** (also called the axial rudiment or somitic blastema, Fig. 23.18) with embryonic ectoderm caps the caudal end of the neural ectoderm, hindgut endoderm, and mesoderm. Growth of the tailbud extends the presumptive spinal cord while giving rise to almost all the tissues of the tail. The exception is the mesenchyme of the anamniotes' dorsal fin made of neural crest cells.

The neural and gut tubes generally remain separate in the tailbud, but in *Xenopus laevis* (see Balinsky, 1981) and in selachians (see Witschi, 1956), closure of superficial ectoderm over the posterior neuropore and blastopore creates a common chamber between the neural canal and the archenteron. Shaped like a U-turn, this **neurenteric canal** (Fig. 23.19) connects the neural ectoderm of the posterior neural tube and endoderm posterior to the hindgut or **postanal** gut. The connection persists, despite the outgrowth of the tail, until fusion of endoderm with ventral tail tissues, regression of the canal, and cell death restrict the endoderm to the hindgut and liberate the tip of the tail's spinal cord.

Heterochronies During Neurulation

Beside the gross absolute difference in early developmental rates in chicks and mammals (i.e., hours versus days), several developmental events are not synchronized in mammals and birds. Heterochronies are apparent in the development of the neural tube and other internal structures (e.g., somites and branchial arches).

The early development of the neural tube is slower in mammals than birds. For example, relative to the number of somites and branchial arches, the closure of the anterior neuropore occurs earlier in the chick than in mammals. The chick has less than half its complement of somites and none of its branchial arches at closure, while mammals have half or more of their somites and two to three branchial arches.

Development of the neural tube speeds up in mammals compared to chicks following the closure of the anterior neuropore. In mammals, the posterior neuropore closes so soon after the anterior neuropore that the number of somites and branchial arches is hardly changed. In contrast, in

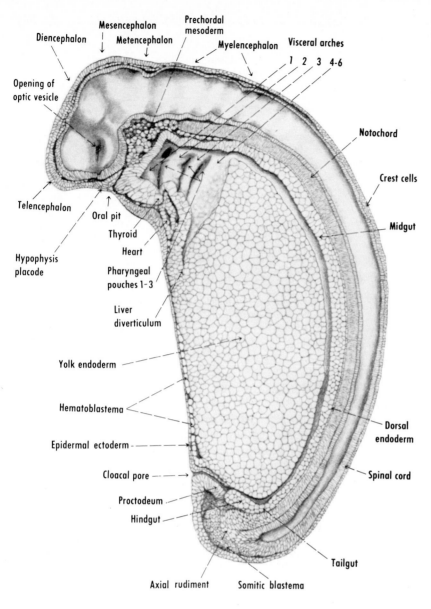

Diencephalon

Mesencephalon
Metencephalon

Prechordal
mesoderm

Myelencephalon

Visceral arches

1 2 3 4-6

Opening of
optic vesicle

Notochord

Crest cells

Midgut

Telencephalon

Oral pit

Thyroid

Hypophysis
placode

Heart

Pharyngeal
pouches 1-3

Liver
diverticulum

Yolk endoderm

Hematoblastema

Epidermal ectoderm

Cloacal pore

Dorsal
endoderm

Spinal cord

Proctodeum

Hindgut

Tailgut

Axial rudiment   Somitic blastema

FIGURE 23.18. Diagrammatic longitudinal section showing medial aspect of the salamander *Taricha torosa* (= *Triturus torosus*) at the early tailbud stage. ×30. From E. Witschi, *Development of vertebrates,* Saunders, Philadelphia, 1956. Used by permission.

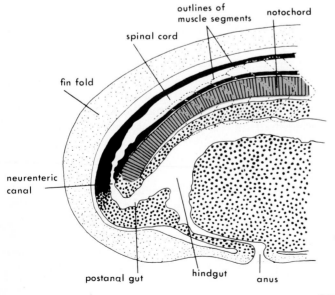

outlines of
muscle segments

notochord

spinal cord

fin fold

neurenteric
canal

postanal gut   hindgut   anus

FIGURE 23.19. Schematic of posterior part of midsagittal section through *Xenopus laevis* showing neurenteric canal connecting canal of spinal cord with lumen of postanal gut. From B. I. Balinksy, *An introduction to embryology,* 5th ed., Saunders, Philadelphia, 1981. Used by permission.

the chick, the number of somites practically doubles and the first two branchial arches are produced in the interval between closing the two neuropores.

Possibly, differences in rates of neural tube development can be attributed to differences in the contents of yolk in mammalian and avian eggs. While the yolk-poor mammalian embryo waits for the aortic arches to carry circulation to the neural tube, yolk in the chick may support a fast rate of early neural tube development. The later acceleration of mammalian neural tube development may be the result of improved access to resources accompanying the establishment of aortic circulation through the branchial arches.

ORGANIZING THE MESODERM

The notochord is generally considered axial mesoderm despite its distribution with endoderm in the surface of the *Xenopus* blastula and its presence in the roof of the archenteron. The early notochord consists of a homogeneous population of cells delineated from surrounding mesoderm by sharp clefts but not by any conspicuous cellular morphology. The "stack of coins" structure of the mature notochord develops from the active insertion (i.e., intercalation) of cells (including some from the surrounding mesoderm) followed by cellular elongation and alignment perpendicular to the notochord's long axis (see Fig. 23.18).

The remainder of the vertebrate mesoderm is divisible into three zones: segmental **paraxial** (also epimeric, axial or segmental), nonsegmental **intermediate** (also mesomeric, stalk of somite), and **lateral plate** (also hypomeric, Fig. 23.20) mesoderm. Initially, the zones are continuous, but eventually they separate and each differentiates into different parts.

Segmentation of the paraxial mesoderm only becomes manifest in the neurula, but rudiments of the paraxial and lateral plate mesoderm are present in the gastrula. Intermediate mesoderm is not distinguishable that early, developing after paraxial segmentation has produced several segments (8–10 in the chick), and may not be present at all in the head (e.g., it is absent in the vicinity of the first five segments of the chick).

Lateral plates are the unsegmented portions of the mesoderm on either side of the body between the dorsal intermediate mesoderm and ventral islands of blood formation and angiogenesis (Figs. 23.10 and 23.20). The **coelom** and **exocoel** split the lateral plates into **somatic** (Gk. *soma* body) and

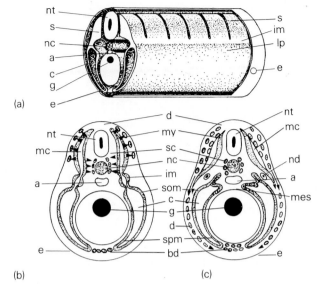

FIGURE 23.20. Stereogram through trunk of a generalized amphibian embryo (ectoderm partially removed). (b) Diagrammatic cross sections at early tailbud stage. Lateral plate mesoderm is still broadly in contact with somite mesoderm through intermediate mesoderm or stalk of somite. (c) Lateral plate mesoderm has lost contact with somite mesoderm. a, dorsal aorta; bd, blood-forming cells; c, coelom; d, dermatome; g, lumen of gut; e, epidermis; im, intermediate mesoderm (= stalk of somite); lp, lateral plate; mc, myocoel; mes, mesentery (suspending gut); my, myotome; nc, notochord; nd, nephric duct; nt, neural tube; s, somite; sc, sclerotome; som, somatic mesoderm (ectoderm + somatopleure); spm, splanchnic mesoderm (endoderm = splanchnopleure). Adapted from V. Blüm, *Vertebrate reproduction,* translated by A. C. Whittle, Springer-Verlag, Berlin, 1986. Used by permission.

splanchnic (Gk. *splanchnos* entrail) mesodermal layers. Fused with the cutaneous ectoderm, the somatic mesoderm forms a **somatopleure** (*soma* + *pleura* side, hence body wall). Similarly, fused with the endoderm, the splanchnic mesoderm forms a **splanchnopleure.** Both layers are lined by smooth, continuous simple **mesothelia** lining the extraembryonic coelom (or exocoel), the embryonic coelom, mesenteries, and viscera (Figs. 23.7 and 23.12).

When the coelomic cavities on each side of the midline approach each other centrally, the sandwich of mesothelia and connective tissue between them becomes a **mesentery.** Dorsal mesenteries (Fig. 23.20) become the supporting structures for the gut and other organs, but, except for mesenteries connected to the liver, ventral mesenteries tend to disappear.

Intermediate mesoderm separates the paraxial and lateral plate mesoderm from the level of the

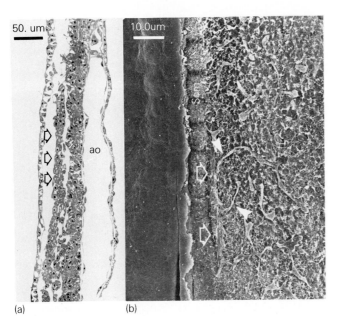

(a)   (b)

FIGURE 23.21. The intermediate mesoderm and the rudiment of the nephric duct seen in the chick embryo of 2 days. (a) Photomicrograph of parasagittal section. (b) Scanning electron micrograph of dorsal view after removal of ectoderm from right side. The rudiment of the nephric duct is a cord of mesenchymal cells (open arrows) that arises from the intermediate mesoderm at the level of the 7–11th somites. The cord moves caudally and hollows out into the nephric duct. ao, Aorta; im, intermediate mesoderm; som, somite; open arrowheads, anlage of nephric duct; closed arrows, blood vessel-forming cells (angioblastic cells). From H. J. Jacob, M. Jacob, and B. Christ, in R. Bellairs, D. A. Ede, and J. W. Lash, eds., *Somites in developing embryos,* Plenum Press, New York, 1986. Courtesy of the authors. Used by permission.

cervical segments (e.g., the sixth segment in the chick) to the proctodeum (except in the larva of primitive myxinoid cyclostomes). The first morphogenic event involving the intermediate mesoderm is its splitting off a cord of mesenchyme adjacent to several cervical segments (7–11 in the chick, Fig. 23.21a,b). This **nephrogenic cord** (Gk. *nephr-* kidney + *genic,* also called nephrotomic plate or cord) bulges against the overlying ectoderm as the **nephrogenic ridge** and tenaciously works its way caudally to the cloaca (Holtfreter, 1943). A lumen gradually opens and the cord becomes the **nephrogenic** or **Wolffian duct** (Jacob et al., 1986).

During its caudal passage, the nephrogenic duct induces the formation of **nephric tubules** or presumptive nephrons at random points of contact with the intermediate mesoderm or **nephrogenic tissue** (sometimes in blocks or **nephrotomes**). The tubules fuse with the duct and their lumina become continuous.

Segmentation is the hallmark of the paraxial mesoderm, and pairs of **somites** separated by deep **clefts** are its most conspicuous component (Fig. 23.22). But paraxial mesoderm does not begin as

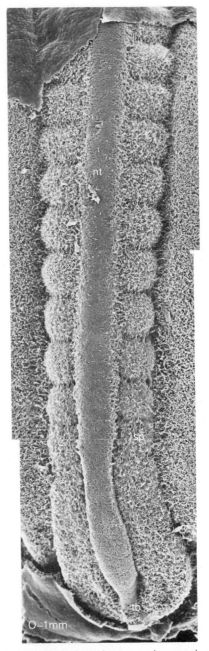

FIGURE 23.22. Scanning electron micrograph of the posterior portion (ectoderm removed) of chick embryo (stage 16 Hamburger & Hamilton, 26–28 somites, 51–56 hours of incubation). Pairs of somites are seen on either side of the neural tube. nt, Neural tube; sp, segmental plate; tb, tailbud. From R. Bellairs, in R. Bellairs, D. A. Ede, and J. W. Lash, eds., *Somites in developing embryos,* Plenum Press, New York, 1986. Courtesy of R. Bellairs. Used by permission.

segmented, and parts of it are never segmented by way of clefts.

Zones of paraxial mesoderm in the head and caudal to the somites lack incisive clefts. Rather, concentric rings of cells in these regions are piled into squat, bilaminar, cylindrical mounds, called **somitomeres.** Discovered serendipitously by Stephen Meier in one of the early applications of the stereo scanning electron microscope to dissected chick embryos, somitomeres are also found in embryos of the mouse, newt, quail, shark, teleost, snapping turtle, and *Xenopus laevis* (Fig. 23.23).

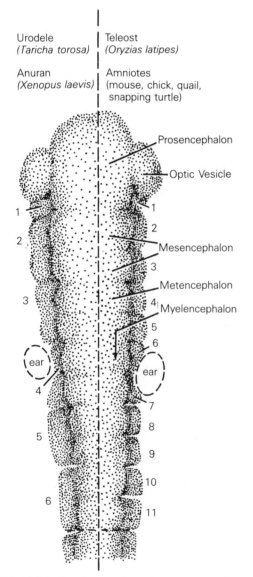

FIGURE 23.23. Somitomeres and neuromeres in amphibians, teleosts, and amniotes. The first somitomere to become a somite is the 5th in amphibians and the 8th in the medaka, a teleost. The 8th somitomere is also the first to become a somite in amniotes. From A. G. Jacobson and S. Neier, *Dev. Biol.,* 106:181 (1984), by permission of Academic Press and courtesy of the authors.

Head somitomeres are **paedomorphic** (i.e., permanently juvenilized). They are normally unable to complete the transition to somites, although their cells differentiate into the same tissues in the head that somite cells differentiate into elsewhere in the organism. Amniotes and teleosts (e.g., the medaka) have seven pairs of head somites, while amphibians have four pairs (Fig. 23.23).

Head somitomeres are formed before the notochordal process appears or the neural plate lifts its edges. The first pair of head somitomeres lies on either side of the **prechordal plate,** a node of condensed mesoderm anterior to the notochord. All other head somitomeres appear on either side of the notochord. They are laid down consecutively just caudal to the amniotic Hensen's node and give rise to skeletal elements, dermis, meninges, and voluntary muscle in the head, including muscles of the jaw and extrinsic muscles of the eye.

Caudally, the **segmental plate** (also called vertebral plate) is an **unsegmented** portion of the paraxial mesoderm (despite its name, sp, Fig. 23.22). Like so many other embryonic structures, the segmental plate is a region in transition.

At the rear, paraxial cells are fed into the segmental plate from the primitive streak and later from the tailbud (tb, Fig. 23.22). At the front, cells leave the segmental plate on their way to become somites. Actually, the cells do not leave the plate so much as the plate leaves them as it moves caudally along with the primitive streak or tailbud.

Results from a host of deletion and *in vitro* experiments with fragments of segmental plates from different animals indicate that caudal somitomeres are **transitional** on their way to become somites (see Jacobson and Meier, 1986). In the first stage of **somitogenesis** (i.e., the production of somites), paraxial mesoderm is condensed or **compacted. Epithelization** (i.e., broadening of intercellular contacts) then changes the somitomere into a somite (Fig. 23.24).

In chicks, somitogenesis begins at about 19 hours. Bilateral pairs of somites are produced at the rate of one pair about every hour until the third day when 50 pairs of somites are completed. Hensen's node seems essential for the production of pairs, since in the absence of the node, a median line of unpaired somites forms (Bellairs, 1986).

Surgically removing the segmental plate has no effect on the formation of somites in the anterior paraxial mesoderm, and transplanted segmental plates continue to produce somites indicative of a high degree of determination in the plate. The plate is not yet determined for polyclonal compartments, however since cells leaving the plate can

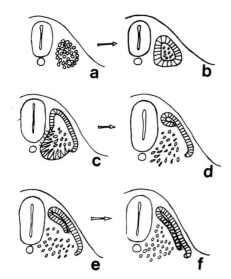

FIGURE 23.24. Somitogenesis in the chick. (A) Compaction; (B) epithelialization; (C–D) sclerotome dispersal and migration; (D–F) folding at dorsomedial lip and migration of myotome beneath dermatome. From D. A. Ede and A. O. A. El-Gadi, in R. Bellairs, D. A. Ede, and J. W. Lash, eds., *Somites in developing embryos,* Plenum Press, New York, 1986. Used by permission.

combine with anterior somites (Tam and Beddington, 1987).

Once somites appear, their anterior and posterior regions are differentially determined. They lose the ability to recombine when brought together via grafting and only rostral portions continue to bind peanut agglutinin, indicative of surface changes (Stern and Keynes, 1986).

The body somites are better organized than head somitomeres. The compact, epithelized body somite is divisible into regions exhibiting different fates. The **dermomyotome** (or myodermatome, Gk. *derma* skin + *mys-* referring to movement, hence muscle + *-tomos* slice or cut) is the outer epithelial cap of the somite. A **myocoel** or somitic cavity develops within the somite, and the dermomyotome's dorsomedial lip folds inward bringing a definitive **myotome** under a **dermatome** (Fig. 23.24). The dermatome provides connective tissue cells to the dermis of the skin, while the myotome provides the voluntary **epaxial** (above the level of ribs) and **hypaxial** (below the level of ribs) body musculature to the body and appendages.

A central **sclerotome** is the first part of the somite to show morphogenic movement. In each somite, the sclerotome **deepithelizes.** It breaks up into **secondarily mesenchymal** cells which migrate toward the notochord (where they will form the centrum of the spinal column), upward around the neural tube (where they will form the neural arches

over the spinal cord), and outward (where they will form ribs and associated structures).

The breakup of the sclerotome does not occur uniformly throughout the somite. The posterior or caudal region of each sclerotome remains dense in contrast to an anterior or rostral region (Fig. 23.25). Neural crest cells and motor axons migrate exclusively through anterior regions of the sclerotome (Fig. 23.16) and the **intervertebral disk** forms in the vicinity. The similarity of somite halves with compartments in *Drosophila* is all the more striking in

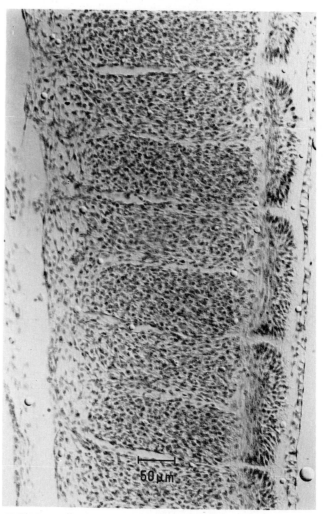

FIGURE 23.25. Photomicrograph of sagittal section through a chick embryo at stage 16 (Hamburger & Hamilton, 51–56 hours, 26–28 somites). Dermomyotomol caps span half-sclerotomes from adjacent segments, a denser posterior or caudal half, and a lighter anterior or rostral half. Von Ebner's fissures lie perpendicular to the dermomyotomal cap between the half-sclerotomes (i.e., at the edges of the resegmented sclerotomes). Anterior or rostral is up. From C. D. Stern and R. J. Keynes, *Development,* 99:261 (1987), by permission of Academic Press and courtesy of the authors.

fish, where virtually no cell mixing occurs between anterior and posterior halves of somites (Thorogood and Wood, 1987).

The anterior and posterior regions of avian sclerotomes separate around **intrasegmental** or **von Ebner's fissures.** While looking like intersegmental borders (Fig. 23.25), von Ebner's fissures lack the rich supply of fibronectin and laminin characteristic of intersegmental borders. Moreover, von Ebner's fissures fail to extend into the dermomyotome, while intersegmental borders traverse the entire somite.

The punctate arrangement of vertebrae seems to result from combining the anterior region of one sclerotome with the posterior region of the next sclerotome. The process, dubbed **resegmentation** by Remak in the mid-18th century, broadly explains how segmental axial muscles span adjacent vertebrae (Fig. 23.25), but the differentiation of vertebrae between successive myotomes seems to be more complex, since the determination of sclerotomal regions is identical (Stern and Keynes, 1986).

EMBRYONIC INTEGUMENT

The embryonic integument or covering layer is formed by a mesenchyme of **mesodermal** and **neural crest** cells fused to an outer **cutaneous ectoderm** (i.e., the ectoderm left on the surface after the withdrawal of the neural ectoderm and neural crest) or **epidermis.** The mesodermal component of the integumentary mesenchyme originates from the somatic layer of the lateral plate and from somite mesoderm (Fig. 23.20).[2] The somatic mesodermal layer continues to form the inner lining of the body wall or mesothelium surrounding the body cavity in the adult, but it is replaced as the lining of the integument by cells of somitic origin.

Adult integumentary mesoderm is provided by the dermatome, the outer portion of the dermomyotome capping the somites (Figs. 23.24 and 23.25). Cells from the dermatome migrate under the cutaneous ectoderm and between it and the somatic mesoderm (Fig. 23.20). The dermatome forms the mesodermal component of the loose and dense connective tissue of the **dermis** (i.e., the leather) of the skin.

The neural crest component of the integumentary mesenchyme, known as **ectomesenchyme** (sometimes mesectoderm), migrates from the neuroepithelial ridges of the neural folds to the **subectodermal space** along the basal lamina of the cutaneous ectoderm. Ectomesenchyme cells also permeate the dermis, and many properties formerly attributed to mesoderm now seem to be due to ectomesenchyme (see Table 25.2).

---

[2]Here is another one of those irritating situations in biology in which important differences in words hinge on single letters. The som*a*tic mesoderm produces the mesothelium and connective tissue surrounding the body cavity; the som*i*tic mesoderm is the source of the dermis.

CENOGENESIS IN VERTEBRATES

*T*he importance of extraembryonic membranes and spaces is sometimes underestimated, possibly out of "adult chauvinism" (i.e., a misguided tendency by adults to place undue emphasis on structures present in adults). For the embryo, the transient structures nourishing and protecting it are just as important as the permanent structures it passes along to adults. Even at birth, the highly vascular human placenta is about one-sixth the weight of the newborn, making it a major organ by size as well as function. Moreover, the placenta, which combines fetal and maternal parts, is as intimately involved in the maternal physiology of pregnancy and parturition as it is involved in fetal development.

Vertebrate embryos may have as many as four extraembryonic membranes and spaces (Fig. 24.1). The membranes are the **yolk sac, amnion** (Gk. *amnos* lamb but referring to membrane surrounding the lamb at birth), **chorion** (Gk. *chori* left apart, referring to the separation of fetal and maternal layers of the eutherian placenta), and **allantois** (Gk. *allantoeides* meaning sausage but referring to the placenta). The mammalian **trophectoderm** (TE) or trophoblast is a precocious form of amnion and chorion. The spaces are the **yolk sac cavity,** filled with yolk except in therian mammals, the **am-** niotic cavity between the embryo proper and amnion, the **exocoel** or **exocoelom** (also extraembryonic coelom) between the amnion and chorion, and the **allantoic cavity** within a vesicular portion of the allantois.

The presence of the extraembryonic amnion around embryonic reptiles, birds, and mammals identifies these embryos as **amniotes.** Lacking amniotic vestments, fish and amphibian embryos are known as **anamniotes.**

The yolk sac, the only extraembryonic membrane appearing in anamniotes, is the oldest of the extraembryonic membranes and the most variable. The allantois is the youngest and the most uniform, differing mainly in the extent of the enclosed allantoic cavity. The amnion and chorion are relatively consistent among amniotes although not necessarily forming in the same way, at the same time, or at the same stage. These membranes may make their appearance before or after gastrulation in the embryo proper and in different orders. The extraembryonic spaces also form via different mechanisms.

As a rule of thumb, extraembryonic membranes make their appearance precociously in eutherians compared to metatherians, prototherians, birds, and reptiles. Extraembryonic develop-

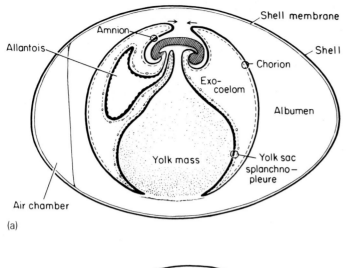

(a)

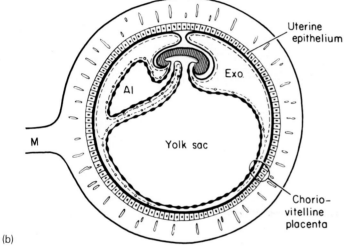

(b)

FIGURE 24.1. Diagrams of extraembryonic membranes in birds and reptiles (a) and prosimians (b). Al, allantois; Exo, extraembryonic coelom or exocoel; M, mesentery supporting uterus (mesometrium). Areas and lines within shell: hatched area, embryo; stippled area, yolk; smooth heavy line, extraembryonic ectoderm; beaded heavy line, extraembryonic endoderm; dashed line, somatic mesoderm lining chorion and amnion; dashes with circles, splanchnic mesoderm lining yolk sac. From W. P. Luckett, in W. P. Luckett and F. S. Szalay, eds., *Phylogeny of the primates,* Plenum Press, New York, 1975. Used by permission.

ment lags in ungulates, prosimians, and many insectivores but reaches extremes in anthropoid primates and rodents (see van Tienhoven, 1983).

DEVELOPMENT OF THE YOLK SAC

The vertebrate yolk sac is typically an appendage of the alimentary tract. It is the only extraembryonic membrane appearing in anamniotes where it ranges from an externally bulging **true yolk sac** (such as that of the marsupial frog, Fig. 23.1) to a mere thickening of the gut wall (e.g., in the lamprey and most amphibians, Fig. 23.18).

In primitive fish, cellular membranes among yolky endodermal cells break down and a yolk mass covered by endoderm forms an internal yolk sac. In the sturgeon and other ganoids, this sac is a dilated portion of the gut in front of the liver (Fig.

24.2) which gives rise to the stomach. In selachians (see Fig. 23.5), as in amniotes (Fig. 23.4), the yolk sac extends from the gut behind the liver.

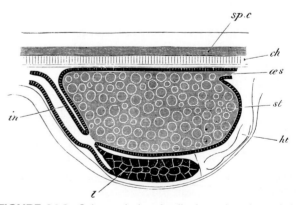

FIGURE 24.2. Schematic longitudinal section through larval *Acipenser,* a ganoid, showing dilated internal yolk sac occupying position of stomach. ch, Notochord; ht, heart; in, intestine; l, liver; oes, esophagus; sp.c, spinal cord; st, stomach. From F. M. Balfour, *Comparative embryology,* Vol. II, Macmillan, London, 1881.

In teleosts, the yolk syncytial layer (YSL) and mesoderm containing the yolk lie within the coelom and outside the gut. Yolk is digested initially by deep strands of cytoplasm and later by folds of tissue (Fig. 24.3).

The amniotic yolk sac is actually a series of yolk sacs which vary considerably, especially among mammals. Ironically, the yolk-poor mammalian embryo forms its primitive yolk sac at an earlier stage of development than the yolk-rich avian embryos.

Although yolk storage and digestion are the most conspicuous functions of the yolk sac, other functions may be more general. In all vertebrates with true yolk sacs, from those with megalecithal eggs (sharks and ganoids, Apoda, some urodeles, large-egg frogs, reptiles, birds, and prototherian mammals) to therian mammals with microlecithal eggs, **angiogenesis** (blood vessel formation) begins in the yolk sac as soon as the mesoderm penetrates. Angiogenesis and early blood cell differentiation then continue in the definitive yolk sac (see Luckett, 1976 e.g., Fig. 18.33).

Avian and reptilian yolk sacs arise as the yolk is covered by extraembryonic ectoderm, mesoderm, and endoderm (Fig. 24.4). At first, only extraembryonic ectoderm and endoderm (i.e., hypoblast) move over the yolk sphere. The early yolk sac, or at least its leading edge, is a **bilaminar** layer of extraembryonic endoderm and ectoderm called an **omphalopleure** (Gk. *omphalos* meaning navel or umbilicus but referring to yolk sac + *pleura* side or rib referring to wall).

The bilaminar condition persists in the yolk sac's cranial region known as the **proamnion.** Elsewhere, mesoderm originating from the primitive streak (see Fig. 18.32) migrates between the extraembryonic ectoderm and endoderm, finally catching up to their leading edge as it moves around the yolk sphere (Fig. 24.4). The single layer of migrating mesoderm cells changes the yolk sac into a **trilaminar yolk sac** or **choriovitelline membrane.**

In viviparous vertebrates (sharks, reptiles, metatherians, and eutherians), a choriovitelline membrane presses against the uterus at some stage of development. The combined embryonic and maternal membranes or **choriovitelline placenta** (Fig. 24.1b) probably serve as the major organ of maternal–embryonic respiratory exchange.

Cavitation in the embryonic mesoderm, which created the coelom between splanchnic and somatic mesoderm (see Fig. 23.7), expands into the extraembryonic mesoderm (see Fig. 23.10), leaving an exocoel between extraembryonic splanchnic and somatic mesoderm. The extraembryonic splanchnic mesoderm and endoderm fuse to form the extraembryonic splanchnopleure of the **definitive** or **splanchnopleuric yolk sac** (Fig. 24.1a). The double layer of fused somatic mesoderm and ectoderm is the **extraembryonic somatopleure** or **chorion.**

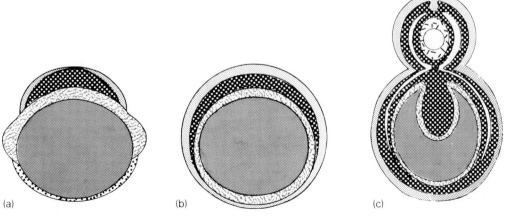

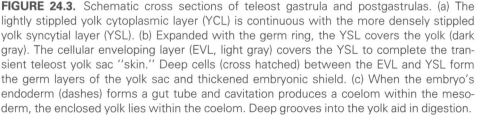

(a)   (b)   (c)

FIGURE 24.3. Schematic cross sections of teleost gastrula and postgastrulas. (a) The lightly stippled yolk cytoplasmic layer (YCL) is continuous with the more densely stippled yolk syncytial layer (YSL). (b) Expanded with the germ ring, the YSL covers the yolk (dark gray). The cellular enveloping layer (EVL, light gray) covers the YSL to complete the transient teleost yolk sac "skin." Deep cells (cross hatched) between the EVL and YSL form the germ layers of the yolk sac and thickened embryonic shield. (c) When the embryo's endoderm (dashes) forms a gut tube and cavitation produces a coelom within the mesoderm, the enclosed yolk lies within the coelom. Deep grooves into the yolk aid in digestion.

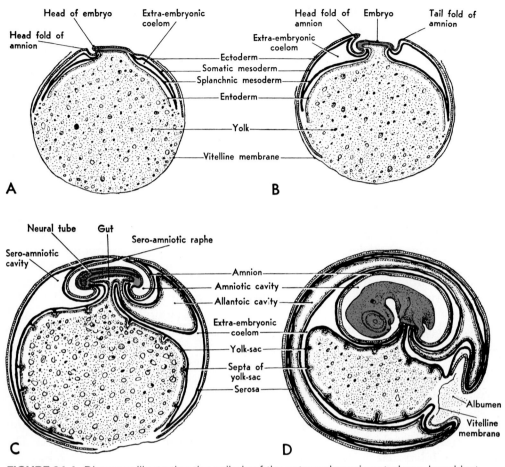

FIGURE 24.4. Diagrams illustrating the epiboly of the extraembryonic ectoderm, hypoblast (extraembryonic endoderm), and extraembryonic mesoderm over the yolk sphere of a chick. From B. M. Carlson, *Patten's foundations of embryology,* 5th ed., McGraw-Hill, New York, 1988. Used by permission.

Prototherian and Metatherian Yolk Sacs

In **prototherians and metatherians,** the **primitive yolk sac** is a bilaminar **omphalopleure** of fused extraembryonic ectoderm and parietal endoderm or hypoblast. A large part of this yolk sac remains an omphalopleure, and the head end of the embryo becomes enshrouded in a bilaminar **proamnion** (Fig. 24.5b).

Both prototherians and metatherians absorb uterine fluids through the shell membranes, but the metatherians lose the shell membrane while still *in utero,* bringing their yolk sac into direct contact with the uterine contents. Metatherians also remain in the uterus longer and take greater advantage of the maternal environment. The thin omphalopleure making up about half of the metatherian yolk sac seems adapted for absorbing ma-

terials from the uterus. Higher concentrations of amino acids and other materials in yolk sac fluid compared to uterine fluid suggest that uptake is active (Renfree, 1973).

Mesoderm penetrates between the extraembryonic endoderm and ectoderm of the prototherian and metatherian yolk sacs, forming a **choriovitelline membrane.** Metatherians also exploit their choriovitelline membrane more than prototherians exploit theirs. The membrane does not necessarily adhere to the uterine lining (e.g., the American opossum), but, when it does (e.g., the Australian native cat), it can be moderately invasive and form a ring-shaped (i.e., annular) placenta. Angiogenesis, which occurs only in the portions of the yolk sac penetrated by mesoderm, provides the yolk sac with a **sinus terminalis** at the border with the omphalopleure. In this position, the large vessel and its drainage system seem ideally adapted

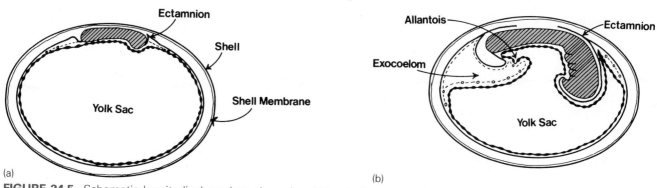

FIGURE 24.5. Schematic longitudinal sections through echidna (spiny anteater) embryos in an early fresh pouch stage with 19 somites (a) and an older pouch stage with 27–30 somites (b). The head end of the embryos (hatched) is toward the right. Yolk is contained in the yolk sac. The amnion develops between the embryo and the inner extraembryonic ectoderm layer of the ectamnion (solid line adjacent to embryo). Smooth solid line, ectoderm; beaded heavy line, extraembryonic endoderm; dashed line, somatic mesoderm lining chorion and amnion; dashes with circles, splanchnic mesoderm lining yolk sac. From W. P. Luckett, in M. K. Hecht, P. C. Goody, and B. M. Hecht, eds., *Major patterns in vertebrate evolution,* Plenum Press, New York, 1976. Used by permission.

for collecting materials absorbed through the omphalopleure and transmitting them to other parts of the embryo.

Cavitation in the posterior yolk-sac mesoderm (Figs. 24.5 and 24.6) provides an exocoel and a splanchnopleure. The completed composite yolk sac consists of an abembryonal bilaminar omphalopleure, cephalic proamnion, anterior choriovitelline membrane, and posterior splanchnopleure. In

the near absence of yolk, the metatherian extraembryonic ectoderm is differentiated as trophectoderm, but otherwise the yolk sac is similar to that of prototherians.

Eutherian Yolk Sac

The **eutherian yolk sac** shows a great deal of variation in representatives of different orders and

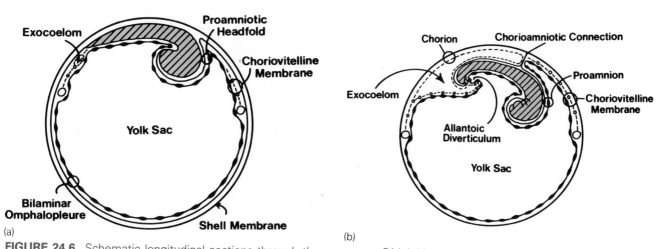

FIGURE 24.6. Schematic longitudinal sections through the opossum, *Didelphis marsupialis,* with 17–19 somites (a) and 30 somites (b). The head end of the embryos (hatched) is toward the right. Yolk is contained in the yolk sac. The amnion develops between the embryo and extraembryonic ectoderm of proamnion and chorioamniotic membranes (solid line adjacent to embryo). Smooth solid line, ectoderm; beaded heavy line, extraembryonic endoderm; dashed line, somatic mesoderm lining chorion and amnion; dashes with circles, splanchnic mesoderm lining yolk sac. From W. P. Luckett, in M. K. Hecht, P. C. Goody, and B. M. Hecht, eds., *Major patterns in vertebrate evolution,* Plenum Press, New York, 1976. Used by permission.

even suborders. Least evolved in insectivores (Fig. 24.7b) and prosimians (Fig. 24.1b), the yolk sac retains portions of the omphalopleure and choriovitelline membrane which, pressed against the uterine lining, form a choriovitelline placenta. Elsewhere on the yolk sac, an extraembryonic splanchnopleure faces the exocoel. Tarsiers, once placed among the prosimians, are now considered full-fledged primates, for one reason, because their definitive yolk sac is composed mainly of splanchnopleure and is totally liberated from attachments with the peripheral trophectoderm.

Cavitation of extraembryonic mesoderm or the expansion of a cavity seems to be the principal force at work in shaping the eutherian yolk sac. Without contact with the chorion, even the largest yolk sac is doomed. In carnivores (Fig. 24.7a), cavitation in the mesoderm of the choriovitelline membrane results in the completion of the splanchnopleuric yolk sac and, at the same time, its undermining. In artiodactyles (Fig. 24.7c), the spread of allantoic mesoderm accomplishes the same thing.

In Old World primates, the yolk sac goes through some puzzling contortions. Typically, the **parietal endoderm** beneath the trophectoderm (i.e., the omphalopleure) and the **visceral endoderm** beneath the epiblast line a continuous **primary yolk sac.** The parietal portion is the more variable.

In the rhesus monkey's yolk sac, the parietal endoderm expands (Fig. 24.8a–c). It bends, branches, and undercuts the embryo with folds, until a smaller **secondary yolk sac** (Fig. 24.9) replaces the larger primary yolk sac. When mesoderm lines the secondary yolk sac and cavitates,

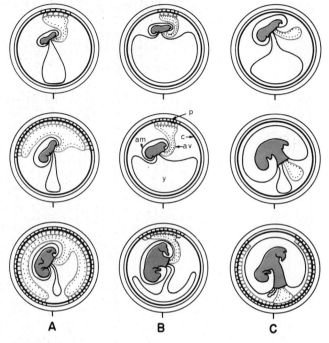

FIGURE 24.7. Diagrams of development in extraembryonic membranes (top to bottom) and their relationships in (a) a carnivore, (b) an insectivore, and (c) an artiodactyl. Differences are seen in (1) the timing and extent of yolk sac involvement, (2) the degree and point of allantoic contact with the chorion, and (3) the position and size of the placenta. The uterus is indicated by the double outer circles and the mesometrium by the dash at the bottom of each figure. The trophoblast (or chorion, c) is indicated by a heavy circle, and the chorioallantoic placenta (p) by outer radial dashes. The presence of allantoic blood vessels (av) is indicated by dots. The amnion (am) surrounds the embryo, and the yolk sac (y) hangs from below. From H. W. Mossman, in R. J. Blandau, ed., *Biology of the blastocyst,* University of Chicago Press, Chicago, 1971. © 1971 by The University of Chicago. Used by permission.

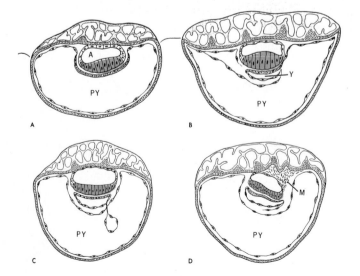

FIGURE 24.8. Diagrams of the rhesus monkey, *Macacca mulatta,* embryo at 11 (a) and 12 (b–d) days after fertilization. The primary yolk sac (PY) is replaced by the secondary yolk sac (Y). The primitive streak is posterior, and extraembryonic mesoderm (M) streams out toward the placenta (above), changing primitive trophoblastic projections into primitive villi with mesenchymal cores. M, mesoderm; PY, primary yolk sac; Y, secondary or definitive yolk sac. From W. P. Luckett, *Am. J. Anat.,* 152:59(1978), Alan R. Liss, Inc. Used by permission.

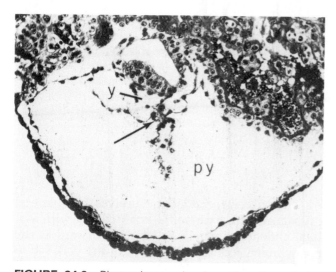

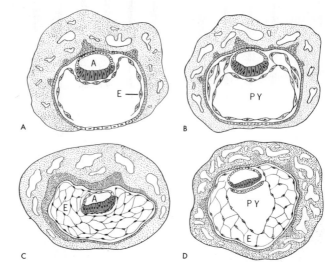

FIGURE 24.9. Photomicrograph of section through implanted rhesus embryo at 12 days. The secondary yolk sac (y) is still continuous with the primary yolk sac (py) at arrows while dislodging and replacing it below the epiblast. A large amniotic cavity is seen above the bilaminar embryonic disk. Thickenings in the peripheral portion of the trophoblast indicate attachment has occurred at a secondary placental site characteristic of Old World monkeys. ×130. From W. P. Luckett, *Am. J. Anat.,* 152:59(1978), Alan R. Liss, Inc. Used by permission.

FIGURE 24.10. Diagrams of human embryos at 9 (a and b), 10.5 (c), and 11 (d) days after fertilization. Expansion of the parietal endoderm (e) into an endoderm meshwork encroaches on the primary yolk sac. The thin amnion roofing the amniotic cavity (a) is completely separated from the well-organized cytotrophoblast. The expanding syncytial trophoblast becomes increasingly lacunary as the cytoytrophoblast undergoes syncytial transformation. A, amniotic cavity; E, extraembryonic endoderm and endodermal meshwork; PY, primary yolk sac. From W. P. Luckett, *Am. J. Anat.,* 152:59(1978), Alan R. Liss, Inc. Used by permission.

the secondary yolk sac becomes a definitive splanchnopleuric yolk sac.

In human beings, the expansion of the parietal endoderm is 3-dimensional. The blastocyst cavity may literally fill up with an **endodermal meshwork** (Fig. 24.10). Lined by a membrane of flattened cells (sometimes called Heuser's exocoelomic membrane), the human **primary yolk sac** varies from a reasonably patent structure (Fig. 24.11) to a vague lineup of vacuoles (sometimes large) within the endodermal meshwork.

When extraembryonic mesoderm migrates, it replaces the endodermal meshwork with a mesenchyme sometimes called the primary mesoblast. The human **secondary yolk sac** forms when the extraembryonic mesoderm fuses with endoderm as a choriovitelline membrane (Figs. 24.12b and 24.13). The human **definitive yolk sac** appears when cavitation within the extraembryonic mesenchyme creates the exocoel and produces a splanchnopleure (Figs. 24.14 and 24.15). Angiogenesis and early blood cell differentiation quickly commence in splanchnopleuric blood islands (Fig. 24.16).

In rodents and rabbits, the visceral and parietal portions of the yolk sac have very different fates.

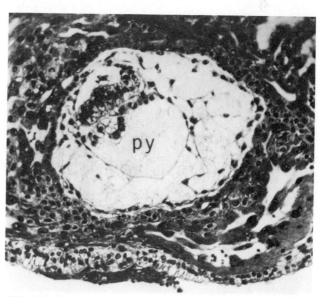

FIGURE 24.11. Photomicrograph of section through implanted human embryo at 11 days after fertilization. The primary yolk sac (py) lies within an extraembryonic endodermal meshwork. ×130. From W. P. Luckett, *Am. J. Anat.,* 152:59(1978), Alan R. Liss, Inc. Used by permission.

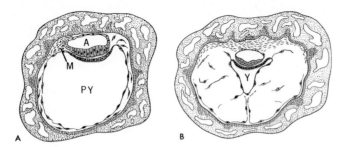

FIGURE 24.12. Diagrams of human embryos at 12 and 13 days after fertilization. A small secondary yolk sac (y) replaces the disintegrated primary yolk sac. Extraembryonic mesoderm (M) from the primitive streak lines about one-third of the trophectoderm. From W. P. Luckett, *Am. J. Anat.,* 152:59(1978), Alan R. Liss, Inc. Used by permission.

The parietal portion of the early endoderm remains an avascular omphalopleure as long as it lasts, but that may not be too long. In the guinea pig (Fig. 24.27), the omphalopleure begins to disintegrate even before it is completely formed.

In mice and rats, a dense basement membrane called Reichert's membrane (R, Fig. 20.14) welds the overlying trophectoderm to the underlying parietal endoderm. The membrane disintegrates on about day 14 postcoitum (p.c.) in mice, while the omphalopleure remains patent until day 17 in rabbits. Ultimately, only a large vein, the sinus terminalis in mice, continues to identify the border of the visceral endoderm and remnants of parietal endoderm (Fig. 24.17).

The visceral endoderm survives the parietal endoderm. The embryonic knob coated with visceral endoderm, which has grown outward as the **egg cylinder,** reduces the lumen of the yolk sac to a narrow U-shaped chamber (Fig. 19.4). This ventral folding of the embryo, or **entypy,** occurs earlier in the mouse (i.e., on day 5 out of a total gestation of 19–32 days) than in the rabbit (i.e., day 12 of a total gestation of 30–35 days), but the results are similar.

The visceral endoderm acquires an undercoating of extraembryonic mesoderm (see Fig. 19.5) and becomes vascularized as a **visceral choriovitelline** membrane. Upon the breakdown of the parietal endoderm, the outer layer of extraembryonic visceral

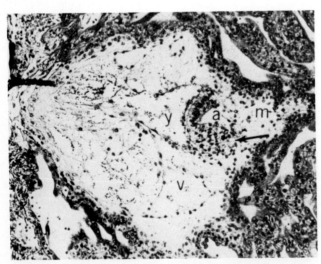

FIGURE 24.13. Photomicrograph of section through early villous human embryo (early stage 6) at 13 days after fertilization. The primary yolk sac has been reduced to vesicular remnants (v) and replaced by the secondary yolk sac (y). A thin-roofed amniotic cavity (a) is seen above the embryonic disk, as extraembryonic mesoderm (m) emerges from its posterior primitive streak (arrow). ×130. From W. P. Luckett, *Am. J. Anat.,* 152:59(1978), Alan R. Liss, Inc. Used by permission.

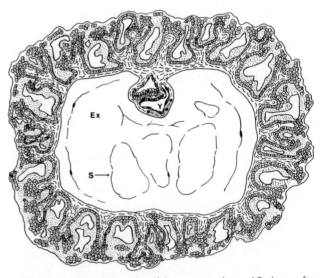

FIGURE 26.14. Diagram of human embryo 16 days after fertilization. Only a few strands (S) of the primary yolk sac remain. The secondary yolk sac (Y) lies within the exocoel (Ex), and angiogenesis occurs in the mesodermal lining of the splanchnopleure. A small allantoic vesicle juts out of the embryo posteriorly, and extraembryonic mesoderm penetrates deeply into the chorionic villi of the highly lacunary placenta. The trilaminar embryo has entered the notochordal process stage. From W. P. Luckett, *Am. J. Anat.,* 152:59(1978), Alan R. Liss, Inc. Used by permission.

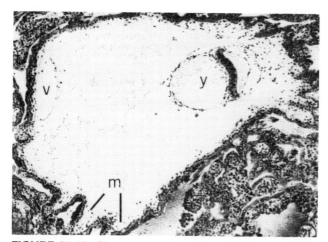

FIGURE 24.15. Photomicrograph of section through early villous human embryo (stage 6) at 14 days after fertilization. Only a vesicular remnant (v) of the primary yolk sac remains as the secondary yolk sac (y) becomes a definitive yolk sac with a splanchnopleure lying in the exocoel. Extra embryonic somatic mesoderm completely lines the cytotrophoblast and villi (m). ×80. From W. P. Luckett, *Am. J. Anat.,* 152:59(1978), Alan R. Liss, Inc. Used by permission.

endoderm covering the everted embryo becomes the inner layer of an open yolk sac. As the embryo withdraws to a more characteristic shape (Fig. 24.17), it leaves an **inverted yolk sac** in place at the surface.

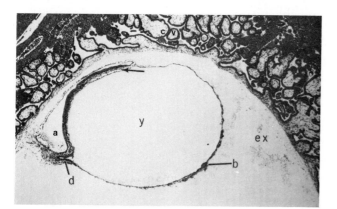

FIGURE 24.16. Photomicrograph of section through presomite human embryo (stage 8) at 19 days after fertilization. The trilaminar embryonic disk and notochordal process (arrow) along with the yolk sac (y) lie between the exocoel (ex) and the amnion (a). The allantoic diverticulum (d) runs out of the hindgut portion of the embryonic disk. Blood islands (b) appear in the splanchnic mesoderm of the yolk sac. An extensive network of chorionic villi (cv) surrounds the embryo. ×25. From W. P. Luckett, *Am. J. Anat.,* 152:59(1978), Alan R. Liss, Inc. Used by permission.

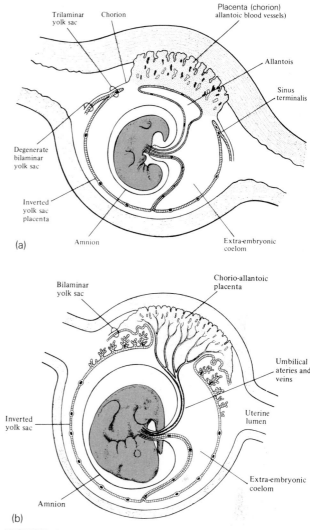

FIGURE 24.17. Diagrams of rabbit (a) and mouse (b) fetuses *in utero* showing arrangements of extraembryonic membranes and placentas. With the breakdown of the bilaminar (omphalopleure) peripheral yolk sac, the inverted yolk sac of splanchnopleure lies open in the uterine lumen and against the uterine wall. From M. B. Renfree, in C. R. Austin and R. V. Short, eds., *Reproduction in mammals,* Book 2, *Embryonic and fetal development,* 2nd ed., Cambridge University Press, Cambridge, 1982. Reprinted with the permission of Cambridge Univ. Press.

AMNIOGENESIS

The trophoblastic shell must be regarded as a peculiar mammalian structure developed solely in response to the demands of viviparity and prolonged gestation in the uterus.

Jack Davies and Hans Hesseldahl (1971, p. 35)

Continuity between the amnion and chorion has led to the tradition of considering amnion and chorion formation under the single title of **amnioge-**

nesis. Amniotes employ folding or various forms of cavitation for amniogenesis.

Folding occurs in reptiles, birds (Figs 24.1a and 24.4), prototherians (Fig. 24.5), metatherians (Fig. 24.6), and eutherians (e.g., prosimians, Fig. 24.1b) whenever amnion and chorion appear late. In reptiles and prototherians, the fold is an **ectamniotic ridge** of extraembryonic ectoderm. The double-layer **ectamnion** slips over the embryo from the anterior end (Fig. 24.5a,b).

Later, the ectamnion is invaded by extraembryonic mesoderm. Expansion of the exocoel within the mesoderm separates an inner amnion from an outer chorion, both of which are somatopleures.

The ectamniotic ridge of birds and metatherians is small. Amniogenesis is by way of circumembryonic **chorioamniotic** or **somatopleuric folds** rather than an ectamniotic ridge. The folds meet at the **chorioamniotic connection** (Fig. 24.6) and seal the embryo within the amniotic cavity. A **raphe** that marks the connection may later disappear, separating the amnion and chorion and opening up the exocoel for expansion or invasion by the allantois (Fig. 24.4).

Eutherians performing amniogenesis by **somatopleuric folding** (Fig. 24.18 upper) tend to attach paraembryonally or equatorially to the uterine lining and exhibit limited invasiveness (see below). The polar trophectoderm (TE) above the ICM disappears, exposing the **embryonic disk** to the uterine environment. The pig (Fig. 24.19), for example, loses its polar TE (also known as Rauber's layer) at the beginning of a period of swelling and elongation. Possibly the swelling and stretching cause the polar trophectoderm to retract to the periphery of the disk, or possibly the TE cells are shed. In either

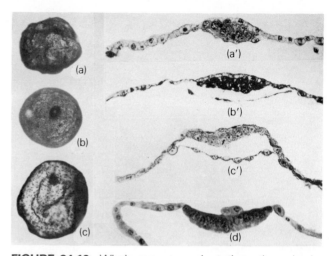

FIGURE 24.19. Whole mounts and sections through pig blastocysts illustrating formation of bilaminar embryonic disk, initial stretching and swelling of blastocyst, and loss of polar trophectoderm. Peripheral trophectodermal cells appear swollen compared to inner cell mass cells and columnar epiblast cells. The early endoderm or hypoblast is a layer of thin cells stretching beneath the epiblast.

Whole mounts	Sections
(a) ×25	(a') ×200. Embryo 0.6 mm in diameter
(b) ×25	(b') ×200. Embryo 0.6 mm in diameter
(c) ×24	(c') ×200. Embryo 0.8 mm in diameter
(d)	×200. Embryo 2–3 mm in diameter

From C. H. Heuser and G. L. Streeter, *Carnegie Inst. Washington Contrib. Embryol.* 20(109):1 (1928). Used by permission.

case, the relationship between the embryonic disk and the extraembryonic membrane is identical to that in metatherians (Fig. 19.1a).

The major difference between extraembryonic somatopleuric folding in birds and in eutherian

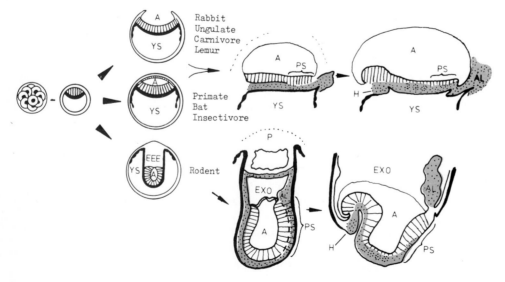

FIGURE 24.18. Modes of amniogenesis in eutherians. A, amniotic cavity or potential cavity; AL, allantois or allantoic stalk; EEE, extraembryonic ectoderm; EXO, exocoel; H, heart; p, placenta; PS, primitive streak; YS, yolk sac. Hatching, epiblast or ectoderm; stippling, mesoderm; solid, endoderm. From M. H. L. Snow, in G. M. Edelman, ed., *Molecular determinants of animal form,* Alan R. Liss, Inc., New York, 1985. Used by permission.

mammals is that the mammalian extraembryonic ectoderm is inherited from the trophectoderm of the blastocysts. The folding and healing into the amnion and chorion seem identical.

Cavitation occurs in the amniogenesis of eutherians in which the polar TE persists. Mammals forming their amnions in this way tend to implant embryonally (at the embryonic pole of the blastocyst, e.g., anthropoid primates).

An amniotic cavity forms within the embryonic ectoderm (Fig. 24.18 middle). Only the chorion inherits the trophectoderm, and it is invasive during implantation (Luckett, 1976). The amniotic ectoderm forms anew from cells in the upper or distal part of the embryonic ectoderm.

In Old World primates, an amniotic cavity forms via cavitation within the embryonic ectoderm very early in development (see Fig. 19.2b). The cavity briefly opens to the extraembryonic ectoderm (see Fig. 19.2c) but quickly forms a thin amniotic roof (see Fig. 19.2d, also Figs. 24.9 and 24.11). The amnion and chorion become somatopleures following the migration of extraembryonic mesoderm (Figs. 24.12–24.16).

Cavitation extending to extraembryonic ectoderm occurs in myomorphic rodents (mice, rats) and hystricomorphs (guinea pigs, porcupines, chinchillas, Fig. 24.18 lower). Embryos implant abembryonally and form invasive placentas.

Cavitation in mice begins in the embryonic knob as a **proamniotic cavity** (sometimes called an amniotic cavity or extraembryonic cavity) at about 5.5 days p.c. (see Fig. 19.4). The cavity expands to the **extraembryonic ectoderm** above the egg cylinder by 6.5 days p.c. As mesoderm emerges from the primitive streak at about 7 days p.c., an **amniotic fold** narrows the cavity midway and, fusing across the cavity at about 7.5 days p.c., separates the amniotic cavity (A) from the exocoel (EXO) (Fig. 24.18).

Embryonic ectoderm furnishes the ectodermal lining of the amniotic somatopleure, while extraembryonic ectoderm furnishes the ectodermal lining of the chorionic somatopleure (Fig. 14.16). The remainder of the chorion is formed by the mural TE and is short-lived. In the Hystricomorpha, the chorion breaks down with the parietal yolk sac.

DEVELOPMENT OF THE ALLANTOIS

The allantois is the membranous **allantoic vesicle** surrounding an **allantoic cavity** in reptiles, birds (Fig. 24.1a), prototherians, metatherians, and some mammals, such as prosimians (Fig. 24.1b) and artiodactyles (Fig. 24.7c). In carnivores (Fig. 24.7a) and rabbits (Fig. 24.17a), the allantois is a vesicle attached to a mesodermal core. In primates, the vesicle is hardly more than a pocket in the hindgut endoderm (Fig. 24.16), and in myomorphic (e.g., rat) and hystricomorphic (e.g., guinea pig) rodents, even a pocket has virtually disappeared. In its place is an **allantoic** or **body stalk** of solid mesoderm (Fig. 24.18), later penetrated by **umbilical vessels** and known as the **umbilical cord** (Fig. 24.17b).

The large saclike allantoic vesicles of birds and reptiles contain allantoic fluid and function in the storage of wastes. The vesicular allantois reaches the chorion to form a highly vascular **chorioallantoic membrane** or CAM (not to be confused with cell adhesion molecules also called CAMs) which functions as a respiratory organ. The prototherian allantois is similar but not as extensive as the chick's. The typical metatherian allantois is a freehanging sac, which is not especially vascularized even where it contacts the chorion, and does not seem to play a major role in exchange.

Wherever the allantois develops as a vesicle, no matter how small, it makes its first appearance as an **allantoenteric diverticulum** from the hindgut region of the embryo (e.g., Fig. 24.16). The allantoic vesicle is therefore a splanchnopleure, and even when no vesicular pocket is seen, the allantoic stalk would seem to be splanchnic mesoderm. Whether in the form of a membrane or a mesodermal core, allantoic mesoderm supports angiogenesis, and the allantois generally becomes highly vascular.

PREIMPLANTATION PERIOD IN THERIAN MAMMALS

The process of implantation takes place when the blastocyst and endometrium are at the right and proper stage of development.

M. C. Chang (1981, p. 32)

In the preimplantation period, the embryo is transported through the oviducts (or uterine tubes); the uterine lining is sensitized (i.e., become receptive to the presence of the embryo), and the blastocyst is brought to the correct stage of development and condition for implantation. All this is ordinarily done on a rigid schedule, the embryo and uterus passing through a **window of synchrony** on their way to successful implantation.

Failure to synchronize the embryo and uterus leads to rejection of the embryo. The rat uterus, for example, becomes unreceptive to embryos 5 days after ovulation. A **blastotoxic substance (blastocidin)**, a low molecular weight polypeptide released into uterine fluid, then dispenses with the blastocyst (Psychoyos and Casimiri, 1981).

Embryo rejection by the uterus is surprising, inasmuch as embryos are sometimes accepted at **ectopic** or **extrauterine** sites (e.g., tubal pregnancy in human beings). Experimental embryo transfer is even successful at sites as unaccustomed to pregnancy as the testis of a different species (e.g., mouse embryo transferred to rat testis) (see Chang, 1981). However, the consequences of implantation at any of these sites is ultimately disastrous, for example, local rupture and severe hemorrhage.

The uterus alone supports a pregnancy to a normal termination and, it would seem, is as finely adapted to pregnancy as is the embryo. Tuning the embryo for implantation occurs in a free preattachment period, while preparations for attachment occur in the uterus.

Free Preattachment Period

A preattachment period consists of a **tubal transport period**, during which the blastocyst moves through the oviducts to the uterus, and an **intrauterine preattachment period** prior to the embryo's adhering to sites on the uterine wall. The embryos of different species spend their free period in different stages, and the period lasts different lengths of time.

Timing of the preattachment period varies, but the different orders of eutherians generally fall into one of four patterns (Table 24.1). In the first pattern, exemplified by the elephant shrew, *Elephantulus myurus* (if not other insectivores, e.g., *Tupaia longipes*), implantation occurs at a relatively early stage following rapid tubal transport and a short intrauterine preattachment period. This pattern is probably primitive, since tubal transport is also 1 day in metatherians (e.g., the American opossum, *Didelphis virginiana*, and Australian native cat, *Dasyurus viverrinus*), and eggs arrive in the uterus still containing pronuclei. But implantation in metatherians does not take place until the embryo has reached the primitive streak stage.

A second pattern is found in bats. A prolonged period of tubal transport is followed by a brief free intrauterine phase. The blastocyst that arrives in the uterus implants after adding only some hypoblast.

In a third preattachment pattern, the periods of tubal transport and intrauterine preattachment are more or less equal. These periods tend to be shorter in primates and rodents (total time about 1 week) than in carnivores (total time about 2 weeks). Implantation occurs with embryos at different ages and at stages ranging from the blastocyst to the complete bilaminar vesicle.

In species with short preattachment periods, such as mice, the window of synchrony is so narrow that blastocysts transported to the uterus too quickly as a result of estrogen administration are generally rejected. Accordingly, estrogenic compounds are generally effective contraceptives when administered during tubal transport (see Chang, 1981).

The ungulates require a category for themselves. A relatively short period of tubal transport is followed by a prolonged free intrauterine period. While the total amount of preattachment time may actually be similar to that of carnivores, the amount of development is profoundly different. The cleaving ungulate embryo arriving in the uterus develops to a multilaminar vesicle with all its extraembryonic membranes in relatively advanced stages of development before finally implanting. The window of synchrony between uterus and embryo is slightly wider in ungulates (e.g., 1–2 days in cows) than in other mammals but still relatively stringent considering the long preattachment period.

Spacing in polyovular species, especially polytocous species (i.e., with large litter sizes), is due to transuterine movements redistributing the embryos along the length of the uterus. Presumably, similar **spacing mechanisms** place the embryo of **monovular** and **monotocous** species (i.e., with single births) in appropriate parts of the uterus (see Böving, 1971).

In most mammals, the entire length of the uterus is capable of being sensitized and accepting an embryo although it does not do so ordinarily. Implantation can occur anywhere in the **simplex uterus** (also called luminal uterus) of bats and primates (Fig. 24.20d), including humans, but normally occurs in the body or middle portion, more often on the posterior wall, and generally near the midline (Boving and Larsen, 1973).[1] In the **bipartite uterus** (b) of rats, mice, and hamsters, the entire endometrium undergoes a decidual transformation

---

[1]Many older accounts incorrectly place the attachment site laterally rather than medially.

TABLE 24.1. Chronologies and stages of preimplantation development for representative eutherians[a]

Order and Species	Days Tubal Transport	Stage Entering Uterus	Total Days Before Attachment	Stage at Attachment
Insectivora				
Elephant shrew, *Elephantulus myurus*	1–2 (?)[b]	1–4 cell	?	Early blastocyst
Tree shrew, *Tupaia longipes*	5–6 (estimate)	Bilaminar vesicle	?	Bilaminar vesicle
Chiroptera				
vampire bat, *Desmodus rotundus*	>16	Blastocyst (some hypoblast)	>16	Blastocyst (partly bilaminar)
Long-tongued bat, *Glossophaga soricina*	12–14	Blastocyst	15	Blastocyst (partly bilaminar)
Little brown bat, *Myotis lucifugus*	?	Morula	>10	Blastocyst or bilaminar vesicle
Primates				
rhesus monkey, *Macacca mulatta*	4	16-cell	8–9	Blastocyst
baboon, *Papio sp.*	4–5	~32-cell	8–9	Blastocyst
human beings, *Homo sapiens*	3–4	?	6–7	Blastocyst
Rodentia				
Rabbit	3–3.5	>20-cell morula	7	Bilaminar vesicle
Mouse	3–3.5	>16-cell morula	6	Blastocyst (some hypoblast)
Rat	4	Morula	6	Blastocyst (some hypoblast)
golden hamster, *Cricetus auratus*	3.5	4–8-cell	4.3	Blastocyst (some hypoblast)
Guinea pig	3.5	8-cell	6	Blastocyst
Carnivora				
Dog	4–5	16-cell to early blastocyst	~11–12	Bilaminar vesicle
Cat	6–7	28–30-cell morula	13–14	Bilaminar vesicle
Ferret	5–6	>32-cell	~12	Bilaminar vesicle
Ungulates, Perissodactyla				
Horse	?	?	49–63	Elongated chorionic vesicle and allantois
Ungulates, Artiodactyla				
Pig	2.5–3.75	3–8-cell	(progressive) 11–20	Elongated chorionic vesicle and allantois
Sheep	2.5–3.5	?	(progressive) 15–17	Elongated chorionic vesicle and allantois
Cow	4–4.5	8–16-cell	(progressive) >40	Elongated chorionic vesicle and allantois

[a]Data from various sources summarized in Wimsatt (1975).
[b]?, unknown.

or thickening. Embryos may implant everywhere along the uterine horns with the exception of the central region above the vagina. Monotocous ungulates (e.g., cow, sheep, horse, see Table 7.11) tend to alternate pregnancies between the horns of their **bicornuate uterus** (c), while polytocous species (e.g., pig) tend to space their embryos evenly along both horns.

Some species express a preference for one horn of the uterus (right horn in certain antelopes and bats, e.g., *Myotis lucifugus*), and some have specialized implantation sites (insectivores and glos-

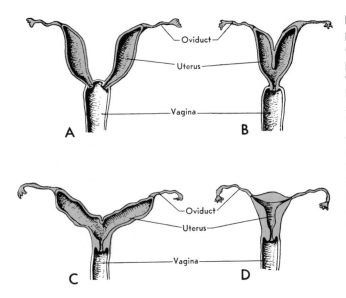

FIGURE 24.20. Types of uterus in therian mammals. (a) Duplex uterus in marsupials. Bifurcation occurs in the vagina. (b) Bipartite uterus of rodents and rabbits. Fusion between proximal portions of horns in rodents produces a single cervix (beginning of uterine horns in vagina). Less fusion in rabbits leaves two cervices entering vagina together (not shown). The shunting of embryos from one horn to the other is blocked, but embryos are evenly spaced throughout a single horn by transuterine movement. (c) Bicornuate (also called bicornate) uterus of ungulates and carnivores. Greater fusion allows shunting of embryos between horns and even spacing throughout uterus. (d) Simplex uterus of primates and bats. Fusion complete. Adapted from H. E. Jordan and J. E. Kindred, *Textbook of embryology,* 4th ed., Appleton-Century, Norwalk, CT, 1942.

sophagine bats). In the bat, *Pipistrellus subflavus,* where ovulation is unequal in the ovaries, spacing mechanisms deliver embryos to the correct site in the uterus.

In rabbits and carnivores, blastocyst expansion sufficient to distend the uterus may be the stimulus for peristaltic-like waves of **uterine squeezing** that space embryos at internodes. Ungulate blastocysts also expand (in the pig, to as much as 5 mm in diameter) and may be spaced by a similar mechanism, but spacing of rat, mouse, and guinea pig blastocysts must involve different mechanisms, since these blastocysts do not expand.

In the absence of confluence between the horns of the bipartite rabbit uterus, embryos are confined to one horn whether crowded or not (Fig. 24.21). This does not necessarily result in an unequal distribution of embryos, at least, in most rodents, ovulation is about equal in both ovaries (as judged by the presence of corpora lutea, see Table 7.11). The movement of blastocysts in the uterine horns is bidirectional, and in carnivores and ungulates where the horns of the uterus are confluent, embryos are shunted from one horn to the other. Spacing mechanisms are coordinated between the horns as well as within the horns, and equal numbers of embryos tend to implant in both horns. Similarly, in sheep, almost all dizygotic twins implant in individual horns even when both eggs come from the same ovary. Cow blastocysts are much less likely to transfer from one horn of the uterus to the other, but, in one recorded case, monozygotic twins implanted in opposite uterine horns (see Wimsatt, 1975).

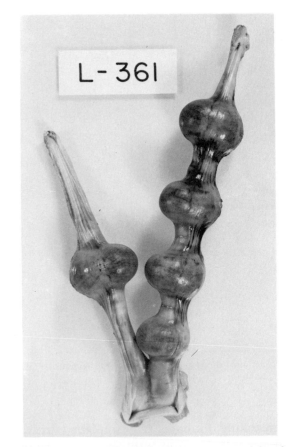

FIGURE 24.21. A rabbit's bipartite uterus containing four evenly spaced implantation domes in one horn and one central dome in the other. From B. G. Boving and J. F. Larsen, in E. S. E. Havez and T. N. Evans, eds., *Human reproduction,* Harper & Row, New York, 1973. Courtesy of B. G. Boving and the Carnegie Institution of Washington.

The resorption of embryos may also be involved in spacing. In the polyovular elephant shrew (releasing 50–120 eggs at a time), only the embryos that attach at four specialized sites develop further. In bats, blastocysts farthest from the oviduct tend to be resorbed, while in the plains viscacha, a polyovular insectivore, blastocysts farthest from the oviducts survive while closer ones are crowded out and resorbed. In the pronghorn antelope, blastocysts closest to the oviduct are also resorbed after being crowded out by the expanding blastocysts farthest from the ovary.

Mechanisms for fixing blastocysts in position vary with the size of the blastocyst and its stickiness. In rabbits, the blastocyst is covered with a nonadhering **extracellular coat,** and blastocysts are arrested in position by the **blastocyst grasp** of the uterus. Possibly, the rabbit's blastocyst enlarges until it literally lodges in the uterine wall (although excess pressure kills the blastocyst).

More likely, the musculature of the myometrium around the blastocyst relaxes, while that between blastocysts contracts, confining each blastocyst to an **implantation dome** (Fig. 24.21). Since progesterone promotes muscle relaxation during pregnancy, local relaxation of the myometrium during implantation may be due to progesterone secretion by the incipient placenta forming at the points of contact between the uterine lining and the blastocyst (Conrad, 1971).

In the rat, mouse, hamster, and guinea pig, swelling due to the decidual reaction and the **attachment reaction,** or the contraction of the uterine surface, obliterates the uterine lumen and locks the previously free blastocysts in position. Hatching and the acquisition of a sticky mural TE then fix the blastocyst in position.

In the pig, the bilaminar blastocyst lengthens to about 1.5 meter at 17 days postfertilization (see Fig. 6.20). The embryo proper or embryonic disk, about 5 mm in length, in the somite stage of embryogenesis, becomes enclosed in its amniotic cavity (Fig. 24.18), and the allantois conveys blood vessels to the chorion. As the trilaminar blastocyst swells (except at its ends) into a **chorionic vesicle,** it fills a **locular enlargement** of the uterus and attaches to the uterine wall (Fig. 24.22).

In carnivores, swelling of the lining of the uterus may arrest the blastocyst (e.g., dogs), and some bat blastocysts are constrained by specialized sites in the wall of the oviduct at the junction with the uterus. In some insectivores (e.g., tupaiids), modified preformed implantation sites alone accept embryos.

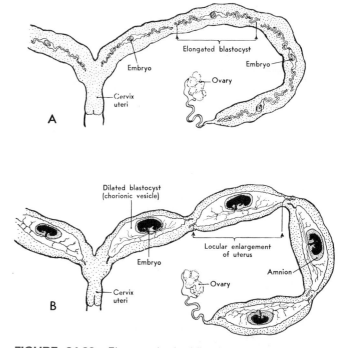

FIGURE 24.22. Elongated pig bilaminar blastocysts in preattachment period (a) and dilated chorionic vesicles within locular enlargements of uterine tube (b). From B. M. Carlson, *Patten's foundations of embryology,* 5th ed., McGraw-Hill, New York, 1988. Used by permission.

Preparations for Attachment

Some of the preparations for attachment are largely maternal, while others are largely embryonal. Most are interactive.

Sensitizing the uterus or rendering it receptive normally involves endogenous hormones. The uterus may also be artificially prepared by exogenous hormones or, in the case of rodents and rabbits, by mating with a vasectomized (sterile) male or by electrical stimulation of the cervix.

A female with an artificially prepared uterus is said to be **pseudopregnant,** and her condition is known as **pseudopregnancy.** In the absence of blastocysts in her uterus, pseudopregnancy recedes after an interval longer than an estrus cycle but shorter than a normal pregnancy. In the presence of blastocysts at the correct stage of development, pregnancy follows pseudopregnancy.

Different hormonal regimes are employed to induce pseudopregnancy in different species. In the rat, the uterus of an ovariectomized animal will be receptive to a transferred embryo 72 hours after providing estrogen and 48 hours after providing progesterone, while in the rabbit, guinea pig, hamster, sheep, rhesus monkey, and wallaby, exogenous estrogen is unnecessary.

The production of steroid hormones in the luteal phase of the estrus cycle and pseudopregnancy (Fig. 24.23), especially progesterone secreted by the corpus luteum, harks back to pituitary control. In general, prolactin and luteinizing hormone (LH) comprise the luteotropic complex of hormones required for luteogenesis and maintenance (see Chapter 7). Prolactin promotes and restores LH receptors in advance of pregnancy and exercises its effect well beyond the point of its secretion or ad-

ministration by stimulating and maintaining luteal activity. LH is essential for implantation (see Amoroso, 1981).

The primary effect of hormones secreted during the ovarian cycle seems to be on the uterus' inner layer, the **mucosa** or **endometrium** (Gk. *endo-* within [at home] + *-metr* womb or uterus). The surrounding muscular **myometrium** (Gk. *mys* referring to muscle + *-metr*) and the outer peritoneal covering, the **perimetrium** (Gk. *peri-* upon + *-metr*), are not conspicuously affected.

In rodents typically used in laboratory experiments, a **decidual transformation,** or **deciduoma formation,** occurs in advance of placental invasion.

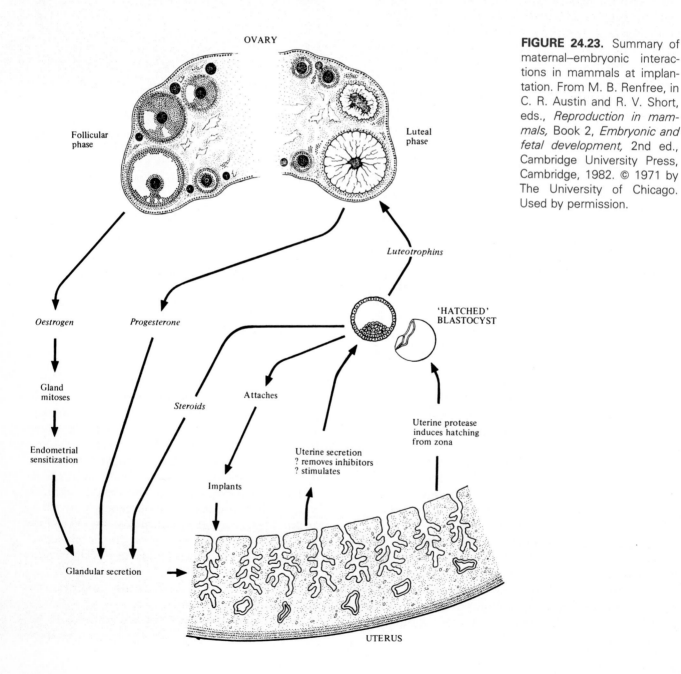

FIGURE 24.23. Summary of maternal–embryonic interactions in mammals at implantation. From M. B. Renfree, in C. R. Austin and R. V. Short, eds., *Reproduction in mammals,* Book 2, *Embryonic and fetal development,* 2nd ed., Cambridge University Press, Cambridge, 1982. © 1971 by The University of Chicago. Used by permission.

Capillary permeability increases; the rate of cell division accelerates, and secretory activity commences in the epithelium lining uterine crypts and glands.

Embryonal effects on the mother are reflected in the maternal recognition of pregnancy (see Heap and Flint, 1984). The blastocyst's ability to promote its survival in the uterus is a function of its ability to prolong luteal function in the mother. The maintenance of the corpus luteum may be due to the **luteotrophic** activity (i.e., luteal stimulation) of luteotrophins (see Fig. 24.23), or the corpus luteum may be rescued from destruction by the suspension of **luteolytic** activity (i.e., lysis or destruction of the corpus luteum) via antiluteolytic agents.

In sheep, the presence of a blastocyst in the uterus results in the inhibition of a luteolytic agent (probably prostaglandin $F_{2\alpha}$) that otherwise causes the destruction of the corpus luteum. Antiluteolytic activity may be due to **trophoblastin** (oTP-1), a member of the interferon family of proteins, produced by trophectodermal cells of the 12–16-day embryo on long-lived stored (maternal) mRNA.

Both proteinaceous and steroidal **luteotrophins** (i.e., hormones that promote the growth and maintenance of the corpus luteum) are frequently produced by trophoectoderm and subsequently by chorionic cells (Fig. 24.23). In metatherians, the maintenance of a functioning corpus luteum depends on an embryonic source of gonadotrophins beginning very shortly after implantation, although luteal progesterone production is dependent on pituitary gonadotrophins.

In human beings, **human chorionic gonadotrophin** (hCG) is synthesized by freshly implanted trophoectoderm beginning at about 8 days after ovulation. The corpus luteum is transformed into the large, stable **corpus luteum of pregnancy** under the influence of hCG, while luteolysis is prevented by the release of trophoblastic estrogen (see van Tienhoven, 1983).

Similar **chorionic** (trophoblastic) **gonadotrophins** are produced in the rat, mouse, rabbit, and pig where they operate much the same way as hCG in human begins. On the other hand, **pregnant mare's serum gonadotrophin** (PMSG), a protein hormone with gonadotrophin activities in other organisms, lacks gonadotrophin activity in mares, and early trophectodermal cells of sheep and cows fail to produce any chorionic gonadotrophins at all.

Estrogen produced by pig and rabbit trophectoderm is luteotrophic. Estrogen may also aid in maintaining the corpus luteum by preventing the release of luteolytic prostaglandin $F_{2\alpha}$.

Maternal effects on the embryo are most conspicuous where embryonic development is delayed at the behest of the mother. In metatherians, bats, laboratory rodents, the European badger, mustilids (skunks and minks), the armadillo, and the roe deer (the sole ungulate in the group), blastocysts of about 100 cells may enter a state of suspended development variously called **delayed implantation** (**obligate** [spontaneous] or **facultative** [induced]), embryonic diapause, or retarded development.

Obligate delayed implantation, lasting as long as a year, takes place in metatherians with young attached to the pouch teat. The African fruit bat, *Eidolon helvum*, has a 3-month period of obligate delayed implantation during the very dry period between rainfall peaks in summer and early fall.

Facultative delayed implantation, known as **lactational delay,** takes place in intensely suckling mice and rats. Early embryos may be retained in the uterus without implanting for as long as 10 days (see Renfree, 1982).

Different hormonal mechanisms seem to be involved in turning blastocyst development on and off in the different species, and a variety of environmental cues, including changing day length and temperature, seem to control the hormonal mechanisms. In general, the uterine environment inducing delayed implantation is nutritionally barren and the dormant embryo virtually starving. The restoration of nutrients (e.g., amino acids, protein-bound glucose and galactose, and probably fructose in the case of roe deer) in uterine secretions as well as changes in circulating hormones reawakens development.

Mouse embryos rendered dormant by ovariectomizing the mother have reduced DNA polymerase activity, RNA polymerase activity, protein synthesis, and metabolism. Injection of exogenous estrogen into the mother quickly restores normal levels of all these activities in the embryos (see Weitlauf and Kiessling, 1981).

Other maternal effects on the preimplantation embryo are subtle. In the rabbit, a uterine-specific progesterone-binding protein known variously as **uteroglobin** or **blastokinin** is secreted in response to progestogens. Uteroglobin stimulates embryonic and fetal growth.

Among the secretions of a sensitized uterus, **uterine protease** appears to aid **hatching,** the release of the blastocyst from the zona pellucida (Fig. 24.23). Alternatively, altered pH in the sensitized uterus or even proteases introduced by sperm may bring about hatching.

The corpus luteum helps maintain pregnancy via progesterone secretion. The relationship of the corpus luteum to the retention of embryos in the uterus seems to be phylogenically quite ancient. Although no bird even approaches an ovoviviparous condition (no less a viviparous condition), the short-lived avian corpus luteum is maintained as long as the egg is present in the oviduct (see Amoroso, 1981).

The immunological problems of pregnancy should be enormous with maternal–fetal interactions at so many levels and over so long a time.

The potential of an immune response to devastate an embryo or fetus is documented in **erythroblastosis fetalis** in human beings and **maternally induced runt disease** in mice where humoral antibodies and sensitized lymphocytes directed against fetal antigens reach the fetus. But even in these cases, damage to fetuses does not extend to the placenta, suggesting that the placenta normally plays some role in regulating maternal immune responses to the fetus.

Earlier, the absence of **major histocompatibility complex** (MHC) antigens on trophoblastic giant cells in rodents (see below) and the syncytiotrophoblast of human beings suggested that the trophectoderm was normally nonantigenic. Recently (see Billington, 1987), however, specific class I MHC antigens have been detected in the placenta with monoclonal antibodies (MAbs), and the hypothesis of nonantigenicity has been rejected.

In general, class I MHC antigens are required for the immunological recognition of any cell-surface antigen. During pregnancy, a maternal immune response might be provoked by class I MHC antigens present on trophectoderm and its derivatives, since only these "foreign" tissues are in direct, unbroken, and continual contact with maternal tissue. Few, if any, maternal lymphocytes traverse the placenta into the fetus, and then only toward the end of pregnancy.

In the case of the mouse and rat, monoclonal antibodies detect class I MHC antigens (so-called H-2 antigens) on diploid trophectoderm cells of the early postimplantation embryo and on some cells in the mature placenta. In particular, paternally inherited **H-2 antigens** appear on compact cellular trophoblastic tissue (known as spongiotrophoblast, see below) adjacent to maternal decidual cells. Comparable antigens (HLA-A, -B, and -C) appear on anatomically analogous extravillus and nonvillus portions (see below) of the human placenta.

Other possibilities must explain the failure of a maternal immune response to normally cause fetal rejection. Possibly, a trophoblastic or placental "barrier" to antibodies keeps fetal antigens away from the mother, or, possibly, the placenta is an immunologically privileged site where immune reactions are inhibited (Davies and Hesseldahl, 1971). In the case of human beings, the trophectoderm is extremely resistant to immune lysis caused by either cell-mediated or antibody mechanisms. The mouse placenta, however, is susceptible to immune lysis.

The transmission of proteins, even antibodies, from mother to fetus is not necessarily damaging. Polypeptide hormones such as LH and ACTH are taken up by the blastocyst, and maternal IgG antibodies are transmitted to the fetus via receptor-mediated transport, in the case of rodents, through the inverted yolk sac. Moreover, the decidual reaction and decidua formation of rodents and a prominent display of leukocytes and development of delayed hypersensitivity at the implantation site in human beings, while like inflammatory responses, are not necessarily pathogenic.

Possibly, maternally produced **enhancing antibodies** promote embryonic placental growth, while **blocking antibodies** frustrate the development or effects of immunity. In rats, mice, and hamsters, lymph nodes are larger when pregnancies involve genetically incompatible offspring rather than compatible ones, and in rats, both the mean placental weight and the mean fetal weight are greater following outbreeding than inbreeding. Moreover, in human beings, failure of immunoregulatory control is correlated with recurrent spontaneous abortions. Maternal immunologic reactivity to the embryo may even make a contribution to hybrid vigor (Beer et al., 1975).

On the other hand, B-cell-depleted mice are able to reproduce normally. Either some other mechanism kicks in when B cells are depleted, or maternal antibodies, whether enhancing or blocking, are not required for pregnancy.

Given that the placenta provides antigenic sites and the maternal immune system is not systemically suppressed, the best guess at present is that soluble **immunomodulating factors** produced by the trophoblast or placental **suppressor lymphocyte factors** produced by maternal lymphocytes act locally to inhibit the induction and expression of antipaternal **cytotoxic lymphocytes** (CTLs). In addition, pregnancy-associated **maternal serum molecules,** such as progesterone, may inhibit the killing of fetal tissue by CTLs (see Billington, 1987).

Autonomous embryonic aspects of implantation are also performed under appropriate in vitro conditions (see Hsu, 1981). The mouse blastocyst *in vitro*, for example, contracts and rotates or expands and relaxes while still enclosed in the zona pellucida. Once hatched, the blastocyst acquires a negative charge and becomes sticky.

The abembryonal **mural trophectoderm** (MT in Fig. 24.24) or circumferential **paraembryonal trophectoderm** attaches randomly to the substratum, resulting in the blastocyst taking on a symmetrical or an asymmetrical orientation. **Polar trophectoderm** (PT) at the embryonal pole continues to identify the original position of the ICM, while the MT migrates centrifugally (Fig. 24.25).

The blastocyst cavity collapses in the process and the embryo, or egg cylinder, grows and develops. Presumably, comparable morphogenic events occur *in utero*, but the interactions of the blastocyst with the uterine lining during implantation are more complex than the mere spreading of mural TE (see Enders et al., 1981).

The relationship of the embryonal pole to the axis of symmetry of the uterus, or orientation, is highly stereotypic and species specific in all eu-

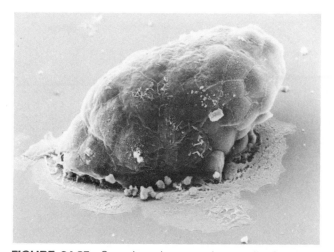

FIGURE 24.25. Scanning electron micrograph of mouse blastocyst *in vitro*. Trophectoderm remains intact over the embryo, but mural cells migrate centrifugally over the surface of the culture dish. From A. C. Enders, D. J. Chávez, and S. Schlafke, in S. R. Glasser and D. W. Bullock, eds., *Cellular and molecular aspects of implantation,* Plenum Press, New York, 1981. Used by permission.

therian orders except the Insectivora and Chiroptera. Using the **mesometrium** (Gk. *mesos* mid- referring to point of suspension + -*metr* womb, hence mesentery of the uterus) or broad ligament, as it is known in human beings, and the position of the **embryonic pole** of the blastocyst or **germ disk** of more advanced vesicles for reference points (Fig. 24.26), orientation may be described as **mesometrial** (top row left in figure, e.g., rabbit, rodents, some insectivores, and bats), **antimesometrial** (center, e.g., armadillo, carnivores, horse, monkeys, apes, human beings,[2] probably cow, pig, deer, and most insectivores, bats, shrews), or **lateral** (right, e.g., a few insectivores and bats).

The orientation of embryos is an especially conservative aspect of mammalian development. Almost every species belonging to a mammalian order orients its embryos toward the uterus in the same way (Luckett, 1976). Even in orders having species differing widely in their mode of placentation (e.g., as in the primates), orientation is relatively constant.

In polytocous species with large litters, the adaptive advantage of orientation may have something to do with spacing, since only one embryo can then occupy a given level of the uterus at a

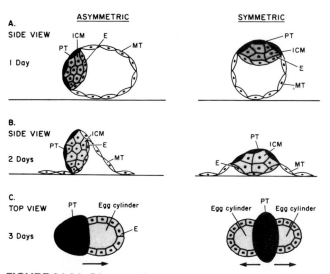

FIGURE 24.24. Diagram of mouse blastocyst "implanting" *in vitro*. The mural trophectoderm (MT) adheres at random, resulting in either an asymmetrical or a symmetrical pattern of attachment. Subsequent migration of the MT leads to the collapse of the blastocyst cavity. The polar trophectoderm (PT) continues to overlie the inner cell mass (ICM), while the embryo (E), or egg cylinder, grows either unidirectionally (left) or bidirectionally (right). From Y-C Hsu, in S. R. Glasser and D. W. Bullock, eds., *Cellular and molecular aspects of implantation,* Plenum Press, New York, 1981.

[2]Disagreement about the orientation of primates is due to uncertainty about what side of the simplex uterus corresponds to the antimesometrial side of other uteri. Inasmuch as the uterine horns seem to have fused along the antimesometrial side, the medial plane should be considered antimesometrial and the orientation classified accordingly (Mossman, 1971).

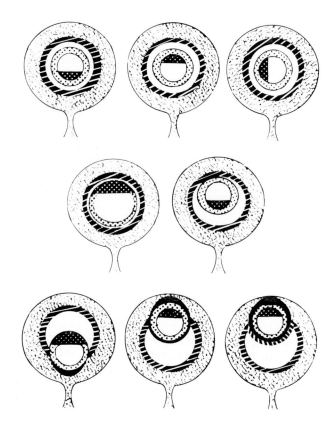

FIGURE 24.26. Diagrams of sections through the uterus and early embryo illustrating common orientations (top row), patterns of primary implantation (middle row), and three types of placenta in primates (bottom row). The mesometrium is at the bottom of each figure. (Top row) Blastocysts or embryonic vesicles may orient with their embryonal pole toward the mesometrium (mesometrial orientation, rodents, e.g., tree shrew, some bats), away from the mesometrium (antimesometrial orientation, e.g., shrews and primates), or toward a side (lateral orientation, e.g., some bats and insectivores). (Middle row) Centric implants (left) occupy the entire lumen of the uterus (e.g., rabbit). Eccentric implants (right) occupy a portion of the surface (e.g., most primates) or a uterine crypt (myomorphic rodents), in which case they are overgrown by the edges of the crypt to become secondarily interstitial. (Bottom row) Three placental types in primates. The tarsier trophoblast attaches abembryonally (left); many New World and Old World monkeys attach embryonally and become partly interstitial (center); hominids attach embryonally and become totally interstitial (right). Diagonal hatching, epithelial lining of uterus; stippling, endometrium (myometrium and perimetrium may be considered the outer line); squares, ICM or epiblast; bricks, cytotrophoblast (trophectoderm or extraembryonic somatopleure); solid, syncytial trophoblast. Adapted from W. A. Wimsatt, *Biol. Reprod.,* 12:41(1975); and W. P. Luckett, in M. K. Hecht, P. C. Goody, and B. M. Hecht, eds., *Major patterns in vertebrate evolution,* Plenum Press, New York, 1976.

time. But orientation is as constant in monotocous species with single births as in polytocous species with multiple births. What adaptive advantage one orientation has over another is not at all obvious.

PERI-IMPLANTATION PERIOD

The peri-implantation period is characterized by labile initial attachments. At the beginning of the period, embryos become apposed to the uterine lining, but initial points of attachment, while specific and direct, may not be permanent. At the end of the period, attachment sites are stabilized for the duration of gestation and implantation commences.

Embryos adhere or attach in one of three **implantation positions: centric** (also central), **eccentric** (toward one side), or **interstitial** (buried within the uterine wall). The initial points of attachment may be distributed **circumferentially** (i.e., diffusely around) or they may be localized, in which case they may be **mesometric** (i.e., toward the mesometrium), **antimesometric** (i.e., away from the mesometrium), or **lateral** (i.e., toward a side).[3] Localized points of attachment may be at the **embryonal** pole, at the opposite **abembryonal** pole, or at intermediate **paraembryonal** (lateral) or circumferential sites (see Blüm, 1986).

Embryos in the centric position (Fig. 24.26 middle left) lie in and may fill the uterine lumen or even press against the uterine wall, as in the case of rabbits. The large expanded blastocysts of rabbits and sciuromorphic rodents (i.e., squirrel-like) attach abembryonally to the antimesometrial side of the uterus. The mural TE may then develop specialized **attachment cones** (Fig. 24.27 top row).

The small centric blastocysts of insectivores and bats attach circumferentially or laterally. In some species, zones of attachment expand circumferentially as growth in the uterine epithelium brings it partially over the embryo.

The elongated blastocysts of ungulates with their prolonged preimplantation periods lie centrically and ultimately form broadly distributed attachments. The noninvasive TE does not penetrate the uterine epithelium, and attachments are superficial.

The attachment of other embryos may involve penetration of the endometrial epithelium and invasion of the uterine mucosa. The trophectoderm of these embryos may form specialized invasive

_____
[3]Note that the same terms are used to identify embryonic orientation as well as attachment sites.

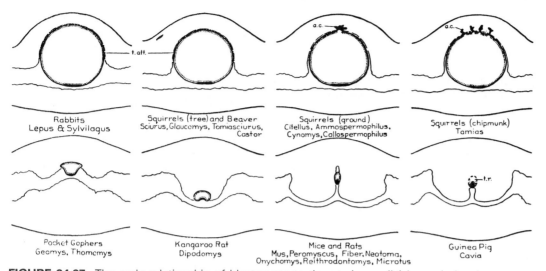

FIGURE 24.27. The early relationship of blastocysts to the uterine wall (shown in longitudinal section) of rabbits and representative rodents. Primary attachments between trophectoderm (omphalopleure) and uterine epithelium are abembryonic and antimesometrial. Final allantoic attachments and definitive placentas (not shown) are embryonal and mesometrial. The mesometrium is at the bottom. Large blastocysts (top row) implant centrally with their embryonal pole oriented away from the antimesometrial side. Small blastocysts (bottom row) implant eccentrically in uterine crypts of the antimesometrial side. In squirrels and chipmunks, the mural trophectoderm develops attachment cones (ac). Dashed line, endoderm. From H. W. Mossman, Comparative morphogenesis of the fetal membranes and accessory uterine structures, *Carnegie Inst. Washington Contrib. Embryol.* 24:129 (1937). Used by permission.

structures and may be overgrown by uterine tissue in a process called **nidation,** resembling wound healing.

Small rodent blastocysts take up eccentric positions (Fig. 24.26 middle right) antimesome-

trically in the uterus. Blastocysts attach within **uterine crypts** or **implantation chambers** (Fig. 24.28) by abembryonic attachments.

The TE of most myomorphic rodents (mice and rats) becomes a prominent **invasive organ** and

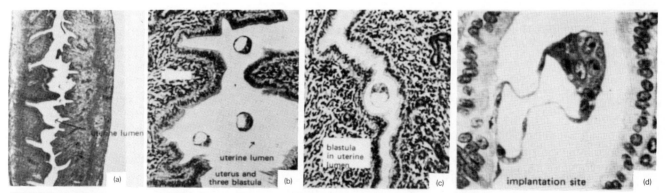

FIGURE 24.28. Photomicrographs of sections through one horn of the mouse's bipartite uterus containing blastocysts during the peri-attachment period (a–c) and initial peri-implantation period (d). (a, b) Longitudinal section through gravid uterus at 3.5 days. Uterine crypts cut deeply into the wall of the uterus. The endometrial epithelium lines the lumen and crypts of the uterus. (c) The blastocyst has entered a crypt where it will subsequently implant. (d) The blastocyst has attached to the crypt epithelium by the mural trophectoderm (compare to Figs. 24.24 and 24.25). From R. Rugh, *Experimental embryology,* Burgess, Minneapolis, 1962. Used by permission.

penetrates the endometrium, while the TE of hystricomorphic rodents (e.g., guinea pigs) survives just long enough to begin the endometrial invasion before disintegrating. In both cases, the embryo proceeds to replace its initial abembryonal points of attachment with **embryonal contacts.**

The new contacts represent the incipient placenta (P, Fig. 24.18). As an invasive organ, it bores a mesometrial path as if trying to escape from the uterine crypt, but the thickened endometrium (i.e., the deciduoma) blocks the retreat. The uterine tissue responds with nidation, enclosing or partially enclosing the embryo in a **trapped** or **secondarily interstitial** position.

In the **partially interstitial** variation of nidation found in many Old and New World monkeys (Fig. 24.26 bottom row center), embryonal invasion erodes the uterine epithelium, but penetration is limited and the embryo comes to lie only partially within the uterine wall. Unlike other primates, the tarsier's initial attachments are abembryonic[4] (Fig. 24.26 bottom row left) and a partially interstitial position is achieved by continued abembryonic invasion.

The interstitial position of hominid embryos (Fig. 24.26 bottom row right) results from rapid penetration of the uterine lining following attachment. After hatching, the embryo exhibits polar embryonal stickiness for a brief adhesive phase, but invasive sites in the polar trophectoderm quickly break through the epithelial lining of the uterus and invade the endometrium. Nidation follows and the entire embryo is enclosed in an interstitial position within uterine tissue (see Figs. 24.10–24.16).

Following the initial penetration of the uterine lining, a decidual reaction begins in the uterine stroma near the blastocyst. Glycogen and lipids accumulate, cells enlarge, vascular permeability increases, and glands become active. The area of penetration and the decidual reaction spread as the blastocyst migrates interstitially, sometimes as far as the myometrium.

IMPLANTATION AND PLACENTATION: CONTACTS FOR THE DURATION

The fetal membranes of monotremes and marsupials exhibit a general pattern of evolutionary

---

[4]The situation with the tarsier is presently ambiguous. The blastocyst's orientation is sometimes described as lateral. The placenta is then considered equatorial or paraembryonal.

conservatism, whereas those of eutherians present considerable variability in their morphogenesis. This appears to be related to the reduction of intraovular yolk and the acquisition of true viviparity in all eutherians.

W. Patrick Luckett (1976, p. 460)

Implantation in therians, or the process of attachment of the embryo to the uterine wall, requires contact, but the degree of intimacy involved varies according to species and regions of the embryo. Contacts may be mere zones of interdigitating surfaces, but frequently contacts are achieved following embryonic penetration of the uterine surface (see below, Fig. 24.31).

Penetration

Various devices of penetration determine the type of contact ultimately achieved. **Intrusion penetration** is the most common type of penetration. Several areas are involved simultaneously depending on the amount of blastocyst surface available, and the penetration can be superficial, as in the ferret and horse, or profound, as in the guinea pig and primates (Fig. 24.29).

In the horse, **trophoblastic tongues** penetrate, engulf, and phagocytize uterine epithelial cells rather than move between them, but surrounding cells are not initially damaged. Trophoblast cells proceed to migrate into the uterine stroma where they differentiate as **cups** of endocrine cells secreting PMSG. Typically, but not necessarily, initial attachments are replaced by more tenacious and intimate **placental bonds** which collectively constitute the placenta.

In the ferret, tongues from **ectoplasmic pads** of the syncytial trophoblast penetrate the uterine epithelium without extensively damaging the tissue. Replacing uterine epithelial cells at junctional complexes and desmosomes, the ectoplasmic pads penetrate to the basal lamina and subsequently to the stroma and vessels (Fig. 24.29c).

In **fusion penetration,** cell membranes between TE cells or syncytial trophoblast and uterine epithelium disappear. The maternal and embryonic nuclei are left in a **merged** but not mixed cytoplasm.

As seen in the rabbit, abembryonal **syncytial trophoblastic knobs** (pegs or processes) penetrate the extracellular coat still present on the trophoblast and fuse with the apical surfaces of uterine epithelial cells (Fig. 24.29b). The knobs reach the uterine basal lamina and, as development proceeds, the underlying stroma and vessels. Later, the uter-

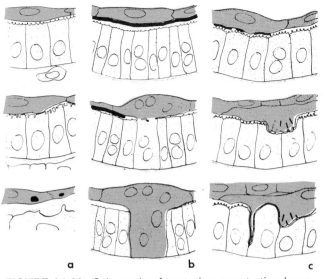

a   b   c

FIGURE 24.29. Schematics for uterine penetration by an adherent trophoblast. The trophoblast is on top. (a) Displacement penetration in rat, mouse, possibly hamster, and vespertilionid bats. Flattened trophoblast cells devour sloughed uterine cells while replacing them along the uterine basal lamina. (b) Fusion penetration in rabbit. Uterine cells merge with syncytial trophoblast of embryo. (c) Intrusion penetration by ferret, etc. Projections from syncytial trophoblast move between uterine epithelial cells without extensive destruction or alteration of adjacent cells of the epithelium. From S. Schlafke and A. C. Enders, *Biol. Reprod.,* 12:41(1975), by permission of the Society for the Study of Reproduction and courtesy of the authors.

ine epithelium fuses into a maternal **symplasma,** and, as the trophoblast's extracellular coat disappears, areas of merging broaden between the syncytial trophoblast and the maternal symplasma.

In **displacement penetration,** seen in the mouse and rat, the invading embryonal trophoblast undermines the adjacent uterine epithelium by sending projections between cells and their basal lamina. The embryonal trophoblast then phagocytizes the dislodged or sloughed uterine epithelial cells (Fig. 24.29a).

Specialized embryonic cells are generally involved in making uterine contacts and penetration. For example, in mice at about 4.5 days p.c. (a day after implantation commences), the mural TE cells no longer divide. Their nuclei undergo DNA endoreduplication and become polyploid. If placed artificially in contact with a 3.5-day ICM, the polyploid cells can still proliferate, but shortly afterward the cells become irreversibly committed to differentiation as nonproliferative **primary giant cells** (GC in Fig. 19.4) (Gardner et al., 1973). Surrounding the egg cylinder, the primary giant cells seem to promote erosion in the adjacent uterine epithelium while stimulating the deciduoma reaction and thickening of the endometrium.

In contrast, the polar TE cells continue dividing and form a thickened **ectoplacental cone** (EPC in Fig. 19.4). Peripherally, EPC cells are transformed into **secondary giant cells** which spread abembryonally and are indistinguishable from primary giant cells of the mural TE, but centrally, the EPC cells undergo a more dramatic change.

In a reversal of morphology, the rodent embryo replaces its initial abembryonal point of uterine attachment with contacts at the embryonal ectoplacental cone. The EPC becomes vascularized following contact with the growing allantoic stalk and, as the new **chorioallantoic membrane,** develops into the embryonic part of the placenta (Fig. 24.17b).

Other specializations occur where maternal–fetal intimacy is most profound in Old and New World primates. Membranes between abembryonic TE cells break down, and a **syncytial trophoblast** (or syncytiotrophoblast) appears with an active **ectoplasm** facing the uterine epithelium (Schlafke and Enders, 1975). The polar embryonal TE of primates probably becomes syncytial at the beginning of implantation. Beneath the invading syncytium, a **cytotrophoblast** of TE cells is also involved in adhesion.

In primates, the syncytiotrophoblast is elevated by vascular projections known as villi (see below), while the cytotrophoblast forms **extravillus attachments** (between villi) with uterine connective tissue cells. Similarly, in rodents, a thin layer of **labyrinthine trophoblast** separates pools of maternal blood from fetal circulation, while a **spongiotrophoblast** or spongy zone of fetal trophoblast cells makes contact with maternal decidual cells. Only the extravillus and nonvillus parts of the cytotrophoblast (i.e., parts where villi fail to form or disappear after an initial appearance) and spongiotrophoblast cells generally express class I MHC antigens (see above), while the syncytiotrophoblast and labyrinthine trophoblast do not.

Mammalian Placentas

Any area of relatively permanent contact between maternal and embryonic or fetal tissue can be called a placenta (Lat. flat cake referring to the human placenta) whether limited to the embryo's yolk sac, chorion, and/or allantois. Metatherians generally employ their trilaminar choriovitelline yolk sac as a **choriovitelline placenta** (Fig. 24.6b), and bandicoots, the koala, and wombats also employ the allantois. The eutherian placenta always begins with the embryo's trophectoderm and somatopleuric chorion (see van Tienhoven, 1983).

Maternal–fetal exchange in eutherians may take place through the allantois alone (Fig. 24.17, e.g., rabbits, rodents, primates), both the yolk sac and allantois (Fig. 24.1, prosimians; Fig. 24.7b, insectivores), or, typically, through the allantois after the yolk sac (Fig. 24.7a,c). The changes require the movement of some attachments and the breakdown of others. In the horse, for example, a trilaminar region of yolk sac combined with a vascular chorion displaces a bilaminar yolk sac combined with an avascular region of chorion only to be displaced by a **chorionic girdle** combined with the vascular allantois (Fig. 24.30).

Three types of chorioallantoic placenta are found among eutherians varying in the degree of uterine penetration (see Amoroso, 1981) (Fig. 24.31). In the **epitheliochorial** or **contact placenta,**

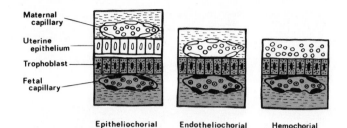

FIGURE 24.31. Diagrams of uterine and chorioallantoic vessels and tissues in three types of chorioallantoic placenta. From W. P. Luckett, in M. K. Hecht, P. C. Goody, and B. M. Hecht, eds., *Major patterns in vertebrate evolution,* Plenum Press, New York, 1976. Used by permission.

a noninvasive chorioallantoic membrane is plastered against the uterus' endometrial epithelium. In the **endotheliochorial placenta,** the endometrial epithelium and part of the endometrial connective tissue is eroded, and the chorioallantoic membrane achieves contact with the endothelial linings of maternal blood vessels. In the **hemochorial placenta,** an initially invasive chorioallantoic membrane erodes the endometrium right through the capillary walls, and **chorionic villi** lie in **lacunas** filled with maternal blood.

The embryonic side of contact placentas is a simple epithelium of cytotrophoblast cells. The chorionic vesicle (i.e., a chorion dilated by the underlying allantois, Figs. 24.7c and 24.22) comprises the embryonic portion of the placenta.

Two types of contact placenta are easily recognized. In the **diffuse placentas** (Fig. 24.32) of horses, pigs, camels, dolphins, whales, and prosimians, points of contact are small and evenly distributed. In the **cotyledonary placentas** of ruminating ungulates such as cows and sheep, the points of contact are concentrated in **placentomes** consisting of fetal **cotyledons** and preformed uterine **caruncles.**

In both endotheliochorial and hemochorial placentas, a syncytial trophoblast may cover a cytotrophoblast, at least in the early placenta. The portion of the chorion most conspicuously specialized for exchange is generally recognized as the placenta while the remainder of the chorion retains the title of chorion.

A **zonary placenta** is the moderately invasive endotheliochorial placenta typical of carnivores (Fig. 24.32) but also found in elephants and related orders. In carnivores, the early yolk sac placenta is gradually replaced by an annular chorioallantoic placenta as the chorionic vesicle is completed (Fig. 24.7a). Blood accumulates centrally or at the edge of the placenta in a **hemophagous** (i.e., blood-

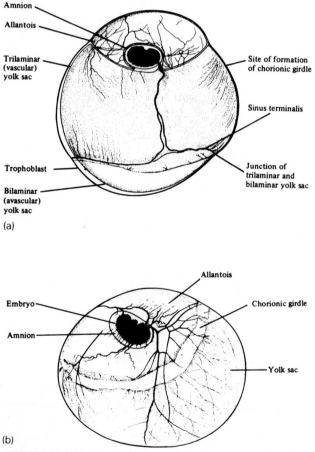

(a)

(b)

FIGURE 24.30. Whole mounts of the horse chorionic vesicle at 25 (a) and 35 (b) days. From M. B. Renfree, in C. R. Austin and R. V. Short, eds., *Reproduction in mammals,* Book 2, *Embryonic and fetal development,* 2nd ed., Cambridge University Press, Cambridge, 1982. Reprinted with the permission of Cambridge University Press.

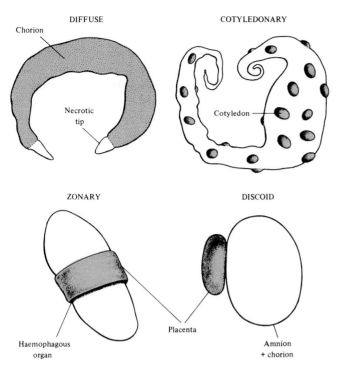

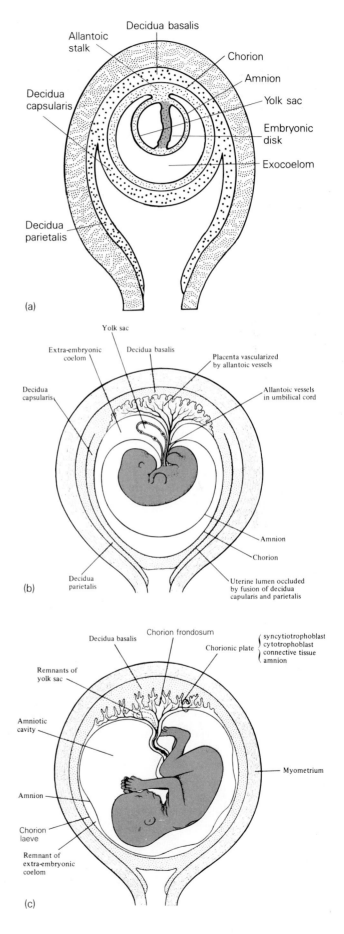

FIGURE 24.32. Diagrams of the four main types of placenta formed by eutherians. From M. B. Renfree, in C. R. Austin and R. V. Short, eds., *Reproduction in mammals,* Book 2, *Embryonic and fetal development,* 2nd ed., Cambridge University Press, Cambridge, 1982. Reprinted with the permission of Cambridge University Press.

eating) organ which, in dogs, has a bright green margin and a red attachment zone.

A **villous placenta** forms in primates as the invasive embryonic surface sends tufts of **chorionic villi** into lacunas excavated in the uterine lining (see Figs. 24.8–24.16). The definitive **discoid placenta** forms as the villi are confined to a **chorionic plate** (Figs. 24.33). Similar discoid placentas are formed by rabbits, rodents (Figs. 24.17), and insectivores (Fig. 24.32).

FIGURE 24.33. A. Diagrams of human embryo (a, b) and fetus (c) showing arrangement of extraembryonic membranes in relation to the decidual portions of the uterus. (a) Chorionic villi form all around the chorion before the decidua capsularis has stretched too far. (b, c) Only the portion of the chorioallantoic placenta nearest the allantoic stalk continues to develop chorionic villi. The decidua capsularis reaches the decidua parietalis and fuses. The amnion expands and covers the umbilical cord (formerly allantoic stalk) and reaches the chorion. The splanchnopleuric yolk sac is gradually reduced and drawn into a remnant of the exocoel within the umbilical cord. Parts (b) and (c) from M. B. Renfree, in C. R. Austin and R. V. Short, eds., *Reproduction in mammals,* Book 2, *Embryonic and fetal development,* 2nd ed., Cambridge University Press, Cambridge, 1982.

In primates, one or more discoid placentas form. The blastocysts of New and Old World primates form one discoid placenta at their invading embryonal edge and possibly an additional discoid placenta on the opposite side. Hominids form only one discoid placenta.

In hominids, almost as soon as the embryo is buried in the uterine wall at the original site of implantation, allantoic mesoderm fuses with chorionic TE and forms solid chorionic villi around the entire chorion (Fig. 24.14). The villi become infiltrated by embryonic blood vessels and elongate as maternal blood fills lacunas.

The chorionic vesicle now grows rapidly, and the protruding part of the uterine wall forms a **decidua capsularis** (Fig. 24.33). Pressure in the underlying chorion presumably prevents the further development of villi and the **chorion laeve** (i.e., the smooth chorion) forms as the initial villi recede. Meanwhile, the chorionic villi near the point of origin of the allantoic stalk flourish and, concentrating in the chorionic plate, form the **chorion frondosum** (i.e., the leafy chorion). The adjoining portion of the uterus thickens as the **decidua basalis.**

The decidua capsularis stretches until it reaches across the entire uterine lumen and fuses with the remaining lining of the uterus known as the **decidua parietalis** (i.e., the decidua wall). Within the chorionic vesicle, the amnion also fuses with the inside of the chorioallantoic membrane. Thus, the membrane whose rupture heralds birth in human beings is actually a combined decidua capsularis, chorion laeve, and amnion. The discoid placenta delivered in the third stage of labor is the decidua basalis and the chorion frondosum, but the fused membranous decidua capsularis and decidua parietalis are also shed at this time.

Placentas in Polyembryony and Twinning

Polyembryony is a mode of modifying reproduction resulting in the development of two or more so-called **identical** offspring from a single egg. The term was coined for plants forming many individuals from the same ovule, but the phenomenon is also found, if rarely, in insects (certain genera of Hymenoptera, e.g., the parasitic cutworm *Copidosoma truncatellum*) and is common in the edentate genus *Dasypus*.

The nine-banded Texas armadillo, *Dasypus novemcinctus*, gives birth to four embryos from a single fertilized egg (Fig. 24.34), and the twelve-banded armadillo, *Dasypus hybridus*, generally forms eight monozygotic offspring (i.e., identical quadruplets and octuplets) with a range of seven to twelve. The offspring have their own allantoic attachments to discrete but fused placentas and each embryo occupies a separate chorionic chamber.

Human beings also give birth to **identical** or **monozygotic twins** (sometimes referred to as MZ twins). Unlike dizygotic twins (DZ, two zygotes produced by separate fertilizations, see Table 7.10),

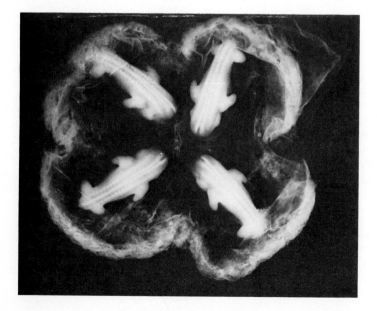

FIGURE 24.34. Dorsal view of polyembryonal (identical) quadruplets of the Texas nine-banded armadillo, *Dasypus novemcinctus.* Each embryo lies within its own chorionic sac (opened for the purpose of photography) and has a discrete portion of a conjoined placenta. From K. Benirschke, in R. M. Wynn, ed., *Fetal Homeostasis, New York Acad. Sci.,* 1:237(1965), by permission and courtesy of K. Benirschke and the New York Academy of Sciences.

monozygotic twins occur with a relatively constant frequency of about 36 incidents per 10,000 births independently of local or ethnic background (MacGillivray et al., 1975). Monozygotic twinning in humans therefore does not seem to be determined genetically or environmentally so much as stochastically (i.e., at random).

Dizygotic twins generally develop separately but may have anastomoses between their placentas and even chorions depending on the degree of fusion occurring *in utero*. The transfusion of blood during development is not uncommon and may result in the development of blood type chimerism as well as birth defects (e.g., free martins in cattle).

Given that the chorion is formed from the TE and the amnion from the ICM, the point in development at which a single embryo becomes doubled can be estimated from the number of chorions, amnions, and the independence of embryos (Fig. 24.35). Some 25–30% of human monozygotic twins are due to early blastomere separation (type 1). Another 70–75% of human monozygotic twins arise from the division of the ICM (type 2). Only about 1% of monozygotic twins are due to later embryonic separation (type 3). Separations occurring later than the primitive streak are likely to be only partial and result in conjoined twins (type 4; see Hamilton and Mossman, 1976).

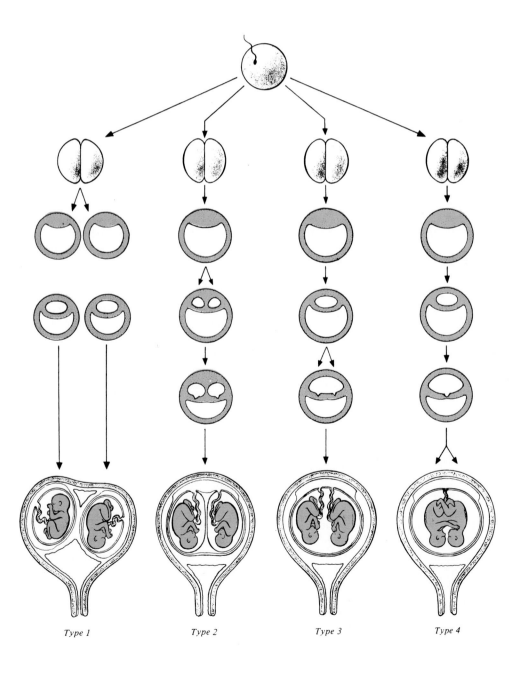

FIGURE 24.35. Mechanisms of monozygotic twinning and the arrangement of amnions, chorions, and discoidal placentas in different types of twins. Type 1: Dichorionic twins. Blastomeres separate 1–3 days after fertilization (also condition of dizygotic twins). Type 2: monochorionic diamniotic twins. The ICM forms two separate axes 3–8 days after fertilization. Type 3: monoamniotic twins. The blastodisk forms two separate axes 8–12 days after fertilization. Type 4: conjoined twins. At later stages, the formation of excess embryonic axes results in only partially separated twins. From M. B. Renfree, in C. R. Austin and R. V. Short, eds., *Reproduction in mammals,* Book 2, *Embryonic and fetal development,* 2nd ed., Cambridge University Press, Cambridge, 1982. Reprinted with permission of Cambridge University Press.

Type 1   *Type 2*   *Type 3*   *Type 4*

PART 8 SUMMARY AND HIGHLIGHTS

Palingenesis is the development of broad taxonomic characteristics, such as body plan, in embryos belonging to the same phylum. Cenogenesis is indicated by embryonic adaptations. Controls over palingenic development, which tend to be systemic, may differ fundamentally from controls over cenogenic development, which tend to be local.

General homologies reflect evolution among species, while serial and sexual homologies reflect evolution in development. Different genetic mechanisms may underlie both types of homology.

Spiralians and arthropods have evolved streamlined development and have specialized in intrinsic controls through truncated cytoplasmic–genetic interactions. The linearly equivalent progenies of founder cells in nematodes and teloblasts in annelids and mollusks generally differentiate along progressively narrowing lines, although positional information may continue to play some role. The rigidity of genetically "hardwired" circuits of determination in spiralian species with direct development may even exceed mechanisms of cytoplasmic determination in species with indirect development.

In *Caenorhabditis elegans*, an initial proliferative phase of about 5 hours provides all the cells but not the form of the round worm, while a second embryogenic phase of the same duration provides form without adding cells. With the exception of the E, D, and P_4 lineages, founder cells give rise to cells of different histotypes, but polyclonal compartments are established early.

The embryo develops folded and twisted around a cleft. The hypodermis, descended from the AB and C founder cells, cements the embryo in its wormlike shape by secreting the cuticle prior to hatching. Differentiation and cell death follow stereotypic lines. Body muscles are generated mainly by the MS, C, and D founder cells, while the AB founder cell generates pharyngeal, intestinal, and rectal muscles. The bulk of nerve cells are from the AB lineage, and all the intestinal precursor cells are from the E lineage. Maternal mutants that disturb the timing of mitosis and heterochronic zygotic mutants that cause temporal displacements of differentiation suggest that development depends on division rather than lineage or regional differences.

In insects, a posterior differentiation center or growth zone may produce segments, or the basic longitudinal body plan, possibly including the imaginal disks of endopterygotes (holometabolous insects), may be established in the germ band. In the fruit fly, *Drosophila melanogaster*, transient intersegmental boundaries appear during gastrulation followed by permanent intersegmental boundaries during the period of germ band shortening. Within 1 day of oviposition, dorsal closure brings the protruding yolk sac under the ectoderm, head involution brings the cephalic segments under the atrium, internal structures differentiate, imaginal disks are segregated from placodes, the cuticle is laid down and embroidered with its denticle pattern, and the first instar hatches.

The concentration of *Antennapedia* and *bithorax* homeotic genes in complexes suggests that genes affecting segmentation and differentiation within segments communicate with each other in genetic networks. Genetically engineered transformants bearing an extra copy of the *Antp* gene mimic the mutant phenotype when the gene is activated during development, suggesting that dominant gain-of-function mutants may compete with recessive loss-of-function mutants for segment determination.

The presence of cross-hybridizing sequences within several homeotic genes in *Drosophila* originally suggested that these homeoboxes were functionally tied to segmentation. A broader interpretation is required to explain the presence of similar sequences in other *Drosophila* genes and related homeobox-encoding domains in the genomes of a wide range of segmented, partially segmented, and unsegmented animals, in yeast genes for mating type antigens, and in viral and bacterial genes for repressor protein. Homeoboxes of about 180 bases encode homeodomains, putative DNA-binding domains of a helix-turn-helix motif.

Found in some mammalian transcription factors and *Drosophila* developmental control proteins, POU domains (named for Pit, Oct, and UNC regulatory proteins) contain POU-related homeobox subdomains, sharing amino acids with prototypic *Drosophila* homeodomains and POU-specific subdomains or POU boxes. The latter subdomain may aid the former in sequence recognition or in the activation of transcription.

Caution should be exercised before identifying similarities among homeodomain-containing proteins and other molecules as examples of general homology. Without independent evidence of evolutionary lineage, attributing molecular similarities to evolutionary inertia is tantamount to molecularizing the biogenetic law.

In contrast to spiralian and arthropod development, vertebrate embryogenesis is an exercise in plasticity, cellular pluripotentiality, extrinsic controls, and induction. While development may be efficient (e.g., in birds, large-egged frogs), heterochrony (retardation and promotion), neoteny, and the development of paedomorphs are widespread.

The vertebrate body takes its shape during the neurular and tailbud stages largely as a result of folding: neural folds bring neural epithelium and neural crest tissue inside; body folds separate the embryo proper from the yolk sac. The closing of neuropores, marking the transition between the two stages, is retarded in mammals. The prolonged presence of a neurenteric canal in *Xenopus laevis* and selachians illustrates continuity among the germ layers in the tailbud.

Mesoderm is distributed in three zones. The lateral plate splits to form the coelom and mesothelia lining the body cavity, mesenteries, and organ systems. Intermediate mesoderm gives rise to nephrogenic cords that induce the formation of nephric tubules as they course caudally and become the nephric ducts. Paraxial mesoderm, on either side of the prechordal plate and notochord, gives rise to somitomeres and, except in the head, to somites. Somitomeres and somites produce the same types of tissue, but somites undergo compaction and epithelialization to become dermomyotomes and sclerotomes.

The dermomyotome separates into the dermatome, which produces much of the cutaneous connective tissue, and myotome, which produces the voluntary epaxial and hypaxial musculature of the body wall and limbs. The sclerotomes of each somite deepithelialize into denser caudal and looser cephalic mesenchymal masses. The arrangement of masses allows neural crest cells to migrate and form dorsal root and sympathetic chain ganglia. Fusion of caudal and cephalic masses from consecutive somites produces spinal centra connected by myotomes.

Extraembryonic membranes generally approach radial symmetry even when produced by bilateral organisms. In the near absence of yolk, the therian mammalian embryo rushes to form extraem-

bryonic membranes and spaces through which it acquires nutrition from uterine contents and maternal tissue. While gastrulation trails in the embryo proper, extraembryonic mesoderm develops rapidly in conjunction with the trophectoderm of the eutherian blastocysts, and splanchnopleuric yolk sacs protrude from the gut.

A yolk sac is identified in anamniotes as well as amniotes, but its strict homology in teleosts and other vertebrates is questionable. Evolutionary novelties in amniotes' yolk sacs include the bilaminar omphalopleure, retained as the proamnion, the trilaminar choriovitelline membrane, and the definitive splanchnopleuric yolk sac. Angiogenesis and blood formation begin in the yolk sac as soon as its endoderm is fused with splanchnic mesoderm.

The amnion and chorion are somatopleures separated by the exocoel. They are made during reptilian and prototherian amniogenesis by ectamnionic folds and cavitation following fusion with somatic mesoderm. Avian and metatherian amniogenesis proceeds from chorioamniotic folds.

The eutherian trophectoderm is the primitive extraembryonic membrane. It becomes the chorion after fusion with somatic mesoderm. In eutherians with massive yolk sacs (e.g., ungulates), an embryonic disk surfaces and the chorion folds over the embryo in a birdlike manner, but in other eutherians, cavitation opens up the amniotic cavity between trophectoderm and embryoblast. A fold between embryonic and extraembryonic portions of ectoderm may separate the amniotic cavity from the exocoel (rodents) or further extraembryonic cavitation may excavate the exocoel between the amnion and chorion (primates).

The amnion is the extraembryonic membrane closest to the embryo proper. Its lining of extraembryonic ectoderm faces the embryo and its layer of extraembryonic somatic mesoderm turns away from the embryo. In reptiles, birds, and metatherians, the amniotic cavity has regions of ectamnion and proamnion depending on the success or failure of mesoderm to penetrate between primary germ layers, cavitate, and expand.

The chorion is farthest from the embryo. Its inner layer of extraembryonic somatic mesoderm faces the exocoel and frequently fuses with allantoic splanchnic mesoderm.

In birds, reptiles, and mammals with dilated chorions (e.g., ungulates), a splanchnopleuric allantoic vesicle arises from an evagination of the hindgut.

The allantoic cavity functions in the storage of metabolic wastes. In other eutherians, the allantois is reduced to a solid stalk of extraembryonic mesoderm. Both the vesicular and solid allantois invade the exocoel and expand circulation in the chorion. The new chorioallantoic membrane functions as a respiratory organ in shelled amniotes and as the chief organ of maternal–fetal exchange in therian mammals.

The integration of the mammalian mother and fetus is profound both in preparation for pregnancy and during pregnancy. The maintenance of the ovarian corpus luteum due to hormonal feedback from the blastocyst is the basis for maternal recognition of pregnancy and the inhibition of embryonal rejection. Luteal activity, especially progesterone production, maintains and supports pregnancy.

The trophectoderm or invading trophoblast may promote capillary permeability, secretory activity, and proliferation in the uterine stroma contributing to the decidua swelling. Despite the intimacy and duration of contact and the presence of major histocompatibility antigens on some embryonic cellular elements of the placenta, maternal immune responses to fetal tissue are not normally detrimental and may be productive.

Implantation in therian mammals is an adaptive, and hence variable, process. It differs in timing as well as pattern, in the duration of the preattachment period, and in the stages of development at attachment. Spacing mechanisms between and within horns of the uterus distribute embryos, and embryos orient to the uterine lining in species-specific patterns. Delayed implantation of new embryos occurs in marsupials with young attached to a pouch teat and in nursing mice and rats, as well as some other mammals.

A placenta is a zone of close and relatively tenacious attachment. Embryos taking up a central position in the uterus and forming epitheliochorial contacts form noninvasive contact placentas of diffuse or cotyledonary varieties (ungulates). Embryos penetrating the uterine lining to form endotheliochorial contacts may form circumferential zonary placentas (carnivores). Embryos taking up eccentric positions may attach to the luminal surface of the uterus (e.g., primates) or to the surface of uterine crypts (e.g., rodents) and form hemochorial placentas. Hominid embryos become interstitial by nidation, while rodent embryos already buried in crypts become secondarily interstitial.

Mechanisms of uterine penetration by embryonic tissue vary from displacement to phagocytosis but frequently involve the formation of syncytia. A syncytial trophectoderm may lead a cytotrophoblast into the endometrium. In rodents, nonproliferative primary giant cells formed by mural trophectodermal cells and polyploid secondary giant cells from the embryonal trophectoderm promote erosion and the decidual reaction in uterine tissue. Polar extraembryonic ectoderm proliferates into an ectoplacental cone, which is invaded by mesoderm to become a new chorion and rudiment of the placenta's embryonic portion.

In hominids, chorionic villi are initially equally well developed around the entire chorion, but gradually a discoidal placenta with a fetal chorion frondosum develops adjacent to a maternal decidua basalis, while elsewhere villi level out and a chorion laeve is surrounded by a decidua capsularis. Dilated by growth and the expansion of the amnion, the decidua capsularis stretches across the uterus and fuses with the decidua parietalis to complete the hominid placenta. The presence of conjoined placentas and chorions with separate amnions in the majority of identical twins and polyembryonal births suggests that embryonic doubling occurs in the embryoblast of a blastocyst.

PART 8 QUESTIONS FOR DISCUSSION

1. Discuss palingenesis as a concept of parallel development and as a concept of recapitulation. What are general, serial, and sexual homologies, and how might they be evaluated? Compare and contrast cenogenesis and palingenesis.

2. Describe intrinsic and extrinsic controls of embryogenesis, and give examples of each. Compare and contrast pluripotentiality and induction in the development of *Caenorhabditis elegans*, *Drosophila melanogaster*, and *Xenopus laevis*.

3. Describe the phenotypes of specific homeotic mutants, and propose a hypothesis about how normal genes might work in determining segmental differentiation.

4. What are a homeobox and the homeodomain? What are a POU domain, a POU-specific subdomain, and a POU-related homeobox subdomain? What is meant when a homeodomain or POU domain is said to be highly conserved? How might they work?

5. Describe the role of folding during vertebrate embryogenesis. How are the vertebrate central nervous system and integument separated? What is the egg cylinder and how is it formed?

6. Distinguish between somites, intermediate mesoderm, and lateral plate. Describe the relationship of somitomeres, somites, and the segmental plate. What are the parts of somites? How are they formed, and what are their fates?

7. What are the hormonal conditions of pregnancy? What does the maternal recognition of pregnancy mean, and what role does the embryo play in the maintenance of the ovarian corpus luteum?

8. What is decidual transformation? What is delayed implantation? What role, if any, does immunity play in pregnancy?

9. How is embryo spacing or positioning accomplished? Describe embryo orientation and initial attachment. Compare and contrast the timing of preimplantation, peri-implantation, and implantation periods in ungulates, rodents, and primates.

10. How is uterine penetration achieved in the horse, rabbit, and mouse? Distinguish between epitheliochorial, endotheliochorial, and hemochorial placentas, and between contact, zonary, and discoidal placentas.

11. Describe the discoidal placenta of hominids during development.

12. Define and describe (a) trophectoderm and chorion and (b) omphalopleure, proamnion, and ectamnion. Distinguish between (a) the choriovitelline yolk sac and the definitive yolk sac and (b) a choriovitelline yolk sac and a choriovitelline placenta. Compare and contrast amniogenesis by folding and by cavitation.

13. Compare and contrast (a) the fates of parietal endoderm and visceral endoderm in rodents and (b) the primary yolk sac and secondary yolk sac of primates. What are giant cells and the ectoplacental cone in rodents?

14. What characteristics of extraembryonic structures separate them from embryonic structures? How does the condition of the placenta reflect the mechanism of identical twinning?

PART 8 RECOMMENDED READING

Austin, C. R. and R. V. Short, eds., 1982. *Reproduction in mammals*, 2nd ed., Book 2, *Embryonic and fetal development*. Cambridge University Press, Cambridge.

Austin, C. R. and R. V. Short, eds., 1984. *Reproduction in mammals*, 2nd ed., Book 3, *Hormonal control of reproduction*. Cambridge University Press, Cambridge.

Bellairs, R., D. A. Ede, and J. W. Lash, eds., 1986. *Somites in developing embryos*. Plenum Press, New York.

Blüm, V., 1986. *Vertebrate reproduction, a textbook*. (Translated from the German by A. C. Whittle.) Springer-Verlag, Berlin.

Cell patterning. Ciba Foundation Symposium 29, new series, 1975. Associated Scientific Publishers, Amsterdam.

Davidson, E. H. and R. A. Firtel, eds., 1984. *Molecular biology of development*. Alan R. Liss, New York.

Edelman, G. M., ed., 1985. *Molecular determinants of animal form*. Alan R. Liss, New York.

Gehring, W. J., 1987. Homeo boxes in the study of development. *Science*, 236:1245–1252.

Glasser, S. R. and D. W. Bullock, eds., 1981. *Cellular and molecular aspects of implantation*. Plenum Press, New York.

Lewis, E. B., 1978. A gene complex controlling segmentation in *Drosophila*. *Nature (London)*, 276:565–570.

McLaren, A. and G. Siracusa, eds. 1987. *Recent advances in mammalian development. Current topics in developmental biology*, Vol. 23. Academic Press, Orlando.

Raff, R. A. and T. C. Kaufmann, 1983. *Embryos, genes, and evolution. The developmental–genetic basis of evolutionary change*. Macmillan, New York.

Schlafke, S. and A. C. Enders, 1975. Cellular basis of interaction between trophoblast and uterus at implantation. *Biol. Reprod.*, 12:41–65.

van Tienhoven, A., 1983. *Reproductive physiology of vertebrates*, 2nd ed. Cornell University Press, Ithaca.

Wimsatt, W. A., 1975. Some comparative aspects of implantation. *Biol. Reprod.*, 12:1–40.

Part 9

ORGANOGENESIS AND SPECIFICATION

Organogenesis (or organogeny) is the development of palingenic larval and adult organs from rudiments laid down in the post-gastrula. **Specification** is the acquisition of functional qualities characteristic of these organs. While the force behind organogenesis and specification appears to start abruptly toward the end of embryonic development, it actually reaches far back into ontogeny.

Ernst Haeckel suggested that this force also reached far back into phylogeny. He proposed that structures in early embryos of advanced species resembled structures in remote adult ancestors, and structures in late embryos of advanced species were comparable to structures in contemporary adults of primitive species. Using evolution as an analogy, he explained development with the concept of **terminal addition,** that ancestral adult stages were incorporated into the preadult stages of descendants.

A different analogy suggests an alternative concept. Like a finely tuned racing car, the embryo may be thought of as revving up its developmental engine during early development. At some unknown signal, the developmental clutch is released, and organogenesis and specification are thrown into gear. Primitive species begin the race at lower rpm than advanced species, but all species run the same race.

According to this alternative concept, the development of adult organs, like the organs themselves, is as adaptive and specialized in primitive species as in advanced species. Differences among species occur because organogenesis and specification take over and transform rudiments into adult organs earlier in primitive species than in advanced species.

In Part 9, organogenesis and specification in vertebrates are scrutinized with an eye toward distinguishing between the two concepts of development. Do palingenic organs in advanced species develop by attaching "add-ons" to organs in primitive species, or do primitive species shift up to organogenesis and specification with less development momentum than advanced species?

VERTEBRATE ORGANOGENESIS

Evolution has probably selected a variety of strategies to orchestrate the organization of embryos—some simple, some more complex—but each decidedly distinct. How much of cellular responses are restricted or preprogrammed into cells of various germ layer origins and how much is left to instructive interactions has yet to be ascertained. What is clear is that instructive induction is phylogenetically ancient and a general principle rather than a specific signal for each inductive event.

EDWARD J. KOLLAR (1983, p. 33)

VERTEBRATE DIGESTIVE SYSTEM AND DERIVATIVES

Vertebrates build the adult alimentary tract and associated organs by fusing ectodermally and endodermally derived parts. The tips of the tract are the ectodermally derived **stomodeum** or mouth and **proctodeum** or anus. The body of the tract has three endodermally derived sections: **foregut, midgut,** and **hindgut** (Figs. 23.4 and 25.1). While all portions of the tract perform alimentary functions, the foregut, especially its anterior end or embryonic **pharynx**[1] (Gk. throat), has journeyed furthest into other roles.

Ectodermally Lined Ends

Ectoderm and endoderm seem to wage a developmental tug-of-war for dominance of the oral

---

[1]Foregut is sometimes equated with the embryonic pharynx. Here, pharynx is equated to the anterior part of the foregut having pharyngeal pouches.

and anal ends of the embryo. In anamniotes, endoderm remains in the oral cavity at sufficiently high levels to become the enamel organs of the more posterior teeth. In amniotes, the endoderm is not as successful in maintaining its position, and no endoderm is left in the oral cavity of mammals.

At the tip of the embryonic mammalian alimentary tract, ectoderm pours over the **first visceral** or **mandibular arch,** around the **maxillary** (Fig. 25.2) and **mandibular processes,** and into the mouth. This ectoderm supplies all the epithelia for the lips, gums, dental ledges, and enamel organs of teeth (see Fig. 25.60). Ventrally, the ectoderm covers the tongue, and dorsally, ectoderm lines the lateral **palatine processes** or **shelves** and forms the secretory portion of parotid salivary glands. The only endodermal holdouts against the incursions of the ectoderm are the secretory epithelia of the sublinguinal and submandibular salivary glands (Fig. 25.3).

The tongue soon drops out of the way, and the palatine processes converge toward the midline where they fuse to form the **palate** (also called sec-

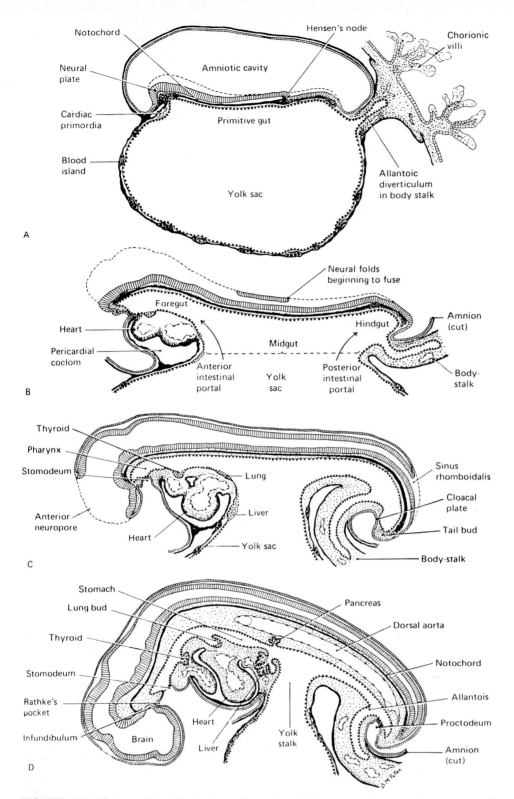

FIGURE 25.1. Schematic midsagittal sections through human embryos at four stages of development. [Stages and ages have been changed to correspond to Carnegie stages (i.e., Streeter's developmental horizons in human embryos) and are only approximate. See G. L. Streeter, *Carnegie Inst. Washington, Contrib. Embryol.* vol. 34, no. 230:167–following 196 (1951).] (a) At 19 days (stage 8), neurula: presomite embryo, neural folds lifting out of neural plate. (b) At 22 days, neurula (stage 10): 4–12 somites. (c) At 24 days, neurula (stage 11): 16 somites, mandibular and hyoid arches, heart prominence, head and tail somewhat bent. (d) Estimated 28 days, tailbud (pharyncula, stage 13). Characteristic **C** shape, about 30+ somites, 4 pairs branchial arches, leg buds appear, lens placode. From B. M. Carlson, *Patten's foundations of embryology,* 5th ed., McGraw-Hill, New York, 1988. Used by permission.

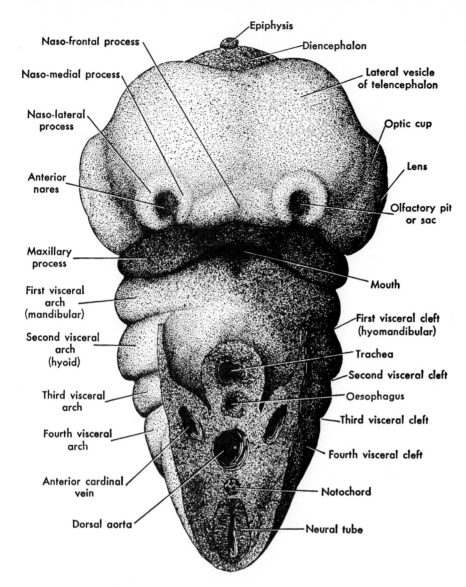

Epiphysis

Naso-frontal process

Diencephalon

Naso-medial process

Lateral vesicle of telencephalon

Naso-lateral process

Optic cup

Lens

Anterior nares

Olfactory pit or sac

Maxillary process

Mouth

First visceral arch (mandibular)

First visceral cleft (hyomandibular)

Second visceral arch (hyoid)

Trachea

Second visceral cleft

Third visceral arch

Oesophagus

Third visceral cleft

Fourth visceral arch

Fourth visceral cleft

Anterior cardinal vein

Notochord

Dorsal aorta

Neural tube

FIGURE 25.2. Cross section cervical region and ventral view of 96-hour chick embryo after removing heart and yolk sac to expose branchial arches. From A. F. Huettner, *Comparative embryology of the vertebrates,* rev. ed. Macmillan, New York, 1949. Used by permission.

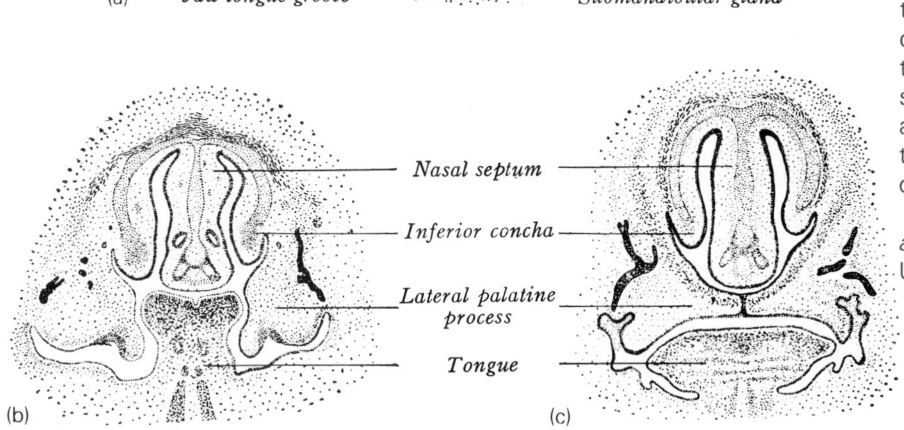

Roof of primitive mouth

Tongue

Palatine process

Parotid gland

Jaw-cheek groove (Vestibule)

Lower gum

Sublingual gland

(a)

Jaw-tongue groove

Submandibular gland

Nasal septum

Inferior concha

Lateral palatine process

Tongue

(b)   (c)

FIGURE 25.3. Diagrammatic frontal sections across jaws of human embryo. (a) The parotid salivary glands at about 2 months developed dorsally in stomodeal ectoderm at the angles of the palatine process and jaw–cheek groove, while the sublinguinal and submandibular glands develop ventrally in foregut endoderm at the angle of the tongue. ×15. (b) The tongue drops down as the nasal septum emerges on the midline. ×11. (c) The palatine shelves converge above the tongue and below the nasal septum to form the palate, dividing the original stomodeal cavity into nasal and oral cavities. ×11. From L. B. Arey, *Developmental anatomy,* Saunders, Philadelphia, 1965. Used by permission.

ondary palate when the inner edge of the maxillary processes is considered the primary palate). The stomodeum is thereafter divided into nasal cavity above the palate and oral cavity below.

Compared to the foregut, the hindgut endoderm is more successful in resisting the incursions of ectoderm. Vertebrate proctodeal ectoderm establishes itself internally only to the extent of invading the common terminal cavity of the hindgut and urogenital systems known as the **cloaca** (Gk. *klyzein* to wash out, hence, sewer) (see Fig. 25.36).

The cloaca is originally separated from the proctodeum by the **cloacal membrane,** consisting of fused hindgut endoderm and proctodeal ectoderm. The adult condition for most vertebrates is achieved when the cloacal membrane breaks down and the proctodeal ectoderm becomes continuous with the rectal endoderm.

In mammals, a **urorectal** or **perineal body fold** partitions the embryonic cloaca into a **urogenital sinus** and a **rectum** (Fig. 25.4). In both female and male mammals, the entire external apparatus of the urogenital system, including most of the vagina in females and the phallic urethra in males, develops beyond the urorectal fold coated with ectoderm. The rectum, or portion of the colon fused with the body wall, is covered by ectoderm only as far as rectal columns just within the anus (i.e., the level of the internal anal sphincter).

Endodermally Lined Center

The simple endodermal epithelium of the gut arises either from the wall of the **archenteron** after separation of mesoderm or from evaginations of the **embryonic endoderm** (also called definitive endoblast and tertiary hypoblast). The foregut and hindgut are defined by the lengths of these evaginations or by the location of organ rudiments.

The tip the foregut may briefly overshoot the mouth as a **preoral gut** (also known as Seessel's pouch in amniotes, SP, Fig. 25.8) which later disappears without trace. Ventrally, the foregut's farthest anterior extension juxtaposes the mouth (oral pit, Fig. 23.18). An **oral membrane** (or stomodeal plate) of fused ectoderm and endoderm initially separates the mouth and pharynx. When the membrane breaks down, the ectoderm and endoderm become continuous, although their fates are quite separate.

When the endodermal portion of the gut is formed by evagination, **intestinal portals** mark the openings (Fig. 25.1). The **anterior intestinal portal** (see Figs. 23.7, 23.10, and 23.12) gapes at the entry to the foregut; the **posterior intestinal portal** yawns at the entry to the hindgut (Fig. 25.1). The midgut contracts, and, as the proximal portion of the yolk sac narrows to a **stalk,** the midgut bends into a loop between the two intestinal portals (Fig. 23.4). The

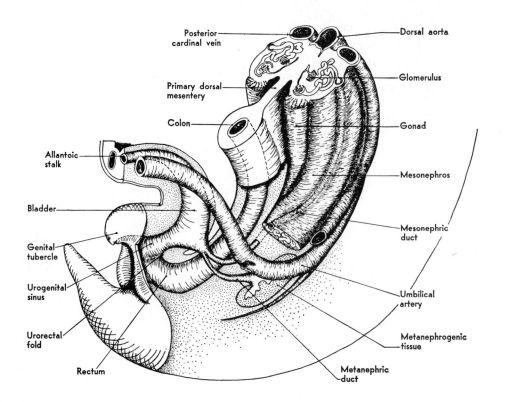

FIGURE 25.4. Reconstruction of posterior end of a generalized mammalian embryo (e.g., a pig at 14–15 mm) following removal of most of body wall to show structures within peritoneal cavity. From B. M. Carlson, *Patten's foundations of embryology,* 4th ed., McGraw-Hill, New York, 1981. Used by permission.

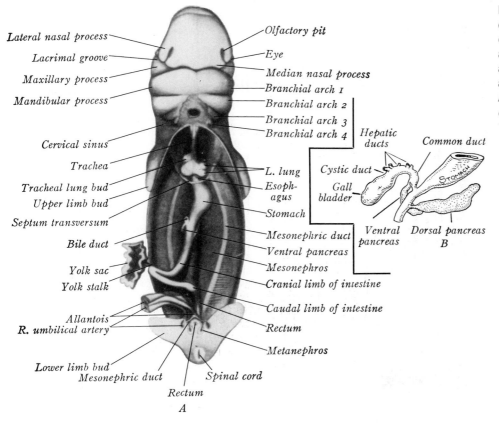

Lateral nasal process
Lacrimal groove
Maxillary process
Mandibular process

Cervical sinus
Trachea
Tracheal lung bud
Upper limb bud
Septum transversum

Bile duct

Yolk sac
Yolk stalk

Allantois
R. umbilical artery

Lower limb bud
Mesonephric duct

Rectum
A

Olfactory pit
Eye
Median nasal process
Branchial arch 1
Branchial arch 2
Branchial arch 3
Branchial arch 4

L. lung
Esoph-
agus
Stomach
Mesonephric duct
Ventral pancreas
Mesonephros
Cranial limb of intestine
Caudal limb of intestine
Rectum
Metanephros

Spinal cord

Hepatic
ducts
Cystic duct
Gall
bladder

Common duct

Ventral
pancreas
Dorsal pancreas
B

FIGURE 25.5. Ventral view of dissected 10 mm pig embryo exposing viscera and branchial arches after removal of heart and deflection of yolk sac and allantois to left. From L. B. Arey, *Developmental anatomy,* Saunders, Philadelphia, 1965. Used by permission.

intestine then develops as a tubular U drawn into the exocoel behind the yolk sac (Fig. 25.5).

The foregut is remarkably plastic (Table 25.1), *and it has a high degree of competence, but induction is required to give competence expression* (Fig. 25.6) (see Spooner, 1974). Foregut endoderm lines or forms the secretory epithelium of a variety of **foregut derivatives** or **appendages,** the thyroid and parathyroid glands, the thymus (but not its lymphocytes), post- (ultimo-) branchial bodies, pharynx beginning at the base of the tongue, trachea, lungs, esophagus, stomach, duodenal portion of the intestine, pancreas, liver, and gallbladder. In addition, sublinguinal and submandibular salivary glands (in mammals) and posterior enamel organs in salamanders are derived from foregut endoderm.

Mesenchyme (both of mesodermal and neurectodermal [i.e., neural crest] origin), which contributes the stroma or connective tissue of the foregut, provides the specific inductive influence directing and restricting the foregut endoderm (Fig. 25.7). As a rule, the stringency of inductive interactions diminishes in time, suggesting that instructive induction precedes permissive induction (see Gurdon, 1987).

TABLE 25.1. Foregut derivatives in mammals (and other vertebrates as indicated) at end of the tailbud stage (listed by branchial arches and somite groups)

Ectodermal	Endodermal	Ectomesenchyme	Mesodermal
	Mandibular (I)		
Pinna, external auditory meatus, external surface of ear drum, lip, body tongue (epidermis), parotid salivary gland, enamel organs, dental ledges	Lining of tympanic cavity, coating of middle ear ossicles and auditory tube, sublinguinal and submandibular salivary glands (enamel organs of posterior teeth in anamniotes)	Mandibular and maxillary cartilage quadrate and articular proc. (= Meckel's cartilage), 1st aortic arch (ventral carotid artery, sharks, degen. elsewhere) dental placodes & odontoblasts	Muscles of mastication, tensor tympani (innervated by V cranial nerve)

TABLE 25.1. (continued)

Ectodermal	Endodermal	Ectomesenchyme	Mesodermal
		Hyoid (II)	
Pinna of ear, neck epidermis	Lining of auditory tube, thyroid (approximately medial ventral diverticulum), palantine fossa (crypt tonsils), epithelium root of tongue, pharynx, epiglottis	Hyomandibular and ceratohyoid cartilage (= columella [amphibians, reptiles, birds], Reichert's cartilage [mammals]), styloid process, hyoid bone, second aortic arch (first afferent in sharks, degenerative elsewhere), GALT lymphocytes of tonsils	Muscles of facial expression (in humans), stapedius (innervated by VII cranial nerve)
		III	
Epidermis of middle neck	Epithelium root of tongue, pharynx, epiglottis Dorsal horn: parathyroid gland (inferior) except fish Ventral horn: thymus (thoracic in mammal)	Hyoid bone (body), third aortic arch (afferent and efferent in aquatic anamniotes, internal carotid elsewhere), paraendocrine and endocrine cells of parathyroid, ultimo-branchial body, lung and gastrointestinal musoca (?)	Muscles of pharynx (innervated by IX cranial nerve), lymphocytes of thymus
		IV	
	Epithelium root of tongue, pharynx, epiglottis Dorsal horn: parathyroid gland (superior) except fish Ventral horn: thymus (cervical portion in mammals)	Cuniform and thyroid cartilage, fourth aortic arch (systemic arches: postmetamorphic amphibians, reptiles; right: aortic arch birds (left: degenerative), left: aortic arch mammals (right: subclavian, otherwise degenerative)	
		V (rudimentary in tetrapods)	
	Laryngotracheal groove (separates into trachea and esophagus) Dorsal horn (does not develop in mammals) Ventral horn: post- (ultimo-) branchial body	Laryngeal cartilages corniculate, arytenoid, cricoid, fifth aortic arch (afferent in shark, degenerative elsewhere except urodeles)	
		VI (fused to body wall in tetrapods)	
		Tracheal cartilages, cricoid cartilage (laryngeal cartilages), sixth aortic arch (ductus arteriosus [disappears in reptiles, birds, and mammals], pulmonocutaneous in postmetamorphic amphibians; pulmonary reptiles, birds, mammals)	
		Occipital region of head (caudal to branchial arches)	
	(+ mesoderm = splanchnopleure) Lung buds and esophagus	Tracheal cartilages	Muscles of tongue (innervated by XII cranial nerve)
		Region of cervical somite	
	Liver and gallbladder, dorsal and ventral pancreas, duodenum	Pancreatic islet cells	Hemopoietic cells of liver

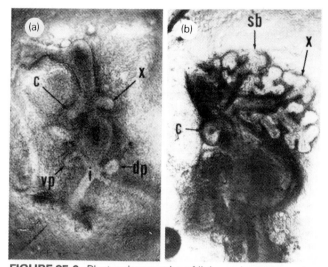

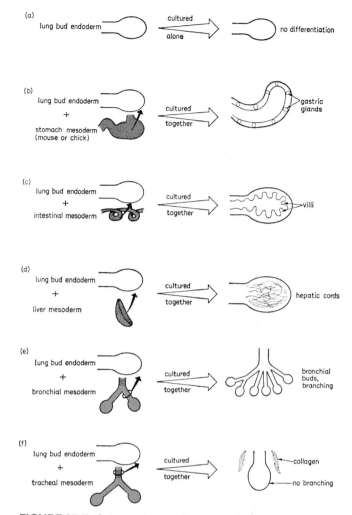

FIGURE 25.6. Photomicrographs of living cultures showing gut explants from 12 somite-stage mouse embryo. Compare to Fig. 25-7 e. (a) After 2 days in culture rudimentary gut and associated organs have appeared: dp: dorsal pancreas; i: intestine; s: stomach (rudimentary); vp: ventral pancreas; C and X: lung buds. (b) Same culture 4 days after replacing mesoderm of X (experimental) lung bud with fresh 11-day bronchial mesoderm. The C (control) lung bud has failed to branch, while the X lung bud has branched extensively. ×80. From B. S. Spooner and N. K. Wessels, *J. Exp. Zool.,* 175:445(1970). Used by permission of Alan R. Liss, Inc.

FIGURE 25.7. Schematic experiments and idealized results with isolated lung bud endoderm cultured alone and in combination with mesoderm from various sources. Endodermal differentiation requires mesodermal induction, but the lung bud endoderm does not differentiate as bronchial buds and branches (i.e., typical of lung) except under the influence of bronchial mesoderm. The endoderm is induced to take a pathway of differentiation corresponding to the mesoderm's source. From E. Deuchar, *Cellular interactions in animal development,* Wiley, New York, 1975. Used by permission.

Neurectodermal cells may also contribute cells to the epithelial components of foregut derivatives. In particular, a neurectodermal source has been hypothesized for paraendocrine (secreting local hormones) and endocrine (secreting systemic hormones) cells in pancreatic islets, parathyroid, ultimobranchial body, lung, and gastrointestinal epithelium or mucosa (Table 25.1) (Pearce, 1976).

The hypothesis of neurectodermal origins for pancreatic islet cells has been supported recently (Albert et al., 1988) by results with transgenic mice harboring molecular markers linked to insulin-promoter sequences. Single and double immunostaining of histological sections through embryos were used to detect the transgenic markers, pancreatic islet hormones (insulin, glucogon, somatostatin, and pancreatic polypeptide), and a **neuronal marker enzyme,** tyrosine hydroxylase (the first enzyme of the catecholamine biosynthetic pathway).

The assumption of the experiment was that common synthetic activities indicate common cellular origins. Not only was tyrosine hydroxylase synthesized in differentiating pancreatic endocrine cells, but the hybrid insulin gene was transcribed by migrating neural crest cells (sympathoblasts or precursors of sympathetic ganglia, see Chapter 26) and by proliferating neuronal cells in the basal portion of the developing neural tube.

These results suggest that pancreatic islet precursor cells and some neurectodermal cells are derived from the same precursor population. The common progenitor would be capable of activating the insulin promoter and specifically transcribing genes encoding the islet-associated hormones and tyrosine hydroxylase.

On the other hand, experiments with explants and transplants seem to rule out the possibility of neurectodermal sources for endocrine cells of the pancreatic islet and gastrointestinal mucosa. In the case of rats, pancreatic rudiments isolated prior to neural crest migration develop islets *in vitro*, and in the case of birds, quail neurectodermal transplants (see Chapter 26) fail to seed chick islets or the gastrointestinal mucosa. These results suggest that the paraendocrine and endocrine cells of foregut derivatives are of endodermal origin (see Le Douarin, 1988).

The midgut develops in conjunction with the yolk sac of amniotes, the yolky endoderm of amphibians (see Fig. 23.18), ***and the epithelial endoderm of fish*** (see Fig. 24.3). Anterior to the point of fusion with the yolk sac, the midgut becomes small intestine (jejunum and ilium). Posterior to the point of fusion with the yolk sac, the midgut quickly anastomoses with the large intestine formed by the hindgut.

Induction in the midgut operates in the direction of mesoderm as well as endoderm. The layer of **splanchnic mesoderm,** which fuses with the endoderm in the splanchnopleure, forms (1) the smooth mesothelium covering the viscera, (2) abundant connective tissue, and (3) layers of circular and longitudinal smooth muscle surrounding the gut. Moreover, beginning in the yolk sac (vascular area, angioblastema, and blood islands, Figs. 23.6–23.12) or in the ventral gut (hematoblastema, Figs. 23.18 and 23.20), endoderm induces **angiogenesis** (formation of blood vessels), **hemopoiesis** (formation of blood cells), and **lymphopoiesis** (the formation of lymphocytes) in the splanchnic mesoderm (Table 25.1).

Hemopoiesis moves to the liver, and angiogenesis spreads to the hindgut and hence to the chorioallantoic membrane or placenta via the allantois (Fig. 25.1). Lymphopoiesis and T-cell production commence in the foregut's thymic rudiments, and in birds, humoral-antibody-producing B cells are generated in a hindgut diverticulum known as the **bursa of Fabricius** (hence the "B" in B cells).

The hindgut is chiefly associated with the colon or large intestine. At its posterior end, the hindgut fuses with the body wall as the **rectum** and opens into the cloaca or, in mammals, through the anus directly to the outside. The hindgut also forms a **cecum** (Lat. *caecus* blind) or cecal diverticulum just posterior to the transition from the small intestine (the iliocolic junction). The cecum may be large (rabbits) or its posterior portion may be reduced to a vestigial **vermiform appendix** (human beings).

In mammals, the hindgut does not form a bursa nor play the role of source for B cells (supplied instead by bone marrow), but the mammalian gut still plays a role in lymphopoiesis. Lymphatic nodules of the **gut-associated lymphatic tissue** (GALT [Tab. 25.1], adenoids, tonsils, Peyer's patches of the small intestine, appendix) seem to be a source for lymphocytes (or plasma cells) that secrete immunoglobulin A (IgA), the antibody of mammalian secretions (i.e., mammary, salivary, tear gland, respiratory and intestinal epithelial secretions).

A receptor polypeptide known as the **secretory component** on the basal side of the secretory epithelia binds IgA from the blood or interstitial space. Endocytosis and transport carry the bound immunoglobulin across the secretory cell to the luminal side. There, the IgA–secretory component complex (SIgA) is released with the gland's major secretory products.

Pharyngeal Pouches and Arches

Behind the mouth, the pharynx swells into a series of **pharyngeal** or **branchial pouches** (Gk. **branchia** gills).[2] Peripherally, the pouches either end at **pouch closing plates,** identified externally by ectodermal **grooves,** or break through into open **pharyngeal clefts** (Fig. 25.2), sometimes called gill slits.

Dorsally and ventrally, horns (or wings) of the third and fourth pouches flare backward. Their walls become rudiments of **parathyroid glands** and **thymus** (Table 25.1). In addition, ventral horns of the fifth pouches provide the rudiments of the **post- (ultimo-) branchial bodies,** the source of thyrocalcitonin, a regulator of the parathyroid glands.

Midventrally, at the level of the second pharyngeal pouch, the pharynx evaginates a **thyroglossal duct** or **thyroid stalk** which sinks below the pharynx and bifurcates into **thyroid lobes,** the rudiments of the thyroid gland (Figs. 25.8–25.10). In mammals, the point of departure of the thyroglossal duct becomes the **foramen caecum** at the base of the tongue, marking the most anterior position occupied by endoderm in the pharynx.

Directly posterior to the thyroid lobes, the **ventral aorta** or **aortic sac,** arising from the truncus arteriosus and heart, breaks up into **aortic arches** (Fig. 25.9). The ultimate position of the thyroid gland and the other pharyngeal derivatives seems

---

[2]**Branchial pouches (sulci branchiales)** is accepted by the International Anatomical Nomenclature Committee, but the pouches are more often called pharyngeal pouches or visceral pouches in the embryological literature.

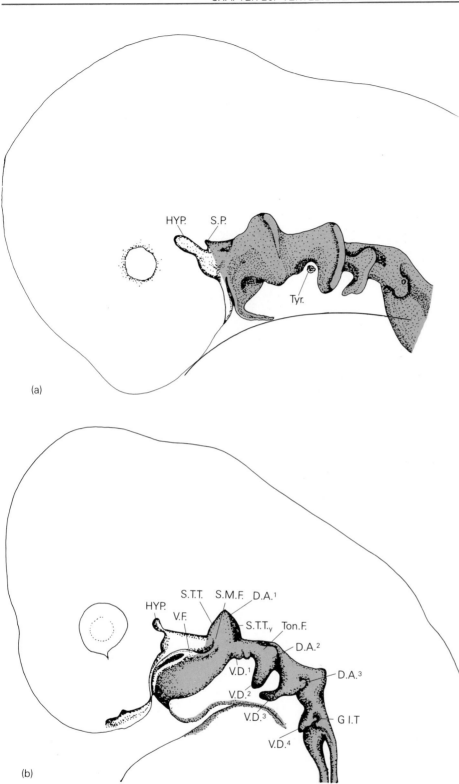

FIGURE 25.8. Lateral view of pharynx reconstructed from serial sections from 6.5-mm (a) and 10-mm pig embryos (b). ×20. Ao.$^{2-6}$, aortic arches 2–6; Ch.Ty., chorda tympani, branch of cranial nerve VII; D.A$^{1-3}$, dorsal apices of first three pouches; D.Ao. dorsal aorta; Gl.T., thymus gland process, dorsal process of fourth pouch; HYP., hypophysis; M, mouth; PH.P.$^{1-4}$, pharyngeal pouches 1–4; P.P.^{4}, posterior process of fourth pouch; Pul., pulmonary artery; S.M.F., submeckelian fold; S.P., Seessel's pouch or preoral gut; S.T.T., sulcus tubotympanicus; S.T.Ty., sulcus tensoris tympani; T.A.o, truncus arteriosus; Ton.F., dorsolateral region of the second pouch, later transformed into tonsillar recess; Tr., trachea or laryngotracheal groove; Tyr., thyroid; V.D.$^{1-4}$, ventral diverticula or process of branchial pouches. V.F., vestibular fold of mouth. From H. Fox, *Am. J. Anat.,* 8:187(1908).

to depend on their anatomical relationship to the final position of the heart.

In mammals, the elongation of the throat and withdrawal of the heart into the thorax draws the thyroid and the pharyngeal glands posteriorly. Those closest to the thyroid move farthest. As a result, the relative positions of the parathyroids and thymic rudiments on the third and fourth pouches are reversed. The third-pouch rudiments migrate posteriorly to become the thoracic or inferior glands, while the fourth-pouch rudiments become the cervical or superior glands.

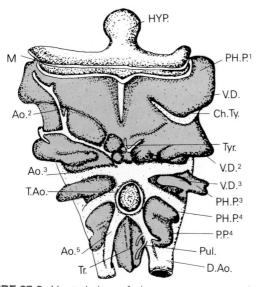

FIGURE 25.9. Ventral view of pharynx reconstructed from serial sections from 6.5 mm (compare to Fig. 25.8). Note interdigitation of aortic arches and pharyngeal pouches. From H. Fox, *Am. J. Anat.,* 8:187(1908).

Thickened mesenchymal columns or **branchial arches**[3] rise between pharyngeal pouches. Mesodermally derived cells are present in the arches, but arch mesenchyme is permeated with ectomesenchymal cells (i.e., of neural crest origin). With the exception of the endothelium lining the blood vessels and most (but not all) of the arches' voluntary (striated) muscle derived from mesoderm, internal arch tissues are formed by ectomesenchymal cells.

[3]**Arcus branchiales** but more often called visceral arches when including the first two (Table 25.1).

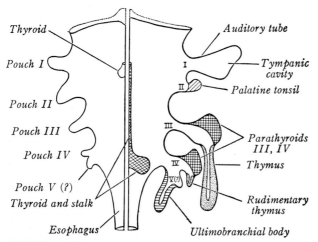

FIGURE 25.10. Scheme for the development of the human pharynx and pharyngeal derivatives. (a) At 4 weeks. ×40. (b) At 6 weeks. ×25. From L. B. Arey, *Developmental anatomy,* Saunders, Philadelphia, 1965. Used by permission.

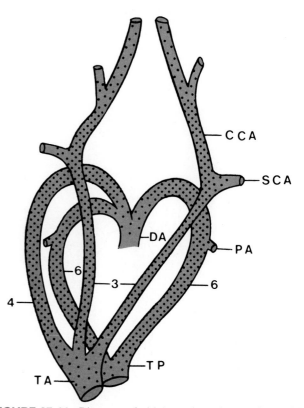

FIGURE 25.11. Diagram of chick aortic arches and carotid arteries summarizing results of experiments with quail–chick chimeras on neural crest involvement in morphogenesis. The intensity of neural crest involvement in structures is indicated by density of stippling. CCA, common carotid artery; DA, dorsal aorta; PA, pulmonary artery; SCA, subclavian artery; TA, aortic trunk; TP, pulmonary trunk; 3–6, arteries derived from third to sixth aortic arches. C. Le Lièvre and N. M. Le Douarin, *J. Embryol. Exp. Morphol.,* 34:125(1975), by permission of the Company of Biologists.

Even the thick connective tissue–smooth muscle walls of the **aortic arches** running through the branchial arches are traceable to the neural crest (Fig. 25.11, Table 25.2) (see Le Douarin, 1984).

The gills of anamniotes are variable and highly adapted structures developing on branchial arches. The simplest form of gills is that of the **external gills** of amphibians. Beginning at the tailbud stage, amphibian **gill buds** or **lobes** appear on branchial arches (now clearly gill arches). The buds elongate into feathery, branched **gill lobules** (Fig. 25.12). Most other gills form in the endodermal pouch epithelium, but external amphibian gills are formed in ectoderm, although urodele gills are later resurfaced with endodermal epithelium.

Urodeles remain at about this stage. Penetrated by accessory capillary loops from the aortic arches,

TABLE 25.2. Contrast between neural crest and mesodermal origins of some cranial and integumentary structures in chick

Neural Crest	Mixed	Mesodermal
	Skull	
Facial part of skull (below notochord), auditory pit	Cartilage and bones of intermediate area	Most of vault and dorsal cranium (above notochord)
Nasal-lateral parts of eye orbit	Otic capsule	
Meninges: leptomeninges (inner coverings of brain and spinal cord) = endomeninx = arachnoid + pia mater		
Prosencephalon	Mesencephalon	Hindbrain and spinal cord
Meninges: pachymeninges (outer covering of brain and spinal cord) = ectomeninx		
Prosencephalon		Mainly mesencephalon, exclusively hindbrain and spinal cord
	Integument	
Dermal papilla, arrector muscle (smooth), endothelial and stromal cells of cornea, orbit, and sclera		Tissue behind eye
Connective tissue of jaw, tongue, and ventral part of neck (floor of mouth), and subcutaneous fat layer		
Connective tissue components of endodermal and ectodermal derived organs		
Pituitary, larcrimal, salivary, thyroid, parathyroid glands, and thymus		
	Arteries	
Muscular-connective tissue wall of branchiocephalic and common carotid arteries		Endothelial lining of all vessels
	Striated muscle	
Ciliary muscles of iris	Core of branchial arches	Extrinsic muscles of eye

Source: Data from Le Douarin (1982).

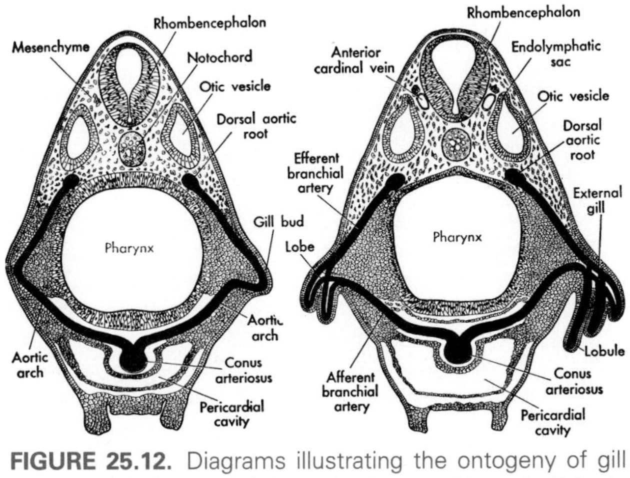

FIGURE 25.12. Diagrams illustrating the ontogeny of gill circulation (left to right) from the interposition of gill buds on the branchial arches and the looping of vessels from the aortic arches through circulation in external gills. From A. F. Huettner, *Comparative embryology of the vertebrates,* rev. ed., Macmillan, New York, 1949. Used by permission.

the external gills function to aerate the blood circulating through them. In frogs, the gill arches are interrupted by gill capillaries, but continuity across the aortic arches is restored at metamorphosis.

Complexity in gill structure emerges with the development of gill slits between the gill arches and the internalization of the gills. The number of arches, on the other hand, seems to have been reduced in the course of evolution from as many as 15 arches found in the hagfish to approximately six found in teleosts and tetrapods.

The separation of gill arches by gill slits seems to be an adaptation toward improved circulation of water around the gills. Eight branchial arches with seven pouches form in the ammocetes larya of cyclostomes (lamprey), and primitive sharks have as many as nine branchial arches with gills and seven slits. In other sharks, the first arch has only temporary gills and its slit remains a small **spiracle.** One or two of the last arches also fail to form and only five to seven arches with four to six slits are usually present.

In teleosts, the number of arches is reduced by fusion from the rear to five. Gills do not form on the first two arches (the mandibular and hyoid), but they appear on each of the four pairs of branchial arches[4] with gill slits between them.

The trend toward reduced numbers of gill arches continues in amphibians and amniotes. Urodeles exhibit reduction of the last pair of gills, usually forming three pairs of external gills (on the third through fifth arches) and four pairs of slits (second to fifth). Anurans form only two pairs of external gills (on the third and fourth arches).

In amniotes, six arches form, but the fifth is rudimentary and the sixth is fused with the posterior body wall (Fig. 25.2). Moreover, the arches form serially rather than simultaneously and are not all present as integral structures at the same time.

Gill covers are commonly formed by anamniotes, and gills are withdrawn into gill chambers. In teleosts, the second pair of branchial arches (the hyoids) send back a stiffened flap of skin (opercula, covers) over the gills. Urodeles partially cover their external gills with **gular folds** (or opercula) of fused lamella from the second arch, and anuran tadpoles draw their gills into a **gill sac** (or opercular chamber). The tadpole's **operculum** arises from the ventral edge of the second branchial arch and fuses, zipper-wise from the right, with a fold on the body wall (the cervical fold) leaving a small vent (also called a spiracle) on the left to drain the gill sac.

---

[4]Because the anterior-most arches do not have gills and hence are not truly "branchial" (gilled), the third "visceral" arch is sometimes called the "first branchial" arch and the sixth is considered the fourth branchial arch. This convention is rejected here in favor of retaining one name and numbering all the arches consecutively.

Inside the anuran gill sac, most of the external gills atrophy, while new **internal gills** or branchial filaments are fashioned from endoderm. Double rows of internal gills develop on the ventral and posterior sides of the third to fifth branchial arches, and single rows of internal gills appear on the anterior surfaces of the paired sixth arches.

The tadpole's remaining gill plates (second, fourth, and fifth) now break open and, joining the third gill slit opened earlier, allow water in the pharynx to flow into the gill sac, past the internal gills, and out the spiracle. Endodermally lined overgrowths (velar plates) on the floor of the pharynx and comblike projections (gill rakers) above the branchial arches protect the openings of the gill slits from clogging with particles (Fig. 25.13).

The anuran gill sac is a unique structure. Under its covering flap of skin, forelimbs as well as internal gills develop, and, at the tadpole's metamorphosis into a frog, the arms break through the cover to emerge externally. The cover, like the gills and the tadpole's tail, degenerates at metamorphosis under the influence of thyroxine.

In fish, gill vessels divide the aortic arches into **afferent branchial arteries** (carrying blood to gills) and **efferent branchial arteries** (carrying blood away from the gills). As the gills develop, one or the other branchial arteries usurps the original arch; the other plows a new path. In elasmobranchs, the afferent branchial artery is original, while in teleosts, the efferent is original.

The patterns of gill arches and aortic arches are not necessarily identical. In fish, as a result of splitting or branching of gills into half gills (hemibranchs), the efferent vessel of one arch drains a gill fed by the afferent vessel of the next arch.

How closely do branchial arches in amniotes resemble gill arches? Three features of the branchial arches of some amniotes make the strongest

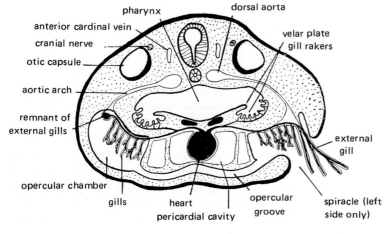

FIGURE 25.13. Diagrammatic cross section through pharyngeal region of 11-mm tadpole showing circulation through internal gills. Slits in the floor of the pharynx allow water entering mouth to rush past the gills on way to spiracle. From R. Rugh, *Guide to vertebrate development,* Burgess, Minneapolis, 1977. Used by permission.

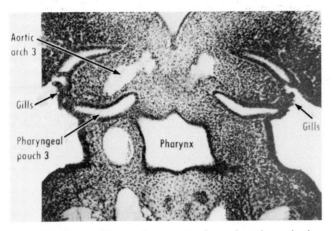

FIGURE 25.14. Photomicrograph of section through the pharyngeal region of a 12-day rabbit embryo showing filamentous ephemeral gill buds in pouch endoderm of the third branchial arch. ×60. From E. Witschi, *Development of vertebrates,* Saunders, Philadelphia, 1956. Used by permission.

case for a resemblance. (1) In mammals, especially human embryos, and in birds, pouch end plates frequently disappear, opening up gill-like slits between the embryonic pharynx and the amniotic cavity. (2) In amniotes, the third, fourth, and rudimentary fifth branchial arches recede within a **cervical sinus** (Fig. 25.5) formed when an **opercular shelf** from the second (hyoid) arch fuses with the **cervical fold** of the body wall over the heart (i.e., the cardiac swelling or prominence). (3) Some mammals, rabbits in particular, develop **ephemeral gill buds** or ripples in the endoderm of the third and fourth pharyngeal pouches (Fig. 25.14). Similar buds appear briefly in birds, especially song birds, as ectodermal folds on the second and third arches within the cervical sinus.

On the other hand, two features of amniotic branchial arches resist comparison with gill arches. (1) The gill-like slits that opened in some amniotes quickly close, and the new pouch end plates as well as persistent old end plates remain closed throughout ontogeny. (2) Unlike the gill sac or gill chamber of anamniotes, the cervical sinus of amniotes is lined with ectoderm, and, instead of protecting gills, the amniotic opercular fold obliterates the arches it covers.

At least in normal development, the opercular fold closes the branchial pouches permanently and seals the future throat. In abnormal development, fistulas remain in the throat, but in no case do arms develop within the cervical sinus nor is its covering eroded by thyroxine.

Finally, what can be said of ephemeral gill buds? Even the most generous interpretation cannot quite puff them up to the level of gills.

All these spurious and short-lived developments are hard to interpret on other than evolutionary principles.

One may, however, admit that the [amniote] embryo at best is a very imperfect fish. (Witschi, 1956, p. 489)

The gills of anamniotes are highly differentiated structures built on branchial arches. While amniotes' ancestors presumably had gills and their embryos presumably developed gills, contemporary amniotes do not have gills, and their embryos do not develop gills. Rather, the branchial arches of amniotes remain undifferentiated well beyond the point contemporary fish's branchial arches develop gills. The further development of the amniote's branchial arch structure depends on elaboration from the undifferentiated, not from the primitive.

The ectomesenchymal cores of the branchial arches form the branchial skeleton and cartilaginous models for several bones in the jaws, throat, and inner ear ossicles of mammals (Fig. 25.15). The fate of the branchial skeleton is different in the various classes of vertebrates, where changes are correlated with the type of jaw suspension.

The most conspicuous changes involve the cartilages of the first and second branchial arches. Arches form upper (epibranchial) and lower (hypobranchial) cartilages. In the case of jawed vertebrates, the first, or **mandibular, arch** bends around the mouth (Fig. 25.2), forming an upper **maxillary process** containing the **palatoquadrate cartilage** and a lower **mandibular process** containing the **mandibular cartilage** (also called Meckel's cartilage). These cartilages form the rudimentary upper and lower jaws, respectively.

The second, or **hyoid, arch** forms an upper **hyomandibular** cartilage and usually fused lower cartilages (ceratohyal and basihyal). In most fish, the hyoid cartilages and the ligaments connecting them fit around the rear of the mandibular cartilages forming a hinge for the jaw. In addition, the hyomandibular cartilage, braced against the otic capsule and the brain case, binds the upper jaw to the skull (i.e., the primitive hyostylic mode of jaw suspension) (Romer, 1955).

In primitive sharks, the upper jaw is both bound to the braincase directly and attached via the hyomandibular cartilage (i.e., the amphistylic mode of jaw suspension). In chimaeras (advanced sharks), the hyomandibular cartilage is nonfunctional by way of supporting the upper jaw (i.e., the primitive autostylic mode of jaw suspension).

Similarly, in lungfish and amphibians, the

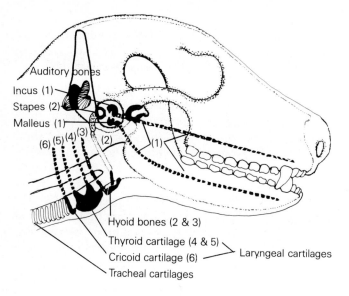

FIGURE 25.15. Schematic view of mammalian skull and throat illustrating derivation of auditory bones (middle ear ossicles), hyoid bones, and cartilage elements from branchial arches (numbers). From H. E. Lehman, _Chordate development,_ 3rd ed., Hunter Textbooks, Winston-Salem, 1987. Used by permission.

palatoquadrate fuses with the base of the skull (at the trabecular cartilages and the otic capsule). The hinge of the jaw is formed posteriorly by the ventrally directed **quadrate process** of the palatoquadrate and the dorsally directed **articular process** of the mandibular cartilage.

In their evolution, tetrapods retained the direct binding of the upper jaw without the aid of the hyomandibular cartilage. The evolution of dermal bones in the jaws and in the palatine structure, moreover, freed the mandibular and hyoid elements for new functions.

Supported by anterior extensions of the lower hyoid cartilage (ceratohyoid) and manipulated by branchial muscles, the **tongue** (not a conspicuous structure in fish) evolved as a major aid to mastication (chewing) in tetrapods. Moreover, the hyomandibular cartilage and other elements fused into the **columnella** of amphibians, reptiles, and birds and functioned as an aid to hearing.

In mammals, the **auditory** or **middle ear ossicles** develop from centers of condrification in direct continuity with the cartilages of the first two arches. The shortened remnant of the columnella (Reichert's cartilage) is transformed into the **stapes** (the stirrup). The dorsal end of the cartilaginous rod in the first arch (Meckel's cartilage) differentiates serially into the **incus** (anvil, corresponding to the quadrate process) and the **malleus** (hammer, corresponding to the articular process, Fig. 25.15).

> These bones, originally gill bar elements, afford a good example of the changes of function which homologous structures can undergo. Breathing aids have become feeding aids and finally hearing aids. (Romer, 1955, p. 522)

In addition, the muscles attached to Meckel's cartilage become the tensor tympani muscle of the malleus, innervated, like other first arch derivatives, by the trigeminal (V) cranial nerve. The muscles of the hyoid arch become the stapedial muscle, innervated, like other second arch derivatives, by the facial (VII) cranial nerve. (See Table 25.1).

Finally, the proximal portion of the first branchial pouch constricts into the **auditory** (eustachian) **tube** rising from the throat. The expanded blind ends of the first and second pouches become the **tympanic cavity** or cavity of the middle ear. The endodermal lining of the pouches coats the middle ear ossicles and forms their suspensory ligaments, called "mesenteries." The end plate of the first pouch provides the primordium of the **tympanic membrane,** the ear drum. The **external acoustic meatus** represents the first pharyngeal groove, and the ectodermal sides of the first and second branchial arches surrounding the groove supply tissue for the auricle or fleshy pinna of the ear.

Here again, the annals of comparative anatomy provide an impressive example of evolutionary resourcefulness, but the comparative embryology of the relevant structures does not suggest terminal additions. At the point in development when a structure in the embryo of a primitive species begins differentiating, the homologous structure in the embryo of an advanced species remains undifferentiated. The material differentiating in the advanced species does not build on the differentiated material of the primitive species but on undetermined resources, especially those with greater developmental plasticity.

In vertebrates, if not other groups, plasticity is characteristic of early stages of development. Contrary to the examples of *Caenorhabditis elegans* and *Drosophila melanogaster* where determination is largely in place prior to morphogenesis, labile determination is characteristic of early development in vertebrates. Progressive induction, requirements for relatively large numbers of cells (i.e., so-called mass effects), and position information predominate in mechanisms of vertebrate determination.

VERTEBRATE CIRCULATORY SYSTEMS

Vertebrate embryos develop their circulatory system from a primitive circulatory loop beginning at the heart as an **arterial system** and ending at the heart as a **venous system** (Fig. 25.16a). The arterial system consists of a **ventral aorta** (or ascending aorta), lateral **aortic arches,** and paired **aortic roots** (or left and right dorsal aortas) that fuse into the

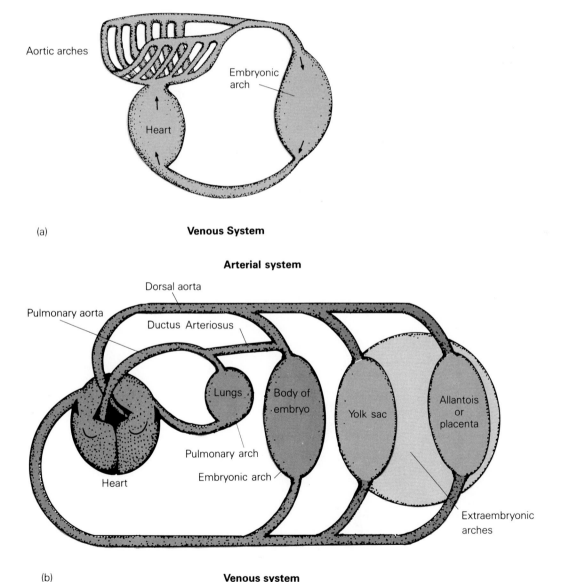

FIGURE 25.16. The primitive circulatory loop of vertebrate embryos. (a) In anamniotes, a single embryonic arch encompasses the yolk. (b) The circulatory loops of vertebrate amniotes embryos. Part b adapted from J. B. Phillips, *Development of vertebrate anatomy,* C. V. Mosby, St. Louis, 1975.

single **dorsal aorta** (or descending aorta). The dorsal aorta gives rise to the capillary networks of the **embryonic arch** through which exchange is achieved with embryonic tissues. The venous system collects blood from the embryonic arch.

Two additional circulatory arches may be added between the arterial and venous systems (Fig. 25.16b). The **extraembryonic arch** consists of a **yolk sac arch,** circulating blood through the yolk sac, with or without an **allantoic arch** circulating blood through the allantois or placenta. In addition, tetrapods and some teleosts have a **pulmonary arch** circulating blood through lungs or related structures. A **ductus arteriosus** (also called the duct of Botalli) may shunt excess pulmonary circulation into the systemic embryonic arch.

Each of the arches is interrupted by one or more sets of **capillaries, sinuses,** or **sinusoids** where exchange occurs between blood and the interstitial environment. In the case of the systemic embryonic arch, exchange involves the delivery of nutrients and the removal of wastes from body tissues. Yolk and yolk sac circulation results in the uptake of nutrients from the yolk or, in the case of early therian mammals and other placental vertebrates with yolk sac placentas (whether omphalopleural or choriovitelline), the uptake of nutrients from the uterine environment. Allantoic circulation in birds and reptiles results in the chorioallantoic exchange of carbon dioxide and oxygen with the ambient environment. Allantoic circulation in therian mammals with some sort of cho-

rionic placenta results in the exchange of nutrients and wastes through the placenta.

The principal parts of the primitive venous system (Fig. 25.17) are the **vitelline veins** (or yolk veins) from the splanchnopleure, the **cardinal veins** from the dorsal somatopleure, and the allied **abdominal veins** from the ventral somatopleure. Terrestrial tetrapods and their relatives add the pulmonary veins, and amniotes take over the abdominal veins for **allantoic** or **umbilical** (Lat. navel, center) **veins.** Although these veins are originally paired, their bilateral symmetry tends to disappear as one or the other side becomes dominant.

Hearts

Early heart development follows the same pattern throughout the vertebrates. **Premyocardial cells** (precursors of myocytes and epimyocardial cells) and **pre-endocardial cells** (precursors of endocardial cells, the endothelial cells lining the heart) are localized as soon as mesoderm makes its appearance (e.g., the primitive streak stage). In the chick, these cells are found in a narrow horseshoe-shaped **heart-forming region** around the anterior neural plate (Fig. 25.18).

Like other **angiogenic** (blood-vessel-producing) cells of the early embryo, the heart-forming cells are part of the splanchnic mesoderm. Endoderm does not contribute heart-forming cells, but, in both _in vitro_ and _in vivo_ studies, endoderm promotes the maturation of **cardiac myoblasts** (heart-

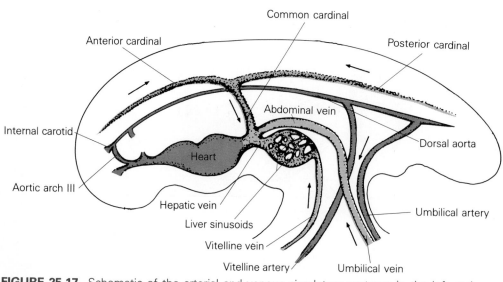

FIGURE 25.17. Schematic of the arterial and venous circulatory systems in the left and central parts in a vertebrate embryo. Adapted from J. B. Phillips, _Development of vertebrate anatomy,_ C. V. Mosby, St. Louis, 1975.

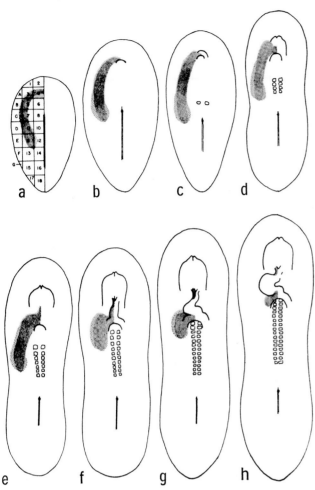

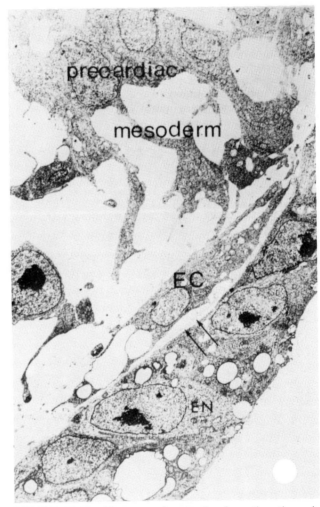

forming cells) into **cardiac myocytes** (heart muscle cells) and induces heart differentiation (see Manasek, 1976).

Premyocardial cells are arranged in a sheet linked by apical junctional complexes. Similarly, in the adult, cardiac muscle cells are linked by intercalated disks. Similar to some other embryonic cell layers (e.g., chick hypoblast), the sheet does not constitute a true epithelium, since it lacks a basal lamina visualizable at the electron microscope level (Fig. 25.19) and does not seem to contain type IV collagen characteristic of basal laminae.

FIGURE 25.18. Diagram of survey for precardiac cells in chick blastoderm. Portions of radioactively labeled embryos were substituted for similar portions in unlabeled embryos (according to a grid, upper left) and the presence of precardiac cells in the transplanted piece ascertained by subsequent autoradiography. The density of stippling is proportional to the frequency with which a portion contributed labeled cells to the heart. (H&H stages from V. Hamburger and H. L. Hamilton, *J. Morphol.,* 88:49, 1951.) Stage 5 H&H: notochordal process, 19–22 h (h = hours of incubation). Stage 6 H&H: anterior body fold (head fold), 23–25 h. Stage 7 H&H: first somite, neural folds, 23–26 h. Stage 8 H&H: 4–6 somites, blood islands, coelom, 26–29 h (approximately comparable with Figs. 23.6 and 23.7). Stage 9 H&H: 7–9 somites, optic vesicle, 29–33 h. Stage 10 H&H: 10–12 somites, anterior amniotic fold, 3 primary brain vesicles, closure anterior neuropore, 33–38 h (approximately comparable with Figs. 23.9 and 23.10). Stage 11 H&H: 13 somites, 5 neuromeres in hindbrain, 40–45 h. Stage 12 H&H: 16 somites, telencephalon, 45–49 h (about 4 hours younger than Figs. 23.11 and 23.12). From G. C. Rosenquist and R. L. De Haan, *Carnegie Inst. Washington Contrib. Embryol.* 38(263):113(1966). Used by permission.

FIGURE 25.19. Electron micrograph of section through avian splanchnopleure of chick embryo (stage 8+ H&H, about 29 hours of incubation, 5 somites). Although a basal lamina is found (arrows) on the basal surface of the endoderm (EN), the precardiac mesoderm and cells assumed to be preendocardial (EC) lack of basal lamina. ×5060. From F. J. Manasek, in G. Poste and G. L. Nicolson, eds., *The cell surface in animal embryogenesis and development,* Elsevier/North-Holland Biomedical Press, Amsterdam, 1976. Used by permission.

With the truly epithelial endoderm leading the way, premyocardial cells move en masse to the region below the foregut as the head rises above the anterior body fold (see Fig. 18.38). The pre-endocardial cells move to the same region individually or in small groups.

Both layers of cells now lift off their endodermal substratum and form thickened, raillike **heart primordia** running parallel on the borders of cavernous coelomic swellings known as the pericardial regions of the coelom or the amniocardiac vesicles. While still attached broadly to the endoderm anteriorly, the heart primordia merge and fuse centrally into a **tubular heart** which, narrowing its attachment to the pharynx above, becomes suspended by a membranous **mesocardium** (see Fig. 23.10).

Initially, a simple cuboidal epithelial **myocar-**

dium comprises the outside of the tube, and a simple squamous epithelial **endocardium** or **endothelium** forms the lining. **Cardiac jelly** or ground substance synthesized by the myocardium separates the two cellular layers.

Cross-banded **myofibrils** appear nearly simultaneously throughout the myocardium at the same time the heart tube becomes discrete (in the chick, stage 9^+–10 Hamburger & Hamilton, 8–10 somites, 33–38 hours of incubation; see Fig. 23.9). Myoblasts are transformed to myocytes, and the heart begins to beat. Contractions begin on the right side and quickly spread to the entire myocardium. Within 4.5 hours of the appearance of the rudimentary tubular heart, blood is entering a beating heart and moving under pressure into the systemic circulation (Fig. 25.20).

Evidence from *in vitro* studies with a strain of

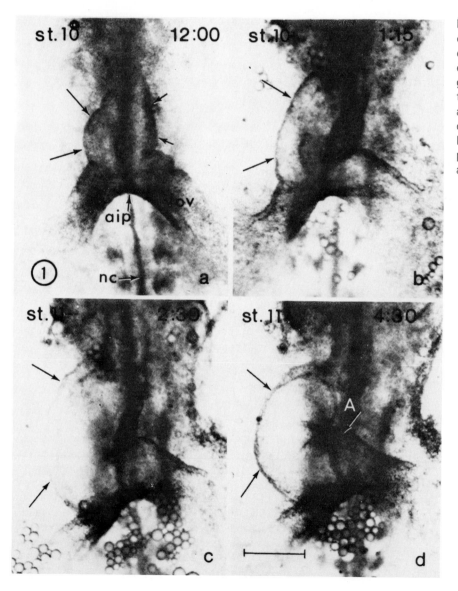

FIGURE 25.20. Photomicrographs of chick heart *in vitro* in ventral view. The corresponding stages of chick development (Hamburger & Hamilton) are given in the upper left; the time in culture is given in the upper right. ×80. aip, Anterior intestinal portal; nc, notochord. From F. J. Manasek and R. G. Monroe, *Dev. Biol.,* 27:584(1972), by permission of Academic Press and the authors.

the axolotl, *Ambystoma* (= *Siredon*) *mexicanum*, carrying the *cardiac lethal c* gene suggests that endodermal induction of cardiac myogenesis is at least partially instructive. In *c/c* animals, a heart appears, but myofibrillogenesis is abnormal. While contractile proteins (actin, alpha actinin, and myosin) are synthesized and accumulate, sarcomeric myofibrils are not organized and the heart fails to beat. Not only are homozygous hearts rescuable by transplantation to normal animals, but the hearts are also rescuable *in vitro* by coculture with anterior endoderm from normal early embryos or incubation in media conditioned by the presence of anterior endoderm.

Ribonuclease abolishes the activity of the endoderm-conditioned medium, while the addition of RNA extracted from the anterior endoderm of normal animals to media stimulates the myofibrillogenesis of sarcomeric myofibrils and the differentiation of beating, functional cardiac muscle.

Other RNA extracts are without effect (Davis and Lemanski, 1987).

Tubular hearts consist mainly of a primitive ventricle (the pulsating muscular portion) whose anterior part swells into a transient bulbus cordis[5] (Lat. bulbus onion + cors heart) or conus arteriosus (Lat. arterial cone) (Table 25.3). Posteriorly, vessels entering the heart are still being **recruited** and incorporated into the primitive **atrium** (Fig. 25.21).

The **sinus venosus** is recruited behind the atrium, and, at the same time, the heart begins the **flexure** that **loops** it into a C-shape, convex side to

---

[5]*Nomina embryologica* recognizes bulbus cordis as official for mammals, and the term is used for birds and reptiles. Conus arteriosus is used for amphibians and some fish where it also identifies the extension of the adult ventricle containing a spiral or semilunar valve.

TABLE 25.3. Names and synonyms for parts of heart at different stages of development and for arterial system

Stage of Heart Development	Part of Heart	
	Name Recognized for Mammals	Synonyms in General Use
Heart primordium	Primordial sinus venosus	
	Primordial atrium	Primordial auricle
	Primordial ventricle	
Tubular heart	Primitive atrium	
	Primitive ventricle	
	Primitive bulbus cordis (distal part ventricle)	Conus arteriosus (bulbus arteriosus)[a]
Sigmoid heart	Sinus venosus	
	Sinoatrial valve	
	Primitive atrium	
	Endocardial cushions	
	Atrioventricular canal	
	Primitive ventricle	
	Bulbus cordis	
	Spiral septum	(In anurans: tend to separate aerated and exhausted blood)
(Subdivisions) 4-chambered heart of mammals and birds		
	Atrium (left/right)	
	Ventricle (left/right)	
	Semilunar valves (at openings systemic and pulmonary trunks)	
	Part of Arterial System	
	Aortic sac	
	Truncus arteriosus	(Bulbus arteriosus?)[a] = ventral aorta (proximal part)
	Aortic Roots	Aortic arches (proximal part = conus arteriosus)
	Pulmonary trunk	

[a]Romer (1955, see footnote, p. 475) objects to equating the conus arteriosus with a bulbus arteriosus, insisting that the latter be reserved for the proximal portion of the ventral aorta. Rugh (1977, p. 318), however, defines bulbus arteriosus as the most anterior division of the early tubular heart (i.e., equivalent to bulbus cordis or conus arteriosus). The truncus arteriosus is equated with the ventral aorta.

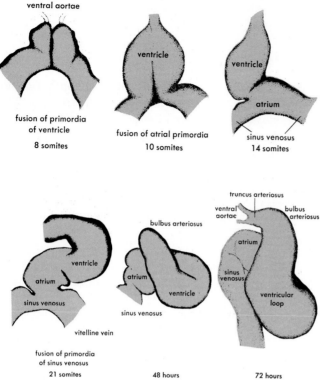

FIGURE 25.21. Diagrams of developing chick heart. Ventral view of heart rudiment (8 somites), tubular heart (10 somites), and during early flexure (14 somites). Dorsal view during late flexure (21 somites), and lateral views of early sigmoid heart (48 hours) and late sigmoid heart (72 hours). From L. G. Barth, *Embryology,* revised and enlarged edition, Holt, Rinehart & Winston, New York, 1953.

the right. Identical looping occurs *in vitro* (Fig. 25.20), indicating that the process is autonomous. Since myocardial cells on the right are columnar, while those on the left are flat, the heart's shape may result from the collective effect of the cells' shapes.

Flexure continues until the heart is bent into an S shape, and the position of the heart's entry point has moved 180° (Fig. 25.21). Instead of blood entering the posterior heart, blood enters the heart at the anterior border. A **sinoatrial valve** (later valvulae venosae in birds and mammals) appears between the sinus venosus and the atrium, guarding against backup, and in dipnoans and anurans, a **spiral septum** curls down the length of the conus arteriosus to the ventricle.

The heart's convolutions have also wrought changes in the **mesocardium** suspending the heart. All but the upper and lower ends have perforated and disappeared, leaving the ventricle entirely free in a common **pericardial cavity** (see Fig. 23.12). A small **atrial mesocardium** serves as a bridge for pulmonary veins, and **lateral mesocardia** provide a causeway for the common cardinals or ducts of Cuvier on their way to the sinus venosus.

At this point, the fish heart begins its final morphological and histological differentiation. The ventricle, above all, thickens conspicuously. Multiple **semilunar valves** appear in the conus arter-

iosus, and an **atrioventricular valve** differentiates between the atrium and ventricle, while the sinoatrial valve occupies the junction of the sinus venosus and atrium (Fig. 25.22).

Sigmoid hearts of tetrapods are mere pulsating coils at the time the fish heart differentiates into a thick, muscular organ. Blood enters the

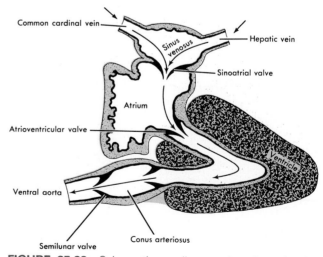

FIGURE 25.22. Schematic median section through sigmoid fish heart. Arrows show direction of flow. Adapted from J. B. Phillips, *Development of vertebrate anatomy,* C. V. Mosby, St. Louis, 1975. Used by permission.

tetrapod heart at the sinus venosus, moves through the atrium, ventricle, and bulbus cordis, and exits into the arterial system.

Differentiation of the tetrapod heart only catches up slowly to the fish heart. Large extracellular spaces disappear; the myoblasts congregate, and the cardiac jelly separating myocardial cells from the endocardium lining is resorbed. The inner surface of the differentiating myocardial cells acquires a relatively smooth basal lamina, and fleshy **trabeculae carnae** begin to differentiate in the ventricle.

Connective tissue elements, blood vessels, and nerve endings move into the thickening heart muscle, and, externally, the heart's cover of **epicardium** differentiates above a **subepicardial** connective tissue layer. By the time the chick embryo has formed 15 somites (stage 12 Hamburger & Hamilton, about 45–49 hours), the heart tube has begun transformation into a chambered organ.

Subdivision of the tetrapod heart involves the development of internal septa, valves, and channels. The heart muscle is not divided by internal walls so much as it is reshaped dynamically in tune with increases in volume and changes in flow patterns.

Subdivision of the primitive atrium and ventricle produces right and left atria and ventricles. As a rule, the venal end of the embryonic heart is absorbed in the adult heart rather than subdivided. In birds and mammals, the sinus venosus is absorbed into the wall of the right atrium. The conus arteriosus is subdivided and split between the left and right ventricles.

In amphibians, subdivision involves the partial partitioning of the atrium by an incomplete **primary interatrial septum** (1° IAS, interatrial septum primum). The sinus venosus moves toward the right, and the minute **pulmonary veins** shift toward the left (Fig. 25.23).

In birds, the onset of subdivision is accelerated compared to mammals, but the routes toward subdivision are parallel. Initially, an **AV constriction** sharply narrows the **AV canal** at the border of the atrium and ventricle, but little is present by way of internal partitions. The only valves are those at the orifice of the shallow sinus venosus into the bulging atrium (Fig. 25.24).

The results of *in vitro* experiments employing

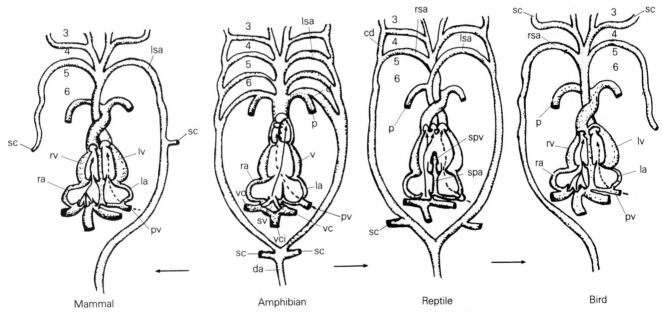

FIGURE 25.23. The fate of aortic arches and the anatomy of the heart of a representative mammal, amphibian, reptile, and bird. The sigmoid-shaped hearts have been straightened for the purposes of illustrating the flow of blood. Each heart is actually folded back on itself. cd, carotid duct; d, Ductus arteriosus; da, aorta; la, left atrium; lsa, left systemic trunk; lv, left ventricle; p, pulmonary artery; pv, pulmonary vein; ra, right atrium; rsa, right systemic trunk; rv, right ventricle; sc, subclavian artery; sv, sinus venosus; spa, interatrial septum; spv, interventricular septum; v, ventricle; vc, common cardinal vein; vci, hepatic vein. Adapted from E. S. Goodrich, *Studies on the structure and development of vertebrates,* Macmillan, London, 1930.

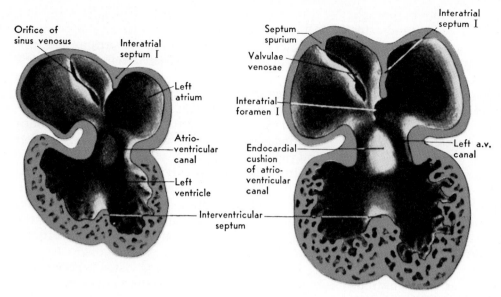

FIGURE 25.24. Stereogram of dorsal half of human hearts. (a) Early in fifth week of development (resembles heart in 3.7-mm pig). (b) Early in sixth week of development (resembles heart in 6-mm pig). Mesenchymal thickenings appear at atrioventricular constriction and in dorsal and ventral endocardial cushions. The interatrial septum I has begun its descent toward the endocardial cushions. From B. M. Carlson, *Patten's foundations of embryology,* 5th ed., McGraw-Hill, New York, 1988. Used by permission.

collagen mats as a substratum suggest that mesenchyme is mobilized from the endothelium under the inductive influence of the myocardium (Markwald et al., 1984). The mesenchymal cells colonize the cardiac jelly of the AV constriction and produce dorsal and ventral **endocardial cushions.** The meeting of these cushions divides the AV canal into right and left channels (Fig. 25.25) and produces the cusps of the two **AV valves,** the left **bicuspid** or mitral valve and the right **tricuspid.** In addition, the endocardial cushions contribute to the second-

ary **interatrial septum** (2° IAS, interatrial septum secundum) and to the **interventricular septum.**

The fused endocardial cushion is pivotal in septal conversion. The 1° IAS closes in, like a diaphragm, around the pupillary **primary interatrial foramen** (interatrial foramen primum) above the endocardial cushions. Below, growth of the cushions and the interventricular septum narrows the **interventricular foramen** (Figs. 25.24 and 25.25).

Ironically, as the 1° IAS closes its primary interatrial foramen centrally, it opens a **secondary**

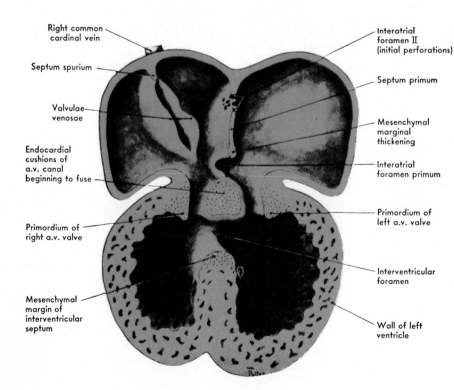

FIGURE 25.25. Stereogram of dorsal half of human heart late in sixth week (i.e., more advanced than in Fig. 25.24), showing narrowing of primary interatrial septum (septum primum) around the interatrial foramen primum (I) and the initial perforation of interatrial foramen secundum (II). Endocardial cushions have fused across the atrioventricular canal forming right and left channels. The muscular interventricular septum is capped by a mesenchymal margin. From B. M. Patten, *Am. J. Anat.,* 107:271–280(1960), Alan R. Liss, Inc. Used by permission.

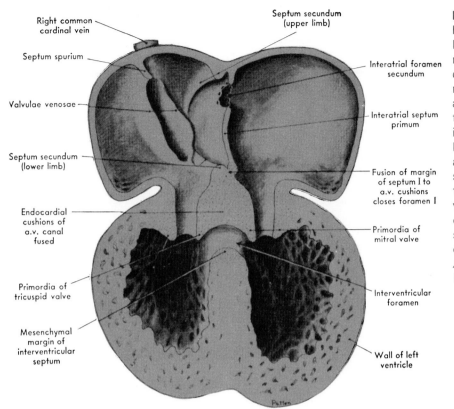

Right common cardinal vein

Septum spurium

Valvulae venosae

Septum secundum (lower limb)

Endocardial cushions of a.v. canal fused

Primordia of tricuspid valve

Mesenchymal margin of interventricular septum

Septum secundum (upper limb)

Interatrial foramen secundum

Interatrial septum primum

Fusion of margin of septum I to a.v. cushions closes foramen I

Primordia of mitral valve

Interventricular foramen

Wall of left ventricle

FIGURE 25.26. Stereogram of dorsal half of human heart slightly later than Fig. 25.25. The interatrial septum primum has fused with the endocardial cushions, closing the interatrial foramen I, only to open the interatrial foramen II. The secondary interatrial septum (septum secundum) now begins its march around the interatrial region. Meanwhile, the sinus venosus is being absorbed into the wall of the atrium. A septum spurium (false septum) marks the sinus' former position. The interventricular canal narrows as the mesenchymal margin of the interventricular septum approaches the endocardial cushions. From B. M. Patten, *Am. J. Anat.,* 107:271–280(1960), Alan R. Liss, Inc. Used by permission.

interatrial foramen (interatrial foramen II) peripherally (Fig. 25.26). Possibly, pressure prevails. The growing volume of blood and the increasingly powerful strokes of the differentiating heart may create the conditions for resorbing the primary membrane at its weakest point.

At the same time, the 2° IAS draws over the primary interatrial foramen like a shade over a window, but the IAS is not sealed. The margin (limbus, Lat. border) of the new septum remains free around the **foramen ovale** (Fig. 25.27), and blood flows from right to left atria.

This relationship between the septa, each overlapping a foramen in the other but allowing flow in one direction, is an engineering and developmental masterpiece. At a time when pulmonary flow on the left side of the heart is low, the double-septa system allows blood to flow from the high-pressure right side to the low-pressure left side. The consequent equalizing of pressure allows both sides of the heart to develop on an equal footing.

Moreover, the arrangement of veins entering the right atrium capitalizes on the position of the septal openings to deliver richer blood to the head region of the developing fetus. The differential distribution of blood in the fetus is possible because the entries of blood from the posterior and anterior

circulations are separate. Blood entering the mammalian heart from the posterior circulation includes the uterine drainage and is therefore nutrient- and oxygen-rich, while blood entering the heart from the anterior circulation is the relatively exhausted blood drained from the head (Fig. 25.27).

Because of the angles of the entry vessels and the positions of the foramen ovale and the secondary foramen, the rich posterior blood moves to the left side and hence to the brain (i.e., via the fourth systemic aortic arch, the brachiocephalic, and common carotid, Fig. 25.31 and see below). On the other hand, relatively depleted anterior blood moves past the tricuspid valve to the right atrium where it is circulated to the body (via the pulmonary trunk of the sixth aortic arch, through the ductus arteriosus, and to the dorsal aorta below the point where the head is supplied)!

Although the interatrial part of this system operates throughout fetal life, the interventricular part does not. At about the third month, in the case of human beings, the interventricular septum closes. Mesenchyme on the margin of the interventricular septum, the endocardial cushions, and ridges in the conus (see below) converge to separate the ventricle into right and left ventricles. The mesenchymatous mass at the top of the ventricle

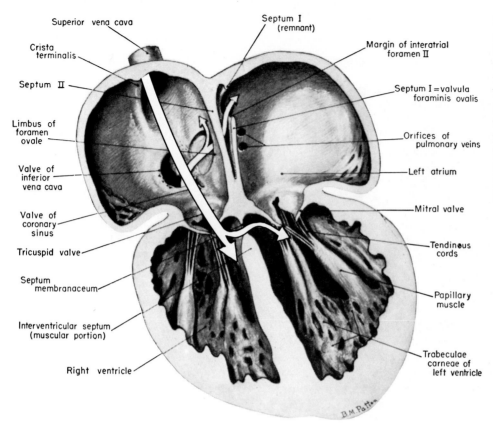

Superior vena cava

Crista terminalis

Septum II

Limbus of foramen ovale

Valve of inferior vena cava

Valve of coronary sinus

Tricuspid valve

Septum membranaceum

Interventricular septum (muscular portion)

Right ventricle

Septum I (remnant)

Margin of interatrial foramen II

Septum I = valvula foraminis ovalis

Orifices of pulmonary veins

Left atrium

Mitral valve

Tendinous cords

Papillary muscle

Trabeculae carneae of left ventricle

B M Patten

FIGURE 25.27. Stereogram of dorsal half of fetal human heart. Lower part of septum primum acts as one-way valve over oval foramen in septum secundum. The arrows illustrate how blood is differentially distributed to the head and body regions of the early fetus. Adapted from B. M. Carlson, *Patten's foundations of embryology,* 5th ed., McGraw-Hill, New York, 1988. Used by permission.

later differentiates into the membranous portion (as opposed to the muscular portion) of the definitive **interventricular septum.**

The heart's conduction system follows the progress of cardiac subdivision. The system imposes the heartbeat on cardiac myocytes, which beat spontaneously from the very beginning of heart formation. The cells are self-excitable and tend to contract as a unit as soon as they are in contact with each other, but the conduction system of modified muscles produces progressive, positive propulsion in a venous-to-atrial direction.

Contraction throughout the embryonic heart muscle is integrated by faster beats initiated at the venous end of the heart imposing their rhythm on slower beats at the atrial end (Patten, 1949). The conduction system in adults consists of two parts. The **sinoatrial node** (SA node, the "pacemaker") generates a rhythmic electrical impulse. The **atrioventricular node** (AV node) conveys the impulse to the ventricle via the **AV bundle** or **Bundle of His** and **conduction myofibers** or **Purkinge fibers.**

The AV node and bundle come into play before the SA node. In the chick, incipient beats begin at about 29 hours of incubation, and electrophysiological properties begin as early as 1½ days of in-

cubation. In human beings, regular heartbeats begin at the end of the third week of pregnancy.

Chick cardiac cells lining the AV canal do not contribute to the conduction system. Precursors of the AV node and bundle are not present before 5½ days of incubation. At 6 days, cells capable of generating slow and long-lasting action potentials, resembling those of the AV node and bundle, are found in the lowest and dorsal segments of the IAS. At this time, separation of the atria by the 2° IAS and the endocardial cushions is completed (Fig. 25.28) (Argüello et al., 1988).

Aortic Systems

A vertebrate is fed by its aortic system. The heart can only push blood, but the arteries must bring it to where it can nourish.

Arteries are large pulsatile vessels carrying blood away from the heart. In amniotes, **internal carotid arteries** combine with the **basilar** and **vertebral** to feed the brain, and the massive dorsal aorta, coursing back between the notochord and gut, feeds almost all the rest of the body and extraembryonic membranes (Fig. 25.29).

In mammals, a central **coeliac** (also celiac) and

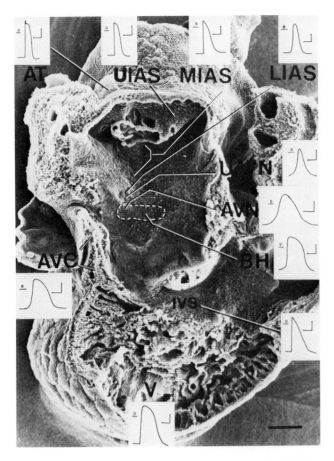

FIGURE 25.28. Scanning electron micrograph of chick heart interior at 6 days of incubation and records of intracellular electrical activity from different sites. The records show a progressive delay in onset in the lower septum (i.e., the UAVN), but only records from the lower and dorsal segment of the interatrial septum (AVC) have the slow, S-shaped rising and decay phases and long duration characteristic of action potential from the adult atrioventricular node. Large-amplitude action potentials with a fast-rising depolarization followed by a slight plateau, recorded from slightly more ventral areas, are characteristic of activity in the atrioventricular bundle of His. Bar = 200 μm. Calibration = 50 mV. 50 msec. Dashed area, area where early atrioventricular node and bundle action potentials are recorded. At, atrial wall; AVC, atrioventricular canal; AVN, atrioventricular node; BH, bundle of His; IVS, interventricular septum; LIAS, lower intratrial septum; MIAS, middle interatrial septum; UAVN, upper atrioventricular node; UIAS, upper interatrial septum; V, ventricle. From C. Argüello, J. Alanis, and B. Valenzuela, *Development,* 102: 623(1988), by permission of the Company of Biologists and the authors.

an **inferior mesenteric artery** run to the foregut and hindgut, respectively, and a **superior mesenteric artery** runs to the midgut and remnant of the yolk sac (Fig. 25.30). The paired **vitelline arteries** of chicks (see Figs. 23.11 and 23.12) running to the yolk sac would seem to be homologs of the single mammalian superior mesenteric artery.

Dorsally, paired **intersegmental arteries** lying between sclerotomes become the **intercostal** and **lumbar arteries** feeding the body wall. Near the border of the neck and thorax, a pair of intersegmentals enlarges and elongates to form the **subclavian arteries** to the forelimb (Fig. 25.31). Posteriorly, another pair of intersegmentals forms the **femoral arteries** (common iliac or ischiadic arteries, Fig. 25.29) of the hind limb. These same arteries provide the trunks of the **allantoic** (birds) or **umbilical arteries** (therian mammals). Laterally, **segmentals** feed the mesonephros, and centrally, paired or fused segmentals feed the adrenals, kidneys (renals), and gonads.

From its inception, the dorsal aorta is paired anteriorly (Fig. 25.15). **Aortic roots** consolidate posteriorly to form the single **dorsal aorta** or descending aorta. The point of fusion is farther posteriorly in amniotes than in anamniotes. In salamanders, aortic roots fuse above the level of the subclavians, in frogs and most reptiles, just below, while in mammals, the subclavians run off right and left aortic roots (Fig. 25.23), and in crocodiles and their avian relatives, the subclavians run off the top of the third aortic arch along with the carotids.

In embryos, the aortic roots arise from sets of right and left **aortic arches** which circle the gut. The aortic arches arise from the **ventral aorta** or **aortic sac** following the **truncus arteriosus.**

The aortic system is conductive and pulsatile (i.e., capable of pulsating when stimulated), but, unlike the heart which is suspended within the pericardial cavity, the aortic vessels are fused to the splanchnopleure or ectomesenchyme. Actually, the results of experiments with chick–quail chimeras (see below, Figs. 26.16, 26.17) indicate that the walls of the aortic arches are derived mainly from neural crest cells (Fig. 25.11) despite the heart's origin from splanchnic mesoderm.

Aortic arches course through the cores of the branchial arches and are identified by corresponding numbers. Typically, the aortic arches develop in the order of their numbering, but the particular aortic arches do not depend on the sequence of development or even on whether arches with lower numbers have been produced.

Development is typically uneven. The first aortic arch may not even form and the first may lose its dorsal connection or disappear before the last emerges (Fig. 25.9). In birds and mammals, the full range of arches is never present at the same time, and in mammals, the fifth and sixth are said

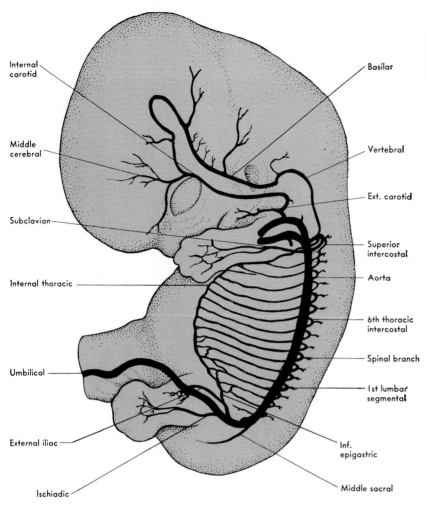

FIGURE 25.29. Major cranial and superficial arteries in human embryo of 7 weeks. From B. M. Patten, _Foundations of embryology,_ 2nd ed., McGraw-Hill, New York, 1984). Used by permission.

to be fused, although one rarely finds even a hint of two posterior aortic arches. Convention and recapitulationism, rather than morphology, require labeling the pulmonary arch of mammals the sixth aortic arch.

The arches are generally presented as variations on a basic vertebrate theme. Represented by cyclostomes (Fig. 25.32), the primitive condition has perfect symmetry, a full spectrum of complete aortic arches, and a patent ventral aorta split only at the level of the first aortic arch (I). The variations found among other vertebrates follow three characteristic paths: the loss of particular arches or connections between them (regression), the disappearance of symmetry (distortion), and the splitting of the ventral aorta (segregation). In elasmobranchs, reduction of gills on the first branchial arch is reflected in the loss of the afferent portion of the first aortic arch (Fig. 25.32, i.e., portion of vessels [stippled] leading toward the gill). Splitting of the ventral aorta accompanies the reduction of the first aortic arch. Teleosts extend both themes, losing the second aortic arch as well as the first

and continuing the splitting of the ventral aorta to the level of the third arch.

The dipnoi (lung fish) form gills on the fourth and fifth arches but the second and third aortic arches are uninterrupted as they run upward to the dorsal aortic roots. Like urodeles and some reptiles (Figs. 25.23 and 25.32), but not like frogs, dipnoi retain the dorsal link between the third and fourth aortic arches (the carotid duct). Dipnoi also supply their lungs, like all tetrapods, with branches from the sixth or last aortic arch, although the branch may be an efferent gill artery or more nearly a branch of the dorsal aorta.

With the reduction of the first and second aortic arches in amphibians, the third becomes the major source of arterial blood to the head region, especially in anurans after the loss of the **carotid duct** linking the third and fourth arches. This situation changes in amniotes when dorsal intersegmental branches fuse to form paired vertebral arteries (Fig. 25.29) and provide additional arterial blood to the head.

Amphibians' **pulmonary arteries,** like those of

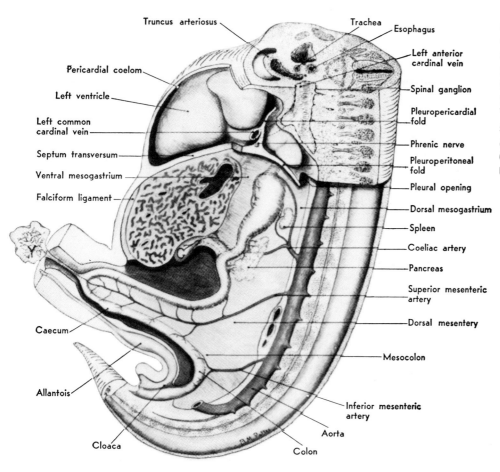

Truncus arteriosus

Pericardial coelom

Left ventricle

Left common cardinal vein

Septum transversum

Ventral mesogastrium

Falciform ligament

Caecum

Allantois

Cloaca

Colon

Aorta

Trachea

Esophagus

Left anterior cardinal vein

Spinal ganglion

Pleuropericardial fold

Phrenic nerve

Pleuroperitoneal fold

Pleural opening

Dorsal mesogastrium

Spleen

Coeliac artery

Pancreas

Superior mesenteric artery

Dorsal mesentery

Mesocolon

Inferior mesenteric artery

FIGURE 25.30. Stereogram of posterior part of human embryo at end of embryonic period showing major internal arteries to the viscera and mesenteries. G, gall bladder; Y, yolk sac. From B. M. Carlson, *Patten's foundations of embryology,* 5th ed., McGraw-Hill, New York, 1988. Used by permission.

other tetrapods, are connected to the sixth aortic arch (Figs. 25.23 and 25.32), but they do not form in the typical vertebrate pattern. Amphibian pulmonary arteries arise separately and fuse with the sixth arch, which may already be complete between the ventral aorta and a dorsal aortic root. The **ductus arteriosus,** or dorsal portion of the arch, allows blood to bypass pulmonary (or pulmonocutaneous) circulation and reach the dorsal aorta directly.

With amphibians, the fourth aortic arches move into the category of **systemic arches.** At metamorphosis, the paired ducti arteriosi shut down and the fifth arches degenerate except in permanently aquatic adults (e.g., *Xenopus laevis*). The fourth aortic arches thereby become the tap roots of the aortic roots.

While amphibians retain the symmetrical development of aortic arches, symmetry breaks down within the ventral aorta. A spiral valve in the truncus arteriosus (Fig. 25.23) directs blood in either of two directions depending on its source. Depleted blood from the general circulation is directed to the sixth aortic arch and hence to the lungs or skin (the alternative respiratory organs of amphibians) and to the dorsal aorta via the ductus arteriosus.

Oxygenated blood returning from the lungs is directed to the head and general body circulation.

With reptiles and mammals, asymmetry reaches the aortic arches. Reptiles favor the right fourth aortic arch, although the left fourth aortic arch is preserved. Birds go further by allowing the connection between the left fourth aortic arch and the dorsal aorta to regress.

Mammals, on the other hand, favor the left fourth aortic arch and allow the right fourth aortic arch to regress. Since it is unlikely that the process of regression begun on the left side in reptiles could have been halted and reversed for the evolution of mammals, the ancestor of mammals might have been an amphibian with aortic symmetry (bottom arrows, Fig. 25.23).

In that event, the four-chambered hearts of birds and mammals would seem to be derived by convergence. Similarities in the respective chambers would then represent analogies, based on function, rather than homologies, based on their presence in a common ancestor. Parallels in the development of chambers may signify requirements of function or extraordinary coincidences rather than palingenesis.

With the loss of the fifth aortic arch, the prox-

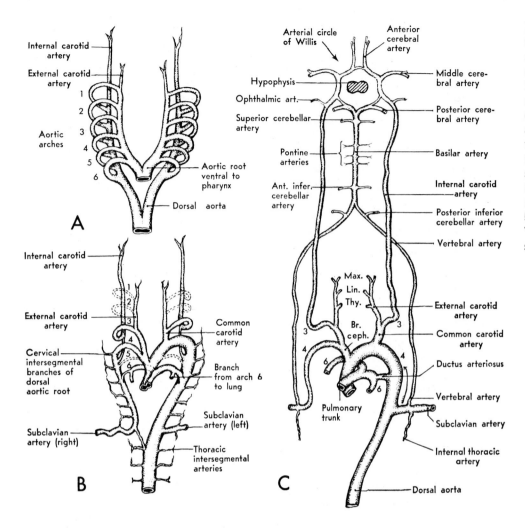

FIGURE 25.31. Diagrams of the major changes in the aortic system of mammalian embryos. (a) Ground plan. (b) Early changes. (c) Adult derivatives. Br. ceph., brachiocephalic artery; Lin., lingual artery; Max., maxillary artery; Thy., thyroid artery. Arrow in c, later change in position of left subclavian. From B. M. Carlson, *Patten's foundations of embryology,* 5th ed., McGraw-Hill, New York, 1988. Used by permission.

imal portion of the right fourth aortic arch becomes the **brachiocephalic** artery (Fig. 25.31). When the distal end of the right fourth aortic arch regresses, the arched portion becomes the root to the right **subclavian artery.** Similarly, the branch of the sixth arch to the right aortic root regresses, leaving only the branch on the left side as a single ductus arteriosus.

Separation in the ventral aorta or truncus arteriosus is especially profound in reptiles where three vessels replace one (i.e., roots of right and left fourth aortic arches and the root of the sixth arch,

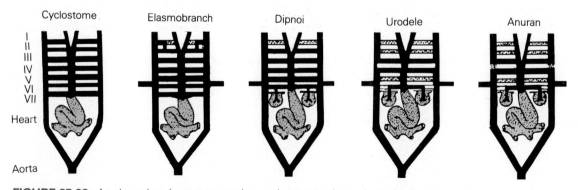

FIGURE 25.32. Aortic arches in representative cyclostome, elasmobranch, dipnoi, urodele (postmetamorphic), and anuran (adult). In permanently aquatic amphibians (e.g., neotenous urodeles and *Xenopus laevis*), the Vth arch remains intact. Roman numerals identify aortic arches. Adapted from H. E. Lehman, *Chordate development,* 3rd ed., Hunter Textbooks, Winston-Salem, 1987.

Fig. 25.23). Birds and mammals, evolving in parallel, show parallel degrees of separation in their ventral aortas.

The mechanism of separation involves paired **ridges** (called truncus ridges) which begin in the wall of the ventral aorta at the level of the pulmonary sixth aortic arches and spiral down to the ventricle. The paired ridges fuse centrally within the ventral aorta and truncus, thereby separating **pulmonary** and **systemic channels** or **trunks.** Within the heart, the ridges line up with the dorsal and ventral endocardial cushions and, fusing with them, match the pulmonary trunk with the right half of the ventricle and the aortic trunk with the left half.

When the interventricular septum and endocardial cushions finally fuse to close the remaining portion of the interventricular foramen (Fig. 25.27), the separation that begins in the ventral aorta extends to the left and right ventricles. In a beautiful display of synchrony and precision in development, pulmonary and systemic arterial functions are separated completely between the right and left ventricles of the heart.

Venous Systems

The venous portions of the embryo's systemic and extraembryonic circulatory arcs (Fig. 25.16) are related developmentally. The vitelline veins, arising in the splanchnopleure, drain the extraembryonic yolk sac and the embryonic intestine. In anamniotes, the vitelline veins branch off anteriorly to the somatopleural cardinal and abdominal veins draining the embryo superficially. The splanchnopleural allantoic or umbilical veins of amniotes usurp the somatopleural abdominal veins of anamniotes for extraembryonic drainage from the allantois or placenta.

The juncture of the splanchnopleure and somatopleure is crucial in bringing these veins together and continues to be a meeting place for the vitelline, cardinal, and allantoic veins during the remainder of embryonic development. The common cardinal veins descend through the lateral mesocardia, and the vitelline derivatives and allantoic veins (rerouted through the liver) move up to the heart through the **septum transversum** (Fig. 25.30).

The initial tendency is for all three venous systems to enter the heart together, but later, with the absorption of the sinus venosus into the atrial wall, anterior and posterior venous systems tend to enter the heart separately (anterior portions of the posterior cardinal veins represent holdouts). With the disappearance of the vitellines and um-

bilicals or allantoics at term or hatching, the last traces of collectivity in the venous system are lost.

Vitelline veins[6] are the first parts of the venous system to appear. Paired but heavily branched, they arise in the splanchnopleure within the embryo and consolidate the myriad of vessels in the extraembryonic yolk sac (e.g., birds). The vessels develop most extensively on the yolk sac but have their greatest developmental impact as the source of the **hepatic portal system** (Fig. 25.33).

The substance of the vitelline veins is broken up by developing liver trabeculae, and the anterior part of the veins form **sinusoids.** From early on, the right side is favored, and as sinusoids meet and anastomose, the right proximal portion of the vein emerges as the definitive **hepatic vein.**

Posteriorly, channeling through sinusoids and anastomoses between right and left veins result in the formation of a single **portal** or **hepatic portal** vein circling the small intestine and collecting branches from other points of abdominal drainage.

Cardinal veins are pairs of anterior (superior or upper) and posterior (inferior or lower) cardinal veins and their common cardinal veins (sometimes called ducts of Cuvier) draining into the sinus venosus (Fig. 23.12). In anamniotes, development of the cardinal system maintains a high degree of bilateral symmetry, while in amniotes, development tends to pare down the paired system to single drainage vessels.

The anamniotic common cardinals tend to be absorbed by the anterior cardinal veins, and the **subclavian veins** draining the fore limbs, including pectoral fins, shift origins anteriorly to the anterior cardinals. A similar shift to the anterior cardinals occurs in amniotes with the formation of **brachiocephalic veins** draining both the anterior cardinals and the subclavian veins (Fig. 25.34), but otherwise, anterior cardinals in amniotes become considerably different from those in anamniotes.

In birds and mammals, the left brachiocephalic vein (formerly innominate vein) is a new formation rather than a derivative of the common or anterior cardinal. By swinging the anterior drainage from left to right sides, the left brachiocephalic changes the right common cardinal into an unpaired **superior vena cava** (Fig. 25.34). Freed of other drainage, the left common cardinal becomes the unpaired **coronary sinus** or sinal horn draining the heart muscle.

[6]The name **vitelline** replaces omphalomesenteric (Gk. *omphal-* navel + mesentery) and removes the distinction formerly made between omphalic portions of the veins on the yolk sac and mesenteric portions within the embryo.

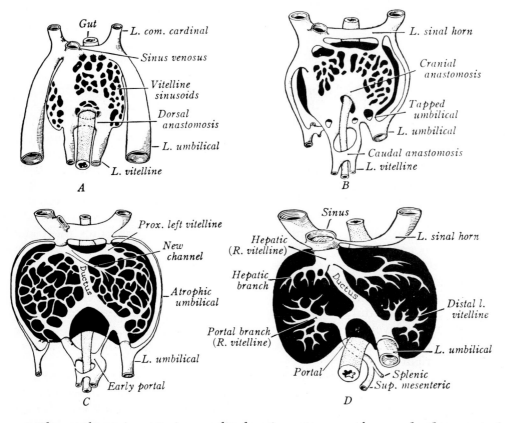

Gut
L. com. cardinal
Sinus venosus
Vitelline sinusoids
Dorsal anastomosis
L. umbilical
L. vitelline

A

L. sinal horn
Cranial anastomosis
Tapped umbilical
L. umbilical
Caudal anastomosis
L. vitelline

B

Prox. left vitelline
New channel
Ductus
Atrophic umbilical
L. umbilical
Early portal

C

Sinus
Hepatic (R. vitelline)
Hepatic branch
Ductus
L. sinal horn
Distal l. vitelline
Portal branch (R. vitelline)
L. umbilical
Portal
Splenic
Sup. mesenteric

D

FIGURE 25.33. Schematic in ventral view of the developing human hepatic portal system and the ductus venosus: (a) 4.5 mm; (b) 5 mm; (c) 6 mm; (d) 9 mm. Liver black; gut stippled. From L. B. Arey, *Developmental anatomy,* 7th ed. Saunders, Philadelphia, 1965. Used by permission.

The embryonic anterior cardinal veins run along the dorsal edges of the pharyngeal pouches draining the brain and cervical region. With the exception of mammals, the anterior cardinal veins give rise to paired **lateral head veins** which receive drainage from the brain of the adult.

In mammals, the anterior cardinals give rise to **internal jugular veins,** draining blood collected from deep in the brain, and to **external jugular veins,** draining blood from more superficial parts of the brain and head (Fig. 25.34). These veins, in turn, deliver blood to the **common jugulars** which pass it along to the brachiocephalic veins.

The posterior cardinal is closely associated with the development of the **renal portal system.** In anamniotes and tetrapods with venous circulation through the kidney, caudal veins (or caudal plus iliac veins in the case of tetrapods) deliver blood to the kidney. The posterior cardinal veins drain the kidney and deliver the filtered blood to the heart. In lung fish and amphibians, the posterior cardinal veins meet centrally to form an **inferior vena cava,** but centralization of posterior drainage is greater in reptiles and birds with the reduction of the renal portal system and in mammals with its elimination.

The inferior vena cava, in amniotes, takes its origins from the posterior cardinal and portions of several other posterior veins. Initially, venous sinuses arising from posterior cardinal veins form plexes and coalesce into ventromesial **subcardinal veins** (Fig. 25.34). Amorphous from their inception, the subcardinal veins meet on the midline and form numerous **intersubcardinal anastomoses.** Presumably, in response to pressure, these sort themselves out into a major channel.

The portion of the subcardinals absorbing the renal veins from the metanephros becomes the **interrenal portion** of the inferior vena cava (lightly stippled in figure). Farther posteriorly, anastomoses with **supracardinal veins** (originally draining the dorsal body wall into the posterior cardinals) result in the formation of the **postrenal portion** of the inferior vena cava (horizontal hatching).

Anteriorly, at the cephalic end of the mesonephros, the subcardinal either makes contact with vessels in a mesentery connected with the liver (e.g., pigs, where the fold is known as the caval plica) or in the dorsal body wall (e.g., humans). Enlarging on the right, the connections between the subcardinals become the **mesenteric portion** of the inferior vena cava (coarse stippling).

Continuing proximally, the inferior vena cava excavates a **hepatic portion** along the dorsal side of the liver. Finally, joining with the right vitelline (hepatic) and the umbilical, the composite inferior vena cava enters the sinus venosus.

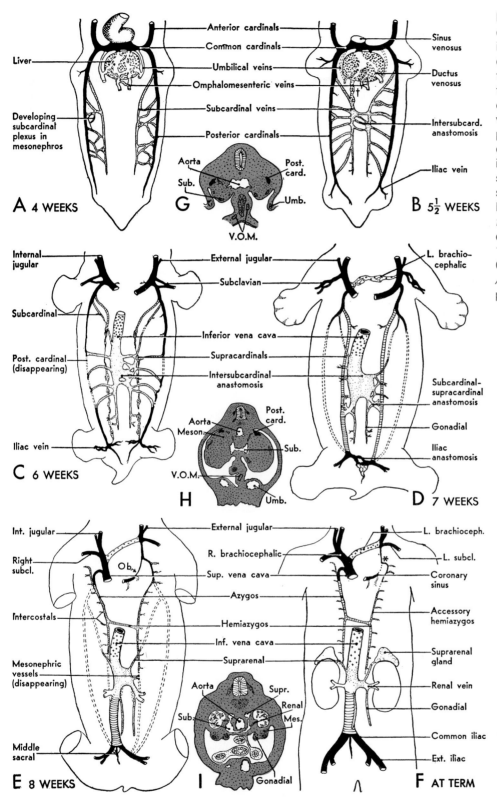

FIGURE 25.34. Schematic for developing human superior and inferior vena cavas. Black, cardinal veins and their derivatives; coarse stippling, mesenteric portion of inferior vena cava; light stippling, subcardinal veins; horizontal hatching, supracardinal veins. L., left; Ext., external; Post., posterior; Int., internal; Intersubcard., intersubcardinal; R., right; Sup., superior; subcl., subclavian. From B. M. Carlson, *Patten's foundations of embryology,* 5th ed., McGraw-Hill, New York, 1988. Based on C. F. W. McClure and E. G. Butler, *Am. J. Anat.,* 35:331(1925). Used by permission.

Allantoic veins take their origins in splanchnopleure along with the allantois. They accompany the allantoic stalk or stalk of the allantoic vesicle to the chorioallantoise in birds, reptiles, and monotremes, or, as **umbilical veins,** they reach the placenta of therian mammals. Later, the same veins are enclosed in the amnion-covered umbilical cord of therian mammals.

The umbilical veins never accompany the hindgut beyond the allantoic vesicle. Instead, at the point where the somatopleure and splanchnopleure meet at the level of posterior and lateral body folds (Fig. 25.17), the umbilical veins link up with the somatopleural homologs of the anamniotic **abdominal veins.** Usurping these superficial channels, the umbilical veins course forward in the body wall.

Originally, the umbilical veins leave the body wall anteriorly. They deliver blood to the common cardinals and hence to the heart (Fig. 25.33), but when the liver makes contact with the anterior abdominal body wall, the umbilical veins begin to exploit the liver as their channel to the heart. Abandoning their more anterior portions to atrophy, the umbilicals take over vitelline sinuses and excavate a channel through the liver.

The umbilical veins may fuse into a single vessel in the umbilical cord (e.g., in primates but not artiodactyles), and only the left umbilical vein may run from the cord to the embryo proper. Responding to increased pressure from the volume of blood passing through this vein, the liver walls off the umbilical channel as the **ductus venosus.** Umbilical venous blood then passes through the anterior remnant of the right vitelline vein (i.e., the hepatic) and hence to the sinus venosus.

Changes at Birth

The heart and several vessels change dramatically at birth in mammals. The changes in eutherians are in response to air breathing and frequently promote air breathing as well.

In embryos and fetuses of tetrapods with four-chambered hearts, the ductus arteriosus aids in equalizing pressure on both sides of the heart as well as in shunting blood to the dorsal aorta. At birth or hatching, the ductus arteriosus is largely shut down, although some backflow to the left pulmonary artery continues for a few days. Closure of the ductus arteriosus is triggered by the high oxygen content of the aortic blood, and by **bradykinin,** the small polypeptide vasodilator, which is released by lungs during their initial inflation.

Bradykinin and high oxygen levels are also responsible for the constriction of the umbilical arteries at birth. Bleeding through the umbilical arteries is thereby minimal, even when cut without benefit of hemostats.

The left umbilical vein of primates remains open for a while after birth, allowing placental blood to reenter neonatal circulation. When completely collapsed and replaced by connective tissue, the intraembryonic portion of the vein becomes the **ligamentum teres hepatius** (ligament Lat. *teres* smooth, round, *hepaticus* liver).

With the ventilation of the lungs, pulmonary vascular flow increases markedly, raising pressure in the left atrium. At the same time, the collapse of placental circulation results in a diminution of blood volume returning to the heart via the posterior vena cava and a drop in pressure in the right atrium.

The same system of double septa and dual foramena which allowed flow between the right and left atria in the fetus (Fig. 25.27) now provides a mechanism for interrupting this flow. Higher blood volumes and venous pressure on the left side of the heart now press the 1° IAS against the foramen ovale like a valve and prevent flow between the atria.

The new equilibrium remains flexible for a while, but a few months after birth, in the case of human beings, the valve heals in place and the IAS is sealed. The process of cardiac subdivision, begun much earlier in the embryo, and the transition to air breathing, begun abruptly at birth, finish in a dead heat.

VERTEBRATE UROGENITAL SYSTEMS

The urogenital system is a composite of urinary and genital systems. Both systems move internally produced materials, from metabolic wastes to new organisms, outside the body. In invertebrates, the movement of these materials is generally done by separate ductworks, but vertebrates combine ducts into a single delivery system.

Urinary Systems

The urinary, excretory, or **nephric** (Gk. *nephros* kidney) part of the urogenital system is derived from **nephrogenic mesenchyme** or **nephrotome** (= intermediate mesoderm) via induction by **nephrogenic ducts** (also of intermediate mesoderm). Beginning as solid nephrogenic cords in the cervical region, hollow nephrogenic ducts pursue superficial paths backward to the cloaca. Wherever the ducts touch down in the adjacent nephrogenic mesenchyme, they induce proliferation and **tubulogenesis,** the differentiation of mesenchyme into a simple epithelium organized as a tubular **nephron** or kidney tubule. Tubules of increasing complexity are induced as the ducts move posteriorly, and, in amniotes, nephrons of even greater complexity are induced in the heels of the nephrogenic mesenchyme by **ureteric buds** sprouted from the ends of the nephric ducts (Jacob et al., 1986).

The results of ablation and *in vitro* experiments suggest that the type of tubules developed is a function of the portion of intermediate mesoderm being induced and not of the nephrogenic duct performing the induction. Moreover, nephrogenic mesenchyme can be induced *in vitro* with spinal cord, suggesting that the induction is per-

missive, but, unlike other permissive systems, cell–cell association seems to be required for tubule formation (Saxén et al., 1980).

The pronephric tubules (i.e., head kidney tubules, Fig. 25.35a) of anamniotes are induced in the pronephrogenic region of intermediate mesoderm. The anterior pronephric tubules open onto the coelom in **nephrostomes** or **peritoneal funnels,**

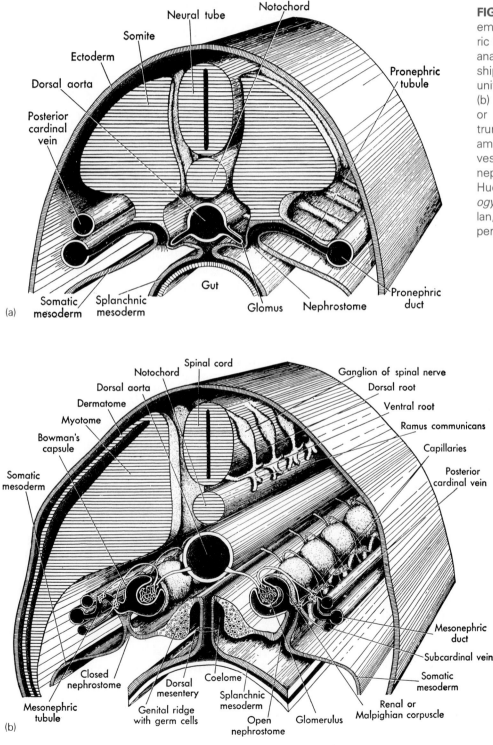

FIGURE 25.35. Stereogram of embryonic kidneys. (a) Pronephric (cervical) area of a generalized anamniote showing relationship of vessels to pronephric units and nephrogenic duct. (b) Opisthonephros (back kidney) or mesonephros (midkidney) in trunk region of generalized anamniote showing relationship of vessels to nephric units and nephrogenic duct. From A. F. Huettner, *Comparative embryology of the vertebrates,* Macmillan, New York, 1949. Used by permission.

and ciliated cells drive coelomic fluid into each tubule's lumen or **nephrocoel.** Posterior pronephric tubules, lacking nephrostomes, are intimately associated with dense capillary networks, or **coelomic glomi,** and function in the filtration of blood. Capillaries or sinuses surrounding the distal portions of the tubules function in collecting the products of tubular filtration while the remaining filtrate in the nephrocoels is passed along to the nephrogenic duct (in its role as urinary collecting duct).

Some pronephric tubules develop in all anamniote embryos (e.g., as many as five in the embryos of freshwater varieties of teleosts and some of their marine descendants). The collective mass of pronephric tubules constitutes the **pronephros** (Gk. *pro-* before + *nephros*). It reaches its fullest development in myxinoid cyclostome embryos (i.e., extending over 70 segments) and persists in adult hagfish and some teleosts, but in sharks the pronephros remaining in the adult is nonfunctional. In other anamniotes, the pronephros degenerates completely. In amniotes, rudimentary pronephric tubules may appear in the cervical region, but they never develop the appurtenances of filtration, and they always degenerate.

Opisthonephric tubules (i.e., back kidney tubules, also called mesonephric tubules or mid-kidney tubules, Fig. 25.35b) are more complex than pronephric tubules and are induced behind the pronephros in the intermediate mesoderm of the opisthonephric region (or mesonephrogenic region) of anamniotes. The typical opisthonephric tubule generally lacks a nephrostome and has an individual **glomerulus** encapsulated in a cup-shaped **glomerular capsule** (also called Bowman's capsule).

Elongated tubules and sometimes branched portions of the nephrogenic duct become wrapped in networks of capillaries or sinuses. The collective mass of tubules, capsules, and blood vessels, known as the **opisthonephric kidney** or **opisthonephros** (Gk. *opisthen* behind or back + *nephros*, also called mesonephros, or mesonephric kidney), lies along the dorsal surface of the body cavity as a pair of ribbons (sharks), rippling masses (amphibians), or bodies in any of a variety of shapes (fish).

Mesonephric kidneys or **mesonephros** (Gk. *mes-* mid- + *nephros*, Fig. 25.36) are the amniotic homologs of anamniotic opisthonephric kidneys.[7] The mesonephros functions in the filtration of

[7]The term **mesonephros** should be reserved for amniotes, since, having only two kidneys, anamniotes do not have a "middle" kidney.

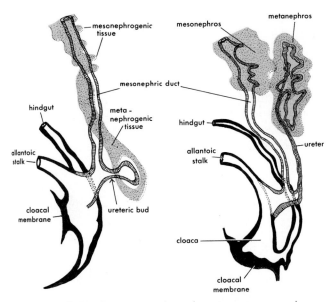

FIGURE 25.36. Reconstruction of ureters, mesonephros, and metanephros in a rabbit embryo. Only the left side of the paired nephric system is shown. From B. I. Balinsky, *An introduction to embryology,* 5th ed. Saunders, Philadelphia, 1981. Used by permission.

blood, in hemopoiesis, and as the site of **adrenal cortex** formation. Among mammals, the mesonephros is especially conspicuous in artiodactyl embryos (e.g., the pig, Fig. 25.4) where the epitheliochorial placenta is relatively inefficient in the removal of metabolic wastes from fetal blood.

As a corollary to the more efficient removal of wastes by the hemochorial placenta, mesonephric development is found to be rudimentary in rodents and poor in primates. Moreover, mesonephric development is followed by early degeneration. In human embryos, five-sixths of the cranial end of the mesonephros is reduced to a suspensory ligament by the second month (Fig. 25.37, compare c to Fig. 25.4 for the pig). By 10 weeks, no unbroken tubules remain in the mesonephros.

In amniotes, the anterior portion of the nephrogenic duct is frequently called the **pronephric cord** or archinephric duct, while the hollow caudal portion is called the mesonephric or Wolffian duct. The duct is typically identified as the **mesonephric duct** when discussed in its capacity as part of a urinary delivery system. The term **Wolffian duct** is used mainly to identify the duct as part of the genital system. Since the same cord or duct seems to induce the formation of nephric units at all levels of the intermediate mesoderm, the duct should probably be identified simply as a nephric duct.

The metanephric kidney or metanephros (Gk. met- *after* **+ nephros) is the last kidney in am-**

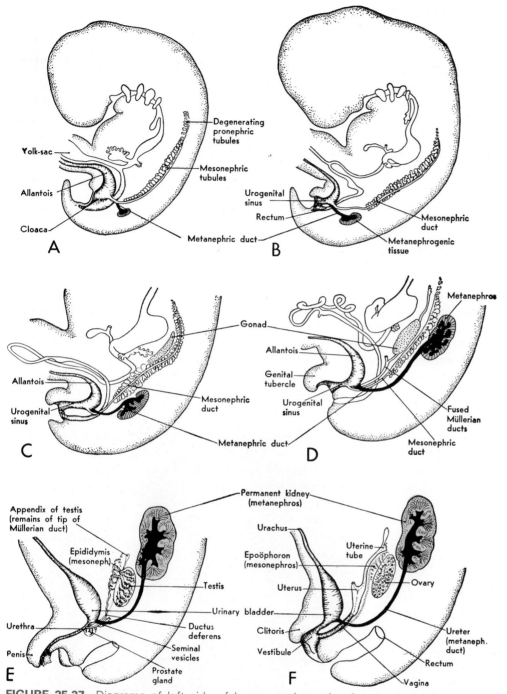

FIGURE 25.37. Diagrams of left side of human embryo showing mesonephric, meta-nephric, and genital development in the indifferent stage (through d), in male (e) and in female (f). (a) Early in fifth week (5–6 mm). (b) Early in sixth week (8 mm). (c) Seventh week (14.6 mm). (d) Eighth week (23–25 mm). (e, f) Male and female about 3 months. (Note: Intestine and colon have been removed for clarity.) From B. M. Carlson, *Patten's foundations of embryology*, 5th ed., McGraw-Hill, New York, 1988. Used by permission.

niotes and the functional kidney of adults. Development begins with the emergence of **ureteric buds** (also called metanephric diverticula) from the caudal ends of the nephric ducts. The buds extend dorsally and reenter the heel of the nephrogenic mes-

enchyme in a region called the **metanephrogenic mesenchyme** or tissue (Fig. 25.36).

Basally, the ureteric bud elongates into the **ureter** (Fig. 25.37). Apically, repeated branching and growth produce a series of metanephric **collecting**

ducts which induce the metanephrogenic mesenchyme to form **metanephric tubules** or **metanephric nephrons**. Proliferation is the first response to induction, followed by enhanced intercellular adhesion, condensation of cells, polarization (i.e., the basal displacement of nuclei and the appearance of apical microvilli), and tubulogenesis.

These metanephrogenic processes can be performed *in vitro* under the influence of either the ureteric bud or the heterologous inducer, embryonic spinal cord (Grobstein, 1956). Some of the metanephrogenic processes can also be isolated *in vitro* with the help of specific antisera.

Monoclonal antibody directed against the cell-surface **disialoganglioside** G_{D3} reduces the branching of the ureter and prevents the growth and condensation of mesenchymal cells. G_{D3} is not expressed in the ureter nor in the mesenchymal cells that can be converted into tubular epithelium but only in the portion of the mesenchyme that will form stroma around the nephric tubules (Fig. 25.38). The antibody seems to work indirectly by interfering with mesenchymal–epithelial interactions (Sariola et al., 1988).

The ureter is not influenced by anti-laminin antibodies, presumably because its basal lamina is already heavily invested with laminin. The neural cell adhesive molecule, N-CAM, and fibronectin are also present before mesenchymal condensation starts, and antibodies against them fail to inhibit further development. Uvomorulin appears during mesenchymal condensation but is not localized, appearing all over the surfaces of condensed cells, and anti-uvomorulin antibodies do not disrupt tubulogenesis.

On the other hand, antisera against laminin inhibit polarization of the mesenchyme. Laminin is a cruciform, trimeric molecule. Its long arm contains an A chain backbone, and its two branches consist of B 1 and B 2 chains. The active antisera are directed against the carboxy-terminal portion of the long arm (i.e., A chain) containing a cell binding site.

The B 1 and B 2 chains are present throughout early development, but the A chain only appears in metanephric mesenchyme cells following induction. Large amounts of the complete trimeric molecule are then deposited in the extracellular

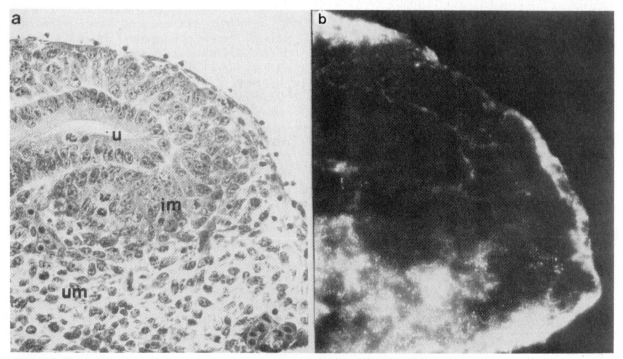

FIGURE 25.38. Section of 12-day embryonic mouse kidney as ureter branches into mesenchyme. (a) Light micrograph. (b) Darkfield indirect immunofluorescence stained for ganglioside G_{D3} antigens. Induction is manifest as the mesenchyme condenses around ureter. The ganglioside is abundant in the uninduced mesenchyme but apparently absent in the ureter and induced mesenchyme. im, induced mesenchyme; u, ureter; um, uninduced mesenchyme. ×500. From H. Sariola et al., *Cell*, 54:235(1988) by permission of Cell Press and the authors.

matrix at the basal end of cells. The induction of laminin A synthesis and the basal localization of the adhesive molecule suggest that the molecule plays a crucial role in polarizing the metanephric mesenchyme prior to tubulogenesis (Klein et al., 1988) (Fig. 25.39).

Resembling the more posterior mesonephric tubules, the metanephric nephrons have glomerular capsules enclosing glomerulae and long metanephric tubules surrounded by capillary networks. Unlike the mesonephros, the metanephros tends to be compact with metanephric glomeruli and glomerular capsules distributed peripherally in a cortex, while tubules and collecting ducts are concentrated centrally in a medulla.

The metanephros is complete when branching, growth, and induction cease, and the proximal portions of some of the collecting ducts regress, thereby forming large **pyramids** draining urine into the ureter. The openings of the ureters move from their origins on the nephric ducts to the base of the allantoic vesicle. The allantoic stalk later degenerates into a ligament (the urachus) while the base of the allantoic vesicle swells into the definitive **urinary bladder.** The ureters then join the bladder to the adult metanephric kidneys (Fig. 25.37).

Genital Systems

Primary sexual structures or the **genital apparatus** includes gonads, or reproductive organs producing sex cells, and the tubes and ducts that deliver these cells or the products of fertilization outside the body. Frequently, vertebrates also have **secondary sexual structures** including a variety of appurtenances that function in sexual behavior to promote fertilization and, after the production of sex cells or offspring, to provide brooding and care (see Chapter 5).

Animals are generally identified by gender on the basis of their genital apparatus: males having testes, females having ovaries, and hermaphrodites having both testes and ovaries. Intersex is the congenitally or experimentally produced condition in which an individual has malformed gonads or combined ovotestis.

Usually two **sexes** are present in vertebrate species, males and females, but some teleosts have hermaphrodites and males, and gynogenic and parthenogenetic species consist of females (see Chapter 5). Gender differences in external appearance and behavior identify **sex morphs** (sometimes called sex phenotypes), and species having two conspicuously different sex morphs are **sexually dimorphic.**

All the many features of genital development are generally harmonized in adults. Developmentally, karyotypic sex (sex-determining chromosomes), genetic sex (sex-determining genes), gonadal sex (ovary versus testis), and somatic sex (nongerminal parts of the sexual apparatus and bodily appearance) have independent features that must be integrated.

Sex determination in vertebrates is usually attributed to particular chromosomes inherited differentially in two sexes (see Chapter 5). In invertebrates, such as *Caenorhabditis elegans* and *Drosophila melanogaster,* chromosomal sex determination harks back to the ratio of the sex-determining chromosome (X) and the population of autosomes, 2:1 determining a female. In vertebrates, a single sex chromosome (e.g., the Y chromosome

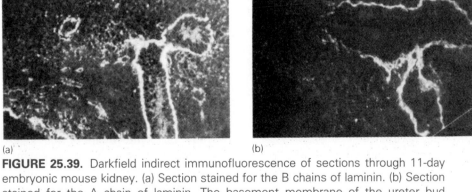

(a)   (b)

FIGURE 25.39. Darkfield indirect immunofluorescence of sections through 11-day embryonic mouse kidney. (a) Section stained for the B chains of laminin. (b) Section stained for the A chain of laminin. The basement membrane of the ureter bud contains laminin with the A chain and both B laminin chains, while the mesenchyme contains laminin B chains exclusively. ×200. G. Klein et al., *Cell,* 55:331(1988) by permission of Cell Press and the authors.

in mammals) determines sex independently of a ratio with autosomes.

Primary germ cells with an XY or XYY karyotype are generally required for testicular development in mammals and, along with cells of the XO karyotype, are incompatible with oocyte differentiation and ovarian development. On the other hand, primary germ cells with an XX or XXX karyotype are generally required for oocyte differentiation and ovarian development and, along with cells of the XXY karyotype, are incompatible with spermatocyte differentiation in testicular development (see McLaren, 1984).

One should note that the concept of "determination" implied by chromosomal sex determination differs sharply from the concept of determination ordinarily implied by the word, namely, invisible differentiation or an ability to self-differentiate. The ability of chromosomes to dictate sex is limited, and sexual structures and the sexual form (or sex morph) of an individual are not generally self-differentiating characteristics.

In male XX/XY chimeric mice (see Chapter 12), for example, XX cells can become part of a seminal vesicle (see below, Fig. 25.41). Despite their "female" karyotype, the XX cells secrete specifically male proteins into semen (Mintz et al., 1972).

Secondary sexual characteristics likewise are generally determined by gonadal hormones rather than by sex chromosomes. If anything, chromosomal genes must supply cells with hormonal receptors, but these genes may not be linked to the organism's sex chromosome constitution. Thus, XY individuals carrying the *testicular feminization (Tfm)* mutation develop testes but fail to become males due to defective androgen receptors. Instead, cells respond to endogenous estrogen and produce a female sex form (Lyon and Hawkes, 1970).

Sex commitment is the first step in determination that drives the organism toward a single sexual form (or morph). In mammals, sex commitment probably depends on the early action of genes on the Y chromosome. Several candidates have been proposed on the basis of antigenic differences attributable to the Y chromosome. Weak histocompatibility antigens (H-Y antigens detected through cellular immunity and the action of T lymphocytes) (Wachtel et al., 1975) have received the most attention, but **serologically detected male** antigens (SDM, detected through humoral immunity and the action of B lymphocytes) (see Stewart, 1983) have also been studied.

Actually, H-Y and SDM antigens may be the same, since female mice sensitized to H-Y antigens by syngeneic grafts of male skin also produce antibodies that are toxic to male cells, including sperm. Similarly, tissue from the heterogametic sex of other vertebrates (e.g., the ZW hen) elicit both cellular and humoral immunity in the homogametic sex (e.g., ZZ rooster).

Serological evidence indicates that the H-Y antigen is expressed in males of virtually all mammalian species. Expression is generally controlled at a locus on the short arm of the Y chromosome, but the situation may be more complex. The mole-vole, *Eliobus lutescens*, for example, lacks the Y chromosome (i.e., having an XO karyotype in males and females) but expresses the H-Y antigen in males. Moreover, ambiguity shades some results (see McLaren, 1987). All-female lines of the wood lemming, *Myopopus schistocolor*, have a Y chromosome but are variously reported as not expressing the H-Y antigen or as expressing it. Likewise, XO female mice and human beings have been reported both positive and negative for the H-Y antigen.

Other genes involved in testis determination are less problematic (while not problem-free). In mice, a *sex-reversal factor (Srf)* gene, which causes XX individuals to develop the male sex form, turns out to be a translocated piece of the Y chromosome. Similarly, the *testis-determining factor (TDF)* gene of human beings is a portion of the Y chromosome.

Individuals whose sexual phenotype conflicted with their karyotype (namely, XX males and XY females) were used to identify the crucial portion of the Y in a DNA library. Most of the XX human males were found to carry a translocated portion of Y on one of their X chromosomes, and most of the XY females were found to have deletions in their Y chromosome. The portion of the Y chromosome most often involved in the translocations and deletions was the 1A2 interval of the short arm. This is the locus of the *TDF* gene.

After narrowing it down and sequencing it, *TDF* was found to have an open-reading frame encoding a zinc finger protein. This protein could be a transcriptional factor with a site-specific DNA binding domain (Page et al., 1987).

TDF genes may be widespread, since DNA from the 1A2 interval of human Y chromosomes cross hybridizes with regions of the Y chromosome from a variety of other mammals. The mechanism through which *TDF* pushes a gonad in the direction of testis determination remains a mystery, however, especially because a cross-hybridizing gene is also found on the X chromosome. This gene may be a nonfunctional pseudogene, an antagonist to the *TDF* on the Y chromosome, or a coordinated gene working in concert with *TDF*.

Although Y-specific antigens appear as early as the 8-cell stage in mice, the crucial event of testis commitment may not occur until 8 days postcoitum when presustentacular cells (see below) begin differentiating as sustentacular cells and are no longer available for differentiation as follicular cells. Presustentacular cells may also produce an inducing signal (via cell contacts or a diffusible "male-determining" substance) capable of recruiting other cells into supporting roles.

Gonadal development in vertebrates begins in paired genital ridges or gonadal blastema on the mesial side of the opisthonephros or mesonephros (Figs. 25.4 and 25.35b). The ridges are never far from the nephric ducts, and, in the male, the ducts are usurped for the transport of sperm (Figs. 25.40b and 25.41a). Other than the covering layer of peritoneal epithelium derived from splanchnic mesoderm, the source of blastemal cells is uncertain. Possibly, they originate from the intermediate mesoderm.

In all vertebrates, the early embryonic gonad is **sterile** (without sex cells) and would remain so if it were not invaded and colonized by **primordial germ cells** (Mintz, 1961). Originating in remote regions of the embryo (the epiblast of birds, hindgut of mammals, marginal mesoderm of urodeles, and vegetal endoderm of most anurans; see Chapter 12), primordial germ cells are either swept into the primitive gonad through the embryonic circulatory system (birds) or migrate to it (mammals) under the direction of stage-specific embryonic antigens (Donovan et al., 1987). The primordial germ cells not reaching gonads presumably die, but those arriving there become the proliferative **primary germ cells** or **gonocytes** of the genital ridge and presumptive gonad.

Early development of the gonad's somatic tissue does not depend on the presence of germ-line cells. The mammalian testis, for example, develops in the absence of germ cells. The ovary also develops in the absence of germ cells but only to a point just prior to birth. Then, prefollicular cells fail to proliferate and disappear, and the ovary degenerates to a **streak gonad.**

Embryonic vertebrate gonads are usually described as "sexually indifferent" and "sexless," but the mammalian embryo's gonad is virtually female

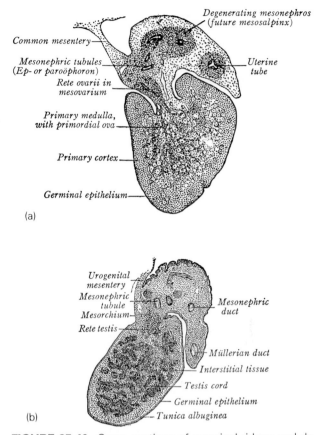

(a)

(b)

FIGURE 25.40. Cross sections of germinal ridges and developing gonads in human embryos. (a) Ovary at 3 months. Oogonia are concentrated peripherally beneath so-called germinal epithelium (actually mesothelium). The Müllerian duct is already transforming into the uterine tube, and the mesonephros is degenerating. ×44. (b) Testis at nearly 8 weeks. Primordial germ cells are developing centrally in testis cords amidst interstitial tissue. The Müllerian duct is well developed and the mesonephric duct is patent. ×70. From L. B. Arey, *Developmental anatomy,* Saunders, Philadelphia, 1965. Used by permission.

inasmuch as it develops into an ovary if left to its own devices (i.e., somatic cells in the gonad are committed to become prefollicular cells, see below). The avian embryo's gonad, on the other hand, is virtually male inasmuch as it develops into a testis if left to its own devices (i.e., somatic cells in the gonad are committed to become presustentacular cells).

The decision to develop as an ovary or testes is probably made in the gonadal somatic tissue. At some point in development, it would seem, a "switch" is thrown in the heterogametic sex to match the gonad's sex with the potential gametes' sex. In mice the switch is thrown at 8 days postcoitum and in human beings at 6–9 weeks when so-

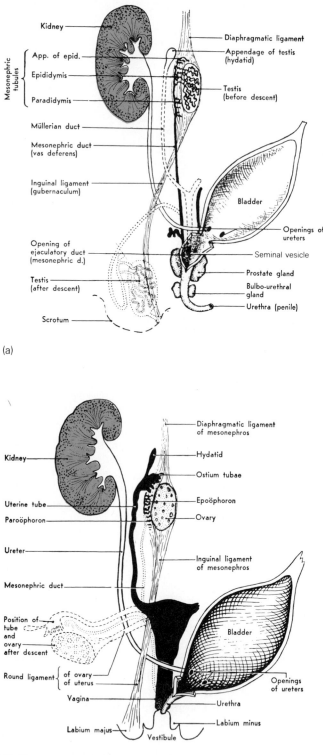

(a)

(c)

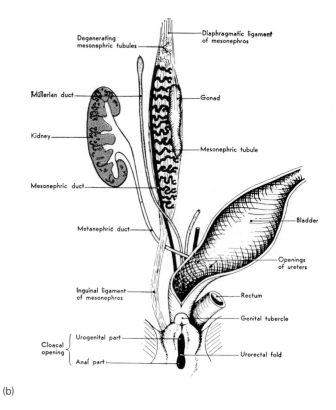

(b)

FIGURE 25.41. Schematics of developing mammalian urogenital system. (a) In male. Dashed lines indicate degenerating Müllerian ducts. Dotted figure at left shows final position of testis after descent into scrotum. (b) In sexually indifferent embryo. (c) In female. Dotted lines show degenerating mesonephric (Wolffian) duct. Dotted figure at left shows final position of ovary after contraction of round ligaments. From B. M. Carlson, *Patten's foundations of embryology,* 5th ed., McGraw-Hill, New York, 1988. Used by permission.

matic XY cells in the gonad are committed to becoming presustentacular cells.

In mammals, the switch not only removes any latent tendency toward ovarian development but may extend to the destruction of XX cells introduced experimentally or as a result of unique con-

genital environments. For example, XX/XY chimeric mice, which develop predominantly as males (in contrast to rats) (Weinberg et al., 1985), have no XX sustentacular cells in their testes, presumably because of the destruction of these cells in the male environment (see McLaren, 1987).

Moreover, karyotypic female calves, twinborn with male calves and sharing their circulation as a result of placental anastomoses, are sterile **free-martins.**

Germ cells taking the male direction become **prospermatogonia** and do not enter meiosis before birth (see Chapter 7). Instead, prospermatogonia proliferate within the confines of **testicular cords** bound by somatic **presustentacular cells** and surrounded by **interstitial endocrinocyte precursors** and mesenchyme. Spermatogenesis beyond the mitotic stage is inhibited until puberty, but the testicular cords become the **seminiferous tubules** of the adult testis supported by sustentacular (Sertoli) cells and enmeshed in interstitial endocrinocytes (Leydig cells) and stroma.

Mammalian gonocytes taking the female direction *in vivo* or *in vitro* (whether XX or XY, in male or female embryos, in ovaries or testes) cease proliferating and begin meiosis, thereby becoming primary oocytes. Becoming invested in a thin layer of follicle cells, the gonocytes become resident oocytes in **primordial follicles.** The layer of follicle cells thickens around some follicles, converting them to **primary follicles,** but further development is suspended until **puberty** (see below and Chapter 7).

Aside from the precocious beginning of meiosis, development in the mammalian fetal ovary is slower than in the testis. Not until shortly before birth is hormonal activity apparent in the ovary.

Ovarian and testicular development also differ topographically. Ovarian development is concentrated in the periphery or cortex of the gonad where oogonia proliferate and accumulate in clusters (nests) (Fig. 25.40a). Testes, on the other hand, form testicular cords in the medullary or central region and confine prospermatogonial proliferation to this region (Fig. 25.40b).

The support of proliferation in gonocytes, as opposed to their entry into meiosis, seems to be at the behest of testicular cords. Genital ridges isolated early from male mice (10.5 days postcoitum), and failing to produce cords, do not support mitosis in gonocytes. Instead, gonocytes enter meiosis. On the other hand, ridges isolated later (11.5 days postcoitum), and developing cords, support mitosis and prevent entry into meiosis. Possibly, a **meiosis-inhibiting factor** produced by testicular cords prevents meiosis while promoting mitosis.

The synthesis of other testicular secretory products does not depend on the morphogenesis of testicular cords. The early genital ridges of males that do not form testicular cords *in vitro* still produce Müllerian inhibiting substance (see below)

and testosterone. Since these products are normally produced by sustentacular cells and interstitial endocrinocytes, respectively, cytodifferentiation must have proceeded in the absence of cords. Furthermore, since these products normally shape the male genital system, cytodifferentiation, rather than cord morphogenesis, would seem to be the key step in testicular determination (see McLaren, 1987).

Linking the Urinal and Genital Systems

Linkage occurs at two points: anteriorly, between the gonad and mesonephros or associated structures, and posteriorly, between urogenital ducts and the cloaca. The linkages diverge considerably in male and female organisms.

Anterior linkage is anticipated by the degeneration of the opisthonephric kidney of anamniotes and the mesonephric kidney of amniotes. In amniotes, the mesonephros continues to degenerate posteriorly as well, but in males, before the latter degeneration starts, the anterior mesonephric tubules are usurped by the gonad as channels for the passage of sperm from the testis. The nephric ducts are then taken over for the delivery of sperm to the cloaca or its derivatives (see Blüm, 1986).

In mammals, testosterone promotes the development of the mesonephric ducts into **vasa deferens** and associated glands (seminal vesicle, prostate, bulbourethral, Fig. 25.41a). Actually, the origins of the connections between the seminiferous tubules in the testis and epididymis (i.e., the rete system and efferent ducts) are ambiguous, and some mesonephric tubules and parts of the nephric duct are not incorporated into the testis proper but remain as a **paradidymis.**

As the mesonephros degenerates in both females and males (Fig 25.40b), a new duct, the **paramesonephric** or **Müllerian duct,** develops laterally and parallel to the nephric ducts. Beginning as grooves in the peritoneal epithelium of the mesonephric ridges, the ducts have open proximal ends. In females, these ends become the funnel-shaped entrances to the **oviducts** (uterine tubes, Fig. 25.40a).

The paramesonephric ducts swing into the midplane along a shelf (the genital cord) and fuse in females to form a rudimentary **uterus** (Fig. 25.37f). The blind end of the fused ducts (Müller's tubercle) becomes the **cervix** of the uterus, while the ducts become the uterine horns (of duplex, bipartite, and bicornuate uteri, Fig. 24.20) or, by zippering backwards, the fundus and corpus of the simplex uterus.

In male mammals, a **Müllerian inhibiting substance** (MIS) checks the development of the Müllerian duct. The remnant of this duct becomes the hydatid or appendage of the testis, while the distal portion of the duct becomes the appendage to the epididymus.

Posterior linkage occurs initially at the cloaca. Except in cyclostomes, some fish, and mammals, the urinary and genital ducts continue to empty into the cloaca along with the colon (rectum) in adults. While the cyclostomes and fish externalize the urogenital openings, mammals partition the cloaca. A **urorectal fold** divides the cloaca into a **urogenital sinus** and rectum (Figs. 25.4 and 25.39b). Later, when the bladder withdraws into the pelvis, the anterior part of the urogenital sinus narrows and becomes the **pelvic portion of the urethra.**

Testosterone produced in rudimentary testes promotes the internalization of the urogenital sinus in males as the **penile urethra** and the elongation of the **genital tubercle** into a penis. Later, with the descent of the testes into the **scrotum,** the male apparatus is complete (Fig. 25.41a).

In females, in the absence of testosterone, the mesonephric tubules and duct degenerate into the vestigial **epoophoron** and **paroophoron** (Fig. 25.41c). The **vagina** is formed in large part (if not entirely) from the invaginated urogenital sinus after linking up with the fused ends of the paramesonephric ducts. The **hymen** lies between the entrance to the vagina and the shallow **vestibule** or urogenital chamber. The **genital tubercle** in females becomes the **clitoris.**

In some female mammals (e.g., spider monkeys, *Ateles*), the urogenital and genital tracts are separated more profoundly. The anterior urinary portion of the vestibule is folded into a **clitoral urethra** identical to the male penile urethra but having no genital function. The vagina then opens into the remaining portion of vestibule entirely on its own.

Male Embryo's Dilemma

Mammalian embryos of both sexes develop in the internal milieu of the mother. The solutes which diffuse from her circulatory system to that of the developing embryo and fetus include her sex hormones. Moreover, the placenta is a major source of steroid hormones, many of which feed

back and forth between fetus and mother. How is it possible to produce a male offspring in this female hormonal environment?

Part of the answer may be that mammalian embryos and fetuses neutralize at least some the steroid hormones in their circulation by binding them to large molecules. Several plasma proteins have the capacity to bind steroid hormones: albumin, α_1-acid glycoprotein, steroid binding proteins (SBP) (see van Tienhoven, 1983), corticosteroid binding globulin (CBG), sex hormone binding globulin (SHBG), and progesterone-binding globulin (PBG) (see Heap and Flint, 1984). Estrogen has an especially high affinity for α-fetoprotein.

If bound estrogen is metabolically inaccessible and enough of it is bound, the embryo and fetus may develop unencumbered by estrogen. Unbound, other hormones (e.g., MIS and testosterone) might then operate on sexual development.

The neutralization of estrogen may have one of its most resounding consequences for the developing brain. The brains of adult female mammals produce gonadotropin releasing hormone (GnRH) cyclically, while those of males produce it noncyclically (continuously). Cyclicity seems inherent (i.e., the default value), while noncyclicity is induced (i.e., must be entered). In rodents, for example, GnRH cyclicity is prevented in females by the injection of testosterone or estrogen early in the neonatal period.

It would seem that in female fetuses, protein-bound estrogen complexes are too large or cumbersome to pass into the fetal brain (i.e., pass a blood–brain barrier). Since the fetal ovary does not produce testosterone, no sex steroid reaches the brain and cyclicity results. Fetal testes, on the other hand, produce testosterone which presumably reaches the brain and destroys its latent capacity for cyclicity.

Actually, testosterone is converted to estrogen in the brain and presumably has its effect as estrogen. While convenient for movement in fetal circulation, testosterone transport turns out to be a cryptic way of delivering estrogen.

Sex steroids may have additional effects on the fetal brain. In primates, mothers receiving testosterone during late pregnancy give birth to females that later display male patterns of behavior (see Goy and McEwen, 1980).

Puberty and Pregnancy

The onset of sexual maturity represents the culmination of many changes, including the commencement of gametogenesis, the maturation of the external and internal genital apparatus, and the enhancement of secondary sexual characteristics (not part of the genital apparatus). Not only must the appropriate hormones be produced at this time, but hormonal receptors must also be available (see Chapter 7).

Briefly, in the mature testis, the gonadotrophic hormone, follicular stimulating hormone (FSH), is bound by receptors on sustentacular (Sertoli) cells in the seminiferous tubules, and luteotrophic hormone (LH, or interstitial cell stimulating hormone, ICSH) is bound by receptors on interstitial endocrinocytes (Leydig cells) at the bases of the seminiferous epithelial cells. In turn, the sustentacular cells and interstitial cells support spermatogenesis, the former while providing a scaffolding for the spermatocytes as they mature and differentiate into sperm, the latter by secreting essential testosterone (Fig. 7.20).

In the mature ovary, FSH is bound by follicle cells or granulosa cells which support the egg's growth and produce large amounts of estrogens. LH is bound by cells of the surrounding theca which transform the follicle into a corpus luteum and produce large amounts of progesterone (Fig. 7.43).

The commencement of gametogenesis in males and females and the resumption of meiosis in females at puberty follows an increase in levels of sex hormones. The mechanism for changing levels of sex hormones is uncertain, but decreased **pineal activity** and serum **melatonin** levels seem to be involved. At least in animals with an annual sex cycle, changes in melatonin in response to changing day length trigger gametogenesis (Waldauser et al., 1984) (see Fig. 7.19). Other physiological parameters must also be involved in the mechanism, since neither changes in exogenous melatonin nor removal of the pineal gland (pinealectomy) triggers or blocks puberty (Klein, 1984).

At a female child's first menstruation, she is said to have reached **menarche,** but she actually may have been ovulating for some time. The age at menarche seems to be heavily dependent on nutrition and is likely to be significantly earlier in places where nutrition is superior. In addition, the age at menarche differs among ethnic groups indicative of sensitivity to hereditary factors. In America, the average age at menarche is between 12 and 14 years. Variation is wide, and delays, especially related to stress, are common.

When a pregnancy begins, the corpus luteum enlarges and continues to produce progesterone. Progesterone (the hormone of pregnancy) promotes gestation and suppresses the development of ad-

ditional eggs. In conjunction with prolactin (PRL), progesterone also promotes the development of mammary tissue in mammals and the crop sac in birds. Milk is ultimately produced in the mammary glands and a product called crop milk (but having nothing to do with mammalian milk) is produced in the crop sac. Brooding behavior in birds is also influenced by PRL.

The high levels of progesterone maintained during pregnancy continue during lactation in some mammals and serve to further suppress egg development. A sterile lactational period results. The ability of progesterone to suppress egg development is the theoretical basis for the use of synthetic steroids with progesterone-like activity as female contraceptives.

At term, or the end of gestation, new hormones take over. The ovary, uterus, and placenta produce **relaxin,** which causes the enlargement of the uterine cervix, birth canal, and pelvic ligaments. The posterior pituitary gland, the neurohypophysis, releases **oxytocin,** actually produced in neurosecretory cells in the hypothalamus, which promotes contractions in the uterus and the beginning of **labor.** Oxytocin later also causes the ejection of milk from the mammary gland during lactation.

When gestation and lactation or brooding are over, diminished LH production dooms the corpus luteum to degenerate, and gradually less progesterone is found in the circulation. Finally, the levels of sex hormones again diminish to the point where their monitor in the hypothalamus cuts in, and the feedback loop regulating their production is again activated. Soon the level of FSH increases and another cycle begins.

Women mark the end of their reproductive lifetime by **menopause,** the last menstruation, but ovulation may continue for a while during their **postmenopausal** lifetime. Usually, menstruation does not stop all at once, but menstrual periods become more irregular and less profound over a period of time. The age at menopause, like the age at menarche, is influenced by nutrition and possibly hereditary factors. In America, menopause is generally reached between the late forties and early fifties.

Reproductive functions come and go therefore, but excretory functions are constant or recurrent. The evolutionary potential for usurping the urinary system for reproductive functions may have been inherent in morphology (i.e., in a ductwork going from inside to outside), but the absence of genital functions at various times during a lifetime may also have relaxed requirements for an exclusively genital system.

Similarly, relaxation in the demand for nephric function in the anterior embryonic kidney may have made the nephric duct available for a genital takeover. Rather than abandoning redundant structures, the embryo, it would seem, appropriates and exploits them.

FINS, LIMBS, INTEGUMENT, AND ALLIED STRUCTURES

Some months ago an undergraduate asked me if there were ways to make a snake or legless lizards' limb bud continue to develop and if so, what would one expect to obtain? Idly turning over in my mind many of the data which other[s] . . . have published, and garnishing this delectable potpourri with a soupçon of Positional information, I realized that Kollar and Fisher's (1980)[8] chickens' teeth were but one manifestation of a fascinating problem which will confront practitioners of paleodevelopmental biology.

Paul F. A. Maderson (1983, p. 231)

Paired pectoral and pelvic fins consist of skeletal elements coupled to abductor and adductor muscle masses. The paired limbs of tetrapods and their supporting girdles are articulated skeletons and musculature. The generalized skeletal elements are **stylopod** (proximal segment, upper arm or thigh, humerus or femur), **zygopod** (yoked, intermediate segment, forearm or leg, radius and ulna, or tibia and fibula), and **autopod** (distal elements, hand or foot, carpus [wrist] or tarsus [ankle], and a series of phalanges arranged in up to five digits). Muscles are combined in complex flexor and extensor groups.

Origins

Primordia of the paired and median fins of fish appear as epithelial folds. Somatic mesoderm (Fig. 25.42) fuses with the fold to form a somatopleural thickening or **provisional** embryonic fin. In the case of paired fins, the somatopleural thickening is invaded by somite processes to form a **definitive** (archetypal) **fin fold**. The **presumptive limb buds** of tetrapods may be located on the neurula fate map (see Fig. 21.21) as **limb disks** in amphibians or circumscribed areas in a ring of thickened somatopleure, known as the **Wolffian ridge** (after Caspar

_____

[8]E. J. Kollar and C. Fisher, 1980. Tooth induction in chick epithelium: Expression of quiescent genes for enamel synthesis. *Science*, 207:993–995.

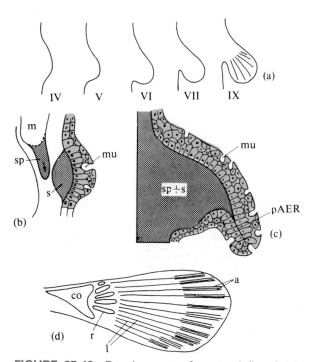

FIGURE 25.42. Development of pectoral fin of teleost. (a) Outline drawing of fin rudiments and stages. (b and c) Cross sections at stage IV (hatching) and V. (d) Complete fin with rays. The base of the fin is made up of combined somitic and somatic mesoderm. m, myotome; mu, mucous cell; pAER, pseudo-apical ectodermal ridge (formed by fold of ectoderm); s, somatic mesoderm; sp, somitic process. From J. R. Hinchliffe and D. R. Johnson, *Development of the vertebrate limb*, Clarendon Press, Oxford, 1980. Used by permission.

Friedrich Wolff who first described the ridge in the chick) (Fig. 25.43).

In fish, early expansion of the somatopleure in one area and regression elsewhere may transpose the fin to a new location, but the site of development is fixed when the fold is invaded by somitic muscle buds from the edges of adjacent myotomes followed by segmentally arranged nerves. Still, the course of segmental nerves at the leading and trailing edges of the fin fold are deflected when the rapid longitudinal growth of the animal's body outstrips the growth of the definitive fin fold (Fig. 25.44).

Somatic and sensory nerves innervating fins (and limbs, see below) come from several axial segments and different numbers of segments (Table 25.4, i.e., a species-specific characteristic). The actual axial segments contributing to fins are not rigidly fixed, however, and may vary within a species and even from one side to the other of the same animal.

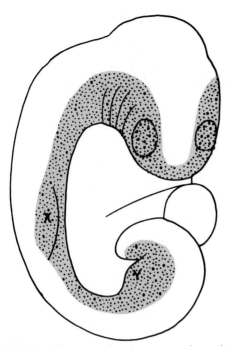

FIGURE 25.43. Diagram of a human embryo (stage 13 Carnegie standard, 4.2 mm, 30 pairs of somites, 31 days) showing the thickened ectodermal ring (stippled), or *Wolffian ridge*, between the upper limb (X) and lower limb (Y) buds. Adapted from R. O'Rahilly and E. Gardner, *Anat. Embryol.*, 148:1 (1975), by permission of Springer-Verlag and the authors.

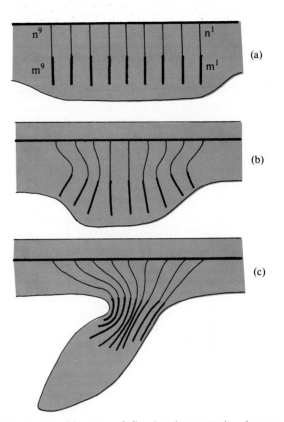

FIGURE 25.44. Diagram of fin development in elasmobranchs. (a) The provisional ridge is invaded by myotomes (m^1-m^9) with their accompanying nerves (n^1-n^9). (b) Differences in relative rates of growth lead to deflection of nerves and angular distortions of myotomes. (c) Contraction at the fin base results in the characteristic posterior notch. From E. S. Goodrich, *Studies on the structure and development of vertebrates,* Macmillan, London, 1930. Used by permission.

Myotomal buds invading the fin folds may be single or paired and many more may invade the ridge than will ultimately become incorporated into the fin. The excluded buds degenerate and disappear, while the included buds separate from the remainder of their myotomes and differentiate into dorsal abductor and ventral adductor muscles (see Hinchliffe and Johnson, 1980).

The evolutionary plasticity of segments associated with limbs is surprising in view of the general homology attributed to limbs. It would seem that different axial segments have produced the same structure in different vertebrates (i.e., are serially homologous).

In tetrapods, limb buds are initially parts of a larger limb field. Any part of a limb field can regenerate a whole limb or cooperate with additional parts in the regulation of a single limb.

Limb fields are the classic example of embryonic fields and can even regenerate limbs following the extirpation of the entire presumptive limb bud (see Stocum and Fallon, 1984). Moreover, a variety of devices for splitting limb buds, from barriers and beads of resin soaked in all-*trans*-retinoic acid (see

Maden, 1984; Tickle et al., 1985) to rotating limb tips and transplanting portions of the posterior bud to the anterior surface (see below), cause the de-

TABLE 25.4. Segments taking part in formation of fin and limb buds based on spinal motor nerves innervating limbs

Group	Pectoral Fin or Forelimb	Pelvic Fin or Hindlimb
Chondrichthyes[a]		
Scyllium	2–13	25–35
Heptanchus	2–19	29–50
Torpedo	4–30	31–42
Amphibians[b]		
Rana catesbeiana	2–4	8–10
Ambystoma	3–5	15–18
Human beings[b]	4–10	20–28
Chick[c]	13–16	25–30

[a]Data from Hinchliffe and Johnson (1980).
[b]Data from Witschi (1956).
[c]Data from Gumpel-Pinot (1984).

velopment of portions of limbs and even complete limbs in the separated parts.

The limbs formed are generally **mirror images** of each other (i.e., face each other, once called Batesian regeneration) even when one limb largely corresponds to the original bud and a **secondary** (or supernumerary) limb is added. It seems that polarity as well as morphogenesis are regulated in the early limb bud (see Holder, 1984).

Limb bud mesoderm is composed of two cell lineages, a **somatic lineage** from the somatopleure and a **somitic lineage** from somites (Christ et al., 1977).[9] In chicks, or at least in quail–chick chimeras in which the cytologically distinguishable quail cells are used to trace cellular origins (see below, Figs. 26.16 and 26.17), all the limb's voluntary musculature is derived from somitic myotomes. With the exception of the scapula (also derived from somites), the skeleton and connective tissue are derived from somatic mesoderm (see Gumpel-Pinot, 1984).

The **motor pools,** or clusters of motor neurons that innervate each limb, extend along the ventral motor horn of the embryo's spinal cord and give rise to axons in several spinal nerves (see Chapter 26). These axons combine in **plexes** at the base of each limb, and their routes become deflected when differential growth ahead of limbs (especially the anterior limbs) drives the nerves (as well as everything else) posteriorly.

Muscles originating in a particular axial segment are generally innervated by motor nerves originating in the same segment, a phenomenon known as **segmental-level association** or **matching.** Estimates of the somitic origins of chick limbs based on quail–chick chimeras roughly agree with estimates based on the motor innervation of the limb (somites 13–16 compared to segmental nerves 12–20, and somites 25–30 compared to segmental nerves 26–32 for the forelimbs and hindlimbs, respectively, i.e., compare Fig. 25.45 and Table 25.4). The faithfulness of segmental-level associations in birds is all the more extraordinary, since the last presumptive muscle cells stop migrating from somites (at stage 20 H&H) a day before motor nerves reach the base of the limb (at stage 23 H&H).

According to **target recognition** (or target segmental matching) **hypotheses,** some sort of **molecular token** originally shared by cells in a segment is required to make the final match between the correct nerve and muscle (Goodrich, 1906). The tokens may leave a trail as they are carried by presumptive muscle cells into the limb; the tokens may provide a bias toward forming particular associations, or the tokens may repulse inappropriate nerves. Associations may also be sorted out by the removal of inappropriate nerve–muscle associations through selective cell death or the elimination of synapses.

The course of nerves may be traced in chick embryos with the aid of the intracellular marker horseradish peroxidase (HRP). The marker is injected into nerves (known as orthograde labeling) or into muscle, in which case it is picked up by nerves through motor end plates and transported backward to the neural cell body (known as retrograde labeling).

Frequently, despite transplantation (i.e., misplacement), motor neurons are found to innervate the correct segmentally associated muscle. For example, craniocaudal rotation of the neural tube, limb bud, or segmental plate does not prevent motor axons from reaching the same muscles they normally innervate (see Summerbell and Stirling, 1982).

Axial segmentation, it would seem, guides axons during an early **emigration stage,** from the neural tube across the neighboring sclerotome. The role of axial segmentation during the later **target recognition stage,** when motor axons innervate their muscles, is less certain.

When the segmental origin of a wing muscle is displaced just one or two segments, motor neurons are not shifted. Similarly, cranial limb axons innervate uncharacteristic caudal limb muscles in secondary wings regenerating from split wing buds. These failures to match nerves and muscles according to segmental-level associations suggest that target recognition is not rigid. Rather than preformed tokens, a more plastic source of guidance may channel axons within the limb. Possibly, the required cues are provided by the lateral plate somatic mesoderm as it differentiates into the limb's connective tissue (Keynes et al., 1987).

Outgrowth

The distal growth of fins is led by an epidermal thickening, the **piscine-apical ectodermal ridge**[10] (pAER, Fig. 25.42), of simple cuboidal epithelium

---

[9]The difference rests with the "a" and "i." One might be helped in remembering which does what by thinking of "mi" as "my" as in myoblast or presumptive muscle.

[10]The pAER is generally known as the **pseudo-apical ectodermal ridge,** but "pseudo-" is neither descriptive nor precise, whereas "piscine-" has the virtue of identifying the group in which this type of ridge appears.

SOMITIC LEVEL OF THE SKELETAL FORMATIONS

MUSCLES OF SOMITIC ORIGIN

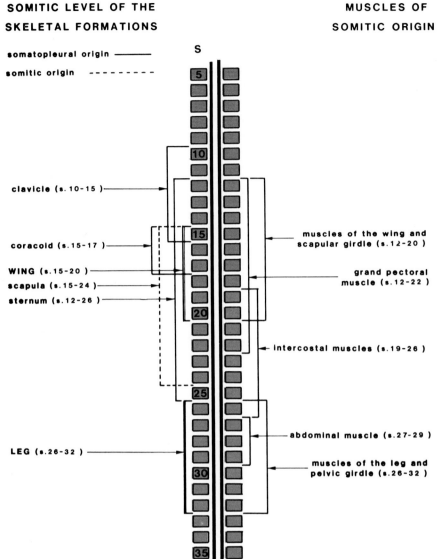

FIGURE 25.45. Summary of results with chick–quail chimeras showing the somitic origins of skeletal muscles (right) and somitic origins of skeletal formations (left). Only the scapula among the skeletal formations is of somitic origin. The somatic mesoderm is identified by corresponding somite numbers. S = somite number. From M. Gumpel-Pinot, in N. Le Douarin and A. McLaren, eds., *Chimeras in developmental biology,* Academic Press, London, 1984. Used by permission.

covered by the enveloping layer or periderm. While the early tetrapod limb typically sports a similar ridge for a considerable time, the pAER is quickly transformed into the flattened **fin fold**. Radial collagenous rays (actinotrichia) soon develop within the fold and support its elongation and rounding into a **fin paddle**. Somatic and somitic mesenchymal cells, driven by population density behind them and directed by contact guidance between collagenous rays, then migrate distally and produce the typical ray pattern of fin connective tissue (see Thorogood and Wood, 1987).

Tetrapod limbs generally form an **apical ectodermal ridge** (AER, Fig. 25.46) prior to mesodermal outgrowth. In amniotes, the AER's epithelium, overlaid with a periderm, is a pseudostratified columnar epithelium (Fig. 25.47) that does not regenerate when surgically removed (although it may

FIGURE 25.46. Scanning electron micrographs of chick hind limb buds showing apical ectodermal ridge (a) at its first appearance (stage 19 H&H) and (b) its maximal elevation (stage 24 H&H). From J. R. Hinchliffe and D. R. Johnson, *Development of the vertebrate limb,* Clarendon Press, Oxford, 1980. Used by permission.

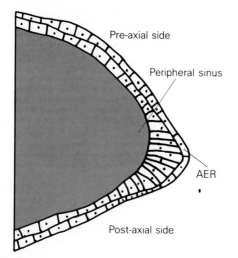

FIGURE 25.47. Schematic longitudinal section through chick hind limb bud (stage 19) showing the stratified columnar epithelium composing the apical ectodermal ridge (AER) in the limb epidermis.

regenerate under other circumstances). The comparable ridge in anurans is multilayered and may regenerate from epidermis. In urodeles, epidermis at the apical edge of limbs is indistinguishable from the two-layered stratified epidermis present elsewhere.

In the poorly vascularized early tetrapod limb, the juncture of the AER and mesoderm is associated with a **peripheral sinus** where venal blood accumulates. Cell division is maintained in the mesoderm beneath the sinus, and cells accumulate basally.

The outgrowth of the limb (Fig. 25.48) is not dependent on an increase in mitotic activity in the limb, although mitotic activity diminishes in the areas surrounding the limb. Division in the limb remains relatively constant if slightly skewed toward the postaxial side (i.e., beyond the middle digit).

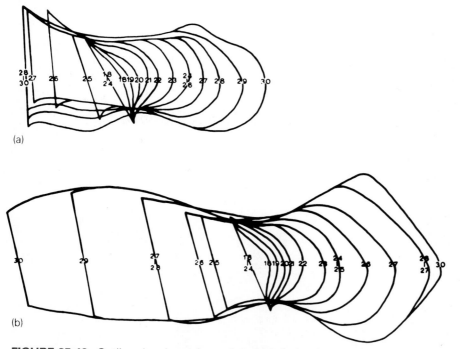

FIGURE 25.48. Outline drawings of growing chick limbs. Numbers refer to Hamburger & Hamilton stages. Note the disproportional amount of growth due to elongation of the proximal portion of the limbs after stage 25. 16 (not shown), Wing represented by thickened ridge, 51–53 hours after laying; 18, wing and leg bud definitive, 52–64 h; 24, toe (digital) plate appears in leg bud, 4.5 days after laying; 25, elbow and knee joints appear, 4.5–5.0 days; 26, first three toes appear, 5 days; 28, three wing digits, four toes; 29, middle wing digit distinctly longer, rudiment fifth toe, 6.0–6.5 days; 30, 6.5–7.0 days. From D. A. Ede, in G. Poste and G. L. Nicolson, eds., _The cell surface in animal embryogenesis and development,_ Elsevier, Amsterdam, 1976. Used by permission.

The growing limb is not shaped by points of increased mitosis. Mitosis is maintained toward the limb's periphery from which cells move to central **blastemata** (pl., **blastema** sing.) of high cell density and diminished mitosis. Cells remain distinct in **chondrogenic blastemata** where abundant collagen and the amorphous deposits of cartilage matrix accumulate, but myoblasts fuse into myotubules and postmitotic sarcomeres differentiate in **myoblastic blastemata.**

Cellular condensations and blastemata appear in a proximodistal order. Blastemata associated with limb-girdle elements appear before those associated with stylopodial elements, and these elements, in turn, are laid down before zygopodial and autopodial elements. Myogenic blastemata also appear before chondrogenic blastemata, and all blastemata seem to be determined before they make their appearance, since interference with the dif-ferentiation of early blastemata does not prevent the appearance of later blastemata.

In newts and chicks, limb cartilages are laid down in solid rods and then divided by joints into separate digits. Chick limb cartilages acquire their elementary form in organ culture, demonstrating that intrinsic factors are largely responsible for the form and architecture of skeletal elements and joints. On the other hand, the elongation of rudiments and maintenance of formed joint cavities fail in culture, suggesting that they require environmental factors and movement (see Shubin and Alberch, 1986).

Differential molecular changes accompany the interruption of the cartilage anlagen by joints in chick limbs. For example, collagen type II (i.e., the collagen of cartilage, see Table 2.3) first appears throughout the cartilage anlage, even across the presumptive joint region (Fig. 25.49), but soon dis-

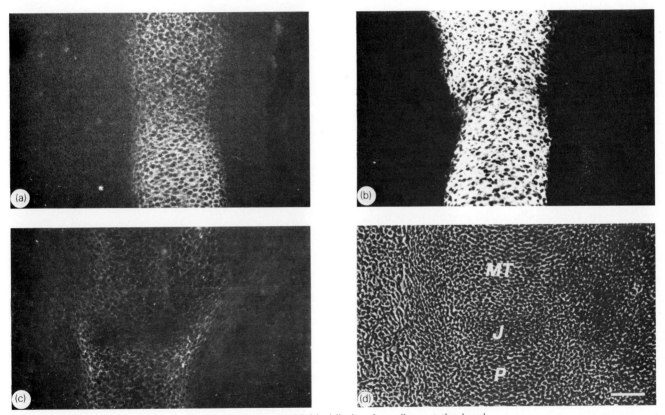

FIGURE 25.49. Longitudinal sections through stage 28 hind limb primordium at the level of the developing joint (J) between metatarsal (MT) and phalangeal (P) elements. Sections (a–c) are stained with monoclonal antibodies for (a) collagen type II, (b) keratan sulfate-containing proteoglycan (KSPG), and (c) collagen type I. (d) Phase contrast of (a). Bar = 61 μm. Collagen type II staining is less in presumptive joint but, along with KSPG staining, occurs across the joint (J) and elsewhere in cartilage anlage. Collagen type I staining is absent in joint. From F. M. Craig, G. Bentley, and C. W. Archer, *Development,* 99:383 (1987), by permission of the Company of Biologists and the authors.

appears from the articular surfaces of the presumptive joint (Fig. 25.50). Keratan sulfate-containing proteoglycan (KSPG) also appears throughout the early matrix, but it disappears specifically in areas of the joint undergoing cavitation. Collagen type I (general body collagen and collagen of bone) is absent at first from joint regions but later appears in the interzone region of the presumptive joint between centers of chondrification and in the perichondrium surrounding the cartilage anlage (Craig et al., 1987).

The shape of the limb outgrowth is also influenced by **differential cell death.** In the chick wing bud, a **posterior necrotic zone** (PNZ) running along the posterior edge suffers a cataclysmic necrosis that frees the humerus from the body and gives the shoulder liberty to rotate. Cell death in an **anterior necrotic zone** (ANZ) helps to shape the shoulder, and necrosis in an **opaque patch** separates the radius and ulnar rudiments.

In the leg, a zone of necrotic activity runs two-thirds along the length of the preaxial margin, and **interdigital necrotic zones** carve out the chick's toes. In contrast, areas of less intensive necrosis in the duck leg bud leave toes joined by webbing (Fig. 25.51).

A cellular **death clock** seems to be running in the PNZ, since even freshly erupted buds from stage 17 embryos (52–64 hours) transplanted to a "neutral" site, such as the somite region, develop PNZs and cell death occurs on schedule. The PNZ is not irreversibly "programmed" to die, however. As late as stage 21 (between 3 and $3\frac{1}{2}$ days of incubation, when bud length is less than half the bud width, see Fig. 25.48), cells in the presumptive PNZ can be rescued by transplantation to the dorsal side of the wing bud.

At stage 22 (when the bud is not quite as tall as it is wide, about $3\frac{1}{2}$ days), the death clock would seem to be set, since the PNZ is beyond recall. Its cells will die at stage 24 (when bud length distinctly exceeds width, about 4 days) even if transplanted to the dorsal side of the wing bud (see Saunders and Fallon, 1966).

Polarity

In the axolotl, the determination of polarity occurs in a definite order. In a series of limb bud transplantation and rotation experiments, Ross Harrison and colleagues demonstrated that the type of limb (i.e., forelimb versus hindlimb) is determined first (see Harrison, 1969). The large areas of the gastrula capable of forming both forelimbs

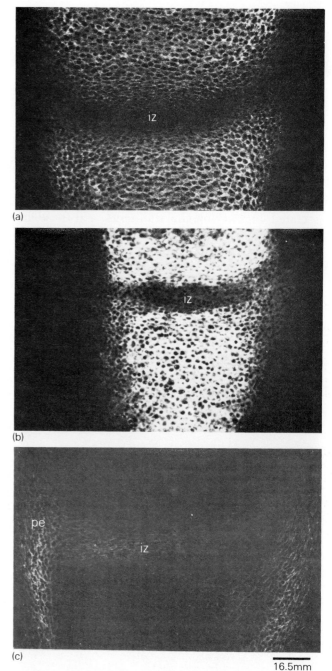

FIGURE 25.50. Longitudinal sections through stage 30 hind limb primordium at the level of the metatarsophalangeal joint showing the developing interzone (iz) and perichondrium (pe). Sections (a–c) are stained with monoclonal antibodies for (a) collagen type II, (b) keratan sulfate-containing proteoglycan (KSPG), and (c) collagen type I. Collagen type II and KSPG staining are reduced or absent in the highly cellular interzone (iz), although both secretions occur throughout the cartilage anlage matrix. Collagen type I staining is largely restricted to the interzone and the perichondrium surrounding the cartilage anlage matrix. From F. M. Craig, G. Bentley, and C. W. Archer, *Development,* 99:383(1987), by permission of the Company of Biologists and the authors.

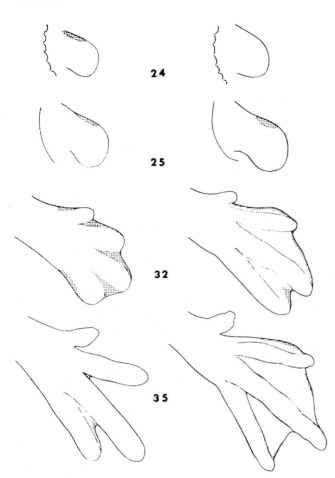

FIGURE 25.51. Pattern of necrosis (stippling) in hind limb bud of chick (left) and duck (right). From J. W. Saunders, Jr. and J. F. Fallon, in M. Locke, ed., *Major problems in developmental biology,* 25th Symposium of the Society for Developmental Biology, Academic Press, Orlando, 1966. Used by permission.

or hindlimbs become restricted to areas forming either one or the other type of limb.

The anterior–posterior plane of the limb is determined next. In the neurula, neither rotation by 180° nor mixing of parts of presumptive limb bud areas in abnormal combinations influences the symmetry of the future limb. The limb that grows out takes its symmetry entirely from the animal. But immediately after neurulation when the bud first appears as a rounded outgrowth, only the dorsal–ventral axis continues to escape the effects of rotation (i.e., takes its cue from the animal). The anterior–posterior axis is fixed and appears backward in the rotated limb. Finally, the limb's dorsoventral axis becomes fixed in embryos at the elongating tailbud stage when the limb bud is longer than it is wide. Its planes of symmetry are no longer malleable under the influence of the animal (see Stocum and Fallon, 1984).

Axes of symmetry in the chick's limb are determined much earlier than in the axolotl. Limb development following rotation of presumptive limb bud areas fails to show any influence from the surrounding tissue. The limbs' axes of symmetry are those of their sites of origin (see Javois, 1984).

The results of separating limb bud epithelium and mesodermal core, coupled with transplantation to the dorsal side of the chick, suggest that the limb bud mesodermal component alone conveys the limb's anterior–posterior axis, but the epithelial component conveys the dorsoventral axis (see Saunders, 1982). Specifically, the chick limb bud's proximodistal axis is determined by the relationship of the AER and mesodermal core.

Hypotheses Concerned with the AER's Effect on Chick Limb Outgrowth

Mesodermal outgrowth may stem directly from action of the AER. Alternatively, mesodermal outgrowth may be molded by elasticity of the AER–mesodermal boundary in conjunction with physical constraints present elsewhere (Amprino, 1977).

The Saunders–Zwilling hypothesis advanced by John Saunders and Edgar Zwilling proposes a "two way street" of ectodermal–mesodermal interactions (closed arrows in Fig. 25.52). The AER is thought to promote the mesoderm's outward growth and its production of an **apical ectodermal maintenance factor** (AEMF) that feeds back on the AER inducing it to further promote mesodermal growth.

The hypothesis was deduced largely from the results of removing and grafting chick AERs. Initially, mesodermal outgrowth was shown to be inhibited by the removal of the AER (Fig. 25.53). Similar results were obtained with frog and salamander limbs deprived of their epidermis and transplanted to the peritoneal cavity to prevent epidermal regeneration from surrounding epidermis.

Mesodermal outgrowth was restored to the denuded chick core by grafting an AER. Similarly, growth was restored to the amphibian core by allowing the apical ectoderm to regenerate (see Goss, 1969). In addition, supernumerary outgrowths were induced in chick limbs by dorsal grafts of AERs.

Since chick AERs at different stages of development provoked the same type and degree of out-

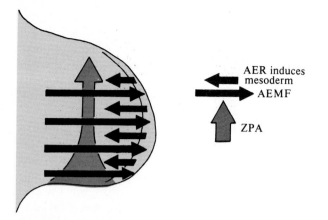

FIGURE 25.52. Schematic of the Saunders–Zwilling hypothesis for the mutual interactions of the apical ectodermal ridge (AER) and mesoderm. The AER stimulates mesodermal outgrowth (short closed arrows), while an apical ectodermal maintenance factor (AEMF, long closed arrows) from mesoderm promotes the health and well being of the AER. The zone of polarizing activity (ZPA, open arrow) may influence anterior–posterior determination. From J. R. Hinchliffe and D. R. Johnson, *Development of the vertebrate limb,* Clarendon Press, Oxford, 1980. Used by permission.

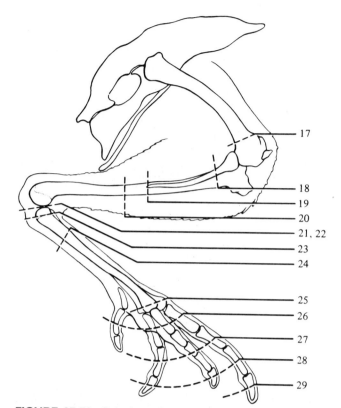

FIGURE 25.53. Drawing of right chick leg in lateral view illustrating the results of removing the apical ectodermal ridge (AER). Numbers identify stages at which the AER was removed. Dashed lines indicate levels at which limb ceased developing. From D. A. Rowe and J. F. Fallon, *J. Embryol. Exp. Morphol.,* 65(Suppl.):309(1981), by permission of the Company of Biologists.

growth, the ability of the AER to promote growth in the mesoderm had to be quantitative, not qualitative (Zwilling, 1968). The AER was imagined to interact **permissively** with the core mesoderm, promoting growth and setting up a chain reaction leading to differentiation.

The absence of **instructive** information in the putative AER signal was also implicit in results of joining ectoderm and mesoderm components from different sources. Forelimb and hindlimb grafts took on the character of the mesoderm as did chick and duck combinations. Rat AER promoted growth in chick wing mesodermal cores transplanted to a chick flank and occasionally yielded winglike limbs covered with ratlike skin. The foreign AER promoted growth in the mesodermal core, but the type of limb formed was determined by the core.

Abnormally high amounts of AEMF may be responsible for **talpid** mutants having excess amounts of chondrogenic condensations (see Ede, 1976) and **polydactylous** mutants having redundant digits. In the **eudiplodia** mutant with double legs, excess AEMF may be produced by the double AERs found on the early leg buds. Abnormally low amounts of AEMF, on the other hand, may be responsible for **wingless** and **chondrodystrophic** mutants lacking wings or having incomplete limbs.

Alternatives to the Saunders–Zwilling hypothesis are frequently versions of positional in-

formation models in which some sort of gradient determines linear patterns of differentiation (if not outgrowth, see Fig. 25.54) (Wolpert, 1969). Gradients are variously proposed in the concentration of morphogens or in some physical aspect of the tissue.

The **morphogen profile hypothesis** explains the results of various deletion and transplantation experiments with the help of a morphogen released from a posterior **zone of polarizing activity** (ZPA, open arrows in Fig. 25.52). First of all, amputation of the posterior half of the wing bud between stages 17 and 22 results in extensive cell death in the distal anterior necrotic zone mesenchyme and in developmental failure in the entire anterior half. Incomplete posterior amputation (i.e., leaving some ZPA *in situ*), on the other hand, results in distal survival and the complete development of a wing. Moreover, grafts of ZPAs to the distal ends (but not the proximal ends) of anterior bud halves **rescue** the halves. The grafts inhibit death in the anterior necrotic zone and promote regulation in

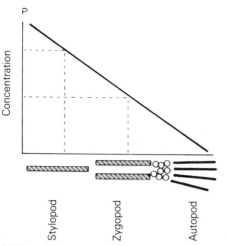

FIGURE 25.54. A concentration gradient of a diffusible morphogen may convey positional information and influence differentiation if threshold values (vertical dashes) direct cellular differentiation. Adapted from M. Maden, in G. M. Malacinsky, *Pattern formation,* 1984.

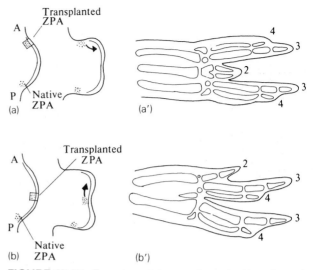

FIGURE 25.55. Experimental procedures (a, b) and results (a', b') of grafting additional putative zones of polarizing activity (ZPA) to different sites on chick limb buds. The ZPA grafted to a preaxial site (a) induces additional digits with reversed polarity, while the ZPA grafted to a midaxial site (b) induces additional digits with the same polarity as the host limb. From J. R. Hinchliffe and D. R. Johnson, *Development of the vertebrate limb,* Clarendon Press, Oxford, 1980. Used by permission.

the anterior mesenchyme to the extent of its forming a complete set of small digits (Wilson and Hinchliffe, 1987).

Furthermore, transplantation of the ZPA elsewhere on a complete limb induces the formation of an ectopic limb (McCabe and Parker, 1976). Inserted in the presumptive preaxial edge of the limb (i.e., before the site of the middle digit), the ZPA induces additional digits with reversed polarity, while inserted in the presumptive midaxial position (i.e., site of the middle digit), the ZPA induces additional digits with the same polarity as the host limb (Fig. 25.55).

A simpler explanation for these results than the morphogen profile hypothesis and ZPA is provided by the **polar coordinate model** of French, Bryant, and Bryant (1976). This model (illustrated by conical projections, Fig. 25.56I) proposes that limbs grow whenever possible until a set of **coordinate values** normally inscribed in a whole limb is completely elaborated and integrated with adjacent values.

When transplantation disrupts a coordinate system capable of regeneration, such as the early tetrapod limb, tissue representing additional coordinates may be **intercalated,** or grow between existing tissue. According to the **shortest intercalation rule,** gaps in polar coordinates are filled with the fewest possible additions required to reintegrate adjacent coordinate values.

If the ZPA is assigned a coordinate value of 4 (corresponding to the 4th digit), transplantation to

a preaxial position (Figs. 25.55a and 25.56IIa) requires intercalating digits 3 and 2 to fill in the blanks between 3 in the host and 4 in the graft. Transplantation to a midaxial position (Figs. 25.55b and 25.56IIb) requires no intercalation between the graft and the host tissue lying to the right but requires regeneration on the preaxial side of the graft to complete the polar coordinates of a complete structure.

Stage 18 buds (see Fig. 25.48) and, to a lesser degree, stage 20–22 buds can also fill in missing coordinates (i.e., regulate) in the axial direction (A–C, Fig. 25.56). Regulation occurs following both the extirpation and the addition of bud tissue (Hampé, 1959). Results with quail–chick chimeras indicate that regulation proceeds by the conversion of presumptive distal structures to proximal ones (Kiny, 1977).

Finally, Wolpert's **progress zone model** is essentially a positional information model that does not deal with tissue regulation. Since regulation is not observed in a large number of grafts between limb bud tips and bases of different ages (Summerbell et al., 1973), the progress zone model proposes that postmitotic cells are irreversibly committed to particular lines of differentiation.

According to the model, only cells within the mitotically active sub-AER zone can progress

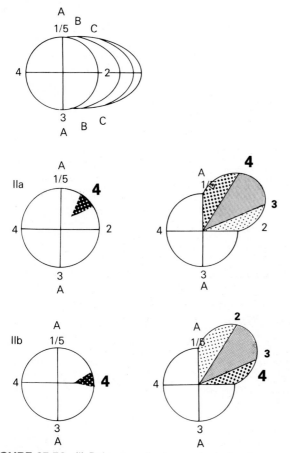

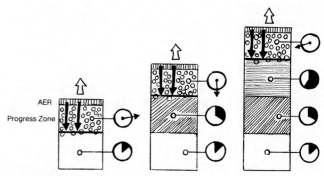

FIGURE 25.56. (I) Polar coordinates projected onto a conical solid representing a chick limb bud. The longitudinal axis is marked by rings labeled with letters. The cross section is marked by numbered coordinates. The high and low numbers (1 and 5 in this case) are considered identical values and assigned to the same position. Development brings all the coordinates to their fullest expression and fills in all the blanks between both the letters and numbers. (II) Two illustrations of how disruptions caused by transplantation are resolved by the shortest intercalation rule. In both cases, the transplanted tissue has a coordinate value of 4. (a) When the transplant is placed between 2 and $\frac{1}{5}$, tissue corresponding to coordinates 3 and 4 is intercalated. (b) When the transplant is placed at the level of 2, tissue corresponding to coordinates 2 and 3 is intercalated.

(hence the name progress zone) and ultimately differentiate as more distal structures. Cells leaving the progress zone early differentiate into proximal limb elements; cells leaving the zone late differentiate into distal limb elements. Support for the hypothesis comes from the ability of transplants to give rise to tissue independently of underlying host tissue or even in the absence of proximal tissue (see Maden, 1982)

Possibly, a mitotic clock ticks off a cell's time in the progress zone and inscribes cells with posi-

tional information. The cell's clock would stop ticking once the cell leaves the progress zone, and the cell would become determined for its ultimate course of differentiation (Fig. 25.57).

Compound Ectodermal Appendages

Compound structures contain cells of more than one germ layer or portion of a layer. A variety of compound structures attached to the epidermis are sometimes grouped as ectodermal appendages. They frequently involve neural crest cells, especially in the form of ectomesenchyme, but epithelial placodes may also take part. Cooperation between layers generally requires cell movement and inductive interactions.

Epidermis and neural crest derivatives are frequently so intimately related that they are indistinguishable except for products. As pigment cells or melanocytes, neural crest cells inject melanin into the feathers and hairs of birds and mammals and help to create and change, by expanding and contracting, the colors of reptiles, amphibians, and fish. Ectomesenchyme also organizes the corneal stroma and provides corneal fibroblasts and the smooth internal corneal endothelium (Fig. 25.58).

Ectomesenchyme cells are concentrated in the **dermal papilla** of hair and feather follicles, in condensations beneath scales, and in the **dental papilla**

FIGURE 25.57. Illustration of the progress zone model of limb determination. Cells in the progress zone (small circles) keep "ticking" and counting the number of times they pass through mitosis (clock faces). The positional values of cells determined by the number of passes a cell has made through mitosis fixes the cell's fate (frozen hands of clocks) and commits it to differentiate in a particular way once beneath the progress zone. White box, stylopodial differentiation; diagonal hatching, zygopodial differentiation; horizontal hatching, autopodial differentiation; open arrow, axial growth of limb bud; closed arrows, cells leaving the progress zone. From D. A. Ede, *Introduction to developmental biology*, Wiley, New York, 1978. Used by permission.

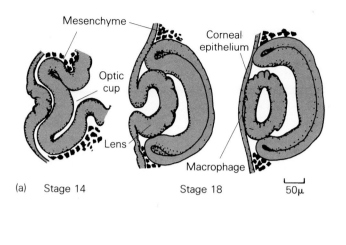

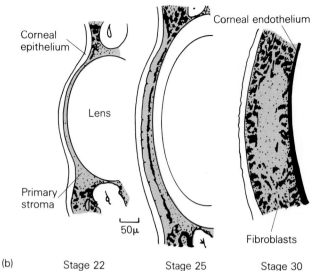

FIGURE 25.58. Cross sections through optic area of chick embryo showing formation of the cornea. Hamburger & Hamilton stages. Stage 14: lens placode begins invaginating. Presumptive corneal epithelium lies at edges of optic cup. Stage 18 (3 days of incubation): lens vesicle separates from corneal epithelium. Macrophages remove debris. Stage 22 (4 days): corneal epithelium secretes primary stroma and presumptive corneal endothelium begins to invade area (arrow near top). Stage 25 (4.5–5 days): corneal endothelium almost complete, followed by swelling of primary stroma and invasion by corneal fibroblasts. Stage 30 (6.5–7 days): fibroblasts occupy all of stroma except for narrow subepithelial zone (curved arrow). Bar = 50 μm. Adapted from E. D. Hay and J.-P. Revel, *Fine structure of the developing avian cornea,* Karger, Basel, 1969, by permission of Karger AG and courtesy of the authors.

of teeth, all of which are interchangeable to an amazing degree. The types of integumentary appendage (scales, feathers, or hairs) and teeth formed at the sites of **heterotypic grafts** (between tissues from different structures in the same animal), **heterospecific grafts** (i.e., between tissues from organisms of different species), and **xenografts** (i.e., between tissues from organisms of different taxa above the species level), whether cultured *in vivo* or *in vitro*, are dependent on the epidermis. Their distribution and pattern of structures are dictated by the mesenchymal component (see Fig. 25.59) (Sengel and Dhouailly, 1977). For example, epidermis from a featherless lizard does not produce feathers, and epidermis induced in the chorioallantoic membrane of chicks does not produce hairs despite the source of induction, but each epidermis produces appendages within its competence according to the pattern of the inducing substratum.

The interaction of papillae with ectoderm seems to be a two-way street for the transfer of developmental information, both to and from both tissues, but the transfer begins in the ectoderm. In the case of scales, feathers, and hair, epithelial

thickenings (sometimes called placodes) appear in the developing ectoderm before dermal papillae appear in the underlying mesenchyme. The order of appearance suggests that the epidermal component of composite structures triggers the formation of the dermal counterpart (see Dhouailly, 1984).

Similarly, ridges of ectoderm called **dental ledges** (or laminae) sink inward before dental papillae appear in the jaw. The **dental organ** or **tooth primordium** forms where mesenchyme cells condense beneath the ledge (see Fig. 25.60), again indicative of a role for epidermis in the induction of mesenchyme.

The formation of dermal condensations may not require further epidermal participation beyond initial permissive induction. Quail epithelium, for example, can support the differentiation of dental papillae into dentine-producing odontoblasts without forming corresponding enamel-producing ameloblasts. On the other hand, dental papillae seem to transmit low information containing triggers that shape microscopic prisms in the enamel matrix of teeth.

Instructive induction, on the other hand, seems to control tooth shape (molar shape rather

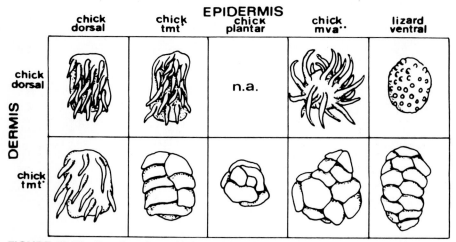

FIGURE 25.59. Results of combining epidermis and dermis from various sources. Chick dorsal dermis induces feather formation. A feather tract pattern and hexagonal distribution of buds are elicited in competent epidermis including lizard ventral epidermis. Similarly, tarsometatarsal dermis induces scales in competent epidermis or a quadrangular pattern of widely dispersed feather primordia. tmt, Anterior tarsometatarsal; mva, midventral apterium (bare area having neither feathers nor scales), n.a., not applicable. From D. Dhouailly, in G. M. Malacinski, ed., *Pattern formation,* Macmillan, New York, 1984. Used by permission.

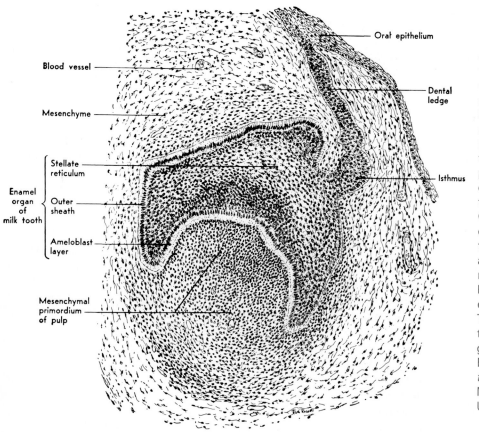

FIGURE 25.60. Section through lower jaw of human embryo (11th week, about 58-mm crown–rump length) containing a tooth primordium. Epidermal cells from the oral epithelium grow into the jaw as a dental ledge and fuse with a condensation of mesenchyme. The mesenchyme becomes the dental papilla or primordium of the pulp and will later give rise to an odontoblast layer that produces the dentine matrix of the tooth. Overlying the dental papilla, the epidermal component swells into the enamel organ consisting of an outer sheath, a central stellate reticulum, and an inner ameloblast layer that lays down the enamel matrix capping the tooth. The isthmus is the point where the dental ledge and enamel organ later separate. ×150. From B. M. Carlson, *Patten's foundations of embryology,* 5th ed., McGraw-Hill, New York, 1988. Used by permission.

than incisor shape). The source of the dermal papilla determines the shape of the **enamel organ** and the **amelogenic** (i.e., enamel-producing) **ameloblast layer.** Moreover, chick pharyngeal epithelium may even respond to mouse dental papilla by forming an enamel organ capable of amelogenesis (see Kollar, 1983; Maderson, 1983), something that no other avian epithelium has done since the time of *Hesperornis regalis* and the last of the toothed birds of the Cretaceous period (about 100,000,000 years ago)!

The ability of epidermal–mesenchymal composites to produce structures is indicative of homologies in the inductive systems whether the terms of induction are dictated by the inducing tissue or merely subscribed to by the reacting tissue (Slavkin et al., 1984). The success of so many combinations of mesenchyme and epithelia to produce patent structures suggests that developmental signals are highly conserved in the course of evolution.

Homologies in permissive induction were discovered by Oscar Schotté, in Spemann's laboratory, following the exchange of frog and salamander ectoderm from presumptive mouth regions (see Spemann, 1938). Host frog and salamander dermis induced oral structures in the grafted ectoderm, but the structures formed corresponded to those normally produced in the donor. The frog ectoderm developed typical frog adhesive glands and horny keratinized mouth parts (Fig. 25.61), while the salamander ectoderm developed typical salamander balancers and calcified teeth.

Finally, neuro-ectodermal interactions operate in the development in the pituitary gland although the direction (or directions) and the type (or types) of interaction are unknown. A compound structure (see Fig. 7.4), the pituitary is composed of a posterior lobe (or neurohypophysis) protruding from the brain on an infundibular (funnel-shaped) stem (or stalk) and an anterior lobe (or adenohypophysis) attached to the posterior lobe.

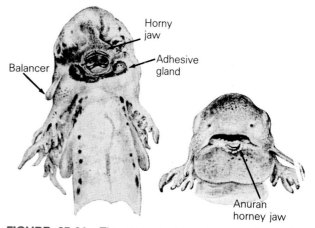

FIGURE 25.61. The ventral side of two salamanders, which received grafts of embryonic frog oral epidermis as tailbud embryos. The animal receiving the larger graft developed one of its own balancers (left), typical frog adhesive glands, and horny jaws. The animal receiving the smaller graft developed an anuran horny jaw in the transplanted frog tissue. From H. Spemann, *Embryonic development and induction,* Yale University Press, New Haven, 1938; reprinted by Hafner, New York, 1962. Used by permission.

The pituitary arises from two separate sources. A region of ectoderm called the **hypophyseal placode,** located just below the curvature of the neural groove, invaginates as the **hypophysis** or **Rathke's pouch** (see Figs. 23.18 and 26.9) from the roof of the stomodeum. Folding inward and lengthening in conjunction with movements in the floor of the brain (Figs. 25.1, 25.8, and 25.9), the hypophysis attaches to the **infundibulum,** a ventral evagination from the brain (see Fig. 26.11) and breaks its ectodermal attachment. The hypophysis forms the anterior lobe of the pituitary (and a portion of the posterior lobe in the adult called the pars intermedia), while the infundibulum gives rise to most of the posterior lobe.

Chapter

26

SPECIFICATION OF THE NERVOUS SYSTEM

*E*mbryology should tell something about how a complex system, such as the nervous system, works, and embryology does. As a matter of fact, many features of the nervous system are shaped in the remote environment of the embryo. Some contortions of the central nervous system and convolutions in the peripheral nervous system, for example, make sense primarily (if not only) because of early embryonic segmentation (e.g., segmental-level association, Chapter 25) and the organization of the rudimentary nervous system (Fig. 26.1).

For example:

> Since in early embryos the hind-brain lies directly above the pharynx, fore-gut and heart, . . . it is natural that the centers concerned with the regulation of chewing, tasting, swallowing, digestion, respiration and circulation remain located in the hind-brain, even though the organs innervated become considerably dislocated in position. (Leslie Brainerd Arey, 1965, p. 482)

In the course of morphogenesis and embryogenesis, the neuraxis of the early embryo gives rise to the axial neural tube, the rudiment of the vertebrate central nervous system (CNS) (see below Table 26.1). The paraxial neural crest and surface neural placodes provide cells to ganglia in the peripheral nervous system (PNS). Diversification separates the nervous tissue from other tissues; differentia-

tion provides it with more types of cell than any other tissue; and specification brings the nervous system to its functional maturity.

Still, embryology has not fulfilled its potential. Something is missing. Embryology has yet to explain what makes the nervous system a fast acting, precise system of information acquisition, transfer, and analysis in the immediate environment of the adult. How does development turn on and tune in the long distance transmission system, the repository for vast amounts of specific information, the modulators of internal and external stimuli, the decoders of positional signals, the transmitters of sensory and motor functions, and so on?

This is a great deal to demand of embryology, and possibly embryology is not yet ready to respond. Then again, possibly it is.

Currently, an overarching principle organizing a great deal of embryological research (or possibly a cloud hanging over embryology) puts the cell surface in charge of cell proliferation, movement, dispersion, congregation, shape, and death. According to the **morphoregulator hypothesis** (Edelman, 1988),[1] resident cell adhesion molecules (CAMs,

---

[1]G. M. Edelman and R. R. Porter shared the Nobel Prize in Physiology or Medicine in 1972 for elucidating the nature and structure of antibodies.

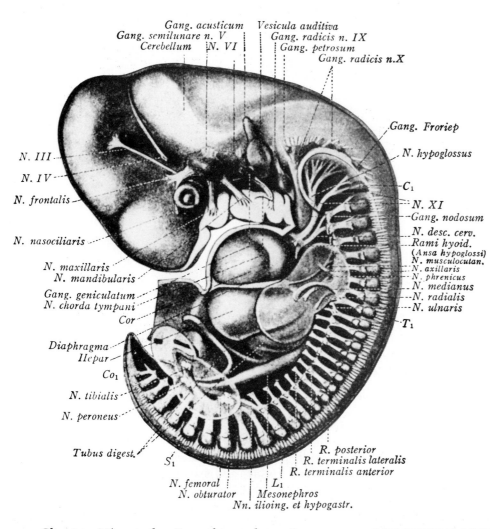

Gang. acusticum
Gang. semilunare n. V
Cerebellum
N. VI
Vesicula auditiva
Gang. radicis n. IX
Gang. petrosum
Gang. radicis n.X

Gang. Froriep

N. hypoglossus

N. III

N. IV

N. frontalis

N. nasociliaris

N. maxillaris
N. mandibularis

Gang. geniculatum
N. chorda tympani

Cor

Diaphragma
Hepar

Co₁

N. tibialis

N. peroneus

Tubus digest.

S₁

N. femoral
N. obturator

L₁
Mesonephros
Nn. ilioing. et hypogastr.

R. posterior
R. terminalis lateralis
R. terminalis anterior

C₁

N. XI
Gang. nodosum
N. desc. cerv.
Rami hyoid.
(Ansa hypoglossi)
N. musculocutan.
N. axillaris
N. phrenicus
N. medianus
N. radialis
N. ulnaris

T₁

FIGURE 26.1. Lateral view of the rudimentary human nervous system. The superficial ectoderm and overlying mesoderm are removed or rendered transparent. ×12. Adapted from G. L. Streeter, *Am. J. Anat.*, 8:285(1908).

see Chapter 15) are the "morphoregulators" responsible for morphology and are regulated by it. The cell's environment transmits information to the cell through CAMs, the cell's cytoskeleton processes information gained from CAMs, and the cell's tissue-specific gene expression depends on CAMs. In turn, the cell regulates its CAMs.

Other hypotheses place equal responsibility on substrate adhesion molecules (SAMs, matrix-, basal lamina-, and membrane-associated), junctional adhesion molecules (JAMs), complex networks of extracellular material, position-specific microenvironments, instructional inductors and permissive inductors released by remote sources, paracrine and endocrine hormones, and a variety of cellular receptors. Still other hypotheses place responsibility on the shoulders of genetic programs, developmental clocks, selector genes, morphoregulatory genes, and historegulatory genes.

The challenge of contemporary embryology is to sort out all these possibilities. The field of developmental neurobiology is currently meeting this challenge.

VERTEBRATE CENTRAL NERVOUS SYSTEM

The functional properties of the vertebrate nervous system depend critically on the intricate network of neuronal connection that is generated during development.

Jane Dodd and Thomas M. Jessell (1988, p. 692)

Cytogenesis in the Neural Tube

Cytogenesis provides the cells required for differentiation. When cytogenesis begins, the wall of the neural tube is a pseudostratified columnar epithelium, or **neuroepithelium**, of **matrix cells** spanning the wall of the neural tube. A **basal lamina** and **external limiting membrane** surround the **outer** or pial surface (named for the pia mater [Lat. tender mother], the nourishing, highly vascularized connective tissue layer surrounding the brain and spinal cord). An **apical** or **ventricular limiting membrane** (formerly the ependymal limiting membrane) lines the central **ventricle** or **neural canal**.

TABLE 26.1. Neural tube derivatives

Primary Divisions	Subdivisions	Constituent Parts	Cavities
Prosencephalon	Telencephalon	Rhinencephalon Corpora striata Cerebral cortex	Lateral ventricles Rostral portion of the third ventricle
	Diencephalon	Epithalamus Thalamus (including Metathalamus) Hypothalamus	Most of the third ventricle
Mesencephalon	Mesencephalon	Colliculi Tegmentum Crura cerebri	Cerebral aqueduct
Rhombencephalon	Metencephalon	Cerebellum Pons	Fourth ventricle
	Myelencephalon	Medulla oblongata	
Spinal cord	Spinal cord	Spinal cord	Central canal

Source: Arey (1965, p. 475).

When cytogenesis is complete, the neural tube is a thickened, multilayered rudiment of the spinal cord and brain (see Table 26.1).

The spinal cord is established through three stages of cytogenesis. In **stage I** (the proliferative stage), nuclei move on a cytological "elevator" within fixed matrix cells (Fig. 26.2). The nuclei undergo DNA synthesis in an S zone toward the outer surface. The nuclei move centrally while in the G-2 period of interphase (I) and undergo mitosis (M) in an apical **ventricular zone** (i.e., the zone surrounding the ventricle; also called the ventricular germinal zone and formerly the ependymal zone). Cytokinesis follows, and cells repeat the cycle of synthesis and division.

In **stage II** of cytogenesis (the stage of neuron production or neurogenesis), cells undergo a **terminal mitosis** and leave the matrix cell population as postmitotic (G-0) **neuroblasts** (nb). The nucleus-containing body of the neuroblast follows an elongating process through the neuroepithelium the same way neurons will later follow their processes in movement outside the CNS (called perikaryal translocation). The neuroblast then detaches from the ventricular layer and begins its differentiation.

Basally, neuronal processes establish the **marginal zone** at the outer level of the tube. The moving neuroblasts establish the **intermediate zone** (formerly the mantle layer) between the inner ventricular and the outer marginal zones (see below, Fig. 26.8).

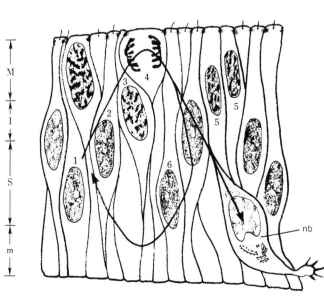

FIGURE 26.2. The "elevator" hypothesis. While so-called matrix cells remain attached at the lumen, their nuclei migrate through zones of DNA synthesis (S), and late interphase (I). When the cells begin differentiation into neuroblast (nb), they enter the basal portion of the future intermediate zone (formerly called the mantle zone, m). From S. Fujita, *Curr. Top. Dev. Biol.,* 20:223(1986), by permission of the Yamada Science Foundation, Academic Press Japan, and the author.

The organization of the early neuroblast's cytoskeleton may set the initial orientation of axon extension, and early axons may take their direction passively in the midst of the abundant substrate available within and outside the neural tube. Substrates are provided by the calcium-independent neural adhesion molecule (N-CAM) (Fig. 20.8d) and Ca^{2+}–dependent adhesion molecule (*N*-cadherin) present on cell surfaces in neural epithelia, and by laminin in the extracellular matrix of the neural epithelium and surrounding mesenchyme. Axonal interaction with laminin requires laminin receptors (i.e., membrane-bound glycoproteins of the integrin family), while N-CAM and *N*-cadherin are cohesive (i.e., homophilic), requiring no intermediate for cell-to-cell adhesion (see Dodd and Jessell, 1988).

The first neuroblasts to differentiate in the spinal cord are large, possibly because they have had few divisions. Present in the ventral portion of the neural tube, these neurons become the segmentally arranged motor neurons of the spinal cord. Dorsally, large ganglion cells lay out the sensory pathways that convey peripheral sensory information to the brain.

Reflex arches may thereby be established early in development. In amphibian tadpoles and other chordate larvas, for example, the large Rohon–Beard cells of the dorsal spinal cord function in the reflex arch that couples tactile stimulation of the tail with flexure of the tail.

The remaining matrix cells continue to proliferate, but, gradually, the duration of the cell cycle increases. In the chick and mouse, the increase is from about 5 hours to about 10 hours from early to late periods of neuroembryogenesis. Possibly, the cycle is extended as cells accumulate irreversibly inactivated replicons.

Later, in **stage III** of cytogenesis, when all the cells that are about to become neuroblasts have entered a permanent postmitotic period, other cells become determined for **glioblast** and for **neuroglial** differentiation. The remaining matrix cells are transformed into **ependymoglioblasts** (stem sustentacular cells lining the canal and ventricles of the CNS), but immunological and cytochemical evidence suggests that no transdifferentiation occurs between neuroblasts and glioblasts (Fujita, 1986).

Neurogenesis in the Rudimentary Central Nervous System

Neurogenesis leads to regional specificity in the central nervous system. One part of the complexity of the CNS is created by neuroblasts dif-

ferentiating into nerve cells. Another part is created by glioblasts differentiating into neuroglia or simply glial cells.

Glial cells come in a variety of different sorts and sizes. Long **radial glial cells,** for example, stretching between the ventricular and pial limiting membranes may guide or direct the neuroblasts' radial migration, while **oligodentroglial cells** (i.e., glial cells with few processes) myelinate axonal processes.

Nerve cells are vastly more complicated. They probably constitute the most heterogeneous population of cell types going under the name of a single tissue. Nevertheless, they are all built of similar parts.

The **neuron** or **perikaryon** (i.e., cytoplasm around a nucleus) is the nerve cell body containing the nucleus. Nerve **processes** or **fibers** are cytoplasmic extensions leaving the neuron (Fig. 26.3).

Two types of nerve cell processes are typically recognized. The **axon** arises from a cone, the **axon hillock,** having a dense undercoating and a funnel of microfilaments. Axons usually have a uniform diameter throughout their lengths although they may give off branches. In addition, axons usually are covered by layers of **myelin** comprising a **myelin sheath.** Other nerve cell processes, called **dendrites** or **neurites,** do not arise from a hillock, may not be of uniform thickness, and are not coated by a myelin sheath.

In the early spinal cord (Fig. 26.4), neuroblasts completely abandon the ventricular zone (VZ). Neurons accumulate in the intermediate zone (IZ), while processes dominate a marginal zone (MZ). Myelination turns the marginal zone into the **white matter,** while the increased cellularity of the intermediate zone turns it into the **gray matter** (Fig. 26.5).

The marginal and intermediate zones do not develop uniformly around the neural tube. Originally, a **roof plate** constitutes the dorsal half of the neural tube, a **floor plate** the ventral half. These plates are arrested as ventricular zones in the midsagittal plane of the spinal cord. In the region derived from the notoplate (see Fig. 23.13, i.e., neural ectoderm above the notochord), the floor plate is reduced to a seam, the **raphe** (Gk. seam), containing no neuroblasts.

Lateral expansion of the original roof plate gives rise to thick dorsolateral **alar plates** (Lat. *alaris* wing), while expansion of the original floor plate gives rise to ventrolateral **basal plates.** Internally, right and left portions of these plates are separated by the dorsoventrally elongated **central canal.** As development continues, the alar and

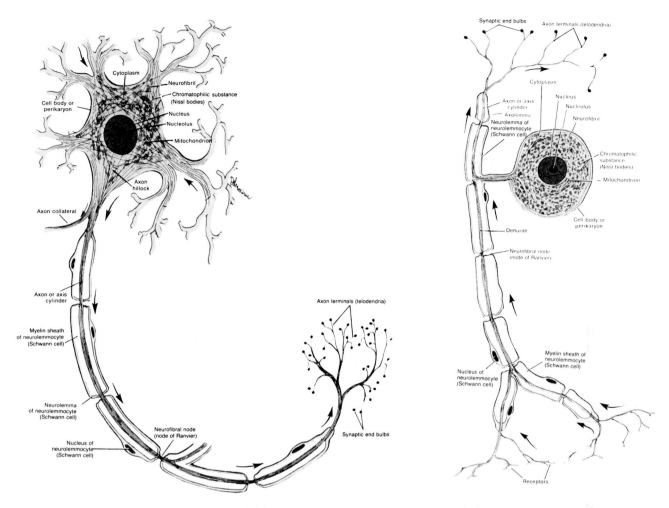

FIGURE 26.3. The conventional nerve cells. (a) Motor nerve (efferent, somatic type). (b) Sensory nerve (afferent). From G. J. Tortora and N. P. Anagnostakos, *Principles of anatomy and physiology,* Harper & Row, New York, 1987. Used by permission.

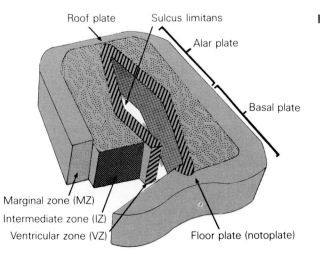

FIGURE 26.4. Schematic of primitive neural tube.

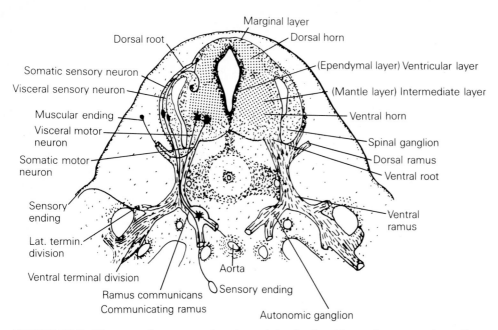

Marginal layer
Dorsal root
Dorsal horn
Somatic sensory neuron
Visceral sensory neuron
(Ependymal layer) Ventricular layer
(Mantle layer) Intermediate layer
Muscular ending
Visceral motor neuron
Ventral horn
Somatic motor neuron
Spinal ganglion
Dorsal ramus
Ventral root
Sensory ending
Lat. termin. division
Ventral ramus
Ventral terminal division
Aorta
Sensory ending
Ramus communicans
Communicating ramus
Autonomic ganglion

FIGURE 26.5. Diagram of cross section through back of a 10-mm human embryo illustrating idealized relationships of spinal nerves and their functional elements. ×30. Adapted from L. B. Arey, *Developmental anatomy,* 7th ed. Saunders, Philadelphia, 1965.

basal plates are separated on each side by a **sulcus limitans** (Lat. limiting furrow).

Most of the subsequent development of the neural tube involves the further elaboration of the intermediate and marginal zones in the alar and basal plates. An exception is the anterior portions of the brain where the basal plates disappear.

In the spinal cord, the gray matter of the basal plates expands into the **ventral columns** or **horns** (actually longitudinally oriented ridges). **Motor nuclei,**[2] especially those in the ventral horn, send out axons (called fibers) in segmentally arranged **ventral roots.**

The alar plates expand into the **dorsal columns** or **horns** (also longitudinally oriented ridges) containing many **association** or **adjustatory neurons** (i.e., sensory relay neurons) that pass afferent impulses along to the sites of reflex arcs and higher ordered functions at cranial levels of the CNS. Ultimately, dendrites from afferent nerves in segmentally arranged **dorsal roots** penetrate the alar plate and either continue through the white matter to higher levels or make local connections with nuclei in the dorsal gray matter.

The emergence of the "butterfly" pattern of gray matter in cross sections of the spinal cord (Fig. 26.6) results from the dorsolateral and ventrolateral expansions of the intermediate zone into columns. The circle of white matter around the butterfly results from the expansion of the marginal zone around the gray columns.

The white matter forms bands or **funiculi** (pl., funiculus sing., Lat. rope): lateral funiculi between dorsal and ventral gray columns, dorsal funiculi between right and left dorsal columns, and ventral funiculi between right and left ventral columns. Externally, right and left funiculi are separated by furrows or grooves (called the dorsal median sulcus and ventral median fissure) that become increasingly deep as expansion continues. Fibers cross in every direction along the length of the neural tube, including the intermediate zone of the spinal cord, comprising a diffuse, reticular (Lat. little net) organization.

In the early brain, areas with crossing fibers give rise to reticular formations. Within these areas, gray matter is preserved in **nuclear regions** (e.g., sensory relays and motor nuclei), while white areas form **tracts** connecting to other parts of the brain, but the regional separation of gray and white matter generally breaks down.

More conspicuous structures develop elsewhere in the brain where gray matter moves to the outside. In **cortical** or **laminated** (i.e., layered) parts of the brain (e.g., cerebellum, cerebral cortices, and optic tectum), the outer **gray matter** or **cortex** is formed when neuroblasts congregate in **cortical**

_____

[2]Motor nuclei refers to aggregates of neurons and not to cellular nuclei as such.

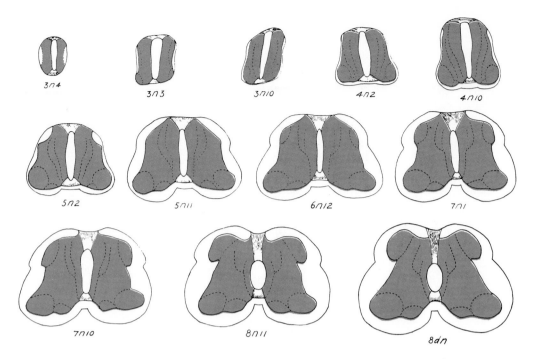

FIGURE 26.6. Sections through the spinal cord of chick embryos from 3 to $8\frac{1}{2}$ days of incubation. Note change in overall cross-sectional shape from oval to pear shape and in spinal canal from slitlike to oval. At first, the entire thickness of the cord is a homogeneous matrix layer. An intermediate zone and ventral horns of gray matter do not appear before day 3. At 4 days, the ventral horns are obvious and white matter has begun to accumulate peripherally. Dorsal horns are only vaguely indicated at $6\frac{1}{2}$ days but clearly outlined at 7 days. By $8\frac{1}{2}$ days, the "butterfly" appearance of the gray matter is unmistakable and mitotic activity has all but ceased in the ventricular layer. From V. Hamburger, *J. Comp. Neurol.,* 88:221(1948), by permission of Alan R. Liss, Inc., and the author.

plates from the "inside out." In **noncortical** or **nuclear** parts of the brain, neuroblasts accumulate from the "outside in." Other regions of the brain accumulate cells and processes by variations on both themes.

In **corticogenic** (i.e., cortex-producing) parts of the brain, the first neuroblasts to migrate out of the matrix layer become embedded in a **molecular layer** composed of the fine terminal processes of **matrix cells. Matrix processes** from these cells stretch across the thickness of the brain wall and, in the course of development, become **fascicled** or bunched. Neuroblasts migrate along the fascicled processes and thereby migrate through layers of earlier neuroblasts. Presumably, neural contacts are established between newly migrating cells and already established cells during migration. (Fig. 26.7a)

In highly cellular **noncorticogenic** regions (e.g., hypothalamus and thalamus), matrix cell processes do not form bundles. Processes fail to maintain contact with the external membrane as the brain thickens, and a molecular layer is not formed. Neuroblasts accumulate in the "outside in" order

corresponding to the order of division and migration into the intermediate zone (Fig. 26.7b).

Development of the Brain

The brain is the cranial portion of the CNS anterior to the first segmental nerves. The embryonic brain rudiment is broader than the remainder of the neuraxis even as an open neural plate (Figs. 23.17 and 26.8), and following formation of the neural tube, the brain swells into three primary **vesicles** with cavities called **ventricles**—the forebrain, midbrain, and hindbrain. Further localized swellings in the forebrain and hindbrain produce a total of five secondary brain vesicles (Table 26.1; Fig. 26.9).

The forebrain or prosencephalon (Gk. pro before + kephal head) is the most anterior brain vesicle. Its cavity is the **prosocoel.**

Originally, the anterior tip of the prosencephalon surrounds the anterior neuropore. The neuropore leaves no trace in the brain, but the

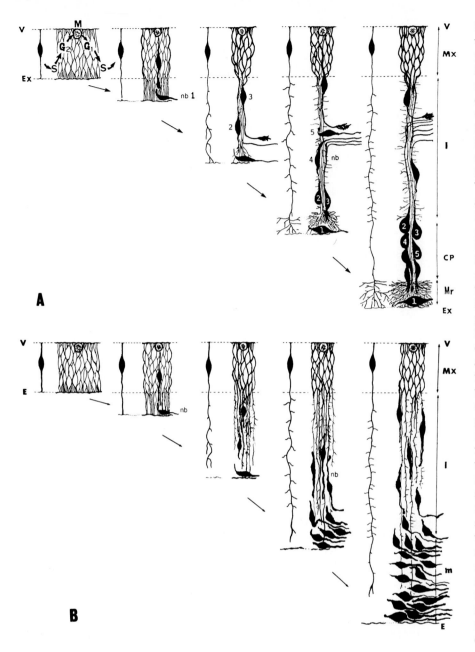

A

B

FIGURE 26.7. Schema for the development of corticogenic (a) and noncorticogenic (b) portions of the brain. The regions begin with similar matrix layers (Mx) in which cells cycle between DNA synthesis (S), G_2, mitosis (M), and G_1. Fascicles of intertwined cell processes span the brain wall, and the first neuroblasts (nb) to migrate out of the matrix layer enter the intermediate layer (I). The first cells to migrate are generally larger than those that migrate later. At this point differences appear between regions that will later become cortical and those that will be noncortical. (a) The corticogenic regions are formed primarily from the "inside out." The first neuroblasts migrate before matrix cell processes fasciculate. Later neuroblasts (2–5) migrate along the fascicled processes and tend to pass one another, but they cannot displace the first neuroblasts embedded in the external molecular layer. (b) The noncorticogenic regions are formed primarily from the "outside in." Processes originally reaching the external surface (E) fail to be maintained as the brain thickens, and a molecular layer fails to form. Instead, the mantle (m) acquires neuroblasts consecutively on the inside surface. Cp, cortical plate; E, Ex, external limiting membrane; I, intermediate zone; m, mantle zone; Mr, molecular layer; Mx, matrix layer; nb, neuroblast; V, ventricle. From S. Fujita, *Curr. Top. Dev. Biol.*, 20: 223(1986), by permission of the Yamada Science Foundation, Academic Press Japan, and the author.

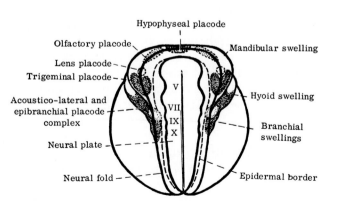

FIGURE 26.8. Surface view of early amphibian neurula (stage 13) showing locations of ectodermal placodes, neural plate, and neural folds. Dashed line = epidermal border: limits of neural infolding. Mandibular swelling: rudiment of first visceral arch. Hyoid swelling: rudiment of second visceral arch. Branchial swellings: rudiments of remaining visceral arches. From E. Witschi, *Development of vertebrates,* Saunders, Philadelphia, 1956. Used by permission.

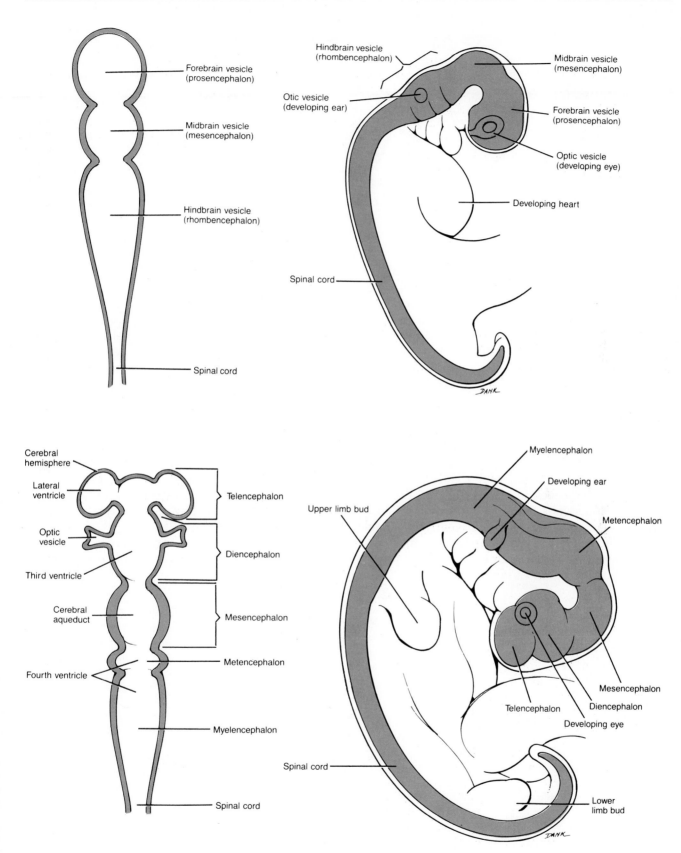

FIGURE 26.9. Conventional diagrams of three-vesicle and five-vesicle brain.

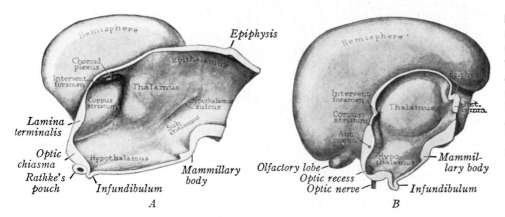

FIGURE 26.10. Drawings of right halves of late embryonic (a) and early fetal (b) human telencephalon and diencephalon. (a) At 7 weeks. ×10. (b) At 10 weeks. ×4.5. From L. B. Arey, *Developmental anatomy,* 4th ed. Saunders, Philadelphia, 1965. Used by permission.

horseshoe of tissue reaching below the neuropore (i.e., the crosswise frontal fold) becomes the **lamina terminalis,** a median band later associated with **commissures** or fiber tracts reaching across the brain (see below). Because the greatest part of forebrain growth is lateral, the band is left behind, buried deep in the longitudinal fissure between cerebral hemispheres (Fig. 26.10).

Shortly after emerging, the forebrain is separated into a **diencephalon** (i.e., second brain) and an anterior **telencephalon** (i.e., end brain). The divide is marked above by the **velum** (or velum transversum, Fig. 26.10b, a depression in front of the epiphysis) and below by the **optic recess.** The diencephalon retains a central cavity called the **third ventricle** (also diocoel, Figs. 26.11 and 26.12). The telencephalon swells laterally into **cerebral hemispheres,** each containing a **lateral ventricle** (sometimes identified as I and II) connected by an **interventricular foramen** (or foramen of Monroe) to the anterior portion of the third ventricle.

The forebrain grows rapidly during neurulation and quickly overshoots the tip of the notochord. Cell kinetics and the movement of labeled cells *in vitro* indicate that some forebrain cells are gained at the expense of the midbrain and rostral hindbrain. The originally posterior cells move anteriorly without altering the positional relationship between the brain regions.

Mitotic spindle orientation reflects the direction of cell movement but probably does not "pump" the cells, since microfilament-dominated events such as cell movement can generally proceed without cell division. **Cell displacement** or **cell flow** (in which cells exchange neighbors) within an intact epithelium may provide the mechanism of forebrain expansion (see Keller and Hardin, 1987; Morris-Kay and Tuckett, 1987).

The diencephalon does not have a basal plate. Only the alar and roof plates expand over the floor of the third ventricle.

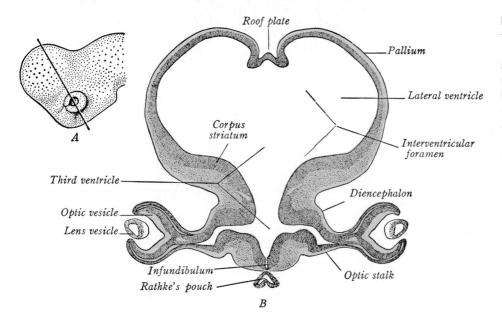

FIGURE 26.11. Drawing of anterior end of human embryo at 10 mm (a) showing angle of sectioning and idealized section showing main divisions of forebrain, optic vesicle, and vesicles (b). From L. B. Arey, *Developmental anatomy,* Saunders, Philadelphia, 1965. Used by permission.

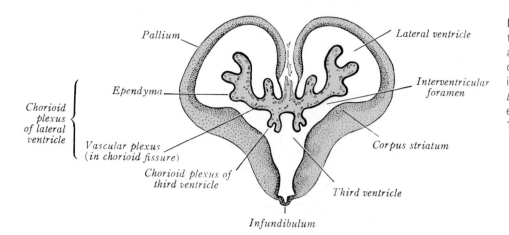

FIGURE 26.12. Idealized section of human embryo forebrain at 3 months showing incursion of anterior choroid plexuses into ventricles. From L. B. Arey, *Developmental anatomy,* 7th ed. Saunders, Philadelphia, 1965. Used by permission.

Anteriorly, the roof plate thins to a differentiated ventricular layer that fuses with a portion of the pia mater (the vascular innermost covering of the brain) known as the **tela choroidea** (Lat. web + chorion-like, hence a highly vascular membrane). Wrinkling and folding into the ventricle, the fused layers form the **anterior choroid plexus** (i.e., anterior vascular network) bringing circulation to the forebrain (Fig. 26.12).

Dorsally, the small medial rudiment of the **pineal gland,** known as the **epiphysis,** emerges from the roof plate (Fig. 26.10). Another **median parietal eye** of some fish, amphibians, and reptiles emerges from yet another rudiment. Each may later function in the perception and measurement of day length and hence sexual development via hormones.

Internally, the walls of the diencephalon produce a series of thickenings. Dorsal **epithalami** thicken the junctions between roof and alar plates. Lateral **thalami** (Gk. woman's apartment, possibly rotunda) swell the upper portion of the alar plate, and **hypothalami** bulge from the lower portion of the plate. **Hypothalamic sulci** separate the hypothalamus from the thalamus or its lower portion called the **subthalamus** (or ventral thalamus, Fig. 26.10).

The hypothalamus is the brain's center for the autonomic nervous system (see below) with nuclei concerned with many visceral functions. It is the coordinating center for regulating digestion, sleep, internal temperature, and some emotions. In addition, the optic chiasma, the infundibulum, and the mammillary bodies (i.e., nipple-shaped bodies, also called corpora albicans) are derivatives of the hypothalamus (Fig. 26.10).

The **infundibulum** (Fig 26.10) at the tip of the funnel-shaped floor of the hypothalamus, forms the major part of the **posterior lobe** (neurohypophysis) of the pituitary gland. Tracts from nuclei

in the hypothalamus deliver hormones to the posterior lobe, and portal vessels of the **hypothalamo–hypophyseal portal system,** traversing the distance between the hypothalamus and the anterior lobe, deliver gonadotrophin releasing hormone (GnRH) to the anterior lobe (see Fig. 7.4). This intimate relationship between the anterior and posterior lobes is built in developmentally by the formation of the anterior lobe from Rathke's pouch following contact with the infundibulum (see Chapter 25).

The thalamus contains the major tracts from the mesencephalon, and its nuclei are association and relay centers between the cerebral hemispheres. In mammals, almost all nervous impulses to the cerebral cortex pass through centers of correlation and tracts in the thalamus (the exception being impulses related to smell).

Growth in the thalamus originally lags behind that in the epithalamus and hypothalamus, but ultimately thalamic growth exceeds that of the other diencephalic regions, and two thalami unite at the **massa intermedia** (intermediate mass). In mammals, some thalamic centers will later operate reflexes involving pleasurable and painful sensations.

The optic vesicles or bulbs are early diverticula (outpocketings or evaginations) of the diencephalon (Figs. 23.9 and 26.13). Their cavities, called **opticoels,** are continuous with the third ventricle. In human beings, **optic pits** or depressions mark the open neural plate even before it becomes a tube. Optic vesicles or pits may induce lens placodes in overlying cutaneous ectoderm to form lens vesicles and finally lenses (see Chapter 21).

Beginning at the ventral surface, the edges of the optic vesicles fold inward, and the opticoel collapses except at the base of the future eye. When the walls of the vesicle meet, they fuse and the

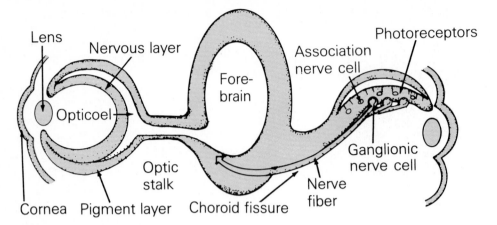

FIGURE 26.13. Diagram of section through optic cups and forebrain. The two eyeballs are not cut in precisely the same plane. (Left) Entire circumference of optic cup shown. (Right) Section passes through choroid fissure in ventral rim of optic cup. Optic nerves leave inner surface of eyeball through fissure. Major circulatory vessels also traverse the fissure.

optic vesicle is transformed to a two-layered **optic cup.** The layer lining the concavity of the cup becomes the neural retina or nervous layer; the layer facing the surrounding tissue becomes the pigmented retina or pigmented layer. The degree to which the development of the mammalian visual apparatus lags behind that of chicks can be appreciated by comparing the size of the eye vesicles or cups in Figs. 26.11 and 23.12.

As the optic vesicle rounds up, it retains a ventral indentation known as the **choroid fissure** that later conveys the major circulatory vessels (hence choroid) to the lining of the eye. The same fissure provides the route through which the **optic nerve** or tract, consisting of the collection of nerve processes arising from the neural retina, exits the eyeball and moves to the brain.

The points of attachment of diencephalon and optic cups move ventrally, narrow, and lengthen into the hollow **optic stalks.** The optic nerve follows the path of the stalk to the base of the diencephalon. There, at the **optic chiasma,** some or all the nerves cross and move through tracts to opposite sides of the brain.

Because the retina is invaginated (Fig. 26.13), the **inner retinal region** surrounding the cavity of the eye corresponds to the outer region of the spinal cord. The eye's equivalent to the spinal cord's canal is the narrow **opticoel** between the nervous and the pigmented layers of the retina.

When differentiated, processes of the optic nerve or tract will arise at the peak of sensory-**neuronal pyramids** spanning the thickness of the nervous layer. Above the peripheral **photoreceptors** (rods and cones), the retina is structured in layers by the processes and cell bodies of **association neurons:** horizontal, bipolar, amacrine (with one process), and ganglion neurons. Each layer collects and transmits signals centripetally until ganglion cells send the collected impulses down processes of the optic nerve (Fig. 26.13).

Mirroring its structure, the differentiation of the retina proceeds centrifugally _in vivo:_ the differentiation of association neurons precedes that of photoreceptors. Determination, however, seems to proceed in the opposite direction, at least in the chick, since cells dissociated from retinas early (embryonic day-6 [ED-6]) and cultured _in vitro_ differentiate predominantly as photoreceptors, while those dissociated later (ED-8 through ED-10) differentiate predominantly as association neurons (Adler and Hatlee, 1989).

Retinal differentiation occurs in postmitotic cells that had previously undergone their terminal mitosis as shown by their failure to incorporate tritiated thymidine. Cells that become postmitotic early (before ED-5) are not necessarily determined at that time. When dissociated a day later and cultured _in vitro,_ they differentiate into both photoreceptor cells and association neurons. The same postmitotic cells seem to become determined a few days later if left in the retinal environment, since, when dissociated at that time (ED-8), the cells differentiate predominantly into association neurons.

If retinal cells are not determined upon becoming postmitotic, how do they acquire their determination postmitotically? Association neurons may acquire their determination while migrating centripetally through the retinal environment. Photoreceptors, on the other hand, may acquire their determination by default, becoming photo-receptors as a consequence of having failed to migrate.

The telencephalon also develops without benefit of basal plates. Most conspicuously, the alar plates swell laterally into the cerebral hemispheres. An olfactory portion of each hemisphere, the **rhinencephalon** (Gk. *rhin* nose + *kephal*), enlarges ventrally into an **olfactory lobe** (later separating into an olfactory bulb and stalk) which is entered by the olfactory cranial nerve (n I)[3] and penetrated by the **olfactory nerve tract.** The non-olfactory portion of each hemisphere, or **neopallium** (Gk. new cloak or mantle), undergoes the greatest amount of growth in mammals, especially

[3]Following convention, the cranial nerves are identified by the letter "n" and a Roman number.

in primates and most especially in human beings (Fig. 26.14).

At early stages of development, the cerebral hemispheres of amphibians and chicks are relatively large compared to mammals but do not truly deserve the title of neopallium assigned to them, since they do not cover anything. Only the cerebral hemispheres of mammals at later stages of development cover the diencephalon and overshadow the rest of the brain, thereby earning the title neopallium. During the second half of fetal life, mammals add the characteristic **gyri** or convolutions, **furrows** (relatively large fissures), and **sulci** (relatively small fissures). Smaller sulci peculiar to the human brain are only added during the last fetal months.

The marginal zone of the telencephalon's alar plates gives rise to a V-shaped **internal capsule** or lamina of tracts running between the neopallium

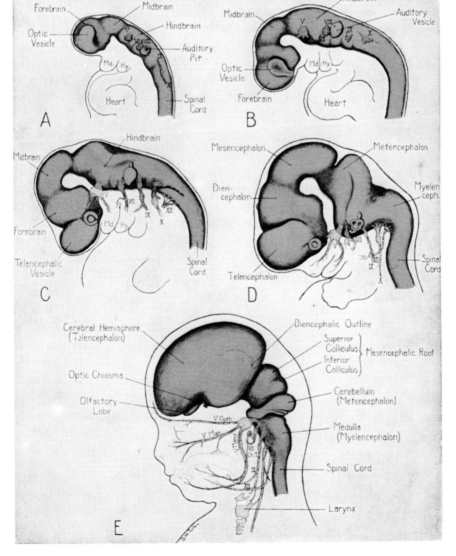

FIGURE 26.14. Schematic stages in the development of the human brain, cranial nerves, and visceral arches: (a) 3 mm; (b) 4 mm; (c) 8 mm; (d) 7 weeks; (e) 3 months. From B. M. Carlson, *Patten's foundations of embryology*, 5th ed., McGraw-Hill, New York, 1988. Used by permission.

and the thalamus. Cellular proliferation in the intermediate zones provides the nuclei of the **corpus striatum's** (layered body, Figs. 26.10, 26.11, 26.12) cellular portions (the adult caudate nucleus and lentiform nucleus). Later, when fused with the thalamus, these regions are the centers for the high-order reflex coordination, coupling sensory input with motor output.

In mammals, the lamina terminalis is the site of several **commissures** connecting the cerebral hemispheres. At the dorsal border of the lamina terminalis, the **hippocampus** (Gk. mythical seahorse, named for fanciful resemblance, also archipallium) fuses into a **dorsal commissure** (also called the hippocampal commissure). An anterior thickening (the torus transversus) becomes the **anterior commissure,** and later the large **corpus callosum** arches over the laminar terminalis (Fig. 26.10).

The midbrain or mesencephalon remains undivided. It is separated from the diencephalon by thickenings (ventrally, the tuberculum posterius or mammillary bodies, and dorsally, the posterior commissure) and from the hindbrain by a constriction, the **isthmus.** Initially, the midbrain is the retarded portion of the original neural tube connecting the rapidly changing forebrain and hindbrain. Later, growth in the midbrain thickens its walls and narrows its ventricle into the **cerebral aqueduct** (or aqueduct of Sylvius, Fig. 26.9).

Mesencephalic morphogenesis begins with the disappearance of the floor plate. Nuclei of the **oculomotor** cranial nerve (n III) and **trochlear** (n IV), which control some of the eyeball's extrinsic musculature, appear in the basal plate in the midst of a reticulum formation (later to become the **tegmentum**). Motor fibers from the oculomotor exit from the floor of the mesencephalon, but those from the trochlear migrate up the wall of the mesencephalon and then backward to exit at the dorsal isthmus.

The mesencephalic alar plates form one or two pairs of bulbous swellings that thicken into the adult **tectum.** In most vertebrates (mammals being the exception), the anterior bilobed **optic tectum** is the center of visual ordering. In amphibians, the undivided posterior part of the tectum receives nerves from the ear, while in amniotes, the tectum has four swellings, the **corpora quadrigemina.** The anterior two swellings, the **superior colliculi,** receive fibers from the **optic nerves** (or tracts) that arise in the **neural retina** of the eye. The posterior **inferior colliculi** become acoustic reflex centers receiving auditory and balance information from nerves of the ear (Fig. 26.14).

In mammals, the superior colliculi provide optic reflex centers associated with moving the head and eyes in response to visual stimuli (i.e., in coordination with the oculomotor, trochlear, and other motor nuclei). Centers of visual sensibility are in the cortex of the cerebral hemispheres. Acoustic sensibility has also moved to the cerebral cortex.

The hindbrain or rhombencephalon (Lat. *rhombus* spinning top + *kephal*) ***is separated from the mesencephalon by the narrow isthmus and from the spinal cord by the first pair of spinal ganglia.*** The thickened lateral walls swing out on the hingelike floor plate giving the **fourth ventricle** (the ventricle of the hindbrain) a characteristic V shape (Fig. 23.12a–d).

The hindbrain's medial roof plate and part of its alar plates thin and stretch across the opening of the V while fusing with the overlying tela chorioidea (vascular membrane). Folding into the fourth ventricle, the fused membrane forms the **posterior choroid plexus** (posterior vascular network) nourishing posterior portions of the brain.

Posteriorly, the rhombencephalon continues to have the basic structure of the spinal cord, but anteriorly, things change. Expansion of the marginal layers on either side gives rise to lateral thickenings (the rhombic lips). The ventricle becomes compressed dorsoventrally into a diamond shape. A rhomboid fossa creases the floor; the sulcus limitans represents the edges of the diamond, and the roof plate represents its upper edge.

The hindbrain is initially undivided, but seven segmental constrictions (rhombic grooves) mark off a series of six **neuromeres** (or rhombomeres, see Fig. 23.18). Four neuromeres form the **myelencephalon** of the adult **brain stem** before narrowing into the spinal cord. Anteriorly, especially in anamniotes, two neuromeres swell into the **metencephalon.**

The metencephalon gives rise to the ventral **pons** and to the dorsal **cerebellum** (enlarged especially in primates but also in flying and swimming vertebrates). A huge shifting of cells and growth of fibers occur in the formation of these areas.

Flexure and Torsion

The brain is especially affected by bends and twists along the embryo's axis. The **cranial flexure** (also cephalic flexure or flexure of the midbrain, Fig. 26.15) occurs in anamniotes and amniotes. Pivoting on mesencephalon, the flexure brings the embryonic forehead of amniotes to rest against the

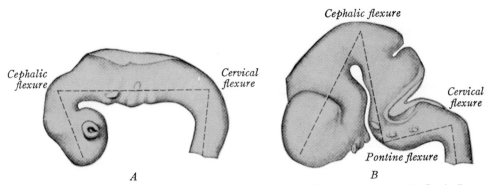

FIGURE 26.15. Flexures in human brain: (a) 6 mm, ×13; (b) 14 mm, ×7. Cephalic or midbrain flexure is widespread among vertebrates and produces permanent change. Cervical flexure has no lasting effect. Pontine flexure places metencephalon temporarily above myelencephalon but straightens as the brain stem develops. From L. B. Arey, *Developmental anatomy,* 7th ed. Saunders, Philadelphia, 1965. Used by permission.

heart (Fig. 26.1) and sets the forebrain permanently at an angle with the rest of the brain.

Other flexures are characteristic of amniotes but are not as uniform or as permanent. In amniotes, a downward **cervical flexure** forms at the juncture of brain and spinal cord, and in mammals, an upward **pontine flexure** (Lat. *pons* bridge) forms near the metencephalon–myelencephalon junction. The developing **cerebellum** of mammals thereby comes to overlie the **medulla oblongata,** but both the cervical and pontine flexures ultimately straighten as the brain stem thickens (Fig. 26.14).

A **caudal flexure** in amniotes tucks the tail under, and a **dorsal flexure** straightens the thorax just below the neck. In many mammals, a **lumbosacral flexure** helps to fold the embryo into a C (Fig. 26.9), while in other amniotes, the posterior spinal cord remains more nearly straight.

In birds, the **torsion** (or rotation around the central axis) lays the head on its left side and pushes the heart into the right exocoel (Fig. 23.11). A small **caudal torsion** in mammals may twist the tail off to the right, and when the connection with the yolk sac has narrowed sufficiently, mammals also turn right and wind up with their left side toward the yolk.

PERIPHERAL NERVOUS SYSTEM

Anatomy of the PNS

The PNS includes all the nerve cell bodies and associated nerve processes or fibers outside the CNS. While visceral motor cells synapse with other motor cells in ganglia of the PNS, sensory and somatic motor cells do not. Those connections are made via the CNS.

PNS nerves are classified hierarchically (Table 26.2). The first level specifies the direction of signals relative to the central nervous system (afferent versus efferent). The second level stipulates the location of nerve endings (somatic versus visceral), and the third level reflects the complexity of sensory organs giving rise to nerve fibers (special versus general).

In the adult, **afferent** or **sensory nerves** deliver impulses from sensory organs to the CNS, while **efferent** or **motor nerves** deliver impulses from the CNS to effector organs. The sensory fibers tend to make their functional synapses in the dorsal part of the CNS, while motor fibers tend to arise from the ventral part (Fig. 26.5).

The **general somatic** or **somaesthetic nerves**

TABLE 26.2. Classification of fibers of the peripheral nervous system

A. Afferent or sensory
  1. Somatic
    a. General (or somaesthetic): mostly from the skin
    b. Special: cranial nerves from sensory layers of eye and ear
  2. Visceral
    a. General: from viscera
    b. Special: cranial nerves from sensory layers of nose and taste buds
B. Efferent or motor
  1. Somatic: to striated muscle
  2. Visceral[a]
    a. General (autonomic): smooth muscle, cardiac muscle, and glandular tissue
    b. Special: cranial nerves to striated muscle of visceral arch origin

[a]Not all these fibers actually go to the viscera proper. Some actually go to the body wall.

and the **general visceral afferent nerves** begin in free nerve endings or in relatively simple sense organs distributed in the skin and viscera, respectively (Fig. 26.5). The cell bodies of both types of afferent nerve are located in **spinal ganglia** (also known as **dorsal root ganglia**) and in ganglia of the n V, n VII, n IX, n X, and, for a transient period in the embryo, the n XI and n XII (Fig. 26.1).

Cell bodies of the **special afferent** nerves lie in **special** sense organs. The **special somatic afferent nerves** are the **optic** (n II) and **acoustic** (n VIII, also called auditory or otic). The **special visceral afferent** nerves are the **olfactory** (n I) from the nose, and portions of the **facial** (n VII), the **glossopharyngeal** (n IX) from taste buds in the tongue, and the **vagus** from taste buds in the region of the epiglottis. The cell bodies of gustatory nerves lie in the **geniculate ganglion** of the n VII and in the **inferior ganglia** of the n IX and n X, the **petrosal ganglion** of the n IX and the **nodose ganglion** of the n X (Fig. 26.1).

In mammals, somatic sensory nerves in the n IX and n X ganglia innervate small areas of the external ear. Most of the nerves in these ganglia are visceral sensory nerves concerned with taste, and in anamniotes, the n X receives impulses from the lateral line organ.

The **somatic efferent nerves** have their nerve cell bodies within the central nervous system and send out fibers directly to skeletal muscles. Motor nuclei in the basal plate of the mesencephalon supply fibers of the oculomotor (n III) and trochlear (n IV). The metencephalon sends fibers into the trigeminal (n V), abducens (n VI), and facial (n VII). The myelencephalon provides motor nuclei for the n IX–n XII.

The **visceral efferent nerves** comprise the **autonomic nervous system.** Two types of nerve are linked in this system. (1) Nerves with cell bodies within the CNS send out **preganglionic** or **presynaptic fibers** through **ventral roots** (Fig. 26.5) to synapses in peripheral ganglia or plexi. (2) Nerves with cell bodies within these ganglia send out **postganglionic** or **postsynaptic fibers** through **communicating rami** (sometimes called gray rami communicans) primarily to smooth muscle and glandular

tissue. Only the postsynaptic fibers have motor end plates capable of exciting nonnervous cells, but only presynaptic fibers are myelinated. Bundles (or fascicles) of myelinated axons in the peripheral nervous system are also invested in an impermeable cellular sheath called the **perineurium.**

The autonomic nervous system is further subdivided into **parasympathetic** and **sympathetic** nerves whose functions are sometimes antagonistic. The criteria for this subdivision of the autonomic nervous system are anatomical and biochemical.

The presynaptic fibers of the parasympathetic nerves are found at the level of the head and pelvis (i.e., "cranial–sacral" nerves). Postsynaptic parasympathetic neurons lie in **terminal ganglia** or scattered in **nervous plexi** close to or in the innervated organs (e.g., heart, lungs, pelvic viscera, and digestive tube).

The presynaptic fibers of the sympathetic nerves are found at the level of the chest and abdomen (i.e., "thoraco–lumbar" nerves). The postsynaptic sympathetic neurons lie in regularly distributed interconnected ganglia on the ventral surface of the vertebral column, the **prevertebral** or **orthosympathetic trunk ganglia.** In addition, **collateral ganglia** (coeliac, superior, and inferior mesenteric ganglia) formed near the aorta and not in the orthosympathetic chain are part of the sympathetic system.

The preganglionic and postganglionic fibers of the parasympathetic nerves are cholinergic, employing acetylcholine as their terminal neurotransmitter. The presynaptic sympathetic fibers are also cholinergic, but postsynaptic sympathetic nerves are adrenergic, employing norepinephrine as their neurotransmitter.

Sources of Sensory Nerves

With the exception of the optic nerve derived from the eye, special sensory neurons of the cranial nerves are derived from **neural placodes** (Fig. 26.8), **neural crest cells,** or both (Table 26.3). Generally,

TABLE 26.3. Sources of somatic and visceral sensory components of cranial nerves

Cranial Nerves					
Olfactory I	Optic II	Facial VII	Acoustic VIII	Glossopharyngeal IX	Vagus X
Dorsolateral placode	Brain	Neural crest and epibranchial placode	Acousticolateral placode	Neural crest and epibranchial placode	Neural crest and epibranchial placode

Source: Adapted from Witschi (1956, p. 284, Table 11).

sensory **ganglioblasts,** or cells that become general sensory neurons of dorsal root ganglia, are derived from the neural crest.

In the trunk region, neural crest cells migrate either along a dorsal pathway, leading to the epidermis and pigment cell differentiation, or along ventral pathways (1–3, Fig. 23.16), leading, among other places, to presumptive ganglia and neuronal differentiation.

The vertebrate optic nerve (n II) arises from **ganglion cells** on the inner surface of the neural retina (Fig. 26.13). Initially running through the choroid fissure that still creases the ventral side of the eye, the optic nerve runs along the length of the optic stalk to the brain. Healing of the optic vesicle and stalk over the choroid fissure brings the optic nerve into the CNS.

Other special sensory nerves arise either from nerve cells in or near the sensory organ or from sensory ganglioblasts in more centrally located ganglia. Anteriorly, paired **olfactory placodes** sink into **olfactory** or **nasal pits** (see Fig. 25.2) and send olfactory nerve fibers (n I) directly into the olfactory bulbs of the telencephalon (Fig. 26.10). All the other placodes and aggregates of sensory ganglioblasts send fibers to the hindbrain.

The placodes play a more prominent role in forming cranial nerve ganglia in amphibians, while neural crest cells are implicated more in amniote development. The rate of ganglia development is also variable. The thickening of placodes and gathering of neural crest cells lag, like the rest of neural development, especially in birds and mammals.

Generalities about the neural crest origins of ganglioblasts are based on results of a variety of experiments: deficiencies developing following the removal of embryonic tissue, and differentiation observed *in vitro* and in intracellularly labeled tissue transplants. Experiments with amphibians preceded those with birds, but, except for a greater latitude in the differentiation of anterior amphibian neural crest, results with representatives of the two orders are similar (see Le Douarin, 1982).

Neural crest from tritiated-thymidine-labeled chick embryos (see Weston, 1983) and from quail with a dense nucleolar marker have been transplanted (Fig. 26.16) and traced in chimeras (Fig. 26.17). In the trunk region, labeled cells bunch up behind the sclerotomes of somites and differentiate as dorsal root ganglia (see Fig. 23.16, pathway 2) (Le Douarin et al., 1984).

Differentiating sensory ganglioblasts become **bipolar** by sending out separate nerve fibers in two directions. Soon these fibers are rearranged in a T shape. The ganglioblast is then **unipolar** with only one process arising from it. The long neurite branch moves peripherally, while the axonal branch enters the CNS (Fig. 26.3b).

Sensory neurites are myelinated, and peripheral bundles of fibers (fascicles) are invested with a perineurium. To ascertain the source of perineurial cells, neural crest cells or fibroblasts were infected with a recombinant retrovirus carrying the *Escherichia coli lacZ*-β-galactosidase gene and combined *in vitro* with a purified neuronal population (dissociated dorsal root ganglia from 15-day rat fetuses). After 6–8 weeks, when extensive myelination and perineurium formation had taken

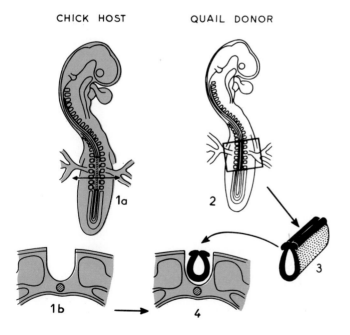

CHICK HOST   QUAIL DONOR

1a   2   3

1b   4

FIGURE 26.16. Schematic for exchange of portion of neural tube between chick and quail embryos. In this case the host and donor are of the same age, and the tissues involved come from corresponding segments. (1a) Dorsal view of chick. Dashes indicate portion of neural tube and crest to be removed. (1b) Schematic cross section showing condition of host prior to grafting. (2) Dorsal view of quail. Lines indicate area from which graft tissue is taken. (3) Neural tube and folds containing neural crest after isolation with aid of trypsin. (4) Schematic cross section showing condition after grafting. From N. Le Douarin, *The neural crest,* Cambridge University Press, Cambridge, 1982. Used by permission.

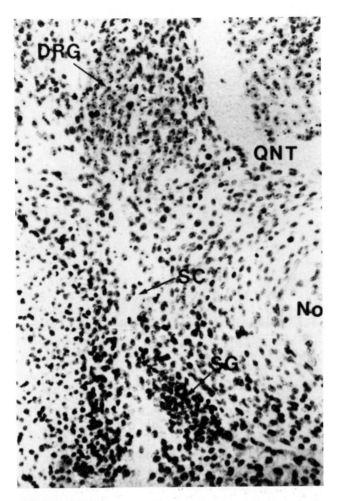

FIGURE 26.17. Photomicrograph of section through chick–quail chimera 5 days after grafting. The host chick cells in the notochord (NO) and surrounding mesenchyme contain typically pale nuclei, whereas the quail neural tube (QNT) is labeled with dense quail nucleoli. The dorsal root ganglia (DRG), sympathetic ganglia (SC), and neurolemmocytes (i.e., Schwann cells, SC) are of quail origin and therefore from the neural crest. ×300. From N. Le Douarin, *The neural crest,* Cambridge University Press, Cambridge, 1982. Used by permission.

place, the cultures were fixed and lacZ-β-galactosidase detected histochemically.

The perineurium was found to be labeled only in cultures containing infected fibroblasts, while the myelin sheaths were labeled only in cultures containing infected neural crest cells. Unlike myelin sheaths derived from neural crest cells, the perineurium seems to be derived from fibroblasts (Bunge et al., 1989).

Bilateral **trigeminal placodes** in amphibians or comparable ectodermal thickenings and aggregates of neural crest cells in birds and mammals form

the **semilunar ganglia** (also called the trigeminal or Gasserian ganglia) of n V. In mammals, central and rostral movement of sensory processes will later bring them from the semilunar ganglia into the brain.

Axons of n V sensory nerves move to **sensory relay nuclei** in the alar plate of the metencephalon. Sensory neurites move peripherally in a three-way split (hence the name **trigeminal** for this cranial nerve). A large body of neurites moves toward the first visceral arch (Figs. 25.2 and 26.14). As the arch bends around the mouth and forms the maxillary process, the nerve supplies the upper embryonic jaw with a **maxillary branch** and lower jaw with a **mandibular branch.** The third, **ophthalmic branch** moves toward the eye.

The second placode or complex of placodes, the **acoustico-lateral** and **epibranchial** (i.e., dorso-pharyngeal) **placode complex** plus aggregated neural crest cells, forms the the sensory parts of ganglia for the n VII, n VIII, n IX, and n X. All these sensory cells send axons into the myelencephalon.

Initially combined in the **acousticofacialis** ganglion, the VIII's **acoustic** (also auditory or otic) **ganglion** and the VII's **geniculate ganglion** gradually separate (Fig. 26.14). The geniculate's sensory neurites join those of the n V in the lower jaw. The acoustic enlarges and, in amniotes, separates again into the **spiral ganglion** associated with the part of the ear specialized for hearing and the **vestibular ganglion** associated with the part of the ear specialized for balance and equilibrium.

Sources of Peripheral Motor Nerves

Postsynaptic motor nerves are derived from motor **ganglioblasts** of neural crest origin. Neural crest cells giving rise to orthosympathetic chain ganglia arise below the level of the fifth somite (Fig. 26.18 right). The cells migrate between somites to the vicinity of the aorta where they differentiate as chain ganglia (see Fig. 23.16, pathway 1), or they take an intermediate route and contribute to the chain (pathway 3). The related adrenergic adrenal medullary cells and paraganglia are derived from neural crest between the 18th and 24th somites.

The parasympathetic **ciliary ganglia** of the iris arise from mesencephalic neural crest. Other parasympathetic terminal ganglia in the head (e.g., lingual, submandibular, otic, sphenopalatine, and ethnoid) are formed chiefly by motor ganglioblasts of neural crest origin migrating out of the **semi-**

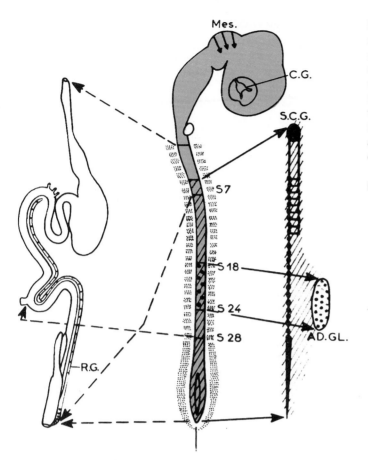

FIGURE 26.18. Summary of neural crest origins of autonomic ganglia and related structures based on heterospecific grafting and tracing nucleolar-labeled cells in chick–quail chimeras. (Left) Parasympathetic ganglia including ciliary ganglion of iris (C.G.) and Remak's ganglion (R.G.) found along the intestine of birds. (Right) Sympathetic ganglia include the superior cervical ganglion (S.C.G.) and the orthosympathetic chain. The associated adrenal medulla (AD.GL) is derived from neural crest cells originating in an overlapping area. From N. Le Douarin, *The neural crest,* Cambridge University Press, Cambridge, 1982. Used by permission.

lunar ganglion (n V). Additional cells are from the **geniculate ganglion** of the facial (n VII) and the **petrosal** (or inferior) **ganglion** of the **glossopharyngeal** (n IX).

Terminal parasympathetic ganglia in the remainder of the body are formed by motor ganglioblasts from two sources. Cells from a source cephalad to the seventh somite migrate along the route of the **vagus** (n X) and its associated ganglia to the viscera (Fig. 26.18 left). In addition, many of the parasympathetic cells of Remak's ganglion in birds migrate out of the neural crest posterior to the 28th lumbosacral somite and then along the length of the intestine (see Le Douarin, 1982).

The motor fibers from n V, n VII, n IX, and n X have strikingly stereotypic relationships to the first to fourth visceral arches. The n V (trigeminal) innervates the first visceral or mandibular arch and hence the lower jaw. The n VII (facial) innervates the second visceral or hyoid arch and, in primates, the facial musculature that spreads from this arch. Motor fibers from the n IX innervate the third visceral arch, and hence the parotid salivary gland and muscles at the base of the tongue. Motor fibers from the n X innervate the fourth arch from which they spread backward along the course of the lat-

eral line to muscles of the pharynx and larynx and even further to the heart and viscera.

ORIGINS OF SENSATION AND RESPONSIVENESS

General Sensory Organs

In addition to the special sensory organs on the head (eye, ear, taste buds, and nose), a variety of general sense organs are spread throughout the vertebrate body. Touch receptors, for example, are the most abundant receptors in the human body and universal among vertebrates. Other sensory organs provide some vertebrates with sensations completely unknown to us. Birds may have electromagnetic receptors in their brains, and fish are equipped with chemical and electrical sensors associated with lateral line organs and the vagus (n X).

The type of sensory organ stimulating a sensory nerve determines the types of information transmitted by the nerve. General somatic nerves or somaesthetic nerves carry impulses from relatively simple sense organs in the skin. These impulses are interpreted as pain, heat, touch, and light pressure. General visceral nerves similarly transmit from equally simple sense organs in blood vessels, membranes of the body cavities, and internal organs. Both types of sensory nerve include **proprioceptive nerves** whose signals arise from sensory cells in joints, tendons, and muscles.

The simplest and most numerous sensory structures are the free nerve endings of general somatic sensory cells inserted among epithelial cells or connective tissue cells of the integument. The nerve endings appear to be sensitive to chemical stimulation and in the adult are associated with pain.

Like other cells of neural crest origin, differentiation of sensory cells lags behind the differentiation of central nervous system cells. In the case of the human fetus, for example, somatic sensory cells do not form free nerve endings in the epidermis until the end of three months.

Tactile receptors such as corpuscles of touch (Meissner's corpuscles, Fig. 26.19) begin to form at 4 months when terminal nerve endings just below the epidermis become encapsulated by mesenchymal cells. The corpuscles are not completely differentiated until a year after birth, while Merkel's tactile disks only begin to form at the end of 4 months neonatal.

Lamellated corpuscles (Pacinian corpuscles), looking like tiny onions buried in the skin (Fig. 26.19), are sensitive to deep pressure. These organs

start from mesenchymal cells clustered around nerve endings. The mesenchymal cells progressively flatten into concentric lamellae, increasing the size of the corpuscle from the outside. These corpuscles begin differentiating in fetuses of 4 months and are complete by 8 months.

Terminal processes from other sensory cells induce small groups of muscle fibers to form **muscle spindles** which function as **stretch receptors** or **proprioceptive units** during the third month (Zelená, 1964). **Neurotendinous end organs,** which later provide proprioceptive information from tendons and joints, begin to form at the same time.

Reflex Functions

Reflex arcs are classified by the level of the central nervous system through which nervous impulses are funneled. In **simple reflexes** a sensory "input" is connected through **association** or **adjustatory neurons** of the gray matter to a motor "output" or response. **Intrasegmental reflexes** are automatic local responses involving sensory and motor nerves of the same body segment. **Intersegmental reflexes** are integrated between sensory nerves of one segment and motor nerves of a few adjacent segments.

More complex responses involve sensory transmissions to the involuntary reflex centers in the cerebellum, brain stem, and cerebral cortex. Somaesthetic nerve fibers reaching these cephalad cen-

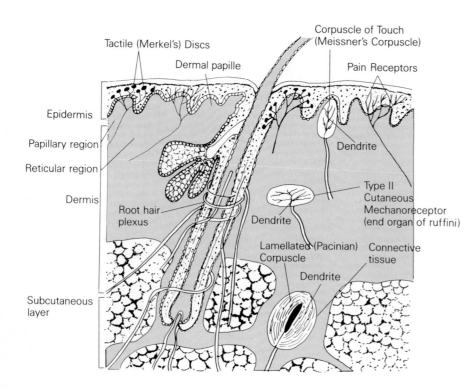

FIGURE 26.19. Structure and location of cutaneous receptors. From G. J. Tortora and N. P. Anagnostakos, *Principles of anatomy and physiology,* Harper & Row, New York, 1987. Used by permission.

ters may send branches to the generalized **reticular formation** before terminating in dorsal thalamic nuclei of sensory projection.

The ability of an organism to perform reflex functions does not depend solely on the formation of nerves. Reflexes depend on the completion of nervous circuits between sensory and terminal organs, and the parts of these circuits do not develop simultaneously. Sensory nerves frequently reach their sensory organs before these are differentiated and functional, and, although motor nerves may develop in advance of sensory nerves, the organs innervated by motor nerves may not be responsive to coordinated impulses until later.

Furthermore, the ability of investigators to detect a reflex may be hampered by **spontaneous activities** which do not depend on nervous control. For example, the motor end plates of somatic efferent nerves begin differentiating in the human fetus at 4 months, but striated and cardiac muscle become contractile upon differentiating and long before they are innervated.

Determining when reflex arcs are completed and when sensations, for example, pain, exist cannot be done with certainty, but tentative conclusions are available. The first functional reflex arcs seem to be laid down in human embryos between the mid- to late second month. At that time the embryo becomes capable of weak twitches of the neck in response to being struck on the lips or nose with a fine bristle (Hooker, 1952). This simple reflex competence spreads caudally. By the end of the second month, the first spontaneous side to side oscillations of the body wall occur.

At the start of the third month, facial movements start and the fetus' hands are capable of grasping reflex movements. A week later, the foot is also capable of grasping reflex movements. By the 10th to 11th weeks, swallowing and occasional rhythmical movements of the chest have begun, and at the end of the third month, almost all the skin is sensitive to touch. Spontaneous fetal movements are weak and irregular, and responses to stimulation often exceed the strength of the stimulus indicative of incomplete coordination through higher brain functions (Bradley and Mistretta, 1975; Gottlieb, 1976).

With the end of fourth month the mother feels fetal movements ("quickening"). From then on the fetus alternates periods of activity with periods of rest. Infants born prematurely late in the fifth month exhibit spontaneous but not sustained breathing. Sensory functions expand after this time, following the pattern of myelination of nerves. General cutaneous sensation, taste, balance (vestibular system), hearing, and vision mature in this order.

Specificity in Connections

Reflex and voluntary behaviors, like other features of organisms, display a species-specific component indicative of hereditary control. Possibly, neurons are inscribed with genetically encoded surface labels whose **chemoaffinities** (chemical attraction) specify hook ups among nervous pathways.

Roger Wolcott Sperry (born 1913), who first suggested this possibility,[4] made the surprising discovery that severed optic nerves of the newt, *Notophthalmus* (= *Triturus*) *viridescens*, regenerated and restored **optokinetic responses** (or visuomotor behavior, i.e., movement triggered by optic stimulation). Instead of confused or random movement, animals moved adaptively in the direction of a visual stimulus. In order to test the generality of his result, he performed the operation on tadpoles and on adults of six species of frogs and toads, all of which responded the same as had the newt. Enough, if not all, the optic nerve's 29,000 fibers had regenerated and had reestablished functional contacts in the optic tectum.

To ascertain if the regenerated nerve's contacts were molded somehow by functional rewards, Sperry rotated the eyeballs at the ends of the severed nerves. The surviving animals subsequently responded to visual stimuli directed toward the rotated eye by moving opposite the normal direction (e.g., Fig. 26.20) (Sperry, 1944). Tadpoles swam away from an orienting stimulus instead of toward it, and adults presented with a fly below eye level snapped at the air above.

Since this misdirected behavior was nonfunctional, in the sense of not bringing the animal closer to a reward, the contacts established by the regenerating nerve had not been reinforced. Furthermore, these behaviors were not reversed with experience although some adults inhibited their inappropriate responses.

Particular parts of the neural retina, it would seem, fed fibers to particular sites in the optic tectum independent of functional reward. Possibly, the fibers were **anatomically specified,** determined by inherent properties capable of selectively influencing central synaptic connections.

The same experimental paradigm has now been repeated and extended to autonomic reflex

---

[4]R. W. Sperry shared the Nobel Prize in Physiology or Medicine in 1981 for his contribution to neurophysiology.

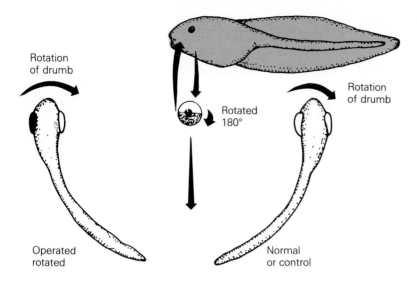

FIGURE 26.20. Sperry's experiment with tadpoles. After severing the optic nerve and rotating the eyeball, regeneration seems to reestablish anatomically specific contacts.

Rotation
of drumb

Rotation
of drumb

Rotated
180°

Operated
rotated

Normal
or control

arcs. Severed sensory nerves regenerate and, it would seem, undergo respecification to central connections with the result that motor reflexes are sometimes misdirected. Moreover, although motor axons can innervate inappropriate muscles, such "foreign" nerves are displaced by appropriate nerves, at least in frogs and neonatal rodents (see Hopkins and Brown, 1984).

The optic nerve processes of amphibian embryos do not seem to be initially specified in the retinal epithelium, since the experimental rotation of optic cups at early tailbud stages does not alter the response of adults. **Retinal specification** takes place at later tailbud stages but prior to the optic nerve establishing functional connections in the tectum.

Research on neural specificity rose to a new plateau when the courses of nerve processes in the neural retina of an adult _Xenopus laevis_ were mapped in the optic tectum. The technique, known as **retinotectal projection,** employs point sources of light directed at the neural retina and recordings of action potentials of neural impulses made in the optic tectum.

When optic vesicles were rotated in early tailbud stages and allowed to heal and regenerate, recordings from the tectum were normal (Fig. 26.21a), but when the rotation was performed at slightly later stages, the anterior–posterior axis of the projection was reversed (Fig. 26.21b). When the operation was performed on tadpoles at still later stages, the dorsoventral axis was also reversed (Fig. 26.21c). Instead of specificity matching ganglion cells in the neural retina with cells in the tectum, it would seem, specificity proceeded epigenically according to geometric coordinates or axes of symmetry.

The possibility that retinal neurons were not targeted for specific connections in the optic tectum was soon tested through a variety of surgical interventions involving the tectum and eyeball (see Easter et al., 1985). Surprisingly, the results showed considerable potential for plasticity among the nervous connections. New hypotheses are needed to account for the specificity that undoubtedly appeared under some circumstances in light of the plasticity that occurred under other circumstances.

Imprinting

The ontogeny of learning is no less subject to genetic control than the ontogeny of reflex behavior. While learning _in utero_ or _in ovo_ is difficult to study, because the environmental experience of embryos and fetuses is restricted and experimenting with it is difficult, **imprinting** on some signals may occur prior to birth or hatching.

Imprinting is the capacity to acquire particular kinds of information during a brief **sensitive period** (or critical phase) with subsequent consequences for behavior. It occurs prior to hatching in birds and may occur prior to birth in human beings. Infants may imprint on the sound of their mother's heartbeat while _in utero_. Later, when a parent holds an infant in the usual manner over the left side, the sound of the heartbeat may trigger placid infant behavior (see Salk, 1973).

In theory, ontogeny brings the brain to a point where it is capable of being imprinted. A **template** or selective system for acquiring information from the environment is laid down in specific neuronal and hormonal conditions. The imprinting stimulus must be available during the sensitive period, but

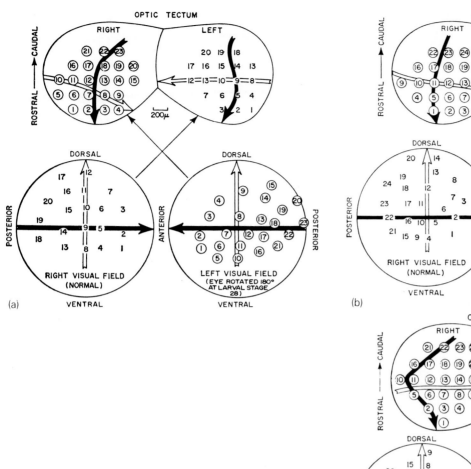

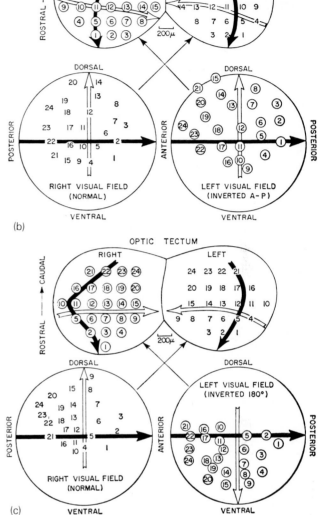

FIGURE 26.21. Maps of retinotectal projections in adult *Xenopus laevis*. In each case, a normal projection is on the left. The projection on the right represents the results following the rotation of the eye rudiment at early to later tailbud stages. (a) Rotation of initial optic cup in early tailbud (stage 28–29 Nieuwkoop and Faber). Projection of rotated eye normal. (b) Rotation in mid-tailbud (stage 30). Projection of rotated eye reversed in anterior–posterior axis. (c) Rotation in late tailbud (stage 32). Projection of rotated eye reversed in dorsal–ventral axis as well (5–10 hours later; embryo complete and elongating). From M. Jacobson, *Science,* 155:1106(1967), by permission of the American Association for the Advancement of Science and the author. Copyright 1967 by the AAAS.

reinforcement or repetition may not be required at that time. The behavior that is later triggered may require reinforcement or improve with practice and environmental feedback, but without successful imprinting, the behavior does not even appear.

The ontogeny of bird song in the white-crowned sparrow, *Zonotrichia leucophrys*, requires both early imprinting and later fine tuning (Konishi and Nottebohm, 1969; Marler and Mundinger,

1971). An acoustic template for the mating song seems to be prepared in the male under the influence of testosterone, since females and castrated males do not acquire the song except under the influence of testosterone implants.

Imprinting is prevented by hand-rearing fledglings in acoustically isolated chambers. Under these conditions, only a **subsong** (visualized as a sound spectrogram, Fig. 26.22 middle) develops

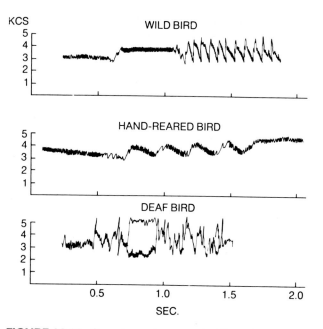

FIGURE 26.22. Sound spectrograms in kilocycles per second (KCS) as a function of time in seconds (SEC.) are the "signatures" of different songs generated by male white-crowned sparrows reared under different conditions. (Top) Birds reared in the wild produce a full song. (Middle) Hand-reared birds raised in acoustic isolation as fledglings produce a subsong that is simple and incomplete compared to the full song. (Bottom) Hand-reared birds deafened after initial exposure to their species-specific bird song produce an abnormal song. Male white-crowned sparrow would seem to require both exposure to their species-specific song and auditory feedback as fledglings to produce their full song. From M. Konishi, _Z. Tierpsychol.,_ 22:770(1965), by permission of the author.

which is incomplete and simpler than the **full song** (top) present in wild birds or those exposed during rearing to the song of wild males. The importance of additional learning is illustrated by the more complex but abnormal song produced by birds deafened prior to performing the song themselves (Fig. 26.22 bottom) (Konishi, 1965).

The ages of birds and mammals at which they experience critical periods for imprinting differ considerably. In **precocial** species, such as chickens and herd ungulates, young are precociously competent with regard to many aspects of behavior and learning. In **altricial** species, such as songbirds, rodents, and human beings, hatching or birth occurs at a relatively early developmental age (i.e., prior to locomotory competence). Parental care and the intensity of interactions with parents is greater in altricial species and, it would seem, so are the opportunities to be imprinted with a variety of stimuli.

CELLULAR ACTIVITIES SPECIFY THE NERVOUS SYSTEM

The Neuron Doctrine

The analysis of mechanisms specifying behaviors could not begin until the role of nerves in building the nervous system was clarified. Nineteenth century concepts of the developing nervous tissue were dominated by the **reticular theory**— that nerve fibers formed by the fusion of cell processes. The consequence of fusion was a continuous network or reticulum, and the consequence of the reticular theory was confusion about nervous integration.

The alternative **neuron doctrine,** or cell theory of the nervous system, was rarely heard prior to the 20th century. Jan E. Purkinje (1787–1869) advanced the idea that the nervous system consisted of a network of individual cells as early as 1836, and in 1880, Wilhelm His (1831–1904) described the extensions of axons and dendrites on developing nerve cells.

When the neuron doctrine was finally promulgated by Santiago Ramón y Cajal (1852–1934), it was based on histological studies utilizing the silver impregnation staining method discovered by Camillo Golgi (1844–1926). Ironically, when Ramón y Cajal and Golgi shared the Nobel Prize in Physiology or Medicine in 1906 for studies on the nervous system, Golgi still espoused the reticular theory.

The reticular theory only declined in importance after Ross Granville Harrison (1870–1959) (see Fig. 21.3) invented **tissue culture** to test the ability of embryonic frog neurons to produce fibers without the aid of other cells (Harrison, 1907). He thereby moved analysis of nerve growth from histology and the study of static (i.e., dead) tissues to the study of dynamic (living) cells. Like Roux who had grappled experimentally with the embryo, Harrison grappled experimentally with the nerve cell and, in doing so, observed living cells differentiate.

Building on earlier experiments by Leo Loeb, and working under aseptic conditions, Harrison placed embryonic frog tissue inside a drop of frog lymph on a coverglass slip. As the lymph clotted, the coverglass was inverted over a shallow depression previously ground into a microscope slide. The coverglass and slide were then sealed together.

Amazingly, the tissue remained alive for weeks and processes grew out of individual nerve cells! Their differentiation was, in today's terms, autonomous, that is, not dependent on other cells.

Harrison's success was recognized when he was voted the Nobel Prize in 1917, but the prize was not awarded due to World War I. Ironically, tissue culture received little additional attention. The technique that was to become molecular biology's stock in trade was too difficult and cumbersome to attract many early buyers.

Originally, tissue culture depended entirely on aseptic technique. A modern tissue culturist would hardly recognize tissue culture laboratories of half a century ago. Procedures intended to promote asepsis, such as changing clothing and passing through sterile anterooms, were performed with the intensity of religious rituals. Experiments were actually performed in the "holy of holies" or "inner sanctum." Arcane precautions excluded most bacteria and mold, but failure to exclude them all was common, and death in the culture was frequent. The advent of bacterial filters and antibiotics after World War II made most of the "hocus pocus" unnecessary and turned tissue culture into a feasible technique for research and industry.

By the early 1960s, the electron microscope had also advanced to the point where cells could be examined routinely. Each nerve cell was seen to be enclosed in a cell membrane, and nerve processes were seen to meet at synapses or junctions rather than points of fusion. The reticular theory of the nervous system was finally laid to rest.

Cellular Studies of Neural Specificity

Neural outgrowths are generally hidden within organisms, but in the *in vitro* environment, **growth cones** are seen at the ends of growing nerve fibers. A thin **lamellipodium** or ruffled membrane caps the growth cone and long **filopodia** or **microspikes** reach out from the leading edge (Fig. 26.23). Microspikes seem to compete at the tip of the growth cone, until one or another microspike prevails. The others withdraw, and the growth cone follows the successful microspike into new territory (see Smith, 1988).

At least three structurally distinct types of fine fiber occur in nerve cell bodies and neuronal processes: neurotubules or microtubules, neurofilaments or neuronal-type intermediate filaments,

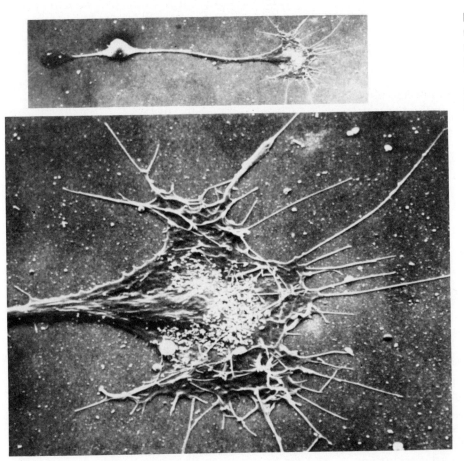

FIGURE 26.23. Scanning electron micrograph of a parasympathetic nerve showing growth cone and microspikes. From N. K. Wessels, *Tissue interactions and development,* Benjamin/Cummings, Menlo Park, 1977. Used by permission.

and microfilaments or actin. Cultures of nerve cells have been treated with drugs known to affect these cytoskeletal elements (among other things) in order to ascertain the actions of each type of fiber on elongation.

Colchicine, which disrupts microtubules, caused the withdrawal of cytoplasmic processes into neurons, suggesting that the length of processes depended on microtubules. Similar cultures of nerve cells treated with cytochalasin B, which disrupts actin, did not show process withdrawal, but the microspikes present on growth cones disappeared, and the cytoplasmic processes ceased to make any further movements. Upon removal of the cytochalasin, actin filaments and protrusions reappeared at the leading edge, suggesting that extension depended on actin (Spooner, 1974).

Neurotubules are present in the neuron either singly or in loose groups that stretch out into the central region of axons and dendrites. These "tracks" of neurotubules probably control **fast axonal transport** (at several hundred millimeters a day) of cytoplasm including organelles or more general **orthograde movement** from the neuron to the tip of the axon.

Fast movement is presumably involved in neurophysiology. In addition, **slow** (at about 1 mm per day) **movement** is involved in growth and maintenance of fiber dimensions, probably including the transport of actin to the active edge of the growth cone (see Vasiliev, 1987).

The movements of materials within nerves can also be **retrograde,** in the direction of the nerve cell body. Nerve growth factor (NGF, see below) may be transported in this way.

Neurofilaments are found in nerve cell bodies, concentrated in large axons but rare in dendrites. Neurofilaments may transport substances such as peptides along nerve processes to their destinations at synapses.

Microfilaments or actin are highly concentrated in growth cones and their radiating lamellipodia and microspikes. The filaments are also abundant in growing dendrites, especially just beneath the plasma membrane in an "actin cortex." The ability of actin to polymerize and depolymerize is probably most directly responsible for the motility of growth cone spikes.

Growth cone morphology is probably determined by localized **actin binding proteins** (ABPs) and their mechanical and transducing linkages to cell-surface receptors. Chief among the ABPs are profilins, which bind and sequester actin monomers, fimbrin, spectrin, and vinculin, which crosslink actin filaments into bundles and attach them

to membranes, and myosin, which in conjunction with actin produces sliding forces.

Receptors may generate second messengers such as Ca^{2+} that regulate actin or have local effects on the cytoskeleton or on other cell-surface receptors. The sensitivity of growth cones' morphology and behavior to neurotransmitters, for example, may be regulated by receptors acting on second messengers acting on protein kinases acting on ABPs (Smith, 1988). Several protein kinases are also activated by NGF.

The undifferentiated neuroblast seems to have an "actinoplastic" organization (i.e., a discrete adhesive and motile system organized by the actin cortex). In contrast, the differentiated neuron has a "tubuloplastic" organization (i.e., dominated by microtubules) with no lamellipodium but long narrow processes filled with microtubules and intermediate filaments. Morphogenesis and differentiation depend, in many respects, on competition and interaction between the actin cortex and the cytoskeletal content, leading to the transition from one form of organization to the other (see Vasiliev, 1987).

Chemoaffinity theories of growth cone guidance depend on the cone's recognition of molecular cues in its environment. Antibodies that disturb the adhesion of neural cells *in vitro* have led to the identification of several cell adhesive molecules, but whether these provide sufficient specificity to direct growth cones remains to be seen (see Dodd and Jessell, 1988).

Vertebrate integrins, N-CAM, and *N*-cadherin and insect amalgam are expressed on all axons and presumably play general roles in neural cell adhesion. Other glycoproteins (e.g., Ng-CAM in chicks, L1, F11, neurofascin, and contactin in mammals, fasciclin II in insects) are restricted to some axonal surfaces; still others are restricted to functional subsets of developing axons (vertebrate TAG-1, RB-8, and TRAP, insect fasciclin III), and some are dynamic, changing from general expression to expression on individual neurons and axon bundles (e.g., insect fasciclin II) (see Harrelson and Goodman, 1988).

Laminin, a substrate adhesive molecule (SAM), or other heterophilic adhesive molecules (i.e., requiring a cell-surface receptor) may guide growth cones especially during early axonal outgrowth when axons are absent. On the other hand, guidance in the presence of axons may be led by homophilic (i.e., self-sticking or cohesive) cell adhesive molecules (e.g., vertebrate N-CAM, *N*-cadherin, L1, and G4) (see Fig. 20.8).

These molecules may be responsible for the **selective fasciculation** (Fig. 26.7a) and tight cohesion of growing axonal bundles (e.g., spinal nerves emerging from the spinal cord). In insects, similar glycoproteins (amalgam and fasciclins) play the role of homophilic cell adhesive molecules (see Harrelson and Goodman, 1988).

Other cues may **defasciculate** or break up axonal bundles. In the case of motor nerves entering limbs, for example, spinal nerves defasciculate at the level of the limb plexes before moving to their target muscle. Cues for fasciculated movement seem to be independent of cues for defasciculated movement, since nerves forced to take novel routes to limbs (e.g., following neural tube rotation, see below) are still able to find their target muscles.

The inhibition (or paralysis) of growth cone extension by nonpermissive substrates, or **contact-mediated inhibition of movement,** may passively redirect growth cones to permissive substrates. For example, axons leaving the neural tube never probe the nonpermissive posterior half of the adjacent somite but only the permissive anterior half. Growth cones may also be directed away from unrelated axons and toward related axons, since growth cones from unrelated nerves (e.g., CNS versus PNS) collapse and retract upon encountering each other *in vitro.*

The small neuroglial cells of the CNS known as oligodendrocytes (cells with few and slender processes) do not permit axonal overgrowth over their surfaces *in vitro.* Two cell-surface proteins, of 35 and 250 kd, may be responsible for the cell's nonadhesiveness, since axons can move over oligodendrocytes treated with antibodies to these proteins. The failure of nerves to regenerate following trauma to the CNS may be due to the presence of these proteins on oligodendrocytes.

Chemotropic guidance or direction by gradients of diffusible chemotropic factors secreted by some cells within a target organ was suggested by Ramón y Cajal. A **maxillary factor,** for example, may guide trigeminal sensory neurites to the maxillary arch, since maxillary epithelium initiates an oriented outgrowth of trigeminal sensory axons *in vitro.*

In the vertebrate embryo, diffusion could create a gradient over a few hundred millimicrons (Crick, 1970), and chemotropism may proceed through a series of **intermediate targets** (also called landmarks or guideposts in invertebrates). A **basal plate factor,** for example, may guide the growth cones of nonfasciculated commissural (crossing) neurons in rats from the embryonic dorsal spinal cord toward the embryonic floor plate. Having reached this intermediate target, the axons fasciculate and turn sharply toward a longitudinal trajectory.

Graded **positional cues** generated by target organs may generate the highly ordered topographical projections of axons seen in the vicinity of target organs and direct connections on a local level within **target fields.** Projections from retinal ganglion cells, for example, may connect to the optic tectum (see above) by recognizing preformed positional cues.

Other "negative" chemotropic factors and positional cues may have an aversive effect on growth cones. For example, axons from retinal ganglion cells on the temporal side of the eye reject posterior optic tectum membranes in favor of anterior optic tectum membranes *in vitro,* their normal substratum (see Dodd and Jessell, 1988)

Competitive interaction theories provide an alternative to chemoaffinity theories. Rather than specificity arising from affinities and restrictive capacities, specificity is attributed to competitive interactions and a physiological competitive edge in relation to other similar cells. According to competitive interaction theories, under particular circumstances or in the event of experimental intervention, one cell's competitive advantage allows it to make a connection and survive, while other cells are weeded out.

The possibility of competitive interactions in the nervous system was first suggested by Viktor Hamburger (born 1900) and colleagues, especially S. R. Detwiler (1936) and R. Levi-Montalcini (1987), on the basis of histological observations and experimental results. First, cell death was seen to be widespread within the developing nervous system. The number of nerve cells in vertebrates was actually higher at the beginning of neurogenesis than at the end (e.g., Fig. 26.24). Second, cell death was frequently localized and seemed to aid in shaping neural structures.

Some experimental results, such as the dramatic reduction in size of the spinal cord and sensory ganglion following the amputation of a wing bud (Fig. 26.25a), can be explained by selective cell death, while other experimental results suggest that functional connections preserve cells that would otherwise have died. For example, when portions of a newt's spinal cord are reversed at the tailbud stage, the number of nerve cells surviving remains high in segments taking part in forming

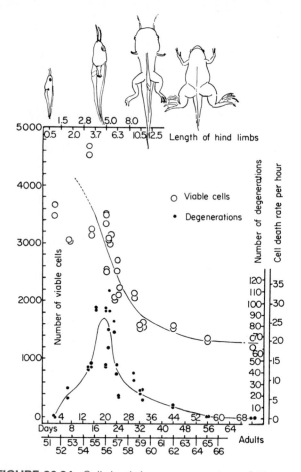

FIGURE 26.24. Cell death in nervous system of *Xenopus laevis* tadpole and metamorphic adult. The numbers of viable cells (open circles, left axis), degenerating cells, and the rate of cell death (closed circles, right axes) in the ventral horns of the spinal cord at lumbar level are plotted against time in days and stages of tadpole and adult life and length of hind limbs. From A. Hughes and J. A. Fozzard, *Br. J. Radiol.,* 34:302(1961). Used by permission.

the brachial plexus (Fig. 26.25b). Similarly, nerves are preserved and an additional brachial plexus is established when transplanted newt forelimbs are innervated (Fig. 26.25c).

Today, the importance of functional nerve ending in the maintenance of adult muscle (known as neurotrophic effects) is well known especially from the tragic sequelae to viral diseases of nerves, such as poliomyelitis. Fifty years ago, the requirement of nerves for trophic reinforcement was surprising, but its experimental pursuit led to even more surprises.

Nerve growth factor (NGF) was discovered as a result of serendipity and shrewd judgments made by Rita Levi-Montalcini and Stanley Cohen,

a contribution for which they were awarded the Nobel Prize in Physiology or Medicine for 1986. Not only did these investigators bring a new focus to the problem of neural specificity, but they started the current rage for studies on circulating **polypeptide growth factors** and cellular **growth factor receptors.**

NGF is now only one among many **neurotrophic factors** capable of promoting nerve growth, but growth factors, in general, do more than promote growth. They also modulate cellular differentiation during development (see Thorburn et al., 1986) and transform tissue culture cells to neoplastic states.

NGF has a broad range of activities. It promotes survival and development of cultured embryonic sensory neurons and certain embryonic or neonatal sympathetic ganglia (e.g., the embryonic chick superior cervical ganglion, Fig. 26.26), although neonatal sensory neurons and parasympathetic neurons at any stage of development do not respond to NGF.[5] Other neurotrophic factors may affect neurons not affected by NGF in general or at stages in which they are insensitive to NGF.

Administered to young animals, NGF causes outgrowth of adrenergic neurons and of fibers. In addition, the growth and development of neuron-like cells of the adrenal medulla and extraadrenal **chromaffin tissue** (i.e., tissue staining heavily with chromium salts) associated with sympathetic ganglia (i.e., paraganglia) are promoted by NGF administration. As a corollary, injections of antibodies raised against NGF (anti-NGF) destroy postganglionic sympathetic neurons, thereby **immunosympathectomizing** the animal.

What is NGF? The neurotrophic factor purified from the submaxillary gland of mature male mice (where it probably functions as an ectopic promoter of wound healing for animals licking wounds) is a high molecular weight complex (140,000 daltons), containing two pairs of α and γ peptide subunits, and one β subunit, all held together with zinc atoms. The β subunit (β-NGF) is a dimer whose noncovalently linked identical chains contain 118 amino acids. β-NGF alone has all the neurotrophic activity of the macromolecule.

How does NGF work? The ability of NGF to direct neural outgrowth may simply depend on local differences in NGF's concentration caused by interaction with the cellular NGF receptor. If the receptors are equally distributed along the length of microspikes arising from a growth cone, then

[5]Hence, NGF is sometimes called **sympathetic neuron survival factor.**

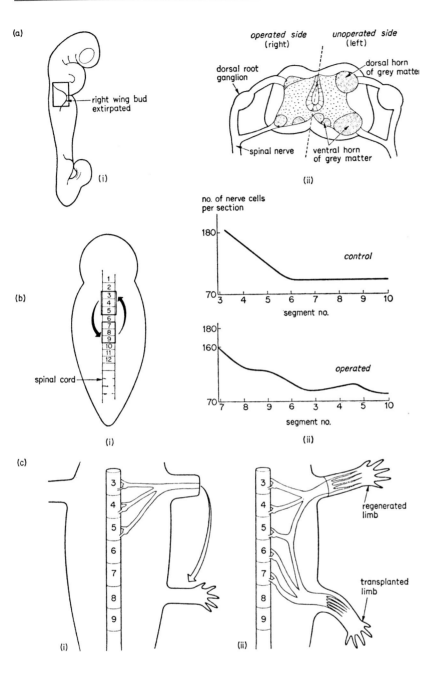

FIGURE 26.25. Summary of the experiments and results of Detwiler and Hamburger. (a) Effect of removing limb bud from chick embryo. Spinal ganglia and dorsal and ventral horns of the nearby spinal cord become smaller than normal. (b) Effect of reversing portions of the neural tube in a newt tailbud. At the end of neurogenesis, the number of nerves in the original third to fifth segments has declined to levels resembling those normally outside the brachial region. Larger numbers of cells are preserved at the level of segments 3–5 whose processes innervate the limb. (c) The transplantation of a new limb results in the development of a second brachial plexus. From E. M. Deuchar, *Cellular interactions in animal development,* Wiley, New York, 1975. Used by permission.

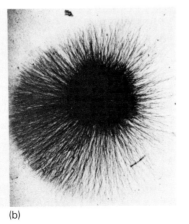

FIGURE 26.26. Photomicrograph of sensory ganglion from a 7-day chick *in vitro* in absence (a) and presence (b) of nerve growth factor (NGF) for 24 hours. Note the apparent preferential growth of neurites (b) toward source of NGF in lower left. From R. Levi-Montalcini and R. Amprino, *Arch. Biol.,* 58:265(1947), by permission of Imprimerie Vaillant-Carmanne and the authors.

the greater surface area of the cylindrical portion of the spike will present more receptors than the hemispheric tip (see Meinhardt, 1982). Assuming that NGF concentration is reduced proportionally, the highest concentration of NGF will be at the tip of the microspike farthest from the growth cone. Growth in response to NGF concentration would then follow the direction of the longest microspike.

Other NGF activities are more difficult to explain even hypothetically. NGF protects nerve cells against the lethal effects of some noxious treatments (e.g., transection). Does NGF in peripheral tissue promote the survival of connected nerves while allowing unconnected nerves to die? Transmitted in the retrograde direction, NGF increases neurotransmitter synthesis. Is NGF genotropic?

cDNA encoding β-NGF from mouse, human, bovine, and chick sources has now been isolated, with the help of synthetic oligonucleotides, and cloned. The 307 amino acids encoded by the precursor transcripts of humans and mice are not especially similar, but the 118 amino acids of the β-NGF domains differ in only 13 amino acids.

> **Studies on the immunological and biological relatedness of NGFs purified from different species strongly support the hypothesis that the site (or sites) of interaction with their receptors has remained structurally more constant than is the case for other epitopes [determinants of an antigen's immunological specificity], probably free to mutate in view of their less fundamental biological functions.** (Rita Levi-Montalcini, 1987, p. 1159)

Possibly, conserved circulating growth factors, such as NGF, represent biochemical signals that call up different cell lineages and link different developmental responses. Extrinsic controls, such as those working in neural development, may operate in the realm of computer networking rather than individual programming. NGF may influence the development of the vertebrate nervous system by accessing modified programs and data banks within cells rather than directing cells to specific information. The future will tell.

PART 9   *SUMMARY AND HIGHLIGHTS*

Vertebrate organogenesis pits development against differentiation. Prolonged morphogenic plasticity rather than terminal addition is characteristic of evolved structures, while early determination and rapid differentiation are exhibited by primitive structures. Induction, rather than intrinsic controls of determination, is the rule.

In a tug-of-war between ectoderm and endoderm in the oral cavity, endoderm holds its own in amniotes, while ectoderm wins in amniotes. Some parts can be formed by either embryonic epithelium.

The foregut splanchnopleure is a remarkably plastic embryonic organ. It forms the sublinguinal and submandibular salivary glands (in mammals), thyroid and parathyroid glands, the thymus (but not the lymphocytes), post- (ultimo-) branchial bodies, pharynx, trachea, lungs, esophagus, stomach, duodenal portion of the intestine, pancreas, liver, and gallbladder under inductive influences from mesenchyme. Midgut and hindgut are not well known for plasticity, but their relationships with hematopoietic and lymphopoietic tissue are indicative of developmental interactions.

In the embryonic pharyngeal region, branchial pouches alternate with branchial arches with their core of ectomesenchyme and central aortic arch. Anamniotes develop gills on their branchial arches but amniotes do not. Instead, they utilize some of the cartilaginous elements of the branchial arches to form parts of their middle ear.

The rudimentary heart becomes tubular as it begins beating and circulating blood. In anamniotes, the heart tissue differentiates valves as it becomes sigmoid shaped, while amniotes only gradually add valves and subdivisions.

Regression, distortion, and segregation characterize the development of the aortic arches in amniotes. Segmental arteries tend to fuse, and paired arteries tend to form single medial trunks. In mammals and birds, ridges that aid in subdividing the ventricle also subdivide the ventral aorta into systemic and pulmonary trunks leading to the fourth and sixth aortic arches.

In contrast to the symmetry of veins in amniotes, condensation and reduction, especially posteriorly, dramatically restructure veins in amniotes. The umbilical veins are the most opportunistic, carving their way through the body wall and liver.

A variety of cenogenic adaptations in mammals and birds equalize pressure in the right and left sides of the embryonic heart and distribute the rich umbilical or allantoic blood preferentially to the head. At birth or hatching, air breathing promotes changes that rectify patterns of blood flow and complete the subdivision of the heart into chambers.

In response to induction by a nephrogenic cord, nephrogenic mesenchyme produces nephric tubules in a cephalocaudal gradient of complexity. The early differentiation of the pronephros and opisthonephros in anamniotes contrasts with the undifferentiated pronephros and the unevenly differentiated mesonephros of amniotes, especially those with efficient hemochorial placentas. The metanephros develops from the metanephrogenic mesenchyme under the inductive influence of the ureteric bud.

The genital ridge is a relatively plastic presumptive ovary lying mesially to the opisthonephros or mesonephros. Primordial germ cells seed the ridge, those of the heterogametic sex directing or redirecting its development possibly via weak histocompatibility antigens. The testis determining factor on the Y chromosome of mammals seems to encode a zinc finger protein with a DNA binding domain.

Gonads develop in conjunction with the anterior end of the nephrogenic mesenchyme. Male vertebrates usurp nephric tubules to connect the genital ridge with the nephric or Wolffian duct, and testosterone promotes its growth into the vas deferens. Müllerian inhibitory substance (MIS) produced in males undermines the development of the paramesonephric or Müllerian duct. The Wolffian ducts do not grow in females due to the absence of testosterone, while the paramesonephric ducts grow into the oviducts and uterus in the absence of MIS.

Openings to the urogenital system and colon are usually recessed in the vertebrate cloaca. In mammals, urogenital folds separate a urogenital sinus from the rectum. Further separation accompanies the invasion of ectoderm to form the urethra and vagina.

Just prior to birth, or soon thereafter, males produce enough testosterone to reach the brain and inhibit the development of cyclic patterns of gonadotropin releasing hormone. Melatonin, produced by the pineal gland, inhibits gonadal function through a juvenile stage, and a diminution in melatonin release brings about puberty.

Fins and limbs are joint projects of somitic myotomal buds and somatopleural thickenings. In fish, a piscine-apical ectodermal ridge (pAER) flattens and elongates, supported by collagenous rays and penetrated by mesenchymal cells. In tetrapods, an apical ectodermal ridge (AER) atop an ectodermal cap typically promotes the outgrowth of a mesenchymal core.

The suspension of distal development resulting from removal of the AER suggests that limb structures are determined progressively in a proximodistal direction. Outgrowth of the underlying mesenchyme may be promoted by the AER, which in turn is supported by an apical ectodermal maintenance factor (AEMF) released from the mesenchyme. Alternatively, determination may result from the release of morphogens from a zone of polarizing activity (ZPA), positional information built into mesenchyme, or a biological clock measuring the time cells spend in a progress zone beneath the AER. Differential cell death also contributes to distinctive patterns.

Reiterative pairs of ventral motor roots emerge from the spinal cord and converge in plexi at the bases of limbs. These segmental nerves bring efferent capabilities to parts of the body occupied by subgroups of skeletal muscle originating in corresponding somites. Target recognition theories attempt to explain segmental-level association via molecular tokens, possibly cell or substrate adhesion molecules.

The vertebrate integument contains an epidermis derived from ectoderm and a dermis containing neural crest, dermatomal, and lateral plate components. Neural crest contributes to connective tissue and forms the dermal papillae of cutaneous appendages while exercising inductive potential over patterns of scales, feathers, and hair. Neural crest cells participate in forming the cornea. In dental papillae, they become dentine-producing odontoblasts capable of inducing the differentiation of dental ridges and the formation of the ameloblast and the enamel organ in oral epithelia.

Adultlike functional qualities in the nervous system result from the integration of neural epithelium, neural crest, and ectodermal placodes, followed by tissue diversification and cellular differentiation. In general, the longer the period of developmental plasticity the greater the final neural complexity.

Plasticity is reflected in process as well as product. During cytogenesis in the neural tube's pseudostratified epithelium or matrix layer, nuclei move on a proliferative elevator between inner zones of mitosis and outer zones of DNA synthesis. Cells move out of the matrix layer as neuroblasts and differentiate. Sustentacular glioblasts follow neuroblasts, abandoning the inner ventricular layer to ependymoglioblast cells.

Gray matter of nerve cell bodies accumulates in an intermediate zone, while white matter of myelinated nerve cell processes accumulates in the marginal zone. Cortical plates are built from the inside out with the help of fascicled matrix processes, while noncortical regions, lacking similar processes, are built from the outside in.

The spinal cord, hindbrain, midbrain, and forebrain of early vertebrate embryos are built by the differential development of alar and basal plates, separated by the sulcus limitans. The embryonic roof and floor (noto-) plates are reduced along the midline.

In the spinal cord, gray matter of the basal plate becomes ventral horns involved with motor functions, while that of the alar plate becomes dorsal horns involved with reflex integration and the transmission of sensory functions. Commissural fibers run downward and turn abruptly in the ventral horn.

In the hindbrain, the alar plate expands into the cerebellum; the basal plate thickens into the posterior portion of the brain stem, and the thin roof plate, fused with vascular coverings, becomes a posterior choroid plexus. In the midbrain, the basal plate includes the major sources of motor nerves to the eyeball, while the alar plate provides the optic tectum and, in tetrapods, centers for the reflex integration of special sensory information— hearing and seeing. In the forebrain, the floor and basal plates are absent anteriorly, and the thin roof plate becomes an anterior choroid plexus. Alar plates alone expand into the telencephalon and cerebral hemispheres, and into the eyes and divisions of the thalamic region of the diencephalon.

The neural layer of the eye's retina exhibits a hierarchical structure of layered photoreceptors and association neurons. While differentiating from the bottom up, cells become committed from the top down, possibly as a result of migration.

In the fetal stage, the lagging cerebral hemispheres of mammals finally surpass those of birds. In Old World primates, the olfactory bulbs are outstripped by the nonolfactory dorsolateral wall whose folds, convolutions, and fissures announce the formation of the neopallium.

The sensory portion of the peripheral nervous system is built largely by neural crest cells with the addition of ectodermal placode cells. Developmentally plastic, these cells migrate to locations where they form sensory ganglia or sensory portions of the I, V, VII, VIII, IX, and X cranial nerves. Postganglionic fibers of both sympathetic and parasympathetic autonomic motor nerves are also built by neural crest cells, but the sources of preganglionic and voluntary motor fibers are cells within the central nervous system.

Receptors for the nervous system are either free afferent nerve endings or specialized transducer cells capable of responding to environmental stimuli and stimulating afferent nerve endings. In mammals, the differentiation of sensory organs and sensory cells lags, and spontaneous activity and reflex functions only become apparent in the fetus. Malleable or critical periods of imprinting and learning may occur before nervous pathways become fixed.

While connections between the frog embryo's central nervous system and periphery may be determined in part by cleavage division, connections in salamander and chick embryos are more plastic and are specified epigenically. The behavior of amphibians with regenerated peripheral and optic nerves indicates that connections of sensory nerves with the central nervous system are precise, but some latitude is also revealed by retinotectal projections following regeneration of *Xenopus laevis* optic nerves. Matching ganglion and tectum cells proceeds epigenically and by axial coordinates rather than by cell-to-cell affinities.

The discovery of nerve outgrowth in tissue culture established the neuronal or cellular doctrine of the nervous system and suggested that nerves trace their courses through growth. Integrin and substrate adhesion molecules may contribute to directing early nerve processes, while axons and homophilic neural cell adhesion molecules may direct later nerve processes.

A nerve growth factor that may be instrumental in directing nerves serves as a model for maturation and growth factors in many systems. The nerve growth factor's activities range from promoting outgrowth, directing differentiation, and selectively protecting nerves with central connections from cell death. The similarity in nerve growth factor-encoding domains of mammals and birds suggests that the factor may key cells into control systems rather than determine their fates.

PART 9 QUESTIONS FOR DISCUSSION

1. Compare and contrast the concepts of prolonged morphogenic plasticity and terminal additions. How do data on organogenesis stack up for and against the two concepts? What adaptive advantages might induction have for integrating prolonged development?

2. What is the role of induction in foregut organogenesis? What are the prospects for growing teeth in birds? For growing gills in mammals?

3. Describe the stages of heart organogenesis. How do these stages differ in anamniotes and amniotes? What adaptations to fetal circulation are changed at birth or hatching and how are the changes brought about?

4. Describe the fates of the aortic arches in representative vertebrates. Discuss regression, distortion, and segregation in the development of arteries and plasticity and opportunism in the development of veins.

5. Describe the development of the pronephros and opisthonephros, the mesonephros and metanephros. Formulate a theory to explain their competitive exclusion.

6. How and where are the urogenital and genital systems paired? Compare and contrast the cellular and hormonal mechanisms involved in the sex determination of gonads.

7. What are the embryonic origins of different fin and limb tissues, and how are these integrated in the course of development? What is the role of cell death in limb development? Suggest how death might be controlled.

8. Describe the proximodistal determination of limbs, and the role of the apical ectodermal ridge in limb outgrowth. Describe the results of transplantation, ablation, and addition experiments on limbs and propose hypotheses to explain these results.

9. Describe the activities and fates of neural crest cells. What do pigment cells, dermal papillae, sensory nerves, and peripheral motor ganglia have in common? Describe the contributions of neural epithelium, neural crest, and ectodermal placodes to the nervous system.

10. What is the escalator concept of neuroepithelial growth? Describe the inside out and outside in theories of cortical and noncortical growth. What are the implications of glioblast differentiation following neuroblast differentiation?

11. Compare and contrast amniote and anamniote development of the forebrain, midbrain, and hindbrain. Describe the fates of the roof, alar, basal, and floor plates in each region of the brain and in the spinal cord.

12. What impact does segmentation have on nervous integration? What are the cranial nerves and how do they develop? Compare and contrast the development of the CNS and PNS.

13. What is the neuron doctrine, and what impact did it have on research in neuroembryology? Describe the ontogeny of a receptor. How is specificity built into the nervous system?

14. What is a critical period for imprinting and what does it imply about plasticity in the nervous system? In your opinion, what is the likelihood of *in utero* learning? Of teaching intellectual or motor skills to fetuses? To neonates?

15. What is nerve growth factor and how might it work? What effects does it have on neural differentiation? On differentiation of neural crest cells? On nerve cell survival?

PART 9 RECOMMENDED READING

Bradley, R. M. and C. M. Mistretta, 1975. Fetal sensory receptors. *Physiol. Rev.*, 55:352–382.

Easter, S. S. Jr., D. Purves, P. Rakic, and N. C. Spitzer, 1985. The changing view of neural specificity. *Science*, 230:507–511.

Gordon, R., 1985. A review of the theories of vertebrate neurulation and their relationship to the mechanics of neural tube birth defects. *J. Embryol. Exp. Morphol.*, 89, Suppl:229–255.

Gottlieb, G., 1976. Conceptions of prenatal development: Behavioral embryology. *Psychol. Rev.*, 83:215–234.

Goy, R. and B. McEwen, 1980. *Sexual differentiation of the brain*. MIT Press, Cambridge.

Guroff, G., 1983. *Growth and maturation factors*, Vol. 1. Wiley, New York.

Heaysman, J. E., C. A. Middleton, and F. M. Watt, eds., 1987. *Cell behaviour: Shape, adhesion and motility*, Second Abercrombie Conference. *J. Cell Sci.*, Suppl. 8. The Company of Biologists, Cambridge.

Hopkins, W. G. and M. C. Brown, 1984. *Development of nerve cells and their connections*. Cambridge University Press, Cambridge.

Le Douarin, N., 1982. *The neural crest*. Cambridge University Press, Cambridge.

Le Douarin, N. and A. McLaren, eds., 1984. *Chimeras in developmental biology*. Academic Press, London.

Malacinski, G. M., 1984. *Pattern formation. A primer in developmental biology*. Macmillan, New York.

Moscona, A. A. and A. Monroy, eds., 1987. *Neural development*, Part IV, *Current topics in developmental biology*, Vol. 21. Academic Press, Orlando.

Raff, R. A. and E. C. Raff, eds., 1987. *Development as an evolutionary process*. Alan R. Liss, New York.

Sawyer, R. H. and J. F. Fallon, eds., 1983. *Epithelial–mesenchymal interactions in development*. Praeger Scientific, New York.

Science, 1988, Vol. 242. *Frontiers in neuroscience*.

REFERENCES

Abercrombie, M., 1961. The bases of the locomotory behaviour of fibroblasts. *Exp. Cell Res.*, 8:188–198.

Abisogun, A. O., P. Braquet, and A. Tsafriri, 1989. The involvement of platelet activating factor in ovulation. *Science*, 243:381–383.

Adamson, E. D. and H. R. Woodland, 1974. Histone synthesis in early amphibian development: Histone and DNA syntheses are not coordinated. *J. Mol. Biol.*, 88:263–285.

Adelmann, H. B., 1942. *The embryological treatices of Hieronymus Fabricius of Aquapendente.* Cornell University Press, Ithaca.

Adelmann, H. B., 1965. *Marcello Malpighi and the evolution of embryology.* Cornell University Press, Ithaca.

Adler, R. and M. Hatlee, 1989. Plasticity and differentiation of embryonic retinal cells after terminal mitosis. *Science*, 243:391–393.

Afzelius, B. A., 1972. Sperm morphology and fertilization biology. In R. A. Beatty and S. Gluecksohn-Waelsch, eds., *Edinburgh symposium on the genetics of the spermatozoon, 1971.* Department of Genetics, University of Edinburgh, Edinburgh, pp. 131–143.

Agrell, I., 1960. Detection of a graded mitotic activity within the sea urchin embryo through the use of oestradiol. *Ark. Zool.*, 12:411–414.

Albert, S., D. Hanahan, and G. Teitelman, 1988. Hybrid insulin genes reveal a developmental lineage for pancreatic endocrine cells and imply a relationship with neurons. *Cell*, 53:295–308.

Alberts, B. M., 1986. The function of the hereditary materials: Biological catalyses reflect the cell's evolutionary history. *Am. Zool.*, 26:781–796.

Allen, G. E., 1983. T. H. Morgan and the influence of mechanistic materialism on the development of the gene concept 1910–1940. *Am. Zool.*, 23:829–843.

Altman, P. L. and D. S. Dittmer, eds., 1962. *Growth including reproduction and morphological development.* Biological Handbooks Federation of American Societies for Experimental Biology, Bethesda.

Altman, P. L. and D. S. Dittmer, eds., 1972. *Biology data book*, 2nd ed. Biological Handbooks, Federation of American Societies for Experimental Biology, Bethesda.

Amoroso, E. C., 1952. Placentation. In A. S. Parkes, ed., *Marshall's physiology of reproduction*, Vol. 2. Longmans Green, London, pp. 127–311.

Amoroso, E. C., 1981. Viviparity. In S. R. Glasser and D. W. Bullock, eds., *Cellular and molecular aspects of implantation.* Plenum, New York, pp. 3–25.

Amprino, R., 1977. Morphogenetic interrelationships between ectoderm and mesoderm in chick embryo limb development. In D. A. Ede, J. R. Hinchliffe, and M. Balls, eds., *Vertebrate limb and somite morphogenesis.* Cambridge University Press, London, pp. 245–255.

Ancel, P. and P. Vintenberger, 1948. Recherches sur le determinisme de la symmetrie bilatérale dans l'oeuf des amphibiens. *Arch. Anat. Microsc. Morphol. Exp.*, 31:1–182.

Ancel, P. and P. Vintenberger, 1949. La rotation de symétrisation, facteur de la polarisation dorso-ventrale des ébauches primordiales, dans l'oeuf des amphibiens. *Arch. Anat. Microsc. Morphol. Exp.*, 38:167–183.

Anderson, D. T., 1962. The embryology of *Dacus tryoni* (Frogg.) [Diptera, Trypetidae (= Tephritidae)], the Queensland fruit-fly. *J. Embryol. Exp. Morphol.*, 10:248–292.

Anderson, D. T., 1966. The comparative early embryology of the Oligochaeta, Hirudinea, and Onychorphora. *Proc. Linnean Soc. New South Wales*, 91:10–43.

Anderson, D. T., 1969. On the embryology of cirripede crustaceans *Tetraclita rosea* (Krauss), *Tetraclita puysurascens* (Wood), *Chthamalus antennatus* (Darwin) and *Chthamalus columna* (Spengler) and some consideration of crustacean phylogenetic relationships. *Philos. Trans. R. Soc. London Ser. B*, 256:183–235.

Anderson, D. T., 1972a. The development of hemimetabolous insects. In S. J. Counce and C. H. Waddington, eds., *Developmental systems: Insects*, Vol. 1. Academic Press, London, pp. 96–163.

Anderson, D. T., 1972b. The development of holometabolous insects. In S. J. Counce and C. H. Waddington, eds., *Developmental systems: Insects*, Vol. 1. Academic Press, London, pp. 166–242.

Anderson, D. T., 1973. *Embryology and phylogeny in annelids and arthropods*, Pergamon Press, Oxford.

Anderson, E. and H. W. Beams, 1960. Cytological observations on the fine structure of the guinea pig ovary with special reference to the oogonium, primary oocyte and associated follicle cells. *J. Ultrastruct. Res.*, 3:432–446.

Arey, L. B., 1965. *Developmental anatomy, a textbook and laboratory manual of embryology*, 7th ed. Saunders, Philadelphia.

Argüello, C. J. Alanis and B. Valenzuela, 1988. The early development of the atrioventricular node and bundle of His in the embryonic chick heart, an electrophysiological and morphological study. *Development*, 102:623–637.

Arion, D., L. Meijer, L. Brizuela, and D. Beach, 1988. *cdc2* is a component of the M phase-specific histone H1 kinase: Evidence for identity with MPF. *Cell*, 55:371–378.

Aristotle, 1984. *Generation of animals*. In J. Barnes, ed., *The complete works of Aristotle*, revised Oxford translation. Bollingen series LXXI. 2. Princeton University Press, Princeton.

Arnold, J. M. and L. D. Williams-Arnold, 1976. The egg cortex problem as seen through the squid eye. *Am. Zool.*, 16:421–446.

Artz, K. P. Dubois, D. Bennett, H. Condamine, C. Babinet, and F. Jacob, 1973. Surface antigens common to mouse cleavage embryos and primitive teratocarcinoma cells in culture. *Proc. Natl. Acad. Sci. USA*, 70:2988–2992.

Astrow, S., B. Holton, and D. Weisblat, 1987. Centrifugation redistributes factors determining cleavage patterns in leech embryos. *Devel. Biol.*, 120:270–283.

Atchison, M. L., 1988. Enhancers: Mechanisms of action and cell specificity. *Annu. Rev. Cell Biol.*, 4:127–153.

Austin, C. R., 1962. Sex chromatin in embryonic and fetal tissue. *Acta Cytol.*, 6:61–65.

Austin, C. R., 1974. Principles of fertilization. *Proc. R. Soc. Med.*, 67:925–927.

Austin, C. R., 1978. Patterns in metazoan fertilization. *Current topics in developmental biology*, Vol. 12. Academic Press, Orlando, pp. 1–9.

Austin, C. R., 1982. The egg. In C. R. Austin and R. V. Short, eds., *Reproduction in mammals, Book 1, Germ cells and fertilization*, 2nd. ed. Cambridge University Press, Cambridge, pp. 46–62.

Avise, J. C. and R. A. Lansman, 1983. Polymorphism of mitochondrial DNA in populations of higher animals. In M. Nei

and R. K. Hoehn, eds., *Evolution of genes and proteins*. Sinauer, Sunderland, pp. 147–164.

Bachvarova, R. Gene expression during oogenesis and oocyte development in mammals. In L. W. Browder, ed., *Developmental biology: A comprehensive synthesis*, Vol. 1, Oogenesis. Plenum Press, New York, pp. 453–524.

Baer, K. E. v., 1827. *Epistola de ovi mammalium et hominis genesi*. Lipsiae.

Baer, K. E. v., 1828. *Uber Entwickelungsgeschichte der Thiere*. Königsberg.

Baird, D. T., 1984. The ovary. In C. R. Austin and R. V. Short, eds., *Reproduction in mammals, Book 3, Hormonal control of reproduction*, 2nd ed. Cambridge University Press, Cambridge, pp. 91–114.

Bajer, A. S. 1977. Interaction of microtubules and the mechanism of chromosome movements (zipper hypothesis). II. Dynamic architecture of the spindle. In T. L. Rost and E. M. Gifford, Jr., eds., *Mechanisms and control of cell division*. Dowden, Hutchinson & Ross, Stroudsburg, pp. 233–258.

Bajer, A. S. and J. Molé-Bajer, 1985. Drugs with colchicine-like effects that specifically disassemble plant but not animal microtubules. *Ann. N.Y. Acad. Sci.*, 466:767–784.

Baker, P. F., 1988. Exocytosis in electropermeabilized cells: Clues to mechanism and physiological control. In N. Düzgünes and F. Bronner, eds., *Current topics in membranes and transport*, Vol. 32, *Membrane fusion in fertilization, cellular transport, and viral infection*. Academic Press, Orlando, pp. 115–138.

Baker, T. G., 1963. A quantitative and cytological study of germ cells in human ovaries. *Proc. R. Soc. London Ser. B*, 158:417–433.

Baker, T. G., 1982. Oogenesis and ovulation. In C. R. Austin and R. V. Short, eds., *Reproduction in mammals, Book 1, Germ cells and fertilization*, 2nd ed. Cambridge University Press, Cambridge, pp. 17–45.

Balfour, F. M., 1880. *A treatise on comparative embryology*, Vol 1. Macmillan, London.

Balinsky, B. I., 1981. *An introduction to embryology*, 5th ed. Saunders, Philadelphia.

Ballard, W. W., 1981. Morphogenetic movements and fate maps of vertebrates. *Am. Zool.*, 21:391–399.

Ballard, W. W., 1984. Morphogenetic movements in embryos of the holostean fish, *Amia calva*: A progress report. *Am. Zool.*, 24:539–543.

Barg, P. E., M. Z. Wahrman, B. E. Talansky, and J. W. Gordon, 1986. Capacitated, acrosome reacted but immotile sperm, when microinjected under the mouse zona pellucida will not fertilize the oocyte. *J. Exp. Zool.*, 237:365–374.

Barra, J. and J.-P. Renard, 1988. Diploid mouse embryos constructed at the late s-cell stage from haploid parthenotes and androgenotes can develop to term. *Development*, 102:773–779.

Barth, L. G. and L. J. Barth, 1954. *The energetics of development: A study of metabolism in the frog egg*. Columbia University Press, New York.

Bayna, E. M., J. H. Shaper, and B. D. Shur, 1988. Temporally specific involvement of cell surface β-1,4 galactosyltransferase during mouse embryo morula compaction. *Cell*, 53:145–157.

Beadle, G. W. and E. L. Tatum, 1941. Genetic control of biochemical reactions in *Neurospora*. *Proc. Natl. Acad. Sci.*

USA, 27:499–506. Reprinted in H. O. Corwin and J. B. Jenkins, eds., 1976. *Conceptual foundations of genetics.* Houghton Mifflin, Boston.

Beams, H. W. and R. G. Kessel, 1976. Cytokinesis: A comparative study of cytoplasmic division in animal cells. *Am. Sci.*, 64:279–290.

Beato, M., 1989. Gene regulation by steroid hormones. *Cell*, 56:335–344.

Beatty, R. A., 1957. *Parthenogenesis and polyploidy in mammalian development.* Cambridge University Press, Cambridge.

Beckerle, M. C., K. Burridge, G. N. DeMartino, and D. E. Croall, 1987. Coloralization of calcium-dependent protease II and one of its substances and sites of adhesion. *Cell*, 51:569–577.

Beddington, R., 1983. The origin of the foetal tissues during gastrulation in the rodent. In M. H. Johnson, ed., *Development in mammals*, Vol. 5. Elsevier Science Publishers, Amsterdam, pp. 1–32.

Bedford, J. M., 1982. Fertilization. In C. R. Austin and R. V. Short, eds., *Reproduction in Mammals, Book 1, Germ cells and fertilization*, 2nd ed. Cambridge University Press, Cambridge, pp. 128–163.

Bedford, J. M. and G. W. Cooper, 1978. Membrane fusion. In G. Poste and G. L. Nicolson, eds., *Cell Surface Review*, Vol. 5. Elsevier/North-Holland Biomedical Press, Amsterdam, pp. 66–125.

Beer, A. E., R. E. Billingham, and J. R. Scott, 1975. Immunogenetic aspects of implantation, placentation and fetoplacental growth rates. *Biol. Reprod.*, 12:176–189.

Bell, E., A. B. Chepelinsky, and J. Brown, 1972. I-DNA—illusion or reality. In R. Harris, P. Allin, and D. Viza, eds., *Cell differentiation*, Proceedings of the First International Conference on Cell Differentiation. Scandinavian University Books, Munksgaard, Copenhagen, pp. 10–16.

Bell, G., 1982. *The masterpiece of nature: The evolution and genetics of sexuality.* University of California Press, Berkeley.

Bellairs, R., 1971. *Developmental processes in higher vertebrates.* University of Miami Press, Coral Gables.

Bellairs, R., 1986. The tail bud and cessation of segmentation in the chick embryo. In R. Bellairs, D. A. Ede, and J. W. Lash, eds., *Somites in developing embryos.* Plenum Press, New York, pp. 161–178.

Bellairs, R. and M.-C. van Peteghem, 1984. Gastrulation: Is it analogous to malignant invasion? *Am. Zool.*, 24:563–570.

Bellairs, R., F. W. Lorenz, and T. Dunlap, 1978. Cleavage in the chick embryo. *J. Embryol. Exp. Morphol.*, 43:55–69.

Bellvé, A. R. and D. A. O'Brien, 1983. The mammalian spermatozoon: Structure and temporal assembly. In J. F. Hartmann, ed., *Mechanism and control of animal fertilization.* Academic Press, Orlando, pp. 55–137.

Benbow, R. M., 1985. Activation of DNA synthesis during early embryogenesis. In C. B. Metz and A. Monroy, eds., *Biology of fertilization*, Vol. 3, *The fertilization response of the egg.* Academic Press, Orlando, pp. 299–345.

Bendori, R., D. Salomon, and B. Geiger, 1987. Contact-dependent regulation of vinculin expression in cultured fibroblasts: A study with vinculin-specific cDNA probes. *EMBO J.*, 6:2897–2905.

Berg, J. M., 1986. Potential metal-binding domains in nucleic acid binding proteins. *Science*, 232:485–487.

Berg, W. E. and W. J. Humphreys, 1960. Electron microscopy of four-cell stages of the ascidian *Ciona* and *Steyla*. *Dev. Biol.*, 2:42–60.

Bernstein, H., H. C. Byerly, F. A. Hopf, and R. E. Michod, 1985. Genetic damage, mutation, and the evolution of sex. *Science*, 229:1277–1281.

Berrill, N. J., 1950. Size and organization in the development of ascidians. In P. Medawar, ed., *Essays on growth and form.* Oxford University Press, Oxford.

Berrill, N. J., 1961. *Growth, development, and pattern.* Freeman, San Francisco.

Berry, S. J., 1985. RNA synthesis and storage during insect oogenesis. In L. W. Browder, ed., *Developmental biology: A comprehensive synthesis*, Vol. 1, *Oogenesis.* Plenum Press, New York, pp. 351–384.

Bertin, L., 1952. Oviparité, ovoviviparité, viviparité. *Soc. Zool. Fr. Bull.*, 77:84–88.

Betchaku, T. and J. P. Trinkaus, 1978. Contact relations, surface activity, and cortical microfilaments of marginal cells of the enveloping layer and of the yolk syncytial and yolk cytoplasmic layers of *Fundulus* before and during epiboly. *J. Exp. Zool.*, 206:381–426.

Bier, K., 1963a. Autoradiographische Untersuchungen über die Leistungen des Follikelpithels und der Näzellen bei der Dotterbildung und Eiweissynthese im Fliegenovar. *Wilhelm Roux' Arch. Entwicklungsmech. Org.*, 154:552–575.

Bier, K, 1963b. Syntheses, interzellulärer transport, und abbau von ribonucleinsäure im ovar der stubenfliege *Musca domestica*. *J. Cell Biol.*, 16:436–440.

Bier, K., W. Kunz, and D. Ribbert, 1972. Insect oogenesis with and without lampbrush chromosomes. In M. C. Gould et al., *Invertebrate oogenesis. II.* MSS Information Corporation, New York, pp. 123–133.

Billington, W. D., 1987. Immunological aspects of implantation and fetal survival: The central role of trophoblast. In A. McLaren and G. Siracusa, eds., *Recent advances in mammalian development*, Academic Press, San Diego. *Current topics in developmental biology*, 23:209–232.

Bishop, J. M., 1985. Viral oncogenes. *Cell*, 42:23–38.

Blackler, A. W., 1962. Transfer of primordial germ cells between two subspecies of *Xenopus laevis*. *J. Embryol. Exp. Morphol.*, 10:641–651.

Blackler, A. W., 1970. The integrity of the reproductive cell line in the amphibia. In A. A. Moscona and A. Monroy, eds., *Current topics in developmental biology*, Vol 5. Academic Press, Orlando, pp. 71–87.

Blackler, A. W. and M. Fischberg, 1961. Transfer of primordial germ cells in *X. laevis*. *J. Embryol. Exp. Morphol.*, 9:634–641.

Blackler, A. W. and C. A. Gecking, 1972. Transmission of sex cells of one species through the body of a second species in the genus *Xenopus*. I. Intraspecific matings. II. Interspecific matings. *Dev. Biol.*, 27:376–384, 385–394.

Bleil, J. D. and P. M. Wassarman, 1980. Structure and function of the zona pellucida: Identification and characterization of the proteins of the mouse oocyte's zona pellucida. *Dev. Biol.*, 76:185–202.

Bloom, W. and D. W. Fawcett, 1986. *A textbook of histology.* 11th ed., Saunders, Philadelphia.

Blüm, V., 1986. *Vertebrate reproduction, a textbook.* Translated from the German edition by A. C. Whittle. Springer-Verlag, Berlin.

Bodemer, C. W., 1971. The biology of the blastocyst in historical perspective. In R. J. Blandau, ed., *Biology of the blastocyst.* University of Chicago Press, Chicago, pp. 1–25.

Bodenstein, D., 1955. Insects. In B. H. Willier, P. A. Weiss, and V. Hamburger, eds., *Analysis of development.* Saunders, Philadelphia, pp. 337–345.

Bold, H. C., C. J. Alexopoulous, and T. Delevoryas, 1987. *Morphology of plants and fungi,* 5th ed. Harper & Row, New York.

Boletzky, S. V., 1988. Characteristics of cephalopod embryogenesis. In J. Wiedmann and J. Kullmann, eds., *Cephalopods—Present and past.* Schweizerbart'sche Verlagsbuchhandlung, Stuttgart, pp. 167–179.

Bonner, J. T., 1962. *Ideas of biology.* Harper & Row, New York.

Bonner, W. M., 1975. Protein migration into nuclei. I. Frog oocyte nuclei in vivo accumulate microinjected histones, allow entry to small proteins, and exclude large proteins. II. From oocyte nuclei accumulate a class of microinjected oocyte nuclear proteins and exclude a class of microinjected oocyte cytoplasmic proteins. *J. Cell Biol.,* 64:421–430, 431–437.

Bonner-Fraser, M. E., 1987. Adhesive interactions in neural crest morphogenesis. In P. R. A. Maderson, ed., *Developmental and evolutionary aspects of the neural crest.* Wiley, New York, pp. 11–38.

Borum, K., 1966. Oogenesis in the mouse. A study of the origin of the mature ova. *Exp. Cell Res.,* 45:39–47.

Boucaut, J. C., T. Darribère, H. Boulekbache, and J.-P. Thiery, 1984. Prevention of gastrulation but not neurulation by antibodies to fibronectin in amphibian embryos. *Nature (London),* 307:364–376.

Boucaut, J. C., T. Darribère, S. De Li, H. Boulekbache, K. M. Yamada, and J.-P. Thiery, 1985. *J. Embryol. Exp. Morphol. Suppl.,* 89:211–227.

Bouillon, J. and B. Werner, 1965. Production of medusae buds by the polyps of Rathkea octopunctata (M. Sars) (Hydroida Athecata). *Helgol. Wiss. Meeresunters.,* 12:137–148.

Bounoure, L., 1934. Recherches sur la lignée germinale chez la grenouille rousse aux premiers stades du développement. *Ann. Sci. Natur. Zool.,* 10(Ser. 17):67–248.

Bounoure, L., R. Aubry, and M.-L. Huck, 1954. Nouvelles recherches experiméntales sur les origines de la lignée reproductrice chez la grenouille rousse. *J. Embryol. Exp. Morphol.,* 2:245–263.

Boveri, T., 1902. On multipolar mitosis as a means of analysis of the cell nucleus. In B. H. Willier and J. M. Oppenheimer, eds., *Foundations of experimental embryology.* Prentice-Hall, Englewood Cliffs, NJ, pp. 74–97.

Böving, B. G., 1971. Biomechanics of implantation. In R. J. Blandau, ed., *The biology of the blastocyst.* University of Chicago Press, Chicago, pp. 423–442.

Böving, B. G. and J. F. Larsen, 1973. Implantation. In E. S. E. Havez and T. N. Evans, eds., *Human reproduction.* Harper & Row, New York, pp. 133–156.

Bowler, P. J., 1988. *Non-Darwinian revolution.* Johns Hopkins University Press, Baltimore.

Brachet, J., 1945. *Embryologie chimique,* 2nd ed. Desoer, Liège.

Brachet, J., 1950. *Chemical embryology,* translated from the French by L. G. Barth. Interscience, New York.

Brachet, J., 1957. *Biochemical cytology.* Academic Press, Orlando.

Brachet, J., 1977. An old enigma: The gray crescent of amphibian eggs. In A. A. Moscona and Alberto Monroy, eds., *Current topics in developmental biology,* Vol. 11, *Pattern development.* Academic Press, Orlando, pp. 133–186.

Brachet, J. and S. Donini-Denis, 1978. Studies on maturation and differentiation without cleavage in *Chaeptopterus variopedatus.* Effects of ions, ionophores, sulphydril reagents, colchicine and cytochalasin B. *Differentiation,* 11:19–37.

Bradley, R. M. and C. M. Mistretta, 1975. Fetal sensory receptors. *Physiol. Rev.,* 55:352–382.

Bramhill, D. and A. Kornberg, 1988. Duplex opening by dnaA protein at novel sequences in initiation of replication at the origin of the E. coli chromosome. *Cell,* 52:743–755.

Brandhorst, B. P., P.-A. Bédard, F. Tufaro, and R. Conlon, 1986. The persistent role of maternal information during embryogenesis of the sea urchin. In J. G. Gall, ed., *Gametogenesis and the early embryo,* 44th Symposium Society for Developmental Biology. Alan R. Liss, New York, pp. 283–303.

Bravo, R. and J. Knowland, 1979. Classes of proteins synthesized in oocytes, eggs, embryos, and differentiated tissues of *Xenopus laevis. Differentiation,* 13:101–108.

Breathnach, R., C. Benoist, K. O'Hara, F. Gannon, and P. Chambon, 1978. Ovalbumin gene: Evidence for a leader sequence in mRNA and DNA sequences at the exon–intron boundaries. *Proc. Natl. Acad. Sci. USA,* 75:4853–4857.

Breckenridge, L. J., R. L. Warren, and A. E. Warner, 1987. Lithium inhibits morphogenesis of the nervous system but not neuronal differentiation in *Xenopus laevis. Development,* 99:353–370.

Brenner, S., 1974. The genetics of *Caenorhabditis elegans. Genetics,* 77:71–94.

Brick, I. and C. Weinberger, 1984. Introduction to the symposium: Gastrulation. *Am. Zool.,* 24:537–538.

Brien, P., 1968. Blastogenesis and morphogenesis. *Adv. Morphog.,* 7:151–204.

Brien, P. and M. Reniers-Decoen, 1950. Etude d'Hydra viridis (Linnaeus). *Ann. Soc. R. Zool. Belgium,* 81:33–108.

Briggs, R., 1979. Genetics of cell type determination. In J. F. Danielli and M. A. DiBerardino, eds., *Nuclear transplantation. Int. Rev. Cytol. Suppl. No. 9.* Academic Press, Orlando, pp. xx–xx.

Briggs, R. and T. J. King. 1952. Transplantation of living nuclei from blastula cells into enucleated frogs' eggs. *Proc. Nat. Acad. Sci. USA,* 38:455–463.

Brink, N. C., 1968. Protein synthesis during spermatogenesis in *Drosophila melanogaster. Mutat. Res.,* 5:192–194.

Brinkley, B. R., 1985. Microtubule organizing centers. *Annu. Rev. Cell Biol.,* 1:145–172.

Britten, R. J. and E. H. Davidson, 1969. Gene regulation for higher cells: A theory. *Science,* 165:349–357.

Brodsky, F. M., 1988. Living with clathrin: Its role in intracellular membrane traffic. *Science,* 242:1396–1402.

Brodsky, V. Ya. and I. V. Uryvaeva, 1985. *Genome multiplication in growth and development.* Cambridge University Press, Cambridge.

Brody, E. and J. Abelson, 1985. The splicosome—yeast pre-messenger RNA associates with a 40S complex in a splicing dependent reaction. *Science,* 228:963–967.

Brothers, A. J., 1979. A specific case of genetic control of early development: The *o* maternal effect mutation of the Mexican axolotl. In S. Subtelny, ed., *Determinants of spatial organization*, 37th Symposium of the Society for Developmental Biology. Academic Press, Orlando, pp. 167–183.

Brown, C. R. and R. Jones, 1987. Binding of zona pellucida proteins to a boar sperm polypeptide of M_r 53000 and identification of zona moieties involved. *Development*, 99:333–339.

Brown, D. D. and I. B. Dawid, 1968. Specific gene amplification in oocytes. *Science*, 160:272–280.

Brown, P. M. and K. Kalthoff, 1983. Inhibition by ultraviolet light of pole cell formation in *Smittia* sp. (Chironomidae Diptera): Action spectrum and photoreversibility. *Dev. Biol.*, 97:113–122.

Brun, R. B., 1978. Developmental capacities of *Xenopus* eggs provided with erythrocyte or erythroblast nuclei from adults. *Dev. Biol.*, 65:271–284.

Brunson, R. B., 1963. Aspects of the natural history and ecology of the Gastrotricha. In E. C. Dougherty, ed., *The lower metazoa: Comparative biology and phylogeny.* University of California Press, Berkeley, pp. 473–478.

Brushkin, A. M., P.-A. Bedard, A. L. Tyner, R. M. Showman, B. P. Brandhorst, and W. H. Klein, 1982. A family of proteins accumulating in ectoderm of sea urchin embryos specified by two related cDNA clones. *Dev. Biol.*, 91:317–324.

Buck, C. A. and A. F. Horwitz, 1987. Cell surface receptors for extracellular matrix molecules. *Annu. Rev. Cell Biol.*, 3:179–205.

Buck, C. A., K. A. Knudsen, C. H. Damsky, C. L. Decker, R. R. Greggs, K. E. Duggan, D. Bozyczko, and A. F. Horwitz, 1985. Integral membrane protein complexes in cell–matrix adhesion. In G. M. Edelman and J.-P. Thiery, eds., *Cell in contact: Adhesions and junctions as morphogenetic determinants.* A Neurosciences Institute Publication, Wiley, New York, pp. 179–205.

Buehr, M. and A. W. Blackler, 1970. Sterility and partial sterility in the South African clawed toad following pricking of the egg. *J. Embryol. Exp. Morphol.*, 23:375–484.

Buehr, M. and A. McLaren 1974. Size regulation in chimaeric mouse embryos. *J. Embryol. Exp. Morphol.*, 31:229–234.

Bunge, M. B., P. M. Wood, L. B. Tynan, M. L. Bates, and J. R. Sanes, 1989. Perineurium originates from fibroblasts: Demonstration in vitro with a retroviral marker. *Science*, 243:229–231.

Burch, J. B. E. and H. Weinbraub, 1983. Temporal order of chromatin structural changes associated with activation of the major chicken vitellogenin gene. *Cell*, 33:65–76.

Burdick, M. L., 1970. Cell sorting out according to species in aggregates containing mouse and chick embryonic limb mesoblast cells. *J. Exp. Zool.*, 175:357–368.

Burgeson, R. E., 1988. New collagens, new concepts. *Annu. Rev. Cell Biol.*, 4:551–577.

Bürglin, T. R., 1988. The yeast regulatory gene *PH02* encodes a homeo box. *Cell*, 53:339–340.

Bürki, K., 1986. *Experimental embryology of the mouse, monographs in developmental biology*, Vol. 19. Karger, Basel.

Burridge, K., L. Molony, and T. Kelly, 1987. Adhesion plaques: Sites of transmembrane interaction between the extracellular matrix and the actin cytoskeleton. *J. Cell Sci. Suppl.*, 8:211–229.

Burridge, K., K. Fath, T. Kelly, G. Nuckolls, and C. Turner, 1988. Focal adhesions: Transmembrane junctions between the extracellular matrix and the cytoskeleton. *Annu. Rev. Cell Biol.*, 4:487–525.

Bursa, W. B. and R. L. Gimlich, 1989. Lithium-induced teratogenesis in frog embryos prevented by a polyphosphoinositide cycle intermediate or a diaglycerol analog. *Dev. Biol.*, 132:315–324.

Butler, H. and B. H. J. Juurlink, 1987. *An atlas for staging mammalian and chick embryos.* CRC Press, Boca Raton, FL.

Byrd, E. W. Jr. and F. D. Collins, 1975. Absence of ast block to polyspermy in eggs of sea urchin *Strongylocentrotus purpuratus. Science*, 257:675–677.

Byskov, A. G., 1982. Primordial germ cells and regulation of meiosis. In C. R. Austin and R. V. Short, eds., *Reproduction in mammals, Book 1, Germ cells and fertilization*, 2nd ed., Cambridge University Press, Cambridge, pp. 1–16.

Cairns, J., 1963. The bacterial chromosome and its manner of replication as seen by autoradiography. *J. Mol. Biol.*, 6:208–213.

Callan, H. G., 1972. Replication of DNA in the chromosomes of eukaryotes. *Proc. R. Soc. London Ser. B*, 181:19–41.

Callan, H. G., 1973. DNA-replication in chromosomes of eukaryotes. *Cold Spring Harbor Symp. Quant. Biol.*, 38:195–203.

Callan, H. G., 1982. Lampbrush chromosomes. *Proc. R. Soc. London Ser. B*, 214:417–448.

Callard, G. V., 1983. Androgen and estrogen actions in the vertebrate brain. *Am. Zool.*, 23:607–620.

Calow, P., 1978. *Life cycles: An evolutionary approach to the physiology of reproduction, development and ageing.* Chapman and Hall, London.

Campos-Ortega, J. A. and V. Hartenstein, 1985. *The embryonic development of Drosophila melanogaster.* Springer-Verlag, Berlin.

Carlson, B. M., 1988. *Patten's foundations of Embryology*, 5th ed. McGraw Hill, New York.

Carroll, S. B., G. M. Winslow, V. J. Twombly, and M. P. Scott, 1987. Genes that control dorsoventral polarity affect gene expression along the anteroposterior axis of the *Drosophila* embryo. *Development*, 99:327–332.

Caspersson, T. O., 1950. *Cell growth and cell function.* Chapman and Hall, London.

Cather, J. N., 1971. Cellular interactions in the regulation of development in annelids and molluscs. *Adv. Morphog.*, 9:67–125.

Cather, J. N., N. H. Verdonk, and M. R. Dohmen, 1976. Role of the vegetal body in the regulation of development in *Bithynia tentaculata* (Prosobranchia, Gastropoda). *Am. Zool.*, 16:455–468.

Chandler, D. E., 1988. Exocytosis and endocytosis: Membrane fusion events captured in rapidly frozen cells. In N. Düzgünes and F. Bronner, eds., *Current topics in membranes and transport*, Vol. 32, *Membrane fusion in fertilization, cellular transport, and viral infection.* Academic Press, Orlando, pp. 169–202.

Chang, M. C., 1981. My life with mammalian eggs. In S. R. Glasser and D. W. Bullock, eds., *Cellular and molecular aspects of implantation.* Plenum Press, New York, pp. 27–36.

Chapman, D. M., 1966. Evolution of the scyphistoma. In W. J. Rees, ed., *The Cnidaria and their evolution.* Academic Press, Orlando, pp. 51–75.

Chargaff, E., 1950. Chemical specificity of nucleic acid and mechanism of their enzymatic degradation. *Experientia*, 6:201–209.

Chen, P. S., 1967. Biochemistry of nucleo-cytoplasmic interactions in morphogenesis. In R. Weber, ed., *The biochemistry of animal development*, Vol. 2. Academic Press, Orlando, pp. 115–191.

Chen, W.-T., E. Hasegawa, T. Hasegawa, C. Weinstock, and K. M. Yamada, 1985. Development of cell surface linkage complexes in cultured fibroblasts. *J. Cell Biol.*, 100:1103–1114.

Chen, W.-T., J. Wang, T. Hasegawa, S. S. Yamada, and K. M. Yamada, 1986. Regulation of fibronectin receptor distribution by transformation, exogenous fibronectin, and synthetic peptides. *J. Cell Biol.*, 103:1649–1661.

Child, C. M., 1914. Susceptibility gradients in animals. *Science*, 39:73–76. Reprinted in B. H. Willier and J. M. Oppenheimer, eds., 1964. *Foundations of experimental embryology*. Prentice-Hall, Englewood Cliffs, NJ, pp. 129–134.

Child, C. M., 1941. *Patterns and problems of development*. University of Chicago Press, Chicago.

Child, C. M., 1946. Organizers in development and the organizer concept. *Physiol. Zool.*, 19:89–148.

Christ, B., H. J. Jacob, and M. Jacob, 1977. Experimental analysis of the origin of the wing musculature in avian embryos. *Anat. Embryol.*, 150:171–186.

Chung, A. E., R. Jaffe, I. L. Freeman, J. P. Vergnes, J. E. Braginski, and B. Carlin, 1979. Properties of a basement membrane-related glycoprotein synthesized in culture by a mouse embryonal carcinoma-derived cell line. *Cell*, 16:277–287.

Churchill, F. B., 1984. Weismann, hydromedusae, and the biogenetic imperative: A reconsideration. In T. J. Horder, J. A. Witkowski, and C. C. Wylie, eds., *A history of embryology*, 8th Symposium of the British Society for Developmental Biology. Cambridge University Press, Cambridge, pp. 7–33.

Clark, P., P. Connolly, A. S. G. Curtis, J. A. T. Dow, and C. D. W. Wilkinson, 1987. Topographical control of cell behavior: 1. Simple step cues. *Development*, 99:439–448.

Clayton, A. B. and A. F. Horwitz, 1987. Integrin, a transmembrane glycoprotein complex mediating cell–substratum adhesion. *J. Cell Sci. Suppl.*, 8:231–250.

Clayton, R. M., 1953. Distribution of antigens in the developing newt embryo. *J. Embryol. Exp. Morphol.*, 1:25–43.

Clegg, K. B. and L. Pikò, 1982. RNA synthesis and cytoplasmic polyadenylation in the one-cell mouse embryo. *Nature (London)*, 295:342–345.

Clement, A. C., 1967. The embryonic value of the micromeres in *Ilyanassa obsoleta* as determined by deletion experiments. I. The first quartet cells. *J. Exp. Zool.*, 166:77–88.

Clement, A. C., 1976. Cell determination and organogenesis in molluscan development: A reappraisal based on deletion experiments in *Ilyanassa*. *Am. Zool.*, 16:447–453.

Clerc, R. G., L. M. Corcoran, J. H. LeBowitz, D. Baltimore, and P. A. Sharp, 1988. The B-cell-specific Oct-2 protein contains POU box- and homeo box-type domains. *Genes & Dev.*, 2:1570–1581.

Clermont, Y., 1962. Quantitative analysis of spermatogenesis in the rat. A revised model for renewal of spermatogonia. *Am. J. Anat.*, 111:111–129.

Clutton-Brock, T. H., 1988. Introduction. In T. H. Clutton-Brock, ed., *Reproductive success: Studies of individual variation in contrasting breeding systems*. University of Chicago Press, Chicago, pp. 1–6.

Coffino, P., 1981. Hormonal regulation of cloned genes. *Nature (London)*, 292:492–493.

Coggins, L. W., 1973. An ultrastructural and radio-autographic study of early oogenesis in the toad *Xenopus laevis*. *J. Cell Sci.*, 12:71–93.

Cohen, J. and B. Massey, 1982. *Living embryos*, 3rd ed. Pergamon Press, Oxford.

Cole, C. J., 1975. Evolution of parthenogenetic species of lizards. In R. Reinboth, ed., *Intersexuality*. Springer-Verlag, Berlin, pp. 340–355.

Colwin, A. L. and L. H. Colwin, 1963. Role of the gamete membranes in fertilization in *Saccoglossus kowalevskii* (Enteropneusta). I. The acrosomal region and its changes in early stages of fertilization. *J. Cell Biol.*, 19:477–500.

Commoner, B., 1964. Roles of deoxyribonucleic acid in inheritance. *Nature (London)*, 202:960–968.

Conklin, E. G., 1896. Cleavage and differentiation (paper presented at the Marine Biological Laboratory. Reprinted in J. Maienschein, 1986. *Defining biology: Lectures from the 1890s*. Harvard University Press, Cambridge, pp. 151–177.

Conklin, E. G., 1905. Organization and cell lineage of the ascidian egg. *J. Acad. Nat. Sci. Philadelphia Ser. 2*, 13:1–119.

Conklin, E. G., 1932. The embryology of Amphioxus. *J. Morphol.*, 54:69–151.

Conrad, J. T., 1971. The biophysics of nidation. In R. J. Blandau, ed., *The biology of the blastocyst*. University of Chicago Press, Chicago, pp. 443–462.

Cook, P. R. and I. A. Brazell, 1980. Mapping sequences in loops of nuclear DNA by their progressive detachment from the nuclear cage. *Nucleic Acids Res.*, 18:2895–2906.

Cooke, J., 1987. Dynamics of the control of body pattern in the development of *Xenopus laevis*. IV. Timing and pattern in the development of twinned bodies after reorientation of eggs in gravity. *Development*, 99:417–427.

Cooke, J. and J. C. Smith, 1987. The midblastula cell cycle transition and the character of mesoderm in u.v.-induced nonaxial *Xenopus* development. *Development*, 99:197–210.

Costa, M., M. Weir, A. Coulson, J. Sulston, and C. Kenyon, 1988. Posterior pattern formation in *C. elegans* involves position-specific expression of a gene containing a homeobox. *Cell*, 55:747–756.

Costello, D. P., 1945. Experimental studies of germinal localization in Nereis. I. The development of isolated blastomeres. *J. Exp. Zool.*, 100:19–66.

Costello, D. P. and C. Henley, 1976. Spiralian development: A perspective. *Am. Zool.*, 16:277–291.

Cowden, R. R., 1976. Cytochemistry of oogenesis and early embryonic development. *Am. Zool.*, 16:363–374.

Cowin, P. W., W. Franke, C. Grund, H.-P. Kapprell, and J. Kartenbeck, 1985. The desmosome-intermediate filament complex. In G. M. Edelman and J.-P. Thiery, eds., *Cell in contact: Adhesions and junctions as morphogenetic determinants*. A Neurosciences Institute Publication, Wiley, New York, pp. 427–460.

Craig, F. M., G. Bentley, and C. W. Archer, 1987. The spatial and temporal pattern of collagens I and II and keratan sulphate in the developing chick metatarsophalangeal joint. *Development*, 99:383–391.

Crick, F. H. C., 1958. On protein synthesis. *Symp. Soc. Exp. Biol.*, 12:138–163.

Crick, F. H. C., 1966. Codon-anticodon pairing: The wobble hypothesis. *J. Mol. Biol.*, 19:548–555.

Crick, F., 1970. Diffusion in embryogenesis. *Nature (London)*, 225:420–422.

Crick, F., 1981. *Life itself, its origin and nature.* Simon and Schuster, New York.

Crick, F. H. C. and P. A. Lawrence, 1975. Compartments and polyclones in insect development. *Science*, 189:340–347.

Critchley, D. R., M. A. England, J. Wakely, and R. O. Hynes, 1979. Distribution of fibronectin in the ectoderm of gastrulating chick embryos. *Nature (London)*, 280:498–500.

Cruz, Y. P. and R. A Pedersen, 1985. Cell fate in the polar trophectoderm of mouse blastocysts as studied by microinjection of cell lineage tracers. *Dev. Biol.*, 112:73–83.

Cunningham, B. A., 1985. Structures of cell adhesion molecules. In G. M. Edelman and J.-P. Thiery, eds., *Cell in contact: Adhesions and junctions as morphogenetic determinants.* A Neurosciences Institute Publication, Wiley, New York, pp. 197–217.

Curtis, A. S. G., 1962. Morphogenetic interactions before gastrulation in the amphibian, *Xenopus laevis*—the cortical field. *J. Embryol. Exp. Morphol.*, 10:410–423.

Dabauvalle, M. C., M. Doree, R. Bravo, and E. Karsenti, 1988. Role of nuclear material in the early cell cycle of *Xenopus* embryos. *Cell*, 52:525–533.

Dahl, G., T. Miller, D. Paul, R. Voellmy, and R. Werner, 1987. Expression of functional cell–cell channels from cloned rat liver gap junction complementary DNA. *Science*, 236:1290–1293.

Dalcq, A. and J. Pasteels, 1938. Potentiel morphogénétique, régulation et "Axial gradients" de Child. Mireau point des bases physiologiques de la morphogénése. *Bull. Acad. R. Méd. Belg. Sér. 3*, 6:261–308.

Dale, B., 1987. Mechanism of fertilization. *Nature (London)*, 325:762–763.

Dale, B. and A. DeSantis, 1981a. The effects of cytochalasin B and D on the fertilization of sea urchins. *Dev. Biol.*, 83:232–237.

Dale, B. and A. DeSantis, 1981b. Maturation and fertilization of the sea urchin oocyte: An electrophysiological study. *Dev. Biol.*, 85:474–484.

Dale, L. and J. M. W. Slack, 1987. Fate map for the 32-cell stage of *Xenopus laevis. Development*, 99:527–551.

Dalton, A. J. and M. D. Felix, 1953. Phase contrast and electron micrography of the cloudman S91 mouse melanoma. In M. Gordon, ed., *Pigment cell growth*, Conference on the Biology of normal and atypical pigment cell growth. Academic Press, Orlando, pp. 267–276.

Daly, M. and M. Wilson, 1979. Sex and strategy. *New Sci.*, 81:15–17.

Damsky, C. H., J. Richa, M. Wheelock, I. Damjanov, and C. A. Buck, 1985. Two cell adhesion molecules: characterization and role in early mouse embryo development. In G. M. Edelman and J.-P. Thiery, eds., *Cell in contact: Adhesions and junctions as morphogenetic determinants.* A Neurosciences Institute Publication, Wiley, New York, pp. 233–254.

Dan, J. C., 1967. Acrosome reaction and lysins. In C. B. Metz and A. Monroy, eds., *Fertilization*, Vol. I. Academic Press, London, pp. 237–294.

Dan, K., 1952. Cyto-embryological studies of sea urchins. II. blastula stage. *Biol. Bull.*, 102:74–89.

Dan, K. and M. Ikeda, 1971. On the system controlling the time of micromere formation in sea urchin embryos. *Dev. Growth Differ.*, 13:285–301.

Dan, K. and T. Ono, 1952. Cyto-embryological studies of sea urchins. I. The means of fixation of the mutual positions among the blastomeres of sea urchin larvae. *Biol. Bull.*, 102:58–73.

Danilchik, M. V. and S. D. Black, 1988. The first cleavage plane and the embryonic axis are determined by separate mechanisms in *Xenopus laevis*: I. Independence in undisturbed embryos. *Dev. Biol.*, 128:58–64.

Darewskii (or Darevski), I. S. and W. N. Kulidowa, 1961. Natürliche Parthenogenese in der polymorphen Gruppe der kaukasischen Felseidechse (*Lacerta saxicola* Eversmann). *Zool. Jb. Syst.*, 89:119–176.

Darlington, C. D., 1981. Genetics and plant breeding, 1910–80. *Philos. Trans. R. Soc. London Ser. B*, 292:401–405.

Darwin, C., 1872. *The origin of species by means of natural selection or the preservation of favoured races in the struggle for life*, 6th ed. First published by John Murray, 1859. The Mentor Edition, 1958, New American Library, New York.

Davidson, E. H., 1986. *Gene activity in early development*, 3rd ed. Academic Press, Orlando.

Davidson, E. H., 1985. Genome function in sea-urchin embryos: Fundamental insights of Th. Boveri reflected in recent molecular discoveries. In T. J. Horder, J. A. Witkowski, and C. C. Wylie, eds., *History of embryology*, 8th Symposium of the British Society for Developmental Biology. Cambridge University Press, Cambridge.

Davidson, E. H. and R. J. Britten, 1979. Regulation of gene expression: Possible role of repetitive sequences. *Science*, 204:1052–1059.

Davies, J. and H. Hesseldahl, 1971. Comparative embryology of mammalian blastocysts. In R. J. Blandau, ed., *Biology of the blastocyst.* University of Chicago Press, Chicago, pp. 27–48.

Davis, L. A. and L. F. Lemanski, 1987. Induction of myofibrillogenesis in cardiac lethal mutant axolotl hearts rescued by RNA derived from normal endoderm. *Development*, 99:145–154.

Davis, L. E., 1973. Ultrastructural changes during dedifferentiation and redifferentiation in the regenerating, isolated gastrodermis. In A. L. Burnett, ed., *Biology of Hydra.* Academic Press, Orlando, pp. 171–219.

Davis, R. L., H. Weintraub, and A. B. Lassar, 1987. Expression of a single transfected cDNA converts fibroblasts to myoblasts. *Cell*, 51:987–1000.

Dawid, I. G. and A. W. Blackler, 1972. Maternal and cytoplasmic inheritance of mitochondrial DNA in *Xenopus. Dev. Biol.*, 29:152–161.

de Beer, G. R., 1951. *Embryos and ancestors.* Clarendon Press, Oxford.

de Crombrugghe, B., S. Busby, and H. Buc, 1984. Cyclic AMP receptor protein: Role in transcription activation. *Science*, 224:831–838.

Deka, N., K. E. Paulson, C. Willard, and C. W. Schmid, 1986.

Repetitive human DNA sequences. II. Properties of a transposon-like human element. *Cold Spring Harbor Symp. Quant. Biol.*, 51:472–477.

Denis, H., 1972. RNA synthesis during oogenesis of *Xenopus laevis.* In R. Harris, P. Allin, and D. Viza, eds., *Cell differentiation*, Proceedings of the 1st International Conference on Cell Differentiation. Scandinavian University Press, Munksgaard, Copenhagen.

DePamphilis, M. L., 1987. Replication of simian virus 40 and polyoma virus chromosomes. In Y. Aloni, ed., *Molecular aspects of papovaviruses.* Martinas Nijhoff, Boston.

DePamphilis, M. L., 1988. Transcriptional elements as components of eukaryotic origins of DNA replication. *Cell,* 52:635–638.

De Pomerai, D., 1985. *From gene to animal, an introduction to molecular biology of animal development.* Cambridge University Press, Cambridge.

Detwiler, S. R., 1936. *Neuroembryology: An experimental study.* Macmillan, New York. (Reprinted by Hafner, New York, 1964.)

Dhouailly, D., 1984. Specification of feather and scale patterns. In G. M. Malacinski, ed., *Pattern formation.* Macmillan, New York, pp. 581–601.

DiBerardino, M. A., 1979. Nuclear and chromosomal behavior in amphibian nuclear transplants. In J. F. Danielli and M. A. DiBerardino, eds., *Nuclear transplantation. Int. Rev. Cytol. Suppl. No. 9.* Academic Press, Orlando, pp. 129–160.

DiBerardino, M. A., 1988. Genomic multipotentiality of differentiated somatic cells. In G. Eguchi, T. S. Okada, and L. Saxén, eds., *Regulatory mechanisms in developmental processes.* Elsevier Scientific Publishers, Ireland Ltd., pp. 129–136.

DiBerardino, M. A. and T. J. King, 1965. Transplantation of nuclei from the frog renal adenocarcinoma. II. Chromosomal and histogenic analysis of tumor nuclear-transplant embryos. *Dev. Biol.,* 15:102–128.

DiBerardino, M. A., N. J. Hoffner, and L. D. Etkin, 1984. Activation of dormant genes in specialized cells. *Science,* 224:946–952.

DiBerardino, M. A., N. H. Orr, and R. G. McKinnell, 1986. Feeding tadpoles cloned from *Rana* erythrocyte nuclei. *Proc. Natl. Acad. Sci. USA,* 83:8231–8234.

Dimino, M. J. and M. D. Campbell, 1980. The function and differentiation of ovarian mitochondria. In P. M. Motta and E. S. E. Hafez, *Biology of the ovary.* Martinus Nijhoff, The Hague, pp. 191–195.

Dodd, J. and T. M. Jessell, 1988. Axon guidance and the patterning of neuronal projections in vertebrates. *Science,* 242:692–699.

Dodson, M., F. B. Dean, P. Bullock, H. Echols, and J. Hurwitz, 1987. Unwinding of duplex DNA from the SV40 origin of replication by T antigen. *Science,* 238:964–967.

Doe, C. Q., Y. Hiromi, W. J. Gehring, and C. S. Goodman, 1988. Expression and function of the segmentation gene *fushi tarazu* during *Drosophila* neurogenesis. *Science,* 239:170–175.

Doetschman, T. C., H. Eistetter, M. Katz, W. Schmidt, and R. Kemler, 1985. The *in vitro* development of blastocyst-derived embryonic stem cell lines: Formation of visceral yolk sac, blood islands and myocardium. *J. Embryol. Exp. Morphol.,* 87:27–45.

Dohmen, M. R., 1983. The polar lobe in eggs of molluscs and annelids: Structure, composition, and function. In W. R.

Jeffery and R. A. Raff, eds., *Time, space, and pattern in embryonic development.* Alan R. Liss, New York, pp. 197–220.

Donovan, P. J., D. Stott, I. Godin, J. Haesman, and C. C. Wylie, 1987. Studies on the migration of mouse germ cells. *J. Cell Sci. Suppl.,* 8:359–367.

Dow, J. A. T., P. Clark, P. Connolly, A. S. G. Curtis, and C. D. W. Wilkinson, 1987. Novel methods for the guidance and monitoring of single cells and simple networks in culture. *J. Cell Sci. Suppl.,* 8:55–79.

Draetta, G. and D. Beach, 1988. Activation of *cdc2* protein kinase during mitosis in human cells: Cell cycle-dependent phosphorylation and subunit rearrangement. *Cell,* 54:17–26.

Dreyfuss, G., 1986. Structure and function of nuclear and cytoplasmic ribonucleoprotein particles. *Annu. Rev. Cell Biol.,* 2:459–498.

Driesch, H., 1892. Entwicklungsmechanische Studien. I. *Z. Wiss. Zool.,* 53:160–178, 183–184. The potency of the first two cleavage cells in echinoderm development. Experimental production of partial and double formations. Abridged and translated by L. Mezger, M. Hamburger, V. Hamburger, and T. S. Hall with references, plate VII, and explanation of figures. In B. H. Willier and J. M. Oppenheimer, eds., 1964. *Foundations of experimental embryology.* Prentice-Hall, Englewood Cliffs, NJ, pp. 40–50.

Driever, W. and C. Nüsslein-Volhard, 1988a. A gradient of *bicoid* in *Drosophila* embryos. *Cell,* 54:83–93.

Driever, W. and C. Nüsslein-Volhard, 1988b. The *bicoid* protein determines position in the *Drosophila* embryo in a concentration-dependent manner. *Cell,* 54:95–104.

Ducibella, T., 1977. Surface charges of the developing trophoblast cell. In M. H. Johnson, ed., *Development in mammals,* Vol. 1. North-Holland, Amsterdam, pp. 5–30.

Duesberg, P. H., 1985. Activated proto-onc genes: Sufficient or necessary for cancer? *Science,* 228:669–677.

Dumont, J. N. and A. R. Brummett, 1985. Egg envelopes in vertebrates. In L. W. Browder, ed., *Developmental biology: A comprehensive synthesis,* Vol. 1, *Oogenesis.* Plenum Press, New York, pp. 235–288.

Dunn, B. E. and T. P. Fitzharris, 1987. Endocytosis in the embryonic chick chorionic epithelium. *J. Exp. Zool. Suppl.,* 1:75–79.

Dunphy, W. G., L. Brizuela, D. Beach, and J. Newport, 1988. The Xenopus *cdc2* protein is a component of MPF, a cytoplasmic regulator of mitosis. *Cell,* 54:423–431.

Du Pasquier, L. and M. R. Wabl, 1977. Transplantation of nuclei from lymphocytes of adult frogs into enucleated eggs. Special focus on technical parameters. *Differentiation,* 8:9–19.

Duprey, P., K. Chowdhury, G. R. Dressler, R. Balling, D. Simon, J.-L. Guenet, and P. Gruss, 1988. A mouse gene homologous to the *Drosophila* gene *caudal* is expressed in epithelial cells from the embryonic intestine. *Genes & Dev.,* 2:1647–1654.

Dustin, P., 1978. *Microtubules.* Springer-Verlag, Berlin.

Dvorak, M. and J. Tesarik, 1980. Ultrastructure of human ovarian follicles. In P. M. Motta and E. S. E. Hafez, eds., *Biology of the ovary.* Martinus Nijhoff, The Hague, pp. 121–137.

Dworkin, M. B. and E. Dworkin-Rastl, 1985. Changes in RNA titers and poly-adenylation during oogenesis and oocyte maturation in *Xenopus laevis. Dev. Biol.,* 112:451–457.

Dyban, A. P and V. S. Baranov, 1987. *Cytogenetics of mam-*

malian embryonic development. Clarendon Press, Oxford.

Easter, S. S. Jr., D. Purves, P. Rakic, and N. C. Spitzer, 1985. The changing view of neural specificity. *Science,* 230:507–511.

Eckstein, P., 1977. Endocrine activities of the ovary. In S. Zuckerman and B. J. Weir, eds., *The ovary,* 2nd ed., Vol. 2. Academic Press, Orlando, pp. 275–313.

Eddy, E. M., 1975. Germ plasm and the differentiation of the germ cell line. *Int. Rev. Cytol.,* 43:229–280.

Eddy, E. M., and A. C. Hahnel, 1983. Establishment of the germ cell line in mammals. In A. McLaren and C. C. Wylie, eds., *Current problems in germ cell differentiation,* 7th Symposium of the British Society for Developmental Biology Cambridge University Press, Cambridge, pp. 41–69.

Eddy, E. M. and S. Ito, 1971. Fine structural and radioautographic observations on dense perinuclear cytoplasmic material in tadpole oocytes. *J. Cell Biol.,* 49:90–108.

Eddy, E. M., J. M. Clark, D. Gong, and B. A. Fenderson, 1981. Origin and migration of primordial germ cells in mammals. *Gamete Res.,* 4:333–362.

Ede, D. A., 1976. Cell interactions in vertebrate limb development. In G. Poste and G. L. Nicolson, eds., *Cell surface in animal embryogenesis and development.* Elsevier/North-Holland Biomedical Press, Amsterdam, pp. 495–543.

Edelman, G. M., 1985. Specific cell adhesion in histogenesis and morphogenesis. In G. M. Edelman and J.-P. Thiery, eds., *Cell in contact: adhesions and junctions as morphogenetic determinants.* A Neurosciences Institute Publication, Wiley, New York, pp. 139–168.

Edelman, G. M., 1986. Cell adhesion molecules in the regulation of animal form and tissue pattern. *Annu. Rev. Cell Biol.,* 2:81–116.

Edelman, G. M., 1988. *Topobiology: An introduction to molecular embryology.* Basic Books, New York.

Edelman, G. M., S. Hoffman, C.-M. Chuong, and B. A. Cunningham, 1985. The molecular bases and dynamics of cell adhesion in embryogenesis. In G. M. Edelman, ed., *Molecular determinants of animal form.* Alan R. Liss, New York, pp. 195–221.

Edidin, M., 1976. The appearance of cell-surface antigens in the development of the mouse embryo: A study of cell-surface differentiation. In *Embryogenesis in mammals,* CIBA Foundation Symposia No. 40 (new series). Elsevier/Excerpta Medica/North-Holland, Amsterdam, pp. 177–197.

Edwards, C., 1973. Contributor thoughts on form, function, habitat and classification of hydroids and hydromedusae. In T. Takioka and S. Nishimura, eds., *Second International Symposium on Cnidaria,* Seto Marine Biology Laboratory, Shirahama, pp. 11–22.

Edwards, R. G., 1974. Fertilization of human eggs in vitro: Morals, ethics and the law. *Q. Rev. Biol.,* 49:3–26.

Elinson, R. P., 1980. The amphibian egg cortex in fertilization and early development. In S. Subtelny, ed., *The cell surface: Mediation of developmental processes,* 38th Symposium of the Society for Developmental Biology. Academic Press, Orlando, pp. 217–234.

Elinson, R. P., 1986. Fertilization in amphibians: The ancestry of the block to polyspermy. *Int. Rev. Cytol.,* 101:59–100.

Elinson, R. P., 1987. Change in developmental patterns: Embryos of amphibians with large eggs. In R. A. Raff and E. C. Raff, eds., *Development as an evolutionary process.* Alan R. Liss, New York, pp. 1–21.

Elinson, R. P., 1988. Parallel tracks of microtubules in the cortex of the frog egg. A potential mechanism for grey crescent formation. *Cell Motil.,* 10:342 (abstract).

Elinson, R. P. and B. Rowning, 1988. A transient array of parallel microtubules in frog eggs: Potential tracks for a cytoplasmic rotation that specifies the dorso-ventral axis. *Dev. Biol.,* 128:185–197.

Emanuelsson, H., 1973. Karyomeres in early cleavage embryos of *Ophryotrocha labronica* La Creca and Bacci. *Wilhelm Roux' Arch. Dev. Biol.,* 173:27–45.

Emlen, S. T. and L. W. Oring, 1977. Ecology, sexual selection, and evolution of mating systems. *Science,* 197:215–223.

Emmons, S. W., 1987. Mechanisms of *C. elegans* development. *Cell,* 51:881–883.

Enders, A. C., D. J. Chávez, and S. Schlafke, 1981. Comparison of implantation in utero and in vitro. In S. R. Glasser and D. W. Bullock, eds., *Cellular and molecular aspects of implantation.* Plenum Press, New York, pp. 365–382.

Epel, D., 1978. Mechanisms of activation of sperm and egg during fertilization of sea urchin gametes. In A. A. Moscona and A. Monroy, eds., *Current topics in developmental biology,* Vol. 12, *Fertilization.* Academic Press, Orlando, pp. 246–285.

Epel, D., 1980. Experimental analysis of the role of intracellular calcium in the activation of the sea urchin egg at fertilization. In S. Subtelny, ed., *Cell surface: Mediation of developmental processes,* 38th Symposium of the Society for Developmental Biology. Academic Press, Orlando, pp. 169–185.

Epel, D., 1989. Arousal of activity in sea urchin eggs at fertilization. In G. Schatten, ed., *Cell biology of fertilization,* Vol. 2. Academic Press, Orlando, pp. 361–385.

Epel, D. and V. D. Vacquier, 1978. Membrane fusion events during invertebrate fertilization. In G. Poste and G. L. Nicolson, eds., *Membrane fusion.* Elsevier/North-Holland Biomedical Press, Amsterdam, pp. 1–63.

Epel, D., R. Steinhardt, T. Humphreys, and D. Mazia, 1974. An analysis of the partial metabolic derepression of sea urchin eggs by ammonia: The existence of independent pathways. *Dev. Biol.,* 40:245–255.

Epel, D., N. L. Cross, and N. Epel, 1977. Flagellar motility is not involved in the incorporation of the sperm into the egg at fertilization. *Dev. Growth Differ.,* 19:15–21.

Ephrussi, A., G. M. Church, S. Tonegawa, and W. Gilbert, 1985. B lineage-specific interactions of an immunoglobulin enhancer with cellular factors in vitro. *Science,* 227:134–140.

Ephrussi, B., 1942. Chemistry of "eye-color hormones" of *Drosophila melanogaster. Q. Rev. Biol.,* 17:327–338.

Eppig, J. J., 1985. Oocyte-somatic cell interaction during oocyte growth and maturation in the mammal. In L. W. Browder, ed., *Developmental biology,* Vol. 1, *Oogenesis.* Plenum Press, New York, pp. 313–347.

Erickson, H. P. and H. C. Taylor, 1987. Hexabrachion proteins in embryonic chicken tissues and human tumors. *J. Cell Biol.,* 105:1387–1394.

Ettensohn, C. A. and D. R. McClay, 1988. Cell lineage conversion in the sea urchin embryo. *Dev. Biol.,* 125:396–409.

Evans, H. J., 1972. Properties of human X and Y sperm. In R. A. Beatty and S. Gluecksohn-Waelsch, eds., *Proceedings of the International Symposium on the Genetics of the Spermatozoon.* Department of Genetics, University of

Edinburgh, Edinburgh, and Albert Einstein College of Medicine, New York, pp. 144–159.

Evans, R. M. and S. M. Hollenberg, 1988. Zinc fingers: Gilt by association. *Cell*, 52:1–3.

Eyal-Giladi, H. and B. Fabian, 1980. Axis determination in uterine chick blastoderms under changing positions during the sensitive period for polarity. *Dev. Biol.*, 77:228–232.

Eyal-Giladi, H., M. Ginsburg, and A. Forbarox, 1981. Avian primordial germ cells are of epiblastic origin. *J. Embryol. Exp. Morphol.*, 65:139–147.

Fawcett, D. W., 1970. A comparative view of sperm ultrastructure. *Biol. Reprod. Suppl.*, 2:90–127.

Fawcett, D. W., 1972. Observations on cell differentiation and organelle continuity in spermatogenesis. In R. A. Beatty and S. Gluecksohn-Waelsch, eds., *Edinburgh symposium on the genetics of the spermatozoon, 1971*. Department of Genetics, University of Edinburgh, Edinburgh, pp. 37–68.

Fawcett, D. W., 1975. Review article: The mammalian spermatozoon. *Dev. Biol.*, 44:394–436.

Feldherr, C., 1965. The effect of the electron-opaque pore material on exchanges through the nuclear annuli. *J. Cell Biol.*, 25:43–53.

Fernández, J. and N. Olea, 1982. Embryonic development of Glossiphoniid leeches. In F. W. Harrison and R. R. Cowden, eds., *Developmental biology of freshwater invertebrates*. Alan R. Liss, New York, pp. 317–361.

Finney, M., G. Ruvkun, and H. R. Horvitz, 1988. The *C. elegans* cell lineage and differentiation gene *unc-86* encodes a protein with a homeodomain and extended similarity to transcription factors. *Cell*, 55:757–769.

Fischberg, M. and A. W. Blackler, 1963. Nuclear changes during the differentiation of animal cells. In *Cell differentiation*, Symposia of the Society for Experimental Biology, Vol. 17. Academic Press, Orlando, pp. 138–156.

Fischer, E. A. and C. W. Petersen, 1987. The evolution of sexual patterns in the seabasses. *BioScience*, 37:482–489.

Fishel, S., 1986. IVF—Historical perspective. In S. Fishel and E. M. Symonds, eds., *In vitro fertilisation: Past, present and future*. IRL Press, Oxford, pp. 1–16.

Fleming. T. P. and M. H. Johnson, 1988. From egg to epithelium. *Annu. Rev. Cell Biol.*, 4:459–485.

Flickinger, R. A., 1976. Replication-dependent transcription in eukaryotes. *Int. J. Biochem.*, 7:85–93.

Florman, H. M. and P. M. Wassarman, 1985. O-linked oligosaccharides of mouse egg ZP3 account for its sperm receptor activity. *Cell*, 41:313–324.

Franchi, L. L. and T. G. Baker, 1973. Oogenesis and follicular growth. In E. S. E Hafez and T. N. Evans, eds., *Human reproduction, conception and contraception*. Harper & Row, New York, pp. 53–83.

Fraser, R. C., 1954. Studies on the hypoblast of the young chick embryo. *J. Exp. Zool.*, 126:349–399.

Freeman, G., 1979. The multiple roles which cell division can play in the localization of developmental potential. In S. Subtelny, ed., *Determinants of spatial organization*, 37th Symposium of the Society for Developmental Biology. Academic Press, Orlando, pp. 53–76.

Freeman, G., 1980. The role of cleavage in the establishment of the anterior–posterior axis of the hydrozoan embryo. In P. Tardent and R. Tardent, eds., *Developmental and cellular biology of coelenterates*. Elsevier/North-Holland Biomedical Press, Amsterdam, pp. 97–108.

Freeman, G. and J. W. Lundelius, 1982. The developmental genetics of dextrality and sinistrality in the gastropod *Lymnaea peregra*. *Wilhelm Roux' Arch. Dev. Biol.*, 191:69–83.

French, V., P. J. Bryant, and S. V. Bryant, 1976. Pattern regulation in epimorphic fields. *Science*, 193:969–981.

Fujita, S., 1986. Transitory differentiation of matrix cells and its functional role in the morphogenesis of the developing vertebrate CNS. In A. A. Moscona and A. Monroy, eds., *Current topics in developmental biology*, Vol. 20. Academic Press, Orlando, pp. 223–241.

Fullilove, S. L. and A. G. Jacobson, 1971. Nuclear elongation and cytokinesis in *Drosophila montana*. *Dev. Biol.*, 26:560–577.

Fulton, A., 1984. *The cytoskeleton*: cellular architecture and choreography. Chapman and Hall (Methuen), New York.

Gachelin, G., R. Kemler, F. Kelly, and F. Jacob, 1977. PCC4, a new cell surface antigen common to multipotential embryonal carcinoma cells, spermatozoa, and mouse early embryos. *Dev. Biol.*, 57:199–209.

Gaddum-Rosse, P., 1985. Mammalian gamete interactions: What can be gained from observations on living eggs? *Am. J. Anat.*, 174:347–356.

Gall, J. G., 1969. The genes for ribosomal RNA during oogenesis. *Genetics, Suppl.*, 61:121–132.

Gall, J. G., M. O. Diaz, E. C. Stephenson, and K. A. Mahon, 1983. The transcription unit of lampbrush chromosomes. In S. Subtelny, ed., *Gene structure and regulation in development*. Alan R. Liss, New York, pp. 137–146.

Gallera, J., 1971. Primary induction in birds. *Adv. Morphogen.*, 9:149–180.

Gans, M., C. Audit, and M. Masson, 1975. Isolation and characterization of sex-linked female sterile mutants in *Drosophila melanogaster*. *Genetics*, 81:683–704.

Garcia, R. B., S. Pereyra-Alfonso, and J. R. Sotelo, 1979. Protein-synthesizing machinery in the growing oocyte of the cyclic mouse. *Differentiation*, 14:101–106.

Garcia-Bellido, A., 1975. Genetic control of wing disc development in *Drosophila*. In *Cell patterning, Ciba Foundation Symposium No. 29*. Elsevier/Excerpta Medica/North-Holland, Amsterdam, pp. 161–182.

Garcia-Bellido, A. and J. R. Merriam, 1969. Cell lineage of the imaginal discs in *Drosophila* gynandromorphs. *J. Exp. Zool.*, 170:61–75.

Gardner, E. J., 1965. *History of biology*, 2nd ed. Burgess, Minneapolis.

Gardner, R. L., 1975. Analysis of determination and differentiation in the early mammalian embryo using intra- and interspecific chimeras. In C. L. Markert and J. Papaconstantinou, eds., *The developmental biology of reproduction*, 33rd Symposium of the Society for Developmental Biology. Academic Press, Orlando, pp. 207–236.

Gardner, R. L., 1978. Developmental potency of normal and neoplastic cells of the early mouse embryo. In J. W. Littlefield and J. Grouchy, eds., *Birth defects*, Excerpta Medica International Congress Series No. 432. Excerpta Medica, Amsterdam.

Gardner, R. L., 1982. Investigation of cell lineage and differentiation in the extraembryonic endoderm of the mouse embryo. *J. Embryol. Exp. Morphol.*, 68:175–198.

Gardner, R. L. and J. Rossant, 1979. Investigation of the fate of 4.5 day *post-coitum* mouse inner cell mass cells by blastocyst injection. *J. Embryol. Exp. Morphol.*, 52:141–152.

Gardner, R. L., V. E. Papaioannou, and S. C. Barton, 1973. Origin of the ectoplacental cone and secondary giant cells in mouse blastocysts reconstituted from isolated trophoblast and inner cell mass. *J. Embryol. Exp. Morphol.*, 30:561–572.

Gartler, S. M. and A. D. Riggs, 1983. Mammalian X-chromosome inactivation. *Am. Rev. Genet.*, 17:155–190.

Gaunt, S. J., 1983. Spreading of a sperm surface antigen within the plasma membrane of the egg after fertilization in the rat. *J. Embryol. Exp. Morphol.*, 75:259–270.

Gautier, J., C. Norbury, M. Lohka, P. Nurse, and J. Maller, 1988. Purified maturation-promoting factor contains the product of a *Xenopus* homolog of the fission yeast cell cycle control gene *cdc2*⁺. *Cell*, 54:433–439.

Gehring, W. J., 1973. Genetic control of determination in the *Drosophila* embryo. In F. R. Ruddle, ed., *Genetic mechanisms of development*, 31st Symposium of the Society for Developmental Biology. Academic Press, Orlando, pp. 103–128.

Gehring, W. J., 1984. Homeotic genes and the control of cell determination. In E. H. Davidson and R. A. Firtel, eds., *Molecular biology of development*. Alan R. Liss, New York, pp. 3–22.

Gehring, W. J., 1987. Homeo boxes in the study of development. *Science*, 236:1245–1252.

Gehring, W. J. and Y. Hiromi, 1986. Homeotic genes and the homeobox. *Annu. Rev. Genet.*, 20:147–173.

Geiger, B., Z. Avnur, T. Volberg, and T. Volk, 1985. Molecular domains of adherens junctions. In G. M. Edelman and J.-P. Thiery, eds., *Cell in contact: Adhesions and junctions as morphogenetic determinants*. Wiley, New York, pp. 461–489.

Geiger, B., T. Volk, T. Volberg, and R. Bendori, 1987. Molecular interactions in adherens-type contacts. *J. Cell Sci. Suppl.*, 8:251–272.

Gerace, L., 1988. Functional organization of the nuclear envelope. *Annu. Rev. Cell Biol.*, 4:335–374.

Gerhart, J. C., 1980. Mechanisms regulating pattern formation in the amphibian egg and early embryo. In R. F. Goldberger, ed., *Biological regulation and development*, Vol. 2, *Molecular organization and cell function*. Plenum Press, New York.

Gerhart, J. C., 1987. Determinants of early amphibian development. *Am. Zool.*, 27:593–605.

Gerhart, J. and R. Keller, 1986. Region-specific cell activities in amphibian gastrulation. *Annu. Rev. Cell Biol.*, 2:201–229.

Gerhart, J., G. Ubbels, S. Black, K. Hara, and M. Kirschner, 1981. A reinvestigation of the role of the grey crescent in axis formation in *Xenopus laevis*. *Nature (London)*, 292:511–516.

Gerhart, J., S. Black, R. Gimlich, and S. Scharf, 1983. Control of polarity in the amphibian egg. In W. R. Jeffery and R. A. Raff, eds., *Time, space, and pattern in embryonic development*. Alan R. Liss, New York, pp. 261–286.

Gibson, G. and W. J. Gehring, 1988. Head and thoracic transformations caused by ectopic expression of *Antennapedia* during *Drosophila* development. *Development*, 102:657–675.

Gilbert, S. F., 1985. Genes-in-pieces revisited. *Science*, 228:823–824.

Gilbert, S. F., 1988. Cellular politics: Ernest Everett Just, Richard B. Goldschmidt, and the attempt to reconcile embryology and genetics. In R. Rainger, K. R. Benson, and J. Maien-schein, eds., *The American development of biology*. University of Pennsylvania Press, Philadelphia, pp. 311–346.

Gilkey, J. C., L. F. Jaffe, E. B. Ridgway, and G. T. Reynolds, 1978. A free calcium wave traverses the activating egg of the medaka, *Oryzias latipes*. *J. Cell Biol.*, 76:448–466.

Gilula, N. B., 1985. Gap junctional contact between cells. In G. M. Edelman and J.-P. Thiery, eds., *Cell in contact: Adhesions and junctions as morphogenetic determinants*. A Neurosciences Institute Publication, Wiley, New York, pp. 395–409.

Gimlich, R. L., 1985. Cytoplasmic localization and chorda-mesoderm induction in the frog embryo. *J. Embryol. Exp. Morphol. Suppl.*, 89:89–111.

Giudice, G., 1973. *Developmental biology of the sea urchin embryo*. Academic Press, Orlando.

Giudice, G., 1986. *The sea urchin embryo: A developmental biological system*. Springer-Verlag, New York.

Godsave, S. F., H. V. Isaacs, and J. M. W. Slack, 1988. Mesoderm-inducing factors: A small class of molecules. *Development*, 102:555–566.

Golden, L., U. Schafer, and M. Rosbash, 1980. Accumulation of individual poly(A) plus RNAs during oogenesis of *Xenopus laevis*. *Cell*, 22:835–844.

Goldman, M. A., G. P. Holmquist, N. C. Gray, L. A. Caston, A. Nag, 1984. Replication timing of genes and middle repetitive sequences. *Science*, 224:686–692.

Goldschmidt, R. B., 1955. *Theoretical genetics*. University of California Press, Berkeley.

Goldschmidt, R. B., 1956. *Portraits from Memory*. University of Washington Press, Seattle.

Gondos, B., 1978. Oogonia and oocytes in mammals. In R. E. Jones, ed., *The vertebrate ovary: Comparative biology and evolution*. Plenum, New York, pp. 83–120.

Goodrich, E. W., 1906. On the development of the fins of fish. *Q. J. Microsc. Sci.*, 50:333–376.

Gordon, R., 1985. A review of the theories of vertebrate neurulation and their relationship to the mechanics of neural tube birth defects. *J. Embryol. Exp. Morphol. Suppl.*, 89:229–255.

Goss, R. J., 1969. *Principles of regeneration*. Academic Press, Orlando.

Goss, R. J., 1978. *Physiology of growth*. Academic Press, Orlando.

Gossler, A., A. L. Joyner, J. Rossant, W. C. Skarnes, 1989. Mouse embryonic stem cells and reporter constructs to detect developmentally regulated genes. *Science*, 244:463–465.

Gottlieb, G., 1976. Conceptions of prenatal development: Behavioral embryology. *Psychol. Rev.*, 83:215–234.

Gould, M., J. L. Stephano, and L. Z. Holland, 1986. Isolation of protein from urechis sperm acrosomal granules that binds sperm to egg and initiates development. *Dev. Biol.*, 117:306–318.

Gould, S. J., 1977. *Ontogeny and phylogeny*. Belknap Press of the Harvard University Press, Cambridge.

Gould, S. J., 1985. Geoffrey and the homeobox. *Natural History*, 94:12–23.

Goy, R. and B. McEwen, 1980. *Sexual differentiation of the brain*. MIT Press, Cambridge.

Graham, C. F., 1966. The regulation of DNA synthesis and mitosis in multinucleate frog eggs. *J. Cell Sci.*, 1:363–374.

Graham, C. F., 1971. Virus assisted fusion of embryonic cells. In E. Diczfalusy, ed., *In vitro methods in reproductive biology*, Karolinska Symposium on Research Biology, Vol. 3. Stockholm, pp. 154–167.

Graham, C. F., 1977. Teratocarcinoma cells and normal mouse embryogenesis. In M. J. Sherman, ed., *Concepts in mammalian embryogenesis*. MIT Press, Cambridge, pp. 315–394.

Grant, P., 1953. Phosphate metabolism during oogenesis in *Rana temporaria. J. Exp. Zool.*, 124:513–544.

Gray, P., 1970. *The encyclopedia of the biological sciences*, 2nd ed. Van Nostrand Reinhold, New York.

Grell, R. F., 1976. Distributive pairing. In M. Ashburner and E. Novitski, eds., *Genetics and biology of Drosophila*, Vol. 1a. Academic Press, London, pp. 435–486.

Grey, R. D., D. P. Wolf, and J. L. Hedrick, 1974. Formation and structure of the fertilization envelope in *Xenopus laevis. Dev. Biol.*, 36:44–71.

Grey, R. D., P. K. Working, and J. L. Hedrick, 1976. Evidence that the fertilization envelope blocks sperm entry in eggs of *Xenopus laevis:* Interaction of sperm with isolated envelope. *Dev. Biol.*, 54:52–60.

Grobstein, C., 1956. Trans-filter induction of tubulin in mouse meta-morphogenic mesenchyme. *Exp. Cell Res.*, 10:424–440.

Gross, K. W., J. Jacobs-Lorena, C. Baglioni, and P. R. Gross, 1973. Cell-free translation of maternal messenger RNA from sea urchin eggs. *Proc. Natl. Acad. Sci. USA*, 70:2614–2618.

Gross, P. R., L. I. Malkin, and W. A. Moyn, 1964. Template for the first proteins of embryonic development. *Proc. Natl. Acad. Sci. USA*, 61:414–447.

Groudine, M. and K. F. Conklin, 1985. Chromatin structure and de novo methylation of sperm DNA: Implications for activation of the paternal genome. *Science*, 228:1061–1068.

Gulyas, B. J., 1971. A reexamination of cleavage patterns in eutherian mammalian eggs: Rotation of blastomere pairs during second cleavage in the rabbit. *J. Exp. Zool.*, 193:235–248.

Gumpel-Pinot, M., 1984. Muscle and skeleton of limbs and body wall. In N. Le Douarin and A. McLaren, eds., *Chimeras in developmental biology*. Academic Press, London, pp. 281–310.

Guraya, S. S., 1978. Maturation of the follicular wall of non-mammalian vertebrates. In R. E. Jones, ed., *The vertebrate ovary: Comparative biology and evolution*. Plenum Press, New York, pp. 261–329.

Guraya, S. S., 1985. *Biology of ovarian follicles in mammals*. Springer-Verlag, Berlin.

Gurdon, J. B., 1963. Nuclear transplantation in amphibia and the importance of stable nuclear changes in promoting cellular differentiation. *Q. Rev. Biol.*, 38:54–78.

Gurdon, J. B., 1982. Amphibian oocytes and gene control in development. In M. M. Burger and R. Weber, eds., *Embryonic development, Part A, Genetic aspects*. Alan R. Liss, New York, pp. 211–225.

Gurdon, J. B., 1987. Embryonic induction—molecular prospects. *Development*, 99:285–306.

Gurdon, J. B. and D. D. Brown, 1965. Cytoplasmic regulation of RNA synthesis and nucleolus formation in developing embryos of *Xenopus laevis. J. Mol. Biol.*, 12:27–35.

Gurdon, J. B. and D. A. Melton, 1981. Gene transfer in amphibian eggs and oocytes. *Annu. Rev. Genet.*, 15:189–218.

Gurdon, J. B., R. A. Laskey, and O. R. Reeves, 1975. The developmental capacity of nuclei transplanted from keratinized skin cells of adult frogs. *J. Embryol. Exp. Morphol.*, 34:93–112.

Gurdon, J. B., E. M. DeRobertis, and G. Partington, 1976. Injected nuclei in frog oocytes provide a living cell system for the study of transcriptional control. *Nature (London)*, 260:116–120.

Gurdon, J. B., S. Brennan, S. Fairman, N. Dathan, and T. J. Mohun, 1984. The activation of actin genes in early *Xenopus* development. In E. H. Davidson and R. A. Firtel, eds., *Molecular biology of development*. Alan R. Liss, New York, pp. 109–118.

Gustafson, T., 1965. Morphogenetic significance of biochemical patterns in sea urchin embryos. In R. Weber, ed., *The biochemistry of animal development*, Vol. 1, *Descriptive biochemistry of animal development*. Academic Press, Orlando, pp. 139–202.

Gustafson, T., 1975. Cellular behavior and cytochemistry in early stages of development. In G. Czihak, ed., *The sea urchin embryo: Biochemistry and morphogenesis*. Springer-Verlag, Berlin, pp. 233–266.

Gustafson, T. and L. Wolpert, 1967. Cellular movement and contact in sea urchin morphogenesis. *Biol. Rev.*, 42:442–498.

Gwatkin, R. B. L., 1977. *Fertilization mechanisms in man and mammals*. Plenum Press, New York.

Hadorn, E., 1966. Dynamics of determination. In M. Locke, ed., *Major problems in developmental biology*, 25th Symposium of the Society for Developmental Biology. Academic Press, Orlando, pp. 85–104.

Hadzi, J., 1963. *Evolution of the Metazoa*. Macmillan, New York.

Haeckel, E., 1874. Die gastraea-theorie, die phylogenetische Klassification des Tierreiches und homologie der Keimblätter. *Z. Naturwiss.(Jena)*, 8:1–55.

Haeckel, E., 1892. *The history of creation*, 2 vols., tr. E. R. Lankester from 8th ed. of *Natürliche Schöpfungsgeschichte*, Kegan Paul, Trench, Trubner & Co., London.

Hamburger, V., 1985. Hans Spemann, Nobel Laureate 1935. *Trends Neurosci.*, 1:385–387.

Hamburger, V., 1988. *The heritage of experimental embryology: Hans Spemann and the organizer*. Oxford University Press, New York.

Hamburger, V. and H. L. Hamilton, 1951. A series of normal stages in the development of the chick embryo. *J. Morphol.*, 88:49–92.

Hamilton, W. J. and H. W. Mossman, 1976. *Human embryology, prenatal development of form and function*, 4th ed. Macmillan, London.

Hampé, A., 1959. Contribution a l'étude du dévelopment et de la régulation des déficiences et des excédents dans la patte de l'embryon de Poulet. *Arch. Anat. Microsc. Morphol. Exp.*, 48:345–478.

Hanawalt, P. C., 1982. Processing of damaged DNA in mammalian cells. In R. T. Schimke, ed., *Gene amplification*. Cold Spring Harbor Laboratory, Cold Spring Harbor, NY, pp. 257–262.

Hannun, Y. A. and R. M. Bell, 1989. Functions of sphingolipids and sphingolipid breakdown products in cellular regulation. *Science*, 243:500–507.

Hara, K., 1977. The cleavage pattern of the axolotl egg studied by cinematography and cell counting. *Wilhelm Roux' Arch. Dev. Biol.,* 181:73–87.

Hara, K., P. Tydeman, and M. Kirschner, 1980. A cytoplasmic clock with the same period as the division cycle in *Xenopus* eggs. *Proc. Natl. Acad. Sci. USA,* 77:462–466.

Hardisty, M. W., 1978. Primordial germ cells and the vertebrate germ line. In R. E. Jones, ed., *Vertebrate ovary: Comparative biology and evolution.* Plenum Press, New York, pp. 1–45.

Hardy, D. E., 1983. The fruit flies of the tribe Euphrantini of Indonesia, New Guinea, and adjacent islands (Tephritidae Diptera). *Int. J. Entomol.,* 25:152–205.

Hargitt, C. W., 1911. Some problems of coelenterate ontogeny. *J. Morphol.,* 22:493–549.

Harper, M. J. K., 1982. Sperm and egg transport. In C. R. Austin and R. V. Short, eds., *Reproduction in mammals, Book 1, Germ cells and fertilization,* 2nd ed. Cambridge University Press, Cambridge, pp. 102–127.

Harrelson, A. L. and C. S. Goodman, 1988. Growth cone guidance in insects: Fasciclin II is a member of the immunoglobulin superfamily. *Science,* 242:700–708.

Harrington, R. E. and D. P. Easton, 1982. A putative precursor to the major yolk protein of the sea urchin. *Dev. Biol.,* 94:505–508.

Harris, A. K., 1982. Cell migration and its directional guidance. In K. M. Yamada, ed., *Cell interactions and development, molecular mechanisms.* Wiley, New York, pp. 123–151.

Harrison, R. G., 1907. Observations on the living developing nerve fiber. *Anat. Rec.,* 1:116–118. Reprinted in B. H. Willier and J. M. Oppenheimer, eds. *Foundations of experimental embryology.* Prentice-Hall, Englewood Cliffs, NJ, pp. 100–103.

Harrison, R. G., S. Wilens, ed., 1969. *Organization and development of the embryo.* Yale University Press, New Haven.

Harrison, R. G., D. M. Rand, and W. C. Wheeler, 1985. Mitochondrial DNA size variation within individual crickets. *Science,* 228:1446–1448.

Hartman, W. D. and H. M. Reiswig, 1971. In R. S. Boardman, A. H. Cheetham, and W. A. Oliver, Jr., eds., *Animal colonies: Development and function through time.* Dowden, Hutchinson, and Ross, Stroudsburg, PA, pp. 567–584.

Harvey, E. B., 1940. A comparison of the development of nucleate and non-nucleate eggs of *Arbacia punctulata. Biol. Bull.,* 79:166–187.

Harvey, W., 1651. *Generatione animalium: Anatomical exercieses on the generation of animals.* Translated by R. Willis. Reprinted 1952, Encyclopedia Brittanica, Inc., Great Books of the Western World, Chicago.

Harvey, R. P. and D. A. Melton, 1988. Microinjection of synthetic Shos-1A homeobox mRNA disrupts somite formation in developing *Xenopus* embryos. *Cell,* 53:687–697.

Hauswirth, W. W. and P. J. Laipis, 1982. Mitochondrial DNA polymorphism in a maternal lineage of Holstein cows. *Proc. Natl. Acad. Sci. USA,* 79:4686–4690.

Heap, R. B. and A. P. F. Flint, 1984. Pregnancy. In C. R. Austin and R. V. Short, eds., *Reproduction in mammals, Book 3, Hormonal control of reproduction,* 2nd ed. Cambridge University Press, Cambridge, pp. 153–194.

Heap, R. B. and D. V. Illingworth, 1977. The mechanisms of action of estrogens and progesterone. In S. Zuckerman and B. J. Weir, eds., *Ovary,* 2nd ed., Vol. 3. Academic Press, Orlando, pp. 59–150.

Hedrich, P. W., G. Thomson, and W. Klitz, 1986. Evolutionary genetics: HLA as an exemplary system. In S. Karlin and E. Nevo, eds., *Evolutionary processes and theory.* Academic Press, Orlando, pp. 583–606.

Helfand, M. S., 1977. T. H. Huxley's "Evolution and ethics": The politics of evolution and the evolution of politics. *Victorian Stud.,* 20:159–177.

Herbert, J., 1977. External factors and ovarian activity in mammals. In S. Zuckerman and B. J. Weir, eds., *Ovary,* 2nd ed., Vol. 2. Academic Press, Orlando, pp. 458–505.

Herbert, M. C. and C. R. Graham, 1974. Cell determination and biochemical differentiation of the early mammalian embryo. In A. A. Moscona and A. Monroy, eds., *Current topics in developmental biology,* Vol. 8. Academic Press, Orlando, pp. 151–178.

Herr, W., R. A. Sturm, R. G. Clerc, L. M. Corcoran, D. Baltimore, P. A. Sharp, H. A. Inhraham, M. G. Rosenfeld, M. Finney, G. Ruvkun, and H. R. Horvitz, 1988. The POU domain: A large conserved region in the mammalian *pit-1, oct-1, oct-2,* and *Caenorhabditis elegans unc-86* gene products. *Genes & Dev.,* 2:1513–1516.

Heuser, C. H. and G. L. Streeter, 1928. Early stages in the development of pig embryos from the period of initial cleavage to the time of the appearance of limb-buds. Carnegie Institution of Washington, Pub. No. 109. *Contrib. Embryol.,* 20:1–30.

Hinchliffe, J. R. and D. R. Johnson, 1980. *Development of the vertebrate limb: An approach through experiment, genetics, and evolution.* Clarendon Press, Oxford.

Hiromi, Y. and W. J. Gehring, 1987. Regulation and function of the *Drosophila* segmentation gene *fushi tarazu. Cell,* 50:963–974.

Hirose, G. and M. Jacobson, 1979. Clonal organization of the central nervous system of the frog. I. Clones stemming from individual blastomeres of the 16-cell and earlier stages. *Dev. Biol.,* 71:191–202.

Hirsh, D., D. Oppenheim, and M. Klass, 1976. Temperature-sensitive developmental mutants of *Caenorhabditis elegans. Dev. Biol.,* 49:220–235.

His, W., 1874. *Unsere Körperform und das physiologische Problem ihrer Entstehung.* Vogel, Leipzig. Quoted from R. W. Oppenheim, 1983. Preformation and epigenesis in the origins of the nervous system and behavior: Issues, concepts, and their history. In P. P. G. Bateson and P. H. Klopfer, eds., *Perspectives in ethology,* Vol. 5, *Ontogeny.* Plenum Press, New York, pp. 1–100.

Hoekstra, D., K. Klappe, T. Stegmann, and S. Nir, 1987. Parameters affecting the fusion of viruses with artificial and biological membranes. In S. Ohki, D. Doyle, T. D. Flanagan, S. W. Hui, and E. Mayhew, eds., *Molecular mechanisms of membrane fusion.* Plenum Press, New York, pp. 399–412.

Hoffner, N. J. and M. A. DiBerardino, 1980. Developmental potential of somatic nuclei transplanted into meiotic oocytes of *Rana pipiens. Science,* 209:517–519.

Hofstetter, H., A. Kressmann, and M. L. Birnstiel, 1981. A split promoter for a eucaryotic tRNA gene. *Cell,* 24:573–585.

Hogue, M., 1910. Über die Wirkung der Zentrifugalkraft auf die Eier von Ascaris megalocephala. *Arch. Entwicklungsmech. Org.,* 29:109–145.

Holder, N., 1984. Regeneration of the axolotl limb: Patterns and polar coordinates. In G. M. Malacinski, ed., *Pattern formation.* Macmillan, New York, pp. 521–537.

Holland, P. W. H. and B. L. M. Hogan, 1986. Phylogenetic distribution of *Antennapedia*-like homeoboxes. *Nature (London)*, 321:251–253.

Holtfreter, J., 1939. Tissue affinity, a means of embryonic morphogenesis. English translation by Konrad Keck, amended by Professor Holtfreter. Reprinted in B. H. Willier and J. M. Oppenheimer, eds., *Foundations of experimental embryology*. Prentice-Hall, Englewood Cliffs, NJ, pp. 186–225.

Holtfreter, J., 1943. Experimental studies on the development of the pronephros. *Rev. Can. Biol.*, 3:220–250.

Holtfreter, J., 1944. A study of the mechanics of gastrulation. II. *J. Exp. Zool.*, 95:171–212.

Holtfreter, J. and V. Hamburger, 1955. Amphibians. In B. H. Willier, P. A. Weiss, and V. Hamburger, eds., *Analysis of development*. Saunders, Philadelphia, pp. 230–296.

Honda, B. M. and R. G. Roeder, 1980. Association of a 5S gene transcription factor with 5S RNA and altered levels of the factor during cell differentiation. *Cell*, 22:119–126.

Hooker, D., 1952. *The prenatal origin of behavior*, Porter Lectures, Series 18. University of Kansas Press, Lawrence.

Hopkins, W. G. and M. C. Brown, 1984. *Development of nerve cells and their connections*. Cambridge University Press, Cambridge.

Hoppe, P. C. and K. Illmensee, 1977. Microsurgically produced homozygous-diploid uniparental mice. *Proc. Natl. Acad. Sci. USA*, 74:5657–5661.

Hopper, A. F. and N. H. Hart, 1985. *Foundations of animal development*, 2nd ed. Oxford University Press, New York.

Horder, T. J. and P. J. Weindling, 1985. Hans Spemann and the organiser. In T. J. Horder et al., eds., *A history of embryology*, 8th Symposium of the British Society for Developmental Biology. Cambridge University Press, Cambridge, pp. 183–242.

Hörstadius, S., 1939. The mechanics of sea urchin development, studied by operative methods. *Biol. Rev.*, 14:132–179.

Hörstadius, S., 1973. *Experimental embryology of echinoderms*. Clarendon Press, Oxford.

Horvitz, H. R., P. W. Steinberg, I. S. Greenwald, W. Fixsen, and H. M. Ellis, 1983. Mutations that affect neural cell lineages and cell fates during the development of the nematode *Caenorhabditis elegans*. *Cold Spring Harbor Symp. Quant. Biol.*, 48:453–463.

Hotter, Y. and S. Benzer, 1973. Mapping of behavior in *Drosophila* mosaics. In F. R. Ruddle, ed., *Genetic mechanisms of development*, 31st Symposium of the Society for Developmental Biology. Academic Press, Orlando, pp. 129–167.

Hough-Evans, B. R., S. G. Ernst, R. J. Britten, and E. H. Davidson, 1979. RNA complexity in developing sea urchin oocytes. *Dev. Biol.*, 69:258–269.

Howard, A. and S. R. Pelc, 1951. Nuclear incorporation of P^{32} as demonstrated by autoradiographs. *Exp. Cell Res.*, 2:178–187.

Howlett, S. K., 1986. A set of proteins showing cell cycle dependent modification in the early mouse embryo. *Cell*, 45:387–396.

Hozumi, N. and S. Tonegawa, 1976. Evidence for somatic rearrangement of immunoglobulin genes coding for variable and constant regions. *Proc. Natl. Acad. Sci. USA*, 73:3628–3632.

Hsu, Yu-C., 1981. Time-lapse cinematography of mouse embryo development from blastocysts to early somite stage. In S. R. Glasser and D. W. Bullock, eds., *Cellular and mo-

lecular aspects of implantation*. Plenum Press, New York, pp. 383–392.

Huang, T. T. F. Jr. and R. Yanagimachi, 1985. Inner acrosomal membrane of mammalian spermatozoa: Its properties and possible functions in fertilization. *Am. J. Anat.*, 174:249–268.

Hubbard, R., 1982. The theory and practice of genetic reductionism—from Mendel's laws to genetic engineering. In S. Rose, ed., *Towards a liberatory biology*. Allison and Busby, London, pp. 62–78.

Hubert, J., 1962. Étude histologique des jeunes stades du développement embryonnaire du lézard vivipare *(Lacerta vivipara jacquin.)* Arch. Anat. Microsc. Morphol. Exp., 55:11–26.

Hugo, I., 1966. *Life of Mendel*. Hafner, New York.

Humphreys, T., 1963. Chemical dissolution and *in vitro* reconstruction of sponge cell adhesions. 1. Isolation and functional demonstration of the components involved. *Dev. Biol.*, 8:27–47.

Hunter, R. H. F., B. Cook, and T. G. Baker, 1976. Dissociation of response to injected gonadotropin between the Graafian follicle and oocyte in pigs. *Nature (London)*, 260:156–158.

Hutchinson, J. S. M. and P. J. Sharp, 1977. Hypothalamus—pituitary control of the ovary. In S. Zuckerman and B. J. Weir, eds., *Ovary*, Vol. III, *Regulation of oogenesis and steroidogenesis*, 2nd ed. Academic Press, Orlando, pp. 237–303.

Hyman, L. H., 1940. *Invertebrates*, Vol. 1, *Protozoa through ctenophora*. McGraw-Hill, New York.

Illmensee, K. and A. P. Mahowald, 1974. Transplantation of posterior polar plasm in *Drosophila*. Induction of germ cells at the anterior pole of the egg. *Proc. Natl. Acad. Sci. USA*, 71:1016–1020.

Imoh, H., 1985. Formation of the neural plate and the mesoderm in normally developing embryos of *Xenopus laevis*. *Dev. Growth Differ.*, 27:1–11.

Ingraham, H. A., R. Chen, H. J. Mangalam, H. P. Elsholtz, S. E. Flynn, C. R. Lin, D. M. Simmons, L. Swanson, and M. G. Rosenfeld, 1988. A tissue-specific transcription factor containing a homeodomain specifies a pituitary phenotype. *Cell*, 55:519–529.

Inoué, S., 1981. Video image processing greatly enhances contrast, quality, and speed in polarization-based microscopy. *J. Cell Biol.*, 89:346–356.

Ivanova-Kasas, O. M., 1972. Polyembryony in insects. In S. J. Counce and C. H. Waddington, eds., *Developmental systems: Insects*, Vol. 1. Academic Press, London, pp. 243–271.

Jacob, F., 1976. *Logic of life: A history of heredity* (translated by B. E. Spillman). Vintage Books, New York.

Jacob, F. 1979. Cell surface and early stages of mouse embryogenesis. In A. A. Moscona and A. Monroy, eds., *Current topics in developmental biology*, Vol. 13. Academic Press, Orlando, pp. 117–137.

Jacob, F. and J. Monod, 1961. On the regulation of gene activity. *Cold Spring Harbor Symp. Quant. Biol.*, 26:193–211.

Jacob, F. and J. Monod, 1963. Genetic repression, allosteric inhibition and cellular differentiation. In M. Locke, ed., *Cytodifferential and macromolecular synthesis*, 21st Symposium of the Society for the Study of Development and Growth. Academic Press, Orlando, pp. 30–64.

Jacob, H. J., M. Jacob, and B. Christ, 1986. The early development of the intermediate mesoderm in the chick. In R. Bellairs, D. A. Ede, and J. W. Lash, eds., *Somites in developing embryos*. Plenum Press, New York, pp. 61–68.

Jacobson, A. G., 1984. Further evidence that formation of the neural tube requires elongation of the nervous system. *J. Exp. Zool.*, 230:23–28.

Jacobson, A. G., 1985. Adhesion and movement of cells may be coupled to produce neurulation. In G. M. Edelman and J.-P. Thiery, eds., *Cell in contact: Adhesions and junction as morphogenetic determinants*. A Neurosciences Institute Publication, Wiley, New York, pp. 49–65.

Jacobson, A. G. and R. Gordon, 1976. Changes in the shape of the developing vertebrate nervous system analyzed experimentally, mathematically and by computer simulation. *J. Exp. Zool.*, 197:191–246.

Jacobson, A. G. and S. Meier, 1986. Somitomeres: The primordial body segments. In R. Bellairs, D. A. Ede, and J. W. Lash, eds., *Somites in developing embryos*. Plenum Press, New York.

Jacobson, A. G., G. M. Odell, and G. F. Oster, 1985. The cortical tractor model for epithelial folding: Application to the neural plate. In x.xxxx, ed., *Molecular determinants of animal form*. Alan R. Liss, New York, pp. 143–166.

Jacobson, M. 1982. Origins of the central nervous system in amphibians. In N. C. Spitzer, ed., *Neuronal development*. Plenum Press, New York, pp. 45–99.

Jacobson, M., 1983. Clonal organization of the central nervous system of the frog. III. Clones stemming from individual blastomeres of the 128-, 256-, and 512-cell stages. *J. Neurosci.*, 3:1019–1038.

Jacobson, M. and G. Hirose, 1981. Clonal organization of the central nervous system of the frog. II. Clones stemming from individual blastomeres of the 32- and 64-cell stages. *J. Neurosci.*, 1:271–284.

Jacobson, M. and S. L. Klein, 1985. Analysis of clonal restriction of cell mingling in *Xenopus*. *Philos. Trans. R. Soc. London, Ser. B*, 312:57–65.

Jaffe, L. A. 1976. Fast block to polyspermy in sea urchin eggs ie electrically mediated. *Nature (London)*, 261:68–71.

Jaffe, L. F., 1980. Calcium explosions as triggers of development. *New York Acad. Sci.*, 339:86–101.

Jaffe, L. F., 1985. The role of calcium explosions, waves, and pulses in activating eggs. In C. B. Metz and A. Monroy, eds., *Biology of fertilization*, Vol. 3, *Fertilization response of the egg*. Academic Press, Orlando, pp. 127–165.

Jaffe, L. A. and M. Gould, 1985. Polyspermy-preventing mechanisms. In C. B. Metz and A. Monroy, eds., *Biology of fertilization*, Vol. 3, *Fertilization response of the egg*. Academic Press, Orlando, pp. 223–250.

Jamieson, B. G. M., 1987. *Ultrastructure and phylogeny of insect spermatozoa*. Cambridge University Press, Cambridge.

Javois, L. C., 1984. Pattern specification in the developing chick limb. In G. M. Malacinski, ed., *Pattern formation*. Macmillan, New York, pp. 557–579.

Johnson, K. E., 1970. The role of changes in cell contact behavior in amphibian gastrulation. *J. Exp. Zool.*, 175:391–428.

Johnson, M. H., 1985. Three types of cell interaction regulate the generation of cell diversity in the mouse blastocyst. In G. M. Edelman and J.-P. Thiery, eds., *Cell in contact: Adhesion and junction as morphogenetic determinants*. A Neurosciences Institute Publication, Wiley, New York, pp. 27–48.

Johnson, M. H. and H. P. M. Pratt, 1983. Cytoplasmic localizations and cell interactions in the formation of the mouse blastocyst. In W. R. Jeffery and R. A. Raff, eds., *Time, space, and pattern in embryonic development*. Alan R. Liss, New York, pp. 287–312.

Jones, R., C. R. Brown, and R. T. Lancaster, 1988. Carbohydrate-binding properties of boar sperm proacrosin and assessment of its role in sperm–egg recognition and adhesion during fertilization. *Development*, 102:781–792.

Judson, H. F., 1978. *Eighth day of creation: Makers of the revolution in biology*. Simon and Schuster, New York.

Jura, Cz., 1972. Development of apterygote insects. In S. J. Counce and C. H. Waddington, eds., *Developmental systems: Insects*, Vol. 1. Academic Press, London, pp. 49–94.

Jura, Cz., A. Krzysztofowicz, and E. Kisiel, 1987. Embryonic development of *Tetrodontophora bielanensis* (Collembola): Descriptive, with scanning electron micrographs. In H. Ando and Cz. Jura, eds., *Recent advances in insect embryology in Japan and Poland*. Arthropod, Embryology Society of Japan, ISEBU Co. Ltd., Tsukuba, pp. 77–124.

Just, E. E., 1939. *Biology of the cell surface*. P. Blakiston's Son & Co., Philadelphia.

Kageura, H. and K. Yamana, 1983. Pattern regulation in isolated halves and blastomeres of early *Xenopus laevis*. *J. Embryol. Exp. Morphol.*, 74:221–234.

Källén, B., 1961. Embryological aspects of the concept of homology. *Ark. Zool.*, 12:137–142.

Kallman, K. D., 1968. Evidence for the existence of transformer genes for sex in the telesot *Xiphophorus maculatus*. *Genetics*, 60:811–828.

Kalt, M. R., 1973. Ultrastructural observations on the germ line of *Xenopus laevis*. *Z. Zellforsch.*, 138:41–62.

Kalt, M. R., 1976. Morphology and kinetics of spermatogenesis in *Xenopus laevis*. *J. Exp. Zool.*, 195:393–408.

Kalthoff, K., 1979. Analysis of a morphogenetic determinant in an insect embryo (*Smittia* spec., Chironomidae, Diptera). In S. Subtelny, ed., *Determinants of spatial organization*, 37th Symposium of the Society for Developmental Biology. Academic Press, Orlando, pp. 97–126.

Kalthoff, K., 1983. Cytoplasmic determinants in dipteran eggs. In W. R. Jeffery and R. A. Raff, eds., *Time, space, and pattern in embryonic development*. Alan R. Liss, New York, pp. 313–348.

Kao, K. R. and R. P. Elinson, 1988. The entire mesodermal mantle behaves as Spemann's organizer in dorsoanterior enhanced *Xenopus laevis* embryos. *Dev. Biol.*, 127:64–77.

Kao, K., R. Elinson, and Y. Masui, 1985. Lithium chloride (LiCl) rescues ultraviolet light (UV)-induced axis-deficient *Xenopus laevis* embryos. *Am. Zool.*, 25:A14.

Karsch, F. J., 1984. The hypothalamus and anterior pituitary gland. In C. R. Austin and R. V. Short, eds., *Reproduction in mammals, Book 3, Hormonal control of reproduction*, 2nd ed. Cambridge University Press, Cambridge.

Katagiri, C., 1974. A high frequency of fertilization in premature and mature coelomic toad eggs after enzymic removal of vitelline membrane. *J. Embryol. Exp. Morphol.*, 31:573–587.

Kaufman, M. H., 1983. *Early mammalian development: Parthenogenetic studies*. Cambridge University Press, Cambridge.

Kaulenas, M. S. and D. Fairbairn, 1968. RNA metabolism of

fertilized *Ascaris lumbroicoides* eggs during uterine development. *Exp. Cell Res.,* 52:233–251.

Kedes, L. H. and P. R. Gross, 1969. Synthesis and function of messenger RNA during early embryonic development. *J. Mol. Biol.,* 42:559–575.

Keller, R. E., 1975. Vital dye mapping of the gastrula and neurula of *Xenopus laevis.* I. Prospective areas and morphogenetic movements of the superficial layer. *Dev. Biol.,* 42:222–241.

Keller, R. E., 1976. Vital dye mapping of the gastrula and neurula of *Xenopus laevis.* II. Prospective areas and morphogenetic movements of the deep layer. *Dev. Biol.,* 51:118–241.

Keller, R. and J. Hardin, 1987. Cell behaviour during active cell rearrangement: Evidence and speculations. *J. Cell Sci. Suppl.,* 8:369–393.

Keller, R. E. and J. P. Trinkaus, 1987. Rearrangement of enveloping layer cells without disruption of the epithelial permeability barrier as a factor in *Fundulus* epiboly. *Dev. Biol.,* 120:12–24.

Kemphues, K. J., J. R. Priess, D. G. Morton, and N. Cheng, 1988. Identification of genes required for cytoplasmic localization in early *C. elegans* embryos. *Cell,* 52:311–320.

Kessel, E. L., 1955. The mating activities of balloon flies. *Sys. Zool.,* 4:97–104.

Kessel, R. G., 1985. Annulate lamellae (porous cytomembranes): With particular emphasis on their possible role in differentiation of the female gamete. In L. W. Browder, ed., *Developmental biology: A comprehensive synthesis,* Vol 1, *Oogenesis.* Plenum Press, New York, pp. 179–233.

Keynes, R. J., R. V. Stirling, C. D. Stern, and D. Summerbell, 1987. The specificity of motor innervation of the chick wing does not depend upon the segmental origin of muscles. *Development,* 99:565–575.

Kimelman, D. and M. Kirschner, 1987. Synergistic induction of mesoderm by FGF and TFG-β and the identification of an mRNA coding for FGF in the early *Xenopus* embryo. *Cell,* 51:869–877.

Kimmel, C. B. and R. D. Law, 1985. Cell lineage of zebrafish blastomeres: I. Cleavage pattern and cytoplasmic bridges between cells. *Dev. Biol.,* 108:78–85.

Kimmel, C. B. and R. M. Warga, 1986. Tissue-specific cell lineages originate in the gastrula of the zebrafish. *Science,* 231:365–368.

King, C. R. and J. Piatigorsky, 1983. Alternative RNA splicing of the murine alpha A-crystallin gene: Protein-coding information within an intron. *Cell,* 32:707–712.

Kintner, C. R. and D. A. Melton, 1987. Expression of *Xenopus* N-CAM RNA in ectoderm is an early response to neural induction. *Development,* 99:311–325.

Kiny, M., 1977. Regulation of proximo-distal pattern formation in the developing limb. In M. Karkinen-Jääskeläinen and L. Saxén, eds., *Cell interactions in differentiation.* Academic Press, London, pp. 125–140.

Kirschner, M., J. C. Gerhart, J. Hara, and G. A. Ubbels, 1980. Initiation of the cell cycle and establishment of bilateral symmetry in *Xenopus* eggs. In S. Subtelny, ed., *Cell surface: Mediator of developmental processes,* 38th Symposium of the Society for Developmental Biology. Academic Press, Orlando, pp. 187–215.

Kirschner, M., J. Newport, and J. Gerhart, 1985. The timing of early developmental events in *Xenopus. Trends Genet.,* 1:41–47.

Klee, C. B., T. H. Crouch, and P. G. Richman, 1980. Calmodulin. *Annu. Rev. Biochem.,* 49:489–515.

Klein, D. C., 1984. Melatonin and puberty. *Science,* 224:6.

Klein, G., M. Langegger, R. Timpl, and Ekblom, 1988. Role of laminin A chain in the development of epithelial cell polarity. *Cell,* 55:331–341.

Klein, S. L., 1987. The first cleavage furrow demarcates the dorsal-ventral axis in *Xenopus* embryos. *Dev. Biol.,* 120:299–304.

Klug, A. and A. A. Travers, 1989. The helical repeat of nucleosome-wrapped DNA. *Cell,* 56:10–11.

Ko, H.-S., P. Fast, W. McBride, and L. M. Staudt, 1988. A human protein specific for the immunoglobulin octamer DNA motif contains a functional homeobox domain. *Cell,* 55:135–144.

Kochav, S. and H. Eyal-Giladi, 1971. Bilateral symmetry in chick embryo determination by gravity. *Science,* 171: 1027–1029.

Kochav, S., M. Ginsburg, and H. Eyal-Giladi, 1980. From cleavage to primitive streak formation: A complementary normal table and a new look at the first stages of development of the chick. II. Microscopic anatomy and cell population dynamics. *Dev. Biol.,* 79:298–308.

Kollar, E. J., 1983. Interspecific epithelio-mesenchymal interaction in the mammalian integument: Tooth development as a model for instructive induction. In R. H. Sawyer and J. F. Fallon, eds., *Epithelial–mesenchymal interactions in development.* Praeger, New York, pp. 27–50.

Koller, P. C., 1964. Chromosomes in neoplasia. In P. Emmelot and O. Mühlbock, eds., *Cellular control mechanisms and cancer.* Elsevier, Amsterdam, pp. 174–189.

Konischi, M., 1965. The role of auditory feedback in the control of vocalization in the white-crowned sparrow. *Z. Tierpsychol.,* 22:770–783.

Konishi, M. and F. Nottebohm, 1969. Experimental studies on the ontogeny of avian vocalization. In R. A. Hinde, ed., *Bird vocalization.* Cambridge University Press, New York.

Kornberg, A., 1982. *Supplement to DNA replication.* W. H. Freeman, San Francisco.

Korschelt, E. and K. Heider, 1895. *Text-book of the embryology of invertebrates.* Macmillan, New York.

Kozak, M., 1987. An analysis of 5'-noncoding sequences from 699 vertebrate messenger RNAs. *Nucleic Acids Res.,* 15:8125–8148.

Krämer, A., 1985. 5 S ribosomal gene transcription during *Xenopus* oogenesis. In L. W. Browder, ed., *Developmental biology: A comprehensive synthesis,* Vol. 1, *Oogenesis.* Plenum Press, New York, pp. 431–451.

Krumlauf, R., P. W. H. Holland, J. H. McVey, and B. L. M. Hogan, 1987. Developmental and spatial patterns of expression of the mouse homeobox gene, *Hox2.1. Development,* 99:603–617.

Kühn, A., 1971. *Lectures on developmental physiology* (translated by Roger Milkman), 2nd ed. Springer-Verlag, New York.

Kumé M. and K. Dan, 1968. *Invertebrate embryology* (translated from the Japanese by J. C. Dan). Originally published by Bai Fu Kan Press, Tokyo, 1957. Reprinted by Garland Publishing, New York, 1988.

Kunkel, T. A., 1988. Exonucleolytic proofreading. *Cell,* 53: 837–840.

Lallier, R., 1975. Animalization and vegetalization. In G. Czihak, ed., *The sea urchin embryo: Biochemistry and morphogenesis.* Springer-Verlag, Berlin, pp. 473–507.

Lane, C. D. and J. Knowland, 1975. The injection of RNA into living cells: The use of frog oocytes for the assay of mRNA and the study of the control of gene expression. In R. Weber, ed., *The biochemistry of animal development*, Vol. 3. Academic Press, New York, pp. 145–181.

Lange, R. H., 1985. The vertebrate yolk-platelet crystal: Comparative analysis of an *in vivo* crystalline aggregate. *Int. Rev. Cytol.*, 97:133–181.

Lansing, A. I., 1952. Biological and cellular problems of ageing: I. General physiology. In A. I. Lansing, ed. *Cowdry's Problems of Ageing*, 3rd ed. Williams and Wilkins, Baltimore, pp. 14–19.

Lawrence, P. A., 1981. The cellular basis of segmentation in insects. *Cell*, 26:3–10.

Lawrence, P. A., 1988. Background to *bicoid. Cell*, 54:1–2.

Leblond, C. P., 1972. Growth and renewal. In R. J. Goss, ed., *Regulation of organ and tissue growth.* Academic Press, Orlando, pp. 13–39.

Leblond, C. P. and B. E. Walker, 1956. Renewal of cell populations. *Physiol. Rev.*, 36:255–276.

Le Douarin, N.M., 1976. Cell migration in early vertebrate development studied in interspecific chimaeras. In *Embryogenesis in mammals*, Ciba Foundation Symposia No. 40 (new series). Elsevier/Excerpta Medica/North-Holland, Amsterdam, pp. 71–101.

Le Douarin, N. M., 1982. *The neural crest.* Cambridge University Press, Cambridge.

Le Douarin, N. M., 1984. A model for cell line divergence in the ontogeny of the peripheral nervous system. In I. Black, ed., *Cellular and molecular biology of neuronal development.* Plenum Press, New York, pp. 3–28.

Le Douarin, N. M., 1988. On the origin of pancreatic endocrine cells. *Cell*, 53:169–171.

Le Douarin, N. M. and J. Smith, 1988. Development of the peripheral nervous system from the neural crest. *Annu. Rev. Cell Biol.*, 4:375–404.

Le Douarin, N. M., P. Cochard, M. Vincent, J.-L. Duband, G. C. Tucker, M.-A. Teillet, and J.-P. Thiery, 1984. Nuclear, cytoplasmic, and membrane markers to follow neural crest cell migration: A comparative study. In R. L. Trelstad, ed., *The role of extracellular matrix in development.* Alan R. Liss, New York, pp. 373–398.

Lentz, T. L. and J. P. Trinkaus. 1967. A fine structural study of cytodifferentiation during cleavage, blastula and gastrula stages of *Fundulus heteroclitus. J. Cell Biol.*, 32:121–138.

Levi-Montalcini, R., 1987. The nerve growth factor 35 years later. *Science*, 237:1154–1162.

Levine, M., 1988. Molecular analysis of dorsal–ventral polarity in *Drosophila. Cell*, 52:785–786.

Levine, M. and T. Hoey, 1988. Homeobox proteins as sequence-specific transcription factors. *Cell*, 55:537–540.

Lewin, R., 1984. Why is development so illogical? *Science*, 224:1327–1329.

Lewis, E. B., 1964. Genetic control of regulation of developmental pathways. In M. Locke, ed., *The role of chromosomes in development.* Academic Press, Orlando, pp. 231–252.

Lewis, E., 1978. A gene complex coordinating segmentation in *Drosophila. Nature (London)*, 276:565–570.

Lewis, S., A. Gifford, B. Baltimore, 1985. DNA elements are

asymmetrically joined during the site-specific recombination of kappa immunoglobulin genes. *Science*, 228:677–685.

Libby, W., 1922. *History of medicine, in its salient features.* Houghton Mifflin, Boston.

Licht, P., 1983. Evolutionary divergence in the structure and function of pituitary gonadotropins of tetrapod vertebrates. *Am. Zool.*, 23:673–683.

Lillie, F. R., 1902. Differentiation without cleavage in the egg of the annelid Chaetopterus pergamentaceus. *Arch. Entwicklungsmech. Org.*, 14:477–499.

Lillie, F. R., 1907. Observations and experiments concerning the elementary phenomena of embryonic development in Chaetopterus. *J. Exp. Zool.*, 3:154–269.

Lillie, F. R., 1913. The mechanism of fertilization. *Science*, 38:524–528. Reprinted 1964 in B. H. Willier and J. M. Oppenheimer, eds. *Foundations of experimental embryology.* Prentice-Hall, Englewood Cliffs, pp. 120–126.

Lillie, F. R., 1916. The theory of the free-martin. *Science*, 43:611–613. Reprinted 1964 in B. H. Willier and J. M. Oppenheimer, eds., *Foundations of experimental embryology.* Prentice-Hall, Englewood Cliffs, NJ, pp. 138–142.

Lima-de-Faria, A., M. Birnstiel, and H. Jaworska, 1972. Amplification of ribosomal cistrons in the heterochromatin of acheta. In M. C. Gould et al., eds., *Invertebrate oogenesis. II.* MSS Information Corporation, New York, pp. 72–86.

Lincoln, G. A. and R. V. Short, 1980. Seasonal breeding: Nature's contraceptive. *Recent Prog. Hormone Res.*, 36:1–43.

Lindenmayer, A., 1982. Developmental algorithms: Lineage versus interactive control mechanisms. In S. Subtelny, ed., *Developmental order: Its origin and regulation*, 40th Symposium of the Society for Developmental Biology. Alan R. Liss, New York, pp. 219–245.

Liotta, L. A., U. M. Wewer, C. N. Rao, and G. Bryant, 1985. Laminin receptor. In G. M. Edelman and J.-P. Thiery, eds., *Cell in contact: Adhesions and junctions as morphogenetic determinants.* A Neurosciences Institute Publication, Wiley, New York, pp. 333–344.

Littlefield, C. L., 1985. Germ cells in *Hydra oligactis*. I. Isolation of a subpopulation of interstitial cells that is developmentally restricted to sperm production. *Dev. Biol.*, 112:185–193.

Loeb, J., 1913. *Artificial parthenogenesis and fertilization.* University of Chicago Press, Chicago.

Lohka, M. J. and Y. Masui, 1984. Role of cytosol and cytoplasmic particles in nuclear envelope assembly and sperm pronuclei formation in cell-free preparations from amphibian eggs. *J. Cell Biol.*, 98:1222–1230.

Longo, F. J., 1973. Fertilization: A comparative ultrastructural review. *Biol. Reprod.*, 9:149–215.

Longo, F. J., 1976. Derivation of the membrane comprising the male pronuclear envelope in inseminated sea urchin eggs. *Dev. Biol.*, 49:347–368.

Longo, F. J., 1985. Pronuclear events during fertilization. In C. B. Metz and A. Monroy, eds., *Biology of fertilization*, Vol. 3, *The fertilization response of the egg.* Academic Press, Orlando, pp. 251–298.

Longo, F. J., 1987. *Fertilization.* Chapman and Hall, New York.

Longo, F. J. and E. Anderson, 1968. The fine structure of pronuclear development and fusion in the sea urchin, *Arbacia punctulata. J. Cell Biol.*, 39:339–368.

Longo, F. J. and M. Kunkle, 1978. Transformations of sperm nuclei upon insemination. In A. A. Moscona and A. Monroy, eds., *Current topics in developmental biology*, Vol. 12, *Fertilization*. Academic Press, Orlando, pp. 149–184.

Lopez, L. C. and B. D. Shur, 1987. Redistribution of mouse sperm surface galactosyltransferase after the acrosome reaction. *J. Cell Biol.*, 105:1663–1670.

Lopo, A. C. 1983. Sperm–egg interactions in invertebrates. In J. F. Hartmann, ed., *Mechanism and control of animal fertilization*. Academic Press, Orlando, pp. 270–324.

Lopo, A. C. and V. D. Vacquier, 1981. Gamete interaction in the sea urchin: A model for understanding the molecular details of animal fertilization. In L. Mastroianni, Jr. and J. D. Biggers, eds., *Fertilization and embryonic development in vitro*. Plenum Press, New York, pp. 199–232.

Løvtrup, S., 1974. *Epigenetics, a treatise on theoretical biology*. Wiley, London.

Løvtrup, S., 1975. Fate maps and gastrulation in Amphibia—A critique of current views. *Can. J. Zool.*, 53:473–479.

Løvtrup, S., U. Landström, and H. Løvtrup-Rein, 1978. Polarities, cell differentiation and primary induction in the amphibian embryo. *Biol. Rev.*, 53:1–42.

Luckett, W. P., 1976. Ontogeny of amniote fetal membranes and their application to phylogeny. In M. K. Hecht, P. C. Goody, and B. M. Hect, eds., *Major patterns in vertebrate evolution*. Plenum Press, New York, pp. 439–516.

Lyon, M. F., 1961. Gene action in the X chromosome of the mouse (*Mus musculus* L.) *Nature (London)*, 190:372–373.

Lyon, M. F. and S. G. Hawkes, 1970. X-linked gene for testicular feminisation in the mouse. *Nature (London)*, 227:1217–1219.

Mabuchi, I., 1986. Biochemical aspects of cytokinesis. *Int. Rev. Cytol.*, 101:175–213.

MacGillivray, I., P. P. S. Hylander, and G. Corney, 1975. *Human multiple reproduction*. Saunders, London.

Maden, M., 1982. Supernumerary limbs in amphibians. *Am. Zool.*, 22:131–142.

Maden, M., 1984. Retinoids as probes for investigating the molecular basis of pattern formation. In G. M. Malacinski, ed., *Pattern formation*. Macmillan, New York, pp. 539–555.

Maderson, P. F. A., 1983. An evolutionary view of epithelial-mesenchymal interactions. In R. H. Sawyer and J. F. Fallon, eds., *Epithelial–mesenchymal interactions in development*. Praeger, New York, pp. 215–242.

Magnuson, T. and C. J. Epstein, 1981. Use of concanavalin A to monitor changes in glycoprotein synthesis during early mouse development. In S. R. Glasser and D. W. Bullock, eds., *Cellular and molecular aspects of implantation*. Plenum Press, New York, pp. 409–411.

Mahon, K. A. and J. G. Gall, 1984. The expression of repetitive sequences on amphibian lampbrush chromosomes. In E. H. Davidson and R. A. Firtel, *Molecular biology of development*. Alan R. Liss, New York, pp. 227–239.

Mahowald, A. P., 1977. The germ plasm of *Drosophila*: A model system for the study of embryonic determination. *Am. Zool.*, 17:551–563.

Mahowald, A. P., 1983. Genetic analysis of oogenesis and determination. In W. R. Jeffery and R. A. Raff, *Time, space, and pattern in embryonic development*. Alan R. Liss, New York, pp. 349–363.

Mahowald, A. P. and S. Hennen, 1971. Ultrastructure of the "germ plasm" in eggs and embryos of *Rana pipiens*. *Dev. Biol.*, 24:37–53.

Mahowald, A. P. and R. E. Boswell, 1983. Germ plasm and germ

cell development in invertebrates. In A. McLaren and C. C. Wylie, eds., *Current problems in germ cell differentiation*. Cambridge University Press, Cambridge, pp. 3–17.

Mahowald, A. P., C. D. Allis, K. M. Karrer, E. M. Underwood, and G. L. Waring, 1979. Germ plasm and pole cells of *Drosophila*. In S. Subtelny, ed., *Determinants of spatial organization*, 37th Symposium of the Society for Developmental Biology. Alan R. Liss, New York, pp. 127–146.

Mahowald, A. P., K. Konrad, L. Engstrom, and N. Perrimon, 1984. Genetic approach to early development. In E. H. Davidson and R. A. Firtel, eds., *Molecular biology of development*. Alan R. Liss, New York, pp. 185–197.

Maienschein, J., 1985. Preformation or new formation—or neither or both? In T. J. Horder, J. A. Witkowsky, and C. C. Whylie, eds., *A history of embryology*, 8th Symposium of the British Society for Developmental Biology. Cambridge University Press, Cambridge, pp. 73–108.

Maienschein, J., ed., 1986. *Defining biology: Lectures from the 1890s*. Harvard University Press, Cambridge.

Malacinski, G. M., 1984. Axis specification in amphibian eggs. In G. M. Malacinski, ed., *Pattern formation: A primer in developmental biology*. Macmillan, New York, pp. 435–456.

Maller, J. L., 1985. Oocyte maturation in amphibians. In L. W. Browder, *Developmental biology: A comprehensive synthesis*, Vol. 1, *Oogenesis*. Plenum Press, New York, pp. 289–311.

Manasek, F. J., 1976. In G. Poste and G. L. Nicolson, eds., *Cell surface in animal embryogenesis and development*. Elsevier/North-Holland Biomedical Press, Amsterdam.

Manes, M. E. and R. P. Elinson, 1980. Ultraviolet light inhibits grey crescent formation in the frog egg. *Wilhelm Roux' Arch. Dev. Biol.*, 189:73–76.

Manes, C., M. J. Byers, and A. S. Carver, 1981. Mobilization of genetic information in the early rabbit trophoblast. In S. R. Glasser and D. W. Bullock, eds., *Cellular and molecular aspects of implantation*. Plenum Press, New York, pp. 113–124.

Maniatis, T., S. Goodbourn, and J. A. Fischer, 1987. Regulation of inducible and tissue-specific gene expression. *Science*, 236:1237–1245.

Mann, J. R. and R. H. Lovell-Badge, 1987. The development of XO gynogenetic mouse embryos. *Development*, 99:411–416.

Manton, S. M., 1949. Studies on the onychophora. VII. The early embryonic stages of *Peripatopsis* and some general considerations concerning the morphology and phylogeny of the Arthropoda. *Philos. Trans. R. Soc. London Ser. B*, 233:343–363.

Margulis, L., 1981. *Symbiosis and cell evolution*. W.H. Freeman, San Francisco.

Markert, C. L. and R. M. Petters, 1977. Homozygous mouse embryos produced by microsurgery. *J. Exp. Zool.*, 201:295–302.

Markert, C. L. and R. M. Petters, 1978. Manufactured hexaparental mice show that adults are derived from three embryonic cells. *Science*, 202:56–58.

Markwald, R. R., R. B. Runyan, G. T. Kitten, F. M. Funderburg, D. H. Bernanke, and P. R. Brauer, 1984. Use of collagen gel cultures to study heart development: Proteoglycan and glycoprotein interactions during the formation of endocardial cushion tissue. In R. L. Trelstad, ed., *The role of extra-*

cellular matrix in development. Alan R. Liss, New York, pp. 323–350.

Marler, P. and P. Mundinger. 1971. Vocal learning in birds. In H. Moltz, ed., *The ontogeny of vertebrate behavior.* Academic Press, Orlando, pp. 389–450.

Martindale, M. Q., 1986. The ontogeny and maintenance of adult symmetry properties in the ctenophore, *Mnemiopsis mccradyl. Dev. Biol.,* 118:556–576.

Martinez-Arias, A. and P. A. Lawrence, 1985. Parasegments and compartments in the *Drosophila* embryo. *Nature (London),* 313:639–642.

Marx, J. L., 1988. Homeobox linked to gene control. *Science, (Res. News)* 242:1008–1009.

Masui, Y. and H. J. Clarke, 1979. Oocyte maturation. *Int. Rev. Cytol.,* 57:185–282.

Mayr, E., 1963. *Animal species and evolution.* Harvard University Press, Cambridge.

Mayr, E., 1982. *Growth of biological thought, diversity, evolution, and inheritance.* Belknap Press of Harvard University Press, Cambridge.

Mazia, D., 1963. Synthetic activities leading to mitosis. *J. Cell. Comp. Physiol. Suppl. 1,* 62 (Symposium on macromolecular aspects of the cell cycle): 123–140.

McCabe, J. A. and B. W. Parker, 1976. Evidence for a gradient of a morphogenetic substance in the developing limb. *Dev. Biol.,* 54:297–303.

McClarin, J. A., C. A. Frederick, B.-Cheng Wang, P. Greene, H. W. Boyer, J. Grable, and J. M. Rosenberg, 1986. Structure of the DNA-Eco RI endonuclease recognition complex at 3 Å resolution. *Science,* 234:1526–1541.

McClay, D. R., 1979. Surface antigens involved in interactions of embryonic sea urchin cells. In A. A. Moscona and A. Monroy, eds., *Current topics in developmental biology,* Vol. 13. Academic Press, Orlando, pp. 199–214.

McClay, D. R. and C. A. Ettensohn, 1987. Cell adhesion in morphogenesis. *Annu. Rev. Cell Biol.,* 3:319–345.

McClintock, B., 1956. Controlling elements and the gene. *Cold Spring Harbor. Symp. Quant. Biol.,* 21:197–216.

McClintock, B., 1984. The significance of responses of the genome to challenge. *Science,* 226:792–801.

McGregor, D. D., 1968. Bone marrow origin of immunologically competent lymphocytes in rats. *J. Exp. Med.,* 127:953–966.

McKinnell, R. G., 1972. Nuclear transfer in *Xenopus* and *Rana* compared. In R. Harris, P. Allin, and D. Viza, eds., *Cell differentiation.* Munksgaard International, Copenhagen, pp. 61–64.

McKinnell, R. G., B. A. Deggins, and D. D. Labat, 1969. Transplantation of pluripotential nuclei from triploid frog tumors. *Science,* 165:394–396.

McLaren, A., 1976. Growth from fertilization to birth in the mouse. *Ciba Foundation Symposium* No. 40 (new series). Elsevier/Excerpta Medica/North-Holland, Amsterdam, pp. 47–51.

McLaren, A. 1981. *Germ cells and soma: A new look at an old problem.* Yale University Press, New Haven.

McLaren, A., 1982. The embryo. In C. R. Austin and R. V. Short, eds., *Reproduction in mammals, Book 2, Embryonic and fetal development,* 2nd ed. Cambridge University Press, Cambridge, pp. 1–25.

McLaren, A., 1984. Germ cell lineages. In N. Le Douarin and A. McLaren, eds., *Chimeras in developmental biology.* Academic Press, London, pp. 111–129.

McLaren, A., 1987. Testis determination and the H-Y hypothesis. *Curr. Topics Dev. Biol.,* 23:163–183.

Meinhardt, H., 1982. *Models of biological pattern formation.* Academic Press, London.

Meinhardt, H., 1986. Hierarchical inductors of cell states: A model for segmentation in *Drosophila. J. Cell Sci. Suppl.,* 4:357–381.

Melton, D. A., 1987. Translocation of a localized maternal mRNA to the vegetal pole of *Xenopus* oocytes. *Nature (London),* 328:80–82.

Mercola, M. and C. D. Stiles, 1988. Growth factor superfamilies and mammalian embryogenesis (review article). *Development,* 102:451–460.

Meselson, M., 1965. The duplication and recombination of genes. In J. A. Moore, ed., *Ideas in modern biology,* Proceedings of the XVI International Congress on Zoology, Vol. 6. Natural History Press, New York, pp. 3–16.

Meselson, M. and F. W. Stahl, 1958. The replication of DNA in *Escherichia coli. Proc. Natl. Acad. Sci. USA,* 44:671–682.

Meshcheryakov, V. N. and L. V. Beloussov, 1975. Asymmetrical rotations of blastomeres in early cleavage of Gastropoda. *Wilhelm Roux' Arch. Dev. Biol.,* 177:193–203.

Miller, J., A. D. McLuchlan, and A. Klug, 1985. Repetitive zinc-binding domains in the protein transcription factor IIIA from *Xenopus* oocytes. *EMBO J.,* 4:1609–1614.

Miller, O. L. and B. A. Hamkalo, 1972. Visualization of RNA synthesis on chromosomes. *Int. Rev. Cytol.,* 33:1–25.

Miller, R. L., 1980. Species-specificity of sperm chemotaxis in the hydromedusae. In P. Tardent and R. Tardent, eds., *Development and cellular biology of coelenterates.* Elsevier/North-Holland Biomedical Press, Amsterdam, pp. 89–94.

Mintz, B., 1960. Formation and early development of germ cells. *Symposium on the germ cells and earliest stages of development.* Fondazione A. Baselli, Milano, pp. 1–24 II E/A.

Mintz, B., 1962. Experimental recombination of cells in the developing mouse egg: Normal and lethal mutant genotypes. *Am. Zool.,* 2:541–542.

Mintz, B., 1964. Synthetic processes and early development in the mammalian egg. *J. Exp. Zool.,* 157:85–100.

Mintz, B., 1970. Gene expression in allophenic mice. In H. A. Padykula, ed., *Control mechanisms in the expression of cellular phenotypes,* Symposia of the International Society for Cell Biology, Vol. 9. Academic Press, Orlando, pp. 15–42.

Mintz, B. and K. Illmensee, 1975. Normal genetically mosaic mice produced from malignant teratocarcinoma cells. *Proc. Natl. Acad. Sci. USA,* 72:3585–3589.

Mintz, B., M. Domon, D. A. Hungerford, and J. Morrow, 1972. Seminal vesicle formation and specific male protein secretion by female cells in allophenic mice. *Science,* 175:657–659.

Mirsky, A. E. and H. Ris., 1949. Variable and constant components of chromosomes. *Nature (London),* 163:666–667.

Misevic, G. N. and M. M. Burger, 1982. The molecular basis of species specific cell–cell recognition in marine sponges, and a study on organogenesis during metamorphosis. In M. M. Burger and R. Weber, eds., *Embryonic development, Part B, Cellular aspects.* Alan R. Liss, New York, pp. 193–209.

Mitchison, J. M. and M. M. Swann, 1955. The mechanical porperties of the cell surface. III. The sea urchin egg from fertilization to cleavage. *J. Exp. Biol.*, 32:734–750.

Mitchison, T. J., 1988. Microtubule dynamics and kinetochore function in mitosis. *Annu. Rev. Cell Biol.*, 4:527–549.

Mittwoch, U., 1978. Parthenogenesis: Review article. *J. Med. Genet.*, 15:165–181.

Modlinski, J. A., 1975. Haploid mouse embryos obtained by microsurgical removal of one pronucleus. *J. Embryol. Exp. Morphol.*, 33:897–905.

Monesi, V., R. Geremia, A. D'Agostino, and C. Boitani, 1978. Biochemistry of male germ cell differentiation in mammals: RNA synthesis in meiotic and postmeiotic cells. In A. A. Moscona and A. Monroy, eds., *Current topics in developmental biology*, Vol. 12. Academic Press, Orlando, pp. 11–36.

Monk, M., M. Boubelik, and S. Lehnert, 1987. Temporal and regional changes in DNA methylation in the embryonic, extraembryonic and germ cell lineages during mouse embryo development. *Development*, 99:371–382.

Monroy, A. and A. Tyler, 1963. Formation of active ribosomal aggregates (polysomes) upon fertilization and development of sea urchin eggs. *Arch. Biochem. Biophys.*, 103:431–435.

Monroy, A. and A. A. Moscona, 1979. *Introductory concepts in developmental biology.* University of Chicago Press, Chicago.

Monroy, A. and F. Rosati, 1979. Cell surface differentiations during early embryonic development. In A. A. Moscona and A. Monroy, eds., *Current topics in developmental biology*, Vol. 13. Academic Press, Orlando, pp. 45–69.

Moody, W. J. and M. M. Bosma, 1985. Hormone-induced loss of surface membrane during maturation of starfish oocytes: Differential effects on potassium and calcium channels. *Dev. Biol.*, 112:396–404.

Moon, R. T., M. V. Danilchik, and M. B. Hille, 1982. An assessment of the masked messenger hypothesis: Sea urchin egg messenger ribonucleoprotein complexes are efficient templates for *in vitro* protein synthesis. *Dev. Biol.*, 93:389–402.

Moore, H. D. M. and J. M. Bedford, 1978. Ultrastructure of the equatorial segment of hamster spermatozoa during penetration of oocytes. *J. Ultrastruct. Res.*, 62:110–117.

Moore, J. A., 1960. Serial back-transfers of nuclei in experiments involving two species of frogs. *Dev. Biol.*, 2:535–550.

Moore, J. A., 1983. Thomas Hunt Morgan—the geneticist. *Am. Zool.*, 23:855–865.

Moore, J. A., 1986. Science as a way of knowing—genetics. *Am. Zool.*, 26:583–747.

Moore, J. A., 1987. Science as a way of knowing—developmental biology. *Am. Zool.*, 27:415–573.

Moore, K. L., 1988. *Developing human*, 4th ed., W. B. Saunders, Philadelphia.

Morgan, T. H., 1901. *Regeneration.* Macmillan, New York.

Morgan, T. H., 1927. *Experimental embryology.* Columbia University Press, New York.

Morishita, K., D. S. Parker, M. L. Mucenski, N. A. Jenkins, N. G. Copeland, and J. N. Ihle, 1988. Retroviral activation of a novel gene encoding a zinc finger protein in IL-3-dependent myeloid leukemia cell lines. *Cell*, 54:831–840.

Moritz, C., T. E. Dowling, and W. M. Brown, 1987. Evolution of animal mitochondrial DNA: Relevance for population biology and systematics. *Annu. Rev. Ecol. Syst.*, 18:269–292.

Morrill, J. B., 1982. Development of the pulmonate gastropod, *Lymnaea*. In F. W. Harrison and R. R. Cowden, eds., *Developmental biology of freshwater invertebrates*. Alan R. Liss, New York, pp. 399–483.

Morrill, J. B. and L. L. Santos, 1985. Scanning electron microscopical overview of cellular and extracellular patterns during blastulation and gastrulation in the sea urchin, *Lytechinus variegatus*. In R. H. Sawyer and R. M. Showman, eds., *The cellular and molecular biology of invertebrate development.* University of South Carolina Press, Columbia, pp. 3–33.

Morrill, J. B., C. A. Blair, and W. J. Larsen, 1973. Regulative development in the pulmonate gastropod, *Lymnaea palustris*, as determined by blastomere deletion experiments. *J. Exp. Zool.*, 183:47–56.

Morris-Kay, G. and R. Tuckett, 1987. Fluidity of the neural epithelium during forebrain formation in rat embryos. *J. Cell Sci. Suppl.*, 8:433–449.

Morse, R. H. and R. T. Simpson, 1988. DNA in the nucleosome. *Cell*, 54:285–287.

Mountain, I. M., 1983. An introduction to Thomas Hunt Morgan and Lilian Vaughan Morgan. *Am. Zool.*, 23:825–827.

Müller, F., P. Walker, P. Aeby, H. Neuhaus, E. Back, H. Felder and H. Tobler, 1982. Molecular cloning and sequence analysis of highly repetitive DNA sequences contained in the eliminated genome of *Ascaris lumbricoides. Biol. Cell*, 45:78.

Murray, A. W. and J. W. Szostak, 1985. Chromosome segregation in mitosis and meiosis. *Annu. Rev. Cell Biol.*, 1:289–315.

Nagl, W., 1978. *Endopolyploidy and polyteny in differentiation and evolution.* Elsevier/North-Holland, Amsterdam.

Nakamura, O., Y. Hayashi, and M. Asashima, 1978. A half-century from Spemann—historical review of studies on the organizer. In O. Nakamura and S. Toivonen, eds., *Organizer—a milestone of a half-century from Spemann.* Elsevier/North-Holland, Amsterdam, pp. 1–48.

Nakauchi, M., 1982. Asexual development of ascidians: Its biological significance, diversity, and morphogenesis. *Am. Zool.*, 22:753–763.

Needham, A. E., 1964. *Growth process in animals.* Pitman & Sons, London.

Needham, J., 1931. *Chemical embryology.* University Press, Cambridge.

Needham, J., 1942. *Biochemistry and morphogenesis.* University Press, Cambridge.

Needham, J., 1959. *A history of embryology.* University Press, Cambridge.

Nemer, M., E. C. Travaglini, E. Rondinelli, and J. D'Alonzo, 1984. Developmental regulation, induction, and embryonic tissue specificity of sea urchin metallothionein gene expression. *Dev. Biol.*, 102:471–482.

Newport, J., 1987. Nuclear reconstitution in vitro: Stages of assembly around protein-free DNA. *Cell*, 48:205–217.

Newport, J. and M. Kirschner, 1982. A major developmental transition in early *Xenopus* embryos: I. Characterization and timing of cellular changes at the midblastula stage. *Cell*, 30:675–686. II. Control of the onset of transcription. *Cell*, 30:687–696.

Nicolet, G., 1971. Avian gastrulation. *Adv. Morphogen.*, 9:221–262.

Nieuwkoop, P. D., 1973. The "organization center" of the amphibian embryo: Its origin, spatial organization, and morphogenetic action. In M. Abercrombie, J. Brachet, and T. J. King, eds., *Advances in morphogenesis*, Vol. 10. Academic Press, Orlando, pp. 1–39.

Nieuwkoop, P. D., 1977. Origin and establishment of embryonic polar axes in amphibian development. In A. A. Moscona and A. Monroy, eds., *Current topics in developmental biology*, Vol. 11. Academic Press, Orlando, pp. 115–132.

Nieuwkoop, P. D. and L. A. Sutasurya, 1979. *Primordial germ cell in the chordates: Embryogenesis and phylogenesis.* Cambridge University Press, Cambridge.

Nieuwkoop, P. D. and L. A. Sutasurya, 1981. *Primordial germ cell in the invertebrates: From epigenesis to preformation.* Cambridge University Press, Cambridge.

Nieuwkoop, P. D., A. G. Johnen, and B. Albers, 1985. *The epigenetic nature of early chordate development: Inductive interaction and competence.* Cambridge University Press, Cambridge.

Nieuwkoop, P. D., and others, 1952. Activation and organization of the central nervous system in amphibians. Part I. Induction and activation. *J. Exp. Zool.*, 120:1–31. Part II. Differentiation and organization. *J. Exp. Zool.*, 120:33–81. Part III. Synthesis of a new working hypothesis. *J. Exp. Zool.*, 120:83–108.

Nirenberg, M. W. and P. Leder, 1964. RNA codewords and protein synthesis: The effect of trinucleotides upon the binding of sRNA to ribosomes. *Science*, 145:1399–1407.

Nobil, E., 1980. The neuroendocrine control of the menstrual cycle. *Recent Prog. Hormone Res.*, 36:53–88.

Noble, R. C., 1987. Lipid metabolism in the chick embryo: Some recent ideas. *J. Exp. Zool. Suppl.*, 1:65–73.

Nordenskiöld, E. 1929. *The history of biology.* Kegan Paul, Trench, Trubner & Co., London.

Nørrevang, A., 1968. Electron microscopic morphology of oogenesis. *Int. Rev. Cytol.*, 23:113–186.

Nüsslein-Volhard, C., 1979. Maternal effect mutations that alter the spatial coordinates of the embryo of *Drosophila melanogaster*. In S. Subtelny, ed., *Determinants of spatial organization*, 37th Symposium of the Society for Developmental Biology. Alan R. Liss, New York, pp. 185–211.

Nüsslein-Volhard, C. and E. Wieschaus, 1980. Mutations affecting segment number and polarity in *Drosophila. Nature (London)*, 287:795–799.

Ochoa, S., 1963. Synthetic polynucleotides and the genetic code. *Fed. Proc. (Symposium on genetic mechanics)* 22:62–74.

Okazaki, K., 1975. Normal development to metamorphosis. In G. Czihak, ed., *The sea urchin embryo: Biochemistry and morphogenesis.* Springer-Verlag, Berlin, pp. 177–232.

Okazaki, R., T. Okazaki, K. Sakabe, K. Sugimoto, and A. Sugino, 1968. Mechanism of DNA chain-growth. I. Possible discontinuity and unusual secondary structure of newly synthesized chains. *Proc. Natl. Acad. Sci. USA*, 59:598–605.

Oliphant, G. and L. A. Eng, 1981. Collection of gametes in laboratory animals and preparation of sperm for in vitro fertilization. In L. Mastroianni, Jr. and J. D. Biggers, eds., *Fertilization and embryonic development in vitro.* Plenum Press, New York, pp. 11–26.

Oliphant, G., A. B. Reynolds, and T. S. Thomas, 1985. Sperm surface components involved in the control of the acrosome reaction. *Am. J. Anat.*, 174:269–283.

Oliver, G., C. V. E. Wright, J. Hardwicke, and E. M. DeRobertis, 1988. A gradient of homeodomain protein in developing forelimbs of *Xenopus* and mouse embryos. *Cell*, 55:1017–1024.

Olsen, M. W., 1960a. Nine-year summary of parthenogenesis in turkeys. *Proc. Soc. Exp. Biol. Med.*, 105:279–381.

Olsen, M. W., 1960b. Performance record of a parthenogenetic turkey male. *Science*, 132:1661.

Oppenheim, R. W., 1982. Preformation and epigenesis in the origins of the nervous system and behavior: Issues, concepts, and their history. In P. P. G. Bateson and P. H. Klopfer, eds., *Perspectives in ethology*, Vol. 5. Plenum Press, New York, pp. 1–100.

Oppenheimer, J. M., 1940. The non-specificity of the germ-layers. *Q. Rev. Biol.*, 15:1–27. Reprinted in J. M. Oppenheimer, 1967. *Essays in the history of embryology and biology.* MIT Press, Cambridge, pp. 256–294.

Oppenheimer, J. M., 1967. *Essays in the history of embryology and Biology.* MIT Press, Cambridge.

Oppenheimer, J. M., 1981. Walter Landauer and developmental genetics. In S. Subtelny, ed., *Levels of genetic control in development.* Alan R. Liss, New York, pp. 1–13.

Oppenheimer, J. M., 1983. Thomas Hunt Morgan as an embryologist: The view from Bryn Mawr. *Am. Zool.*, 23: 845–854.

Ortalani, G., L. Tosi, M. Branno, and E. Patricolo, 1982. Development of animal halves of sea urchin eggs after trypsin treatment. *Acta Embryol. Morphol. Exp.*, No. 3: XVII–XVIII (abstract).

Padgett, R. A., M. M. Konarska, P. J. Grabowski, S. F. Hardy, and P. A. Sharp, 1984. Lariat RNAs as intermediate and products in the splicing of messenger RNA precursors. *Science*, 225:898–903.

Page, D., R. Mosher, E. M. Simpson, E. M. C. Fisher, G. Mardon, J. Pollack, B. McGillivray, A. de la Cahpelle, and L. G. Brown, 1987. The sex-determining region of the human Y chromosome encodes a finger protein. *Cell*, 51: 1091–1104.

Palade, G. E. and K. R. Porter, 1954. *J. Exp. Med.*, 100:641.

Palmiter, R. D., T. M. Walkie, H. Y. Chen, and R. L. Brinster, 1984. Transmission distortion and mosaicism in an unusual transgenic mouse pedigree. *Cell*, 36:869–877.

Papaioannou, V. E., 1982. Lineage analysis of inner cell mass and trophectoderm using microsurgically reconstituted mouse blastocysts. *J. Embryol. Exp. Morphol.*, 68:199–209.

Papaioannou, V. E. and K. M. Ebert, 1988. The preimplantation pig embryo: Cell number and allocation to trophectoderm and inner cell mass of the blastocyst *in vivo* and *in vitro. Development*, 102:793–803.

Pardee, A. B., R. Dubrow, J. L. Hamlin, and R. F. Kletzien, 1978. Animal cell cycle. *Annu. Rev. Biochem.*, 47:715–750.

Parisi, E., S. Filosa, B. De Petrocellis, and A. Monroy, 1978. The pattern of cell division in the early development of the sea urchin, *Paracentrotus lividus. Dev. Biol.*, 65:43–49.

Parisi, E., S. Filosa, and A. Monroy, 1979. Actinomycin D disruption of the mitotic gradient in the cleavage stages of the sea urchin embryo. *Dev. Biol.*, 72:167–174.

Partanen, C. R., 1965. On the chromosomal basis for cellular differentiation. *Am. J. Bot.*, 52:204–209.

Pasteels, J., 1964. The morphogenetic role of the cortex of the amphibian egg. *Adv. Morphogen.*, 3:363–388.

Patten, B. M., 1949. Initiation and early changes in the character of the heart beat in vertebrate embryos. *Physiol. Rev.*, 29:31–47.

Patten, B. M., 1958. _Foundations of embryology._ McGraw-Hill, New York.

Pauling, L. and M. Delbrück, 1940. The nature of the intermolecular forces operative in biological processes. _Science,_ 92:77–79.

Pauly, P. J., 1988. Summer resort and scientific discipline: Woods Hole and the structure of American biology, 1882–1925. In R. Rainger, K. R. Benson, and J. Maienschein, eds., _American development of biology._ University of Pennsylvania Press, Philadelphia, pp. 121–150.

Pearce, A. G. E., 1976. Peptides in brain and intestine. _Nature (London),_ 262:92–94.

Pederson, T., 1972. Chromatic structure and the cell cycle. _Proc. Natl. Acad. Sci. USA,_ 69:2224–2228.

Perris, R., Y. von Boxberg, and J. Löfberg, 1988. Local embryonic matrices determine region-specific phenotypes in neural crest cells. _Science,_ 241:86–89.

Perry, J. S. and I. W. Rowlands, 1962. Early pregnancy in the pig. _J. Reprod. Fertil.,_ 4:175–188.

Peter, R. E., 1983. Evolution of neuro-hormonal regulation of reproduction in lower vertebrates. _Am. Zool.,_ 23:685–695.

Peters, H., 1978. Folliculogenesis in mammals. In R. E. Jones, ed., _The vertebrate ovary: Comparative biology and evolution._ Plenum Press, New York, pp. 121–144.

Pethica, B. A., 1961. The physical chemistry of cell adhesion. _Exp. Cell Res.,_ Suppl. 8 (Symposium on cell movement and cell contact): 123–140.

Petraglia, F., P. Sawchenko, A. T. W. Lim, J. Rivier, and W. Vale, 1987. Localization, secretion, and action of inhibin in human placenta. _Science,_ 237:187–189.

Petzoldt, U. and A. Muggleton-Harris, 1987. The effect of the nucleocytoplasmic ratio on protein synthesis and expression of a stage-specific antigen in early cleaving mouse embryos. _Development,_ 99:481–491.

Phillips, D. M., 1974. _Spermiogenesis._ Academic Press, Orlando.

Phillips, R. G. and K. Kalthoff, 1986. An immunological technique for isolation of RNA involved in UV-induced and photorepairable effects on development. In H. C. Slavkin, ed., _Progress in developmental biology, Part B._ Alan R. Liss, New York, pp. 365–368.

Phillips, W. H., J. A. Winkles, and R. M. Grainger, 1984. A method for analyzing oocyte messenger RNAs which persist in the embryo of _Drosophila melanogaster._ In E. H. Davidson and R. A. Firtel, eds., _Molecular biology of development._ Alan R. Liss, New York, pp. 241–252.

Pienta, K. J. and D. S. Coffey, 1985. The nuclear matrix: An organizing structure for the interphase nucleus and chromosome. In C. Nicolini and P. O. P. Ts'o, eds., _Structure and function of the genetic apparatus._ Plenum Press, New York, pp. 83–98.

Platt, J. R., 1964. Strong inference: Certain systematic methods of scientific thinking may produce much more rapid progress than others. _Science,_ 146:347–353.

Poccia, D., 1986. Remodeling of nucleoproteins during gametogenesis, fertilization, and early development. _Int. Rev. Cytol.,_ 105:1–65.

Podell, S. B. and V. D. Vacquier, 1984. Wheat germ agglutinin blocks the acrosome reaction in _Strongylocentrotus purpuratus_ sperm by binding a 210,000-mol-wt membrane protein. _J. Cell Biol.,_ 99:1598–1604.

Policansky, D., 1987. Evolution, sex, and sex allocation. _BioScience,_ 37:466–468.

Porter, R. R., 1973. Structural studies of immunoglobulins. _Science,_ 180:713–716.

Poste, G. and A. C. Allison, 1973. Membrane fusion. _Biochim. Biophys. Acta,_ 300:421–465.

Postlethwait, J. H. and F. Giorgi, 1985. Vitellogenesis in insects. In L. W. Browder, ed., _Developmental biology, a comprehensive synthesis,_ Vol. 1, Oogenesis. Plenum Press, New York, pp. 85–126.

Prescott, D. M., 1976. _Reproduction of eukaryotic cells._ Academic Press, Orlando.

Prescott, D. M. and M. A. Bender, 1963. Synthesis and behavior of nuclear proteins during the cell life cycle. _J. Cell. Comp. Physiol. Suppl. 1,_ 62 (Symposium on macromolecular aspects of the cell cycle): 175–194.

Psychoyos, A. and V. Casimiri, 1981. Uterine blastotoxic factors. In S. R. Glasser and D. W. Bullock, eds., _Cellular and molecular aspects of implantation._ Plenum Press, New York, pp. 327–334.

Quastler, H. and F. G. Sherman, 1959. Cell population kinetics in the intestinal epithelium of the mouse. _Exp. Cell Res.,_ 17:420–438.

Queen, C. and M. Rosenberg, 1981. Differential translation efficiency explains discoordinate expression of the galactose operon. _Cell,_ 25:241–249.

Raff, R. A. and E. C. Raff, eds., 1987. _Development as an evolutionary process._ MBL Lectures in Biology, Vol. 8. Alan R. Liss, New York.

Raikova, E. V., 1973. Life cycle and systematic position of _Polypodium hydriforme_ Ussov (Coelenterata), a cnidarian parasite of the eggs of Acipenseridae. In T. Takioka and S. Nishimura, eds., _Second International Symposium on Cnidaria._ Seto Marine Biology Laboratory, Shirahama, pp. 165–174.

Rappaport, R., 1961. Experiments concerning the cleavage stimulus in sand dollar eggs. _J. Exp. Zool.,_ 161:81–89.

Raven, C. P., 1961. _Oogenesis: The storage of developmental information._ Pergamon Press, Oxford.

Raven, C. P., 1966. _Morphogenesis: An analysis of molluscan development._ Pergamon Press, Oxford.

Raven, C. P., 1976. Morphogenetic analysis of spiralian development. _Am. Zool.,_ 16:395–403.

Redi, F., 1668. _Experiments on the generation of insects._ Translated by M. A. B. Bigelow. Excerpt reproduced in A. Rook, ed., 1964. _The origins and growth of biology._ Penguin Books, Baltimore, pp. 75–79.

Reeck, G. R., 1987. Homology in proteins and nucleic acids: A terminology muddle and a way out of it. _Cell,_ 50:667.

Reeder, R. H., S. Busby, M. Dunaway, S. C. Pruitt, G. Morgan, P. Labhart, A. H. Bakken, and B. Sollner-Webb, 1984. Nucleolar dominance and the developmental regulation of RNA polymerase I promoter in _Xenopus._ In E. H. Davidson and R. A. Firtel, eds., _Molecular biology of development._ Alan R. Liss, New York, pp. 199–212.

Regier, J. C., G. D. Mazur, and F. C. Kafatos, 1980. The silkmoth chorion. Morphological and biochemical characterization of four surface regions. _Dev. Biol.,_ 76:286–304.

Regulski, M., K. Harding, R. Kostriken, F. Karch, M. Levine, and W. McGinnis, 1985. Homeo box genes of the Antennapedia and Bithorax complexes of _Drosophila. Cell,_ 43:71–80.

Renfree, M. W., 1973. The composition of fetal fluids of the marsupial *Macropus eugenii. Dev. Biol.*, 33:63–79.

Renfree, M. B., 1982. Implantation and placentation. In C. R. Austin and R. V. Short, eds., *Reproduction in mammals, Book 2, Embryonic and fetal development*, 2nd ed. Cambridge University Press, Cambridge, pp. 26–69.

Reverberi, G., 1972. *Experimental embryology of marine and fresh-water invertebrates*. North-Holland, Amsterdam.

Reynaud, G., 1969. Transfert de cellules germinales primordiales de dindon à l'embryon de poulet par injection intravasculaire. *J. Embryol. Exp. Morphol.*, 21:485–507.

Reynaud, G., 1976. Capacités reproductrices et descendance de poulets ayant subi un transfert de cellules germinales primordiales durant la vie embryonnaire. *Wilhelm Roux' Arch. Dev. Biol.*, 179:85–110.

Rice, M. E., 1976. Larval development and metamorphosis in sipuncula. *Am. Zool.*, 16:563–571.

Rinaldi, A. M. and A. Monroy, 1969. Polyribosome formation and RNA synthesis in the early post-fertilization stages of the sea urchin egg. *Dev. Biol.*, 19:73–86.

Rivier, C., J. Rivier, and W. Vale, 1986. Inhibin-mediated feedback control of follicle-stimulating hormone secretion in the female rat. *Science*, 234:205–208.

Roberts, D. B., 1971. Antigens of developing *Drosophila melanogaster. Nature (London)*, 233:394–397.

Robertson, D. J., 1959. The ultrastructure of cell membranes and their derivatives. *Biochem. Soc. Symp. (Cambridge, England)*, 16:3–43.

Rochaix, D.-D., A. Bird, and A. Bakken, 1974. Ribosomal RNA gene amplification by rolling circles. *J. Mol. Biol.*, 87:473–487.

Roeder, K. D., 1963. *Nerve cells and insect behavior*. Harvard University Press, Cambridge.

Roeder, R. G., L. B. Schwartz, and V. E. F. Sklar, 1976. Function, structure, and regulation of eukaryotic nuclear RNA polymerases. In J. Papaconstantinou, ed., *The molecular biology of hormone action*. Academic Press, Orlando, pp. 29–52.

Romanoff, A. L., 1967. *Biochemistry of the avian embryo*. Wiley, New York.

Romanoff, A. L. and A. J. Romanoff, 1949. *The avian egg*. Wiley, New York.

Romer, A. S., 1955. *The vertebrate body*, 2nd edition, Saunders, Philadelphia.

Roosen-Runge, E. C., 1977. *The process of spermatogenesis in animals*. Cambridge University Press, London.

Rosa, F., A. B. Roberts, D. Danielpour, L. L. Dart, M. B. Sporn, and I. B. Dawid, 1988. Mesoderm induction in amphibians: The role of TGF-β_2-like factors. *Science*, 239:783–785.

Ross, M. H. and E. J. Reith, 1985. *Histology, a test and atlas*. Harper & Row, New York.

Rossant J. and V. E. Papaioannou, 1977. The biology of embryogenesis. In M. I Sherman, ed., *Concepts in mammalian embryogenesis*. MIT Press, Cambridge, pp. 1–36.

Roth, S., 1979. Plasma membrane glycosyltransferases and morphogenesis. In J. D. Ebert and T. S. Okada, eds., *Mechanisms of cell change*. Wiley, New York, pp. 215–223.

Roth, S., J. E. McGuire, and S. Roseman, 1971. Evidence for cell-surface glycosyltransferases. Their potential role in cellular recognition. *J. Cell Biol.*, 51:536–547.

Rothschild, Lord, 1946. Physiology of fertilization. *Nature (London)*, 157:720–722.

Rothschild, Lord and M. M. Swann, 1952. The fertilization reaction in the sea urchin. The block to polyspermy. *J. Exp. Biol.*, 29:469–483.

Roux, W., 1888. Contributions to the developmental mechanics of the embryo. On the artificial production of half-embryos by destruction of one of the first two blastomeres, and the later development (postgeneration) of the missing half of the body. Translation by Hans Laufer in B. H. Willier and J. M. Oppenheimer, eds., 1964. *Foundations of experimental embryology*. Prentice-Hall, Englewood Cliffs, NJ, pp. 2–37.

Roux, W., 1894. The problems, methods, and scope of developmental mechanics: An introduction to the *Archiv für Entwicklungsmechanik der Organismen*. Translated from the German by William Morton Wheeler in J. Maienschein, ed., 1986. *Defining biology: Lectures from the 1890s*. Harvard University Press, Cambridge, pp. 107–148.

Ruddle, F. H., C. P. Hart, and W. McGinnis, 1985. Homeo-box sequences—relevance to vertebrate developmental mechanisms. In F. Costantini and R. Jaenisch, eds., *Genetic manipulation of the early mammalian embryo*, Banbury report 20. Cold Spring Harbor Laboratory, Cold Spring Harbor, NY, pp. 169–177.

Rudnick, D., 1955. Teleosts and birds. In B. H. Willier, P. A. Weiss, and V. Hamburger, eds., *Analysis of development*. Saunders, Philadelphia, pp. 297–314.

Rugh, R., 1977. *Guide to vertebrate development*. Burgess, Minneapolis.

Runnström, J., 1933. Zur Kenntnis der Stoffwechselvorgänge bei der Entwicklungserregung des Seeigeleises. *Biochem. Z.*, 258:257–279.

Runnström, J., 1966. The vitelline membrane and cortical particles in sea urchin eggs and their function in maturation and fertilization. *Adv. Morphogen.*, 5:222–325.

Runnström, J., 1975. Integrating factors. In G. Czihak, ed., *The sea urchin embryo: Biochemistry and morphogenesis*. Springer-Verlag, Berlin, pp. 646–670.

Russell, E. S., 1930. *The interpretation of development and heredity: A study in biological method*. Clarendon Press, Oxford.

Sakai, H., 1968. Contractile properties of protein threads from sea urchin eggs in relation to cell division. *Int. Rev. Cytol.*, 23:89–112.

Salk, L., 1973. The role of the heartbeat in the relations between mother and infant. *Sci. Am.*, 228(May):24–29.

Sander, K., 1975. Pattern specification in the insect embryo. In *Cell patterning*, Ciba Foundation Symposium No. 29 (new series). Elsevier Excerpta Medica North-Holland, Amsterdam, pp. 241–263.

Sander, K., 1984. Embryonic pattern formation in insects: Basic concepts and their experimental foundations. In G. M. Malacinski, ed., *Pattern formation: A primer in developmental biology*. Macmillan, New York, pp. 245–268.

Sanders, E. J., 1982. Ultrastructural immunocytochemical localization of fibronectin in the early chick embryo. *J. Embryol. Exp. Morphol.*, 71:155–170.

Sang, J. H., 1984. *Genetics and development*. Longman, New York.

Sariola, H., E. Aufderheide, H. Bernhard, S. Henke-Fahle, W. Dippold, and P. Ekblom, 1988. Antibodies to cell surface ganglioside G_{D3} perturb inductive epithelial–mesenchymal interactions. *Cell*, 54:235–245.

Sarvella, P., 1973. Adult parthenogenetic chickens. *Nature (London)*, 243:171.

Sato, E., H. N. Wood, D. G. Lynn, and S. S. Koide, 1985. Modulation of oocyte maturation by cyclic adenosine 3', 5'-pyrophosphate. *Cell Differ.*, 17:169–174.

Satoh, N., 1982. Timing mechanisms in early embryonic development. *Differentiation*, 22:156–163.

Saunders, J. W. Jr., 1982. *Developmental biology, patterns, problems, principles*. Macmillan, New York.

Saunders, J. W. Jr. and J. F. Fallon, 1966. Cell death in morphogenesis. In M. Locke, ed., *Major problems in developmental biology*. Academic Press, Orlando, pp. 289–314.

Sawada, I. and C. W. Schmid, 1986. Repetitive human DNA sequences. I. Evolution of the primate α-globin gene cluster and interspersed Alu repeats. *Cold Spring Harbor Symp. Quant. Biol.*, 51:471–472.

Sawicki, J. A., T. Magnuson, and C. J. Epstein, 1981. Evidence for expression of the paternal genome in the two-cell mouse embryo. *Nature (London)*, 294:450–451.

Sawyer, R. H. and R. B. Shames, 1987. Expression of β-keratin genes during development of avian skin appendages. In R. H. Sawyer, ed., *Molecular and developmental biology of keratins. Current topics in developmental biology*, Vol. 22. Academic Press, Orlando, pp. 235–253.

Saxén, L. and S. Toivonen, 1962. *Primary embryonic induction*. Prentice-Hall, Englewood Cliffs, NJ.

Saxén, L. and S. Toivonen, 1985. Primary embryonic induction in retrospect. In T. J. Horder et al., eds., *A history of embryology*, 8th Symposium of the British Society for Developmental Biology. Cambridge University Press, Cambridge, pp. 261–274.

Saxén, L., S. Toivonen, and O. Nakamura, 1978. Concluding remarks—primary embryonic induction: an unsolved problem. In O. Nakamura and S. Toivonen, eds., *Milestone of a half-century from Spemann*. Elsevier/North-Holland Biomedical Press, Amsterdam.

Saxén, L., M. Karkinen-Jääskeläinen, E. Lehtonen, S. Nordling, and J. Wartiovaara, 1977. Inductive tissue interactions. In G. Poste and G. L. Nicolson, eds., *The cell surface in animal embryogenesis and development*. Elsevier/North-Holland Biomedical Press, Amsterdam, pp. 331–407.

Saxén, L., P. Ekblom, and I. Thesleff, 1980. Mechanisms of morphogenetic cell interactions. In M. H. Johnson, ed., *Development in mammals*, Vol. 4. Elsevier/North-Holland, Amsterdam, pp. 161–201.

Schatten, G., 1981. Sperm incorporation, the pronuclear migrations, and their relation to the establishment of the first embryonic axis: Time-lapse video microscopy of the movements during fertilization of the sea urchin *Lytechinus variegatus*. *Dev. Biol.*, 86:426–437.

Schatten, G., H. Schatten, T. H. Bestor, and R. Balczon, 1982. Taxol inhibits the nuclear movements during fertilization and induces asters in unfertilized sea urchin eggs. *J. Cell Biol.*, 94:455–465.

Scheer, U. and M.-C. Dabauvalle, 1985. Functional organization of the amphibian oocyte nucleus. In L. W. Browder, ed., *Developmental biology: A comprehensive synthesis*, Vol. 1, *Oogenesis*. Plenum Press, New York, pp. 385–430.

Schierenberg, E., 1982. Development of the nematode *Caenorhabditis elegans*. In F. W. Harrison and R. R. Cowden, eds., *Developmental biology of freshwater invertebrates*. Alan R. Liss, New York, pp. 249–281.

Schimke, R. T., 1982. Studies on gene duplications and amplifications—an historical perspective. In R. T. Schimke, ed., *Gene amplification*. Cold Spring Harbor Laboratory, Cold Spring Harbor, NY, pp. 1–6.

Schlafke, S. and A. C. Enders, 1975. Cellular basis of interaction between trophoblast and uterus at implantation. *Biol. Reprod.*, 12:41–65.

Schleif, R., 1988. DNA binding by proteins. *Science*, 241:1182–1187.

Schliwa, M., J. van Blerkom, and K. B. Pryzwansky, 1981. *Cold Spring Harbor Symp. Quant. Biol.*, 46:51.

Schmahl, G., 1980. Morphogenesis in isolated apical and basal parts of strobilating scyphistomae in *Aurelia aurita* and *Chrysaora sp.* (Scyphozoa). In P. Tardent and R. Tardent, eds., *Developmental and cellular biology of coelenterates*. Elsevier/North-Holland Biomedical Press, Amsterdam, pp. 251–262.

Schneider, D., 1969. Insect olfaction: Deciphering a system for chemical messages. *Science*, 163:1031–1036.

Schor, S. L. and A. M. Schor, 1987. Foetal-to-adult transitions in fibroblast phenotype: Their possible relevance to the pathogenesis of cancer. *J. Cell Sci. Suppl.*, 8:165–180.

Schrödinger, E., 1967. *What is life?* and *Mind and matter*, combined reprint. Cambridge University Press, Cambridge.

Schroeder, T. E., 1980a. The jelly canal marker of polarity for sea urchin oocytes, eggs, and embryos. *Exp. Cell Res.*, 128:490–494.

Schroeder, T. E., 1980b. Expressions of the prefertilization polar axis in sea urchin eggs. *Dev. Biol.*, 79:428–443.

Schroeder, T. E., 1985. Physical interactions between asters and the cortex in echinoderm eggs. In R. H. Sawyer and R. M. Showman, eds., *The cellular and molecular biology of invertebrate development*. University of South Carolina Press, Columbia, pp. 69–89.

Schroeder, T. E., 1986. The origin and action of the contractile ring. In N. Akkas, ed., *Biomechanics of cell division*. Plenum Press, New York, pp. 209–230.

Schuel, H., 1985. Functions of egg cortical granules. In C. B. Metz and A. Monroy, eds., *Biology of fertilization*, Vol. 3, *Fertilization response of the egg*. Academic Press, Orlando, pp. 1–43.

Schuetz, A. W., 1985. Local control mechanisms during oogenesis and folliculogenesis. In L. W. Browder, ed., *Developmental biology: A comprehensive synthesis*, Vol. 1. *Oogenesis*. Plenum Press, New York, pp. 3–83.

Schultz, R. J. and K. D. Kallman, 1968. Triploid hybrids between the all-female teleost *Poecilia formosa* and *Poecilia sphenops*. *Nature (London)*, 219:280–282.

Schwann, Th., 1839. Mikroscopische Untersuchungen über die Uebereinstimmung in der Structur und dem Wachsthum der Thiere und Planzen. Berlin. Translated by H. Smith, 1847, in *Sydenham Soc.*, XII. London. Reprinted 1955 in part in M. L. Gabriel and S. Fogel, eds., *Great experiments in biology*. Prentice-Hall, Englewood Cliffs, NJ.

Scott, M. P. and P. H. O'Farrell, 1986. Spatial programming of gene expression in early *Drosophila* embryogenesis. *Annu. Rev. Cell Biol.*, 2:49–80.

Scully, N. F. and B. D. Shur, 1988. Stage-specific increase in cell surface galactosyltransferase activity during spermatogenesis in mice bearing *t* alleles. *Dev. Biol.*, 125:195–199.

Searle, R. F., M. H. Sellens, J. Elson, E. J. Jenkinson, and W. D.

Billington, 1985. Detection of alloantigens during preimplantation development and early trophoblast differentiation in the mouse by immunoperoxidase labeling. *J. Exp. Med.*, 143:348–359.

Seidel, F., 1929. Untersuchungen über das Bildungsprinzip der Keimanlage im Ei der Libelle *Platycnemis pennipes. Arch. Entwicklungsmech. Org.*, 119:322–440.

Sellos, D., 1985. The histones isolated from the sperm of the oyster *Crassostrea gigas. Cell Differ.*, 17:183–192.

Sengel, P. and D. Dhouailly, 1977. Tissue interaction in amniote skin development. In M. Karkinen-Jaaskelainen, L. Saken, and L. Weiss, eds., *Cell interaction in differentiation.* Academic Press, Orlando, pp. 549–554.

Senger, D. R. and P. R. Gross, 1978. Macromolecule synthesis and determination in sea urchin blastomeres at the sixteen-cell stage. *Dev. Biol.*, 65:404–416.

Serras, F. and J. A. M. van den Biggelaar, 1987. Is a mosaic embryo also a mosaic of communication compartments? *Dev. Biol.*, 120:132–138.

Setchell, B. P., 1982. Spermatogenesis and spermatozoa. In C. R. Austin and R. V. Short, eds., *Reproduction in mammals, Book 1, Germ cells and fertilization*, 2nd ed. Cambridge University Press, Cambridge, pp. 63–101.

Shapiro, B. M., 1981. Awakening of the invertebrate egg at fertilization. In L. Mastroianni and J. D. Biggers, eds., *Fertilization and embryonic development in vitro.* Plenum Press, New York, pp. 233–255.

Shapiro, D. Y., 1987. Differentiation and evolution of sex change in fishes. *BioScience*, 37:490–497.

Sheinin, R., J. Humbert, and R. E. Pearlman, 1978. Some aspects of eukaryotic DNA replication. *Annu. Rev. Biochem.*, 47:277–316.

Shen, S. S., 1983. Membrane properties and intracellular ion activities of marine invertebrate eggs and their changes during activation. In J. F. Hartmann, ed., *Mechanisms and control of animal fertilization.* Academic Press, Orlando, pp. 213–267.

Shen, S. S. and R. A. Steinhardt, 1978. Direct measurement of intracellular pH during metabolic depression at fertilization and ammonia activation of the sea urchin egg. *Nature (London)*, 272:253–254.

Shepherd, G. W., E. Rondinelli, and M. Nemer, 1983. Differences in abundance of individual RNAs in normal and animalized sea urchin embryos. *Dev. Biol.*, 96:520–528.

Shimizu, T., 1982. Development in the freshwater oligochaete *Tubifex.* In F. W. Harrison & R. R. Cowden, eds., *Developmental biology of freshwater invertebrates.* Alan R. Liss, New York, pp. 283–316.

Short, R. V., 1977. The discovery of the ovaries. In S. Zuckerman and B. J. Weir, eds., *The ovary*, Vol. 1. Academic Press, Orlando, pp. 1–67.

Short, R. V., 1982. Sex determination and differentiation. In C. R. Austin and R. V. Short, eds., *Reproduction in mammals, Book 2, Embryonic and fetal development.* Cambridge University Press, Cambridge, pp. 70–113.

Short, R. V., 1984. Oestrous and menstrual cycles. In C. R. Austin and R. V. Short, eds., *Reproduction in mammals, Book 3, Hormonal control of reproduction*, 2nd ed. Cambridge University Press, Cambridge, pp. 115–152.

Shostak, S., 1977. Vegetative reproduction by budding in Hydra: A perspective on tumors. *Prosp. Biol. Med.*, 20:545–568.

Shott, R. J., J. J. Lee, R. J. Britten, and E. H. Davidson, 1984.

Differential expression of the actin gene family of *Stongylocentrotus purpuratus. Dev. Biol.*, 101:295–306.

Shott-Akhurst, R. J., F. J. Calzone, R. J. Britten, and E. H. Davidson, 1984. Isolation and characterisation of a cell lineage-specific cytoskeletal actin gene family of *Strongylocentrotus purpuratus.* In E. H. Davidson and R. A. Firtel, eds., *Molecular biology of development.* Alan R. Liss, New York, pp. 119–128.

Shubin, N. H. and P. Alberch, 1986. A morphogenetic approach to the origin and basic organisation of the tetrapod limb. *Evol. Biol.*, 20:319–387.

Shur, B. D., 1983. Embryonal carcinoma cell adhesion: The role of surface galactosyltransferase and its 90 kD lactosaminoglycan substrate. *Dev. Biol.*, 99:360–372.

Shur, B. D., 1989. Calactosyltransferase as a recognition molecule during fertilization and development. In H. Schatten and G. Schatten, eds., *Molecular biology of fertilization.* Academic Press, Orlando, pp. 37–71.

Siekevitz, P. and G. E. Palade, 1958. A cytochemical study on the pancreas of the guinea pig. 1. Isolation and enzymatic activities of cell fractions. *J. Biophys. Biochem. Cytol.*, 4:203–218. 2. Functional variations in the enzymatic activity of microsomes. *J. Biophys. Biochem. Cytol.*, 4:309–318. 3. *In vivo* incorporation of leucine-1-^{14}C into the proteins of cell fractions. *J. Biophys. Biochem. Cytol.*, 4:557–566.

Simoncini, L., M. L. Block, and W. J. Moody, 1988. Lineage-specific development of calcium currents during embryogenesis. *Science*, 242:1572–1575.

Singer, C., 1959. *A short history of scientific ideas to 1900.* Oxford University Press, London.

Singer, S. J. and G. L. Nicolson, 1972. Fluid mosaic model of the structure of cell membranes. *Science*, 175:720–731.

Škreb, N., A. Švajger, and B. Levak-Švajger, 1976. Developmental potentialities of the germ layers in mammals. In *Embryogenesis in mammals*, Ciba Foundation Symposium No. 40. Elsevier, Amsterdam, pp. 27–45.

Slack, J. M. W., 1983. *From egg to embryo: Determinative events in early development.* Cambridge University Press, Cambridge.

Slack, J. M. W., 1984. The early amphibian embryo—a hierarchy of developmental decisions. In G. M. Malacinski, ed., *Pattern formation, a primer in developmental biology.* Macmillan, New York, pp. 457–480.

Slack, J. M. W., B. G. Darlington, J. K. Heath, and S. F. Godsave, 1987. Mesoderm induction in early *Xenopus* embryos by heparin-binding growth factors. *Nature (London)*, 326: 197–200.

Slavkin, H. C., M. L. Snead, M. Zeichner-David, P. Bringas, Jr., and G. L. Greenberg, 1984. Amelogenin gene expression during epithelial–mesenchymal interactions. In R. L. Trelstad, ed., *The role of extracellular matrix in development.* Alan R. Liss, New York, pp. 221–253.

Smith, J. C., 1987. A mesoderm-inducing factor is produced from a *Xenopus* cell line. *Development*, 99:3–14.

Smith, J. C., L. Dale, and J. M. W. Slack, 1985. Cell lineage labels and region-specific markers in the analysis of inductive interactions. *J. Embryol. Exp. Morphol. Suppl.*, 89:317–331.

Smith, J. Maynard, 1978. *The evolution of sex.* Cambridge University Press, Cambridge.

Smith, L. Dennis, 1964. A test of the capacity of presumptive somatic cells to transform into primordial germ cells in the Mexican axolotl. *J. Exp. Zool.*, 156:229–242.

Smith, L. Dennis, 1966. The role of a "germinal plasm" in the formation of primordial germ cells in *Rana pipiens*. *Dev. Biol.*, 14:330–347.

Smith, L. Dennis, 1975. Molecular events during oocyte maturation. In R. Weber, ed., *The biochemistry of animal development*, Vol. III, *Molecular aspects of animal development*. Academic Press, Orlando, pp. 1–46.

Smith, L. Dennis, 1986. Regulation of translation during amphibian oogenesis and oocyte maturation. In R. G. Gall, ed., *Gametogenesis and the early embryo*, 44th Symposium of the Society for Developmental Biology. Alan R. Liss, New York, pp. 131–150.

Smith, L. Dennis and M. A. Williams, 1975. Germinal plasm and determination of the primordial germ cells. In C. L. Markert and J. Papaconstantinou, eds., *The developmental biology of reproduction*. Academic Press, Orlando, pp. 3–24.

Smith, L. Dennis, P. Michael, and M. A. Williams, 1983. Does a predetermined germ line exist in amphibians. In A. McLaren and C. C. Wylie, *Current problems in germ cell differentiation*. Cambridge University Press, Cambridge, pp. 19–39.

Smith, S. J., 1988. Neuronal cytomechanics: The actin-based motility of growth cones. *Science*, 242:708–715.

Snow, M. H. L., 1976. Embryo growth during the immediate postimplantation period. *Ciba Found. Symp.*, 40 (new series):53–70.

Snow, M. H. L., 1981. Autonomous development of parts isolated from primitive-streak-stage mouse embryos. Is development clonal? *J. Embryol. Exp. Morphol. Suppl.*, 65:269–287

Snow, M. H. L., 1985. The embryonic cell lineage of mammals and the emergence of the basic body plan. In G. M. Edelman, ed., *Molecular determinants of animal form*. Alan R. Liss, New York, pp. 73–98.

Snow, M. H. L., P. P. L. Tam, and A. McLaren, 1981. On the control and regulation of size and morphogenesis in mammalian embryos. In S. Subtelny and U. K. Abbot, eds., *Levels of genetic control in development*. Alan R. Liss, New York, pp. 201–217.

Sogin, M. L., J. H. Gunderson, H. J. Elwood, R. A. Alonso, and D. A. Peattie, 1989. Phylogenetic meaning of the kingdom concept: An unusual ribosomal RNA from *Giardia lamblia*. *Science*, 243:75–77.

Sollner-Webb, B., 1988. Surprises in polymerase III transcription. *Cell*, 52:153–154.

Solter, D. and B. B. Knowles, 1979. Developmental stage-specific antigens during mouse embryogenesis. In A. A. Moscona and A. Monroy, ed., *Current topics in developmental biology*, Vol. 13. Academic Press, Orlando, pp. 139–165.

Soltynska, M. S., 1982. The possible mechanism of cell positioning in mouse morulae: an ultrastructural study. *J. Embryol. Exp. Morphol.*, 68:137–147.

Sonneborn, T. M., 1954. Patterns of nucleo-cytoplasmic integration in *Paramecium*. *Caryologia Suppl.*, 6:307–325.

Sonneborn, T. M., 1967. The evolutionary integration of the genetic material into genetic systems. In R. A. Brink, ed., *Heritage from Mendel*, Proceedings of the Mendel Centennial Symposium of the Genetics Society of America, 1965. University of Wisconsin Press, Madison, pp. 375–401.

Spemann, H., 1938. *Embryonic development and induction.* Yale University Press, New Haven. (Reprinted by Hafner, New York, 1962.)

Spemann, H. and H. Mangold, 1924. Über Induktion von Embryonalanlagen durch Implantation artfremder Organisatoren. *Wilhelm Roux' Arch. Entwicklungsmech. Org.*, 100:599–638. Reprinted in B. H. Willier and J. M. Oppenheimer, eds., *Foundations of experimental embryology*. Prentice-Hall, Englewood Cliffs, NJ, pp. 144–185.

Speroff, L. and R. L. VandeWiele, 1971. Regulation of the human menstrual cycle. *Am. J. Obstet. Gynecol.*, 109:234–247.

Sperry, R. W., 1944. Optic nerve regeneration with return of vision in anurans. *J. Neurophysiol.*, 7:57–69. Reprinted in C. Fulton and A. O. Klein, 1976. *Explorations in developmental biology*. Harvard University Press, Cambridge, pp. 630–642.

Spiegel, M., 1954. The role of specific surface antigens in cell adhesion. Part II. Studies on embryonic amphibian cells. *Biol. Bull.*, 107:149–155.

Spooner, B. S., 1974. Morphogenesis of vertebrate organs. In J. Lash and J. R. Whittaker, eds., *Concepts of development*. Sinauer, Sunderland, MA, pp. 213–240.

Spradling, A. C. and A. P. Mahowald, 1980. Amplification of genes for chorion proteins during oogenesis in *Drosophila melanogaster*. *Proc. Natl. Acad. Sci. USA*, 77:1096–1100.

Spratt, N. Jr., 1946. Formation of the primitive streak in the explanted chick blastoderm marked with carbon particles. *J. Exp. Zool.*, 103:259–304.

Spratt, N. T. Jr., 1952. Location of prospective neural plate in the early chick blastoderm. *J. Exp. Zool.*, 120:109–130.

Srivastava, P. N., R. G. Sheikhnejad, R. Fayrer-Hosken, H. Malter, and B. G. Brackett, 1986. Inhibition of fertilization of the rabbit ova *in vitro* by the antibody to the inner acrosomal membrane of rabbit spermatozoa. *J. Exp. Zool.*, 238:99–102.

Steel, G. G., 1977. *Growth kinetics of tumours, cell population kinetics in relation to the growth and treatment of cancer.* Clarendon Press, Oxford.

Steinberg, M. S., 1964. The problem of adhesive selectivity in cellular interactions. In M. Locke, ed., *Cellular membranes in development*. Academic Press, Orlando, pp. 321–366.

Steinberger, A. and E. Steinberger, 1973. Hormonal control of mammalian spermatogenesis. In S. J. Segal, R. Crozier, P. A. Corfman, and P. G. Condliffe, eds., *Regulation of mammalian reproduction*. Charles C. Thomas, Springfield, IL, pp. 139–150.

Steinhardt, R. A. and D. Epel, 1974. Activation of sea urchin eggs by calcium ionophore. *Proc. Natl. Acad. Sci. USA*, 71:1915–1919.

Steitz, T. A., T. J. Richmond, D. Wise, and D. Engelman, 1974. The *lac* repressor protein: Molecular shape, subunit structure, and proposed model for operator interaction based on structure studies of microcrystals. *Proc. Natl. Acad. Sci. USA*, 71:593–597.

Stent, G. S., 1985. Thinking in one dimension: The impact of molecular biology on development. *Cell*, 40:1–2.

Stern, C. D. and R. J. Keynes, 1986. Cell lineage and the formation and maintenance of half somites. In R. Bellairs, D. A. Ede, and J. W. Lash, eds., *Somites in developing embryos*. Plenum Press, New York, pp. 147–159.

Stern, H., 1977. DNA synthesis during microsporogenesis. In L. Bogorad and J. H. Weil, eds., *Nucleic acid and protein synthesis in plants*. Plenum Press, New York, pp. 1–13.

Stevens, L. C., 1960. Embryonic potency of embryoid bodies derived from a transplantable testicular teratoma of the mouse. *Dev. Biol.*, 2:285–297.

Stewart, A. D., 1983. The role of the Y-chromosome in mammalian sexual differentiation. In M. H. Johnson, ed., *Development in mammals*, Vol. 5. Elsevier/North-Holland, Amsterdam, pp. 321–367.

Stocum, D. L. and J. F. Fallon, 1984. Mechanisms of polarization and pattern formation in urodele limb ontogeny: A polarizing zone model. In G. M. Malacinski, ed., *Pattern formation*. Macmillan, New York, pp. 507–520.

Streeter, G. L., 1938. Characteristics of the primate egg immediately preceding its attachment to the uterine wall. *Carnegie Inst. Wash. Pub.*, No. 501:397–414.

Strickberger, M. W., 1986. The structure and organization of genetic material. *Am. Zool.*, 26:769–780.

Strome, S. and W. B. Wood, 1983. Generation of asymmetry and segregation of germ-line granules in early *C. elegans* embryos. *Cell*, 35:15–25.

Sturm, R. A., G. Das, and W. Herr, 1988. The ubiquitous octamer-binding protein Oct-1 contains a POU domain with a homeo box subdomain. *Genes & Dev.*, 2:1582–1599.

Sturtevant, A. H., 1929. The claret mutant type of *Drosophila simulans*: A study of chromosomal elimination and cell lineage. *Z. Wiss. Zool.*, 135:323–356.

Sturtevant, A. H., 1965. *A history of genetics*. Harper & Row, New York.

Suarez, S. S., D. F. Katz, and S. Meizel, 1984. Changes in motility that accompany the acrosome reaction in hyperactivated hamster spermatozoa. *Gamete Res.*, 10:253–265.

Sugiyama, M., 1951. Re-fertilization of the fertilized eggs of the sea urchin. *Biol. Bull.*, 101:335–344.

Sulston, J. E. and J. G. White, 1980. Regulation and cell autonomy during postembryonic development in *Caenorhabditis elegans*. *Dev. Biol.*, 78:577–598.

Sulston, J. E., E. Schierenberg, J. G. White, and H. N. Thomson, 1983. The embryonic cell lineage of the nematode *Caenorhabditis elegans*. *Dev. Biol.*, 100:64–119.

Summerbell, D. and R. V. Stirling, 1982. Development of the pattern of innervation of the chick limb. *Am. Zool.*, 22:173–184.

Summerbell, D., J. H. Lewis, and L. Wolpert, 1973. Positional information in chick limb morphogenesis. *Nature (London)*, 244:492–496.

Suomalainen, E., A. Saura, and J. Lokki, 1987. *Cytology and evolution in parthenogenesis*. CRC Press, Boca Raton, FL.

Swenson, K. I., K. M. Farrell, and J. V. Ruderman, 1986. The clam embryo protein cyclin A induces entry into M phase and the resumption of meiosis in *Xenopus* oocytes. *Cell*, 47:861–870.

Szarski, H., 1976. Cell size and nuclear DNA content in vertebrates. *Int. Rev. Cytol.*, 44:93–111.

Szollosi, D., P. Calarco, and R. P. Donahue, 1972. Absence of centrioles in the first and second meiotic spindles of mouse oocytes. *J. Cell Sci.*, 11:521–541.

Takata, K. and T. Hama, 1978. Primary induction in cyclostomes and prochordates. In O. Nakamura and S. Toivonen, eds., *Organizer—a milestone of a half-century from Spemann*. Elsevier/North-Holland Biomedical Press, Amsterdam, pp. 267–282.

Takeichi, M., 1988. The cadherins: Cell–cell adhesion molecules controlling animal morphogenesis. *Development*, 102:639–655.

Takeichi, M., K. Hatta, and A. Nagafuchi, 1985. Selective cell adhesion mechanism: Role of the calcium-dependent cell adhesion system. In G. M. Edelman, ed., *Molecular determinants of animal form*. Alan R. Liss, New York, pp. 223–233.

Talbot, P., 1985. Sperm penetration through oocyte investments in mammals. *Am. J. Anat.*, 174:331–346.

Tam, P. P. L. and R. S. P. Beddington, 1987. The formation of mesodermal tissues in the mouse embryo during gastrulation and early organogenesis. *Development*, 99:109–126.

Tarkowski, A. K., 1959. Experimental studies on regulation in the development of isolated blastomeres of mouse eggs. *Acta Theriol.*, 3:191–267.

Tarkowski, A. K., 1961. Mouse chimeras developed from fused eggs. *Nature (London)*, 190:857–860.

Tarkowski, A. K., 1975. Induced parthenogenesis in the mouse. In C. L. Markert and J. Papaconstantinou, eds., *The developmental biology of reproduction*, 23rd Symposium of the Society for Developmental Biology. Academic Press, Orlando, pp. 107–129.

Tarkowski, A. K., 1977. *In vitro* development of haploid mouse embryos produced by bisection of one-celled fertilized eggs. *J. Embryol. Exp. Morphol.*, 55:319–330.

Tarkowski, A. K., 1982. Nucleo-cytoplasmic interactions in oogenesis and early embryogenesis in the mouse. In M. M. Burger and R. Weber, eds., *Embryonic development, Part A*. Alan R. Liss, New York, pp. 407–416.

Tarkowski, A. K. and J. Wroblewska, 1967. Development of blastomeres of mouse eggs isolated at the 4- and 8-cell stage. *J. Embryol. Exp. Morphol.*, 18:155–180.

Tattersall, P. and D. C. Ward, 1976. Rolling hairpin model for replication of parvovirus and linear chromosomal DNA. *Nature (London)*, 263:106–109.

Tegner, M. J. and D. Epel, 1976. Scanning electron microscope studies of sea urchin fertilization. I. Eggs with vitelline layers. *J. Exp. Zool.*, 197:31–58.

Telfer, W. H., 1954. Immunological studies of insect metamorphosis. II. The role of a sex limited blood protein in egg formation of the cecropia silkworm. *J. Gen. Physiol.*, 37:539–558.

Telfer, W. H., 1965. The mechanism and control of yolk formation. *Annu. Rev. Entomol.*, 10:161–184.

Terranova, V. P., D. H. Rohbach, and G. R. Martin, 1980. Role of laminin in the attachment of PAM 212 (epithelial) cells to basement membrane collagen. *Cell*, 22:719–726.

Thibault, C., 1973. *In vitro* maturation and fertilization of rabbit and cattle oocytes. In S. J. Segal, R. Crozier, P. A. Corfman, and P. G. Condliffe, eds., *Regulation of mammalian reproduction*. Charles C. Thomas, Springfield, IL, pp. 231–246.

Thiery, J.-P., J. C. Boucaut, and K. M. Yamada, 1985a. Cell migration in the vertebrate embryo. In G. M. Edelman, ed., *Molecular determinants of animal form*. Alan R. Liss, New York, pp. 167–193.

Thiery, J.-P., J.-L. Duband, and A. Delouvée, 1985b. The role of cell adhesion in morphogenetic movements during early embryogenesis. In G. M. Edelman and J.-P. Thiery, eds., *Cell in contact: Adhesion and junctions as morphogenetic determinants*. A Neurosciences Institute Publication, Wiley, New York, pp. 169–196.

Thompson, D. W., 1963. *On growth and form.* Cambridge University Press, London.

Thorburn, G. D., I. R. Young, M. Dolling, D. W. Walker, C. A. Browne, and G. G. Carmichael, 1986. Growth factors in fetal development. In G. Guroff, ed., *Growth and maturation factors,* Vol. 3. Wiley-Interscience, New York, pp. 177–201.

Thorogood, P. and A. Wood, 1987. Analysis of *in vivo* cell movement using transparent tissue systems. *J. Cell Sci. Suppl.,* 8:395–413.

Tickle, C. J. Lee and G. Eichele, 1985. A quantitative analysis of the effect of all-*trans*-retenoic acid on the pattern of chick wing development. *Dev. Biol.,* 109:182–195.

Tiedemann, H., 1975. Substances with morphogenetic activity in differentiation of vertebrates. In R. Weber, ed., *The biochemistry of animal development,* Vol. III, *Molecular aspects of animal development.* Academic Press, Orlando, pp. 257–292.

Tiedemann, H., 1978. Chemical approach to the inducing agents. In O. Nakamura and S. Toivonen, eds., *Organizer—a milestone of a half-century from Spemann.* Elsevier/North-Holland Biomedical Press, Amsterdam, pp. 91–117.

Tijssen, P., 1985. *Laboratory techniques in biochemistry and molecular biology,* Vol. 15, *Practice and theory of enzyme immunoassays.* Elsevier, Amsterdam.

Tobler, H., K. D. Smith, and H. Ursprung, 1972. Molecular aspects of chromatin elimination in *Ascaris lumbricoides. Dev. Biol.,* 27:190–203.

Toivonen, S., 1978. Regionalization of the embryo. In O. Nakamura and S. Toivonen, eds., *Organizer—a milestone of a half-century from Spemann.* Elsevier/North-Holland Biomedical Press, Amsterdam, pp. 119–156.

Townes, P. L. and J. Holtfreter, 1955. Directed movements and selective adhesion of embryonic amphibian cells. *J. Exp. Zool.,* 128:53–120.

Trimmer, J. S. and V. D. Vacquier, 1986. Activation of sea urchin gametes. *Annu. Rev. Cell Biol.,* 2:1–26.

Trinkaus, J. P., 1984a. *Cells into organs: Forces that shape the embryo,* 2nd ed. Prentice-Hall, Englewood Cliffs, NJ.

Trinkaus, J. P., 1984b. Mechanism of *Fundulus* epiboly—a current view. *Am. Zool.,* 24:673–688.

Tsafriri, A., 1978. Oocyte maturation in mammals. In R. E. Jones, ed., *Vertebrate ovary.* Plenum Press, New York, pp. 409–442.

Tung, T. C., S. C. Wu, and Y. Y. F. Tung, 1962a. The presumptive areas of the egg of *Amphioxus. Scientia Sinica,* 11: 629–644.

Tung, T. C., S. C. Wu and Y. Y. F. Tung, 1962b. Experimental studies on the neural induction in *Amphioxus. Sci. Sin.,* 11:805–820.

Turing, A., 1952. The chemical basis of morphogenesis. *Philos. Trans. R. Soc. Ser. B,* 237:37–72.

Turner, E., L. J. Hager, and B. M. Shapiro, 1988. Ovothiol replaces glutathione peroxidase as a hydrogen peroxide scavenger in sea urchin eggs. *Science,* 242:939–941.

Turner, F. R. and A. P. Mahowald, 1977. Scanning electron microscopy of *Drosophila melanogaster* embryogenesis. II. Gastrulation and segmentation. *Dev. Biol.,* 57:403–416.

Turner, F. R. and A. P. Mahowald, 1979. Scanning electron microscopy of *Drosophila melanogaster* embryogenesis. III. Formation of the head and caudal segments. *Dev. Biol.,* 68:96–109.

Turner, R. S. Jr., 1987. Cell–cell associations in echinoderm embryos. In A. H. Greenberg, ed., *Invertebrate models: Cell receptors and cell communication.* Karger, Basel, pp. 143–171.

Tyler, A. and H. Bauer, 1937. Polar body extrusion and cleavage in artificially activated eggs of *Urechis caupo. Biol. Bull.,* 73:164–180.

Vacquier, V. D., 1975. The isolation of intact cortical granules from sea urchin eggs: Calcium ions trigger granule discharge. *Dev. Biol.,* 43:62–74.

Vacquier, V. D., M. J. Tegner, and D. Epel, 1973. Protease release from sea urchin eggs at fertilization alters the vitelline membrane layer and aids in preventing polyspermy. *Exp. Cell Res.,* 80:111–119.

Vakaet, L., 1984. The initiation of gastrular ingression in the chick blastoderm. *Am. Zool.,* 24:555–562.

van Blerkom, J. and R. W. McGaughey, 1978. Molecular differentiation of the rabbit ovum. II. During the preimplantation development of in vivo and in vitro matured oocytes. *Dev. Biol.,* 63:151–164.

van den Biggelaar, J. A. M., 1971a and b. Timing of the phases of the cell cycle with tritiated thymidine and Feulgen cytophotometry during the period of synchronous division in *Lymnaea. J. Embryol. Exp. Morphol.,* 26:351–366, 367–391.

van de Vyver, G., 1980. A comparative study of the embryonic development of hydrozoa athecata. In P. Tardent and R. Tardent, eds., *Developmental and cellular biology of coelenterates.* Elsevier/North-Holland Biomedical Press, Amsterdam, pp. 109–120.

van Tienhoven, A., 1983. *Reproductive physiology of vertebrates.* Cornell University Press, Ithaca.

Varmus, H. E., 1985. Reverse transcriptase rides again. *Nature* (London), 314:583–584.

Vasiliev, J. M., 1987. Actin cortex and microtubular system in morphogenesis: Cooperation and competition. *J. Cell Sci. Suppl.,* 8:1–18.

Vendrely, R. and C. Vendrely, 1956. The results of cytophotometry in the study of the deoxyribonucleic acid (DNA) content of the nucleus. *Int. Rev. Cytol.,* 5:171–197.

Verdonk, N. H. and J. A. M. van den Biggelaar, 1983. Early development and the formation of the germ layers. In N. H. Verdonk, J. A. M. van den Biggelaar, and A. S. Tompa, eds., *The mollusca,* Vol. 3, *Development.* Academic Press, Orlando, pp. 91–122.

Virchow, R., 1858. *Die Cellularpathologie in ihrer Begründung auf physiologische und pathologische Gewebelehre.* Berlin. 1860. *Cellular pathology: As based upon physiological and pathological histology* (translated from the 2nd edition by F. Chance), Robert M. De Witt, New York. 1978. Classic of Medicine Library, Birmingham.

Vogt, W., 1925. Gestaltungsanalyse am Amphibienkeim mit örtlicher Vitalfärbung. Vorwort über Wege und Ziele. I. Methodik u. Wirkungsweise der örtlicher Vitalfärbung. *Wilhelm Roux' Arch. Entwicklungsmech. Org.,* 106:542–610.

Wachtel, S. S., S. Ohno, G. C. Koo, and E. A. Boyse, 1975. Possible role for H-Y antigen in the primary determination of sex. *Nature (London),* 257:235–236.

Waddington, C. H., 1952. *The epigenetics of birds.* Cambridge University Press, Cambridge.

Waddington, C. H., 1962. *New patterns in genetics and development.* Columbia University Press, New York.

Wahli, W. and I. B. Dawid, 1979. Vitellogenin in *Xenopus laevis* is encoded by a small family of genes. *Cell*, 16:535–549.

Wahli, W., I. B. Dawid, G. U. Ryffel, and R. Weber, 1981. Vitellogenesis and the vitellogenin gene family. *Science*, 212:298–306.

Wakahara, M., 1978. Induction of supernumerary primordial germ cells by injecting vegetal pole cytoplasm into *Xenopus* eggs. *J. Exp. Zool.*, 203:159–164.

Wall, R., 1973. Physiological gradients in development—a possible role for messenger ribonucleoprotein. *Adv. Morphogen.*, 10:41–114.

Wallace, H. and T. R. Elsdale, 1964. Effects of actinomycin D on amphibian development. *Acta Embryol. Morphol. Exp.*, 6:275–282.

Wallace, R. A., 1963. Studies on amphibian yolk. IV. An analysis of the main-body component of yolk platelets. *Biochim. Biophys. Acta*, 74:505–518.

Wallace, R. A., 1978. Oocyte growth in nonmammalian vertebrates. In R. E. Jones, ed., *Vertebrate ovary: Comparative biology and evolution*. Plenum Press, New York, pp. 469–502.

Walter, P. and G. Blobel, 1983. Disassembly and reconstitution of signal recognition particle. *Cell*, 34:525–533.

Warburg, O., 1908. Über die Oxydationen in lebenden Zellen nach Versuchen am Seeigelei. *Z. Physiol. Chem.*, 66:305–340. Reprinted in B. H. Willier and J. M. Oppenheimer, eds., 1964. *Foundations of experimental embryology*. Prentice-Hall, Englewood Cliffs, NJ, pp. 104–116.

Warner, A. E., S. C. Guthrie, and N. B. Gilula, 1984. Antibodies to gap junctional protein selectively disrupt junctional communication in the early amphibian embryo. *Nature (London)*, 311:127–131.

Wartiovaara, J. and A. Vaheri, 1980. Fibronectin and early mammalian embryogenesis. In M. H. Johnson, ed., *Development in mammals*, Vol. 4. Elsevier/North-Holland Biomedical Press, Amsterdam, pp. 233–266.

Wassarman, P. M., 1983. Fertilization. In K. M. Yamada, ed., *Cell interactions and development: Molecular mechanisms*. Wiley-Interscience, New York, pp. 1–27.

Wassarman, P. M., 1987. Early events in mammalian fertilization. *Annu. Rev. Cell Biol.*, 3:109–142.

Wassarman, P. M., J. M. Greve, R. M. Perona, R. J. Roller, and G. S. Salzmann, 1984. How mouse eggs put on and take off their extracellular coat. In E. H. Davidson and R. A. Firtel, eds., *Molecular biology of development*. Alan R. Liss, New York, pp. 213–225.

Wasserman, S. A., J. M. Dungan, and N. R. Cozzarelli, 1985. Discovery of a predicted DNA knot substantiates a model for site-specific recombination. *Science*, 229:171–174.

Wasylyk, B., R. Derbyshire, A. Guy, D. Molko, A. Roget, R. Téoule, and P. Chambon, 1980. Specific *in vitro* transcription of conalbumin gene is drastically decreased by single-point mutation in TATA box homology sequence. *Proc. Natl. Acad. Sci. USA*, 77:7024–7028.

Watson, J. D., 1968. *The double helix*. Norton, New York.

Watson, J. D., 1976. *Molecular biology of the gene*, 3rd rev. ed. With N. H. Hopkins, J. W. Roberts, J. A. Steitz, and A. M. Weiner, 1987. Vol. 1, *General principles*. Vol. 2, *Specialized aspects*, 4th ed. Benjamin-Cummings, Menlo Park.

Watson, J. D. and F. H. C. Crick, 1953a. Molecular structure of nucleic acids: A structure for deoxyribonucleic acids. *Nature (London)*, 171:737–738.

Watson, J. D. and F. H. C. Crick, 1953b. The structure of DNA. *Cold Spring Harbor Symp. Quant. Biol.*, 18:123–131.

Watterson, R. L., 1955. Selected invertebrates. In B. H. Willier, P. A. Weiss, and V. Hamburger, eds., *Analysis of development*. Saunders, Philadelphia, pp. 315–336.

Way, J. C. and M. Chalfie, 1988. *mec-3*, a homeobox-containing gene that specifies differentiation of the touch receptor neurons in C. elegans. *Cell*, 54:5–16.

Weeks, D. L. and D. A. Melton, 1987. A maternal messenger RNA localized to the vegetal pole in *Xenopus* eggs codes for a growth factor related to TGF-β. *Cell*, 51:861–867.

Weinberg, W. C., J. C. Howard, and P. M. Iannaccone, 1985. Histological demonstration of mosaicism in a series of chimeric rats produced between congenic strains. *Science*, 227:524–527.

Weiner, A. M., 1988. Eukaryotic nuclear telomeres: Molecular fossils of the RNP world? *Cell*, 52:155–157.

Weir, B. J. and I. W. Rowlands, 1977. Ovulation and atresia. In S. Zuckerman and B. J. Weir, eds., *Ovary*, 2nd ed., Vol. 1. Academic Press, Orlando, pp. 265–301.

Weir, M. P. and C. W. Lo, 1984. Gap-junctional communication compartments in the *Drosophila* wing imaginal disk. *Dev. Biol.*, 102:130–146.

Weir, M. P. and C. W. Lo, 1985. An anterior/posterior communication compartment border in *engrailed* wind discs: Possible implications for *Drosophila* pattern formation. *Dev. Biol.*, 110:84–90.

Weismann, A., 1885. *Die Continuität des Keimplasmas als Grundlage einer Theorie der Vererbung*. Fischer, Jena.

Weismann, A., 1892. *Das Keimplasma. Eine Theorie der Vererbung*. Fischer, Jena.

Weiss, P., 1939. *Principles of development: A text in experimental embryology*. Holt, New York.

Weiss, P., 1958. Cell contact. *Int. Rev. Cytol.*, 7:391–423.

Weitlauf, H. M. and A. A. Kiessling, 1981. Activity of RNA and DNA polymerases in delayed-implanting mouse embryos. In S. R. Glasser and D. W. Bullock, eds., *Cellular and molecular aspects of implantation*. Plenum Press, New York, pp. 125–136.

Wellauer, P. K., R. H. Reeder, I. B. Dawid, and D. D. Brown, 1976. The arrangement of length heterogeneity in repeating units of amplified and chromosomal ribosomal DNA from *Xenopus laevis*. *J. Mol. Biol.*, 105:487–505.

Werren, J. H., 1987. Labile sex ratios in wasps and bees. *BioScience*, 37:498–506.

Weston, J. A., 1983. Regulation of neural crest cell migration and differentiation. In K. M. Yamada, ed., *Cell interactions and development: Molecular mechanisms*. Wiley, New York, pp. 153–184.

Whitaker, M. and R. F. Irvine, 1984. Inositol 1,4,5-trisphosphate microinjection activates sea urchin eggs. *Nature (London)*, 312:636–639.

Whitaker, M. J. and R. A. Steinhardt, 1985. Ionic signaling in the sea urchin egg at fertilization. In C. B. Metz and A. Monroy, eds., *Biology of fertilization*, Vol. 3, *Fertilization response of the egg*. Academic Press, Orlando, pp. 167–221 and 453–455.

White, J. H. and W. R. Bauer, 1989. The helical repeat of nucleosome-wrapped DNA. *Cell*, 56:9–10.

Whitman, C. O., 1899. Animal behavior. In J. Maienschein, ed., 1986. *Defining biology: Lectures from the 1890s*. Harvard University Press, Cambridge, pp. 285–338.

Whittaker, J. R., 1979. Cytoplasmic determinants of tissue differentiation in the ascidian egg. In S. Subtelny, ed., *Determinants of spatial organization*, 37th Symposium of the Society for Developmental Biology. Academic Press, Orlando, pp. 29–51.

Wiley, L. M., 1979. Early embryonic cell surface antigens as developmental probes. In A. A. Moscona and A. Monroy, eds., *Current topics in developmental biology*, Vol. 13. Academic Press, Orlando, pp. 167–197.

Wilkins, M. H. F., A. R. Stokes, H. R. Wilson, 1953. Molecular structure of deoxypentose nucleic acids. *Nature* (London), 171:738–740.

Willardsen, S. M., 1979. A method for culture of micromanipulated sheep embryos and its use to produce monozygotic twins. *Nature (London)*, 277:298–300.

Williams, G. C., 1975. *Sex and evolution.* Princeton University Press, Princeton.

Williams, J., 1965. Chemical constitution and metabolic activities of animal eggs. In R. Weber, ed., *The biochemistry of animal development*, Vol. 1. Academic Press, Orlando, pp. 14–71.

Willier, B. H., 1937. Experimentally produced sterile gonads and the problem of the origin of germ cells in the chick embryo. *Anat. Rec.*, 70(Suppl. 1):89–112.

Willier, B. H. and J. M. Oppenheimer, eds., 1964. *Foundations of experimental embryology.* Prentice-Hall, Englewood Cliffs, NJ.

Wilson, A. C., R. L. Cann, S. M. Carr, M. George, U. B. Byllensten, K. M. Helm-Bychowski, R. G. Higuchi, S. R. Palumbi, E. M. Prager, R. D. Sage, and M. Stoneking, 1985. Mitochondrial DNA and two perspectives on evolutionary genetics. *Biol. J. Linnean Soc.*, 26:375–400.

Wilson, D. J. and J. R. Hinchliffe, 1987. The effect of the zone of polarizing activity (ZPA) on the anterior half of the chick wing bud. *Development*, 99:99–108.

Wilson, E. B., 1896. *The cell in development and inheritance.* Macmillan, London. 1928, definitive 3rd ed., reprinted 1987, Garland Publishing, New York.

Wilson, E. B., 1898. Cell lineage and ancestral reminiscence. Biology Lectures from the Marine Biological Laboratory, Woods Holl, MA, pp. 21–42. Reprinted in B. H. Willier and J. M. Oppenheimer, eds., 1964. *Foundations of experimental embryology.* Prentice-Hall, Englewood Cliffs, NJ, pp. 55–72.

Wilson, E. B., 1904a and b. Experimental studies on germinal localization. I. The germ-region in the egg of dentalium. II. Experiments on the cleavage-mosaic in Patella and Dentalium. *J. Exp. Zool.*, 1:1–72, 197–268.

Wilson, H. V., 1907. On some phenomena of coalescence and regeneration in sponges. *J. Exp. Zool.*, 5:245–258.

Wilt, F. H. and C. R. Phillips, 1984. Localization of poly(A) in Xenopus embryos. In E. H. Davidson and R. A. Firtel, eds., *Molecular biology of development.* Alan R. Liss, New York, pp. 23–36.

Wily, H. S. and R. A. Wallace, 1981. The structure of vitellogenin. *J. Biol., Chem.*, 256:8626–8634.

Wimsatt, W. A., 1975. Some comparative aspects of implantation. *Biol. Reprod.*, 12:1–40.

Wischnitzer, S., 1966. The ultrastructure of the cytoplasm of the developing amphibian egg. *Adv. Morphogen.*, 5:131–179.

Witschi, E., 1956. *Development of vertebrates.* Saunders, Philadelphia.

Wolcott, D. L., 1981. Effect of potassium and lithium ions on protein synthesis in the sea urchin embryo. *Exp. Cell Res.*, 132:464–468.

Wolcott, D. L., 1982. Does protein synthesis decline in lithium-treated sea urchin embryos because RNA synthesis is inhibited? *Exp. Cell Res.*, 137:427–431.

Wolf, D. P., 1974. The cortical granule reaction in living eggs of the toad, *Xenopus laevis. Dev. Biol.*, 36:62–71.

Wolf, D. P., 1981. The mammalian egg's block to polyspermy. In L. Mastroianni and J. D. Biggers, eds. *Fertilization and embryonic development in vitro.* Plenum Press, New York, pp. 183–197.

Wolffe, A. P. and D. D. Brown, 1988. Developmental regulation of 2 5S ribosomal RNA genes. *Science*, 241:1626–1632.

Wolgemuth, D. J., 1983. Synthetic activities of the mammalian early embryo: Molecular and genetic alterations following fertilization. In J. F. Hartmann, ed., *Mechanism and control of animal fertilization.* Academic Press, Orlando, pp. 415–452.

Wolpert, L., 1969. Positional information and the spatial pattern of cellular differentiation. *J. Theoret. Biol.*, 25:1–47.

Wolpert, L., 1985. Gradients, position and pattern: A history. In T. J. Horder, J. A. Witkowski, and C. C. Wylie, eds., *History of embryology*, 8th Symposium of the British Society for Developmental Biology, Cambridge University Press, Cambridge, pp. 347–361.

Wood, A. and L. P. M. Timmermans, 1988. Teleost epiboly: A reassessment of deep cell movement in the germ ring. *Development*, 102:575–585.

Wood, W. B., ed., 1988. The nematode *Caenorhabditis elegans.* Cold Spring Harbor Laboratory, Cold Spring Harbor, NY.

Wylie, C. C., J. Heasman, A. Snope, M. O'Driscoll, and S. Hollwill, 1985. Primordial germ cells of *Xenopus laevis* are not irreversibly determined early in development. *Dev. Biol.*, 112:66–72.

Wyllie, A. H., R. A. Laskey, J. Finch, and J. B. Gurdon, 1978. Selective DNA conservation and chromatin assembly after injection of SV40 DNA into *Xenopus* oocytes. *Dev. Biol.*, 64:178–188.

Xia, L., Y. Clermont, M. Lalli, and R. B. Buckland, 1986. Evolution of the endoplasmic reticulum during spermiogenesis of the rooster: An electron microscopic study. *Am. J. Anat.*, 177:301–312.

Yamada, K. M., M. J. Humphries, T. Hasegawa, E. Hasegawa, K. Olden, W.-T. Chen, and S. K. Akiyama, 1985. Fibronectin: Molecular approaches to analyzing cell interactions with the extracellular matrix. In G. M. Edelman and J.-P. Thiery, eds., *Cell in contact: Adhesion and junctions as morphogenetic determinants.* A Neurosciences Institute Publication, Wiley, New York, pp. 303–332.

Yan, S., M. Du, N. Wu, J. Yan, G. Jin, Y. Qin, and X. Zhang, 1986. Identification of intergeneric nucleo-cytoplasmic hybrid fish obtained from the combination of carp nucleus and crucian cytoplasm. In H. C. Slavkin, *Progress in developmental biology, Part A*, 10th International Congress of the International Society of Developmental Biologists. Alan R. Liss, New York, pp. 35–38.

Yanagimachi, R., 1978. Sperm–egg association in mammals. In A. A. Moscona and A. Monroy, eds., *Current topics in de-*

velopmental biology, Vol. 12, *Fertilization*. Academic Press, Orlando, pp. 83–105.

Yanagimachi, R., 1981. Mechanisms of fertilization in mammals. In L. Mastroianni and J. D. Biggers, eds., *Fertilization and embryonic development in vitro*. Plenum Press, New York, pp. 81–182.

Yanagimachi, R., 1988. Sperm–egg fusion. In N. Düzgünes and F. Bronner, *Current topics in membranes and transport, Vol. 32, Membrane fusion in fertilization, cellular transport, and viral infection*. Academic Press, Orlando, pp. 3–43.

Yoshida-Noro, C., N. Suzuki, and M. Takeichi, 1984. Molecular nature of the calcium-dependent cell–cell adhesion system in mouse teratocarcinoma and embryonic cells studied with a monoclonal antibody. *Dev. Biol.*, 101:19–27.

Zagris, N., 1980. Erythroid cell differentiation in unincubated chick blastoderm in culture. *J. Embryol. Exp. Morphol.*, 58:209–216.

Zamboni, L., 1970. Ultrastructure of mammalian oocytes and ova. *Biol. Reprod. Suppl.*, 2:44–63.

Zaug, A. and T. R. Cech, 1986. The intervening sequence RNA of *Tetrahymena* is an enzyme. *Science*, 231:470–475.

Zelená, J., 1964. Development, degeneration and regeneration of receptor organs. *Prog. Brain Res.*, 13:175–213.

Zust, B. and K. E. Dixon, 1977. Events in the germ cell lineage after entry of the primordial germ cells into the genital ridge in normal and U.V.-irradiated *Xenopus laevis*. *J. Embryol. Exp. Morphol.*, 41:33–46.

Zwilling, E., 1968. Morphogenetic phases in development. In M. Locke, ed., *The emergence of order in developing systems*. Academic Press, Orlando.

INDEX

Note: Definitions of terms are identified by page numbers in **boldface** type. Illustrations are identified by page numbers in *italic* type. Tables are identified by a *t* following the page number.